A. GREIDANUS

Encyclopedia of Composite Materials and Components

Encyclopedia of Composite Materials and Components

ENCYCLOPEDIA REPRINT SERIES

Editor: Martin Grayson

A WILEY-INTERSCIENCE PUBLICATION

JOHN WILEY & SONS

NEW YORK • CHICHESTER • BRISBANE • TORONTO • SINGAPORE

Library of Congress Cataloging in Publication Data:

Main entry under title:

Encyclopedia of composite materials and components.

(Encyclopedia reprint series)
Reprint of articles taken from the Encyclopedia of chemical technology.
"A Wiley-Interscience publication."
Includes index.
1. Composite materials—Dictionaries. I. Grayson, Martin, II. Encyclopedia of chemical technology.

TA418.9.C6E5 1983 620.1'18'0321 82-23823

ISBN 0-471-87357-8

Printed in the United States of America

10 9 8 7 6 5 4 3 2 1

CONTENTS

EDITORIAL STAFF

CONTRIBUTORS

A. F. Alciatore, *Grefco, Inc., New Orleans, Louisiana,* Diatomite

David G. Bailey, *United States Department of Agriculture, Philadelphia, Pennsylvania,* Leather

Marilyn Bakker, *Business Communications Co., Inc., Stamford, Connecticut,* Engineering plastics

G. A. Baum, *The Institute of Paper Chemistry, Appleton, Wisconsin,* Paper

John G. Blumberg, *PQ Corporation, Lafayette Hill, Pennsylvania,* Fillers

N. L. Bottone, *Union Carbide Corporation, Cleveland, Ohio,* Carbon, carbon and artificial graphite

J. P. Brazel, *General Electric Company, Philadelphia, Pennsylvania,* Ablative materials

Peter R. Buechler, *United States Department of Agriculture, Philadelphia, Pennsylvania,* Leather

Rointan F. Bunshah, *University of California, Los Angeles, California,* Refractory coatings

R. M. Bushong, *Union Carbide Corporation, Cleveland, Ohio,* Carbon, carbon and artificial graphite

C. C. Chamis, *National Aeronautics and Space Administration, Cleveland, Ohio,* Laminated and reinforced metals

F. P. Civardi, *Inmont Corporation, Clifton, New Jersey,* Leatherlike materials

Eli M. Dannenberg, *Cabot Corporation, Billerica, Massachusetts,* Carbon, carbon black

Robert F. Davis, *North Carolina State University, Raleigh, North Carolina,* Ceramics, scope

Robert Desch, *Armstrong World Industry, Lancaster, Pennsylvania,* Plastic building products

Arthur Drelich, *Chicopee Division, Johnson and Johnson, Milltown, New Jersey,* Nonwoven textile fabrics, staple fibers

H. S. Dugal, *The Institute of Paper Chemistry, Appleton, Wisconsin,* Paper

Lawrence H. Dunlap, *Consultant, Lancaster, Pennsylvania,* Plastic building products

D. B. Easty, *The Institute of Paper Chemistry, Appleton, Wisconsin,* Paper

Alfred L. Everett, *United States Department of Agriculture, Philadelphia, Pennsylvania,* Leather

James S. Falcone, Jr., *PQ Corporation, Lafayette Hill, Pennsylvania,* Fillers

Stephen H. Feairheller, *United States Department of Agriculture, Philadelphia, Pennsylvania,* Leather

A. L. Friedberg, *University of Illinois, Urbana, Illinois,* Enamels, porcelain or vitreous

N. R. Greening, *Portland Cement Association, Skokie, Illinois,* Cement

Ralph E. Grim, *University of Illinois, Urbana, Illinois,* Clays, uses

Thomas M. Hare, *North Carolina State University, Raleigh, North Carolina,* Ceramics, properties and applications

Charles A. Harper, *Westinghouse Electric Corporation, Baltimore, Maryland,* Embedding

Joseph S. Hayes, Jr., *American Kynol, Inc., New York, New York,* Novoloid fibers

Richard A. Helmuth, *Portland Cement Association, Skokie, Illinois,* Cement

William F. Herbes, *American Cyanamid Company, Bound Brook, New Jersey,* Amino resins and plastics

W. B. Hillig, *General Electric Co., Schenectady, New York,* Composite materials; High temperature composites

Arnold J. Hoiberg†, *Manville Corporation, Denver, Colorado,* Roofing materials

R. H. Hutchinson, *Armstrong Cork Co., Lancaster, Pennsylvania,* Cork

G. Frederick Hutter, *Inmont Corporation, Clifton, New Jersey,* Leatherlike materials

M. G. Jacko, *Bendix Corporation, Southfield, Michigan,* Brake linings and clutch facings

B. R. Joyce, *Union Carbide Corporation, Cleveland, Ohio,* Carbon, carbon and artificial graphite

P. W. Juneau, Jr., *General Electric Company, Philadelphia, Pennsylvania,* Ablative materials

Fred A. Keimel, *Bell Telephone Laboratories, Berkeley Heights, New Jersey,* Adhesives

W. D. Keller, *University of Missouri, Columbia, Missouri,* Clays, survey

R. C. Krutenat, *Exxon Research and Engineering Company,* Linden, New Jersey, Metallic coatings, survey

H. D. Leigh, *C-E Basic, Inc., Bettsville, Ohio,* Refractories

J. D. Litvay, *The Institute of Paper Chemistry, Appleton, Wisconsin,* Paper
Ernest G. Long, *Manville Corporation, Denver, Colorado,* Roofing materials
J. C. Long, *Union Carbide Corporation, Cleveland, Ohio,* Carbon, carbon and artificial graphite
E. W. Malcom, *The Institute of Paper Chemistry, Appleton, Wisconsin,* Paper
Frederick J. McGarry, *Massachusetts Institute of Technology, Cambridge, Massachusetts,* Laminated and reinforced plastics
John N. McGovern, *The University of Wisconsin, Madison, Wisconsin,* Fibers, vegetable
J. T. Meers, *Union Carbide Corporation, Cleveland, Ohio,* Carbon, carbon and artificial graphite
W. C. Miiller, *Manville Corporation, Denver, Colorado,* Refractory fibers
F. M. Miller, *Portland Cement Association,* Skokie, Illinois, Cement
Sewell T. Moore, *American Cyanamid Company, Stamford, Connecticut,* Amino resins and plastics
G. A. Morneau, *Borg-Warner Chemicals, Washington, West Virginia,* Acrylonitrile polymers, ABS resins
David I. Netting, *ARCO Chemical Co., Glenolden, Pennsylvania,* Fillers
E. L. Neu, *Grefco, Inc., Los Angeles, California,* Diatomite
T. R. O'Connor, *Portland Cement Association, Skokie, Illinois,* Cement
W. A. Pavelich, *Borg-Warner Chemicals, Washington, West Virginia,* Acrylonitrile polymers, ABS resins
Fred M. Peng, *Monsanto Co., Indian Orchard, Massachusetts,* Acrylonitrile polymers, survey and styrene–acrylonitrile copolymers
E. L. Piper, *Union Carbide Corporation, Cleveland, Ohio,* Carbon, carbon and artificial graphite
Andrew Pocalyko, *E. I. du Pont de Nemours & Co., Inc., Coatesville, Pennsylvania,* Metallic coatings, explosively clad
K. Porter, *ICI Fibres, Harrogate, North Yorkshire, England,* Nonwoven textile fabrics, spunbonded
Mary K. Porter, *Consultant, Troy, New York,* Felts
J. Preston, *Monsanto Triangle Park Development Center, Inc., Research Triangle Park, North Carolina,* Aramid fibers
R. L. Reddy, *Union Carbide Corporation, Cleveland, Ohio,* Carbon, carbon and artificial graphite
S. K. Rhee, *Bendix Corporation, Southfield, Michigan,* Brake linings and clutch facings
John A. Roberts, *Arco Ventures Co., Troy, Michigan,* Metal fibers
William J. Roberts, *Consultant, Bernardsville, New Jersey,* Fibers, chemical
L. G. Roettger, *Borg-Warner Chemicals, Washington, West Virginia,* Acrylonitrile polymers, ABS resins
Philip B. Roth, *American Cyanamid Company, Bound Brook, New Jersey,* Amino resins and plastics
R. Russell, *Union Carbide Corporation, Cleveland, Ohio,* Carbon, carbon and artificial graphite
P. M. Scherer, *Union Carbide Corporation, Cleveland, Ohio,* Carbon, carbon and artificial graphite
Raymond B. Seymour, *The University of Southern Mississippi, Hattiesburg, Mississippi,* Sealants
R. E. Skochdopole, *The Dow Chemical Company, Midland, Michigan,* Foamed plastics

Leonard H. Smiley, *PQ Corporation, Lafayette Hill, Pennsylvania,* Fillers

R. W. Soffel, *Union Carbide Corporation, Cleveland, Ohio,* Carbon, carbon and artificial graphite

Robert M. Sowers, *Ford Motor Company, Lincoln Park, Missouri,* Laminated materials, glass

Robert E. Sparks, *Washington University, St. Louis, Missouri,* Microencapsulation

Daniel R. Stewart, *Owens-Illinois, Toledo, Ohio,* Glass-ceramics

E. R. Stover, *General Electric Company, Philadelphia, Pennsylvania,* Ablative materials

William C. Streib, *Johns-Manville Corporation, Denver, Colorado,* Asbestos

K. W. Suh, *The Dow Chemical Company, Granville, Ohio,* Foamed plastics

J. W. Swanson, *The Institute of Paper Chemistry, Appleton, Wisconsin,* Paper

Fred N. Teumac, *Uniroyal, Inc., Mishaka, Indiana,* Coated fabrics

Ivor H. Updegraff, *American Cyanamid Company, Stamford, Connecticut,* Amino resins and plastics

H. F. Volk, *Union Carbide Corporation, Cleveland, Ohio,* Carbon, carbon and artificial graphite

D. Wahren, *The Institute of Paper Chemistry, Appleton, Wisconsin,* Paper

J. H. Westbrook, *General Electric Company, Schenectady, New York,* Materials standards and specifications

L. L. Winter, *Union Carbide Corporation, Cleveland, Ohio,* Carbon, carbon and artificial graphite

John A. Youngquist, *United States Department of Agriculture, Madison, Wisconsin,* Laminated wood-based composites

† Deceased.

PREFACE

This volume is one of a series of carefully selected reprints from the world-renowned Kirk-Othmer *Encyclopedia of Chemical Technology* designed to provide specific audiences with articles grouped by a central theme. Although the 25-volume Kirk-Othmer Encyclopedia is widely available, many readers and users of this key reference tool have expressed interest in having selected articles in their specialty collected for handy desk reference or teaching purposes. In response to this need, we have chosen all of the original, complete articles related to composite materials and components to make up this new volume. The full texts, tables, figures and reference materials from the original work have been reproduced here unchanged. All articles are by industrial or academic experts in their field and the final work represents the result of careful review by competent specialists and the thorough editorial processing of the professional Wiley staff. Introductory information from the Encyclopedia concerning Chemical Abstracts Registry Numbers, nomenclature, SI units and conversion factors and related information has been provided as a further guide to the contents for those concerned with the materials aspects of composites.

The contents of this volume include coverage of nearly every aspect of composites and provide detailed information on methods of manufacture, properties and uses of composites and their components. Alphabetical organization, extensive cross references and a complete index further enhance the utility of this Encyclopedia. The approximately 50 main entries in this Encyclopedia each average 10,000 words in length and have been prepared by 75 leading authorities from industries, universities and research institutes. The contents should be of interest to all those engaged in the manufacture and use of modern, lightweight, tough engineering materials for use in consumer goods, transportation, aerospace, communications and related industrial activities. The contents range over such diverse fields as metals, plastics, coatings and fibers, natural products, and standards and specifications for materials. The book should be an important research reference tool, desk-top information resource and supplementary reading asset for teaching professionals and their students.

M. GRAYSON

NOTE ON CHEMICAL ABSTRACTS SERVICE REGISTRY NUMBERS AND NOMENCLATURE

Chemical Abstracts Service (CAS) Registry Numbers are unique numerical identifiers assigned to substances recorded in the CAS Registry System. They appear in brackets in the *Chemical Abstracts* (CA) substance and formula indexes following the names of compounds. A single compound may have many synonyms in the chemical literature. A simple compound like phenethylamine can be named β-phenylethylamine or, as in *Chemical Abstracts*, benzeneethanamine. The usefulness of the *Encyclopedia* depends on accessibility through the most common correct name of a substance. Because of this diversity in nomenclature careful attention has been given the problem in order to assist the reader as much as possible, especially in locating the systematic CA index name by means of the Registry Number. For this purpose, the reader may refer to the CAS Registry Handbook-Number Section which lists in numerical order the Registry Number with the *Chemical Abstracts* index name and the molecular formula; eg, **458-88-8,** Piperidine, 2-propyl-, (*S*)-, $C_8H_{17}N$; in the *Encyclopedia* this compound would be found under its common name, coniine [*458-88-8*]. The Registry Number is a valuable link for the reader in retrieving additional published information on substances and also as a point of access for such on-line data bases as Chemline, Medline, and Toxline.

In all cases, the CAS Registry Numbers have been given for title compounds in articles and for all compounds in the index. All specific substances indexed in *Chemical Abstracts* since 1965 are included in the CAS Registry System as are a large number of substances derived from a variety of reference works. The CAS Registry System identifies a substance on the basis of an unambiguous computer-language description of its molecular structure including stereochemical detail. The Registry Number is a machine-checkable number (like a Social Security number) assigned in sequential order to each substance as it enters the registry system. The value of the number lies in the fact that it is a concise and unique means of substance identification, which is independent of, and therefore bridges, many systems of chemical nomenclature. For polymers, one Registry Number is used for the entire family; eg, polyoxyethylene (20) sorbitan monolaurate has the same number as all of its polyoxyethylene homologues.

Registry numbers for each substance will be provided in the third edition cumulative index and appear as well in the annual indexes (eg, Alkaloids shows the Registry Number of all alkaloids (title compounds) in a table in the article as well, but the intermediates have their Registry Numbers shown only in the index). Articles such as Analytical methods, Batteries and electric cells, Chemurgy, Distillation, Economic evaluation, and Fluid mechanics have no Registry Numbers in the text.

Cross-references are inserted in the index for many common names and for some systematic names. Trademark names appear in the index. Names that are incorrect, misleading or ambiguous are avoided. Formulas are given very frequently in the text to help in identifying compounds. The spelling and form used, even for industrial names, follow American chemical usage, but not always the usage of *Chemical Abstracts* (eg, *coniine* is used instead of (*S*)-*2-propylpiperidine*, *aniline* instead of *benzenamine*, and *acrylic acid* instead of *2-propenoic acid*).

There are variations in representation of rings in different disciplines. The dye industry does not designate aromaticity or double bonds in rings. All double bonds and aromaticity are shown in the *Encyclopedia* as a matter of course. For example, tetralin has an aromatic ring and a saturated ring and its structure appears in the

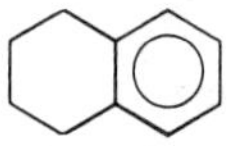

Encyclopedia with its common name, Registry Number enclosed in brackets, and parenthetical CA index name, ie, tetralin, [*119-64-2*] (1,2,3,4-tetrahydronaphthalene). With names and structural formulas, and especially with CAS Registry Numbers the aim is to help the reader have a concise means of substance identification.

CONVERSION FACTORS, ABBREVIATIONS, AND UNIT SYMBOLS

SI Units (Adopted 1960)

A new system of measurement, the International System of Units (abbreviated SI), is being implemented throughout the world. This system is a modernized version of the MKSA (meter, kilogram, second, ampere) system, and its details are published and controlled by an international treaty organization (The International Bureau of Weights and Measures) (1).

SI units are divided into three classes:

BASE UNITS

length	meter[†] (m)
mass[‡]	kilogram (kg)
time	second (s)
electric current	ampere (A)
thermodynamic temperature[§]	kelvin (K)
amount of substance	mole (mol)
luminous intensity	candela (cd)

[†] The spellings "metre" and "litre" are preferred by ASTM; however "-er" are used in the Encyclopedia.

[‡] "Weight" is the commonly used term for "mass."

[§] Wide use is made of "Celsius temperature" (t) defined by

$$t = T - T_0$$

where T is the thermodynamic temperature, expressed in kelvins, and T_0 = 273.15 K by definition. A temperature interval may be expressed in degrees Celsius as well as in kelvins.

SUPPLEMENTARY UNITS

plane angle	radian (rad)
solid angle	steradian (sr)

DERIVED UNITS AND OTHER ACCEPTABLE UNITS

These units are formed by combining base units, supplementary units, and other derived units (2–4). Those derived units having special names and symbols are marked with an asterisk in the list below:

Quantity	*Unit*	*Symbol*	*Acceptable equivalent*
*absorbed dose	gray	Gy	J/kg
acceleration	meter per second squared	m/s^2	
*activity (of ionizing radiation source)	becquerel	Bq	1/s
area	square kilometer	km^2	
	square hectometer	hm^2	ha (hectare)
	square meter	m^2	
*capacitance	farad	F	C/V
concentration (of amount of substance)	mole per cubic meter	mol/m^3	
*conductance	siemens	S	A/V
current density	ampere per square meter	A/m^2	
density, mass density	kilogram per cubic meter	kg/m^3	g/L; mg/cm^3
dipole moment (quantity)	coulomb meter	C·m	
*electric charge, quantity of electricity	coulomb	C	A·s
electric charge density	coulomb per cubic meter	C/m^3	
electric field strength	volt per meter	V/m	
electric flux density	coulomb per square meter	C/m^2	
*electric potential, potential difference, electromotive force	volt	V	W/A
*electric resistance	ohm	Ω	V/A
*energy, work, quantity of heat	megajoule	MJ	
	kilojoule	kJ	
	joule	J	N·m
	electron volt†	eV†	
	kilowatt-hour†	kW·h†	

† This non-SI unit is recognized by the CIPM as having to be retained because of practical importance or use in specialized fields (1).

Quantity	*Unit*	*Symbol*	*Acceptable equivalent*
energy density	joule per cubic meter	J/m^3	
*force	kilonewton	kN	
	newton	N	$kg{\cdot}m/s^2$
*frequency	megahertz	MHz	
	hertz	Hz	1/s
heat capacity, entropy	joule per kelvin	J/K	
heat capacity (specific), specific entropy	joule per kilogram kelvin	J/(kg·K)	
heat transfer coefficient	watt per square meter kelvin	$W/(m^2{\cdot}K)$	
*illuminance	lux	lx	lm/m^2
*inductance	henry	H	Wb/A
linear density	kilogram per meter	kg/m	
luminance	candela per square meter	cd/m^2	
*luminous flux	lumen	lm	cd·sr
magnetic field strength	ampere per meter	A/m	
*magnetic flux	weber	Wb	V·s
*magnetic flux density	tesla	T	Wb/m^2
molar energy	joule per mole	J/mol	
molar entropy, molar heat capacity	joule per mole kelvin	J/(mol·K)	
moment of force, torque	newton meter	N·m	
momentum	kilogram meter per second	kg·m/s	
permeability	henry per meter	H/m	
permittivity	farad per meter	F/m	
*power, heat flow rate, radiant flux	kilowatt	kW	
	watt	W	J/s
power density, heat flux density, irradiance	watt per square meter	W/m^2	
*pressure, stress	megapascal	MPa	
	kilopascal	kPa	
	pascal	Pa	N/m^2
sound level	decibel	dB	
specific energy	joule per kilogram	J/kg	
specific volume	cubic meter per kilogram	m^3/kg	
surface tension	newton per meter	N/m	
thermal conductivity	watt per meter kelvin	W/(m·K)	
velocity	meter per second	m/s	
	kilometer per hour	km/h	
viscosity, dynamic	pascal second	Pa·s	
	millipascal second	mPa·s	
viscosity, kinematic	square meter per second	m^2/s	

Quantity	*Unit*	*Symbol*	*Acceptable equivalent*
	square millimeter per second	mm^2/s	
volume	cubic meter	m^3	
	cubic decimeter	dm^3	L(liter) (5)
	cubic centimeter	cm^3	mL
wave number	1 per meter	m^{-1}	
	1 per centimeter	cm^{-1}	

In addition, there are 16 prefixes used to indicate order of magnitude, as follows:

Multiplication factor	*Prefix*	*Symbol*	*Note*
10^{18}	exa	E	
10^{15}	peta	P	
10^{12}	tera	T	
10^{9}	giga	G	
10^{6}	mega	M	
10^{3}	kilo	k	
10^{2}	hecto	h[a]	[a] Although hecto, deka, deci, and centi are SI prefixes, their use should be avoided except for SI unit-multiples for area and volume and nontechnical use of centimeter, as for body and clothing measurement.
10	deka	da[a]	
10^{-1}	deci	d[a]	
10^{-2}	centi	c[a]	
10^{-3}	milli	m	
10^{-6}	micro	μ	
10^{-9}	nano	n	
10^{-12}	pico	p	
10^{-15}	femto	f	
10^{-18}	atto	a	

For a complete description of SI and its use the reader is referred to ASTM E 380 (4) and the article Units and Conversion Factors which will appear in a later volume of the *Encyclopedia*.

A representative list of conversion factors from non-SI to SI units is presented herewith. Factors are given to four significant figures. Exact relationships are followed by a dagger. A more complete list is given in ASTM E 380-79(4) and ANSI Z210.1-1976 (6).

Conversion Factors to SI Units

To convert from	*To*	*Multiply by*
acre	square meter (m^2)	4.047×10^3
angstrom	meter (m)	1.0×10^{-10}†
are	square meter (m^2)	1.0×10^{2}†
astronomical unit	meter (m)	1.496×10^{11}
atmosphere	pascal (Pa)	1.013×10^5
bar	pascal (Pa)	1.0×10^{5}†
barn	square meter (m^2)	1.0×10^{-28}†

† Exact.

To convert from	*To*	*Multiply by*
barrel (42 U.S. liquid gallons)	cubic meter (m^3)	0.1590
Bohr magneton (μ_β)	J/T	9.274×10^{-24}
Btu (International Table)	joule (J)	1.055×10^3
Btu (mean)	joule (J)	1.056×10^3
Btu (thermochemical)	joule (J)	1.054×10^3
bushel	cubic meter (m^3)	3.524×10^{-2}
calorie (International Table)	joule (J)	4.187
calorie (mean)	joule (J)	4.190
calorie (thermochemical)	joule (J)	4.184†
centipoise	pascal second (Pa·s)	1.0×10^{-3}†
centistoke	square millimeter per second (mm^2/s)	1.0†
cfm (cubic foot per minute)	cubic meter per second (m^3/s)	4.72×10^{-4}
cubic inch	cubic meter (m^3)	1.639×10^{-5}
cubic foot	cubic meter (m^3)	2.832×10^{-2}
cubic yard	cubic meter (m^3)	0.7646
curie	becquerel (Bq)	3.70×10^{10}†
debye	coulomb·meter (C·m)	3.336×10^{-30}
degree (angle)	radian (rad)	1.745×10^{-2}
denier (international)	kilogram per meter (kg/m)	1.111×10^{-7}
	tex‡	0.1111
dram (apothecaries')	kilogram (kg)	3.888×10^{-3}
dram (avoirdupois)	kilogram (kg)	1.772×10^{-3}
dram (U.S. fluid)	cubic meter (m^3)	3.697×10^{-6}
dyne	newton (N)	1.0×10^{-5}†
dyne/cm	newton per meter (N/m)	1.0×10^{-3}†
electron volt	joule (J)	1.602×10^{-19}
erg	joule (J)	1.0×10^{-7}†
fathom	meter (m)	1.829
fluid ounce (U.S.)	cubic meter (m^3)	2.957×10^{-5}
foot	meter (m)	0.3048†
footcandle	lux (lx)	10.76
furlong	meter (m)	2.012×10^{-2}
gal	meter per second squared (m/s^2)	1.0×10^{-2}†
gallon (U.S. dry)	cubic meter (m^3)	4.405×10^{-3}
gallon (U.S. liquid)	cubic meter (m^3)	3.785×10^{-3}
gallon per minute (gpm)	cubic meter per second (m^3/s)	6.308×10^{-5}
	cubic meter per hour (m^3/h)	0.2271
gauss	tesla (T)	1.0×10^{-4}
gilbert	ampere (A)	0.7958
gill (U.S.)	cubic meter (m^3)	1.183×10^{-4}
grad	radian	1.571×10^{-2}
grain	kilogram (kg)	6.480×10^{-5}
gram force per denier	newton per tex (N/tex)	8.826×10^{-2}

† Exact.

‡ See footnote on p. xvi.

To convert from	*To*	*Multiply by*
hectare	square meter (m^2)	1.0×10^{4}†
horsepower (550 ft·lbf/s)	watt (W)	7.457×10^{2}
horsepower (boiler)	watt (W)	9.810×10^{3}
horsepower (electric)	watt (W)	7.46×10^{2}†
hundredweight (long)	kilogram (kg)	50.80
hundredweight (short)	kilogram (kg)	45.36
inch	meter (m)	2.54×10^{-2}†
inch of mercury (32°F)	pascal (Pa)	3.386×10^{3}
inch of water (39.2°F)	pascal (Pa)	2.491×10^{2}
kilogram force	newton (N)	9.807
kilowatt hour	megajoule (MJ)	3.6†
kip	newton (N)	4.48×10^{3}
knot (international)	meter per second (m/s)	0.5144
lambert	candela per square meter (cd/m^2)	3.183×10^{3}
league (British nautical)	meter (m)	5.559×10^{3}
league (statute)	meter (m)	4.828×10^{3}
light year	meter (m)	9.461×10^{15}
liter (for fluids only)	cubic meter (m^3)	1.0×10^{-3}†
maxwell	weber (Wb)	1.0×10^{-8}†
micron	meter (m)	1.0×10^{-6}†
mil	meter (m)	2.54×10^{-5}†
mile (statute)	meter (m)	1.609×10^{3}
mile (U.S. nautical)	meter (m)	1.852×10^{3}†
mile per hour	meter per second (m/s)	0.4470
millibar	pascal (Pa)	1.0×10^{2}
millimeter of mercury (0°C)	pascal (Pa)	1.333×10^{2}†
minute (angular)	radian	2.909×10^{-4}
myriagram	kilogram (kg)	10
myriameter	kilometer (km)	10
oersted	ampere per meter (A/m)	79.58
ounce (avoirdupois)	kilogram (kg)	2.835×10^{-2}
ounce (troy)	kilogram (kg)	3.110×10^{-2}
ounce (U.S. fluid)	cubic meter (m^3)	2.957×10^{-5}
ounce-force	newton (N)	0.2780
peck (U.S.)	cubic meter (m^3)	8.810×10^{-3}
pennyweight	kilogram (kg)	1.555×10^{-3}
pint (U.S. dry)	cubic meter (m^3)	5.506×10^{-4}
pint (U.S. liquid)	cubic meter (m^3)	4.732×10^{-4}
poise (absolute viscosity)	pascal second (Pa·s)	0.10†
pound (avoirdupois)	kilogram (kg)	0.4536
pound (troy)	kilogram (kg)	0.3732
poundal	newton (N)	0.1383
pound-force	newton (N)	4.448
pound per square inch (psi)	pascal (Pa)	6.895×10^{3}
quart (U.S. dry)	cubic meter (m^3)	1.101×10^{-3}

† Exact.

To convert from	*To*	*Multiply by*
quart (U.S. liquid)	cubic meter (m^3)	9.464×10^{-4}
quintal	kilogram (kg)	$1.0 \times 10^{2\dagger}$
rad	gray (Gy)	$1.0 \times 10^{-2\dagger}$
rod	meter (m)	5.029
roentgen	coulomb per kilogram (C/kg)	2.58×10^{-4}
second (angle)	radian (rad)	4.848×10^{-6}
section	square meter (m^2)	2.590×10^{6}
slug	kilogram (kg)	14.59
spherical candle power	lumen (lm)	12.57
square inch	square meter (m^2)	6.452×10^{-4}
square foot	square meter (m^2)	9.290×10^{-2}
square mile	square meter (m^2)	2.590×10^{6}
square yard	square meter (m^2)	0.8361
stere	cubic meter (m^3)	$1.0^{\dagger}$
stokes (kinematic viscosity)	square meter per second (m^2/s)	$1.0 \times 10^{-4\dagger}$
tex	kilogram per meter (kg/m)	$1.0 \times 10^{-6\dagger}$
ton (long, 2240 pounds)	kilogram (kg)	1.016×10^{3}
ton (metric)	kilogram (kg)	$1.0 \times 10^{3\dagger}$
ton (short, 2000 pounds)	kilogram (kg)	9.072×10^{2}
torr	pascal (Pa)	1.333×10^{2}
unit pole	weber (Wb)	1.257×10^{-7}
yard	meter (m)	$0.9144^{\dagger}$

Abbreviations and Unit Symbols

Following is a list of commonly used abbreviations and unit symbols appropriate for use in the *Encyclopedia*. In general they agree with those listed in *American National Standard Abbreviations for Use on Drawings and in Text* (*ANSI Y1.1*) (6) and *American National Standard Letter Symbols for Units in Science and Technology* (*ANSI Y10*) (6). Also included is a list of acronyms for a number of private and government organizations as well as common industrial solvents, polymers, and other chemicals.

Rules for Writing Unit Symbols (4):

1. Unit symbols should be printed in upright letters (roman) regardless of the type style used in the surrounding text.

2. Unit symbols are unaltered in the plural.

3. Unit symbols are not followed by a period except when used as the end of a sentence.

4. Letter unit symbols are generally written in lower-case (eg, cd for candela) unless the unit name has been derived from a proper name, in which case the first letter of the symbol is capitalized (W,Pa). Prefix and unit symbols retain their prescribed form regardless of the surrounding typography.

5. In the complete expression for a quantity, a space should be left between the numerical value and the unit symbol. For example, write 2.37 lm, *not* 2.37lm, and 35 mm, *not* 35mm. When the quantity is used in an adjectival sense, a hyphen is often used, for example, 35-mm film. *Exception:* No space is left between the numerical value and the symbols for degree, minute, and second of plane angle, and degree Celsius.

6. No space is used between the prefix and unit symbols (eg, kg).

7. Symbols, not abbreviations, should be used for units. For example, use "A," not "amp," for ampere.

8. When multiplying unit symbols, use a raised dot:

N·m for newton meter

In the case of W·h, the dot may be omitted, thus:

Wh

An exception to this practice is made for computer printouts, automatic typewriter work, etc, where the raised dot is not possible, and a dot on the line may be used.

9. When dividing unit symbols use one of the following forms:

$$\mathrm{m/s} \text{ or } \mathrm{m{\cdot}s^{-1}} \text{ or } \frac{\mathrm{m}}{\mathrm{s}}$$

In no case should more than one slash be used in the same expression unless parentheses are inserted to avoid ambiguity. For example, write:

$$\mathrm{J/(mol{\cdot}K)} \text{ or } \mathrm{J{\cdot}mol^{-1}\cdot K^{-1}} \text{ or } \mathrm{(J/mol)/K}$$

but *not*

J/mol/K

10. Do not mix symbols and unit names in the same expression. Write:

joules per kilogram *or* J/kg *or* $\mathrm{J{\cdot}kg^{-1}}$

but *not*

joules/kilogram *nor* joules/kg *nor* $\mathrm{joules{\cdot}kg^{-1}}$

ABBREVIATIONS AND UNITS

A	ampere
A	anion (eg, H*A*); mass number
a	atto (prefix for 10^{-18})
AATCC	American Association of Textile Chemists and Colorists
ABS	acrylonitrile–butadiene–styrene
abs	absolute
ac	alternating current, *n.*
a-c	alternating current, *adj.*
ac-	alicyclic
acac	acetylacetonate
ACGIH	American Conference of Governmental Industrial Hygienists
ACS	American Chemical Society
AGA	American Gas Association
Ah	ampere hour
AIChE	American Institute of Chemical Engineers
AIME	American Institute of Mining, Metallurgical, and Petroleum Engineers
AIP	American Institute of Physics
AISI	American Iron and Steel Institute
alc	alcohol(ic)
Alk	alkyl
alk	alkaline (not alkali)
amt	amount
amu	atomic mass unit
ANSI	American National Standards Institute
AO	atomic orbital
AOAC	Association of Official Analytical Chemists
AOCS	American Oil Chemist's Society
APHA	American Public Health Association

API	American Petroleum Institute
aq	aqueous
Ar	aryl
ar-	aromatic
as-	asymmetric(al)
ASHRAE	American Society of Heating, Refrigerating, and Air Conditioning Engineers
ASM	American Society for Metals
ASME	American Society of Mechanical Engineers
ASTM	American Society for Testing and Materials
at no.	atomic number
at wt	atomic weight
av(g)	average
AWS	American Welding Society
b	bonding orbital
bbl	barrel
bcc	body-centered cubic
BCT	body-centered tetragonal
Bé	Baumé
BET	Brunauer-Emmett-Teller (adsorption equation)
bid	twice daily
Boc	*t*-butyloxycarbonyl
BOD	biochemical (biological) oxygen demand
bp	boiling point
Bq	becquerel
C	coulomb
°C	degree Celsius
C-	denoting attachment to carbon
c	centi (prefix for 10^{-2})
c	critical
ca	circa (approximately)
cd	candela; current density; circular dichroism
CFR	Code of Federal Regulations
cgs	centimeter–gram–second
CI	Color Index
cis-	isomer in which substituted groups are on same side of double bond between C atoms
cl	carload
cm	centimeter
cmil	circular mil
cmpd	compound
CNS	central nervous system
CoA	coenzyme A
COD	chemical oxygen demand
coml	commercial(ly)
cp	chemically pure
cph	close-packed hexagonal
CPSC	Consumer Product Safety Commission
cryst	crystalline
cub	cubic
D	Debye
D-	denoting configurational relationship
d	differential operator
d-	dextro-, dextrorotatory
da	deka (prefix for 10^1)
dB	decibel
dc	direct current, *n.*
d-c	direct current, *adj.*
dec	decompose
detd	determined
detn	determination
Di	didymium, a mixture of all lanthanons
dia	diameter
dil	dilute
DIN	Deutsche Industrie Normen
dl-; DL-	racemic
DMA	dimethylacetamide
DMF	dimethylformamide
DMG	dimethyl glyoxime
DMSO	dimethyl sulfoxide
DOD	Department of Defense
DOE	Department of Energy
DOT	Department of Transportation
DP	degree of polymerization
dp	dew point
DPH	diamond pyramid hardness
dstl(d)	distill(ed)
dta	differential thermal analysis
(*E*)-	entgegen; opposed
ϵ	dielectric constant (unitless number)
e	electron

ECU	electrochemical unit
ed.	edited, edition, editor
ED	effective dose
EDTA	ethylenediaminetetraacetic acid
emf	electromotive force
emu	electromagnetic unit
en	ethylene diamine
eng	engineering
EPA	Environmental Protection Agency
epr	electron paramagnetic resonance
eq.	equation
esp	especially
esr	electron-spin resonance
est(d)	estimate(d)
estn	estimation
esu	electrostatic unit
exp	experiment, experimental
ext(d)	extract(ed)
F	farad (capacitance)
F	faraday (96,487 C)
f	femto (prefix for 10^{-15})
FAO	Food and Agriculture Organization (United Nations)
fcc	face-centered cubic
FDA	Food and Drug Administration
FEA	Federal Energy Administration
fob	free on board
fp	freezing point
FPC	Federal Power Commission
FRB	Federal Reserve Board
frz	freezing
G	giga (prefix for 10^9)
G	gravitational constant = 6.67×10^{11} N·m²/kg²
g	gram
(g)	gas, only as in $H_2O(g)$
g	gravitational acceleration
gem-	geminal
glc	gas-liquid chromatography
g-mol wt; gmw	gram-molecular weight
GNP	gross national product
gpc	gel-permeation chromatography

GRAS	Generally Recognized as Safe
grd	ground
Gy	gray
H	henry
h	hour; hecto (prefix for 10^2)
ha	hectare
HB	Brinell hardness number
Hb	hemoglobin
hcp	hexagonal close-packed
hex	hexagonal
HK	Knoop hardness number
HRC	Rockwell hardness (C scale)
HV	Vickers hardness number
hyd	hydrated, hydrous
hyg	hygroscopic
Hz	hertz
^{i}(eg, Pr^i)	iso (eg, isopropyl)
i-	inactive (eg, *i*-methionine)
IACS	International Annealed Copper Standard
ibp	initial boiling point
IC	inhibitory concentration
ICC	Interstate Commerce Commission
ICT	International Critical Table
ID	inside diameter; infective dose
ip	intraperitoneal
IPS	iron pipe size
IPTS	International Practical Temperature Scale (NBS)
ir	infrared
IRLG	Interagency Regulatory Liaison Group
ISO	International Organization for Standardization
IU	International Unit
IUPAC	International Union of Pure and Applied Chemistry
IV	iodine value
iv	intravenous
J	joule
K	kelvin
k	kilo (prefix for 10^3)
kg	kilogram
L	denoting configurational relationship
L	liter (for fluids only)(5)
l-	*levo*-, levorotatory
(l)	liquid, only as in $NH_3(l)$

LC_{50}	conc lethal to 50% of the animals tested
LCAO	linear combination of atomic orbitals
LCD	liquid crystal display
lcl	less than carload lots
LD_{50}	dose lethal to 50% of the animals tested
LED	light-emitting diode
liq	liquid
lm	lumen
ln	logarithm (natural)
LNG	liquefied natural gas
log	logarithm (common)
LPG	liquefied petroleum gas
ltl	less than truckload lots
lx	lux
M	mega (prefix for 10^6); metal (as in MA)
M	molar; actual mass
$\overline{M}_w$	weight-average mol wt
$\overline{M}_n$	number-average mol wt
m	meter; milli (prefix for 10^{-3})
m	molal
m-	meta
max	maximum
MCA	Chemical Manufacturers' Association (was Manufacturing Chemists Association)
MEK	methyl ethyl ketone
meq	milliequivalent
mfd	manufactured
mfg	manufacturing
mfr	manufacturer
MIBC	methyl isobutyl carbinol
MIBK	methyl isobutyl ketone
MIC	minimum inhibiting concentration
min	minute; minimum
mL	milliliter
MLD	minimum lethal dose
MO	molecular orbital
mo	month
mol	mole
mol wt	molecular weight
mp	melting point
MR	molar refraction
ms	mass spectrum

mxt	mixture
μ	micro (prefix for 10^{-6})
N	newton (force)
N	normal (concentration); neutron number
N-	denoting attachment to nitrogen
n (as n_D^{20})	index of refraction (for 20°C and sodium light)
^{n} (as Bu^n), n-	normal (straight-chain structure)
n	neutron
n	nano (prefix for 10^9)
na	not available
NAS	National Academy of Sciences
NASA	National Aeronautics and Space Administration
nat	natural
NBS	National Bureau of Standards
neg	negative
NF	*National Formulary*
NIH	National Institutes of Health
NIOSH	National Institute of Occupational Safety and Health
nmr	nuclear magnetic resonance
NND	New and Nonofficial Drugs (AMA)
no.	number
NOI-(BN)	not otherwise indexed (by name)
NOS	not otherwise specified
nqr	nuclear quadruple resonance
NRC	Nuclear Regulatory Commission; National Research Council
NRI	New Ring Index
NSF	National Science Foundation
NTA	nitrilotriacetic acid
NTP	normal temperature and pressure (25°C and 101.3 kPa or 1 atm)
NTSB	National Transportation Safety Board
O-	denoting attachment to oxygen

o-	ortho
OD	outside diameter
OPEC	Organization of Petroleum Exporting Countries
o-phen	*o*-phenanthridine
OSHA	Occupational Safety and Health Administration
owf	on weight of fiber
Ω	ohm
P	peta (prefix for 10^{15})
p	pico (prefix for 10^{-12})
p-	para
p	proton
p.	page
Pa	pascal (pressure)
pd	potential difference
pH	negative logarithm of the effective hydrogen ion concentration
phr	parts per hundred of resin (rubber)
p-i-n	positive-intrinsic-negative
pmr	proton magnetic resonance
p-n	positive-negative
po	per os (oral)
POP	polyoxypropylene
pos	positive
pp.	pages
ppb	parts per billion (10^9)
ppm	parts per million (10^6)
ppmv	parts per million by volume
ppmwt	parts per million by weight
PPO	poly(phenyl oxide)
ppt(d)	precipitate(d)
pptn	precipitation
Pr (no.)	foreign prototype (number)
pt	point; part
PVC	poly(vinyl chloride)
pwd	powder
py	pyridine
qv	quod vide (which see)
R	univalent hydrocarbon radical
(*R*)-	rectus (clockwise configuration)
r	precision of data
rad	radian; radius
rds	rate determining step

ref.	reference
rf	radio frequency, *n.*
r-f	radio frequency, *adj.*
rh	relative humidity
RI	Ring Index
rms	root-mean square
rpm	rotations per minute
rps	revolutions per second
RT	room temperature
^{s} (eg, Bu^s); *sec*-	secondary (eg, secondary butyl)
S	siemens
(*S*)-	sinister (counterclockwise configuration)
S-	denoting attachment to sulfur
s-	symmetric(al)
s	second
(s)	solid, only as in $H_2O(s)$
SAE	Society of Automotive Engineers
SAN	styrene–acrylonitrile
sat(d)	saturate(d)
satn	saturation
SBS	styrene–butadiene–styrene
sc	subcutaneous
SCF	self-consistent field; standard cubic feet
Sch	Schultz number
SFs	Saybolt Furol seconds
SI	Le Système International d'Unités (International System of Units)
sl sol	slightly soluble
sol	soluble
soln	solution
soly	solubility
sp	specific; species
sp gr	specific gravity
sr	steradian
std	standard
STP	standard temperature and pressure (0°C and 101.3 kPa)
sub	sublime(s)
SUs	Saybolt Universal seconds
syn	synthetic

^{t} (eg, Bu^{t}), *t*-, *tert*-	tertiary (eg, tertiary butyl)	Twad	Twaddell
T	tera (prefix for 10^{12}); tesla (magnetic flux density)	UL	Underwriters' Laboratory
t	metric ton (tonne); temperature	USDA	United States Department of Agriculture
TAPPI	Technical Association of the Pulp and Paper Industry	USP	*United States Pharmacopeia*
tex	tex (linear density)	uv	ultraviolet
T_g	glass-transition temperature	V	volt (emf)
tga	thermogravimetric analysis	var	variable
THF	tetrahydrofuran	*vic*-	vicinal
tlc	thin layer chromatography	vol	volume (not volatile)
TLV	threshold limit value	vs	versus
trans-	isomer in which substituted groups are on opposite sides of double bond between C atoms	v sol	very soluble
TSCA	Toxic Substance Control Act	W	watt
TWA	time-weighted average	Wb	Weber
		Wh	watt hour
		WHO	World Health Organization (United Nations)
		wk	week
		yr	year
		(*Z*)-	zusammen; together; atomic number

Non-SI (Unacceptable and Obsolete) Units		*Use*
Å	angstrom	nm
at	atmosphere, technical	Pa
atm	atmosphere, standard	Pa
b	barn	cm^2
bar†	bar	Pa
bbl	barrel	m^3
bhp	brake horsepower	W
Btu	British thermal unit	J
bu	bushel	m^3; L
cal	calorie	J
cfm	cubic foot per minute	m^3/s
Ci	curie	Bq
cSt	centistokes	mm^2/s
c/s	cycle per second	Hz
cu	cubic	exponential form
D	debye	C·m
den	denier	tex
dr	dram	kg
dyn	dyne	N
dyn/cm	dyne per centimeter	mN/m
erg	erg	J
eu	entropy unit	J/K
°F	degree Fahrenheit	°C; K
fc	footcandle	lx
fl	footlambert	lx
fl oz	fluid ounce	m^3; L
ft	foot	m
ft·lbf	foot pound-force	J

† Do not use bar (10^5Pa) or millibar (10^2Pa) because they are not SI units, and are accepted internationally only for a limited time in special fields because of existing usage.

Non-SI (Unacceptable and Obsolete) Units		*Use*
gf den	gram-force per denier	N/tex
G	gauss	T
Gal	gal	m/s^2
gal	gallon	m^3; L
Gb	gilbert	A
gpm	gallon per minute	(m^3/s); (m^3/h)
gr	grain	kg
hp	horsepower	W
ihp	indicated horsepower	W
in.	inch	m
in. Hg	inch of mercury	Pa
in. H_2O	inch of water	Pa
in.-lbf	inch pound-force	J
kcal	kilogram-calorie	J
kgf	kilogram-force	N
kilo	for kilogram	kg
L	lambert	lx
lb	pound	kg
lbf	pound-force	N
mho	mho	S
mi	mile	m
MM	million	M
mm Hg	millimeter of mercury	Pa
mμ	millimicron	nm
mph	miles per hour	km/h
μ	micron	μm
Oe	oersted	A/m
oz	ounce	kg
ozf	ounce-force	N
η	poise	Pa·s
P	poise	Pa·s
ph	phot	lx
psi	pounds-force per square inch	Pa
psia	pounds-force per square inch absolute	Pa
psig	pounds-force per square inch gauge	Pa
qt	quart	m^3; L
°R	degree Rankine	K
rd	rad	Gy
sb	stilb	lx
SCF	standard cubic foot	m^3
sq	square	exponential form
thm	therm	J
yd	yard	m

BIBLIOGRAPHY

1. The International Bureau of Weights and Measures, BIPM (Parc de Saint-Cloud, France) is described on page 22 of Ref. 4. This bureau operates under the exclusive supervision of the International Committee of Weights and Measures (CIPM).
2. *Metric Editorial Guide* (*ANMC-78-1*) 3rd ed., American National Metric Council, 1625 Massachusetts Ave. N.W., Washington, D.C. 20036, 1978.
3. *SI Units and Recommendations for the Use of Their Multiples and of Certain Other Units* (*ISO 1000-1981*), American National Standards Institute, 1430 Broadway, New York, N. Y. 10018, 1981.
4. Based on *ASTM E 380-82* (*Standard for Metric Practice*), American Society for Testing and Materials, 1916 Race Street, Philadelphia, Pa. 19103, 1982.
5. *Fed. Regist.*, Dec. 10, 1976 (41 FR 36414).
6. For ANSI address, see Ref. 3.

R. P. LUKENS
American Society for Testing and Materials

ABLATIVE MATERIALS

Ablation

The word ablation is used to describe the erosion and disintegration of meteors resulting from the intense heat generated by passage through the atmosphere at high velocities. During development of the first operational ballistic missiles at Peenemunde, Germany, aerodynamic heating of the missile skin caused many to disintegrate during test; even with redesign, about 650 of the 3550 operational V-2s launched failed to reach their targets because of such air disintegration (1). The development and study of materials which protect the payload and structure of missiles, earth satellites, or space probes from damage during ablation is a relatively new area of chemical technology. The same principles of thermal protection also apply to other applications, such as the protection of rocket nozzles from attrition by propellant gases and protection from laser beams.

The Ablation Environment. The selection of materials to resist ablation is highly dependent on the intensity and duration of heating, which can be much more severe than the hot environments encountered in other technologies. Figure 1 illustrates schematically some typical reentry velocities and heating rates at different altitudes, representative of different types of reentry bodies (2).

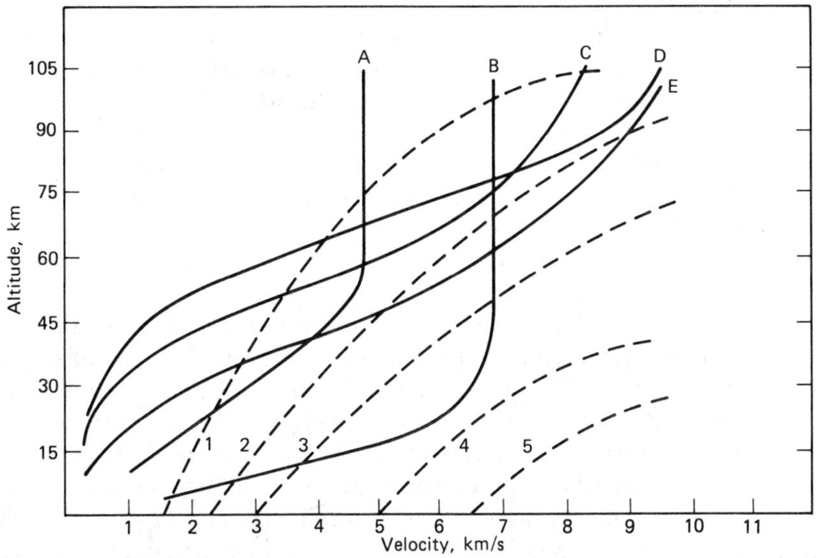

Figure 1. Hypervelocity flight environments (2). A, intermediate range vehicle; B, intercontinental range vehicle; C, manned orbital vehicle; D, satellite; E, space probe. The dashed lines represent calculated heating rates for laminar flow about a body with a nose radius of 15.24 cm, and corresponding radiation equilibrium temperatures (calculated for an emissivity of 0.9) for each numbered curve are shown below.

Curve	Heating rate, W/cm^2	Radiation equilibrium temperature, K
1	57	1830
2	230	2590
3	570	3260
4	2300	4590
5	5700	5810

Near the leading point, or stagnation point, of a hypervelocity vehicle, gas molecules are abruptly accelerated to the velocity of the vehicle. This translational energy appears as a temperature rise, although dissociation and reactions of the gas molecules also occur; the specific energy content of the gas is approximately proportional to the square of the velocity. The gross heating rate depends not only on this energy but upon the local pressure, which varies with velocity, altitude (gas density), and position on the vehicle, decreasing from stagnation point to trailing edge. The heat-transfer rate from gas-boundary layer to surface also depends on whether the stream is in laminar flow or is turbulent; in the latter case, heating is more severe. The duration of intense heat is longer if the flight path has a low angle with the horizontal, although the rate of heating is less because the body decelerates at high altitudes where pressure is low. Heating rate also depends on the vehicle design, since a light blunt body decelerates at a higher altitude than a heavy pointed body having a higher ratio of weight to drag.

In Figure 1, the numbered curves are lines of equal heating rate calculated for laminar heating of a blunt body, as described in the caption, at the different combinations of altitude and velocity. The lettered curves show velocity–altitude trajectories for typical reentry conditions. Note that the most severe condition B develops temperatures over 4000 K for radiation equilibrium of the surface, which limits the temperature of the material.

Simulation of the more severe environments for laboratory tests has been the subject of considerable development. Various chemical flames, plasma arcs, solar furnaces, lasers, and electric arc air heaters have been used as heat sources, often in combination with wind tunnels. Feeding devices have often been incorporated to maintain samples in a controlled or constant heat flux while they recede. In some tests, samples have been fired down a ballistics range and profiles photographed at successive stations while the surface recedes. In almost all cases, some aspects of the real environment are lacking, however, so that full-scale tests are eventually necessary to confirm the selection. A reliable prediction of performance under conditions which cannot be adequately simulated usually requires study of the mechanisms involved in ablation.

Because of the diversity in environments, many types of materials have been utilized where ablation is involved. During reentry, a protective heat shield must perform as an aerodynamic body, a predictably decomposing ablative material, a structural component and an insulator (3–4). In missile nose tips, where shape change may affect the trajectory, a material such as graphite may be selected primarily to minimize recession under very high heat fluxes (5–6). If the environment involves erosion by snow or ice particles, a refractory metal may be considered (7). On the other hand, ability to insulate or protect the payload from the heat generated, with minimum weight, is ordinarily the primary consideration, and a variety of ablative plastics and fiber-reinforced plastic composites have been developed (8–10). This article reviews the principles involved in material selection for the most common areas of application.

Metals and Ceramics as Ablative Materials

If one surface of a large slab of solid metal is exposed to a very hot stream of inert gas, its surface temperature rises at a rate proportional to the difference between the rate at which heat reaches the surface and the rate of heat conduction into the material. Eventually, surface melting will occur, and hot liquid may drip or be blown off the surface. After this, the face of the solid will recede at a steady rate which depends on the latent heat of melting of the metal, its heat capacity, and its thermal conductivity.

The initial, transient phase of this process is the basis for protection against heat pulses of relatively short duration by using a heat sink. Beryllium [*7440-41-7*], copper [*7440-50-8*], tungsten [*7440-33-7*], molybdenum [*7439-98-7*], and even steel have functioned in this way. Refractory carbides, borides, nitrides, and oxides, beryllium oxide [*1304-56-9*] in particular, have also been considered because of high melting point, high thermal conductivity, and high specific heat. In some designs, fluid is forced through the pores of such a material to provide transpiration cooling.

Tungsten and molybdenum have been evaluated as nose tip materials, and recession rates of samples ablated in both flight and laboratory tests have been correlated by analysis of the mechanisms involved (7). The formation of volatile oxide species is significant with these materials, and it could be assumed that the melt layers on the surface, either of oxide at low temperatures or of the metal in the steady-state region, were infinitesimally thin.

The Ablation of Graphite and Other Carbons

The element carbon [*7440-44-0*], due to its high heat capacity per unit weight, high energy of vaporization, and high temperature and pressure required for melting, has the highest "heat of ablation" (energy absorbed per mass lost) of any material, provided that mechanical removal of particulates does not occur. Factors involved in predicting recession rates, which usually must be minimized, are reviewed.

The melting temperature and pressure of the solid–liquid–vapor triple point is important because reradiation of energy absorbed from the gas stream is limited by the surface temperature. Several studies have shown that wetting droplets of liquid form only if the pressure of the surrounding gas exceeds 10 MPa (100 atm); solidified liquid is identified by graphite [*7782-42-5*] crystals having basal planes which grow in the direction of solidification (11). The most recent determination, by laser heating of pyrolytic graphite, indicated a triple point at 4130 ± 30 K and 12 ± 1 MPa (120 atm), with little change in temperature up to 21.5 MPa (215 atm) (12).

However, other studies of laser-heated spinning-pyrolytic graphite rods (13–15) have shown evidence of liquid formation at a lower temperature, 3780 ± 30 K, at gas pressures as low as 20 kPa (0.2 atm); the phenomena were associated with the formation of a variety of nongraphitic carbon structures, primarily linear polymorphs, above about 2600 K with vapor pressures different from that of graphite. Similar studies on industrial graphites (15) indicated that liquid formation at grain boundaries during ablation may contribute to the ejection of particulate material observed in many types of graphite heated to vaporization temperatures (15–17) and postulated as an important mechanism in carbon arcs (18). Spontaneous emission of particulates may also result from the expansion of gas trapped in closed pores during manufacture of graphite (19–20).

Such mechanisms of ablation inmaterials composed of pure carbon have introduced uncertainties into calculations based on much of the vaporization data in the literature (21). Although modified data for the free energy functions and vapor pressures of the molecular products of vaporization (primarily C_3) have appeared recently (22–25), the JANAF tables (26) have been used in most studies and have permitted good agreement with vaporization rates in arcs and laser beams (16,27). Calculated rates have also generally matched tests intended to simulate reentry conditions, where chemical reaction with dissociated species in the boundary layer of gas is important (28–30). The calculated heating of carbon–carbon heat shields during flight has provided good agreement with temperatures measured by thermocouples on the backface (30). However, these comparisons have not covered some conditions of heat flux and pressure which may be encountered under more severe reentry trajectories.

Surface roughening of graphite affects the surface heating, if flow changes from laminar to turbulent conditions, and thus can increase the recession rate. Preferential vaporization in subsurface porosity is a mechanism for such roughening (31), and the proportion of large pores in the material affects ablation rate (32). In fiber reinforced composites, spalling of bundles from the surface or large porosity can affect ablation (33), but performance is otherwise similar to graphite (32). Bonding of the constituents can be important, and a carbon–carbon composite processed at low temperatures has shown resistance to laser beam penetration superior to industrial graphites (34). Pyrolytic graphite, which is free of open porosity, is superior to the other types in resistance to laser penetration (34).

Ablation in environments like the combustion gases in rocket nozzle throats can be predicted, as in reentry, by analysis of the concentration of reacting species available by diffusion through a boundary layer of reaction products (35). The relative temperatures of surface and gas are important, because the proportion of reactive, dissociated species among the molecules which reach the surface is affected (36). For example, reaction rates of pyrolytic graphite plates heated in a cold gas stream *decreased* with increasing temperature in the range 1800–2500 K for CO_2 at 110 kPa (1.09 atm), and in the range 2400–2700 K for steam at 130 kPa (1.28 atm) (37). Rotating pyrolytic graphite discs, inductively heated in oxygen–argon mixtures, showed more rapid reaction on cut surfaces (edges of graphite basal planes exposed) than on as-deposited surfaces, which had fewer reacting sites (38). In contrast, tests by one of the authors (Stover) demonstrate that pyrolytic graphite heated in hydrogen–oxygen flames, where reactive species are present in the gas initially, instead of being created on the surface by dissociation of molecules, shows no decrease with increasing temperature, and no effect of AB (edge) versus C (basal plane) orientations (Fig. 2). The tests, in which penetration rates of slabs heated in opposed flames were measured, showed a higher rate for a porous industrial graphite (ATJ) susceptible to loss of particles, and a lower rate for tungsten, where more oxygen is required to remove an atom (W reacting to WO_3 [*1314-35-8*]). Rate also varied with flame composition at the same surface temperature. Thus, proper simulation of the environment is essential in tests of graphite ablation below temperatures where vaporization is predominant.

Ablative Plastics and Organic Insulations

Gas evolution from an ablative material is important in reentry, because it thickens the boundary layer and interferes with the convective transfer of heat to the surface. Such blocking action can reduce the net heating of the solid by more than 50% in many practical situations.

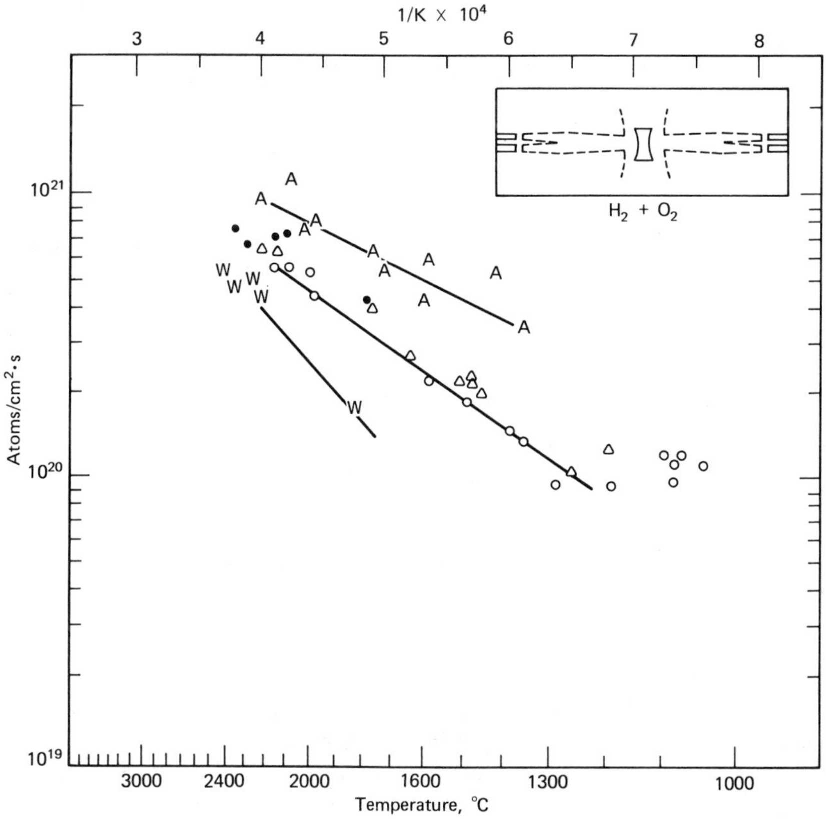

Figure 2. Rate of penetration of plates by flames from opposing hydrogen–oxygen welding torches, multiplied by bulk density and atomic weight, and plotted vs surface temperature at the point of impingement, corrected for emissivity; rates for pyrolytic graphite (PG) slabs cut from thick plate (A direction), ●, are compared with plates with the deposition surface exposed (C direction) without boron, ○, and with the addition of 0.6 wt % boron, △, which improves resistance to oxidation in air; A, ATJ graphite (Union Carbide Corp.); and W, tungsten data under the same conditions.

This principle is utilized with thermoplastic ablative materials such as polytetrafluoroethylene [*9002-84-0*] or Teflon, which decomposes above about 500°C to form the volatile monomer without leaving a solid residue. This is useful where uniform, clean removal of the ablative material is desired. A low weight heat shield for a Venus entry probe has been designed from Teflon with a reflective coating on the back face for further insulation (39).

In most cases, however, formation of a porous solid at the surface is desired. Elastomeric shield materials (ESM) were developed for applications where heat flux and aerodynamic shear forces are relatively low and thermal insulation is a primary requirement. For example, General Electric Co. RTV 560 is a foamed silicone elastomer loaded with silicon dioxide [*7631-86-9*] and iron oxide [*1309-38-1*] (Fe_2O_3) particles, which decomposes to a similar foam of SiO_2, SiC, and $FeSiO_3$ (40). Elastomers may contain hollow microspheres of silicon dioxide or phenolic resin to control void size, and are often contained in fiber-reinforced phenolic honeycombs (41). As shown in Figure 3, silicone resins are relatively resistant to decomposition, and the silicon dioxide forms a viscous liquid when molten.

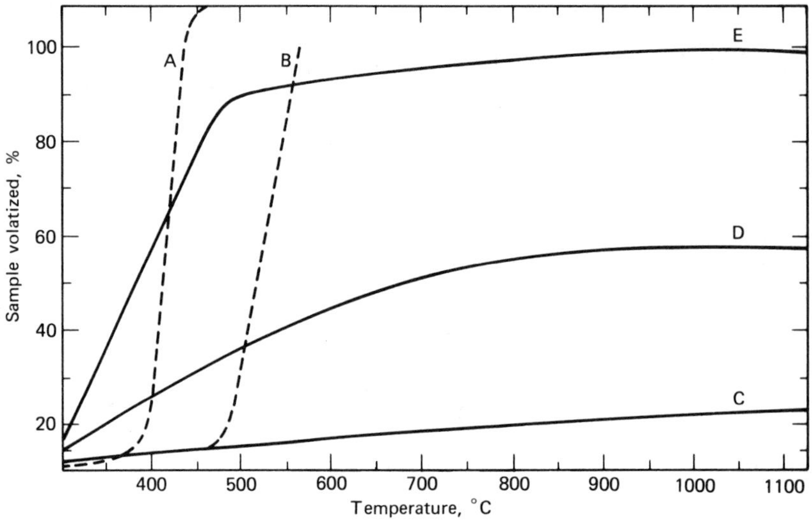

Figure 3. Decomposition of polymers as a function of temperature during heating (42). A, polymethylene; B, polytetrafluoroethylene; C, silicone; D, phenolic; E, epoxy.

A low cost, thermally-insulating ablative material, which can be used in mild environments, is cork; the porous carbon char permits high surface temperatures, and the gaseous products provide transpiration cooling of the outer layers.

Ablative Plastic Composites. The phenolic resins give the highest yield of carbon during pyrolysis, among the common plastics, and they have been widely used as surface charring ablative materials. Since a char is relatively weak, and removed mechanically by high shear forces associated with reentry, fibers of carbon, silicon dioxide, refractory oxides, mineral asbestos, or even glass have been added to assist the char retention.

When fiber-reinforced plastic ablators are exposed to ablative environments, they first act as heat sinks (see also Laminated and reinforced plastics). As heating progresses, the outer layer of polymer may become viscous and then begins to degrade, producing a foaming carbonaceous mass and ultimately a porous carbon char. The char is a thermal insulation; the interior is cooled by volatile material percolating through it from the decomposing polymer. The process is schematically illustrated in Figure 4. During percolation, the volatile materials are heated to very high temperatures with decomposition to low molecular weight species, which are injected into the boundary layer of gases. This mass injection creates a blocking action which reduces the heat transfer to the material. Thus, a char-forming resin acts as a self-regulating ablation radiator, providing thermal protection through transpiration cooling and insulation. The high efficiency, in terms of heat absorbed per weight of material lost of many such materials, is about forty times that of a copper heat sink.

As the layer of char thickens, the heat reaching the moving interface, in Figure 4, is reduced by the insulating effect of the porous char combined with the cooling effect of the diffusing gases. The rate of decomposition and gas formation is also reduced. Ablation of the char occurs by sublimation of carbon and oxidation in the boundary layer with concomitant vaporization. The rate of ablation slows as the rate of char-surface recession equals or exceeds the recession rate of the interface between the char and the virgin polymer.

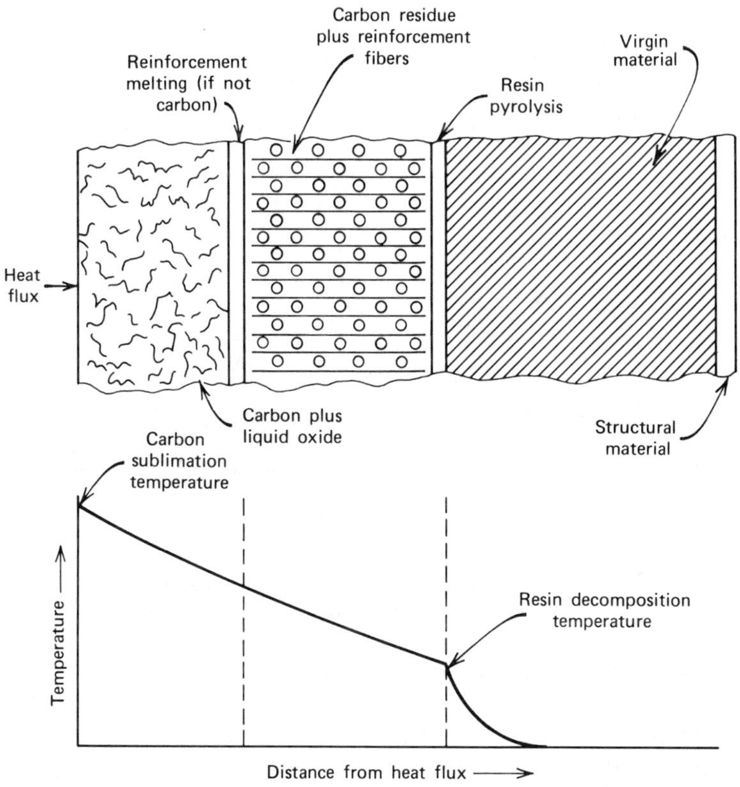

Figure 4. Schematic diagram of the ablation process (43).

An example of the response of a carbon fiber-reinforced phenolic composite to heat fluxes typical of reentry environments is shown by the test data in Figure 5 (44). A slab of the composite was exposed to arc-heated air in a graphite channel in such a way that the surface was first exposed to a low heat flux, 495 kW/m^2 [44 $Btu/(ft^2 \cdot s)$], for 230 s, and then to 4950 kW/m^2 [440 $Btu/(ft^2 \cdot s)$] for about 100 s. The temperatures recorded by thermocouples at various depths from the surface are shown; the deepest thermocouple, 10.67 mm, indicated that decomposition was just beginning at the end of test. Separately measured surface temperatures were approximately 1200 K before the increase in heat flux and 2000 K afterwards.

The resinous component of a typical reinforced plastic ablator comprises about 35% of the total weight. Since this part provides the essential gas for transpiration cooling of the char, the decomposition products are important. In phenolic resins, water is the principal product up to about 350°C; depolymerization then occurs, with release of phenol and larger species; however, above 530°C, lower molecular weight products are evolved, with H_2 as the primary product (in the absence of oxygen) at 700°C and above (45). Several phenolic compounds have the combination of high carbon content to provide high char yield, low oxygen to minimize CO and CO_2 formation, and many carbon–hydrogen bonds which provide the evolution of H_2 and CH_4 for transpiration cooling (46). The plastic must also have good forming characteristics, good adherence to the fibers, strength in the composite, and form a char which adheres well to the fiber reinforcements. During ablation, some pyrolytic deposition of carbon from the CH_4

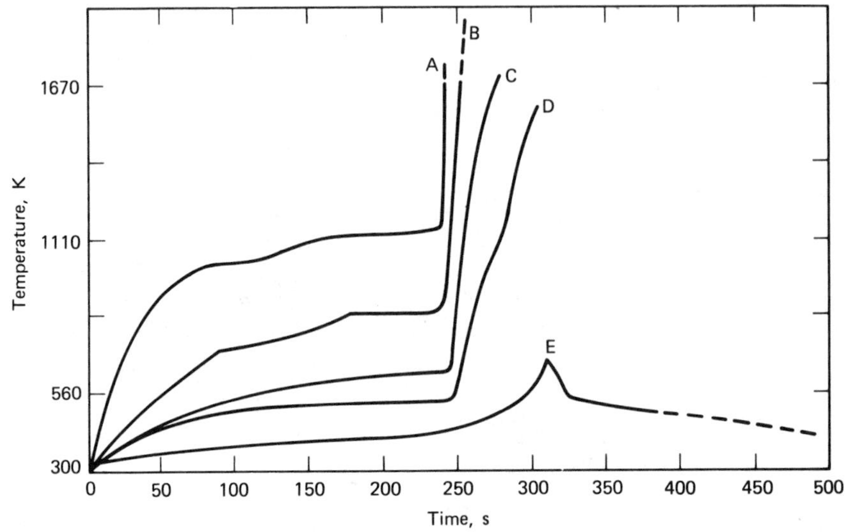

Figure 5. Subsurface temperature response of carbon–phenolic composite during ablation test. Thermocouple depth in mm, A, 2.26; B, 4.07; C, 6.20; D, 7.82; E, 10.67.

or other gases in the outer portions of the char can contribute to the strength and resistance to mechanical removal from the flow field. Addition of particulate carbon or other additives to the resin can also contribute to strength of the char. Very good char retention has been obtained with carbon phenolic heat shield materials (47).

Other resins besides the phenolics have even higher char yields and higher decomposition temperatures, and, hence, have been evaluated for ablative composites. A polyimide resin, eg, was found to show very little degradation at 400°C for two hours, and rapid degradation at 600°C, with evolution of low molecular weight gases (44). Other resins considered have been polybenzimidazole [*26986-65-9*] (PBI), poly(*p*-diphenylene oxide) [*32033-78-6*] (DPO), polyphenylenes, *p*-phenylphenols, and substituted heterocyclic polymers. Composite processing is a limitation in some cases.

When silica fiber reinforcements are used, as in less severe reentry trajectories, the mechanisms of ablation are somewhat more complex, as indicated in Figure 4. For example, the Apollo heat shield material (48) was a low density ablator consisting of an epoxy–novalac resin containing phenolic microballoons and a silica fiber reinforcement in the cavities of a fiberglass-reinforced phenolic honeycomb. Upon heating to about 1600 K agglomerated silica fibers first appeared on the surface. Silica globules then partially covered the surface, decreasing as silicon monoxide vapor formed on further heating to about 2100 K. Above 2100 K, silica melt was no longer found on the surface, and the material suffered mechanical removal of particles at pressures over 101 kPa (1 atm).

Thermal-Structural Considerations in Application of Ablators

The mechanical and thermal properties of the ablative materials can limit their application, due to structural requirements or stresses resulting from the thermal gradients. The plastics and plastic composites are relatively simple to attach to structures because the back face remains at low temperatures. The fiber reinforcement must provide good mechanical properties, however. Table 1 lists typical properties in the fiber directions of cloth-reinforced phenolic plates (49).

Table 1. Typical Properties of Ablative Composites Used for Heat Shields [a,b]

Property	Carbon phenolic	Silica phenolic
density, g/cm^3	1.4–1.5	1.7
tensile strength, MPa (psi)	138 (20,000)	103 (15,000)
elastic modulus, GPa (psi)	19 (2.8 × 10^6)	17 (2.48 × 10^6)
compressive strength, MPa (psi)	359 (52,000)	214 (31,000)

[a] Ref. 49.
[b] At room temperature.

Laminated composites of this type have usually been made as heat shields by tape-wrapping cloth impregnated with the resin, molding, and curing. For example, an 8-harness satin cloth may be layed up with fibers at 45° to the wrapping direction (bias orientation) and then layed up on a cylinder or frustum so that the plane of the tape makes an angle of 20°, eg, to the tangent plane. This arrangement provides resistance to delamination, which may occur with a simple cylindrical (scroll) wrap, and allows the fibers to make a low angle to the ablating surface, thus minimizing thermal conduction.

As previously discussed, pyrolytic graphite has shown the best resistance to surface recession under very high heat-flux conditions. This material is made as flat plates up to 3 cm thick or monolithic shapes by chemical vapor deposition (CVD) of carbon from methane or other carbonaceous gases onto industrial graphite substrates, usually at 2100–2600 K (50). Unlike "aerospace grades" of industrial graphites, which are processed from granular material and pitch and have densities in the range 1.7–1.9 g/cm^3, pyrolytic graphite is impervious to gas and has densities of 2.1–2.2 g/cm^3 as deposited; upon "graphitizing" at 2500°C or above, densities approaching the graphite single crystal, 2.26 g/cm^3, are obtained.

However, the high degree of preferred orientation of graphite basal planes parallel to the substrate results in a severe anisotropy in mechanical and thermal properties. These can vary widely, depending on the microstructure resulting from different deposition conditions (51), but tensile strength is typically 100–200 MPa in the plane of deposit (AB orientation) and 5–10 MPa through the thickness (C), compared with 30–60 MPa for aerospace graphites (see Carbon and artificial graphite). Thermal conductivity is only about 1% as high through the thickness as in the plane of deposit. The most critical property affecting thermal stresses, however, is the thermal expansion, compared in Table 2 (52–53).

Table 2. Typical Thermal Expansion of Graphite from 300 to 2500 K

Orientation	Pyrolytic graphite, %	Union Carbide ATJ-S, %	Poco AXF-5Q, %
radial or AB direction	0.5	0.9	1.9
axial or C direction	6	1.2	1.9

The high thermal expansion anisotropy of pyrolytic graphite means that any curvature in the deposit will either distort (if free) or develop bending stresses (if restrained) during heating or cooling. Such stresses place a limit on the thickness-to-radius ratio which can be deposited without cracking during cooling after deposition (typically 0.1 thickness-to-radius). Further distortion will also occur if the deposit graphitizes at higher temperatures, since not only does the anisotropy increase, but the deposit shrinks through the thickness (usually 10% or more) and expands radially (typically 3–8%, depending on initial microstructure). This has limited the use of pyrolytic graphite coatings on rocket nozzles (53). Designs with minimum recession throats have often utilized stacks of plates with the C direction (thickness) parallel to the nozzle axis. Although the anisotropy can be greatly reduced by codeposition of 14 vol % SiC [*409-21-2*] (54), such "alloying" generally results in an increase in vaporization rate.

The thickness limitations inherent to pyrolytic graphite are absent in industrial graphites; both ATJ-S (Union Carbide Corp.) and AXF-5Q (Poco Graphite Inc.), in Table 2, have been used in reentry protection systems (5–6). However, the biaxial stresses developed in heated rocket nozzles or nosetips cause catastrophic cracking in all particulate-processed graphites, if the tensile strain at failure is exceeded. Some special grades have been made with improvements in this property, combined with adequately low thermal expansion coefficients and elastic moduli, by modifications and quality control of the constituents and processing, together with nondestructive evaluation to eliminate crack initiators in the material (55).

Carbon–Carbon Composites

The thermal-stress limitations of industrial graphites have been overcome without significant degradation in ablation performance, although at significantly higher cost, by incorporating 40–50 vol % of high strength graphite fibers as a three-dimensional reinforcement (56–59). Woven preforms, usually having 5–30 cm dimensions with center-to-center bundle spacings of 0.3–3 mm (60) have been made with several types of geometries; examples are shown in Figure 6 (61). The 3-directional packing has been made as cylinders or frusta with polar coordinate geometry, axial, radial, and circumferential reinforcements (62). It is also made, in principle, by laminating woven cloth and injecting rods or yarn perpendicular to the "pierced fabric" with 2.5 mm or less between centers (63).

In all cases, the fiber assemblies are impregnated with resin or pitch and carbonized in several cycles to reduce porosity to levels where acceptable ablation performance is obtained. Usually, graphitizing temperatures (normally 2700–2800°C) are employed after all or most of the cycles. In some cases, gaseous infiltration and deposition of pyrolytic carbon is employed. The 4-directional and 7-directional geometries (Fig. 6) permit better processing, in this way, because of the porosity continuity (61).

The properties of the 3-directional composites on a macroscopic scale are determined primarily by small unidirectional composites which fill the reinforcement prism positions in Figure 6. About 60–65% of the volume of the fiber bundles is filled by filaments of the yarn or tow, with graphite or carbon obtained from repeated im-

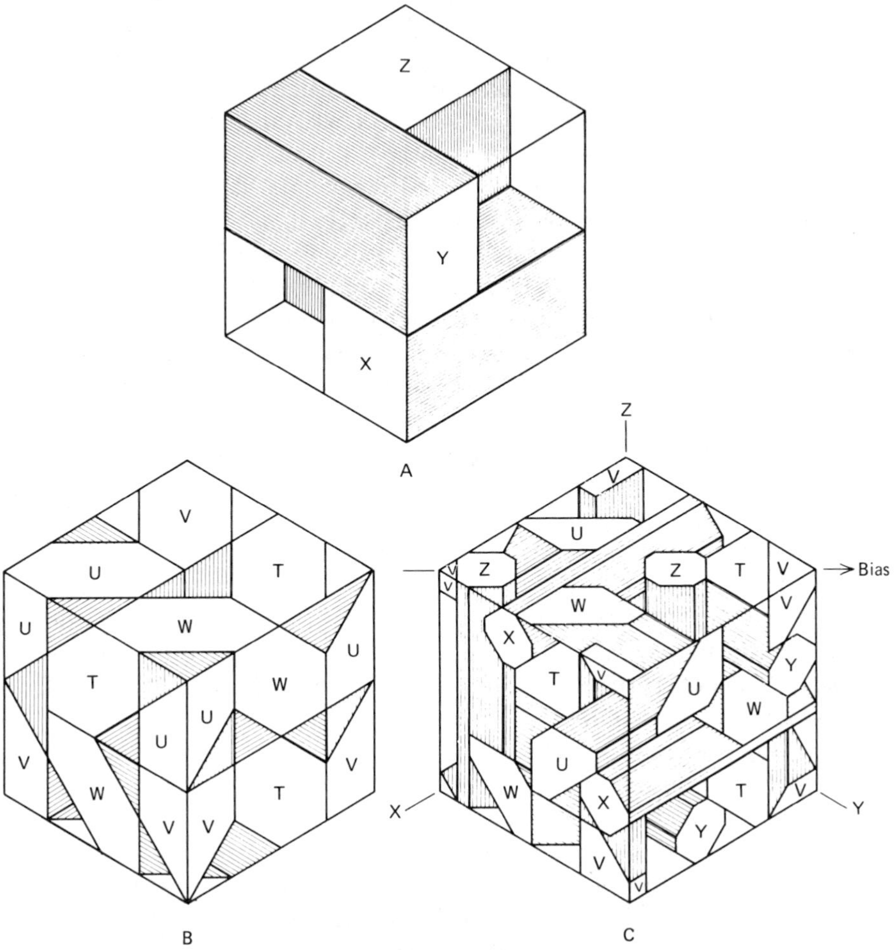

Figure 6. Three-dimensional packing geometries used in carbon–carbon composites; note the shape of the continuous-prism positions filled by fibers (61). A, 3-directional geometry, parallel to cube sides (space filled ≥ 75%); B, 4-directional geometry, parallel to cube diagonals (space filled = 75%); C, 7-directional geometry, 3-directional plus 4-directional combination (space filled = 60%).

pregnation–graphitization cycles comprising the rest of the material. The fibers, prepared from rayon, acrylic, or pitch precursors, are usually processed to high modulus of 300–500 GPa ($44–73 \times 10^6$ psi) and strength (1–3 GPa) (145,000–435,000 psi) and have a high preferred orientation of the graphite layers. This results in a low thermal expansion coefficient in the bundle filament direction (about 0.5% from 300 to 3000 K) and a high expansion coefficient transverse to the bundle (about 3% from 300 to 3000 K). When the material is heated to ca 3000 K at the conclusion of processing, the transverse-bundle contraction upon cooling is resisted by bundles in opposing directions, so that fissures tend to open around or within the bundles due to cracking in tension across the layers. Upon reheating, these spaces accommodate the high transverse expansions, so that the composite as a whole has a low coefficient of expansion, similar to that of the high modulus fibers. The fissures are probably the primary reason for resistance to crack propagation, which prevents thermal-stress failure even though tensile strain to failure (0.1–0.4%) is less than that of the brittle aerospace graphites (0.3–0.7% at low temperatures). Strength and modulus are higher than the graphites and depend on the proportion of fibers reinforcing the direction of test; 100–300 MPa (15,000–45,000 psi) tensile strengths are found in the 3-directional carbon–carbon composites (56–59).

Most published literature on carbon–carbon composites is concerned with 2-dimensional reinforcements made by laminating cloth or unidirectional tape oriented in different directions, or by filament winding, and with relatively isotropic, low filament-density composites made by chemical vapor deposition (CVD)-infiltrating carbonized felt which has been formed to shape (64–66). In the former case, a resin matrix is often prepared initially, and the carbon fiber–resin composite is precharred or carbonized. This processing is often followed by CVD infiltration to increase density near the surface and interlaminar shear strength (67–68). Such materials have been evaluated for rocket nozzles. With additions of silicon carbide to improve low temperature oxidation resistance, they have been utilized as the leading edge-thermal protection for the space shuttle (69). In the CVD-felt composites, the microstructure of the deposit may be varied by changing deposition conditions to control the properties after graphitization (70). Both CVD-felt and CVD-infiltrated filament-wound frusta have been flight tested successfully (71); the properties can vary considerably, depending on the filament-winding orientation or degree of needling the felt (which pushes some filaments perpendicular to the layers); but both materials have a fracture toughness superior to graphite (72).

One example of a successful "2-D" composite was the protective container around the radioactive-fuel capsule used to power the Apollo lunar geophysical stations. High temperatures generated by the fuel during storage prevented use of a carbon–phenolic ablative material. The composite was fabricated from phenolic resin-impregnated cloth, made from low modulus carbon fibers, layed in a rosette pattern 6–9° to the tangent of the cylinder (Fig. 7). After molding and carbonizing, it was heated to about 2500 K and then CVD-infiltrated at about 1300 K to densify the outer layers to about 1.5 g/cm^3; the center remained at about 1.3 g/cm^3. The lower elastic modulus associated with the lower central density contributed to a reduction of thermal stresses. Although relatively low in tensile strength (ca 90 MPa axial, ca 30 MPa circumferential, and ca 6 MPa through the thickness), this material had much better resistance to crack propagation than graphite, thus providing protection against fuel dispersal upon impact as well as surviving the ablation and thermal stress environment during reentry (73). Such reentry occurred during the aborted Apollo 13 mission.

Composite Dielectric Ablative Materials

As in the case of carbon heat shields, the antenna window portion of a reentry thermal shield has been subject to development to improve mechanical properties by incorporating fiber reinforcements. In these cases, either high purity silicon dioxide or boron nitride [*10043-11-5*] (BN) are used to avoid forming a conducting char.

Table 3 summarizes the pertinent properties of both composite grades and typical ceramic materials which have been considered. Fused silica has been used in optically transparent and porous slip-cast grades. Hot-pressed boron nitride has higher resis-

Table 3. Properties of Ablative Materials for Antenna Windows at ca 300 K [a]

Material description[b]	Bulk density, g/cm^3	Thermal conductivity[c] MW/(m·K)	Dielectric constant	Loss tangent, 10^{-3}	Tensile properties: Modulus GPa[d]	Tensile properties: Strength kPa[e]	Tensile properties: Strain, %	Relative impact strength[f]	Heat of ablation, MJ/kg
fused SiO_2, transparent (Corning 7940)	2.2	1.4	3.8	1.6	69	47		0.5–1.5	7–12
slip cast SiO_2 (Corning 7941)	1.9–2.1	0.9	3.3	0.2	58	37–62		0.5–1.5	7–12
3-directional SiO_2–SiO_2 (A-F AS-3DX)	1.7	0.7	2.9	8.0	28	30 (35 flexure)	0.2	4.0–6.0	10–26
4-directional SiO_2–SiO_2 (GE ADL-4D6)	1.6	0.7	2.8–3.1	6.0	(14 flexure)	(35 flexure)	1	4.0–8.0	14–26
4-directional SiO_2–silicone (GE ADL-10)	1.8	0.6	3.0–3.3	2–5	7–14	70–140	1–2	5.0	
boron nitride, hot pressed (UCC HBC)	1.9–2.05	62 AB 44 C-direction	4	0.26	41–76	41		0.9 max	26
pyrolytic (UCC Boralloy)	2.1	62 AB 1.6 C-direction	5.1 AB 3.4 C	0.15 0.10 max	21 AB	83 AB 41 C	0.4 AB		HBC-AB lower-C
0.4 BN–0.6 SiO_2 (Carborundum M)	2.12	26 AB 10 C-direction	3.7 AB 4.2 C	0.5 1.0	106	97			
3-directional BN–BN (A-F BN-3DX)	1.5–1.6	8.7	2.9–3.2	3.0	15	25 (58 flexure)	0.2	3.0	26–56

[a] Ref. 74.
[b] Manufacturer: AF, Aeronutronic-Ford; UCC, Union Carbide Corporation; GE, General Electric Co., Re-entry & Environmental Systems Division.
[c] Measured at 340 K.
[d] 1.00 GPa = 14.5×10^4 psi.
[e] 1.00 kPa = 14.5×10^{-2} psi.
[f] Relative spall threshold in plate-slap tests.

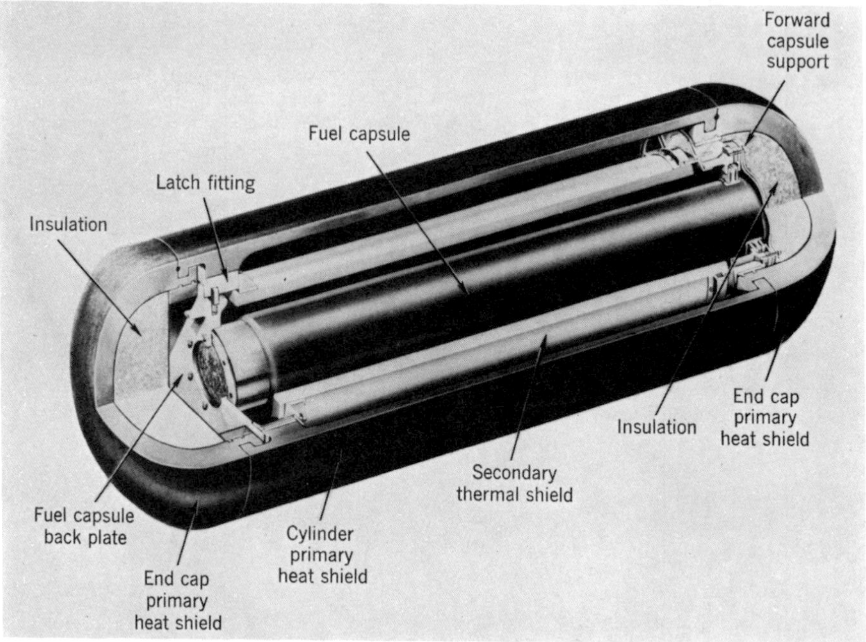

Figure 7. Design of the carbon–carbon reentry primary heat shield used for protection of the radioactive fuel source carried on the Apollo lunar excursion modules.

tance to ablation but the disadvantage of high thermal conductivity; the low interlayer shear strength of pyrolytic (CVD) BN contributes to more rapid ablation, under high shear environments.

The fiber-reinforced composites in Table 3 were developed for improved mechanical toughness. Significant improvements in tensile strain to failure and relative impact resistance have been obtained, as shown in the table. The Aeronutronic-Ford AS-3DX utilizes a 3-directional orthogonal arrangement of silica fibers, densified by repeated impregnation with an aqueous suspension of colloidal silica and heating to 900 K, until about 1.7 g/cm^3 bulk density is obtained (75). A 4-directional composite, having the geometry in Figure 6, utilizes the same high purity fibers and a similar densification process (76); it is known as General Electric ADL-4D6. A 3-directional-BN composite, Aeronutronic-Ford BN-3DX (77) is based on high purity Carborundum Co. boron nitride fibers and the formation of boron nitride from boron oxide and ammonia in the pores. A reinforced 4-directional-SiO_2 containing a silicone-resin matrix (74) has also been tested, but is limited to 1800 K surface temperature to avoid forming an electrically conducting char.

BIBLIOGRAPHY

"Ablation" in *ECT* 2nd ed., Vol. 1, pp. 11–21, by I. J. Gruntfest, General Electric Company.

1. D. K. Huzel, *Pennemunde to Canaveral,* Prentice-Hall, Inc., Englewood Cliffs, N. J., 1962.
2. D. L. Schmidt, "Ablative Plastics for Re-entry Thermal Protection," *WADD TR 60-862,* USAF ASD, Wright-Patterson AFB, Ohio, Aug. 1961.
3. H. J. Allen, *J. Aeronaut. Sci.* **25,** 217 (1958).
4. M. C. Adams, *Jet Propul.* **29,** 625 (1959).
5. A. M. Morrison, *J. Spacecr. Rockets* **12,** 633 (1975).
6. P. J. Schneider and co-workers, *J. Spacecr. Rockets* **10,** 592 (1973).
7. H. L. Moody and co-workers, *J. Spacecr. Rockets* **13,** 746 (1976).

8. D. L. Schmidt in D. V. Rosato and R. T. Schwartz, eds., *Environmental Effects on Polymeric Materials,* Interscience Publishers, a division of John Wiley & Sons, Inc., New York, 1968, pp. 487–587.
9. D. L. Schmidt in G. F. D'Alelio and J. A. Parker, eds., *Ablative Plastics,* Marcel Dekker, Inc., New York, 1971, pp. 1–39; *J. Macromol. Sci. Chem.* **3,** 327 (1969).
10. M. L. Minges in G. F. D'Alelio and J. A. Parker, eds., *Ablative Plastics,* Marcel Dekker, Inc., New York, 1971, pp. 287–313.
11. N. S. Diaconis, E. R. Stover, J. Hook, and G. J. Catalano, "Graphite Melting Behavior," *AFML-TR-71-119,* ASD, Wright-Patterson AFB, Ohio, July 1971.
12. N. A. Gokcen and co-workers, *High Temp. Sci.* **8,** 81 (1976).
13. L. S. Nelson, A. G. Whittaker, and B. Tooper, *High Temp. Sci.* **4,** 445 (1972).
14. A. G. Whittaker and P. L. Kintner, "Carbon Solid–Liquid–Vapor Triple Point and the Behavior of Superheated Liquid Carbon," *12th Biennial Conference on Carbon, Extended Abstracts,* University of Pittsburgh, Pittsburgh, Pa., 1975, pp. 45–47.
15. A. G. Whittaker and P. L. Kintner, *Carbon* **14,** 257 (1977).
16. J. H. Lundell and R. R. Dickey, *AIAA J.* **11,** 216 (1973).
17. J. H. Lundell and R. R. Dickey, *AIAA J.* **13,** 1079 (1975).
18. J. Abrahamson, *Carbon* **12,** 111 (1974).
19. M. L. Lieberman, *Carbon* **9,** 345 (1971).
20. R. T. Meyer and A. W. Lynch, *High Temp. Sci.* **4,** 283 (1972).
21. H. B. Palmer and M. Shelef, "Vaporization of Carbon," in P. L. Walker, Jr., ed., *Chemistry and Physics of Carbon,* Vol. 4, Marcel Dekker, Inc., New York, 1968, pp. 85–135.
22. H. R. Leider, O. H. Krikorian, and D. A. Young, *Carbon* **11,** 555 (1973).
23. W. E. Pearson and W. C. Davy, *AIAA J.* **11,** 1207 (1973).
24. R. T. Meyer and A. W. Lynch, *High Temp. Sci.* **5,** 195 (1973).
25. P. D. Zavitsanos and G. A. Carlson, *J. Chem. Phys.* **59,** 2966 (1973).
26. *NSRDS-NBS-37,* 2nd ed., National Bureau of Standards, Washington, D. C., 1971.
27. J. H. Lundell and R. R. Dickey, "Vaporization of Graphite in the Temperature Range of 4000 to 4500 K," *AIAA Paper No. 76-166, 14th Aerospace Sciences Meeting, January 1976,* American Institute of Aeronautics and Astronautics, New York.
28. S. M. Scala and L. M. Gilbert, *AIAA J.* **3,** 1625, 2124 (1965).
29. J. A. Segletes, *J. Spacecr. Rockets* **12,** 251 (1975).
30. K. E. Putz and E. P. Bartlett, *J. Spacecr. Rockets* **10,** 15 (1973).
31. K. M. Kratsch and co-workers, *AFML-TR-70-307,* Vol. IV, ASD, Wright-Patterson AFB, Ohio, May 1973.
32. I. Auerbach and co-workers, *J. Spacecr. Rockets* **14,** 19 (1977).
33. E. Frye, *Nucl. Technol.* **12,** 93 (1971).
34. J. H. Lundell, R. R. Dickey, and J. T. Howe, "Simulation of Planetary Entry Radiative Heating with a CO_2 Gasdynamic Laser," *ASME Conference on Environmental Systems, San Francisco CA, July 1975,* American Society of Mechanical Engineers, New York.
35. J. W. Schaefer, H. Tong, and R. J. Bedard, *J. Spacecr. Rockets* **12,** 552 (1975).
36. L. M. Gilbert, *J. Spacecr. Rockets* **12,** 184 (1975).
37. J. C. Lewis, I. J. Floyd, and F. C. Cowlard, "Laboratory Investigation of Carbon-Gas Reactions of Relevance to Rocket Nozzle Erosion," *Third Conference on Industrial Carbons and Graphite (1970),* Society of Chemical Industry, London, 1971, pp. 282–296.
38. T. R. Acharya and D. R. Olander, *Carbon* **11,** 7 (1973).
39. D. L. Petersen and W. E. Nicolet, *J. Spacecr. Rockets* **11,** 382 (1974).
40. T. F. McKeon, "Ablative Degradation of a Silicone Foam," in G. F. D'Alelio and J. A. Parker, eds., *Ablative Plastics,* Marcel Dekker, Inc., New York, 1971, pp. 259–286.
41. S. S. Tompkins and W. P. Kabana, "Effects of Material Composition on the Ablation Performance of Low-Density Elastomeric Ablators," *NASA TND-7246,* NASA Langley Research Center, Hampton, Va., 1973; *Chem. Abstr.* **79,** 32555G.
42. P. W. Juneau, Jr., *Third Annual Polymer Conference Series, Program VII, 1972,* University of Detroit, Detroit, Mich.
43. P. W. Juneau, Jr., Paper No. XVII-3, *22nd Annual Technology Conference, Society of Plastics Engineers,* Vol. 12, Mar. 1966.
44. R. W. Farmer, "Extended Heating Ablation of Carbon Phenolic and Silica Phenolic," *AFML-TR-74-75,* ASD, Wright-Patterson AFB, Ohio, Sept. 1974.
45. G. F. Sykes, *NASA TN D-3810,* National Aeronautics and Space Administration, Feb. 1967.
46. R. P. Rastigi and D. Deepak, *AIAA J.* **12,** 1146 (1974).
47. J. W. Metzer, "The Behavior of Ablating Carbon Phenolic," *AIAA–AIME–SAE 13th Structures, Structural Dynamics, and Materials Conference, San Antonio, Tex., Apr. 1972.*

48. E. P. Bartlett and L. W. Anderson, *J. Spacecr. Rockets* **8,** 463 (1971).
49. *Ablative Materials Handbook,* U.S. Polymeric, Inc., Santa Ana, Calif., 1964.
50. W. H. Smith and D. H. Leeds, "Pyrolytic Graphite," *Modern Materials,* Vol. 7, Academic Press, Inc., New York, 1970, pp. 139–221.
51. J. C. Bokros, "Deposition, Structure and Properties of Pyrolytic Carbon," in P. L. Walker, Jr., ed., *Chemistry and Physics of Carbon,* Vol. 5, Marcel Dekker, Inc., New York, 1969, pp. 1–118.
52. C. D. Pears, "Characterization of Several Typical Polygraphites With Some Convergence on Solid Billet ATJ-S," *Proceeding of the Conference on Continuum Aspects of Graphite Design, 1970, CONF-701105* National Technology Information Service (NTIS), Springfield, Va., 1972, pp. 115–136.
53. J. G. Baetz, "Characterization of Advanced Solid Rocket Nozzle Materials," *SAMSO-TR-75-301,* Air Force Rocket Propulsion Laboratories, Edwards AFB, Calif., Dec. 1975.
54. R. H. Singleton, E. L. Olcott, and C. Pears, "Thermal and Mechanical Properties of a Codeposited Pyrographite Silicon Carbide Composite Material," *11th Biennial Conference on Carbon, 1973, CONF-730601,* NTIS, Springfield, Va., p. 192.
55. C. A. Pratt, Jr., "Petroleum Derived Carbons" in M. L. Deviney and T. M. O'Grady, eds., *ACS Symposium Series,* No. 21, American Chemical Society, Washington, D. C., 1976, pp. 203–211.
56. A. R. Taverna and L. E. McAllister in J. Buckley, ed., *Advanced Materials, Composites and Carbon, 1971 Symposium,* American Ceramic Society, Columbus, Ohio, 1972, pp. 203–211.
57. K. M. Kratsch, J. C. Schutzler, and D. A. Eitman, "Carbon–Carbon 3D Orthogonal Material Behavior," *AIAA Paper No. 72365, AIAA–ASME–SAE 13th Structural Dynamics and Materials Conference, 1972,* American Institute of Aeronautics and Astronautics, New York.
58. E. R. Stover and co-workers, *11th Biennial Conference on Carbon, 1973, CONF-730601,* National Technology Information Service, Springfield, Va., pp. 277, 335–336.
59. J. L. Perry and D. F. Adams, *Carbon* **14,** 61 (1976).
60. Product data, Fiber Materials, Inc., Biddeford, Maine, Jan. 1975.
61. J. J. Gebhardt and co-workers in M. L. Deviney and T. M. O'Grady, eds., *ACS Symposium Series,* No. 21, American Chemical Society, Washington, D. C., 1976, pp. 212–227.
62. U.S. Pat. 3,904,464 (Sept. 9, 1975), R. W. King.
63. P. G. Rolincik, *11th Biennial Conference on Carbon, 1973, CONF-730601,* National Technology Information Service, Springfield, Va., pp. 343–344.
64. D. L. Schmidt, *SAMPE J.* 9 (May–June, 1972).
65. H. M. Stoller and co-workers in J. Weeton and E. Scala, eds., *Composites: State of the Art,* Metropolitan Society AIME, New York, 1974, p. 69.
66. W. V. Kotlensky in P. L. Walker, Jr. and P. A. Thrower, eds., *Chemistry and Physics of Carbon,* Vol. 9, Marcel Dekker, Inc., New York, 1973, pp. 173–261.
67. H. O. Davis, *J. Spacecr. Rockets* **13,** 456 (1976).
68. L. Boyne, J. Hill, and K. Turner in J. M. Blocher and co-workers, eds., *Proceedings of the 5th International Conference on Chemical Vapor Deposition,* The Electrochemical Society, Princeton, N. J., 1975, pp. 577–588.
69. J. M. Williams and R. J. Imprescia, *J. Spacecr. Rockets* **12,** 151 (1975).
70. H. O. Pierson and D. A. Northrup, *J. Compos. Mater.* **9,** 118 (1975).
71. H. M. Stoller and co-workers, "Properties of Flight-Tested CVD-Felt and CVD-FW Composites," *SC-DC 71-4046,* Sandia Laboratories, Albuquerque, N. M., 1971; *Summary of Papers, Tenth Biennial Conference on Carbon, 1971,* pp. 90–91.
72. T. R. Guess and W. P. Hoover, *J. Compos. Mater.* **7,** 12 (1973).
73. R. Gavert, *SAMPE Q.* **1**(1), 56 (1969).
74. J. P. Brazel, *AMMRC CR-76-4,* U.S. Army Materials and Mechanics Research Center, Watertown, Mass., Feb. 1976.
75. T. M. Place, *Proceedings of the 12th Symposium on Electromagnetic Windows,* Georgia Institute of Technology, Atlanta, Ga., 1974, pp. 47–51.
76. J. P. Brazel and R. Fenton, *Proceedings of the 13th Symposium on Electromagnetic Windows,* Georgia Institute of Technology, Atlanta, Ga., 1976, pp. 9–16.
77. T. M. Place, *Proceedings of the 13th Symposium on Electromagnetic Windows,* Georgia Institute of Technology, Atlanta, Ga., 1976, pp. 17–22.

E. R. STOVER
P. W. JUNEAU, JR.
J. P. BRAZEL (Composite Dielectric Ablative Materials)
General Electric Company

ACRYLONITRILE POLYMERS

SURVEY AND STYRENE–ACRYLONITRILE COPOLYMERS

Acrylonitrile has been established as one of the most important building blocks of the plastics industry (see Acrylonitrile). Its unique properties have resulted in broad applications for a number of important market areas. Acrylonitrile was discovered in 1893 (1), but had limited industrial application until the development of nitrile rubbers during World War II. Later in the 1950s, acrylonitrile found its greatest use in synthetic fibers, and to this day synthetic fibers still claim the largest portion of acrylonitrile consumption. In the late 1950s. a family of plastics, deriving superior properties of toughness, rigidity, and resistance to chemicals and solvents from acrylonitrile, appeared in the market. These plastics are styrene–acrylonitrile copolymers [*9003-54-7*] (SAN) and acrylonitrile–butadiene–styrene terpolymers [*9003-56-9*] (ABS). In the late 1960s, acrylonitrile polymers were introduced into the packaging industry in a new class of plastics called barrier resins (qv). These resins have started to penetrate into the packaging market area, competing with glass containers and other plastic films. A substantial growth is anticipated for these barrier resins.

Acrylonitrile homopolymer has little application outside fibers because the combination of high melting point, poor thermal stability, and high melt viscosity make melt processing very difficult. However, by copolymerizing acrylonitrile with other monomers, these deficiencies have been tempered and a large measure of the unusual properties of acrylonitrile have been incorporated into melt-processable resins. Some of the special properties of polyacrylonitrile that can contribute to copolymers are hardness, heat resistance, slow burning, resistance to most solvents and chemicals, resistance to sunlight and outdoor exposure, ability to form oriented fibers and films, reactivity of the nitrile group, compatibility with certain polar substances, and low permeability of gases (2).

By 1975, the production of acrylonitrile grew to over a half million metric tons in the United States (about 5% of all the plastics and resin materials produced in that year). The production and price history of acrylonitrile (3–4) is shown in Figure 1. The broad licensing of Sohio's propylene–ammonia process in the 1960s greatly increased the United States acrylonitrile capacity, and resulted in a significant price reduction. The distribution of acrylonitrile usage over the years (5–6) is shown in Table 1. The miscellaneous portion encompasses very broad and important applications. They are acrylamide and water-soluble polymers, acrylic esters, adhesives, adiponitrile, alkyd resins, antioxidants, coatings, chemotherapy, cyanoethylated natural fibers and paper, dielectric paper, dyes, electrically conductive rubber, emulsifying agents, foam, glutamic acid, glutaronitrile, insecticides, latex paints, photographic emulsions, plasticizers, soil aggregating agents, synthetic leathers, and wire insulation, etc (7).

Synthetic fibers contain 20–90 wt % of acrylonitrile; ABS and SAN, up to about 35%; barrier resins, 60% or higher; and nitrile rubbers, 18–50%. Synthetic fibers and ABS/SAN are expected to remain as the major usages of acrylonitrile. Barrier resins are still new in the market, but because of their potential for various packaging applications, they could become very important in the plastic industry.

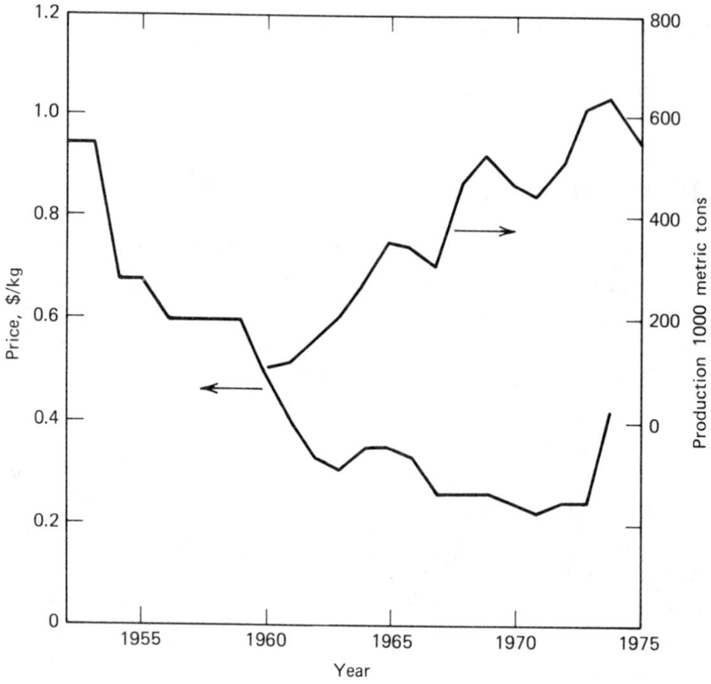

Figure 1. United States price (3–4) and production history (3) of acrylonitrile.

When SAN grafted-rubber is blended with SAN, it becomes a two-phase system and is called ABS. The amount of rubber in ABS varies from about 5 to 30 wt % with the remainder being SAN copolymers. ABS has the useful properties of SAN, eg, rigidity and resistance to chemicals and solvents, etc, and its rubber additive imparts toughness. Because ABS is a two-phase system and the two phases have different indexes of refraction, ABS is opaque. In contrast, SAN is optically clear. A clear ABS product can be made by adding a monomer such as methyl methacrylate which adjusts the refractive indexes of both phases. ABS is a very versatile material which can be tailored to meet specific needs. Modifications have brought out specialty grades of ABS, such as clear ABS, high-heat, and flame retardant ABS, etc.

Since the introduction of SAN and ABS about 15 to 20 years ago, their United States market has grown at a very rapid rate, with ABS outperforming SAN, to about 360,000 metric tons of production in 1975, or about 3.2% of all plastics and resins produced in that year in the United States. Very few polymeric products have enjoyed this kind of growth. The following discussions are directed to the manufacturing processes, applications, economics, and some other aspects of SAN copolymers and ABS terpolymers (see Styrene plastics, also Styrene).

SAN Properties and Test Methods

The incorporation of acrylonitrile into polymeric systems imparts some important properties. Table 2 shows some properties of SAN copolymers with various amounts of acrylonitrile (8–9). Methods for testing the chemical composition of SAN copolymers and some typical physical properties are also listed in Table 3. In addition to these

Table 1. Acrylonitrile Usage Distribution

	1975[a]			1970[b]			1965[c]		
Use	U.S.A.	Europe	Japan	U.S.A.	Europe	Japan	U.S.A.	Europe	Japan
synthetic fibers, %	55	81	54	56	83	71	52	75	80
ABS/SAN, %	20	8	11	16	11	15	9	12	10
nitrile rubbers, %	5	5	1	6	3	3	6	7	3
miscellaneous, %	15	6	19	11	3	11	11	6	7
export, %	5		15[d]	11			22		

[a] From Ref. 5.
[b] From Ref. 6.
[c] From Ref. 6.
[d] 11% to Far East and 4% to Europe.

properties listed, SAN copolymers have optical clarity, good processability, and are relatively inexpensive. All these factors have made SAN copolymers useful products with broad applications.

Table 2. Properties of SAN Copolymers

	Acrylonitrile content, wt %						
Properties	5.5	9.8	14.0	21.0	27.0	23–24	27–28
Physical[a]							
tensile strength, MPa	42.3	54.6	57.4	63.8	72.5		
(psi)	(6130)	(7922)	(8321)	(9259)	(10511)		
elongation, %	1.6	2.1	2.2	2.5	3.2		
notched impact,[b] J/cm	0.67	0.65	0.67	0.67	0.67		
Chemical[c]							
water absorption, %							
at 60°C, 2 weeks						+0.54[d]	+0.57
at RT, 15 days in							
10% NaOH						+0.42	+0.53
40% NaOH						+0.07	+0.05
50% H_2SO_4						+0.01	+0.01
Solvent resistance[c]							
soluble in	acetone, chloroform, dioxane, methyl ethyl ketone, pyridine						
swells in	benzene, ether, toluene						
insoluble in	carbon tetrachloride, ethyl alcohol, gasoline, lubricating oil, Solvesso (trade name of solvents)						

[a] From Ref. 8.
[b] To convert J to cal divide by 4.184.
[c] From Ref. 9.
[d] As change in weight.

Table 3. SAN Copolymer Properties and Test Methods

Properties	Analytical, test methods	Typical values
Chemical		
SAN composition	carbon–hydrogen–nitrogen analysis	20–35 AN wt %
copolymer molecular weights	osmotic pressure, light scattering, gel-permeation chromatography, solution viscosity	$\overline{M}_n$ = 30,000–150,000 $\overline{M}_w$ = 150,000–400,000
residual SAN monomers	gas chromatography	0.05–1.0 wt %
Physical[a]		*Lustran SAN 31*
tensile strength	ASTM D-638-72	72.4 MPa (10,497 psi)
elongation	ASTM D-638-72	3.0% at fail
tensile modulus	ASTM D-638-72	3276 MPa (475,054 psi)
Izod impact strength	ASTM D-256-72a	0.62 J/cm, 1.27 × 1.27 cm bar, 0.0254 cm notch radius
deflection temperature under load	ASTM D-648-72	102°C at 1.82 MPa (265 psi), injection molded, annealed 1.27 × 1.27 cm bar
Vicat softening point	ASTM D-1525-70	110°C
deformation under load	ASTM D-621-64	1.5% at 27.6 MPA (3997 psi), 50°C, 24 h
coefficient of linear thermal expansion	ASTM D-696-70	0.000068/°C
specific gravity	ASTM D-792-66, A	1.08
Rockwell hardness	ASTM D-785-65	83, "M" scale
water absorption	ASTM D-570-63	0.25%, in 24 h
burning rate	ASTM D-635-72	1.27–2.54 cm/min, 3.18 cm bar
refractive index	ASTM D-542-50	1.57
melt flow	ASTM D-1238-70	8.0 g/10 min, condition I
bulk factor	ASTM D-1132	1.6, lubricated pellets
mold shrinkage		0.002–0.006 cm/cm
dielectric strength	ASTM D-149-64	181 × 10^3 V/cm, short time 165 × 10^3 V/cm, step by step
dielectric constant	ASTM D-150-70	2.95 at 10^3 Hz 2.86 at 10^6 Hz
dissipation factor	ASTM D-150-70	0.0062 at 10^3 Hz 0.0078 at 10^6 Hz
volume resistivity	ASTM D-257-66	4.6 × 10^{16} Ω·cm at 50°C

[a] From Ref. 5.

SAN Manufacture

General Considerations. Because the reactivities of acrylonitrile and styrene radicals towards their monomers are not the same, SAN copolymer compositions vary from their monomer compositions (10). Such variations are illustrated in Figure 2. Acrylonitrile is soluble in water (11), and the slight difference in behavior between the mass and the emulsion copolymerizations evident in the figure is attributed to this phenomenon (12). SAN copolymer compositions can be calculated for given monomer compositions from the copolymerization equations (13) using compiled reactivity ratios (14). The temperature dependence of the reactivity ratios has been determined by Goldfinger and Steidlitz (15). Because of the difference in reactivity ratios, the copolymer composition drifts away as copolymerization proceeds (16), as illustrated in Figure 3, except at the azeotrope composition where the monomer and copolymer compositions remain the same throughout the entire course of the copolymerization. When the SAN copolymer compositions vary significantly, the copolymers become incompatible (17). This incompatibility causes the loss of optical clarity, mechanical strength, heat, solvent and chemical resistance, modability, and other important properties (18). Therefore, it is important to maintain the compositional homogeneity in the manufacture of SAN copolymers, by techniques described later.

Both the rates (19–20) and heat (21) of SAN copolymerization vary with acrylonitrile monomer concentration, as shown in Figures 4 and 5. The solid curves are the fittings of Walling's and Alfrey's equations, respectively, and the broken curve in Figure 5 assumes a simple ideal case, namely, the sum of the products of mole fraction and its heat of homopolymerization of styrene and acrylonitrile, as detailed in the referenced works.

Commercially, SAN copolymers are manufactured by three processes. They are emulsion, suspension, and continuous mass. Each process is discussed in the following sections.

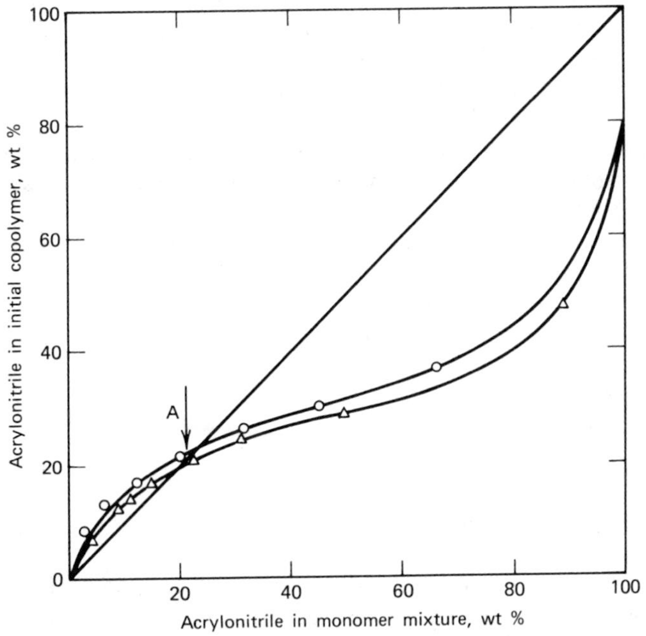

Figure 2. Styrene–acrylonitrile monomer–polymer compositions (10). A, azeotrope composition; O, mass polymerization; Δ, emulsion polymerization.

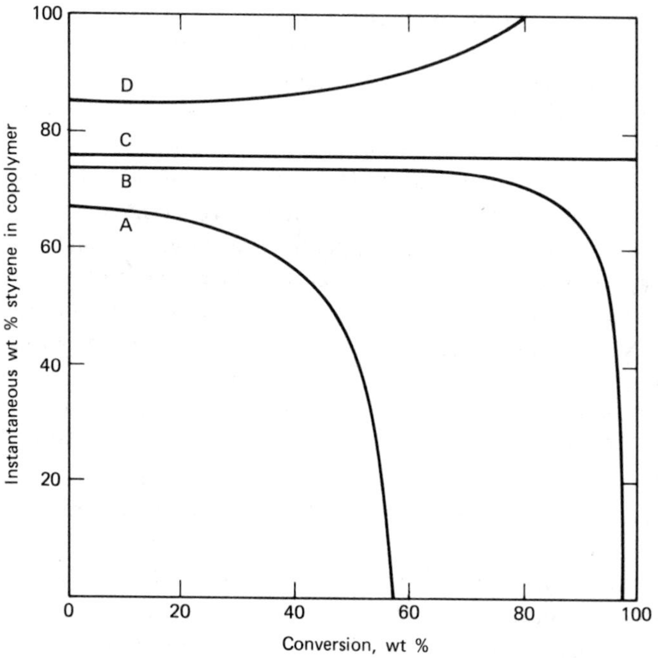

Figure 3. Approximate compositions of styrene–acrylonitrile copolymers formed at different conversions starting with various monomer mixtures (16). A, 35/65; B, 70/30; C, 76/24; D, 90/10 (S/AN).

Emulsion Process. In the emulsion copolymerization (22–23) of SAN, both batch and continuous processes are used. Generally, the copolymerizing system contains emulsifier, initiator, chain-transfer agent, monomers, and water. The copolymerization is carried out in the temperature range of 70–100°C, and to a conversion of 97% or higher. With redox catalysis systems (24), the copolymerization can be carried out at temperatures down to 38°C. The copolymer latex may be used directly to blend with SAN grafted-rubber latex to make ABS, or it may be coagulated, washed, and dried to recover the SAN copolymers.

Figure 6 shows the conditions and the process diagram of a typical batch emulsion process (25). The recipe illustrated is:

Initiator–emulsifier solution:	
Dresinate (rosin soap, Hercules)	3 parts
$K_2S_2O_8$	0.06 part
desalted water	146 parts
Monomer solution:	
styrene	68.5 parts
acrylonitrile	31.5 parts
tert-dodecyl mercaptan,	0.4 part

Portions of both solutions are charged initially and purged with N_2; the remaining portions are added continuously. The reaction is controlled at 80°C through reflux cooling of a fraction of the polymerization mixture from the reactor. After the continuous charges are completed, the reaction is allowed to continue for another hour to complete the polymerization. The cycle time is claimed to be about 1–3 h.

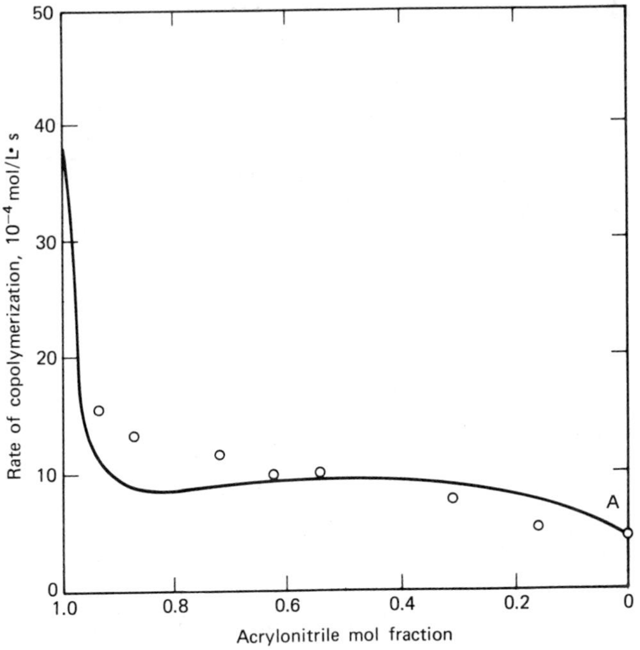

Figure 4. Rate of styrene–acrylonitrile copolymerization as function of acrylonitrile monomer concentration (19). A, Walling's equation.

A continuous emulsion process has been described (26–27), using two stirred-tank reactors in series followed by a large hold-tank. In a continuous stirred-tank reactor, the theoretical emulsion particle residence time in the reactor varies from zero to infinity (28). Thus, process kinetics become more complicated.

The polymer is generally recovered by coagulation of the emulsion latex, either by addition of electrolyte or by freezing, then the coagulum is washed and dried. Because fine particles cause difficulties in the washing and drying steps, process refinements have been developed to reduce the formation of such particles during coagulation (29–31).

Suspension Process. In the suspension process, the reaction system contains monomers, chain-transfer agent, initiator, suspending agent, and water. Copolymerization is carried out from temperatures as low as 60°C to as high as 150°C. A typical recipe is:

styrene	70 parts
acrylonitrile	30 parts
dipentene	1.2 parts
di-*tert*-butyl peroxide	0.03 part
acrylic acid–2-ethylhexyl acrylate (90:10)	0.03 part
water	100 parts

In the emulsion process emulsifier levels are typically 1–5% of the monomer weight. Suspending agent levels, however, are quite low in the suspension process. Generally, the amounts used are approximately 0.01–0.05% of the monomer weight. The resulting

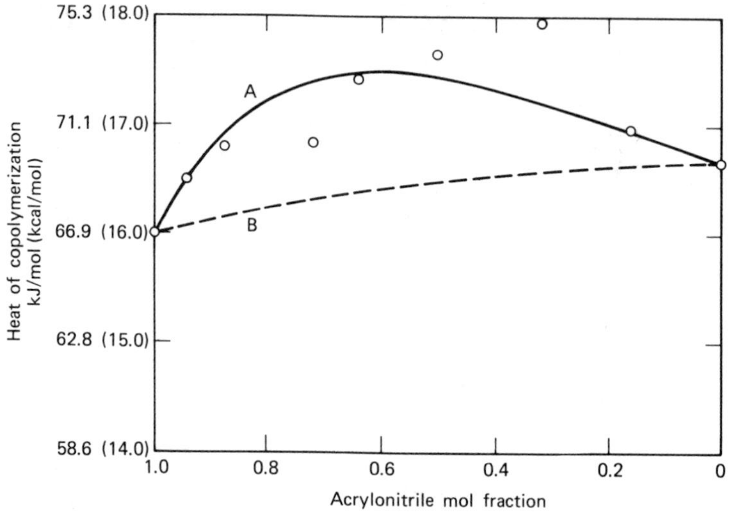

Figure 5. Heat of styrene–acrylonitrile copolymerization as function of acrylonitrile monomer concentration (21). A, Alfrey's equation; B, simple equation.

product is small polymer spheres of approximately 0.06 cm in diameter as opposed to the very fine particles of submicrometer size obtained from the emulsion process. This greatly increases the ease of polymer recovery and water removal. Figure 7 shows a typical suspension process (32). All the components are charged into a pressure reactor and purged with N_2. The mixture is agitated and heated to 128°C for 3 h, then to 150°C for 2 h. At the end of the copolymerization, the unreacted monomers can be distilled out of the polymeric beads. The slurry is then transferred to a centrifuge for washing and dewatering. Due to the relatively large size and smooth surface of the suspension polymer beads, surface moisture is quite low, and can readily be removed by drying.

Several variations from the above described suspension process have been claimed to generate good quality products of low haze level (33–36) and good color (37–41). To avoid compositional drift in nonazeotrope copolymerization, special techniques are employed (42–43). Other process refinements have been described which produce larger particles (44) and improved reaction conditions (45–46).

Continuous Mass Process. SAN continuous mass copolymerization is conceptually simple but practically complicated. It can be initiated either thermally or catalytically and a chain-transfer agent may be used. Copolymerization is carried out between 100 and 200°C. Solvents can be used to reduce the viscosity or the copolymerization can be conducted at a low conversion level (40–70%) followed by devolatilization to remove unreacted monomers and solvent. Devolatilization is carried out from 120 to 260°C under vacuum lower than 20 kPa (2.9 psi). The devolatilized polymeric melt is then fed through a strand die, cooled, and pelletized. Because of the high viscosity of the copolymerizing medium, it requires sophisticated design of equipment to handle the highly viscous material, to remove the heat of copolymerization, the unreacted monomers and solvent, and to keep the composition uniform. A general copolymerization process will be illustrated first, and then various designs of equipment will be discussed.

As shown in Figure 8, the monomers are fed continuously into the screw reactor

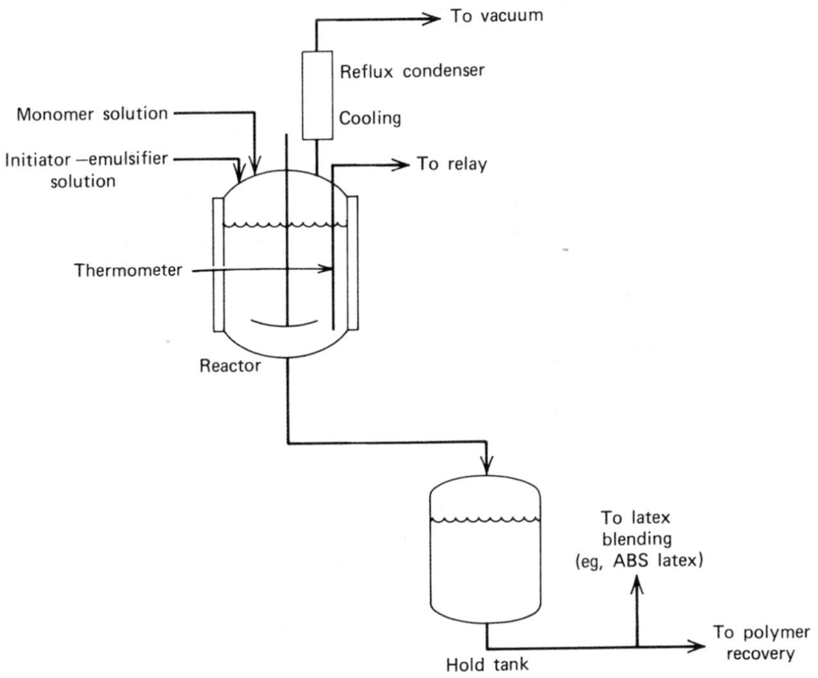

Figure 6. Styrene–acrylonitrile batch emulsion process (25).

(47). Copolymerization is carried out at 150°C to 73% conversion in 55 min. The heat of copolymerization is removed from both the barrel and the screw surfaces. The polymeric melt is withdrawn from the reactor and fed into the devolatilizer to remove the unreacted monomers under low pressure (4 kPa or 30 mm Hg) and high temperature (220°C). The final product is claimed to contain less than 0.7% volatiles. Two devolatilizers in series are claimed to yield better quality product, and better operational control than a single devolatilizer (48–49).

The two basic reactors used in the continuous mass process are the stirred-tank reactor (50) and the linear-flow reactor. The stirred-tank reactor consists of a horizontal cylindrical chamber with external cooling jacket. Various designs of agitators are used for mixing the viscous melt (51–52) in these reactors. The two types of linear-flow reactors are the screw reactor (47) and the tower reactor (52). A screw reactor is composed of two concentric cylinders. The mixture is polymerized while it is conveyed towards the outlet by rotating the inner cylinder which has helical threads. Heat of polymerization is removed from the surface of both cylinders. A tower reactor with three separate temperature zones has been claimed. The upper region has a scraper agitator and the lower portion generates plug flow. In a stirred-tank reactor with adequate mixing the composition of the melt inside the reactor is homogeneous. Therefore, operating at a fixed conversion with make-up monomers added at an amount and ratio equal to the amount and composition of copolymer withdrawn, the copolymer composition will become constant. However, in a linear-flow reactor the conversion varies along the axial direction of the reactor, as does the copolymer composition, except when operating at the azeotrope composition. Consequently, streams of monomer added along the reactor will be required to maintain SAN co-

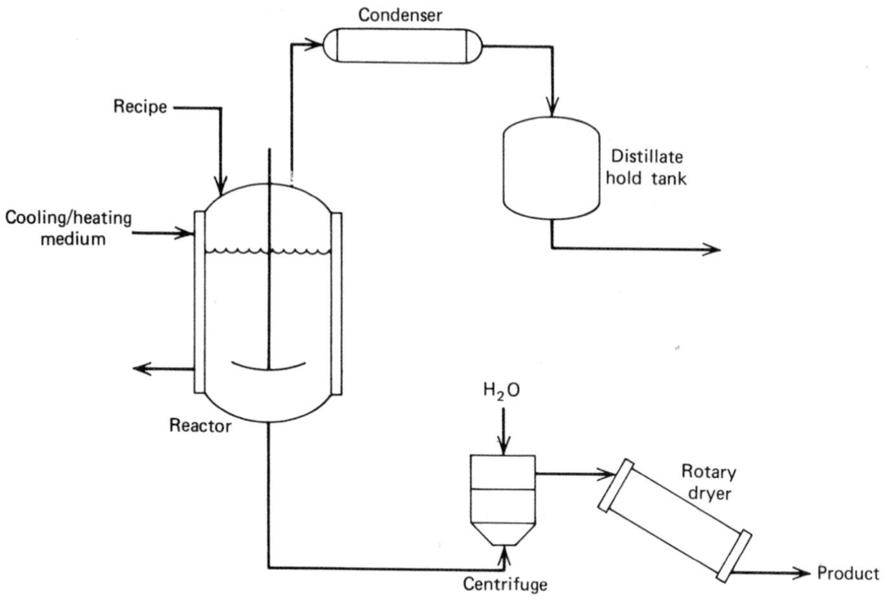

Figure 7. Styrene–acrylonitrile suspension process (32).

polymer compositional homogeneity if the copolymerization is carried out to a high conversion. A two-stage combination of a stirred-tank reactor followed by a linear-flow reactor has been claimed (52). It has been claimed (53) that through continuous recycle copolymerization, a copolymer of the same composition as the monomer feed can be achieved, regardless of disproportionate monomer reactivities. In addition to the different reactors used in this polymerization process, various devolatilizer designs have also been described (48–49,51,54–56).

Process Comparisons. Because there is no need for emulsifiers, suspending agents, salts, or water in the continuous mass process, it is a self-contained system without waste treatment or environmental problems. The other two processes do require additional waste treatment steps. In addition, since the SAN copolymers manufactured by the continuous mass process do not contain residual emulsifying or suspending agents, they generally have superior color and haze in molded form, and are preferred for applications requiring high degree of optical transparency. Both continuous mass and suspension SAN copolymers can be used for either molding applications directly or blending with other resins such as ABS to dilute its rubber concentration. Emulsion SAN copolymers are uniquely suited for manufacture of ABS by virtue of the ease of obtaining uniform dispersion of the grafted rubber in the SAN matrix through blending and co-coagulation of the SAN and the grafted-rubber latexes.

The continuous mass process does not use a large quantity of water as do both the emulsion and suspension processes. It has very efficient space and time utilization. Because there is no need to use water in this process, it does not require a polymer drying operation. Therefore, it consumes less energy, and is more economical than the other processes. However, because of the residence time distribution and the backmixing in the continuous stirred-tank reactor, and the lateral mixing in the linear-flow reactor, it requires some time to reach a steady state. Hence, during start-up and shutdown operation, some off-grade materials will be produced. In addition, when

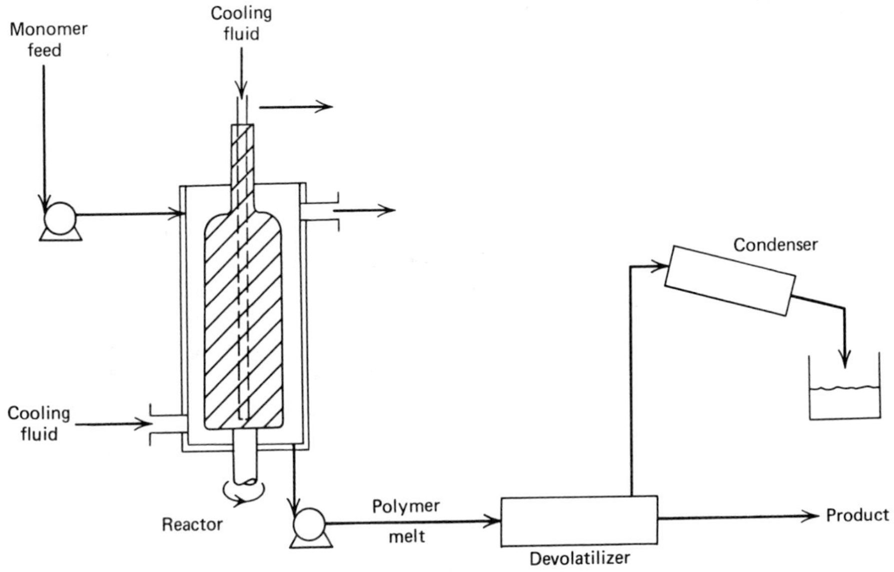

Figure 8. Styrene–acrylonitrile continuous mass process (47).

several different grades of product are produced in a production line, during the transition period of product grade change-over, some intermediate grade of material is produced which may not be desirable. Due to the highly viscous polymeric melt of the continuous mass process, good mixing and adequate heat removal are very difficult to achieve. Hence, in order to overcome these problems, the process operation and equipment design are complicated. In both emulsion and suspension processes, the polymer particles are dispersed and suspended in the low viscosity aqueous medium. Mixing and heat transfer are not a problem.

Economic Aspects

A major portion of SAN produced is incorporated in ABS. Their markets (particularly ABS) have been growing steadily since they were introduced in the 1950s. Figure 9 shows the United States sales history of SAN (3,57), and Figure 10 shows its price history (3–4,58). These histories for acrylonitrile monomer and for all plastic and resin materials are also included in these two figures. The price reductions in the 1960s resulted from the acrylonitrile capacity increase during this period. Since then, the sales increased steadily while prices held firmly or slightly reduced. In 1974, the price increased drastically and sales decreased slightly. This was because of the oil embargo and the double digit inflation in the United States. In 1975, sales continued to drop due to the continued 1974 recession and the reduction of the excessive build-up of inventory prompted by the oil embargo. In 1975, there were four producers of SAN copolymers in the United States (59).

Manufacturers	*Trade names*
The Dow Chemical Company	Tyril
B. F. Goodrich Chemical Co.	

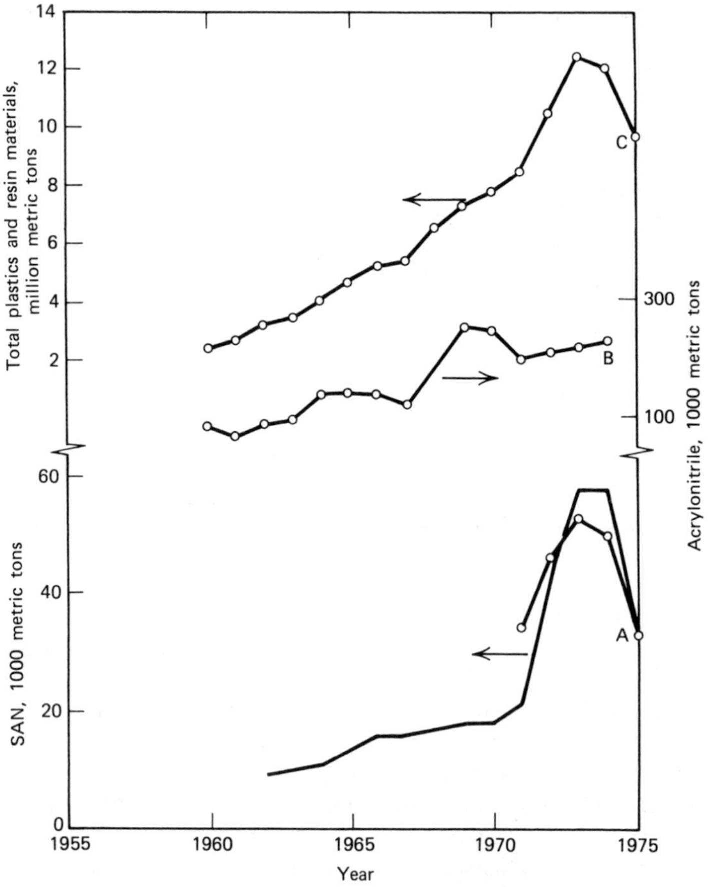

Figure 9. United States sales history of styrene–acrylonitrile copolymers. A, SAN (3,57); B, acrylonitrile (3); and C; total (3).

Monsanto Company	Lustran SAN
Union Carbide Corporation	Bakelite

The first three are also ABS producers.

Health and Toxicology

Acrylonitrile monomer is a toxic and flammable substance which can create an explosion hazard, but it has been handled satisfactorily on an industrial scale. Although there is no evidence of cumulative effect, its toxic action is believed to be similar to hydrogen cyanide (enzyme inhibition of cellular metabolism) (60) (see Acrylonitrile). The toxicity of styrene monomer is believed to be relatively low. It is an irritant to the eyes and respiratory tract. Inhalation of high concentrations produces an anesthetic effect (61). The odor and irritation of the vapor make voluntary exposure to high concentrations rather unlikely. It has been reported that exposure to 10,000 ppm for 30–60 min, or 2500 ppm for 8 h is dangerous to life (62–63) (see Styrene).

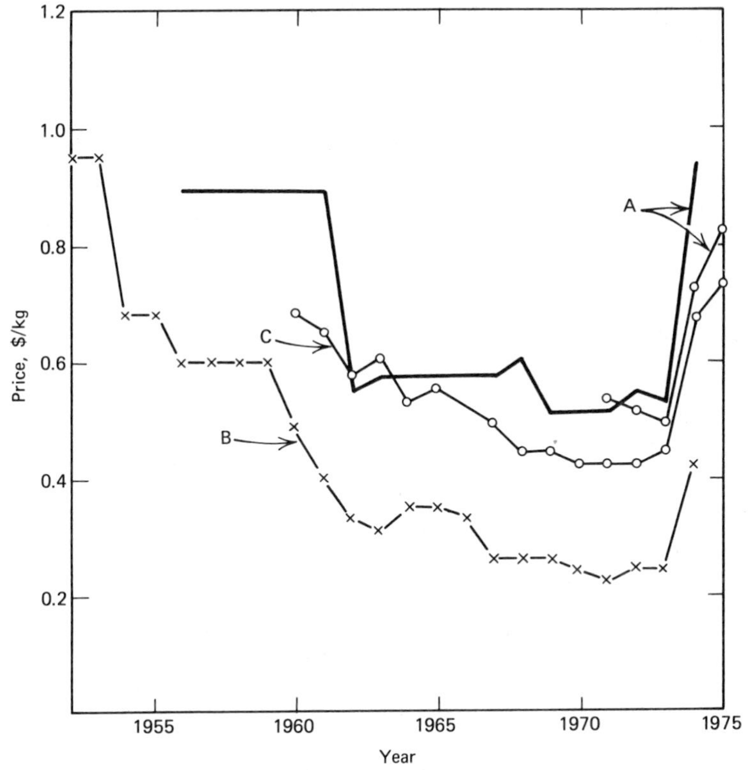

Figure 10. United States price history of styrene–acrylonitrile copolymers. A, SAN (58); B, acrylonitrile (3–4); and C, total plastics and resin materials (3).

Table 4 lists the maximum allowable acrylonitrile and styrene vapor concentrations in the working environment according to OSHA. The SAN copolymers are not considered toxic, but in food contact applications the amounts of extractable acrylonitrile monomer must not exceed the maximum set forth by the regulations of the FDA. Currently, the regulations are found in Title 21 CFR 181.32 and Title 21 CFR 181.22.

Uses

The useful properties of SAN copolymers, such as rigidity, resistance to chemicals and solvents, and toughness, etc, are due to the acrylonitrile content in the copolymers. Hence, the choice of the styrene and acrylonitrile composition of the copolymers is dictated by their particular applications. The well balanced and unique properties possessed by SAN copolymers have led to broad application in the market, and swift growth in the plastic industry. Table 5 lists the general uses of SAN copolymers and their United States market breakdown in 1974 (5,65).

Table 4. Maximum Allowable Acrylonitrile and Styrene Vapor Concentrations in Air [a]

Substance	8-h time weighted average	Acceptable ceiling concn	Acceptable maximum peak above the acceptable ceiling concn for an 8-h shift	
			Concentration	Maximum duration
acrylonitrile	20 ppm			
styrene	100 ppm	200 ppm	600 ppm	5 min in any 3 h

[a] Ref. 64.

Table 5. SAN Copolymer Market [a]

Applications	1974 United States market share, %	Articles
houseware	18	brush blocks and handles, broom and brush bristles, cocktail glasses, disposable dining utensils, hangers, ice buckets, jars, mugs, soap containers, and tumblers
appliances	13.5	air conditioner parts, decorated escutcheons, washer and dryer instrument panels, washing machine filter bowls, refrigerator shelves and crisper pans, blenders, mixers, lenses, knobs, and covers
packaging	13.5	bottles, closures, containers, display boxes, and films
industrial	10.8	batteries, business machines, medical apparatus and equipment, and tape reels
automotive	9	batteries, bezels, lenses, signals, and interior trim
general (custom molding)	35.2	aerosol nozzles, bottle sprayers, camera parts, dentures, pen and pencil barrels, sporting goods, toys, telephone parts, filter bowls, tape dispensers, terminal boxes, etc

[a] Refs. 5 and 65.

BIBLIOGRAPHY

1. American Cyanamid Company, *The Chemistry of Acrylonitrile,* The Beacon Press, Inc., New York, 1957, p. 7.
2. *Ibid.,* p. 56.
3. *Synthetic Organic Chemicals, U.S. Production and Sales,* United States International Trade Commission, Washington, D. C., 1961–1976.
4. J. D. Idol, Jr., *Appl. Polym. Symp.* **25,** 6 (1973).
5. Courtesy of Monsanto Company.
6. Ref. 4, pp. 3–4.
7. P. L. Morse, *Process Economics Program, Report No. 17, Acrylonitrile,* Stanford Research Institute, Menlo Park, Calif., 1966, p. 21.
8. A. W. Hanson and R. L. Zimmerman, *Ind. Eng. Chem.* **49,** 1803 (1957).
9. Brit. Pat. 590,247 (July 11, 1947), (to Bakelite Ltd.).
10. R. G. Fordyce and E. C. Chapin, *J. Am. Chem. Soc.* **69,** 581 (1947).
11. H. S. Davis and O. F. Wiedeman, *Ind. Eng. Chem.* **37**(5), 482 (1945).
12. W. V. Smith, *J. Am. Chem. Soc.* **70,** 2177 (1948).
13. P. J. Flory, *Principles of Polymer Chemistry,* Cornell University Press, Ithaca, New York, 1957, Chapt. V.
14. L. J. Young, in J. Brandrup and E. H. Immergut, eds., *Polymer Handbook,* 2nd ed., John Wiley & Sons, Inc., New York, 1975, pp. II-137 to 138, II-306.

15. G. Goldfinger and M. Steidlitz, *J. Polym. Sci.* **3,** 786 (1948).
16. C. H. Basdekis, *ABS Plastics,* Reinhold Publishing Corp., New York, 1964, p. 47.
17. G. E. Molau, *J. Polym. Sci. B* **3,** 1007 (1965).
18. Brit. Pat. 1,328,625 (Aug. 30, 1973), (to Daicel Ltd.).
19. M. Suzuki, H. Miyama, and S. Fujimoto, *Bull. Chem. Soc. Jpn.* **35,** 60 (1962).
20. G. Mino, *J. Polym. Sci.* **22,** 369 (1956).
21. H. Miyama and S. Fujimoto, *J. Polym. Sci.* **54,** S32 (1961).
22. C. A. Uraneck, in J. P. Kennedy and E. G. M. Törnqvist, eds., *Polymer Chemistry of Synthetic Elastomers,* Pt I, Vol. 23, of *High Polymers,* Interscience Publishers, a division of John Wiley & Sons, Inc., New York, 1968, pp. 127–178.
23. F. A. Bovey, I. M. Kolthoff, A. I. Medalia, and E. J. Meehan, *Emulsion Polymerization,* Vol. 9, of *High Polymers,* Interscience Publishers, a division of John Wiley & Sons, Inc., New York, 1965.
24. Brit. Pat. 1,093,349 (Nov. 29, 1967), T. M. Fisler (to Ministerul Industriei Chimice).
25. U.S. Pat. 3,772,257 (Nov. 13, 1973), G. K. Bochum, P. K. Hurth Efferen, J. M. Liblar, and H. J. K. Hurth (to Knapsack Aktiengesellschaft).
26. U.S. Pat. 3,547,857 (Dec. 15, 1970), A. G. Murray (to Uniroyal, Inc.).
27. Brit. Pat. 1,168,760 (Oct. 29, 1969), (to Uniroyal, Inc.).
28. O. Levenspiel, *Chemical Reaction Engineering,* John Wiley & Sons, Inc., New York, 1964, Chapt. 9.
29. U.S. Pat. 3,248,455 (Apr. 26, 1966), J. E. Harsch, C. Moruska, A. M. Smith, T. G. Lainas, and A. G. Murray (to U.S. Rubber Co.).
30. Brit. Pat. 1,034,228 (June 29, 1966), (to U.S. Rubber Co.).
31. U.S. Pat. 3,249,569 (May 3, 1966), J. Fantl (to Monsanto Co.).
32. Brit. Pat. 971,214 (Sept. 30, 1964), (to Monsanto Co.).
33. U.S. Pat. 3,198,775 (Aug. 3, 1965), R. E. Delacretaz, S. P. Nemphos, and R. L. Walter (to Monsanto Co.).
34. U.S. Pat. 3,258,453 (June 28, 1966), H. K. Chi (to Monsanto Co.).
35. U.S. Pat. 3,287,331 (Nov. 22, 1966), Y. C. Lee and L. P. Paradis (to Monsanto Co.).
36. U.S. Pat. 3,681,310 (Aug. 1, 1972), K. Moriyama and T. Moriwaki (to Daicel Ltd.).
37. U.S. Pat. 3,243,407 (Mar. 29, 1966), Y. C. Lee (to Monsanto Co.).
38. U.S. Pat. 3,331,810 (July 18, 1967), Y. C. Lee (to Monsanto Co.).
39. U.S. Pat. 3,331,812 (July 18, 1967), Y. C. Lee and S. P. Nemphos (to Monsanto Co.).
40. U.S. Pat. 3,356,644 (Dec. 5, 1967), Y. C. Lee (to Monsanto Co.).
41. U.S. Pat. 3,491,071 (Jan. 20, 1970), R. Lanzo (to Montecatini Edison S.p.A.).
42. U.S. Pat. 3,738,972 (June 12, 1973), K. Moriyama and T. Osaka (to Daicel Ltd.).
43. Brit. Pat. 1,328,625 (Aug. 30, 1973), (to Daicel Ltd.).
44. U.S. Pat. 3,682,873 (Aug. 8, 1972), Y. Hozumi, Y. Sonoyama, and M. Omata (to Daicel Ltd.).
45. U.S. Pat. 3,444,270 (May 13, 1969), V. A. Aliberti (to Monsanto Co.).
46. U.S. Pat. 3,444,271 (May 13, 1969), V. A. Aliberti (to Monsanto Co.).
47. U.S. Pat. 3,141,868 (July 21, 1964), E. P. Fivel (to Resines et Verms Artificiels).
48. U.S. Pat. 3,201,365 (Aug. 17, 1965), R. K. Charlesworth, W. Creck, S. A. Murdock, and K. G. Shaw (to The Dow Chemical Co.).
49. U.S. Pat. 2,941,985 (June 21, 1960), J. L. Amos and C. T. Miller (to The Dow Chemical Co.).
50. U.S. Pat. 3,031,273 (Apr. 24, 1962), G. A. Latinen (to Monsanto Co.).
51. U.S. Pat. 2,745,824 (May 15, 1956), J. A. Melchore (to American Cyanamid Co.).
52. Jpn. Pat. 48-21783 (Mar. 19, 1973), H. Sato, I. Nagai, T. Okamoto, and M. Inoue (to Toray, K. K. Ltd.).
53. R. L. Zimmerman, J. S. Best, P. N. Hall, and A. W. Hanson, in R. F. Gould, ed., *Adv. Chem. Ser.* **34,** 225 (1962).
54. U.S. Pat. 2,849,430 (Aug. 26, 1958), J. A. Amos and A. F. Roche (to The Dow Chemical Co.).
55. U.S. Pat. 3,211,209 (Oct. 12, 1965), G. A. Latinen and R. H. M. Simon (to Monsanto Co.).
56. U.S. Pat. 3,067,812 (Dec. 11, 1962), G. A. Latinen and R. H. M. Simon (to Monsanto Co.).
57. *Mod. Plast. Materials and Market Statistics,* Jan. issues.
58. H. E. Frey, *Chemical Economics Handbook,* Stanford Research Institute, Menlo Park, Calif., Sept. 1975, p. 580.1503C.
59. *SPI Committee on Resin Statistics, Section I, Monthly Statistical Report,* compiled by Ernst & Ernst, New York, March 26, 1976, pp. 9–16.
60. *Acrylonitrile, Hygienic Guide Series,* American Industrial Hygiene Association, Detroit, Mich., Mar. 1957.
61. *Styrene Monomer, Hygienic Guide Series,* American Industrial Hygiene Association, Detroit, Mich., Sept.–Oct. 1968.

62. H. W. Gerarde, in F. A. Patty, ed., *Industrial Hygiene and Toxicology*, Vol. II, 2nd ed., Interscience Publishers, a division of John Wiley & Sons, Inc., New York, 1963, p. 1230.
63. *Occupational Safety and Health Reporter, Tab Section Contents 74, Toxic Substances*, The Bureau of National Affairs, Inc., Washington, D. C., 1971, pp. 74:2113, 74:2327.
64. *Occupational Safety and Health Standards, Subpart Z—Toxic and Hazardous Substances, Section 1910.1000 Air Contaminants*, U. S. Department of Labor, Washington, D. C., May 28, 1975, pp. 642.2 and 642.6.
65. G. P. Ziemba, *Acrylonitrile–Styrene Copolymers*, under "Acrylonitrile Polymers," in N. M. Bikales, ed., *Encyclopedia of Polymer Science and Technology*, Vol. 1, Interscience Publishers, a division of John Wiley & Sons, Inc., New York, 1964, p. 434.

FRED M. PENG
Monsanto Co.

ABS RESINS

The preparation and properties of poly(styrene-*co*-acrylonitrile) (SAN) have been discussed above. Because this copolymer is relatively brittle, it has been modified in different ways to form engineering thermoplastics with greater impact strength. Initially nitrile rubber was blended in mechanically. In the next development, a discrete elastomeric phase was uniformly dispersed in the copolymer. The interfacial behavior was modified by grafting polymer or copolymer onto the elastomeric phase. The resulting rubber and thermoplastic composite is known as ABS, based on its common monomers: acrylonitrile–butadiene–styrene [*9003-56-9*]. This *is not* a random terpolymer. Several manufacturing processes have been developed and ABS is produced in a wide range of compositions. Because of the composition range available, ABS exhibits a wide range of properties. This flexibility of composition and structure allows the use of ABS in automobiles, refrigerators, telephones, and other areas.

Properties

The physical properties of ABS plastics vary somewhat with their method of manufacture but more so with their composition. In general, emulsion processes are used to make materials of higher impact strength, and bulk or suspension processes are preferred for materials with less impact strength. The range of properties typically available is presented in Table 1. In this table, different classifications of impact strength and heat resistance are shown to illustrate the balance of properties that results from designing materials for these properties. Several conclusions can be drawn from Table 1. First, because higher impact strength is usually achieved by increased rubber content; tensile strength, modulus, hardness, and deflection temperature are generally lower for materials having higher impact strength. Also, elongation, specific gravity, and coefficient of expansion vary directly with rubber content.

Table 1. Typical Properties of ABS Plastics[a]

Properties	ASTM test method	Units	Ranges within major grade classifications			
			High impact	Medium impact	Low impact	Heat resistant
Mechanical, at 23°C						
impact strength, Izod	D256	J/m	375–640	215–375	105–215	105–320
	3.2 mm notched	(ft-lb/in.)	(7–12)	(4–7)	(2–4)	(2–6)
tensile strength	D638	MPa	33–41	41–48	41–52	41–52
		(psi)	(4800–6000)	(6000–7000)	(6000–7500)	(6000–7500)
elongation	D638	%	15–70	10–50	5–30	5–20
tensile modulus	D638	GPa	1.7–2.1	2.1–2.5	2.1–2.6	2.1–2.6
		(psi × 10^5)	(2.5–3.0)	(3.0–3.6)	(3.0–3.8)	(3.0–3.8)
hardness	D785	HRC (Rockwell)	88–90	95–105	105–110	105–110
specific gravity	D792		1.02–1.04	1.04–1.05	1.05–1.07	1.04–1.06
Thermal						
deflection temperature, annealed samples	D648	°C, 1820 kPa (264 psi)	93–99	96–102	96–104	102–112
linear coefficient of thermal expansion	D696	$°C^{-1} \times 10^{-5}$	9.5–11.0	7.8–8.8	7.0–8.2	6.5–9.3

[a] Manufacturers' literature for nonpigmented material.

Heat resistant grades thus tend to have lower impact strength as a consequence of reducing the rubber content to maintain stiffness at elevated temperatures. The most heat resistant grades are made by incorporating a third monomer, α-methyl styrene, into the matrix phase to raise its glass transition temperature (T_g).

Not shown in the table, but important in the use of ABS plastics, is the processability of these materials. Generally, lower impact grades are the most easily processed. Higher impact grades are somewhat more difficult because their higher rubber content makes them more viscous. Heat resistant grades, designed to be stiff at elevated temperatures, require somewhat higher processing temperature and pressures.

Physicochemical Aspects

ABS can be considered a blend of a glassy copolymer and a rubbery domain. The glassy copolymer can be varied in composition and becomes tougher as the acrylonitrile content is increased. Simple blending of rubber with this glassy copolymer does not lead to optimal impact properties. Free rubber phases tend to separate into large aggregates that are inefficient impact modifiers. Rubbery copolymers that are compatible with the styrene–acrylonitrile copolymer matrix can improve impact resistance, but properties such as tensile strength, hardness, and melt flow are then adversely influenced. Such compatibility allows the formation of uniformly distributed small rubbery domains. Compatibility with the glassy matrix is achieved by grafting styrene–acrylonitrile to the rubber molecules. The exact mechanism has not been elucidated. A reasonable hypothesis is that polybutadiene, or a butadiene-containing copolymer, has a labile hydrogen that can form an allylic free radical on abstraction by a radical, R•:

$$-\!\!(CH_2-CH=CH-CH_2)\!\!- + R\cdot \longrightarrow -\!\!(\dot{C}H-CH=CH-CH_2)\!\!- + RH$$

This allylic radical can now initiate polymerization:

$$-\!\!(\dot{C}H-CH=CH-CH_2)\!\!- + CH_2=CHC\equiv N \longrightarrow -\!\!(\underset{\underset{\underset{\cdot CH_2-CN,\ etc}{|}}{CH_2}}{\underset{|}{CH}}-CH=CH-CH_2)\!\!-$$

or can combine with a growing polymer chain, P•:

$$-\!\!(\dot{C}H-CH_2=CH-CH_2)\!\!- + P\cdot \longrightarrow -\!\!(\underset{\underset{P}{|}}{CH}-CH=CH-CH_2)\!\!-$$

In either case, a grafted side chain is formed. An interface between the rubber domain and the free copolymer is created by this grafted material:

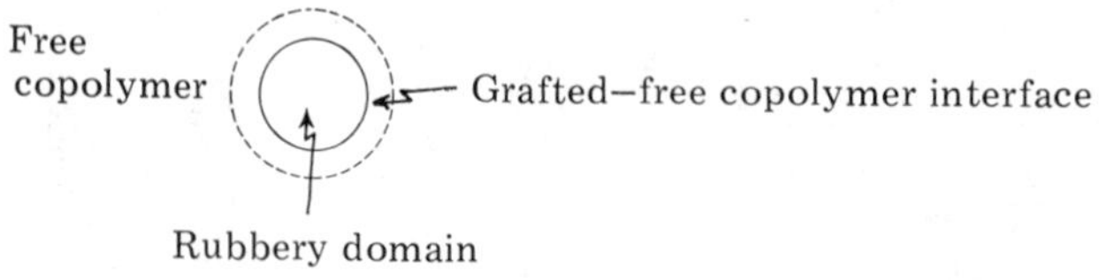

The nature and influence of the grafted phase is discussed in several articles (1–3).

In addition to the compatibility introduced by the grafted interface, some integrity must be given to the rubbery domain. This is achieved by crosslinking the rubber before or during the grafting reaction. This can be visualized as a combination of two polybutadienyl radicals:

$$2\!-\!\!(\dot{C}H\!-\!CH\!=\!CH\!-\!CH_2)\!\!- \longrightarrow \begin{array}{l} -\!\!(CH\!-\!CH\!=\!CH\!-\!CH_2)\!\!- \\ \quad\; | \\ -\!\!(CH\!-\!CH\!=\!CH\!-\!CH_2)\!\!- \end{array}$$

This introduces two elements of structure: crosslink density and domain size. Although a three-dimensional rubbery network throughout the polymer might be considered optimal for impact resistance, such a material (gel) would not have the melt flow required for processing. Typical domains found in commercial systems range from 1×10^{-5} to 100×10^{-5} cm.

Although infrared spectroscopy can give the overall composition of an ABS, it does not tell what rubber was used when a copolymer, such as styrene–butadiene, was employed. To better assess this element, the T_g can be measured (4) using torsion pendulum or thermoanalytical methods. One can also gain insights into the nature of the interaction between the grafted and free copolymer in this way.

Because the two phases are interacting, the rubbery phase modifies the melt flow of the free copolymer. The amount of rubber also contributes to the hardness and impact behavior of the composite (5). Increasing the amount of rubber raises the impact strength but decreases the hardness. The impact resistance is also influenced by the particle size distribution of the rubbery phase (6–7). The failure mechanism has not been defined but is presently being discussed in the literature (8–10). Through the controlled variation of molecular weight of the rigid and grafted copolymers, and the composition, particle size distribution and crosslink density of the rubbery phase, a wide variety of end use properties is created under the label "ABS."

By using monomers other than the three basic monomers of ABS, additional properties are introduced into this system. For example, the replacement of part of the styrene with α-methylstyrene raises the T_g of the thermoplastic. This allows end uses requiring higher heat distortion temperatures. The melt flow (processability) is adversely affected, however, because the same molecular characteristics that resist local response to thermal energy also create a stiffer, less fluid chain.

The use of a styrene–butadiene rubbery phase with a styrene–acrylonitrile–methyl methacrylate terpolymer thermoplastic phase yields a composite with a uniform refractive index. Although most ABS is translucent, this formulation is a clear impact-resistant material.

Manufacture

Three types of polymerization processes are used for the commercial production of ABS plastics, emulsion, suspension, and bulk. In addition, there are numerous commercial ABS materials which are physical polyblends of SAN with one or more types of ABS graft polymers made by the above processes. Historically, emulsion and suspension processes have dominated the field of ABS manufacture. Recently, however, the bulk process has achieved commercial importance. Because bulk polymerization does not proceed in water, it has two inherent advantages over suspension and emulsion polymerization. First, wastewater treatment is minimal. Second, less energy per kilogram of product is consumed since dewatering, drying, and compounding steps are not necessary. Disadvantages of the bulk process include less product flexibility, greater mechanical complexity, and less complete conversion of monomer to polymer. This means that most ABS materials made by bulk require devolatization to remove residual monomers prior to compounding of the final product.

Emulsion Process. The emulsion process consists of three distinct polymerizations. A polybutadiene substrate latex is prepared; styrene and acrylonitrile are grafted onto the polybutadiene substrate; and styrene–acrylonitrile copolymer is formed. The latter two reactions may take place simultaneously in the same vessel, or in separate vessels and then latex blended. A flow sheet for the emulsion process is shown in Figure 1.

The polybutadiene substrate used is generally prepared in emulsion batch reactions either as the homopolymer or as a copolymer with up to 35% styrene or acrylonitrile. Reaction temperatures may be varied from 5 to 70°C depending upon the desired structure of the polymer (11). Water-soluble free-radical initiators such as potassium persulfate (11) or redox initiation systems may be used. The following formulation is a typical redox system (12) for a polybutadiene substrate.

Material	*Parts by weight*
butadiene	175.00
cumene hydroperoxide	0.30
sodium pyrophosphate	2.50
sodium oleate	4.00
dextrose	1.00
ferrous sulfate	0.05
water	200.00

The initiator (cumene hydroperoxide), activators (sodium pyrophosphate, dextrose, and ferrous sulfate), and emulsifier (sodium oleate) solutions are prepared in separate vessels and added to a reactor which has been purged of oxygen. Then, the demineralized water and butadiene are added, the temperature is increased, and the reaction cycle begins. Reactions are run in vessels equipped with a water circulating jacket to remove the heat of polymerization, and designed to withstand pressures up to 1000

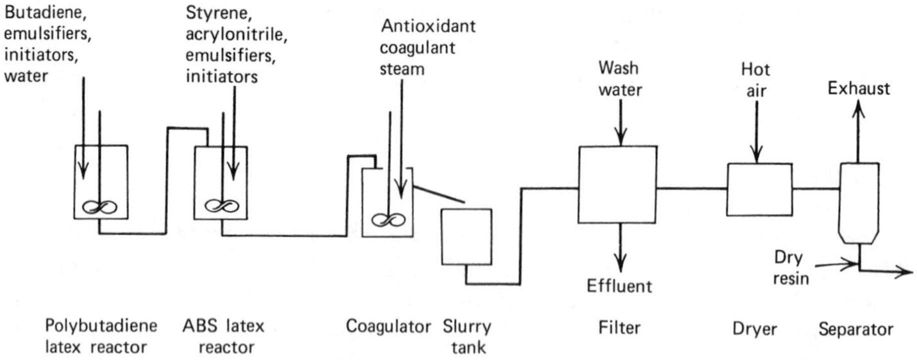

Figure 1. Emulsion ABS process.

kPa (145 psi). Reactors range in capacity from 13 to 30 m^3 (3400–7900 gal) and can be either stainless steel or glass-lined carbon steel. The reaction rate is limited by the ability of the cooling jacket to remove the heat of polymerization of 1278 J/g (550 Btu/lb) (13), and reaction times usually range from 12 to 24 h. The average latex particle size is determined by the emulsifier and monomer quantities and the ionic strength of the reaction medium (14). One can also vary the size of the polybutadiene domains by growing the latex in stages (15) using the "seed latex" technique. To prevent extensive crosslinking and ring-formation at higher conversions, in some processes the polymerization is stopped at 70–90% conversion and a shortstop such as sodium dithiocarbamate (11,13) is added. In these cases, the latex must be stripped to remove residual butadiene before it is stored for later use in ABS graft reactions.

In the next step in the emulsion process, styrene and acrylonitrile are grafted into the polybutadiene substrate. The amount of substrate used in the graft reaction is determined by the physical properties desired in the final polymer and usually ranges from 10 to 60 wt % of the total polymer. Whereas grafted ABS polymers with higher polybutadiene contents are generally blended with SAN copolymers, for grafts with lower polybutadiene contents the free SAN copolymer is generated *in situ*. A recipe for a 40% rubber graft reaction is given below (16).

Material	*Parts by weight*
polybutadiene–acrylonitrile (93:7) latex (50% solids)	900.0
water	1055.0
rubber reserve soap	2.0
2% aqueous solution potassium persulfate	240.0
styrene	455.0
acrylonitrile	235.0
terpinolene	4.8

In this reaction, potassium persulfate is used as the free-radical initiator and terpinolene as the chain-transfer agent. The product has a grafted SAN to substrate ratio of 1.0 to 1.0. The remainder may be considered to be free (nongrafted) SAN copolymer. This graft latex, containing 40% rubber substrate, is blended with emulsion SAN copolymer latex prepared separately to produce typical ABS resins containing 10–30% rubber. The thermal stability of the resin may be improved by adding such antioxidants as di-*tert*-butyl-*p*-cresol and tris(nonylphenyl)phosphite to the latex (17). Emulsion graft reactions are run at temperatures in the range of 55–75°C at atmospheric pressure in stainless steel or glass-lined vessels of up to 20 m^3 (5300 gal).

The graft reaction may be either a batch process or semibatch. In the latter case, the polybutadiene latex, the initiator solution, and the emulsifier initially are charged to the reactors. The reactor is heated to reaction temperature, and styrene and acrylonitrile monomers are pumped in over a period of 1–6 h. The batch is then cooled and pumped to the coagulation system.

Emulsion ABS resins are recovered from latex by coagulation with dilute salt or acid solutions. Calcium chloride, sodium chloride, sulfuric acid, and hydrochloric acid are commonly used for soap emulsions. Detergent emulsions can only be coagulated with salt (13). Coagulation is carried out at elevated temperatures (80–100°C) to promote agglomeration of the resin particles. The resin particle size can be controlled by salt or acid concentration, temperature, and slurry concentration. The slurry is then dewatered by filtration and/or centrifuging and dried in common hot air dryers such as a rotary fluid-bed, or flash dryer. The dry resin may be pneumatically conveyed to storage silos.

Suspension Process. In contrast to the emulsion process, the suspension process begins with a polybutadiene rubber which is so lightly crosslinked, it is soluble in the monomers.

Various techniques including emulsion or bulk polymerization may be used to make the polybutadiene, but it must be recovered and dried for use in the suspension process. Stereospecific polybutadiene and block styrene–butadiene copolymers have also been used for suspension polymerization (18–20). The following formulation is representative of suspension ABS (21).

Material	*Parts by weight*
soluble polybutadiene rubber	14.00
styrene	62.00
acrylonitrile	26.00
tert-butyl peracetate	0.07
di-*tert*-butyl peroxide	0.05
terpinolene	0.90
water	120.00
acrylic acid-2-ethylhexyl acrylate copolymer	0.30

The suspension ABS process is an extension of the suspension high-impact polystyrene process. A flow sheet is given in Figure 2. The first step in the suspension process, the "prepoly" step, is carried out in mass. The polybutadiene is dissolved in the monomers to produce a solution free of crosslinked rubber gels. A free-radical initiator such as an organic peroxide or an azo compound is added to the solution along with chain-transfer agents such as mercaptans or terpenes. It is then heated to 80–120°C for a period of 2–8 h with shearing agitation sufficient to prevent crosslinking and maintain the desired polymer particle size (18,21). The mass polymerization is carried out until 25–35% conversion of monomer to polymer is achieved and phase inversion has taken place. SAN is the continuous phase. This polymer syrup is then transferred via a high viscosity gear pump to a suspension reactor where it is dispersed in water with agitation. Many common suspension aids such as poly(vinyl alcohol), carboxymethylcellulose, or water soluble acrylic polymers are used to maintain the stability of the dispersion. The reactor is heated to 100–170°C depending upon the half-life of the ini-

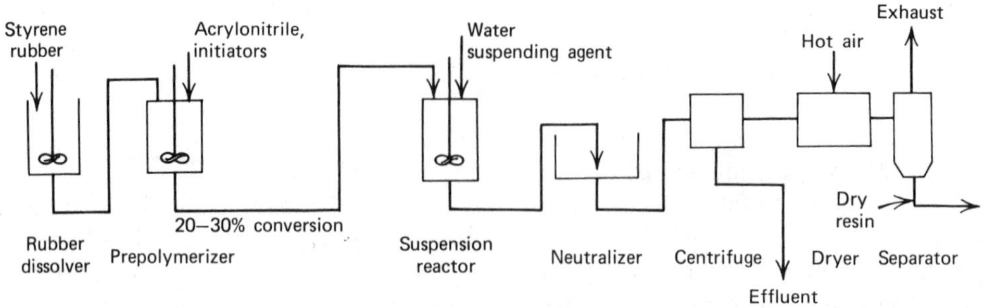

Figure 2. Suspension ABS process.

tiator used, until polymerization is essentially completed (6–8 h). Suspension reactors have capacities of up to m^3 (10,600 gal) may be either stainless steel, stainless-clad carbon steel, or glass-coated carbon steel, and must withstand the autogenous pressures of the monomers at the reaction temperature of 350 kPa (3.5 atm).

Reactor cycle time is limited by the heat transfer capacity of the reactor. The heat of polymerization of typical ABS reactions ranges from 810 to 930 J/g (350–400 Btu/lb) of monomer converted to polymer depending upon the ratio of styrene to acrylonitrile used.

When the suspension batch has reached the desired conversion, it is cooled, and the slurry is pumped to the dewatering system. There is sufficient post reactor surge capacity so that the dewatering system may be operated continuously. The slurry is pumped from a feed tank to a continuous centrifuge where the beads are washed and dewatered to less than 10% moisture. The wet product is then conveyed into a conventional hot air dryer where it is dried to less than 1% moisture. Suspension ABS beads range in size from 0.4 to 1.2 mm in diameter. The dry beads are stored in silos prior to compounding.

Bulk Process. The first step in the bulk ABS process is essentially the same as in the suspension process. Figure 3 shows a flow sheet for the bulk process. A monomer soluble polybutadiene rubber or butadiene copolymer is dissolved in the styrene and acrylonitrile with initiators and modifiers. This mixture is polymerized through phase inversion to approximately 30% conversion under sufficient shearing conditions to prevent the rubber from crosslinking. Instead of dispersing the prepolymerized syrup in an aqueous medium, as in the suspension process, the syrup is pumped into a specially designed bulk polymerizer where conversion is taken to from 50 to 80%. Several types of these polymerizers may be used interchangably for high impact polystyrene or ABS (21–23). Other bulk processes more recently described in the patent literature (24–30) are designed specifically for ABS polymerization. Bulk polymerizers are operated continuously at 120–180°C with residence times of 1–5 h. The heat of polymerization is removed by evaporation cooling of the monomers, heat transfer through the reactor wall, and heating of the fresh monomers being charged. The monomer vapors are condensed and recycled with the monomer feed stream. After the reaction, the polymer is pumped to a devolatizer where unreacted monomers are removed under vacuum at temperatures in excess of 150°C. Normally, about 5–30% of the feed stream is removed as unreacted monomer and recycled. The ABS polymer is removed from the devolatizer via melt pump or extruder and chopped into pellets.

An organic solvent may be added to the feed stream to reduce the viscosity of

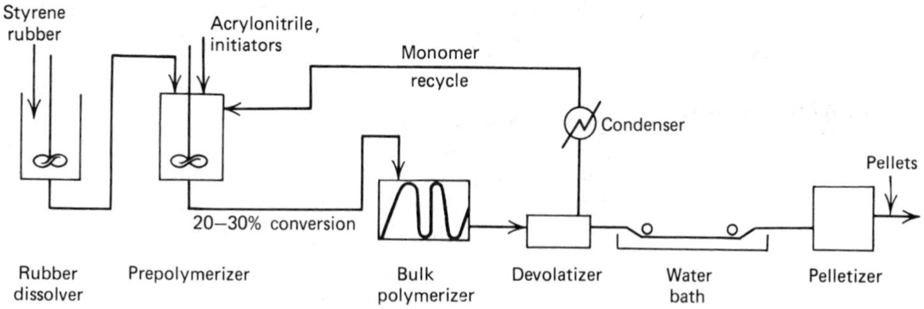

Figure 3. Bulk ABS process.

the polymerizing mass and aid in control of this polymerizing mass and the reaction rate (24). These diluents are used in amounts of up to 20% of the feed, and they are recovered in the devolatilization zone and recycled with the unreacted monomers.

Resin Blending and Compounding. Each process for ABS polymers produces a material with unique characteristics in the grafted phase. Some ABS processes take advantage of this by blending ABS polymers made by the emulsion process with ABS polymers made by the suspension or bulk process to achieve a material with some of the desirable characteristics of each (21). Those materials may be mechanically blended using conventional melt-mixing equipment. One process, however, incorporates a dewatered emulsion ABS graft into the polymerization zone of a bulk process (25–26). Several ABS polymers are also mechanical blends of high rubber content ABS grafts made by emulsion with SAN copolymer made by suspension or bulk polymerization. This enables the manufacturer to vary independently the structure of the graft and the rigid phases.

The dry resin made by either the emulsion or suspension process is usually compounded into pellet form before being sold to plastic processors. High shear devices such as Banbury mixers, single-screw extruders, or twin-screw extruders may be used. In these machines, the polymer is subjected to high shear conditions which create sufficient frictional heat to flux the polymer and disperse pigments, lubricants, stabilizers, or other additives (150–250°C). Also, ABS resins may be blended with other ABS resins, SAN copolymers, PVC, polycarbonates, acrylics, or other materials in the compounding step to achieve a product with the desired properties.

Banbury mixers are batch operated. The polymer melt is dropped from the Banbury onto a conventional two-roll mill which provides enough surge capacity so that the melt may be continuously diced or chopped into pellets. Extruders operate continuously, and the pellets are chopped directly from the extruded melt.

Quality Control. Quality control specifications are as numerous as the grades of ABS that are produced. However, the same properties are of key importance in most grades: impact strength, melt flow, tensile strength, and residual monomer content. These properties are monitored periodically by sampling the resin at the first point in the process where no additional polymerization occurs. The polymer is recovered from the aqueous phase in the case of emulsion and suspension processes. Residual monomer content is then determined by vapor-phase chromatography; the melt index of the material is measured; and molded samples are measured for impact and tensile strength by the appropriate ASTM method.

Materials which are off-quality may often be blended with first grade materials to make an acceptable product. After being compounded, each lot of compounded pellets is checked for a more complete series of physical properties before shipment to the customer.

Analytical Methodology

Analysis of ABS is undertaken to ascertain the overall composition of the sample, the composition and morphology of the grafted rubber phase, and the nature and amount of nonpolymeric materials present. The methods employed are similar to those used in the impact polystyrene field and are described by Moore and co-workers (31).

The initial step is to run a general infrared spectroscopic scan on a thin film. The overall monomer composition is then calculated. Elemental analysis can be used to supplement this data. In addition, the nature of the additive package present sometimes can be inferred from the spectral data.

The free copolymer and the grafted rubber phase are separated by centrifugation after dispersing the ABS in a solvent such as acetone. The supernatant contains the copolymer and is decanted from the grafted phase. After drying, the amounts of the two phases are determined gravimetrically. The composition of each phase can be determined spectroscopically. The copolymer molecular weight distribution is assessed using gel-permeation chromatography. Appropriate column and carrier solvent systems have been suggested by Dubin (32). Typically, one finds number-average molecular weights of 20×10^3, with weight- to number-average molecular weight ratios of 3 to 10. The polybutadiene can be oxidized to leave the grafted copolymer. This can then be evaluated for composition and molecular weight distribution.

The morphology of the grafted phase is qualitatively defined by transmission electron microscopy of thin sections that have been stained with OsO_4. Two general types of morphology are discerned. Figure 4 shows typical rubber domains of a grafted, emulsion polymerized rubber substrate. Figure 5 shows the rubber domains created in the suspension process. The size distribution of these domains can be inferred by measuring the photomicrographs.

The nonpolymer components consist of unpolymerized monomers and additives such as antioxidants and lubricants. Monomers can be vaporized from solid samples directly into a gas chromatograph, or can be freed from polymer by dispersing a sample in a solvent and injecting the dispersion into a gas chromatograph. Lubricants, plasticizers, and antioxidants are determined by liquid chromatography. If necessary, preparative liquid chromatography and infrared spectroscopy can be combined for positive identification of these additives.

Economics

Reliable worldwide statistics are virtually unavailable because most reporting agencies include ABS with styrene or styrene copolymers. Table 2 shows United States production and consumption of ABS plastics for the years 1971 through 1975. Prior to 1971, statistics are unavailable. Consumption in Western Europe was estimated to be 160,000 metric tons in 1975 (33).

Table 2. Production, Sales, and Use of ABS Plastics [a] in the United States, 1971–1975 [b]

Year	Production	Sales and use
1971	309	296
1972	389	365
1973	405	409
1974	386	367
1975		294

[a] In thousands of metric tons.

[b] Sources: various, includes trade magazines, *Synthetic Organic Chemicals Report*, and *Society of the Plastics Industry Reports.*

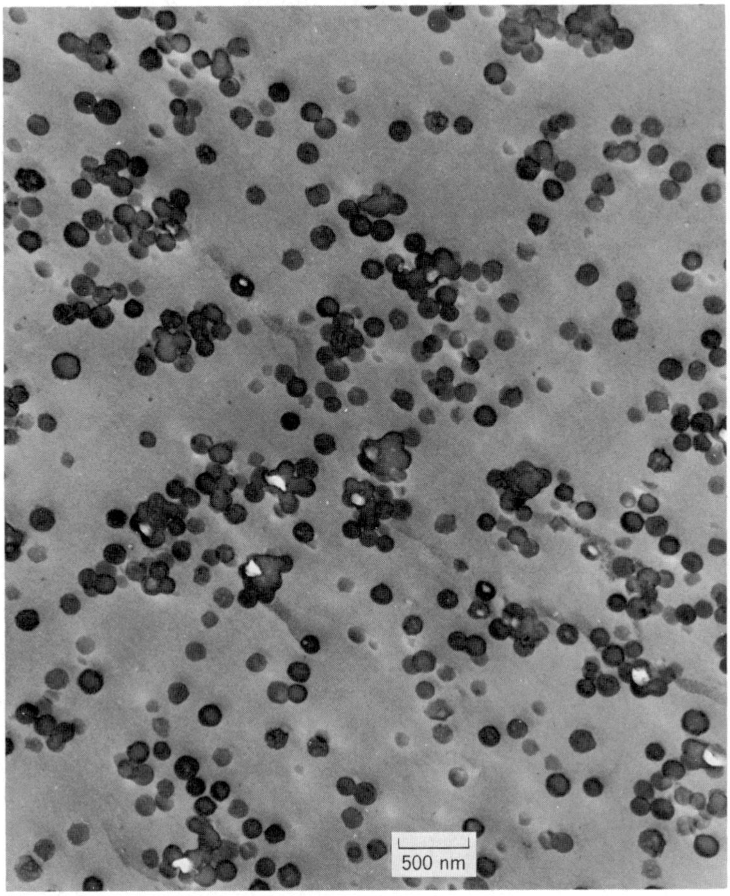

Figure 4. Electron photomicrograph showing emulsion rubber domains.

The variations in the volume of ABS sold and used indicate the general flow of ABS into "durable goods" markets. Construction (pipe), automotive, and appliances are three major markets for ABS plastic that were vigorous in 1973 and severely depressed in 1975. The price history of ABS in the United States is presented in Table 3. For the years 1952 through the early 1970s, economies achieved due to the increased volume allowed lower prices. However, since 1970, growth in volume has been less dramatic and increased costs, particularly of fuel, labor, and feedstocks has forced the price of ABS upwards. In 1975, the total estimated production capacity for ABS plastics in the United States was 625,000 t. This capacity is shared by six producers as shown in Table 4.

End Use Processing and Applications

ABS plastics are fabricated into end products by virtually all of the common thermoplastic processes including injection and compression molding, extrusion, blow molding, and calendering. Important secondary operations include vacuum forming, vapor metalizing, plating, hot stamping, painting, ultrasonic welding and staking, and solvent and adhesive bonding (see Plastics technology).

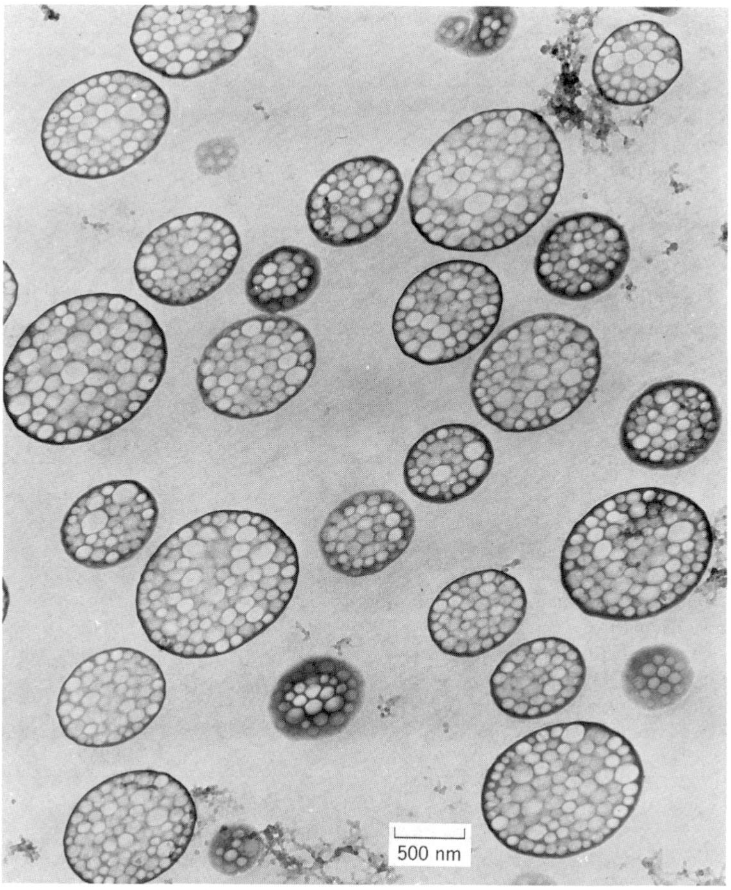

Figure 5. Electron photograph showing suspension prepared domains.

Because ABS plastics are mildly hygroscopic, these pellets must be dried before melt processing to avoid surface defects. Hot air dryers or dehumidifiers are used. Generally, subjecting pellets to temperatures of 85°C for 2–3 h is sufficient to reduce the moisture in the pellets to a level (<0.1%) which is adequate for most applications.

Design of tooling for injection molding and extrusion for ABS plastics must take into account their relatively high melt viscosities. Runners in molds should be large and full round; extrusion dies should be streamlined. Stock temperatures ranging from 200 to 270°C are generally employed.

Table 5 shows a breakdown of the estimated current and projected major end use markets for ABS in the United States. The largest single market is pipe and fittings. While this market is expected to grow in absolute terms, the percentage of ABS used for piping will decrease slightly. Drain, waste, and vent (DWV) pipe consumes the principal share of this market but other applications including mine drainage pipe and conduit are emerging. The next largest end use market for ABS is automotive. This market is expected to grow significantly partially due to design emphasis on weight and cost reduction and partially because of increased use of special grades such

Table 3. Price History of ABS Plastics in the United States [a]

Year	Price range, \$/kg[b]
1952	1.28–1.43
1955	1.17–1.28
1960	1.08–1.12
1970	0.73–0.93
1973	0.62–0.90
1976	1.15–1.43

[a] According to industry price schedules.
[b] Basis: truckload black or nonpigmented. Range is due to variations in properties available.

Table 4. Major Producers of ABS Plastics in the United States [a]

Producer	Estimated capacity, 1975, thousands of metric tons
Borg Warner Chemicals Division, Borg-Warner Corp.	234
Monsanto Co.	200
Uniroyal Chemical Division, Uniroyal Inc.	90
The Dow Chemical Co.	65
Rexene Polymers Co., Division Dart Ind. Inc.	22
B. F. Goodrich Co.	14
Total	*625*

[a] According to Industry estimates from trade publications.

as more heat resistant grades for paint curing and platable grades for trim. ABS is suited to these and other markets because of its combination of toughness, rigidity, appearance, processability, and moderate cost.

Outside the United States, consumption patterns are different. In Western Europe, as shown in Table 6, furniture is a significant market for ABS, but pipe is not. Consumer taste and styling account for the use of ABS in furniture in Europe, whereas in the United States, consumers generally prefer more traditional furniture designs to the contemporary "plastics" look. In the European pipe and fittings market, PVC

Table 5. Major Applications for ABS Plastics in the United States [a]

Application	1975, %	1980, %
pipe and fittings	30.5	26.0
automotive	13.5	15.5
recreation	12.4	15.5
appliances	11.2	12.5
business machines and telephones	5.3	6.5
luggage and cases	2.8	2.0
electrical and electronic	2.3	3.0
other[b]	17.9	15.0

[a] According to industry estimates.
[b] Includes packaging, toys, furniture, shoe heels, and other minor uses.

Table 6. Major Applications for ABS Plastics in Western Europe, 1975 [a]

Market	%
appliances	30
automotive	21
electrical/electronics [b]	13
furniture	9
pipe and fittings	4
recreation	2
other	21

[a] According to *Modern Plastics International,* January, 1976.
[b] Includes telephones and business machines.

is used far more widely than ABS. This accounts for the relatively small fraction (4%) of ABS used for piping in Europe compared with the fraction used in United States (30%).

BIBLIOGRAPHY

1. L. Bohn, *Angew. Makromol. Chem.* **20,** 129 (1971).
2. G. Kampf and H. Schuster, *Angew. Makromol. Chem.* **27,** 81 (1972).
3. J. Stabenow and F. Haaf, *Angew. Makromol. Chem.* **29/30,** 1 (1973).
4. L. Bohn, *Angew. Makromol. Chem.* **29/30,** 25 (1973).
5. W. J. Frazer, *Chem. Ind.* 1399 (1966).
6. R. P. Kambour, *J. Polym. Sci. Macromol. Revs.* **1,** 1 (1973).
7. L. Morbitzer, D. Kranz, G. Humme, and K. H. Ott, *J. Appl. Polym. Sci.* **20,** 2691 (1976).
8. E. H. Hagerman, *J. Appl. Polym. Sci.* **17,** 2203 (1973).
9. S. L. Rosen, *J. Elastoplast.* **2,** 105 (1970).
10. P. Zitek, S. Myvic, and J. Zelinger, *Angew. Makromol. Chem.* **6,** 116 (1969).
11. L. F. Albright, *Processes for Major Addition-Type Plastics and Their Monomers,* McGraw-Hill, New York, 1974, pp. 365–370.
12. U.S. Pat. 3,238,275 (Mar. 1, 1966), W. C. Calvert (to Borg-Warner Corporation).
13. W. M. Smith, *Manufacture of Plastics,* Vol. I, Reinhold, New York, 1964, pp. 446–454.
14. B. Vollmert, *Polymer Chemistry,* Springer-Verlag, New York, 1973, pp. 152 ff.
15. C. F. Parsons and E. L. Suck, Jr., *Adv. Chem. Ser.* **99,** 340 (1971).
16. U.S. Pat. 3,509,238 (Apr. 28, 1970), N. E. Aubrey and M. B. Jastrzebski (to Monsanto Company).
17. U.S. Pat. 3,222,422 (Dec. 7, 1965), L. A. Cohen (to Monsanto Company).
18. U.S. Pat. 3,278,642 (Oct. 11, 1966), L. Lee (to The Dow Chemical Company).
19. U.S. Pat. 3,442,981 (May 6, 1969), O. L. Stafford, D. V. Wing, and D. E. Stolsmark (to The Dow Chemical Company).
20. U.S. Pat. 3,346,520 (Oct. 10, 1967), L. Lee (to The Dow Chemical Company).
21. U.S. Pat. 3,509,237 (Apr. 28, 1970), N. E. Aubrey (to Monsanto Company).
22. U.S. Pat. 3,243,481 (Mar. 29, 1966), N. R. Ruffing, B. A. Kozakiewicz, B. B. Cave, and J. L. Amos (to The Dow Chemical Company).
23. U.S. Pat. 2,694,692 (Nov. 16, 1954), J. L. Amos, J. L. McCurdy, and O. R. McIntire (to The Dow Chemical Company).
24. U.S. Pat. 3,511,895 (May 12, 1970), N. Kydonieus, S. P. Sence, R. F. Blanks, and D. E. James (to Union Carbide Corporation).
25. U.S. Pat. 3,903,199 (Sept. 2, 1975), W. O. Dalton (to Monsanto Company).
26. U.S. Pat. 3,903,200 (Sept. 2, 1975), D. L. Cincera, W. O. Dalton, M. B. Jastrzebski, and C. E. Wyman (to Monsanto Company).
27. U.S. Pat. 3,903,202 (Sept. 2, 1975), D. E. Carter and R. H. M. Simon (to Monsanto Company).
28. U.S. Pat. 3,928,495 (Dec. 23, 1975), W. O. Dalton (to Monsanto Company).
29. U.S. Pat. 3,751,526 (Aug. 7, 1973), H. Okasaka, T. Okamoto, T. Hirai, and M. Inoue (to Toray Industries, Inc.).

30. U.S. Pat. 3,950,455 (Apr. 13, 1976), T. Okamoto, A. Kishimoto, M. Inoue, I. Nagui, and M. Otani (to Toray Industries, Inc.).
31. L. D. Moore, W. W. Moyer, and W. J. Frazer, *Appl. Polym. Symp.* **7**, 67 (1968).
32. P. L. Dubin, *Ind. Res.* **18**(7), 55 (1976).
33. *Modern Plastics International,* Jan. 1976, p. 9.

G. A. MORNEAU
W. A. PAVELICH
L. G. ROETTGER
Borg-Warner Chemicals

ADHESIVES

Adhesives, often referred to as cements, glues, or pastes, have been used since antiquity. But only since the advent of modern polymer technology have they become a significant item of commerce in manufacture and construction (Fig. 1).

The ancients used waxes, natural resins, gums, and asphaltic pitches as "hot melt adhesives" to join a variety of substrates, and many of these materials are still used successfully today. These compositions had limited heat, moisture, and biological resistance. Modern adhesives made from synthetic resins are more durable.

Phenolic resins were used in the earliest synthetic adhesives for heat, moisture, and biological resistance but they were often too brittle. The first adhesives that were both strong and tough under thermal stress were made by modifying the high tensile

Figure 1. Adhesives are used to bond the 3224 hollow concrete segments of this 14-k bridge, the President Costa e Silva in Rio de Janeiro, one of the longest in the world. Each segment is 12.8 m wide and weighs from 80–110 metric tons. The Campenon Bernard technique of segmented concrete construction (1) was used with quick-setting epoxy-based mortar. The adhesive-bonded joints carry the vertical shear loads between the segments, and stressed steel cables passed horizontally through the segments take all the bending tensile loads in the beam which they form. The epoxy mortars have proved themselves more than equal to the shear forces encountered. They also help to secure the steel cables within the concrete, ensuring an even distribution of stress and providing maximum protection from the environment. The construction technique has been proved in about 100 bridges, viaducts, and other major construction projects (2). (**a**) Scene showing two rolling jib cranes used to lift prefabricated deck segments which will later be adhesive bonded. (**b**) Prefabricated concrete segments used to construct the deck beams. Courtesy of CIBA-GEIGY.

strength phenolics with high modulus synthetic rubbers and flexible resins. Such heating and cooling stresses often occur during transportation and use of commercial products.

The most successful widely known product of the new technology was the automotive bonded brake lining first introduced in 1947, and now regarded as a symbol of quality and integrity (see Brake linings).

The phenolics are condensation resins that liberate water when formed. If trapped and heated, the steam can drastically weaken or literally blow a joint of impermeable surfaces apart. Early adhesives based on condensation resins required bonding under compression to contain the steam. This limited their usefulness to flat objects bonded in presses, curved objects in specially designed pressure fixtures, or more complex shapes bonded in autoclaves. Newer resins, such as epoxy, urethane, and silicone are addition polymers and can be used without expensive presses. The adhesive industry grew with the advent of these new resins because no special equipment was needed to make useful joints. Most commercial bonding is done without even a rudimentary knowledge of adhesive technology. The high technology of adhesive bonding has originated from high-volume low-cost items (plywood, laminated plastics, and wet and dry abrasive paper) and low-volume, high-cost items (aircraft and aerospace systems, electronic equipment, and sophisticated plastic items).

Definition of an Adhesive

There is no class of materials known as adhesives. Rather, an adhesive is defined by usage. The industry-accepted (3) definition is "a substance capable of holding materials together by surface attachment." Substances may attach to surfaces and develop the internal or cohesive strength necessary to hold materials together while cooling from the liquid to the solid state, while losing solvent, or during chemical reaction. Substances called pressure-sensitive adhesives generally do not undergo a phase change in order to hold materials together.

Not all such substances are commonly thought of as adhesives. For example, water frozen between two surfaces may serve as an adhesive when performing a useful function, but is considered a nuisance when it joins a frozen food package to the freezer shelf.

Waxes, gums, tars, natural resins, and thermoplastic synthetic resins cooled from the liquid state, between substrates in thin layers called bondlines, may be considered adhesives if they can transfer loads from one substrate to another. Substances that lose solvent and surface-attach, such as sodium silicate (waterglass), or natural and synthetic resins and elastomers when dissolved or dispersed in water or organic solvents, may also be adhesives. Substances that surface-attach and then chemically combine in thin joints, such as chain lengthening or crosslinking synthetic resins and inorganic metal oxide–phosphoric acid dental cements, may also be designated as adhesives. Solvent-free natural and synthetic resins having a viscoelastic property termed tack belong to that special group of materials called pressure-sensitive adhesives and are commonly used on surgical, household, and industrial tapes.

What is Not an Adhesive. Many of the same substances designated above as adhesives, when applied in thin films to only one surface, may be called paints, finishes, or coatings, and when employed in thick masses, may be called caulking, potting, casting, or encapsulating compounds (see Chemical grouts; Sealants; Embedding).

Low-melting glasses and metal alloys, such as lead-tin solders (qv), high-melting welding filler metals, and thick layers of inorganic or polymer mortars and concretes are not usually termed adhesives, even though they may satisfy the accepted definition (see Cement; Welding). Adhesive technologists do not all agree as to what is and what is not an adhesive (4–5).

Borderline Materials. High tensile strength elastomeric sealants, sulfur melted and cooled in a joint, and polymer concrete used to join fresh, wet concrete to old concrete, are often termed adhesives when used in thin bondlines. Hot tar (qv) used to join roofing papers, waxed paper used in packaging, materials used to bond particles into abrasive wheels and particle board, so-called plastic wood and plastic solder, vulcanizing rubber tie-coats, nonwoven fiber binders and inorganic dental and refractory joining compounds, may not be considered adhesives by persons using the same materials in different industries.

To be termed an adhesive, a substance must be a liquid or tacky semisolid, at least for an instant to contact and wet a surface, and be used in a relatively thin layer to form a useful joint capable of transmitting stresses from one substrate to another.

The Adhesives Industry

Adhesives must compete with nonadhesive joining techniques, such as welding, and the use of mechanical fasteners which are used predominantly, eg, bolts, nails, rivets, screws, and staples. Adhesive joining is used to replace or supplement these more traditional joining methods only where joint quality, performance, or economics of the former are in question except in the aerospace and laminating industries where adhesive joining is often the technique of choice. Most of the literature on adhesive bonding was spawned in the aerospace industry. Where human life and/or multimillion dollar hardware items and mission success may depend upon an adhesive joint, there can be no compromise of quality. In down-to-earth applications, economic considerations may determine alternative joining techniques.

Adhesive Industry Economics. The 1972 census of manufacturers by the United States Department of Commerce assigns Standard Industrial Classification Number (S.I.C.) 2891 to adhesives and sealants. The group includes cement (qv), glue (qv), paste and waxes (qv). The census detailing adhesive and sealant industry general statistics (in 1972) reports 463 establishments employing approximately 15,000 persons primarily engaged in adhesives and sealants manufacture (6). There were fewer than 50 employees in each of 81% of those establishments. The industry shipped products valued at $954,300,000 in 1972. This has been estimated to include 6.6% of all polymers shipped, with an estimated 11.4% annual growth rate (7). The next set of figures, due in late 1977, should be well above $1.5 billion. There is every reason to believe that this growth rate will be sustained.

The value contributed to products through use of adhesives can be measured directly in the plywood industry. But in automotive and electronic uses where adhesives contribute only a small part of the final value of the product there is no way to measure the contribution of adhesives.

The same problem as that in defining what is an adhesive exists in defining many of the products shipped to or by the industry. For example, a drum of toluene may be used by a sign manufacturer to join molded polystyrene letters. In this application, toluene is a true adhesive but would not be included in the "value added" statistics. A tank car of methylene chloride may be used by another manufacturer in a paint remover or as the major ingredient in an acrylic adhesive formulation. A quantity of epoxy resin might be used by an electrical equipment manufacturer for in-house use to formulate wire coatings, potting compounds, and adhesives. It is extremely unlikely that each use would be reflected in the economic statistics.

Industry Problem Areas. The adhesives industry, as it exists today, has been described as "fragmented and having a low degree of sophistication" (8). The major firms in the adhesives business are part of giant conglomerates operating in many areas including raw material production and manufacture of adhesives. A number of these firms compete with their customers by selling raw materials and finished intermediates and also by manufacturing, and sometimes wholesaling and retailing, adhesive products to industrial and home consumers. Adhesive volumes are usually too small to devote even limited research funds to the search for new polymers just for adhesives, and even when such polymers are found, their development into marketable products is hindered by ignorance of user needs.

Formerly high strength adhesives required pressure-producing apparatus to make bonds. Expensive equipment, such as large heated and cooled platen presses and huge autoclaves, took many years to amortize and the users of adhesives did not welcome new technology that required replacement of equipment and facilities. Surface preparation chemicals and equipment come from firms oriented to the surface coatings industry. Thus, chemical solutions must often be made up by the adhesive bonder; and time and temperature ranges different from those used in the finishes industry must be established. A Swedish adhesive manufacturer has taken the systems approach and one result has been unique adhesives and application equipment for the wood working industry (9).

Theories of Adhesion

No one knows why adhesives join objects together although there are many theories, each with its supporters and detractors. It is at least agreed that adhesion is an interfacial phenomenon, and that wetting of substrates is essential to adhesion. Substantial loads may be carried across properly designed adhesive bonded joints, certain substrate materials require particular adhesives, and many substrates can produce stronger joints with special surface treatment procedures.

Electrical Theory. The electrical theory presumes that the adhesive and substrates are like the plates of a capacitor that become charged due to the contact of two substances. The theory fails to predict the strong joints that result when a layer of water is frozen to join two blocks of ice, or when an epoxy adhesive is used to join two previously cured blocks of cast epoxy.

Diffusion Theory. The diffusion theory (10) calls for the penetration of the substrates by the adhesive prior to the solidification of the adhesive. The diffusion of a solvent into many porous plastics is readily visualized; however, diffusion of an adhesive into metal, glass, or glazed ceramic is difficult to accept.

Adsorption Theory. The adsorption theory depends upon the concept of molecular forces, such as van der Waals forces, acting across the space between molecules in a material. The adhesive, being a liquid at some stage of the bonding process, supposedly displaces gases so as to fill the voids and make intimate contact with the substrates so that such forces can take effect.

Rheological Theory. The rheological theory suggests that the removal of weak boundary layers in plastics leaves the mechanical properties of the joint to be determined by the mechanical properties of the materials making up the joint and by local stresses. In the absence of weak boundary layers, joint failure must be cohesive within the bondline or the substrates.

Each of these theories has contributed to the beginnings of a science of adhesion and has stimulated the gathering of a mass of experimental data. The works of Zisman (11), from a surface chemistry and wetting viewpoint, and Bikerman (12), for a theory of weak boundary layers, give insights into the origins of the science of adhesion. Schonhorn (13) gives a detailed exposition of the various theories and more recent articles may be found in the general references of the bibliography.

The Question of the Ideal Adhesive

There is no ideal adhesive and it is doubtful that there will ever be such a product because of conflicting requirements. Many bonding requirements are for permanent structural adhesives, but many temporary or demountable adhesives, such as masking tapes, must be easily removed. For more than two decades, there have been hundreds of technical papers in which the terms "structural and nonstructural adhesives" have been used. Yet no one has any precise idea of the difference.

Low viscosity is required when the adhesive must penetrate small cracks or be used in very thin bondlines, but heavy bodied or thixotropic adhesives are required to fill gaps and resist sag in vertical applications. The terms thin and thick carry no precise definition. In capacitor adhesives 1.0 μm is a thin bondline and 0.1 mm is a standard and 1.0 mm is a thick bondline, but in bonding highway dividers, 1.0 mm is a thin bondline and 10 mm is a thick one.

Many adhesives must have fast grab or tack, but must be easily repositionable for a controlled time prior to the tacky stage. Adhesives must readily wet the bonding area, but not bleed out and contaminate areas not to be bonded. Heat-curing adhesives should cure at low temperatures, but have unlimited shelf stability at temperatures normally attained in transportation and storage. They should remain flexible at cryogenic temperatures, but not creep at elevated temperatures. Low, normal, and high temperature ranges vary. Adhesives should be easy to apply and require no special application or fixturing equipment. They should also be able to bond through oil or other contamination to all solid materials, and be available at low cost.

Types and Forms of Adhesives

Just as it has been stated that adhesives belong to no specific class of materials, there is no best way to classify the various types of adhesives. They could be listed by chemical type, but to complete such a listing would include almost all polymers, plastics, homopolymers, and copolymers, including most thermoplastics, thermosets, and alloys of both, most inorganic silicates and phosphates, most natural gums, resins, tars, animal and vegetable proteins, starches and dextrins, many solvents, and some organic–inorganic hybrids. Shields (14) devotes 50 pages to a detailed listing and description of 89 different types of adhesives. Cagle (15) gives a full chapter to such a listing and Houwink and Salomon (16), their entire first volume.

Adhesives have been categorized by suitability for bonding various substrates, and in one excellent reference by Rider (17), taken one step further by temperature resistance recommendations for many substrates when using a particular type bonding agent. Adhesives have been listed by physical form, by application method, eg, spatula, brush, roller, spray, curtain coater, notched trowel, etc, by economic factors, such as cost or tonnage, and by end-product usage.

Adhesives are available in many forms, including liquids, pastes, and solids. The encapsulated adhesive patented by National Cash Register, Inc. (18) is a liquid form of adhesive that is dry to the touch. NCR's "carbonless carbon paper" in which encapsulated toluene is released when pressure or heat ruptures the microballons, is based on the same principle. In both cases, the solvent reactivates a dry film. The 3M Company encapsulates epoxy hardener and mixes it with high molecular weight epoxy resins which are solid at normal temperatures. This mixture dispersed in a fluid carrier, is coated on bolts or screw threads. The dry coated screws are shipped to the user. When the screw is used, the mating screw threads rupture the capsules of hardener and mix them with the resin forming a room temperature curing epoxy adhesive (19) (see Microencapsulation).

Solid adhesives are made by removing most or all of the solvent from liquid adhesives, by using higher molecular weight resins, or by partial reaction of resins to an advanced state of cure termed a B-stage. These solid adhesives may be produced in the form of thin or thick film or tape, either unsupported or supported on a woven or nonwoven carrier. The films may be die-cut to various configurations and in this form the adhesive is placed only where it is required. The thickness of the film allows precise control of bondline thickness in designs where tolerances are critical. Some epoxy adhesives may be purchased as powders and the powder may be preformed into shapes such as washers, rods, tubes, etc. Many types of reactive resins and curing agents are accurately proportioned, mixed, and quick frozen and then shipped worldwide by air in specially insulated containers. The package should be warmed above the dew point before opening. Liquid, paste, film, and specially shaped forms of adhesive are available in frozen form.

One manufacturer makes sticks of rubber cement which melts from the frictional heat generated as it is rubbed across a sheet of paper (20). Another has patents on epoxy, polysulfide, curing type neoprenes, and other resins in which a base and a catalyzed hardener system are extruded in correctly proportioned layered ribbons. When a length of ribbon is torn off and mixed by hand or machine a paste adhesive results (21). Another familiar form of adhesive is dry film adhesive coated on a textile material that is reactivated by heat (eg, iron-on patches). General purpose rubber adhesives are impregnated into nonwoven fabric and paper carriers to form a dry film adhesive used to fasten items like nameplates, soap dishes, paper cup and towel dispensers to building walls. These may be reactivated with heat or the consumer obtains a gelatin capsule which is punctured to spread the solvent contents over the adhesive surface. The item with the wet solvent-reactivated adhesive backing is pressed against the wall for a few seconds and a strong joint is formed in a few minutes. The advantage of the adhesive in this form is that nonporous substrates may be bonded as the small amount of solvent used diffuses rapidly into the thick adhesive bondline instead of migrating to the edges before it evaporates.

Special Properties

The adhesive formulator may build-in many properties by controlling the molecular weight of resins or the addition of various constituents.

With epoxy adhesives, even the user may combine epoxy resins of various molecular weights with dozens of amines, polyamines, anhydrides, boron trifluoride complexes or other curing agents. The resulting compositions may have viscosities varying from those suitable for fine needle hypodermic syringe use to highly viscous compounds applicable by a spatula. Potlife, useful application life, can vary from almost instant cure to several days at room temperature. Heat resistance from 70 to 200°C may be obtained by sacrificing flexibility and low-temperature impact strength. Pigments and thixotropic agents ground in epoxy resins are available so that color, gloss, and opacity may be controlled along with sag and gap filling properties. Proprietary formulations contain other fillers, such as quartz or metal powders to match the expansion properties of substrates, or to contribute special thermal or electrical properties. In one user-formulated epoxy adhesive, Seger and Sharpe (22), incorporated a thixotropic agent to prevent flow of adhesive into a porous ceramic and a silane coupling agent to ensure long-term bond durability.

Other adhesive materials are not as amenable to formulation changes as the epoxies. With some materials, such as cyanoacrylates, urethanes, phenolics, and silicones, slight changes in pH or moisture content of additives may cause shelf stability problems or even spontaneous gelling. Fillers containing active metal ions will cause similar problems with anaerobics. The expert formulator may take advantage of these limitations by including such destabilizing compounds in materials called primers or accelerators. These are often applied to substrates by the user prior to adhesive application. Upon later contact with the adhesive, accelerated curing takes place. An entirely new type of acrylic adhesive has recently evolved using this technique. The two-part system is applied, one part to each substrate; and within minutes after assembling the substrates, the adhesive cures to form a high strength joint.

In only one area, very high temperature adhesives for aerospace, have really significant sums of money been spent for research and development of sophisticated new adhesive systems. The resulting adhesives include polyimides, aramid–imides, polybenzimidazoles, silicones, and a number of organic–inorganic hybrids. In this case, the necessary funds came from government agencies, as commercial use of large quantities of such adhesives is many years in the future.

Design for Adhesive Bonding

All solid materials may be adhesive bonded, and many may be joined with strong durable joints approaching the strength of the substrates themselves. However, when adhesives are used to join high strength metal-to-metal, metal-to-nonmetal, and nonmetal-to-nonmetal composites, they are, by comparison, very weak materials. Useful joints of such materials may be made if proper consideration is given to the design techniques that have been developed over the years.

When properly selected high strength adhesives are used to bond materials listed below the epoxies in Table 1, no special design precautions must be taken to assure a break in the substrate in assemblies that are loaded in tensile or shear modes. Only if cleavage or peel are factors must special designs be used to compensate for the weakness of the adhesive.

Table 1. Approximate Tensile Strength of Materials

Material	Tensile strength, MPa[a]
glass fibers	4000
alloy steels	2000
titanium	1400
nylon fibers	800
carbon steels	700
high strength aluminum	650
magnesium	380
medium strength aluminum	280
alumina ceramics	200
glass-filled nylon	170
low strength aluminum	100
EPOXIES	90
nylon	70
acrylics	70
hard rubber	55
phenolics	55
wood composition board	35
black rubber	30
gum rubber	10
foam plastics	2
concrete	1
foam rubber	0.4

[a] 1 MPa = 145 psi.

With the high strength materials listed above the epoxies, special designs must be used. In tension these designs may include a reinforcing coupling (Fig. 2). If the coupling materials are strong enough, the cross-section thick enough, and if the overlap length is sufficient, failure in a tensile mode will come outside the adhesive joint. The strength of the adhesive in the butt joint area will play only a small part. This joint will be seen as similar to that made by the pipefitter who joins two pipes with a similar collar, nipple, or coupling. The same reinforcing principle can be used to make high strength joints between rods, tubes, flat plates, and more complex structures.

High strength metal bonding has been mostly to the light metals such as aluminum, titanium, and, to a lesser extent, magnesium. With these metals and their alloys, welding problems occur unless highly skilled labor and exotic techniques are used. With steel and other ferrous alloys, reliable welds with joint strength equivalent to the parent materials are achieved at low cost, so there has been no incentive to develop adhesive bonding extensively.

It is rarely possible to substitute directly an adhesive for a design planned for welding or a mechanical fastener and achieve satisfactory results. This seems to account for a major portion of the bond failures experienced in adhesive-bonded consumer products.

The size and shape of a bonded structure are major determinants of its stress distribution. Changing the shape to reduce stress may make the parts lighter yet surprisingly stronger. Inadequate fillets or abrupt changes in section thickness may

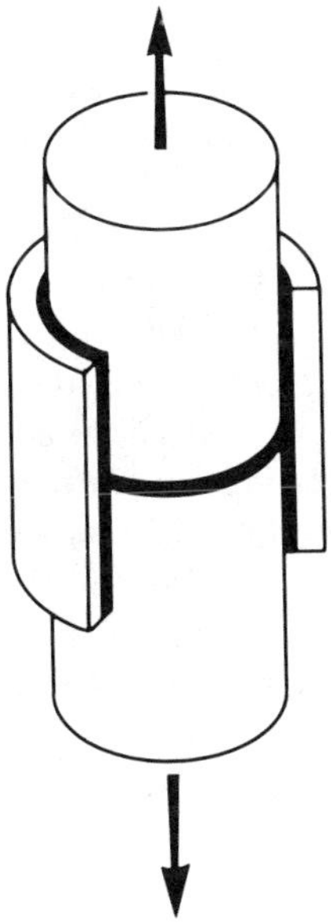

Figure 2. Adhesive bonded reinforcing coupling for butt joints.

concentrate stress. Stress concentrations in adverse environments, especially high humidity, can lead to displacement or desorption, stress relief within the adhesive, or corrosion of metallic substrates; any of which reduces joint strength in an unpredictable fashion. Some adhesive joints, which in the absence of high humidity might be expected to retain their integrity under load indefinitely, have been shown to fail in just a few hours when subjected to high humidity at normal temperatures (23). Durability test results are just beginning to be included in technical data sheets of adhesive suppliers.

The design of the joint may dictate the form of the adhesive or vice versa. If any adhesive is to be introduced into a small gap that is already assembled and fixtured, it must be a low viscosity liquid. Not so obvious is the requirement for specific thermal, electrical, or mechanical properties which may only be achieved with a carefully controlled bondline thickness where a film adhesive must be used.

Analytical approaches to joint design have received comparatively little attention. Analysis is complicated by the complexity of structures containing an adhesive layer with totally different strength properties than the substrates. Testing the number of assemblies required for statistical treatment is seldom practical where high cost items like aircraft structures are involved.

Surface Preparation

With paper and wood, the surfaces may ordinarily be used in the "as received" condition. But with ceramics, metals, and plastics, specific procedures must be used to enhance the end-product quality. Even though initial adhesion may be excellent, if the joint is subjected to adverse environments, bond permanence or durability may be poor. Chemically active metals, eg, magnesium, cannot simply be solvent cleaned or abraded, but must receive chemical or anodic surface treatments. Aluminum seems to require that the surface be micro-rough and this condition is best obtained with a sulfuric acid–sodium dichromate treatment often termed "sulfochromate" by the aerospace industry which has been using such a procedure for several decades. Literally hundreds of articles have been written on the surface preparation of titanium metal. Titanium was the leading contender as a heat-resistant skin material for the supersonic transport aircraft (SST) and adhesive bonding was the only practical means of attachment. But for many years titanium was considered almost impossible to bond. Most of the proposed surface treatments resulted in impressive improvements in joint shear strength, but almost no peel strength or bond durability. Vazirani (24), in 1970, patented a procedure giving dramatic improvements in both categories, but because his procedures were developed outside the aerospace industry they have received little attention. His procedure calls for the addition of concentrated phosphoric acid, 60 vol %, to replace part of the hydrofluoric acid and oxidizing acids, such as chromic and nitric acid, that make up most of the surface treatment solutions found in the adhesive bonding literature. Most procedures call for a preliminary vapor degreasing in a chlorinated solvent in spite of the fact that such solvents have been shown to cause stress cracking of titanium (25).

With plastics, surface preparation includes the removal of mold-release agents used in the molding process by solvent wiping or abrasion. This may be followed by a chemical treatment or other special surface preparation procedures. These may include the substitution of one chemical species for another on the surface through use of strong acid oxidation or gas plasma impingement. Even difficult to bond surfaces, such as polytetrafluoroethylene, ie, Teflon and other fluorinated polymers, may be made bondable when treated with commercially available sodium metal preparations. Polyolefins, such as polyethylene and polypropylene, may be joined in low strength joints with pressure-sensitive adhesives, but for higher strength bonds, treatment by oxidizing flames, Tesla coils, plasma etching, or concentrated sulfuric acid–sodium dichromate is required. Ceramics and glass also require rigorous cleaning in so-called glass cleaning solutions made up of the same concentrated sulfuric acid–sodium or potassium dichromate solutions used for plastics, if long-term durability is to be achieved. Snogren (26), has written a book on surface preparation prior to adhesive bonding. Specific cleaning and surface treatment procedures may also be found in references 27–28, and most books on adhesives technology devote a full chapter to the subject.

Adhesive Selection Process

Adhesive selection has been almost entirely empirical. Most design guides emphasize that selection of a specific adhesive for a specific application should be left to the experts. The factors to be taken into account, and the weights to be assigned to each factor, are the essence of engineering judgement. There is no one adhesive best suited for joining any two substrates. The same properties that may be cited in one application as an advantage may be a limitation in another. Each adhesive expert may recommend a different adhesive, depending upon his analysis of the factors and weights given each factor. Small, seemingly insignificant changes in substrate, joint design, environment, application or processing equipment or technique, and even the mode of transportation of the finished product may negate an earlier recommendation. In fact, many of the failures of adhesives in product are the result of such changes introduced without reappraisal of the adhesive selection. Among those properties not already mentioned, which must be considered, are design life, loading factors, effect on nearby materials, economic factors, such as quantity of assemblies, cost of materials, and labor, and often health and safety factors, such as toxicity, flammability, odor, etc.

Table 2 lists what are often cited as advantages of adhesive bonding. Each listing has one or more examples of an actual application where the advantage is critical or important. Table 3 shows offsetting limitations, with examples.

Application of Adhesives

Much of the application equipment for adhesives is the same equipment used to apply surface coatings such as roller, flow, curtain and knife coaters (see Coating processes). Hot and cold spray systems include air and airless, electrostatic liquid and powder, and the ubiquitous hand-held pressurized cans. Special spray units may be equipped with mixing heads for multicomponent formulations. Spray guns may be hand held or included in fixed or traveling fixtures as part of an automated production or packaging line (see Coating processes).

Roller coaters use pressure, dip or kiss rolls of smooth or grooved rubber or steel to apply continuous films or interrupted areas of adhesives. Application speeds may vary from a few meters to a hundred or more meters per minute. Application equipment unique to the adhesive industry includes gravity, capillary wick, pressure or vacuum feeding of adhesives between prepositioned substrates. Such equipment has been used for the bonding of extruded aluminum frames to cast aluminum corners of electronic equipment cabinets (29).

Dispensing equipment for solvents, solvent dispersions, aqueous emulsions or latexes and reactive materials, that will handle single component or premixed adhesives varying in viscosity from 1 mPa·s to 5 kPa·s (1 cP–5 $\times$ 10^6 cP), are applicators often used to fill containers with precise amounts of nonadhesive materials. Heat may be used to reduce the viscosity of liquids or pastes and compressed air, inert gases or pumps are then used to extrude the adhesive. Applicator tips may be metal, glass, or plastic, and are often hypodermic needles. Dispensing equipment is available to apply spots of adhesive as small as 0.1 mm in diameter. Milligram spots to continuous streams of adhesive may be applied with the same equipment. In the electronics industry these applicators may be used in conjunction with binocular microscopes and micrometer-positioned X, Y axis substrate-positioning devices.

Similar dispensers may also proportion, mix, and degas multicomponent adhesives. Some of these adhesives cure within seconds after being mixed with a curing agent and cannot be used without this special mixing equipment. Ratios, from 1:1 to 1:100, can be controlled by gear pumps or by adjustable or fixed ratio piston pumps. Mixing is done by moving vane or motionless mixers. Fast-curing adhesives will cure in the machines unless precautions are taken. When thin liquids are being dispensed, the unit can be equipped with a vacuum system to draw back the last drop of material to prevent dripping and assure deposition of precisely reproducible volumes.

Hot-melt adhesive applicators are really miniature thermostatically controlled extruders similar to those used to extrude thermoplastics. The adhesive may be supplied in the form of blocks, chips, granules, ribbons, or even bulk solids packaged in pails and drums. In the latter case, they may be melted in the container and forced

Table 2. Advantages of Adhesive Bonding (Selected Examples)

1. Allows any solid material or combination of materials to be joined (human bone and tissue, metal hypodermic needles to glass syringes).
2. Permits joining of thin materials not suitable for mechanical fasteners or welding (metal foil and paper or plastic film labels, panel skins to honeycomb cores).
3. Can bond massive structures defying conventional fastening techniques (laminate wood timbers, concrete bridge sections, precast highway dividers).
4. Tolerates materials with vastly different thermal coefficients of expansion (windows to frames in high rise buildings, metal rearview mirror to windshield).
5. Makes joints nearly impermeable to liquids and gases (submarine cable repeater seals, plastic water and drain pipe, balloon seams).
6. Makes thermally and electrically isolated joints when required (insulated wall panels, laminated motor cores, multilayer printed circuits).
7. Eliminates corrosion due to metal fasteners.
8. Simplifies disassembly where required (abrasive paper on sanding discs, masking tape, medical sensors to skin tissue).
9. Makes optically transparent joints possible (laminated safety glass, plastic sun screen films to windows, microscope lenses).
10. Provides transparency to other electromagnetic radiation (solar cell covers, radar and microwave antenna covers, x-ray film holders).
11. Produces joints resistant to high temperature (ceramic spark plugs to metal base, ablative surfaces on reentry vehicles).
12. Makes joints resistant to low temperatures (arctic shelters, cryogenic container walls such as bulkhead for Saturn vehicle).

Table 3. Limitations of Adhesive Bonding (Selected Examples)

1. Many combinations require costly equipment or involved procedures (autoclaves for high-pressure aerospace bonding, sodium etching of fluorocarbons).
2. Thin materials often require special high-speed laminating equipment (plywood presses, bottle and can label application equipment).
3. Massive structures often require special equipment (single use fixtures required for building one-of-a-kind wood beams).
4. Warping due to "bimetallic spring" effect with temperature cycling (curling of plastic coated metal foil and other labels and nameplates).
5. Impermeability can deteriorate with time and temperature (oil and gas additives often attack adhesives, liquids can undercut adhesive bondlines).
6. Humidity may change thermal and electrical properties of adhesives over time (porous insulating ceramic adhesives become thermally and electrically conductive).
7. Adhesives may cause corrosion of adjacent metals (neoprene adhesives break down and release acids, room-curing epoxies corrode copper).
8. Many adhesives are stronger than substrates and disassembly destroys part (wood adhesives stronger than wood, foam rubber gaskets often tear on removal).
9. Adhesives may change in light transmission over time (organics often yellow in ultraviolet light, moisture absorption may cause haze).
10. Radiation may cause internal heating and loss of adhesion (nuclear radiation destroy polyolefin hot-melt adhesives).
11. Organic adhesives have limited temperature resistance (overheated clutch and brake adhesives fail, auto gaskets fall off in summer).
12. Many adhesives become brittle at low temperatures (labels fall off in freezers, fiberglass boat patches fail upon impact).

through heated hoses by pressure plates fitted within the container. The hoses may be heated by steam, hot oil, or by embedded electric wires.

Excessive heating of adhesive can cause degradation by oxidation or polymerization, and this is minimized by temperature control, nitrogen blankets, or by heating the material only at the point of application.

Manufacturers of adhesive equipment may sell off-the-shelf equipment or supply complete custom made systems tailored to specific adhesives or specific end uses. The costs associated with adhesive placement including application equipment, labor and equipment maintenance are often a significant factor to be considered when selecting a joining method.

Setting and Curing of Adhesives

Although pressure-sensitive, solvent releasing and hot-melt adhesives have been formulated to cure chemically, most do not cure. The term "cure" is a misnomer. The correct term is set. In the case of solvent-releasing adhesives setting is a function of loss of solvent by diffusion into some substrates, migration through other porous ones, and solvent or vapor travel through the bondline to the exposed edge of impermeable substrates such as metals and ceramics. Hot-melt adhesives set immediately upon cooling from the liquid to the solid state, but may develop improved properties over minutes or days with crystallization. Pressure-sensitive adhesives usually require only an initial application of pressure to complete the joint. They may increase substantially in bond strength over time, as a result of increased wetting, especially if exposed to heat during or after lamination.

It is only the reactive adhesives which truly cure. There are a number of curing mechanisms. Here are brief comments on some typical cures. Acrylics, unsaturated polyesters and other monomer adhesives containing ethylenic unsaturation cure by formation of free radicals when catalyzed by peroxides and accelerated by metal-ion donators such as cobalt and other metallic naphthenates. The base monomers usually contain inhibitors, such as hydroquinone or materials classed as stabilizers, to prevent premature cure in the container. Epoxies cure by addition mechanisms when basic materials, such as amines, or acidic materials, such as anhydrides, open the reactive oxirane ring structure. The curing agent, often called an epoxy hardener, may take part in the reaction or, as a tertiary amine, act as a catalyst to join the short monomer or polymer molecules to form large crosslinked structures. The curing reaction is exothermic and with amines may take place at normal room temperatures, or may require heat input to thermoset the monomers or prepolymers into insoluble and infusible polymers. Acidic curing agents require elevated temperature (and sometimes catalysts) to complete the reaction in a reasonable time. Polysulfides react with active oxidizing agents (such as lead or manganese oxides) to form rubbery polymers, especially useful in joining low expansion glasses to metals, as in glazing sealants for aircraft and high-rise buildings (see Sealants). Curing type neoprene rubber adhesives depend upon similar reactions. Resorcinol, phenol, and phenol–resorcinol adhesives, with all their modifications, cure by condensation reactions with formaldehyde, in the presence of acid or base.

Many urethane and silicone adhesives cure in the presence of moisture. The primary reaction yielding urethanes is that of isocyanate and hydroxyl groups, the moisture in the air furnishing the hydroxyl in the so-called one-part moisture cured systems. The two-part systems may be cured with amines or amines may be used as catalysts in hydroxyl reactions. High molecular weight linear-thermoplastic urethanes dispersed in solvent and used as noncuring adhesives are often supplied commercially with a liquid component that the user may add to obtain a cured elastomer with increased heat and chemical resistance. The liquid component may be an isocyanate, peroxide, or epoxy–amine combination. All of the above reactive adhesive cures are accelerated by heat. Heat may be supplied by platen presses, conventional ovens, infrared radiation and by more complex means, such as microwaves, electromagnetic excitation of metallic particles incorporated in the adhesive, ultrasonic vibration, and laser beams.

Cyanoacrylate adhesives cure only in thin bondlines, ie, 1.0 μm to 0.1 mm, without the addition of catalyst or hardener. The basicity present on many substrates initiates cure when inhibitors present in the adhesive are overcome. Curing times as short as a few seconds are possible on substrates such as alkaline glasses. Acrylic acid diesters, such as those produced by the Loctite Corporation for thread-locking applications, polymerize anaerobically. These anaerobic adhesives must be packaged in small volumes in low molecular weight polyethylene containers. The free passage of oxygen through the container walls inhibits the cure during shipping and storage. When air is excluded the material will cure into a useful adhesive. Metal ions present upon many surfaces accelerate the cure mechanism. Adhesive outside the bondline does not cure and must normally be removed. However, with some of the newer compounds, heat will cause even the exuded material to cure.

Many adhesives are combinations of the materials mentioned and cure by combinations of the mechanisms described. Some pressure-sensitive adhesives crosslink when subjected to heat. Unlike dry film adhesives, they have the advantage of being stuck in place without the need of fixtures, but have the disadvantage of being difficult to reposition if not affixed perfectly. Reactive adhesives are available in premixed frozen form and are shipped by air. Many solvent-based adhesives are of the reactive type which cure with heat after the solvent has been driven off. Curing type adhesives available in dry film form may have the properties of thermoplastic adhesives and may be reactivated with solvent. In the bonding of vinyl copolymer tops on round vinyl storage battery cases for the telephone system, the adhesive is a film of the same vinyl incorporating carbon black. This film is placed between the substrates and the joint area is exposed to infrared radiation. The radiation penetrates the clear vinyl without heating it, but the black vinyl film absorbs the heat and liquefies to wet the substrates and act as a heat reactivated adhesive (30).

Testing of Adhesives

The American Society for Testing and Materials (ASTM) Committee D-14 on Adhesives, has been instrumental in preparing and distributing specifications on adhesives, including chemical test methods for quality control of raw materials, processing solutions, finished adhesives, and physical test methods for evaluating adhesive performance in joints (31). The test data, as shown in adhesive manufacturers' data sheets, cannot be used directly in design. The test results, usually based on ASTM test procedures, are only meaningful on a comparative basis between adhesive systems or adhesive products within a system. One adhesive reported in the aerospace literature as a room temperature curing system with excellent elevated temperature properties, proved upon close examination to be in reality a system that cured at high temperatures in a very few minutes. The discrepancy came from a standardized testing procedure that required a five minute soak time at test temperature. This five minute soak was actually sufficient time to cure the adhesive to the point that it exhibited excellent properties at the test temperature. In laboratory experiments where this adhesive was heated to the use temperature in a few seconds with an infrared source, it showed no strength whatever at the elevated temperature. For the proposed use on a reentry vehicle, the standard derived test data were useless and misleading. For use on a friction braking system with almost instantaneous heat-up, it would be dangerous. Test methods must be tailored to the specific usage.

Uses

Automotive. One of the most significant areas of growth has been in the manufacture of automobiles, mobile homes, and recreational vehicles (8). Anaerobic thread-locking sealants and adhesives are being used today in almost every vehicle (32). Adhesive-based body solder to replace lead solder is eimated to be the largest volume use in the future (8). Pressure-sensitive adhesive tapes with a neoprene foam inner lamination are used to hold decorative strips on some car bodies (33). They eliminate a major source of corrosion where holes used to be punched in the body and mechanical fasteners in the form of spring clips used for decorative strip attachment. Vinyl car tops are bonded to metal roofs. Sound, heat, and vibration damping insulation are bonded to car hoods, firewalls, and "headliners," foam crash padding is laminated to an outer covering and bonded to dashboards. Weldbonding, where a spotweld is made through uncured adhesive, is a rapidly growing technique. The final adhesive cure takes place while the paint finish is being baked (see also Welding).

Construction. It has been estimated that by 1985, one-half of all architects will be specifying builder-applied high strength adhesives (8). Today the major drawbacks are lack of performance specifications in building codes. The mobile home and recreational vehicle industry have led the way by using adhesives to provide the added strength and stiffness necessary to withstand wind damage. For field applications, new neoprene adhesives for wood bond plywood to wet and frozen wood framing even at subfreezing temperatures and have gap filling properties to compensate for poorly mated surfaces (34). Rubber-based adhesives are used to join building wallboard and paneling to studs, and hot-melt adhesives serve the same purpose in automated factory assembly of panels for mobile homes and prefabricated homes.

Concrete patches, especially where feathered edges are required, have developed failures in flexure due to heavy loads or thermal cycling. Epoxy adhesives with flexibility modifiers that will cure underwater have allowed the patching of potholes in bridges and highways with inexpensive concrete, after application of an epoxy tie-coat to the damaged concrete (35). Polymer concrete is usually an acrylic or epoxy compound containing sand and aggregate but no portland cement (36). Epoxy adhesive has been used to repair earthquake damaged structures. In one such repair, 76 m^3 (20×10^3 gal) of foamed epoxy was used to bond the masonry walls of the 28-story Los Angeles City Hall (37).

The advantages of bonding anisotropic materials to make plywood and glass fiber-reinforced laminates in selected patterns for dimensional stability and multidirectional strength has long been recognized. New uses will be found for the newer fibers of boron and graphite, such as surfacing (or cores) for less expensive or weaker materials of construction, and high strength adhesives will be used to assemble the structures (38) (see Laminated and reinforced plastics).

Electronic Bonding. New applications include either the ability to make adhesives conduct electricity almost as well as metals or insulate as well as plastics (which most electronic adhesives are). Table 4 shows the properties available in comparison to some other materials. Aluminum heat sinks are being bonded to active and inactive micro and macro electronic elements such as resistors, transistors, and entire integrated circuit chips (39). The chips are also being interconnected electrically with conductive adhesive bonded solid or stranded wires where soldering might damage heat-sensitive components. Other similar but less conductive adhesives are used to bond conductive rubber floors to drain off static charges in hospital operating rooms, and to bond braided radio frequency shielding gaskets to electronic equipment enclosure doors (40).

Flexible and rigid circuitry is made possible by bonding metallic conductors to insulative backing. The Western Electric Company plant in Richmond, Virginia, the world's largest producer of printed circuit boards (PCBs), turns out about 250,000 PCBs weekly, worth about $1,200,000 (41). In most of the rigid PCBs, the copper conductor layer is bonded to a B-staged epoxy backing, the interface of which acts as the adhesive. In older PCBs a separate adhesive layer was used (see Electrostatic sealing).

New computer memories, such as those used in the electronic switching systems by the telephone industry, consume millions of dollars worth of adhesives each year. In one application, millions of square meters of a 25 μm thick nylon–epoxy adhesive film are used to join a similar thickness magnetic alloy to an aluminum sheet twenty times as thick. By photoetching techniques, most of the magnetic alloy is removed, leaving behind thousands of tiny discrete rectangular magnetic areas. Here the chemical resistance of the adhesive film allows immersion in the many strong chemical solutions necessary for the etching processes. Figure 3 shows one of the finished memory cards (42).

Table 4. Approximate Thermal and Electrical Properties of Some Common Materials

Medium	Thermal conductivity, W/(m·K)	Electrical resistivity, $\mu\Omega$/m
air	2.6×10^{-2}	
aluminum	2.0×10^{2}	6×10^{-2}
ceramic (aluminum oxide)	3.6×10^{1}	1×10^{14}
copper	3.8×10^{2}	2×10^{-2}
epoxy (unfilled)	2.0×10^{-1}	1×10^{18}
epoxy (average conductive)	1.7×10^{0}	1×10^{1}
epoxy (dielectric filled)	1.0×10^{0}	1×10^{12}
epoxy (high-grade silver or gold filled)	2.4×10^{0}	5×10^{-1}
graphite	2.1×10^{2}	1×10^{1}
silicone RTV	2.0×10^{-1}	1×10^{18}
solder	3.5×10^{1}	2×10^{-1}

Figure 3. Electronic switching system—twistor memory plane and card. Courtesy of Bell Telephone Laboratories.

Packaging. Corrugated and paperboard products have used sodium silicate-based adhesives for many years. More limited use of synthetic adhesives for waterproofed paperboard is a growing market. Fiber tubes and cores for rolled paper products, pressure-sensitive tapes, and shipping containers use similar adhesives. Composite cans and adhesive-seamed metal cans are used to package dry foods, motor oil, beverages, and sanitary household products. Dextrin and animal glues were formerly used for many of these applications, but they are being replaced by synthetic adhesives (43). Large volumes of adhesives are used for liners for bottle and jar caps, seams on paper cups, and closures for cereal, cake mixes, and dessert boxes (see Packaging materials, consumer products).

Plastics Joining. Plastics joined with adhesives have made possible huge structures that could not be fabricated in a single piece. The adhesives used in low cost plastic piping for residential hot and cold water, sanitary lines, irrigation, and acid lines in chemical plants, range from simple solvents which are bodied to fill voids in poorly dimensioned pipes to epoxies used in thermoset glass-fiber piping. Couplings similar to those used by the pipefitter or plumber are used to obtain high-strength bonds to high-strength materials. An adhesive can be substituted for solder in areas where sparks and flame cannot be tolerated, eg, copper tubing is often joined with adhesives. Plastic piping uses couplings similar to those used in joining copper tubing.

The increasing use of plastics in furniture, such as high-pressure laminates for cabinet door and countertop facings, and low-pressure lamination of vinyl decorative films to wood, other plastics, and metals is based upon new moisture-resistant adhesives that are taking the place of the familiar poly(vinyl acetate) "white glues" (see Laminated and reinforced wood).

Today, all women's shoes manufactured in the United States, and a growing share of children's and men's shoes, are adhesive bonded without supplemental mechanical fasteners or sewing (see also Leather-like materials).

Textile Products and Apparel. Today, almost all tufted carpet production in the United States uses adhesive back coatings based upon carboxylated styrene–butadiene rubber latexes, and much of this in commercial installations is adhesive bonded to the floor (44). Flocking applications for use in shoes, wall coverings, shades, instrument cases and other lining applications, decorative applications, and even the pages of magazine advertisements, are making a comeback after going through a period where poor adhesive performance gave consumers a connotation of inferiority (45–46). Adhesive bonding, along with heat sealing, together referred to as stitchless seaming, of new synthetic fiber textiles and nonwoven textile and paper used in disposable apparel, such as hospital gowns and even swim suits, is a major industry (see Nonwoven textiles). Adhesive laminating of backing materials and adhesives to supplement stitching is found in sporting goods (waterproof hiking and ski boots, backpacks, and sleeping bags), watch bands, shower curtains, drapes, lamp shades, beach and regular umbrellas, women's pocketbooks, and babies' diapers (see Coated fabrics).

Energy, Safety, and Environmental Considerations

The adhesives industry has felt the impact of the recent energy shortage caused by the petroleum crisis, the new emphasis on worker and consumer safety, and the concern for environmental protection, to a greater extent than many other industries. Recent budgets for research and development and much of the time of technical personnel throughout the industry have been devoted to a sometimes frantic search for alternative materials and less energy-intensive procedures.

Energy. The production of many adhesives requires the use of petroleum-derived chemicals and the use of large amounts of energy. Hot-melt adhesives, as an example, must be heated and extruded, then chopped into pellets, slugs, or formed into ribbons for use in special application equipment. In the applicator, the adhesive must again be heated to a fluid consistency, and if it is to be applied to metal substrates, they too should be heated to facilitate wetting action. Room temperature curing reactive adhesives and water- and solvent-based adhesives require the least energy input. But, where fast set of these adhesives is required, heat must be supplied to evaporate the solvent. Solvent recovery systems require energy, usually in the form of steam, and solvent incineration disposal systems also require large energy inputs because of the usual low concentration of even the highest energy solvents. Solvent-based, heat-activated, crosslinking adhesives must first have the solvent evaporated, usually with added heat to speed up the process, and then be heated again for extended time periods

With large structures made from two different metals, or the same metal with thin and thick sections, the assemblies must often be "heat staged," ie, heated and held at one temperature until the most massive part has reached the same temperature as the thinnest section. The entire assembly is then raised in temperature, or staged again. Staging may take place in several steps, and is necessary to prevent adhesive cure in low mass areas before the full effects of thermal expansion have taken place in the thick sections. Without staging, distortion will take place (47). Most of the energy consumed by the adhesive user is used to heat conventional ovens. A recent survey predicts that by 1980 only 10% of the energy used in adhesive bonding will be by nonconventional sources such as infrared, ultraviolet or microwave radiation, radio frequency, or induction heating or by ultrasonic vibration (8).

Safety. In the application of adhesives, the user may be exposed to a variety of hazards. High strength bonding to ceramics, metals, and plastics may require substrate preparation with strong acid or base solutions. Fluorocarbon plastics require surface chemical oxidation with metallic sodium formulations. Other plastics may require plasma or other radiation treatment or oxidizing in a flame. Bonding usually requires abrasive cleaning or surface cleaning with solvents in the liquid or vapor phase.

A basic chemical product 4,4′-methylenebis(2-chloroaniline), used for crosslinking urethane resins in an extraordinarily diverse group of applications, was listed among the carcinogens by the Occupational Safety and Health Administration (OSHA) in May 1973 (48–49). To date, no satisfactory substitute has been found for this important product. "Glue sniffing" has made national headlines, and for a time the Consumer Product Safety Commission banned a number of spray adhesives that were suspected of causing an undue risk of chromosomal damage (50). New envelope flap and remoistenable label adhesives and adhesives in contact with food products require extensive testing prior to marketing.

Pollution. The controls by the Environmental Protection Administration (EPA) and the National Institute for Occupational Safety and Health (NIOSH) are continuing to bring about major changes in production techniques by both suppliers and large-scale users of solvent-based adhesives. Toluene is probably the most widely used solvent in the industry, and it continues to receive the attention of the safety standards writers (51). Because of the small volumes of adhesives in the usually thin bondlines, flammability and smoke generation have only become a concern in a small number of applications such as those in manned space vehicles.

Sources of Information

There is an impressive mass of literature, including many books, monographs, periodicals, a monthly magazine, *Adhesives Age,* and a growing flood of technical conference proceedings, all designed to further the education of the expert and inform the uninitiated, *The Journal of Adhesion* is the only regularly issued English language scientific journal in the field.

The most completely indexed and cross-referenced literature on adhesive bonding is contained in a series of documents prepared by PLASTEC (Plastics Technical Evaluation Center) of Picatinny Arsenal, Dover, N. J. (52). Included is information on many proprietary adhesives, adhesive systems, surface preparation procedures and durability data. Unfortunately, much of this information, while not security classified, is restricted to government agencies or to individuals approved by Picatinny. Much of what is public may be obtained from the National Technical Information Service (NTIS) (53). The National Aeronautics and Space Administration (NASA) and the Forest Products Laboratories at Madison, Wisconsin have published extensively on adhesives.

The American Society of Metals (ASM) periodically publishes a bibliography on adhesives and has an abstracting service that references many foreign as well as English language publications. The ASM also provides photocopies of the material referenced.

BIBLIOGRAPHY

"Adhesives" in *ECT* 1st ed., Vol. 1, pp. 191–206, by V. N. Morris, C. L. Weidner, and N. St. Landau, Industrial Tape Corporation; "Adhesives" in *ECT* 1st ed., Suppl. 1, pp. 18–32, by R. F. Blomquist, Forest Products Laboratory, Forest Service, U.S. Department of Agriculture; "Adhesives" in *ECT* 2nd ed., Vol. 1, pp. 371–405, by R. F. Blomquist, Forest Products Laboratory, Forest Service, U.S.D.A.; "Adhesion" in *ECT* 2nd ed., Suppl. Vol., pp. 16–27, by H. Schonhorn, Bell Telephone Laboratories.

1. J. Muller, *J. Prestressed Concrete Instit.* **20**(1), 28 (1975).
2. Araldite Mortar Secures South America's Longest Bridge, Plastics Division, CIBA-GEIGY, Cambridge, Eng., 1974.
3. *Terms Relating to Adhesives, D 907,* American Society for Testing and Materials, Philadelphia, Pa.
4. G. Salomon in R. Houwink and G. Salomon, eds., *Adhesion and Adhesives,* Vol. 1, Elsevier Scientific Publishing Co., Inc., New York, 1965, p. 2.
5. W. R. Lewis, in Ref. 4, Chapt. 10 is devoted to welding and soldering.
6. U.S. Bureau of the Census, *Census of Manufacturers 1972 Industry Series: Miscellaneous Chemical Products, MC72(2)-28H,* U.S. Government Printing Office, Washington, D.C., 1974.
7. I. Skeist, *Adhes. Age* **19**(4), 41 (1976).
8. G. M. Estes, ed., *A Delphi Forecast of the Future of the Adhesives Industry,* E. I. du Pont de Nemours & Co., Inc., 1975, 6 volume unpaged study by 382 individuals in 132 organizations.
9. O. Mattsson and P. Zaunschirm, *Adhes. Age* **18**(10), 25 (1975).
10. S. S. Voyutskii, *Autohesion and Adhesion of High Polymers* (*Polymer Reviews, No. 4*), Interscience Publishers, a division of John Wiley & Sons, Inc., New York, 1963.
11. W. A. Zisman, in P. Weiss, ed., *Adhesion and Cohesion,* Elsevier Scientific Publishing Co. Inc., New York, 1962.
12. J. J. Bikerman, *The Science of Adhesive Joints,* Academic Press, Inc., New York, 1961; 2nd ed., 1968.
13. H. Schonhorn, "Adhesion," in A. Standen, ed., *Kirk-Othmer Encyclopedia of Chemical Technology,* Suppl. Vol., 2nd ed., Interscience Publishers, a division of John Wiley & Sons, Inc., New York, 1970, pp. 16–27.
14. J. Shields, *Adhesives Handbook,* CPR Press, Cleveland, Ohio, 1970, Chapt. 4.
15. C. V. Cagle, *Adhesive Bonding,* McGraw-Hill, Inc., New York, 1968, Chapt. 2.
16. Ref. 4, Vol. 1.
17. D. K. Rider, in I. Skeist, ed., *Handbook of Adhesives,* Reinhold Publishing Corp., New York, Chapt. 59, and 2nd ed. 1977.
18. U.S. Pat. 2,986,477 (1961), H. J. Eichel (to National Cash Register Co.).
19. U.S. Pat. 3,179,143 (1965), B. W. Schultz and A. F. Thomson (to 3M Company); U.S. Pat. 3,642,937 (1972), F. W. Deckert and G. Matson (to 3M Company).
20. U.S. Pat. 3,267,052 (1966), G. A. Brennan.
21. U.S. Pat. 3,708,379 (1973) and 3,837,981 (1974), T. L. Flint.

22. S. G. Seger and L. H. Sharpe, in L. H. Lee, ed., *Adhesion Science and Technology,* Vol. 9B, Plenum Press, 1975, pp. 577–595.
23. L. H. Sharpe in M. J. Bodnar, ed., *Structural Adhesives Bonding,* Interscience Publishers, a division of John Wiley & Sons, Inc., New York, 1966, pp. 353–359.
24. U.S. Pat. 3,676,223 (July 11, 1972), H. N. Vazirani (to Bell Telephone Laboratories).
25. H. L. Logan, *The Stress Corrosion of Metals,* John Wiley & Sons, Inc., New York, 1966, p. 239.
26. R. C. Snogren, *Handbook of Surface Preparation,* Palmerton Publishing Co., New York, 1974.
27. ASTM Specification, *D 2651, Preparation of Metal Surfaces for Adhesive Bonding; D 2093, Preparation of Surfaces of Plastics Prior to Adhesive Bonding.*
28. A. H. Landrock, *Processing Handbook on Surface Preparations for Adhesive Bonding, AD B0104531,* National Technical Information Service, Springfield,Va., 1975, specific procedures for 80 substrates with extensive references.
29. A. J. Setzer and M. A. Graybeal, *Assem. Eng.* **8**(7) 24 (1965).
30. D. W. Dahringer and J. R. Schroff, *Bell Syst. Tech.* **49**(7), 1403 (1970).
31. *Part 22 Adhesives,* American Society for Testing and Materials, 1976.
32. J. L. Claflin, *Adhes. Age* **19**(4), 28 (1976).
33. C. V. Cagle, *Handbook of Adhesive Bonding,* McGraw-Hill, Inc., New York, 1973, Chapt. 26, pp. 26–27.
34. *APA Glued Floor System,* American Plywood Association, Tacoma, Wash., Form U405, 1974.
35. G. B. Lowe, *Adhes. Age* **16**(12), 41 (1973).
36. *Introductory Course on Concrete–Polymer Materials,* Brookhaven National Laboratory, New York, BNL 1952S, 1974.
37. *Mod. Plast.* **52**(2), 32 (1975).
38. I. Crivelli-Visconti, *Polym. Eng. Sci.* **15**(3), 167 (1975).
39. J. Shields, *J. Phys. E.* **5**(2), 109 (1972).
40. *Mach. Des.* **47**(19), 74 (1975).
41. *Bell Labs News* (Aug. 16, 1976).
42. *Bell Lab. Rec.* **43**(8), 207 (1965).
43. P. Lambert, *Adhes. Age* **16**(7), 22 (1973).
44. M. Stone, *Adhes. Age* **12**(12), 23 (1969).
45. G. C. Kantner, *Adhes. Age* **10**(6), 22 (1967).
46. U. Maag, *Adhes. Age* **18**(9), 23 (1975).
47. Ref. 29, pp. 24–29.
48. *Fed. Reg.* **38**(85), (1973).
49. H. E. Schroeder, *Res. Manage.* **18**(5), 11 (1975).
50. *Fed. Reg.* **38**(162), (1973).
51. H. M. D. Utidjian, *J. Occup. Med.* **16**(2), 107 (1974).
52. *Plast. World* **30**(1), 52 (1972).
53. R. E. Herzog, *Mach. Des.* **45**(23), 132 (1973).

General References

"Adhesives," Vol. 1 and "Applications," Vol. 2 in R. Houwink and G. Salomon, eds., *Adhesion and Adhesives,* Elsevier Scientific Publishing Co., Inc., New York, 1965, 1967.

L. H. Lee, ed., *Adhesion Science and Technology,* in two volumes, Plenum Press, 1975.

C. V. Cagle, *Adhesive Bonding,* McGraw-Hill, Inc., New York, 1968.

I. Katz, *Adhesive Materials,* Foster Publishing Co., Long Beach, Calif., 1971.

N. J. DeLollis, *Adhesives for Metals: Theory and Technology,* Industrial Press, Inc., New York, 1970.

Adhesives Guidebook and Directory, Noyes Data Corp., Park Ridge, N. J., 1972. Information on selected adhesives taken from manufacturers' literature.

J. Shields, *Adhesives Handbook,* CRC Press, Cleveland, Ohio, 1970.

Adhesives Redbook, Palmerton Publishing Co., New York, issued annually since 1968 as a directory of the adhesives industry.

D. J. Alner, ed., *Aspects of Adhesion,* in seven volumes, University of London Press Ltd., 1965–1973.

J. P. Cook, *Construction Adhesives and Sealants,* Interscience Publishers, a division of John Wiley & Sons, Inc., New York, 1970.

Effect of Varying Processing Parameters in the Fabrication of Adhesive Bonded Structures, an 18 part series detailing work supported by the U.S. Government, Jan. 1970 to July 1973, National Technical Information Service, Springfield, Va.

M. W. Ranney, *Epoxy and Urethane Adhesives,* Noyes Data Corp., Park Ridge, N. J., 1971, abridgment of patents from 1960 thru early 1971.
C. V. Cagle, *Handbook of Adhesive Bonding,* McGraw-Hill, Inc., New York, 1973.
I. Skeist, ed., *Handbook of Adhesives,* Reinhold, New York, 1977, 2nd ed.
H. Lee and K. Neville, *Handbook of Epoxy Resins,* McGraw-Hill, Inc., 1967.
R. C. Snogren, *Handbook of Surface Preparation,* Palmerton Publishing Co., New York, 1974.
M. J. Satriana, *Hot Melt Adhesives,* Noyes Data Corp., 1974, abridgment of patents from 1955 thru early 1974.
D. H. Kaelble, *Physical Chemistry of Adhesion,* Interscience Publishers, a division of John Wiley & Sons, Inc., New York, 1971.
L. F. Martin, *Pressure Sensitive Adhesives,* Noyes Data Corp., Park Ridge, N. J., 1974, abridgment of patents middle 1960s thru 1973.
J. J. Bikerman, *The Science of Adhesive Joints,* Academic Press, Inc., New York, 1961, 2nd ed., 1968.
R. L. Patrick, ed., *Treatise on Adhesion and Adhesives,* in four volumes, Marcel Dekker, Inc., New York, 1966, 1969, 1973, 1976.
K. Johnson, *Vinyl and Acrylic Adhesives,* Noyes Data Corp., Park Ridge, N. J., 1971, abridgment of patents from 1960 thru early 1971.

FRED A. KEIMEL
Bell Telephone Laboratories

AMINO RESINS AND PLASTICS

Amino resins are manufactured throughout the industrialized world to provide a wide variety of useful products. Adhesives (qv), representing the largest single market, are used to make plywood, chipboard, and sawdust board. Other types are used to make laminated wood beams, parquet flooring, and for furniture assembly (see Laminated and reinforced wood).

Some amino resins are used as additives to modify the properties of other materials. For example, a small amount of amino resin added to textile fabric imparts the familiar wash-and-wear qualities to shirts and dresses. Automobile tires are strengthened by amino resins which improve the adhesion of rubber to tire cord. A racing sailboat may have a better chance to win because the sails of Dacron polyester have been treated with an amino resin (1). Amino resins can improve the strength of paper even when it is wet. Molding compounds based on amino resins are used for parts of electrical devices, bottle and jar caps, molded plastic dinnerware, and buttons.

Amino resins are also often used for the cure of other resins such as alkyds and reactive acrylic polymers. These polymer systems may contain 5–50% of the amino resin and are commonly used in the flexible backings found on carpets and draperies, as well as in protective surface coatings, particularly the durable baked enamels of appliances, automobiles, etc.

The term amino resin is usually applied to the broad class of materials regardless of application, whereas the term aminoplast or sometimes amino plastic is more commonly applied to thermosetting molding compounds based on amino resins. Amino plastics and resins have been in use for the past fifty years. Compared to other segments of the plastics industry, they are mature products, and their growth rate is now only about half of that of the plastics industry as a whole. At present they account for about 3% of the United States plastics and resins production.

Most amino resins are based on the reaction of formaldehyde with urea (qv) or melamine (see Cyanamides). Although formaldehyde combines with many other amines, amides, or aminotriazines to give useful products, only a few have found commercial utility, and they are of minor importance compared to the major products based on urea and melamine.

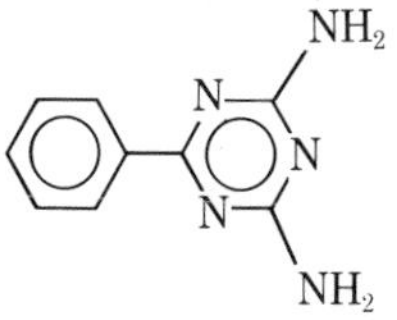

benzoguanamine

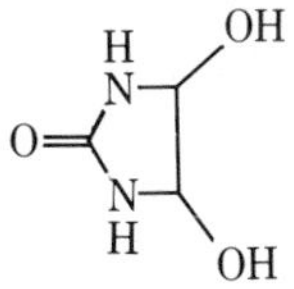

dihydroxyethyleneurea

Benzoguanamine, eg, is used in amino resins for coatings because it provides excellent resistance to laundry detergent, a definite advantage in coatings for automatic washing machines. Dihydroxyethyleneurea is used for making amino resins that provide wash-and-wear properties to clothing.

Aniline–formaldehyde resins [*25214-70-4*] were formerly important because of their excellent electrical properties, but have been supplanted by newer thermoplastics (see Amines, aromatic–aniline). Nevertheless, some aniline resins are still used as modifiers for other resins. Acrylamide (qv) occupies a unique position in the amino resin field since it not only contains a formaldehyde-reactive site but also a polymerizable double bond. Thus it forms a bridge between the formaldehyde condensation polymers and the versatile vinyl polymers and copolymers.

Formaldehyde links two molecules together and is hence difunctional. Each amino group has two replaceable hydrogens that can react with formaldehyde and thus is also difunctional. Since urea and melamine, the amino compounds commonly used for making amino resins, contain two and three amino groups, they react polyfunctionally with formaldehyde to form three-dimensional, cross-linked polymers. Compounds with a single amino group, such as aniline or toluenesulfonamide, can react with formaldehyde to form only linear polymer chains. This is true under mild conditions, but in the presence of an acid catalyst at higher temperatures, the aromatic ring of aniline, eg, may react with formaldehyde to produce a cross-linked polymer.

The basic chemistry of amino resins was established as early as 1908 (2), but the first commercial product, a molding compound, was patented in England by Rossiter (3) only in 1925. It was based on a resin made from an equimolar mixture of urea and thiourea and reinforced with purified cellulose fiber and was trademarked Beetle (indicating it could "beat all" others). Patent rights were acquired by the American Cyanamid Company along with the Beetle trademark, and by 1930 a similar molding compound was being marketed in the United States. The new product was hard and not easily stained and was available in light, translucent colors; furthermore, it had no objectionable phenolic odor.

The use of thiourea improved gloss and water resistance, but stained the steel molds. As amino resin technology progressed the amount of thiourea in the formulation could be reduced and finally eliminated altogether.

In the early 1920s, Goldschmidt and Neuss (4) were experimenting with urea–formaldehyde resins [*9011-05-6*] in Germany, while Pollak (5) and Ripper (6) were making similar amino resins in Austria. The latter team discovered that urea–formaldehyde resins might be cast into beautiful clear transparent sheets, and it was proposed that this new synthetic material might serve as an organic glass. In fact, an experimental product called Pollopas was introduced, but lack of sufficient water resistance prevented commercialization. Melamine–formaldehyde resin [*9003-08-1*] does have better water resistance but the market for synthetic glass was taken over by new thermoplastic materials such as polystyrene and polymethylmethacrylate (see Methacrylic polymers; Styrene plastics).

Melamine resins were introduced about ten years after Rossiter's Beetle molding compound. They were very similar to those based on urea but had superior qualities. Henkel in Germany was issued a patent for a melamine resin in 1936 (7). Melamine resins rapidly supplanted urea resins and were soon used in molding, laminating, and bonding formulations, as well as for textile and paper treatments. The remarkable stability of the symmetrical triazine ring made these products resistant to chemical change once the resin had been cured to the insoluble, cross-linked state.

Before the rapid expansion of thermoplastics following World War II, amino plastics served a broad range of applications in molding, laminating, and bonding. As the newer and more versatile thermoplastic materials moved into these markets, amino resins became more and more restricted to applications demanding some specific property best offered by the thermosetting amino resins. Current sales patterns are very specific. Urea molding powders find application in moldings for electrical devices and jar and bottle caps. Urea resins have retained their use in electrical wiring devices because of good electrical properties and heat resistance and availability of colors not obtainable with phenolics. The preference for making caps is based on the excellent resistance of urea resins to oils, fats, and waxes often found in cosmetics, as well as to the availability of a broad range of colors. Melamine molding compound is now used mainly for molded dinnerware primarily because of outstanding hardness, and water and stain resistance. Melamine–formaldehyde is the hardest commercial plastic material.

Molding practice for amino resins has changed during the past ten years. Aminoplasts and other thermosetting plastics are now being molded by an automatic injection molding process similar to that used for thermoplastics, but with an important difference (8). Instead of being plasticized in a hot cylinder and then injected into a much cooler mold cavity, the thermosets are plasticized in a warm cylinder and then injected into a hot mold cavity where the chemical reaction of cure sets the resin to the solid state. The process is best applied to relatively small moldings. Melamine plastic dinnerware is still molded by standard compression-molding techniques. The great advantage of injection molding is that it reduces costs by eliminating manual labor, thereby placing the amino resins in a better position to compete with thermoplastics (see Plastics technology).

Future markets for amino resins and plastics appear to be secure because they provide unusual qualities. New developments will probably occur in the areas of more highly specialized materials for treating textiles, paper, etc, and for use with other resins in the formulation of surface coatings where a small amount of an amino resin can significantly increase the value of the basic material.

Looking farther into the future, the fact that amino resins are largely based on nitrogen may put them into a position to compete with other plastics as raw materials based on fossil fuels become more costly.

Raw Materials

Urea. Urea (carbamide) is the most important building block for amino resins because urea–formaldehyde is the largest selling amino resin, and urea is the raw material for melamine, the amino compound used in the next largest selling type of amino resin. Urea is also used to make a variety of other amino compounds, such as ethyleneurea, and other cyclic derivatives used for amino resins for treating textiles. They are discussed below.

Urea is soluble in water, and the crystalline solid is somewhat hygroscopic, tending to cake when exposed to a humid atmosphere. For this reason, urea is frequently pelletized or prilled (formed into little beads) to avoid caking and making it easy to handle (see Pelleting).

Only about 10% of the total urea production is used for amino resins, which thus appear to have a secure source of low-cost raw material. Urea is made by the reaction of carbon dioxide and ammonia at high temperature and pressure to yield a mixture of urea and ammonium carbamate; the latter is recycled.

$$CO_2 + 2\,NH_3 \rightarrow NH_2CONH_2 + H_2O \rightleftarrows H_2NCOONH_4$$

Melamine. Melamine (cyanurotriamide, 2,4,6-triamino-*s*-triazine) is a white crystalline solid, melting at approximately 350°C with vaporization, only slightly soluble in water, commercial product, recrystallized grade, is at least 99% pure. Melamine was synthesized early in the development of organic chemistry, but it remained of theoretical interest until it was found to be a useful constituent of amino resins. Melamine was first made commercially from dicyandiamide (see Cyanamides), but is now made from urea, a much cheaper starting material (9–12) (see also Cyanuric and isocyanuric acids).

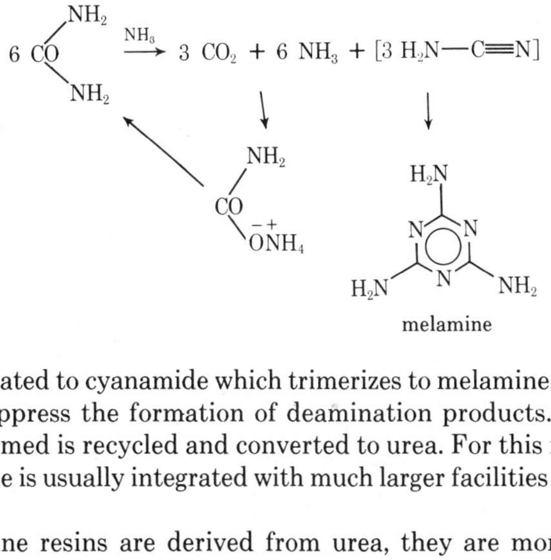

The urea is dehydrated to cyanamide which trimerizes to melamine in an atmosphere of ammonia to suppress the formation of deamination products. The ammonium carbamate also formed is recycled and converted to urea. For this reason the manufacture of melamine is usually integrated with much larger facilities making ammonia and urea.

Since melamine resins are derived from urea, they are more costly and are therefore restricted to applications requiring superior performance. Essentially all of the melamine produced is used for making amino resins and plastics.

Formaldehyde. Pure formaldehyde is a colorless, pungent smelling reactive gas (see Formaldehyde). The commercial product is handled either as solid polymer, paraformaldehyde (13), or in aqueous or alcoholic solutions. Marketed under the trade name Formcel, solutions in methanol, *n*-butanol, and isobutanol, made by the Celanese Chemical Company, are widely used for making alcohol-modified urea and melamine resins for surface coatings and treating textiles.

Aqueous formaldehyde, known as formalin, is usually 37 wt % formaldehyde, though more concentrated solutions are available. Formalin is the general-purpose formaldehyde of commerce supplied unstabilized or methanol-stabilized. The latter may be stored at room temperature without precipitation of solid formaldehyde polymers because it contains 5–10% of methyl alcohol. The uninhibited type must be maintained at a temperature of at least 32°C to prevent the separation of solid formaldehyde polymers. Large quantities are often supplied in more concentrated solutions. Formalin at 44, 50, or even 56% may be used to reduce shipping costs and improve manufacturing efficiency. Heated storage tanks must be used. For example, formalin containing 50% formaldehyde must be kept at a temperature of 55°C to avoid precipitation. Formaldehyde solutions stabilized with urea are used (14), and various other stabilizers have been proposed (15–16). With urea-stabilized formaldehyde the user need only adjust the U/F (urea/formaldehyde) ratio by adding more urea to produce a urea resin solution ready for use.

Paraformaldehyde is a mixture of polyoxymethylene glycols, $HO(CH_2O)_nH$, with n from 8 to as much as 100. It is commercially available as a powder (95%) and as flake (91%). The remainder is a mixture of water and methanol. Paraformaldehyde is an unstable polymer that easily regenerates formaldehyde in solution. Under alkaline conditions, the chains depolymerize from the ends, whereas in acid solution the chains are randomly cleaved (17). Paraformaldehyde is often used when the presence of a large amount of water should be avoided as in the preparation of alkylated amino resins for coatings. Formaldehyde may also exist in the form of the cyclic trimer trioxane. This is a fairly stable compound that does not easily release formaldehyde, hence it is not used as a source of formaldehyde for making amino resins.

Approximately 25% of the formaldehyde produced in the United States is used in the manufacture of amino resins and plastics.

Other Materials. Benzoguanamine and acetoguanamine may be used in place of melamine to achieve greater solubility in organic solvents and greater chemical resistance. Aniline and toluenesulfonamide react with formaldehyde to form thermoplastic resins. They are not used alone, but rather as plasticizers for other resins including melamine and urea–formaldehyde. The plasticizer may be made separately or formed *in situ* during preparation of the primary resin.

Acrylamide is an interesting monomer for use with amino resins; the vinyl group is active in free-radical catalyzed addition polymerizations, whereas the —NH_2 group is active in condensations with formaldehyde. Many patents describe methods of making cross-linked polymers with acrylamide by taking advantage of both vinyl polymerization and condensation with formaldehyde. For example, acrylamide reacts readily with formaldehyde to form N-methylolacrylamide which gives the corresponding isobutyl ether with isobutanol.

$$CH_2{=}CH{-}\overset{\overset{\displaystyle O}{\|}}{C}{-}NH_2 + HCHO \longrightarrow CH_2{=}CH{-}\overset{\overset{\displaystyle O}{\|}}{C}{-}NH{-}CH_2OH$$

$$CH_2{=}CH{-}\overset{\overset{\displaystyle O}{\|}}{C}{-}NH{-}CH_2OH + HO{-}CH_2{-}\underset{\underset{\displaystyle CH_3}{|}}{\overset{\overset{\displaystyle CH_3}{|}}{CH}} \longrightarrow$$

$$CH_2{=}CH{-}\overset{\overset{\displaystyle O}{\|}}{C}{-}NH{-}CH_2{-}O{-}CH_2{-}\underset{\underset{\displaystyle CH_3}{|}}{\overset{\overset{\displaystyle CH_3}{|}}{CH}} + H_2O$$

This compound is soluble in most organic solvents and may be easily copolymerized with other vinyl monomers to introduce reactive side groups on the polymer chain (18). Such reactive polymer chains may then be used to modify other polymers including other amino resins. It may be desirable to produce the cross-links first. Thus *N*-methylolacrylamide can react with more acrylamide to produce methylenebisacrylamide, a tetrafunctional vinyl monomer (see Acrylamide polymers).

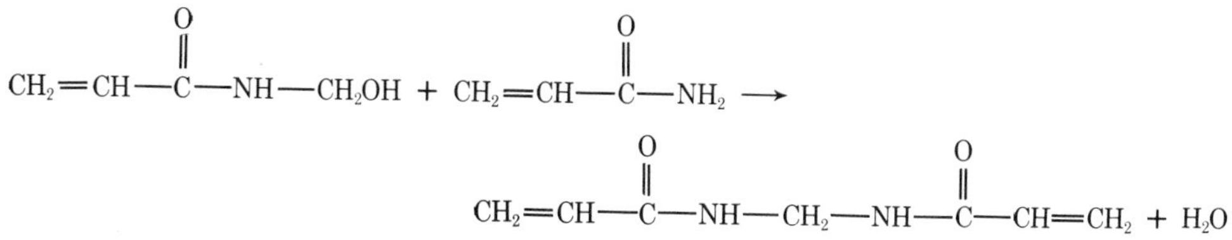

Chemistry of Resin Formation

The first step in the formation of resins and plastics from formaldehyde and amino compounds is the addition of formaldehyde to introduce the hydroxymethyl group, known as methylolation or hydroxymethylation:

$$R{-}NH_2 + HCHO \rightarrow R{-}NH{-}CH_2OH$$

The second step is the condensation of monomer units with the liberation of water to form a dimer, a polymer chain, or a vast network. This is usually referred to as methylene bridge formation, polymerization, resinification, or simply cure.

$$RNH{-}CH_2OH + H_2NR \rightarrow RNH{-}CH_2NH{-}R + H_2O$$

Successful manufacture and utilization of amino resins depend largely on the precise control of these two reactions (19–30).

The first reaction, the addition of formaldehyde to the amino compound, is catalyzed by either acids or bases. The second reaction joins the amino units with methylene links and is catalyzed only by acids. DeJong and DeJonge (28) studied the rates of these reactions over a broad range of pH (Fig. 1). They also examined the formation of more complex urea–formaldehyde condensation products. Rate constants for these reactions at 35°C and pH 4.0 are shown below (U = urea; F = formaldehyde).

Reaction	*Rate constant κ, L/(s·mol)*
$U + F \rightarrow UF$	4.4×10^{-4}
$UF + U \rightarrow U{-}CH_2{-}U$	3.3×10^{-4}
$UF + UF \rightarrow U{-}CH_2{-}UF$	0.85×10^{-4}
$UF_2 + UF \rightarrow FU{-}CH_2{-}UF$	0.5×10^{-4}
$UF_2 + UF_2 \rightarrow FU{-}CH_2{-}UF_2$	$<3 \times 10^{-6}$

The methylol compounds obtained are relatively stable under neutral or alkaline conditions, but form polymeric products under acid conditions. Consequently, the first step in making an amino plastic is usually carried out under alkaline conditions. The amino compound and formaldehyde form a stable resin intermediate that may be used as an adhesive or combined with filler to make a molding compound (see Fillers). The second step is the addition of an acidic substance to catalyze the curing reaction, often with the application of heat. In this reaction, the methylol group is probably protonated and a molecule of water lost giving the intermediate carbo-

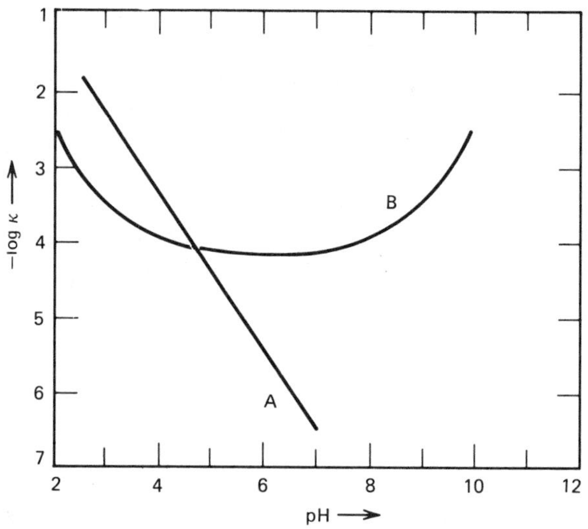

Figure 1. Influence of pH on A, the addition reaction of urea and formaldehyde (1:1); and B, the condensation of methylolurea with the amino hydrogen of a neighboring urea molecule. Temperature = 35°C; 0.1 *M* aq.

nium–imonium ion. This then reacts with an amino group to form a methylene link:

$$RNHCH_2OH + H^+ \rightleftharpoons RNHCH_2OH_2^+$$

$$RNHCH_2OH_2^+ \rightleftharpoons [RNH{-}CH_2]^+ + H_2O$$

$$RNHCH_2^+ + H_2NR' \rightleftharpoons RNHCH_2NH_2R'^+$$

$$RNHCH_2NH_2R'^+ \rightleftharpoons RNHCH_2NHR' + H^+$$

In addition to these two main reactions, methylolation and condensation, a number of other reactions are also important for the manufacture and uses of amino resins. For example, two methylol groups may combine to produce a dimethylene ether linkage and liberate a molecule of water.

$$2\,RNH{-}CH_2OH \rightleftharpoons RNH{-}CH_2{-}O{-}CH_2{-}NHR + H_2O$$

The dimethylene ether so formed is less stable than the diamino–methylene bridge and may rearrange to form a methylene link and liberate a molecule of formaldehyde.

$$RNH{-}CH_2O{-}CH_2NHR \rightarrow RNH{-}CH_2{-}NHR + HCHO$$

The simple methylol compounds and the low molecular weight polymers obtained from urea and melamine are soluble in water and quite suitable for the manufacture of adhesives, molding compounds, and some kinds of textile treating resins. However, amino resins for coating applications require compatibility with the film-forming alkyd or copolymer resins with which they must react. Furthermore, even where compatible, the free methylol compounds are often too reactive and unstable for use in a coating-resin formulation that may have to be stored for some time before use. Reaction of the free methylol groups with an alcohol to convert them to alkoxy methyl groups solves both problems.

The replacement of the hydrogen of the methylol compound with an alkyl group renders the compound much more soluble in organic solvents and more stable. This reaction is also catalyzed by acids and usually carried out in the presence of considerable excess alcohol to suppress the competing self-condensation reaction. After neutralization of the acid catalyst, the excess alcohol may be stripped or left as a solvent for the amino resin.

The mechanism of the alkylation reaction is similar to curing. The methylol group becomes protonated and dissociates to form a carbonium ion intermediate which may react with alcohol to produce an alkoxymethyl group or with water to revert to the starting material. As one might expect, the amount of water in the reaction mixture should be kept to a minimum since the relative amounts of alcohol and water determine the final equilibrium.

Another way of achieving the desired compatibility with organic solvents is to employ an amino compound having an organic solubilizing group in the molecule, such as benzoguanamine. With one of the —NH_2 groups of melamine replaced with a phenyl group, benzoguanamine–formaldehyde resins [*26160-89-4*] have some degree of oil solubility even without additives. Nevertheless, benzoguanamine–formaldehyde resins are generally modified with alcohols to provide a still greater range of compatibility with solvent-based surface coatings. Benzoguanamine resins provide a high degree of detergent resistance together with good ductility and excellent adhesion to metal.

Displacement of a volatile with a nonvolatile alcohol is an important reaction for curing paint films with amino cross-linkers and amino resins on textile fabrics or paper. Following is an example of a methoxymethyl group on an amino resin reacting with a hydroxyl group of a polymer chain.

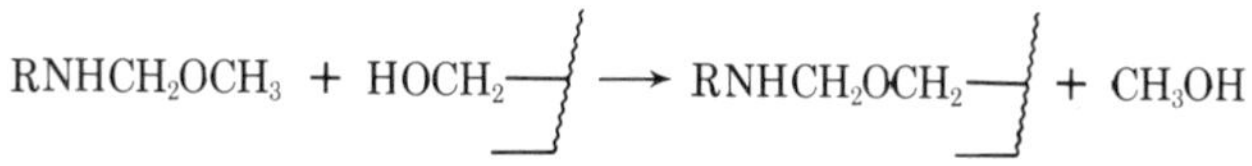

A troublesome side reaction encountered in the manufacture and use of amino resins is the conversion of formaldehyde to formic acid. Often the reaction mixture of amino compound and formaldehyde must be heated under alkaline conditions. This favors a Cannizzaro reaction in which two molecules of formaldehyde interact to yield one molecule of methanol and one of formic acid.

$$2\,HCHO + H_2O \rightarrow CH_3OH + HCOOH$$

Unless this reaction is controlled, the solution may become sufficiently acidic to catalyze the condensation reaction causing abnormally high viscosity or premature gelation of the resin solution.

Manufacture

Precise control of the course, speed, and extent of the reaction is essential for successful manufacture. Important factors are: mole ratio of reactants; catalyst (pH of reaction mixture); and reaction time and temperature. Amino resins are usually made by a batch process. The formaldehyde and other reactants are charged to a kettle, the pH adjusted, and the charge heated. Often the pH of the formaldehyde is adjusted before adding the other reactants. Aqueous formaldehyde is most convenient to handle and lowest in cost.

In general, conditions for the first part of the reaction are selected to favor the formation of methylol compounds. After addition of the reactants, the conditions may be adjusted to control the polymerization. The reaction may be stopped to give a stable syrup. This could be an adhesive or laminating resin and might be blended with filler to make a molding compound. (See also Laminated and reinforced plastics.) It might also be an intermediate for the manufacture of a more complicated product, such as an alkylated amino resin, for use with other polymers in coatings.

The flow sheet, Figure 2, illustrates the manufacture of amino resin syrups, cellulose-filled molding compounds, and spray-dried resins.

In the manufacture of amino resins every effort is made to recover and recycle the raw materials. However, there may be some loss of formaldehyde, methanol, or other solvent as tanks and reactors are vented. Some formaldehyde, solvents, and alcohols are also evolved in the curing of paint films and the curing of adhesives and resins applied to textiles and paper. The amounts of material evolved in curing the resins may be small so that it may be difficult to justify the installation of complex recovery equipment. However, in the development of new resins for coatings and for treating textiles and paper, emphasis is being placed on those compositions that evolve a minimum of by-products on curing.

Uses

Adhesives. Adhesives (qv) based on amino resins can be simple reaction products of urea or melamine with formaldehyde, or they can be complex formulated products that include plasticizers, extenders, stabilizers, curing agents, hardeners, etc.

Large-volume products, such as urea-resin-based plywood and chipboard glues, are often made as needed at the plywood or chipboard mill (see Laminated and reinforced wood). These are simple solutions of methylol compounds containing a fairly high proportion of formaldehyde. For example, a urea adhesive may contain about 1.8 mol of formaldehyde per 1.0 mol of urea. Unlike laminating resins, the adhesive must not saturate the wood chips or veneers, but must remain in the glue line on the surface of the chips or between the plies. In general, the adhesives are made of high viscosity so that they remain in the glue line. Thickeners and extenders, such as finely powdered pecan shells, wheat flour, blood albumen, etc, are often used. Water present in the adhesive formulation need not be removed before curing since it can diffuse through the wood. Laminating resins, on the other hand, must be of low viscosity so that they will quickly saturate the cloth or paper to be laminated. Water or other solvent must then be substantially all removed before the laminate is pressed.

Furniture assembly glues are highly developed compositions that are generally cured by adding a hardener or curing agent just before application. Urea–formaldehyde adhesives may be cured at room temperature by adding a catalyst such as ammonium chloride. A slow-acting basic material, such as tricalcium phosphate or triethanolamine, is often added to neutralize excess catalyst acid that otherwise might damage the wood.

Amino resins may, of course, be cured with heat using less active catalysts. Very rapid cures may be achieved by applying heat in the form of microwave radiation (see Microwave technology). The moist amino resin adhesive absorbs the high-frequency radiation more readily than the dry wood, thereby concentrating the heat in the glue line where it is needed.

Melamine adhesives have excellent water resistance, whereas urea–formaldehyde adhesives are cheaper but more sensitive to water. Often, the best balance between

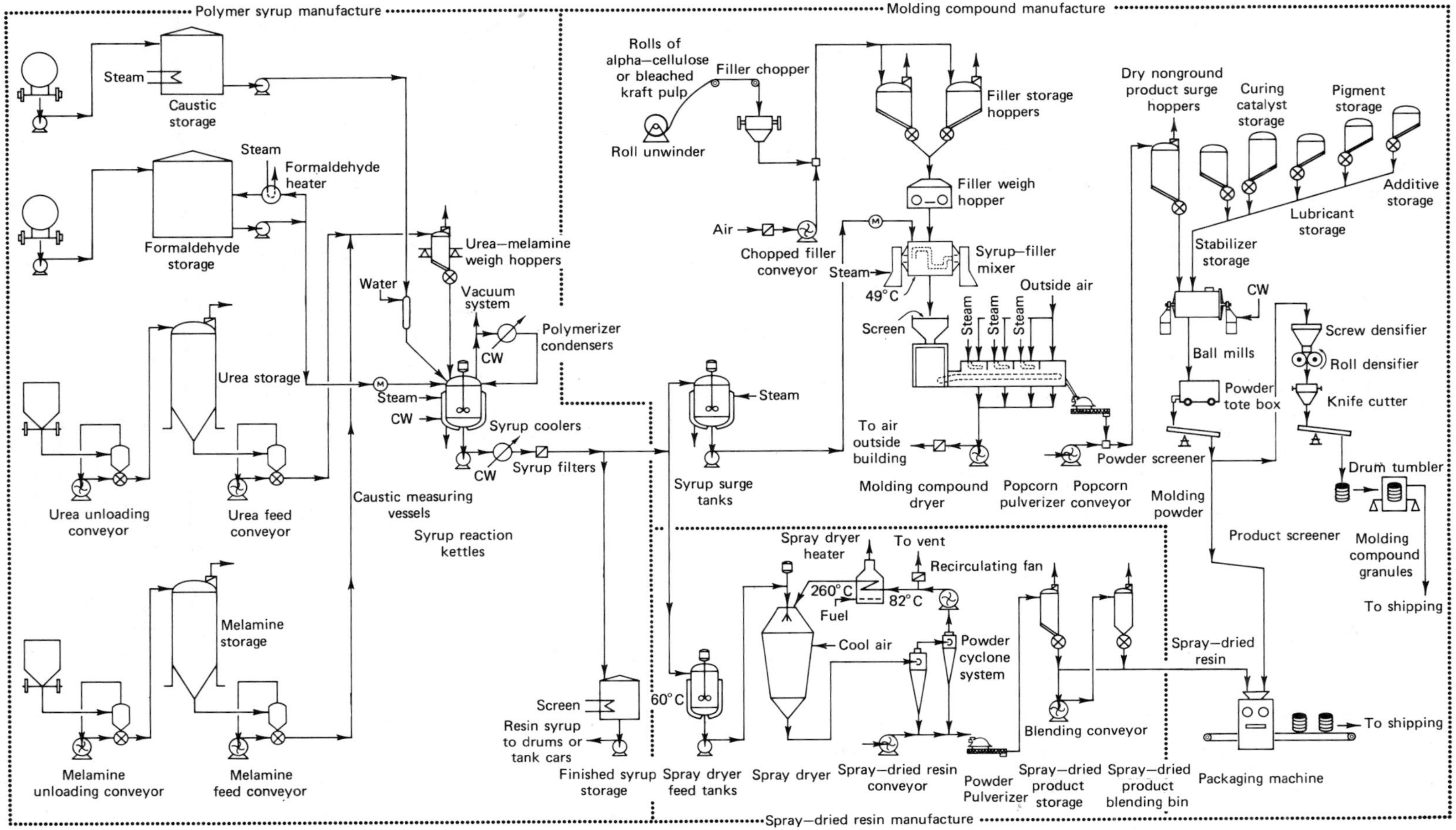

Figure 2. Urea–formaldehyde and melamine–formaldehyde resin manufacture. CW = cold water. Courtesy of Stanford Research Institute.

cost and performance is achieved with a blend of urea– and melamine–formaldehyde resins.

Urea–formaldehyde adhesives may contain modifiers like benzyl alcohol (31) or furfuryl alcohol (32) to make the resin less rigid, ie, less highly cross-linked and better able to relieve stresses imposed by shrinkage due to cure and loss of water in glue lines of nonuniform thickness. These gap-filling adhesives are useful in furniture assembly. Urea–formaldehyde adhesives may also be mixed with poly(vinyl acetate) (PVA) emulsion adhesives. Such a formulation combines the short clamping time of the PVA adhesive with the heat and water resistance of the cross-linked urea–formaldehyde adhesive.

Continuous production of a urea–formaldehyde resin has been described in many patents. In a typical example, urea and formaldehyde are combined and the solution pumped through a multistage unit. Temperature and pH are controlled at each stage to achieve the appropriate degree of polymerization. The product is then concentrated in a continuous evaporator to 58% solids and is recommended for use as a wood adhesive (33).

Laminating Resins. Phenolic and melamine resins are both used in the manufacture of decorative laminated plastic sheets for counters and table tops (see Laminated and reinforced plastics). The phenolic is functional, being used in the backing or support sheets, whereas the melamine resin performs both decorative and functional roles in the print sheet and in the protective overlay. Hardness, transparency, stain resistance, and freedom from discoloration are essential, in addition to a long-lasting working surface. Transparency is achieved because the refractive index of cured melamine–formaldehyde resin approaches that of the cellulose fibers and thus there is little scattering of light. Low-cost and good all-around mechanical properties are provided by the phenolic backing layers. In this instance, the combination of phenolic and amino resins achieves an objective that neither would be capable of performing alone.

Phenolic resins are generally used in alcoholic solution, whereas melamine resins are best handled in water or water–alcohol mixtures. The paper or cloth web is passed through a dip tank containing resin solution, adjusted for pick-up on squeeze rolls, and then passed through a heated drying oven. Once dried, the treated paper or cloth is fairly stable, and, stored in a cool place, it may be kept for several weeks or months before pressing into laminated plastic sheets.

A melamine laminating resin used to saturate the print and overlay papers of a typical decorative laminate might contain two moles of formaldehyde for each mole of melamine. In order to inhibit crystallization of methylol melamines, the reaction is continued until about one-fourth of the reaction product has been converted to low molecular weight polymer. A simple determination of free formaldehyde may be used to follow the first stage of the reaction and the build-up of polymer in the reaction mixture may be followed by cloud-point dilution or viscosity tests.

A particularly interesting and useful test is run at high dilution. One or two drops of resin are added to a test tube half full of water. A cloudy streak as the drop sinks through the water indicates that the resin has advanced to the point where the highest molecular weight fraction of polymer is no longer soluble in water at that particular temperature. At this high dilution, the proportion of water to resin is not critical, hence the only measurement needed is the temperature of the water. The temperature at or below which the drops give a white streak is known as the hydrophobe temperature. This test is particularly useful with melamine resins.

Laminates are pressed in steam-heated, multiple-opening presses. Each opening may contain a book of as many as ten laminates pressed against polished steel plates. Curing conditions are 20–30 min at about 150°C under a pressure of about 6900 kPa (1000 psi).

Molding Compounds. Molding was the first big application for amino resins, although molding compounds are more complex than either laminating resins or adhesives.

A simple amino resin molding compound might be made by combining melamine with 37% formalin in the ratio of two moles of formaldehyde for each mole of melamine at neutral or slightly alkaline pH and a temperature of 60°C. The reaction should be continued until some polymeric product has been formed to inhibit crystallization of dimethylolmelamine upon cooling. When the proper reaction stage has been reached, the resin syrup is pumped to a dough mixer where it is combined with alpha cellulose pulp, approximately one part of cellulose to each three parts of resin solids (see Fillers). The wet, spongy mass formed in the dough mixer is then spread on trays and dried in a humidity-controlled oven to produce a hard, brittle popcorn-like intermediate. This material may be coarsely ground and sent to storage. To make the molding material, the cellulose–melamine resin intermediate is combined in a ball mill with a suitable catalyst, stabilizer, colorants, and mold lubricants. The materials must be ground for several hours to achieve the uniform fine dispersion needed to get the desired decorative appearance in the molded article. The molding compound may be used as a powder or it may be compacted under heat and pressure to a granular product that is easier to handle (34). A urea molding compound might be made in much the same way using a resin made with 1.3–1.5 mol of formaldehyde per 1.0 mol of urea.

To speed up the molding process, the required amount of molding powder or granules is often pressed into a block and prewarmed before placing it in the mold. Rapid and uniform heating is accomplished in a high-frequency preheater, essentially an industrial microwave oven. The prewarmed block is then transferred to the hot mold, pressed into shape, and cured.

Production of decorated melamine plastic dinner plates makes use of molding and laminating techniques. The pattern is printed on the same type of paper used for the protective overlay of decorative laminates, treated with melamine resin and dried, and then cut into disks of the appropriate size.

To make a decorated plate, the mold is opened shortly after the main charge of molding compound has been pressed into shape, the decorative foil is laid in the mold on top of the partially cured plate, printed side down, and the mold closed again to complete the curing process. The melamine-treated foil is thus fused to the molded plate and, as with the decorative laminate, the overlay becomes transparent so that the printed design shows through yet is protected by the film of cured resin.

The excellent electrical properties, hardness, heat resistance, and strength of melamine resins makes them useful for a variety of industrial applications. Some representative properties of amino resin molding compounds, including the industrial-grade melamines, are listed in Table 1.

Industrial-grade urea-resin molding materials in black and brown colors were important in years past. They are made with wood–flour filler, and since they do not have to meet the appearance requirements of decorative grades they are not ground in a ball mill. All the ingredients are added to a mixer. Moldings from these indus-

Table 1. Typical Properties of Filled Amino Resin Molding Compounds

ASTM or UL test	Property	Urea	Melamine			
		Alpha-cellulose	Alpha-cellulose	Macerated fabric	Asbestos	Glass fiber
	Physical					
D792	specific gravity	1.47–1.52	1.47–1.52	1.5	1.7–2.0	1.8–2.0
D570	water absorption, 24 h, 3.2 mm thick, %	0.48	0.1–0.6	0.3–0.6	0.08–0.14	0.09–0.21
	Mechanical					
D638	tensile strength, MPa (10^3 psi)	38–48 (5.5–7)	48–90 (7–13)	55–69 (8–10)	38–45 (5.5–6.5)	35–70 (5–10)
D638	elongation, %	0.5–1.0	0.6–0.9	0.6–0.8	0.3–0.45	
D638	tensile modulus, GPa (10^5 psi)	9–9.7 (13–14)	9.3 (13.5)	9.7–11 (14–16)	13.5 (19.5)	16.5 (24)
D785	hardness, Rockwell M	110–120	120	120	110	115
D790	flexural strength, MPa (10^3 psi)	70–124 (11–18)	83–104 (12–15)	83–104 (12–15)	52–69 (7.4–10)	90–165 (13–24)
D790	flexural modulus, GPa (10^5 psi)	9.7–10.3 (14–15)	7.6 (11)	9.7 (14)	12.4 (18)	16.5 (24)
D256	impact strength, J/m of notch (ft lb/in. of notch)	14–18 (0.27–0.34)	13–19 (0.24–0.35)	32–53 (0.6–1.0)	16–21 (0.3–0.4)	32–1000 (0.6–18)
	Thermal					
C177	thermal conductivity, 10^{-4} (J·cm)/(s·cm^2·°C)[a]	42.3	29.3–42.3	44.3	54.4–71	48.1
D696	coefficient of thermal expansion, 10^{-5} cm/(cm·°C)	2.2–3.6	2.0–5.7	2.5–2.8	2.0–4.5	1.5–1.7
D648	deflection temperature at 1.8 MPa (264 psi), °C	130	182	154	129	204
UL 94	flammability class	VO[b]	VO[b]			VO
	continuous no-load service temperature, °C	77[c]	99[c]	121	149	149–204
	Electrical					
D149	dielectric strength, V/0.00254 cm					
	short time, 3.2 mm thick	330–370	270–300	250–350	410–430	170–300
	step by step	220–250	240–270	200–300	280–320	170–240
D150	dielectric constant, 22.8°C					
	at 60 Hz	7.7–7.9	8.4–9.4	7.6–12.6	6.4–10.2	9.7–11.1
	at 10^3 Hz		7.8–9.2	7.1–7.8	9.0	
D150	dissipation factor, 22.8°C					
	at 60 Hz	0.034–0.043	0.030–0.083	0.07–0.34	0.07–0.17	0.14–0.23
	at 10^3 Hz		0.015–0.036	0.03–0.05	0.07	
D257	volume resistivity, 22.8°C, 50% rh, ohm·cm	$0.5–5.0 \times 10^{11}$	$0.8–2.0 \times 10^{12}$	$1.0–3.0 \times 10^{11}$	1.2×10^{12}	$0.9–2.0 \times 10^{11}$
D495	arc resistance, s	80–100	125–136	122–128	20–180	180–186

[a] To convert J to cal divide by 4.184.
[b] Applies to specimens thicker than 1.6 mm.
[c] Based on no color change.

trial-grade amino resins have good electrical properties but they are not as dimensionally stable or as water resistant as corresponding phenolic molding materials. They may become more important as phenol becomes more costly compared to urea.

Coatings. Cured amino resins are far too hard and brittle to be used alone as coating materials, but in combination with more flexible resins cure or cross-linking is achieved. These resin systems cure rapidly and are ideal for factory-applied industrial coatings that must meet tight production schedules (see Coatings, industrial). Usually the amino resin is in the range of 10–50% of the total resin solids in the formulation. Often the amino resin reacts simultaneously with itself and with the copolymer or alkyd, thus building up hard and rigid amino–formaldehyde polymers in addition to forming cross-links. Newer amino resins for coatings are designed to emphasize the cross-linking function and minimize self-condensation. This provides greatest efficiency for hardening the primary resin with least evolution of volatile reaction products (35–36).

The manufacture and use of protective surface coatings is a highly developed technology, and many different amino resins are needed to meet the varied requirements of the coatings industry. A complete line of amino resins for coatings might include as many as forty different products about equally divided between urea and melamine types. Resins based on benzoguanamine may also be included.

The urea resins are the cheapest and fastest curing. They are used in clear coatings for wood furniture, baked enamels for appliances, and primer coats on automobiles. Melamine resins give better chemical resistance and are preferred for applications involving outdoor exposure or contact with detergents such as automotive topcoats and kitchen appliances. It is not uncommon for a paint formulation to contain a combination of both urea and melamine resins.

The principal factors in the manufacture of alkylated amino resins for coatings are the degree of polymerization and of alkylation, and the type of alcohol used. Resins with a high polymer content cure rapidly since little chemical reaction is needed for cross-linking. Resins of low alkylation are fast curing since they contain a high concentration of reactive methylol groups and replaceable hydrogen atoms that can react to join molecules with methylene linkages. A highly alkylated resin has a greater range of compatibility with other resins and solvents, but may require an acid catalyst for efficient curing. The most commonly used alcohols for alkylation are methanol, *n*-butanol, and isobutanol. Resins alkylated with isobutanol cure somewhat more rapidly than those alkylated with *n*-butanol. Both have good compatibility with most resins and organic solvents. Methylated resins are compatible not only with organic solvents but also with water and are widely used in water-reducible coating formulations. Amino resins for coatings that are deposited from aqueous emulsions by electrodeposition require a precise balance between hydrophilic and hydrophobic properties in order to migrate with the base resin under the influence of the electric current (see Coating processes).

Stability in storage is an important property for coating systems containing amino resins. If the amino resin undergoes self-condensation or reacts at room temperature with the alkyd or other film-forming polymer, the system may become too viscous or thicken to a gel which can no longer be used for coating. Alkyds usually contain sufficient free carboxyl groups to catalyze the curing reaction when the coating is baked but this may also cause the paint to thicken in storage. Partial neutralization of the acid groups with an amine can greatly improve storage stability yet allow the film to cure when baked since much of the amine is vaporized with the solvent during the baking process. 2-Amino-2-methyl-1-propanol, triethylamine, and dimethylaminoethanol are commonly used as stabilizers. Alcohols as solvent also improve storage stability.

Catalyst addition just before the coating is to be applied permits rapid curing and avoids the problem of storage stability. A strong acid soluble in organic solvents such as *p*-toluenesulfonic acid is very effective and may be partially neutralized with an amine to avoid premature reaction.

A butylated urea–formaldehyde resin for use in the formulation of fast-curing baking enamels might be made as follows: Charge urea (1.0 mol), paraformaldehyde (2.12 mol), and butanol (1.50 mol). Add triethanolamine to make the solution alkaline (about 1% of the weight of the urea) and reflux until the paraformaldehyde is dissolved. Then add phthalic anhydride to give a pH of 4.0 and remove the water by azeotropic distillation until the batch temperature reaches 117°C. Cool and dilute to the desired solids content with solvent (37).

A partially alkylated urea resin such as this might be combined with a soya oil–phthalic anhydride alkyd along with suitable solvents and pigmentation to produce a fast-curing appliance finish. During baking, the urea resin self-condenses to insoluble resin and also reacts with the free hydroxyl and carbonyl groups of the alkyd forming a three-dimensional polymer network.

A highly methylated melamine–formaldehyde resin for cross-linking with little or no self-condensation might be made as follows (38).

A solution of formaldehyde in methyl alcohol is charged to a reaction kettle and adjusted to a pH of 9.0–9.5 using sodium hydroxide. Melamine is then added to give a ratio of one mole of melamine for each 6.5 moles of formaldehyde and the mixture refluxed for one half hour. The reaction is then cooled to 35°C and more methanol added to bring the ratio of methanol per mole of melamine up to 11. With the batch temperature at 35°C, enough sulfuric acid is added to reduce the pH to 1.0. After holding the reaction mixture at this temperature and pH for 1 h, the batch is neutralized with 50% sodium hydroxide and the excess methanol stripped to give a product containing 60% solids which is then clarified by filtration. A highly methylated resin, such as this, may be used in water-based (39) or solvent-type coatings. It might also be used to provide crease resistance to cotton fabric.

Textile Finishes. Most amino resins used commercially for finishing textile fabrics are methylolated derivatives of urea or melamine. Although these products are usually monomeric, they may contain some polymer by-product.

Amino resins react with cellulosic fibers and change their physical properties. They do not react with synthetic fibers, such as nylon, polyester, or acrylics, but may self-condense on the surface. This results in a change in the stiffness or resiliency of the fiber. Partially polymerized amino resins of molecular size that prevents them from penetrating the amorphous portion of cellulose also tend to increase the stiffness or resiliency of cellulose fibers.

Monomeric amino resins react predominantly with the primary hydroxyls of the cellulose, thereby replacing weak hydrogen bonds with strong covalent bonds which leads to an increase in fiber elasticity. When an untreated cotton fiber is stretched or deformed by bending, as in forming a crease or wrinkle, the relatively weak hydrogen bonds are broken and then reform to hold the fiber in its new position. The covalent bonds that are formed when adjacent cellulose chains are cross-linked with an amino resin are five to six times stronger than the hydrogen bonds. Covalent bonds are not broken when the fiber is stretched or otherwise deformed. Consequently, the fiber tends to return to its original condition when the strain is removed. This increased elasticity is manifested in two important ways: (*1*) When a cotton fabric is cross-linked while it is held flat, the fabric tends to return to its flat condition after it has been wrinkled during use or during laundering. Garments made from this type of fabric are known as wash-and-wear, minimum care, or no-iron. (*2*) A pair of pants that is pressed to form a crease and then cross-linked tends to maintain the crease through wearing and laundering. This type of garment is called durable press or permanent press (see Textiles).

This increased elasticity is always accompanied by a decrease in strength of the cellulose fiber which occurs even though weak hydrogen bonds are replaced by stronger covalent bonds. The loss of strength is not caused by hydrolytic damage to the cellulose. If the cross-linking agent is removed by acid hydrolysis, eg, the fiber will regain most, if not all, its original strength. The loss in strength is believed to be due to intramolecular reaction of the amino resin along the cellulose chain to displace a larger number of hydrogen bonds, resulting in a net loss in strength. The intramolecular and intermolecular reaction (cross-linking) both occur at the same time.

Although there are many different amino resins used for textile finishing, all of them impart about the same degree of increase in elasticity when applied on an equal molar basis. Elasticity can be measured by determining the recovery from wrinkling. Although all these products impart about the same degree of improvement in elasticity, they also may impart many other desirable or undesirable properties to the fabric. The development of amino resins for textile finishing has been aimed toward maximizing the desirable properties and minimizing the undesirable ones. Most of the resins and reactants used in today's textile market are based on urea as a starting material. However, the chemistry differs considerably from that employed in early textile-finishing operations.

The first amino resins used commercially on textiles were the so-called urea–formaldehyde resins, dimethylolurea, or its mixtures with monomethylolurea. Although their performance falls short of most present finishes, particularly in durability and resistance to chlorine-containing bleaches, urea–formaldehyde resins are still in use. Both urea and formaldehyde are relatively inexpensive, and manufacture is simple; ie, 1–2 mol of formaldehyde as an aqueous solution reacts with 1 mol of urea under mildly alkaline conditions at slightly elevated temperatures.

Since the methylolurea monomers have limited water solubility (about 30%), they are usually marketed in dispersed form as soft pastes containing 55–65% active ingredient in order to decrease container and shipping costs. By increasing the temperature and using slightly acid conditions, dimethylolureas can be made as a series of short polymers that have infinite water solubility and can be marketed at concentrations as high as 85%. However, because these result in increased fabric stiffness, they cannot be used interchangeably with the monomeric materials. Both forms polymerize readily in storage and, unless kept under refrigeration, become water insoluble within a few weeks at ambient temperatures.

To overcome stability and water solubility problems, methylolurea resins are frequently alkylated to block the reactive hydroxyl groups. For reasons of economy the alkylating agent is usually methanol. In this process, 2 mol of aqueous formaldehyde reacts with 1 mol of urea under alkaline conditions to form dimethylolurea. Excess methanol is then added, and the reaction continued under acid conditions to form methoxymethylurea.

Both methylol groups can be methylated by maintaining a low concentration of water and using a large excess of methanol; however, methylation of only one of the methylol groups is sufficient to provide adequate shelf life and water solubility. Upon completion of the methylation reaction, the resin is adjusted to pH 7–10, and excess methanol and water are removed by distillation under reduced pressure to provide syrups of 50–80% active ingredients.

Like the methylol ureas, the cyclic ureas are based on reactions between urea and formaldehyde; however, the amino resin is cyclic rather than linear. Many cyclic urea resins have been used in textile finishing processes, particularly to achieve wrinkle resistance and shrinkage control, but the ones described below are the commercially most important. They are all in use today to greater or lesser extents, depending on specific end requirements (see also Textiles, miscellaneous finishes).

Ethyleneurea. One of the most widely used resins during the 1950s and 1960s was based on dimethylolethyleneurea (1,3-bis(hydroxymethyl)-2-imidazolidinone) commonly known as ethyleneurea. This resin [*28906-87-8*] is most conveniently prepared from urea, ethylenediamine, and formaldehyde. 2-Imidazolidinone is first prepared by the reaction of excess ethylenediamine with urea (40) in an aqueous medium at about 240°C.

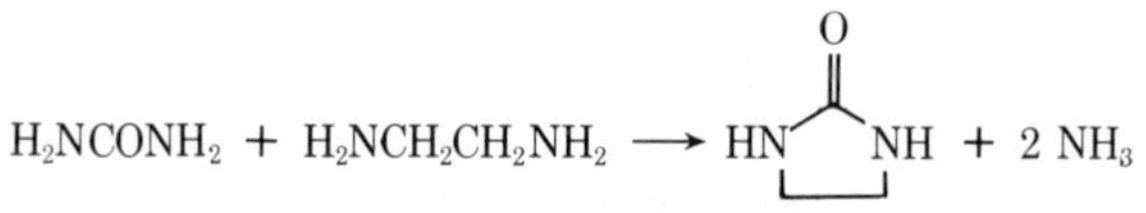

2-imidazolidinone

A fractionating column is required for the removal of ammonia and recycle of ethylenediamine. The molten product (mp 133°C) is then run into ice water to give a solution which is methylolated with 37% aqueous formaldehyde.

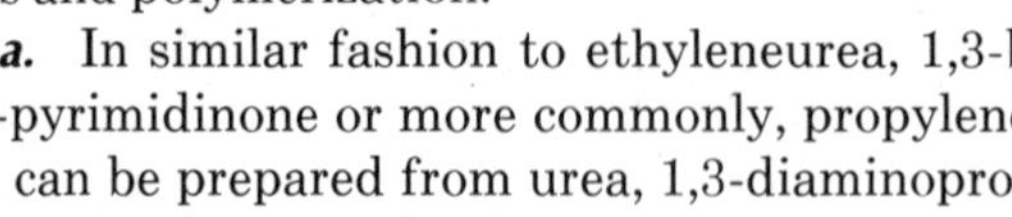

dimethylolethyleneurea

The resin itself, generally a 50% solution in water, has excellent shelf life and is stable to hydrolysis and polymerization.

Propylene Urea. In similar fashion to ethyleneurea, 1,3-bis(hydroxymethyl)-tetrahydro-2-(1*H*)-pyrimidinone or more commonly, propyleneurea–formaldehyde resin [*65405-39-2*], can be prepared from urea, 1,3-diaminopropane, and formaldehyde.

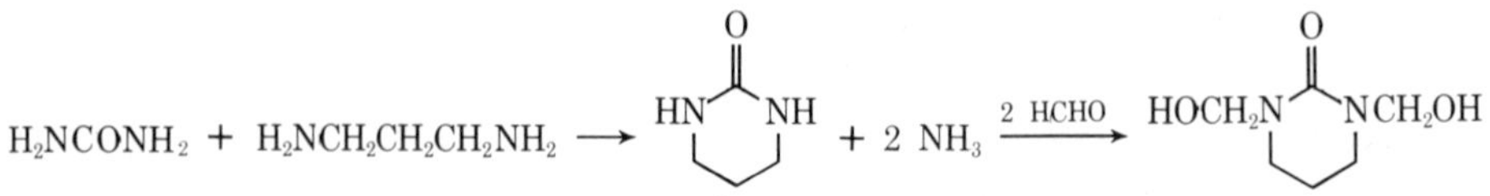

propyleneurea

This resin was temporarily accepted, primarily because of its improved resistance to acid washes. However, the relatively high cost of the diamine precluded widespread commercial acceptance.

Triazone. Triazone is the common name for the class of compounds corresponding to the dimethylol derivatives of tetrahydro-5-alkyl-*s*-triazone. They can be made readily and cheaply from urea, formaldehyde, and a primary aliphatic amine. A wide variety of amines may be used, as first described by Burke (41), to form the six-membered ring; however, for reasons of cost and odor hydroxyethylamine (monoethanolamine) is used preferentially (see Alkanolamines). Since the presence of straight-chain methylol ureas causes no deleterious effects to the fabric finish, the triazones typically are prepared with less than the stoichiometric quantity of the amine. This results not only in a less costly resin but also in improved performance (42).

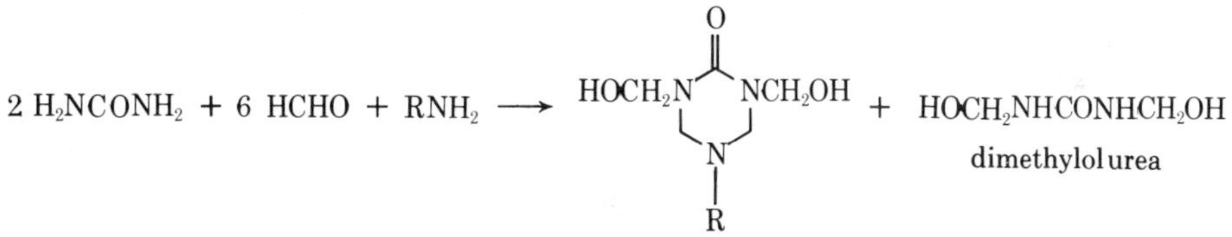

1,3-bis(hydroxymethyl)tetrahydro-5-alkyl-*s*-triazone-2-one

The resin is simply prepared by heating the components together. Usually the urea and formaldehyde are first charged to the kettle and heated under alkaline conditions to give a mixture of polymethylolureas, followed by the slow addition of the amine with continued heating to form the cyclic compound. The order of addition can be varied as can the molar ratios to yield a range of chain–ring compound ratios. The commercial resin is usually sold as a 50% solids solution in water.

Uron Resins. In the textile industry, the term uron resin usually refers to the mixture of a minor amount of melamine resin and so-called uron, which in turn is predominantly *N,N′*-bis(methoxymethyl)uron [*7388-44-5*] plus 15–25% methylated urea–formaldehyde resins, a by-product. *N,N′*-Bis(methoxymethyl)uron was first isolated and described in 1936 (43), but was commercialized only in 1960. It is manufactured (44) by the reaction of 6 mol of formaldehyde with 1 mol of urea at 60°C under highly alkaline conditions to form tetramethylolurea. After concentration under reduced pressure to remove water, excess methanol is charged and the reaction continued under acidic conditions at ambient temperatures to close the ring and methylate the hydroxymethyl groups. After filtration to remove the precipitated salts, the methanolic solution is concentrated to recover excess methanol. The product (75–85% pure) is then mixed with a methylated melamine–formaldehyde resin to reduce fabric strength losses in the presence of chlorine, and diluted with water to 50–75% solids.

N,N′-bis(methoxymethyl)uron

Glyoxal Resins. Since the late 1960s, glyoxal resins have dominated the textile-finishing market for use as wrinkle-recovery, wash-and-wear, and durable-press agents. These resins are based on 1,3-bis(hydroxymethyl)-4,5-dihydroxy-2-imidazolidinone, commonly called dimethyloldihydroxyethyleneurea [*1854-26-8*] (DMDHEU). Several methods of preparation are described in the literature (45). On a commercial scale, DMDHEU can be prepared inexpensively at high purity by a one-kettle process (46): one mole of urea, one mole of glyoxal as 40% solution, and two moles of formaldehyde in aqueous solution are charged to the reaction vessel. The pH is adjusted to 7.5–9.5 and the mixture heated at 60–70°C. The reaction is nearly stoichiometric; excess reagent is not necessary.

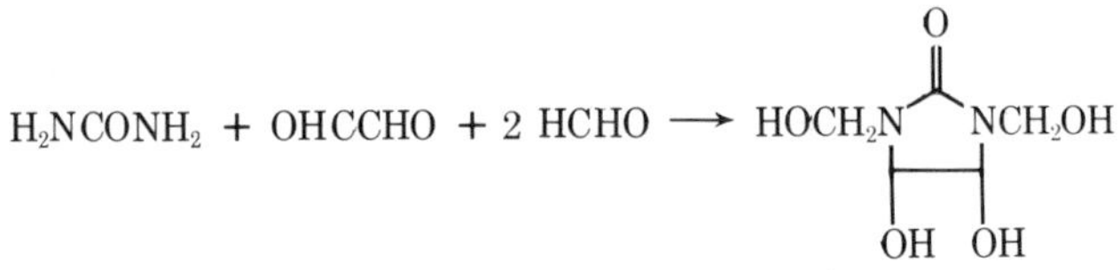

DMDHEU

Glyoxal resins are generally sold at 45% solid solutions in water. Resin usage for crease-resistant fabrics increased to well over 60 million kg by 1974 and over half of this was DMDHEU for durable-press garments.

A less important glyoxal resin is tetramethylolglycoluril [*5395-50-6*] (tetramethylolacetylenediurea) produced by the reaction of one mole of glyoxal with two moles of urea, and four moles of formaldehyde.

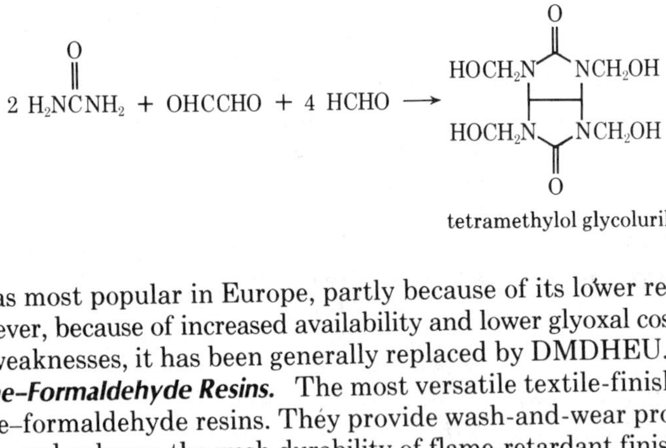

tetramethylol glycoluril

This resin was most popular in Europe, partly because of its lower requirements of glyoxal. However, because of increased availability and lower glyoxal costs plus certain application weaknesses, it has been generally replaced by DMDHEU.

Melamine–Formaldehyde Resins. The most versatile textile-finishing resins are the melamine–formaldehyde resins. They provide wash-and-wear properties to cellulosic fabrics, and enhance the wash durability of flame-retardant finishes. Butylated melamine–formaldehyde resins of the type used in surface coatings may be used in textile printing-ink formulations. A typical textile melamine resin is the dimethyl ether of trimethylolmelamine [*1852-22-8*] which can be prepared as follows:

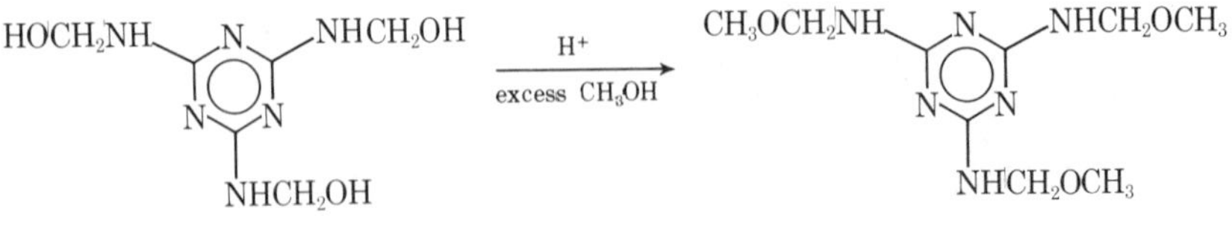

Three mol of formaldehyde react with 1 mol of melamine at elevated temperatures under alkaline conditions. Since water interferes with the methylation, methylolation is carried out in methanol with paraformaldehyde, and by simply adjusting the pH to about 4 with continued heating. After alkylation is complete the pH is adjusted to 8–10 and excess methanol is distilled under reduced pressure. The resulting syrup contains about 80% solids.

Miscellaneous Resins. Next to melamine–formaldehyde and urea–formaldehyde resins, the most important textile finishing resins are the methylol carbamates. They are urea derivatives since they are made from urea and an alcohol (R can vary from methyl to a monoalkyl ether of ethylene glycol).

$$ROH + H_2NCONH_2 \longrightarrow RO\overset{\overset{\displaystyle O}{\|}}{C}NH_2 + NH_3$$

Temperatures in excess of 140°C are required to complete the reaction and pressurized equipment is used for alcohols boiling below this temperature; provision must be made for venting ammonia without loss of alcohol. The reaction is straightforward and, in the case of the monomethyl ether of ethylene glycol, can be carried out at atmospheric pressure using stoichiometric quantities of urea and alcohol (47). Methylolation with aqueous formaldehyde is carried out at 70–90°C under alkaline conditions. The excess formaldehyde needed for complete dimethylolation remains in the resin and prevents more extensive usage because of formaldehyde odor problems in the mill.

Other amino resins used in the textile industry for rather specific properties have included the methylol derivatives of acrylamide (48), hydantoin (49), and dicyandiamide (50).

Textiles are finished with amino resins in four steps: The fabric is (*1*) passed through a solution containing the chemicals; (*2*) through squeeze rolls (padding) to remove excess solution; (*3*) dried; and (*4*) heated (cured) to bond the chemicals with the cellulose or to polymerize them on the fabric surface.

The solution (pad bath) contains one or more of the amino resins described above, a catalyst, and other additives such as a softener, a stiffening agent, or a water repellent. The catalyst may be an ammonium or metal salt, eg, magnesium chloride or zinc nitrate. Synthetic fabrics, such as nylon or polyester, are treated with amino resins to obtain a stiff finish. Cotton or rayon fabrics or blends with synthetic fibers are treated with amino resins to obtain shrinkage control and a durable-press finish.

Normally, fabrics are treated in the sequence outlined above. The temperature of the drying unit is 100–110°C and the temperature of the curing unit can vary between 120 and 200°C but usually ranges from 150 to 180°C. The higher temperatures are employed to polymerize the resins on synthetic fabrics and at the same time to heat-set the fibers. Temperatures up to 180°C are used to allow the amino resins to react with cellulosic fibers alone or blended with synthetic fibers. The fabric is held flat but with minimum tension during drying and curing and always tends to become flat when creased or wrinkled during use or laundering. The resin-treated cellulose absorbs less water and swells less than untreated cellulose. This reduced swelling along with little or no tension induced during drying minimizes shrinkage during laundering.

The steps followed in the precure are repeated in the postcure process, except that after the drying step the goods are shipped to a garment manufacturer who makes garments, presses them into the desired shape with creases or pleats, and then cures the amino resin on the completed garment. It is important that the amino resins used in the postcure process should (*1*) not react with the fabric before it has been fashioned into a garment, and (*2*) release a minimum amount of formaldehyde into the atmosphere, especially while the goods are in storage or during the cutting and sewing operations. These requirements are met, at present, with dimethyloldihydroxyethyleneurea as the amino resin and zinc nitrate as the catalyst.

Amino Resins in the Paper Industry. Paper is a material of tremendous versatility and utility, prepared from a renewable resource. It may be made soft or stiff, dense or porous, absorbent or water repellent, textured or smooth. Some of the versatility originates with the fibers, which may vary from short and supple to long and stiff, but the contribution of chemicals should not be underestimated.

Amino resins are used by the paper industry in large volume for a variety of applications (See Paper). The resins are divided into two classes according to the mode of application. Resins added to the fiber slurry before the sheet is formed are called wet-end additives and are used to improve wet and dry strength and stiffness. Resins applied to the surface of formed paper or board, almost invariably together with other additives, are used to improve the water resistance of coatings, the sag resistance in ceiling tiles, and the scuff resistance in cartons and labels.

The requirements for the two types of resins are very different. Wet-end additives are used in dilute fiber slurries in small amounts. After the sheet is formed, most of the water is drained away and some of the remaining water is pressed out of the sheet before it is dried. The amino resin must be retained (adsorbed) on the surface of the cellulose fibers so that it will not be washed away. On a typical paper machine, fiber concentration in the headbox would be about 1%. If the amount of wet-strength resin used is 1% of the weight of the fiber, the concentration of resin in the headbox would be only 0.01%. If no mechanism for attaching the resin to the fiber is provided, only a trace of the resin added to the slurry would be retained in the finished sheet. Resins for application to the surface of preformed paper are not required to be substantive to cellulose and they may be formulated for adhesion, cure rate, viscosity, compatibility with other materials, etc, without concern for retention.

The integrity of a paper sheet is dependent on the hydrogen bonds which form between the fine structures of cellulose fibers during the pressing and drying operations (see Cellulose). The bonds between hydroxyls of neighboring fibers are very strong when the paper is dry but are severely weakened as soon as the paper becomes wet. Bonding between the hydroxyls of cellulose and water is as energetic as bonding between two cellulose hydroxyls. As a consequence, ordinary paper loses most of its strength when it is wet or exposed to very high humidity. The sheet loses its stiffness and bursting, tensile, and tearing strength.

Many materials have been used over the years in an effort to correct this weakness in paper. If water can be prevented from reaching the sites of the bonding by sizing or coating the sheet, then a measure of wet strength may be attained. Water molecules are so small and cellulose so hydrophilic that this solution usually affords only temporary protection. Formaldehyde, glyoxal, and polyethylenimine have been used to provide temporary wet strength. The former two materials must be applied to the formed paper but the latter is substantitive and may be used as a wet-end additive. Carboxymethyl cellulose–calcium chloride and locust bean gum–borax are examples of two-component systems applied separately to paper that were used to a limited extent before the advent of the amino resins. Today polyamide–polyamine resins treated with epichlorohydrin (51) share the market with conventional amino resins. They are more expensive but very efficient and cure at neutral or alkaline pH values.

The first wet-strength resin to be widely used by paper mills was melamine–formaldehyde. Low molecular weight (or even monomeric) trimethylolmelamine [*1017-56-7*], when dissolved in the proper amount of dilute acid and aged, polymerizes to a colloidal polymer which is retained well by almost all types of paper-making fiber, and produces high wet strength under the mild curing conditions easily attained on a paper machine (52–53). This resin, introduced by American Cyanamid in 1942, is still extensively used when rapid cure, high wet strength, and good dry strength are important. Improvements are still being made in this area. A recent patent (54) describes the formation of a stable melamine resin acid colloid using formic and phosphoric acids. The chemistry of this reaction is quite interesting. Melamine–formaldehyde acts as an amine when dissolved in dilute acid, usually hydrochloric acid. During polymerization, about 20 monomeric units combine to form a polymer of colloidal dimension (6–7 nm) with the elimination of water and HCl (55). The development of cationicity is associated with the loss of HCl, since a unit of charge on the polymer is generated for every mole of acid lost, and the pH decreases steadily during the polymerization. In a typical formulation at 12% solids at room temperature, polymerization is complete in about 3 h. The initially colorless solution develops a light blue haze and shows a strong Tyndall cone.

Such a colloidal sol is highly substantitive to all paper-making fibers, kraft, sulfite, groundwood, and soda. For its successful use in paper mills, the pH must be kept low, both to prevent precipitation of the resin in an unusable form and to promote the curing of the resin, and the concentration of sulfates in the white water on the paper machine must not be allowed to exceed about 100 ppm, again because the resin is precipitated in an inactive form by high concentrations of sulfates. High sulfate concentrations may build up in mills using large amounts of alum for setting size or sulfuric acid for controlling pH.

Maxwell (56) solved the problem of sulfate sensitivity by adding formaldehyde to the aged colloid which improved wet-strength efficiency and reduced sensitivity to sulfates. Later, he obtained equivalent results by adding the extra formaldehyde before the colloid was aged. The additional formaldehyde acts like an acid during the aging process and, unless compensated for by a reduction in the amount of acid charged, lowers the pH to a point where polymerization to the colloid is inhibited. The HE (High Efficiency) resins have been used in mills with sulfate concentrations so high that use of regular trimethylolmelamine (MF_3) colloids would be uneconomical. Sulfate tolerance is a function of the amount of extra formaldehyde present. For best cost–performance, a family of HE colloids is necessary with composition varying from MF_4 for moderate sulfate concentrations to MF_9 for very high sulfates.

Formulations for regular and HE colloids are shown in Table 2 (57). The materials are added in the order listed to a 454 L (120 gal) tank provided with good agitation and ventilation. Formaldehyde fumes are evolved even from the regular colloid. The colloids develop only after aging and freshly prepared solutions are ineffective for producing wet strength. Stability of the colloids depends on temperature and concentration. Colloids at 10–12% are stable at room temperature for at least a week; stability may be extended by dilution after the colloid has aged properly.

Table 2. Formulations for Regular and HE Colloid Resins

	Regular MF_3	HE MF_8
water, 20° ± 10°C, kg	412.0	330.8
HCl, 20° Bé, kg (1.16 g/mL), kg	17.7	14.1
formaldehyde, 37%, kg		84.8
trimethylolmelamine, kg	45.4	45.4
Total	*475.1*	*475.1*

Both regular and HE colloids increase the wet strength of paper primarily by increasing adhesion between fibers; the strength of the fiber itself is unaffected (58). The resin appears to improve the adhesion between the fibers, whether they are wet or dry, by forming bonds that are unaffected by water. The excess formaldehyde in the HE colloid appears to function by increasing the amount of formaldehyde bound in the colloid (57). The regular colloid, starting with about 3 mol of formaldehyde per 1 mol of melamine, has about 2 mol bound in the colloid and 1 mol free. By mass action, the additional formaldehyde increases the amount of bound formaldehyde in the colloid. When an HE colloid is dialyzed or stored at very low concentrations (0.05%), it loses the extra bound formaldehyde and behaves as a regular colloid.

The first urea–formaldehyde resins used to any extent as wet-strength agents were anionic polymers made by the reaction of a urea resin with sodium bisulfite (59). Attempts to use nonionic urea–formaldehyde polymers were unsuccessful because the unmodified polymers were not retained. The sulfomethyl group introduced by reaction with $NaHSO_3$ gave the polymers strong anionicity but substantivity was largely restricted to unbleached kraft pulp. Lignin residues probably provided sites for adsorption of the polymer. The use of alum as a mordant was essential, since both the resin and the fiber were anionic. The reaction of bisulfite with the urea–formaldehyde polymer may be represented as:

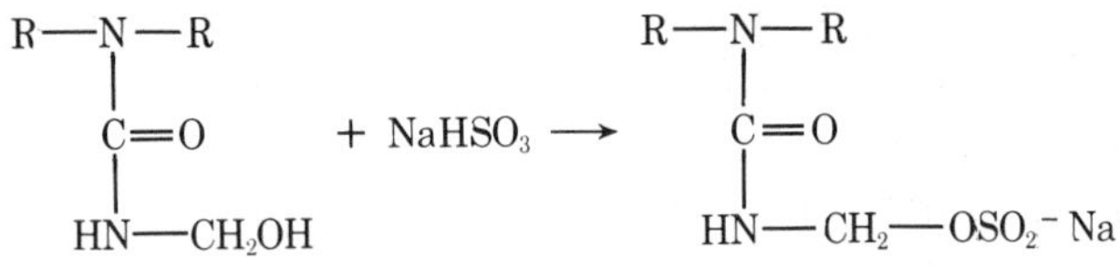

In 1945, cationic urea resins were introduced and quickly supplanted the anionic resins, since they could be used with any type of pulp (60). They have now become commodities and are the most widely used resins in the paper industry. They are commonly made by the reaction of urea and formaldehyde with one or more polyethylenepolyamines. The structure of these resins is very complicated and has not been determined. Ammonia is evolved during the reaction, probably according to the following:

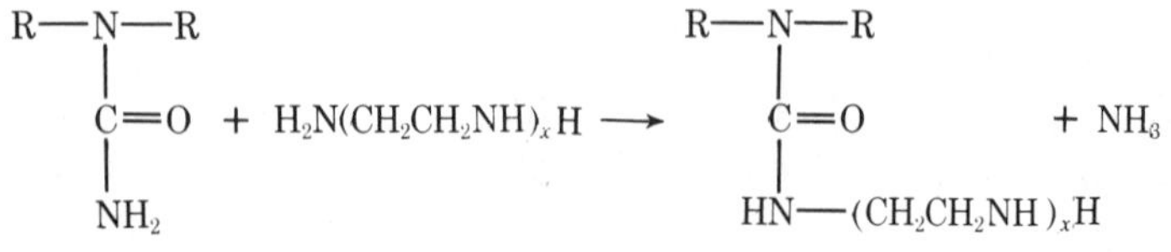

Formaldehyde may react with the active hydrogens on both the urea and amine groups and therefore the polymer is probably highly branched. The amount of formaldehyde (2–4 mol per 1 mol urea), the amount and kind of polyamine (10–15%), and resin concentration are variable and hundreds of patents have been issued throughout the world. Generally, the urea, formaldehyde, polyamine, and water react at 80–100°C. The reaction may be carried out in two steps with an initial methylolation at alkaline pH, followed by condensation to the desired degree at acidic pH, or the entire reaction may be carried out under acidic conditions (61). The product is generally a syrup with 25–35% solids and is stable for up to three months.

The cationic urea resins are added to paper pulp preferably after all major refining operations have taken place. The pH on the paper machine must be acidic for reasonable rates of cure of the resin. Urea resins do not cure as rapidly as melamine–formaldehyde resins and the wet strength produced is not as resistant to hydrolysis. Furthermore, the resins are not retained as well as the melamine resins. On a resin-retained basis, however, their efficiency is as good. The lower retention of the urea–formaldehyde resins is due to their polydisperse molecular weight distribution. High molecular weight species are strongly adsorbed on the fibers and are large enough to bridge two fibers. Low molecular weight species are not retained as well because of fewer charge sites. Attempts to improve the performance of urea–formaldehyde resins by fractionating the syrups by salt or solvent precipitation, or selective freezing or dialysis have been technically successful but economically impractical. The process for production of resins is sufficiently simple so that some paper mills have set up their own production units. With captive production, resins with higher molecular weights and lower stability may be tolerated.

The recovery of fiber from broke (off-specification paper or trim produced in the paper mill) is complicated by high levels of wet-strength resin. The urea resins present a lesser problem than the melamine resins because they cure slower and are not as resistant to hydrolysis. Broke from either resin treatment may be reclaimed by hot acidic repulping. Even the melamine resin is hydrolyzed rapidly under acidic conditions at high temperature. The cellulose is far more resistant and is not harmed if the acid is neutralized as soon as repulping is complete.

The TAPPI monograph (62) is an excellent source of additional information on technical and economic aspects of wet strength. In addition, the Institute of Paper Chemistry, Appleton, Wisc., has published two comprehensive bibliographies on the wet strength of paper with supplements issuing from time to time.

Wet strength applications account for the majority of amino resin sales to the paper industry but substantial volumes are sold for coating applications. The largest use is to improve the resistance of starch–clay coatings to dampness. In off-set printing, which is becoming ever more important in the graphic arts, the printing paper is exposed to both ink and water. If the coating lifts from the paper and transfers to the plate, it causes smears and forces a shutdown for cleaning. A wide variety of materials have been added to the coatings to improve wet rub resistance, including casein, soya protein, poly(vinyl acetate), styrene–butadiene latices, and amino resins. Paper coatings are applied at as high a solids content as possible to ease the problem of drying. Retention is not a problem. The important characteristics for coating resins are high solids at low viscosity, high cure rates, and high wet-rub efficiency. Urea and melamine resins or mixtures are sold as high-solids syrups or dry powders. They are used with starch–pigment coatings with acidic catalysts or with starch–pigment–casein (or protein) coatings usually without catalysts. The syrups are frequently methylated for solubility and stability at high solids. All of the resins are low molecular weight to reduce viscosity (see Coatings, industrial).

Closely allied to resins for treating paper are the resins used to treat regenerated cellulose film (cellophane) which does not have good water resistance unless it is coated with nitrocellulose or poly(vinylidene chloride). Adhesion of the waterproofing coating to the cellophane film is achieved by first treating the cellophane with an amino resin. The cellophane film is passed through a dip tank containing about 1% of a melamine–formaldehyde acid colloid type of resin. Some glycerol may also be present in the resin solution to act as a plasticizer. Resins for this purpose are referred to as anchoring agents.

Other Uses. Water-soluble melamine–formaldehyde resins are used in the tanning of leather in combination with the usual tanning agents (see Leather). By first treating the hides with a melamine–formaldehyde resin, the leather is made more receptive to other tanning agents and the finished product has a lighter color. The amino resin is often referred to as a plumping agent because it makes the finished leather firmer and fuller.

Urea–formaldehyde resins are also used in the manufacture of foams. The resin solution containing an acid catalyst and a surface active agent is foamed with air and cured. The open-cell type of foam absorbs water readily and is soft enough so that the stems of flowers can be easily pressed into it. These features make the urea resin foam ideal for supporting floral displays. Urea–formaldehyde resin may also be foamed in place. A special nozzle brings the resin, catalyst, and foaming agent together. Air pressure is used to deposit the foam where it is desired, eg, within the outside walls of older houses to provide insulation. This application might be expected to grow as energy costs increase.

Urea–formaldehyde resins are also used as the binder for the sand cores used in the molds for casting hollow metal shapes. The amino resin is mixed with moist sand and formed into the desired shape of the core. After drying and curing, the core is assembled into the mold and the molten metal poured in. Although the cured amino resin is strong enough to hold the core together while the hot metal is solidifying, it decomposes on longer heating. Later, the loose sand may be poured out of the hollow casting and recovered.

Tire cord (qv) is normally first treated with rubber latex to improve adhesion to the vulcanized rubber. The latex dip solution often contains additives to further improve adhesion. Resorcinol–formaldehyde resins are commonly used but amino resins are also effective. Both urea and melamine resins are described in the patent literature but melamine resins are preferred (63).

Toxicity

Both urea- and melamine–formaldehyde resins are of low toxicity. In the uncured state, the amino resin contains some free formaldehyde that could be objectionable. However, uncured resins have a very unpleasant taste which would discourage ingestion of more than trace amounts. The molded plastic, or the cured resin on textiles or paper may be considered nontoxic. Combustion or thermal decomposition of the cured resins can evolve toxic gases, such as formaldehyde, hydrogen cyanide, and oxides of nitrogen.

Economic Aspects

Japan produces more amino resin than any other country, the United States is next, with the Union of Soviet Socialist Republics, France, the United Kingdom, and the Federal Republic of Germany following.

Modern Plastics (64) lists United States markets for urea and melamine resins according to uses (see Table 3).

Many large chemical companies produce amino resins and the raw materials needed, ie, formaldehyde, urea, and melamine as well. Some companies may buy raw materials to produce amino resins for use in their own products, such as plywood, chipboard, paper, textiles, or paints, and may also find it profitable to market these resins to smaller companies. The technology is highly developed and sales must be supported by adequate technical service to select the correct resin and see that it is applied under the best conditions.

Following is a list of major manufacturers of amino resins. Because of the highly specialized nature of amino resin technology many of these companies produce only a few types. Allied Chemical Corp., Specialty Chemicals Div., Morristown, N.J.; American Cyanamid Co., Industrial Chemicals and Plastics Div., Wayne, N.J.; Ashland Chemical Co., Div. Ashland Oil, Inc., Columbus, Ohio; BASF Canada Ltd., Station St. Laurent, Canada; Badische Aniline- und Soda-Fabrik A.G., Ludwigshafen, Germany; Berger Chemicals, Resinous Chemicals Div., Newcastle upon Tyne, England; Borden Chemical, Div. Borden, Inc., Columbus, Ohio; British Industrial Plastics, Ltd., Warley, West Midlands, England; British Oxygen Chemicals, Ltd., Chester-le-Street, Durham, England; Cassella Farbwerk Mainkur A.G., Frankfurt-Fechenheim, Ger-

Table 3. Uses of Urea and Melamine Resins, 1976

Resin	Amount, kt
bonding and adhesive	
fibrous and granulated wood	274
laminating	14
plywood	33
molding compounds	43
paper treating and coating resins	24
protective coatings	35
textile treating and coating resins	14
exports	15
other	2
Total	*454*
urea molding powder	
closures	5.8
electrical devices	17.1
other	1.7
Total	*24.6*
Melamine molding powder	
buttons	0.6
dinnerware	16.8
sanitaryware	0.3
other	0.8
Total	*18.6*

many; Celanese Resins, Div. Celanese Coatings Co., Louisville, Kentucky; Commercial Solvents Corp., New York; Cook Paint and Varnish Co., Kansas City, Mo.; DSM, Utrecht, Holland; Dynamit Nobel, A.G., Troisdorf-Koln, Germany; Dynamit Nobel of America, Inc., Northvale, N.J.; Eronel Industries, Hawthorne, Calif.; Ferguson, James, & Sons, Ltd., London, England; Fiberite Corp., Winona, Minn.; Flexcraft Industries, Newark, N.J.; Ford Motor Co., Paint and Vinyl Operations, Mount Clemens, Mich.; George, P. D., Co., St. Louis, Mo.; Georgia-Pacific Corp., Chemical Div., Portland, Oregon; Guardsman Chemical Coatings, Inc., Grand Rapids, Mich.; Gulf Oil Chemicals Co., Gulf Adhesives, Lansdale, Pa.; Hitachi Chemical Co., Ltd., Tokyo 160, Japan; Hoechst Aktiengesellschaft, D-6230 Frankfurt, Germany; Materiales Moldeables, S. A. de C. V., Mexico, D. F., Mexico; Matsushita Electric Works, Ltd., Kadoma-shi, Osaka-fu, Japan; Melamine Chemicals, Inc., Donaldsonville, La.; Monsanto Co., St. Louis, Mo.; Montedison S.p.A., Milano, Italy; Pacific Resins and Chemicals, Inc., Tacoma, Wash.; Patent Plastics Inc., Knoxville, Tenn.; Perstorp AB,

Table 4. United States List Prices of Representative MF and UF Resins, September 1975[a]

Resin	\$/kg
Melamine–formaldehyde	
molding granules for dinnerware, 120 kg fiber drums, such as Cyanamid's Cymel 1077, alpha-cellulose filled	1.17
Type A laminating resin, white, water-soluble, spray-dried powder, 22.7-kg bags such as Monsanto's Resimene 814, 100% resin	1.15
Type B laminating resin, white unfilled powder for laminating or impregnating, such as Cyanamid's Cymel 405	0.63
spray-dried powder, 100% resin, 113-kg drums, such as Borden's MO-608, used for adding to UF syrups	1.39
coating resins, 55 wt % solids	0.88
acid–colloid solution, 10 wt % solids, tanks, for paper treating	0.09
syrup for decorative laminates, 60 wt % solids	0.60
methylated resin, 98 wt % solids such as Monsanto's Resimene 747	1.04
mixed MF–UF spray dried powder, 22.7-kg bag such as Cyanamid's Melurac 400 adhesive	1.01
butylated surface-coating resin, 50 wt % solids, 208 L (55 gal) drums, such as Cyanamid's Cymel 255-10, for mixing with alkyds for paints or baking enamels	1.08
hexamethoxymethylmelamine, drums, such as Cyanamid's Cymel 303, 100% solids	1.15
mixed UMF syrup (60% M–40% U), 65 wt % solids for hot pressing, tank cars or tank trucks, such as Borden's MU-607F	0.62
Urea–formaldehyde	
particle board syrup, 65 wt % solids	0.176
plywood syrup, 60 wt % solids, tank cars	0.185
spray-dried powder, water soluble, 100% resin, in 113-kg drums, such as Borden's Casconite 151 for wood adhesives	0.86
coating resin, general-purpose, 55 wt % solids	0.72
general-purpose molding compound, 90.7-kg drum of granules, such as Allied Chemical's S-447, alpha-cellulose filled	0.97
Type A butylated UF resin, 208 L (55 gal) drums of liquid such as Cyanamid's Beetle 227-8, 50 wt % solids, for mixing with alkyds for baked surface coatings	0.80
butylated UF resin, 15.1-m^3 (4000-gal) tanks, 50 wt % solids in butanol–xylene, such as Cargill's UF 3353 resin, for rapid-curing baking enamels	0.46
powdered UF adhesive, 31.8-kg bags of powder, such as Cyanamid's Urac 110	0.48
adhesive syrup, 250-kg steel drums, chemically modified UF resin in aqueous dispersion, such as Cyanamid's Urac 185	0.73
syrup for paper treating in tank cars, 35 wt % solids in water, such as Borden's Casco PR-335	0.20

[a] Courtesy of Stanford Research Institute.

Perstrop, Sweden; Plastics Engineering Co., Sheboygan, Wis.; Pierrefitte-Auby, Neuilly s/Seine, France; Products Chimiques Ugine Kuhlmann, Div. Plastics, Paris, Fr.; Raymond Chemical Co., Toledo, Ohio; Reichhold Chemicals, Inc., White Plains, N.Y.; Rohm & Haas Co., Philadelphia, Pa.; Sullivan Chemical Coatings, Chicago, Ill.; and Sumitomo Bakelite Co., Ltd., Tokyo, Japan.

Representative prices of amino resins and plastics are shown in Table 4.

Notice

BIBLIOGRAPHY

"Amino Resins and Plastics" in *ECT* 1st ed., Vol. 1, pp. 741–771, by P. O. Powers, Battelle Memorial Institute; *ECT* 2nd ed., Vol. 2, pp. 225–258, by H. P. Wohnsiedler, American Cyanamid Company.

1. R. Bainbridge, *Sail* **8**(1), 142 (1977).
2. A. Einhorn and A. Hamburger, *Ber. Dtsch. Chem. Ges.* **41,** 24 (1908).
3. Brit. Pats. 248,477 (Dec. 5, 1924), 258,950 (July 1, 1925), 266,028 (Nov. 5, 1925), E. C. Rossiter (to British Cyanides Company, Ltd.).
4. Brit. Pats. 187,605 (Oct. 17, 1922), 202,651 (Aug. 17, 1923), 208,761 (Sept. 20, 1922), H. Goldschmidt and O. Neuss.
5. Brit. Pats. 171,096 (Nov. 1, 1921), 181,014 (May 20, 1922), 193,420 (Feb. 17, 1923), 201,906 (July 23, 1923), 206,512 (July 23, 1923), 213,567 (Mar. 31, 1923), 238,904 (Aug. 25, 1924), 240,840 (Oct. 1, 1924), 248,729 (Mar. 3, 1925), F. Pollak.
6. U.S. Pat. 1,460,606 (July 3, 1923), K. Ripper.
7. Ger. Pat. 647,303 (July 6, 1937), Brit. Pat. 455,008 (Oct. 12, 1936), W. Hentrich and R. Köhler (to Henkel and Co., G.m.b.H.).
8. R. Rager, *Mod. Plast.* **49**(4), 67 (1972).
9. E. Drechsel, *J. Prakt. Chem.* [2] **13,** 330 (1876).
10. U.S. Pat. 2,727,037 (Dec. 13, 1955), C. A. Hochwalt (to Monsanto Chemical Company).
11. Ger. Pat. 1,812,120 (June 11, 1970), D. Fromm, K. W. Leonhard, R. Mohr, M. Schwartzmann, and H. Woehrle (to Badische Anilin und Soda-Fabrik A.G.).
12. P. Ellwood, *Chem. Eng.* **77**(23), 101 (1970).
13. J. F. Walker, *Formaldehyde, American Chemical Society Monograph, No. 159,* 3rd ed., Reinhold Publishing Corp., New York, 1964.
14. *U.F. Concentrate-85,* technical bulletin, Allied Chemical Corporation, 61 Broadway, New York, N.Y. 10006.
15. U.S. Pat. 3,129,226 (Apr. 14, 1964), G. K. Cleek and A. Sadle (to Allied Chemical Corp.).
16. U.S. Pat. 3,458,464 (July 29, 1969), D. S. Shriver and E. J. Bara (to Allied Chemical Corp.).
17. Ref. 13, p. 151.
18. *N-(iso-butoxymethyl) acrylamide, Technical Bulletin PRC 126,* American Cyanamid Co., Wayne, N.J., Feb. 1976.
19. M. Gordon, A. Halliwell, and T. Wilson, *J. Appl. Polym. Sci.* **10,** 1153 (1966).
20. J. W. Aldersley, M. Gordon, A. Halliwell, and T. Wilson, *Polymer* **9,** 345 (1968).
21. I. H. Anderson, M. Cawley, and W. Steedman, *Br. Polym. J.* **1,** 24 (1969).
22. K. Sato, *Bull. Chem. Soc. Jpn.* **40**(4), 724 (1967) (in Eng.).
23. K. Sato and T. Naito, *Polym. J.* **5,** 144 (1973).
24. K. Sato and Y. Abe, *J. Polym. Sci. Polym. Chem. Ed.* **13,** 263 (1975).
25. V. A. Shenai and J. M. Manjeshwar, *J. Appl. Polym. Sci.* **18,** 1407 (1974).
26. A. Berge, S. Gudmundsen, and J. Ugelstad, *Eur. Polym. J.* **5,** 171 (1969).
27. A. Berge, B. Kvaeven, and J. Ugelstad, *Eur. Polym. J.* **6,** 981 (1970).
28. J. I. DeJong and J. DeJonge, *Rec. Trav. Chim.* **71,** 643, 661, 890 (1952); **72,** 88, 139, 202, 207, 213, 1027 (1953).
29. R. Steele, *J. Appl. Polym. Sci.* **4,** 45 (1960).
30. G. A. Crowe and C. C. Lynch, *J. Am. Chem. Soc.* **70,** 3795 (1948); **71,** 3731 (1949); **72,** 3622 (1950).

31. U.S. Pat. 2,303,982 (Dec. 1, 1942), A. Brooks (to American Cyanamid Company).
32. U.S. Pat. 2,518,388 (Aug. 8, 1950), W. G. Simons (to American Cyanamid Company).
33. Brit. Pat. 829,953 (Mar. 9, 1960), K. Elbel.
34. U.S. Pats. 3,007,885 (Nov. 7, 1961) 3,114,930 (Dec. 24, 1963), W. N. Oldham, N. A. Granito, and B. Kerfoot (to American Cyanamid Company).
35. U.S. Pat. 3,661,819 (May 9, 1972), J. N. Koral and M. Petschel, Jr. (to American Cyanamid Company).
36. U.S. Pat. 3,803,095 (Apr. 9, 1974), L. J. Calbo and J. N. Koral (to American Cyanamid Company).
37. W. Lindlaw, *The Preparation of Butylated Urea–Formaldehyde and Butylated Melamine Formaldehyde Resins Using Celanese Formcel and Celanese Paraformaldehyde,* technical bulletin, Celanese Chemical Company, 245 Park Ave., New York, N.Y. 10017, Table XIIA.
38. *Technical Bulletin S-23-8,* 1967, *Supplement to Technical Bulletin S-23-8,* 1968, Celanese Chemical Company, Example VIII.
39. W. J. Blank and W. L. Hensley, *J. Paint Technol.* **46,** 46 (1974).
40. U.S. Pat. 2,517,750 (Aug. 8, 1950), A. L. Wilson (to Union Carbide and Carbon Corp.).
41. U.S. Pat. 2,304,624 (Dec. 8, 1942), W. J. Burke (to E. I. du Pont de Nemours & Co., Inc.).
42. U.S. Pat. 3,324,062 (June 6, 1967), G. S. Y. Poon (to Dan River Mills Inc.).
43. H. Kadowaki, *Bull. Chem. Soc. Jpn.* **11,** 248 (1936).
44. U.S. Pat. 3,089,859 (May 14, 1963), T. Oshima (to Sumitomo Chemical Company, Ltd.).
45. U.S. Pats. 2,731,472 (Jan. 17, 1956) 2,764,573 (Sept. 25, 1956), Bruno V. Reibnitz and co-workers (to Badische Anilin-und Soda-Fabrik); U.S. Pat. 2,876,062 (March 3, 1959), Erich Torke (to Phrix-Werke A.G.).
46. U.S. Pat. 3,487,088 (Dec. 30, 1969), K. H. Remley (to American Cyanamid Company).
47. U.S. Pat. 3,524,876 (Aug. 18, 1970), J. E. Gregson (to Dan River Mills, Inc.).
48. U.S. Pat. 3,658,458 (Apr. 25, 1972), D. J. Gale (to Deering Milliken Research Corporation).
49. U.S. Pats. 2,602,017; 2,602,018 (July 1, 1952), L. Beer.
50. C. Hasegawa, *J. Soc. Chem. Ind. Jpn.* **45,** 416 (1942).
51. U.S. Pats. 2,926,116; 2,926,154 (Feb. 23, 1960), G. L. Keim (to Hercules Powder Company).
52. U.S. Pat. 2,345,543 (Mar. 28, 1944), H. P. Wohnsiedler and W. M. Thomas (to American Cyanamid Company).
53. C. G. Landes and C. S. Maxwell, *Pap. Trade J.* **121**(6), 37 (1945).
54. Ger. Pat. 2,332,046 (Jan. 23, 1975), W. Guender and G. Reuss (to Badische Anilin-und Soda-Fabrik A.G.).
55. J. K. Dixon, G. L. M. Christopher, and D. J. Salley, *Pap. Trade J.* **127**(20), 49 (1948).
56. U.S. Pat. 2,559,220 (July 3, 1951), C. S. Maxwell and C. G. Landes (to American Cyanamid Company).
57. C. S. Maxwell and R. R. House, *TAPPI* **44**(5), 370 (1961).
58. D. J. Salley and A. F. Blockman, *Pap. Trade J.* **121**(6), 41 (1945).
59. U.S. Pat. 2,407,599 (Sept. 10, 1946), R. W. Auten and J. L. Rainey (to Resinous Products and Chemical Company).
60. U.S. Pat. 2,742,450 (Apr. 17, 1956), R. S. Yost and R. W. Auten (to Rohm and Haas Company).
61. U.S. Pat. 2,683,134 (July 6, 1954), J. B. Davidson and E. J. Romatowski (to Allied Chemical and Dye Corporation).
62. J. P. Weidner, ed., *Wet Strength in Paper and Paper Board, Monograph Series, No. 29,* Technical Association of Pulp and Paper Industry, New York, 1965.
63. U.S. Pat. 3,212,955 (Oct. 19, 1965), S. Kaizerman (to American Cyanamid Company).
64. *Mod. Plast.* **54**(1), 52 (1977).

General References

T. J. Suen in "Condensations with Formaldehyde," C. E. Schildknecht, ed., *Polymer Processes,* Interscience Publishers, New York, 1956, Chapter VIII, a comprehensive survey of amino and phenolic resin technology up to 1956, and contains detailed directions for making a number of different kinds of amino resins.

C. P. Vale and W. G. K. Taylor, *Amino Plastics,* Iliffe Books, Ltd., London, 1964, a very comprehensive review of amino resin chemistry and technology.

G. Widmer in N. Bikales, ed., *Encyclopedia of Polymer Science and Technology,* Vol. 2, Interscience Publishers, a division of John Wiley & Sons, Inc., New York, 1965, pp. 1–94, an excellent discussion of amino resin chemistry and technology with major emphasis on the development of the technology.

S. A. Heap, R. E. Hunt, P. A. Rennison, R. Tattersall, W. F. Herbes, S. J. O'Brien, and R. G. Weyker in H. Mark, N. Wooding, and S. M. Atlas, eds., *Chemical Aftertreatment of Textiles,* Interscience Publishers, a division of John Wiley & Sons, Inc., New York, 1971, Chapter VI, it contains much detailed information on the manufacture and application of all kinds of amino resins used in the treatment of textiles.
J. C. Petropoulos, I. H. Updegraff, and L. L. Williams in J. K. Craver and R. W. Tess, eds., *Applied Polymer Science,* Organic Coatings and Plastics Chemistry Division of the American Chemical Society, Washington, D.C., 1975, Chapter 46, a recent review with emphasis on the development of amino resins for surface coatings.

IVOR H. UPDEGRAFF
SEWELL T. MOORE
WILLIAM F. HERBES
PHILIP B. ROTH
American Cyanamid Company

ARAMID FIBERS

Aromatic polyamides are formed by reactions that lead to the formation of amide linkages between aromatic rings. In practice this generally means the reaction of aromatic diamines and aromatic diacid chlorides in an amide solvent. From solutions of these polymers it is possible to produce fibers of exceptional heat and flame resistance and fibers of good to quite remarkable tensile strength and modulus. Because the physical property differences between fibers of aromatic and aliphatic polyamides are greater than those between other existing generic classes of fibers, a new generic term for fibers from aromatic polyamides was requested by the DuPont Company in 1971. Subsequently, the generic term aramid was adopted (1974) by the United States Federal Trade Commission for designating fibers of the aromatic polyamide type: aramid—a manufactured fiber in which the fiber-forming substance is a long-chain synthetic polyamide in which at least 85% of the amide (—CO—NH—) linkages are attached directly to two aromatic rings.

At the same time that the new generic term became effective, the generic term for nylon was amended as follows: nylon—a manufactured fiber in which the fiber-forming substance is a long-chain synthetic polyamide in which less than 85% of the amide (—CO—NH—) linkages are attached directly to two aromatic rings.

Aromatic polyamide–imides have also been considered as aramids by the ISO which has proposed the definition: synthetic linear macromolecules made from aromatic groups joined by amide linkages, in which at least 85% of the amide linkages are joined directly to two aromatic rings and in which imide groups may be substituted for up to 50% of the amide groups. In this article, amide–imides are not discussed as aramid fibers and only those polymers having amide condensation linkages are reviewed (see also Polyamides; Heat-resistant polymers; Flame-retardant textiles).

There is no recognized systematic nomenclature for aromatic polyamides as exists for aliphatic polyamides. However, acronyms or the initial letters of the various monomers are frequently combined (see equations 1 and 3), with the initials of the diamine moieties preceding those of the diacid moieties. The letters T and I have the commonly-accepted significance of standing for, respectively, terephthalamide and isophthalamide.

The first aramid fiber to be developed, known experimentally as HT-1 (1) and almost certainly based on poly(*m*-phenyleneisophthalamide), was commercialized under the trademark Nomex by the DuPont Company in 1967. This fiber was introduced for applications requiring heat resistance far in excess of that possessed by conventional synthetic fibers. In Japan, Teijin introduced Conex in the early 1970s and has disclosed (2) the composition to be poly(*m*-phenyleneisophthalamide), ie, apparently the same as that used for Nomex. Conex production, however, appears to be quite low at present, and is probably only a fraction of that of Nomex. Fenilon (or Phenylone), also based on MPD-I is produced in the Soviet Union for both civilian and military as well as space exploration needs. Although the quantity of Fenilon produced has not been announced, it is almost certain that production is less than 4500 metric tons per year.

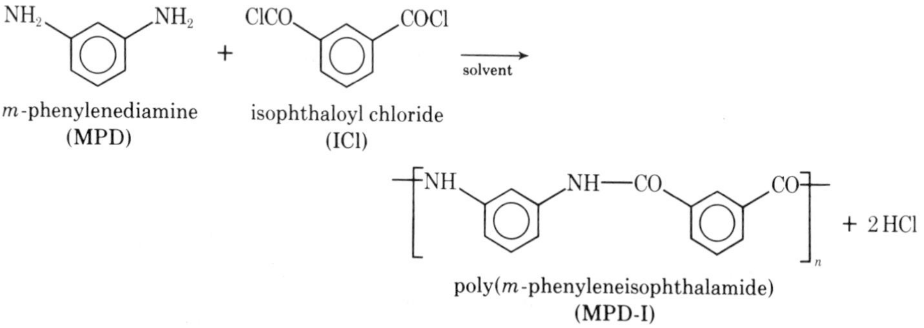

(1)

An aramid fiber of high strength and exceptionally high modulus was introduced by the DuPont Company in 1970 for use in tires under the experimental name of Fiber B (3). This early high modulus aramid tire cord (qv) appears to have been based on poly(*p*-benzamide) (PPB) spun from an organic solvent. Later, another version of

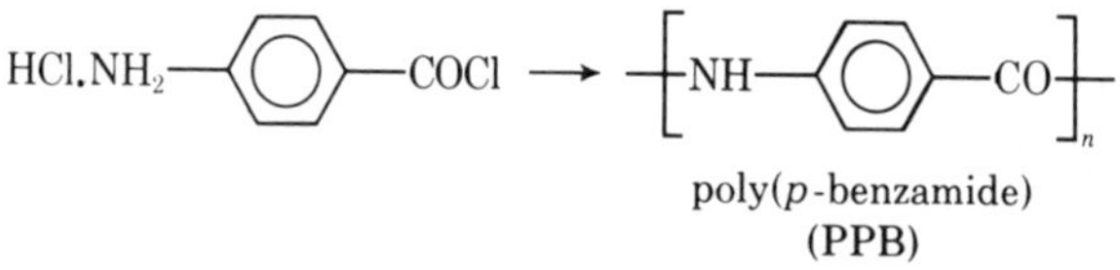

(2)

Fiber B, apparently based on poly(*p*-phenyleneterephthalamide) (PPD-T) and probably spun from sulfuric acid, was introduced (4). This second generation fiber was of considerably higher strength (by nearly two-fold) than that of Fiber B.

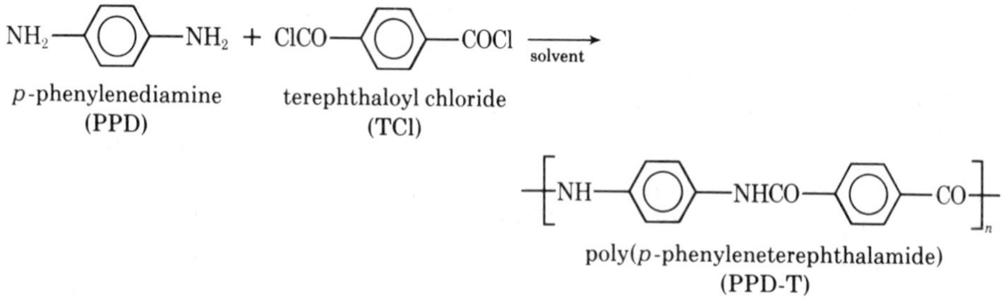

(3)

An even higher modulus fiber intended for use in rigid composites was introduced under the experimental name of PRD-49; it too is believed to be composed essentially of PPD-T, the higher modulus of PRD-49 being achieved through hot-drawing of this fiber. Upon announcement of construction of commercial facilities for these fibers, the trademarks Kevlar-29 and Kevlar-49, respectively, were introduced. In early 1975, the Akzo Company of the Netherlands announced their intention to commercialize an aramid fiber under the name Arenka; this fiber too is probably based on PPD-T.

Properties

Aramid fibers do not melt in the conventional sense because decomposition generally occurs simultaneously. An endothermic peak in the differential thermal analysis (DTA) thermogram for aramid fibers can be obtained and the values are generally >400°C and can be as high as 550°C. Glass transitions range from ca 250 to >400°C. Weight loss, as determined by thermogravimetric analysis (tga) in an inert gas, begins at about 425°C for most aramids although some of the rod-like polymers do not lose substantial weight to 550°C.

Aramid fibers are characterized by medium to ultra-high tensile strength, medium to low elongation, and moderately high modulus to ultra-high modulus. Most of these fibers have been reported to be either highly crystalline (5–6) or crystallizable (7) and densities for crystalline fibers range from about 1.35 to 1.45 g/cm^3. Poly(*m*-phenyleneisophthalamide) fiber of low orientation has a density of ca 1.35 g/cm^3 and hot-drawn fiber has a density of ca 1.38 g/cm^3; Kevlar, poly(*p*-phenyleneterephthalamide), has a density of ca 1.45 g/cm^3.

Since most of the work on aramid fibers has been directed toward the development of heat and flame-resistant fibers, and ultra high-strength–high-modulus fibers, it is possible to categorize them in these terms. In general, polymers useful for the former contain a high portion of meta-oriented phenylene rings and polymers useful for the latter fibers contain principally para-oriented phenylene rings. The meta-oriented polymers are generally considered to be chain-folding polymers. On the other hand, small angle x-ray diffraction studies of the rod-like para-oriented polymers show no evidence of chain-folding for this class of polymers. However, for purposes of discussion, the two classes of polymers will be referred to here as heat and flame-resistant fibers and high-strength–high-modulus fibers, respectively.

Heat and Flame-Resistant Aramid Fibers. The fiber properties of some aramid fibers having medium strength and elongation-to-break (expressed in %) are given in Table 1. No attempt has been made to present data for all of the fibers reported in the literature. Only the commercially important or potentially important compositions and fibers of those compositions that show significantly different properties from the simple aramids are considered. Although random copolyamides (eg, mixed isophthalamides and terephthalamides) have been reported, no clearcut advantage has been demonstrated for fibers from such polymers. On the contrary, lower softening points and lower crystallinity are usually observed.

Tensile Properties at Elevated Temperatures. The tensile properties at elevated temperatures of selected aramid fibers are given in Table 2. These fibers have tensile strengths at 250°C that are characteristic of conventional textile fibers at room temperature, and have useful tenacities to >300°C. In contrast, nylon-6,6 fiber loses almost all of its strength at about 205°C. The tensile moduli of the aramid fibers fall off considerably at 300–350°C; the modulus values are still substantial, however, even

Table 1. Tensile Properties of Selected Heat and Flame Resistant Aramid Fibers

Structure	CAS Registry No.	Mol formula	η_{inh}[a]	Method of prep[b]	Spun from[c]	Spinning method[d]	T/E/M_i[e]	Ref.
AB Type								
(1) 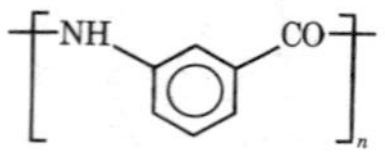	[*25735-77-7*]	$(C_7H_5NO)_n$	1.25	S S	DMAC–LiCl DMAC–LiCl	D W	0.51/10/10.4 0.35/14/	8 9
AA–BB Type								
(2) 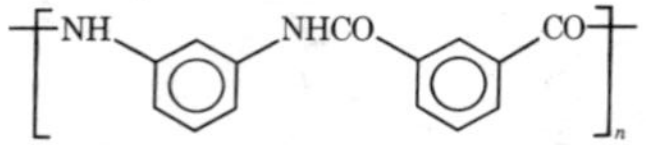	[*24938-60-1*]	$(C_{14}H_{10}N_2O_2)_n$	1.86 2.11	S S	DMAC–$CaCl_2$ DMAC–$CaCl_2$	D W	0.49/28/10.0 0.69/24/10.8	10 11
(3) 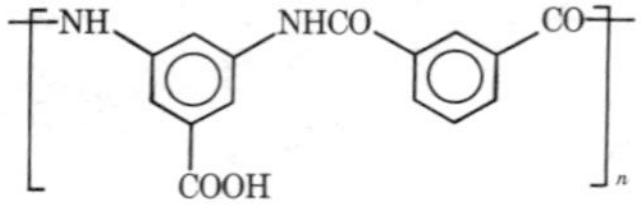	[*31808-02-3*]	$(C_{15}H_{10}N_2O_4)_n$	2.01	S	DMAC	W	0.35/9.0/7.1	12
(4) 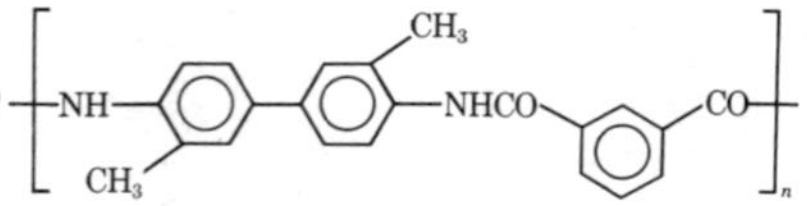	[*25670-09-1*]	$(C_{22}H_{18}N_2O_2)_n$		S	DMSO	D	0.81/6.8/16.1	13
(5) 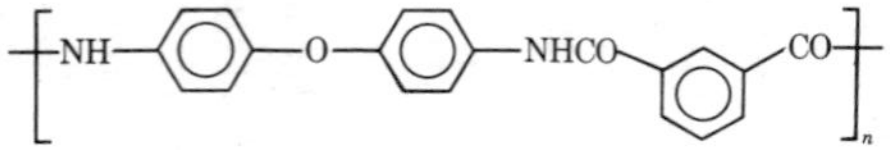	[*26026-92-6*]	$(C_{20}H_{14}N_2O_3)_n$	2.32[f]				0.37/8.5/7.9	14
(6) 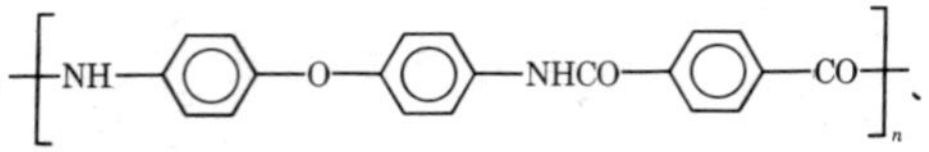	[*26854-93-3*]	$(C_{20}H_{14}N_2O_3)_n$	1.54[f]				0.62/6.0/10.6	14
(7) 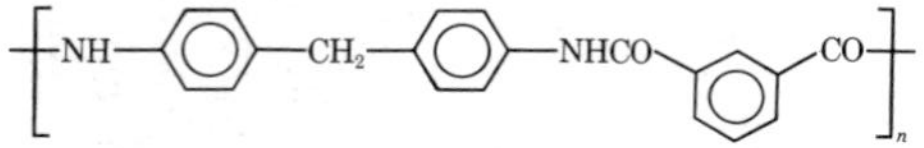	[*25667-73-6*]	$(C_{21}H_{16}N_2O_2)_n$	2.51	S	DMF	D	0.36/20/	15
(8) 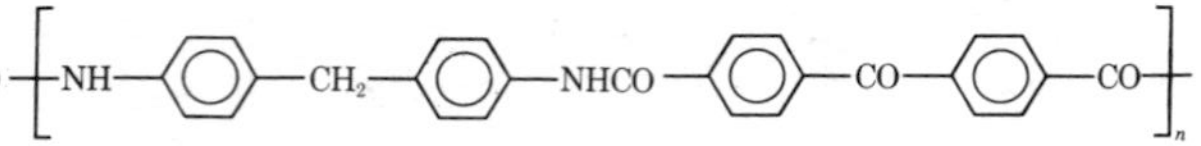	[*32027-57-9*]	$(C_{28}H_{20}N_2O_3)_n$	2.84	S	DMAC–LiCl	D	0.32/22/2.6	16

(9) 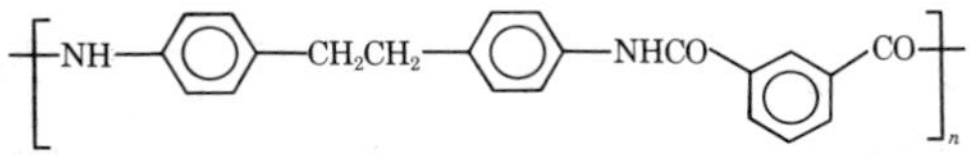	[*31986-64-8*]	$(C_{22}H_{18}N_2O_2)_n$		S	DMAC–LiCl	W	0.33/30/4.6	
(10) 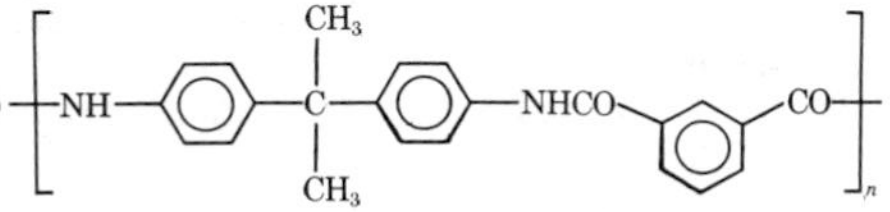	[*27880-27-9*]	$(C_{23}H_{20}N_2O_2)_n$	1.80	I	DMF	D	0.32/28/	15
(11) 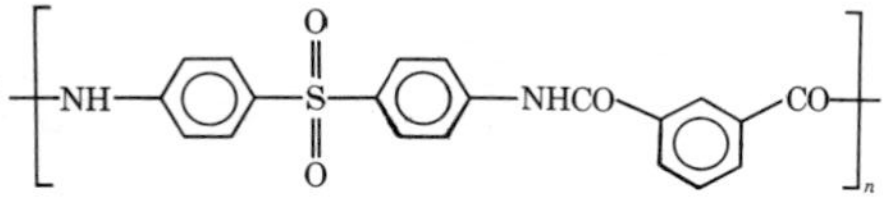	[*26100-95-8*]	$(C_{20}H_{14}N_2O_4S)_n$					0.35/17/4.1	17
(12) 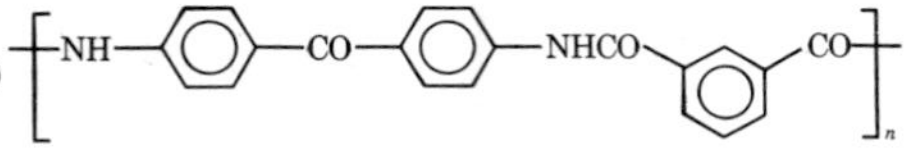	[*26026-95-9*]	$(C_{21}H_{14}N_2O_3)_n$	1.26	S		D	0.27/23/2.5	16
(13) 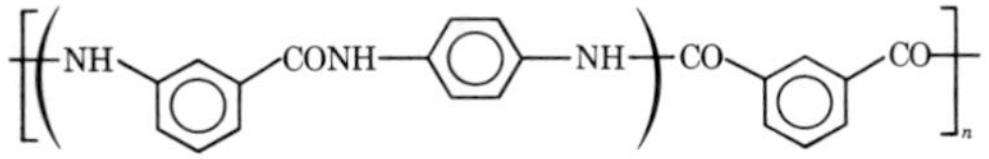	[*29153-49-9*]	$(C_{21}H_{15}N_3O_3)_n$	2.1	S		W	0.50/23/6.8	18
(14) 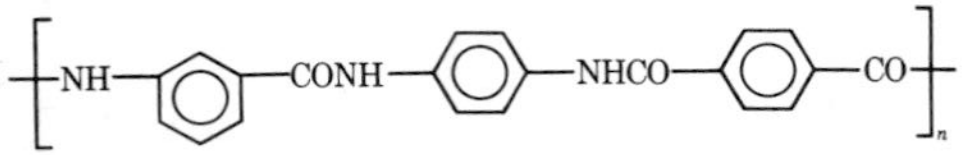	[*52303-33-0*]	$(C_{21}H_{15}N_3O_3)_n$	2.7	S		W	0.47/20/7.4	18
(15) 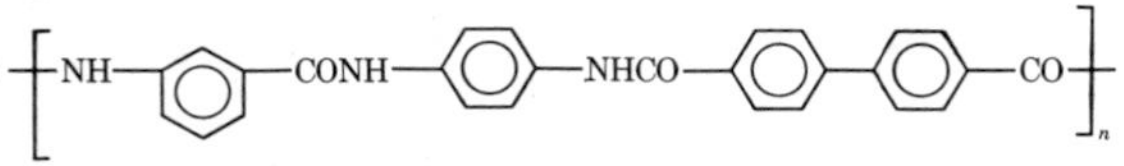	[*52303-35-2*]	$(C_{27}H_{19}N_3O_3)_n$	2.0	S		W	0.24/10/7.5	18
(16) 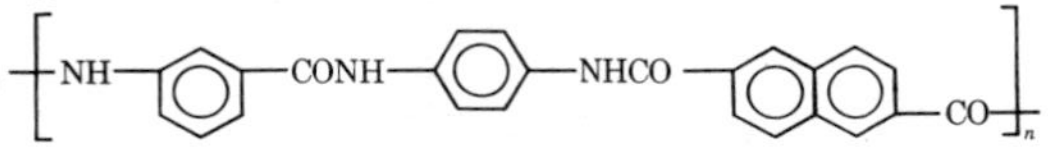	[*52303-34-1*]	$(C_{25}H_{17}N_3O_3)_n$	2.4	S		W	0.38/41/8.6	18
(17) 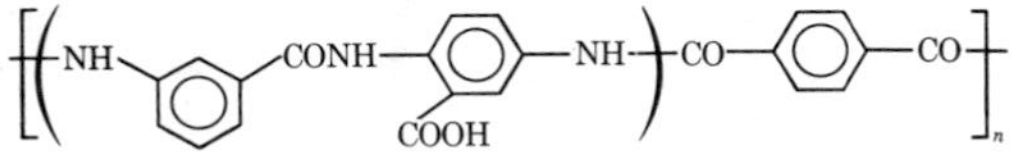	[*65749-41-9*]	$(C_{22}H_{15}N_3O_5)_n$	1.97	S	DMAC	W	0.46/8.8/10.5	19
Ordered copolymers								
(18) 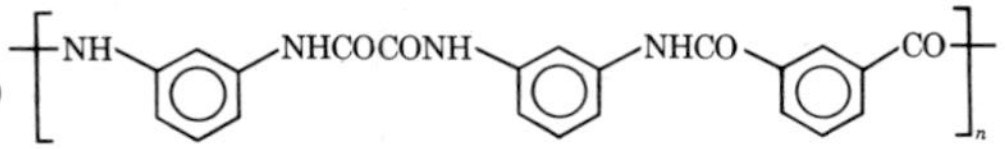	[*65749-42-0*]	$(C_{22}H_{16}N_4O_4)_n$	2.48	S	DMAC–LiCl	W	0.26/24/4.7	

Table 1 *(continued)*

	Structure	CAS Registry No.	Mol formula	η_{inh}[a]	Method of prep[b]	Spun from[c]	Spinning method[d]	T/E/M_i[e]	Ref.
(19)	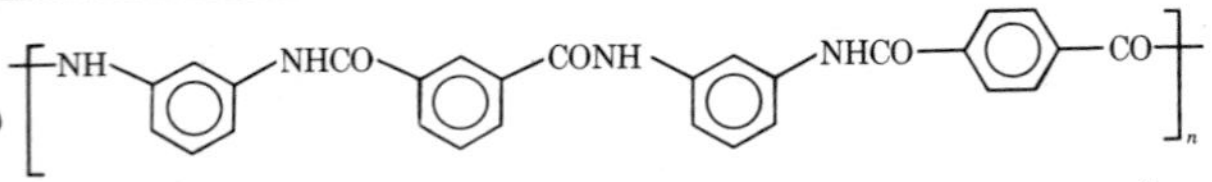	[65749-43-1]	$(C_{28}H_{20}N_4O_4)_n$		S	DMAC–$CaCl_2$	D	0.36/33/7.9	10
(20)	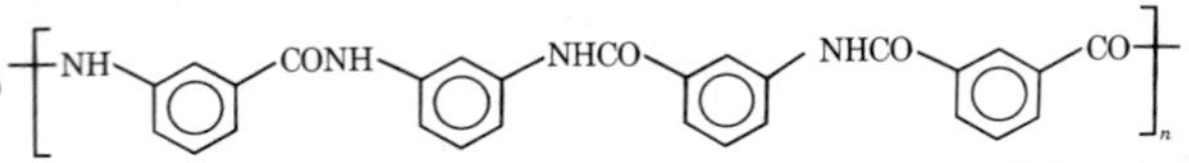	[27290-59-1] and [27307-23-9]	$(C_{28}H_{20}N_4O_4)_n$		S	DMAC–LiCl	W	0.45/37/7.9	20
(21)	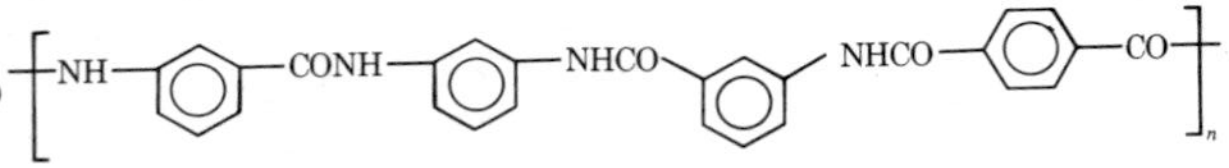	[25035-07-8]	$(C_{28}H_{20}N_4O_4)_n$		S		W	0.53/23/8.9	21
	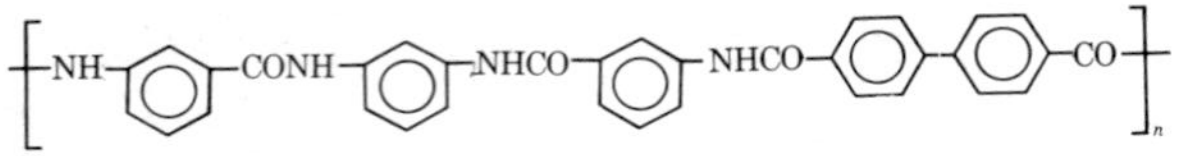	[65749-44-2]	$(C_{34}H_{24}N_4O_4)_n$		S		W	0.52/16/8.2	21
(23)	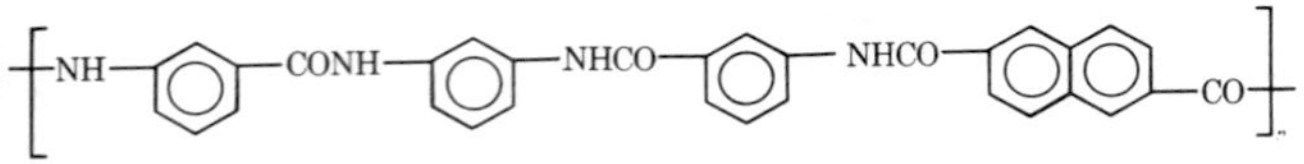	[32195-67-8]	$(C_{32}H_{22}N_4O_4)_n$		S		W	0.56/19/8.3	21
(24)	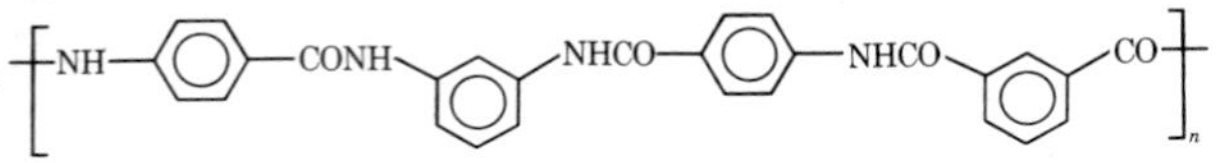	[32195-68-9]	$(C_{28}H_{20}N_4O_4)_n$		S		W	0.55/20/6.2	21
Copolymers of limited order									
(25)	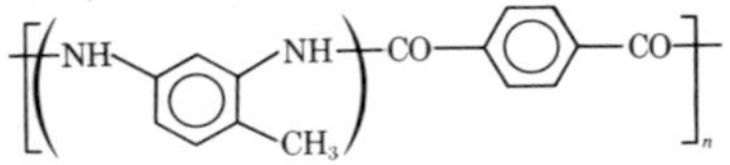	[29153-38-6]	$(C_{15}H_{12}N_2O_2)_n$	1.37	I	DMAC–LiCl		0.12/96/2.8	15
(26)	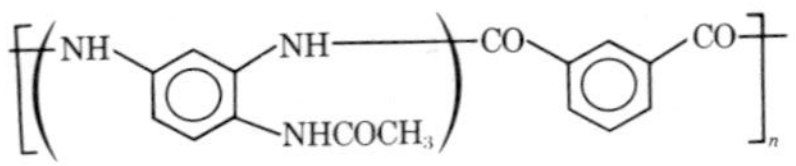	[59789-54-7]	$(C_{16}H_{13}N_3O_3)_n$	1.33	S	DMAC–LiCl	W	0.26/26/4.7	

[a] Determined in sulfuric acid at 30°C on a 0.5 g/100 mL solution. $\eta_{inh} = \ln \eta_{rel}/c$; $\eta_{rel} = \eta_{soln}/\eta_{solvent}$; c = conc of solvent.
[b] S = solution polymerization; I = interfacial polymerization.
[c] DMAC = dimethylacetamide; DMSO = dimethyl sulfoxide; DMF = dimethylformamide.
[d] D = dry spinning; W = wet spinning.
[e] T = tenacity, N/tex (= 11.33 gf/den); E = elongation-to-break, %; M_i = initial modulus, N/tex.
[f] Specific viscosity on 0.5% solution.

Table 2. Tensile Properties of Aramid Fibers at Elevated Temperatures

Structure no.	T/E/M_i [a] at room temp	T/E/M_i [a] at 250°C	T/E/M_i [a] at 300°C	Zero strength temp [b], °C	Ref.
(1)	0.51/10/10.4		0.26/7.5/7.9	480–550	8
(2)	0.49/28/10.0	0.27/22/5.6	0.11/23/6.6	440	10
(4)	0.81/6.8/16.1		0.23/12/2.9		13
(19)	0.36/33/7.9	0.24/48/3.4	0.14/51/2.3	490	10
(20)	0.45/37/7.9		0.13/22/0.97	>400	20
(21)	0.53/23/8.9		0.18/11/4.4	485	21
(22)	0.52/16/8.2	0.14/15/5.1		>450	21
(23)	0.56/19/8.3		0.18/19/	455	21
(24)	0.55/20/6.2	0.29/21/4.1	0.20/19/3.2	470	21
(13)	0.42/27/6.4 [c] 0.76/28/ [d]	0.21/41/2.5			22

[a] T/E/M_i = tenacity, N/tex; elongation-to-break, %; initial modulus, N/tex (= 11.33 gf/den).
[b] Temperature at which the fiber breaks under a load of 8.8 mN/tex (0.1 gf/den).
[c] Tensile properties of polymer of η_{inh} 1.1; fibers prepared by dry spinning.
[d] Tensile properties of polymer of η_{inh} 2.8; fibers prepared by wet spinning.

at these temperatures (Table 2). Surprisingly, in general the elongations of these fibers tend to be less at 300°C than at room temperature.

Tensile Properties after Prolonged Aging at Elevated Temperatures. It is important to be able to predict the long term performance of aramid fibers at elevated temperatures. For this purpose, evaluation of the tensile properties after heat-aging at intermediate temperatures is more meaningful than determining the tensile properties at greatly elevated temperatures. Heat-aging performance in air (thermo-oxidation) is a particularly severe test of stability because of the high surface-to-volume ratio of the fibers.

The effect of heat-aging at 300°C in air on the fibers is given in Table 3. In general, aramid fibers retain useful tensile properties for 1–2 weeks under the severe conditions indicated. Nomex in the form of paper has been reported to have a useful lifetime of only around 40 h at 300°C for a certain electrical application, but a useful lifetime of about 1400 h at 250°C. At 177°C, this fiber retained 80% of its strength after exposure to air for several thousand hours (23).

Table 3. Physical Properties of Aramid Fibers after Prolonged Exposure in Air at 300°C

Structure	T/E/M_i [a], d					
no.	0	1	7	14	35	Ref.
(1)	0.51/10/10.4				0.31/4/11.5	8
(2)	0.49/28/10.0	0.26/9.9/8.6	0.15/5.7/6.8	0.063/1.6/4.1		10
(19)	0.36/33/7.9	0.30/32/6.3	0.088/3.1/4.4	0.063/1.4/5.1		10
(20)	0.45/37/7.9	0.29/25/8.3	0.23/13/7.9			20
(21)	0.53/23/8.9	0.31/9.0/6.5	0.18/3.0/6.6			20
	0.48/15/11.5	0.29/14/8.8		0.16/5/8.8		24
(23)	0.49/25/7.9		0.11/2.5/4.7 [b]			25
(24)	0.57/20/6.9	0.40/17/5.9	0.12/16/			21

[a] T/E/M_i = tenacity, N/tex; elongation-to-break, %; initial modulus, N/tex (= 11.33 gf/den).
[b] Determined after nine days.

Heat-aging almost always results in a decrease in the elongation of the fibers, the loss being roughly proportional to the decrease in tenacity. Frequently, the limit of usefulness of a fiber, such as in flexible products, is even more dependent upon elongation than tenacity, so that loss of elongation is of greater concern.

Flame Resistance. Aramid fibers characteristically burn only with difficulty and they do not melt like nylon-6,6 or polyester fibers. They are useful in a number of applications requiring high flame resistance. Upon burning, the aramid fibers produce a thick char which acts as a thermal barrier and prevents serious burns to the skin. Limiting oxygen index (LOI) values, generally accepted as a measure of flame resistance for polymers, are quite high for aramid fibers (Table 4) (see Flame retardants). For purposes of comparison, Table 4 contains some LOI values for several fibers containing either alkylene chains or heterocycles in combination with arylene rings and amide linkages. From Table 4, it is interesting to compare the LOI values of poly[(5-carboxy-*m*-phenylene)isophthalamide] (**3**) and poly(*m*-phenyleneisophthalamide) (**2**), which have the same structure except for a pendant COOH group in the former. The LOI values given in Table 4 and those usually reported are for ignition from the top of the sample. Bottom ignition of fabric samples generally gives considerably lower LOI values. This fact has been dramatically illustrated by Sprague (27) who reported LOI values for Kevlar-49, Nomex, and Durette (X-400) fabrics (28) of 24.5, 26.0, and 36.0, respectively, when lighted from the top with a wick. LOI values for the same samples ignited from the bottom with a wick were only 16.5, 17.0, and 18.0, respectively.

Some aramid fibers, such as Nomex, shrink away from a flame or a high heat source. Durette fabrics, based on Nomex and treated with hot chlorine gas or other chemical reagents to promote surface cross-linking which stabilizes the fibers, were developed for greater dimensional stability (28). HT-4, an experimental aramid fiber from DuPont, has high LOI values and high dimensional stability in a flame (29). Although the structure of the HT-4 polymer has not been disclosed, patents assigned to the DuPont Company have described the spinning of PPD-T from sulfuric acid followed by incorporation of tetrakis(hydroxymethyl)phosphonium chloride (THPC) and subsequently cross-linking the THPC by means of melamine. PPD-T alone has high dimensional stability but an LOI value comparable to Nomex (ie, 28–30); addition of 1% phosphorus raises the LOI value to 40–42. A process which diffuses THPC and a melamine–formaldehyde resin (Aerotex UM) into never-dried Nomex followed by a heat treatment, has been described in a recent patent (30).

Electrical Properties. Wholly aromatic polyamides have high volume resistivities and high dielectric strengths and, significantly, they retain these properties at elevated temperatures to a high degree. Accordingly, they have considerable potential as high temperature dielectrics particularly for use in motors and transformers. The high temperature electrical properties of aramid fibers are exemplified by breakdown voltages of 76 V/mm at temperatures up to 180°C. By comparison, the breakdown voltage for poly(tetrafluoroethylene) is only 57 V/mm at 150°C and for nylon-6,6 is merely 3 V/mm at 150°C (23) (see Insulation, electric).

Paper of Nomex has almost twice the dielectric strength of high quality rag paper, and retains its useful electrical properties at higher operating temperatures than the upper limit for the rag paper (about 105°C) (see Synthetic paper). In the early years of the commercialization of Nomex it is almost certain that the production of Nomex in the form of paper equaled or outstripped the production of Nomex in the form of fiber.

Table 4. LOI[a] Values for Selected Aramid Fibers[b,c]

	Structure	CAS Registry No.	Mol formula	LOI
(18)				22.0–23.0
(9)				22.5–23.0
(27)	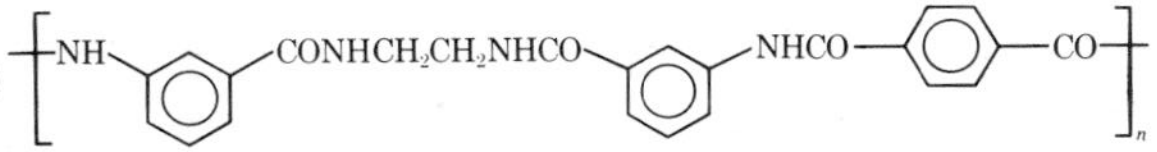	[*65749-40-8*]	$(C_{24}H_{20}N_4O_4)_n$	23.0–23.5
(11)				25.5–26.0
(28)	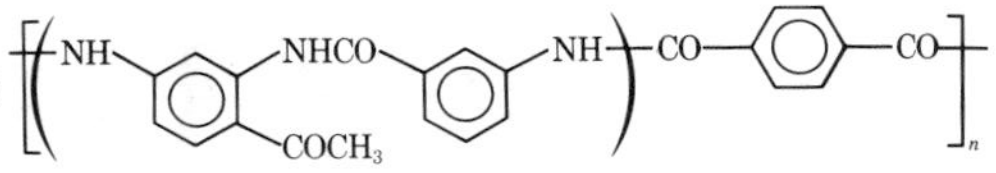	[*59789-57-0*]	$(C_{23}H_{17}N_3O_5)_n$	26.0–26.5
(2)				28.5–29.0
(29)	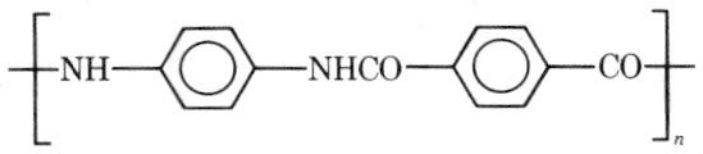	[*24938-64-5*]	$(C_{14}H_{10}N_2O_2)_n$	28.5–29.0
(26)				29.0–30.0
(30)	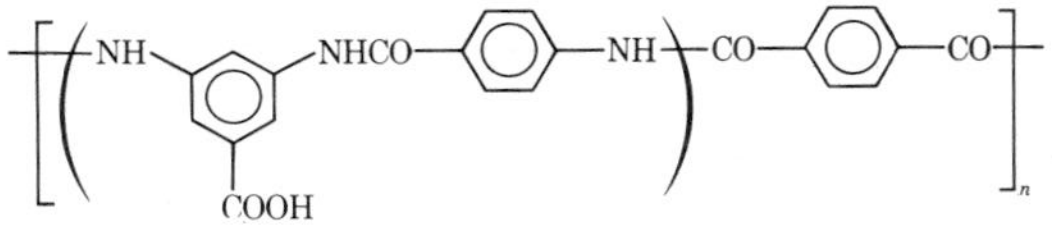	[*59789-55-8*]	$(C_{22}H_{15}N_3O_5)_n$	31.0–31.5
(31)	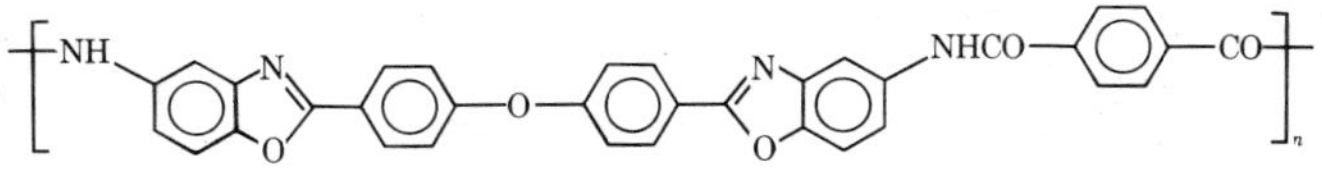	[*25670-03-5*]	$(C_{34}H_{20}N_4O_5)_n$	32.0
(3)				32.0–32.5

[a] Limiting oxygen index.
[b] Determined on thin knits.
[c] Ref. 26.

Resistance to Chemicals, Uv, and Ionizing Radiation. The chemical resistance of aramid fibers is, in general, very good (Tables 5 and 6). Although these fibers are much more resistant to acid than nylon-6,6 fibers, they are not as acid resistant as polyester fibers, except at elevated temperatures, particularly in the range of pH 4–8. Their resistance to strong base is comparable to that of nylon-6,6 fibers and hydrolytic stability of the aramids is superior to that of polyester and comparable to nylon-6,6.

Aramid fibers, like their aliphatic counterparts, are susceptible to degradation by ultraviolet light. Flanking the amide linkage on each side by arylene rings confers no special immunity to degradation by uv irradiation. The mechanism of photodegradation of the aramids is quite different from that of the aliphatic polyamides and presents greater problems as regards the development of color. To date, photo quenchers, which have been highly successful in the stabilization of aliphatic polyamides, have not been found for aromatic polyamides. Nor have uv screening agents been found which are effective and efficient at reasonable levels (= 1%) (see Uv absorbers).

The resistance of aramids to ionizing radiation is greatly superior to that of nylon-6,6 (Table 7). Nomex fiber retained 76% of its original strength after exposure to 155 C/g in a Van de Graff generator, whereas nylon-6,6 fiber was destroyed by the same exposure (23). The greater resistance of Nomex to degradation in comparison with the nylon-6,6 fiber is just as marked upon exposure to gamma radiation and x-rays. That excellent radiation resistance is a characteristic property of the aramid fibers is evidenced by fibers of two ordered aromatic copolyamides that actually gained about 10% strength after exposure to a dose of 30 kGy of gamma radiation.

Dyeability. Aramid fibers are exceedingly difficult to dye, probably as a consequence of their very high glass transition temperatures. A procedure for dyeing Nomex has been published (31) that makes use of a dye carrier (qv) at an elevated temperature (ca 120°C) in pressurized beam and jet-dyeing machines. Cationic dyes, which are used exclusively, must be selected carefully in order to achieve reasonable light fastness; a list of recommended dyes for Nomex has been given in a DuPont technical information bulletin (31).

Table 5. Chemical Resistance of Fibers of Aromatic Ordered Copolyamides[a]

Chemical	Strength	Temp, °C	Time, h	Strength retained, % Polymer (21)[b]	Polymer (23)[b]
dimethylacetamide	100%	21	264	81	99.5
sodium hypochlorite	0.5%	21	8	96	89
			264	55	44
sodium hydroxide	16*N*	21	8	100	84
			264	71	74
	4*N*	100	8	33	38
			22	[c]	19
sulfuric acid	20*N*	21	8	95	70
			264	76	70
	4*N*	100	8	51	51
			96	[d]	[e]

[a] Ref. 25.
[b] From Table 1.
[c] Too weak to test.
[d] Disintegrated.
[e] Weak, but still in fiber form.

Table 6. Chemical Resistance of Nomex High Temperature Resistant Aramid Fiber[a]

Chemical	Concentration, %	Temp, °C	Time, h	Effect on breaking strength[b]
Acids				
formic	90	21	10	none
hydrochloric	10	95	8	none
	35	21	10	appreciable
	35	21	100	appreciable
sulfuric	10	21	100	none
	10	60	1000	moderate
	70	21	100	none
	70	95	8	appreciable
Alkalies				
ammonium hydroxide	28	21	100	none
sodium hydroxide	10	21	100	none
	10	60	100	slight
	10	95	8	appreciable
	40	21	10	none
	50	60	100	degraded
Miscellaneous chemicals				
dimethylformamide	100	70	168	none
perchloroethylene	100	70	168	none
phenol	100	21	10	none
sodium chlorite	0.5	21	10	none
	0.5	60	100	moderate

[a] Ref. 23.
[b] None, 0–9% strength loss; slight, 10–24% strength loss; moderate, 25–44% strength loss; appreciable, 45–79% strength loss; degraded, 80–100% strength loss.

Table 7. Resistance of Fibers to Ionizing Radiation Degradation[a]

Radiation	Tenacity retained, %	
	Nomex	Nylon-6,6
beta radiation (Van de Graff)		
52 C/g	81	29
155 C/g	76	0
gamma radiation (Brookhaven Pile)		
52 C/g	70	32
520 C/g	45	0
x-ray		
50 kV	85	22
100 kV	73	0

[a] Ref. 23.

Ultra High-Strength–High-Modulus Aramid Fibers. Although the rod-like aramids have very high melting points (generally above 500°C with decomposition) and are heat-resistant polymers, they are not particularly useful as heat-resistant fibers. The reason is that fibers from the rod-like polymers have rather low elongations, and they lose elongation rapidly after being heated at an elevated temperature, thereby becoming too brittle to be especially useful. For very short-term exposure, retention of strength at elevated temperatures is outstanding, eg, in the case of Kevlar, tenacity is greater than 0.9 N/tex (10 gf/den) at about 300°C (32). The chemical resistance (32) of Kevlar is similar to that of Nomex (Table 6).

The flame resistance of the aramid fibers from rod-like polymers is high, but a flame tends to propagate along a vertically-held fiber of this class because of the grid formed by the residual char. However, incorporation of phosphorus, as pointed out earlier, has been used to raise the flame resistance to very high levels for fiber from poly(*p*-phenyleneterephthalamide) (PPD-T) which has not been highly drawn during spinning. The high flame resistance of PPD-T fabrics containing phosphorus coupled with the inherent high resistance to shrinkage of PPD-T fabrics can contribute significantly to providing insulation and protection from burns.

The fiber properties of some aramid fibers having ultra-high strength and high modulus are given in Table 8. The rod-like polymers (Table 8) used to prepare these fibers are of the AB, AA–BB, ordered copolymer and limited copolymer types. Terlon, Vnivlon, and SVM, high-strength–high-modulus fibers developed in the Soviet Union, have properties which would suggest that they are poly(*p*-benzamide) (PPB) spun from an organic solvent and poly(*p*-phenyleneterephthalamide) (PPD-T) and a heterocycle–amide copolymer of PPD-T spun from sulfuric acid (33–35).

Apparently the great regularity necessary for the attainment of good tensile properties in homopolymers that chain-fold is not necessary for the rod-like polymers which probably exist in the chain-extended state. In Tables 9 and 10, the tensile properties of selected aramid fibers based on random copolymers are given. These properties show that random copolymers yield fibers having tensile properties comparable to homopolymers (see Copolymers).

As shown for fibers from polyamide–hydrazides (36), the introduction of a few mol % of rings that are not rigid chain-extending types does not necessarily detract from the tensile properties of aramid fibers. In fact, on balance, fiber tensile properties may be enhanced considerably; note, eg, the excellent balance of properties for fiber (**52**), Table 9, and fiber (**61**), Table 10.

Two important features of the ultra high-strength–high-modulus fibers should be noted here: (*1*) the fibers are of fine diameter, usually less than 7×10^{-7} kg/m (6 den) per filament and generally more nearly about $1.7–2.2 \times 10^{-7}$ kg/m (1.5–2 den) per filament, and (*2*) the fibers are produced from rod-like polymers exhibiting extremely high intrinsic and inherent viscosity values although molecular weights of these polymers may be moderate. The situation regarding fine linear densities is not unlike that for glass fibers which must be of very small cross-sectional area in order to exhibit good mechanical properties in flexure.

The stress-strain curves for the ultra high-strength–high-modulus fibers show a strong similarity to those for glass and steel. On a specific basis (ie, taking into account the lower specific gravity of the aramid fibers in comparison to glass and steel) the aramid fibers are stronger and stiffer than glass and steel. These properties suggest that the organic fibers should be quite useful in the reinforcement of rigid and flexible composites. This is indeed the case. Thus Kevlar fiber may be used as a tire cord as a replacement for glass and steel and Kevlar-49 may be used competitively with the lower modulus types of graphite fibers (see Ablative materials; Carbon; Composite materials).

Table 8. Tensile Properties of the Rod-like Aramid Fibers

	Structure	CAS Registry No.	Mol formula	η_{inh}[a]	Spun from[b]	Solids, %	Spinning method[c]	Tex[d]	T/E/M$_i$[e]	W-T-B[f]	Remarks[g]	Ref.
AB Type												
(32)		[*24991-08-0*]	$(C_7H_5NO)_n$	1.67	O (—)		D		0.72/3.1/44.9			37
				2.36	O (A)		W	0.54	0.64/8.1/25.0		spun from anisotropic layer of dope	5
				same dope	O (I)		W	2.53	0.11/9.0/5.6		spun from isotropic layer of dope	5
				3.7	A (A)	18.0	DJ–W	0.11	1.7/4.0/50.3	36.5		6
AA–BB Type												
(33)	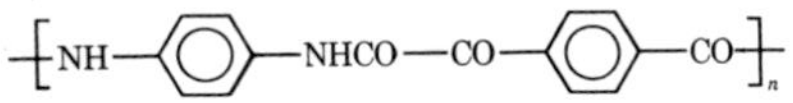	[*24938-64-5*]	$(C_{14}H_{10}N_2O_2)_n$	4.8	A (A)		DJ–W	0.15	2.1/5.0/42.7			6
(34)	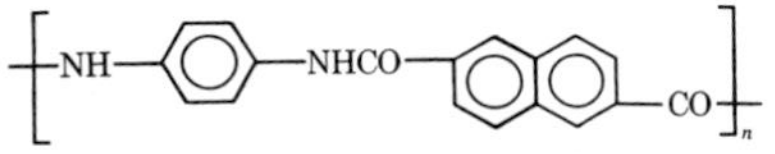	[*27307-20-6*]	$(C_{18}H_{12}N_2O_2)_n$		A (A)		DJ–W	0.20	2.8/3.7/81.2	55.1	hot-drawn	38
(35)	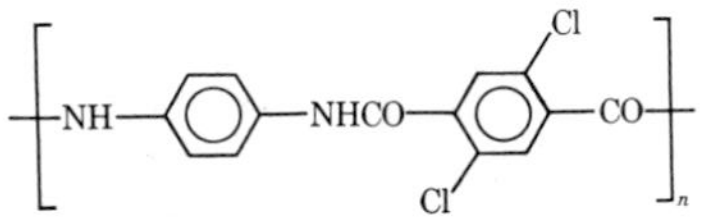	[*26402-76-6*]	$(C_{14}H_8Cl_2N_2O_2)_n$	3.6	A (A)		DJ–W	0.22	15.0/4.9/41.5	35.6		6
				1.59	A (—)	3.9	W	0.10	0.65/2.4/34.1		hot-drawn	5
(36)	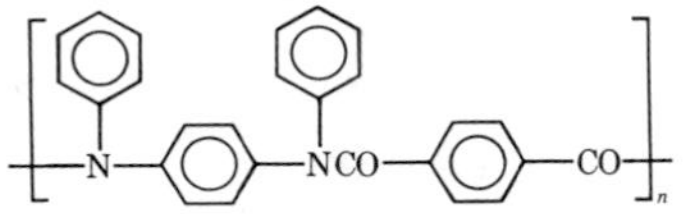	[*28779-61-5*]	$(C_{26}H_{18}N_2O_2)_n$	2.0	O (—)		D		0.24/3.6/8.6		hot-drawn	39
(37)	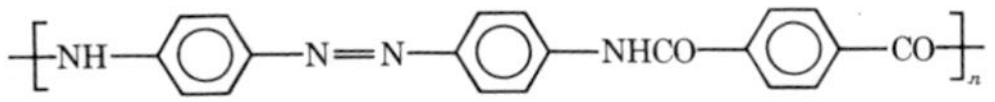	[*27252-16-0*]	$(C_{20}H_{14}N_4O_2)_n$	3.24	A (A)		DJ–W	0.17	0.62/4.9/29.2		hot-drawn	40
	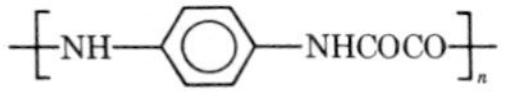	[*65749-45-3*]	$(C_8H_6N_2O_2)_n$	0.9	O (—)		W		0.35/3.0/39.7		hot-drawn	41
(39)		[*37357-28-1*]	$(C_{15}H_{12}N_2O_2)_n$	4.46	O (—)		W		0.86/10.2/25.3			5
				3.1	A (A)		DJ–W	0.24	1.4/2.5/68.8	19.4		6
(40)	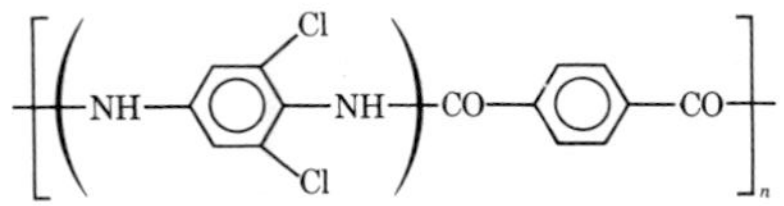	[*26402-76-6*]	$(C_{14}H_8Cl_2N_2O_2)_n$	3.77	O (—)			0.36	0.77/5.6/21.3			5
				same dope				0.30	1.3/3.5/43.2		hot-drawn	5

Table 8 *(continued)*

	Structure	CAS Registry No.	Mol formula	η_{inh}[a]	Spun from[b]	Solids, %	Spinning method[c]	Tex[d]	T/E/M$_i$[e]	W-T-B[f]	Remarks[g]	Ref.
(41)	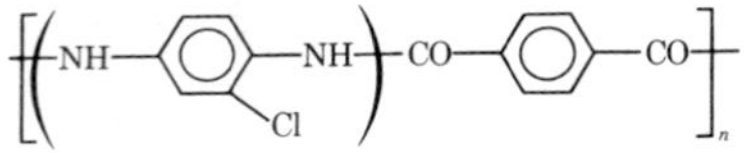	*[51257-61-7]*	$(C_{14}H_9ClN_2O_2)_n$	3.1	A (A)	20.0	DJ–W	3.9	1.6/6.5/32.7	49.4		6
(42)		*[31801-22-6]*	$(C_{14}H_9ClN_2O_2)_n$	3.7	A (A)	20.0	DJ–W	1.9	1.9/4.8/56.5	43.8		6
(43)		*[65749-46-4]*	$(C_{13}H_9N_3O_2)_n$	5.3	A (A)	20.0	DJ–W	3.5	1.6/5.8/41.5	46.2		6
(44)		*[65761-30-0]*	$(C_{28}H_{20}N_4O_4)_n$	2.4	A (A)	3.9	W	6.1	0.42/12.4/14/7			5
				same dope			W	6.3	0.48/0.9/51.5		hot-drawn	5
(45)	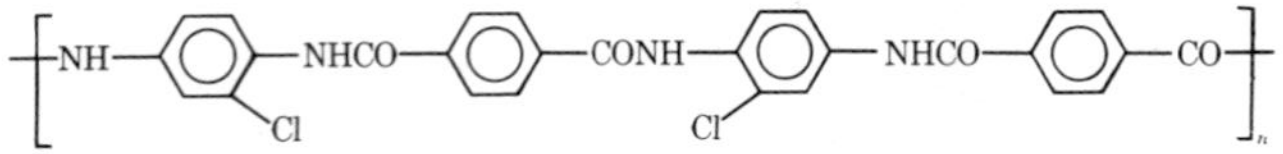	*[65749-48-6]*	$(C_{28}H_{18}Cl_2N_4O_4)_n$	3.25	O (—)		W	1.49	0.94/6.6/30.3			5
				same dope			W	1.36	1.2/1.9/67.3		hot-drawn	5
(46)	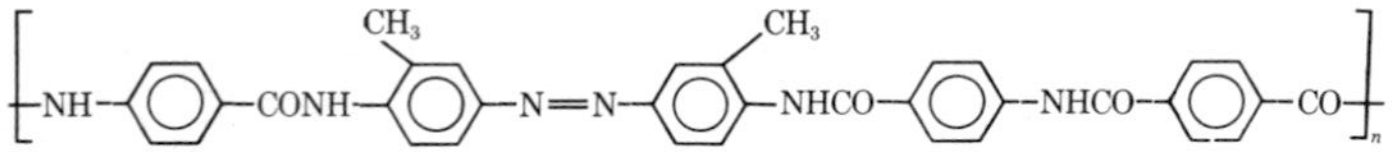	*[65749-49-7]*	$(C_{36}H_{28}N_6O_4)_n$	7.8[h]	O (—)				0.58/18.6/17.4			42
				same dope					0.86/2.4/51.8		hot-drawn	42
(47)	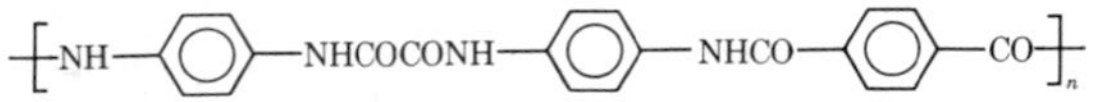	*[52270-04-9]*	$(C_{22}H_{16}N_4O_4)_n$	3.1	O (—)		DJ–W		0.27/7/11.7			43
				same dope			DJ–W		0.89/ /22.1		hot-drawn	43
Copolymers of limited order												
(48)	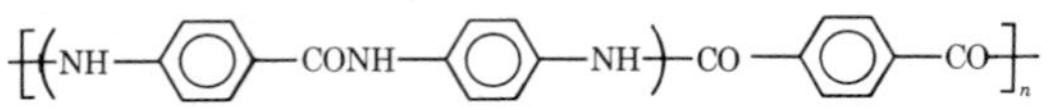	*[29153-47-7]*	$(C_{21}H_{15}N_3O_3)_n$	6.6	O (—)		DJ–W	5.6	0.65/8.3/25.2	43.8	hot-drawn	36
				3.6	A (A)	10.0	W	6.7	0.57/11/15.3			5
				same dope			W	4.2	0.89/1.2/73.0		hot-drawn	5
				4.3	A (A)	20.0	DJ–W	1.6	1.5/3.9/54.7	31.6		6
(49)	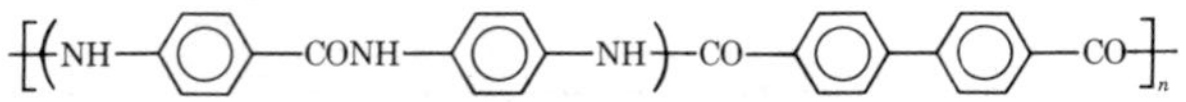	*[65749-50-0]*	$(C_{27}H_{19}N_3O_3)_n$	2.4	A (A)	11.0	DJ–W	7.2	0.49/9/19.1			5
				same dope			DJ–W		1.2/2.4/62.4		hot-drawn	5

[a] Determined in sulfuric acid, 0.5 g/100 mL, at 30°C. $\eta_{inh} = \ln \eta_{rel}/c$; $\eta_{rel} = \eta_{soln}/\eta_{solvent}$; c = conc of solvent.
[b] O = organic solvent; A = acid (H_2SO_4); (I) = isotropic dope; (A) = anisotropic dope.
[c] D = dry spun; W = wet spun; DJ–W = dry-jet–wet spun.
[d] Tex (= 9 den) per filament.
[e] T = tenacity, N/tex; E = elongation-to-break, %; M$_i$ = initial modulus, N/tex (= 11.33 gf/den).
[f] W-T-B = work-to-break, J/g (= 11.33×10^{-3} gf·cm/(den·cm)).
[g] Fiber properties are for as-spun fiber except as noted.
[h] Determined in DMAC–LiCl.

Table 9. Fibers from Random Copolyterephthalamides of *p*-Phenylenediamine (PPD) and Various Diamines[a]

Diamine	Mol formula	Mol %	η_{inh}[b]	Tex[c]	T/E/M_i[d]	W-T-B[e]	Ref.
(50) 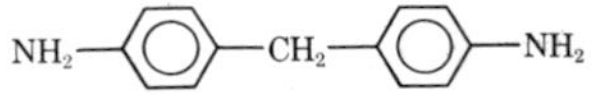	$C_{13}H_{14}N_2$	5	3.8	0.27	1.3/4.1/51.2	30.0	6
(51) 	$C_{14}H_{16}N_2$	7.5	3.4	0.23	1.9/3.8/64.4	38.1	6
(52) 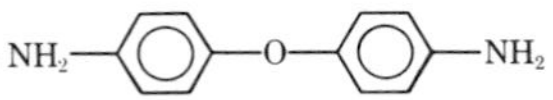	$C_{12}H_{12}N_2O$	5	4.3	0.29	2.1/6.2/45.9	68.9	6
(53) 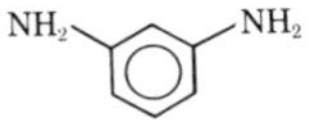	$C_6H_8N_2$	5	4.2	0.39	1.5/5.4/40.6	42.1	6
(54) 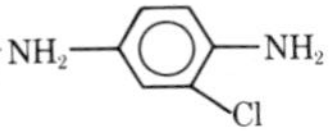	$C_6H_7ClN_2$	25	5.7	0.48	1.9/6.9/30.9	65.6	6
(55) 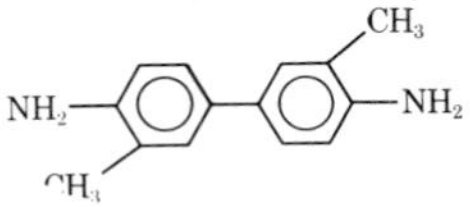	$C_{14}H_{16}N_2$	5	3.4	0.36	1.5/4.9/48.5	39.7	6
(56) 	$C_{14}H_{14}N_4O_2$	50	6.6[f]		0.69/2.4/33.8[g]		44
					1.0/1.8/52.8[h]		44
		60	4.3[f]		1.2/1.4/75.2[g]		44

[a] Spun from sulfuric acid except as noted.
[b] Determined at 30°C in 0.5 g fiber dissolved in 100 mL sulfuric acid. $\eta_{inh} = \ln \eta_{rel}/c$; $\eta_{rel} = \eta_{soln}/\eta_{solvent}$; c = conc of solvent.
[c] Tex (= 9 den) per filament.
[d] T = tenacity, N/tex; E = elongation-to-break, %; M_i = initial modulus, N/tex (= 11.33 gf/den).
[e] W-T-B = work-to-break, J/g (= 11.33 × 10^{-3} gf·cm/(den·cm)).
[f] Determined on polymer in DMAC–LiCl.
[g] Spun from organic solvent and hot drawn.
[h] Double hot-draw.

Table 10. Fibers from Random Copolyterephthalamides of *p*-Phenylenediamine (PPD) and Various Diacid Chlorides[a,b]

Diacid chloride	Mol formula	Mol %	η_{inh}[c]	Tex[d]	T/E/M_i[e]	W-T-B[f]
(57) 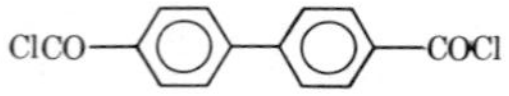	$C_{14}H_8Cl_2O_2$	55	5.3	0.38	1.9/5.5/60.9	56.5
(58) 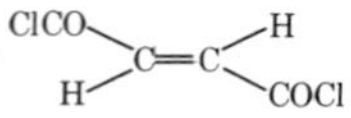	$C_4H_2Cl_2O_2$	40	3.3	0.18	1.6/5.7/49.4	48.5
(59) 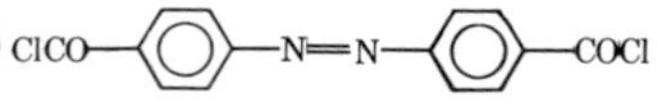	$C_{14}H_8Cl_2N_2O_2$	5	3.4	0.34	1.7/4.6/44.1	39.7
(60) 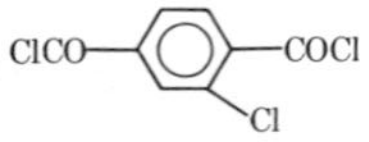	$C_8H_3Cl_3O_2$	5	5.5	0.47	2.0/4.2/51.2	40.6
(61) 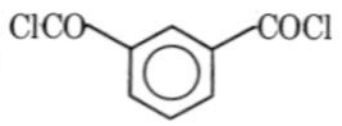	$C_8H_4Cl_2O_2$	5	3.9	0.17	1.8/4.6/51.2	43.2
(62) 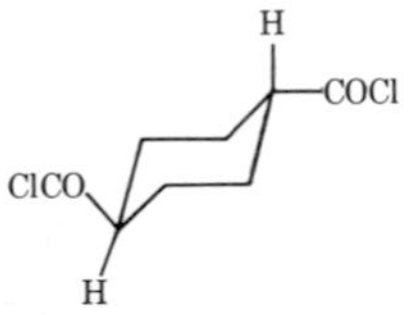	$C_8H_{10}Cl_2O_2$	25 50	4.3 3.4	0.16 0.23	1.9/4.5/47.7 1.6/4.9/43.2	45.0 41.5

[a] Fibers spun from sulfuric acid.
[b] Ref. 6.
[c] Determined at 30°C on 0.5 g of fiber dissolved in 100 mL of sulfuric acid. $\eta_{inh} = \ln \eta_{rel}/c$; $\eta_{rel} = \eta_{soln}/\eta_{solvent}$; c = conc of solvent.
[d] Tex (= 9 den) per filament.
[e] T = tenacity, N/tex; E = elongation-to-break, %; M_i = initial modulus, N/tex (= 11.33 gf/den).
[f] W-T-B = work-to-break, J/g (= 11.33×10^{-3} gf·cm/(den·cm)).

Synthesis

Methods of Preparation. The usual methods for the preparation of aliphatic polyamides are unsuited to the preparation of high molecular weight, wholly aromatic polyamides. However, two general synthetic methods are presently available for the preparation of medium to high molecular weight polymer: low temperature polycondensation and direct polycondensation in solution using phosphites, especially in the presence of metal salts (45).

Low temperature (ie, <100°C) polycondensation procedures developed over the past two decades have led to the preparation of a very large number of high molecular weight polyamides otherwise unobtainable. Morgan, in an excellent text, has reviewed the literature on low temperature polycondensation methods up to 1965 (46). The principal low temperature methods are interfacial polycondensation and solution polycondensation.

Interfacial Polymerization. The interfacial method for the preparation of polyamides is an adaptation of the well-known Schotten-Baumann reaction: the diacid chloride is dissolved or dispersed in an inert, water-immiscible organic solvent which is preferably a swelling agent for the polymer, and the diamine is dissolved or dispersed along with a proton acceptor in the aqueous phase. The two-phase system may be stirred or unstirred, but higher molecular weight polymers are generally obtained when the system is stirred rapidly. In the latter process, the use of an emulsifying agent is usually helpful. The polymer is collected and dried; it can then be dissolved in a suitable solvent and fabricated.

Solution Polymerization. The most significant polymerization process for aromatic polymers is the low temperature polycondensation of diacid chlorides and diamines in amide solvents. This type of polymerization is often more convenient to carry out than interfacial polycondensation and has the further advantage of usually providing a solution of the polymer amenable to direct fabrication of fibers. Nomex is prepared in solution and spun from the solution after neutralization (Fig. 1). Another distinct advantage of solution polycondensation over the interfacial method results from the peculiar solubility characteristics of certain wholly aromatic polyamides: although such polymers are soluble as made in solution by low temperature polycondensation, they often cannot be redissolved in solvents (other than strong acids) once precipitated. Interfacial polymerization gives a precipitated product which cannot be readily fabricated.

In solution polycondensation the polymerization medium is a solvent for at least one of the reactants and is a solvent or swelling agent for the polymer, preferably a solvent. The solvent must be inert or relatively so to the reactants. To ensure completeness of the reaction, an acid acceptor, usually a tertiary amine, is used unless the solvent itself is such an acceptor. The better solvents are amides such as dimethylacetamide (DMAC), *N*-methylpyrrolidinone (NMP), hexamethylphosphoric triamide (HPT), and tetramethylurea (TMU). Dimethylformamide (DMF) is not useful because of its rapid and irreversible reaction with acid chlorides, nor is dimethyl sulfoxide (DMSO) useful because of its sometimes violent reaction with acid chlorides.

For the solution polymerization of many aromatic polyamides, there is no known organic solvent sufficiently powerful to keep the polymer in solution as its molecular weight builds up. The solvating power of many organic solvents is greatly increased, however, by the addition of inorganic salts such as lithium chloride and calcium chloride. Use of these salt-containing solvents has permitted preparation of many high

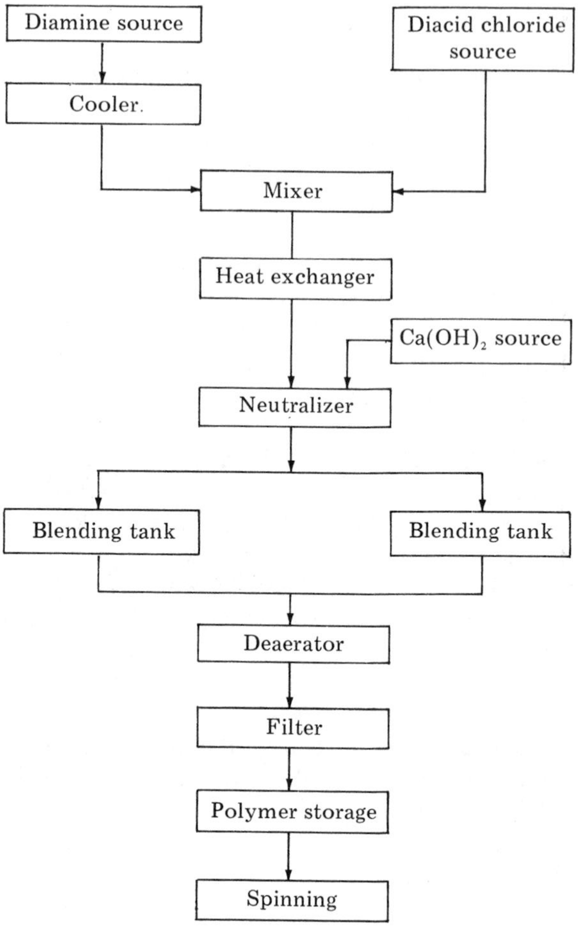

Figure 1. Preparation of Nomex (7). Reactants were *m*-phenylenediamine and isophthaloyl chloride; solvent, dimethyacetamide; polymer(MPD-I) of η_{inh} = 1.65 produced.

molecular weight polymers not obtainable using organic solvents alone. Also, mixtures of the basic amide solvents, such as DMAC with a minor amount of the corresponding hydrochloride, are much better solvents for aromatic polyamides than are the amide solvents alone. This is advantageous since hydrogen chloride is produced in the condensation of aromatic diamines with diacid chlorides and, when amide solvents are used, the desired salt–solvent combination is produced. Moreover, the quantity of the hydrochloride salt increases as the polycondensation proceeds, thereby increasing the solvating power of the solvent at the same time that the molecular weight of the polymer is building up.

Certain of the rod-like aromatic polyamides apparently are best prepared in mixed solvents, especially in specific ratios of these component solvents. Thus it was reported (47) that the highest molecular weight for poly(*p*-phenyleneterephthalamide) had been obtained in a 1:2 weight ratio of HPT to NMP; the optimum ratio of DMAC to HPT was obtained (47) at about 1:1.4. These weight ratios correspond to molar ratios of, respectively, 1:3 and 2:3 (Fig. 2). Preparation in DMAC and NMP or mixtures of

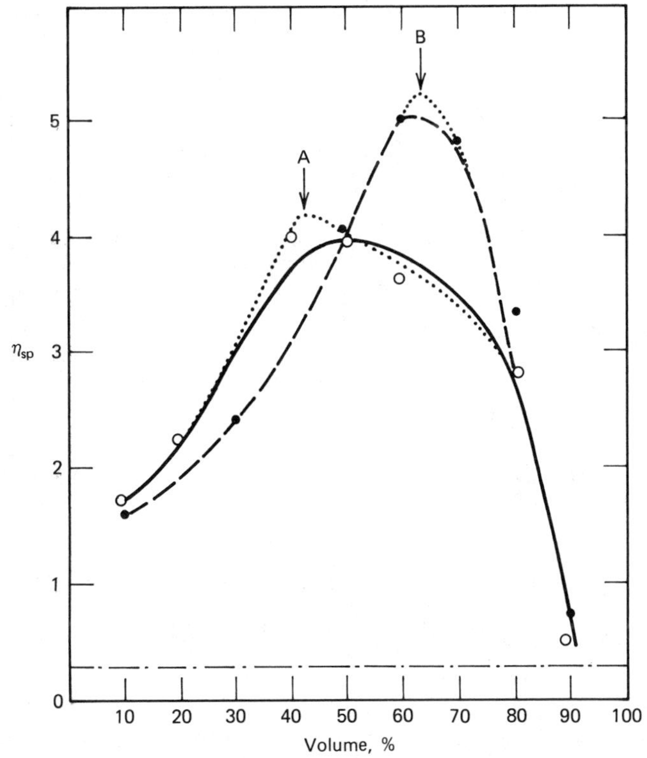

Figure 2. Relationship of molecular weight, as indicated by η_{sp}, for PPD-T prepared in the synergistic solvent mixtures of DMAC–HPT, —; and NMP–HPT, — —. Neither DMAC nor NMP alone or in any combination, · — ·, produce high molecular weight PPD-T. (The original drawing (47) has been modified by the · · · · lines to show the maxima in the curves.) A, the molar ratio of DMAC:HPT, 3:2; B, the molar ratio of NMP:HPT, 3:1.

these solvents at all ratios give low molecular weight polymer, thus showing no synergistic solvent effect (Fig. 2). Addition of salts to the mixtures of DMAC–HPT and NMP–HPT useful for preparation of PPD-T results in lower molecular weight polymer, presumably because the salts complex with the individual solvents, thereby breaking up the complex between the solvent pairs.

For aromatic polyamides that are not of the rod-like type, a solution of 15–25% polymer solids is typically prepared. This level of polymer solids, ie, a concentration of about 0.5 mol/L, yields polymer of optimum molecular weight (Fig. 3). However, for the synthesis of rod-like polymers, eg, PPD-T, it has been pointed out that molecular weight increases (Fig. 3) with concentration only up to about 0.25 mol/L or about 6–7% solids and decreases rapidly above this concentration (48–49). It may be significant that above this concentration rod-like polymers can yield anisotropic solutions in organic solvents but below this level isotropic solutions are obtained (50). In the case of the preparation of polymer at high solids levels, it is possible that the growing closely-packed polymer chains, as in the anisotropic state, may become inaccessible to the monomers, thereby limiting molecular weight.

The preparation of PPD-T, used to spin Kevlar fiber, is shown schematically in Figure 4.

Recently, the direct polycondensation of aromatic dicarboxylic acids and aromatic

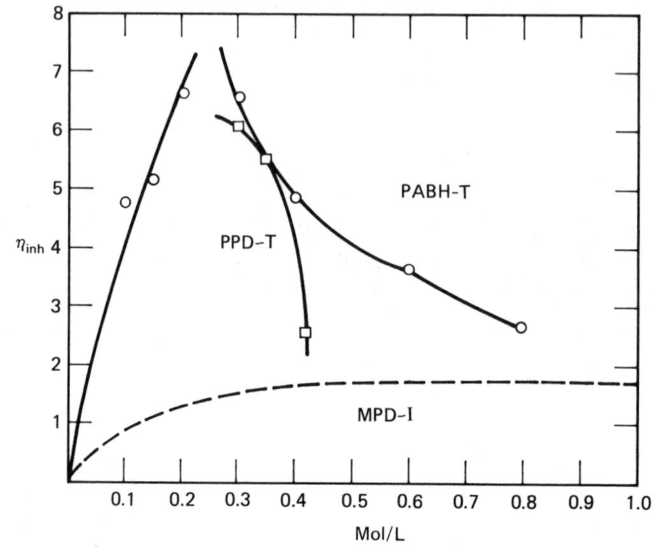

Figure 3. Relationship of mol wt (as indicated by η_{inh}) for aromatic polymers (PABH-T =

; PPD-T data from Ref. 6).

diamines in amide solvents using aryl phosphites in the presence of pyridine has been demonstrated (45). Addition of metal salts, eg, LiCl, to the amide solvents has made it possible to keep the polymer in solution formed during the reaction and thereby allow for the build-up of moderately high molecular weight polymer as indicated by the inherent viscosity values obtained ($\eta_{inh} = 1.0 \pm 0.1$). High molecular weight, wholly aromatic polyamides have not been made using the phosphorylation reaction, presumably because of the reaction of by-product phenol to yield phenyl ester.

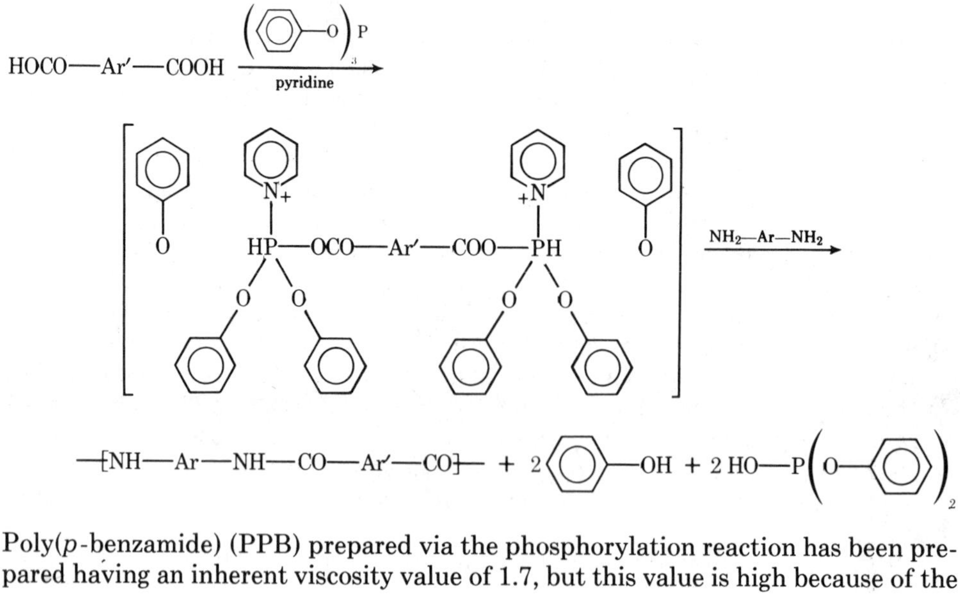

Poly(p-benzamide) (PPB) prepared via the phosphorylation reaction has been prepared having an inherent viscosity value of 1.7, but this value is high because of the rod-like character of PPB; the actual molecular weight of PPB is probably no higher than that of poly(*m*-benzamide) (PMB) having an inherent viscosity of 0.4 (45).

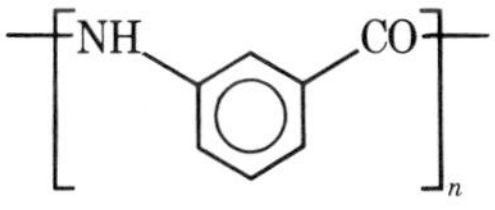

poly(*m*-benzamide)
(PMB)

Preparation of Selected Classes of Polymers. Several classes of fiber-forming wholly aromatic polyamides have been prepared. In addition to the conventional AB and AA–BB classes, some classes of AA–BB polymers of more complex structure have been made. Synthesis of the latter involves the use of diamines or diacids containing preformed amide or heterocyclic units, or pendant functional groups which do not participate in the polycondensation. Only polymers from which fibers have been prepared will be considered here.

A–B Polyamides. Low temperature polycondensation of reactive A–B monomers has been used to obtain high molecular weight wholly aromatic A–B polyamides. Polymers derived from *m*- or *p*-aminobenzoic acid have been prepared from the *m*- or *p*-aminobenzoyl chloride hydrochloride by the addition of a base under special conditions (8,37,52–53). The *m*- and *p*-aminobenzoyl chloride hydrochlorides are synthesized via reaction of the aminobenzoic acids with thionyl chloride to yield the thionylaminobenzoyl chlorides which react with HCl under anhydrous conditions.

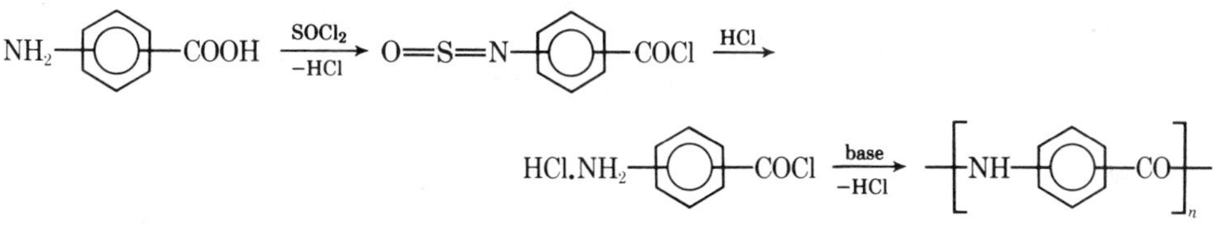

Apparently considerable attention to technique is important in the A–B polyamide polycondensation reaction because one worker reported (8) that interfacial polymerization of *m*-aminobenzoyl chloride hydrochloride yielded a polymer of η_{inh} 0.7 and another reported (52) a polymer of η_{inh} 2.36. The use of solution techniques by the same workers gave polymers of η_{inh} 1.37 and η_{inh} 0.3, respectively.

Use of the interfacial polycondensation method employing *p*-aminobenzoyl chloride hydrochloride yields only *p*-aminobenzoic acid and its dimer (53). PPB of η_{inh} 0.7 is obtained upon the addition of a tertiary organic base, eg, pyridine, to a slurry of monomer in dry dioxane (53), and high molecular weight PPB is obtained by polymerization in an amide solvent at low temperature using either *p*-aminobenzoyl chloride hydrochloride (37) or 4(4′-aminobenzamido) benzoyl chloride hydrochloride (54).

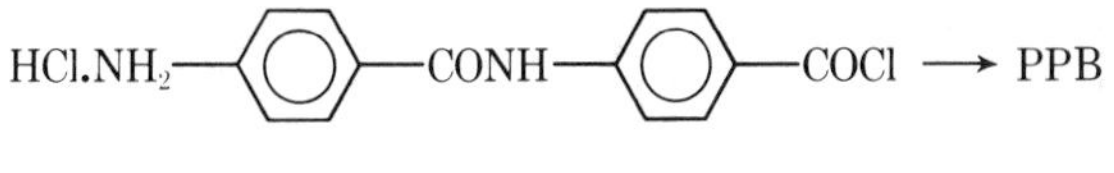

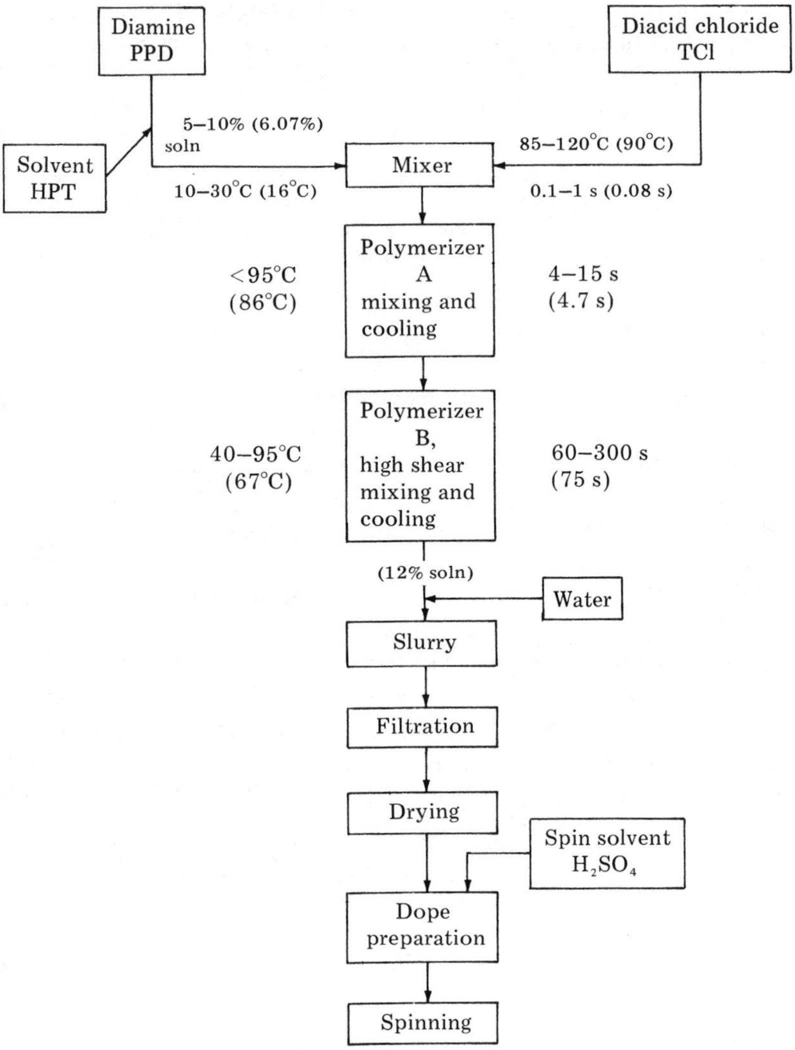

Figure 4. Preparation of PPD-T, used to spin Kevlar fiber (6,51). Values in parentheses are from an actual example in Ref. 51; polymer of $\eta_{inh} = 5.3$ was produced.

AA–BB Polyamides. AA–BB polyamides are readily prepared from aromatic diamines and diacid chlorides.

$$NH_2—Ar—NH_2 + ClCO—Ar'—COCl \longrightarrow —[NH—Ar—NHCO—Ar'—CO]—$$

Ar and Ar′ =

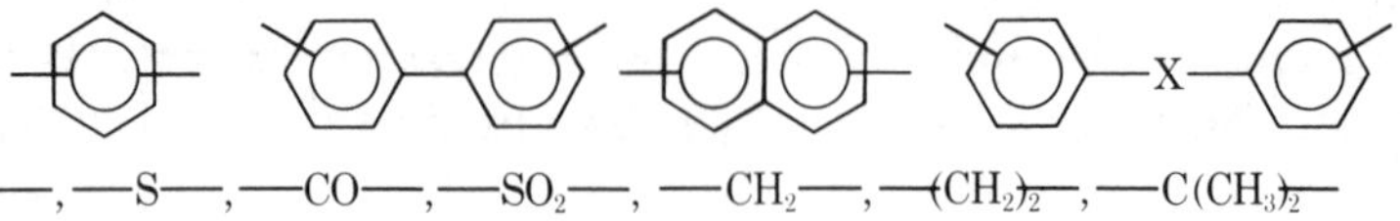

$$X = —O—, —S—, —CO—, —SO_2—, —CH_2—, —(CH_2)_2—, —C(CH_3)_2—$$

Polyamides having phenylene rings bridged by short alkylene chains are included because from a property point of view they closely approach wholly aromatic polymers and because most of the literature concerning wholly aromatic polyamides is interwoven with these polymers. Polymers containing pendant methyl or chloro groups on the phenylene units have also been included.

The interfacial polycondensation technique was the first low temperature process to yield high molecular weight wholly aromatic polyamides. The preparation of poly(benzidine terephthalamide) [*25667-70-3*] prepared via this route, although of relatively low molecular weight, demonstrated that this low temperature route to wholly aromatic polyamides was feasible (55). Although high molecular weight polymer can usually be prepared by the interfacial polycondensation method, doing so is frequently not without difficulty.

Polar cycloaliphatic and heterocyclic solvents that are either miscible or partially miscible with the aqueous phase are most generally useful; consequently weak acid acceptors are preferred to minimize hydrolysis of the acid chloride. The rod-like polymers, because of their low solubility in and resistance to swelling by organic solvents, apparently are not prepared in high molecular weight by the interfacial technique.

In general, high molecular weight wholly aromatic polyamides can be obtained more consistently by low temperature solution polycondensation than by the interfacial method. The great value of the low temperature solution polymerization method has been shown amply by the numerous reports of high molecular weight, fiber-forming polymers of widely varying structure. In general, however, wholly aromatic polyamides will not remain in solution when prepared by solution polymerization unless either inorganic salts, such as lithium chloride or calcium chloride, are present or a basic solvent, such as dimethylacetamide or *N*-methylpyrrolidinone, which forms a salt with hydrogen chloride, is used (56). The products of low temperature solution polycondensation may be used in a convenient, economical, and direct process for the spinning of fibers.

The monomers for rod-like polymers are mixed in solution, but they frequently form insoluble gel structures at high molecular weight (5–6). In order to obtain high molecular weight polymer, shearing action, such as may be obtained with a high speed food blender, is necessary. A process confining mixing of the monomers and the necessary shearing action has been described (51). The polymer then may be washed free of solvent, dried, and redissolved in acids, such as sulfuric, chlorosulfuric, fluorosulfuric or hydrofluoric, for the preparation of fibers by wet-spinning methods (5–6).

AA–BB polyamides having reactive substituents on the rings can be prepared provided that the reactive substituent is either masked during the polycondensation or is unreactive under the conditions of the polycondensation. The polymer from isophthaloyl chloride and 3,5-diaminobenzoic acid, DAB-I, is an example (12).

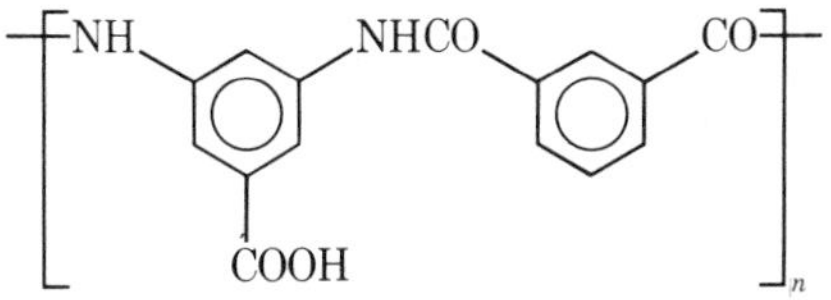

poly[(5-carboxy-*m*-phenylene)isophthalamide]
(DAB-I)

AA–BB Polyamides from Intermediates Containing Preformed Amide Linkages. Ordered copolyamides (10,20) are readily prepared by low temperature polycondensation from diamines or diacid chlorides containing preformed amide linkages because no reorganization of structural units is possible under the conditions of polymerization. Polymers of this type may be prepared by both the interfacial and the solution methods, but more consistent results and polymers of higher inherent viscosities are generally obtained by the solution technique. The preparation of ordered copolyamides may be illustrated by the following examples:

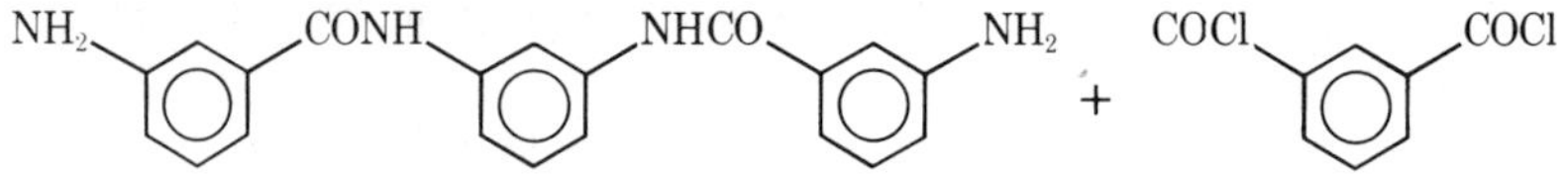

N,N′-m-phenylene-bis(*m*-aminobenzamide) (MPMAB)

isophthaloyl chloride (ICl)

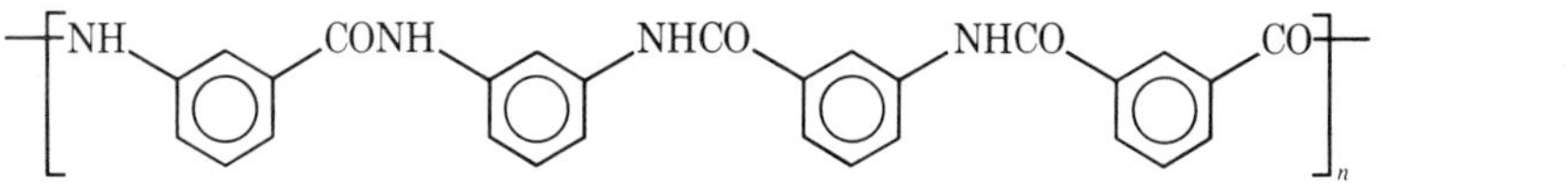

(MPMAB-I)

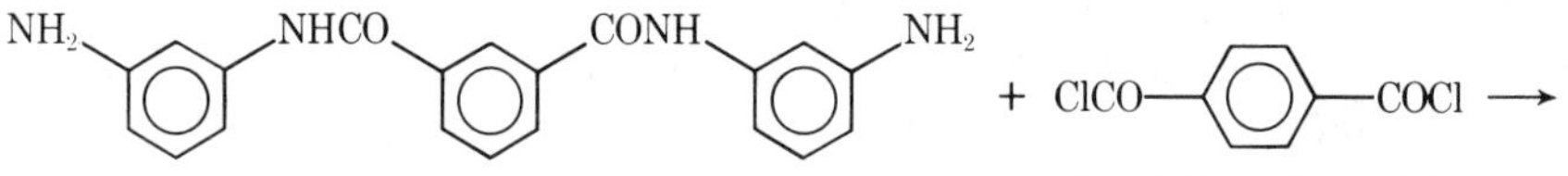

N,N′-bis(*m*-aminophenyl)isophthalamide (MAPI)

terephthaloyl chloride (TCl)

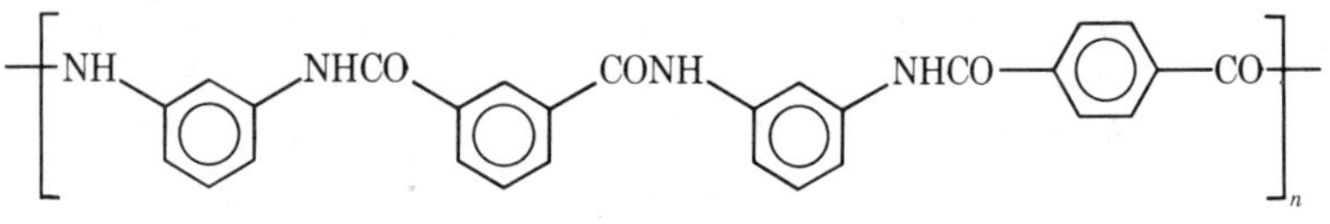

(MAPI-T)

Polycondensation of MPMAB with terephthaloyl chloride yields MPMAB-T and polycondensation of MAPI with isophthaloyl chloride yields MPD-I (Nomex).

Diamines, such as MPMAB, were prepared by reaction of aromatic diamines with nitroaroyl chlorides followed by reduction of the dinitro intermediate. Diamines, such as MAPI, may be prepared by reaction of nitroaniline with a diacid chloride followed by reduction.

A series of copolyamides having somewhat less order (the so-called copolymers of limited order) may be prepared (22) by the low temperature polycondensation of diaminobenzanilides with diacid chlorides; both interfacial and solution methods have been investigated (22). Because the diamine moiety can enter the polymer chain head-to-head or head-to-tail fashion, the diamine moiety is placed in parentheses within the polymer repeat unit.

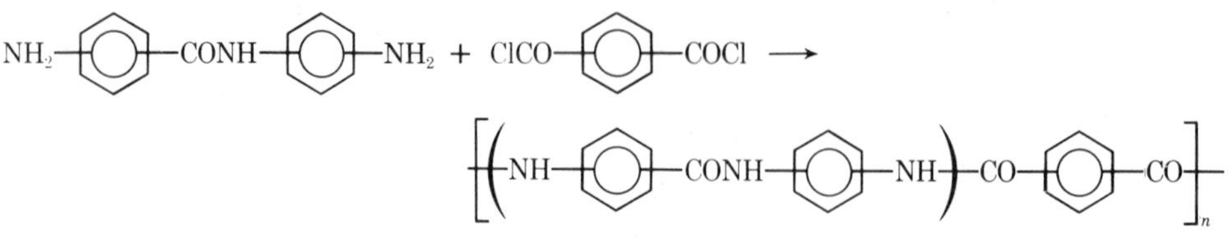

AA–BB Polyamides from Intermediates Containing Preformed Heterocyclic Units. Polymers characterized structurally by an ordered sequence of amide and heterocyclic units in the polymer chain may be prepared by low temperature solution polycondensation of an aromatic diacid chloride and an aromatic diamine, one or both of which contain a heterocyclic unit (57). The preparation of these polymers may be illustrated by the following examples employing heterocyclic diamines:

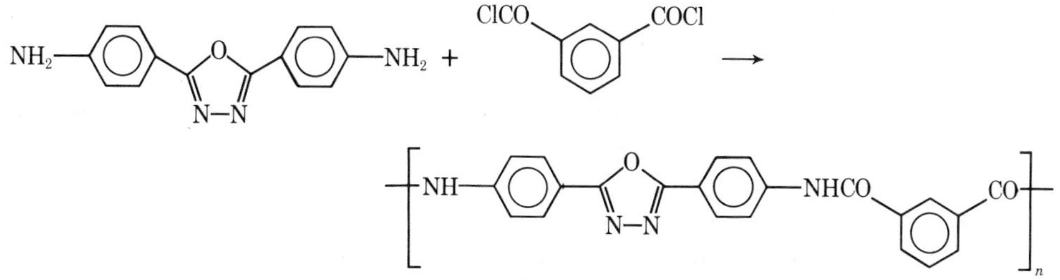

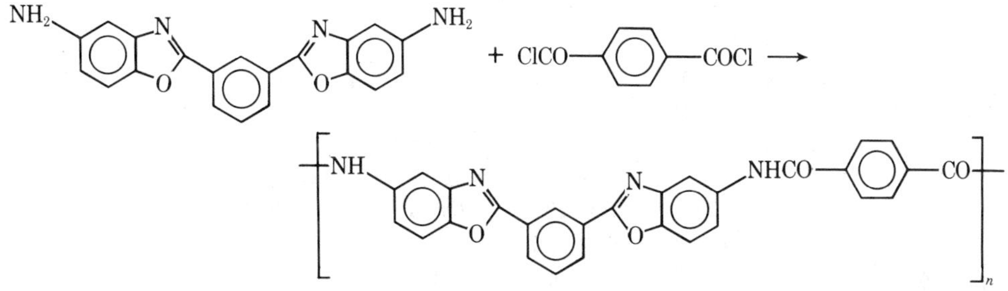

Diacid chlorides containing preformed heterocycles, eg, oxadiazole groups or imide groups, can also be used to prepare amide–heterocyclic copolymers.

A series of copolyamides that probably have some degree of disorder along the chain (amide-heterocyclic polymers of limited order) can be prepared by the low temperature polycondensation of unsymmetrical heterocyclic diamines. The preparation of such polymers may be illustrated by the following example:

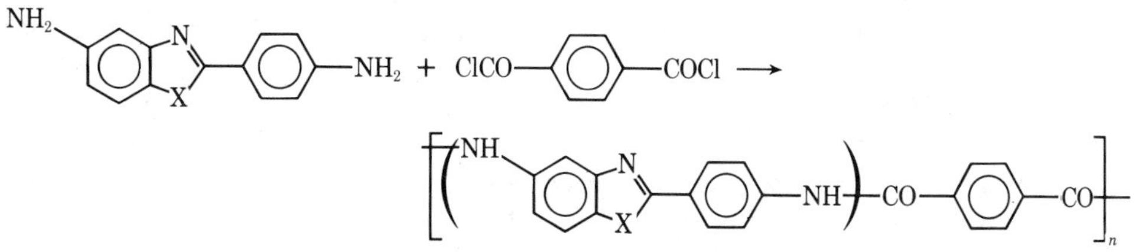

X = —O—, —S—, —NH—

Polymers of this type and copolymers with p-phenylenediamine (PPD) based on a benzimidazolediamine appear to have been employed by Soviet workers to make a high-strength–high-modulus fiber.

The ordered amide–heterocyclic copolymers and the amide–heterocyclic copolymers of limited order appear to fall into the definition of aramids, but they are not discussed further.

Production

All wholly aromatic polyamide fibers reported here have been spun from solutions of the polymers. Both dry-spinning and wet-spinning techniques appear to be equally satisfactory for fiber formation from polymers that are soluble in organic solvents although properties may vary with the method of spinning. The mol wt of a given polymer generally should be higher when spun by the wet method and should be 60,000 or greater for the best balance of initial tensile properties and retention of tensile properties after the fiber is placed in service. Polyamides prepared by solution polycondensation can be conveniently spun from the resulting solution (provided that it is stable), circumventing the need for isolating and redissolving the polymer. The ability to spin the polymer solution as prepared directly is crucial for those polymers that do not redissolve (or redissolve only with difficulty) in organic solvents after isolation in the dry state and that are not stable in strong acid solvents. Polymers prepared by interfacial polymerization must necessarily be dissolved prior to spinning since solutions are not produced directly by this method.

Almost all wholly aromatic polyamides require the presence of a salt, either organic or inorganic, in the amide solvents to dissolve the polymers or prevent them from precipitating if prepared by solution polycondensation. If organic salts such as DMAC hydrochloride are the solvating salts during the course of polymerization, it is common practice to neutralize the hydrogen chloride produced during polycondensation with an inorganic base, eg, lithium hydroxide or calcium oxide, prior to spinning to prevent corrosion of spinning equipment. In the case of dry spinning (58), the salt remains in the spun fibers and must be leached subsequently in order to obtain fiber with optimum mechanical properties. In wet spinning most of the salt diffuses from the fiber during spinning and thus a separate step for leaching the salt is not required. Fiber containing any residual salt has poorer thermal stability and poorer electrical properties than salt-free fiber.

Wet spinning (11,54) in the conventional manner, immersing the jet in the coagulation bath, and the dry-jet or so-called air-gap method (6,59–60), leaving the dry jet above the coagulation bath, have both been described in the patent literature for the spinning of aromatic polyamides. Wet spinning permits the use of both organic and inorganic solvents. With some inorganic solvents, dry spinning would be impossible.

The yarn must be dried and it is usually hot drawn no matter if spun by a wet or dry process. An important exception to the need for a hot-drawing step is described below. In general, the wet-spinning processes lend themselves better to continuous in-line processing than dry spinning when salts are used.

Certain rod-like polymers, such as poly(*p*-phenyleneterephthalamide) (PPD-T) and poly(*p*-benzamide) (PPB), yield anisotropic solutions when the concentration reaches a critical level (5–6,50). The basis for the anisotropy is the formation of liquid crystals (qv), ie, close-packed aggregates of the rod-like molecules mutually aligned within a given packet. Hot drawing of fibers spun from such solutions is not necessary for the attainment of high strength (6); however, hot drawing of these fibers can result in more than a two-fold increase in initial modulus but at a sacrifice in elongation-to-break without an attendant appreciable increase in tenacity.

Because the rod-like aramid polymer molecules have a very high effective volume in solution, it is not surprising that at relatively low concentrations in polar organic solvents and in inorganic solvents (such as sulfuric and hydrofluoric acids), viscous solutions are obtained. For certain of these polymers it has been demonstrated (61) that the viscosity of spinning solutions (dopes) may be *decreased* by *increasing* the polymer solids concentration above a certain critical value by changing the solution from the isotropic to the anisotropic state.

The importance of using anisotropic dopes instead of isotropic ones for the spinning of certain types of polymers has been recognized for some time (62–63). More recently, this principle has been applied to the spinning of aramids.

Economic Aspects

The two types of aramid fibers—heat-resistant fibers, eg, Nomex and Conex, and ultra high-strength–high-modulus fibers, eg, Kevlar and Arenka—are relatively expensive compared to other synthetic fibers, such as nylon and polyester, which are produced in very large volume. The reasons for this are the relatively low volume production of both types of aramid fibers, the high cost of the monomers employed, and high processing costs. The announced annual DuPont capacity for Nomex was about 4500 metric tons; of the actual production, more than one-half went to the manufacture of electrical insulating paper. Expansion of Nomex production to more than 9100 t/yr was completed in 1975. The announced capacity of the Kevlar facilities is 23,000 t/yr and semiworks production was reported to be 27,000 t/yr in 1975.

The price of Nomex bright, 2.2×10^{-7} kg/m (2 den) continuous filament yarns, ie, $2.2–13.3 \times 10^{-5}$ kg/m (200–1200 den), in early 1975 was in the range of $16.28–19.58/kg, and that of $1.7–2.2 \times 10^{-7}$ kg/m (1.5–2.0 den) staple was in the range of $9.24–9.39/kg. The price of 1.1×10^{-6} kg/m (10 den) Nomex staple for use in carpets was $10.45/kg. The price of Kevlar tire cord currently is about $8.69/kg and the prices of Kevlar-29 and Kevlar-49 vary widely according to yarn denier. The price of Kevlar-29 varies from a low of $17.75/kg for 1.6×10^{-4} kg/m (1420 den) yarn to $49.61/kg for 2.1×10^{-5} kg/m (190 den) yarn. For Kevlar-49, the price range is about $19–60/kg for yarns ranging from 1.6×10^{-4} kg/m (1420 den) to 2.1×10^{-5} kg/m (190 den).

Although the annual production of aramid fibers is not great, the total dollar volume is nevertheless substantial and rising. Because of rather high monomer and processing costs coupled with low volume production, aramid fibers are relatively expensive and probably will remain so compared to other synthetic fibers which are produced in very large volume.

The introductory price of Kevlar tire cord was $6.27/kg but was increased in 1975 to $6.93/kg and in early 1977 to $8.69/kg. PRD-49 (now Kevlar-49) was introduced at a price of about $220/kg, but the price was soon cut to $40–60/kg; current (1977) prices for Kevlar-29 and Kevlar-49 have ranged from about $18–60/kg. Further price reductions might be expected with high volume production.

Health and Safety Factors

Toxicology. In recent inhalation studies at the Haskell Laboratory for Toxicology and Industrial Medicine of the DuPont Company, hexamethylphosphoric triamide (HPT) used in the preparation of rod-like polyamide, eg, poly(*p*-phenyleneterephthalamide) (PPD-T), has been found to be carcinogenic in rats (64). Accordingly, any work with HPT should be carried out in an efficient hood to avoid the inhalation of vapors. Since HPT is absorbed by the skin, care should be taken to avoid contact with HPT. Despite the fact that the HPT solvent is washed from the PPD-T prior to drying the polymer and despite the fact that the fibers spun from sulfuric acid are rigorously washed with water, some HPT solvent apparently persists in the fiber. However, because HPT is so tightly bound (most likely by hydrogen bonding to the amide linkage), it seems unlikely that it would be readily lost from the fiber and present an inhalation problem.

Kevlar presents no evidence of skin sensitization when tested on guinea pig skin under occluded conditions or in a standard 200-subject prophetic patch test on human skin, also under occluded conditions (32). In the human patch test, no skin irritation was observed after 48 h of continuous contact. Some irritation, probably from mechanical causes, was observed after 144 h of continuous occluded contact (32). No adverse mechanical or chemical skin effects are to be expected from the usual industrial handling of either Kevlar or Nomex.

For Nomex, and especially for Kevlar, as with any fine, fibrous material, inhalation of the "fly" generated in certain textile handling operations should be avoided.

Safety. Because of its unusual high strength and cutting resistance, caution should be exercised in splicing, handling, or cutting Kevlar (32). Manual cutting and splicing should be attempted only with stationary yarns to avoid possible injury from entanglement in moving yarn or fabric (32).

Applications

Typical applications for aramid fibers are listed according to the distinctive properties of these fibers. A particularly useful discussion of the relationship of Kevlar properties to applications is in a paper by Wilfong and Zimmerman (65).

Heat resistance. Filter bags for hot stack gases (see Air pollution control methods); press cloths for industrial presses (eg, application of permanent press finishes to cotton and cotton–polyester garments); home ironing board covers; sewing thread for high-speed sewing; insulation paper for electrical motors and transformers (see Insulation, electric; Synthetic paper); braided tubing for insulation of wires; paper-makers' dryer belts.

Flame resistance. Industrial protective clothing (eg, pants, shirts, coats, and smocks for workers in laboratories, foundries, chemical plants, and petroleum refineries); welder's clothing and protective shields; fire department turnout coats, pants, and shirts; jump suits for forest fire fighters; flight suits for pilots of the Armed Services (specified by the United States Army, Navy, and Air Force); auto racing drivers' suits; pajamas and robes, particularly for persons who are nonambulatory; mailbags; carpets, upholstery, and drapes (specified on some aircraft and on all ships of the United States Navy); cargo covers, boat covers, and tents (see Flame-retardant textiles).

Dimensional stability. Reinforcement of fire hose and V belts by aramid fibers of moderately high modulus, eg, Nomex; conveyor belts.

Ultra high-strength–high-modulus. Tire cord for use in tire carcasses and as the belt in bias-belted and radial-belted tires (see Tire cords); V belts; cables; parachutes; body armor; and in rigid reinforced plastics in general and more specifically in high performance boats and aircraft (interior trim, exterior fairings, control surfaces, structure) and other vehicles of transportation; in radomes and antenna components; in circuit boards; in filament-wound vessels; in fan blades; in sporting goods (skis, golf clubs, surf boards).

Electrical resistivity. Electrical insulation (paper and fiber).

Chemical inertness. Use in filtration (qv).

Permselective properties. Hollow-fiber permeation separation membranes to purify sea water, brackish water, or to make separation of numerous types of salts and water (see Hollow-fiber membranes) (65–67).

BIBLIOGRAPHY

1. L. K. McCune, *Text. Res. J.* **32**(9), 762 (1962).
2. T. Ono, *Jpn. Text. News* **243,** 71 (Feb. 1975).
3. J. W. Hannell, *Polym. News* **1**(1), 8 (1970).
4. R. E. Wilfong and J. Zimmerman, *J. Appl. Polym. Sci.* **17,** 2039 (1973).
5. U.S. Pat. 3,671,542 (June 20, 1972), S. L. Kwolek (to E. I. du Pont de Nemours & Co., Inc.).
6. U.S. Pat. 3,767,756 (Oct. 23, 1973), H. Blades (to E. I. du Pont de Nemours & Co., Inc.).
7. U.S. Pat. 3,287,324 (Nov. 22, 1966), W. Sweeny (to E. I. du Pont de Nemours & Co., Inc.).
8. U.S. Pat. 3,472,819 (Oct. 14, 1969), C. W. Stephens (to E. I. du Pont de Nemours & Co., Inc.).
9. A. S. Semenova and E. A. Vasil'eva-Sokolova, *Fibre Chem. (USSR)* **5,** 470 (1969).
10. U.S. Pat. 3,049,518 (Aug. 14, 1962), C. W. Stephens (to E. I. du Pont de Nemours & Co., Inc.).
11. U.S. Pat. 3,079,219 (Feb. 26, 1963), F. W. King (to E. I. du Pont de Nemours & Co., Inc.).
12. H. E. Hinderer, R. W. Smith, and J. Preston, *Appl. Polym. Symp.* **21,** 1 (1973).
13. Belg. Pat. 569,760 (1958), (to E. I. du Pont de Nemours & Co., Inc.).
14. E. P. Krasnov and co-workers, *Fibre Chem. (USSR)* **5,** 28 (1972).
15. U.S. Pat. 3,094,511 (June 18, 1963), H. W. Hill, Jr., S. L. Kwolek, and W. Sweeny (to E. I. du Pont de Nemours & Co., Inc.).
16. U.S. Pat. 3,354,123 (Nov. 21, 1967), P. W. Morgan (to E. I. du Pont de Nemours & Co., Inc.).
17. *Khim. Volokna* (4), 78 (1969).
18. J. Preston, R. W. Smith, and S. M. Sun, *J. Appl. Polym. Sci.* **16,** 3237 (1972).
19. H. E. Hinderer and J. Preston, *Appl. Polym. Symp.* **21,** 11 (1973).
20. J. Preston, *J. Polym. Sci. A-1* **4,** 529 (1966).
21. J. Preston, R. W. Smith, and C. J. Stehman, *J. Polym. Sci. C.* **19,** 29 (1967).
22. J. Preston and R. W. Smith, *J. Polym. Sci. B* **4,** 1033 (1966).
23. *Properties of Nomex High Temperature Resistant Nylon Fiber, NP-33 Bulletin,* E. I. du Pont de Nemours & Co., Inc., Wilmington, Del., Oct. 1969.
24. J. O. Weiss, H. S. Morgan, and M. R. Lilyquist, *J. Polym. Sci. C* **19,** 29 (1967).
25. F. Dobinson and J. Preston, *J. Polym. Sci. A-1* **4,** 2093 (1966).
26. J. Preston, *Polym. Eng. Sci.* **16**(5), 298 (1976).
27. B. S. Sprague, *paper presented at the 163rd National ACS Meeting, Boston, Mass., Apr. 1972.*
28. C. E. Hathaway and C. L. Early, *Appl. Polym. Symp.* **21,** 101 (1973).
29. J. C. Shivers and R. A. A. Hentschel, *Text. Res. J.* **44**(9), 665 (1974).
30. U.S. Pat. 3,519,355 (Apr. 13, 1976), B. R. Baird (to E. I. du Pont de Nemours & Co., Inc.).
31. *Dyeing and Finishing Nomex Type 450 Aramid, Bulletin NX-1,* E. I. du Pont de Nemours & Co., Inc., Wilmington, Del., May 1976.
32. *Properties of Industrial Filament Yarns of Kevlar Aramid Fiber for Tires and Mechanical Rubber Goods, Bulletin K-1,* E. I. du Pont de Nemours & Co., Inc., Wilmington, Del., Dec. 1974.
33. *Khim. Volokna* (6), 20 (1972).
34. T. S. Sokolova and co-workers, *Khim. Volokna* (3), 25 (1974).
35. G. I. Kudryavtsev and co-workers, *Khim. Volokna* (6), 70 (1974).
36. J. Preston, H. S. Morgan and W. B. Black, *J. Macromol. Sci. Chem.* **A7**(1), 325 (1973).
37. U.S. Pat. 3,600,350 (Aug. 17, 1971), S. L. Kwolek (to E. I. du Pont de Nemours & Co., Inc.).
38. U.S. Pat. 3,869,429 (Mar. 4, 1975), H. Blades (to E. I. du Pont de Nemours & Co., Inc.).
39. U.S. Pat. 3,296,201 (Jan. 3, 1967), C. W. Stephens (to E. I. du Pont de Nemours & Co., Inc.).
40. U.S. Pat. 3,804,791 (Apr. 16, 1974), P. W. Morgan (to E. I. du Pont de Nemours & Co., Inc.).
41. U.S. Pat. 3,932,365 (Jan. 13, 1976), R. Penisson (to Rhône-Poulenc-Textile).
42. H. C. Bach and H. E. Hinderer, *Polym. Prepr. Am. Chem. Soc. Div. Polym. Chem.* **11**(1), 334 (1970).
43. U.S. Pat. 3,770,704 (Nov. 6, 1973), F. Dobinson (to Monsanto Company).
44. U.S. Pat. 3,738,964 (June 12, 1973), F. Dobinson and F. M. Silver (to Monsanto Company).

45. N. Yamazaki, M. Matsumoto, and F. Higashi, *J. Polym. Sci. Polym. Chem. Ed.* **13,** 1373 (1975).
46. P. W. Morgan, *Condensation Polymers: By Interfacial and Solution Methods,* Interscience Publishers, a division of John Wiley & Sons, Inc., New York, 1965.
47. A. A. Federov and co-workers, *Viskomol Soyedin Ser. B* **15**(1), 74 (1973).
48. J. Preston, *Polym. Eng. Sci.* **15**(3), 199 (1975).
49. T. I. Bair, P. W. Morgan, and F. L. Killian, *Polym. Prepr. Am. Chem. Soc. Div. Polym. Chem.* **17**(1), 59 (1976).
50. S. P. Papkov and co-workers, *J. Polym. Sci. Polym. Phys. Ed.* **12,** 1753 (1974).
51. U.S. Pat. 3,850,888 (Nov. 26, 1974), J. A. Fitzgerald and K. K. Likhyani (to E. I. du Pont de Nemours & Co., Inc.).
52. U.S. Pat. 3,203,933 (Aug. 31, 1965), W. A. Huffman, R. W. Smith, and W. T. Dye, Jr., (to Monsanto Company).
53. U.S. Pat. 3,225,011 (Dec. 21, 1965), J. Preston and R. W. Smith (to Monsanto Company).
54. U.S. Pat. 3,541,056 (Nov. 17, 1970), J. Pikl (to E. I. du Pont de Nemours & Co., Inc.).
55. U.S. Pat. 2,831,834 (Apr. 22, 1958), E. E. Magat (to E. I. du Pont de Nemours & Co., Inc.).
56. U.S. Pat. 3,063,966 (Nov. 13, 1962), S. L. Kwolek, P. W. Morgan, and W. R. Sorensen (to E. I. du Pont de Nemours & Co., Inc.).
57. J. Preston and W. B. Black, *J. Polym. Sci. B* **4,** 267 (1966).
58. U.S. Pat. 3,360,598 (Dec. 26, 1967), C. R. Earnhart (to E. I. du Pont de Nemours & Co., Inc.).
59. U.S. Pat. 3,414,645 (Dec. 3, 1968), H.S. Morgan (to Monsanto Company).
60. U.S. Pat. 3,642,706 (Feb. 15, 1972), H. S. Morgan (to Monsanto Company).
61. S. W. Kwolek and co-workers, *Polym. Prepr. Am. Chem. Soc. Div. Polym. Chem.* **17**(1), 53 (1976).
62. U.S. Pat. 3,089,749 (May 14, 1963), D. G. H. Ballard (to Courtaulds, Ltd.).
63. U.S. Pat. 3,121,766 (Feb. 18, 1964), D. G. H. Ballard and J. D. Griffths (to Courtaulds, Ltd.).
64. J. A. Zapp, Jr., *Science* **190,** 422 (1975).
65. R. E. Wilfong and J. Zimmerman, *Appl. Polym. Symp.* **31,** 1 (1977).
66. U.S. Pat. 3,567,632 (Mar. 2, 1971), J. W. Richter and H. H. Hoehn.
67. U.S. Pat. 3,775,361 (Nov. 27, 1973), J. H. Jensen (to E. I. du Pont de Nemours & Co., Inc.).

General References

O. E. Snider and R. J. Richardson, "Polyamide Fibers" in N. Bikales, ed., *Encyclopedia of Polymer Science and Technology,* Vol. 10, Interscience Publishers, a division of John Wiley & Sons, Inc., New York, 1969, pp. 347–460.

W. Sweeny and J. Zimmerman, "Polyamides" in N. Bikales, ed., *Encyclopedia of Polymer Science and Technology,* Vol. 10, Interscience Publishers, a division of John Wiley & Sons, Inc., New York, 1969, pp. 483–597.

W. B. Black and J. Preston, "Fiber-Forming Aromatic Polyamides" in H. F. Mark, S. M. Atlas, and F. Cernia, eds., *Man-Made Fibers: Science and Technology,* Vol. 2, Interscience Publishers, a division of John Wiley & Sons, Inc., New York, 1968, pp. 297–364.

W. B. Black, "Wholly Aromatic High-Modulus Fibers" in C. E. H. Bawn, ed., *MTP Int. Rev. Sci., Phys. Chem. Series 2, Macromol. Sci. Vol.,* Chapt. 2, Butterworths, London, 1975, pp. 34–122.

W. B. Black and J. Preston, eds., *High-Modulus Wholly Aromatic Fibers,* Marcel Dekker, Inc., New York, 1973.

G. B. Carter and V. T. J. Schenk, "Ultra-High Modulus Organic Fibres" in I. M. Ward, ed., *Structure and Properties of Oriented Polymers,* Halsted Press, a division of John Wiley & Sons, Inc., New York, 1975, pp. 454–491.

J. Preston, "Synthesis and Properties of Rod-Like Condensation Polymers" in A. Blumstein, ed., *Liquid Crystalline Order in Polymers,* Academic Press, Inc., New York, 1978.

J. PRESTON
Monsanto Triangle Park Development Center, Inc.

ASBESTOS

Asbestos is a generic term describing a variety of naturally formed hydrated silicates that, upon mechanical processing, separate into mineral fibers. There are two fundamental varieties of asbestos: serpentine and the amphiboles. Serpentine asbestos is known as chrysotile and the amphiboles include five species identified as anthophyllite, amosite, crocidolite, actinolite, and tremolite. Each of these varieties of asbestos differ from each other chemically as illustrated in Table 1.

Asbestos fibers are unique minerals combining unusual physical and chemical properties which make them useful in the manufacture of a wide variety of residential and industrial products. Of mineral origin, asbestos does not burn, does not rot, and, dependent on variety, possesses extremely high tensile strength as well as resistance to acids, bases, and heat. Similarly, when processed into long, thin fibers, asbestos is sufficiently soft and flexible to be woven into fire-resistant fabrics.

Historical records show that asbestos has been known for more than 2000 years. Applications of this noncombustible fiber are mentioned by Plutarch and Pliny, particularly with reference to asbestos textiles used for cremation cloths, oil lamp wicks, etc.

The asbestos industry *per se* had its inception in the 18th century in the Russian Ural mountains and by the mid-19th century both Italian chrysotile and tremolite varieties were mined and processed into commercial products. At the same time asbestos was discovered and mined on a commercial scale at Thetford Asbestos in Quebec, Canada. To this day, these Canadian and Russian locations are the major producers of chrysotile asbestos.

The amphibole asbestos industry is of more recent origin. Blue asbestos, crocidolite, was discovered in South Africa about 1803 to 1806, but it was not until 1893 that this variety was commercially exploited. Production, sale, and use of the amosite species from the Transvaal followed in the early 20th century (1).

Origin and Occurrence

Through the years, the origin of asbestos has been the subject of extensive geological research. Serpentine asbestos occurs under widely differing geological conditions from the amphiboles. Similarly, the modes of occurrence or the manner in which the fibers are physically imbedded in the host rock also differ widely. The current opinion is that chrysotile fiber resulted from two separate metamorphic reactions in ultrabasic

Table 1. Asbestos

Species	CAS Registry Number	Variety	Chemical composition
chrysotile[a]	[*12007-29-5*]	serpentine	$3MgO.2SiO_2.2H_2O$
anthophyllite	[*17068-78-9*]	amphibole	$7MgO.8SiO_2.H_2O$
amosite[a]	[*12172-73-5*]	amphibole	$11FeO.3MgO.16SiO_2.2H_2O$
actinolite	[*13768-00-8*]	amphibole	$2CaO.4MgO.FeO.8SiO_2.H_2O$
tremolite	[*14567-73-8*]	amphibole	$2CaO.5MgO.8SiO_2.H_2O$
crocidolite[a]	[*12001-28-4*]	amphibole	$Na_2O.Fe_2O_3.3FeO.8\ SiO_2.H_2O$

[a] Asbestos species of major commercial significance.

rocks of igneous origin. The initial hydrothermal reaction altered the olivines and pyroxenes to serpentine. At a subsequent point the serpentine was redissolved and the mineral-rich solutions flowed into cracks and crevices in the host rock where chrysotile fiber was reprecipitated.

In this reprecipitation process, asbestos fiber was usually deposited in a cross-vein mode of occurrence; ie, the fiber is arranged perpendicular to the wall rock as illustrated in Figure 1.

In some cases, chrysotile was either deposited or affected by earth movements such that the fibers lie principally parallel to the wall rock as illustrated in Figure 2. This mode of occurrence is referred to as slip fiber.

The third and unusual mode of occurrence of chrysotile is referred to as a massive or agglomerated form wherein the fibers have been deposited as platelets having no specific fiber orientation. This unique formation has been found and commercially mined in the New Idria serpentine deposits of California (2) and at Stragari, Yugoslavia. In these cases the asbestos content of the ore is abnormally high, but the fiber length is very short as compared to commercially useful cross-vein or slip fiber deposits. This mode of occurrence is illustrated in Figure 3.

The origin of the amphibole varieties of asbestos are not as clearly defined as those of chrysotile. The two commercially significant amphibole fibers, crocidolite and amosite, occur in metamorphosed sedimentary strata known as banded ironstones. These sedimentary formations vary considerably in composition which accounts for the compositional variations of the associated amphibole fibers (1).

Figure 1. Cross-vein chrysotile (Courtesy of Johns-Manville Research Center).

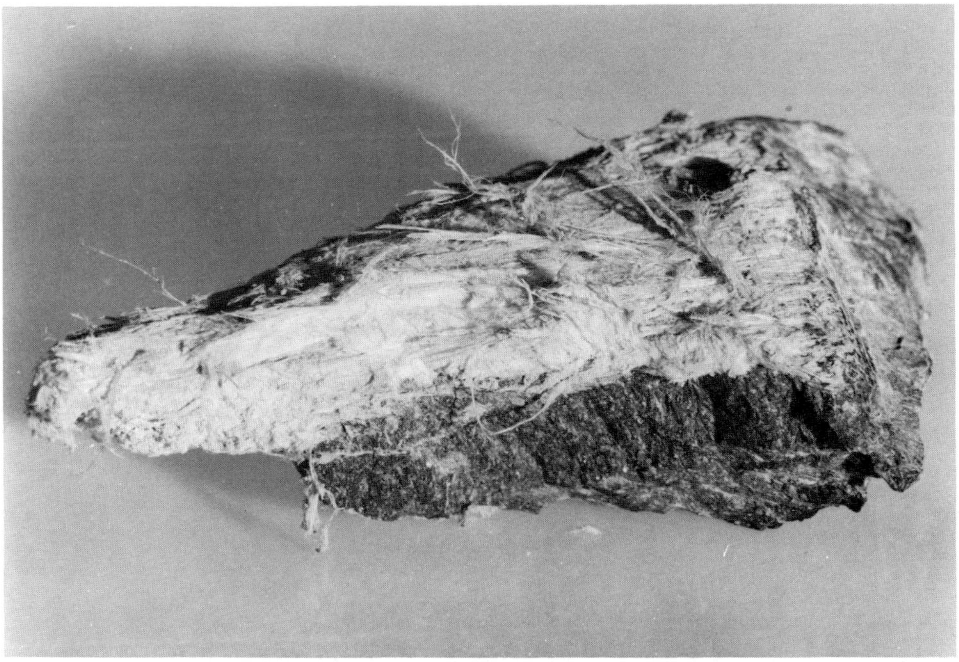

Figure 2. Slip fiber chrysotile (Courtesy of Johns-Manville Research Center).

Crocidolite and amosite species of asbestos occur in cross-vein modes and the anthophyllite and tremolite species often occur in slip modes although both of the latter can also occur in a massive mode wherein the fibers are nonoriented. Actinolite, a noncommercial species of amphibole asbestos, is most often found as brittle, acicular, or bladed forms in a massive mode.

In its generic connotation, asbestos is a mineral found all over the world. In most cases, however, the asbestos species most easily located are those having limited or no commercial utility such as actinolite, anthophyllite, or tremolite. In 1977, Canada and Russia are the major producers of chrysotile asbestos, and Africa, China, and the United States provide modest quantities. The major producing area of the important amphiboles, crocidolite, and amosite is Africa. In fact, the only deposits of amosite known to date are in the Eastern Transvaal.

Chrysotile Asbestos

Crystal Structure. The mineral species associated with the serpentine group, serpentine [*12168-92-2*], chrysotile, lizardite [*12161-84-1*], and antigorite [*12135-86-3*], although differing structurally, are compositionally almost identical. All have the approximate chemical composition of $Mg_3(Si_2O_5)(OH)_4$. The crystal structure of chrysotile is layered or sheeted similarly to the kaolinite group [*1318-74-7*] (3–4). It is based on an infinite silica sheet $(Si_2O_5)_n$ in which all the silica tetrahedra point one way (see Silica). On one side of the sheet structure, and joining the silica tetrahedra, is a layer of brucite, $Mg(OH)_2$, in which two out of every three hydroxyls are replaced by oxygens at the apices of the tetrahedra. The result is a layered structure illustrated in Figure 4 (1,5).

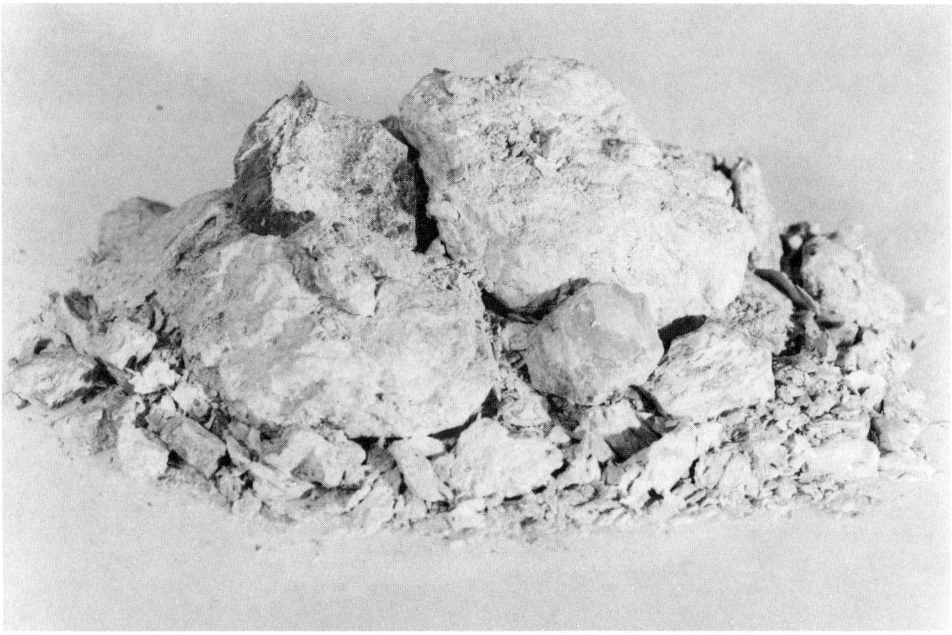

Figure 3. New Idria (Coalinga) chrysotile platelets (Courtesy of Johns-Manville Research Center).

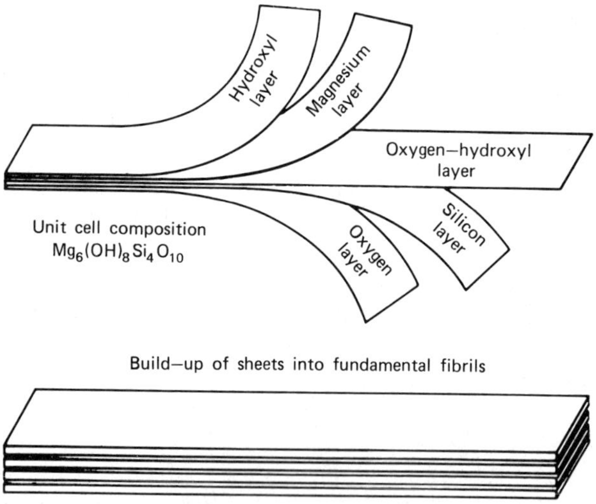

Figure 4. Fundamental sheet of a chrysotile structure.

Mismatches and strains between the layers cause the structure to curve and form cylinders or fibers (6–7). Individual chrysotile fibers have ultimate diameters of 0.02–0.03 μm. X-ray and electron microscope studies have confirmed this cylindrical form and diameter range. In fact, the first electron microphotographs indicated a tubular structure as illustrated in Figure 5 (8–10). This characteristic appearance has now become one of the more definitive identification techniques for chrysotile.

Figure 5. Electron photomicrograph (80,000×) of chrysotile (Courtesy of Johns-Manville Research Center).

Electron microphotographs have shown most chrysotile fibers with a hollow cylindrical form and a single magnesia–silica sheet rather than the earlier double-layer concept (11–12). The lattice planes have a multispiral arrangement as suggested by earlier x-ray studies (13).

Chemical and Surface Properties. Chrysotile asbestos is a naturally formed mineral. The chemical compositions vary somewhat, depending on deposit location, from the idealized composition of $Mg_3(Si_2O_5)(OH)_4$. Chemical analyses of chrysotile range approximately as follows: SiO_2, 37–44%; MgO, 39–44%; FeO, 0–6.0%; Fe_2O_3, 0.1–5.0%; Al_2O_3, 0.2–1.5%; CaO, trace–5.0%; H_2O, 12.0–15.0%. Variations in chemical analyses may be due to either associated mineral impurities or to isomorphic substitutions in the crystal lattice. Common mineral impurities found in commercial grades of chrysotile from various locations include magnetite, chromite, brucite, calcite, dolomite, and awaruite. Within the chrysotile lattice, nickel and iron can occur as minor isomorphic substitutions for magnesium (14–17). Chrysotile, a hydrated silicate, is subject to thermal decomposition at elevated temperatures. This thermal decomposition is a two-stage reaction consisting first of a dehydroxylation phase and then a structure phase change. Dehydroxylation or the loss of water occurs at 600–780°C. At 800–850°C the anhydride breaks down to forsterite and silica. These reactions are irreversible and are illustrated by the typical differential thermal analysis shown in Figure 6.

Structural changes in chrysotile can occur under conditions of intense grinding. As a result of these effects the structure can become amorphous and no longer identifiable by either x-ray diffraction or electron microscopy. These structural changes apparently occur because of localized temperature surges in the fibrils with accompanying dehydroxylation as they absorb the tremendous impact energies, eg, extensive dry ball milling results in an amorphous mass and wet ball milling results in short fibrils that retain their crystallinity (16).

Because of its hydroxyl outer layer, chrysotile is readily attacked by acid and will, ultimately, completely dissolve the magnesium component, leaving essentially a fibrous, but fragile, silica structure. Similarly, because of its alkaline surface, chrysotile is not readily attacked by caustic solutions except under conditions of extreme alkali concentration and elevated temperatures (18–19).

Dispersions of chrysotile fiber in carbon dioxide-free distilled water exhibit al-

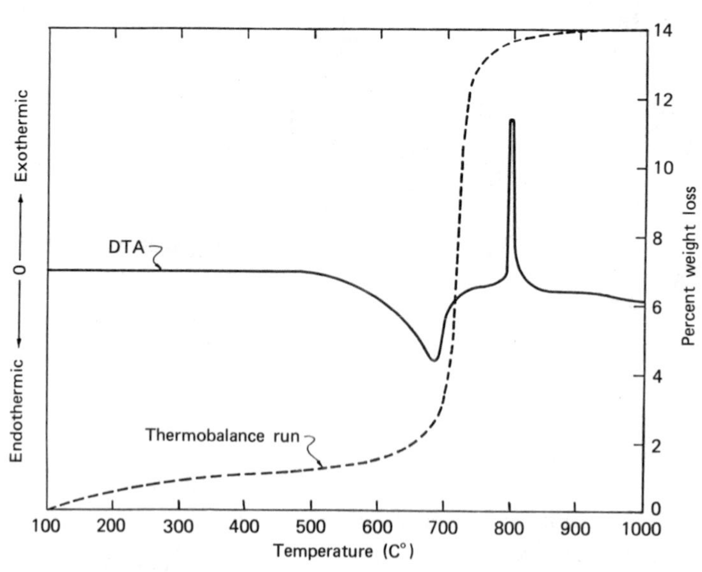

Figure 6. Typical differential thermal analysis of chrysotile.

kaline properties. Such suspensions will reach a pH of 10.33 as with magnesium hydroxide suspensions tested under the same conditions. Solubility product constants for chrysotile fibers range from 1.0×10^{-11} to 3×10^{-12}, indicative of the magnesium hydroxide outer layer (20–22).

The electrokinetic behavior of chrysotile is also related to its surface characteristics. Normally, below its isoelectric point of approximately pH 11–12, chrysotile exhibits a positive charge. Above this pH range, it demonstrates a negative charge. Exceptions to this general behavior have been noted with chrysotile fibers from certain locations (23).

Because of its very small fiber diameter, its high specific surface area and its relatively reactive surface, chrysotile is a selectively adsorptive material. Commercial grades of chrysotile adsorb as much as 2–3 wt % moisture from saturated air. Adsorption studies of a variety of organic compounds from both vapor and liquid media show that chrysotile has a greater affinity for polar molecules. Heats of adsorption have been measured ranging from 38 kJ/g (9 kcal/g) for hexane to 67 kJ/g (16 kcal/g) for water (5,24–27). Chrysotile also adsorbs iodine from solutions in a manner similar to magnesium hydroxide or brucite. This adsorption characteristic is often used as a staining technique for the detection of chrysotile asbestos (28).

Physical Properties. The common physical properties of chrysotile asbestos are given in Table 2.

Asbestos fibers are used in composite materials (qv) to provide reinforcement. Tensile strength of the fiber is, therefore, an important and highly significant physical property. Unfortunately, because of their extremely fine diameter and the complicating factor of the effect of sample length on strength determinations, it is extremely difficult to measure the tensile strengths of asbestos with precision. Most recent information indicates typical chrysotiles have tensile strengths in the order of 3727 MPa (5.4×10^5 psi) which exceeds corresponding values for steel piano wire and fiber glass. A comparison of typical strength values for the different asbestos varieties is given in Table 3. Since all these measured values are far less than the theoretical value of over 10,000 MPa (1.45×10^6 psi) attributable to silicate chain structures, the values given should be considered as relative for the different varieties rather than specific (1,29–31).

Physical strengths of asbestos are adversely affected by elevated temperatures. Table 4 shows decreasing tensile strength as dehydroxylation takes place (1).

What visually appears to be a single fiber in commercial grades of asbestos is in actuality a bundle of a large number of individual fibrils. These bundles can be subdivided into a multitude of finer bundles, but only with special processing can a large portion of fiber mass be divided to its ultimate fibril diameter. The specific surface areas of commercial asbestos fibers vary with the extent of mechanical defibrillation. Surface areas by nitrogen adsorption tests on samples teased by hand from chrysotile crude are 4–12 m^2/g; however, when aggressively milled or fiberized, surface areas of 30–50 m^2/g result (32). Chrysotile asbestos can be separated into smaller diameter fibrils (higher specific area) more readily under wet processing conditions than under dry mechanical milling. For this reason many asbestos product manufacturing processes utilize wet opening techniques to provide improved fiber reinforcing efficiencies.

The term harshness in the asbestos industry refers to the fiber's brittleness, flexibility, form, and modulus of elasticity. Commercial grades of chrysotile are usually

Table 2. Properties of Asbestos Fibers

	Chrysotile	Anthophyllite	Amosite (ferroanthophyllite)	Crocidolite	Tremolite	Actinolite
structure	in veins of serpentine, etc	lamellar, fibrous asbestiform	lamellar, coarse to fine fibrous and asbestiform	fibrous in iron-stones	long, prismatic and fibrous aggregates	reticulated long prismatic crystals and fibers
mineral association	in altered peridotite adjacent to serpentine, and limestone near contact with basic igneous rocks	in crystalline schists and gneisses	in crystalline schists, etc	in iron-rich silicious argillite in quartzose schists	in Mg limestones as alteration product of highly magnesian rocks, metamorphic and igneous rocks	in limestone and in crystalline schists
origin	alteration and metamorphism of basic igneous rocks rich in magnesium silicates	metamorphic, usually from olivine	metamorphic	regional metamorphism	metamorphic	results of contact metamorphism
veining	cross and slip fibers	slip, mass fiber unoriented and interlacing	cross fiber	cross fiber	slip or mass fiber	slip or mass fiber
essential composition	hydrous silicates of magnesia	Mg silicate with iron	silicate of Fe and Mg, higher iron than anthophyllite	silicate of Na and Fe water	Ca and Mg silicate with some water	Ca, Mg, Fe, silicates, water up to 5%
crystal structure	fibrous and asbestiform	prismatic, lamellar to fibrous	prismatic, lamellar to fibrous	fibrous	long and thin columnar to fibrous	long and thin columnar to fibrous
crystal system	monoclinic (pseudoorthorhombic?)	orthorhombic	monoclinic	monoclinic	monoclinic	monoclinic

color	white, gray, green, yellow	gray–white, brown, gray, or green	ash gray, green, or brown	lavender, blue	gray–white, green, yellow, blue	green
luster	silky	vitreous to pearly	vitreous, somewhat pearly	silky to dull	silky	silky
hardness, Mohs	2.5–4.0	5.5–6.0	5.5–6.0	4	5.5	ca 6
specific gravity	2.4–2.6	2.85–3.1	3.1–3.25	3.2–3.3	2.9–3.2	3.0–3.2
cleavage	010 perfect	110 perfect	110 perfect	110 perfect	110 perfect	110 perfect
optical properties	biaxial positive extinction parallel	biaxial positive extinction parallel	biaxial positive extinction parallel	biaxial extinction inclined	biaxial negative extinction inclined	biaxial negative extinction inclined
refractive index	1.50–1.55	ca 1.61	ca 1.64	1.7 pleochroic	1.61	1.63 weakly pleochroic
fusibility, Seger cones	fusible at 6, 1190–1230°C	infusible or difficult to fuse	fusible at 6, loses water at moderate temperatures	fusible at 3, 1145–1170°C	fusible at 4, 1165–1190°C	fusible at 4, 1165–1190°C
flexibility	very flexible	very brittle, nonflexible	good, less than chrysotile	fair to good	generally brittle, sometimes flexible	brittle and nonflexible
length	short to long	short	5–28 cm	short to long	short to long	short to long
texture	soft to harsh, also silky	harsh	coarse but somewhat pliable	soft to harsh	generally harsh, sometimes soft	harsh
acid resistance	soluble up to approximately 57%	fairly resistant to acids	fairly resistant to acids	fairly resistant to acids	fairly resistant to acids	relatively insoluble in HCl
spinnability	best	poor	fair	fair	generally poor, some are spinnable	poor
specific heat, J/(kg·K) [or Btu/(lb·°F)]	1113 [0.266]	879 [0.210]	908 [0.217]	841 [0.201]	887 [0.212]	908 [0.217]

Table 3. Tensile Strength of Asbestos[a]

	Tensile strength, MPa (psi × 10^3)	Young's modulus, GPa (psi × 10^6)
chrysotile, Arizona, USA	3780 (548)	145 (21.1)
chrysotile, Thetford, Canada	3640 (528)	146 (21.2)
crocidolite, Koegas, Cape Province, Africa	2840 (413)	147 (21.3)
crocidolite, Koegas, Cape Province, Africa	3090 (448)	151 (21.9)
crocidolite, Pomfret, Cape Province, Africa	4660 (675)	169 (24.5)
crocidolite, Pomfret, Cape Province, Africa	3550 (515)	175 (25.3)
crocidolite, Cochabąmbo, Bolivia	1440 (209)	170 (24.6)
amosite, Penge, Transvaal, Africa	2580 (374)	143 (20.8)
amosite, Penge, Transvaal, Africa	1980 (287)	143 (20.8)
anthophyllite, Paakilla, Finland	2450 (356)	156 (22.6)

[a] See ref. 1.

Table 4. Effect of Temperature on Tensile Strength of Asbestos[a]

	Percentage of original tensile strength		
Temperature, °C	Chrysotile	Crocidolite	Amosite
200		100	100
320	91.6	70	58
430	73.3	32	32
550	59.5	20	15
670	32.0		
period of heat soak	3 min	4 h	4 h

[a] See ref. 1.

classed as soft or nonharsh. Commonly, they are silky, of fine diameter, and extremely flexible. Contrary to chrysotile, the commercial amphiboles are harsh fibers. They are relatively stiff, brittle, coarser in diameter, and rodlike in appearance under the microscope. These differing physical characteristics account for the different operating characteristics exhibited in the manufacture of various asbestos-containing products. For example, soft chrysotiles can be more readily spun into textiles than the amphiboles but they have poorer drainage properties when used in wet manufacturing processes. However, several fiber treatments have been developed to improve the filtration or drainage properties of chrysotile and thereby increase the production rate of wet machines manufacturing asbestos–cement products (33–37).

Amphibole Asbestos

Crystal Structure. The structure of all the amphiboles consists of two chains or ribbons based on Si_4O_{11} units separated by a band of cations. Seven cations form the basal unit. Two hydroxyl groups are attached to the central cation in each unit cell. These hydroxyls, unlike the chrysotile structure, are contained entirely within the amphibole structure. The final structure is composed of stacks of these sandwich ribbons as illustrated in Figure 7 (1,5). The bonding between these ribbons is rather weak and the crystals are easily cleaved parallel to the ribbons along cleavage line A-A. If the cleavage is very facile, the result is an asbestiform mineral (1,5). Amphiboles

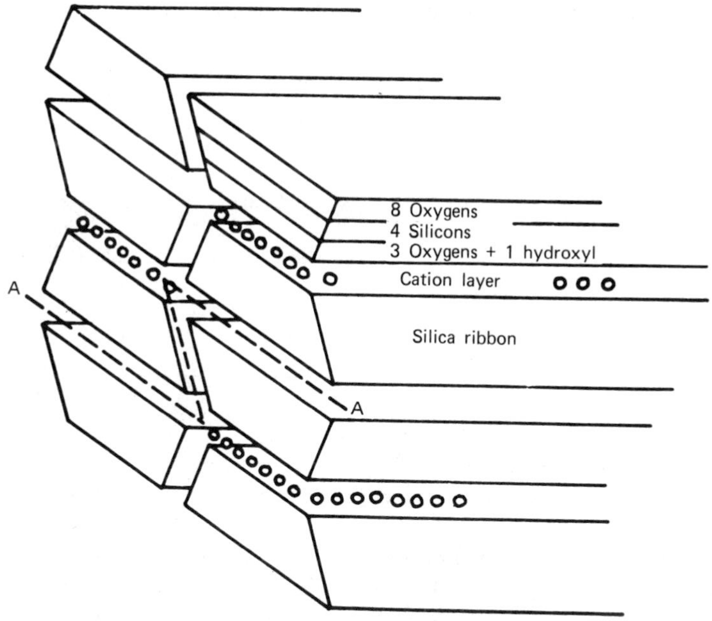

Figure 7. Amphibole structure.

can also occur in nonfibrous forms (38) which may result because of structural disorder (39). The dominant cations are Mg^{2+}, Fe^{2+}, Fe^{3+}, Na^{+}, and Ca^{2+} (see Table 1). Minor isomorphic substitutions of Al^{3+}, Ti^{4+}, K^{+}, and Li^{+} also occur. Because of the wide compositional range, the amphiboles are often assigned to three generic series; ie, the anthophyllite–cummingtonite [*17499-08-0*] series, the calcic amphiboles and the soda amphiboles.

Chemical and Surface Properties. The chemical compositions of the amphibole fibers are more complex and variable than chrysotile. Typical compositions are shown in Table 5.

Like chrysotile, the amphibole asbestos fibers dehydroxylate and decompose at elevated temperatures. The presence of large quantities of iron (particularly ferrous iron) makes the decomposition or thermal analysis determinations particularly complex and very dependent on the composition of the atmosphere. Table 6 is an

Table 5. Typical Chemical Compositions of Amphibole Asbestos

	Crocidolite, %	Amosite, %	Anthophyllite, %	Actinolite, %	Tremolite, %
SiO_2	49–53	49–53	56–58	51–52	55–60
MgO	0–3	1–7	28–34	15–20	21–26
FeO	13–20	34–44	3–12	5–15	0–4
Fe_2O_3	17–20			0–3	0–0.5
Al_2O_3	0–0.2		0.5–1.5	1.5–3	0–2.5
CaO	0.3–2.7			10–12	11–13
K_2O	0–0.4	0–0.4		0–0.5	0–0.6
Na_2O	4–8.5	trace		0.5–1.5	0–1.5
H_2O	2.5–4.5	2.5–4.5	1–6	1.5–2.5	0.5–2.5

Table 6. Decomposition Reactions of Amphiboles Under Neutral Conditions[a]

Amphibole variety	Dehydroxylation DTA peak, °C	Structural breakdown, °C	Decomposition products
crocidolite	610	800	Na–Fe pyroxene, magnetite, silica
amosite	780	600–900	Fe–Mg pyroxene, silica
anthophyllite	950	950	Mg–Fe pyroxene, magnetite, silica
actinolite	1040	1040	CaMgFe pyroxene, silica

[a] See ref. 1.

oversimplification of the thermal decomposition reactions of the amphibole fibers (1,40).

Compared to chrysotile, the amphibole fibers are relatively acid resistant. However, under boiling conditions and high acid concentrations the amphiboles can exhibit weight losses of approximately 2–20%. Relative order of acid resistance is:

tremolite > anthophyllite > crocidolite > actinolite >

amosite >>> chrysotile

Amphibole fibers have a negative charge as contrasted to chrysotile's usual positive charge. The magnitude of the charge exhibited by the amphiboles is substantially lower than chrysotile's.

Physical Properties. See Table 2 for the more common physical properties and characteristics of the various amphibole fibers. In general, amphibole fibers are harsh, springy, and brittle as compared to the chrysotile variety. These physical properties make the amphiboles fast draining and bulky when used in manufacturing processes.

As illustrated in Table 3, the tensile strengths of amphibole asbestos fibers differ widely. The typical tensile strengths of asbestos fibers have the order:

crocidolite > chrysotile > amosite > anthophyllite >

tremolite > actinolite

As shown in Table 4, the amphiboles also lose their strength with increasing temperature.

Amphibole fibers do not divide into fibrils as fine in diameter or as symmetrical as the chrysotile variety. Ultimate diameters of amphiboles have been reported to be about 0.1 μm (1) and the surface areas of amphibole asbestos are considerably smaller than chrysotile. Fully fiberized commercial grades of crocidolite, for example, have surface areas by gas adsorption of 3–15 m^2/g compared to the 30–50 m^2/g values of chrysotile (2,41).

Milling

Imbedded asbestos fibers are removed from the ore by a repeated series of crushing, fiberizing, screening, aspirating, and grading operations (milling). A typical, greatly simplified asbestos mill is shown in Figure 8. The ore is crushed, dried, and fiberized in a variety of impact mills. The short fiber and granular material is removed by screening the fiberized mass. The oversized fractions are stratified on a screen where the spherical, granular material of high density seeks the screen's surface and the fluffy, low-density fiber rises to the top of the bed. At the end of the screen, the fiber is separated from the rock by an aspirating hood. The coarse granular fractions that still contain veins of chrysotile fiber, are refiberized and rescreened to recover shorter fibers. Fibers recovered from these primary screening operations are rescreened to remove entrapped granular material and classified into grades by fiber length.

The recovery of milled asbestos fiber from ore is fairly low. In general, a 5% recovery is typical. The chrysotile mines in California are notable exceptions where 50% recoveries are common from agglomerated ores (Fig. 3). The fibers, however, are very short and are normally sold as reinforcing fillers (qv).

Conventional asbestos milling uses large quantities of air both for separating the fibers as they are freed from the ore and for dust control. Approximately 130 m^3/s (275,000 cfm) are used to process one metric ton of ore (42). Preconcentration of ore includes selective grinding, screening, and magnetic techniques (43–45).

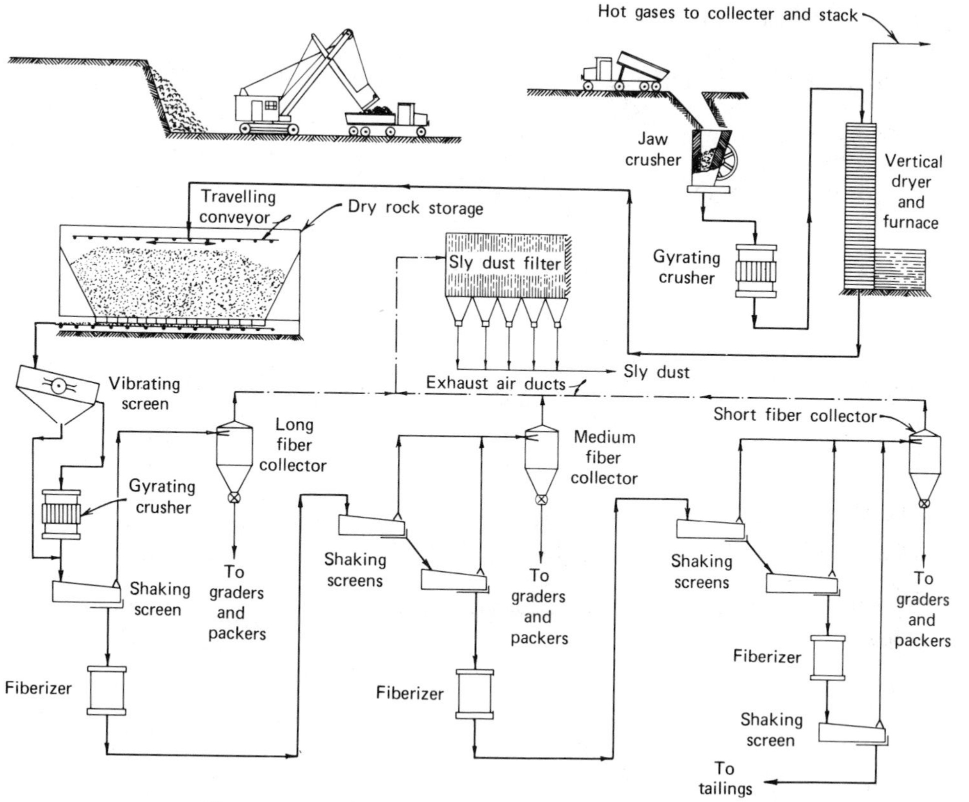

Figure 8. Schematic of a typical asbestos milling flowline.

Production and Usage

In 1977, 80% of the world's chrysotile was produced by Canada and Russia. Most of the Canadian production is from Quebec and the balance from Newfoundland, British Columbia, and the Yukon Territory. Russian production is predominantly in the Bazhenovo District in the Central Urals, the Dzhetygara area northwest of Kazakhstan, and the Aktourak deposits near the Yenisei River.

Crocidolite and Amosite are produced in significant quantities only in South Africa. Main producing areas are at Bosrand, Cornheim, Ouplaas, Owendale, the Kuruman area in the Cape Province, and the Lydenburg District in the Transvaal (46).

Production statistics vary widely with the source of information. Table 7 summarizes ranges given by various authorities (47–53).

The largest use of asbestos is in asbestos–cement for products such as pipes, ducts, and flat and corrugated sheets. Pipe products find use in water supply, sewage disposal, and irrigation systems. Asbestos–cement sheets are used in a wide variety of construction applications. Other uses of asbestos include fire-resistant textiles, friction materials (see Brake linings), underlayment and roofing papers, and floor tiles. Table 8 shows the uses of asbestos in the United States (the largest consumer) and the world. The United States usage patterns differ considerably from the rest of the world (54–56).

Standards and Test Methods

Classification. Canadian chrysotile crudes are classified as follows:

crude #1:	1.9 cm staple and longer
crude #2:	0.95 cm to 1.9 cm staple
crude run of mine:	unsorted crudes

Table 7. 1976 Asbestos Production

Type and location	Thousands of metric tons
Chrysotile	
Canada	1537
USSR	2285
South Africa	111
Rhodesia (estimate)	239
China (estimate)	150
Europe	321
USA	104
South America	69
Australia	66
others	45
Total chrysotile	*4927*
Amphiboles (South Africa)	
crocidolite	178
amosite	79
anthophyllite	2
Total amphiboles	*259*
Total asbestos	*5186*

Table 8. Asbestos Usage by Product Line Estimated Percent of Production

Product line	United States consumption, %	World consumption, %
asbestos–cement	23–30	65–70
asbestos papers	26–38	7–8
floor tile	13–21	4–7
friction materials	6–8	2–3
roof coatings	4–7	2–3
textiles	1–2	1–4
plastics	1–2	trace–1
miscellaneous	12–20	13–15

Milled Canadian fibers sold from Quebec are classified by a dry screen technique known as the Quebec Standard Asbestos Test. This test method grades fibers roughly by fiber or staple length. Minimum test values are guaranteed for each grade and a numerical classification system has been established for fibers ranging from Group 3, the longest grade, to Group 7, the shortest grade.

Chrysotile fibers produced outside Quebec are graded or controlled by screening test methods differing from the Quebec Screen Test; however, these basically identify grade by staple length.

Test Methods. The major properties of concern are length, granular content, degree of openness or effective surface area, drainage or filtration rate, color, absorption, electrical resistivity, bulk density, and strength-giving properties (57).

Fiber length is most reliably measured by wet screening techniques. Fiber openness is commonly measured by air permeability surface area methods. A variety of filtration tests are used to estimate a fiber's operating performance in asbestos–cement wet machines or paper machines. Granular content is often determined by a dry screening test where the fibrous material is aspirated from the granular impurity remaining on the various screen meshes.

The most complex measurement is the determination of the strength-giving property of a fiber for use in asbestos–cement products. In this case, small pressed sheets are made from a slurry of fiber and cement. After curing, the sheet strength is determined and the reinforcing effectiveness or value of the fiber is calculated.

Health and Safety Factors

The inhalation of excessive quantities of free asbestos fibers over prolonged periods of time can increase the risk of developing certain diseases of the lung within 20 or 30 years. The three diseases associated with the inhalation of asbestos are: asbestosis, a nonmalignant fibrotic lung condition; bronchogenic (lung) carcinoma; and mesothelioma, a rare cancer of the lining of the chest or abdominal cavities (58).

Reduction of asbestos dust exposure is at present the only known method of preventing disease among asbestos industry workmen. When dust levels are low, risk to employees and the incidence of asbestos-related disease drop sharply. Cigarette smoking greatly increases the risk of developing bronchogenic cancer among persons encountering heavy asbestos exposure. Nonsmoking asbestos workers show no greater incidence of bronchogenic cancer than the average nonsmoker.

There are governmental regulations (OSHA) that describe the allowable airborne fiber levels in work areas. Extensive dust control together with corrective work practices are used to implement these regulations and medical examinations are regularly provided to assure worker protection (see Air pollution control methods; Industrial hygiene and toxicology).

BIBLIOGRAPHY

"Asbestos" in *ECT* 1st ed., Vol. 2, pp. 134–142, by M. S. Badollet, Johns-Manville Research Center; "Asbestos" in *ECT* 2nd ed., Vol. 2, pp. 734–747 by M. S. Badollet, Consultant.

1. A. A. Hodgson, "Fibrous Silicates," *The Royal Institute of Chemistry, London, Lecture, Series No. 4 1965.*
2. F. A. Mumpton, H. W. Jaffe, and C. S. Thompson, *Am. Mineral.* **50,** 1893 (1965).
3. B. E. Warren and W. L. Bragg, *Krist.* **76,** 201 (1930).
4. B. E. Warren and K. W. Hearing, *Phys. Rev.* **59,** 925 (1941).
5. S. Speil and J. P. Leineweber, *Environ. Res.* **2**(3), (1969).
6. F. J. Wicks and E. J. W. Whittaker, *Can. Mineral.* **13,** 227 (1975).
7. F. J. Wicks and J. Zussman, *Can. Mineral.* **13,** 244 (1975).
8. T. F. Bates, L. P. Sand, and J. F. Mink, *Science* **111,** 512 (1950).
9. W. Noll and H. Kercher, *Naturwissenschaften* **37,** 540 (1950).
10. J. Turkevich and J. Hillier, *Anal. Chem.* **21,** 475 (1949).
11. M. Maser, R. V. Rice, and H. P. Klug, *Am. Mineral.* **45,** 680 (1960).
12. K. Yada, *Acta Crystallogr.* **23,** 704 (1967).
13. E. J. W. Whittaker, *Acta Crystallogr.* **6,** 747 (1953); **8,** 571 (1955); **9,** 855 (1956); **10,** 149 (1957).
14. H. Berman, *Am. Mineral.* **17,** 313 (1932).
15. E. H. Nickel, *Can. Mineral.* **6,** 307 (1959).
16. G. P. Reimschussel, unpublished information, Johns-Manville Research Center, Denver, Colo.
17. P. Hahn-Weinheimer and A. Hirner, *3rd International Conference on the Physics and Chemistry of Asbestos Minerals,* Quebec, Canada, 1975.
18. M. S. Badollet, *Can. Min. Metall. Bull.* **1** (1951).
19. D. G. Hiscock, unpublished information, Cape Asbestos Ltd., 1964.
20. F. L. Pundsack, *J. Phys. Chem.* **59,** 892 (1955).
21. F. L. Pundsack and G. P. Reimschussel, *J. Phys. Chem.* **60,** 1218 (1956).
22. J. W. Ryznar, J. Greon, and M. G. Winterstein, *Ind. Eng. Chem.* **38,** 1057 (1946).
23. E. Martinez and G. L. Zucker, *J. Phys. Chem.* **64,** 924 (1960).
24. L. M. Pidgeon and A. van Wissen, *Can. J. Res.* **9,** 153 (1933).
25. G. J. Young and F. H. Healy, *J. Phys. Chem.* **58,** 881 (1954).
26. A. C. Zettlemoyer, G. J. Young, and F. H. Healey, *J. Phys. Chem.* **57,** 649 (1953).
27. C. H. Gorski and L. E. Stettler, *Amer. Ind. Hyg. Assoc. J.* **36,** (1975).
28. M. Morton and W. G. Baker, *Can. Inst. Min. Metall. Trans.* **XLIV,** (1941).
29. M. S. Badollet, *Can. Min. Metall. Bull.* **1** (1951).
30. R. Zukowski and R. Gaze, *Nature (London)* **183,** 35 (1959).
31. D. R. Burman, *Paper No. 2–8, 1st International Conference on the Physics and Chemistry of Asbestos Minerals,* Oxford, Eng., 1967.
32. F. L. Pundsack, *J. Phys. Chem.* **65,** 30 (1961).
33. M. S. Badollet and W. C. Streib, *Can. Min. Metall. Bull.* **48,** (1955).
34. U.S. Pat. 2,616,801 (Nov. 1952), M. S. Badollet and W. C. Streib (to Johns-Manville).
35. U.S. Pat. 3,173,831 (Mar. 1965), F. L. Pundsack and G. P. Reimschussel (to Johns-Manville).
36. U.S. Pat. 3,891,498 (June 24, 1975), T. H. Sadler (to Johns-Manville).
37. U.S. Pat. 3,715,230 (Feb. 6, 1973), T. H. Sadler (to Johns-Manville).
38. J. E. Chisholm, *J. Mater. Sci.* **8,** 475 (1973).
39. D. R. Veblen, R. R. Buseck, and C. W. Burnhem, *Science* **198,** 359 (1977).
40. J. J. Cilliers and co-workers, *Econ. Geol.* **56,** (1961).
41. J. H. Patterson and R. L. Thompson, *Paper No. 2–5, 1st International Conference on the Physics and Chemistry of Asbestos Minerals,* Oxford, Eng., 1967.
42. J. Goldfield, *Min. Eng.* **7,** (Nov. 1955).

43. H. Berger, *CIM Bull.* **68,** (Feb. 1975).
44. R. G. Evans, *CIM Bull.* **68,** (July 1975).
45. E. Martinez, *Trans. Soc. Min. Eng.* **255,** (1974).
46. H. H. Gossling, *2nd Industrial Minerals International Conference,* Munich, Ger., 1976.
47. *Asbestos* **58,** (Dec. 1976).
48. *Asbestos* **58,** (Mar. 1977).
49. *Asbestos* **59,** (Nov. 1977).
50. *Mineral Industries of the USSR,* U.S. Bureau of Mines, MP2, July 1977.
51. *Asbestos-1977,* U.S. Bureau of Mines, MCP6, Sept. 1977.
52. *Mineral Trade Notes,* **74**(6), (June 1977).
53. *World Mining Magazine* **30,** June 25, 1977.
54. *Ind. Min.* (June 1975).
55. B. Lincoln, *Ind. Min.,* (Oct. 1975).
56. N. W. Hendry, unpublished information, Johns-Manville Corporation, Denver, Colo.
57. *Chrysotile Asbestos Test Manual,* 3rd ed., Asbestos Textile Institute and the Quebec Asbestos Mining Association, 1974.
58. *Chem. Week,* 16 (May 3, 1978).

WILLIAM C. STREIB
Johns-Manville Corporation

BRAKE LININGS AND CLUTCH FACINGS

This review concerns the current state of friction materials technology with regard to types of friction materials, their applications, friction and wear characteristics, raw materials, manufacturing methods, evaluation and test methods, environmental considerations, and current trends.

During a stop, a brake converts the kinetic energy of the moving vehicle into heat, absorbs the heat, and gradually dissipates it into the atmosphere. The brake is a sliding friction couple consisting of a rotor connected to the wheel and a stator on which is mounted the friction material (pad, lining, or block). The friction material is considered to be the expendable portion of the brake couple which, over a long period of use, is converted to debris and gases (1).

During engagement a clutch transfers the kinetic energy of a rotating crankshaft (coupled to a power source) to the transmission and wheels. Any slippage results in the generation of heat, which is absorbed and eventually dissipated to the atmosphere by the clutch. Thus the clutch is basically a static friction couple which momentarily slides during gear shifts. The clutch friction material is considered to be expendable.

Brakes and clutches operate both dry and wet. In dry friction couples, the heat is removed by conduction to the surrounding air and structural members. Wet friction couples operate within a fluid (usually an oil) which absorbs the heat and maintains the couple at relatively low temperatures (below 200°C). The fluid also traps the wear debris (see Hydraulic fluids).

Friction materials serve in a variety of ways to control the acceleration and deceleration of a variety of vehicles and machines which may be as small as a clutch in a business machine or a brake on a bicycle to as large as jumbo-aircraft brakes. The brakes on bicycles may have friction couples of iron sliding against iron in the coaster brake or rubber-bound composite pads sliding against a chromium-plated wheel rim in hand-activated brakes. Passenger cars may have drum brakes or disk brakes exclusively or a combination of disk fronts and drum rears. The friction materials may be resin- or rubber-bound composites based on asbestos or metallic fibers. Trucks and off-highway vehicles usually have very large drum brakes; only a few now have disk brakes, with friction couples that usually operate at higher friction levels and temperatures than passenger cars. Large aircraft are equipped exclusively with disk brakes which contain multiple rotor and stator arrangements with the most popular friction couple consisting of a sintered friction material sliding against a high-temperature resistant steel. The newer aircraft brakes consist of carbon composites serving as both the rotor and the stator (see Carbon).

Friction and Wear

Friction. A qualitative analysis of frictional force suggests that a frictional force is likely to consist of several components such as adhesion-tearing, ploughing (or abrasion), elastic and plastic deformation, and asperity interlocking, all occurring at the sliding interface. These mechanisms presumably depend upon the temperature as well as the normal load and sliding speed, since material properties are known to be dependent upon these variables. In the case of automotive friction materials, the coefficient of friction is usually found to decrease with increasing unit pressure and sliding speed at a given temperature, contrary to Amontons' law (2–4). This decrease in friction is controlled by the composition and microstructure of friction materials. As the temperature of the sliding interface increases, the coefficient of friction varies. This variation is unpredictable, and there exists no general trend except that at extremely high temperatures the coefficient may become very low (less than 0.1). This temporary loss in friction is referred to as fade (5). Like automotive friction materials, aircraft cermet friction materials exhibit decreasing coefficient of friction with increasing unit pressure (6).

Terminology. *Effectiveness*, essentially a measure of the stopping efficiency, can be expressed in a number of different ways: as the coefficient of friction, ie, hydraulic or air line pressure required, torque developed, or distance required to stop a vehicle. The various temperatures are identified for passenger cars as *cold* (under 100°C), *normal* (150–250°C), or *hot* (above 300°C). Effectiveness can be measured *new* or *off-rack* (without any prior use), *preburnished* (after little prior use), *burnished* (after moderate use), and *faded* (after use at elevated temperatures). Although the same terms are used for all friction materials, for large aircraft materials the temperatures are: cold, under 300°C; normal, 400–600°C; and hot, above 700°C. The normal fade-free maximum operating temperatures of various friction materials are: drum linings and clutch friction materials, 250°C; Class A organic disk pads, 300°C; Class B organic materials and blocks, 350°C; semimetallics, 400°C; and ceremetallics and carbon composites, 700°C.

Friction peaking is an increase in friction known to occur during or after prior high temperature operation. *Unbalance* occurs when friction peaking or fade causes one wheel or axle to change in friction, yielding side-to-side or front-to-rear unbalance. *Friction stability* is the ability of the friction materials to produce similar friction or friction changes at all wheels through all duty-cycles and especially during a return to normal operation after a temporary severe duty. *Recovery* from fade is the ability of the friction material to return to its prefade friction level. *Speed sensitivity* is the ability to maintain effectiveness at varying surface or rubbing speeds. Most materials show losses in effectiveness at higher speeds, with semimetallics being the notable exception. *Load sensitivity* is the ability to maintain effectiveness at various weight loadings. The ability of a friction material to recover from loss in effectiveness due to exposure to water is called *water recovery*.

Wear. For a fixed amount of braking the amount of wear of automotive friction materials tends to increase slightly or remain practically constant with respect to brake temperature, but once the brake rotor temperature reaches about 204°C the wear of resin-bonded materials increases exponentially with increasing temperature (7–10). This exponential wear is due to thermal degradation of organic components. At low temperatures the practically constant wear rate is primarily controlled by abrasion and adhesion (11). The wear, W, of friction materials can best be described by the following wear equation (12–13):

$$W = KP^aV^bt^c$$

where K is the wear coefficient, P the normal load, V the sliding speed, t the sliding time, and a, b, and c are a set of parameters for a given friction material–rotor pair at a given temperature.

Wear is an economic consideration. Wear resistance generally (not always) is inversely related to friction level and other desirable performance characteristics within any class of friction material. The formulator's objective is to provide the highest level of wear resistance in the normal use temperature range, a controlled moderate increase at elevated temperatures, and a return to the original lower wear rate when temperatures again return to normal. Contrary to common belief, maximum wear life does not require maximum physical hardness.

There are four possible mechanisms of wear at sliding surfaces: (*1*) Adhesion of the two materials, followed by cohesion failure of one as the two slide past each other. (*2*) Ploughing or gouging of one material by a fragment of another (harder) material. (*3*) Thermal or mechanical fatigue or melting, which permits solid pieces to become detached from the surface. (*4*) Oxidation and pyrolysis, leading to gas formation (see Ablative materials).

In the low temperature range Class A organic materials and semimetallic materials wear at a substantially lower rate than Class B organic materials. At extremely high temperatures the wear rate of Class B organic materials is the best, that of semimetallic materials next-best, and that of Class A organic materials the worst. In the intermediate temperature range, semimetallic materials are the best, followed by Class B organic materials.

Cermet or carbon friction materials operate at substantially higher temperatures than normal automotive or truck friction materials. Still the wear rates of these materials increase with brake temperature. One unique feature of these two materials is the formation of a glazed friction layer which reduces the wear rate. Without this glazed layer the wear rate is usually very high.

Types of Friction Materials

Organic Materials. The most common type of friction materials used in brakes and clutches for normal duty are termed organics. These materials usually contain about 30–40 wt % of organic components (14).

The primary applications of organic frictional materials and their requirements are as follows: (*1*) primary drum brake linings must provide high and stable friction at all temperatures and pressures; (*2*) secondary drum brake linings must provide stable friction and wear resistance; (*3*) Class A disk pads provide nonabrasive friction and wear properties, quiet operation, and rotor compatibility; (*4*) Class B disk pads provide higher friction and high temperature wear resistance at the expense of some low temperature wear resistance, noise properties, and rotor compatibility; (*5*) Class C friction materials consist of both disk pads and block-type friction materials for extremely heavy-duty operations where high friction, minimal fade, and wear resistance are essential at the expense of other brake characteristics; and (*6*) clutch friction materials provide stable friction, good wear properties, quiet operation, and rotor compatibility combined with high strength properties.

The major constituent of practically all organic friction materials is asbestos fiber, although small quantities of other fibrous reinforcement may be used. Asbestos is chosen because of its thermal stability, its relatively high friction, and its reinforcing properties (see Asbestos; Composite materials; Laminated and reinforced plastics). Since asbestos alone does not offer all of the desired properties, other materials called property modifiers are added to provide desired levels of friction, wear, fade, recovery, noise, and rotor compatibility. A resin binder holds the other materials together. This binder is not completely inert and makes contributions to the frictional characteristics of the composite. More commonly used ingredients can be found in various patents (15–17); see also Table 1 (18).

Table 1. Organic Friction Materials, Wt %[a]

	Copoly-mers[b]	5K asbestos	Sulfur	Zinc oxide	Cardolite no. 753	Resin	Additive 1	Additive 2
X-1	22.00	49.00	2.00	4.00			11.50[c]	11.50[d]
X-2	11.00	63.00	1.00	2.00			11.50[c]	11.50[d]
X-3	17.00	54.30	1.60	3.10			11.50[c]	11.50[d]
								1.00[e]
X-4	15.00	57.80	1.48	2.72			11.50[c]	11.50[d]
X-5	15.00	57.80		2.72			12.98[c]	11.50[d]
X-6	15.00	78.80	1.48	2.72		2.00		
X-7	15.00	68.80	1.48	2.72	10.00	2.00		
X-8	15.00	68.80	1.48	4.72	10.00			
X-9	15.00	78.80	1.48	2.72				
X-10	15.00	68.80	1.48	2.72		2.00	10.00[f]	
X-11	13.30	55.17	1.31	2.42	8.85	1.75	17.20[g]	
X-12	13.00	64.20	1.30	3.00	10.00	6.50	1.00[h]	1.00[e]
X-13	15.00	15.00	0.30	6.00		5.00	40.00[i]	18.70[j]
X-14	15.00	68.80	1.48	2.72		2.00	10.00[k]	
X-15	15.00	68.80		4.20	10.00	2.00		
X-16	17.00	48.80		4.20	30.00			
X-17	17.00	64.80		4.20	10.00		4.0[l]	
X-18	17.00	68.80		4.20	10.00			

[a] Ref. 18.
[b] Acrylonitrile–butadiene copolymer.
[c] Barytes.
[d] Rottenstone.
[e] Alundum (600X).
[f] B-196 friction particle.
[g] Zinc dust.
[h] Paraformaldehyde (curing agent).
[i] Steel wool.
[j] Iron oxide.
[k] B-221 friction particle.
[l] Orlon fiber.

Metallic Materials. Gray cast iron is of reasonably low cost, provides good wear resistance and damping characteristics, and has been in long time use as a brake drum or disk material for passenger cars and trucks. However, under severe operating conditions, gray cast iron can undergo phase transformations leading to heat checking of brake drums or disks (19). Copper or aluminum rotors have been evaluated experimentally (20–22) and alloy steel rotors are being used for certain nonautomotive brakes, including aircraft and trains. In the case of carbon-composite aircraft brakes the rotor material is the same as the stator material.

Semimetallic Materials. Semimetallics were introduced in the late 1960s and gained widespread usage in the mid 1970s. These materials usually contain more than 50 wt % metallic components. They are primarily used as disk pads and blocks for heavy-duty operation (23).

The major constituent of practically all semimetallics is iron powder in conjunction with a small amount of steel fiber. Various property modifiers are added to enhance performance to desired levels, and a resin binder—necessary to hold the materials together in a mass—is also added. Semimetallic friction materials may therefore consist of metallic powder, sponge iron particles, ceramic powder, steel fiber, rubber particles, graphite powder, and phenolic resin (24).

Sintered Materials or Cermets. Heavy weights and high landing speeds of modern aircraft or high-speed trains require friction materials that are extremely stable thermally. Organic or semimetallic friction materials are frequently unsatisfactory for these applications. Cermet friction materials are metal-bonded ceramic compositions (25–26). The metal matrix may be copper or iron. Several typical compositions are given in Table 2 (27) (see Glassy metals; Ceramics).

Carbon Composites. Cermet friction materials tend to be heavy, thus making the brake system less energy-efficient. Compared with cermets, carbon (or graphite) is a thermally stable material of low density and reasonably high specific heat. A combination of these properties makes carbon attractive as a brake material. Currently several companies are actively developing and field-testing lighter brake materials based on carbon fiber-reinforced carbon-matrix composites, which are to be used primarily for aircraft brakes (see Carbon and artificial graphite; Ablative materials).

Raw Materials

Binders. In selecting a resin binder system the processing characteristics must be considered along with the final frictional and physical properties. Two types of systems are used. In wet processing, the binder is a viscous liquid (usually a resole) having characteristics suited to thermoplastic processing techniques (see Plastics technology). In dry processing, the binder is a powdered material (usually a novolac) that is mixed directly with the other materials; it does not cross-link until heat and pressure are applied.

Synthetic resins such as phenolic and cresylic resins are most commonly used

Table 2. Cermet Friction Materials, Wt % [a]

Material	1	2	3	4	5	6	7	8	9	10
Matrix										
copper	47.1	49.6	44.6	44.7	44.9	46.2	52.4	43.6	18.6	46.2
zinc										5.5
tin								7.4		3.3
nickel	7.1	7.1	7.1			7.0	6.6	6.6	7.1	
titanium	3.6	3.6	3.6			3.5	3.3	3.3	3.6	
brass chips									28.6	
iron				15.0	15.0					8.8
Total	57.8	60.3	55.3	59.7	59.9	56.7	62.3	60.9	57.9	63.8
Friction material										
calcined kyanite	26.1	26.1	26.1	24.9	19.9	25.6	24.0	24.2	26.1	22.0
silica	4.8	4.8	4.8	4.6	4.6	4.8	4.4	4.5	4.8	4.4
Total	30.9	30.9	30.9	29.5	24.5	30.4	28.4	28.7	30.9	26.4
Lubricant										
graphite	1.2	1.2	1.2	1.1	6.1	1.1	1.1	1.1	1.2	8.8
hard coal										
lead						2.0	3.7			
Total	1.2	1.2	1.2	1.1	6.1	3.1	4.8	1.1	1.2	8.8
Antioxidant										
molybdenum	10.0	7.5	12.5	9.5	9.5	9.8	4.6	9.3	10.0	1.0
Total	10.0	7.5	12.5	9.5	9.5	9.8	4.6	9.3	10.0	1.0

[a] Ref. 27.

friction material binders, and are usually modified with drying oils, rubber, cardanol, or an epoxy (see Phenolic resins). They are prepared by the condensation of the appropriate phenol with formaldehyde in the presence of an acidic or basic catalyst. Polymerization takes place at elevated temperatures. Other resin systems are based on elastomers, drying oils, or combinations of the above or other polymers.

Metals such as copper, iron, tin, and lead are used as binders of sintered friction materials where deformation under the high forming pressure is required to lock together the property modifiers within a matrix.

Fibrous Reinforcements. The asbestos usually used in friction materials is chrysotile (mined in Quebec and Vermont). Chrysotile is the principal mineral of the serpentine group ($3MgO.2SiO_2.2H_2O$) (28). Long-fiber asbestos (eg, Grade 5) is generally used when dry processing techniques are employed in manufacturing. Shorter-fiber asbestos (eg, Grade 7) is used with wet processing techniques. The two grades differ considerably in length of fiber, bulk, absorptiveness, cost, and reinforcing value. Longer fibers permit the bending of secondary linings from flat blanks to curved segments.

Steel fiber may have a variety of aspect ratios, cleanliness (straight versus curved fibers), surface roughness, and chemical compositions. Fibers with tight specifications in terms of cleanliness, chemical composition, and aspect ratio are necessary for sintered friction materials. The fibers are usually machined from larger metallic forms.

Carbon fibers are produced by graphitization of organic or pitch fibers by techniques that provide parallel alignment of the carbon chain to the fiber length for maximum tensile strength.

Organic clutch materials contain continuous-strand reinforcements in addition to fibrous reinforcements. These include cotton (primarily for processing) and asbestos yarn, brass wire, or copper wire for high burst strength.

Property Modifiers. Property modifiers can, in general, be divided into two classes: nonabrasive and abrasive. Nonabrasive modifiers can be further classified as high friction and low friction.

The most frequently used nonabrasive modifier is a cured resinous friction dust derived from cashew nutshell liquid. Ground rubber is used in particle sizes similar to or slightly coarser than those of the cashew friction dusts for noise, wear, and abrasion control. Carbon black, petroleum coke flour, natural and synthetic graphite, or other carbonaceous materials are used to control the friction and improve wear (when abrasives are used) or to reduce noise. The above-mentioned modifiers are primarily used in organic and semimetallic materials except for graphite which is used in all friction materials.

Abrasive modifiers are used in several types of friction materials. Very hard materials such as alumina, silicon carbide, and kyanite are used in fine particle sizes in organic, semimetallic, and cermet materials (generally less than 74 μm or 200 mesh). Particle size is limited by the fact that large particles of such hard materials would groove cast-iron mating surfaces. Larger particle sizes are possible for harder mating surfaces in the special steel rotors used with sintered materials.

Minerals are generally added to improve wear resistance at minimum cost. The most commonly used are ground limestone (whiting) and barytes, although various types of clay, finely divided silicas, and other inexpensive or abundant inorganic materials may also perform this function.

Metal or metal oxides may be added to perform specific functions. Brass chips and copper powder are frequently used in heavy-duty organics where they act as scavengers to break up undesirable surface films. Zinc and aluminum are also used. Zinc chips used in Class A organics contribute significantly to recovery of normal performance following fade. Most of these inorganic materials tend to detract from anti-noise properties and mating surface compatability.

Manufacturing

An important relation exists between composition, including performance requirements and ease of manufacture. Organic linings which must bend also require higher resin contents and longer fibers. Heavy-duty materials with reduced resin loading for improved performance require molding-to-shape. Sintered and carbon friction materials require high pressure forming and high temperature treatment in inert atmospheres. Woven and some clutch materials require special fiber-forming methods.

Linings. Most linings are produced from resin wet-mix by either an extrusion or a rolling process. Initially, the fibrous reinforcement and the friction modifiers are mixed with a liquid resin at approximately 50°C. The binder solvent serves as a plasticizer to yield a dense putty-like mass with good wet strength. In the extrusion process, the mix is heated to approximately 90°C and extruded at 13.8–27.6 MPa (2000–4000 psi) as a flat, pliable tape which is dried for 2 h at 80°C. In the rolling process, the mix is partially dried, sized into particles, then fed between two rolls of slightly different speeds to align the fibers in the flat, pliable tape or green lining which is formed.

The green lining is then cut to length, formed into an arcuate segment at 150°C, then placed in curved mold cavities and cured 4–8 h at 180–250°C. Final grinding produces the finished brake lining.

Linings for heavy-duty use such as for medium-sized trucks are produced by a dry-mix process. The fiber, modifiers, and a dry novolac resin are mixed in a Littleford mixer. The blend is then formed into about a 60 by 90 cm briquet at 2.8–4.1 MPa (400–600 psi). The briquets are hot-pressed for 3–10 min at 140–160°C and then cooled. The resin is only partially cured at this point and will be thermoplastic when subsequently reheated for bending. The hot-pressed briquets are then cut to desired size and bent at 170–190°C and cured in curved molds for 4–8 h at 220–280°C. Final grinding produces the finished segments.

Disk Pads. Organic and semimetallic disk pads are produced by somewhat similar processes after the mixes are formed. The mix for organics is prepared as for heavy-duty linings. The mix for semimetallics is produced in a less intensive blender. The mix is then formed into briquets at room temperature and 27.6–41.4 MPa (4000–6000 psi). These briquets are then hot-pressed at 160–180°C for 5–15 min at 27.6–55.2 MPa (4000–8000 psi). The pads are then cured at 220–300°C for 4–8 h and ground to produce the finished disk pads.

In many instances the friction material mix is integrally molded into holes within the backing plate or shoe. Painting of the final assembly is rare in the United States but it is the rule in Europe.

Blocks. The mix for organic blocks is prepared as for heavy-duty segments and the mix for semimetallic blocks is prepared as for semimetallic disk pads. Briquets are formed at 10.3–17.2 MPa (1500–2500 psi). To reduce blisters in hot pressing, the briquets may be heated to 90°C for 15–30 min. The blocks are formed at 130–150°C at 13.8–20.7 MPa (2000–3000 psi) for periods of 10–30 min. After slitting to width, the blocks undergo grinding of internal and external radii. Final cure may be in an unconfined form at temperatures as low as 180°C for 15 h or in confined form at temperatures as high as 280°C for 6 h. Grinding, drilling, and chamferring produce the final block.

Clutch Materials. Methods for producing clutch friction materials are concerned with the placement of the reinforcement strand or wire within the matrix. Processing includes: molding of mix without strands or wire; molding of mix around strand or wire preforms; and weaving curable preforms. In the first two cases, a dry mix is used. In the latter case, a wet mix is prepared and a strand is run through the premix to pick up the viscous mass along the strand which can be woven after drying. After hot-pressing and curing, the surface is ground to final shape.

For trucks and heavy off-the-road equipment, cermet friction materials (see below) are also used. Sintered cermet segments are attached to a metal plate to form the clutch facing.

Woven Bands. Woven bands for heavy-duty operation are produced by an expensive process which begins with an asbestos cord, which may be reinforced with wire, being passed through a wet-mix to pick up resin and modifiers. The saturated cord is then woven into tapes which pass through heated rolls to partially cure the resin. The material can be post-cured at low temperatures (ca 160°C) to remain as a flexible roll lining or post-cured at higher temperature (180–230°C) to form rigid segments. Such materials are used in large band brakes used to control large machinery.

Cermets. Cermet materials are manufactured using the powder metallurgy (qv) technique. Desired amounts of individual ingredients are weighed, mixed, compacted, sintered, and coined (or recompacted). The sintering is performed in a reducing or neutral atmosphere, and the sintering temperature has to be high enough so that the metal ingredients will adhere to each other.

Carbon Composites. In this class of materials, carbon or graphite fibers are embedded in a carbon or graphite matrix. The matrix can be formed by two methods: chemical vapor deposition and coking. In the case of chemical vapor deposition a hydrocarbon gas is introduced into a reaction chamber in which carbon formed from the decomposition of the gas condenses on the surface of carbon fibers. An alternative method is to mold a carbon fiber–resin mixture into shape and coke the resin precursor at high temperatures. In both of the methods the process has to be repeated until a desired density is obtained.

Evaluation Methods

Chemical, Physical, and Mechanical Tests. Manufactured friction materials are characterized by various chemical, physical, and mechanical tests in addition to friction and wear testing. The chemical tests include tga (thermogravimetric analysis), dta (differential thermal analysis), pgc (pyrolysis gas chromatography), acetone extraction, lc (liquid chromatography), ir analysis, and x-ray or sem (scanning electron microscope) analysis (see Analytical methods). Physical and mechanical tests determine properties such as thermal conductivity, specific heat, tensile or flexural strength, and hardness.

Dynamometer and Vehicle Testing. Friction materials are evaluated in the laboratory by a great variety of tests and equipment. Friction and wear characteristics can be evaluated with sample dynamometers such as the Chase machine. In the most reliable sample test machines the output torque is controlled so that different materials all do the same amount of work. One disadvantage of sample test machines is that the ratios of friction-material area to rotor area and friction-material mass to rotor mass are quite different from the ratios used on vehicles. The heat generation, storage, and rejection conditions are therefore quite different. A second disadvantage is that only one material is tested, whereas in vehicles with drum brakes two types of friction materials are used together and there are interaction effects. The advantage is mainly one of economics—more tests at less cost.

The full brake dynamometer, when properly instrumented and controlled, reflects with reasonable accuracy the actual brake performance in a vehicle. The high initial investment is recovered through operation independent of the climatic conditions and by a fully automatic operation for extended periods, thus minimizing manpower costs.

No attempt will be made here to detail the numerous vehicle test procedures used by different organizations. Performance tests are essentially designed to appraise initial effectiveness, burnish and normal effectiveness, fade and recovery, and final effectiveness. Side-to-side and front-to-rear balance can also be determined. Only vehicle tests can determine noise properties accurately. Wear measurements are generally made in accelerated performance tests, but the results are a reflection of high-temperature wear properties and are usually not valid for predicting normal driving wear. More valid predictions of normal wear life result from specifically designed extended road traffic wear measurement tests involving a great number of stops with restricted maximum temperatures.

Vehicle tests are considered the ultimate in friction material evaluation, but to be accurate they must be carefully designed to eliminate variations caused by changing weather. Controlled-temperature tests and parallel-test control vehicles normally perform the function satisfactorily, but at increased cost.

Environment and Health

Manufacturing. Organic friction materials have been in increasing jeopardy because OSHA regulations have limited the exposure of workers to airborne asbestos fibers. High performance friction materials can only be produced by the dry-mix process and this tends to be dustier than other processes. Prior to any regulations, typical fiber concentrations were in the 10–30 fibers/cm^3 range. In 1974 the time-weighted average was set at 5 fibers/cm^3 and in 1976 it was reduced to 2 fibers/cm^3. The cost of capital equipment to effect these improvements is extremely high. OSHA has been considering a further reduction to 0.5 or 0.1 fibers/cm^3.

Because of the wide variety of asbestos bundle sizes and even individual fibrils in commercial asbestos, the handling of asbestos causes degradation of the larger fiber bundles to fibers with diameters less than 2 μm which remain airborne for extended periods of time. These airborne fibers are prone to inhalation and lung entrapment which may lead to disabling diseases. The exact definition of harmful fibers and the mechanism by which they affect the body is not accurately known.

Friction materials are known to contain other harmful materials. Lead is found in secondary linings, Class A, B, and C organic disk pads, and other friction materials. At least one original equipment and after-market supplier is known to have a policy against incorporation of lead in its products.

Wear Products. Friction material and rotor emissions are generated by normal wear. Because of large-scale usage and the potential health hazard of asbestos, organic friction materials and their wear debris have been extensively studied (1,29–30). Below 250°C, abrasive and adhesive wear are considered to be the most important mechanisms and the wear rates are low. Above 250°C, organic friction materials begin to pyrolyze or oxidize such that both gases and particulates are released.

In order to define the extent of emissions from automotive brakes and clutches, a study was carried out in which specially designed wear debris collectors were built for the drum brake, the disk brake, and the clutch of a popular United States vehicle (1). The vehicle was driven through various test cycles to determine the extent and type of brake emissions generated under all driving conditions. Typical original equipment and after-market friction materials were evaluated. Brake relinings were made to simulate consumer practices. The wear debris was analyzed by a combination of optical and electron microscopy to ascertain the asbestos content and its particle size distribution. It was found that more than 99.7% of the asbestos was converted to a nonfibrous form and that only 3.2% of the total asbestos was emitted to the atmosphere. A second study of brake emissions adjacent to a city freeway exit ramp on the downwind side indicated that the asbestos emissions were so low they could not be distinguished from the background on the upwind side (30).

Future Prospects

The trend toward more energy-efficient passenger cars and trucks will put an increased demand on friction materials. There is already a trend toward smaller and lighter vehicles. More efficient vehicles have manual transmissions with smaller brakes. Organic friction materials will continue to serve the drum brake industry. In the past 10 years, more and more vehicles were equipped with ventilated disk brakes. In time these may be replaced by solid disk brakes for weight savings. When the brakes become smaller, thus producing higher brake temperatures, the Class A organic materials will become less suitable and Class B organic and semimetallic materials will fill the gap. Trucks and other heavy vehicles are moving toward more efficient disk brakes. More sintered friction materials will appear in the heavy-vehicle clutch market. At the same time, aircraft will move toward light carbon brakes.

Future brakes will be required to satisfy health standards. Some passenger car manufacturers are known to be in favor of removing all asbestos from passenger cars. One large automotive manufacturer has a policy for removal of asbestos-based friction materials by model year 1981. The only proven asbestos-free friction materials at this time are the semimetallics. In some cases, current vehicles already have front disk brakes equipped with semimetallics (15–20% of the original equipment market in model year 1977). Equipping rear disk brakes with semimetallic materials also would be one way to achieve a vehicle with asbestos-free friction materials before 1980.

Current friction materials are relatively noise-free. The major problem with new heavier-duty friction materials will be their noise properties. Noise-free friction materials will continue to be used in the heavy-truck and bus markets (see Noise pollution).

BIBLIOGRAPHY

"Brake Linings and Other Friction Facings" in *ECT* 1st ed., Vol. 2, pp. 622–628, by F. C. Stanley, The Raybestos Division, Raybestos-Manhattan, Inc.; "Friction Material" in *ECT* 2nd ed., Vol. 10 pp. 124–134, by Clyde S. Batchelor, The Raybestos Division, Raybestos-Manhattan, Inc.

1. M. G. Jacko, R. T. DuCharme, and J. H. Somers, *SAE Trans.* **82,** 1813 (1973).
2. S. K. Rhee, *SAE Trans.* **83,** 1575 (1974).
3. S. K. Rhee, *Wear* **28,** 277 (1974).
4. W. R. Tarr and S. K. Rhee, *Wear* **33,** 373 (1975).
5. J. M. Herring, *Mechanisms of Brake Fade in Organic Brake Linings, SAE Paper No. 670146,* SAE, New York, Jan. 1967.
6. N. A. Hooton, *Bendix Tech. J.* **2,** 55 (1969).
7. S. K. Rhee, *SAE Trans.* **80,** 992 (1971).
8. S. K. Rhee, *Wear* **29,** 391 (1974).
9. T. Liu and S. K. Rhee, *Wear* **37,** 291 (1976).
10. T. Liu and S. K. Rhee, "High-Temperature Wear of Semimetallic Disc Brake Pads," in K. C. Ludema, W. A. Glaeser, and S. K. Rhee, eds., *Wear of Materials—1977,* American Society of Mechanical Engineers, New York, 1977.
11. S. K. Rhee, *Wear* **23,** 261 (1973).
12. S. K. Rhee, *Wear* **16,** 431 (1970).
13. S. K. Rhee, *Wear* **18,** 471 (1971).
14. F. W. Aldrich and M. G. Jacko, *Bendix Tech. J.* **2**(1), 42 (Spring 1969).
15. U.S. Pat. 2,428,298 (Sept. 30, 1947), R. E. Spokes and E. C. Keller (to American Brake Shoe Co.).
16. U.S. Pat. 2,685,551 (Aug. 3, 1954), R. E. Spokes (to American Brake Shoe Co.).
17. U.S. Pat. 3,007,549 (Nov. 7, 1961), B. W. Klein (to Bendix Corporation).
18. U.S. Pat. 3,007,890 (Nov. 7, 1961), S. B. Twiss and E. J. Sydor (to Chrysler Corporation).
19. S. K. Rhee, *Characterization of Cast Iron Friction Surfaces, SAE Paper No. 720056* SAE, New York, Jan. 1972.
20. S. K. Rhee, R. M. Rusnak, and W. M. Spurgeon, *SAE Trans.* **78,** 1031 (1969).
21. S. K. Rhee, J. L. Turak, and W. M. Spurgeon, *SAE Trans.* **79,** 503 (1970).
22. S. K. Rhee and J. E. Byers, *SAE Trans.* **81,** 2085 (1972).
23. B. W. Klein, *Bendix Tech. J.* **2**(3), 109 (Autumn 1969).
24. U.S. Pat. 3,835,118 (Sept. 10, 1974), S. K. Rhee and J. P. Kwolek (to Bendix Corporation).
25. N. A. Hooton, *Bendix Tech. J.* **2,** 55 (1969).
26. K. Aoki and J. Shirotani, *Bendix Tech. J.* **6,** 1 (1973–1974).
27. U.S. Pat. 2,948,955 (Aug. 16, 1960), A. W. Allen and R. H. Herron.
28. A. A. Hodgson, "Fibrous Silicates," *Lecture Series No. 4,* Royal Institute of Chemistry, London, Eng., 1965.
29. A. E. Anderson and co-workers, "Asbestos Emissions from Brake Dynamometer Tests," *SAE Paper No. 730549,* SAE, New York, May 1973.
30. J. C. Murchio, W. C. Cooper, and A. DeLeon, "Asbestos Fibers in Ambient Air of California," *University of California (Riverside) Report EHS No. 73-2,* Mar. 1973.

M. G. Jacko
S. K. Rhee
Bendix Corporation

CARBON

CARBON AND ARTIFICIAL GRAPHITE

STRUCTURE, TERMINOLOGY, AND HISTORY

Carbon [*7440-44-0*] occurs widely in its elemental form as crystalline or amorphous solids. Coal, lignite, and gilsonite are examples of amorphous forms; graphite and diamond are crystalline forms. Carbon forms chemical bonds with other elements, but is capable of forming compounds in which carbon atoms are bound to carbon atoms and, as such, is the chemical element that is the basis of organic chemistry. Carbon and graphite can be manufactured in a wide variety of products with exceptional electrical, thermal, and physical properties. They are a unique family of materials in that properties of the final product can be controlled by changes in the manufacturing processes and by raw material selection.

Crystallographic Structure

Elemental carbon exists in nature as two crystalline allotropes: diamond [*7782-40-3*] and graphite [*7782-42-5*]. The diamond crystal structure (see page 683) is face-centered cubic with interatomic distances of 0.154 nm. Each atom is covalently bonded to four other carbon atoms in the form of a tetrahedron. Diamond is transformed to graphite in the absence of air at temperatures above 1500°C.

The accepted ideal structure for graphite was proposed by Bernal in 1924 (1). This structure is described as infinite layers of atoms of carbon which are arranged in the form of hexagons lying in planes (see page 691). The stacking arrangement is ABAB so that the atoms in alternate planes align with each other. The spacing between the layers is 0.3354 nm, the interatomic distance within the planes 0.1415 nm, and the crystal density 2.266 g/cm^3. A less frequently occurring structure is rhombohedral with a stacking arrangement of ABCABC in which the atoms of every fourth layer align with each other (2). This rhombohedral form, which occurs only in conjunction with the hexagonal form, is less stable, and converts to the hexagonal form at 1300°C.

There are six electrons in the carbon atom with four electrons in the outer shell available for chemical bonding. Three of the four electrons form strong covalent bonds with the adjacent in-plane carbon atoms. The fourth electron forms a less-strong bond of the van der Waals type between the planes. Bond energy between planes is 17 kJ/mol (4 kcal/mol) (3) and within planes 477 kJ/mol (114 kcal/mol) (4). The weak forces between planes account for such properties of graphite as good electrical conductivity, lubricity, and the ability to form interstitial compounds.

Terminology

Although the terms carbon and graphite are frequently used interchangeably in the literature, the two are not synonymous. The terms carbon, formed carbon, manufactured carbon, amorphous carbon, or baked carbon refer to products that result from the process of mixing carbonaceous filler materials such as petroleum coke, carbon blacks, or anthracite coal with binder materials of coal tar or petroleum pitch, forming these mixtures by molding or extrusion, and baking the mixtures in furnaces at temperatures from 800–1400°C. The term carbon–graphite designates a formed, composite product in which the filler materials, in addition to a carbon, such as petroleum coke, contain a graphitized carbon that has been heat-treated to high temperatures. The filler is material that makes up the body of the finished product (see Fillers). Green carbon refers to formed carbonaceous material that has not been baked.

Graphite, also called synthetic or artificial graphite, electrographite, manufactured graphite, or graphitized carbon refers to a carbon product that has been further heat-treated at a temperature exceeding 2400°C. This process of graphitization, described in a later section, changes not only the crystallographic structure, but also the physical and chemical properties.

With the development of nuclear and aerospace technologies, several new forms of carbon and graphite are being commercially produced. Products that are deposited on a heated graphite substrate by vapor phase decomposition of gaseous hydrocarbons, usually methane, at 1800–2300°C, are termed pyrolytic carbons (5). Chemical vapor deposited (CVD) carbon or pyrocarbon are other terms used to designate pyrolytic carbons. Pyrolytic graphite is a product resulting from high temperature annealing, and has a crystallite interlayer spacing similar to that of ideal graphite (6) (see Ablative materials; Nuclear reactors).

The term polymeric carbon is a generic term for products that result when high polymers with some degree of cross-linking are heated in an inert atmosphere. These organic starting materials do not coke; instead, they result in chars that represent a distinct group of materials possessing a graphite ribbon-network structure rather than extensive graphite sheets (7). Vitreous or glassy carbon prepared by carbonization of cellulose and phenolic or polyfurfuryl resins is an example of a polymeric carbon. Carbonization of acrylic fibers to produce carbon fibers is another example.

Because of the diversity of carbon and graphite forms, British, French, and German carbon groups have agreed to cooperate in improving the characterization of carbon solids. The ultimate aim is to prescribe standard primary and secondary methods of characterization as well as making available standard samples (8). In the United States, the ASTM has issued standard definitions of terms relating to manufactured carbon and graphite (9).

There is no standard industry-wide system for designating the various grades of carbon and graphite that are commercially available. Each of the more than 20 manufacturers has its own nomenclature to describe grades, sizes, and shapes available for specific purposes. The *Directory of Graphite Availability* (10) characterizes many of the materials that were available in 1967, but the catalogues and technical literature issued by carbon and graphite manufacturers must be consulted for current grade and property data. The data cited in the following sections are representative average values for commercially available materials.

History of the Industry

Graphite as it is found in nature has been known for several centuries for its use in making clay–graphite crucibles and its lubricating properties. The first known use was for drawing or writing, and it was because of this attribute that the German mineralogist A. G. Werner named graphite after the Greek word *graphein,* which means to write (11).

Manufacture of artificial graphite did not come about until the end of the 19th century. Its manufacture was preceded by developments mainly in the fabrication

Table 1. Applications for Manufactured Carbon and Graphite [a]

Aerospace	*Metallurgical*
nozzles	electric furnace electrodes for the production of iron and steel, ferroalloys, and nonferrous metals
nose cones	furnace linings for blast furnaces, ferroalloy furnaces, and cupolas
motor cases	aluminum pot liners and extrusion tables
leading edges	run-out troughs for molten iron from blast furnaces and cupolas
control vanes	metal fluxing and inoculation tubes for aluminum and ferrous furnaces
blast tubes	ingot molds for steel, iron, copper, and brass
exit cones	extrusion dies for copper and aluminum
thermal insulation	*Nuclear*
Chemical	moderators
heat exchangers and centrifugal pumps	reflectors
electrolytic anodes for the production of chlorine, aluminum, and other electrochemical products	thermal columns
electric furnace electrodes for making elemental phosphorus	shields
activated carbon	control rods
porous carbon and graphite	fuel elements
reaction towers and accessories	*Other*
Electrical	motion picture projector carbons
brushes for electrical motors and generators	turbine and compressor packing and seal rings
anodes, grids, and baffles for mercury arc power rectifiers	spectroscopic electrodes and powders for spectrographic analyses
electronic tube anodes and parts	structural members in applications requiring high strength-to-weight ratios
telephone equipment products	
rheostat disks and plates	
welding and gouging carbons	
electrodes in fuel cells and batteries	
contacts for circuit breakers and relays	
electric discharge machining	

[a] Ref. 15.

and processing of carbon electrodes. H. Davy is credited with using the first fabricated carbon in his experiments on the electric arc in the early 1800s. During the 19th century, several researchers received patents on various improvements in carbon electrodes. The invention of the dynamo and its application to electric current production in 1876 in Cleveland, Ohio, by C. F. Brush, provided a market for carbon products in the form of arc-carbons for street lighting. The work of a Frenchman, F. Carré, in the late 19th century, established the industrial processes of mixing, forming, and baking necessary for the production of carbon and graphite (12).

A significant development occurred when E. G. Acheson patented an electric resistance furnace capable of reaching approximately 3000°C, the temperature necessary for graphitization (13). This development was the beginning of a new industry where improved carbon and graphite products were used in the production of alkalies, chlorine, aluminum, calcium and silicon carbide, and for electric furnace production of steel and ferroalloys. In 1942 a new application for graphite was found when it was used as a moderator by E. Fermi in the first self-sustaining nuclear chain reaction (14). This nuclear application and subsequent use in the developing aerospace industries opened new fields of research and new markets for carbon and graphite. Carbon and graphite fibers are an example of a new form and a new industry. A list of major applications is shown in Table 1.

BIBLIOGRAPHY

"Active Carbon" under "Carbon" in *ECT* 1st ed., Vol. 2, pp. 881–899, by J. W. Hassler, Nuclear Active Carbon Division, West Virginia Pulp and Paper Company, and J. W. Goetz, Carbide and Carbon Chemicals Corporation; "Arc Carbon" under "Carbon" in *ECT* 1st ed., Vol. 2, pp. 899–915, by W. C. Kalb, National Carbon Company, Inc.; "Baked and Graphitized Products" under "Carbon" in *ECT* 1st ed., Vol. 3, pp. 1–34, by H. W. Abbott, Speer Carbon Company; "Carbon Black" under "Carbon" in *ECT* 1st ed., Vol. 3, pp. 34–65, and Suppl. 1, pp. 130–144, by W. R. Smith, Godfrey L. Cabot, Inc.; "Acetylene Black" under "Carbon" in *ECT* 1st ed., Vol. 3, pp. 66–69, by B. P. Buckley; Shawinigan Chemicals Ltd.; "Diamond" under "Carbon" in *ECT* 1st ed., Vol. 3, pp. 69–80, by C. V. R.; "Lampblack" under "Carbon" in *ECT* 1st ed., Vol. 3, pp. 80–84, by W. J. Colvin and W. E. Drown; Monsanto Chemical Company; "Natural Graphite" under "Carbon" in *ECT* 1st ed., Vol. 3, pp. 84–104, by S. B. Seeley and E. Emendorfer, Joseph Dixon Crucible Co.; "Structural and Specialty Carbon" under "Carbon" in *ECT* 1st ed., Vol. 3, pp. 104–112, by F. J. Vosburgh, National Carbon Company; "Activated Carbon" under "Carbon" in *ECT* 2nd ed., Vol. 4, pp. 149–158, by E. G. Doying, Union Carbide Corporation; "Baked and Graphitized Products, Manufacture" under "Carbon" in *ECT* 2nd ed., Vol. 4, pp. 158–202, by L. M. Liggett, Speer Carbon Company; "Baked and Graphitized Products, Uses" under "Carbon" in *ECT* 2nd ed., Vol. 4, pp. 202–243, by W. M. Gaylord, Union Carbide Corporation; "Carbon Black" under "Carbon" in *ECT* 2nd ed., Vol. 4, pp. 243–282, by W. R. Smith, Cabot Corporation, and D. C. Bean, Shawinigan Chemicals Ltd.; "Diamond, Natural" under "Carbon" in *ECT* 2nd ed., Vol. 4, pp. 283–294, by H. C . Miller, Super-Cut, Inc.; "Diamond, Synthetic" under "Carbon" in *ECT* 2nd ed., Vol. 4, pp. 294–303, by R. H. Wentorf, Jr., General Electric Research Laboratry; "Natural Graphite" under "Carbon" in *ECT* 2nd ed., Vol. 4, pp. 304–335, by S. B. Seeley, The Joseph Dixon Crucible Company.

1. J. D. Bernal, *Proc. R. Soc. (London) Ser. A* **106,** 749 (1924).
2. H. Lipson and A. R. Stokes, *Proc. R. Soc. (London) Ser. A* **181,** 101 (1942).
3. G. J. Dienes, *J. Appl. Phys.* **23,** 1194 (1952).
4. M. A. Kanter, *Phys. Rev.* **107,** 655 (1957).
5. D. B. Fischbach, *Chem. Phys. Carbon* **7,** 28 (1971).
6. A. W. Moore, *Chem. Phys. Carbon* **8,** 71 (1973).
7. G. M. Jenkins and K. Kawamura, *Polymeric Carbons—Carbon Fibre, Glass and Char,* Cambridge University Press, New York, 1976.
8. International Cooperation on Characterisation and Nomenclature of Carbon and Graphite, *Carbon* **13,** 251 (1975).

9. *Standard Definitions of Terms Relating to Manufacturing Carbon and Graphite, ASTM Standard C 709-75,* American Society for Testing and Materials, Philadelphia, Pa.
10. J. Glasser and W. J. Glasser (Chemical and Metallurgical Research, Inc.), *Directory of Graphite Availability, Report AFML-TR-67-113,* 2nd ed., U.S. Air Force Materials Laboratory, 1967.
11. F. Cirkel, *Graphite; its Properties, Occurrence, Refining and Use,* Department of Mines, Montreal, Canada, 1906.
12. F. Jehl, *The Manufacture of Carbons for Electric Lighting and Other Purposes,* "The Electrician" Printing and Publishing Co., Ltd., London, Eng., 1899.
13. U.S. Pat. 568,323 (Sept. 28, 1896), E. G. Acheson.
14. E. Fermi, *Collected Papers of Enrico Fermi,* Vol. 2, University of Chicago Press, Chicago, Ill., 1965.
15. E. L. Piper, *Preprint Number 73-H-14,* Society of Mining Engineers of AIME, 1973.

General References

Periodicals

Carbon, Pergamon Press, New York, 1963.
Tanso (Carbons), Tanso Zairyo Kenkyukai, Tokyo, Japan, 1949.

Conferences

Proceedings of the First (1953) and Second (1955) Conferences on Carbon, Waverly Press, Baltimore, Md., 1956.
Proceedings of the Third Conference (1957) on Carbon, Pergamon Press, New York, 1959.
Proceedings of the Fourth Conference (1959) on Carbon, Pergamon Press, New York, 1961.
Proceedings of the Fifth Conference (1961) on Carbon, Vol. 1, 1962, and Vol. 2, 1963, Pergamon Press, New York.
M. L. Deviney and T. M. O'Grady, *Petroleum Derived Carbons,* American Chemical Society, Washington, D.C., 1976.
Carbon Society of Japan, *Symposium on Carbon, July 20–23, 1964,* Carbon Society of Japan, Tokyo, 1964.
Symposium on Carbonization and Graphitization, 1968, Societé de Chimie Physique, Paris, Fr., 1968.
Industrial Carbon and Graphite, 1st, 1957, Society of Chemical Industry, London, Eng., 1958.
Industrial Carbon and Graphite, 2nd, 1965, Society of Chemical Industry, London, Eng., 1966.
Industrial Carbons and Graphite, 3rd, 1970, Society of Chemical Industry, London, Eng., 1971.
Carbon and Graphite Conference, 4th, 1974, Society of Chemical Industry, London, Eng., 1976.
Carbon '72, 1972, Deutsche Keramische Gesellschaft, Badhonnef, Ger., 1972.
Carbon '76, 1976, Deutsche Keramische Gesellschaft, Badhonnef, Ger., 1976.

Books

L. C. F. Blackman, *Modern Aspects of Graphite Technology,* Academic Press, New York, 1970.
R. L. Bond, *Porous Carbon Solids,* Academic Press, New York, 1967.
H. W. Davidson and co-workers, *Manufactured Carbon,* Pergamon Press, New York, 1968.
Carbones, Par le Groupe Francais d'Etude des Carbones, Masson et Cie, Paris, Fr., 1965.
C. L. Mantell, *Carbon and Graphite Handbook,* 3rd ed., Interscience Publishers, a division of John Wiley & Sons, Inc., New York, 1968.
Nouveau Traite de Chimie Minerale, Vol. 8, Part 1, Masson et Cie, Paris, Fr., 1968.
A. R. Ubbelohde and F. A. Lewis, *Graphite and its Crystal Compounds,* The Clarendon Press, Oxford, 1960.
P. L. Walker, Jr., ed., *Chemistry and Physics of Carbon; a Series of Advances,* Vol. 1, Marcel Dekker, New York.

J. C. LONG
Union Carbide Corporation

ACTIVATED CARBON

Activated carbon, a microcrystalline, nongraphitic form of carbon, has been processed to develop internal porosity. Activated carbons are characterized by a large specific surface area of 300–2500 m^2/g which allows the physical adsorption of gases and vapors from gases and dissolved or dispersed substances from liquids. Commercial grades of activated carbon are designated as either gas-phase or liquid-phase adsorbents. Liquid-phase carbons are generally powdered or granular in form; gas-phase, vapor-adsorbent carbons are hard granules or hard, relatively dust-free pellets. Activated carbons are widely used to remove impurities from liquids and gases and to recover valuable substances from gas streams (see Adsorptive separation).

Physical Properties

Surface area is the most important physical property of activated carbon. For specific applications, the surface area available for adsorption depends upon the molecular size of the adsorbate and the pore diameter of the activated carbon. Generally, liquid-phase carbons are characterized as having a majority of pores of 3 nm diameter and larger (1), whereas most of the pores of gas-phase adsorbents are 3 nm in diameter and smaller. Liquid-phase adsorbents require larger pores because of the need for rapid diffusion in the liquid and because of the large size of many dissolved adsorbates. Methods of testing adsorbency employ substances having a range of molecular sizes. Liquid-phase carbons are usually characterized by phenol, iodine, and molasses numbers. Gas-phase carbons are characterized by carbon tetrachloride and benzene adsorption activities.

The bulk density or apparent density of an activated carbon, together with its specific adsorptive capacity for a given substance, can be used to determine bed capacity in the design of an adsorption system or to determine grades of carbon required for an existing system.

The range of particle sizes of activated carbon is important: the rate of adsorption has been shown to depend inversely upon the particle size, (small particles having the fastest rates); however, in fixed beds pressure drop increases as particle size decreases.

The mechanical strength or hardness and the attrition resistance of the particles are important where pressure drop and carbon losses are a concern.

The kindling point of the carbon must be high enough to prevent excessive carbon oxidation in gas-phase adsorption where high heats of adsorption, particularly of ketones, are involved.

Chemical Properties

The most important chemical properties of activated carbon are the ash content, ash composition, and pH of the carbon. Discrepancies between the expected performance of an activated carbon, based upon surface area and pore-size distribution data, and actual adsorptive capacity can often be explained by oxygen-containing groups on the surfaces of the carbon. The pH or pK_a of the carbon, as a measure of surface acidity or basicity of the oxygen-containing groups, assists in predicting hydrophilicity and anionic or cationic adsorptive preferences of the carbon (2–5).

Manufacture and Processing

Almost any carbonaceous material of animal, plant, or mineral origin can be converted to activated carbon if properly treated. Activated carbon has been prepared from the blood, flesh, and bones of animals; it has been made from materials of plant origin, such as hardwood and softwood, corncobs, kelp, coffee beans, rice hulls, fruit pits, nutshells, and wastes such as bagasse and lignin. Activated carbon has also been made from peat, lignite, soft and hard coals, tars and pitches, asphalt, petroleum residues, and carbon black. However, for economic reasons, lignite, coal, bones, wood, peat, and paper mill waste (lignin) are most often used for the manufacture of liquid-phase or decolorizing carbons, and coconut shells, coal, and petroleum residues are used for the manufacture of gas-adsorbent carbons.

Activation of the raw material is accomplished by two basic processes, depending upon the starting material and whether a low or high density, powdered or granular carbon is desired: (*1*) Chemical activation depends upon the action of inorganic chemical compounds, either naturally present or added to the raw material, to degrade or dehydrate the organic molecules during carbonization or calcination. (*2*) Gas activation depends upon selective oxidation of the carbonaceous matter with air at low temperature, or steam, carbon dioxide, or flue gas at high temperature. The oxidation is usually preceded by a primary carbonization of the raw material.

Decolorizing carbons are coal- and lignite-based granules, or light, fluffy powders derived from low density starting materials such as sawdust or peat. Many decolorizing carbons are prepared by chemical activation. Some raw materials, such as bones, contain inorganic salts that impart some degree of activity to the carbon when the raw material is simply carbonized or heated in an inert atmosphere (2–3). Decolorizing carbons are usually prepared by admixing or impregnating the raw material with chemicals that yield oxidizing gases when heated or that degrade the organic molecules by dehydration. Compounds used successfully are alkali metal hydroxides, carbonates, sulfides, and sulfates; alkaline earth carbonates, chlorides, sulfates, and phosphates; zinc chloride; sulfuric acid; and phosphoric acid (2).

Gas- and vapor-adsorbing carbons may be prepared by the chemical activation process by using sawdust or peat as raw material and phosphoric acid, zinc chloride, potassium sulfide, or potassium thiocyanate as the activator (2–3). In some cases, the chemically activated carbon is given a second activation with steam to impart physical properties not developed by chemical activation.

Processes involving selective oxidation of the raw material with air or gases are also used to make both decolorizing- and gas-adsorbing carbons. In both instances, the raw material is activated in granular form. The raw material is carbonized first at 400–500°C to eliminate the bulk of the volatile matter and then oxidized with gas at 800–1000°C to develop the porosity and surface area. Some decolorizing carbons are oxidized with air at low temperature; however, because this reaction is exothermic, it is difficult to control and is suitable only for low activity carbon. The high temperature oxidation process with steam, carbon dioxide, or flue gas is endothermic, easier to control, and generally used more often (6–7).

Some gas-adsorbing carbons are made from hard, dense starting materials such as nutshells and fruit pits. These are carbonized, crushed to size, and activated directly to give hard, dense granules of carbon. In other cases, it is advantageous to grind the charcoal, coal, or coke to a powder, form it into briquettes or pellets with a tar or pitch binder, crush to size, calcine at 500–700°C, and then activate with steam or flue gas at 850–950°C. This method gives more easily activated particles because they possess more entry channels or macropores for the oxidizing gases to enter and for the reaction products to leave from the center of the particles.

The production of activated carbon from granular coal (8) as well as the activation of briquetted coal (9–10) has been described. In commercial plants, the raw material is carbonized in horizontal tunnel kilns, vertical retorts, or horizontal rotary kilns. Activation is accomplished in continuous internally or externally fired rotary retorts, or in large, cylindrical, multiple hearth furnaces where the charge is stirred and moved from one hearth to the next lower one by rotating rabble arms, or in large vertical retorts where the charge cascades over triangular ceramic forms as it moves downward through the furnace.

Fibrous activated carbon has been made by carbonizing infusible, cured phenolic fibers, air oxidizing, and then steam activating the fibers. The activated carbon fibers of 300–1000 m^2/g surface area are relatively strong and flexible with tensile strengths of at least 103 MPa (ca 15,000 psi). The manufacture of the activated fibers and of activated quilted fabrics has been reported (11).

Antipollution laws have increased the sales of activated carbon for control of air and water pollution (see Air pollution control methods). Pollution control regulations also affect the manufacture of activated carbons. Chemical activating agents have a large potential for the emission of corrosive acid gases during the activation process. If the emissions from chemical activation processes cannot be economically controlled, alternative selective oxidation methods will dominate. Selection of raw materials favoring low-sulfur materials will depend upon the local standards for permissible emission levels of sulfur oxides.

Economic Aspects

Activated carbon usage in the United States in 1976 was ca 90,000 metric tons valued at 9×10^7 dollars, or \$0.99/kg. Of the total activated carbon usage, 45,000 t was powdered carbon at an average price of \$0.62/kg and 45,000 t was granular carbon at an average price of \$1.32/kg (12–13). The capacities of activated carbon production of United States companies in 1976 are shown below (14–16):

Company	*Location*	*Capacity, metric tons*
Westvaco Corp.	Covington, Ky.	38,500
ICI United States	Marshall, Tex.	34,000
Calgon Corp.	Catlettsburg, Ky.	39,000
	Pittsburgh, Pa.	
Husky Industries	Romeo, Fla.	10,000
Barnebey-Cheney Co.	Columbus, Ohio	5,500
Union Carbide	Fostoria, Ohio	2,250
Witco Chemical	Petrolia, Pa.	2,250
Total		*131,500*

The 1976 production of gas- and vapor-adsorbent carbon in the United States is estimated to be 14,000 t. Gas-phase carbon consumption is much lower than liquid-phase carbon consumption, as there are fewer large volume applications and, in most applications, the carbon is regenerated repeatedly, remaining in service for several years. Prices were \$1.10–4.08/kg with special grades running as high as \$5.50/kg.

Since activated carbon manufacture is energy-intensive, prices increase with increases in the costs of fuels used to supply heat for calcination and activation of the carbon.

Regeneration of activated carbons is a major factor in the cost-effectiveness of the use of carbon (13). Liquid treatment costs can be minimized best by using the lowest effective carbon dosage of a carbon that retains its adsorptive capacity and mechanical strength after many thermal regeneration cycles (17).

Gas-phase activated carbons used for recovery of valuable process gases and solvents are responsible for the continued cost-effectiveness of many industrial processes, particularly solvent-based fiber and tape manufacture, dry cleaning, and rotogravure printing. Regeneration methods used are steam, thermal-vacuum, and pressure-swing desorption. Regeneration, pressure drop through carbon beds, carbon life, and system capital cost are the principal economic factors in gas and vapor recovery by activated carbon (18–19).

Specifications

Activated carbon is supplied on the basis of minimum values (eg, bulk density), maximum values (eg, moisture and ash content), and ranges of properties. In many cases, setting a narrow specification range for a property may considerably increase the price of the carbon. Specifications are based upon estimates and tests that indicate those properties of the carbon yielding the lowest cost of use or the maximum return on investment.

Table 1 shows some typical physical properties of activated carbons. These are approximate values; properties vary by grade and manufacturer within each raw material type.

Analytical and Test Methods

The American Society for Testing and Materials (ASTM), Committee D-28 on activated carbon, has defined several common property tests (20). These standard tests are listed in Table 2.

Many industrial activated carbon tests have not yet been standardized by ASTM but are in common use. The following are standard industrial tests: shaking a sample with several different sized steel balls for a standard time period reveals its hardness or strength; the percent hardness reported is based upon the decrease in average

Table 1. Physical Properties of Typical Activated Carbon Grades

	Liquid-phase carbon			Gas-phase carbon		
Property	Lignite base	Wood base	Bituminous coal base	Granular coal	Pelleted coke	Granular coconut
mesh (Tyler)	−100	−100	8–30	−4, +10	−6, +8	−6, +14
(mm)	(0.15)	(0.15)	(2.38–0.59)	(4.76, 1.70)	(3.36, 2.38)	(3.36, 1.18)
CCl_4 activity, % min	30	40	50	60	60	60
iodine no., min	500	700	950	1000	1000	1000
bulk density, g/mL, min	0.48	0.25	0.50	0.50	0.52	0.53
ash, % max	18	7	8	8	2	4

Table 2. ASTM Test Methods for Activated Carbon

Test	ASTM test no.
definition of terms relating to activated carbon	D2652
apparent density of activated carbon	D2854
liquid-phase evaluation of activated carbon	D2355
moisture in activated carbon	D2867
particle size distribution of granular activated carbon	D2862
total ash content of activated carbon	D2866

particle size of the carbon calculated from screen analyses before and after shaking with the steel balls.

Gas adsorptive properties of activated carbon may be statically determined by exposure to the desired substance at constant temperature and humidity levels in a bell jar. Adsorptive capacity is reported as weight percent pickup of the adsorbate by the carbon adsorbent. Gas adsorptive properties can be determined dynamically by blowing a test gas through a carbon bed or tube of a standardized depth and diameter. The equilibrium amount of adsorbate per unit weight of carbon is determined when the carbon bed becomes saturated with adsorbate. The most commonly applied gas adsorption tests are carbon tetrachloride and benzene activity tests. The carbon tetrachloride activity of a sample is its weight percent pickup, at equilibrium, of carbon tetrachloride from a dry air stream saturated with carbon tetrachloride. Benzene activity is based on the same principle, except that the benzene-saturated air stream is usually diluted to 10 vol % saturation (21). Another common gas-phase carbon test is retentivity. Both benzene and carbon tetrachloride retentivities measure the weight percent of adsorbate retained by a carbon sample first saturated with adsorbate and then subjected to air blowing for 6 h (2).

Activated carbon adsorption tests are useful in classifying carbons for general application areas and in manufacturing quality control. Although actual process fluids might appear more useful for testing the adsorptive powers of activated carbon than standard gas or liquid methods, process stream adsorbate and other impurity concentrations often vary within such wide limits that process fluid tests may be less reliable and less convenient than standard gas and liquid tests (2,22).

Storage

Activated carbon packaging includes bags, drums, cartons, railroad cars, and tank trucks. To avoid contamination and possible loss of properties, gas-phase activated carbon should not be exposed to vapors or moisture. Liquid-phase activated carbon may be stored as a water slurry. For storage of specially impregnated grades, activated carbon may be supplied in vapor-barrier drums.

Health and Safety Factors

Although activated carbon is not an easily combustible material (kindling points in pure oxygen are usually above 370°C), stored carbon should be kept away from heat, electricity, and flames. Dry carbon should not be allowed to contact strong oxidizing agents. Electrical controls used on or near activated carbon should be dust-tight and explosion-proof, and electrical motors should be totally enclosed. Foam or a fine water spray can extinguish activated carbon fires. When organic gases or vapors are adsorbed from an air stream, the carbon should be wet, steamed, or purged with inert gas prior to the first adsorption cycle since the initial heat of adsorption, especially for ketones, is sufficiently high to represent a fire hazard.

Use of activated carbon in protective masks does not protect the wearer from an oxygen depleted atmosphere's asphyxiation hazard (23). Also, an asphyxiation hazard exists inside activated carbon vessels because of adsorption of oxygen from the air onto the carbon. Proper ventilation and breathing equipment should be used by persons entering activated carbon vessels. Since flammable solvent vapors may be present, only explosion-proof lamps and spark-proof tools should be used inside the adsorber vessels. Dust-free loading of activated carbon into adsorbers may be accomplished by using special dry-loading equipment or by slurrying the carbon and wet-loading the adsorbers.

Uses

Liquid-Phase Carbon. Approximately 60% of the activated carbon manufactured for liquid-phase applications is used in powdered form. The two principal uses of powdered activated carbon are: (*1*) removal from solution of color, odor, taste, or other objectionable impurities such as those causing foaming or retarding crystallization, and (*2*) concentration or recovery from solution of a solute. Usually liquid treatment with powder is a batch process. The liquid to be treated is mixed with an appropriate amount of carbon predetermined by laboratory tests, heated if necessary to reduce viscosity, and agitated for 10–60 min. When the batch process is complete, the carbon is separated from the liquid by settling or filtration and is either discarded or eluted. The use of powdered carbons has been declining because granular carbons offer advantages in improved handling and reduced need for final filtering. The possibility of regenerating the carbon makes granular carbon more cost effective than powdered carbon. Increased exports to sugar-producing countries in Latin America mask the decline in domestic use of powdered carbon.

Granular carbon can be used in a continuous process where the liquid is slowly percolated through fixed beds of carbon until the carbon becomes saturated with adsorbate. The liquid stream is then diverted to a second adsorber, allowing the carbon in the first to be regenerated by hot gas or by solvent extraction. In applications where modest amounts of carbon are used, carbon saturated with adsorbed impurities is sent back to the manufacturer for regeneration in a multiple-hearth furnace.

Activated carbon is widely used to remove color from sugar (qv) (2). The use of carbon in glucose manufacture removes protein and hydroxymethylfurfural from the syrup, rendering it colorless and stable.

Although there are many chemical, physical, and biological methods of treating water contaminated with industrial and municipal wastes to produce safe and palatable drinking water, none have the potential of activated carbon treatment (24). Most methods require careful control for the effective removal of taste and odor contaminants. Activated carbon is a broad-spectrum agent that effectively removes toxic or biorefractive substances (25). Insecticides, herbicides, chlorinated hydrocarbons, and phenols, typically present in many water supplies, can be reduced to acceptable levels by activated carbon treatment. Powdered activated carbon is most effectively used in smaller water treatment systems where the equipment investment required for granular carbon would be prohibitive. Granular carbon systems may consist of moving beds, beds in series, beds in parallel, or expanded beds (26). Expanded-bed systems require water flow rates sufficient to expand the bed at least to 115% of static bed volume. Removal of organics by expanded-bed systems is comparable to fixed beds without the need for frequent backwashing required by fixed beds (27). Granular water-treatment carbons are thermally regenerated for reuse, thus reducing the cost of treatment.

Industrial and municipal waste water treatment will be an expanding market for liquid-phase activated carbons (14,28). *The 1972 Federal Water Pollution Control Act* expanded the authority of the EPA to conduct research for the improvement of municipal and industrial water treatment (29). Thus government and industry may cooperate through an EPA cost-sharing grant system in the improvement and application of activated carbon for effective, economical waste water treatment. The EPA program for advanced industrial waste water treatment will be especially concerned with activated carbon regeneration technologies (16) (see Water).

The dry cleaning industry uses powdered and granular carbons for reclaiming liquid solvent that has become contaminated with dyes and rancid extracted oils and grease. Carbon use is cheaper than distillation, and carbon more completely removes odor-causing substances.

Activated carbon has many applications in food and pharmaceutical manufacturing. Such uses include purification and color removal during the processing of fruit juices, honey, maple syrup, candy, soft drinks, and alcoholic beverages (2). Pharmaceutical uses include removal of pyrogens from solutions for injections, vitamin decolorizing and deodorizing, and insulin purification. Activated carbon (drug-pure grade) can be administered orally to poison victims (30). Research is being conducted for the development of activated carbon filters for artificial kidney devices and for the dialysis (qv) of poisons and drugs (31–32).

Activated carbon is used in electroplating (qv) to remove organic impurities from the bath. Carbon may also be used to allow the recovery of gold from cyanide-leach pulps (33). Gold may be recovered by burning the carbon or by eluting the gold (34). Activated carbon is not commonly used for the purification of inorganic chemicals (1,35).

Miscellaneous uses for liquid-phase carbons include laboratory uses, reaction catalysis, and aquarium water filters.

Gas-Phase Carbon. Gas-phase carbons typically have a surface area of 1000–2000 m^2/g and are made in larger particle sizes of greater strength and density than liquid-phase carbons. The small (3 nm and less) pores of gas-phase carbons provide high adsorptive capacity and selectivity for gases and organic vapors. Desirable characteristics of a good gas adsorbent carbon are: high adsorptive capacity per unit volume; high retentive capacity; high preferential adsorption of gases in the presence of moisture; low resistance to gas flow; high strength or breakage resistance; and complete release of adsorbates at increased temperatures and decreased pressures.

The largest single application for gas-phase carbon is in gasoline vapor emission control cannisters on automobiles. Evaporative emissions from both fuel tank and carburetor are adsorbed on the carbon (33,36). An evaporative control carbon should have good hardness, a high vapor working capacity, and a high saturation capacity (37). The working capacity of a carbon for gasoline vapor is determined by the adsorption–desorption temperature differential, the rate at which purge air flows through the carbon cannister, and by the extent to which irreversibly adsorbed, high molecular weight gasoline components accumulate on the carbon (15).

Another application for gas-phase carbons is the purification and separation of gases. Many industrial gas streams are treated with activated carbon to remove impurities or recover valuable constituents. Odors are removed from air in activated carbon air conditioning systems, and gas masks containing activated carbon are used to protect individuals from breathing toxic gases or vapors (38).

The purification of industrial gases is usually performed with gas under pressure in a two-bed system. The carbon is supported in tall vertical towers having one on stream as the other is regenerated, dried, or cooled. Regeneration is accomplished by temperature or pressure change or both, with or without a purge gas.

Activated carbon is used in air conditioning (qv) systems to remove industrial odors and irritants from building inlet air and to remove body, tobacco, and cooking odors, etc from recirculated building air (39). The use of carbon to remove odors allows more economical building operation by permitting higher percentages of recycled air to be used, thus minimizing the need for heating or cooling large amounts of fresh inlet air (40). Manufacturers use carbon to remove objectionable odors from process exhaust gases, and to remove corrosive gases and vapors, eg, sulfur compounds, from intake air to protect electrical switching equipment. Activated carbon also prolongs the storage life of fruits and flowers by adsorbing gases that contribute to odor and deterioration of these products (2).

Since air is circulated by low-pressure fans in air conditioning systems, activated carbon is always used in thin beds from 13–26 mm thick. The carbon is usually supported in cannisters of perforated sheet metal that are either flat, cylindrical, or corrugated. The effective life of the carbon depends on the application but may range from several months to 1.5 yr (41).

Activated carbon is used in nuclear reactor systems to adsorb radioactive gases in carrier or coolant gases, and from the air in reactor emergency exhaust systems (42). Radioactive gas adsorbers are very deep beds of carbon used to adsorb and retain radioactive materials long enough for isotopes to decay to safe levels of radioactivity before they are discharged into the atmosphere. Radioactive iodine and organic iodides are adsorbed by specially impregnated grades of high-activity carbon (43). The radioactive fission products, krypton and xenon, are adsorbed for decay to safe isotopes before they are released to the atmosphere (44–45).

The use of solvent recovery-grade activated carbon for the recovery of volatile organic compounds from process air streams is important to the operating economics of several industries (18,46). A simple activated carbon recovery system consists of an air filter, a blower, two horizontal adsorbers (each containing a bed of carbon 0.3–0.6 m deep), a vapor condenser, and a solvent decanter or continuous still (19).

Some solvent recovery plants are manually operated, but most are automatic, using a time–cycle controller. The newer, larger industrial units allow adsorption cycle times to be controlled according to variable vapor loadings in the process air stream (47).

Specially impregnated grades of activated carbon in cigarette filters adsorb some of the harmful components of tobacco smoke (48). Other applications for activated carbon include use in kitchen range hoods, gas sampling tubes (49), refrigerator deodorizers, and as getters used in pill bottles and vacuum equipment for adsorption of harmful contaminant vapors or to aid in attaining high vacuum.

The catalytic oxidation of compounds by activated carbon has been recognized for many years. It is generally conceded that this action results from the presence of irreversibly adsorbed oxygen on the carbon surface. Many salts, such as ferrous sulfate, sodium arsenite, potassium nitrite, and potassium ferrocyanide, are oxidized in solution by activated carbon; the reaction can be made continuous by passing both solution and air over granular carbon (50).

Hydrogen sulfide in air is oxidized to sulfur when passed over activated carbon. It can be removed from manufactured gas by adding a small amount of air to the gas before bringing it in contact with the carbon (51).

Activated carbon is used as the catalyst in the manufacturing of phosgene (qv) from carbon monoxide and chlorine, and sulfuryl chloride is produced by the reaction of sulfur dioxide with chlorine in the presence of activated carbon.

BIBLIOGRAPHY

"Active Carbon" under "Carbon" in *ECT* 1st ed., Vol. 2, pp. 881–899, by J. W. Hassler, Nuclear Active Carbon Division, West Virginia Pulp and Paper Company, and J. W. Goetz, Carbide and Carbon Chemicals Corporation; "Activated Carbon" under "Carbon" in *ECT* 2nd ed., Vol. 4, pp. 149–158, by E. G. Doying, Union Carbide Corporation, Carbon Products Division.

1. A. J. Juhola, *Carbon* **13,** 437 (Nov. 5, 1975).
2. J. W. Hassler, *Purification with Activated Carbon,* Chemical Publishing Co., Inc., New York, 1974.
3. M. Smisek and S. Cerny, *Active Carbon,* American Elsevier Publishing Company, Inc., New York, 1970.
4. Y. Matsumura, *J. Appl. Chem. Biotechnol.* **25,** 39 (Jan. 1975).
5. R. Prober and co-workers, *AIChE J.* **21,** 1200 (Nov. 1975).
6. N. K. Chaney, *Trans. Am. Electrochem. Soc.* **36,** 91 (1919); U.S. Pat. 1,497,544 (June 10, 1924), N.-K. Chaney.
7. A. B. Ray, *Chem. Metall. Eng.* **30,** 977 (1923).
8. U.S. Pat. 3,876,505 (Apr. 8, 1975), G. R. Stoneburner (to Calgon Corp.).
9. A. C. Fieldner and co-workers, *U.S. Bur. Mines Tech. Paper* **479,** (1930).
10. A. E. Williams, *Min. J. (London)* **227,** 1026 (Dec. 28, 1946).
11. J. Economy and R. Y. Lin, *Appl. Polym. Symp.,* 199 (Nov. 29, 1976); U.S. Pats. 3,769,144 (1973), and 3,831,760 (1974), J. Economy and R. Y. Lin (to Carborundum Co.).
12. *Chem. Eng. News* **52,** 7 (July 22, 1974); W. J. Storck, *Chem. Eng. News* **55,** 10 (Apr. 18, 1977).
13. *Chem. Week* **118,** 44 (Jan. 14, 1976).
14. *Chem. Eng. News,* **55,** 10 (Apr. 18, 1977).
15. U.S. Dept. of Health, Education, and Welfare, *Fed. Reg.* **33**(108), (June 4, 1968).
16. *Chem. Eng. News,* **56,** 10 (Apr. 3, 1978).
17. *Chem. Eng. (N. Y.)* **82,** 113 (Apr. 28, 1975).
18. J. C. Enneking, *Air Pollution Control—Part II, ASHRAE Bulletin CH 73-2,* American Society of Heating, Refrigerating & Air Conditioning Engineers, July, 1973, p. 20.
19. R. R. Manzone and D. W. Oakes, *Pollut. Eng.* **5,** 23 (Oct. 1973).
20. *Annual Book of ASTM Standards, Part 30,* American Society for Testing and Materials, Philadelphia, Pa., 1974.
21. *Japanese Industrial Standard K 1412-1958* (reaffirmed 1967).
22. J. J. Kipling, *Adsorption from Solutions of Non-Electrolytes,* Academic Press, New York, 1965.
23. C. L. Mantell, *Adsorption,* McGraw-Hill Book Company, New York, 1951.
24. *Ind. Res.,* 30 (Jan. 1976).
25. *Chem. Eng. (N. Y.)* **77,** 32 (Sept. 7, 1970).
26. *Pollut. Eng.* **8,** 24 (July 1976).
27. L. D. Friedman and co-workers, *Improving Granular Carbon Treatment,* U.S. Environmental Protection Agency, Grant 17020 GDN, July 1971.
28. *Chem. Week,* 40 (Feb. 16, 1977).
29. G. Rey and co-workers, *Chem. Eng. Prog.* **69,** 45 (Nov. 1973).
30. U.S. Pat. 3,917,821 (Nov. 4, 1975), M. Manes.

31. J. H. Knepshield and co-workers, *Trans. Am. Soc. Artif. Int. Organs* **19,** 590 (1973).
32. T. A. Davis, *Activated Carbon Filters for Artificial Kidney Devices,* Southern Research Institute, Report No. AK-2-72-2208, 1973.
33. J. B. Zadra, *U.S. Bur. Mines Rep. Invest.,* 4672 (1950).
34. U.S. Pat. 3,935,006 (Mar. 19, 1975), D. D. Fischer (to U.S. Department of Interior).
35. D. N. Strazhesko, ed., *Adsorption and Adsorbents,* John Wiley & Sons, Inc., New York, 1973.
36. J. O. Sarto and co-workers, *Society of Automotive Engineers Paper No. 700150,* New York, 1970.
37. R. S. Joyce and co-workers, *Society of Automotive Engineers Paper No. 690086,* New York, 1969.
38. *Am. Ind. Hyg. Assoc. J.* **32,** 404 (June 1971).
39. H. Sleik and A. Turk, *Air Conservation Engineering,* 2nd ed., Connor Engineering Co., Danbury, Ct., 1953.
40. M. Beltran, *Chem. Eng. Prog.* **70,** 57 (May 1974).
41. *ASHRAE Guide and Data Book,* American Society of Heating, Refrigerating & Air Conditioning Engineers, Inc., New York, 1970.
42. R. J. Bender, *Power* **116,** 56 (Sept. 1972).
43. A. G. Evans, *paper presented at 13th AEC Air Cleaning Conference,* 1974, Report DP-MS-74-2 (NSA, 30, 23732).
44. P. J. Geue, *Australian Atomic Energy Commission Research Establishment,* Report AAEC LIB/BIB 401 (1973) (NSA 29,18310).
45. M. N. Myers, *Literature Survey: Sampling, Plate-Out and Cleaning of Gas-Cooled Reactor Effluents,* General Electric Co., Report XDC 61-4-702 (TID 14,469) (NSA 16,6114), 1961.
46. C. R. Wherry, *Activated Carbon, Report 731-2020,* in *Chemical Economics Handbook,* Stanford Research Institute, Menlo Park, Calif., 1969.
47. B. Dundee, *Gravure,* (June 1972).
48. U.S. Pat. 3,355,317 (Mar. 18, 1966), C. H. Keith II and V. Norman (to Liggett and Myers Tobacco Co.).
49. A. Turk and co-workers, eds., *Human Responses to Environmental Odors,* Academic Press, New York, 1974.
50. U.S. Pat. 2,365,729 (Dec. 26, 1944), E. A. Schumacher and G. W. Heise (to Union Carbide Corporation).
51. H. Krill and K. Storp, *Chem. Eng.* (*N. Y.*) **80,** 84 (July 23, 1973).

General Bibliography

D. G. Hager, *Chem. Eng. Prog.* **72,** 57 (Oct. 1976).
R. A. Hutchins, *Chem. Eng.* (*N. Y.*) **80,** 133 (Aug. 20, 1973).
W. G. Timpe and co-workers, *Kraft Pulping Treatment and Reuse—State of the Art,* U. S. Environmental Protection Agency, Grant 12040 EJU, Dec. 1973 (EPA-660/2-75-004).
E. W. Long and co-workers, *Activated Carbon Treatment of Unbleached Kraft Effluent for Reuse,* U.S. Environmental Protection Agency, Grant 12040 EJU, Dec. 1973 (EPA-R2-73-164).

R. W. SOFFEL
Union Carbide Corporation

BAKED AND GRAPHITIZED CARBON

Raw Materials

The raw materials used in the production of manufactured carbon and graphite largely determine the ultimate properties and practical applications of the finished products. This dependence can be attributed to the nature of carbonization and graphitization processes.

Throughout the entire process of the thermal conversion of organic materials to carbon and graphite, the natural chemical driving forces cause the growth of larger and larger fused-ring aromatic systems, and ultimately result in the formation of the stable hexagonal carbon network of graphite. Differences in the final materials depend upon the ease and extent of completion of these overall chemical and physical ordering processes.

The first few steps in the carbonization process can be considered as a dehydrogenative polymerization where hydrocarbon molecules lose hydrogen and combine to form larger planar molecular networks. As the process moves toward carbon, solid-state reorganization and recrystallization processes begin to take place. In these processes, a gradual improvement in both in-plane and stacking perfection occurs; the result is the three-dimensionally ordered graphite structure with ABAB stacking (see page 691).

The starting materials in the carbonization process may be partly or entirely aromatic hydrocarbons or heterocyclics derived from coal or petroleum. After some heat treatment, the partially-polymerized products constitute a pitch, which is a complex mixture of many hundreds or even thousands of aromatic hydrocarbons with 3–8 condensed rings and an average molecular weight of 300. Although the individual, pure compounds may melt at fairly high temperatures (>300°C), the pitch mixture acts as a eutectic and softens at a much lower temperature, eg, 50–180°C. Both liquid and solid pitches are isotropic materials, ie, perfectly random in their mutual molecular orientation.

Coke is the next stage in the process of carbonization. Coke consists of aromatic polymers of much higher molecular weights, >3000. In contrast to pitch, coke is a totally infusible solid and is generally anisotropic, ie, the flat molecules within fairly large domains can all have nearly the same orientation.

Coke is well-oriented compared with pitch because an easily oriented liquid crystal or mesophase of aromatic molecules forms during the pitch-to-coke transformation. As isotropic pitch is heated above 400°C, small anisotropic mesophase spheres appear which grow, coalesce, and finally form large anisotropic regions (1). As polymerization continues, these anisotropic regions become very viscous or semisolid and are essentially coke. Such well-oriented coke retains this ordered structure and can be converted to crystalline graphite by high temperature heat treatment. Fine insoluble solids in the pitch are not incorporated within the growing mesophase sphere but tend, instead, to aggregate on the surfaces of the spheres and may modify the coalescence, usually resulting in a less anisotropic coke. Pitches that contain this mesophase material are now generally known as mesophase pitches; the term mesophase is derived from the Greek mesos or intermediate and indicates the pseudo-crystalline nature of this highly oriented anisotropic material.

In order to produce useful carbon and graphite bodies, filler and binder materials are mixed, formed by molding or extrusion, and finally baked, or baked and graphitized, to yield the desired shaped carbon or graphite bodies. More than 30 different raw materials are used in the manufacture of carbon and graphite products. The primary materials, in terms of tonnage consumed, are the petroleum coke or anthracite coal fillers and the coal-tar or petroleum pitch binders and impregnants. Other materials, called additives, are often included to improve processing conditions or to modify certain properties in the finished products.

Filler Materials. ***Petroleum Coke.*** Petroleum coke is produced in large quantities in the United States as a by-product of the petroleum cracking process used to make gasoline and other petroleum products. It is the largest commercially available source of synthetically produced carbon that can be readily graphitized by heating above 2800°C.

Most of the petroleum coke is produced by the delayed coking process (2). The properties of cokes vary according to the feedstock and to operating variables in the delayed coker unit such as operating pressure, recycle ratio, time, and coker heater outlet temperature.

The cokes to be used in any one product are selected to provide optimum quality. The major quality factors are sulfur content, volatiles, ash, and thermal expansion. High sulfur content is undesirable because of environmental concerns and because it leads to uncontrolled expansion (puffing) during graphitization, resulting in cracking of the product. The volatiles in raw cokes affect the quality of the calcined coke; high levels of volatiles lead to low density and low strength in the calcined product. High ash content is usually undesirable and leads to contamination of the finished product. Also, if these ash contaminants are present in the feedstock from which the coke is made, they impede the formation of desired crystal structure during the coking and lower the strength and density of the final coke product.

High-quality or needle cokes have low thermal expansion, low ash content, high density, and good crystal structure as indicated by x-ray analysis. Such needle cokes are used primarily in the manufacture of large graphite products such as electrodes for electric steel-making furnaces. Cokes of lower quality are known as metallurgical-grade cokes.

Needle cokes are produced from aromatic feedstocks such as decant oils from a refinery catalytic cracking unit or tars made by thermally cracking gas oils. Other feedstocks, such as atmospheric or vacuum residues from a refinery, usually produce sponge cokes. Usually the more aromatic the feedstock, the higher the coke quality will be, since larger, more highly ordered domains of oriented molecules will form during the mesophase stage of the coking reaction.

Sponge cokes are used primarily as fuel and in the manufacture of anodes for the aluminum industry. For every kilogram of aluminum produced, approximately 0.4 kg of petroleum coke is consumed (3).

The properties required in cokes for the aluminum industry are different from those required in needle cokes. Metallic impurities in the coke are more critical and a low ash content is essential since impurities from the coke tend to concentrate in the aluminum as the anode is consumed.

Particularly important impurities are vanadium, nickel, iron, and silicon; other impurities such as sulfur are slightly less critical. Since the carbon consumption per kilogram of aluminum produced affects the economics of the process, bulk density is also an important property (see Aluminum).

Table 1 shows a comparison of properties of typical sponge- and needle-grade cokes.

Natural Graphite. Natural graphite is a crystalline mineral form of graphite occurring in many parts of the world (see Carbon, natural graphite).

Carbon Blacks. Carbon blacks are commonly used as components in mixes to make various types of carbon products (see Carbon, carbon black).

Table 1. Typical Properties of Sponge and Needle Cokes

Property	Aluminum anode grade sponge coke		Graphite electrode-grade needle coke	
	Raw	Calcined[a]	Raw	Calcined[a]
sulfur, wt %	2.5	2.5	1.0	1.0
ash, wt %	0.25	0.50	0.10	0.15
vanadium, ppm	150[b]	200[b]	10	10
nickel, ppm	150[b]	200[b]		20–40
silicon, wt %	0.02	0.02	0.04	0.04
volatile matter, wt %	10–12		8	
resistivity, μΩ·m (particles, −35 mesh to +65 mesh Tyler, 0.49 to 0.23 mm)		890		965
real density, g/cm^3		2.06		2.12
bulk density, g/cm^3		0.80		0.88
coefficient of thermal expansion of graphite per °C (30–100°C)		20×10^{-7}		5×10^{-7}

[a] Calcining, discussed later in this section, is a thermal treatment which removes volatiles from the raw materials and shrinks the particles.
[b] Ref. 4.

Anthracite. Anthracite is preferred to other forms of coal in the manufacture of carbon products because of its high carbon-to-hydrogen ratio, its low volatile content, and its more ordered structure. It is commonly added to carbon mixes that are used for fabricating metallurgical carbon products to improve specific properties and reduce cost. Anthracite is used in mix compositions for producing carbon electrodes, structural brick, blocks for cathodes in aluminum manufacture, and in carbon blocks and brick used for blast furnace linings (see Coal).

Pitches. Carbon articles are made by mixing a controlled-size-distribution of coke filler particles with a binder such as coal tar or petroleum pitch. The mix is then formed by molding or extruding and is heated in a packed container to control the shape and set the binder. This brief description of the carbon-making process serves to identify the second most important raw material for making a carbon article, the pitch binder. The pitch binder preserves the shape of the green carbon and also fluidizes the carbon particles, enabling them to flow into an ordered alignment during the forming process. During the subsequent baking steps, the pitch binder is pyrolyzed to form a coke that bridges the filler particles and serves as the permanent binding material. These carbon bridges provide the strength in the finished article and also provide the paths for energy flow through thermal and electric conductance (see Tar and pitch).

A binder used in the manufacture of electrodes and other carbon and graphite products must (*1*) have high carbon yield, usually 40–60 wt % of the pitch; (*2*) show good wetting and adhesion properties to bind the coke filler together; (*3*) exhibit acceptable softening behavior at forming and mixing temperature, usually in the range of 90–180°C; (*4*) be low in cost and widely available; (*5*) contain only a minor amount of ash and extraneous matter that could reduce strength and other important physical properties; and (*6*) produce binder coke that can be graphitized to improve the electrical and thermal properties.

The primary binder material, coal-tar pitch, is produced primarily as a by-product of the destructive distillation of bituminous coal in coke ovens during the production of metallurgical coke. The second largest source of binder is petroleum pitch obtained from the cracking of petroleum in refinery processes. Manufacturers of carbon products require pitches with various softening points, depending upon the products to be made and the processing conditions required in forming and baking. The softening behavior and rheology of the pitch are important factors in establishing the forming conditions of the carbon product (see Rheological measurements). Baking schedules are then developed to improve the carbon yield and to reduce energy input to the furnace.

Petroleum Pitch. Although coal-tar pitch is a main source of binder pitch for the carbon industry, pitches from other precursors are becoming increasingly important. Petroleum pitch has been investigated extensively as a potential source of aluminum cell anode pitch (5). These studies have shown that pitch from petroleum sources, if produced from the proper charging stock, can be the approximate equivalent in quality of coal-tar pitch for manufacture of carbon products. Petroleum pitch is produced by cracking petroleum liquids at 450–500°C, 690–1380 kPa (100–200 psi). The cracking step produces both light and heavy feedstocks. The heavy feedstock is then processed by heat soaking at ca 350°C to form a viscous pitch.

The primary differences (shown in Table 2) between petroleum pitches and coal-tar pitches are in viscosity, benzene insolubles (BI), and quinoline insolubles (QI). In the graphite industry, petroleum pitch is used as an impregnant to increase the density of carbon and graphite products because of its low QI content.

Certain compounds found in some coal-tar and petroleum pitches are carcinogenic. Individuals working with pitches or exposed to fumes or dust should wear protective clothing to avoid skin contact. Respirators should be worn when pitch dust or fume concentrations in the air are above established limits (see Industrial hygiene and toxicology).

Additives. In addition to the primary ingredients in the mixes (the fillers and binders), minor amounts of other materials are added at various steps in the process. Although the amounts of these additives are usually in small percentages, they play an important role in the economics of the process and determine the quality of the final products. Light extrusion oils and lubricants, including light petroleum oils, waxes, and fatty acids and esters, are often added to the mix to improve the extrusion rates and structure of extruded products. Inhibitors are used to reduce the detrimental effects of sulfur in high sulfur cokes. Iron oxide is often added to high sulfur coke to

Table 2. Comparison of Typical Physical Properties of Petroleum and Coal-Tar Pitches

Property	Petroleum	Coal tar
softening point, cube-in-air, °C	120	110
sp gr, 15.5–26.7°C	1.22	1.33
coking value, wt %	51	58
benzene insolubles (BI), wt %	3.6	33
quinoline insolubles (QI), wt %	none	14
ash, wt %	0.16	0.10
sulfur, wt %	1.0	0.8
viscosity in Pa·s[a] at		
160°C	0.8	1.4
177°C	0.3	0.4
199°C	0.1	0.2

[a] To convert Pa·s to poise, multiply by 10.

prevent puffing, the rapid swelling of the coke caused by volatilization of the sulfur at 1600–2400°C. Iron from Fe_2O_3 or other iron compounds prevents this action by forming a more stable iron sulfide, which reduces the gas pressure in the coke particles. Other sulfide-forming compounds such as sodium, nickel, cobalt, and vanadium may also be used. Sodium carbonate is often used in applications requiring a low ash product.

Calcining

Coke and pitch purchased by carbon companies from suppliers must meet rigid specifications. Some materials, such as pitch and natural graphite, may be used as received from suppliers; other materials, such as raw coke and anthracite, require calcining, a thermal treatment to temperatures above 1200°C.

Calcining consists of heating raw filler to remove volatiles and to shrink the filler to produce a strong, dense particle. Raw petroleum coke, eg, has 5–15% volatile matter; when the coke is calcined to 1400°C, it shrinks approximately 10–14%. Less than 0.5% of volatile matter in the form of hydrocarbons remains in raw coke after it is calcined to 1200–1400°C. During calcination the evolving volatiles are primarily methane and hydrogen which burn during the calcining process to provide much of the heat required. The calcining step is particularly important for those materials used in the manufacture of graphite products, such as electrodes, since the high shrinkages occurring in raw coke during the baking cycle of large electrodes would cause the electrode to crack. To prevent partial fusion of the coke during calcining, the volatile content of the green coke is kept below 12%.

Anthracite is calcined at appreciably higher temperatures (1800–2000°C). The higher calcining temperatures for anthracite are necessary to complete most of the shrinkage and to increase the electrical conductivity of the product for use in either Soderberg or prebaked carbon electrodes for aluminum or phosphorus manufacture. Some other forms of carbon used in manufacturing of carbon products, such as carbon black, are also calcined.

The selection of calcining equipment depends upon the temperatures required and the materials to be calcined. Two major types of calcining units are used for cokes: the horizontal rotary drum-type calciner and the vertical rotary hearth-type calciner. Modifications of these have been developed for particular applications and to reduce energy input and product loss.

Rotary-type calciners for calcining raw petroleum coke are similar in design to those used for calcining limestone and cement (6–7).

The conventional rotary calciner is energy-intensive, and 10–15% loss of coke is experienced through oxidation. Newer calciners have been designed to prevent excessive loss of product and to reduce fuel consumption.

Since anthracite must be calcined at higher temperatures than can be reached reasonably in conventional gas-fired kilns, an electrically heated shaft kiln is used to calcine the coal (qv) at temperatures up to 2000°C (8).

BIBLIOGRAPHY

"Baked and Graphitized Products" under "Carbon" in *ECT* 1st ed., Vol. 3, pp. 1–34, by H. W. Abbott, Speer Carbon Company; "Baked and Graphitized Products, Manufacture" under "Carbon" in *ECT* 2nd ed., Vol. 4, pp. 158–202, by L. M. Liggett, Speer Carbon Company.

1. J. D. Brooks and G. H. Tayler, *Chem. Phys. Carbon* **4,** 243 (1968).
2. K. E. Rose, *Hydrocarbon Process,* **50,** 85 (Nov. 7, 1971).
3. C. B. Scott, *Chem. Ind. London,* 1124 (July 1, 1967).
4. C. B. Scott and J. W. Connors, *Light Metals, 1971,* American Institute of Mining, Metallurgical, and Petroleum Engineers, Inc., New York, 1972, p. 277.
5. L. F. King and W. D. Robertson, *Fuel* **47,** 197 (1968).
6. R. F. Wesner, P. T. Luckie, and E. A. Bagdoyan, *Light Metals, 1973,* Vol. 2, American Institute of Mining, Metallurgical, and Petroleum Engineers, Inc., New York, 1973, p. 629.
7. V. D. Allred in ref. 4, p. 313.
8. M. M. Williams, *Light Metals, 1972,* American Institute of Mining, Metallurgical, and Petroleum Engineers, Inc., New York, 1972, p. 163.

General References

E. Wege, *High Temp. High Pressures* **8,** 293 (Nov. 3, 1976).
J. M. Hutcheon in L. C. F. Blackman, ed., *Modern Aspects of Graphite Technology,* Academic Press, New York, 1970, Chapt. 2.
M. L. Deviney and T. M. O'Grady, eds., *Petroleum Derived Carbons,* American Chemical Society, Washington, D.C., 1976.

L. L. WINTER
Union Carbide Corporation

PROCESSING OF BAKED AND GRAPHITIZED CARBON

Raw Material Preparation

Crushing and Sizing. Calcined petroleum coke arrives at the graphite manufacturer's plant in particle sizes ranging typically from dust to 5–8 cm dia. In the first step of artificial graphite production the run-of-kiln coke is crushed, sized, and milled to prepare it for the subsequent processing steps. The degree to which the coke is broken down depends on the grade of graphite to be made. If the product is to be a fine-grained variety for use in aerospace, metallurgical, or nuclear applications, the milling and pulverizing operations are used to produce sizes as small as a few micrometers in diameter. If, on the other hand, the product is to be coarse in character for products like graphite electrodes used in the manufacture of steel, a high yield of particles up to 1.3 cm dia is necessary.

The wide variety of equipment available for the crushing and sizing operations is well-described in the literature (1–2). Roll crushers are commonly used to reduce the incoming coke to particles that are classified in a screening operation. The crushed coke fraction, smaller than the smallest particle needed, is normally fed to a roll or hammer mill for further size reduction to the very fine (flour) portion of the carbon mix. A common flour sizing used in the graphite industry contains particles ranging from 149 μm (100 mesh) to a few micrometers, with about 50% passing through a 74-μm (200-mesh) screen.

For a coarse-grained (particle-containing) graphite, the system depicted in Figure 1 is typical. The run-of-kiln coke is brought in on railroad cars and emptied into pits where the coke is conveyed to an elevator. The elevator feeds a second conveyer which empties the coke into any one of a number of storage silos where the coke is kept dry. The manufacturer usually specifies a maximum moisture content in the incoming coke, at about 0.1–0.2%, to ensure that mix compositions are not altered by fluctuations in moisture content.

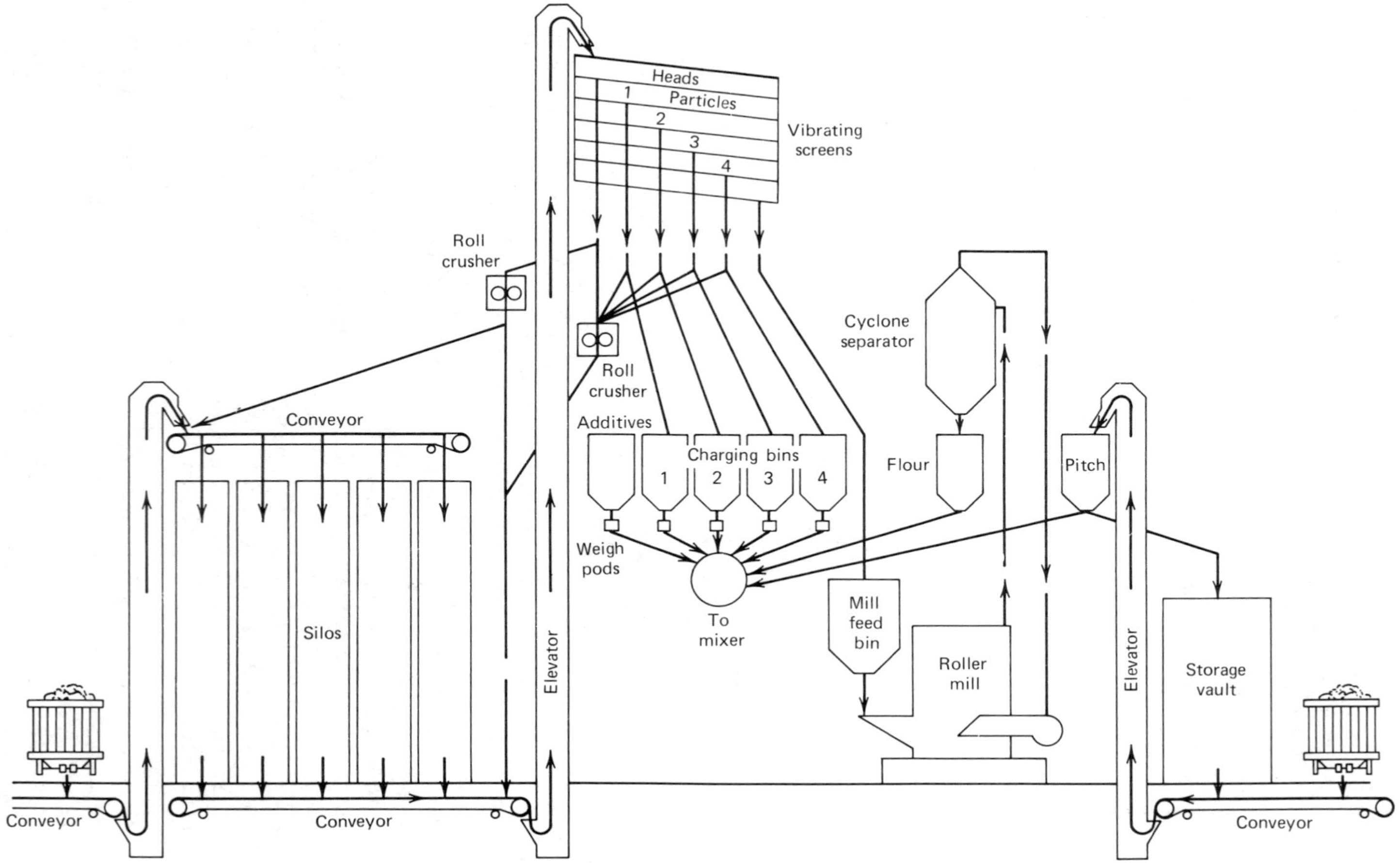

Figure 1. Raw materials handling system.

In the system shown in Figure 1, the oversized coke particles (heads) are diverted to a roll crusher. Most raw materials systems provide the option of further reducing the sizes of particles by passing them through a second crusher directly from the screens and recycling the resulting fractions through the screening system. The undersized coke fractions are transferred to a bin that supplies a mill for production of the flour portion of graphite composition. The mills used in this application may be of impact (hammer) variety or of roller variety. A commonly-used mill consists of a rotating roller operating against a stationary steel ring. The coke is crushed to very fine sizes that are air-classified by a cyclone separator. The sizes larger than those desired in the flour are returned to the mill and the acceptable sizes are fed to a charge bin.

The coal-tar-pitch binder used in graphite manufacture also arrives in railroad cars. If the pitch is shipped in bulk form, the large pieces must be crushed to ca 3 cm and smaller to facilitate uniform melting in the mixer and control of the weighing operation. Many vendors of binder pitches now form their product either by prilling, extruding, or flaking to ensure ease of handling and storage.

The pitch system shown in Figure 1 conveys the incoming pitch through a crusher to an elevator that deposits it into a charging bin. The graphite manufacturer tries to avoid long-term storage of 100°C-softening point pitch because of its tendency to congeal at ambient conditions into masses extremely difficult to break up and handle. Thus, whenever possible, cars of pitch are ordered and used as needed at the carbon plant.

In some plants the pitches are delivered in heated tank cars as liquids.

Proportioning. The size of the largest particle is generally set by application requirements. For example, if a smoothly-machined surface with a minimum of pits is required, as in the case of graphites used in molds, a fine-grained mix containing particles no larger than 0.16 cm with a high flour content is ordinarily used. If high resistance to thermal shock is necessary (eg, in graphite electrodes used in melting and reducing operations in steel plants), particles up to 1.3 cm are used to act as stress absorbers in preventing catastrophic failures in the electrode.

Generally, the guiding principle in designing carbon mixes is the selection of the particle sizes, the flour content, and their relative proportions in such a way that the intergranular void space is minimized. If this condition is met, the volume remaining for binder pitch and the volatile matter generated in baking are also minimized. The volatile evolution is often responsible for structural and property deterioration in the graphite product. In practice, most carbon mixes are developed empirically with the aim of minimizing binder demand and making use of all the coke passed through the first step of the system. From an economic standpoint, accumulation of one size component cannot be tolerated in making mixes for commonly used graphite grades since this procedure will amount to a loss of relatively costly petroleum coke. Typically, a coarse-grained mix may contain a large particle (eg, 6 mm dia), a small particle half this size or smaller, and flour. In this formulation, approximately 25 kg of binder pitch would be used for each 100 kg of coke.

Although binder levels increase as particle size is reduced, and they are greatest in all-flour mixes where surface area is very high, the principle of minimum binder level still applies. The application of particle packing theory to achieve minimum binder level in all-flour mixes is somewhat more complex because of the continuous gradation in sizes encountered (3).

For some carbon and graphite grades, particle packing and minimum pitch level concepts are not used in arriving at a suitable mix design. For relatively small products, eg, where large dimensional changes can be tolerated during the baking and graphitizing operations, high binder levels are often used. Increased pitch content results in greater shrinkage which gives rise to high density and strength in the finished products.

Mixing. Once the raw materials have been crushed, sized, and stored in charging bins and the desired proportions established, the manufacturing process begins with the mixing operation. The purpose of mixing is to blend the coke filler materials and to melt and distribute the pitch binder over the surfaces of the filler grains. The intergranular bond ultimately determines the property levels and structural integrity of the graphite. Thus the more uniform the binder distribution is throughout the filler components, the greater the likelihood for a structurally sound product.

The degree to which mixing uniformity is accomplished depends on factors such as time, temperature, and batch size. However, a primary consideration in achieving mix uniformity is mix design. A number of mechanically agitated, indirectly-heated mixer types are available for this purpose (4–5). Each mixer type operates with a different mixing action and intensity. Ideally, the mixer best suited for a particular mix composition is one that introduces the most work per unit weight of mix without particle breakdown. In practice, only a few mixer types are used in graphite manufacture.

The cylinder mixer is commonly used for coarse-grained mixes. It is equipped with an axial rotating shaft fitted with several radial arms where paddles are attached. The intensity of this mixer is relatively low to avoid particle breakdown, and long mixing times, such as 90 min, are therefore needed to complete the mixing operation. With fine-grained compositions, more intensive mixers may be used with a corresponding reduction in mixing time. Bread or sigma-blade mixers and the high intensity twin-screw mixers of the Werner-Pfleiderer and Banbury variety are examples of the equipment that can be used on fine-grained compositions. For both mixers, temperatures at the time of discharge are 160–170°C.

Following the mixing operation, the hot mix must be cooled to a temperature slightly above the softening point of the binder pitch. Thus the mix achieves the proper rheological consistency for the forming operation and the formed article is able to maintain its shape better as it cools to room temperature. At the end of the cooling cycle, which typically requires 15–30 min, the mix is at 100–110°C and is ready to be charged into an extrusion press or mold.

Forming. One purpose of the forming operation is to compress the mix into a dense mass so that pitch-coated filler particles and flour are in intimate contact. For most applications, a primary goal in the production of graphite is to maximize density; this goal begins by minimizing void volume in the formed (green) product. Another purpose of the forming step is to produce a shape and size as near that of the finished product as possible. This reduces raw material usage and cost of processing graphite that cannot be sold to the customer and must be removed by machining prior to shipment.

The two important methods of forming are extrusion and molding.

Extrusion. The extrusion process is used to form most carbon and graphite products. In essence, the various extrusion presses comprise a removable die attached by means of an adapter to a hollow cylinder, called a mud chamber. This cylinder is charged with mix that is extruded in a number of ways depending on the press design. For one type of press, the cooled mix is introduced into the mud chamber in the form of plugs which are molded in a separate operation. A second type of extruder, called a tilting press, makes use of a movable mud chamber–die assembly to eliminate the need for precompacting the cooled mix. Loading occurs directly from coolers with the assembly in the vertical position; the mixture is extruded with the assembly in the horizontal position. A third type of extrusion press makes use of an auger to force mix through the die. This press is used principally with fine-grained mixes because of its tendency to break down large particles.

The basic steps in the extrusion operation when a tilting press is used are depicted in Figure 2. The cooled mix is usually fed to the press on a conveyor belt where it is discharged into the mud chamber in the vertical position. A ram descends on the filled chamber, tamping the mix to compact the charge. A closing plate located in the pit beneath the press is often used to seal off the die opening thereby preventing the mix from extruding during the application of high tamping pressures. The filling and tamping procedures are repeated until the mud chamber is filled with tamped mix and then rotated back to the horizontal position. The extrusion ram then enters the mud cylinder forcing the mix through the die at 7–15 MPa (69–148 atm). A guillotine-like knife located near the die outlet cuts the extruded stock to the desired length. Round products are rolled into a tank of water where the outer portions are quickly cooled to prevent distortion of the plastic mass. Products having large rectangular cross sections may be transferred from the press to the cooling tank by means of an overhead crane. Water temperatures are regulated to avoid cracking as a result of too rapid cooling. Products with smaller cross sections, such as the 3.2 × 15.2 × 81 cm plates used as anodes in chlorine cells, may be cooled in air on steel tables. Bulk densities of green products range typically from 1.75–1.80 g/cm^3.

The anisotropy, usually observed in graphite products, is established in the forming operation. In extruded products, the anisotropic coke particles orient with

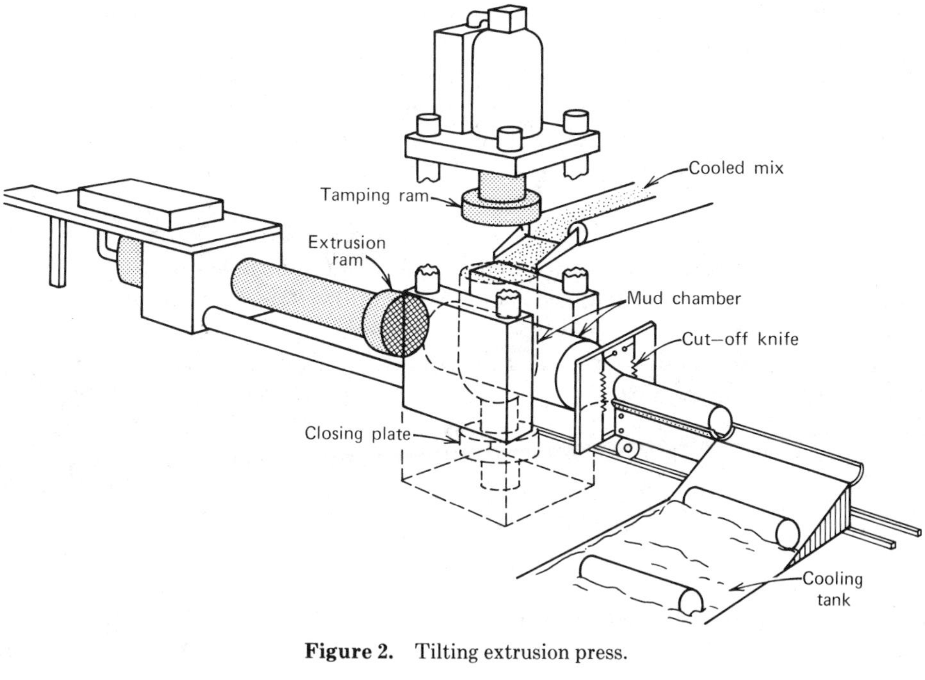

Figure 2. Tilting extrusion press.

their long dimensions parallel to the extrusion direction. The layer planes of the graphite crystals are predominantly parallel to the long dimension of the coke particle. Accordingly, the highly anisotropic properties of the single crystal are translated, to a greater or lesser degree depending on several factors, to the graphite product. The most important of these factors are coke type, particle size, and the ratio of die-to-mud chamber diameters. The more needlelike the coke particle, the greater the difference is between properties with-grain (parallel to the extrusion direction) and cross-grain. The use of smaller particles in the mix design also increases this property difference; the presence of large particles interferes with the alignment process. As the ratio of mud cylinder to die diameter increases, the with-grain to cross-grain ratios of strength and conductivity increase, while the with-grain to cross-grain ratios of resistivity and expansion coefficient decrease. Thus anisotropy is increased for the same coke type and mix design when going from a 0.60 m dia die to a 0.40 m dia die on the same extrusion press. As a result of particle orientation in extruded graphitized products, strength, Young's modulus, and thermal conductivity values are greater, whereas electrical resistivity and coefficient of thermal expansion are smaller in the with-grain direction than in the two cross-grain directions.

Molding. Molding is the older of the two forming methods and is used to form products ranging in size from brushes for motors and generators to billets as large as 1.75 m dia by 1.9 m in length for use in specialty applications.

Several press types are used in molding carbon products. The presses may be single-acting or double-acting, depending on whether one or both platens move to apply pressure to the mix through punched holes in either end of the mold. The use of single-acting presses is reserved for products whose thicknesses are small compared with their cross-sections. As thickness increases, the acting pressure on the mix diminishes with distance from the punch because of frictional losses along the mold wall. Acceptable thicknesses of molded products can be increased by using double-acting presses which apply pressures equally at the top and bottom of the product.

Jar molding is another method used to increase the length of the molded piece and keep nonuniformity within acceptable limits. By this technique the heated mold is vibrated as the hot mix is introduced, thus compacting the mix during the charging operation. Pieces as large as 2.5 m in diameter and 1.8 m in length have been molded in this way; the green densities are comparable with those obtained in extruded materials.

Smaller products, such as brushes and seal rings, are often molded at room temperature from mix that is milled after cooling. When binder levels exceed approximately 30% of the mix, the compacted milled mix has sufficient green strength to facilitate handling in preparation for the baking operation.

In a typical hot-molding operation to form a 1.7 m dia billet 1.3 m long, approximately 7200 kg of mix at 160°C is introduced into a steam-heated mold without cooling. The platens of the press compact the mix at ca 5 MPa (ca 50 atm), holding this pressure for 15–30 min. The cooling step for pieces of this size is the most critical part of the forming operation. Owing to the low thermal conductivity of pitch, 0.13 W/(m·K) (6), and its relatively high expansion coefficient (4.5×10^{-4}/°C at 25–200°C) (7), stresses build rapidly as the outer portions of the piece solidify. If cooling is too rapid, internal cracks are formed which are not removed in subsequent processing steps. As a result, a cooling schedule is established for each product size and is carefully followed by circulating water of various temperatures through the mold for specified time periods.

When the outside of the piece has cooled sufficiently, it is stripped from the mold and the cooling operation continued by direct water spray for several hours. If cooling is stopped too soon, heat from the center of the piece warms the pitch binder to a plastic state, resulting in slumping and distortion. The cooled piece is usually stored indoors prior to baking in order to avoid extreme temperature changes which may result in temperature gradients and damage to the structure. Bulk density of the green billet is usually 1.65–1.70 g/cm^3.

As with extruded products, molded pieces have a preferred grain orientation. The coke particles are aligned with their long dimensions normal to the molding direction. Thus the molded product has two with-grain directions, and one cross-grain direction which coincides with the molding direction. Strength, modulus, and conductivity of molded graphites are higher in both with-grain directions, and expansion coefficient is higher in the cross-grain direction.

Isostatic molding is a forming technique used to orient the coke filler particles randomly, thereby imparting isotropic properties to the finished graphite. One approach to isostatic molding involves placing the mix or blend into a rubber container capable of withstanding relatively high molding temperatures. The container is evacuated, then sealed and placed in an autoclave which is closed and filled with heated oil. The oil is then pressurized to compact the mass which may then be processed in the usual way to obtain isotropic graphite.

Baking. The next stage is the baking operation during which the product is fired to 800–1000°C. One function of this step is to convert the thermoplastic pitch binder to solid coke. Another function of baking is to remove most of the shrinkage in the product associated with pyrolysis of the pitch binder at a slow heating rate. This procedure avoids cracking during subsequent graphitization where very fast firing rates are used. The conversion of pitch to coke is accompanied by marked physical and chemical changes in the binder phase, which if conducted too rapidly, can lead to serious quality deficiencies in the finished product. For this reason, baking is generally regarded as the most critical operation in the production of carbon and graphite.

Several studies discuss the kinetics of pitch pyrolysis and indicate, in detail, the weight loss and volatile evolution as functions of temperature (8–9). During this process weight losses of 30–40% occur, indicating that for every 500 kg of green product containing 20% pitch, 30–40 kg of gas must escape. In terms of gas volume, approximately 150 cm^3 of volatiles at standard conditions must be evolved per g of pitch binder during the baking operation. The product in the green state is virtually impermeable, and the development of a venting porosity early in the bake must be gradual to avoid a grossly porous or cracked structure. The generation of uniform structure during the bake is made more difficult by the poor thermal conductivity of pitch. Long firing times are usually needed to drive the heat into the center of the product which is necessary for pitch pyrolysis and shrinkage. If the heating rates exceed a value which is critical for the size and composition of the product, differential shrinkage leads to splitting scrap. Shrinkage during baking is of the order of 5% and increases with increasing pitch content. Added to these difficulties is the complete loss of mechanical strength experienced by the product in the 200–400°C range where the pitch binder is in a liquid state. To prevent slumping and distortion during this period, the stock must be packed in carefully sized coke or sand which provides the necessary support and is sufficiently permeable to vent the pitch volatiles.

A variety of baking furnaces are in use to provide the flexibility needed to bake a wide range of product sizes and to generate the best possible temperature control. One common baking facility is the pit furnace, so named because it is positioned totally or partially below ground level to facilitate improved insulation. In essence, the pit furnace is a box with ceramic brick walls containing ports or flues through which hot gases are circulated. Traditionally, natural gas has been the fuel used to fire pit furnaces; however, because of natural gas shortages, pit furnaces are being converted to use fuel oil (see Furnaces, fuel-fired).

Another common baking facility is the so-called ring furnace; one form of this is depicted in Figure 3. Two equal rows of pit furnaces are arranged in a rectangular ring. Ports in the furnace walls permit the heated gases from one furnace to pass to the next until the cooled gases are exhausted by a movable fan to a flue leading to a stack. A movable burner, in this case located above one furnace, fires it to a predetermined off-fire temperature. The firing time per furnace is 18–24 h. When the desired temperature has been reached, the burner is moved to the adjacent furnace which has been heated by gases from the most recently fired pit. At the same time, the fan is moved to a furnace that has just been packed. This process continues, with packing,

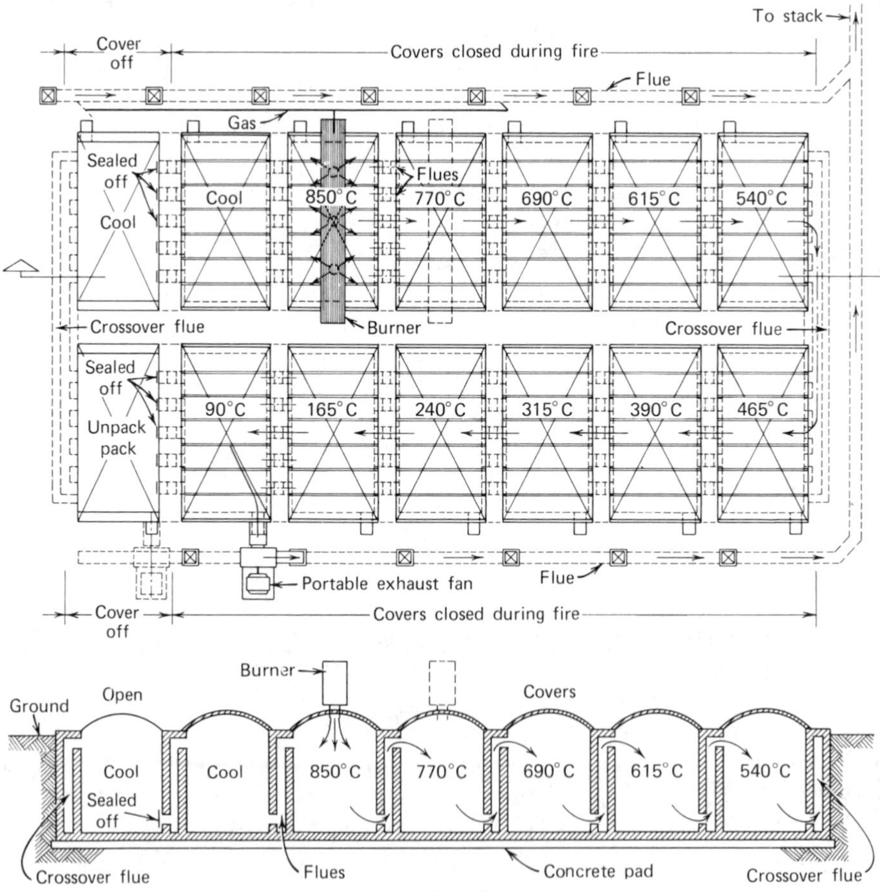

Figure 3. Ring furnace system.

unloading, and cooling stages separating the fan and the burner. Cycle times in this furnace are 3–4 wk. Thermally, the ring furnace is highly efficient but it has the disadvantage that very little control can be exercised over heating rates.

The firing schedules used in the baking operation vary with furnace type, product size, and binder content. A bulk furnace packed with 61 × 81 × 460 cm pieces of specialty graphite may require 6 wk to fire and an additional 3–4 wk to cool. In contrast, very small products, such as seal rings, may be baked in tunnel kilns in a few h. A sagger furnace containing electrodes may require 12–14 d to reach final temperature with an additional 3–5 d to cool. Firing rates early in the baking schedule are reduced to permit pitch volatiles to escape slowly, minimizing damage to the structure. For most carbon products, temperatures must be well below 400°C prior to unpacking to avoid cracking due to thermal shock. The product is scraped or sanded to remove adhering packing materials and is then weighed, measured, and inspected prior to being stored for subsequent processing. Some products that are sold in the baked state are machined at this stage. Baked products include submerged arc furnace electrodes, cathode blocks for the electrolytic production of aluminum, and blast furnace lining blocks.

Impregnation. In some applications the baked product is taken directly to the graphitizing facility for heat treatment to 3000°C. However, for many high performance applications of graphite, the properties of stock processed in this way are inadequate. The method used to improve those properties is impregnation with coal-tar or petroleum pitches. The function of the impregnation step is to deposit additional pitch coke in the open pores of the baked stock, thereby improving properties of the graphite product. Table 1 lists the graphite properties of unimpregnated and impregnated stock 15–30 cm dia and containing a 1.5 mm particle.

Further property improvements result from additional impregnation steps separated by rebaking operations. However, the gains realized diminish quickly, for the quantity of pitch picked up in each succeeding impregnation is approximately half of that in the preceding treatment. Many nuclear and aerospace graphites are multiply pitch-treated to achieve the greatest possible assurance of high performance.

Table 1. Effect of One Pitch Impregnation on Graphite Properties

Property		Unimpregnated	Impregnated
bulk density, g/cm^3		1.6	1.7
	wg[b]	7.4	11.0
Young's modulus[a], GPa	ag[c]	4.4	6.3
	wg	10,000	17,000
flexural strength[a], kPa	ag	7,100	13,000
	wg	5,000	8,100
tensile strength[a], kPa	ag	4,400	7,300
	wg	21,000	34,000
compressive strength[a], kPa	ag	21,000	33,000
	wg	0.40	0.19
permeability, Darcys	ag	0.35	0.16
	wg	1.3	1.5
coefficient of thermal expansion, 10^{-6}/°C	ag	2.7	3.1
	wg	8.8	7.6
specific resistance, $\mu\Omega\cdot m$	ag	13	11

[a] To convert Pa to atm, divide by 101×10^3.
[b] With the grain.
[c] Across the grain.

During the baking operation, binder pitch exuding the product surface creates a dense impermeable skin. In addition, the exuding pitch causes packing material to adhere to the baked stock. The skin and the packing material must be removed by sanding, scraping, or machining before the stock can be impregnated on a reasonable time cycle. Unless this operation is properly performed, the impregnant may not reach the center of the product and a so-called dry core will result. When this condition exists, the product usually splits during graphitization as a consequence of the greater concentration of pitch and greater shrinkage in the outer portions of the stock. The likelihood of a dry core increases with the quinoline-insoluble solids content of the impregnant. During the impregnation process, the insolubles form a filter cake of low permeability on the stock surface, reducing the penetrability of the impregnant. Quinoline insolubles significantly greater than 5% reduce the penetration rate and increase the incidence of dry cores.

A schematic diagram of the pitch impregnation process is shown in Figure 4. Before it is placed in an autoclave, the skinned baked stock is preheated to 250–300°C to thoroughly dry it and to facilitate free flow of the molten impregnant into the open pores. The first step in the impregnation process is to evacuate the stock to pressures below 3.5 kPa (26 mm Hg) for a period of 1 h or more depending on the size and permeability of the stock. Unless the stock is adequately evacuated, the remaining air prevents thorough penetration of the impregnant to the center of the product. Heated pitch is then introduced by gravity flow into the autoclave from a holding tank until the charge is completely immersed. The system is then subjected to pressures of 700–1500 kPa (6.9–14.8 atm) for several hours to shorten the time for pitch penetration. When the pressure cycle has been completed, the pitch is blown back to the holding tank by means of compressed air. The autoclave is then opened, and the stock is transferred to a cooler where water and circulating air accelerate the cooling process. After cooling, the stock is weighed to determine the quantity of pitch picked up. If the pickup is below a specified limit, the stock is scrapped. Depending on the density of the baked stock, the pickup is 14–16% on the first impregnation and 7–8% on the second impregnation.

If the stock is to receive a second impregnation, it must be rebaked. In the past, stock containing raw impregnating pitch could be graphitized directly. However, the air polluting effect caused by this practice has made rebaking a necessary preliminary step to graphitization in order to meet the requirements of the EPA.

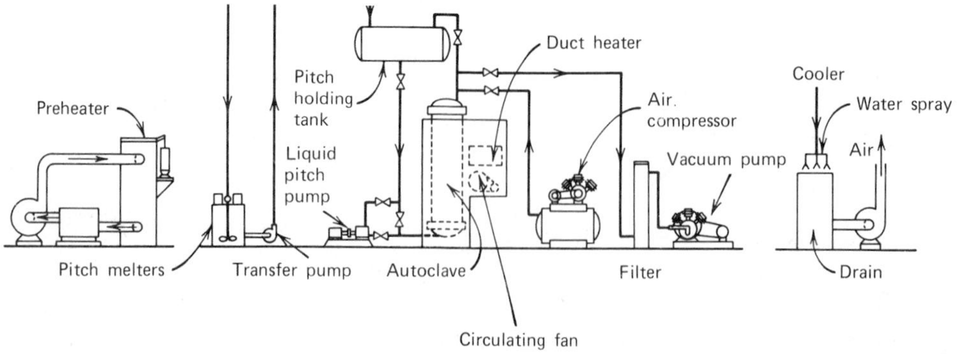

Figure 4. Pitch impregnation system.

Graphitization. Graphitization is an electrical heat treatment of the product to ca 3000°C. The purpose of this step is to cause the carbon atoms in the petroleum coke filler and pitch coke binder to orient into the graphite lattice configuration. This ordering process produces graphite with the intermetallic properties that make it useful in many applications.

Very early in the carbonization of coker feeds and pitch, the carbon atoms are present in distorted layers of condensed benzene ring systems formed by the polymerization of the aromatic hydrocarbons in these materials. The x-ray studies of raw coke, for example, show that two-dimensional order exists at that early stage of graphite development (10). As the temperature of coke increases, the stack height of the layer planes increases. The layers are skewed about an axis normal to them, however, and it is not until a temperature of ca 2200°C is reached that three-dimensional order is developed. As the graphitizing temperature is increased to 3000°C, the turbostratic (see pp. 690–691) arrangement of the layer planes is effectively eliminated, and the arrangement of the carbon atoms approaches that of the perfect graphite crystal. Depending on the size and orientation of these crystals, the properties of manufactured graphites can be varied controllably to suit a number of critical applications.

The furnace that made the graphite industry possible was invented in 1895 by Acheson (11) and is still in use today with only minor modifications. It is an electrically-fired furnace capable of heating tons of charge to temperatures approaching 3000°C. The basic elements of the Acheson furnace are shown in Figure 5. The furnace bed is made up of refractory tiles supported by concrete piers. The furnace ends are

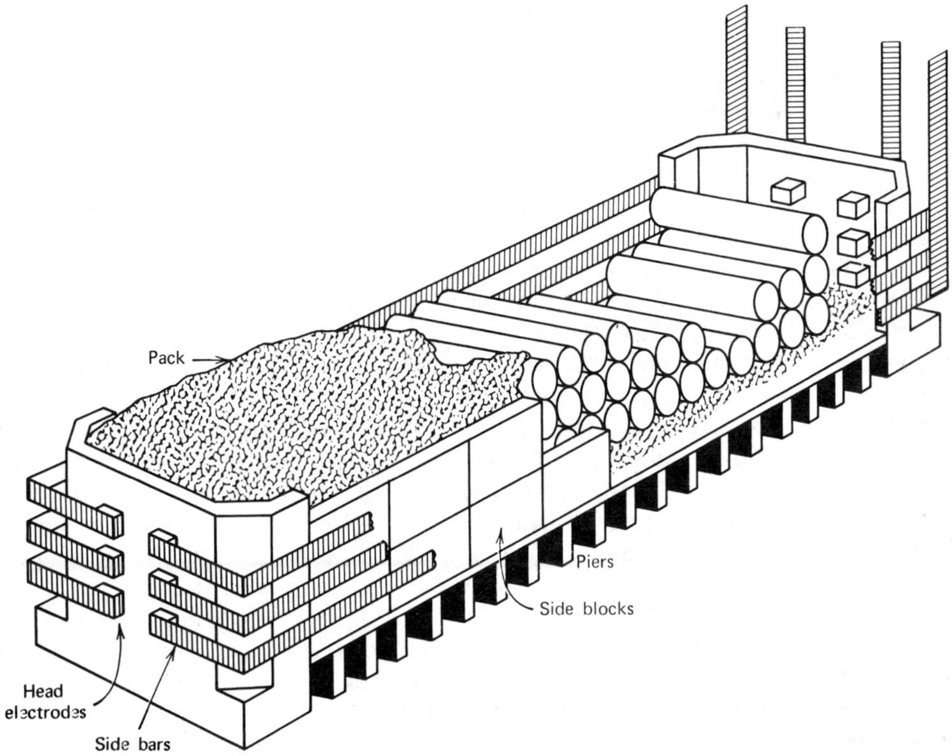

Figure 5. The Acheson furnace.

U-shaped concrete heads through which several graphite electrodes project into the pack. These electrodes, which are water-cooled during operation, are connected by copper bus work to the secondary of a transformer. The product is placed on a layer of metallurgical coke with its long axis transverse to current flow. Although a cylindrical product is shown in Figure 5, any product shape can be graphitized in the Acheson furnace so long as the product pieces are carefully spaced. This feature of the Acheson furnace makes it extremely versatile. The spacing between pieces may vary from less than a centimeter to several centimeters, depending on the shape and size of the product. With the product in place, a coarsely-sized metallurgical coke, called resistor pack, is used to fill the interstices between pieces; most of the heat needed to reach graphitizing temperatures is generated in the resistor material. Once the charge and resistor material are loaded, the furnace is covered with a finer blend of metallurgical coke, sand, and silicon carbide to provide thermal and electrical insulation. Concrete side blocks, usually 0.5–1 m from the charge ends, are used to retain the insulation. The procedure for loading a furnace usually requires one day.

Acheson furnace sizes may vary, depending on the product size and the production rate desired. Typically, the furnace may be 12–15 m by 3–3.5 m. Loads ranging from 35–55 metric tons of product are charged to these furnaces. The transformers used are rated 4000–6000 kW and are capable of delivering up to 60,000 A to the charge. Heating rates are usually 40–60°C/h, the total firing time being approximately three days. At the end of this time, the product temperatures are 2800–3000°C. Total power input varies, depending on the product and load size; for graphite electrodes, total power (energy) inputs average 4.5 kW·h/kg, and total power inputs in excess of 9 kW·h/kg may be used in the thermal purification of nuclear graphites. Following the heating cycle, 8–10 d are needed to cool and unload the furnace. The total cycle time on an Acheson furnace is ca two weeks. The cooling procedure is hastened by the gradual removal of pack with care to leave sufficient cover to prevent oxidation of the product. The insulation and resistor materials are screened to specified sizes and proportions for reuse, and new materials added as necessary. The product is cleaned and inspected prior to being measured and weighed for bulk density and resistivity determinations. If the properties are within specified limits, the product is stored and is ready for machining.

Furnaces other than the Acheson furnace are used commercially, but on a much smaller scale and usually for smaller products. For example, electrographitic brushes are graphitized in tube furnaces, wherein a current-carrying graphite tube is the heating element. These furnaces are particularly useful in the laboratory because of the ease with which they can be loaded and unloaded without the need for handling large quantities of packing material. Inductively-heated furnaces are also used commercially to graphitize a limited number of products, such as some aerospace grades and graphite fibers. These furnaces, also popular in the laboratory, consist basically of a cylindrical graphite shell susceptor positioned inside a water-cooled copper coil. High frequency power supplied to the coil induces current to flow in the susceptor, heating it and causing it to radiate heat to the contained charge (see Furnaces, electric).

More recently, several patents (12–15) have been issued describing a process for graphitization where the carbon charge to be heated is placed in a longitudinal array and covered with insulation to prevent heat losses and oxidation of the charge. An electric current is passed directly through the carbon array, generating within the carbon the heat required to raise the carbon to the graphitization temperatures. These patents describe reduced energy requirements and reductions in process time.

Puffing. In the temperature range of 1500–2000°C, most petroleum cokes undergo an irreversible volume increase known as puffing. This effect has been associated with thermal removal of sulfur from coke and increases with increasing sulfur content. Because of the recent emphasis on the use of low sulfur fuels, many of the sweet crudes that had been used as coker feeds are now being processed as fuels. Desulfurization of the sour crudes available for coking is possible but expensive. The result is an upward trend in the sulfur content of many petroleum cokes, leading to greater criticality in heating rate in the puffing temperature range during graphitization.

Many studies of the puffing phenomenon and of means for reducing or eliminating it have been made (16–18). As a general rule, puffing increases as particle size increases and is greater across the product grain. Depending on particle size and on the product size, heating rates must be adjusted in the puffing range to avoid splitting the product. Fortunately, the use of puffing inhibitors (discussed in the previous section) has eased the problem and has permitted the use of graphitization rates greater than would otherwise be possible.

BIBLIOGRAPHY

"Baked and Graphitized Products" under "Carbon" in *ECT* 1st ed., Vol. 3, pp. 1–34, by H. W. Abbott, Speer Carbon Company; "Baked and Graphitized Products, Manufacture" under "Carbon" in *ECT* 2nd ed., Vol. 4, pp. 158–202, by L. M. Liggett, Speer Carbon Company.

1. F. J. Hiorns, *Br. Chem. Eng.* **15,** 1565 (Dec. 1970).
2. A. Ratcliffe, *Chem. Eng. (N.Y.)* **79,** 62 (July 10, 1972).
3. A. E. Goldman and H. D. Lewis, *U.S. Los Alamos Scientific Laboratory, Report LA 3656,* 1968.
4. W. L. Root and R. A. Nichols, *Chem. Eng. (N.Y.)* **80,** 98 (Mar. 19, 1973).
5. V. W. Uhl and J. B. Gray, *Mixing,* Vol. II, Academic Press, New York, 1967, Chapt. 8.
6. D. McNeil and L. J. Wood, *Industrial Carbon and Graphite; Papers Read at the Conference Held in London, Sept. 24–26, 1957.* Society of Chemical Industry, London, Eng., 1958, p. 162.
7. R. E. Nightingale, *Nuclear Graphite,* Academic Press, New York, 1962, Chapt. 2.
8. M. Born, *Fuel* **53,** 198 (1974).
9. A. S. Fialkov and co-workers, *J. Appl. Chem. USSR* **35,** 2213 (1964).
10. R. E. Franklin, *Acta Cryst.* **4,** 253 (1951).
11. E. G. Acheson, *Pathfinder,* Acheson Industries, Inc., 1965.
12. Ger. Pat. 2,018,764 (Oct. 28, 1971), (to Sigri Elektrographit GmbH).
13. Ger. Pat. 2,316,494 (Oct. 2, 1974), (to Sigri Elektrographit GmbH).
14. Jpn. Kokai 75-86494 (July 11, 1975), (to Toyo Carbon Co. Ltd.).
15. Ger. Pat. 2,623,886 (Dec. 16, 1976), (to Elettrocarbonium SP).
16. M. P. Whittaker and L. I. Grindstaff, *Carbon* **7,** 615 (1969).
17. I. Letizia, *Paper No. CP-69,* presented at the 10th Biennial Conference on Carbon, Lehigh University, Bethlehem, Pa., June 1971.
18. H. F. Volk and M. Janes, *Paper No. 172,* presented at 7th Biennial Conference on Carbon, Cleveland, Ohio, June 1965.

E. L. Piper
Union Carbide Corporation

PROPERTIES OF MANUFACTURED GRAPHITE

Physical Properties

The graphite crystal, the fundamental building block for manufactured graphite, is one of the most anisotropic bodies known. Properties of graphite single crystals illustrating this anisotropy are shown in Table 1 (1). Anisotropy is the direct result of the layered structure with extremely strong carbon–carbon bonds in the basal plane and weak bonds between planes. The anisotropy of the single crystal is carried over in the properties of commercial graphite, although not nearly to the same degree. By the selection of raw materials and processing conditions, graphites can be manufactured with a very wide range of properties and degree of anisotropy. The range of room temperature properties, attainable for various forms of graphite, is shown in Figures 1 and 2 (1). The range extremities represent special graphites having limited industrial utility, whereas the bulk of all manufactured graphites fall in the bracketed areas marked conventional.

The directional properties of graphite arise in the following way. When the coke aggregate is crushed and sized, the resulting coke particles tend to have one axis longer than the other two. As the plastic mix of particles and binder pitch is formed into the desired shape, the long axis of particles tends to align perpendicularly to the molding force in molded graphites and parallel to the extrusion force in extruded graphites. The particle alignment is preserved during the subsequent processing so that properties of the finished graphite have an axis of symmetry that is parallel to the forming force. Properties in the plane perpendicular to the axis of symmetry are essentially independent of direction. Samples cut parallel to the molding force for molded graphites or perpendicular to the extrusion force for extruded graphites are designated as cross-grain. Samples cut parallel to the molding plane of molded graphites or parallel to the extrusion axis for extruded graphites are designated as with-grain. A number of special test procedures for determining the properties of carbon and graphite have been adopted by ASTM (2).

Manufactured graphite is a composite of coke aggregate (filler particles), binder carbon, and pores. Most graphites have a porosity of 20–30%, although special graphites can be made that have porosity well outside this range. Manufactured graphite is a highly refractory material that has been thermally stabilized to as high as 3000°C. At atmospheric pressure, graphite has no melting point but sublimes at 3850°C, the triple point being approximately 3850°C and 12.2 MPa (120 atm) (3). The strength of graphite increases with temperature to 2200–2500°C; above 2200°C, graphite becomes

Table 1. Room Temperature Properties of Graphite Crystals[a]

Property	Value in basal plane	Value across basal plane
resistivity, Ω·m	40×10^{-4}	ca 6000×10^{-4}
elastic modulus[b], TPa	0.965	0.034
tensile strength (est)[b], TPa	0.096	0.034
thermal conductivity, W/(m·K)	ca 2000	10
thermal expansion, $°C^{-1}$	-0.5×10^{-6}	27×10^{-6}

[a] Ref. 1.
[b] To convert TPa to psi, multiply by 1.45×10^8.

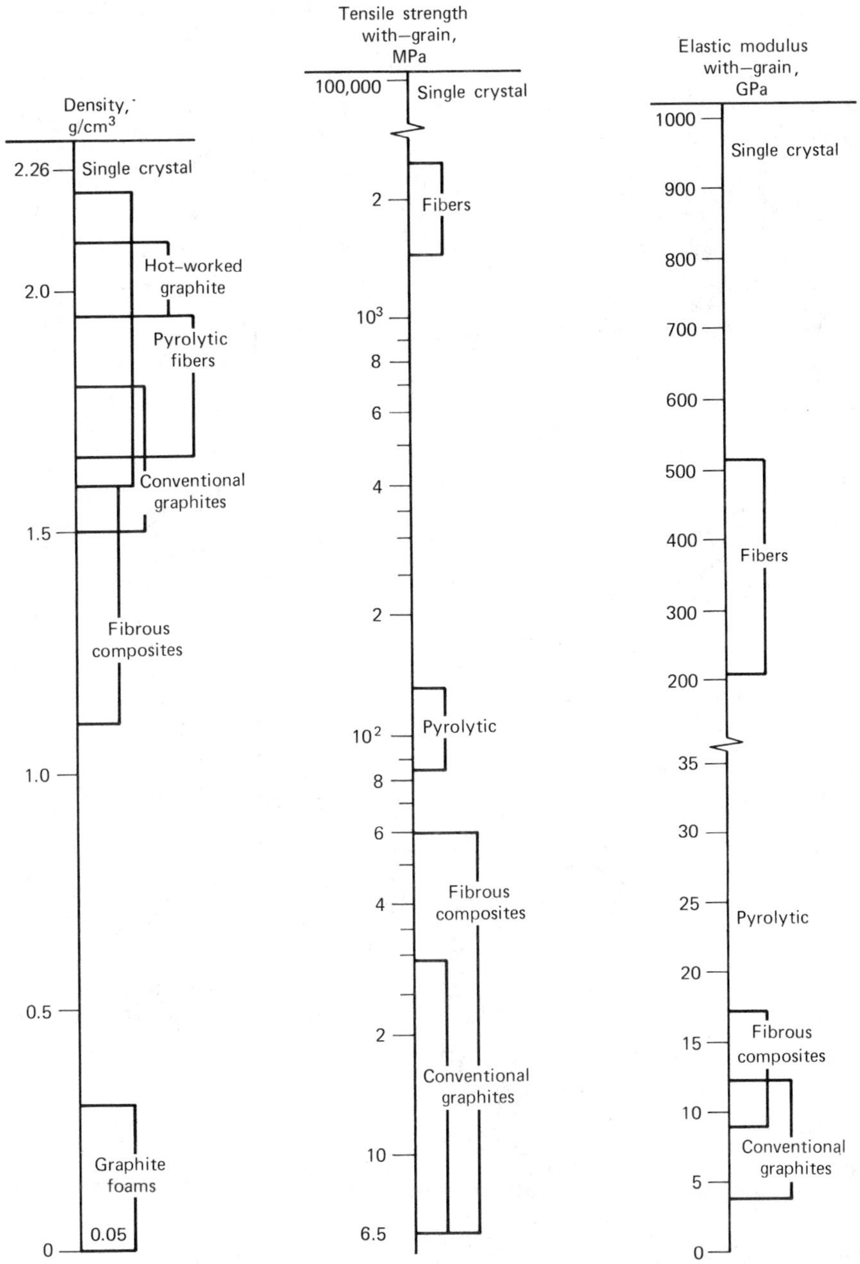

Figure 1. Mechanical properties of artificial graphite (1). To convert Pa to psi, multiply by 145 $\times 10^{-6}$.

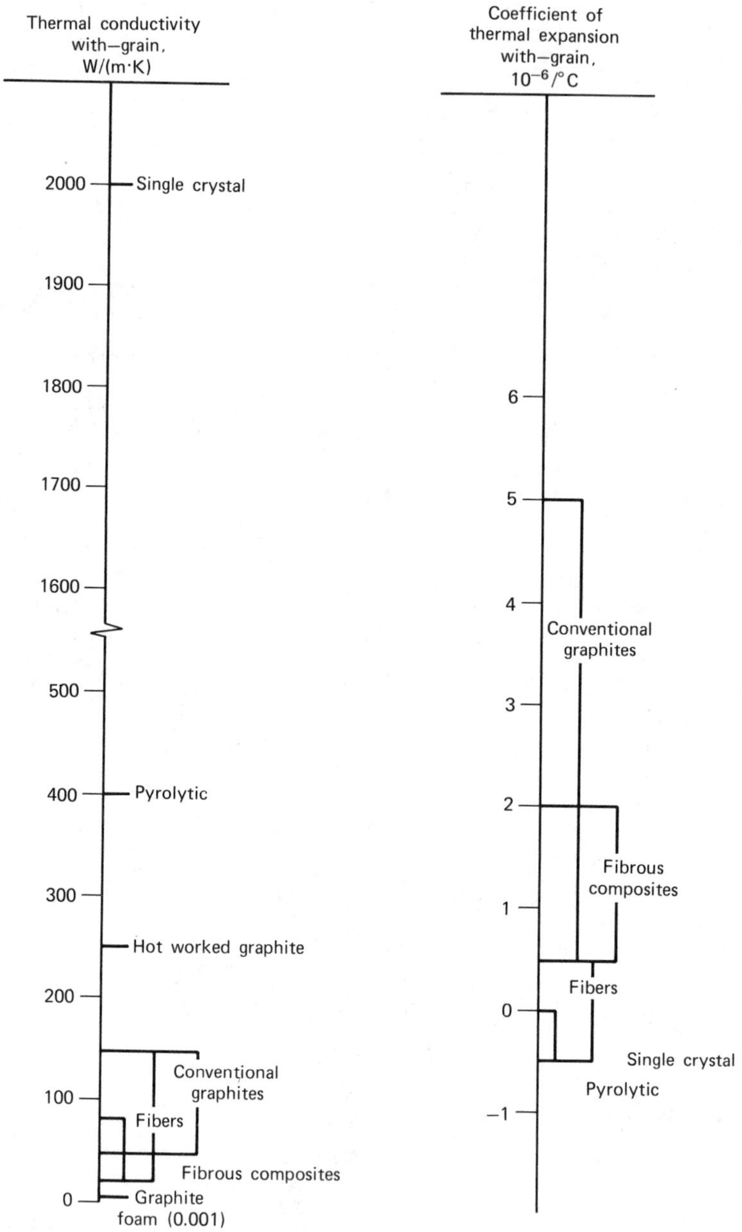

Figure 2. Thermal properties of artificial graphites (1).

plastic and exhibits viscoelastic creep under load (4). Graphite has high resistance to thermal shock, a property that makes it a valuable structural material at higher temperatures than most metals and alloys. For many applications of graphite, one or more of the following characteristics are important: density, elastic modulus, mechanical strength, electrical and thermal conductivity, and thermal expansion.

Electrical Properties. Manufactured graphite is semimetallic in character with the valence and conduction bands overlapping slightly (5–7). Conduction is by means of an approximately equal number of electrons and holes that move along the basal planes. The resistivity of single crystals as measured in the basal plane is approximately 0.4 μΩ·m; this is several orders of magnitude lower than the resistivity across the layer planes (8–10). Thus the electrical conductivity of formed graphite is dominated by the conductivity in the basal plane of the crystallites and is dependent on size, degree of perfection, and orientation of crystallites and on the effective carbon–carbon linkages between crystallites. Manufactured graphite is strongly diamagnetic and exhibits a Hall effect, a Seebeck coefficient, and magnetoresistance. The green carbon body is practically nonconductive; however, heat treatment at 1000°C decreases the resistivity by several orders of magnitude, and thereafter resistivity decreases slowly. After graphitization to over 2500°C, the room temperature electrical resistivity may range from a few hundredths to a few tenths Ω·m, depending upon the type of raw materials used. Graphites made from petroleum coke usually have a room temperature resistivity range of 5–15 μΩ·m and a negative temperature coefficient of resistance to about 500°C, above which it is positive. Graphites made from a carbon black base have a resistivity several times higher than those made from petroleum coke, and the temperature coefficient of resistance for the former remains negative to at least 1600°C.

Thermal Conductivity. Compared with other refractories, graphite has an unusually high thermal conductivity near room temperature (11); above room temperature, the conductivity decreases exponentially to approximately 1500°C and more slowly to 3000°C (12). With the grain the thermal conductivity of manufactured graphite is comparable with that of aluminum; against the grain it is comparable to that of brass. However, graphite is similar to a dielectric solid in that the principal mechanism for heat transfer is lattice vibrations. The electronic component of thermal conductivity is less than 1%. Graphite does not obey the Wiedemann-Franz Law; however, at room temperature the ratio of thermal and electrical conductivities is equal to approximately 0.126 when the thermal conductivity is in W/(m·K) and the electrical conductivity is in S(=1/Ω) (13–14). For most graphites, a value of thermal conductivity at room temperature accurate to ±5% can be obtained from the measured value of the electrical conductivity.

Coefficient of Thermal Expansion. The volumetric thermal expansion of manufactured graphite is anomalously low when compared to that of the graphite single crystal. At room temperature, the volume coefficient of thermal expansion of a single crystal is approximately 25×10^{-6}/°C (15–16), whereas those of many manufactured graphites fall in the range of $4–8 \times 10^{-6}$/°C. There are exceptions, and some commercially available, very fine-grain, near-isotropic graphites have a volumetric expansion as high as two thirds the value for the single crystal. The low value of volume expansion of most manufactured graphite has been related to the microporosity within the coke particles. The microcracks within the coke particle accommodate the large *c*-axis expansion of graphite crystallites (17–19) and effectively neutralize it. The coefficient of thermal expansion (CTE) is somewhat sensitive to the filler particle sizing and to the method of processing, but the anisotropy and perfection degree of filler carbon particles largely determine the expansion characteristics of the finished graphite. Except for differences in absolute values, plots of the CTEs of manufactured graphite vs temperature are essentially parallel to each other, showing that the change in CTE with temperature is approximately the same for all graphites at high temperatures. The mean linear coefficient of thermal expansion between room temperature and any final temperature can be obtained by adding the value of CTE for the temperature interval 20–100°C to the appropriate factor which varies from 0 at 100°C to 2.52×10^{-6} at 2500°C (20). This method is valid for stock of any grain orientation.

Mechanical Properties. The hexagonal symmetry of a graphite crystal causes the elastic properties to be transversely isotropic in the layer plane; only five independent constants are necessary to define the complete set. The self-consistent set of elastic constants given in Table 2 has been measured in air at room temperature for highly ordered pyrolytic graphite (21). With the exception of c_{44}, these values are expected to be representative of those for the graphite single crystal. Low values of shear and cleavage strengths between the layer planes compared with very high C–C bond strength in the layer planes, suggest that graphite always fails through a shear or cleavage mechanism. However, the strength of manufactured graphite depends upon the effective network of C–C bonds across any stressed plane in the graphite body. Until these very strong bonds are broken, failure by shear or cleavage cannot take place. Porosity affects the strength of graphite by reducing the internal area over which stress is distributed and by creating local regions of high stress. Because of the complexity of the graphite structure, a simple analytical model of failure has not been derived (22). The stress–strain relation for bulk graphite is concave toward the strain axis. The relaxation of the stress leads to a small residual strain; repeated stressing to larger loads followed by gradual relaxation leads to a set of hysteresis loops contained within the stress–strain envelope (4,23–26). Each successive load causes an increase in the residual strain and results in a decreased modulus for the sample. The residual strain can be removed by annealing the sample to the graphitizing temperature after which its original stress–strain response is restored. In the limit of zero stress, the elastic modulus of graphite is the same in compression and tension, and is equal to the modulus derived from dynamic measurements (27). The modulus of graphite is weakly dependent upon temperature, increasing with temperature to approximately 2000°C and decreasing thereafter. The strain at rupture of most graphites is 0.1–0.2%; however, values of strain at rupture approaching 1.0% have been obtained for specially-processed, fine-grain graphites (28). Graphite exhibits measurable creep under load and at temperatures above 1600°C, but for most applications creep can be neglected below 2200°C. As the temperature is increased above 2500°C, the creep rate increases rapidly and the short-time strength decreases rapidly.

Thermal shock resistance is a primary attribute of graphite and a number of tests have been devised in attempts to establish a quantitative method of measurement (29–30). These tests, which establish very large thermal gradients in small specially shaped samples, continue to give only qualitative data and permit one to establish only the relative order of shock resistance of different graphites. A commonly used thermal shock index is the ratio of the thermal conductivity and strength product to the expansion coefficient and modulus product (31). At high temperatures, values of this index for graphite are higher than for any other refractory material. To show the range of property values of graphite, several properties for a very coarse grain graphite and a very fine grain graphite are given in Table 3 (27,32).

Table 2. Elastic Constants of Graphite[a]

$c_{11} = 1.06 \pm 0.002$	$s_{11} = 0.98 \pm 0.03$
$c_{12} = 0.18 \pm 0.02$	$s_{12} = 0.16 \pm 0.06$
$c_{13} = 0.015 \pm 0.005$	$s_{13} = 0.33 \pm 0.08$
$c_{33} = 0.0365 \pm 0.0010$	$s_{33} = 27.5 \pm 1.0$
$c_{44} = 0.00018 - 0.00035$	$s_{66} = 2.3 \pm 0.2$

[a] Units c_{ij} (stiffness constant) in TPa, s_{ij} (compliance constant) in $(\text{TPa})^{-1}$. To convert TPa to psi, multiply by 1.45×10^8.

Table 3. Properties of Fine and Coarse Grain Graphites[a]

Temperature, °C	Thermal cond., W/(m·K)		CTE[b], cm/cm ×10⁶/°C		Specific heat[c] kJ/(kg.K)	Tensile[d]				Compression[d]			
						Modulus, GPa		Strength, MPa		Modulus, GPa		Strength, MPa	
	wg	cg	wg	cg		wg	cg	wg	cg	wg	cg	wg	cg
Fine-grained graphite, 180 μm maximum grain size													
21.1	150	114	2.15	3.10	0.63	11.5	7.9	17.4	15.0	9.7	7.2	26.6	20.1
260	117	93	2.50	3.46	1.30	11.6	8.0	19.3	17.2	10.0	7.4	27.9	21.7
538	91	72	2.82	3.84	1.63	11.7	8.1	21.7	19.7	10.3	7.6	29.3	23.4
816	73	57	3.16	4.12	1.80	11.9	8.3	24.1	22.1	10.6	7.9	30.9	25.2
1093	60	46	3.45	4.45	1.95	12.1	8.6	26.0	24.3	11.4	8.3	32.4	26.9
1371	52	40	3.70	4.69	2.03	12.5	9.0	28.3	26.2	12.4	9.0	35.2	29.3
1649	46	35	3.95	4.91	2.11	13.2	9.6	29.9	27.9	13.4	9.7	38.1	31.6
1927	42	32	4.17	5.16	2.16	13.7	10.5	31.0	29.3	13.4	9.7	32.2	37.2
2204	40	29	4.35	5.39	2.18	11.5	8.4	31.7	30.1	12.1	9.0	37.9	32.6
2482	38	28	4.58	5.71	2.20	8.0	5.9	31.0	29.3	10.0	7.9	32.4	26.6
2760	36	28	4.83	6.04	2.20	5.2	4.3	26.9	24.8	7.9	6.2	26.6	19.1
Coarse-grained graphite, 6400 μm maximum grain size													
21.1	156	108	0.46	1.03		4.2	2.6	3.75	2.91	3.0	2.6	9.3	12.1
1371	30	22	2.4	3.2		5.8	2.9	5.34	4.54	3.4	2.8	12.0	14.7
1927	24	19	2.7	2.85		6.5	3.7	5.39	4.36	4.3	3.3	14.1	17.4
2427	24	20	3.0	4.2		5.6	3.0	7.32	5.17				

[a] wg = with-grain; cg = cross-grain.
[b] CTE = coefficient of thermal expansion.
[c] To convert J to cal, divide by 4.184.
[d] To convert Pa to psi, multiply by 145×10^{-6}.

Chemical Properties

The impurity (ash) content of all manufactured graphite is low, since most of the impurities originally present in raw materials are volatilized and diffuse from the graphite during graphitization. Ash contents vary from 1.5% for large diameter graphites to less than 10 ppm for purified graphites. Iron, vanadium, calcium, silicon, and sulfur are major impurities in graphite; traces of other elements are also present (33). Through selection of raw materials and processing conditions, the producer can control the impurity content of graphites to be used in critical applications. Because of its porosity and relatively large internal surface area, graphite contains chemically and physically adsorbed gases. Desorption takes place over a wide temperature range, but most of the gas can be removed by heating in a vacuum at approximately 2000°C.

Graphite reacts with oxygen to form CO_2 and CO, with metals to form carbides, with oxides to form metals and CO, and with many substances to form laminar compounds (34–35). Of these reactions, oxidation is the most important to the general use of graphite at high temperatures. Oxidation of graphite depends upon the nature of the carbon, the degree of graphitization, particle size, porosity, and impurities present (36). These conditions may vary widely among graphite grades. Graphite is less reactive at low temperatures than many metals; however, since the oxide is volatile, no protective oxide film is formed. The rate of oxidation is low enough to permit the effective use of graphite in oxidizing atmospheres at very high temperatures when a modest consumption can be tolerated. A formed graphite body alone will not support combustion. The differences in oxidation behavior of various types of graphite are greatest at the lowest temperatures, tending to disappear as the temperature increases. If an oxidation threshold is defined as the temperature at which graphite oxidizes at 1% per day, the threshold for pure graphite lies in the range of 520–560°C. Small amounts of catalyst, such as sodium, potassium, vanadium, or copper, reduce this threshold temperature for graphite by as much as 100°C but greatly increase the oxidation rate in the range of 400–800°C (33). Above 1200°C, the number of oxygen collisions with the graphite surface controls the oxidation reaction. Oxidation of graphite is also produced by steam and carbon dioxide; general purpose graphite has a temperature oxidation threshold of approximately 700°C in steam and 900°C in carbon dioxide. At very low concentrations of water and CO_2, there is also a catalytic effect of impurities on the oxidation behavior of graphite (37).

BIBLIOGRAPHY

"Baked and Graphitized Products" under "Carbon" in *ECT* 1st ed., Vol. 3, pp. 1–34, by H. W. Abbott, Speer Carbon Company; "Baked and Graphitized Products, Manufacture" under "Carbon" in *ECT* 2nd ed., Vol. 4, pp. 158–202, by L. M. Liggett, Speer Carbon Company.

1. E. L. Piper, *Soc. Min. Eng. AIME,* Preprint Number 73-H-14 (1973).
2. *Annual Book of ASTM Standards,* Part 17, American Society for Testing and Materials, Philadelphia, Pa., 1976.
3. N. A. Gokcen and co-workers, *High Temp. Sci.* **8,** 81 (June 1976).
4. E. J. Seldin, *Proceedings of the 5th Conference on Carbon,* Vol. 2, Pergamon Press, New York, 1963, p. 545.
5. B. D. McMichael, E. A. Kmetko, and S. Mrozowski, *J. Opt. Soc. Am.* **44,** 26 (1954).
6. G. A. Saunders in L. C. F. Blackman, ed., *Modern Aspects of Graphite Technology,* Academic Press, New York, 1970, p. 79.

7. J. A. Woollam in M. L. Deviney and T. M. O'Grady, eds., *Petroleum Derived Carbons,* American Chemical Society, Washington, D. C., 1976, p. 378.
8. N. Ganguli and K. S. Krishnan, *Nature (London)* **144,** 667 (1939).
9. A. K. Dutta, *Phys. Rev.* **90,** 187 (1953).
10. D. E. Soule, *Phys. Rev.* **112,** 698 (1958).
11. Y. S. Touloukian and co-eds., *Thermophysical Properties of Matter,* Vol. 2, IFI/Plenum Press, New York, 1970, p. 5.
12. B. T. Kelley, *Chem. Phys. Carbon* **5,** 128 (1969).
13. R. W. Powell and F. H. Schofield, *Proc. Phys. Soc. London* **51,** 170 (1939).
14. T. J. Neubert, private communication quoted by L. M. Currie and co-workers in *Proceedings of the International Conference on the Peaceful Uses of Atomic Energy, Geneva, 1955,* Vol. 8, United Nations, New York, 1956, p. 451.
15. J. B. Nelson and D. P. Riley, *Proc. Phys. Soc. London* **57,** 477 (1945).
16. B. T. Kelley and P. L. Walker, Jr., *Carbon* **8,** 211 (1970).
17. S. Mrozowski, *Proceedings of the 1st and 2nd Conferences on Carbon,* University of Buffalo, Buffalo, N.Y., 1956, p. 31.
18. A. L. Sutton and V. C. Howard, *J. Nucl. Mater.* **7,** 58 (1962).
19. W. C. Morgan, *Carbon* **10,** 73 (1972).
20. *Industrial Graphite Engineering Handbook,* Union Carbide Corporation, Carbon Products Division, New York, 1970, Section 5B.02.03.
21. O. L. Blakslee and co-workers, *J. Appl. Phys.* **41,** 3380 (1970).
22. W. L. Greenstreet, *U.S. Oak Ridge National Laboratory Report ORNL-4327* (Dec. 1968).
23. C. Malmstrom, R. Keen, and L. Green, *J. Appl. Phys.* **22,** 593 (1951).
24. P. P. Arragon and R. Berthier, *Industrial Carbon and Graphite,* Society of Chemical Industry, London, Eng., 1958, p. 565.
25. H. H. W. Losty and J. S. Orchard in *Proceedings of the 5th Conference on Carbon,* Vol. 1, Pergamon Press, New York, 1962, p. 519.
26. G. M. Jenkins, *Br. J. Appl. Phys.* **13,** 30 (1962).
27. E. J. Seldin, *Carbon* **4,** 177 (1966).
28. H. S. Starrett and C. D. Pears, *Southern Research Institute Technical Report, AFML-TR-73-14,* Vol. 1, 1973.
29. J. J. Gangler, *Am. Ceram. Soc. J.* **33,** 367 (1950).
30. E. A. Carden and R. W. Andrae, *Am. Ceram. Soc. J.* **53,** 339 (1970).
31. L. Green, Jr., *J. Appl. Mech.* **18,** 346 (1951).
32. J. K. Legg and S. G. Bapat, *Southern Research Institute Technical Report, AFML-TR-74-161,* 1975.
33. L. M. Currie, V. C. Hamister, and H. G. MacPherson, *Proceedings of the International Conference on the Peaceful Uses of Atomic Energy, Geneva, 1955,* Vol. 8, United Nations, New York, 1956, p. 451.
34. R. E. Nightingale, ed., *Nuclear Graphite,* Academic Press, New York, 1962, p. 142.
35. M. C. Robert, M. Aberline, and J. Mering, *Chem. Phys. Carbon* **10,** 141 (1973).
36. P. L. Walker, Jr., M. Shelef, and K. A. Anderson, *Chem. Phys. Carbon* **4,** 287 (1968).
37. M. R. Everett, D. V. Kinsey, and E. Romberg, *Chem. Phys. Carbon* **3,** 289 (1968).

J. T. MEERS
Union Carbide Corporation

APPLICATIONS OF BAKED AND GRAPHITIZED CARBON

Aerospace and Nuclear Reactor Applications

Graphite is an important material for aerospace and nuclear reactor applications because of a unique combination of thermal, chemical, and mechanical properties that enable its survival under the extremely hostile environments encountered.

Graphite is a lightweight structural material that retains good mechanical strength to extremely high temperatures and is readily machinable and commercially available. It also demonstrates good neutron interaction characteristics and stability under irradiation. The most troublesome problem is oxidation at high temperatures.

Aerospace and nuclear reactor applications of graphite demand both high reliability and reproducibility of properties, and physical integrity of the product. The manufacturing processes require significant additional quality assurance steps that result in high cost.

Aerospace. Graphite has long been employed in rocket nozzles, as wing leading edges, as nose cones, and as structural members for both ballistic and glider types of reentry vehicles (see Ablative materials). Graphite is unique in that it can be used both as a heat sink and as an ablation–sublimation material.

The erosion of graphite in nozzle applications is the result of both chemical and mechanical factors. Changes in temperature, pressure, or fuel-oxidizing ratio markedly affect erosion rates. Graphite properties affecting the erosion resistance include density, porosity, and pore size distribution.

The entrance cap, throat, and exit cone sections in a typical nozzle are frequently made or lined with conventional bulk graphite, especially in small nozzles because a small change in dimension causes a relatively large change in performance. In other designs, the throat may be made of conventional graphite with the entrance cap and exit cone molded of carbon or graphite fibrous materials that serve as reinforcement in conjunction with high-temperature plastic resins. In larger nozzles, all three sections might be made of fibrous, reinforced material owing to the ease of construction as well as the entire assembly being lighter in weight.

Nose cones and wing leading-edge components fabricated of graphite are used on both ballistic and glider types of reentry vehicles. Ballistic missiles are subjected to short-duration and extremely severe friction heating and oxidizing conditions when reentering the atmosphere, whereas glider-type reentry vehicles are exposed to less severe conditions for longer periods. Design technology has overcome any adverse effects of high anisotropy relative to thermal stress.

Nuclear Reactors. Manufactured graphite is the most extensively used material for moderator and reflector materials in thermal reactors. Since its use in the first reactor, CP-1, constructed in 1942 at Stagg Field, University of Chicago, many thousands of metric tons of graphite have been used for this purpose. Recently, great interest has been shown in the use of graphite as a construction material in the High-Temperature Gas-Cooled Reactor (HTGR) system.

Graphite is chosen for use in nuclear reactors because it is the most readily available material with good moderating properties and a low neutron capture cross section. Other features that make its use widespread are its low cost, stability at elevated temperatures in atmospheres free of oxygen and water vapor, and its good heat transfer characteristics, good mechanical and structural properties, and excellent machinability.

Neutron economy in graphite occurs since pure graphite has a neutron capture cross section of only $0.0032 \pm 0.0002 \times 10^{-24}$ cm^2. Taking into account the density of reactor grade graphite (bulk density 1.71 g/cm^3), the bulk neutron absorption coefficient is 0.0003/cm. Thus a slow neutron may travel >32 m in graphite without capture.

The purity of reactor-grade graphite is controlled by raw material selection and subsequent processing and purification. Although high temperature purification is most commonly used, some moderator applications require considerably higher purity levels. This objective is accomplished by halogen purification to remove extremely stable carbides, especially of boron, as volatile halides. The actual purity requirements are determined by the reactor design.

The major effect of high temperature radiation (1) over a long period of time is to produce dimensional changes in the graphite involved. When graphite initially contracts upon exposure to fast neutron doses, the rate of contraction decreases with exposure until it reaches a minimum volume; further exposure causes volume expansion, with the rate of expansion increasing rapidly at neutron doses above 3×10^{22} neutrons/cm^2 (>50 keV) in all bulk graphite tested to date. This behavior is caused by atomic displacements that take place when graphite is exposed to fast neutrons, resulting in anisotropic crystallite growth rates. The crystal expands in the *c*-axis direction and contracts in the *a*-axis directions (see page 691). The bulk dimensional change depends upon the geometrical summation of the individual crystallite changes and, hence, is dependent upon the starting materials and the method of fabrication. The extent of radiation damage is also strongly dependent upon the temperature of the graphite during irradiation. The severity of graphite radiation damage at high temperatures was underestimated since the magnitude of this temperature dependence was not recognized until about 1965.

Figure 1 shows the volume change in a conventional nuclear graphite during irradiations at various temperatures of relatively high fluxes. Figure 2 shows the length change in an isotropic nuclear graphite during irradiations at various temperatures at relatively high fluxes. The actual changes in dimensions are, of course, different from grade to grade and depend largely on the degree of anisotropy present in the graphite (1).

Table 1 (2) lists some useful properties of several graphites used for moderators or reflectors in nuclear reactors.

Reactor designers have taken advantage of graphite's properties in applying the material to other than moderator and reflector components, usually in conjunction with some other material.

Combined as an admixture with some forms of boron or other high-neutron-absorbing elements, graphite offers advantages as a neutron shield, control rod, or secondary shutdown material of high temperature stability, without danger of meltdown. In fast reactors, where high-energy neutrons reach the shield region, the presence of carbon atoms slows these neutrons down to energy levels where the probability of capture in the neutron absorber greatly increases. Graphite also serves as a stable matrix for the neutron absorber because it is able to withstand neutron and localized alpha recoil damage, offering protection against gross shield degradation.

Bulk graphites are also used in the HGTR concept to support and surround the active fuel core. These components tend to be large, complex-shaped blocks and have been produced from commercial grades of molded graphites.

In combination with compounds of uranium or thorium, graphite offers advantages as a matrix for fissile or fertile reactor fuel in thermal reactors. In this instance, the graphite serves a dual purpose, as a moderator and as a stable disbursing phase for fuel. Its stability under irradiation and at high temperature aids in minimizing fuel degradation and permits longer useful fuel life. Because of its excellent thermal properties and mechanical integrity, graphite offers an exceptional heat transfer medium for heat removal and also resists thermal shock.

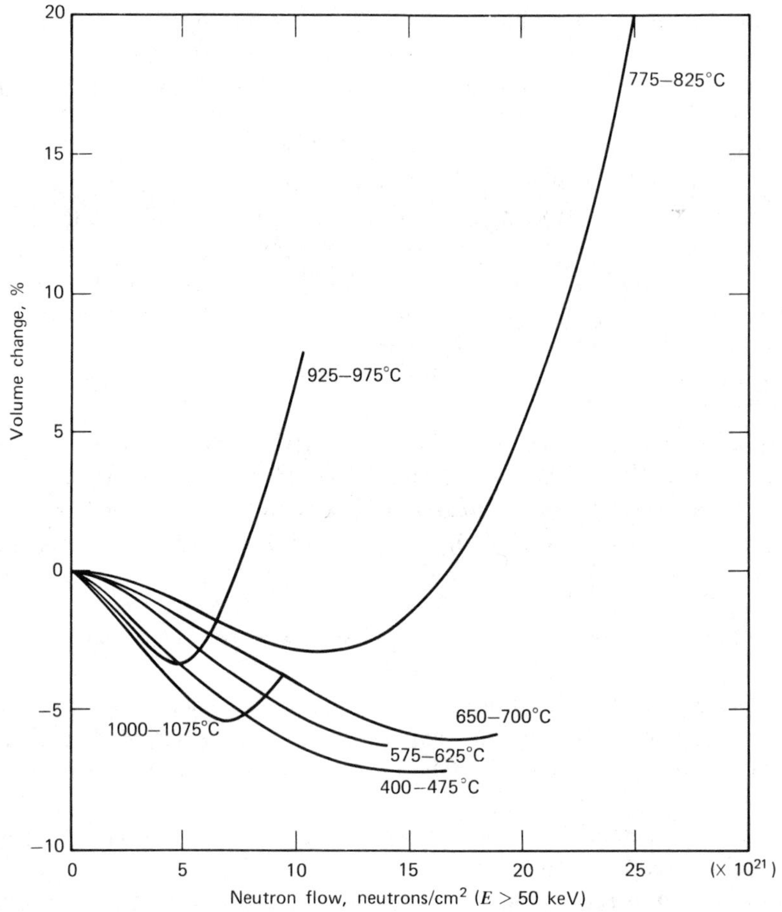

Figure 1. Volume change in anisotropic graphite during GETR (General Electric Test Reactor) irradiations. Courtesy of Oak Ridge National Laboratory, operated by Union Carbide Corporation for the DOE, former Energy Research and Development Administration.

Chemical Applications

The excellent corrosion resistance of carbon and graphite (3) and that of impervious carbon and graphite to acids, alkalies, organics, and inorganic compounds has led to the use of these materials in process equipment where corrosion is a problem. Most of these applications are in chemical process industries but many are in steel, food, petroleum, pharmaceutical, and metal finishing industries.

Other properties, such as the high thermal conductivity of graphite, excellent high temperature stability, and immunity to thermal shock, make these materials useful in applications involving combinations of heat and corrosion, such as heat exchange and high-temperature gas-spray cooling.

Carbon and graphite exhibit varying degrees of porosity, depending on grade, and equipment fabricated of these materials must be operated essentially at atmospheric pressure; otherwise, some degree of leakage must be tolerated. A good example is the carbon brick used to line tanks and vessels handling acids such as hydrofluoric,

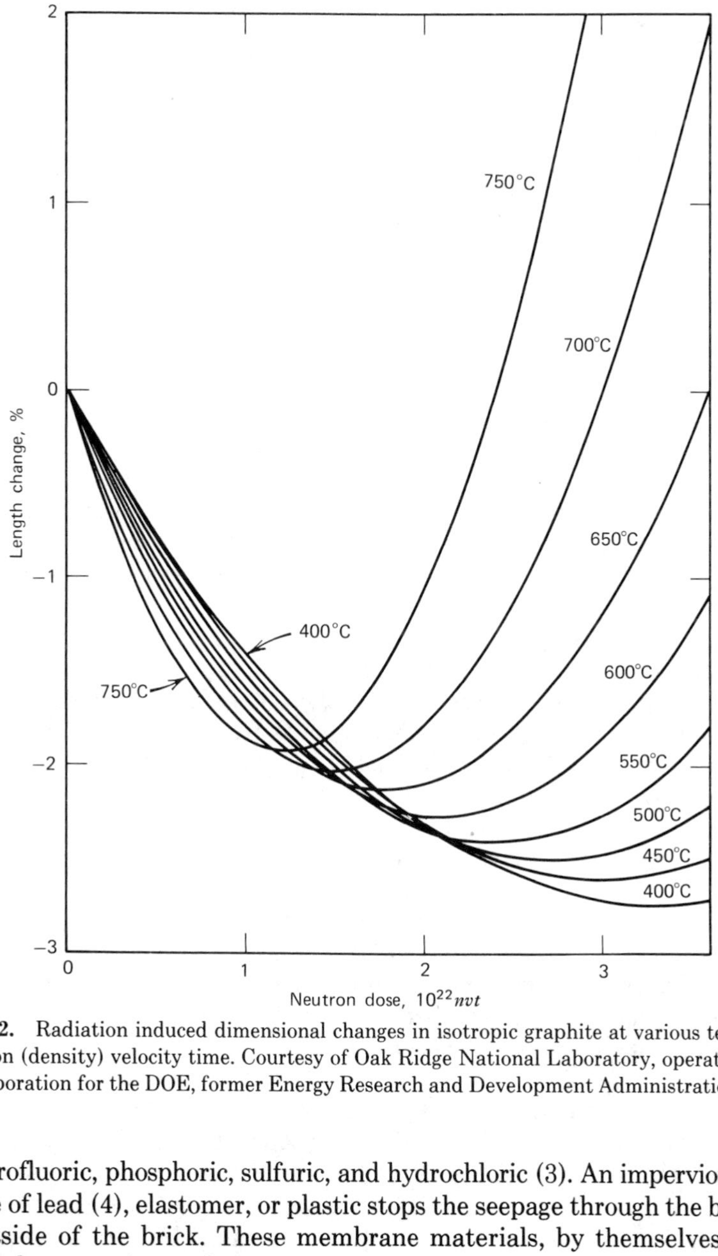

Figure 2. Radiation induced dimensional changes in isotropic graphite at various temperatures. *nvt* = neutron (density) velocity time. Courtesy of Oak Ridge National Laboratory, operated by Union Carbide Corporation for the DOE, former Energy Research and Development Administration.

nitric–hydrofluoric, phosphoric, sulfuric, and hydrochloric (3). An impervious backing membrane of lead (4), elastomer, or plastic stops the seepage through the brick lining at the outside of the brick. These membrane materials, by themselves, may not withstand the corrosive and temperature conditions in the vessel; however, as a membrane behind the carbon lining, they are protected from adverse temperature and abrasion effects. Carbon linings have provided indefinite life with a minimum of maintenance.

Self-Supporting Structures. Self-supporting structures of carbon and graphite are used in a variety of ways. Water-cooled graphite towers serve as chambers for the burning of phosphorus in air or hydrogen in chlorine. The high thermal conductivity of graphite allows rapid heat transfer to the water film, thus maintaining inside wall temperatures below the graphite threshold oxidation temperature of ca 500°C.

Table 1. Properties of Nuclear Graphites[a]

Property	Anisotropic graphite	Isotropic graphite
density, g/cm^3	1.71	1.82
resistance, $\mu\Omega$·cm	735	1,000
tensile strength, kPa[b]	9,930	16,550
coefficient of thermal expansion (CTE), $10^{-6}/°C$		
with grain	2.2	4.8
against grain	3.8	4.6
ansiotropy ratio (CTE ratio)	1.73	0.96
total ash, ppm	740	500
boron content, ppm	0.4	0.3

[a] Ref. 2.
[b] To convert kPa to psi, multiply by 0.145.

Phosphorus combustion chambers for the production of thermal phosphoric acid using cemented graphite block construction have been built 6 m in diameter and 11 m high (5). In other thermal phosphoric acid systems, cement–carbon block structures are used as spray-cooling, hydrating, and absorbing towers.

The immunity of graphite to thermal shock, its high temperature stability, and its corrosion resistance, permit its use for the fabrication of high temperature (800–1650°C), self-supporting reaction vessels such as those used in direct chlorination of metal and alkaline earth oxides. Cemented joints are vulnerable in the units, and it is necessary to machine the graphite components to close tolerances (±0.10 mm) to minimize the cement joint thickness. Graphite is easily machined on commercial metal-machining equipment.

Carbon Raschig-ring tower packing is available in sizes of 10–75 mm dia. Bubble-cap trays up to 3 m dia for hydrochloric–organic stripping towers and packing support structures up to 5.5 m dia for scrubbing towers in pulp and paper mill liquor recovery processes have been installed. Because none of these components requires complex machining or a high degree of imperviousness, carbon rather than graphite is often used in these applications because of its lower cost.

Impervious Graphite. For those applications where fluids under pressure must be retained, impregnated materials are available (3). Imperviousness is attained by blocking the pores of the graphite or carbon material with thermosetting resins such as phenolics, furans, and epoxies. Because the resin pickup is relatively small (usually 12–15 wt %), the physical properties exhibited by the original graphite or carbon material are retained. However, the flexural and compressive strengths are usually doubled. Graphite is also made impervious in a vacuum impregnation process.

Because carbon is difficult to machine, very little impervious carbon equipment is made. However, impervious graphite has been accepted as a standard material of construction by the chemical process industry for the fabrication of process equipment, such as heat exchangers, pumps, valves, towers, pipe, and fittings (6–7).

Many types of impervious graphite shell and tube, cascade, and immersion heat exchangers are in service throughout the world (8). The most common is the shell and tube design where an impervious graphite tube bundle with fixed and floating covers is employed in combination with a steel shell. Whenever parts must be joined, such as the tube to the tube sheet in a shell and tube heat exchanger, very thin resin cement joints are used. These resin cements have the same corrosion-resistant characteristics as the resins used to impregnate the graphite. Because of the high thermal conductivity of graphite, heat exchangers fabricated of impervious graphite have thermal efficiencies equal to metal heat exchangers of equivalent heat transfer area. Heat exchangers up to 1.8 m dia with areas up to 1300 m^2 are commercially available with operating pressures to 690 kPa (100 psi) and temperatures to 170°C (9–11).

Impervious graphite shell and tube heat exchangers are used in boiling, cooling, condensing, heating, interchanging, and cooling–condensing. Large units are used extensively for cooling–condensing wet sulfur dioxide gas in sulfuric acid production plants that burn sludge acid (4). The operation of six units was analyzed to compare actual operation with the calculated design using the Colburn-Hougen analogy; good agreement was reported (12).

These heat exchangers are also used for evaporation of phosphoric acid and rayon spin bath solution; cooling electrolytic copper cell liquor; heating pickle liquor used for descaling sheet steel; boiling, heating, cooling, and absorbing hydrochloric acid and hydrogen chloride; and in many heating and cooling applications involving chlorinated hydrocarbons and sulfuric acid.

Impervious graphite heat exchangers machined from solid blocks are also available (13–14). The solid block construction is less susceptible to damage by mechanical shock, such as steam and water hammer, than are shell and tube exchangers. Block exchangers are limited in size and cost from 50–100% more than shell and tube units on an equivalent area basis.

For in-tank heat exchange, a variety of immersion heat exchangers, such as plate, coil, and steam injection types, are available in impervious graphite. These units are highly efficient, and the heat transfer area required for most services can be easily determined from manufacturers' monographs.

Impervious graphite centrifugal pumps, pipe fittings, and valves were developed because most chemical processes require the movement of liquids.

Impervious graphite pipe and fittings of 25–635 mm ID are used to convey corrosive fluids.

Towers, entrainment separators, thermowells, and rupture disks are fabricated of impervious graphite material. Many equipment items are available from stock. Special equipment can be custom-designed and built, and both standard and special items can be integrated to handle a complete process step. Systems for the absorption of hydrogen chloride in water to produce hydrochloric acid use impervious graphite equipment throughout. Usually, absorption is done in a falling-film absorber (15), a special design adaption of the shell and tube heat exchanger. This approach to absorption of hydrogen chloride (16) was developed and expanded in the United States and is now accepted as the standard.

Stripping hydrogen chloride (13–19) from aqueous hydrochloric acid, and the subsequent production of anhydrous hydrogen chloride, can be efficiently and economically achieved with a series of impervious graphite shell and tube heat exchangers that operate as falling-film reboiler, water and brine-cooled condensers, and bottoms acid cooler. In plants with available chlorine and hydrogen, the production of hydrogen chloride in any form or concentration can be achieved in a system that combines the burning of hydrogen in chlorine in a water-cooled graphite combustion chamber; absorption is carried out in an impervious graphite falling-film absorber, and a train of impervious graphite exchangers is used for stripping and drying (20).

Low Permeability Graphite. Most resin-impregnated impervious graphite materials have a maximum operating temperature limit of 170°C because of resin breakdown above this temperature. Certain special grades with a temperature limitation of 200°C are on the market (21). The chemical industry has developed high temperature processes (370°C and above) where equipment corrosion is a serious problem. Graphite equipment could solve the corrosion problem, but complete fluid containment is usually needed. To meet this need, graphite manufacturers have developed low permeability graphite materials where permeability is reduced by deposition of carbon and graphite in the pores of the base material (21). This material is not limited in its operating temperature, except in oxidizing conditions, and it is used to fabricate high-temperature interchanger ejectors, fused salt cells, fused salt piping systems, and electric resistance heaters.

Several grades of carbon and graphite are commercially available. A controlled combination of high permeability and porosity characterizes these materials. Average pore diameters for typical grades are 0.03–0.12 mm with a total porosity of 48%. Porous graphite is manufactured by graphitizing the amorphous material.

Porous carbon and graphite are used in filtration of hydrogen fluoride streams, caustic solutions, and molten sodium cyanide; in diffusion of chlorine into molten aluminum to produce aluminum chloride; and in aeration of waste sulfite liquors from pulp and paper manufacture and sewage streams.

Electrical Applications

Arc Carbons. A current of electricity produces carbon arc light by leaping across a gap between two carbon electrodes. The typical arc light for visible radiation uses direct current which produces a relatively uniform ball of light across the crater of the positive electrode. The core of the positive electrode contains rare earth compounds to produce white light in the visible portion of the radiation spectrum.

The carbon arc can produce the highest useful brightness of any known artificial light source and provides a color quality matching that of sunlight. The size and shape of the carbon arc light source and the distribution of brilliancy bear an important relation to the optical problems involved in its use. The flat light source at the crater of the positive electrode produces a relatively even brilliancy distribution over a considerable portion of the area. This characteristic permits the light to be collected and directed with a simple reflector or lenses.

The carbon rods used in carbon arc lighting usually consist of two parts, the shell and the core. The shell forms the outer wall and has a central longitudinal hole throughout its length where the core is inserted. The core hole is formed during the extrusion process. The core is usually inserted into the shell after the shell is baked, or alternatively, the shell and core may be extruded simultaneously (22). The core contains, in addition to the conventional carbon filler–binder mix, small percentages of an arc-supporting material, such as potassium, and a flame material such as the metals of the cerium group. For some applications, the carbons are copper-plated to increase electrical conductivity.

Carbon arcs have been used for motion picture projection since the earliest days of the motion picture industry. Owing to the simplicity and reliability of the carbon arc system, thousands of carbon arc lamps are still in use after 20–30 yr. The quantity of light projected by the arc lamps is 10,000–50,000 lm. The amount of light depends on the carbon size and required brightness.

Projector carbon electrodes are manufactured in 7–13.6 mm dia. The diameter of the light ball at the positive crater is proportional to the carbon diameter. Carbon diameters are selected so that the light image at the projector aperture has a side to center brightness of at least 80%.

Peak crater brightness of the d-c high-intensity arc is near 1.8 ncd/m^2. The luminous efficiency is on the order of 25–35 lm/W. Color temperature of the carbon arc light is ca 4000–7000 K, and is dependent on the composition of the core of the positive electrode.

Carbon arc lighting is also used for spotlighting, for searchlights, and in the graphic arts industry for photoengraving, photolithography, and other mechanical reproduction processes. The light sensitized materials, such as dichromates, used for transferring the photographic negative image for subsequent graphic arts operations are affected by radiation in the 0.3–0.5 μm range. Carbon electrodes are specially designed to give a high spectral response in this spectral range.

Light from the carbon arc is used to simulate outdoor exposure for measuring the relative lightfastness, fading, and deterioration of materials such as clothing, paint, plastics, etc. Materials can be tested under precisely controlled conditions which may be repeated often (23–24).

Simulation of the solar radiation in space, by means of the carbon arc, is used for environmental testing of space vehicles, missiles, and their components (25).

Brushes for Motors, Generators, and Slip Rings. Electric motors and generators employ carbon and graphite brushes to maintain electrical contact between an external circuit and the windings of the machine's rotating element. Brushes have several functions, each being essential to the satisfactory operation of the particular machine (26). The brushes must provide electrical contact: they must be good conductors of electricity. The brushes must maintain firm mechanical contact with the rapidly revolving surface of the commutator and function as a bearing material, and they must control the currents resulting from commutation (27). These functions must be fulfilled without destructive sparking, serious commutator wear, or excessive electrical or frictional losses. The brushes must also maintain a smooth commutator surface and a commutator film of suitable characteristics. Only carbon and graphites are capable of meeting all these requirements.

The brushes currently manufactured may be classified into four general types: electrographitic, carbon–graphite, resin-bonded graphite, and metal–graphite. Electrographitic brushes are generally made from calcined blacks and coal-tar pitch. These composites may be supplemented by the addition of graphites and other additives. Mixing and forming follow conventional carbon processing procedures. The final baking temperature approaches 3000°C. Electrographitic brushes have exceptionally good commutating properties and low friction coefficients; they are also efficient in high speed, high current density applications. Carbon–graphite brushes are made from calcined carbon black, calcined coke, graphite, or any combination of the three. They are more abrasive than electrographitic brushes and are generally adapted to lower speed and lower current density applications. Resin-bonded graphite brushes are made from graphite flours and polymeric resins. These brushes have high electrical resistance and are used in applications where very low current densities are required. Metal–graphite brushes can be classified into two general types: those where powdered metal and graphite are mixed and bonded mechanically or chemically and those where carbon–graphite base stock is impregnated with molten metal. Because of the lowered electrical resistance of metal–graphite brushes, they are usually used where high current carrying capacity is required. Table 2 lists typical physical properties of each type of brush.

The performance of brushes in terms of wear and commutation is dependent upon environmental, electrical, and mechanical aspects of each particular application. The environmental aspects which most affect the commutator surface film and consequent performance are temperature and atmosphere. In the absence of water vapor or oxygen, the brushes lose their self-lubricating quality and some impregnant is generally employed to make up for this deficiency. Performance is also dependent on the electrical design of the commutating machine: the location of the brushes with respect to the electrical neutrality of the field poles, the current density required by the application, and the number of commutator bars bridged by the brush face. The mechanical features that affect brush performance include commutator design, brush configuration, holder design, and wear of the moving parts on the commutating machine.

Spectroscopic Electrodes and Powders. Electrodes made of some form of carbon or graphite are employed extensively in optical emission spectrochemical analysis (28–35). The sample to be analyzed (in powder, solid, or liquid form) is excited by means of an electric arc at 5,000–8,000°C or by a spark source at 7,000–20,000°C between two electrodes. At these temperatures, most chemical compounds are dissociated and elements are excited. The chemical elements present are determined qualitatively, semiquantitatively, or quantitatively by measurements of the wavelength and intensity of spectral lines produced by such excitation and dispersed by a suitable optical device.

Graphite and carbon electrodes are used in spectrochemical analysis for several reasons: they are available in high purity grades at reasonable cost; graphite electrodes are easily machinable; they have a uniform structure; they have reasonable electrical and thermal properties; they emit few spectral lines or bands; they do not absorb appreciable moisture; they are chemically inert at room temperature and not wetted even at arc temperatures by most materials; they are porous; they produce an arc with a high excitation potential; they produce a chemically reducing atmosphere in the arc discharge; and they sublime rather than melt.

Graphite is the most universally employed material for spectroscopic electrodes because it is easier than carbon to machine into desired shapes and ship without breakage. Graphite electrodes are available in rods of different diameters and in a variety of preforms, whereas carbon is offered only in rod form. Spectroscopic electrodes are usually supplied in several grades of graphite and one grade of carbon. The graphite grades are of various densities and of different purity levels. The carbon grade is a low crystallinity, high purity material of moderate density. A given spectroscopic grade of graphite or carbon is determined by selection of raw materials and processing conditions.

High purity graphite and carbon powders are also used extensively in spectroscopic techniques as matrix material, either in powder form or as spectroscopic pellets. Graphite carries the sample into the arc thus eliminating excessively long burn times; in addition, fractional distillation of elements from the sample is reduced during exposure. Artificial or synthetic graphite has excellent mixing properties because its almost spherical particles flow more easily and can be transferred and mixed thoroughly with other powders. Natural graphite is specially suited for pelletizing applications because of its flat platelets or needles (depending on whether it is of the

Table 2. Typical Physical Properties of Brushes

Type	Resistivity, μΩ·m	Apparent density, g/cm³	Scleroscope hardness	Flexural strength, MPa[a]	Current capacity, kA/m²	Contact drop[b]	Coefficient of friction[b]	Abrasiveness[b]
electrographitic	10	1.72	35	14	93	*M*	*M*	*VL*
	63	1.43	60	34	124	*H*	*L*	*L*
carbon–graphite	20	1.80	40	25	62	*M*	*M*	*M*
resin-bonded	1400	1.60	9	6	47	*VH*	*L*	*VL*
graphite	13	1.90	30	31	101	*L*	*M*	*L*
metal–graphite	0.05	6.60	8	103	232	*VL*	*VL*	*VL*
	8	2.50	45	21	116	*L*	*M*	*M*

[a] To convert MPa to psi, multiply by 145.

[b]

	VL = very low	*L* = low	*M* = medium	*H* = high	*VH* = very high
contact drop, V	<0.05	ca 0.2	ca 0.4	ca 0.6	≥0.8
coefficient of friction	≤0.15	≤0.22	≤0.3	≥0.3	

Madagascar or Sri Lanka (Ceylon) type) which compress along their long slippage plane and cohere to a large contact area.

The areas where optical emission spectrochemical analysis and graphite or carbon spectroscopic products are employed are varied; they include: metallurgy, geology and prospecting, agriculture, wear, metal analysis, the petroleum industry, medicine, forensics, customs work, and astronomy (see Analytical methods).

The temperature of the electrode, the temperature distribution, and the processes occurring in the arc are of concern to the spectroscopist. Vaporization of the electrode material into the arc is a very complex mechanism affected by particle size, particle shape, particle bonding, the degree of crystallization, and other factors. It is possible to control these properties by manufacturing techniques. Selected raw materials are blended under special conditions to prevent contamination. The blend is extruded into rods and subsequently graphitized. From the graphitized rod stock, preformed electrodes are machined to specific dimensions with precision tools and then purified by a halogen gas process that removes the last traces of nearly all contaminants. After the purified electrodes are tested by means of a very sensitive cathode layer spectrographic technique to make sure they meet the purity requirements, they are packed in spectroscopically clean containers. Each step must be carefully executed and controlled to assure a consistent and uniform product, with a total impurity level of less than 6 ppm.

Miscellaneous Electrical Applications. Carbon and graphite can be processed to have the electrical, mechanical, chemical, and thermal properties required for many electrical and electronic applications. Graphite is an ideal electrode material for electrical discharging machining (EDM). Ultrafine grain, high strength grades that are readily machinable to complex shapes are used as electrodes in the EDM process. Fine-grain graphites are used as EDM electrodes in roughing applications where rapid metal removal and electrode cost are of primary concern (see Electrolytic machining methods).

Graphite is employed in a variety of electronic tube applications. The low inherent gas evolution of graphite makes it easily outgassed; it is pure, free from melting and distortion at high temperatures, and conforms closely in radiating characteristics to the theoretical black body (36). Graphite is used for many electronic tube applications (37). Combinations of the electrical and mechanical properties of carbon and graphite make these materials useful in such diverse applications as pantograph contacts and collector shoes, rheostat disks and plates, high-temperature furnace heating elements, telephone and microphone components, and lightning arrestor parts (27).

Electrode Applications

With the exception of carbon use in the manufacture of aluminum, the largest use of carbon and graphite is as electrodes in electric-arc furnaces. In general, the use of graphite electrodes is restricted to open-arc furnaces of the type used in steel production, whereas carbon electrodes are employed in submerged-arc furnaces used in phosphorus, ferroalloy, and calcium carbide production.

Graphite Electrodes. Graphite electrodes are commercially produced in many sizes ranging from 32 mm dia by 610 mm length to 700 mm dia by 2800 mm length. Such electrodes are used in open-arc furnaces for the manufacture of steel (38), iron and steel castings, brass, bronze, copper and its alloys, nickel and its alloys, fused cast refractories, and fused refractory grain. By far the largest use of graphite electrodes is in the manufacture of steel and, as a consequence, the growth of graphite production has been closely related to the growth in electric furnace steel production. A cutaway sketch of an open-arc furnace is shown in Figure 3. Steel is produced by filling the cylindrical shell with ferrous scrap, metallized iron ore, or occasionally, molten pig iron, then melting and refining the metallic charge with the heat derived from the electric arc generated at the tips of the electrodes.

Prior to the mid 1940s, the arc furnace was used almost exclusively for the production of low tonnage, high quality steels such as stainless and alloy steels. Since then its use has been extended to production of the more common high-tonnage steel grades (see Steel). Domestic growth of arc furnace steel production has been dramatic, rising

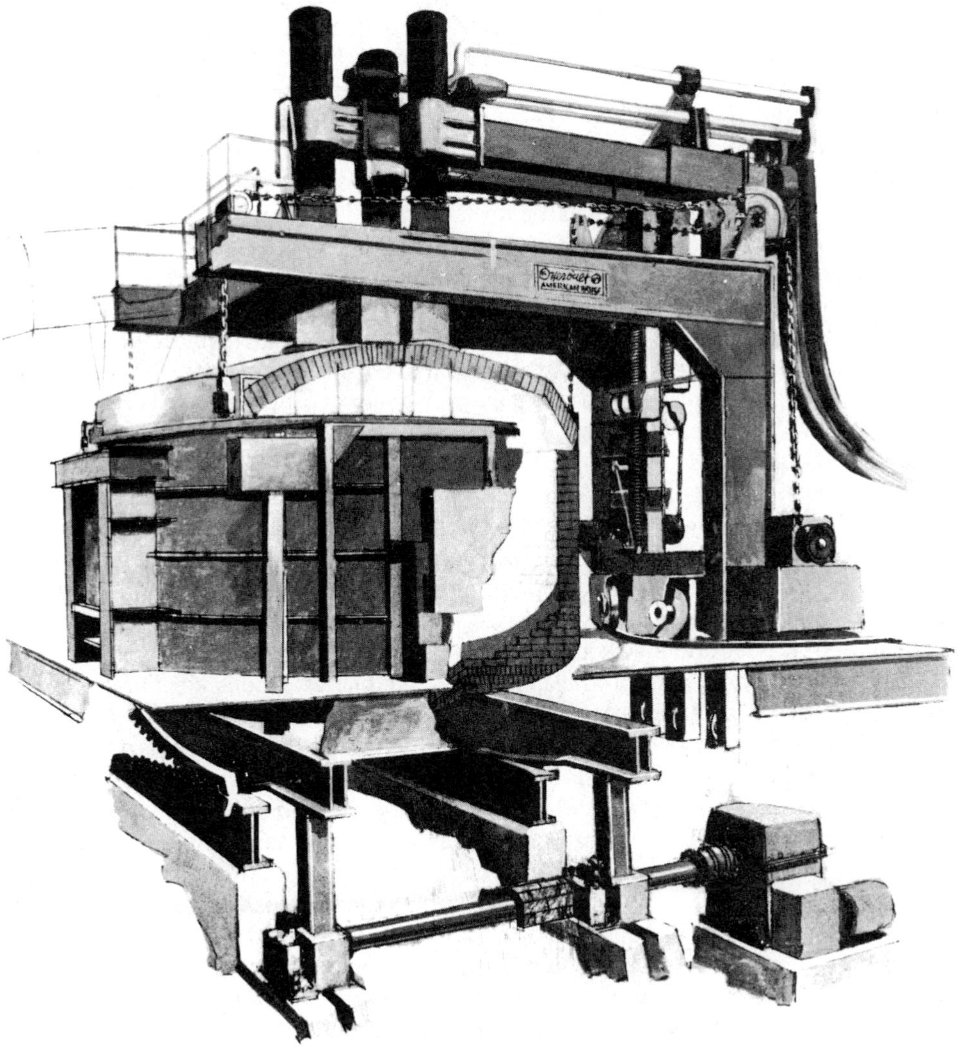

Figure 3. Overall sketch of an electric-arc furnace, cut away to show sections of bottom, sidewall, and roof. Courtesy of American Bridge Division, U.S. Steel Corporation.

from 6% of total domestic steel production in 1950 to 20% in 1975. Over 100 million metric tons of steel were produced in 1975 in electric-arc furnaces in the free world, approximately 20% of total world steel production (39), and these furnaces consumed over 600,000 metric tons of graphite electrodes.

Graphite electrodes are consumed in the melting process. For iron and steel production, the average consumption is ca 5–8 kg/t, depending upon the quality of charge material, the quality of electrodes, and numerous factors related to the operation of the furnace (22). Electrode consumption can be classified into three broad categories: tip consumption, sidewall consumption, and breakage. Roughly half of the observed consumption occurs at the electrode tip where the intensely hot and rapidly moving arc spot produces both vaporization of the graphite and some ejection of small graphite particles. In addition, the electrode tip is eroded by contact with the liquid metal and slag. The rate of incremental tip consumption generally increases when operating currents or power are increased. However, since increased current or power levels generally result in higher productivity, the electrode consumption when expressed in terms of kg/t of metal produced, may exhibit little or no increase. The periphery or sidewall of the hot electrode is slowly consumed by reaction with oxidizing atmospheres both inside and outside the furnace, resulting in a tapering of the electrode toward the arc tip. Sidewall consumption is time dependent and is greatest for low productivity furnaces (40). It is also increased by the use of many fume removal systems and by the use of oxygen in the furnace for assisting melting or refining. Since sidewall consumption may account for 40% or more of total electrode consumption, extensive efforts have been made to reduce this component of consumption through the use of oxidation retardants and electrode coatings. Such efforts have had little success to date, primarily because of the extreme thermal and chemical environment to which the electrode is exposed. A third form of consumption consists primarily of electrode breakage resulting from excessive movement of large masses of scrap during melting or the presence of nonconductors in the charge. Although such breakage usually accounts for less than 10% of net electrode consumption, excessive thermal shock, improper joining practice, and incorrect phase rotation can magnify this form of electrode consumption (41). Although attempts have been made to correlate electrode consumption with relatively small changes in electrode properties, it is apparent that charge quality and furnace operating practice exert a more profound influence on electrode performance. Most notable is the established inverse relationship between furnace productivity and electrode consumption (40).

Graphite electrodes are produced in two broad classifications, regular grade and premium grade. Typical properties of these grades are given in Table 3.

The major differences between the two grades are that the premium grade is made from needle-grade coke and is pitch-impregnated prior to graphitization. The premium grade electrode is used where very high performance is required, such as in the high current operation typical of ultrahigh-powered arc furnaces. The current carrying capacity of an electrode column depends on many characteristics of the furnace operation as well as the characteristics of the electrode and electrode joint. Over the years, significant progress has been achieved in improving the current carrying capacity of electrode columns. For example, the 510 mm dia electrode first introduced in 1938 was designed to carry 26,000 A; by 1961, this same size electrode carried ≤45,000 A and, by 1975, permitted >55,000 A. Such improvements stem primarily from improved raw materials and process technology advancements that are not fully reflected in changes in electrode properties (42).

Table 3. Typical Properties of Regular and Premium Grade Graphite Electrodes[a]

Property	Regular grade	Premium grade
bulk density, g/cm^3	1.58	1.68
resistivity, $\mu\Omega\cdot m$	7.7	6.0
flexural strength, kPa[b], wg	6900	10350
cg	5865	8300
elastic modulus, GPa[b], wg	6.2	7.6
cg	3.5	5.5
coefficient of thermal expansion (CTE), $10^{-6}/°C$		
wg	0.8	0.9
cg	1.7	1.5
thermal conductivity W/(m·K), wg	134	168
cg	67	101
thermal shock parameter[c], wg	186×10^3	254×10^3
cg	66×10^3	102×10^3

[a] wg = with-grain; cg = cross-grain.
[b] To convert Pa to psi, multiply by 1.45×10^{-4}.
[c] Thermal shock parameter $= \dfrac{\text{thermal conductivity} \times \text{strength}}{\text{CTE} \times \text{elastic modulus}}$.

In service, graphite electrodes operate at up to 2500 K and are subject to large thermal and mechanical stresses and extreme thermal shock. Graphite is unique in its ability to function in this extreme environment. The relatively low electrical resistance along the length of the electrode minimizes the power loss owing to resistance heating and helps keep the electrode temperature as low as possible. This characteristic is most important in ultrahigh-power furnaces. For such furnaces approximately 30% lower electrode resistivity of premium grade electrodes is usually essential. A high value of the thermal shock parameter is also important (see Table 3); this is enhanced by high strength and high thermal conductivity combined with low elastic modulus and low coefficient of thermal expansion.

The joints between electrodes are an extremely important part of the electrode system, both from the standpoint of resisting the mechanical forces of scrap caves and of carrying high current density without localized overheating (42a). Such joints should possess high strength, especially in flexure, and possess low electrical resistance. Careful assembly and proper torque are vital to good performance (41).

Carbon Electrodes. Carbon electrodes are used primarily in submerged-arc furnaces for the manufacture of ferroalloys, phosphorus, silicon metal, calcium carbide, pig iron, and fused refractory grain. There are two broad types of carbon electrodes. The self-baking Soderberg type consisting of carbon paste is used extensively in the production of calcium carbide and electric-furnace pig iron; and the prebaked electrode finds its major use in production of phosphorus, silicon metal, and several ferroalloys. Prebaked carbon electrodes are produced in sizes ranging from 250 mm dia by 1520 mm length to 1400 mm dia by 2800 mm length. In general, the sizes up to 1140 mm dia are extruded; the larger sizes are molded.

Submerged-arc furnaces (see Fig. 4) differ significantly from open-arc furnaces in both function and operation. The name submerged-arc furnace is derived from the fact that the high temperature reaction zone in such furnaces is always isolated from the walls and roof by cooler charge material, ie, the reaction zone is submerged. As

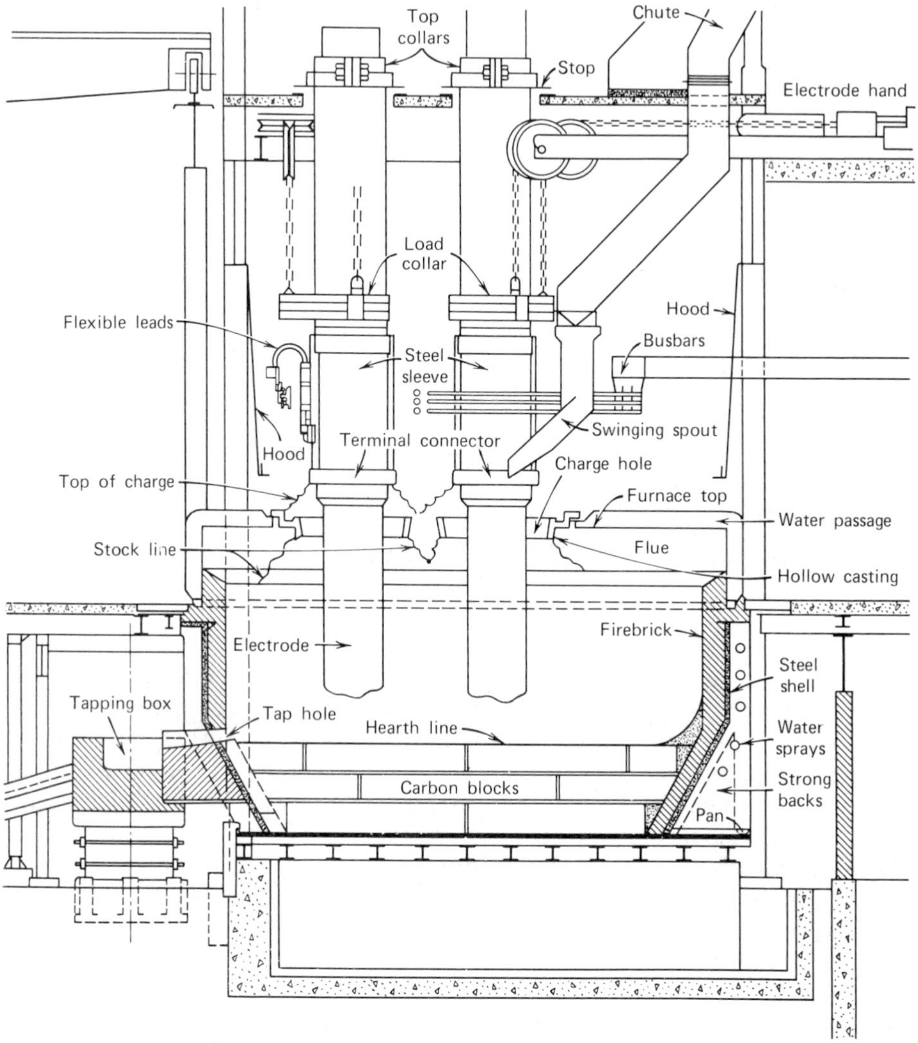

Figure 4. Design of a submerged-arc ferroalloy furnace. Courtesy of Union Carbide Corporation (3).

charge material is consumed in the reaction zone, fresh charge material is added to keep the furnace nearly full. In addition to melting the charge, the furnace also performs the function of a reaction vessel where reduction of oxides is achieved, usually by reaction with carbon. Electrode carbon consumption by reduction reactions varies greatly with the type of product produced and the amount of reducing agent added to the charge. Thus a broad range of specific electrode consumption rates is found in commercial practice, extending from approximately 20 kg/t product for phosphorus or standard ferromanganese to more than 140 kg/t for silicon metal. Unlike open-arc furnaces, the tip of the electrodes in submerged-arc furnaces is often a few meters above the hearth floor so that the volume beneath and around the electrode tip is heated by a combination of resistance heating and arcing (43). Temperatures in this region may reach 2200 K. The volume of this heated reaction zone largely determines furnace productivity. Therefore, to attain the largest possible reaction zone, large diameter electrodes are employed. In many cases, owing to the relatively low electrical currents involved, relatively low cost carbon electrodes may be used rather than graphite which might be prohibitively expensive in the sizes required. However, in certain cases, such as the modern phosphorus furnaces operating at up to 90,000 kW (see Phosphorus), semigraphite electrodes are required. This type of electrode is produced primarily from graphite raw materials bonded with a carbonized pitch binder. In addition to possessing significantly lower electrical resistance, semigraphite electrodes are more easily machined than carbon and have superior thermal shock resistance.

Typical properties of carbon and semigraphite electrodes are shown in Table 4. Strength is an important factor, especially in the larger size electrodes, because of the massive weight of the electrode columns used on many furnaces. Low resistance is an advantage in reducing joule heating losses, although such losses are relatively minor except for very highly powered furnaces. Since relatively large thermal gradients can exist in the submerged electrode, thermal shock or thermal stress resistance is also important in cases of furnaces subjected to transient or intermittent operation and especially where large diameter electrodes are used.

Anode Applications

Graphite has served as the primary material for electrolysis anodes in which chlorine or chlorates are produced at the anode (see Alkali and chlorine products). Recent technological advances, however, have resulted in a dimensionally stable anode (DSA) consisting of precious metal oxides deposited on a titanium substrate that is replacing graphite as the primary anode in mercury cells used in the chlor–alkali industry (40–46).

Although the particles and binder of graphite anodes are subject to oxidative attack under anodic conditions, graphite anodes can provide 270 days of service in diaphragm cell applications. Diaphragm cells account for 75% of the chlor–alkali capacity in the United States (47–48).

Graphite anodes meet the requirements for electrolytic cell applications which include: high degree of insolubility, low initial cost, availability in almost unlimited quantities, few limitations as to size and shape, good electrical conductivity, and high purity to prevent contamination of cell products.

The two basic types of graphite anodes used are plain and impregnated. The purpose of impregnation is to prevent anolyte penetration of the pores in graphite and attendant corrosion inside the anode. For impregnated anodes, base graphite with an initial porosity of 20–30% is given a vacuum-pressure impregnation usually with an oil such as linseed to fill or coat the accessible pores. Proper impregnation provides a 20–50% increase in anode life over unimpregnated graphite.

The electrolysis of fused magnesium chloride for the production of magnesium metal is the second largest use for graphite anodes. The electrolytic process is the current principal method of magnesium production although developments in the metallothermic process and the carbothermic process show potential for economical alternatives (49).

Other applications of graphite anodes include electrolysis of fused chlorides for the production of sodium, lithium, tantalum, and columbium. Graphite anodes are also used for electrolysis of aqueous manganese sulfate and for the anodic deposition of manganese dioxide used as dry battery depolarizer. Carbon anodes are utilized in the electrolytic production of fluorine.

Table 4. Average Properties of Carbon and Semigraphite Electrodes (Measured With the Grain at Room Temperature)

Electrode diameter range, mm	Bulk density, g/cm^3	Compressive strength[a], kPa	Flexural strength[a], kPa	Elastic modulus[a], GPa	Specific resistance, $\mu\Omega\cdot m$	Coefficient of thermal expansion (CTE), $10^{-6}/°C$	Thermal conductivity, W/(m·K)
				Carbon			
425–1125	1.63	11700	4830		50	3.2	16
1250–1375	1.62	11700	3860	3.3	46	3.2	12
				Semigraphite			
875–1125	1.64	12400	5400		24	3.0	34
1250–1375	1.64	12400	4660	3.7	30	3.0	34

[a] To convert Pa to psi, multiply by 1.45×10^{-4}.

Graphite anodes are also used in cathodic protection applications for corrosion prevention of underground and underwater metal structures where low cost, light weight, and excellent electrical conductivity are required. Life of treated graphite ranges from 3–30 yr in cathodic protection applications (50).

Mechanical Applications

Carbon–graphite finds use in many mechanical applications where its natural lubricity, high-temperature mechanical strength, and corrosion resistance give it important advantages over other materials. This lubricity, strength, and corrosion resistance, together with ease of machining, dimensional stability, and high thermal conductivity, make carbon–graphite the material of choice for mechanically supporting loads in sliding or rotating contact. The mechanical applications are: mechanical seals—face, ring, and circumferential types; bearings—carbon cages for roller and ball bearings, carbon sleeve bearings and bushings, carbon thrust bearings or washers, and combination sleeve and thrust bearings; packing rings—steam and water shaft packing rings, and compressor tail-rod packing rings; nonlubricated compressor parts—piston rings, wear rings, segments, scuffer shoes, shaft tail-rod packing rings, pistons, and piston skirts; and miscellaneous applications—flat plate slider parts for support of apparatus and facilitating sliding movement under load, rotor vanes, and metering device parts such as the metering ball and plates.

Carbon–graphite materials employed for mechanical applications are prepared by mixing selected sizes and types of carbon and graphite with binder materials such as pitches and resins. The mixtures are formed into compacts and baked to temperatures of ca 1000–3000°C. Specific raw materials and processing techniques are employed to obtain desired properties for the finished carbon–graphite materials (51).

The successful application of carbon–graphite as a sliding contact is dependent upon the proper use of proprietary additives or impregnants, or both, in the carbon–graphite materials. Carbon–graphite, long considered to be self-lubricating, depends on the presence of adsorbed films of water vapor and/or oxygen for its low friction and low wear properties. This adsorbed boundary layer is soon lost when the operation is conducted at high altitude, high temperature, or in cold, dry air. A substitute boundary layer can be formed by incorporating certain additives or impregnants, or both, such as thermoplastic or thermosetting resins, or metallic sulfides, oxides, or halides. In addition to reducing the friction and wear of the carbon–graphite materials, the additives and impregnants can serve to improve oxidation resistance, provide impermeability to high pressure gases and liquids, and even permit operation under high vacuum conditions (52). Successful operation under high vacuum conditions is a primary requirement of equipment used for exploring outer space.

Carbon–graphite materials will not gall or weld even when rubbed under excessive load and speed. Early carbon materials contained metal fillers to provide strength and high thermal conductivity, but these desirable properties can now be obtained in true carbon–graphite materials that completely eliminate the galling tendency and other disadvantages of metals.

Maintenance of flat faces in rotating and stationary mating seals is important for successful operation. Dimensional stability is necessary in high speed, high-load face seal applications, and carbon–graphite materials with high elastic moduli have been developed to meet the requirements. Distortion of face seals caused by unequal loading or thermal stress leads to high unit loads at contact points, causing high localized friction and additional heat which produces further distortion (53–54).

Carbon–graphite materials are compatible with a wide range of mating materials, such as chrome-plated steel, 440-C stainless, and 300 series stainless (for corrosion resistance) steel, fine-grained cast iron, flame-plated oxides and carbides, ceramics, cermets, and at times, with themselves. However, the importance of rubbing contact has been minimized for certain applications that employ face seals equipped with self-acting lift augmentation (55). For this new generation of seals, pads are machined on the seal face which, during operation, act as a thrust bearing and cause the seal to lift from the counter face and ride on a thin gas film. Ideally, the self-acting seal will experience mechanical wear only during startup and shutdown of the equipment on which it is installed. The advent of this new seal design will enhance the ability of carbon–graphite materials to meet the ever increasing speed, pressure, and temperature requirements for certain applications (56–57).

Metallurgical Applications

Because of their unique combination of physical and chemical properties, manufactured carbons and graphites are widely used in several forms in high temperature processing of metals, ceramics, glass, and fused quartz. A variety of commercial grades is available with properties tailored to best meet the needs of particular applications (58). Industrial carbons and graphites are also available in a broad range of shapes and sizes.

Structural Graphite Shapes. In many metallurgical and other high temperature applications, manufactured graphite is used because it neither melts nor fuses to many common metals or ceramics, exhibits increasing strength with temperature, has high thermal shock resistance, is nonwarping, has low expansion, and possesses high thermal conductivity. However, because of its tendency to oxidize at temperatures above 750 K, prolonged exposure at higher temperatures frequently necessitates use of a nonoxidizing atmosphere. In addition, prolonged contact both with liquid steel and with liquid metals that rapidly form carbides should be avoided.

Some of the more common applications for structural graphite shapes are: (*1*) hot-pressing molds and dies (59) for beryllium at 1370 K and 6.9 MPa (1000 psi); diamond-impregnated drill bits and saw tooth segments at 1250 K and 13.8 MPa (2000 psi); tungsten and other refractory metals and alloys up to 2370 K and 6.9 MPa (1000 psi); and boron nitride and boron carbide up to 2060 K; (*2*) molds for metal casting steel railroad car wheels made by the controlled-pressure pouring process (60); steel slabs and billets made by the controlled-pressure pouring process (61); continuous casting of copper and its alloys, aluminum and its alloys, bearing materials, zinc, and gray iron (62–63); centrifugal casting of brasses, bronzes, steels, and refractory metals (64); nickel anodes; welding rods and thermite welding molds; shapes of refractory metals (Ti, Zr, Mo, Nb, W) and carbides; and shapes of gray, ductile and malleable irons (65); (*3*) foundry accessories including: mold chill plates, core rods, and riser rods; crucible skimmer floats; plunging bells for magnesium additions to ductile iron and desulfurization of blast-furnace hot metal (66–67); stirring rods for nonferrous metals; and railroad brake shoe inserts; (*4*) injection tubes and nozzles for purifying molten aluminum (68) and other nonferrous metals, desulfurization of blast furnace and foundry iron with calcium carbide or magnesium, and carbon raising of foundry iron with graphite powders; (*5*) aluminum extrusion components including dies, guides from die openings, run-out table boards, and cooling-rack inserts; (*6*) rolls for handling metal sheets are used in certain processes because they are self-lubricating and reduce surface marring; (*7*) immersion thermocouple protection tubes for nonferrous metals; (*8*) welding electrodes for welding, gouging, and cutting iron and steel, particularly with the aid of an air blast (69); (*9*) crucibles, either induction or resistance heated, for producing tungsten carbide, beryllium fluoride and beryllium, titanium and zirconium fluoride, semiconductor crystals of germanium and silicon, and for laboratory chemical analysis equipment; (*10*) ceramic and glass production including: casting molds for fused cast refractories of alumina, magnesia, and chrome–magnesite compositions up to 2650 K (70); mold susceptors for fabricating fused magnesia crucibles; susceptors, electrical resistor elements, fusion crucibles, molds and dies for the production of fused quartz (71); linings for float-glass tanks (72); take-out pads for automatic glass-blowing machines; diablo-wheels for glass tubing production; and linings for hydrofluoric acid tanks for glass etching; (*11*) boats, trays, and plates for sintering clutch plates, brake disks, and cemented carbides and for the manufacture of semiconductor material and transistors; and (*12*) furnace jigs for brazing honeycomb panels, automotive ignition points and arms, automotive radiator cores, transistor junction assembly, and glass-to-metal seals.

Electric Heating Elements. Machined graphite shapes are widely used as susceptors and resistor elements to produce temperatures up to 3300 K in applications utilizing nonoxidizing atmospheres. The advantages of graphite in this type of application include its very low vapor pressure (lower than molybdenum), high black body emissivity, high thermal shock resistance, and increasing strength at elevated temperatures with no increase in brittleness. Graphites covering a broad range of electrical resistivity are available and can be easily machined into complex shapes at lower cost than refractory metal elements. Flexible graphite cloth is also used widely as a heating element since its low thermal mass permits rapid heating and cooling cycles. Porous carbon or graphite, and flexible carbon or graphite felts are used for thermal insulation in many high temperature furnaces. Typical applications include molten-iron or steel-holding furnaces, continuous-casting tundishes, liquid-steel degassing units, chemical-reaction chambers, quartz-fusion apparatus, zinc-vaporization chambers, sintering furnaces, vapor deposition units; the felts are also used in the manufacture of semiconductors (qv) (73–74).

Graphite Powder and Particles. Manufactured graphite powders and particles are used extensively in metallurgical applications where the uniformity of physical and chemical characteristics, high purity, and rapid solubility in certain molten metals are important factors (75). The many grades of graphite powders and particles are classified on the basis of fineness and purity. Applications for these materials include facings for foundry molds and steel ingot molds, additives to molten iron to control carbon level and chill characteristics, covering material for molten nonferrous metals and salt baths to prevent oxidation, additives to sintered materials to control carbon level and frictional characteristics of oil-less bearings, and as charge-carbon in steels made in electric arc furnaces.

Refractory Applications

Various forms of carbon and graphite materials have found wide application in the metals industry, particularly in connection with the production of iron and aluminum. Carbon has been used as a refractory material since 1850, although full commercial acceptance and subsequent rapid increase in use has taken place only since 1945 (see Refractories).

Carbon as a Blast Furnace Refractory. The first commercial use of carbon as a refractory for a blast furnace lining took place in France in 1872, followed in 1892 by a carbon block hearth for a blast furnace of the Maryland Steel Company at Sparrows Point, Md. After a period of abated interest, the excellent results obtained with several carbon hearths in Germany and the United States during the late 1930s and early 1940s renewed enthusiasm for this approach. Although initially used only for the hearth floor of blast furnaces, carbon and graphite refractories have been applied successfully to hearth walls, lower and upper boshes, and most recently, to the lower stack of modern high-performance blast furnaces throughout the world (76). More than 360 individual carbon or graphite blast furnace linings have been installed in North America through the end of 1975. Carbon is also used extensively for blast furnace slag troughs and iron runners.

Experience has shown that carbon and graphite refractories have several characteristics contributing to their successful use in blast furnaces. (*1*) They show no softening or loss of strength at operating temperatures. (*2*) They are almost immune to attack by either blast furnace slags or molten iron. (*3*) Their relatively high thermal conductivity, when combined with adequate external cooling, assures solidification of iron and slag far from the furnace exterior, thus promoting long life while maintaining safe lining thicknesses (77). (*4*) A high level of thermal shock resistance avoids cracking or spalling in service. (*5*) A positive, low thermal expansion provides both dimensional stability and a tightening of joints in the multiblock linings. However, because of their poor oxidation resistance such refractories must not be exposed to air, carbon dioxide, or water vapor at elevated temperatures (78–79).

The prime requirement for linings in blast furnace hearths or hearth walls is to contain liquid iron and slag safely within the crucible throughout extended periods of continuous operation. This requirement is most readily achieved by providing sufficient cooling of the lining to assure solidification of penetrating iron and slag far from the furnace exterior (80). Where only peripheral cooling is used, a thick carbon bottom is required, the thickness of which is approximately one-quarter of the furnace diameter between the cooling system (77). In North and South America, such hearth bottoms are made from 3 or 4 horizontal layers of long carbon beams, each of which may be up to 6.5 m long and weigh as much as 5 metric tons. Anchoring the ends of these long beams beneath the furnace sidewall prevents flotation of the blocks in the denser molten iron, a phenomenon that has been observed in other parts of the world where long carbon beams are not used. Where short carbon blocks are used in the hearth, whether in a vertical or horizontal orientation, extensive interlocking, keying, and cementing must be employed to prevent loss by flotation. The application of carbon hearth and hearth wall refractories has virtually eliminated the once-prevalent danger of breakouts, a phenomenon that represents a serious threat to both life and property. As of 1976, carbon hearth walls were used in virtually all blast furnaces in North America, and of these, 80 had carbon hearth floors. One such furnace remained in operation after 22 years of service on the original carbon hearth, and another produced over 13 million metric tons of iron on its carbon hearth. In recent years, there has been an increasing tendency to incorporate some form of underhearth cooling in the larger furnaces to enhance hearth life. Medium size furnaces generally use a layer of high conductivity graphite beneath the carbon hearth for improved cooling. Larger furnaces incorporate tightly sealed steel plenums beneath the hearth through which air (or infrequently water) is passed to effect cooling of the hearth (81).

During 1960–1975 the use of carbon and graphite as bosh refractories has tripled. In 1976 ninety North American blast furnaces were operating with carbon boshes, and record performance on a single carbon bosh has exceeded 8 million metric tons of pig iron. As in the hearth and hearth wall, long life is assured by efficient cooling of the carbon bosh lining. 100% External shower cooling is used, and special cements are employed to assure that the high conductivity refractory is in good thermal contact with the cooled bosh shell. Long life of carbon bosh refractories also necessitates a high level of resistance to abrasion and alkali attack (82).

The success of carbon bosh linings has prompted several operators in Europe and Japan to extend the use of this refractory into the lower stack of several large modern blast furnaces where design permits this feature (76).

Refractories in the Aluminum Industry. The Hall-Heroult aluminum cell uses carbon materials for the anode, cathode, and sidewall, since carbon is the only material able to withstand the corrosive action of the molten fluorides used in the process (see Aluminum). The aluminum industry uses more carbon per metric ton of virgin metal than any other industry. Production of 1 t of molten aluminum requires about 500 kg of anode carbon and 7.5–10 kg of cathode carbon.

Because of the very large consumption of anode carbon, aluminum plants usually contain an on-site carbon plant for the manufacture of anodes.

Two types of cathodes are used in aluminum cells: both are produced from blends of calcined anthracite, metallurgical coke, and pitch. Compared with a tamped lining, prebaked carbon block linings provide higher operating strength, higher density, lower porosity, and longer life. Their lower resistance results in a lower voltage drop through the lining that improves the overall electrical efficiency of the cell (83). Prebaked cathodes also exhibit more predictable starting characteristics and possess greatly improved ability to restart successfully after a shutdown. This characteristic helps increase production and service life and usually more than compensates for the higher initial cost of prebaked cathode blocks. The service life of a prebaked cathode is usually 3–4 yr.

Refractories for Cupolas. In many ways, the use of carbon cupola linings has paralleled the application of carbon in the blast furnace. Carbon brick and blocks are used to form the cupola well (84), or crucible, up to the tuyeres. When properly installed and cooled, carbon linings last for many months, or even years of intermittent operation. Their resistance to molten iron and both acid and basic slags provides not only insurance against breakouts but also operational flexibility to produce different iron grades without the necessity of changing refractories. Carbon is also widely used for tap holes, breast blocks, slagging troughs, and dams.

Refractories for Electric Reduction Furnaces. Carbon hearth linings are used in submerged-arc electric reduction furnaces producing phosphorus, calcium carbide, all grades of ferrosilicon, high carbon ferrochromium, ferrovanadium, and ferromolybdenum. They are also used in the production of beryllium oxide and beryllium copper, where temperatures up to 2273 K are required.

Most of the principles pertaining to carbon blast furnace hearths apply as well to hearths for submerged-arc furnaces, although the fact that carbon is an electrical conductor is of importance in the case of electric reduction furnaces. The very long life of carbon linings in this application is attributable to their exceptional resistance to corrosive slags and metals at relatively high temperatures.

BIBLIOGRAPHY

"Baked and Graphitized Products, Uses" under "Carbon" in *ECT* 2nd ed., Vol. 4, pp. 202–243, by W. M. Gaylord, Union Carbide Corporation.

1. P. R. Kasten and co-workers, *U.S. Oak Ridge National Laboratory, ORNL-TM 2136,* Feb. 1969.
2. J. T. Meers and co-workers, *Am. Nucl. Soc. Trans.* **21,** 185 (1975); A. E. Goldman, H. R. Gugerli, and J. T. Meers, *paper presented at NUCLEX 75* Meeting, Basel, Switz., Oct. 6–10, 1975.
3. M. R. Hatfield and C. E. Ford, *Trans. Am. Inst. Chem. Eng.* **42,** 121 (1946).
4. W. M. Gaylord, *Ind. Eng. Chem.* **51,** 1161 (1959).
5. N. J. Johnson, *Ind. Eng. Chem.* **53,** 413 (1961).
6. S. H. Friedman, *Chem. Eng.* (*N.Y.*) **69**(14), 133 (1962).
7. J. R. Schley, *Chem. Eng.* (*N.Y.*) **81,** 144 (Feb. 18, 1974); **81,** 102 (Mar. 18, 1974).
8. D. Hills, *Chem. Eng. NY* **81,** 80 (Dec. 23, 1974); **82,** 116 (Jan. 20, 1975).
9. F. L. Rubin, *Chem. Eng.* (*N.Y.*) **60,** 201 (1953).
10. W. W. Palmquist, *Chem. Eng. Costs Q.* **4,** 111 (1954).
11. C. H. Baumann, *Ind. Eng. Chem.* **54,** 49 (1962).
12. J. F. Revilock, *Chem. Eng.* (*N.Y.*) **66,** 77 (Nov. 19, 1959).
13. W. M. Gaylord, *Ind. Eng. Chem.* **49,** 1584 (1957).
14. W. S. Norman, A. Hilliard, and C. H. Sawyer, *Materials of Construction in the Chemical Process Industries,* Society of Chemical Industry, London, Eng., 1950, p. 239.
15. J. Coull, C. A. Bishop, and W. M. Gaylord, *Chem. Eng. Prog.* **45,** 525 (1949).
16. W. M. Gaylord and M. A. Miranda, *Chem. Eng. Prog.* **53,** 139 (Mar. 1957).
17. T. F. Meinhold and C. H. Draper, *Chem. Process. Chicago* **23**(8), 92 (1960).
18. C. C. Brumbaugh, A. B. Tillman, and R. C. Sutter, *Ind. Eng. Chem.* **41,** 2165 (1949).
19. C. W. Cannon, *Chem. Ind.* (*N.Y.*) **65,** 3554 (1949).
20. R. W. Naidel, *Chem. Eng. Prog.* **69,** 53 (Feb. 1973).
21. J. F. Revilock and R. P. Stambaugh, *Chem. Eng.* (*N.Y.*) **69,** 148 (June 25, 1962).
22. S. Chari and J. Bohra, *Carbon* **10,** 747 (1972).
23. L. I. Nass, *Plast. Tech.* **17,** 91 (Oct. 1971); **18,** 31 (Mar. 1972).
24. R. E. Harrington, *Inst. Environ. Sci. Proc.,* 501 (1968).
25. Institute of Environmental Sciences, *Space Simulation; Proceedings of a Symposium Held in New York City, May 1972,* NASA Special Publication NASA-SP-298, 1972.
26. F. K. Lutz and W. C. Kalb, *Carbon Brushes for Electrical Equipment,* Union Carbide Corporation, New York, 1966.
27. R. Holm, *Electric Contacts; Theory and Application,* 4th ed., Springer Verlag, New York, 1967.
28. E. L. Grove, *Analytical Emission Spectroscopy,* Vol. 1, Marcel Dekker, Inc., New York, 1971.
29. L. H. Ahrens, and S. R. Taylor, *Spectrochemical Analysis,* Addison-Wesley, Reading, Mass., 1961.
30. G. L. Clark, *The Encyclopedia of Spectroscopy,* Reinhold Publishing Corp., New York, 1960.
31. ASTM Committee E-2, *Methods for Emission Spectrochemical Analysis,* 6th ed., American Society for Testing and Materials, Philadelphia, Pa., 1971.
32. ASTM Committee E-2, *Annual Book of ASTM Standards,* Part 42, American Society for Testing and Materials, Philadelphia, Pa.
33. N. H. Nachtrieb, *Principles and Practice of Spectrochemical Analysis,* McGraw-Hill Book Co., New York, 1950.
34. G. R. Harrison, R. C. Lord, and J. R. Loofbourow, *Practical Spectroscopy,* Prentice-Hall, Inc., New York, 1948.
35. E. L. Grove and A. J. Perkins, *Developments in Applied Spectroscopy,* Vol. 9, Plenum Press, New York-London, 1971.
36. G. A. Beitel, *J. Vac. Sci. Technol.* **8,** 647 (Sept.–Oct. 1971).

37. W. A. Kohl, *Handbook of Materials and Techniques for Vacuum Devices,* Reinhold Publishing Corp., New York, 1967, p. 137.
38. C. E. Sims, ed., *Electric Furnace Steelmaking,* John Wiley & Sons, Inc., New York, 1962–1963.
39. *World Steel in Figures,* International Iron and Steel Institute, Brussels, Belg., 1976.
40. W. E. Schwabe, *J. Met.* **24,** 65 (Nov. 1972).
41. *Electric Arc Furnace Digest,* Carbon Products Division, Union Carbide Corporation, 1975.
42. A. Ince, *Ironmaking Steelmaking* **3**(6), 310 (1976).
42a. J. S. Davis and P. Schroth, *AIME Electr. Furn. Steel Proc.* **29,** 145 (1971).
43. V. Paschkis and J. Persson, *Industrial Electric Furnaces and Appliances,* Interscience Publishers Inc., New York, 1960, p. 245.
44. *Chem. Process Chicago* **39**(9), 60 (1976).
45. V. H. Thomas, *J. Electrochem. Soc.* **74,** 618 (1974).
46. S. Puschaver, *Chem. Ind. London,* 236 (Mar. 15, 1975).
47. *Chem. Week* **113,** 32 (Oct. 1973).
48. J. C. Davis, *Chem. Eng. (N.Y.)* **81,** 84 (Feb. 1974).
49. B. S. Gulyanitskii, *Itogi Nauki Tekh.,* (5), 5 (1972).
50. W. W. Palmquist, *Pet. Eng. Los Angeles* **22,** D22 (Jan. 1950).
51. N. J. Fechter and P. S. Petrunich, *Development of Seal Ring Carbon–Graphite Materials,* NASA Contract Reports CR-72799, Jan. 1971; *CR-72986,* Aug. 1971; *CR-120955,* Aug. 1972; and *CR-121092,* Jan. 1973.
52. D. H. Buckley and R. L. Johnson, *Am. Soc. Lubr. Eng. Trans.* **7,** 91 (1964).
53. M. J. Fisher, *Paper D4 presented at International Conference on Fluid Sealing, British Hydromechanics Research Association, Apr. 1961.*
54. C. F. Romine and J. P. Morley, *Mach. Des.* **40,** 173 (Dec. 5, 1968).
55. R. L. Johnson and L. P. Ludwig, *NASA TN-D-5170* (Apr. 1969).
56. L. P. Ludwig, *NASA Contract Report TM-X71588,* 1974.
57. A. Zobens, *Lubr. Eng.* **31,** 16 (Jan. 1975).
58. *Industrial Graphite Engineering Handbook,* Union Carbide Corporation, Carbon Products Div., New York, 1970.
59. R. M. Spriggs in A. M. Alper, ed., *High Temperature Oxides,* Vol. V-3, Academic Press, New York, 1970, p. 183.
60. *J. Met.* **24,** 50 (Nov. 1972).
61. E. A. Carlson, *Iron Steel Eng.* **52,** 25 (Dec. 1975).
62. R. Thomson, *Am. Foundrymen's Soc. Trans.* **79,** 161 (1971).
63. H. A. Krall and B. R. Douglas, *Foundry* **98,** 50 (Nov. 1970).
64. *Foundry* **90,** 63 (Feb. 1962).
65. C. A. Jones and co-workers, *Am. Foundrymen's Soc. Trans.* **79,** 547 (1971).
66. *Foundry* **93,** 132 (Feb. 1965).
67. W. H. Duquette and co-workers, *AIME Open Hearth Proc.* **56,** 79 (1973).
68. *33 Magazine* **13,** 64 (Aug. 1975).
69. L. J. Christensen, *Welding J.* **52,** 782 (Dec. 1973).
70. A. M. Alper and co-workers in A. M. Alper, ed., *High Temperature Oxides,* Vol. V-1, Academic Press, New York, 1970, p. 209.
71. U.S. Pat. 2,852,891 (Sept. 23, 1958), H. J. C. George (to Quartz & Silica, S.A.).
72. U.S. Pat. 3,486,878 (Dec. 30, 1969), R. J. Greenler (to Ford Motor Co.).
73. H. G. Carson, *Ind. Heat.,* (Nov. 1962) and (Jan. 1963).
74. J. G. Campbell, *Second Conference on Industrial Carbon and Graphite,* Society of Chemical Industry, London, Eng., 1966, p. 629.
75. A. T. Lloyd, *Mod. Cast.* **64,** 46 (Dec. 1974).
76. G. Kahlhofer and D. Winzer, *Stahl Eisen* **92**(4), 137 (1972).
77. L. W. Tyler, *Blast-Furnace Refractories,* The Iron and Steel Institute, London, Eng., 1968.
78. F. K. Earp and M. W. Hill, *Industrial Carbon and Graphite,* Society of Chemical Industry, London, Eng., 1958, p. 326.
79. S. Ergun and M. Mentser, *Chem. Phys. Carbon* **1,** 203 (1965).
80. R. D. Westbrook, *Iron Steel Eng.* **30,** 141 (Mar. 1953).
81. S. A. Bell, *J. Met.* **18,** 365 (Mar. 1966).
82. R. J. Hawkins, L. Monte, and J. J. Waters, *Ironmaking Steelmaking* **1,** 151 (Nov. 3, 1974).
83. L. E. Bacon in G. Gerard, ed., *Extractive Metallurgy of Aluminum,* Vol. 2, John Wiley & Sons, Inc., New York, 1969, p. 461.

84. *The Cupola and Its Operation,* American Foundrymen's Society, Des Plaines, Ill., 1965.

General References

Aerospace and Nuclear Reactor Applications

R. M. Bushong, *Aerosp. Eng.* **20,** 40 (Jan. 1963).
C. E. Ford, R. M. Bushong, and R. C. Stroup, *Met. Prog.* **82,** 101 (Dec. 1962).
M. W. Riley, *Mater. Des. Eng.* **56,** 113 (Sept. 1962).
S. Glasstone, *Principles of Nuclear Engineering,* D. Van Nostrand Co., Princeton, N. J., 1955.
R. E. Nightingale, *Nuclear Graphite,* Academic Press, New York, 1962.

Arc Carbons

J. E. Kaufman, ed., *IES Lighting Handbook,* 5th ed., Illuminating Engineering Society, New York, 1972.
G. Kirschstein and co-workers, *Gmelins Handbuch der Anorganischen Chemie, 8th ed., System Number 14, Carbon, Teil B, Lieferung 1,* Verlag Chemie GMBH, Weinheim, Ger., 1967, p. 207.

Chemical Applications

J. R. Schley, *Mater. Prot. Perform.* **9,** 11 (Oct. 1970).
A. Hilliard, *Chem. Ind. London,* 40 (Jan. 10, 1970).
A. R. Ford and E. Greenhalgh in L. C. F. Blackman, ed., *Modern Aspects of Graphite Technology,* Academic Press, London, Eng., 1970, p. 272.

Electrical Applications

F. P. Bowden and D. Tabor, *The Friction and Lubrication of Solids,* Oxford University Press, London, Eng., 1958.
V. Berger and U. Schroeder, *ETZ-A* 8(4), 91 (1962).
W. T. Clark, A. Conolly, and W. Hirst, *Br. J. Appl. Phys.* **14,** 20 (1963).
H. M. Elsey and C. Lynn, *Electr. Eng. Am. Inst. Electr. Eng.* **68,** 106 (June 1949).
J. K. Lancaster, *Br. J. Appl. Phys.* **13,** 468 (1962).
J. K. Lancaster, *Wear* **6,** 341 (Nov. 5, 1963).
J. W. Midgley and D. G. Teer, *ASME Trans. Ser. D* **85,** 488 (1963).
E. I. Shobert, *Carbon Brushes,* Chemical Publishing Co., Inc., New York, 1965.
W. J. Spry and P. M. Scherer, *Wear* **4,** 137 (Nov. 2, 1961).
K. Binder, *ETZ-B* **86,** 285 (1965).

Electrode Applications

J. R. Bello, *AIME Electr. Furn. Steel Proc.* **29,** 219 (1971).
J. A. Persson, *AIME Electr. Furn. Steel Proc.* **21,** 131 (1973).
W. M. Kelly, *Carbon and Graphite News,* Vol. 5, No. 1, Union Carbide Corp., 1958, p. 1.

Anode Applications

L. E. Vaaler, *Electrochem. Technol.* **5**(5–6), 170 (1967).
J. P. Randin in A. J. Bard, ed., *Encyclopedia of Electrochemistry of the Elements,* Vol. VII, C, V, Marcel Dekker Inc., New York, 1976, Chapt. VII–I.
F. L. Church, *Mod. Met.* **23,** 90 (Aug. 1967).
Eur. Chem. News, (Oct. 1969).
R. R. Irving, *Iron Age* **210,** 64 (Nov. 1972).
V. A. Kolesnikov, *Tsvetnye Met.,* 53 (Nov. 7, 1975).
W. A. Rollwage, *Paper Presented at AIME Annual Meeting,* (Feb. 1967).
V. de Nora, *Chem. Ing. Tech.* **47,** 125 (Feb. 1975).
D. Bergner, *ibid.,* p. 136.

Mechanical Applications

J. W. Abar, *Lubr. Eng.* **201,** 381 (October 1964).
G. P. Allen and D. W. Wisander, *NASA-TN-D-7381* (Sept. 1973).
G. P. Allen and D. W. Wisander, *NASA-TN-D-7871* (Jan. 1975).
P. F. Brown, N. Gordon, and W. J. King, *Lubr. Eng.* **22,** 7 (Jan. 1966).
L. J. Dobek, *NASA Contract Report CR-121177,* (Mar. 1973).
Crane Packing Co., *Packing and Mechanical Seals,* 2nd ed., Morton Grove, Ill., 1966.
J. P. Giltrow, *Composites* **4,** 55 (Mar. 1973).
W. R. Lauzau, B. R. Shelton, and R. A. Waldheger, *Lubr. Eng.* **19,** 201 (May 1963).
G. Oley, *Mech. Eng.* **94,** 18 (Apr. 1972).
R. R. Paxton, *Electrochem. Tech.* **5,** 174 (May–June 1967).
V. P. Povinelli, Jr., *J. Aircr.* **13,** 266 (Apr. 1975).
F. F. Ruhl, A. B. Wendt, and P. N. Dalenberg, *Lubr. Eng.* **23,** 241 (June 1967).
A. G. Spores, *Lubr. Eng.* **31,** 248 (May 1975).
R. D. Taber, J. H. Fuchsluger, and M. L. Rutherber, *Lubr. Eng.* **31,** 565 (Nov. 1975).

R. M. BUSHONG (Aerospace and Nuclear Applications)
R. RUSSELL (Chemical Applications)
B. R. JOYCE (Electrical Applications, Arc Carbons)
P. M. SCHERER (Electrical Applications, Brushes; Anode Applications; and Mechanical Applications)
N. L. BOTTONE (Electrical Applications; Spectroscopic Electrodes)
R. L. REDDY (Electrode Applications; Metallurgical Applications; and Refractory Applications)
Union Carbide Corporation

CARBON FIBERS AND FABRICS

Carbon fibers are filamentary forms (fiber dia 5–15 μm) of carbon (carbon content exceeding 92 wt %) and are characterized by flexibility, electrical conductivity, chemical inertness except to oxidation, refractoriness, and in their high performance varieties, high Young's modulus and high strength. Considerable confusion exists over the terms carbon and graphite fibers. The term graphite fibers should be restricted to materials with the three-dimensional order characteristic of polycrystalline graphite; essentially all commercial fibers are carbon fibers.

Carbon fibers were first made intentionally in 1878 by Edison (1) who pyrolyzed cotton to make incandescent lamp filaments. Interest in carbon fibers remained at a very low level until the late 1950s when commercially useful products, made by carbonizing rayon cloth and felt, were introduced (2). These relatively low strength, low modulus products were followed in 1964 by the development of high modulus, high strength, rayon-based carbon yarns (3) and of intermediate modulus, high strength, polyacrylonitrile (PAN)-based carbon yarns (4). High modulus, intermediate strength fibers made from a mesophase (liquid crystal) pitch precursor were introduced in 1974 as a mat product (5) and in 1975 as continuous yarns (6–7). Carbon fibers can also be made from ordinary (nonmesophase) pitch, but they generally have low strength and low modulus. Other precursor materials, including phenolics, polyacetylene, poly(vinyl alcohol), and polybenzimidazole, have been investigated (8–9). As of early 1977 no evidence was available to indicate that fibers from these latter precursors would become commercially significant.

Most desirable properties of carbon fibers, eg, Young's modulus, electrical and thermal conductivity, and to a lesser extent tensile strength, depend on the degree of preferred orientation. In a high modulus fiber, the carbon layer planes are predominantly parallel to the fiber axis; however, when viewed in cross section, the layers in most carbon fibers are oriented in all directions, although mesophase pitch-based fibers with concentric layers (onionskin structure) and with layers radiating from the fiber axis (radial structure) have been made (7). The high strength, high modulus properties are found only along the fiber axis. Very highly oriented carbon fibers possess a Young's modulus of nearly 900 GPa (130×10^6 psi), very close to that of the graphite single crystal, and tensile strength of over 3.4 GPa (500,000 psi). In commercial production, Young's moduli are usually held below 550 GPa (80×10^6 psi) for three principal reasons: the tensile strength does not always increase proportionally with the modulus, leading to unacceptably low strain-to-failure ratios; the interlamellar shear strength (adhesion between the fiber and the matrix resin) decreases with increasing modulus; and the production costs for very high modulus fibers are high.

Detailed discussions of the carbon fiber structure and the methods of structure determination are reported (10–13). The Young's modulus of carbon fibers depends entirely upon the degree of preferred orientation, as shown in Figure 1. Since carbon fibers of different origin can vary greatly in density, the data in Figure 1 have been normalized to correct for these differences.

The electrical and thermal conductivity also increase with increased orientation (11). The elastic constants of rayon- and PAN-based carbon fibers with Young's moduli ranging from 41–551 GPa ($6–80 \times 10^6$ psi) and of epoxy-matrix composites made from

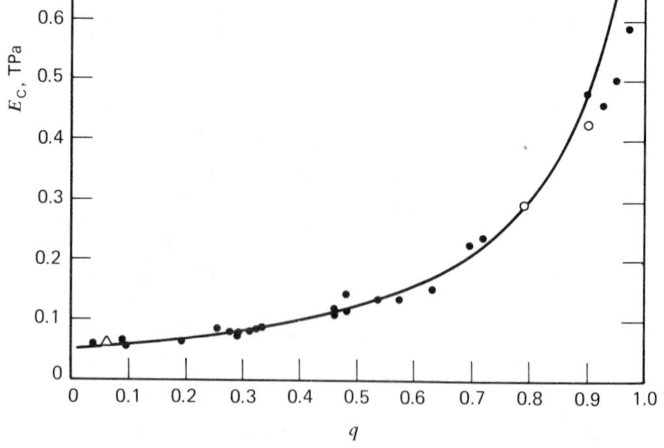

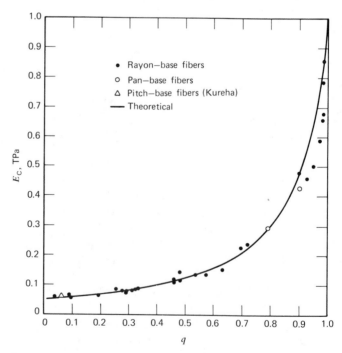

Figure 1. Young's modulus corrected for porosity E_c as a function of preferred orientation q; curve is based on theoretical model (10). To convert TPa to psi, multiply by 145×10^6.

these fibers have been determined (14). The relation between the axial thermal expansion of a variety of commercial carbon fibers to their preferred orientation was measured (15) and the transverse thermal expansion was determined (16).

The tensile strength of highly oriented carbon fibers should, in theory, be approximately 5% of the Young's modulus. Graphite whiskers with tensile strength of over 21 GPa (3×10^6 psi) have been made (17). For commercially produced carbon fibers, the tensile strength is controlled by defects in the fiber structure; these include stress risers in the form of surface irregularities, voids, and particulate inclusions (18–25). The nature and frequency distribution of these limiting defects vary for different carbon fibers, but all fibers show significant increases in single-filament tensile strength (frequently over 100%) when the test-gage length is reduced from 5 cm to 1 mm. The useful tensile strength (translatable into composite properties) of commercial carbon yarn rarely exceeds 3.8 GPa (550,000 psi). In 1977 the standard, low cost, PAN-based carbon fibers had a modulus of 207 GPa (30×10^6 psi) and tensile strengths of 2.4–2.9 GPa (350,000–425,000 psi).

The interlamellar shear strength of carbon fiber composites tends to decrease with increasing fiber modulus. This property is important because a low shear strength results in low transverse composite properties and possibly low compressive strength (26–27). A large number of surface treatments to improve shear strength have been published. Most processes (28) depend on some form of oxidative etching in air, nitric acid, or other oxidizing liquids, anodic oxidation, or the deposition of a low temperature carbon char (29) on the fiber surface. The actual processes used by various producers are closely guarded trade secrets.

There are considerable differences in the process chemistry, production economics, and properties of rayon-, PAN-, and pitch-based carbon fibers. These processes and products are, therefore, discussed separately.

Rayon-Based Carbon Fibers

The production process for rayon-based carbon yarn and cloth involves three distinct steps: preparation and heat treating, carbonization, and optional high temperature heat treatment. Heat treating at 200–350°C to form a char with thermal stability is required for rapid subsequent carbonization. Principally, water evolves during this stage although complex thermal tars also form. These tars can redeposit on the yarn or cloth and render the material brittle upon carbonization. To reduce tar formation and to quicken the process, the heat treatment is frequently carried out in a reactive atmosphere, such as air, or with the aid of chemical impregnants (9). The weight loss during this step is approximately 50–60% and any structure or preferred orientation that may have existed in the starting material is destroyed during the formation of the amorphous char. The yarn or cloth is carbonized next at 1000–2000°C during which additional weight is lost and an incipient carbon layer structure is formed. The material may then be heat treated at temperatures approaching 3000°C. The overall carbon yield is approximately 20–25%. The products have densities of 1.43–1.7 g/cm^3, filament tensile strengths of 345–690 MPa ($5–10 \times 10^4$ psi), and Young's moduli of 21–55 GPa ($3–8 \times 10^6$ psi). They are usually used in the form of cloth for aerospace applications, eg, phenolic impregnated heat shields, and in carbon–carbon composites for missile parts and aircraft brakes. Production statistics are not available; estimated consumption for 1970–1976 was 100,000–250,000 kg/yr. Through 1975 the volume of rayon-based, low modulus yarn and cloth probably exceeded that of all other carbon fibers combined. The price for cloth is \$60–120/kg (see Ablative materials).

To produce high modulus, high strength rayon-based fiber, it is necessary to stretch the yarn during the heat treatment process at 2700–3000°C. The high temperature increases the size, perfection, and parallel stacking of the carbon layers, and stretching orients them parallel to the fiber axis. The tensile strength of rayon-based yarn also increases with increasing stretch and Young's modulus, resulting in an almost constant strain-to-failure of approximately 0.5%. The density of the yarn is 1.65 g/cm^3 at a modulus of 345 GPa (50×10^6 psi) or 1.82 g/cm^3 at a modulus of 517 GPa (75×10^6 psi). There are comprehensive reviews of production processes (10), and the structure and properties of rayon-based yarn (8).

The hot-stretching process is very costly, and the price of high modulus rayon-based yarn is \$600–1300/kg. Production volume is relatively low and applications are confined to aerospace uses.

The long range future availability of all rayon-based carbon fiber products depends on the continued production of suitable continuous filament rayon yarn. The rapid decline in the tire cord use of rayon has caused several major rayon manufacturers to discontinue production. Many applications of rayon-based carbon fibers depend on their stability at very high temperatures and the high carbon content of over 99%. In these areas, they are not readily replaceable by the standard PAN-based fibers that have a lower carbon content and change when heated beyond the original process temperature. Mesophase pitch-based fibers can be made with properties very similar to those of rayon-based products and may be expected to replace them in many applications.

PAN-Based Carbon Fibers

Production of PAN-based carbon fibers also involves the three steps of low temperature heat treatment, carbonization, and optional high temperature heat treatment. The principal differences from the rayon-based fiber process are that a well-oriented ladder polymer structure is developed during oxidative heat treatment under tension, and the orientation is essentially maintained through carbonization.

Most producers probably use copolymer PAN fibers containing at least 95% acrylonitrile units. Textile PAN yarns with a high percentage of copolymer content and those containing appreciable quantities of brighteners, such as TiO_2, are not considered suitable precursors. To improve the alignment in the polymer structure and reduce the fiber diameter, the as-spun PAN fibers are frequently stretched by 100–500% at approximately 100°C. Stretching of the polymer also increases the tensile strength and Young's modulus of the resulting carbon fibers (30).

Heat treatment is carried out in an oxidizing atmosphere (usually air) at 190–280°C for 0.5–5 h. Sufficient tension must be applied at this stage to prevent the yarn from shrinking and to develop a high degree of alignment of the ladder polymer chain. Heat treatment is an exothermic process, and precautions against a runaway reaction are required. After heat treatment, the yarn is black, infusible, and can be rapidly carbonized in an inert atmosphere, usually at 1000–1300°C, where the tensile strength reaches a maximum. Hydrogen cyanide is evolved during carbonization, and suitable precautions must be taken for combustion of volatiles. The yield from raw to carbonized yarn is approximately 45–50%. At this stage, the material consists of 92–95%

carbon and most of the remainder is nitrogen. This yarn has a Young's modulus of ca 210 GPa (30×10^6 psi). The modulus can be further increased to 350–520 GPa (50–75 $\times 10^6$ psi) by heating to above 2500°C, but the tensile strength tends to decrease. In at least one type of PAN-based fiber this decrease in strength is caused by impurities in the raw PAN fibers (24). Fibers spun under clean-room conditions increased in strength as well as modulus upon heating above 2500°C.

The tensile strength of some fibers with 210 GPa (30×10^6 psi) Young's modulus is apparently limited by stress-rising surface irregularities (24), and etching to remove or blunt these flaws can increase the strength (31). PAN-based fibers can also be hot-stretched, with a resulting increase in both tensile strength and Young's modulus. However, because of the high cost, this process is not the usual commercial practice. Detailed reviews of the PAN-based fiber process and fiber characteristics have been published (8–9,32).

Most PAN-based fibers produced in 1976 had a tensile strength of ca 2.4–2.75 GPa (350,000–400,000 psi) and a Young's modulus of 193–241 GPa (28–35 $\times 10^6$ psi), although fibers with moduli up to 482 GPa (70×10^6 psi) and generally lower strength are available. Fiber densities are usually 1.7–1.8 g/cm^3. Product forms include yarns of 1,000, 3,000, 6,000, and 10,000 filaments, tows up to 500,000 filaments, and fabrics of various styles and weave construction.

Although official production statistics are not available, it is estimated that the combined production in the three principal producing countries (United States, Japan, and England) may have risen from 10 metric tons in 1970 to 250 t in 1976. With increased production, the price of the standard 207 GPa (30×10^6 psi) Young's modulus fibers has decreased from approximately $1000/kg before 1970 to $40–80/kg in 1977, although higher modulus fibers and low denier yarns, such as 1000-filament products, are sold at premium prices.

Until about 1972 PAN-based carbon fibers were used almost exclusively in epoxy "prepregs" where structures for the aerospace industry were fabricated. The higher costs had to be justified by weight savings and performance improvements. Long-time reliability experience was meager and aircraft structures were mostly limited to noncritical parts. Fiber price reductions and significant advances in composite fabrication technology had, by 1976, made carbon fiber composites cost competitive with many metal parts in aircraft construction. Confidence in design and reliability has been established; composites are currently used in major flight-critical structures. During the same time period, a second major market for PAN-based carbon fibers developed in the sporting goods industry, first in golf clubs, then in fishing rods, tennis rackets, bows and arrows, skis, and sailboat masts and spars. Applications for textile and computer machinery, automotive and general transportation, and musical instruments are rapidly emerging (see Composite materials).

Mesophase Pitch-Based Fibers

The process for these fibers starts with commercial coal tar or petroleum pitch which is converted through heat treatment into a mesophase or liquid crystal state. Since the mesophase is highly anisotropic, the shear forces acting on the pitch during spinning and drawing result in a highly oriented fiber, with the essentially flat aromatic polymer molecules oriented parallel to the fiber axis. The spun yarn is then thermoset in an oxidizing atmosphere which renders the fibers infusible and amenable to rapid carbonization. The original preferred orientation is maintained during thermosetting, which does not require tensioning, and is further enhanced during carbonization and graphitization. Mesophase pitch-based fibers have a Young's modulus over 690 GPa (100×10^6 psi) when heated (without stretching) to 3000°C. These are the only fibers that can truly be called graphite. The carbon yield from spun to carbonized fiber is 75–85%, depending on the molecular weight of the mesophase pitch and the heat treatment conditions. Fibers with a Young's modulus over 345 GPa (50×10^6 psi) are almost pure (over 99.5%) carbon. Lower modulus fibers may contain small amounts (ca 1%) of sulfur. The fiber density is approximately 2 g/cm^3.

The tensile strength of mesophase pitch-based fibers increases with increasing carbonization temperatures until a modulus of 345 GPa (50×10^6 psi) is reached. It then remains essentially flat to 550 GPa (80×10^6 psi) and rises again with further heat treatment. Single-filament tensile strengths (2 cm gage length) over 3.5 GPa (5×10^5 psi) have been measured. In 1977 the commercial yarn with a 345 GPa (50×10^6 psi) modulus had a tensile strength of approximately 2.0 GPa (3×10^5 psi). The strength-limiting defects consist primarily of voids of 1–2 μm dia and, at this stage of development, of particulate inclusions similar to but more severe than those found in PAN-based fibers. Considerable process details and the microstructure of mesophase pitch-based fibers are described in the literature (7,33).

In 1977 mesophase pitch-based fibers were produced in three forms. One form is a low modulus (138 GPa or 20×10^6 psi), low cost ($17/kg) mat product that is not recommended for structural reinforcement. This material is useful as a veil mat for sheet molding to provide an electrically conductive surface for electrostatic spraying or improved surface appearance or both. Milled mat with a fiber length of ca 1.0 mm is used in injection molding to achieve electrical conductivity, resistance to heat distortion, and improved wear characteristics (for bearing applications).

The second form consists of pitch-based carbon fiber fabrics (12) that have filament moduli of 138–517 GPa ($20–75 \times 10^6$ psi), depending on process temperature, and present an attractive, low cost alternative to rayon-based cloth.

The third form, a continuous filament yarn with a Young's modulus of 345 GPa (50×10^6 psi) and 2.0 GPa (3×10^5 psi) tensile strength, is by far the lowest priced high modulus carbon yarn. Initial uses are in stiffness-critical applications.

Health and Environmental Considerations

Carbon fibers, like other forms of carbon, are very inert. For this reason they are now used in medical research for body implant studies such as pacemaker leads (see Prosthetic and biomedical devices). The only known medical hazard is a possible skin irritation caused by broken filaments, similar to that caused by glass fibers. Carbon fibers are good electrical conductors. Broken filaments can remain airborne over considerable distances and can cause short circuits in electrical equipment.

BIBLIOGRAPHY

"Baked and Graphitized Products, Uses" under "Carbon" in *ECT* 2nd ed., Vol. 4, pp. 202–243, by W. M. Gaylord, Union Carbide Corporation.

1. U.S. Pat. 223,898 (Jan. 27, 1880), T. A. Edison.
2. U.S. Pat. 3,107,152 (Sept. 12, 1960), C. E. Ford and C. V. Mitchell (to Union Carbide).

3. R. Bacon, A. A. Pallozzi, and S. E. Slosarik, *Technical and Management Conference of the Reinforced Plastics Division, Proceedings of the 21st Annual, Feb. 8–10, 1966, Chicago, Ill.*, Society of the Plastics Industry, Inc., New York, 1966, Sec. 8-E.
4. Brit. Pat. 1,110,791 (Apr. 24, 1968), W. Johnson, L. N. Phillips, and W. Watt (to National Research Corp.).
5. H. F. Volk, *Carbon and Graphite Conference, 4th London International; extended abstracts of paper presented at Imperial College, Sept. 23–27, 1974,* Society of Chemical Industry, London, Eng., 1974, Session IV, Paper 102.
6. H. F. Volk, *Proceedings of the 1975 International Conference on Composite Materials,* Vol. I, Metallurgical Society of AIME, New York, 1976, p. 64.
7. J. B. Barr and co-workers, *Appl. Polym. Symp.* **29,** 161 (1976).
8. P. J. Goodhew, A. J. Clarke, and J. E. Bailey, *Mater. Sci. Eng.* **17,** 3 (1975).
9. H. M. Ezekiel, *U. S. Air Force Materials Laboratory Report AFML-TR-70-100,* Jan. 1971.
10. R. Bacon, *Chem. Phys. Carbon* **9,** 1 (1973).
11. R. Bacon and W. A. Schalamon, *Appl. Polym. Symp.* **9,** 285 (1969).
12. W. Ruland, *Appl. Polym. Symp.* **9,** 293 (1969).
13. B. Harris in M. Langley, ed., *Carbon Fibers in Engineering,* McGraw-Hill Book Co., London, Eng., 1973.
14. R. E. Smith, *J. Appl. Phys.* **43,** 2555 (1972).
15. B. Butler, S. Duliere, and J. Tidmore, *10th Biennial Conference on Carbon,* American Carbon Committee, 1971, Abstract FC-29, p. 45.
16. R. C. Fanning and J. N. Fleck, *10th Biennial Conference on Carbon,* American Carbon Committee, 1971, Abstract FC-30, p. 47.
17. R. Bacon, *J. Appl. Phys.* **31,** 283 (1960).
18. W. N. Reynolds and J. V. Sharp, *Carbon* **12,** 103 (1974).
19. S. G. Burnay and J. V. Sharp, *J. Micros.* **97,** 153 (1973).
20. D. J. Thorne, *Nature (London)* **248,** 754 (1974).
21. W. S. Williams, D. A. Steffens, and R. Bacon, *J. Appl. Phys.* **41,** 4893 (1970).
22. J. W. Johnson, *Appl. Polym. Symp.* **9,** 229 (1969).
23. J. W. Johnson and D. J. Thorne, *Carbon* **7,** 659 (1969).
24. R. Moreton and W. Watt, *Nature (London)* **247,** 360 (1974).
25. W. R. Jones and J. W. Johnson, *Carbon* **9,** 645 (1971).
26. H. M. Hawthorne and E. Teghtsoonian, *J. Mater. Sci.* **10,** 41 (1975).
27. N. L. Hancox, *J. Mater. Sci.* **10,** 234 (1975).
28. D. W. McKee and V. J. Mimeault, *Carbon* **8,** 151 (1973).
29. J. V. Duffy, *U.S. Naval Ordnance Laboratory Report NOLTR-73-153,* Aug. 6, 1973, A.D. 766,782.
30. R. Moreton, *Industrial Carbons and Graphite; 3rd Conference; papers read at the conference held at Imperial College of Science and Technology, London, 14th–17th April, 1970,* Society of Chemical Industry, London, Eng., 1971.
31. J. W. Johnson, *Appl. Polym. Symp.* **9,** 229 (1969).
32. W. Watt, *Proc. R. Soc. London Ser. A* **319,** 5 (1970).
33. U.S. Pat. 4,005,183 (Jan. 25, 1977), L. S. Singer (to Union Carbide).

H. F. VOLK
Union Carbide Corporation

OTHER FORMS OF CARBON AND GRAPHITE

The many forms and applications of graphite already mentioned illustrate the versatility and unique character of this material. Several other forms of graphite with specialized properties have been developed during recent years (1), and although all are characterized by relatively high cost, the specialized properties are sufficiently intriguing that applications have been found resulting in sales of a few millions of dollars in each category.

Carbon and Graphite Foams

The earliest foamed graphite was made from exfoliated small crystals of graphite bound together and compacted to a low density (2–4). This type of foam is structurally weak and will not support loads of even a few newtons per square meter. More recently, carbon and graphite foams have been produced from resinous foams of phenolic or urethane base by careful pyrolysis to preserve the foamed cell structure in the carbonized state. These foams have good structural integrity and a typical foam of 0.25 g/cm^3 apparent density has a compressive strength of 9,300–15,000 kPa (1350–2180 psi) with thermal conductivity of 0.87 W/(m·K) at 1400°C. These properties make the foam attractive as a high-temperature insulating packaging material in the aerospace field and as insulation for high temperature furnaces (see Insulation, thermal). Variations of the resinous-based foams include the syntactic foams where cellular polymers or hollow carbon spheres comprise the major volume of the material bonded and carbonized in a resin matrix.

Pyrolytic Graphite

The carbon commercially produced by the chemical vapor deposition (CVD) process is usually referred to as pyrolytic graphite (5–6). The material is not true graphite in the crystallographic sense, and wide variations in properties occur as a result of deposition methods and conditions; nevertheless, the term pyrolytic graphite is generally accepted and used extensively. Pyrolytic graphite was first produced in the late 1800s for lamp filaments yet received little further attention until the 1950s. Since then pyrolytic graphite has been studied extensively and has been commercially produced in massive shapes by several companies. Commercial applications for pyrolytic graphite, limited by the price of $18 or more per kg, include rocket nozzle parts, nose cones, laboratory ware, and pipe liners for smoking tobacco. Pyrolytic graphite coated on surfaces or infiltrated into porous materials is also used in other applications such as nuclear fuel particles and prosthetic devices (qv).

The greatest quantity of commercial pyrolytic graphite is produced in large, inductively-heated, low pressure furnaces for which natural gas is used as the carbon source. Temperature of deposition is ca 1800–2200°C on a carefully prepared substrate of fine-grained graphite. The properties of pyrolytic graphite are highly anisotropic. The ultimate tensile strength in the *ab* direction is five to ten times greater than that of conventional graphite, and the *c* direction strength is proportionally weaker. The thermal conductivity is even more anisotropic. In the *ab* direction, pyrolytic graphite is one of the very best conductors among elementary materials, whereas in the *c* direction conductivity is quite low. At room temperature the thermal conductivity values are approximately three hundred times as high in the *ab* direction as in the *c* direction. The commercially produced pyrolytic graphite is quite dense, usually 2.0–2.1 g/cm^3, and is quite low in porosity and permeability.

A special form of pyrolytic graphite is produced by annealing under pressure at temperatures above 3000°C. This pressure-annealed pyrolytic graphite exhibits the theoretical density of single crystal graphite; and although the material is polycrystalline, the properties of the material are close to single-crystal properties. The highly reflective, flat faces of pressure-annealed pyrolytic graphite have made the material valuable as an x-ray monochromator (see X-ray techniques).

Glassy Carbon

When carbon is produced from certain nongraphitizable carbonaceous materials, the material resembles a black glass in appearance and brittleness; hence, the terms glassy carbon or vitreous carbon (7–8). Nongraphitizing carbons are obtained from polymers that have some degree of cross-linking, and it is believed that the presence of these cross-linkages inhibits the formation of crystallites during subsequent heat treatments. By pyrolyzing polymers such as cellulosics, phenol–formaldehyde resins, and poly(furfuryl alcohol) under closely controlled conditions, glassy carbon is produced. These carbons are composed of random crystallites, of the order of 5.0 nm across, and are not significantly altered by ordinary graphitization heat treatment to 2700°C.

The properties of glassy carbon are quite different from those of conventional carbons and graphites. The density is low (1.4–1.5 g/cm^3), but the porosity and permeability are also quite low. The material has approximately the same hardness and brittleness as ordinary glass, whereas the strength and modulus are higher than for ordinary graphite. The extreme chemical inertness of glassy carbon, along with its impermeability, make it a useful material for chemical laboratory glassware, crucibles, and other vessels. It has found use as a susceptor for epitaxial growth of silicon crystals and as crucibles for growth of single crystals. Applications have been limited because of high cost owing to technological difficulties in manufacturing, although recently billets of reasonable size and with good properties have become available (9).

Carbon Spheres

Carbon and graphite have been produced for many years in roughly spherical shape in small sizes (1–2 mm dia). Globular carbon of this type was used in telephone receivers in the mid 1930s. Carbon of nearly spherical shape can be produced by mechanically tumbling or rolling a mixture of finely divided carbon or graphite with a resinous binder, screening to select the desired size balls, and carefully carbonizing the selected material. Graphite balls in small sizes have also been made from manufactured graphite by crushing, milling in a chopping-type mill, sizing, and tumbling to abrade away the irregular corners. Usage of these nearly spherical carbon balls has been limited to specialized applications where flowability was at a premium or uniform contact area was important. The telephone resistor already mentioned and shim particles for use in nuclear reactor fuel applications are examples.

Recently, a new type of carbon sphere, usually hollow, has been made from pitch. This technology, developed in Japan (10), produces carbon spheres available in 40–400 μm dia. The spheres are of uniform, very nearly spherical shape and are graphitizable. The particle density is usually 0.20–0.25 g/cm^3, and consequently, attention has been given to application in syntactic carbon foam for high temperature insulation and for lightweight composite structures. The spheres can be activated to produce an adsorptive material of high flow-through capacity. At present, cost has limited the application of these spheres. Although the spheres can be produced for about \$1/kg, the cost is greater than for most starting materials for carbon.

BIBLIOGRAPHY

"Baked and Graphitized Products, Uses" under "Carbon" in *ECT* 2nd ed., Vol. 4, pp. 202–243, by W. M. Gaylord, Union Carbide Corporation.

1. R. W. Cahn and B. Harris, *Nature (London)* **221,** 132 (Jan. 11, 1969).
2. R. A. Mercuri, T. R. Wessendorf, and J. M. Criscione, *Am. Chem. Soc. Div. Fuel Chem. Prepr.* **12**(4), 103 (1968).
3. C. R. Thomas, *Mater. Sci. Eng.* **12,** 219 (1973).
4. S. T. Benton and C. R. Schmitt, *Carbon* **10,** 185 (1972).
5. J. C. Bokros, *Chem. Phys. Carbon* **5,** 1 (1969).
6. A. W. Moore, *Chem. Phys. Carbon* **11,** 69 (1973).
7. F. C. Cowtard and J. C. Lewis, *J. Mater. Sci.* **2,** 507 (1967).
8. G. M. Jenkins and K. Kannenmura, *Polymeric Carbons—Carbon Fibre, Glass and Char,* Cambridge University Press, New York, 1976, p. 178.
9. C. Nakayama and co-workers, *Proccedings of the Carbon Society of Japan, Annual Meeting, 1975,* p. 114; C. Nakayama, M. Okawa, and H. Nagashima, *13th Biennial Conference on Carbon, 1977, Extended Abstracts and Program,* American Carbon Society, 1977, p. 424.
10. Y. Amagi, Y. Nishimura, and S. Gomi, *SAMPE 16th National Symposium 1971,* p. 315.

R. M. BUSHONG
Union Carbide Corporation

CARBON BLACK

Carbon black is an important member of the family of industrial carbons. Its various uses depend on chemical composition, pigment properties, state of subdivision, adsorption activity and other colloidal properties. The basic process for manufacturing carbon black has been known since antiquity. The combustion of fuels with insufficient air produces a black smoke containing extremely small carbon black particles which, when separated from the combustion gases, comprise a fluffy powder of intense blackness. The term carbon black refers to a wide range of such products made by partial combustion or thermal decomposition of hydrocarbons in the vapor phase, in contrast to cokes and chars which are formed by the pyrolysis of solids.

This printed page was made with an ink containing a pigment grade of carbon black. Prehistoric cave wall paintings and objects from ancient Egypt were decorated with paints and lacquers containing carbon black. Carbon black was made in China about 3000 BC and records show that it was exported to Japan about 500 AD. The original process consisted of the partial burning of specially purified vegetable oils in small lamps with ceramic covers. The smoke impinged on the covers from which the adhering carbon black was painstakingly removed. Thus carbon black is one of the oldest industrial products. In the United States carbon black has been manufactured for over 100 years. Its use as a pigment continues as an important and growing application, but the major market for the last 50 years has been as a strengthening or reinforcing agent for rubber products, particularly tires.

Relationship of Carbon Black to Other Forms of Industrial Carbons

Carbon exists in two crystalline forms, and numerous so-called amorphous, less-ordered forms. The crystalline forms are diamond and graphite, and the less-ordered forms are mainly cokes and chars. Diamond has a cubic structure in which every carbon atom in the space lattice is bonded by its four valences to adjacent atoms situated in the apexes of a regular tetrahedron around the central atom. The carbon–carbon distance is 0.154 nm, similar to that in aliphatic hydrocarbons. The diamond structure is shown in Figure 1**a**. Diamond has a specific gravity of 3.54 (high for such a low atomic weight element), high refractive index, brilliance and clarity, electrical insulating properties, and a hardness greater than any other known material (see Hardness). The graphite form of carbon has a completely contrasting set of properties: lower density, a grayish-black appearance, softness, slipperiness, and electrical conductivity.

In Figure 1**b** the carbon atoms of the graphite structure are strongly bonded forming large sheets of hexagonal rings in which the electrons are quite mobile. The layer planes are stacked on each other in an ABAB arrangement, ie, layers in which every atom has an atom directly above it separated by one layer. Within a layer plane the carbon atoms are separated by 0.142 nm, comparable to the aromatic carbon separation distance of 0.139 nm in benzene. The mobility of electrons within the layer planes results in a 250-fold higher electrical conductivity in the planar direction than perpendicular to the planes. The bond strengths within the layer planes are over one hundred times stronger than the interlayer van der Waals bonding. The large graphite interplanar distance of 0.335 nm results in a specific gravity of 2.26.

All forms of industrial carbons other than diamond and graphite, including carbon black, can be classified as amorphous carbons characterized by degenerate or imperfect graphitic structures. The parallel layer planes in these carbons are not perfectly oriented with respect to their common perpendicular axis; the angular displacement of one layer with respect to another is random and the layers overlap one another irregularly. This arrangement has been termed turbostratic structure. In this arrangement a separation distance of 0.350–0.365 nm is found for the layer planes. A schematic model of the short-range crystalline order in carbon black is shown in Figure 1**c**. Table 1 lists some of the basic structural and crystallite properties of the various carbons.

Microstructure

X-ray diffraction patterns of carbon black show two or three diffuse rings similar to the more intense rings of natural graphite. Analysis of these patterns by Warren (1) yielded the first model of carbon black microstructure consisting of a random arrangement of crystallites within the particles. A diffuse (002) reflection band provided an estimate of the interplanar spacing and the thickness of the crystallites in the direction perpendicular to the planes. The weak (10) and (11) bands yielded values of the average diameter of the parallel layers. For most furnace and channel-type carbon blacks these values are about 0.35 nm for interplanar spacing (d), 1.2–1.5 nm for average crystallite thickness (L_c), and 1.7 nm for average diameter (L_a). Later studies of carbon black graphitization (2), electron diffraction analysis (3), oxidation studies (4), and high resolution diffracted-beam electron microscopy (5) have led to the conclusion that the microstructure of carbon black consists of a more concentric ar-

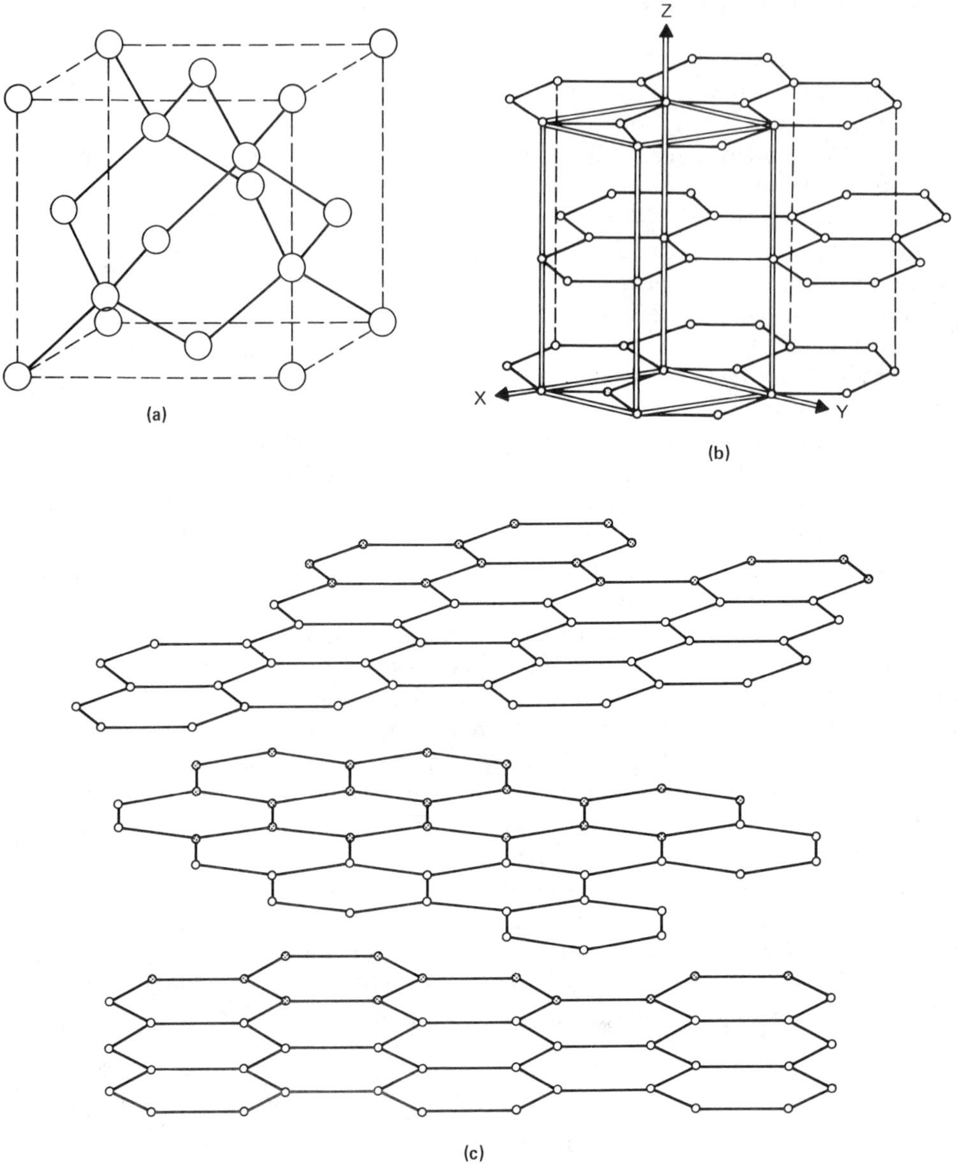

Figure 1. Crystallographic arrangement of carbon atoms in: (**a**) diamond, face-centered cubic structure; (**b**) graphite, parallel layers of hexagons in ordered positions (every third layer is in the same position as the first layer, the second layer corresponds with the fourth, etc); (**c**) carbon black, hexagonal layers are parallel but farther apart and arranged without order—turbostratic arrangement.

rangement of layer planes. Ordinary transmission electron microscopy provides information on the size and shape of carbon black aggregates from their two-dimensional projections as shown in Figure 2 for a rubber-reinforcing grade (N339). High resolution electron microscopy provides information on the microstructure of the aggregates from interference patterns caused by electron diffraction. These patterns are due to a concentric-layer configuration, with an interplanar distance of 0.35 nm, somewhat

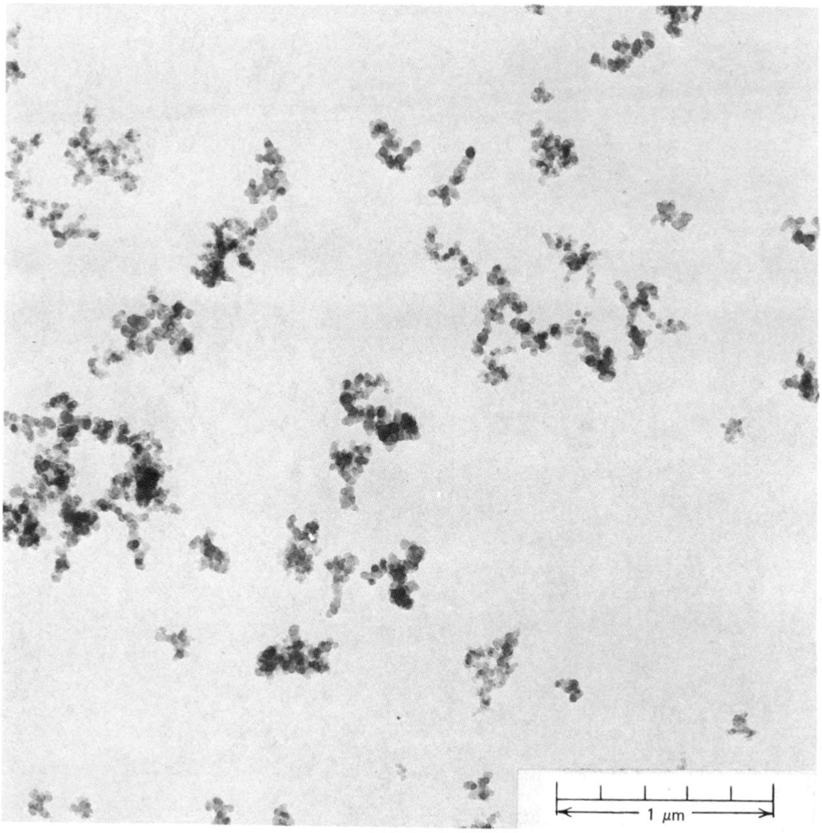

Figure 2. Electron micrograph of furnace black N339 (HAF-HS).

Table 1. Forms of Carbon and Characteristics

Form	Crystal system	Sp gr	C—C distance, nm	Layer distance, nm
diamond	cubical	3.52	0.155	
graphite	hexagonal	2.27	0.142	0.335
carbon black	hexagonal-turbostratic[a]	1.86–2.04	0.142	0.365
cokes (oven and calcined)	hexagonal-turbostratic[a]	1.3–2.1		
chars and activated carbons	hexagonal-turbostratic[a]	1.1–1.3		
fibrous carbon	hexagonal-turbostratic[a]	1.65		
vitreous carbon	hexagonal-turbostratic[a]	1.47		
pyrolytic graphite	hexagonal-turbostratic[a]	1.2–2.2		

[a] Turbostratic crystals have randomly oriented layer planes, see pp. 690–691.

larger than in graphite. Figure 3 shows a high resolution electron micrograph of a portion of a carbon black aggregate. A structural model with concentric-layer orientation proposed by Harling and Heckman (6) is shown in Figure 4. According to this model the interior of the aggregate is less ordered than the surface, has a lower density,

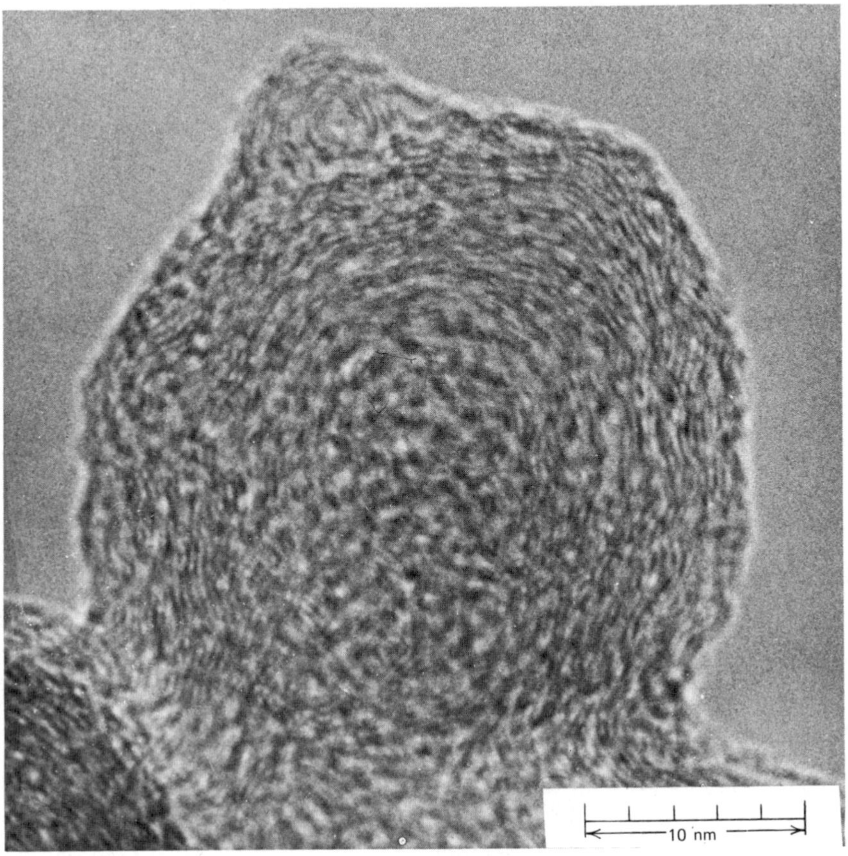

Figure 3. High resolution electron micrograph of carbon black primary aggregate nodule showing concentric orientation of layer planes.

and is chemically more reactive. Graphitization and oxidation studies confirm this view as illustrated by electron micrographs of oxidized and graphitized carbon blacks. Figure 5 shows that during oxidation the more reactive interior carbon is burned away leaving hollow shells of the less reactive surface layers. Figure 6 shows that graphitization causes increased crystallinity. The aggregates take on a capsule-like appearance with hollow centers. These changes are consistent with a concentric-layer structure.

High resolution electron microscopy also reveals that the interiors of some aggregates contain smaller concentric-layer regions which may have been separate entities at some stage of the carbon formation process. This can be seen in Figure 3. These have been called growth centers, and it is assumed that early in the carbon formation process they agglomerated, and continued to grow. The surface layers of these aggregates are distorted and folded in order to conform to aggregate geometry. This structure is clearly evident in the high resolution electron micrograph of graphitized N330 (HAF) in Figure 7.

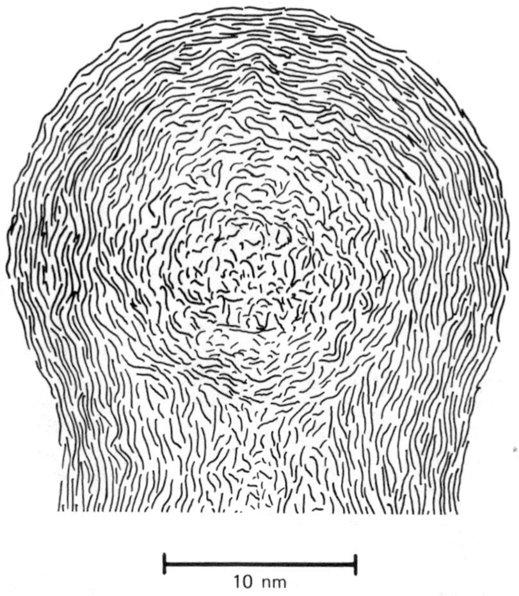

Figure 4. Concentric layer plane model of carbon black microstructure.

Characterization

Carbon blacks differ in particle size or surface area, average aggregate mass, particle and aggregate mass distributions, morphology or structure, and chemical composition. The form of these products, loose or pelleted, is another feature of some special grades. The ultimate colloidal units of cabon black, or the smallest dispersible entities in elastomer, plastic, and fluid systems are called aggregates, fused assemblies of particles. The particle size is related to the surface area of the aggregates. The oil furnace process produces carbon blacks with particle diameters of 10–250 nm and the thermal black process, 120–500 nm.

A primary aggregate is further characterized by its size, the volume of carbon comprising the aggregate, and its morphology. Aggregates have a variety of structural forms. Some are clustered like a bunch of grapes, whereas others are more open, branched, and filamentous.

Figure 8 illustrates the wide range of particle and aggregate sizes of commercial carbon blacks used for rubber reinforcement and for pigment applications.

Surface Area. The three most important properties used to identify and classify carbon blacks are surface area, structure, and tinting strength. Surface areas are measured by both gas- and liquid-phase adsorption techniques and depend on the amount of adsorbate required to produce a monolayer. If the area occupied by a single adsorbate molecule is known, a simple calculation will yield the surface area. The most familiar method is the procedure developed by Brunauer, Emmett, and Teller (BET). All commercial carbon blacks provide well-defined S-shaped isotherms with nitrogen as adsorbate at $-193°C$; the break in the isotherm (B point) corresponding to monolayer coverage V_m is determined and used to calculate the surface area. A useful extension of the BET method, called the "t" (adsorbed layer thickness) method, distinguishes between internal (or porous) surface area and external surface area (8). Most

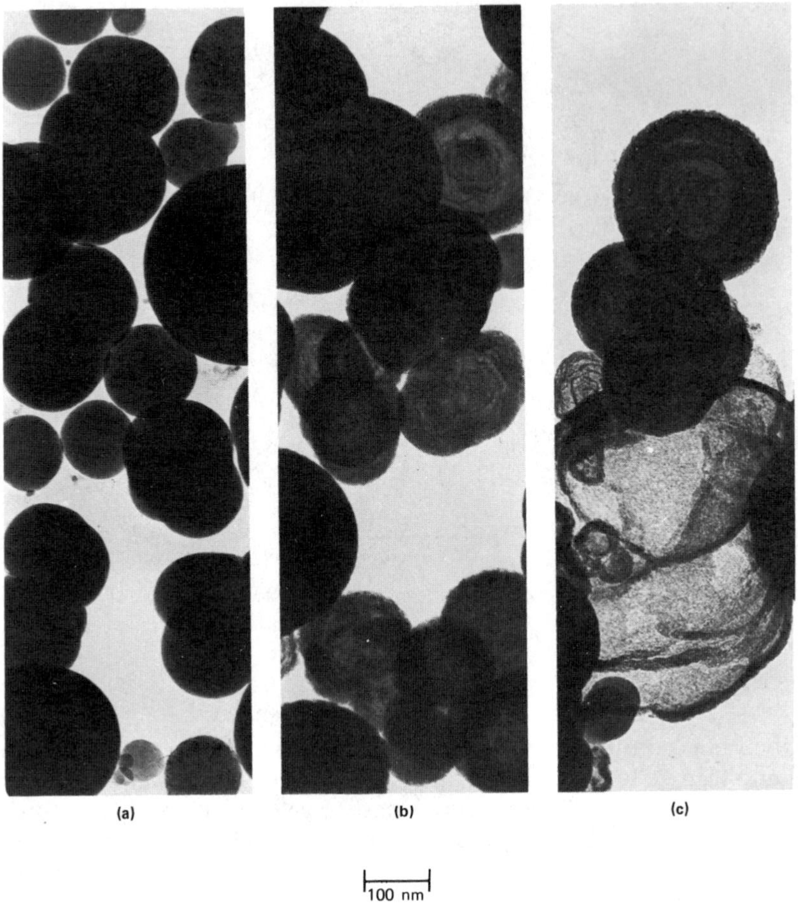

Figure 5. Electron micrographs of medium thermal carbon black particles: (**a**) original; (**b**) air-oxidized to 33% weight loss; (**c**) air-oxidized to 51% weight loss.

rubber-grade carbon blacks are nonporous, so that the BET and "t" methods give identical results. For some porous carbon blacks used for pigments and electrical applications, the "t" method is useful to assess their internal surface areas.

Liquid-phase adsorption methods are widely used for product control and specification purposes. The adsorption of iodine from potassium iodide solution is the standard ASTM method for classification of rubber-grade carbon blacks (9). The method is simple and precise, but its accuracy is affected by the presence of adsorbed hydrocarbons and the degree of surface oxidation. More recently developed surface area methods based on the adsorption of cetyltrimethylammonium bromide (CTAB) or Aerosol OT from aqueous solution are not affected by the same factors as the iodine adsorption test (10). These adsorbate molecules are so large that they cannot penetrate into micropores, and therefore provide an estimate of external surface area. The CTAB surface area method gives values in good agreement with the BET method for nonporous rubber grade blacks.

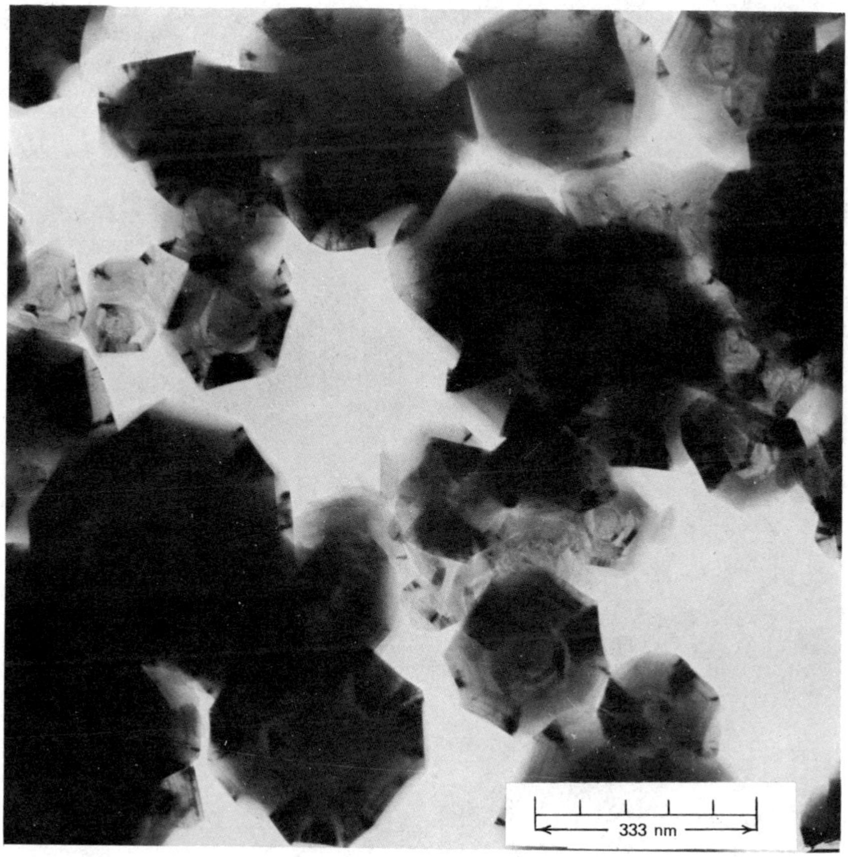

Figure 6. Electron micrograph of graphitized medium thermal carbon black.

Structure. The second most important property of carbon blacks is structure. Structure is determined by aggregate size and shape, the number of particles per aggregate, and their average mass. These characteristics affect aggregate packing and the volume of voids in the bulk material. The measurement of void volume, a characteristic related to structure, is a useful practical method for assessing structure. It is measured by adding linseed oil or dibutyl phthalate (DBP) to carbon black until the consistency of the mixture suddenly changes, at which point almost complete absorption or filling of the voids has occurred. Automated laboratory equipment is used to measure void volume by the DBP absorption (DBPA) method, a standard industry procedure (11). A modification of this procedure uses samples that have been preconditioned by four successive compression and grinding steps to fracture the weaker aggregate structures and more nearly represent the state of carbon black in rubber mixtures (12).

Tinting Strength. Tinting strength is another test method widely used to classify carbon blacks. In this test, a small amount of carbon black is mixed with a larger amount of white pigment, usually zinc oxide or titanium dioxide, in an oil or resinous vehicle to give a gray paste (13) and its diffuse reflectance measured. A carbon black with high tinting strength has a high light absorption coefficient and a low reflectance value. Tinting strength is related to average aggregate solid volume. Centrifugal sed-

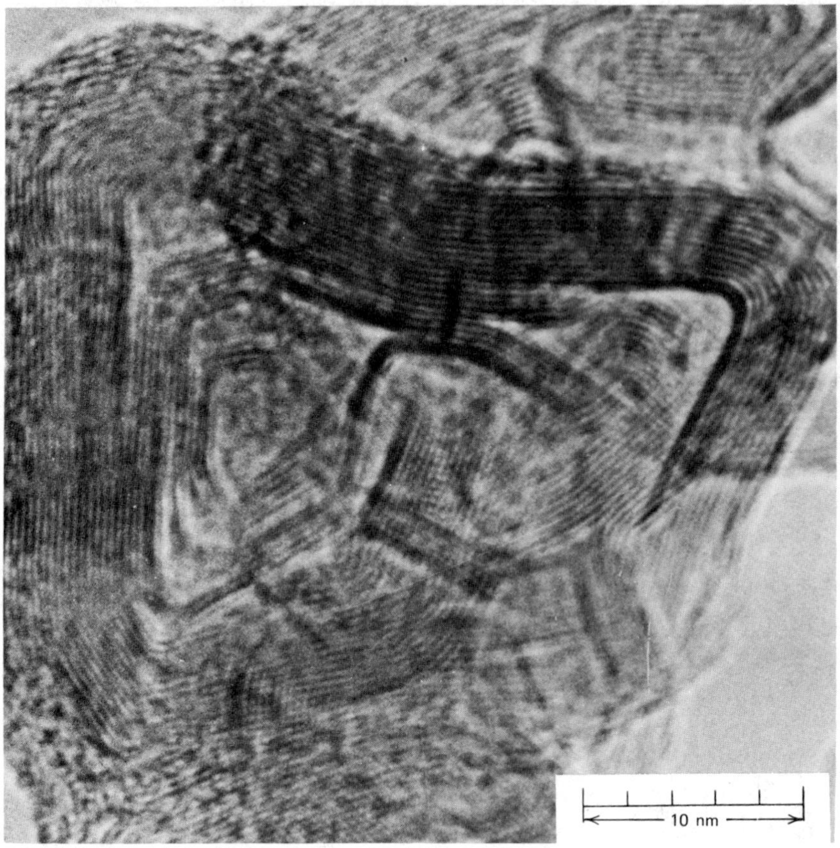

Figure 7. High resolution electron micrograph of graphitized N330 (HAF) carbon black.

imentation of aqueous carbon black suspensions has been used to estimate average aggregate volume and volume distributions (14). Figure 9 shows the correlation of tinting strength with average aggregate Stokes diameter, D_{St}. The Stokes diameter defines a sphere which sediments at the same rate as the actual aggregate.

Properties. Most commercial rubber-grade carbon blacks contain over 97% elemental carbon. A few special pigment grades have carbon contents below 90%. In addition to chemically combined surface oxygen, carbon blacks contain varying amounts of moisture, solvent-extractable hydrocarbons, sulfur, hydrogen, and inorganic salts. Hygroscopicity is increased by surface activity, high surface area, and the presence of salts. Extractable hydrocarbons result from the adsorption of small amounts of incompletely burned hydrocarbons. The combined sulfur content of carbon black has its origin in the sulfur content of the feedstocks. Most of the inorganic salt content comes from the water used for quenching and pelletizing. Table 2 lists the chemical composition of a few typical carbon blacks. Table 3 lists other analytical properties for rubber-grade carbon blacks. Table 4 lists analytical properties of carbon blacks for the ink, paint, plastics, and paper industries.

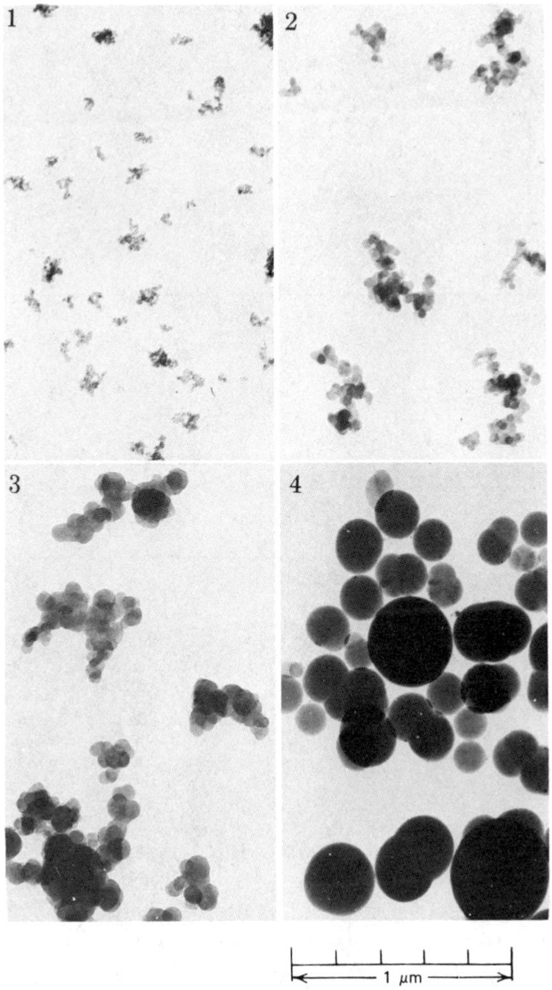

Figure 8. Electron micrographs of a range of carbon blacks: (1) high color pigment grade, 240 m^2/g; (2) rubber reinforcing grade, 90 m^2/g; (3) general purpose rubber grade, 36 m^2/g; (4) medium thermal grade, 7 m^2/g.

Formation Mechanisms

Carbon black is made by partial combustion processes involving flames, and by thermal decomposition processes in the absence of flames. Whether partial combustion or thermal decompsition methods are used, the basic reaction is represented by:

$$C_xH_y \xrightarrow{\Delta} x\,C + \frac{y}{2}H_2$$

With some hydrocarbon gases, such as methane, the reaction is endothermic, whereas with acetylene it is highly exothermic. The decomposition of heavy aromatic feedstocks in the furnace process is slightly endothermic, so that more energy is required than necessary to vaporize and heat the feedstock to reaction temperatures. The actual energy requirement will vary according to the process used, the average temperature of the carbon black formation reaction, the heat losses from the reactors, the degree

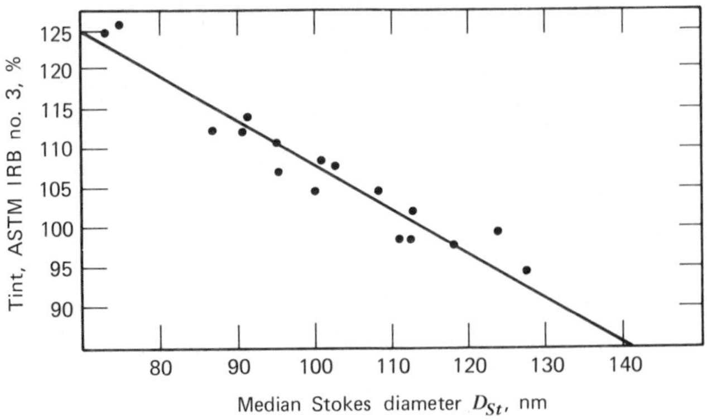

Figure 9. Tinting strength vs Stokes diameter D_{St}.

Table 2. Chemical Composition of Carbon Blacks

Symbol	Carbon, %	Hydrogen, %	Oxygen, %	Sulfur, %	Ash, %
medium thermal (N990) MT	99.4	0.3	0.1	0.0	0.3
semireinforcing furnace (N770) SRF	98.6	0.4	0.2	0.6	0.2
general purpose furnace (N660) GPF	98.6	0.4	0.2	0.6	0.2
fast extruding furnace (N550) FEF	98.4	0.4	0.4	0.7	0.2
high abrasion furnace (N330) HAF	98.0	0.3	0.8	0.6	0.3
superconducting furnace SCF	97.4	0.2	1.2	0.6	0.6
acetylene	99.8	0.1	0.1	0.0	0.0

of preheating of the input streams, the amount of heat recovery from the flue gases, and the extent to which the flue gases are used as auxiliary fuel. Modern furnace carbon black manufacture is quite efficient, reaching levels of 50–70% carbon recovery. The energy required to produce a kilogram of carbon black by the furnace process is in the range of $9–16 \times 10^7$ J/kg ($4–7 \times 10^4$ Btu/lb) depending on the grade of carbon black (see Burner technology).

The mechanism of carbon formation is not well defined. It is likely that no single mechanism can explain carbon black formation from different raw materials and by different processes. The manufacturing process has more of an influence on product quality than the raw material from which it is made. Thus natural gas or liquid aromatic hydrocarbons used in the channel black process produce the same type of carbon black. The decomposition of natural gas and benzene in a thermal process also will produce similar products. Thus a mechanism is called for in which different hydrocarbons in the vapor phase break down into smaller fragments at high temperature and recombine quickly to form particulate carbon. The temperatures and concentrations of reacting species determine the types of carbon black formed. There are several reviews of carbon formation theories (15–18).

Many of the proposed mechanisms of carbon black formation do not account for the microcrystalline morphology of the products. A model reaction mechanism of carbon formation involves sequentially: (*1*) the raw material hydrocarbon molecules decompose into smaller units or radicals with a loss of hydrogen; (*2*) these units, which may be monatomic carbon, C_2, C_3, C_6, or C_xH_y radicals, recombine into nuclei or

Table 3. Properties of Rubber-Grade Carbon Blacks

ASTM[a]	Type	Typical I_2 adsptn no.[b] D1510, mg/g	Surface area (BET), m^2/g	Particle size, nm[c]	Typical DBPA, mL/100 g D2414	Volatile content, %	Typical pour density, kg/m^3 (lb/ft^3) D1513	Tinting strength[d], % IRB 3 D3265
N110	SAF	145			113		336 (21.0)	128
N121	SAF–HS	120			130		320 (20.0)	
N219	ISAF–LS	118			78		441 (27.5)	124
N220	ISAF	121	115	22	114	1.5	344 (21.5)	114
N231	ISAF–LM	125			91		392 (24.5)	
N234		118			125		320 (20.0)	130
N242	ISAF–HS	123			126		328 (20.5)	119
N293	CF	145			100		376 (23.5)	
N294	SCF	205			106		368 (23.0)	
N326	HAF–LS	82	80	27	71	1.0	465 (29.0)	109
N330	HAF	82	80	27	102	1.0	376 (23.5)	104
N339	HAF–HS	90			120		344 (21.5)	114
N347	HAF–HS	90	90	26	124	1.0	336 (21.0)	104
N351		67			120		344 (21.5)	100
N358	SPF	84			150		288 (18.0)	
N363		66			68		481 (30.0)	
N375	HAF	90			114		344 (21.5)	116
N440	FF	50			60		481 (30.0)	
N472	XCF	270	254	31	178	2.5	256 (16.0)	
N539		42				9		384 (24.0)
N542		44			67		505 (31.5)	
N550	FEF	43	42	42	121	1.0	360 (22.5)	63
N568	FEF–HS	45			132		336 (21.0)	
N650	GPF–HS	36			125		368 (23.0)	
N660	GPF	36			91		424 (26.5)	51
N683	SPF	30			132		336 (21.0)	
N741		20			105		368 (23.0)	
N762		26			62		505 (31.5)	
N765	SRF–HS	31			111		376 (23.5)	54
N774		27			70		497 (31.0)	
N880	FT	13	12	200	35	0.5	673 (42.0)	29
N990	MT	7	8	450	35	0.5	673 (42.0)	17
acetylene black			67	40	260	0.3		

[a] ASTM designations are determined according to *Recommended Practice D2516, Nomenclature for Rubber-Grade Carbon Blacks.*

[b] In general, Method D1510 can be used to estimate the surface area of furnace blacks but not channel, oxidized, and thermal blacks.

[c] Approximate number-average particle size d_n in nm, based on electron microscopy.

[d] Industry reference black, IRB 3, an HAF grade, is taken as 100.

droplets by polymerization and condensation from the vapor phase; (*3*) the droplets continue to lose hydrogen and grow by further condensation on their surfaces of polymerized species, or the carbon nuclei grow into particles by further condensation of carbon from the vapor phase; (*4*) carbon particles or droplets flocculate; and (*5*) further carbon deposition on the carbon particle flocculates or continued loss of hydrogen from the agglomerated droplets result in the formation of primary aggregates. Higher reaction temperatures favor the formation of nuclei, smaller particle size within

Table 4. Properties of Furnace Process Carbon Blacks for Inks, Paints, and Plastics

	Nigrometer index[a]	Surface area (BET), m^2/g	Particle size, nm	Oil (DBP) absorption, mL/100 g		Tinting strength index	Volatile content, %	Fixed carbon, %	pH	Toluene extract, %
				Fluffy	Pellets					
High Color Furnace										
HCF-1	64	560	13	121	105	100	9.5	90.5	3.3	0.08
HCF-2	65	240	14	65	50	115	2.0	98.0	7.0	0.08
HCF-3	69	230	15	70	65	120	2.0	98.0	7.0	0.08
Medium Color Furnace										
MCF-1	74	220	16	120	110	122	1.5	98.5	8.0	0.08
MCF-2	74	210	17	7.5	68	120	1.5	98.5	8.0	0.08
MCF-3	78	200	18	122	117	118	1.0	99.0	8.0	0.08
Long Flow Furnace	83	138	24	60	55	112	5.0	95.0	3.4	0.10
Medium Flow Furnace	84	96	25	72	70	112	2.5	97.5	4.5	0.10
Conductive Furnace	87	254	30	185	178	82	2.0	98.0	5.0	0.10
Regular Color Furnace										
RCF-1	84	140	19		114	114	1.5	98.5	7.0	0.10
RCF-2	83	112	24	62	60	116	1.0	99.0	7.5	0.10
RCF-3	83	86	25	65		112	1.0	99.0	8.0	0.30
RCF-4	84	94	25	62	70	110	1.0	99.0	8.5	0.10
RCF-5	87	80	27		72	104	1.0	99.0	7.5	0.10
RCF-6	90	46	36		60	92	1.0	99.0	7.0	0.10
RCF-6A	90	85	27	103		100	1.0	99.0	9.0	0.10
RCF-7	93	45	37	95		73	1.0	99.0	9.0	0.10
Low Color Furnace										
LCF-1	94	30	60		64	59	1.0	99.0	8.5	0.10
LCF-2	95	42	41		120	61	1.0	99.0	7.5	0.10
LCF-3	96	35	50		91	49	1.0	99.0	7.5	0.10
LCF-4	99	25	75	71	70	49	1.0	99.0	8.5	0.10

[a] A method for measuring the diffuse reflectance from a black paste with a black tile standard. The low numbers represent the jettest or most intense black grades.

the aggregates, and higher surface area. Aggregate formation is favored by increasing feedstock aromaticity. Particles are formed in close proximity, agglomerate, and form aggregates. The particles within an aggregate are remarkably monodisperse compared to the wide range of sizes apparent for aggregates (Figs. 2 and 8). In practice the various steps described above do not occur in an ideal sequence. For this reason carbon blacks are characterized by broad particle and aggregate size distributions.

Manufacture

Carbon black manufacture has evolved from primitive methods to continuous, high capacity systems using sophisticated designs and control equipment. About 1870 channel black manufacture began in the United States in natural gas producing areas to replace lampblack for pigment use. The discovery of rubber reinforcement by carbon black about 1912 and the growth of the automobile and tire industries transformed carbon black from a small-volume specialty product to a large-volume basic industrial raw material. Today's carbon black production is almost entirely by the oil furnace black process, which was introduced in the United States during World War II. However, the unique relationship of manufacturing process to special performance features of products has prevented the total abandonment of the older processes. The thermal, lamp black, channel black, and acetylene black processes account for less than 10% of the world's production. The easy sea and land transportation of feedstocks for the oil furnace process has made possible worldwide manufacturing facilities convenient to major consumers.

There are currently 30 furnace plant locations in the United States, 25 in Western Europe, 12 in the Eastern bloc countries, and 32 in the rest of the world. Eight United States companies produce carbon black. A recent estimate of their capacities (19) is shown in Table 5. Distribution of world carbon black manufacturing capacities is shown in Table 6.

Oil Furnace Process. The oil furnace process was developed during World War II by the Phillips Petroleum Company. The first grade of black produced was FEF (N500 type), followed by an HAF grade (N300 type) a few years later. The first successful carbon black reactor (Fig. 10) was introduced in the early fifties (20).

The oil furnace process was preceded by the gas furnace process. Gas furnace blacks were manufactured from about 1922 to the 1960s. In this process gas underwent partial combustion in refractory-lined retorts or furnaces, and carbon black was separated from the combustion gases with electrostatic precipitators and cyclone separators. Temperatures in the range of 1200–1500°C were used to produce carbon black grades designated as semireinforcing (SRF), high modulus furnace (HMF), fine furnace (FF), and high abrasion furnace (HAF). Some of these letter designations have been retained for oil furnace blacks. The gas furnace blacks were characterized by low structure levels and low-to-medium reinforcing performance. The carbon recovery of the process depended on the grade of black. The low-surface-area semireinforcing grades had yields of 25–30%, and the yields of the higher-surface-area reinforcing grades were 10–15%. The remarkable success of the oil furnace process was due to its improved yield performance in the range of 50–70%, higher capacities, and its ability to produce a wide range of products.

The use of petroleum oil for the manufacture of carbon black was patented as early as 1922 (21). The reactor contained many of the basic features of present oil

Table 5. United States Carbon Black Capacity, 1976

Manufacturer	Capacity, thousands of metric tons Furnace	Thermal
Ashland Chemical	*304*	
Arkansas Pass, Tex.	68	
Baldwin, La.	116	
Belpre, Ohio	45	
Mojave, Calif.	27	
Shamrock, Tex.	48	
Cabot Corp.	*425*	
Big Spring, Tex.	113	
Franklin, La.	91	
Pampa, Tex.	29	
Parkersburg, W. Va.	74	
Ville Platte, La.	118	
Cities Service	*343*	
Conroe, Tex.	44	
El Dorado, Ark.	37	
Eola, La.	32	
Franklin, La.	71	25
Hickok, Kan.	23	
Mojave, Calif.	24	
Moundsville, W. Va.	71	
Seagraves, Tex.	41	
Continental Oil	*194*	
Bakersfield, Calif.	35	
Ponca City, Okla.	61	
Sunray, Tex.	43	
Westlake, La.	54	
J. M. Huber Corp.	*179*	*19*
Baytown, Tex.	117	
Borger, Tex.	62	
Phillips Petroleum	*219*	
Borger, Tex.	130	
Orange, Tex.	52	
Toledo, Ohio	36	
Richardson Co.	*95*	
Addis, La.	45	
Big Spring, Tex.	50	
Thermatomic Carbon		
Sterlington, La.[a]		59
Total	*1759*	*94*

[a] Plant closed in 1977.

furnace processes including feedstock preheating and atomization, and water quenching. Another important oil furnace process was patented in 1942 (22). Table 7 compares yields for various processes. Energy requirements for each process are also shown.

Feedstocks. The feedstocks (qv) used by the carbon black industry are viscous, residual aromatic hydrocarbons consisting of branched polynuclear aromatic types mixed with smaller quantities of paraffins and unsaturates. Feedstocks are preferred that are high in aromaticity, free of coke or other gritty materials, and contain low levels of asphaltenes, sulfur, and alkali metals. For a particular plant location, other limi-

Table 6. Distribution of World Carbon Black Capacity[a]

Country	Number of plants	Estimated annual capacity, thousands of metric tons
North America		
Canada	3	164
Mexico	2	88
Total	*5*	*252*
South America		
Argentina	1	54
Brazil	3	152
Colombia	2	32
Peru	1	15
Venezuela	1	23
Total	*8*	*276*
Europe		
Great Britain	5	300
France	3	197
West Germany	5	269
Holland	2	116
Italy	4	175
Spain	3	91
Sweden	1	33
Yugoslavia	1	20
Total	*24*	*1201*
Australia, S. E. Asia		
Australia	3	89
India	5	111
Indonesia	1	3
Japan	9	479
Korea	1	23
Malaysia	1	13
Philippines	1	12
Taiwan	1	15
Total	*22*	*745*
Africa	1	60
Middle East		
Iran	1	15
Israel	1	11
Turkey	2	30
Total	*4*	*56*
United States	32	1888
remainder of the world[b]		680
Total		*5158*

[a] Including announced expansions through 1978.
[b] Estimated capacity.

tations usually involve the quantities available on a long-term basis, uniformity, ease of transportation, and cost.

In 1975 over 1.9 GL (5×10^8 gal) of feedstock was used for carbon black production in the United States. Sources of feedstock include petroleum refineries, ethylene plants, and coal coking plants. As a result of the introduction of more efficient petroleum cracking catalysts, decant oil from gasoline production has now become

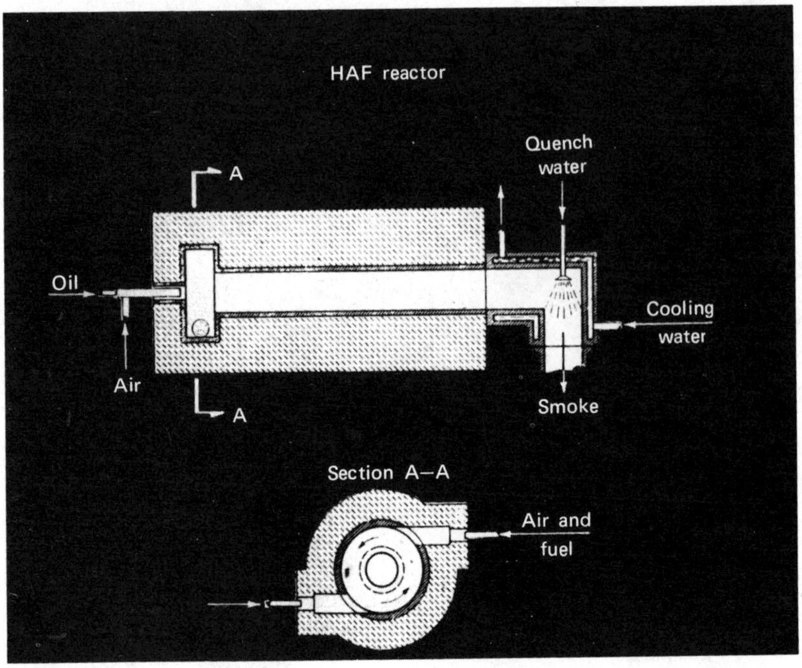

Figure 10. Phillips HAF black reactor (20).

Table 7. Yields from Different Processes

Process	Raw material	Commercial yields, g/m³ [a]	Yield % of theoretical carbon content	Energy utilization in manufacture, J/kg [b]
channel	natural gas	8–32	1.6–6.0	$1.2–2.3 \times 10^{9}$
gas-furnace	natural gas	144–192	27–36	$2.3–3.0 \times 10^{8}$
thermal	natural gas	160–240	30–45	$2.0–2.8 \times 10^{8}$
oil-furnace	liquid aromatic hydrocarbons	300–660[c]	23–70	$9.3–16 \times 10^{7}$

[a] To convert g/m³ to lb/1000 ft³, divide by 16.
[b] To convert J/kg to Btu/lb, multiply by 0.43×10^{-3}.
[c] In kg/m³ (2.5–5.5 lb/gal).

the most important feedstock in the United States, displacing residual thermal tars, and catalytic cycle stock extracts. Another important raw material is the steam pyrolysis residual tar from the manufacture of ethylene based on naphtha or gas oil. It is expected that ethylene tars will become more important as ethylene capacity increases. In Europe more ethylene process pyrolysis tars are used than in the United States. Coal tars, naphthalene and anthracene oils from coal tars, are also used for carbon black manufacture, but quantities are limited and prices are high.

Equipment. Figure 11 is a flow diagram (23) of an oil furnace process. The reactor feeds are preheated feedstock, preheated air, and gas. The carbon-containing combustion products are quenched with water sprays and pass through heat exchangers which preheat the primary combustion air. The product stream is again cooled by a secondary water quench in vertical towers. Carbon black in light fluffy form is separated in bag filters and conveyed to micropulverizers discharging into a surge tank. From the surge tank the carbon black is fed to dry drums for dry pelletization (not shown in Fig. 11) or pin-type wet pelletizers, followed by dryers, to produced pelleted products (see Pelleting).

Different grades of carbon black are made by changing reactor temperatures and residence times as shown in Table 8. The long residence time for thermal black, the largest particle size grade, is obtained with large volume reactors and relatively low gas velocities. The shortest residence times for the highest-surface-area reinforcing grades are obtained using small, high velocity reactors.

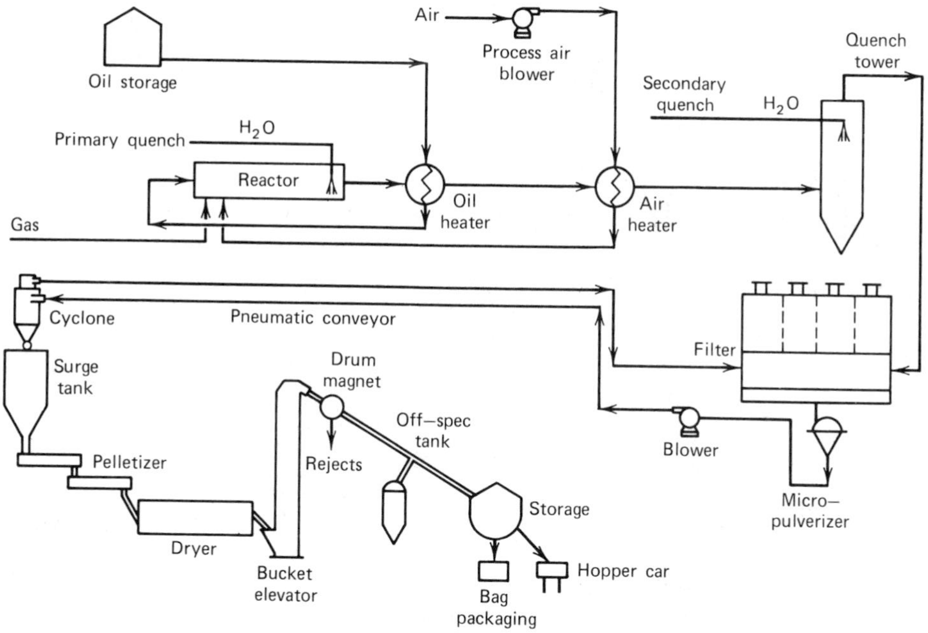

Figure 11. Flow diagram of oil furnace process.

Table 8. Time–Temperature Conditions in Carbon Black Reactors

Black grade	Residence time, s	Temperature, °C
thermal (N990)	10	1200–1350
SRF (N700 series)	0.9	1400
HAF (N300 series)	0.031	1550
SAF (N100 series)	0.008	1800

Reactors. The design and construction features of reactors are determined by economics and product specifications. Some reactors can produce over 45 metric tons of carbon black per day. At least two different types of reactors are required to make all the furnace grades for the rubber industry: one for the reinforcing and another for the less reinforcing grades. Additional types are used to produce special pigment grades. During the last decade reinforcing furnace black reactors have undergone substantial design changes to give higher gas velocities and turbulence, more rapid mixing of reactant gases and feedstock, and increased capacity (24–27). There have been fewer advances in the design of the larger reactors used for the less reinforcing, lower-surface-area grades. A notable exception has been the commercial use of huge vertical reactors having a more precise control of turbulence (28–29). Some of the latest reactor designs have refractory-lined metal construction, increased thermal efficiency, more efficient feedstock atomization, and improved quenching.

Carbon black properties, yield, and production rate can be adjusted by controlling the reactor feed variables. Combustion air and primary fuel rate are set to provide 20–50% excess air in the primary combustion flame. The desired temperature determines the feedstock rate. Lowering the feedstock rate produces higher temperatures, lower yields, and higher surface area products. In the mixing zone of the reactor a portion of the feedstock reacts with the excess air to give the required temperature. From 30 to 40% of the feedstock is burned to raise the temperature from 1450°C to 1500–1600°C when producing the most reinforcing types of carbon black. The temperature declines from this level as energy is used in the slightly endothermic hydrocarbon decomposition reactions.

In addition to the control of surface area, it is necessary to control average aggregate size, assessed by the dibutyl phthalate adsorption method (DBPA). Increasing the aromaticity of the feedstock increases the degree of aggregation. One of the most effective methods for reducing carbon black aggregation is by the injection of trace quantities of alkali metal salts into the carbon formation zone (30). This technique is useful for maintaining product DBPA specifications and for the production of low aggregation grades. The effectiveness of alkali metal addition is in the order of ionization potentials. The most commonly used alkali metal is potassium.

Post-reactor processing. The newly-formed carbon black aggregates are protected from attack by the water vapor in the hot reactor stream by rapidly reducing the temperature from 1300–1600°C to about 1000°C with a water spray. The primary water quench represents an unrecoverable energy loss and has the disadvantage of diluting the tail gas with moisture and lowering its heating value. Attempts to use indirect cooling methods to conserve heat energy have not been successful. The product stream is passed through heat exchangers to preheat the primary combustion air. The product is further cooled by a secondary water quench in a vertical tower, lowering its temperature to about 270°C before entering the bag filters. Glass cloth is commonly used in the bag filters. The product stream enters the inside of the bag filter tubes depositing carbon black and the tail gas is redistributed for use as fuel, or is flared or vented to the atmosphere. Periodically a bag filter compartment of tubes is repressurized with tail gas from the outside, releasing the carbon black.

The carbon black is collected in hoppers at the bottom of the bag filter unit, conveyed to grinders or micropulverizers which break up lumps, then conveyed to a cyclone and into the surge tank. The surge tank accumulates and partially compacts the fluffy carbon black and provides a reservoir for maintaining a constant flow to the pelletizers.

Most of the carbon black used by the rubber industry is wet pelletized. Dry pelletization is practiced for some rubber applications and for pigmentation. In the dry process the slightly compacted loose black is conveyed to large rotating steel drums whose rolling motion causes a steady increase in bulk density and the formation of pellets. A fraction of the pelletized product is recirculated and added to the loose black at the drum entry to increase the density of the bed and provide seed material to increase the formation rate.

The wet pellet process uses pin-type agitators enclosed in tubes about 0.5 m dia and about 3 m in length. The pins are arranged as a double helix, or other configuration, around a rotating shaft extending through the center of the tube. Water is added to the loose black through sprays downstream of its entry, resulting in a rapid increase in density while forming wet pellet beads. The water–carbon black ratio is critical; an insufficient amount of water results in a dusty product, and an excess causes a sticky paste which can clog the equipment. The optimum water–black ratio increases with increasing aggregation. For normal aggregation blacks about equal amounts by weight of water and black are used. There are many different types of wet pelletizers. Their water spray pressures, nozzle locations, and rotating shaft speeds influence the characteristics of the pellets.

The wet pellets are conveyed to a dryer which consists of a long, slowly-revolving cylindrical drum heated by natural gas or tail gas burners located under its entire length. A fraction of the heater exhaust gases may be passed over the moving bed of pellets in the drier drum to purge liberated moisture and remove small amounts of carbon black dust formed by pellet breakdown. The wet pellets have an initial moisture content of about 50% on entry to the drier, and after about one hour residence time exit with less than 0.5% moisture. The dried product is screened to remove oversized pellets which are recycled through the pelletization system. The product is conveyed to storage bins for bulk shipment or for bag packing. In the United States as much as 90% of carbon black production is shipped in bulk.

Good pellet quality is necessary for easy loading and unloading of bulk cars, for conveying without appreciable pellet breakdown and dustiness, and for automatic weighing in rubber factories. High bulk density is preferred for economical transportation, cleanliness, and rapid incorporation into rubber mixtures. The maximum bulk density is however limited by the requirement for good dispersion in rubber. Bulk densities of commercial furnace carbon black grades are 256–513 kg/m^3 (16–32 lb/ft^3). Their pellet size distributions are in the range of 125–2000 μm.

New Process Technology. During the early 1970s the industry adopted improved reactor and burner designs which provided more rapid and uniform feedstock atomization, more rapid mixing, and shorter residence times (27). Higher capacities and increases of 6–8% in yields for products rated at equivalent roadwear levels have been reported (31). These new carbon blacks have lower surface areas than the standard grades. Electron microscopy shows fewer large aggregates, a smaller average aggregate size, a narrower aggregate size distribution, and the individual aggregates appear to be more open, branched, and bulky. In addition to these changes in aggregate morphology, the new carbon blacks have higher tinting strengths and give evidence of higher surface activity as shown by an increase in bound rubber formation (14). In rubber, enhanced surface-polymer interaction results in higher tensile strength, modulus, hysteresis, and treadwear. The improvements in pigment and rubber performance are described in recent patents (32) (see Rubber compounding).

Thermal Decomposition Process. The high temperature decomposition of hydrocarbons in the absence of air or flames is the basis for the manufacture of thermal blacks and acetylene black. Thermal black is made by a strongly endothermic reaction requiring a large heat energy input, whereas the acetylene black decomposition reaction is strongly exothermic.

In the thermal process two refractory-lined cylindrical furnaces, or generators, alternate on about a 5-minute cycle between black production and heating, making the overall production continuous from the two generators. The generators are about 4 m in diameter and 10 m high and are nearly filled with an open checkerwork of silica brick. During operation one generator is fired with a stoichiometric ratio of air and fuel, while the other generator, heated in the previous cycle to an average temperature of 1300°C, is fed with natural gas. The product stream contains suspended carbon black, hydrogen, methane, and other hydrocarbons. This stream is cooled with water sprays and the black removed by a bag filter. Carbon recovery is about 30–45% of the total carbon content of the gas used for the heating and production cycles. The gas from the bag filter containing about 90% hydrogen is cooled, dehumidified, compressed, and used as fuel for reheating the generators. It is also used to dilute the natural gas feed for the production of a smaller particle size grade known as Fine Thermal (N880), as well as to fire boilers for plant steam and electricity. The fluffy black is passed through a magnetic separator, screened, and micropulverized. Most of the production is pelletized, packed in paper bags, or loaded directly into bulk hopper cars. Table 9 lists commercial grades of thermal blacks. The N900 series of medium thermal grades are the most widely used. The nonstaining grades N907 and N908 are made at higher average generator temperatures and longer contact times. This virtually eliminates residual hydrocarbon residues which cause staining by diffusion to the surface of many rubber products. A comparison of the analytical properties of medium and fine thermal blacks with carbon blacks made by other processes is shown in Table 10. Thermal blacks have the lowest surface area and lowest aggregation of the commercial carbon blacks. Medium thermal black particles are 400–500 nm and fine thermal blacks 120–150 nm in diameter. The electron micrograph of medium thermal black in Figure 8 shows a predominance of spherical particles in marked contrast to the aggregate nature of furnace black.

A medium thermal carbon black variety is produced in England by a cyclic process similar to the thermal process. This process uses oil for both the heating and production cycles. During the production cycle steam and oil are fed to the hot generator producing a gas containing suspended carbon black. The gas contains about 60% H_2, 25% CH_4, and 15% CO. After water cooling, about 85–90% of the black is removed with cyclones,

Table 9. Thermal Black Grades

ASTM classification	Industry type	Description
N880	FT–FF	fine thermal black, free flowing
N881	FT	fine thermal black
N990	MT–FF	medium thermal black, free flowing
N907	MT–NS–FF	medium thermal, nonstaining, free flowing
N908	MT–NS	medium thermal, nonstaining
N991	MT	medium thermal black

Table 10. Typical Analyses of Carbon Black Grades from Five Different Processes

Property	*Type:* Furnace	Thermal		Acetylene	Channel	Lampblack
Symbol:	HAF	MT	FT	Shawinigan	EPC	Lb
ASTM No.:	N-330	N-990	N-880		S300	
average particle diameter, nm	28	500	180	40	28	65
surface area (BET), m^2/g	75	47	13	65	115	22
DBPA, mL/100 g	103	36	33	250	100	130
tinting strength, % SRF	210	35	65	108	180	90
benzene extract, %	0.06	0.3	0.8	0.1	0.00	0.2
pH	7.5	8.5	9.0	4.8	3.8	3.0
volatile material, %	1.0	0.5	0.5	0.3	5	1.5
ash, %	0.4	0.3	0.1	0.0	0.02	0.02
Composition, %						
C	97.9	99.3	99.2	99.7	95.6	98
H	0.4	0.3	0.5	0.1	0.6	0.2
S	0.6	0.01	0.01	0.02	0.20	0.8
O	0.7	0.1	0.3	0.2	3.5	0.8

and the remainder is removed as a wet slurry used in the manufacture of carbon electrodes (33).

Thermal black has had wide use as a low-cost extender for rubber, as well as a functional filler for specially engineered rubber products (see Fillers). Substantial price increases during the last few years, caused by the rise in natural gas prices, have discouraged the extender use. It remains an essential ingredient in many specialty rubber products. The market share of thermal black has been as high as 11%, but in 1976 had declined to less than 5% due to the closing of some production facilities.

Acetylene Black Process. The high carbon content of acetylene (qv) (92%) makes it attractive for conversion to carbon. It decomposes exothermically at high temperatures, a property which was the basis of an explosion process initiated by electrical discharge (34). Acetylene black is made by a continuous decomposition process at 800–1000°C in water-cooled metal retorts lined with a refractory. The process is started by burning acetylene and air to heat the retort to reaction temperature, followed by shutting off the air supply to allow the acetylene to decompose to carbon and hydrogen in the absence of air. The large heat release requires water cooling in order to maintain a constant reaction temperature. The high carbon concentration, high reaction temperature and relatively long residence time produce a unique type of carbon black. After separation from the gas stream it is very fluffy with a bulk density of only 19 kg/m^3 (1.2 lb/ft^3). Acetylene black is difficult to compact by compression and resists pelletization. Commercial grades are compressed to various bulk densities up to a maximum of 200 kg/m^3 (12.5 lb/ft^3).

Table 10 lists the properties of acetylene black in comparison with carbon blacks from other processes. It is the purest form of carbon black listed in Table 10 with a carbon content of 99.7%, and a hydrogen content of 0.1%. It has the highest aggregation with a DBPA value of 250 cm^3/100 g. X-ray analysis indicates that it is the most crystalline or graphitic of the commercial blacks (35). These features result in a product with low surface activity, low moisture adsorption, high liquid adsorption, and high electrical and thermal conductivities.

A major use for acetylene black is in dry cell batteries because it contributes low electrical resistance and high capacity. In rubber it gives electrically conductive properties to heater pads, heater tapes, antistatic belt drives, conveyor belts, and shoe soles. It is also used in electrically conductive plastics. Some applications of acetylene black in rubber depend on its contribution to improved thermal conductivity, such as rubber curing bags for tire manufacture.

Lampblack Process. Early lampblack processes used large open shallow pans from 0.5 to 2 m in diameter and 16 cm deep for burning various oils in an enclosure with a restricted air supply. The smoke from the burning pans was allowed to pass at low velocities through settling chambers from which it was cleared by motor-driven ploughs. Yields of ca 480–600 kg/m^3 (4–5 lb/gal) of oil were obtained when making rubber grades. Today lampblack substitutes are made by the furnace process. Traditional lampblack manufacture is still practiced, but the quantities produced are small.

Channel Black Process. The channel black process has had a long and successful history, beginning in 1872 and ending in the United States in 1976. Small quantities are still produced in a few scattered plants operating in Germany (roller-process using oil), Eastern Europe, and Japan. Rising natural gas prices, smoke-pollution, low yield, and the rapid development of furnace process grades caused the termination of channel black production in the United States.

The name channel black came from the use of steel channel irons whose flat side was used to collect carbon black deposited from many small flames in contact with its surface. The collecting channels and thousands of flames issuing from ceramic tips were housed in sheet metal buildings, each 35–45 m long, 3–4 m wide, and about 3 m high. The air supply came from the base of these buildings; the waste gases, containing large quantities of undeposited product, were vented to the atmosphere as a black smoke. Carbon black was removed from the channels by scrapers and fell into hoppers beneath the channels. Yields were very low, in the range of 1–5%. The blackest pigment grades had the lowest yields. The product was conveyed from the hot houses to a processing unit where grit, magnetic scale, coke, and other foreign material were removed. From an initial bulk density of 80 kg/m^3 (5 lb/ft^3) it was compacted and pelletized to over 400 kg/m^3 (25 lb/ft^3) for use in rubber. Lower bulk densities were used for pigment applications.

Table 10 lists the properties of easy processing channel black (EPC), a long-time favorite of the rubber industry for tire tread reinforcement. Channel blacks are surface oxidized as a result of their exposure to air at elevated temperatures on the channel irons. Due to surface oxidation, the particles are slightly porous. These features influence performance in most applications. In rubber, the acidity of the oxygen-containing surface groups has a retarding effect on rate of vulcanization. In polyethylene, where carbon black is used as an uv absorber (see Uv absorbers) and to improve weathering resistance, the phenol and hydroquinone surface groups have antioxidant properties (see Antioxidants). In inks, oxygen-containing groups contribute to flow and printing behavior.

Trends in Carbon Black Manufacture (36–37). Since the introduction of the oil furnace process in the 1940s the industry has emphasized the development of new and improved grades to meet customer requirements, and process improvements to increase yields and capacities. Carbon black yields have almost doubled, and reactor capacities have increased about tenfold from 1950 to 1975. The carbon black industry is expected to devote more of its resources in the future to energy conservation, environmental problems, and health and safety aspects of its processes and products. Because of the increasing costs of new production capacity, plant production levels will be maintained closer to rated capacity of existing facilities. The following trends in carbon black manufacture can be expected: increasing use of by-product pyrolysis tars from ethylene manufacture for feedstocks; replacement of natural gas by liquid hydrocarbons for the primary heat source; increased sensible heat recovery from the combustion stream; increased use of tail gas for its heating value; use of higher sulfur content feedstocks; and increasing use of automatic computer control.

Economic Aspects

Carbon black consumption by the rubber industry has grown steadily from the time of the discovery of its outstanding reinforcing properties by S. C. Motte in Great Britain about 1912 (38). The average annual growth rate since 1925 has been about 5.8% in the United States. Figure 12 shows United States consumption from 1887 to 1975 in six distinct growth periods on a semilogarithmic plot.

Outside the United States carbon black consumption has followed a similar pattern with a few notable differences. From 1962–1974, consumption grew at 7.6% for Western Europe and 6.1% for the United States. The United States consumed more carbon black than the rest of the world combined for the period 1915–1965. By 1975 United States consumption accounted for about 42% of world consumption.

The pigment use of carbon black was the only important market until the discovery of rubber reinforcement. The United States market distribution from 1910 to the present is shown in Figure 13 (39). About 75% of the total market depends on motor vehicle applications including tires and automotive products (40). Carbon black prices depend mainly on costs of raw materials, plant equipment, labor, and utilities. Advances in technology and improved process economics had kept prices stable from 1950–1973 but in recent years the cost of feedstock has increased more than fourfold, causing about a twofold increase in the price of the large-volume rubber grades. The bulk price range in the United States for the major rubber grade carbon blacks in 1977 was 28–35¢/kg. Special pigment grades had a very wide price range from 31.4¢/kg for the least expensive low color SRF grades to $4.41/kg for the fluffy high color lacquer and enamel grades. The HAF (N300 series) grade is the most important and now accounts for almost one half the total market. This grade is divided into a number of subgrades with different structure levels. The next important grade is GPF (N660) which was introduced in the late fifties.

Health and Safety Aspects

Carbon is a relatively stable, unreactive element, insoluble in organic or other solvents. Carbon in the form of char has been used in the pharmaceutical industry for many years. Recently, vitreous and pyrolytic carbons have been found to be biocompatible and useful for artificial heart valves, and other medical applications (see Prosthetic and biomedical devices). There is no evidence of carbon black toxicity in humans, despite the fact that it contains trace amounts of some polynuclear aromatic compounds known to be carcinogenic.

During the first years of the carbon black industry, when major production was

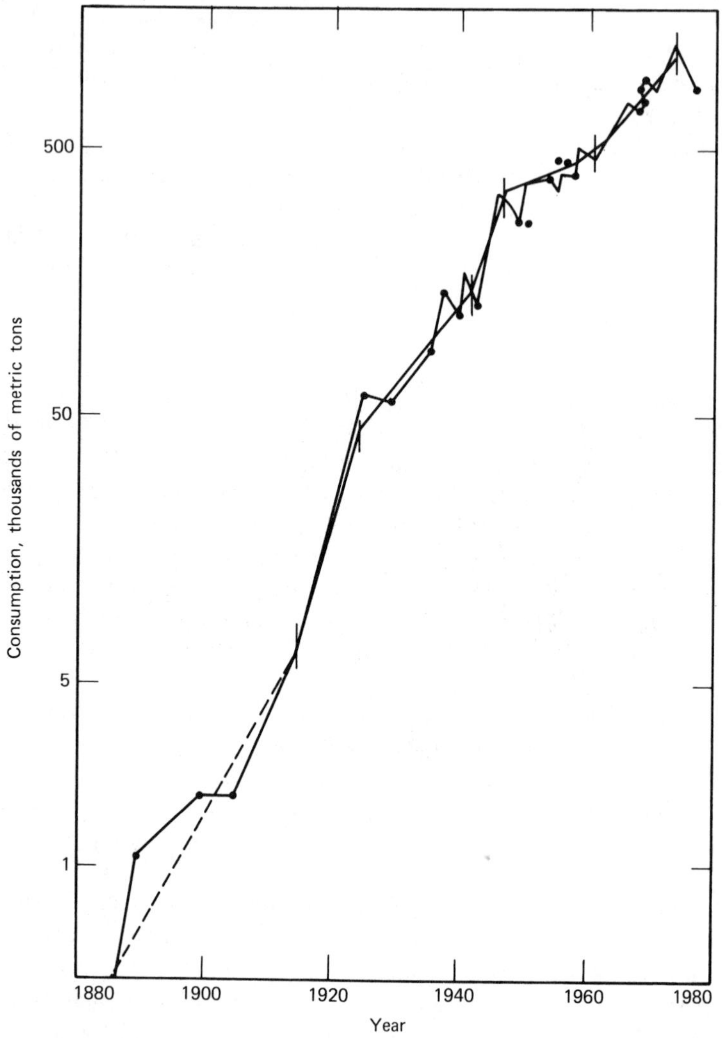

Figure 12. Carbon black consumption in the United States. Growth: 1915–1925, 16%; 1926–1942, 5.2%; 1943–1946, 17%; 1947–1961, 3.3%; 1962–1974, 5.9%.

by the channel black process, workers in the plants were exposed to an atmosphere containing carbon black. Channel black plants in the United States have been replaced by furnace process plants. Because carbon black is fine, light, easily carried by air currents, and jet black, special housekeeping procedures must be used during production and use. In order to reduce dustiness, the carbon black industry supplies its products either in loose compacted forms, or in more highly compacted pelletized forms.

The ingestion, skin contact, subcutaneous injection, and inhalation of channel, furnace, and thermal blacks by various animals causes no significant physiological changes (41–46). Although trace amounts of carcinogens were present, they were at too low a level, or were too inactive in their adsorbed state to cause a health hazard.

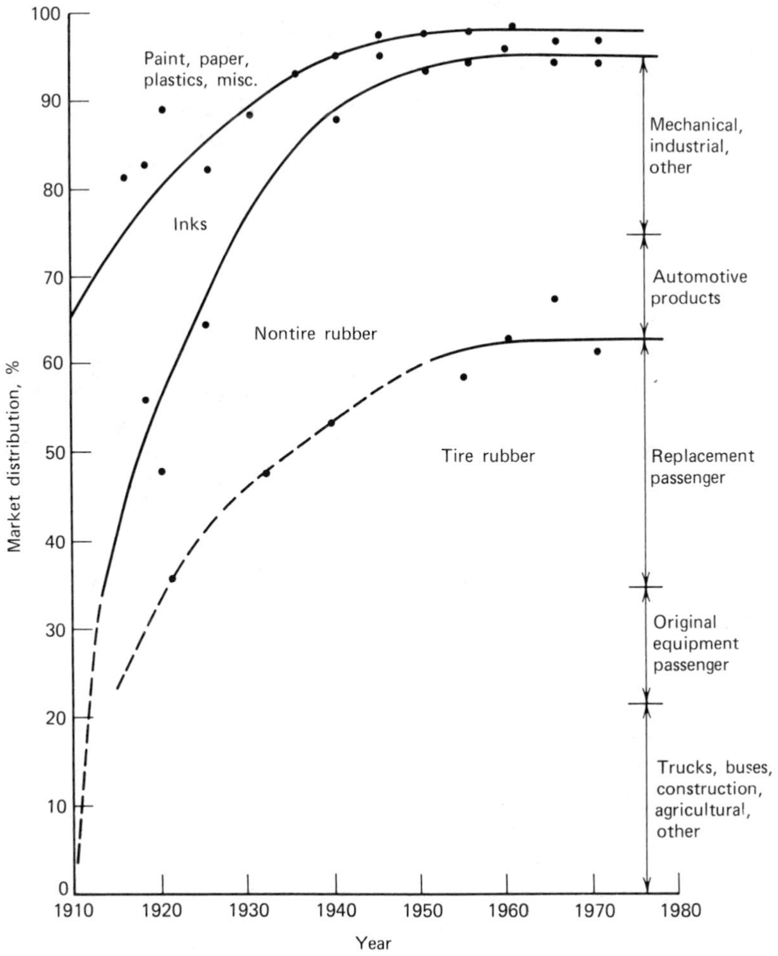

Figure 13. Carbon black market distribution.

Surveys of the health histories of carbon black workers published in 1950 and 1961 (47–48) showed their morbidity rate and observed death rate from cancer to be equal to or lower than those of comparable groups of industrial workers and the general population. Inhalation of high concentrations of carbon black over prolonged periods produced lung changes in animals due primarily to carbon black deposition with minimal or no fibrous tissue proliferation (44).

Carbon black is difficult to ignite, does not undergo spontaneous combustion, and does not produce dust explosions. When ignition does occur from contact with flames, glowing metal, sparks, or lighted cigarettes, it continues slowly with a dull glow. Due to the low conductivity of carbon black, storage fires may go undetected for some time. Fires may be controlled by purging with carbon dioxide. When this is done respirators must be used to avoid breathing carbon monoxide. Even in the absence of a fire, an air line or adequate respirator should be used when entering any confined carbon black storage facilities to avoid breathing possible toxic gases or a low oxygen atmosphere.

The level of toxic substances associated with carbon black is very low. Analytical values for a typical rubber grade semireinforcing carbon black (N774) showed 0.6 ppm phenols, 2.7 ppm lead, less than 0.05 ppm for cyanides, and less than 0.5 ppm for all other heavy metals (49). The metals have their origin in the petroleum feedstocks which have undergone various treatments and catalytic processes. The inorganic content of carbon black also includes such salts as sodium, potassium, calcium, and magnesium sulfates, chlorides, carbonates, and silicates which are contained in the water used for quenching and for wet pelletization.

The combustion of fuels can cause serious problems of atmospheric pollution from the emission of particulate material, sulfur compounds, nitric oxides, hydrocarbons, and other gases. Because of the intense blackness of carbon black it cannot be allowed to escape to the atmosphere even in minute quantities (50). Huge bag filters for complete recovery were put in operation over 30 years ago (see Air pollution control methods).

In the furnace process, tail gas burning to convert CO and H_2S to less noxious CO_2 and SO_2 has been practiced in some plants since 1950. This trend is increasing, stimulated by the necessity to recover the energy content of the tail gases. Not all the sulfur contained in carbon black feedstocks is converted to gases. One third to one half of the sulfur content is incorporated as nonextractable sulfur in the products.

Uses

Rubber. Table 11 lists the major grades used by the rubber industry, the general mechanical properties they contribute to rubber, and some examples of typical uses. A review of carbon black compounding and the types used in rubber products has been published (51) (see Rubber compounding).

Classification. The classification of carbon black grades for rubber was originally based on various performance or property characteristics, including: (*1*) levels of abrasion resistance—high abrasion furnace (HAF), intermediate super abrasion furnace (ISAF), and super abrasion furnace (SAF); (*2*) level of reinforcement—semireinforcing furnace (SRF); (*3*) a vulcanizate property—high modulus furnace (HMF); (*4*) a rubber processing property—fast extrusion furnace (FEF); (*5*) utility—general purpose furnace (GPF) and all-purpose furnace (APF); (*6*) particle size—fine furnace (FF), and large particle size furnace (LPF), fine thermal (FT), and medium thermal (MT); and (*7*) electrically conductive properties (XCF). Within some of these grades there were a variety of subgrades having different aggregate levels, eg, HAF with a high aggregate subgrade (HAF-HS), and a low aggregate subgrade (HAF-LS). The obvious inadequacies of this unwieldy classification procedure led ASTM Committee D-24 on carbon black to establish a letter and number system, shown in Table 3. In the ASTM system the N-series numbers increase as iodine adsorption values or surface areas decrease. The SAF grades have designated numbers from N100 to N199; the ISAF grades, N200 to N299; the HAF grades, N300 to N399; the FF and XCF (with this grade surface areas may be out of order with respect to the ASTM number) grades, N400 to N499; the FEF grades, 500 to N599; the HMF, GPF, and APF grades, N600 to N699; the SRF grades, N700 to N799; fine thermal (FT) has been designated as N880, medium thermal (MT) N990, and nonstaining medium thermal (MT-NS), N907. Aggregate levels or other quality variations within a grade are given arbitrary numbers. Table 3 lists a selected group of carbon black grades, their ASTM number,

Table 11. Applications of Major Rubber Grade Carbon Blacks

ASTM N-Type	Designation	General rubber properties	Typical uses
N990	medium thermal (MT)	low reinforcement, modulus, hardness, hysteresis, tensile strength; high loading capacity and high elongation	wire insulation and jackets, mechanical goods, footwear, belts, hose, packings, gaskets, O-rings, mountings, tire innerliners
N880	fine thermal (FT)	low reinforcement, modulus, hardness, hysteresis, tensile strength; high elongation, tear strength, and flex resistance	mechanical goods, gloves, bladders, tubes, footwear uppers
N700 Series	semireinforcing (SRF)	medium reinforcement, high elongation, high resilience, low compression set	mechanical goods, footwear, inner tubes, floor mats
N660	general purpose (GPF)	medium reinforcement, medium modulus, good flex and fatigue resistance, low heat buildup	standard tire carcass black; tire innerliners and widewalls; sealing rings, cable jackets, hose, soling, and extruded goods; EPDM compounds
N650	general purpose–high structure (GPF–HS)	medium reinforcement, high modulus and hardness, low die swell, smooth extrusion	tire innerliners, carcass, radial belt and sidewall compounds; extruded goods and hose
N550	fast extrusion (FEF)	medium-high reinforcement; high modulus and hardness; low die swell and smooth extrusion	tire innerliners, carcass, and sidewall compounds; innertubes, hose and extruded goods
N326	high abrasion–low structure (HAF–LS)	medium-high reinforcement; low modulus, high elongation, good fatigue resistance, flex resistance, and tear strength	tire belt, carcass, and sidewall compounds
N330	high abrasion (HAF)	medium-high reinforcement; moderate modulus, good processing	tire belt, sidewall, and carcass compounds; retread compounds, mechanical and extruded goods
N339, N347, N375	high abrasion–high structure (HAF–HS)	high reinforcement, modulus, and hardness; excellent processing	standard tire tread blacks
N220	intermediate super abrasion (ISAF)	high reinforcement, tear resistance; good processing	passenger and off-the-road tire treads; special service tires
N110	super abrasion (SAF)	high reinforcement	special tire treads, airplane, off-the-road, racing tires; products for highly abrasive service

their category, and typical analytical properties. Table 12 is a breakdown of the United States carbon black market according to consumption by type in 1975. Almost 80% of the market consisted of three types, HAF, GPF, and FEF.

Figure 14 is a chart of the carbon black product spectrum illustrating performance differences due to surface area, as assessed by iodine absorption, and aggregation or structure, as assessed by DBP absorption.

Properties of Carbon Black–Rubber Compounds. The general effects of different carbon blacks on rubber properties are dominated mainly by surface area and aggregation or structure. High surface area and small particle size impart higher levels of reinforcement as reflected in tensile strength, tear resistance, and resistance to abrasive wear with resulting higher hysteresis and poorer dynamic performance. Higher aggregation gives improved extrusion behavior, higher stock viscosities, improved green strength, and higher modulus values. A summary of the effects of carbon black structure and particle size on rubber processing and vulcanizate properties appears in Table 13.

Table 12. United States Carbon Black Distribution, 1975

Carbon black grade	Market, %
HAF (N300 series)	46.1
GPF (N600 series	20.3
FEF (N550)	11.5
SRF (N700 series)	8.8
ISAF (N200 series)	8.0
thermal (N800–N900 series)	4.4
SAF (N110)	0.9

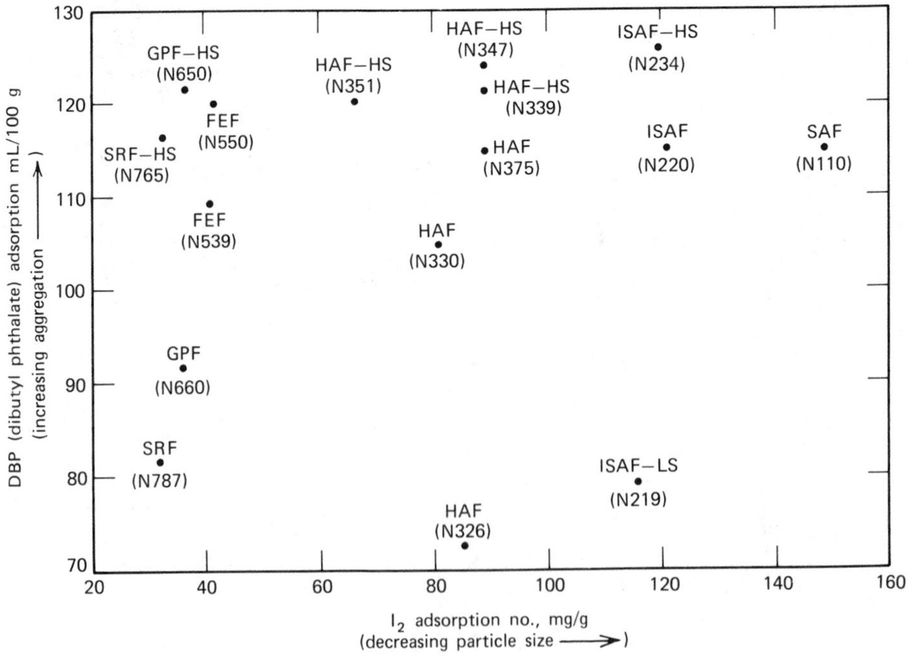

Figure 14. Carbon black grade spectrum.

Table 13. Effect of Carbon Black Colloidal Characteristics on Rubber Processing and Physical Properties

Property	Colloidal characteristics	
	Decreasing particle size	Increasing aggregation
	increasing surface area increasing iodine number increasing tinting strength	increasing aggregate size increasing DBP absorption decreasing bulk density
Rubber processing property		
loading capacity	decreases	decreases
mixing incorporation time	increases	increases
Mooney viscosity	increases	increases
dimension stability (green)	increases	increases
extrusion shrinkage	not significant	decreases
Rubber physical property		
tensile strength	increases	variable
modulus	not significant	increases
elongation	not significant	decreases
hardness	increases	increases
impact resilience	decreases	not significant
abrasion resistance	increases	variable

Rubber property changes produced by carbon black addition depend on loading. Tensile strength and abrasion resistance increase with increased loading to an optimum and then decrease. The optimum is normally in the range of 40–60 phr (parts per hundred of rubber). Increasing the loading level to 35–80 phr produces linear increases in hardness and modulus. The magnitude of these changes increases with structure. These effects are shown in Figure 15 for a group of carbon black grades including HAF (N330) used in tire treads, GPF (N660) and APF (N683) used in tire carcasses, and a medium thermal (MT) (N990) used in mechanical goods. Testing was done in a standard SBR-1500 recipe containing 12 parts of processing oil (52).

Special Carbon Blacks. About 6–7% of the total carbon black production consists of special industrial carbon grades for nonrubber applications (special blacks). Although some of the basic grades produced for the rubber market can be used for special black applications, most of these products are manufactured by methods developed to meet specific use requirements. They sell for a higher average price than the rubber grades. The growth of these markets over the last two decades has been about 4% annually. Of increasing importance in recent years have been applications in plastics where special carbon blacks are required to improve weathering resistance, or to impart antistatic and electrically conductive properties (see Antistatic agents).

About 40% of the special blacks are used in printing inks, 10% in paints and lacquers, 36% in plastics, 2% in paper, and 12% in miscellaneous applications. News inks account for most of the printing ink market. For this use special N300 series (HAF) grades containing 6% mineral oil are made to give rapid dispersion. Medium and high color blacks, the most expensive grades, are used in enamels, lacquers and plastics for their intense jetness. Carbon black grades for these applications are listed in Table 14.

Although the applications for special blacks have not changed greatly from 1965 to 1975, there were major changes in the product line and production technology. During this period the medium and high color grades, which were formerly made by the channel process, were replaced by furnace process grades.

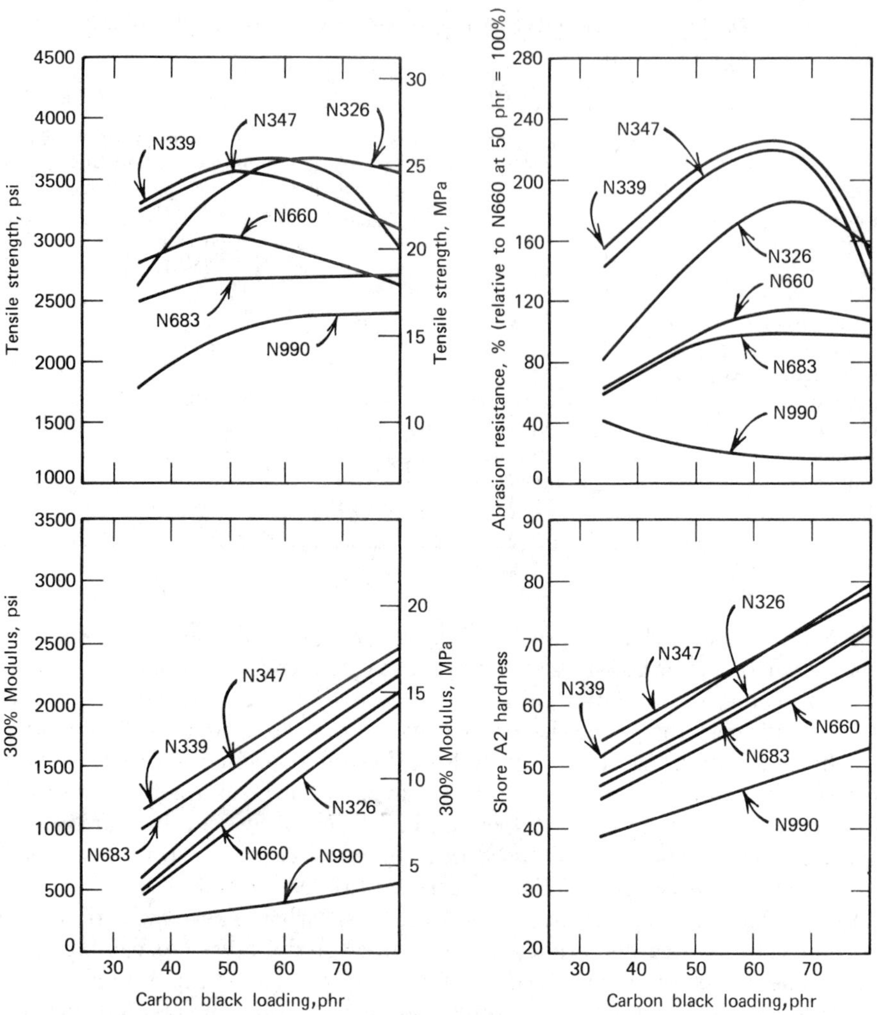

Figure 15. Properties in SBR of N339 (HAF-HS), N347 (HAF-HS), N326 (HAF-LS), N660 (GPF), N683 (APF), N990 (MT). phr = parts per hundred of rubber.

Printing Inks. The printing ink industry uses about 40% of the special blacks production in the United States. The grade and concentration used depend on the type and quality of the ink, and are selected for such factors as degree of jetness, gloss, blueness of tone, viscosity, tack, and rheological properties. Carbon black surface area, structure, degree of surface oxidation, moisture content, and bulk density are the most important factors in performance. Over forty special black grades have been developed having a broad range of properties from 20 m^2/g surface area grades used for inexpensive inks and tinting to oxidized, porous low-aggregation grades of about 500 m^2/g used for high color enamels and lacquers. The color and rheological properties of inks are influenced by both surface area and structure as shown in Table 15. The surface chemistry of carbon blacks affects their behavior in some types of ink. Special flow grades are made by a secondary oxidative aftertreatment process to increase their volatile content, a measure of the chemisorbed surface oxygen complexes. This

Table 14. Types and Applications of Special Carbon Blacks

Type	Surface area, m^2/g	DBP absorption, mL/100 g	Volatile content, %	Uses
high color	230–560	50–120	2–10	high jetness for alkyl and acrylic enamels, lacquers, and plastics
medium color	200–220	70–120	1–1.5	medium jetness and good dispersion for paints and plastics; ultraviolet and weathering protection for plastics
medium color, long flow	138	55–60	5	used in lithographic, letterpress, carbon paper, and typewriter ribbon inks; high jetness, excellent flow, low viscosity, high tinting strength, gloss, and good dispersibility
medium color, medium flow	96	70	2.5	used for gloss printing and carbon paper inks; excellent jetness, dispersibility, tinting strength, and gloss in paints
regular color	80–140	60–114	1–1.5	for general pigment applications in inks, paints, plastics, and paper; gives ultraviolet protection in plastics, and high tint, jetness, gloss, and dispersibility in inks and paints
	46	60	1.0	good tinting strength, blue tone, low viscosity; used in gravure and carbon paper inks, paints, and plastics
	45–85	73–100	1.0	main use is in inks; standard and offset news inks
low color	25–42	64–120	1.0	excellent tinting blacks—blue tone; used for inks—gravure, one-time carbon paper inks; also for paints, sealants, plastics, and cements
thermal blacks	7–15	30–35	0.5–1.0	tinting—blue tone; plastics and utility paints
lamp blacks	20–95	100–160	0.4–9.0	paints for tinting—blue tone
conductive blacks	254	180	2.0	conductivity and antistatic applications in rubber and plastics
acetylene	65	250	0.3	conductive and antistatic applications, dry cells, tire curing bags

treatment improves wetting, dispersion, and prevents pigment reagglomeration. Inks made with flow carbon blacks have lower viscosities and a reduced tendency to body-up or thicken during storage. Another factor influencing ease of dispersion is bulk density. Pelleted products (250–500 g/L) are more difficult to disperse than lower density or fluffy products (100–300 g/L).

Different methods of printing require inks with specially designed rheological and drying properties. Table 16 lists the major types of printing inks, typical carbon black grades and concentrations, and the basic ink vehicle composition. Letterpress news inks normally contain from 9–14% carbon black in a petroleum oil. Such inks are relatively fluid, and dry by penetration of the oil into the fibers of the paper. Cost is a paramount factor in news ink production. Large circulation weekly publications printed on coated papers at very high speeds use inks containing solvent, resin, and 10–22% carbon black. Offset or lithographic gloss inks are used in high quality book printing, illustrations, and other material where excellence in detail and transfer are paramount. Such inks must possess maximum tinting strength and covering power. This is achieved by using carbon black concentrations of 15–22%. Gravure inks vary

Table 15. Surface Area and Aggregation Effects on Ink Performance

Ink property	High surface area	Low surface area
masstone color	darker	lighter
tone	browner	bluer
viscosity	higher	lower
tint strength	stronger	weaker
dispersion	more difficult	easier
wetting	more difficult	easier
Ink property	**Low aggregation**	**High aggregation**
dispersibility	harder	easier
gloss	higher	lower
wetting	faster	slower
viscosity	lower	higher
color	darker	lighter
undertone	browner	bluer
tint strength	higher	lower
thixotropy	lower	higher

Table 16. Types and Composition of Various Printing Inks

Ink	Grades of carbon black	Carbon black concentration, %	Vehicle composition
letterpress news ink	medium color, medium structure; oil pellets	9–14	mineral oil
letterpress gloss ink (nonheat-set, heat-set)	medium color, low structure, oxidized flow; oil pellets	16–22	nonheat-set or heat-set varnish
lithographic news ink	medium color, medium structure; oil pellets	18–20	mineral oil and varnish
lithographic gloss ink (non-heat-set, heat-set)	medium color, low structure, oxidized flow; oil pellets	15–22	nonheat-set or heat-set varnish
gravure (roto)	low-medium color, low structure	5–15	resins, polymers
flexographic	low to medium structure, oxidized flow grades	12–18	resins, polymers

widely in requirements and formulation because of their diversity of applications. These inks must have good strength, low viscosity, and good gloss. Low-to-medium color, low aggregation carbon blacks are used. Highly fluid flexographic inks designed for printing on nonporous surfaces, such as plastics metal foil, and special coated paper, must dry very rapidly. The most widely used carbon blacks for flexographic inks are the medium-to-long flow, medium color grades (see Inks; Printing processes).

Paints, Lacquers, Enamels, and Industrial Finishes. Carbon black pigments for paints, lacquers, enamels, and industrial finishes fall into three categories, classified as high, medium, and standard color (see also Coatings, industrial; Paints; Pigments). In these applications, jetness or masstone is of primary significance. The coating industry uses a wide range of furnace carbon blacks. High color carbon blacks provide exceptional jetness and gloss for automobile finishes and high grade enamels. Medium color carbon blacks are used in industrial enamels, and the standard grades are used in general utility and industrial paints. Large particle size furnace and thermal grades are used to a limited extent in paints and other coatings. Although they are deficient in jetness, they provide a desirable blue tone. Carbon black is sometimes supplied to the lacquer industry in the form of masterbatch chips. These chips consist of a high loading of carbon black dispersed in a resin.

Plastics. Medium and high color carbon blacks are used for tinting and also to provide jetness in plastics (see Plastics technology). Carbon black is used in polyolefins as a protective agent. Polyolefins without carbon black degrade rapidly on exposure to sunlight. Carbon black is an excellent black body, absorbing both ultraviolet and infrared (53). When dispersed in polyethylene it preferentially absorbs and dissipates the incident radiation (54). Since carbon black also appears to be effective in terminating free radicals it provides protection against thermal degradation (see Heat stabilizers). Clear polyethylene cable jackets become brittle after two years of outdoor exposure, but the same compound containing 2% black showed no change after twenty years' exposure (55) (see Insulation, electric). Medium color furnace blacks are preferred in this application as well as for the weathering protection of plastic pipe.

Polyethylene compounds containing up to 50 wt % of furnace or thermal blacks have been developed. The deterioration of physical properties and embrittlement that would normally accompany such high loadings of black is reduced substantially by cross-linking the compound either by radiation or addition of organic peroxides. This provides greatly increased strength, and makes possible strong, tough compounds for cable coatings or pipe (56–57). Electrically conductive compounds for the cable industry depend on the use of about 30% of electrically conductive grades of carbon black (see Polymers, conductive).

Paper. The paper industry employs carbon black to produce a variety of black papers, including album paper, leatherboard, wrapping and bag papers, opaque backing paper for photographic film, highly conducting and electrosensitive paper, and black tape for wrapping high voltage transmission cables. The use of carbon black is essential when a conducting paper is desired. The loading of black will vary with the application in the range of 2–8 wt % of pulp. To obtain adequate dispersion, the black is employed in the fluffy form and added directly to the beater in a dispersible paper bag to avoid dusting. It may also be added in the form of an aqueous slurry. Aqueous dispersions containing 35% of carbon black are available. Reference 58 is a review of the applications of carbon black in the paper industry (see Paper).

Other Applications. The low thermal conductivity of carbon black makes it an excellent high temperature insulating material. For high temperature insulation up to 3000°C (59), it must be maintained in an inert atmosphere to prevent oxidation. Thermal-grade carbon black has been most widely used.

Carbon black is a source of pure carbon both for ore reduction and carburizing. Carbon brushes and electrodes are fabricated from carbon black.

There are a number of other minor applications of carbon black. For example, it is used as a pigment in the cement industry, in linoleum, leather coatings, polishes, and plastic tile.

BIBLIOGRAPHY

"Carbon Black" under "Carbon" in *ECT* 1st ed., Vol. 3, pp. 34–65, and Suppl. 1, pp. 130–144, by W. R. Smith, Godfrey L. Cabot, Inc.; "Acetylene Black" in *ECT* 1st ed., Vol. 3, pp. 66–69, by B. P. Buckley, Shawinigan Chemicals Ltd.; "Carbon Black" under "Carbon" in *ECT* 2nd ed., Vol. 4, pp. 243–282, by W. R. Smith, Cabot Corporation, D.C. Bean (Acetylene Black), Shawinigan Chemicals Limited.

1. B. E. Warren, *Phys. Rev.* **59,** 693 (1941).
2. E. A. Kmetlso, *Proc. 1st and 2nd Conf. on Carbon,* 21 (1956).
3. V. L. Kasatotshkin and co-workers, *J. Chim. Phys.* **52,** 822 (1964).
4. J. B. Donnet and J. C. Bouland, *Rev. Gen. Caoutch. Plast.* **41,** 407 (1964).
5. W. M. Hess and L. L. Ban, *Norelco Rep.* **13**(4), 102 (1966).
6. D. F. Harling and F. A. Heckman, *International Plastics and Elastomers Conference, Milan, Italy, Oct. 1968.*
7. S. Brunauer, P. H. Emmett, and J. Teller, *J. Am. Chem. Soc.* **60,** 310 (1938).
8. J. H. deBoer and co-workers, *J. Catal.* **4,** 649 (1965).
9. *ASTM D1510-76, Annual Book of ASTM Standards,* American Society for Testing and Materials, Philadelphia, Pa.
10. J. Janzen and G. Kraus, *Rubber Chem. Technol.* **44,** 1287 (1971).
11. Ref. 9, *ASTM D2414-76.*
12. R. E. Dollinger, R. H. Kallinger, and M. L. Studebaker, *Rubber Chem. Technol.* **40,** 1311 (1967).
13. Ref. 9, *ASTM D3265-76.*
14. A. I. Medalia and co-workers, *Rubber Chem. Technol.* **46,** 1239 (1973).
15. H. P. Palmer and C. F. Cullis in P. I. Walker, ed., *Chemistry and Physics of Carbon,* Vol. 1, Marcel Dekker, Inc., New York, 1965, p. 265.
16. J. B. Donnet, "Les Carbones," in *Groupe Francais Etude des Carbones,* Vol. 2, Masson, Paris, Fr., 1962, p. 208.
17. A. Feugier, *Rev. Gen. Therm.* **9,** 105, 1045 (1970).
18. J. Abrahamson, *Nature* **266,** 323 (1977).
19. *Chem. Eng. News,* p. 9 (Apr. 5, 1976).
20. U.S. Pat. 2,564,700 (Aug. 21, 1956), J. C. Krejci (to Phillips Petroleum Co.).
21. U.S. Pat. 1,438,032 (Dec. 5, 1922), W. H. Frost (to Wilckes-Martin-Wilckes Co.).
22. U.S. Pat. 2,292,355 (Aug. 11, 1942), J. W. Ayers (to C. K. Williams and Co.).
23. O. K. Austin, "Commercial Manufacture of Carbon Black," in G. Kraus, ed., *Reinforcement of Elastomers,* Interscience Publishers, New York, 1965.
24. E. M. Dannenberg, *J. Inst. Rubber Ind.* **5,** 190 (1971).
25. Can. Pat. 822,024 (Sept. 2, 1969), G. L. Heller (to Columbian Carbon Co.).
26. U.S. Pat. 3,353,915 (Nov. 21, 1967), B. L. Latham (to Continental Carbon Co.).
27. U.S. Pat. 3,619,140 (Nov. 9, 1971); Re. 28,974 (Sept. 21, 1976), A. C. Morgan and M. E. Jordan (to Cabot Corporation).
28. U.S. Pat. 3,003,855 (Oct. 10, 1961), G. L. Heller and C. I. DeLand (to Columbian Carbon Co.).
29. U.S. Pat. 3,253,890 (May 31, 1966), C. L. DeLand, G. L. DeCuir, and L. E. Wiggins (to Columbian Carbon Co.).
30. U.S. Pats. 3,010,794 (Nov. 28, 1961), G. F. Friauf and B. Thorley (to Cabot Corporation).
31. K. R. Dahmen, "The Carbon Black Furnace Process," *paper presented to the Akron Rubber Group, Technical Symposium, Apr. 15, 1977.*
32. U.S. Pats. 3,725,103 (Apr. 3, 1973); 3,799,788 (Mar. 26, 1974), M. E. Jordan, W. G. Burbine, and F. R. Williams (to Cabot Corporation).
33. H. J. Stern, *Rubber-Natural and Synthetic,* Maclaren & Sons, Ltd., London, Eng., 1954, p. 137.
34. Ger. Pat. 103,862 (June 27, 1899), L. J. E. Hubou.
35. A. E. Austin, *Proc. 3rd Conf. on Carbon,* 389 (1958).
36. E. M. Dannenberg, *Paper No. 42,* Rubber Div., American Chemical Society, San Francisco, Calif., 1976.
37. E. M. Dannenberg, *Plast. Rubber Int.* **3,** 11 (1978).
38. H. J. Stern, *Rubber Age Synth.,* 268 (Dec. 1945–Jan. 1946).
39. *Chemical Economics Handbook,* Stanford Research Institute, Menlo Park, Calif., 1975.
40. H. L. Duncombe, *paper presented to RMA Molded and Extruded Products Division, Hot Springs, Va., Rubber and Plastics News, July 12, 1976.*
41. C. A. Nau, J. Neal, and V. Stembridge, *AMA Arch. Ind. Health* **17,** 21 (1958).

42. C. A. Nau, J. Neal, and V. Stembridge, *AMA Arch. Ind. Health* **18,** 511 (1958).
43. C. A. Nau, J. Neal, and V. Stembridge, *Arch. Environ. Health* **1,** 512 (1960).
44. C. A. Nau and co-workers, *Arch. Environ. Health* **4,** 415 (1962).
45. J. Neal, M. Thornton, and C. A. Nau, *Arch. Environ. Health* **4,** 598 (1962).
46. C. A. Nau, G. T. Taylor, and C. Lawrence, *J. Occup. Med.* **18,** 732 (1976).
47. T. H. Ingalls, *Arch. Ind. Hyg. Occup. Med.* **1,** 662 (1950).
48. T. H. Ingalls and R. Risquez-Iribarren, *Arch. Environ. Health* **2,** 429 (1961).
49. H. J. Collyer, *EPA Conference Paper, Akron, Ohio, Mar. 12, 1975.*
50. E. M. Dannenberg, *Rubber Age* **108**(4), 37 (1976).
51. M. Studebaker, "Compounding with Carbon Black," in G. Kraus, ed., *Reinforcement of Elastomers,* Interscience Publishers, New York, 1965.
52. *Technical Report TG-76-1,* Cabot Corporation, Boston, Mass., 1976.
53. A. J. Wells and W. R. Smith, *J. Phys. Chem.* **45,** 1055 (1941).
54. V. T. Wallder and co-workers, *Ind. Eng. Chem.* **42,** 2320 (1950).
55. W. L. Hawkins, *Rubber Plast. Weekly (London)* **142,** 291 (1962).
56. A. Charlesby, *Atomic Radiation and Polymers,* Pergamon Press, Inc., New York, 1960, Chapt. 13.
57. E. M. Dannenberg, M. E. Jordan, and H. M. Cole, *J. Polym. Sci.* **31,** 127 (1958).
58. I. Drogin, *Pap. Trade J.* **147,** 24 (Apr. 1, 1963).
59. W. D. Schaeffer, W. R. Smith, and M. H. Polley, *Ind. Eng. Chem.* **45,** 1721 (1953).

ELI M. DANNENBERG
Cabot Corporation

CEMENT

The term cement is used to designate many different kinds of substances that are used as binders or adhesives (qv). The cement produced in the greatest volume and most widely used in concrete for construction is portland cement. Masonry and oil well cements are produced for special purposes. Calcium aluminate cements are extensively used for refractory concretes (see Refractories). Such cements are distinctly different from epoxies and other polymerizable organic materials. Portland cement is a hydraulic cement, ie, it sets, hardens, and does not disintegrate in water. Hence it is suitable for construction of underground, marine, and hydraulic structures whereas gypsum plasters and lime mortars are not. Organic materials, such as latexes and water soluble polymerizable monomers, are sometimes used as additives to impart special properties to concretes or mortars; furthermore, concretes are sometimes impregnated with liquid organic monomers (or liquid sulfur) and polymerized to produce polymer-impregnated concrete. The term cements as used henceforth will be confined to inorganic hydraulic cements, principally portland and related cements. The essential feature of these cements is their ability on hydration to form with water relatively insoluble bonded aggregations of considerable strength and dimensional stability.

Hydraulic cements are manufactured by processing and proportioning suitable raw materials, burning (or clinkering at a suitable temperature), and grinding the resulting hard nodules called clinker to the fineness required for an adequate rate of hardening by reaction with water. Portland cement consists mainly of tricalcium silicate and dicalcium silicate. Usually two types of raw materials are required: one rich in calcium, such as limestone, chalk, marl, or oyster or clam shells; the other rich in the silica, such as clay or shale. The two other major phases in portland cements are tricalcium aluminate and a ferrite phase. A small amount of calcium sulfate in the form of gypsum or anhydrite is also added during grinding to control the setting time and enhance strength development.

The demand for cement was stimulated by the growth of canal systems in the United States during the 19th century. This led to process improvements in the calcination of certain limestones for the manufacture of natural cements, and to its gradual displacement by portland cement. The latter was named by Aspdin in a 1924 patent because of its resemblance to a natural limestone quarried on the Isle of Portland in England. Research conducted in many parts of the world since that time has provided a clear picture of the composition, properties, and fields of stability of the principal systems found in portland cement. These results led to the widely used Bogue calculation of composition based on oxide analysis (1). Recent research is reported in the *International Symposia on the Chemistry of Cements,* and the annual reviews, beginning in 1974, of the American Ceramic Society in *Cements Research Progress* (see under General References).

Clinker Chemistry

The conventional cement chemists' notation uses the following abbreviations for the most common constituents:

$CaO = C$	$MgO = M$	$K_2O = K$
$SiO_2 = S$	$SO_3 = \overline{S}$	$CO_2 = \overline{C}$
$Al_2O_3 = A$	$Na_2O = N$	$H_2O = H$
$Fe_2O_3 = F$		

Thus tricalcium silicate, Ca_3SiO_5, is denoted by C_3S.

Portland cement clinker is formed by the reactions of calcium oxide with acidic components to give C_3S, C_2S, C_3A, and a ferrite phase approximating C_4AF.

Phase Equilibria. During burning in the kiln, about 20–30% of liquid forms in the mix at clinkering temperatures. Reactions occur at surfaces of solids and in the liquid. The crystalline silicate phases formed are separated by the interstitial liquid. The interstitial phases formed from the liquid in normal clinkers during cooling are also completely crystalline to x-rays, although they may be so finely subdivided as to appear glassy (optically amorphous) under the microscope.

The high temperature phase equilibria governing the reactions in cement kilns were studied, eg, in the $CaO–Al_2O_3–SiO_2$ system illustrated in Figure 1 (2–3). In such a ternary diagram, the primary-phase fields are plotted, ie, the composition regions in which any one solid is the first to separate when a completely liquid mix is cooled (with negligible supercooling). The primary-phase fields are separated by eutectic points on the sides of the triangle such as that at 1436°C between tridymite and α-CS.

In the relatively small portland cement zone almost all modern cements fall in the high lime portion (about 65% CaO). Cements of lower lime content tend to be slow in hardening and may show trouble from dusting of the clinker by transformation of β- to γ-C_2S, especially if clinker cooling is very slow. The zone is limited on the high lime side by the need to keep the uncombined CaO to low enough values to prevent excessive expansion due to hydration of the free lime. Commercial manufacture at compositions near the $CaO–SiO_2$ axis can present difficulties. If the lime content is high, the burning temperatures may be so high as to be impractical. If the lime content is low, the burning temperatures may even be low, but impurities must be present in the C_2S to prevent dusting. On the high alumina side the zone is limited by excessive liquid-phase formation which prevents proper clinker formation in rotary kilns.

The relations between the compositions of portland cements and some other common hydraulic cements are shown in the $CaO–SiO_2–Al_2O_3$ phase diagram of Figure 2 (4), analogous to Figure 1. In this diagram the Fe_2O_3 has been combined with the Al_2O_3 to yield the Al_3O_3 content used. This is a commonly applied approximation that permits a two-dimensional representation of the real systems.

Clinker Formation. Portland cements are ordinarily manufactured from raw mixes including components such as calcium carbonate, clay or shale, and sand. As the temperature of the materials increases during their passage through the kiln, the following reactions occur: (*1*) evaporation of free water; (*2*) release of combined water from the clay; (*3*) decomposition of magnesium carbonate; (*4*) decomposition of calcium carbonate (calcination); and (*5*) combination of the lime and clay oxides. The course of reactions (5) occurring at the high temperature end of the kiln, just before and in the burning zone, is illustrated graphically in Figure 3 (6).

From the phase diagram of the $CaO–SiO_2–Al_2O_3$ system, the sequence of crystallization during cooling of the clinker can be derived if the cooling is slow enough to maintain equilibrium. For example, a mix at 1500°C of relatively low lime content, along the $C_3S–C_2S$ eutectic line in Figure 1, will be composed of solid C_3S and C_2S and a liquid along the $C_3S–C_2S$ eutectic at the intersection with the 1500°C isotherm (to the left of the 1470–1455 line). Upon cooling, this liquid deposits more C_3S and C_2S, moving the liquid composition down to the invariant point at 1455°C, at which C_3A also separates until crystallization is complete. Although real cement clinkers contain

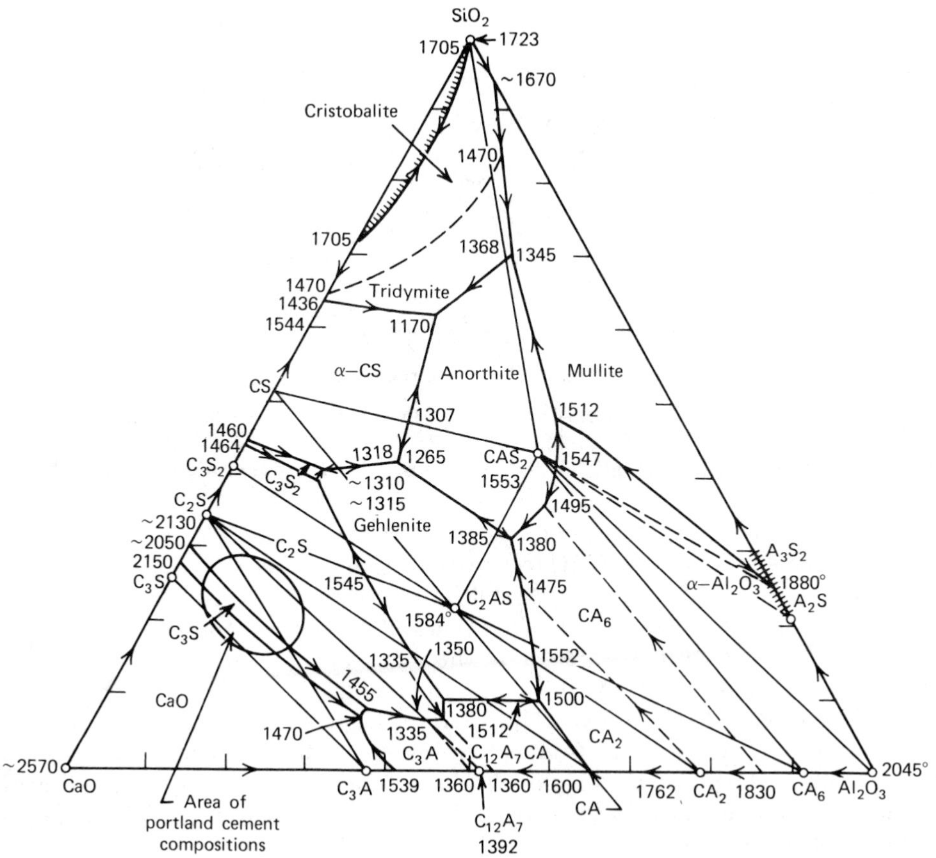

Figure 1. Phase equilibria in the $CaO–Al_2O_3–SiO_2$ system (2–3), in °C. Shaded areas denote two liquids, compositional index marks on the triangle are indicated at 10% intervals.

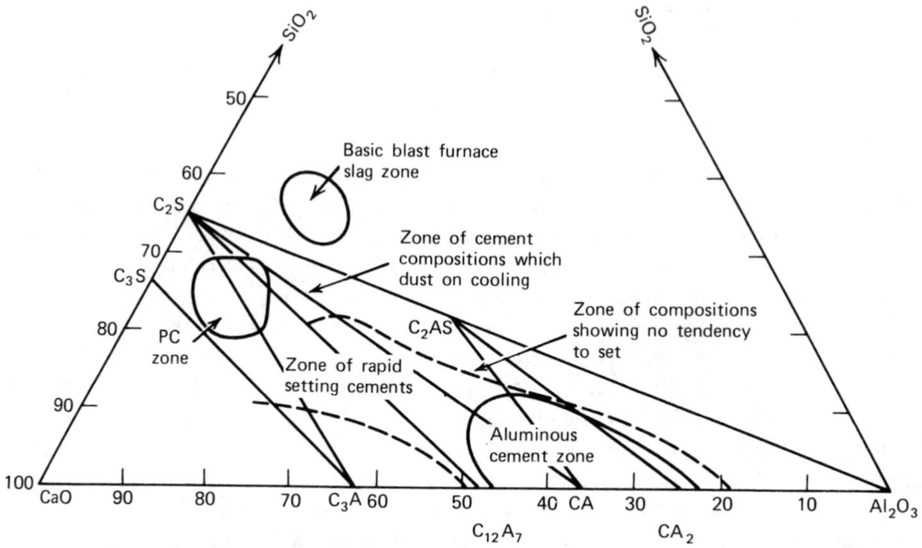

Figure 2. Cement zones in the $CaO–Al_2O_3–SiO_2$ system (4).

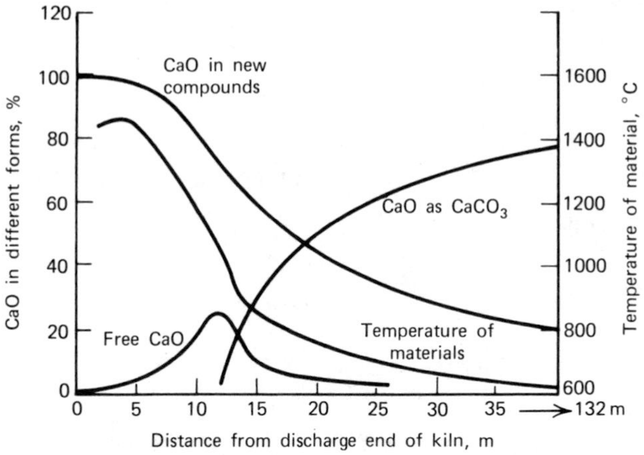

Figure 3. Temperatures and progress of reactions in a 132-meter wet-process kiln.

more components, which alter the system and temperatures somewhat, the behavior is similar.

Cooling is ordinarily too rapid to maintain the phase equilibria. In the above case, the lime-deficient liquid at 1455°C requires that some of the solid C_3S redissolve and that more C_2S crystallize during crystallization of the C_3A. During rapid cooling there may be insufficient time for this reaction and the C_3S content will be higher than when equilibrium conditions prevail. In this event crystallization is not completed at 1455°C, but continues along the $C_3A–C_2S$ boundary until the invariant point at 1335°C is reached. Crystallization of C_2S, C_3A, and $C_{12}A_7$ then occurs to reach complete solidification. Such deviations from equilibrium conditions cause variations in the phase compositions estimated from the Bogue calculation (see below), and cause variations in the amounts of dissolved substances such as MgO, alkalies, and the alumina content of the ferrite phase.

The theoretical energy requirement for the burning of portland cement clinker can be calculated from the heat requirements and energy recovery from the various stages of the process. Knowledge of the specific heats of the various phases, and the heats of decomposition, transformation, and reaction then permits calculation of the net theoretical energy requirement of 1760 kJ (420 kcal) for 1 kg of clinker from 1.55 kg of dry $CaCO_3$ and kaolin (7).

The kinetics of the reactions are strongly influenced by the temperature, mineralogical nature of the raw materials, fineness to which the raw material is ground, percentage of liquid phase formed, and viscosity of the liquid phase. The percentage of liquid formed depends on the alumina and iron oxides (see under Proportioning of Raw Materials). When the sum of these oxides is low, the amount of liquid formed is insufficient to permit rapid combination of the remaining CaO. The viscosity of the liquid at clinkering temperature is reduced by increasing the amounts of oxides such as MnO, Fe_2O_3, MgO, CaO, and Na_2O (8).

The reaction of C_2S with CaO to form C_3S depends upon dissolution of the lime in the clinker liquid. When sufficient liquid is present, the rate of solution is controlled by the size of the CaO particles, which depends in turn on the sizes of the particles of ground limestone. Coarse particles of silica or calcite fail to react completely under commercial burning conditions. The reaction is governed by the rate of solution (9):

$$\log t = \log \frac{D}{A} + 0.43 \frac{E}{RT}$$

t is the time in minutes, D is the particle diameter in mm, A is a constant, T the absolute temperature, and E is the activation energy with a value of 607 kJ/mol (146 kcal/mol). For example, 0.05 mm particles require 59 min for solution at 1340°C but only 2.3 min at 1450°C. A similar relation applies for the rate of solution of quartz grains.

Phases Formed in Portland Cements. Most clinker compounds take up small amounts of other components to form solid solutions. Best known of these phases is the C_3S solid solution called alite. Phases that may occur in portland cement clinker are given in Table 1. In addition, a variety of minor phases may occur in portland cement clinker when certain minor elements are present in quantities above that which can be dissolved in other phases. Under reducing conditions in the kiln, reduced phases, such as FeO and calcium sulfide may be formed.

The major phases all contain impurities. These impurities in fact stabilize the structures formed at high temperatures so that decomposition or transformations do not occur during cooling, as does occur with the pure compounds. For example, pure C_3S exists in (at least) six polymorphic forms each having a sharply defined temperature range of stability, whereas alite exists in three stabilized forms at room temperature depending upon the impurities. Some properties of the more common phases in portland clinkers are given in Table 2.

Structure. Examination of thin sections of clinkers using transmitted light and of polished sections by reflected light reveals details of the structure. A recently developed method (16) employs the polarizing microscope to determine the size and birefringence of alite crystals, and the size and color of the belite to predict later age strength. The clinker phases are conveniently observed by examining polished sections selectively etched with special reagents as shown in Figure 4. The alite appears as clear

Table 1. Phases in Portland Cement Clinker[a]

Name of impure form	CAS Registry No.	Chemical name	Cement chemists' notation
free lime	*[1305-78-8]*	calcium oxide	C
periclase (magnesia)	*[1309-48-4]* and *[1317-74-4]*	magnesium oxide	M
alite	*[12168-85-3]*	tricalcium silicate	C_3S
belite	*[10034-77-2]*	dicalcium silicate	C_2S
C_3A	*[12042-78-3]*	tricalcium aluminate	C_3A
ferrite	*[12612-16-7]*	calcium aluminoferrite[b]	$C_2A_xF_{1-x}$
	[12068-35-8]	tetracalcium aluminoferrite	C_4AF
	[12013-62-6]	dicalcium ferrite[c]	C_2F
mayenite	*[12005-57-1]*	12-calcium-7-aluminate	$C_{12}A_7$
gehlenite	*[1302-56-3]*	dicalcium alumino monosilicate	C_2AS
aphthitalite	*[12274-74-4]* and *[17926-93-1]*	sodium, potassium sulfate[d]	$N_xK_y\overline{S}$
arcanite	*[7778-80-5]* and *[14293-72-2]*	potassium sulfate	$K\overline{S}$
metathenardite	*[7757-82-6]*	sodium sulfate form I	$N\overline{S}$
calcium langbeinite	*[14977-32-8]*	potassium calcium sulfate	$2X\overline{S}.K\overline{S}$
anhydrite	*[7778-18-9]* and *[14798-04-0]*	calcium sulfate	$C\overline{S}$
calcium sulfoaluminate	*[12005-25-3]*	tetracalcium trialuminatesulfate	$C_4A_3\overline{S}$
alkali belite	*[15669-83-7]*	α'- or β-dicalcium (potassium) silicate[e]	$K_xC_{23}S_{12}$
alkali aluminate	*[12004-54-3]*	8-calcium disodium trialuminate	NC_8A_3
	[65430-58-2]	5-calcium disilicate monosulfate	$2C_2S.C\overline{S}$
spurrite	*[1319-44-42]*	5-calcium disilicate monocarbonate	$2C_2S.C\overline{C}$
	[12043-73-1]	calcium aluminate chloride	$C_{11}A_7.CaCl_2$
	[12305-57-6]	calcium aluminate fluoride	$C_{11}A_7.CaF_2$[f]

[a] Refs. 10–12.
[b] Solid solution series ($x = A/(A + F); 0 < x < 0.7$).
[c] End member of series.
[d] Solid solution series ($1/3 \leq x/y$).
[e] Solid solution series ($x \leq 1$).
[f] Mixed notation.

euhedral crystalline grains, the belite as rounded striated grains, the C_3A as dark interstitial material, and the C_4AF as light interstitial material.

Portland cement clinker structures (17–18) vary considerably with composition, particle size of raw materials, and burning conditions, resulting in variations of clinker porosity, crystallite sizes and forms, and aggregations of crystallites. Alite sizes range up to about 80 μm or even larger, most being 15–40 μm.

Raw Material Proportions. The three main considerations in proportioning raw materials for cement clinker are: the potential compound composition, the percentage of liquid phase at clinkering temperatures, and the burnability of the raw mix, ie, the relative ease, in terms of temperature, time, and fuel requirements, of combining the oxides into good quality clinker. The ratios of the oxides are related to clinker composition and burnability. For example, as the CaO content of the mix is increased, more C_3S can be formed, but certain limits cannot be exceeded under normal burning conditions. The lime saturation factor (LSF) is a measure of the amount of CaO that can be combined (19):

$$LSF = \frac{\%\ CaO}{2.8\ (\%\ SiO_2) + 1.1\ (\%\ Al_2O_3) + 0.7\ (\%\ Fe_2O_3)}$$

Table 2. Properties of the More Common Phases in Portland Cement Clinker[a]

Name	Crystal system	Density	Mohs' hardness
alite	triclinic		
	monoclinic	3.14–3.25	ca 4
	trigonal		
belite	hexagonal	3.04	
	orthorhombic	3.40	
	monoclinic	3.28	>4
	orthorhombic	2.97	
C_3A	cubic	3.04	<6
ferrite	orthorhombic	3.74–3.77	ca 5
free lime	cubic	3.08–3.32	3–4
magnesia	cubic	3.58	5.5–6

[a] Refs. 10–15.

An LSF of 100 would indicate that the clinker can contain only C_3S and the ferrite solid solution. Lime saturation factors of 88–94 are frequently appropriate for reasonable burnability; low LSF indicates insufficient C_3S for acceptable early strengths, and higher values may render the mix very difficult to burn. Several other weight ratios such as the silica modulus and the iron modulus are also important (20).

The potential liquid-phase content at clinkering temperatures range from 18 to 25% and can be estimated from the oxide analysis of the raw mix. For example (21), for 1450°C:

$$\%\ \text{liquid phase} = 1.13\ (\%\ C_3A) + 1.35\ (\%\ C_4AF) + \%\ MgO + \%\ \text{alkalies}$$

The potential compound composition of a cement or cement clinker can be calculated from the oxide analyses of any given raw materials mixture, or from the oxide analyses of the cement clinker or finished cement. The simplest and most widely used method is the Bogue calculation (22). The ASTM C150 (23) calculation is somewhat modified.

The techniques of determining the proper proportions of raw materials to achieve a mix of good burnability and clinker composition are readily adaptable to a computer program which uses iterative techniques, starting with raw components of known composition. The concept of targets may be utilized, including fixed values of moduli, compound content, and amount of any raw material element in the final clinker. The number of targets that may be set is one less than the number of raw materials. The fuel ash must be considered as one of the raw materials.

Representative chemical analyses of raw materials used in making portland and high alumina cements are given in Table 3, analyses of cements of various types appear in Table 4, and their potential compound compositions in Table 5.

Hydration

Calcium Silicates. Cements are hydrated at elevated temperatures for the commercial manufacture of concrete products. With low pressure steam curing or hydrothermal treatment above 100°C at pressures above atmospheric, the products formed from calcium silicates are often the same as the hydrates formed from their

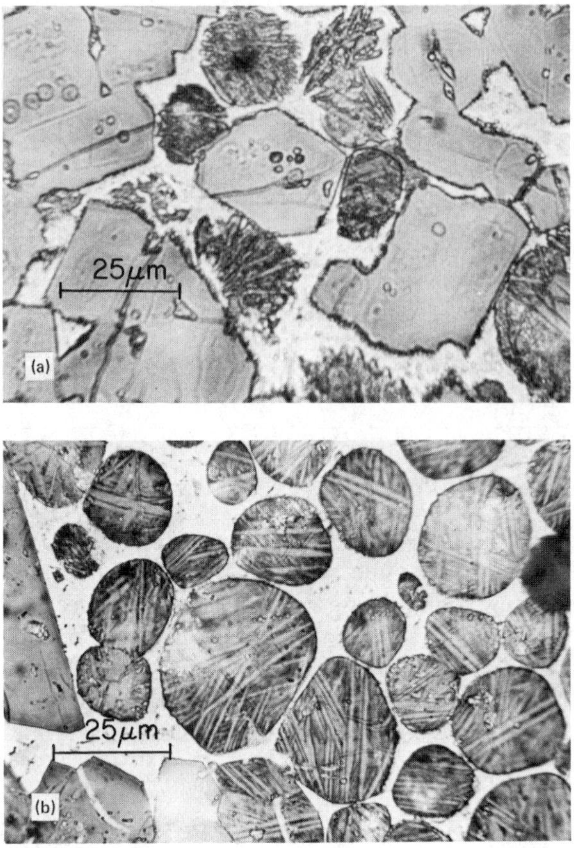

Figure 4. Photomicrograph of polished and etched sections of portland cement clinkers. The C_3A appears as dark interstitial material, the C_4AF as light interstitial material. (**a**) Euhedral and subhedral alite crystals and rounded or ragged belite; (**b**) rounded and striated belite crystals.

oxide constituents. Hence lime and silica are frequently used in various proportions with or without portland cement in the manufacture of calcium silicate hydrate products. Some of these compounds are listed in Table 6.

Although hydration under hydrothermal conditions may be rapid, metastable intermediate phases tend to form, and final equilibria may not be reached for months at 100–200°C, or weeks at even higher temperatures. Hence, the temperatures of formation given in Table 5 indicate the conditions under saturated steam pressure that may be expected to yield appreciable quantities of the compound, although it may not be the most stable phase at the given temperature. The compounds are listed in order of decreasing basicity, or lime/silica ratio. Reaction mixtures with ratios C:S = 1 yield xonotlite at 150–400°C. Intermediate phases of C-S-H (I), C-S-H (II), and crystalline tobermorite are formed in succession. Tobermorite (1.13 nm) appears to persist indefinitely under hydrothermal conditions at 110–140°C; it is a major part of the binder in many autoclaved cement-silica and lime-silica products.

In hydrations at ordinary temperatures (26) pure C_3S and β-C_2S (corresponding to the alite and belite phases in portland cements, respectively) react with water to form calcium hydroxide and a single calcium silicate hydrate (C-S-H), according to the following equations (in cement chemists' notation):

Table 3. Chemical Composition of Raw Materials[a], %

Type	SiO_2	Al_2O_3	Fe_2O_3	CaO	MgO	Ign. loss
cement rock	13.4	3.5	1.7	42.9	1.0	37.2
limestone	1.2	0.2	0.4	53.4	1.3	43.4
limestone	4.5	0.5	1.6	35.0	14.9	44.0
marl	6.0	0.6	2.3	49.1	0.4	40.4
oyster shells	1.5	0.4	1.2	52.3	0.7	41.8
shale	53.8	18.9	7.7	3.2	2.2	8.2
clay	67.8	14.3	4.5	0.9	1.2	8.0
mill scale			ca 100.0			
sandstone	76.6	5.3	3.1	4.7	1.7	6.6
bauxite	10.6	57.5	2.6			28.4

[a] Courtesy of the American Concrete Institute (24).

Table 4. Chemical Composition of Some Typical Cements, %

	SiO_2	Al_2O_3	Fe_2O_3	CaO	MgO	SO_3	Loss	Insoluble residue
Type I	20.9	5.2	2.3	64.0	2.8	2.9	1.0	0.2
Type II	21.7	4.7	3.6	63.6	2.9	2.4	0.8	0.4
Type III	21.3	5.1	2.3	64.9	3.0	3.1	0.8	0.2
Type IV	24.3	4.3	4.1	62.3	1.8	1.9	0.9	0.2
Type V	25.0	3.4	2.8	64.4	1.9	1.6	0.9	0.2
white	24.5	5.9	0.6	65.0	1.1	1.8	0.9	0.2
alumina	5.3	39.8	14.6	33.5	1.3	0.4	0	4.8

Table 5. Potential Compound Composition of Some Typical Cements[a], %

	C_3S	C_2S	C_3A	C_4AF
Type I	55	19	10	7
Type II	51	24	6	11
Type III	56	19	10	7
Type IV	28	49	4	12
Type V	38	43	4	9
white	33	46	14	2

[a] Calculated by the American Society for Testing and Materials C150-76 (23).

$$2\,C_3S + 6\,H \rightarrow C_3S_2H_3 + 3\,CH$$

$$2\,C_2S + 4\,H \rightarrow C_3S_2H_3 + CH$$

These are the main reactions in portland cements since the two calcium silicates constitute about 75% of the cement. The average lime/silica ratio (C:S) may vary from about 1.4 to about 1.7 or even higher, the average value being about 1.5. The water content varies with the ambient humidity, the three moles of water being estimated from measurements in the dry state and structural considerations. As the lime/silica ratio of the C–S–H increases, the amount of water increases on an equimolar basis, ie, the lime goes into the structure as calcium hydroxide, resulting in less free calcium hydroxide.

Calcium silicate hydrate is not only variable in composition, but is very poorly

Table 6. Calcium Silicate Hydrates Formed at Elevated Temperatures[a]

Name	CAS Registry No	Composition[b]	Temperature of formation, °C	Density, kg/m³
tricalcium silicate hydrate	*[54596-90-6]*	$C_6S_2H_3$	150–500	2.61
calciochondrodite	*[12141-47-8]*	C_5S_2H	250–800	2.84
dicalcium silicate hydrates				
α (A)	*[15630-58-7]*	C_2SH	100–200	2.8
β (B) (hillebrandite)	*[18536-02-2]*	C_2SH	140–350	2.66
γ (C) (probably a mixture)	*[15669-77-9]*	$(C_5S_2H + C_3S_2H_x?)$	160–300	2.67
δ (D) (dellaite	*[54694-02-9]*	C_6S_3H	350–800	2.98
afwillite	*[16291-79-5]*	$C_3S_2H_3$	100–160[c]	2.63
foshagite	*[12173-33-0]* and *[62520-56-3]*	C_4S_3H	300–500	2.7
xonotlite	*[12141-77-4]*	C_5S_5H	150–400	2.7
C-S-H (II)	*[18662-40-3]*	$C_{1.5-2.0}SH_x$	<100	2.0–2.2
C-S-H (I)		$C_{0.8-1.5}SH_y$	<130	
1.4 nm tobermorite	*[1319-31-9]* and *[1344-96-3]*	$C_5S_6H_9$	60 (?)	2.2
1.13 nm tobermorite	*[12028-62-5]* and *[12323-54-5]*	$C_5S_6H_5$	110–140	2.44
0.93 nm tobermorite	*[51771-55-2]*	C_5S_6H	250–450	2.7
gyrolite	*[16225-87-9]* *[12141-71-8]* and *[60385-01-5]*	$C_2S_3H_2$	120–200	2.39
truscottite	*[12425-42-2]*	$C_6S_{10}H_3$	200–300	2.36–2.48

[a] Refs. 10–11, and 25.
[b] In cement chemists' notation.
[c] Afwillite can also be formed, and appears to be the thermodynamically stable calcium silicate hydrate in pure systems, at room temperature.

crystallized, and is generally referred to as calcium silicate hydrate gel (or tobermorite gel) because of the colloidal sizes (<0.1 μm) of the gel particles. The calcium silicate hydrates are layer minerals with many similarities to the limited swelling clay minerals found in nature. The layers are bonded together by excess lime and interlayer water to form individual gel particles only 2–3 layers thick. Surface forces, and excess lime on the particle surfaces, tend to bond these particles together into aggregations or stacks of the individual particles to form the porous gel structure.

Significant changes in the structure of the gel continue over very long periods. During the first month of hydration appreciable quantities of the dimer Si_2O_7 are formed, which are reduced by later condensation to higher polysilicates. The amount of the polysilicates and the mean length of the metasilicate chains continues to increase for at least 15 years of moist curing. In one study a mean length of 15.8 silica tetrahedra was found after such prolonged curing (27). These changes appear to have a positive effect on both strength development and reduction of drying shrinkage.

Drying (and other chemical processes) can have significant effects on this structure, there being loss of hydrate water (as well as physically adsorbed water) and collapse of the structure to form more stable aggregations of particles (28–29).

Tricalcium Aluminate and Ferrite. The hydration of the C_3A alone and in the presence of gypsum usually produce well-crystallized reaction products that can be identified by x-ray diffraction and other methods. C_3AH_6 is the cubic calcium aluminate hydrate, C_4AH_{19} and $C_3AC\overline{S}H_{12}$ are hexagonal phases, the latter being commonly referred to as the monosulfate. The highly hydrated trisulfate, ettringite, occurs as needles, rods, or dense columnar aggregations. Its formation on the surfaces of anhydrous grains is responsible for the necessary retardation of hydration of the aluminates in portland cements and the expansion process in expansive cements (30).

The early calcium aluminate hydration reactions in portland cements have been studied in simple mixtures of C_3A, gypsum, calcium hydroxide, and water as shown in Figure 5 (31). The figure shows the progressive reaction of the gypsum, water, and C_3A as ettringite is formed, and then the reaction of the ettringite, calcium hydroxide, and water to form the monosulfate and the solid solution of the monosulfate with C_4AH_{19}. These reactions are important in the portland cements to control the hydration of the C_3A, which otherwise might hydrate so rapidly as to cause flash set, or premature stiffening in fresh concrete.

Other reactions taking place throughout the hardening period are substitution and addition reactions (28). Ferrite and sulfoferrite analogues of calcium monosulfoaluminate and ettringite form solid solutions in which iron oxide substitutes continuously for the alumina. Reactions with the silicate hydrate result in the formation of additional substituted C–S–H gel at the expense of the crystalline aluminate, sulfate, and ferrite hydrate phases

The hydration of the ferrite phase (C_4AF) is of greatest interest in mixtures containing lime and other cement compounds because of the strong tendency to form solid solutions. When the sulfate in solution is very low, solid solutions are formed between the cubic C_3AH_6 and an analogous iron hydrate C_3FH_6. In the presence of water and silica, solid solutions such as $C_3ASH_4.C_3FSH_4$ may be formed (32). Table 7 lists some of the important phases formed in the hydration of mixtures of pure compounds.

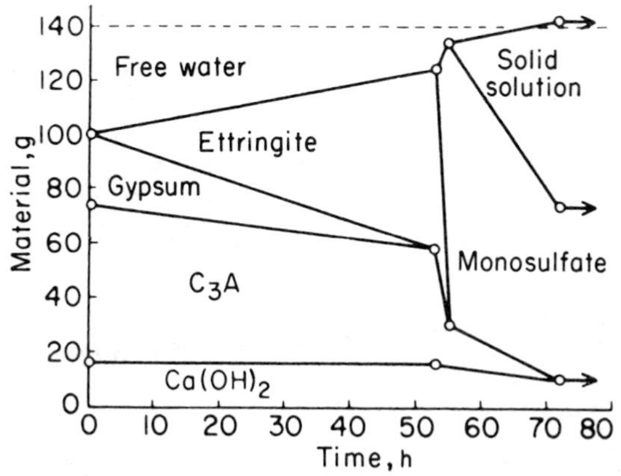

Figure 5. The early hydration reactions of tricalcium aluminate in the presence of gypsum and calcium hydroxide. Initial molar proportions: 1-C_3A; 1-$Ca(OH)_2$; 3/4-$CaSO_4.2H_2O$; 0.4 water–solids ratio (31).

Table 7. Cement Phases Hydrated at Normal Temperatures[a]

Name	CAS Registry No.	Approximate composition[b]	Stability range RH (at 25°C)	Stability range Temp, °C	Crystal system	Density, kg/m^3
calcium sulfate dihydrate (gypsum)	*[10101-41-4]* and *[13397-24-5]*	$C\overline{S}H_2$	100–35	<100	monoclinic	2.32
calcium hydroxide (portlandite)	*[1305-62-0]*	CH	100–0	<512	trigonal–hexagonal	2.24
magnesium hydroxide (brucite)	*[1309-42-8]*	MH	100–0	<350	trigonal–hexagonal	2.37
calcium silicate hydrate gel (C–S–H gel)	*[12323-54-5]*	$C_xS_yH_z$ $1.3 < \frac{x}{y} < 2$; $1 < \frac{z}{y} < 1.5$ (?)	indefinite		indefinite	2.7[c]
tetracalcium aluminate						
19-hydrate	*[12042-86-3]*	C_4AH_{19}	100–85	<15	trigonal–hexagonal	1.80
13-hydrate	*[12042-85-2]*	C_4AH_{13}	81–12		trigonal–hexagonal	2.02
7-hydrate	*[12511-52-3]*	C_4AH_7	2–0	to 120		
tetracalcium aluminate monosulfate						
16-hydrate	*[67523-83-5]*	$C_4A\overline{S}H_{16}$	aq	< 8	trigonal–hexagonal	
14-hydrate	*[12421-30-6]*	$C_4A\overline{S}H_{14}$	100–95	> 9	trigonal–hexagonal	
12-hydrate	*[12252-10-7]*	$C_4A\overline{S}H_{12}$	95–12	> 1	trigonal–hexagonal	1.95
10, 8, x-hydrate	*[12252-09-4]* and *[12445-38-4]*	$C_4A\overline{S}H_x$	< 12			
ettringite (6-calcium aluminate trisulfate, 32-hydrate)	*[12252-15-2]*	$C_6A\overline{S}_3H_{32}$	100–4	< 60	trigonal–hexagonal	1.73–1.79
	[11070-82-9]	$C_6A\overline{S}_3H_8$	4–2	< 110		
garnet-hydrogarnet solid solution series		$C_3(F_{1-x}A_x)\ (S_{1-y}H_{2y})_3$; $x = \frac{A}{A+F}$ and $y = \frac{2\,H}{2\,H+S}$	stable		cubic	
	[12042-80-7]	end member: C_3AH_6	100–0	> 15	cubic	2.52

[a] Refs. 10–11, and 33.
[b] In cement chemists notation.
[c] Wet.

Other Phases in Portland and Special Cements. In cements free lime (CaO) and periclase (MgO) hydrate to the hydroxides. The *in situ* reactions of larger particles of these phases can be rather slow and may not occur until the cement has hardened. These reactions then can cause deleterious expansions and even disruption of the concrete and the quantities of free CaO and MgO have to be limited. The soundness of the cement can be tested by the autoclave expansion test of portland cement ASTM C-151 (23).

The expansive component $C_4A_3\overline{S}$ in expansive cements of type K hydrates in the presence of excess sulfate and lime to form ettringite:

$$C_4A_3\overline{S} + 8\,C\overline{S} + 6\,C + 96\,H \rightarrow 3\,C_6A\overline{S}_3H_{32}$$

The reactions in the regulated-set cements containing $C_{11}A_7.CaF_2$ (mixed notation) as a major phase resemble those in ordinary portland cements. Initial reaction rates are controlled by ettringite formation. Setting occurs with formation of the monosulfate, along with some transitory lower-limed calcium aluminate hydrates that convert to the monosulfate within a few hours.

Pozzolans contain reactive silica which reacts with cement and water by combining with the calcium hydroxide released by the hydration of the calcium silicates to produce additional calcium silicate hydrate. If sufficient silica is added (about 30% of the weight of cement), the calcium hydroxide can be completely combined. Granulated blast-furnace slag is not ordinarily reactive in water, but in the presence of lime reactions occur with the silica framework. This breakdown of the slag releases other components so that a variety of crystalline hydrate phases can also form.

Steam-Curing of Portland Cements. The hydrated silicates formed by portland cements at 100°C are similar to those obtained from lime–silica mixtures or C_3S and C_2S, but with a higher C–S ratio of the C–S–H gel. Some crystalline $C_6S_2H_3$ and $C_5SH(B)$ may also be formed.

Steam curing for 6–12 h at 150–200°C forms C–S–H gel or tobermorite, $C_6S_2H_3$ and $C_2SH(A)$. The formation of the latter is unfavorable for strength development, and silica is often added for its prevention. Small additions result in more $C_2SH(A)$ and no $C_6S_2H_3$. If sufficient silica is added (30–40% of replacement of the cement) formation of 1.1 nm tobermorite is favored, giving optimum strengths. Hydrated calcium aluminates, sulfoaluminates, or hydrogarnets are not usually found (34).

Hydration at Ordinary Temperatures. Portland cement is generally used at temperatures ordinarily encountered in construction (5–40°C). Temperature extremes have to be avoided. The exothermic heat of the hydration reactions can play an important part in maintaining adequate temperatures in cold environments, and must be considered in massive concrete structures to prevent excessive temperature rise and cracking during subsequent cooling.

The initial conditions for the hydration reactions are determined by the concentration of the cement particles (0.2–100 μm) in the mixing water (0.3–0.7 on a weight basis) and the fineness of the cement (2500–5000 cm^2/g). Upon mixing with water, the suspension of particles as shown in Figure 6 (35) is such that the particles are surrounded by films of water with an average thickness of about 1 μm. The anhydrous phases initially react by the formation of surface hydration products on each grain, and by dissolution into the liquid phase. The solution quickly becomes saturated with calcium and sulfate ions, and the concentration of alkali cations increases rapidly. These reactions consume part of the anhydrous grains, but the reaction products tend to fill that space as well as some of the originally water-filled space. The porous gel

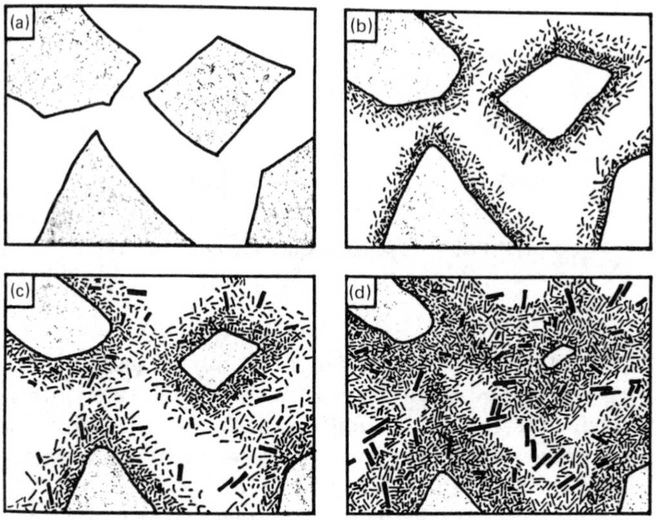

Figure 6. Four stages in the setting and hardening of portland cement: simplified representation of the sequence of changes. (**a**) Dispersion of unreacted clinker grains in water. (**b**) After a few minutes: hydration products eat into and grow out from the surface of each grain; (**c**) After a few hours: the coatings of different clinker grains have begun to join up, the gel thus becoming continuous (setting). (**d**) After a few days: further development of the gel has occurred (hardening). Courtesy of Academic Press Inc. (London) Ltd. (35).

in its most dense configurations occupies about twice the volume of the reacted anhydrous material (36). The hydration products at this stage are mostly colloidal (<0.1 μm) but some larger crystals of calcium aluminate hydrates, sulfoaluminate hydrate, and hydrogarnets form. As the reactions proceed the coatings increase in thickness and eventually form bridges between the original grains. This is the stage of setting. Despite the low solubility and mobility of the silicate anions, growths of the silicate hydrates also form on the crystalline phases formed from the solution and become incorporated into the calcium hydroxide and other phases. With further hydration the water-filled spaces become increasingly filled with reaction products to produce hardening and strength development.

The composition of the liquid phase during the early hydration of portland cements is controlled mainly by the solution of calcium, sulfate, sodium, and potassium ions. Very little alumina, silica, or iron are present in solution. Calcium hydroxide (as calcium oxide) and gypsum (as calcium sulfate) alone have solubilities of about 1.1 and 2.1 g/L at 25°C, respectively. In the presence of alkalies released by the readily soluble alkali sulfate phases in cements (as much as 70–80% may be released in the first 7 min), the composition tends to be governed by the equilibrium:

$$CaSO_4 + 2\,MOH \rightleftharpoons M_2SO_4 + Ca(OH)_2$$

where M represents the alkalies. At advanced stages of hydration of low water–cement ratio pastes, the alkali solution concentration may exceed 0.4 N with a pH above 13. Saturated lime-water has a pH of 12.4 at 25°C.

The exact course of the early hydration reactions depends mainly on the C_3A, ferrite, and soluble alkali contents of the clinker and the amount of gypsum in the cement. Following the rate of reaction by calorimetric measurements, at least two and sometimes three distinct peaks in the rate of heat liberation can be observed (32,37). A large initial peak lasts only a few minutes and may reach 4 J/(g·min) (1 cal/(g·min)) resulting mainly from the solution of soluble constituents and the surface reactions, especially the formation of the sulfoaluminate coating on the highly reactive C_3A phase. After the initial heat peak, the reactions are strongly retarded, producing a 1–2 h delay referred to as the dormant period, during which the cement–water paste remains plastic and the concrete is workable. The C_3A reaction continues slowly to form ettringite. The solution composition remains relatively constant except for a slow increase in supersaturation with respect to calcium hydroxide. Eventually the supersaturation produces nucleation of calcium hydroxide at numerous sites which decreases the calcium concentration in the solution and accelerates the rate of hydration of the alite. This produces the second heat peak which reaches a rate of about 16 J/(g·h) (4 cal/(g·h)), at about six hours of hydration, corresponding to final set. The third peak occurs at a time that depends upon the gypsum and C_3A contents and corresponds to the exhaustion of the solid gypsum, a rapid decrease of sulfate in solution, conversion of the ettringite to monosulfate, and renewed rapid reaction of the remaining C_3A and ferrite phases. At optimum gypsum the third peak ordinarily occurs between 18 and 24 hours with maximum strength and minimum shrinkage.

In these early reactions the reactivities of the individual phases are important in determining the overall reaction rate. However, as the cement particles become more densely coated with reaction products, diffusion of water and ions in solution becomes increasingly impeded. The reactions then become diffusion-controlled at some time depending on various factors such as temperature and water–cement ratio. After about 1 or 2 days (ca 40% of complete reaction) the remaining unhydrated cement phases react more nearly uniformly.

Microscopic examination of sections of hardened cement paste show that the unhydrated cores of the larger cement particles can be distinguished from the hydrated portion or inner product, which is a pseudomorph of the original grain, and the outer product formed in the originally water-filled spaces. Measurements of these cores indicate the depth of penetration of the hydration reactions (38). The overall hydration rate increases with the temperature, the fineness of the cement, and to a lesser extent with the water–cement ratio; measurements of the activation energy indicate that the reaction becomes increasingly diffusion controlled (32). Although more finely ground cements hydrate more rapidly in the first month or more of hydration, these differences gradually disappear at later ages. After one year most portland cements at usual water–cement ratios are more than 90% hydrated if continuously moist-cured. At complete hydration the chemically combined water (the water retained after strong drying) is about 20–25% of the weight of the cement, depending upon its composition. However, a minimum water–cement ratio of about 0.4 is required to provide enough space to permit complete hydration of the cement (36). If moist curing is stopped and the hardened cement is dried sufficiently, say to 80% rh, the hydration process stops.

Cement Paste Structure and Concrete Properties

The properties of both fresh and hardened mortars and concretes depend mainly on the cement–water paste properties. Practical engineering tests are usually made with concrete specimens since their properties also depend on the proportions, size gradation, and properties of the aggregates. Quality control testing and research on cement properties is usually done on cement pastes or cement mortars made with standard sands. The properties of hardened cement pastes, mortars, and concretes are similar functions of the water–cement ratio and degree of hydration of the cement. The properties of fresh concretes that determine the workability, or ease of mixing and placement into forms, also depend strongly on, but are not so simply related to, the cement paste rheological properties (39–40).

The fresh paste even in the dormant period is normally thixotropic, or shear thinning, indicating that the structure is being continuously broken down and reformed during mixing. It is an approximately Bingham plastic body with a finite yield value and plastic viscosity from 5000 to 500 mPa·s (= cP) as the water–cement ratio increases from 0.4 to 0.7 (41). The viscosity and yield values can be greatly reduced by the addition of certain organic water-reducing admixtures especially formulated for this purpose. Workability of concrete is measured by the slump of the concrete determined after removal of a standard slump cone (305 mm high) (42). Workable concretes have slumps of 75 mm or more.

After mixing and casting, sedimentation of the cement particles in the water results in bleeding of water to the top surface and reduction of water–cement ratio in the paste. At high water–cement ratios, some of the very fine particles may be carried with the bleed water to the top resulting in laitance and perhaps the formation of flaws called bleeding channels. In concretes, sedimentation may cause flaws under the larger aggregate particles. If the fresh concrete is not protected from too rapid surface drying, capillary forces cause drying shrinkage which may cause plastic shrinkage cracks. Good construction practices are designed to minimize all of these flaws.

The engineering properties of the concrete, such as strength, elastic moduli, permeability to water and aggresive solutions, and frost resistance, depend strongly on the water–cement ratio and degree of hydration of the cement. A variety of empirical water–cement ratio laws express the strength as functions of water–cement ratio or porosity. The fraction of the original water-filled space which is occupied by hydration products at any stage of hydration, is termed the gel-space ratio X. The compressive strength f_c of hardened cement or mortar then approximately fits the power law:

$$f_c = AX^n$$

where n is about 3.0 and A is the intrinsic strength of the densest ($X = 1$) gel produced by a given cement under normal hydrating conditions. Values of A range upward from ca 100 MPa (15,000 psi), depending on the cement composition (43). Under extreme conditions (hot pressing at very low water–cement ratios) strengths as high as 655 MPa (95,000 psi) have been reported (44). Tensile strengths and elastic moduli are similarly dependent on porosity or gel-space ratio, but the tensile strength is only about one tenth the compressive strength. The Young's modulus of the densest gels produced under normal hydrating conditions is ca 34 GPa (5×10^6 psi) (29).

Under sustained loads hardened cements and concrete creep, or deform continuously with time, in addition to the initial elastic deformation. Under normal working loads this deformation may in time exceed the elastic deformation and must be considered in engineering design. This is especially true in prestressed concrete structural members in which steel tendons under high tensile stress maintain compressive stress in the concrete to prevent tensile cracking during bending. Both creep and drying shrinkage of the concrete may lead to loss of prestress. Some creep in ordinary concrete structures and in the cement paste between the aggregate particles can also be an advantage because it tends to reduce stress concentrations, cracking, and microcracking around aggregate particles.

Drying of hardened cements results in shrinkage of the paste structure and of concrete members. Linear shrinkage of hardened cements is about 0.5% when dried to equilibrium at normal (ca 50%) relative humidities. The cement gel structure is somewhat stabilized during drying so that upon subsequent wetting and drying smaller changes occur. Concretes shrink much less (about 0.05%), depending on the volume fractions of cement paste and aggregates, water–cement ratio and other factors. Drying of concrete structural members proceeds very slowly and results in internal shrinkage stresses because of the moisture gradients during drying. Thick sections continue to dry and shrink for many years. Atmospheric carbon dioxide penetrates the partly dried concrete and reacts with the calcium silicate hydrate gel (as well as with calcium aluminate hydrates) which releases additional water and causes additional shrinkage. The density of the hydration products is increased, however, and the strength is actually increased. This reaction is sometimes used to advantage in the manufacture of precast concrete products to improve their ultimate strength and dimensional stability by precarbonation.

The slowness of drying and the penetration of the hardened cement by carbon dioxide or chemically aggresive solutions (eg, sea water or sulfate ground waters) is a result of the small sizes of the pores. Initially the pores in the fresh paste are the water-filled spaces (capillaries) between cement particles. As these spaces become subdivided by the formation of the hydration products, the originally continuous pore system becomes one of more discrete pores or capillary cavities separated from each other by gel formations in which the remaining pores are very much smaller. These gel pores are so small (ca 3 nm) that most of the water contained in them is strongly affected by the solid surface force fields. These force fields are responsible for a large increase in the viscosity of the water and a decrease in mobility of ionic species in solution. Hence, the permeability of the paste to both water and dissolved substances is greatly reduced as hydration proceeds. This in part accounts for the great durability of concrete, especially when water–cement ratios are kept low and adequate moist curing ensures a high degree of hydration. High water–cement ratios result in large numbers of the capillary spaces (0.1 μm and larger) interconnected through capillaries which are 10 nm or larger. These capillaries not only lower the strength, but also permit the easy penetration of aggresive solutions. Furthermore, these capillary spaces may become filled with water which freezes below 0°C, resulting in destructive expansions and deterioration of the concrete.

To ensure frost resistance, air-entraining admixtures or cements are used to produce a system of small spherical air voids. The amount is adjusted to entrain about 20 vol % of air in the cement paste distributed so that the mean half distance between voids (the void spacing factor) is 0.15–0.20 mm. Such air voids are large enough so that they do not readily fill with water by capillarity. Air entrainment thus ensures the durability of the concrete when exposed to wet and freezing conditions.

Manufacture

PORTLAND CEMENTS

The process of portland cement manufacture consists of (*1*) quarrying and crushing the rock, (*2*) grinding the carefully proportioned materials to high fineness, (*3*) subjecting the raw mix to pyroprocessing in a rotary kiln, and (*4*) grinding the resulting clinker to a fine powder. A layout of a typical plant is shown in Figure 7 (45), which also illustrates differences between wet process and dry process plants, and newer dry process plants shown in Figure 8 (46). The plants outlined are typical of installations producing approximately 1,000 metric tons per day. Modern installations (47–49) are equipped with innovations such as suspension or grate preheaters, roller mills, or precalciner installations.

Because calcium oxide comprises about 65% of portland cement, these plants are frequently situated near the source of their calcareous material. The requisite silica and alumina may be derived from a clay, shale, or overburden from a limestone quarry. Such materials usually contain some of the required iron oxide, but many plants need to supplement the iron with mill scale, pyrite cinders, or iron ore. Silica may be supplemented by adding sand to the raw mix, whereas alumina can be furnished by bauxites and Al_2O_3-rich flint clays.

Industrial by-products are becoming more widely used as raw materials for cement, eg, slags contain carbonate-free lime, as well as substantial levels of silica and alumina. Fly ash from utility boilers can often be a suitable feed component, since it is already finely dispersed and provides silica and alumina. Even vegetable wastes, such as rice hull ash, provide a source of silica. Probably 50% of all industrial by-products are potential raw materials for portland cement manufacture.

Clinker production requires large quantities of fuel. In the United States, coal and natural gas are the most widely used kiln fuels with coal increasing in importance (47). Residual oil furnishes fuel energy for about 9% of United States clinker production, and petroleum coke is also finding increasing application. It is estimated that by 1990 in the United States 90% of the clinker will be produced in pulverized coal-fired kilns. The feasibility of using supplemental refuse-derived fuel (RDF) together with conventional fuels is also being evaluated (see Fuels from waste).

In addition to the kiln fuel, electrical energy is required to power the equipment. This energy, however, amounts to only about one sixth that of the kiln fuel. The cement industry is carefully considering all measures that can reduce this heavy fuel demand.

Raw Materials Preparation. The bulk of the raw material originates in the plant quarry. Control of the clinker composition starts in the quarries with systematic core drillings and selective quarrying in order to utilize the deposits economically.

A primary jaw or roll crusher is frequently located within the quarry and reduces the quarried limestone or shale to about 100 mm top size. A secondary crusher, usually roll or hammer mills, gives a product of about 10–25 mm top size. Clays may require treatment in a wash mill to separate sand and other high silica material. Combination crusher-dryers utilize exit gases from the kiln or clinker cooler to dry wet material during crushing.

Argillaceous, siliceous, and ferriferous raw mix components are added to the crusher product. At the grinding mills, the constituents are fed into the mill separately, using weigh feeders or volumetric measurements. Ball mills are used for wet and dry processes to grind the material to a fineness such that only 15–30 wt % is retained on

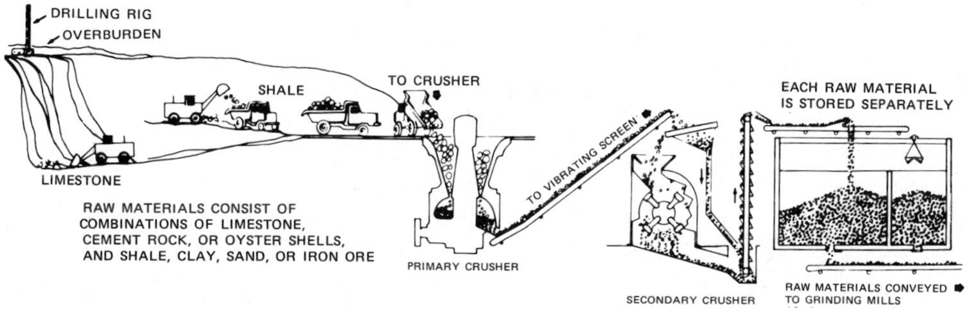

1. Stone is first reduced to 13–cm size, then to 2 cm and stored.

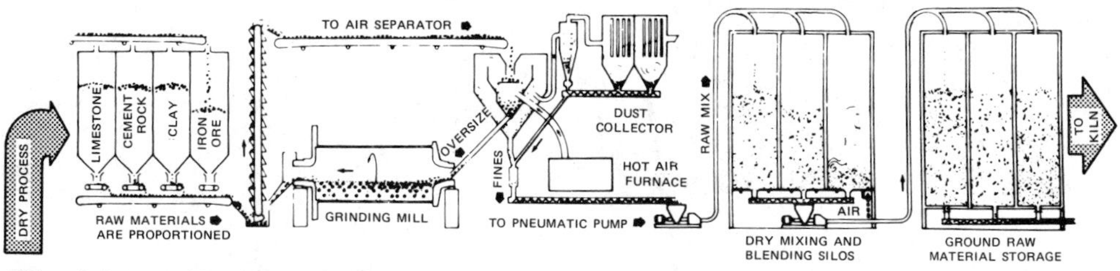

OR

2. Raw materials are ground to powder and blended.

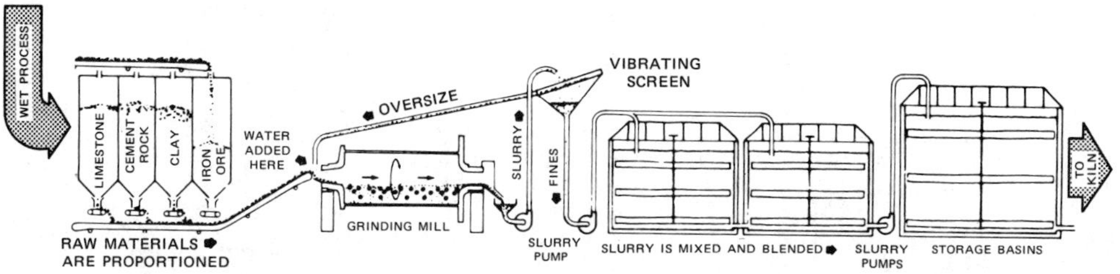

2. Raw materials are ground, mixed with water to form slurry, and blended.

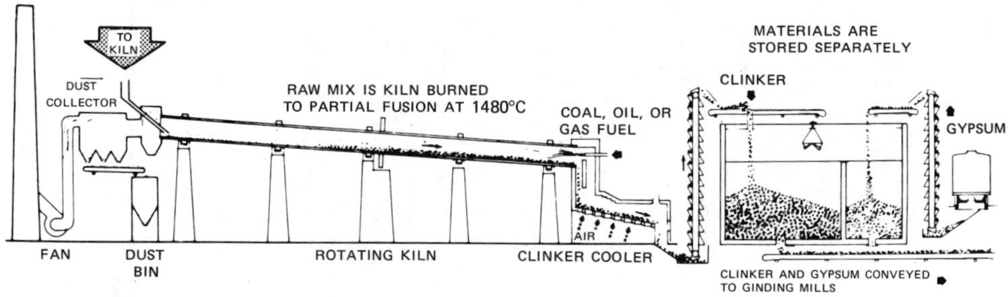

3. Burning changes raw mix chemically into cement clinker.

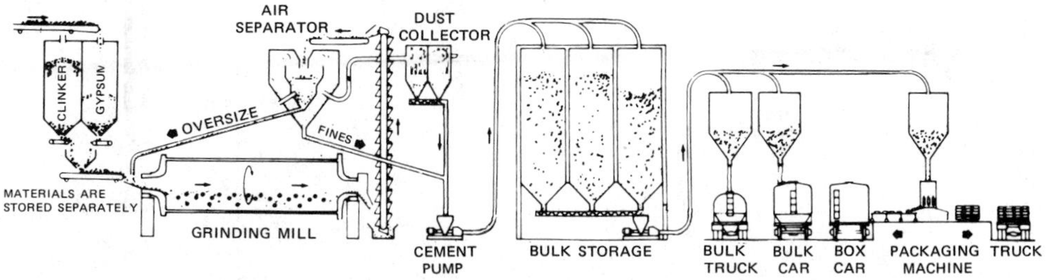

4. Clinker with gypsum is ground into portland cement and shipped.

Figure 7. Steps in the manufacture of portland cement. Courtesy of Portland Cement Association (45).

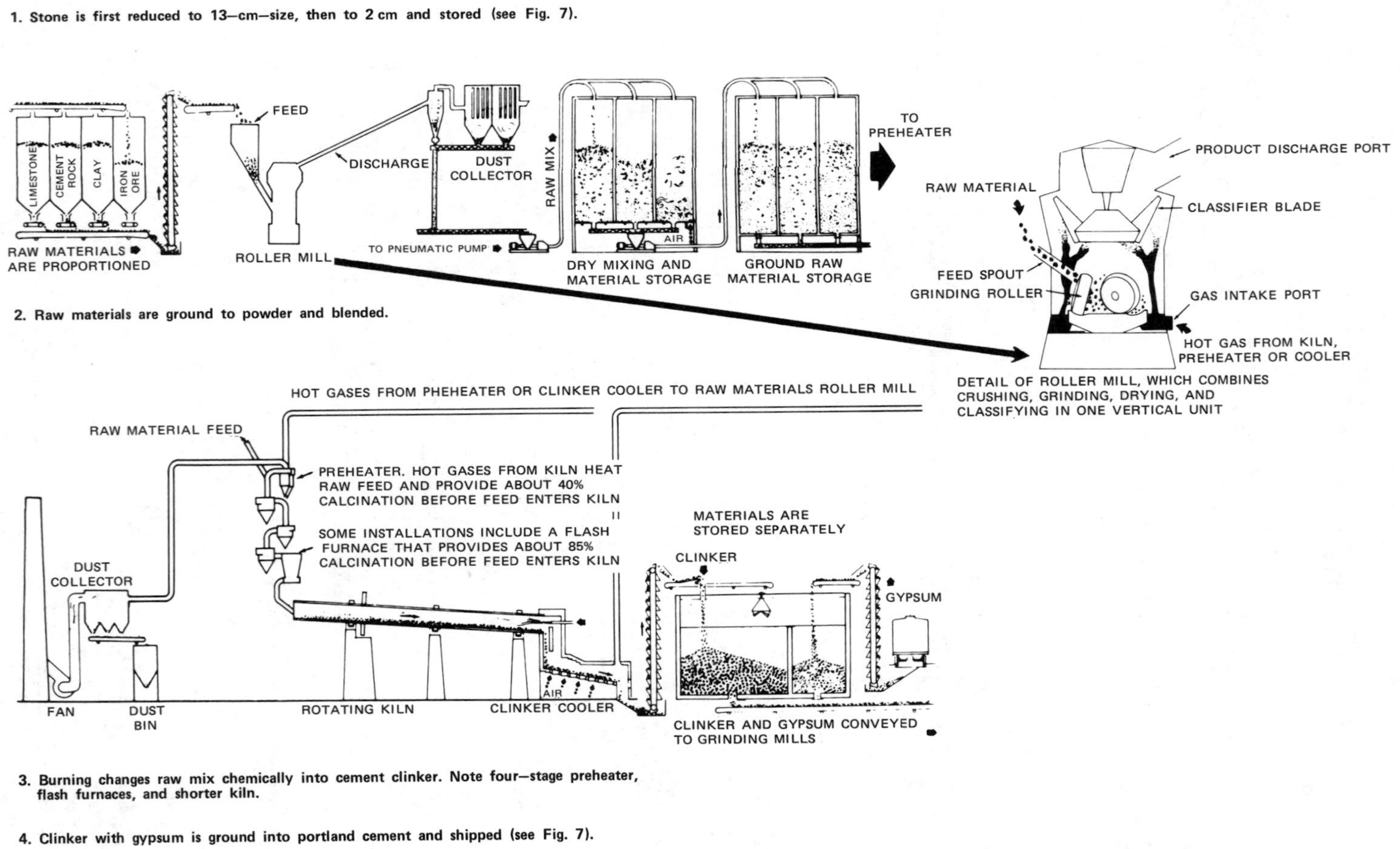

Figure 8. New technology in dry-process cement manufacture. Courtesy of Portland Cement Association (46).

a 74 μm (200 mesh) sieve. In the wet process the raw materials are ground with about 30–40% water, producing a well-homogenized mixture called slurry. Low concentrations of slurry thinners may be added, such as sodium carbonates, silicates, and phosphates, as well as lignosulfonates and modified petrochemicals. Filter presses or other devices remove water from slurries before feeding into the kiln.

Raw material for dry process plants is ground in closed-circuit ball mills with air separators, which may be set for any desired fineness. Drying is usually carried out in separate units, but waste heat can be utilized directly in the mill by coupling the raw mill to the kiln. Autogenous mills, which operate without grinding media are not widely used. For suspension preheater-type kilns, a roller mill utilizes the exit gas from the preheater to dry the material in suspension in the mill.

A blending system provides the kiln with a homogeneous raw feed. In the wet process the mill slurry is blended in a series of continuously agitated tanks in which the composition, usually the CaO content, is adjusted as required. These tanks may also serve as kiln feed tanks, or the slurry after agitation is pumped to large kiln feed basins. Dry-process blending is usually accomplished in a silo with compressed air.

Pyroprocessing. Nearly all cement clinker is produced in large rotary kiln systems. The rotary kiln is a highly refractory-lined cylindrical steel shell (3–8 m dia, 50–230 m long) equipped with an electrical drive to rotate at 1–3 rpm. It is a countercurrent heating device slightly inclined to the horizontal so that material fed into the upper end travels slowly by gravity to be discharged onto the clinker cooler at the discharge end. The burners at the firing end produce a current of hot gases that heats the clinker and the calcined and raw materials in succession as it passes upward toward the feed end (see under Clinker Chemistry). Highly refractory bricks of magnesia, alumina, or chrome–magnesite combinations line the firing end, whereas in the less heat-intensive midsection of the kiln bricks of lower refractoriness and thermal conductivity can be used, changing to abrasion-resistant bricks or monolithic castable lining at the feed end. To prevent excessive thermal stresses and chemical reaction of the kiln refractory lining, it is necessary to form a protective coating of clinker minerals on the hot face of the burning zone brick. This coating also reduces kiln shell heat losses by lowering the effective thermal conductivity of the lining.

It is desirable to cool the clinker rapidly as it leaves the burning zone. This is best achieved by using a short, intense flame as close to the discharge as possible. Heat recovery, preheating of combustion air, and fast clinker cooling are achieved by clinker coolers of the traveling-grate, planetary, rotary, or shaft type. Most commonly used are grate coolers where the clinker is conveyed along the grate and subjected to cooling by ambient air, which passes through the clinker bed in crosscurrent heat exchange. The air is moved by a series of undergrate fans, and becomes preheated to 370–800°C at the hot end of the cooler. It then serves as secondary combustion air in the kiln; the primary air is that portion of the combustion air needed to carry the fuel into the kiln and disperse the fuel.

During the burning process, the high temperatures cause vaporization of alkalies, sulfur, and halides. These materials are carried by the combustion gases into the cooler portions of the kiln system where they condense, or they may be carried out to the kiln dust collector (usually a fabric filter or electrostatic precipitator) together with partially calcined feed and unprocessed raw feed. This kiln dust is reusable. However, ASTM specifications limit the total SO_3 content of the finished cement to 2.3–4.5%, depending upon the cement type and C_3A content. Similarly, an optional ASTM C150 specification limits the total alkali content of the cement to 0.60%, expressed as equivalent

Na_2O. Other potential and actual uses of dust include fertilizer supplements, acid mine waste neutralization, boiler SO_2 control, and soil stabilization.

Wet-Process Kilns. In a long wet-process kiln, the slurry introduced into the feed end first undergoes simultaneous heating and drying. The refractory lining is alternately heated by the gases when exposed and cooled by the slurry when immersed; thus the lining serves to transfer heat, as do the gases themselves. Because large quantities of water (about 0.8 L/kg of clinker product) must be evaporated, most wet kilns are equipped with chains to maximize heat transfer from the gases to the slurry. Large, dense chain systems have permitted energy savings of up to 1.7 MJ/kg (1.6×10^6 Btu/t) clinker produced in exceptionally favorable situations (47). The chain system also serve to break up the slurry into nodules that can flow readily down the kiln without forming mud rings. After most of the moisture has been evaporated, the nodules, which still contain combined water, move down the kiln and are gradually heated to about 550°C where the reactions commence as discussed under Clinker Chemistry. As the charge leaves the burning zone it begins to cool, and tricalcium aluminate and magnesia crystallize from the melt and the liquid phase finally solidifies to produce the ferrite phase. The material drops into the clinker cooler for further cooling by air.

Dry-Process Kilns, Suspension Preheaters, and Precalciners. The dry process utilizes a dry kiln feed rather than a slurry. Early dry process kilns were short, and the substantial quantities of waste heat in the exit gases from such kilns were frequently used in boilers for electric power generation; the power generated was frequently sufficient for all electrical needs of the plant. In one modification, the kiln has been lengthened to nearly the extent of long wet-process kilns, and chains have been added; however, they serve almost exclusively a heat-exchange function. Refractory heat-recuperative devices, such as crosses, lifters, and trefoils, have also been installed. So equipped, the long dry kiln is capable of good energy efficiency. Other than the need for evaporation of water, its operation is similar to that of a long wet kiln.

The second major type of modern dry-process kiln is the suspension preheater system (50). The dry, pulverized feed passes through a series of cyclones where it is separated and preheated several times. The partially calcined feed exits the preheater tower into the kiln at about 800–900°C. The kiln length required for completion of the process is considerably shorter than that of conventional kilns, and heat exchange is very good. Suspension preheater kilns are very energy-efficient (as low as 3.1 MJ/kg or 1334 Btu/lb clinker in large installations).

The intimate mixing of the hot gases and feed in the preheaters promotes condensation of alkalies and sulfur on the feed which sometimes results in objectionably high alkali and sulfur contents in the clinker. To alleviate this problem, some of the kiln exit gases can be bypassed and fewer cyclone stages used in the preheater with some sacrifice of efficiency.

The success of preheater kiln systems, particularly in Europe and Japan where low alkali specifications do not exist, led to precalciner kiln systems. These units utilize a second burner to carry out calcination in a separate vessel attached to the preheater. The flash furnace (51), eg, utilizes preheated combustion air drawn from the clinker cooler and kiln exit gases and is equipped with an oil burner which burns about 60% of the total kiln fuel. The raw material is calcined almost 95%, and the gases continue their upward movement through successive preheater stages in the same manner as in an ordinary preheater.

The precalciner system permits the use of smaller kilns since only actual clinkering is carried out in the rotary kiln. Energy efficiency is comparable to that of a preheater kiln, except that the energy penalty for bypass of kiln exit gases is reduced since only about 40% of the fuel is being burned in the kiln. Precalciner kilns in operation in Japan produce up to 10,000 metric tons of clinker per day; the largest long wet-process kiln, in Clarksville, Missouri, produces only 3270 t/d by comparison.

The burning process and clinker cooling operations for the modern dry-process kiln systems are the same as for long wet kilns.

Finish Grinding. The cooled clinker is conveyed to clinker storage or mixed with 4–6% gypsum and introduced directly into the finish mills. These are large, steel cylinders (2–5 m in dia) containing a charge of steel balls, and rotating at about 15–20 rpm. The clinker and gypsum are ground to a fine, homogeneous powder with a surface area of about 3000–5000 cm^2/g. About 85–96% of the product is in particles less than 44 μm dia. These objectives may be accomplished by two different mill systems. In *open-circuit milling,* the material passes directly through the mill without any separation. A wide particle size distribution range is usually obtained with substantial amounts of very fine and rather coarse particles. In *closed-circuit grinding* the mill product is carried to a cyclonic air separator in which the coarse particles are rejected from the product and returned to the mill for further grinding.

Energy requirements for finish grinding vary from about 33–77 kW·h/t cement, depending also on the nature of the clinker.

Computer Control. Process computer control was introduced to the cement industry in the 1960s and met with varying degrees of success because of complexity of the equipment and control problems. The rotary kiln is the largest and most difficult equipment to operate. Temperature-sensing and gas-analyzing devices present problems. Unless temperature and combustion can be accurately measured, the computer cannot perform its control function. However, progress has been made and a plant of a capacity of 1 million metric tons per year has been built and placed in operation in 1973 with complete computer DDC (direct digital control) process and segmental control (52). Other new plants have been built and some older plants computerized (53). Variables can be measured at intervals of 0.25 s and overall optimum response to operating problems is programmed, not always possible with manual operation.

Quality Control. Beginning at the quarry operation, quality of the end product is maintained by adjustments of composition, burning conditions, and finish grinding. Control checks are made for fineness of materials, chemical composition, and uniformity. Clinker burning is monitored by weighing a portion of sized clinker, known as the liter weight test, a free lime test, or checked by microscopic evaluation of the crystalline structure of the clinker compounds. Samples may be analyzed by x-ray fluorescence, atomic absorption, and flame photometry (see Analytical methods). Wet chemistry is described in ASTM C114 (23). Standard cement samples are available from the National Bureau of Standards. Fineness of the cement is most commonly measured by the air permeability method. Finally, standardized performance tests are conducted on the finished cement (23).

Environmental Pollution Control. With the passage of the *Clean Air Act* and its amendments (54), the cement industry started an intensive program of capital expenditure to install dust collection equipment on kilns and coolers that were not already so equipped. Modern equipment collects dust at 99.8% efficiency. Many smaller dust collectors are installed in new plants, eg, a wet process plant of 430,000 t/yr capacity has 73 collectors connected to points of possible dust emission (55).

The Federal Water Pollution Control Act Amendments of 1972 (56) established limits for cement plant effluents. This includes water run-off from manufacturing facilities, quarrying, raw material storage piles, and waste water. Compliance with these standards has required construction of diversion ditches for surface water, ponds for settling and clarification, dikes and containment structures for possible oil spills, and chemical water treatment in some cases. Since the cement industry obtains most of its raw material by quarrying, the standards for the mineral industry also apply.

SPECIAL PURPOSE AND BLENDED CEMENTS

Special purpose and blended portland cements are manufactured essentially by the same processes as ordinary portland cements but have specific compositional and process differences as noted below.

White cements are made from raw materials of very low iron content. This type is often difficult to burn because almost the entire liquid phase must be furnished by calcium aluminates. As a consequence of the generally lower total liquid-phase content, high burning-zone temperatures may be necessary. Fast cooling and occasionally oil sprays are needed to maintain both quality and color.

Regulated set cements are made with fluorite (CaF_2) additions which also act as fluxing agents, or mineralizers, to reduce burning temperatures. The clinker produced then contains $C_{11}A_7.CaF_2$ (mixed notation) as a major phase.

Expansive cements manufactured in the United States usually depend upon aluminate and sulfate phases that result in more ettringite formation during hydration than in normal portland cements (see under Hydration Chemistry). This can be achieved by three types designated as Type K, Type S, and Type M (57). Type K contains an anhydrous calcium sulfoaluminate, $C_4A_3\overline{S}$, Type S contains a high C_3A content with additional calcium sulfate, and Type M is a mixture of portland cement, calcium aluminate cement, and calcium sulfate. Except for the Type M expansive cement, any of these cements can be made either by integrally burning to produce the desired phase composition, or by intergrinding a special component with ordinary portland cement clinkers and calcium sulfate. Type M can be made by mixing the finished cements in proper proportions, or by intergrinding the clinkers.

Oil well cements are manufactured similarly to ordinary portland cements except that the goal is usually sluggish reactivity. For this reason, levels of C_3A and alkali sulfates are kept low. Hydration-retarding additives are also employed.

The manufacture of blended cements is similar to that of portland cements except for the finish grinding process where the cement clinker is interground with pozzolans, granulated blast-furnace slag, or, in the case of masonry cements, a variety of materials.

Pozzolans include natural materials such as diatomaceous earths, opaline cherts, and shales, tuffs, and volcanic ashes or pumicites, and calcined materials such as some clays and shales. By-products such as fly-ashes and precipitated silica are also employed. In the United States the proportion of pozzolan interground with clinker has varied from 15 to over 30%, whereas in Italy, cements with a 30–40% pozzolan content are produced.

In some European countries portand cement clinker is interground with 10–65%

granulated blast-furnace slag to produce a portland blast-furnace slag cement. The composition of the slag varies considerably but usually falls within the following composition ranges:

CaO	40–50%	MgO	0–8%
SiO_2	30–40%	S (sulfide)	0–2%
Al_2O_3	8–18%	FeO, MnO	0–3%

Most masonry cements are finely interground mixtures with portland cement a major constituent, but also including finely ground limestones, hydrated lime, natural cement, pozzolans, clays, or air-entraining agents. These secondary materials are used to impart the required water retention and plasticity to mortars.

NONPORTLAND CEMENTS

Calcium Aluminate Cements. These cements are manufactured by heating until molten or by sintering a mixture of limestone and bauxite with small amounts of SiO_2, FeO, and TiO_2. In Europe the process is usually carried out in an open-hearth furnace having a long vertical stack into which the mixture of raw materials is charged. The hot gases produced by a blast of pulverized coal and air pass through the charge and carry off the water and carbon dioxide. Fusion occurs when the charge drops from the vertical stack onto the hearth at about 1425–1500°C. The molten liquid runs out continuously into steel plans on an endless belt in which the melt solidifies. Special rotary kilns provided with a tap hole from which the molten liquid is drawn intermittently and electric arc furnaces have also been used.

In a new process called shock sintering (58), finely ground raw materials are pelletized on a disk pelletizer and dried in a drier–preheater. The pellets are heated very rapidly to maximum reaction temperature in the sintering section of a specially designed rotary kiln, and are then ground into cement.

When calcium aluminate cements are made by the fusion process, the solidified melt must be crushed and then ground. The material is very hard to grind and power consumption is high.

Supersulfated Cement. Supersulfated cement contains about 80% slag interground with 15% gypsum or anhydrite and 5% portland cement clinker.

Hydraulic Limes. These materials are produced by heating below sintering temperature a limestone containing considerable clay, during which some combination takes place between the lime and the oxides of the clay to form hydraulic compounds.

Economic Aspects, Production, and Shipment

From the beginning of the United States portland cement industry in 1872, cement consumption grew at an average annual rate of 20% until 1920. As the cement markets matured, the industry grew at an average annual rate of 3% from 1920 to 1975. Annual production and sales tonnages are nearly identical; Table 8 gives United States production figures since 1910, Table 9 the world production.

Since World War II, the cement industry reduced labor and energy costs by increased investment in capital equipment and larger plants to remain competitive with other building materials industries. The average plant size increased more than 65% between 1950 and 1975.

Table 8. United States Portland Cement Production[a]

Year	Production, 1000 metric tons	Number of plants	Average capacity per plant, 1000 metric tons
1910	13,056	111	150
1920	17,059	117	213
1930	27,493	163	281
1940	22,209	152	286
1950	38,550	150	304
1960	54,408	176	417
1970	66,378	181	451
1975	60,597	164	511

[a] Refs. 59–60.

Table 9. World Portland Cement Production[a]**, Million Metric Tons**

	1950	1960	1970	1975
Europe	68	168	322	383
France	7	14	29	31
Germany	11	26	37	33
Italy	5	16	33	35
USSR	10	46	95	122
Spain	2	5	16	24
United Kingdom	10	14	18	18
Africa	4	9	18	24
Western Hemisphere	48	77	111	122
United States	38	53	65	59
Exports	0.413	0.032	0.144	0.331
Imports	0.238	0.699	2.356	3.299
Canada	3	5	7	10
Asia	11	57	121	167
Japan	4	22	57	65
Oceania	2	3	6	6
Total	*133*	*314*	*578*	*702*

[a] Refs. 61–62.

The wet process was used in 60% of the plants in the 1960s because it is less labor intensive than the dry process. However, as energy costs escalated in the early 1970s, dry-process manufacturing was preferred because it is generally more energy efficient.

Energy Usage. In the past 25 years, the cement industry has reduced its unit energy usage by 25.2%, from 9.6 MJ/kg (4131 Btu/lb) in 1950 to 7.2 MJ/kg (3098 Btu/lb) in 1975. In the 1950s and 1960s, the industry had shifted from coal to inexpensive, abundant, and clean natural gas, but by 1975 nearly 80% of the industry's capacity had been converted to permit use of coal as the primary kiln fuel; actual coal usage in 1975 was about 50% of all kiln fuels (see Table 10).

Marketing Patterns. Since 1950 the cement industry has reduced its dependence on bag (container) shipments (54.7% in 1950) and turned to the more labor-efficient bulk transport (92.0% in 1975). In addition, the amount of cement shipped by rail transportation declined from 75% of industry shipments in 1950 to less than 13% in 1975. Table 11 gives the shipment distribution by type.

Table 10. United States Portland Cement Industry Energy Consumption

Year	Coal, 1000 metric tons	Oil, 1000 metric tons	Natural gas, million m³	Power, million kW·h	Energy usage, MJ/t[a]
1950	7206	764	2747	2877	9627
1955	7918	1235	3710	4022	8979
1960	7591	586	4870	5589	8548
1965	8288	649	5635	7485	8222
1970	7227	1455	6003	8717	8008
1975	6866	1065	4531	9315	7201

[a] To convert MJ/t to Btu/lb, divide by 2.32.

Table 11. Portland Cement Shipments by Type[a,b], 1975

Type	Shipments, 1000 metric tons	Average value per metric ton, $
Types I and II	56,987	33.76
Type III	1,911	36.14
oil well	1,016	36.66
white	331	82.51
Type V	314	38.97
portland slag, pozzolan	286	33.54
expansive	83	46.19
miscellaneous	560	41.95
Total	*61,488*	*34.27*

[a] Ref. 59.
[b] See Tables 4–5.

In the past 25 years, the ready-mixed concrete industry became the primary customer for cement manufacturers. In 1975 more than 63% of the cement shipped was sold to the ready-mixed concrete industry, compared with 56% in 1960. The other major uses are in building materials, concrete products, and highway construction.

Specifications and Types

Portland cements are manufactured to comply with the specifications established in each country (63). In the United States, several different specifications are used, including those of the American Society for Testing and Materials, American Association of State Highway and Transportation Officials, and various government agencies. The ASTM annually publishes test methods and standards (23) which are established on a consensus basis by its members, including consumers and producers.

In the United States, portland cement is classified in five general types designated by ASTM Specification C150-76 (23) as follows: Type I, when the special properties are not required; Type II, for general use, and especially when moderate sulfate resistance or moderate heat of hydration is desired; Type III, for high early strength. Type IV, for low heat of hydration, and Type V, for high sulfate resistance. Types I, II, and III may also be specified as air-entraining. Chemical compositional, physical, and performance test requirements are specified for each type; optional requirements for particular uses may also be specified.

Other countries have similar types; some, as in Germany and the Union of Soviet Socialist Republics, are based on age-strength levels by standard tests (63). A product made in Italy and France known as Ferrari cement is similar to Type V and is sulfate resistant. Such cements have high iron oxide and low alumina contents, and harden more slowly.

Uses

Hydraulic cements are intermediate products that are used for making concretes, mortars, grouts, asbestos–cement products, and other composite materials. High early strength cements may be required for precast concrete products or in high-rise building frames to permit rapid removal of forms and early load carrying capacity. Cements of low heat of hydration may be required for use in massive structures, such as gravity dams, to prevent excessive temperature rise and thermal contraction and cracking during subsequent cooling. Concretes exposed to seawater, sulfate-containing ground waters, or sewage require cements that are sulfate resistant after hardening.

Air-entraining cements produce concretes that contain a system of closely spaced spherical voids that protect the concrete from frost damage. They are commonly used for concrete pavements subjected to wet and freezing conditions. Cement of low alkali content may be used with certain concrete aggregates containing reactive silica to prevent deleterious expansions.

Expansive, or shrinkage-compensating cements cause slight expansion of the concrete during hardening. The expansion has to be elastically restrained so that compressive stress develops in the concrete (57,64). Subsequent drying and shrinkage reduces the compressive stresses but does not result in tensile stresses large enough to cause cracks.

Regulated-set cement (or jet cement in Japan) is formulated to yield a controlled short setting time (1 h or less) and very early strength (65).

Natural cements (66) may be regarded as intermediate between portland cements and hydraulic limes (see below) in hydraulic activity.

Blended cements. Portland cement clinker is also interground with suitable other materials such as granulated blast furnace slags and natural or artificial pozzolans (see above). These substances also show hydraulic activity when used with cements, and the blended cements (67) bear special designations such as portland blast-furnace slag cement or portland-pozzolan cement. Pozzolans are used in making concrete both as an interground component of the cement and as a direct addition to the concrete mix. It is only when the two materials are interground that the mixture can be referred to as portland–pozzolan cement. Portland-pozzolan cements (68) were developed originally to provide concretes of improved economy and durability in marine, hydraulic, and underground environments; they also prevent deleterious alkali–aggregate reactions. Blast-furnace slag cements (69) also reduce deleterious alkali–aggregate reactions and can be resistant to seawater if the slag and cement compositions are suitably restricted. Both cements hydrate and harden more slowly than portland cement. This can be an advantage in mass concrete structures where the lower rates of heat liberation may prevent excessive temperature rise, but when used at low temperatures the rate of hardening may be excessively slow. Portland blast-furnace slag cements may be used to advantage in steam-cured products which can have strengths as high as obtained with portland cement. Current interest in the use of blended cements is stimulated by energy conservation and solid waste utilization considerations.

Oil well cements (70) are usually made from portland cement clinker and may also be blended cements. They are specially produced for cementing the steel casing of gas and oil wells to the walls of the bore-hole and to seal porous formations (71). Under these high temperature and pressure conditions ordinary portland cements would not flow properly and would set prematurely. Oil well cements are more coarsely ground than normal, and contain special retarding admixtures.

Masonry cements (72) are cements for use in mortars for masonry construction. They are formulated to yield easily workable mortars and contain special additives that reduce the loss of water from the mortar to the porous masonry units.

Calcium aluminate cement (73) develops very high strengths at early ages. It attains nearly its maximum strength in one day, which is much higher than the strength developed by portland cement in that time. At higher temperatures, however, the strength drops off rapidly. Heat is also evolved rapidly on hydration and results in high temperatures; long exposures under moist warm conditions can lead to failure. Resistance to corrosion in sea or sulfate waters, as well as to weak solutions of mineral acids, is outstanding. This cement is attacked rapidly, however, by alkali carbonates. An important use of high alumina cement is in refractory concrete for withstanding temperatures up to 1500°C. White calcium aluminate cements, with a fused aggregate of pure alumina, withstand temperatures up to 1800°C.

Supersulfated cement (74) has a very low heat of hydration and low drying shrinkage. It has been used in Europe for mass concrete construction and especially for structures exposed to sulfate and seawaters.

Trief cements (75), manufactured in Belgium, are produced as a wet slurry of finely ground slag. When activators (such as portland cement, lime, or sodium hydroxide) are added in a concrete mixer, the slurry sets and hardens to produce concretes with good strength and durability.

Hydraulic limes (76) may be used for mortar, stucco, or the scratch coat for plaster. They harden slowly under water, whereas high calcium limes, after slaking with water, harden in air to form the carbonate but not under water at ordinary temperatures. However, at elevated temperatures achieved with steam curing, lime–silica sand mixtures do react to produce durable products such as sand–lime bricks.

Specialty cements. For special architectural applications, white portland cement with a very low iron oxide content can be produced. Colored cements are usually prepared by intergrinding 5–10% of pigment with white cement.

Numerous other specialty cements composed of various magnesium, barium, and strontium compounds as silicates, aluminates, and phosphates, as well as others, are also produced (77).

BIBLIOGRAPHY

"Cement, Structural" in *ECT* 1st ed., Vol. 3, pp. 411–438, by R. H. Bogue, Portland Cement Association Fellowship (Portland Cement); J. L. Miner and F. W. Ashton, Universal Atlas Cement Company (Calcium–Aluminate Cement); and G. J. Fink, Oxychloride Cement Association, Inc. (Magnesia Cement); "Cement" in *ECT* 2nd ed., Vol. 4, pp. 684–710, by Robert H. Bogue, Consultant to the Cement Industry.

1. R. H. Bogue, *The Chemistry of Portland Cement,* 2nd ed., Rheinhold Publishing Corp., New York, 1955.

2. G. A. Rankin and F. E. Wright, *Am. J. Sci.* **39,** 1 (1915).
3. F. M. Lea, *The Chemistry of Cement and Concrete,* 3rd ed., Edward Arnold (Publishers) Ltd., London, Eng., 1971.
4. Ref. 3, p. 88.
5. Ref. 3, p. 122.
6. P. Weber, *Zem. Kalk Gips* Special Issue No. 9, (1963); ref. 3, p. 130.
7. Ref. 3, p. 126.
8. K. Endell and G. Hendrickx, *Zement* **31,** 357, 416 (1942); ref. 3, p. 128.
9. N. Toropov and P. Rumyantsev, *Zh. Prikl. Khim* **38,** 1614, 2115 (1965); ref. 3, p. 135.
10. *Guide to Compounds of Interest in Cement and Concrete Research, Special Report 127,* Highway Research Board, National Academy of Sciences, Washington, D.C., 1972.
11. H. F. W. Taylor, ed., *The Chemistry of Cements,* Vols. 1 and 2, Academic Press Inc. (London) Ltd., London, Eng., 1964, Appendix 1.
12. Ref. 3, p. 121.
13. A. Guinier and M. Regourd in *Proc. of the 5th Int. Symposium on the Chemistry of Cement, Tokyo, 1968,* The Cement Association of Japan, Tokyo, Japan, 1969.
14. G. Yamaguchi and S. Takagi in ref. 13.
15. Ref. 3, p. 42 ff and p. 270.
16. F. A. DeLisle, *Cement Technol.* **7,** 93 (1976); Y. Ono, S. Kamamura, and Y. Soda in ref. 13.
17. F. Gille and co-workers, *Microskopie des Zementklinkers, Bilderatlas,* Association of the German Cement Industry, Beton-Verlag, Dusseldorf, FRG, 1965.
18. L. S. Brown, *Proc. J. Am. Concr. Inst.* **44,** 877 (1948).
19. H. Kuhl, *Zement* **18,** 833 (1929); ref. 3, p. 164.
20. Ref. 3, p. 166.
21. K. E. Peray and J. J. Waddell, *The Rotary Cement Kiln,* Chemical Publishing Co., New York, 1972, p. 65.
22. R. H. Bogue, *Ind. Eng. Chem. Anal. Ed.* **1,** 192 (1929); ref. 1, p. 246.
23. *1976 Annual Book of ASTM Standards,* Part 13, American Society for Testing and Materials, Philadelphia, Pa., 1976, p. 138.
24. F. R. McMillan and W. C. Hansen, *J. Am. Concr. Inst.* **44,** 553, 564, 565 (1948).
25. Ref. 3, p. 188.
26. S. Brunauer and D. K. Kantro in ref. 11.
27. C. W. Lentz in *Special Report 90, Structure of Portland Cement Paste and Concrete,* Highway Research Board, NRC-NAS, Washington, D.C., 1966.
28. L. E. Copeland and G. Verbeck in *The 6th Int. Congress on the Chemistry of Cement, Moscow, 1974, English Preprints,* The Organizing Committee of the U.S.S.R., 1974.
29. G. Verbeck and R. A. Helmuth in ref. 13.
30. W. C. Hansen in E. G. Swenson, ed., *Performance of Concrete,* University of Toronto Press, Toronto, Can., 1968.
31. G. Verbeck, *Research Department Bulletin 189,* Portland Cement Association, Skokie, Ill., 1965.
32. L. E. Copeland and D. L. Kantro in ref. 11.
33. F. M. Lea, *The Chemistry of Cement and Concrete,* 3rd ed., Edward Arnold (Publishers) Ltd., London, Eng., 1971.
34. Ref. 3, p. 203.
35. H. F. W. Taylor in ref. 11, p. 21.
36. T. C. Powers in ref. 11, Chapt. 10.
37. W. Lerch, *ASTM Proc.* **46,** 1251 (1946).
38. Ref. 3, p. 239.
39. G. H. Tattersall, *The Workability of Concrete, Publ. 11.008,* Cement and Concrete Association, Wexham Springs, Eng., 1976.
40. T. C. Powers, *The Properties of Fresh Concrete,* John Wiley & Sons, Inc., New York, 1968.
41. E. M. Petrie, *Ind. Eng. Chem. Prod. Res. Dev.* **15,** 242 (1976).
42. *1976 Annual Book of ASTM Standards,* Part 14, American Society for Testing and Materials, Philadelphia, Pa., 1976.
43. T. C. Powers in *Proc. of the 4th Int Symposium on the Chemistry of Cement, Washington, D.C., 1960,* U.S. Government Printing Office, Washington, D.C., 1962.
44. D. M. Roy and G. R. Gouda, *Cement Concr. Res.* **5,** 153 (1975).
45. *The U.S. Cement Industry: An Economic Report,* Portland Cement Association, Skokie, Ill., Oct. 1974, p. 7.

46. Ref. 45, p. 8.
47. *Energy Conservation in the Cement Industry, Conservation Paper No. 26,* Federal Energy Administration, Washington, D.C., 1975.
48. W. H. Duda, *Cement Data Book,* Bauverlag G.m.b.H., Wiesbaden, FRG, and Berlin, 1976.
49. K. E. Peray and J. J. Waddell, *The Rotary Cement Kiln,* Chemical Publishing Co., New York, 1972.
50. J. R. Tonry, *Report MP-96,* Portland Cement Association, Skokie, Ill., 1961.
51. *Report: 1973 Technical Mission to Japan,* Portland Cement Association, Skokie, Ill., 1973.
52. D. G. Courteney, *Rock Prod.* **78**(5), 75 (1975).
53. D. Grammes in *Mill Session Papers M-195,* Portland Cement Association, Skokie, Ill., 1969.
54. *Clean Air Act,* Public Law 88-206 (1963); Amendments: *Public Law* 89-675 (1966); 91-604 (1970); and 95-95 (1977).
55. W. E. Trauffer, *Pit Quarry* **67**(8), 52 (1975).
56. *Public Law 92-500.*
57. ACI Committee 223, *Proc. Am. Concr. Inst.* **67,** 583 (1970).
58. *Pit Quarry* **66**(8), 104 (1974).
59. *Minerals Yearbook: Cement,* U.S. Bureau of Mines, Washington, D.C., 1975.
60. *Minerals Yearbook: Cement,* U.S. Bureau of Mines, Washington, D.C., 1972.
61. *Statistical Review No. 33, Production-Trade-Consumption 1974–1975,* Cembureau, Paris, Fr., Oct. 1976.
62. *World Cement Market in Figures, Production-Trade-Consumption 1913–1972,* Cembureau, Paris, Fr., 1973.
63. *Cement Standards of the World,* Cembureau, Paris, Fr., 1968.
64. Ref. 23, ASTM C-806.
65. U.S. Pat. 3,628,973 (Dec. 21, 1971), N. R. Greening, L. E. Copeland, and G. J. Verbeck (to Portland Cement Association).
66. Ref. 23, ASTM C-10.
67. Ref. 23, ASTM C-595.
68. Ref. 3, Chapt. 14; R. Turriziani in ref. 11, Chapt. 14.
69. Ref. 3, Chapt. 15; R. W. Nurse in ref. 11, Chapt. 13.
70. *Specifications for Oil-Well Cements and Cement Additives, API Standards 10A,* 19th ed., API, New York, 1974.
71. D. K. Smith, *Cementing,* Society of Petroleum Engineers of AIME, New York, 1976.
72. Ref. 23, ASTM C-91.
73. Ref. 3, Chapt. 16; T. D. Robson in ref. 11, Chapt. 12.
74. Ref. 3, p. 481.
75. Ref. 3, p. 477.
76. Ref. 23, ASTM C-141.
77. L. Cartz in J. F. Young, ed., *Cements Research Progress 1975,* American Ceramic Society, Columbus, Ohio, 1976, Chapt. 11.

General References

J. F. Young, ed., *Cements Research Progress 1976,* American Ceramic Society, Columbus, Ohio, 1977.
Refs. 1, 11, 13, 17, 23, 28, 33, 40, 47–49, and 63 are also good general references.

RICHARD A. HELMUTH
F. M. MILLER
T. R. O'CONNOR
N. R. GREENING
Portland Cement Association

CERAMICS

SCOPE

"Ceramics comprise all engineering materials or products (or portions thereof) that are chemically inorganic, except metals and alloys, and are usually rendered serviceable through high-temperature processing" (1). Ceramic materials are normally composed of both cationic and anionic species. The primary difference between ceramics and other materials is the nature of their chemical bonding (2–5).

Although there are no distinct boundaries between ceramic and metallic or polymeric materials, it is instructive to compare them in terms of the service requirements in engineering design (3). As a class of materials, ceramics are better electrical and thermal insulators and more stable in chemical and thermal environments than are metals (see Cement). Metals usually have comparable tensile and compressive strengths, whereas ceramics are normally appreciably stronger in compression than in tension. Ceramics exhibit greater rigidity, hardness and temperature stability than polymers; however, polymerization occurs in ceramics, especially in glasses (see Glass; Glass-ceramics).

Modern ceramics encompass a wide variety of materials and products ranging from single crystals and dense polycrystalline materials, through glass-bonded aggregates to insulating foams and wholly vitreous substances (4–7). Such a range of microstructural characteristics allows the considerable versatility evidenced in the range of manufactured industrial products.

On the basis of available statistics, the value of this industrial output, in terms of the value of products shipped during 1975, approximated 25.2 billion dollars. The breakdown by major product classifications is given in Figure 1. Although several of the product areas do not have large dollar values as compared to many industrial commodities, they are nevertheless vital to an industrial economy. Two notable examples are the refractories necessary for the reduction of ores in the metallurgical industries and abrasives (qv) which allow the mass production of machine parts.

The magnitude of the ceramic industry is by no means completely represented by the data in Figure 1. For instance, dielectric and magnetic components in electrical and electronic products, enameled parts of household appliances, refractories in heating systems and fuel materials and other parts of nuclear reactors (qv) are all components of finished goods which should be, but are not currently classified as ceramics (see Enamels; Refractories).

As late as the 1930s, ceramic technology was primarily perceived as applied high-temperature silicate chemistry. Although silicate materials continue to be the inexpensive high-tonnage backbone of the industry, the desire for high performance ceramic materials, particularly those having improved electrical, electronic, piezoelectric, and magnetic and, more recently, electro-optic, pyroelectric, and laser properties has increased steadily in the last ten to twenty years (see Ceramics as

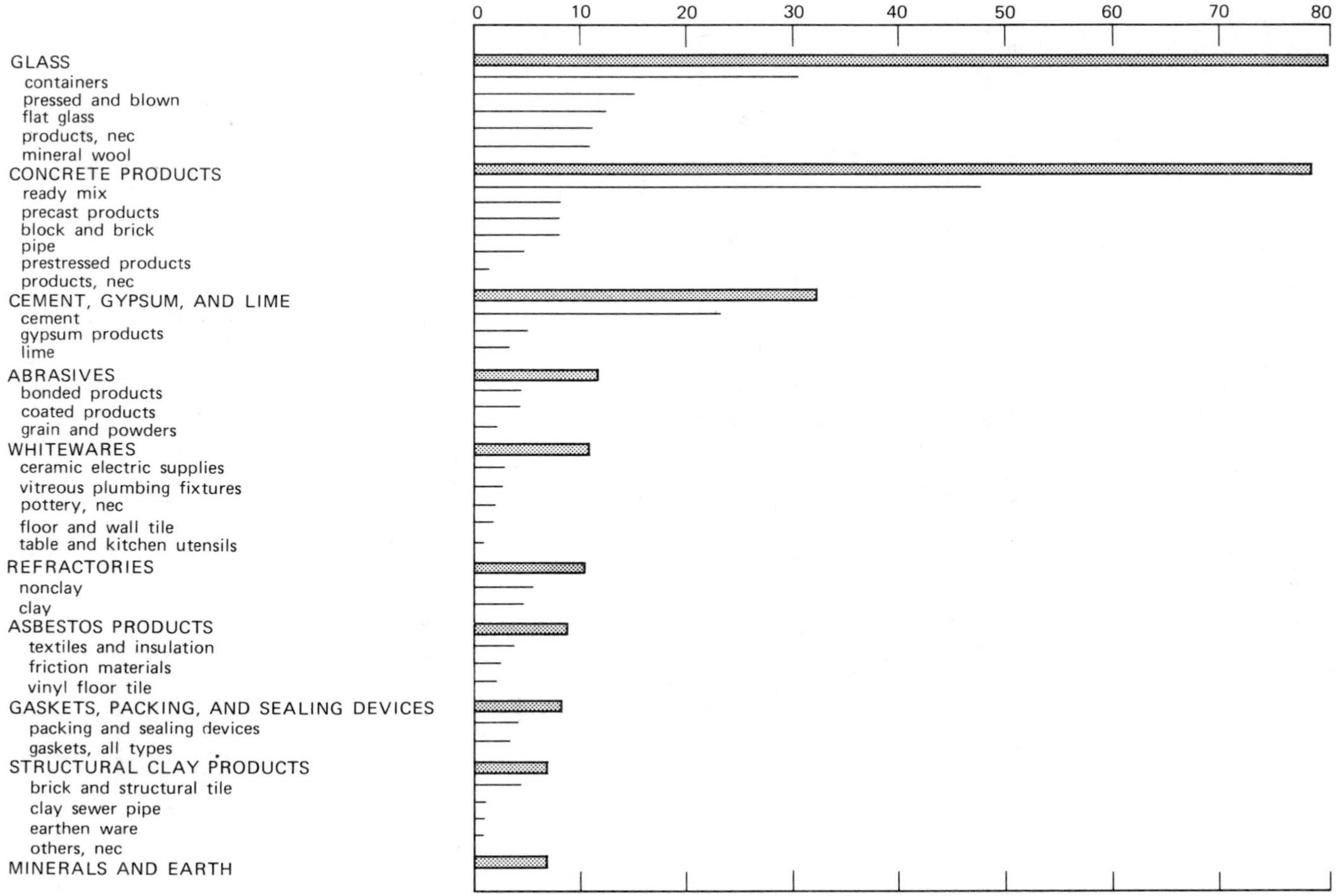

Figure 1. Value of ceramic products shipped in 1975, in billions of dollars; nec = not elsewhere classified. Abstracted from the *1975 Annual Survey of Manufacturers*, U.S. Department of Commerce.

electrical materials). The present urgency to develop materials for energy production, conversion and storage apparatus has stimulated the evolution of solid electrolytes for batteries (qv), refractories for magnetohydrodynamic generators and coal gasification devices (see Coal), strong dense ceramics for high efficiency turbine parts, and new glasses for solar collector panels (see Solar energy). In these newer ceramics, the greatest emphasis has been given to the oxide systems; however, considerable progress has also been made in the synthesis and employment of the borides, carbides, nitrides, and silicides (see Boron compounds, refractory; Carbides; Nitrides; Silicon and silicides).

The development, purification, and utilization of materials often requires the evolution of new processing techniques. Particle preparation, extrusion, dry pressing, slip casting, and sintering remain as important techniques in the ceramic industry, but they have been joined by freeze drying, thermal evaporation or sputtering, and chemical vapor deposition to produce high purity materials or thin films or complex shapes, respectively. Furthermore, the growth of single crystals, improved densification techniques and the glass–ceramic process have led to pore-free crystalline or nearly crystalline ceramics having dramatically improved properties.

Ceramics are frequently termed ionic solids, ie, possessing ionic bonding. In reality the bonding varies as a function of the polarizing power of the cations and the polarizability of the anions and is almost totally ionic in CsF and covalent in SiC. In the layer silicates such as the clays, van der Waals forces also bond the layers together. When these materials are subjected to firing processes, pyrochemical changes result in the formation of new crystalline aggregates dispersed in a vitreous matrix, each having its own ionic-covalent bonding.

Empiricism remains an intrinsic aspect of ceramic technology; however, the latter now involves the cooperative talents of the ceramic scientist, chemist, metallurgist, and solid state physicist in order to effect a more fundamental understanding of these materials. Manufacturing is now highly mechanized with several industries having fully automated, computer-controlled processing (see Instrumentation and control).

The following articles present generalized and unifying characteristics of ceramic technology by means of discussions of raw materials, processing, thermal treatment, and properties.

BIBLIOGRAPHY

"Structural Clay Products" under "Ceramic Industries" in *ECT* 1st ed., Vol. 3, pp. 521–545, by R. M. Campbell, The New York State College of Ceramics; "Whiteware" under "Ceramic Industries" in *ECT* 1st ed., Vol. 3, pp. 545–574, by F. P. Hall, Pass & Seymour, Inc., and Onondaga Pottery Company; "Scope of Ceramics" under "Ceramics" in *ECT* 2nd ed., Vol. 4, pp. 759–762, by W. W. Kriegel, North Carolina State of The University of North Carolina.

1. *Objective Criteria in Ceramic Engineering Education,* American Society for Engineering Education, Urbana, Illinois, 1963.
2. F. H. Norton, *Elements of Ceramics,* 2nd ed., Addison Wesley Publishing Company, Inc., Reading, Mass., 1974.
3. Institute of Ceramics, Textbook Series, published by Maclaren and Sons, Ltd., London: (a) W. E. Worrall, *Raw Materials,* 1964; (b) F. Moore, *Rheology of Ceramic Systems,* 1967; R. W. Ford, *Drying,* 1964; W. F. Ford, *The Effect of Heat on Ceramics,* 1976.

4. W. D. Kingery, H. K. Bowen, and D. R. Uhlmann, *Introduction to Ceramics,* 2nd ed., John Wiley & Sons., Inc., New York, 1976.
5. L. H. Van Vlack, *Physical Ceramics for Engineers,* Addison-Wesley Publishing Company, Inc., Reading, Massachusetts, 1964.
6. J. E. Burke, ed., *Progress in Ceramic Science,* Vols. 1–4, Pergamon Press, Inc., New York, 1962–1966.
7. *Proceedings of the University Conferences on Ceramic Science* (various publishers). This is a series of edited books on cogent topics in the ceramic field beginning in 1964 and continuing to the present (1977 conference at North Carolina State University, Raleigh, in press, Plenum Publishing Co., New York).

ROBERT F. DAVIS
North Carolina State University

PROPERTIES AND APPLICATIONS

Composition and Microstructure

In terms of useful properties and typical applications, there is great diversity among ceramic substances, and indeed between differently processed ceramic items of the same substance. These important differences in materials can, in principle, be described in terms of just two overall parameters, the composition and the microstructure. By composition is meant the chemical and mineralogical makeup of the total substance of the material, including its deliberate and accidental impurities. Microstructure means broadly the crystallographic structure of each component or phase, and in particular, the size, shape, and distribution of each of the phases present, including the unoccupied space, or porosity.

These determinative parameters are fixed by the combined response of the starting materials to all the processing steps. Consequently, microstructural characterization is an important factor in controlling the processing of ceramics. In many applications ceramics are used at service temperatures far below their initial maturing or processing temperature, so that little or no change in composition and microstructure can occur. Thus these two parameters, once fixed during processing, also establish the useful properties of the ceramic product. In service at temperatures near or above the maturing temperature, the parameters are likely to change with temperature, time, environment, etc, so that the useful properties are likely to be transient rather than static.

A ceramic material is a solid and it may be either crystalline or vitreous or of mixed type. It is crystalline if its ions or atoms are in orderly array, row on row, layer upon layer, over long distances compared to one interatomic spacing. Crystalline ceramics include many single oxides, carbides, nitrides, borides, sulfides, etc, as well as their binary and ternary compounds. A material is vitreous if its ions are arranged with some regularity at close range but do not display long-range order. Glass (qv) is the classic vitreous substance. Technically, it can be characterized as an extremely viscous, metastable, supercooled liquid but in practice it behaves in most respects as a solid at room temperature.

A ceramic material may be composed entirely of a single substance and would be described as monophase but it is most likely to be polyphase, containing two or more discrete substances. Monophase crystalline ceramics may occur as monocrystals, having a unique crystallographic orientation throughout, eg, sapphire is the monocrystal form of α-Al_2O_3. Characteristically, most ceramics are polycrystalline masses, with abrupt changes in orientation or composition occurring across each grain or phase boundary. Polyphase materials must, out of necessity, be polycrystalline, and in such materials there may be wide variations in the character of internal boundaries depending on the orientation and composition of the neighboring grain or phase. Boundaries constitute fairly abrupt discontinuities in local atomic arrangements and bonding character, and in properties. Such boundary regions are thermodynamically less stable than the more orderly crystalline material(s) adjoining them.

Another discontinuity exists at the free surface of the ceramic, the final boundary between the ceramic solid and its environment. Many properties of ceramics are surface-sensitive, both with respect to mechanical condition and chemical and atmospheric environment, or both.

In practice, compositional and microstructural parameters are fairly difficult to determine with precision in typically complex ceramic systems. In selected, simple monophase systems, considerable progress has been made in recent years in systematically relating microstructure to processing on the one hand, and in characterizing variations in properties in terms of microstructure on the other. Regardless of whether or not such relationships are amenable to precise measurement, every property of a ceramic is dependent to some degree upon the relative amounts of the phases present, their respective compositions and structures, the spatial disposition of the phases, and the shapes, sizes, and orientations of the grains and pores.

The mechanical behavior of ceramics offers one example of this problem. Whereas it is possible to conceive of the calculation of stress and strain at a particular point within a particular grain of a porous, polyphase polycrystalline ceramic under a particular state of applied stress and strain, it would be almost impossibly difficult to carry out, even if all the boundary conditions were mathematically resolvable. On the other hand, flow or fracture must be related to specific details of stress and strain, in specific directions, on specific planes, within a particular atomic structure if fundamental understanding of the process is to be achieved; similar considerations apply in the case of optical, thermal, and electromagnetic properties. Consequently, there is an important divergence between those properties that must be determined in single crystals so that orientation can be selectively controlled and specifically related to certain discrete atomistic processes, and those properties of polyphase or polycrystalline materials that must of necessity be measured and interpreted in a statistical manner. Experiments of both types are usually required to characterize adequately a particular kind of behavior in ceramic materials.

Many of the elastic, thermal, and optical properties of materials may be quite adequately described by tensor quantities predicated only on an ideal model of the crystal structure. But many of the properties of ceramic materials cannot be so described; mechanical and electromagnetic properties in particular, can be shown to be extremely sensitive to defects existing in the structure. Such properties are termed structure-sensitive; some of the commonly encountered defects in ceramics which account for them are summarized in Table 1.

Chemical Properties of Ceramic Materials

A high degree of chemical stability is characteristic of the oxides, carbides, nitrides, borides, etc, which form the basis for all ceramic materials. In particular, many of the oxides are extremely stable over wide ranges of temperature and environmental conditions, and are resistant to further oxidation and to reduction to suboxides or metals. Some nonoxide ceramics are very stable at extreme temperatures (>2000°C) *in vacuo*, or in neutral or reducing atmospheres, but in general they will not tolerate long-term high temperature exposure to oxidizing environments. The stability of these ceramic compounds is the result of the compactness of the crystal structure, the directed chemical bonding (generally ionic or covalent, or of mixed ionic–covalent character), and of the high field strengths associated with the relatively small, highly charged

Table 1. Types of Defects in Crystalline Ceramics

Defect	Cause	Relative scale
impurity	foreign atom or ion introduced by substitution or interstitially; vacancies may be created to balance valence charges	atomic in three dimensions
vacancy	missing positive or negative ion, or atom	atomic in three dimensions
vacancy pair	missing positive ion coupled with missing negative ion	atomic in three dimensions
vacancy cluster	aggregation of vacancies	microscopic (large clusters can be resolved optically)
color center	anion vacancy plus electron or cation vacancy plus electron hole	atomic in two dimensions, microscopic in length
dislocation	structural misfit, linear character	atomic in two dimensions, microscopic in length
subgrain boundary	structural misfit (at small angles), surface character	atomic in one dimension, microscopic in two dimensions
grain boundary	structural misfit (at large angles), surface character	atomic to microscopic in one dimension, microscopic to macroscopic in two dimensions
phase boundary	compositional or crystallographic discontinuity, surface character	atomic to microscopic in one dimension, microscopic to macroscopic in two dimensions

cations encountered in refractory ceramics. Thermodynamically, the most stable ceramic will be that having the greatest negative free energy of formation from the elements; Kingery (1) points out that of the oxides yttria and thoria are the most stable in this respect. Others widely used for their chemical stability are alumina, beryllia, magnesia, and stabilized zirconia (see Refractories).

Although there are exceptions (SnO_2, ZnO, MgO, etc), most refractory ceramics have very low vapor pressures in the most stable valence state, but the suboxides tend to be volatile (SiO, Al_2O, etc). Limiting conditions for more complex oxides are often established by the volatility of a particular component, such as Na in sodium aluminate, etc.

Interactions between dissimilar ceramics, or between ceramics and metals, becomes increasingly likely as operating temperature is increased, or at low pressure. Heating *in vacuo* favors reaction by facilitating removal of vapor phase products, eg:

$$x\,W + ThO_2 \xrightarrow{\Delta} ThO_{2-2x} + x\,WO_2\uparrow \;(>2000°C) \tag{1}$$

$$3\,C + 2\,Al_2O_3 \xrightarrow{\Delta} 4\,Al\uparrow + 3\,CO_2\uparrow \;(>1950°C) \tag{2}$$

Even a system of materials with very high melting points can form fluid liquids at temperatures much below either melting point. For example, Al_2O_3 melts at 2050°C and CaO at 2500°C, but when in contact, they can form a reactive, eutectic liquid at 1450°C. The presence of a few percent impurities can significantly lower the temperature of liquid formation. Johnson (2), Economos and Kingery (3), and Kingery (1) have discussed the stability of oxide refractories in contact with other materials at high temperatures.

Optical Properties

Many ceramic substances, particularly the oxides, are optically transparent in single crystal or vitreous forms; eg, the transparency of glass is perhaps its most useful and most characteristic feature. Optical transmission occurs in dielectric materials such as ceramics through interactions of the oncoming electromagnetic radiation (photons) with the polarized (or polarizable) electron shells surrounding the nuclei of the constituent ions or atoms. Optical properties are closely linked to composition and structure of the ceramic since the degree of polarization is a function of ion size, bounding energy, and crystallographic direction. The index of refraction is a sensitive quantitative measure of these materials parameters.

A material is often characterized by its transparency in certain wavelength regions (% transmission/mm thickness). Many ceramics and most glasses absorb highly in the infrared and ultraviolet regions of the spectrum. MgF_2 and CaF_2 are used for uv transmission. Other specialty materials, such as CsI and KI, are transparent at very long infrared wavelengths.

Isotropic crystals and vitreous glass display isotropic optical properties, but other crystals are optically anisotropic, the index of refraction being highest when light rays traverse the close-packed crystallographic directions. Dispersion is the term employed to describe the variation of index of refraction with change in wavelength. It results from resonance effects between incoming photons in the visible range and the typical ultraviolet frequencies of oscillating electronic states associated with the constituent ions.

Large ions in crystals or in glass compositions are more readily polarized than small ones and produce a high index of refraction. Most glasses are in the range 1.4–2.0, metals are usually >3.0. The index of refraction of some ceramic materials is shown in Table 2.

Optical absorption measures the loss of intensity as light traverses a translucent medium. In addition to the losses attributable to surface reflection and dispersed-phase scattering, a principal cause of absorption is the presence of unfilled electronic energy bands within the material. One source of such donor or acceptor sites is the color center, which can be produced in normally colorless ionic materials by nonstoichiometric cation–anion ratios. Such defects (F-centers, V-centers) are strongly absorbing at some given wavelength (F-centers, usually in or toward the ir, V-centers in the uv), and so characteristic of the particular defects that optical absorption spectra are used to identify them.

Principal and very practical sources of unfilled energy bands in ceramic materials are impurities, or deliberate colorant oxide additives, which can be selected from any one of a number of transition metal oxides having unfilled $3d$, $4f$, or $5f$ electron shells.

Table 2. Refractive Index of Some Ceramic Crystal Phases[a]

MgO	1.74	TiO_2	2.71
Al_2O_3	1.76	SiC	2.68
$MgAl_2O_4$	1.72	Y_2O_3	1.92
SiO_2 (quartz)	1.55	$BaTiO_3$	2.40
$ZrSiO_4$	1.95		
mullite	1.64	LiF	1.392

[a] Ref. 4.

The $3d$ colorants are especially well known; eg, cobalt blue, chrome green, chrome–alumina pink, ferric iron brick-red, and vanadium yellow are familiar and traditional ceramic colors. The energy band structure responsible for optical effects is quite sensitive to the crystallographic environment of the colorant ion, to the valence state, and a number of other factors. A ceramic (ruby) single crystal, α-Al_2O_3, containing deliberate Cr^{3+} impurities, was the first to be employed as a laser host crystal (other hosts include $CaWO_3$, CaF_2, YIG, YAG, etc) and glasses doped with rare earth and actinide elements are now used in laser devices (see Colorants for ceramics; Lasers).

The microstructure of a polycrystalline ceramic tends to dominate the useful optical properties of the base material. Pores within grains or between boundaries scatter and reflect light and can induce translucency or even opacity in an otherwise transparent medium. Figure 1 shows the decrease in percent transmission of alumina as a function of porosity. Transmission is also reduced with finer pore size, until the pore size approaches the wavelength of light. Materials with extremely small pores <100 nm can be made transparent even at high levels of porosity.

Polyphase polycrystalline materials may develop additional scattering owing to differences in refraction across boundaries. Dispersed phases in particular act as opacifiers, with the degree of opacity depending upon concentration, particle size and

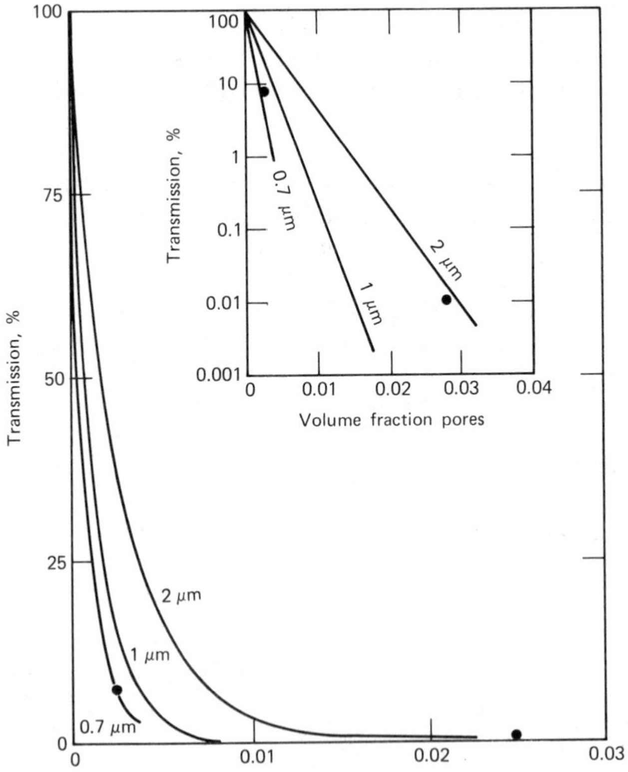

Figure 1. Transmission of polycrystalline alumina containing small amount of residual porosity (equivalent thickness 0.5 mm) (5).

difference in index of refraction of the dispersant phase. Dispersions of SnO_2, $ZrSiO_4$, TiO_2, and/or Sb_2O_3 are commercially employed as opacifiers and whiteners in enamels (qv), glazes, and glasses. They may be inert additions, products of reactions, or nucleated and crystallized from a glass. Dispersed phases are also employed to produce color effects, where the color-absorbing energy bands are not associated directly with the host ceramic matrix but are contained instead in a refractory additive phase called a color stain. In stains, the colorant transition metal oxides have been bound up in a stable crystal structure (the spinel structure is frequently employed) by prior high temperature reaction (calcining), so that it undergoes relatively little change during subsequent thermal processing of the ceramic product. A wide range of calcined, milled color stains for various uses are commercially available.

Just as color and opacity are highly desired in many ceramic applications, the absence of these properties is critically important in others. Highly translucent, almost transparent, pore-free polycrystalline ceramics produced by sintering or the hot-pressing method require careful control of impurities which could jeopardize the desired color-free transparency.

Because of total internal reflection, transparent fibers can transmit light around corners. A bundle of such fibers and its outer sheath can greatly minimize transmission losses over long distances. Proper control of impurities can minimize internal absorption. Other optical devices utilizing important ceramic components are optical wave guides, and electrooptic and acoustioptic materials (4) (see also Fiber optics).

Thermal Properties

The thermal properties of ceramics illustrate the subtle interplay between structure and properties, as well as between properties and applications, which determines the suitability of a particular ceramic material for a given job. The thermal properties of greatest interest include specific heat, thermal conductivity, and thermal expansion. Each is important since ceramics are frequently employed, and are always produced, at some high temperature. In such cases it is useful, or even essential, to know and provide for the quantity of heat stored and/or transferred by the material, and the dimensional changes it undergoes as a result of the added thermal energy. Furthermore, materials with high thermal conductivity and low thermal expansion are most likely to survive severe thermal shock (4).

Thermal energy within a solid takes the form of increased motions of its fundamental particles (electrons, ions, molecules) in a variety of rotational and vibrational modes. The availability of the electronic conduction mode in any oxide materials markedly improves thermal conductivity, eg, in SiC. However, for most ceramics having insulating rather than semiconducting properties, the most important mode involves lattice vibrations; they may be directly correlated with the thermal properties of interest in ceramics. The thermal energy consists of vibrations that are neither random nor independent but take the form of strain waves which move through the lattice with the speed of sound. These strain waves are called phonons. Phonons have quantized energy states and are restricted to the natural vibrational frequencies of the solid, and hence are governed in part by the lattice configuration and in part by the external dimensions of the solid. In essence a phonon is a transient lattice defect and in fact may interact with or be scattered by impurities, grain boundaries, and other defects including other phonons. To avoid such problems in theoretical considerations of the role of phonons in some thermal properties, eg, specific heat, it is convenient to assume an ideal crystal free of defects in which phonons do not interfere with one another.

Specific Heat. At high temperatures, the average vibrational energy of a particle in a solid is $3/2\,kT$ (k = Boltzmann constant), reflecting the three principal orthogonal vibrational modes, and the specific heat of the solid at high temperature becomes a constant, $3\,Nk$ = 24.9 J/(mol·K) [5.96 cal/(mol·°C)] (N = Avogadro's no.). At some lower temperature, however, the vibrational energy decreases and ultimately goes to zero at 0 K. Classical physical concepts are not able to account for this behavior, but the spectral (frequency) distributions of phonons in solids were utilized by Debye in 1912 in arriving at a quantum approach that does correctly predict the variation in thermal properties. His treatment defines a characteristic temperature θ at which the phonons attain their maximum frequency and shortest mean free path in the solid as shown below:

$$\theta = h\nu_{\max}/k \tag{3}$$

where h = Planck's constant and ν = frequency of phonons. Below θ, the specific heat decreases with decreasing temperatures since only lower frequencies of phonons with longer mean free path are permitted; heat is being conducted through rather than stored in the lattice. Above θ, the three principal modes of lattice vibration are in full effect, and the specific heat remains constant. Kingery (4) notes that most ceramic oxides and carbides have Debye temperatures on the order of 1000°C. At higher temperatures, the thermal energy absorbed in processes other than lattice vibration (disordering, creation of defects, electronic effects) causes the measured heat capacity to increase at a modest rate rather than remain constant at $3\,Nk$.

Heat capacity curves for some ceramic materials are shown in Figure 2. Kingery also points out that although the molar heat capacity is essentially a constant, unaffected by the microstructure of the ceramic, the volume heat capacity frequently reported is a function of the porosity in the structure. Much less heat is required to raise the temperature of a given volume of a porous material than of a dense one, and much less must be removed during cooling. Minimum density insulation is therefore much

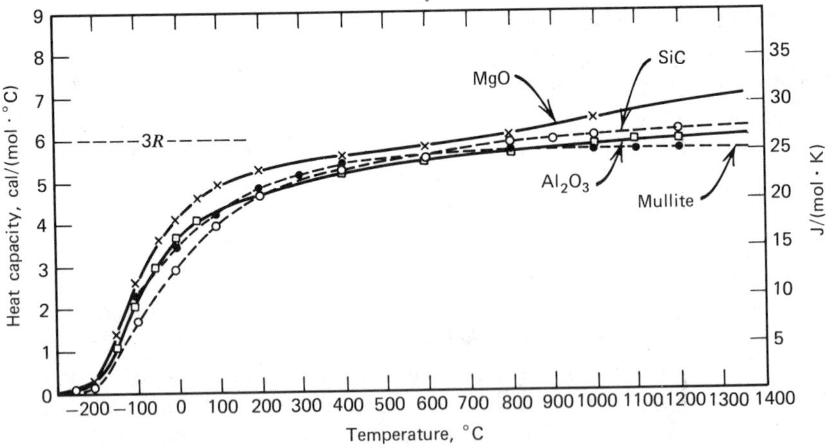

Figure 2. Heat capacity of some ceramic materials at different temperatures (6).

used in the construction of kilns and furnaces which are to be operated in periodic fashion. The selection of minimum density insulation results in energy savings in these kilns.

Modern ZrO_2 fiber insulations for laboratory furnaces have a fractional density of very low thermal conductivity, which allows very light construction and extremely fast heating and cooling rates (10–15 min to 1600°C).

Thermal Conductivity. The mechanism by which thermal energy is transported from a region of higher temperature to one of lower temperature involves a transport of phonons and electrons, or both (7). Electron conduction is of importance in metals, and in semiconductors over some temperature ranges, but in insulators it is phonon transport that is responsible for thermal conductivity:

$$k = pC_p lc \tag{4}$$

where k = coefficient of thermal conductivity, p = density of solid, C_p = heat capacity, l = mean free path of phonons, and c = velocity of propagation of phonons. The mean free path l reflects the effects of scattering of phonons from permanent structural defects l_s as well as from phonon–phonon scattering l_t as shown below:

$$\frac{1}{l} = \frac{1}{l_s} + \frac{1}{l_t} \tag{5}$$

At high temperature $1/l_s$ is essentially a constant, but below 100 K it decreases with decreasing temperature. The number of structural defects decreases with decreasing temperature, therefore the mean path between structural defects becomes larger as the temperature is lowered. At about 20 K, the mean path for the structural scattering factor is quite sensitive to defects and microstructure, being lower in single crystals than in polycrystalline ceramics, etc.

The thermal scattering effect, $1/l_t$, is almost proportional to the absolute temperature, increasing as the vibrational frequency increases and the mean path between phonon–phonon collisions shortens. The combined effects of both types of scattering as a function of temperature of α-Al_2O_3 are shown in Figure 3. The curves in Figure 4 illustrate the wide range of thermal conductivities available in ceramic materials.

The low conductivity of porous refractory ceramics combined with resistance to chemical change at very high temperature make them a valuable insulating material for high temperature applications. It should be noted that an increase in heat flow is indicated at temperatures above about 1500°C. This additive effect is attributable to radiation transfer processes.

Thermal conductivity is a tensor property, strongly dependent upon the crystallographic orientation and bonding character of the solid (7). Kingery (4) discusses the decreased thermal conductivity of binary oxide compounds and solid solutions in comparison to their respective end-member oxides. This behavior is also attributable to an increased likelihood of structural scattering of phonons in the more complex structures that result.

Thermal Expansion. It has been noted earlier that, as the temperature of a solid is increased, vibrations about each lattice point (ion, atom, or molecule) are induced, creating elastic waves or phonons. But the oscillations about the equilibrium lattice point encounter different energy fields in the compressive and expansive half cycles. The differences occur because repulsion forces increase much more rapidly with variations in interatomic distance than do attraction forces. This produces a shift

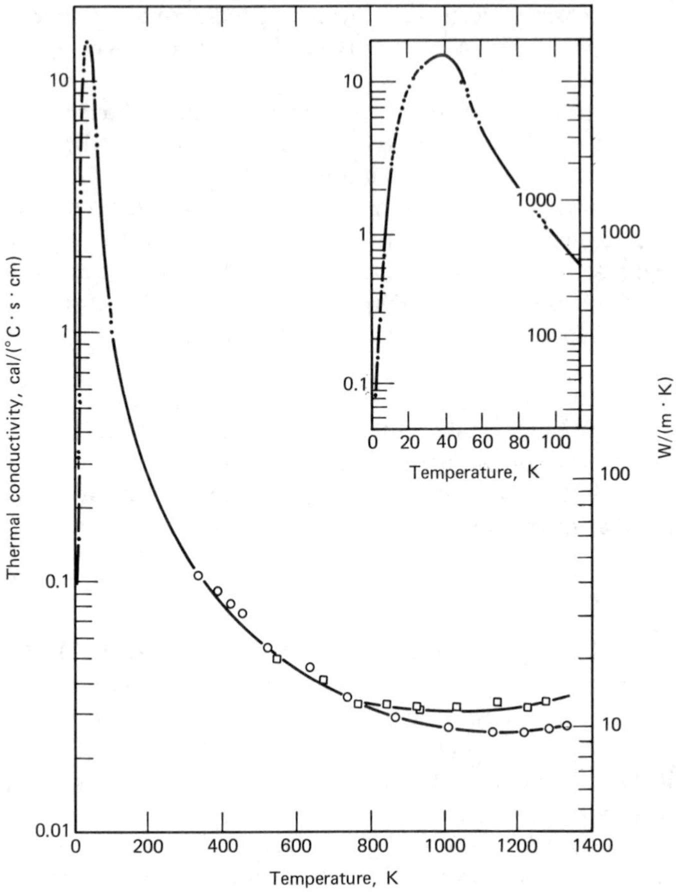

Figure 3. Thermal conductivity of single crystal aluminum oxide (8). ●, Berman (1951); ○, Lee and Kingery (1960) Pt foil interface; □, Lee and Kingery (1960) graphite interface.

toward greater separations of the equilibrium lattice sites as the temperature increases, ie, the crystal expands with increasing temperature (see Table 3).

The expansion for oxide structures with dense packing of oxygen ions is in the range of 6×10^{-6} to 8×10^{-6}/°C at room temperature and increases to 10×10^{-6} to 15×10^{-6}/°C at high temperatures (4). More open structures can have a much lower expansion in some crystallographic direction. The transverse vibrations of ions provide a vibrational energy component which compacts the structure. In some highly anisotropic materials, the thermal expansion coefficient in one direction may be negative, so that the average value of thermal expansion for a randomized polycrystalline material may be very low. This accounts for the excellent thermal shock resistance of ceramics such as cordierite and the lithium alumino–silicates much used as insulators and coil forms in rapid-heating electrical appliances such as soldering irons and radiant heaters.

In an extreme case, these very low or negative coefficients along certain crystallographic axes can induce such severe grain boundary stresses that the body is subject to internal ruptures during cooling from the maturing temperature. On heating, the cracks recombine; this phenomenon has been analyzed in detail for alumino–titanate

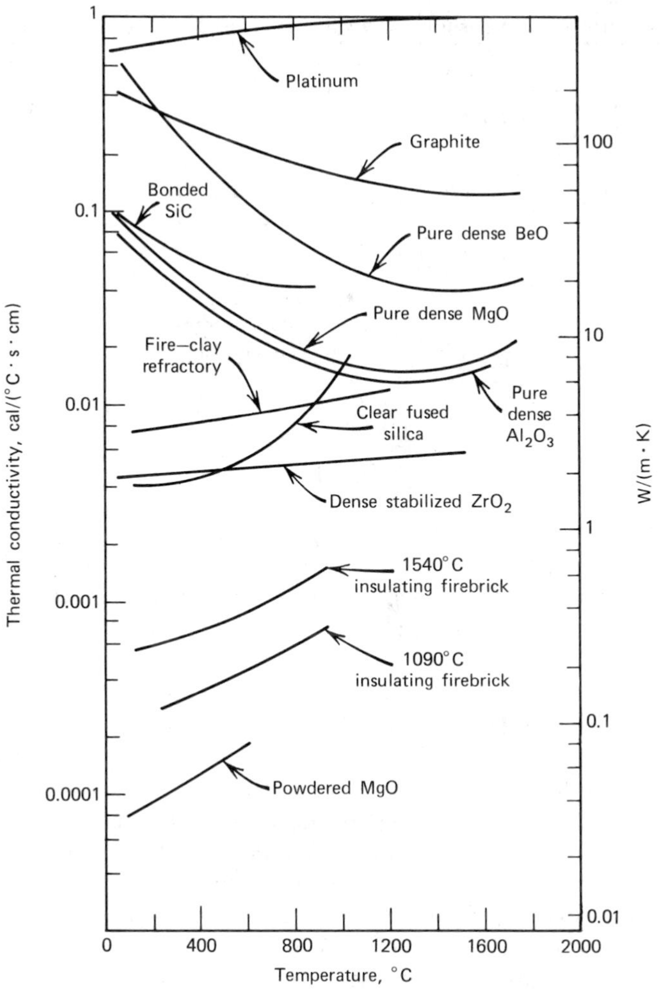

Figure 4. Thermal conductivity as a function of temperature for a variety of ceramic materials (9).

ceramics by Buessem (10). It accounts for the abnormally low strength of such materials at room temperature, and for the typical increases in strength that occur as the temperature is raised. Graphite, some borides, and other highly anisotropic materials also show this anamalous strength–temperature relationship.

Thermal Shock Resistance. The principal properties improving thermal shock resistance are (*1*) low thermal expansion, (*2*) high thermal conductivity, and (*3*) high strength. Ceramic materials in general do not have particularly good thermal shock resistance because of their brittle nature, but are used in many applications where ductile materials could not be used because of other superior high temperature properties. Thermal shock resistance becomes an important property in any application involving rapid changes in temperature or large thermal gradients. Applications such as cookware, spark plug insulators, abrasive wheels, refractory linings, etc, all demand some measure of shock resistance.

Thermal shock resistance is usually measured by quenching from a higher tem-

Table 3. Mean Thermal Expansion Coefficients for Various Ceramic Materials, Range of 0–1000°C

Material	Linear expansion coefficient, $(°C \times 10^6)^{-1}$	Material	Linear expansion coefficient, $(°C \times 10^6)^{-1}$
Al_2O_3	8.8	fused silica glass	0.5
BeO	9.0	soda–lime–silica glass	9.0
MgO	13.5	TiC	7.4
mullite	5.3	porcelain	6.0
spinel	7.6	fire-clay refractory	5.5
ThO_2	9.2	TiC cermet	9.0
zircon	4.2	B_4C	4.5
SiC	4.7	UO_2	10.0
ZrO_2 (stabilized)	10.0	Y_2O_3	9.3

perature and by measuring the strength degradation (as compared to room temperature). Figure 5 shows the effect on strength of quench temperature for aluminum oxide of various grain sizes. Alumina can withstand a quench from 200°C with no effect on strength. From above 200°C, microstructural damage takes place, with a more dramatic effect at finer grain size.

Materials such as fused silica, cordierite, and lithium aluminosilicates have very low thermal expansion and good thermal shock resistance despite being weak when compared to other ceramic materials. Improperly processed low thermal expansion materials such as cordierite can be very weak (owing to thermal expansion anisotropy of individual crystals) and thus have poor thermal shock resistance.

BeO, despite a relatively high thermal expansion, has excellent thermal conductivity and high strength, and thus has good thermal shock resistance. Many nonoxide ceramics also possess high thermal conductivity and can have very high strength. SiC combines high strength with a relatively low thermal expansion and high thermal conductivity. Al_2O_3 is also a good shock resistant material because of its high strength.

A thorough understanding of shock resistance requires a knowledge of microstructural, thermal, and elastic properties, and is discussed in detail by Kingery (4) and Hasselman (12).

Many ceramics contain phases (such as SiO_2 or ZrO_2) that undergo phase transformations accompanied by a volume change. Large particles of quartz, for instance, impart very poor resistance to porcelains in the neighborhood of 573°C. This gives these materials very poor shock resistance. The effect of the transformation can be reduced by additives or with the use of very fine particles.

Elastic Properties

The elastic properties of solids arise from the interaction of mechanical distortion, called strain, with the periodicities of the atomic or ionic structure. Elastic moduli measure the force involved in achieving recoverable, small, unit displacements of atoms or ions from their equilibrium positions. The elastic moduli are consequently strongly dependent upon the type of bonding and upon the electron density configuration of

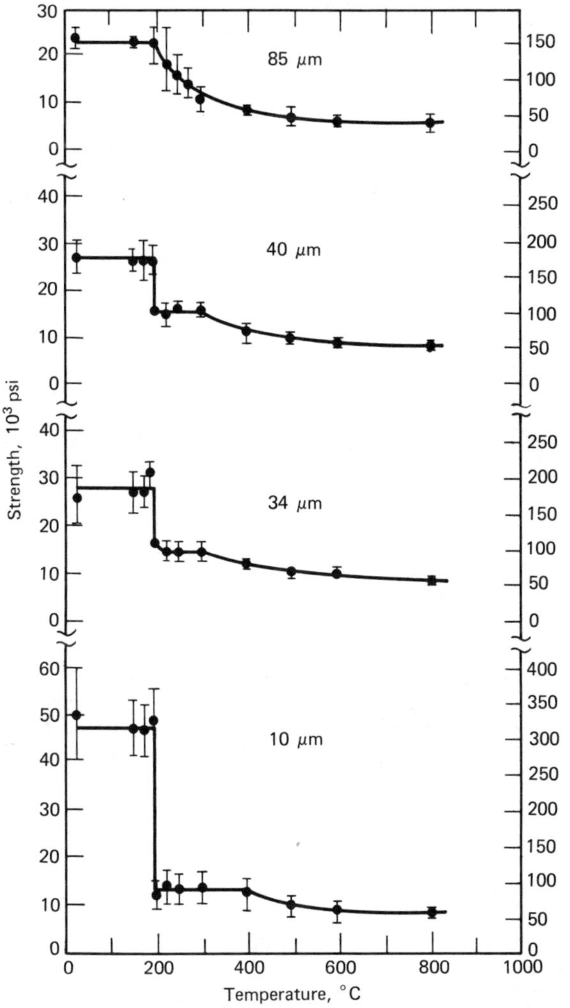

Figure 5. Room temperature strength of Al_2O_3 specimens of various grain sizes as function of quench temperature (11).

the crystal, just as is the melting point of the materials. This relationship is illustrated in Figure 6.

As in the case of optical (photon) waves and thermally induced strain waves (phonons), the elastic properties are directional, and can be described as tensors with respect to an ideal crystal (7):

$$\sigma_{ij} = C_{ijkl}\epsilon_{kl} \quad (6)$$

$$\epsilon_{ij} = S_{ijkl}\sigma_{kl} \quad (7)$$

where C_{ijkl} = elastic rigidity tensor and S_{ijkl} = elastic compliance tensor.

For isotropic substances such as glass, and for randomized polycrystalline materials having pseudoisotropic behavior, the elastic properties can be expressed adequately in terms of the following three quantities:

$$E = \text{Young's modulus of elasticity} = \frac{\sigma}{\epsilon} \quad (8)$$

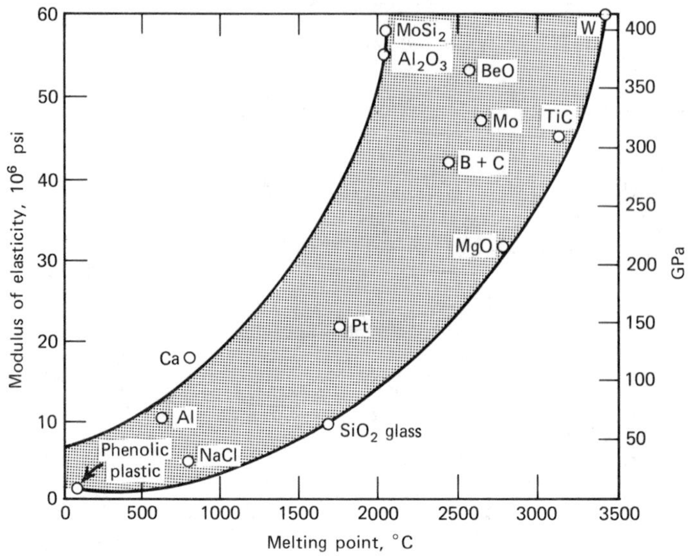

Figure 6. Relationship between elastic modulus and melting point for different materials (4).

$$G = \text{Shear modulus} = \frac{\tau}{\gamma} \tag{9}$$

$$\mu = \text{Poisson's ratio} = \frac{E}{2G} - 1 \tag{10}$$

where σ = normal stress, τ = shear stress, ϵ = normal strain, γ = shear strain. They are strictly valid only in the elastic (ie, recoverable) range of deformation and only when the material is homogeneous and isotropic. Most materials, and certainly most ceramics, are both inhomogeneous and anisotropic, therefore more complete descriptions of elastic behavior are usually required.

The temperature dependence on the elastic moduli for ceramic materials depends in large degree upon the temperature range and upon the types of structural defects present in the material. For monocrystals free from mobile defects within the given temperature span, the temperature dependence over most of the range is almost linear, with the moduli decreasing with increasing temperature. For sapphire (α-Al_2O_3), Wachtman and co-workers, (13) reported the following over the range 77–850 K:

$$E = E_o - bTe^{-T_o/T} \tag{11}$$

where E = Young's modulus of elasticity; T = temperature, K; and E_o, T_o, b = empirical constants for the curve shown in Figure 7. For polycrystalline ceramics, the temperature dependence of elastic moduli is more complex; this is particularly so over the temperature range where inelastic effects occur, as illustrated in Figure 8. In polyphase polycrystalline ceramics, for cases where Poisson's ratio is equal for both phases and where adequate bonding exists across phase boundaries, the elastic moduli are approximately additive in proportion to their respective volume fractions.

The effect of porosity on elastic properties of ceramics has been treated by a number of authors. Mackenzie (15) has derived a quadratic relationship for closed spherical pores:

$$E = E_o\,(1 - 1.9\,P + 0.9\,P^2) \tag{12}$$

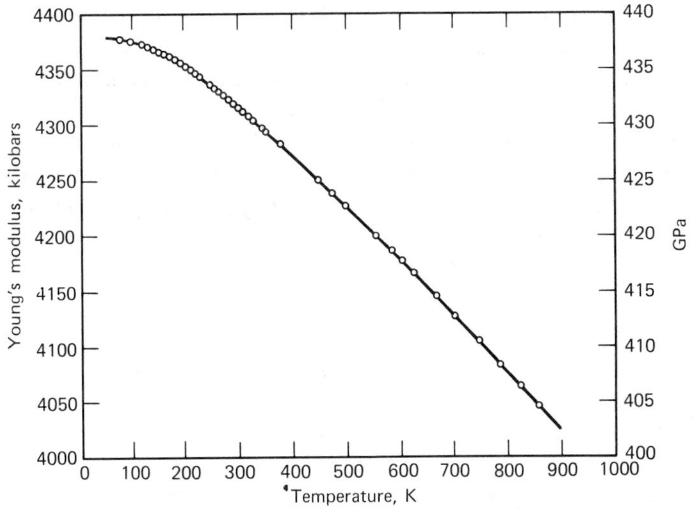

Figure 7. Young's modulus of sapphire as a function of temperature (13).

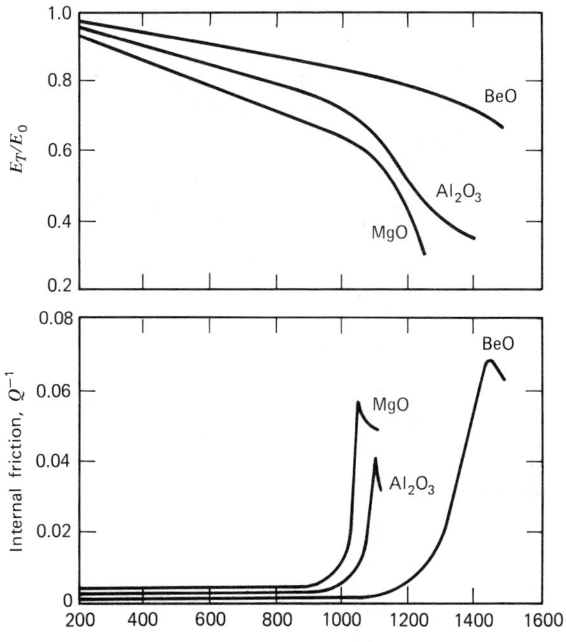

Figure 8. Temperature dependence of Young's modulus and internal friction of polycrystalline BeO, Al_2O_3, and MgO (14).

where E = Young's modulus with porosity, E_o = Young's modulus without porosity, and P = fractional porosity, which reflects the notion that the pore not only fails to contribute its own volume to the load-carrying portion of the structure, but that it shadows an additional, essentially stress-free volume of adjacent good material on either side of the pore along the stress axis. This relationship, and substantiating experimental data for alumina ceramics, are illustrated in Figure 9. Other relationships have been proposed, particularly for low porosity ranges (16).

Inelastic effects are attributable to time-dependent stress–strain relationships. They are particularly evident when the frequency of mechanical oscillation causes resonance of certain structural defects (vacancy pairs, segments of dislocations, grain boundary sliding, etc). At resonance, the defect absorbs energy. Such internal friction peaks are dependent upon the applied frequency, which must be tuned to obtain resonance, and upon the temperature.

Strength

Ceramics are classically considered to be brittle materials. They are strong in compression, but relatively much weaker (frequently by factors of 10 or more) in transverse bending or in tension. In the case of glass at room temperature, it is believed that deformation is entirely elastic up to the point of failure. Brittle fracture always appears to initiate at the specimen surface and can be attributed to the stress-induced catastrophic extension of preexisting defects, the well-known Griffith cracks.

Room Temperature Strengths. The theoretical strengths of ceramic materials, derived from atomic bonding considerations, are far larger than the observed tensile strengths. The presence of flaws, which act as stress concentrators, explain the discrepancy.

The calculation of the stress required to initiate fracture owing to the increased stress in the vicinity of a crack tip was derived by Inglis (17) who derived an expression for stress in the tip vicinity σ_m:

$$\sigma_m = 2\,\sigma \left(\frac{c}{\rho}\right)^{1/2} \tag{13}$$

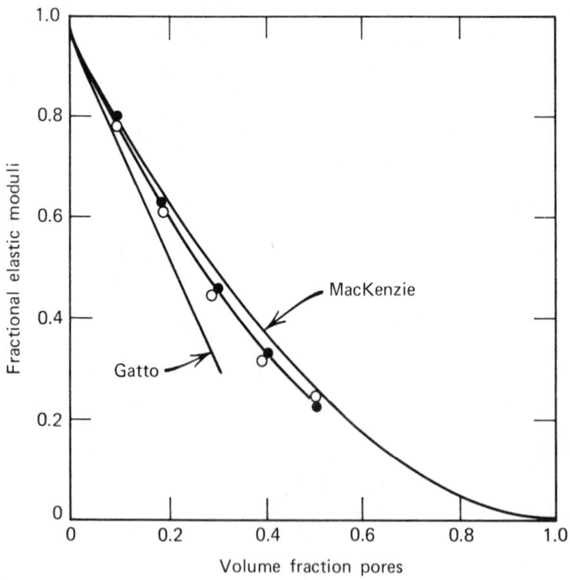

Figure 9. Relative elastic moduli of alumina as a function of porosity. ●, Elastic modulus; ○, rigidity modulus.

where ρ is the crack tip radius, $2c$ is the (elliptical) crack length, and σ is the applied stress.

Using the idea that the crack tip was of atomic spacing dimension and by utilizing the theoretical strength expressed as a function of Young's modulus and surface energy, Orowan (18) expressed the condition for failure as follows:

$$\sigma_f = \left(\frac{E\gamma}{4c}\right)^{1/2} \tag{14}$$

which is similar to the expression derived by Griffith (19) based on elastic and surface energy considerations. The equation assumes that σ_f is smaller than the general plastic yield strength by a factor of two or three. This approach becomes inapplicable when plastic flow becomes possible, which nearly always requires high temperature for ceramic materials.

The crack tip areas, or microflaws from which fracture originates are produced during the fabrication process and can be identified by careful microscopy. An example of such a process-related flaw in alumina is shown in Figure 10.

For many materials the flaw size can be related to the grain size or pore size. It is the largest flaw that affects strength and thus the relative sizes of pores and grains dictates which of these is important (21–22). In fully dense materials there are no large pores and the flaw size is usually related to the grain size. The strength of alumina as a function of grain size is shown in Figure 11. The decreased slope in the smaller region has been attributed to the increased influence of flaws larger than the grain size (21).

High Temperature. At higher temperatures, ceramics can deform plastically by the generation and motion of dislocations. Dislocation nomenclature employs (abc) [abc] or {abc} ⟨abc⟩ Miller indices to denote first the (crystallographic plane) or {family of planes} upon which slip can occur with the dislocations under consideration, and

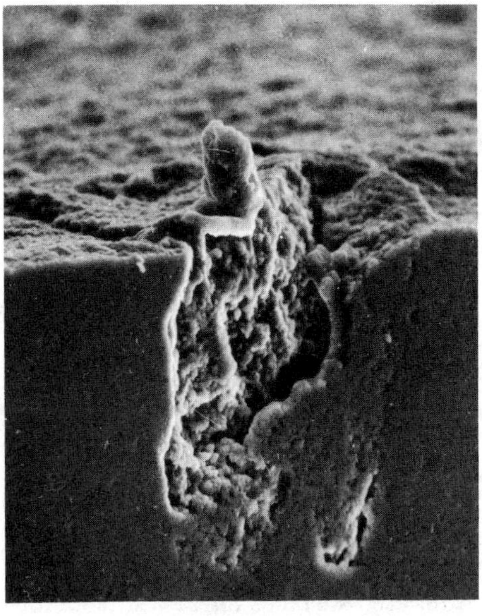

Figure 10. Flaw in well processed Al_2O_3 sintered under rate control (20).

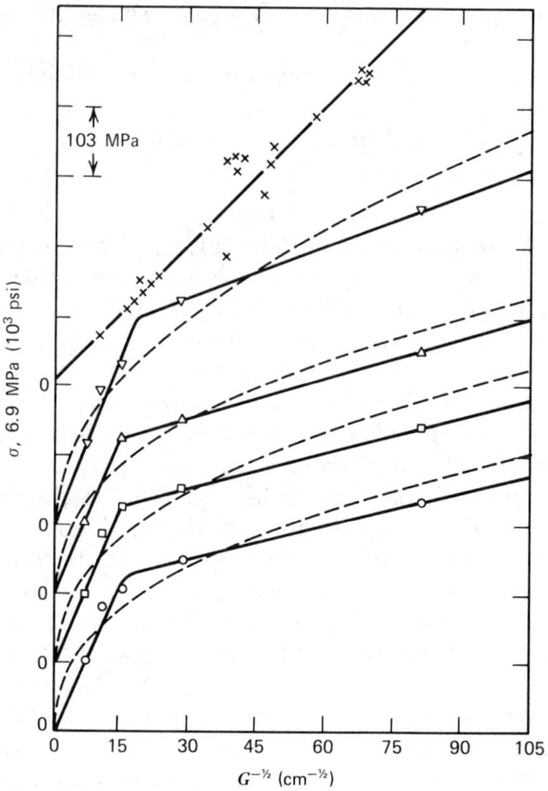

Figure 11. Strength vs grain size in Al_2O_3 (23). x, High pressure, annealed, machine, annealed; σ, room temperature (20). ∇, High pressure, machine, annealed; σ, room temperature (23). Δ, High pressure, machine, annealed; σ, 400°C (23). □, High pressure, machine, annealed; σ, 700°C (23). O, High pressure, machine, annealed; σ, 1000°C (23). To convert MPa to psi, multiply by 145.

second, the [crystallographic direction] or ⟨family of them⟩ of the Burger's vector of the dislocation, which coincides with the slip direction. For further information on the subject of dislocations in ceramics, see ref. 4.

The combined influences of bond character and crystal structure have their greatest effect on the relative mobility of dislocations. When a dislocation moves, bonds must be broken and remade, which is a considerably more difficult process in ionic and covalent-bonded material than in metals having mobile bonding electrons. If the dislocation is a dissociated one, as in alumina or spinel, the partial dislocations and the intervening stacking faults must be moved along together. Only at high temperatures, where ion mobility is high and the several required motions are easy to synchronize, can dislocations be made to move in uniaxially stressed alumina. Even at high temperatures, the stress levels required to induce plastic deformation are high, and are strongly temperature dependent. On the other hand, the simple structure, the predominantly ionic bond, and the relatively simple configuration of the {110} ⟨110⟩ dislocations in magnesium oxide leads to plastic deformation at much lower stress levels, and to a much lower order of temperature dependence (24). Magnesium oxide crystals with properly prepared surfaces may even be deformed in liquid nitrogen, at a temperature where many metals are severely embrittled.

Easy flow requires two particular phenomena; at least five independent slip systems (25) and mobile dislocations. β-SiC has sufficient number of slip systems but does not exhibit ductile behavior until very high temperatures are reached, because of the directionality of covalent bonds. MgO and UO_2 are principally ionic and exhibit flow more readily.

With increases in temperature, each material exhibits first limited and then general plasticity, but the range of behavior is very large. At low temperatures the behavior is as previously discussed: the stresses required for flow are much larger than those required for fracture. As temperature increases, brittle fracture still occurs, but equation 13 no longer strictly applies because flaws are extended by limited plastic processes. At the highest temperature, general plastic flow occurs and the mechanical behavior is similar to metals. Within this framework there are large variations between individual materials.

At high temperatures, ceramics deform in creep, ie, deformation occurs as a function of time under the influence of an applied stress. The general strain rate equation provides an adequate description of such processes:

$$\epsilon = A\sigma n e^{-Q/RT} \qquad (15)$$

where ϵ = strain rate; A = constant; σ = applied stress; n = stress–strain exponent; and Q = activation energy for the flow process. For viscous flow, or grain boundary sliding, the value of the exponent n equals 1: a form of diffusion-controlled deformation, the Nabarro-Herring creep process, also has n equal to 1. With plastic deformation processes, the value of n is much higher, on the order of 4, 4.5, or 5. The creep rate is dependent not only upon the stress level, the stress–strain rate dependence, and the temperature, but also on the average grain size of the material; creep rates are higher for fine-grained material than for coarse-grained. It is important to note that this grain size effect at high temperatures is opposite to the effect of grain size on strength at low temperatures. A ceramic material selected for resistance to creep at high temperatures may be rather weak and prone to brittle fracture at low temperatures, but subject to excessive deformation at high temperatures, creeping even under its own weight. Other microstructural considerations are important to creep behavior; Figure 12 illustrates the marked effect of porosity on creep rates and flow stresses in alumina at high temperature. In many common ceramics a vitreous phase is present and usually is localized at the grain boundary. At elevated temperatures, the strength decreases rapidly, an effect attributable to lowered viscosity and increased lubricity of the intergranular glassy phase.

Engineering with Ceramics. The strength of a polycrystalline ceramic is a very complex quantity, strongly influenced by temperature and environment, the size, shape, and loading geometry of the specific part being tested, its composition, microstructure, and surface condition, as well as its prior thermal and mechanical history.

In structural design with brittle materials, the average breaking strength is of little value. Often a particular sample will break at a considerably lower stress level than the average. Microscopic fracture analysis usually reveals that a large flaw in the form of an impurity crystal, pore or large grain is located at the fracture origin. Failure analyses of ceramic materials, particularly at room temperature, is nearly always concerned with flaw identification.

By subjecting an individual piece to a stress higher than will be seen in service

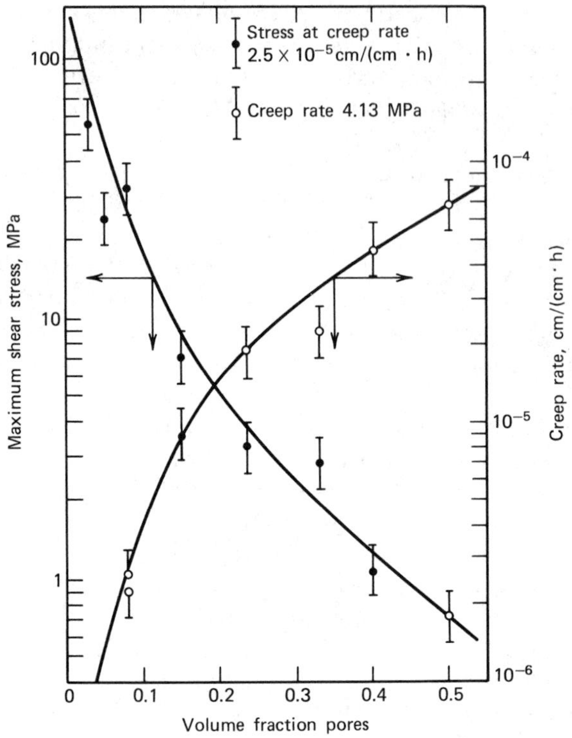

Figure 12. Effect of porosity on torsional creep of Al_2O_3 at 1275°C. To convert MPa to psi, multiply by 145.

(proof testing), the upper limit to the size of flaws which can be present can be calculated. A minimum service life guarantee can be determined from knowledge of crack propagation behavior (11,26).

The contributions of mechanical surface damage to loss of strength and increased likelihood of fracture in ceramics have been described by Westwood as an extreme form of notch sensitivity (27). The action of chemical polishing in reducing notch sensitivity is probably not entirely dependent on complete removal of a crack, but on a blunting of its tip to acceptable proportions. For example, in alkali halides Pulliam (28) found that a 60 s exposure to H_2O caused the radius of curvature of a crack tip to increase from an initial value of 0.2–1 nm to about 200 nm. Chemical polishing and etching operations are essential in detailed studies of plasticity in ceramic materials. Polishing is employed to remove cracks and cold-worked surface layers; etching is used to reveal grain boundaries as grooves and individual dislocations by etch pits at their surface termini.

The value obtained for strength in a test is highly dependent on how the tests are made, particularly with respect to surface preparation. The type of flaw present at the surface can dominate if these flaws are larger than those inherent in the bulk materials.

Strength values of ceramics are also dependent on the time of load, being lower over a longer time as cracks can propagate and cause delayed failure at a load which does not cause catastrophic failure for short loading times. The rate of crack propagation under the influence of stress and chemical environment is very important and is a factor in design criteria. See reference 29 for more information on stress corrosion and time dependence of failure.

The range of transverse bending strengths at room temperature for a variety of ceramic materials is indicated in Table 4. Additional information on strength of ceramic systems can be found in references 4 and 24.

Many ceramics in common usage are not simple, monophase materials, but frequently are quite complex proprietary compositions which have been developed to afford some suitable compromise between a number of conflicting property requirements, together with acceptable workability and firing behavior. The best sources of information on strength and other properties of such ceramic compositions are often to be found in the technical brochures issued by individual ceramic manufacturers.

Electrical and Magnetic Properties

An increasing emphasis on ceramics as electronic and magnetic components in a host of applications has been apparent since World War II, and it continues to intensify (see Ceramics as electrical materials). Ceramic components, ceramic substrates, and ceramic thin films are major constituents in the present development of microcircuits for electronics. Most oxide ceramics in their normal valence states are insulators, and serve very important roles in electronic and electrical devices purely for that reason (see Insulation, electric). Most nonoxide ceramics, and some oxides, are

Table 4. Strength Values for Some Ceramic Materials

Material	Modulus of rupture, MPa[a]
aluminum oxide crystals	345–1034
sintered alumina (ca 5% porosity)	207–345
alumina porcelain (90–95% Al_2O_3)	345
sintered beryllia (ca 5% porosity)	138–276
hot-pressed boron nitride (ca 5% porosity)	48–103
hot-pressed boron carbide (ca 5% porosity)	345
sintered magnesia (ca 5% porosity)	103
sintered molybdenum silicide (ca 5% porosity)	690
sintered spinel (ca 5% porosity)	90
dense silicon carbide (ca 5% porosity)	172
sintered titanium carbide (ca 5% porosity)	1100
sintered stabilized zirconia (ca 5% porosity)	83
silica glass	107
vycor glass	69
pyrex glass	69
mullite porcelain	69
steatite porcelain	138
superduty fire-clay brick	5.2
magnesite brick	27.6
bonded silicon carbide (ca 20% porosity)	13.8
1090°C insulating firebrick (80–85% porosity)	0.28
1430°C insulating firebrick (ca 75% porosity)	1.17
1650°C insulating firebrick (ca 60% porosity)	2.0

[a] To convert MPa to psi, multiply by 145.

semiconductors (qv) with very interesting and very useful properties, eg, SiC, a semiconductor, is much used as high temperature heating element, and as a voltage-dependent resistor to ground (conducting only at high voltages), providing lightning protection to TV antennae, etc.

Those ceramics that contain transition metal ions may display pronounced magnetic effects associated with the spins of unpaired electron orbital states. The magnetic ceramics, notably the ferrites (qv) and the synthetic garnets, constitute a significant segment of technical ceramic endeavor, and are of major importance in computer memory circuits, delay lines, and Faraday rotators in microwave electronic systems, and in a variety of permanent magnet applications which range from TV tube deflection yokes to refrigerator door closures. Magnetic ceramics, in contrast to magnetic metals, do not act as conductors of electricity (see Magnetic materials; Glassy metals; Amorphous magnetic materials; Microwave technology).

In these specialized ceramics, the matter of principal concern is the mobility of electrons, electron holes, and ions under the driving force of alternating electrical and/or magnetic fields. Interactions with these materials occur between nonthermodynamic defects, particularly dislocations, and the electromagnetic properties of ceramics, so that the thermo–mechanical history of the specimens may have a pronounced influence on the observed electrical or magnetic behavior. Many of the interactions of charge carriers with structural defects are analogous to the behavior of photons and phonons with structural defects discussed earlier.

Magnetization. Ceramic materials respond to magnetic fields in a way which is analogous to inelastic and dielectric behavior. The magnetic susceptibility is as follows:

$$\chi_m = M/H = N\alpha_m \tag{16}$$

where M = magnetic moment; H = magnetizing field; N = number of elementary magnetic dipoles per unit volume; and α_m = magnetizability of the elementary dipoles.

Only those electrons having unpaired spin orbitals can contribute to magnetization; they are associated with conduction electrons, with atoms and molecules having an odd number of electrons, and with atoms and ions having partially filled inner electron shells (mainly the $3d$, $4f$, and $5f$ transition series). Magnetic ceramics rely principally upon the last category, and are usually based on compounds involving one or more of those atoms or ions.

Magnetic materials are generally classified as hard or soft; the hard materials are permanent magnets, soft are materials that can be magnetized and demagnetized readily, as required in electronic applications. The main types are outlined below:

Spinel ferrites have a wide range of composition, based on the spinel structure AB_2O_4, which contains both tetrahedral and octahedral sites for the ions, including those with unpaired spin orbitals. Spinels can be either hard or soft and cover a wide range of magnetic properties.

Rare earth garnets have the general formula $A_3Fe_5O_{12}$, with the iron in two valence states, and are used in magnetic bubble memories (see Magnetic materials).

Hexagonal ferrites are related to the spinel structure with a hexagonally packed structure of the formula $AB_{12}O_{19}$, where A is usually divalent Ba, Sr, or Pd and B is trivalent Al, Ga, Cr, or Fe. Hexagonal ferrites are primarily used for permanent magnets.

Ceramic magnets are of increasing importance in the electronics industry particularly in high frequency devices. For more detail on magnetic properties see reference 30.

Composites and Cermets

An important segment of the ceramic industry is concerned with complex materials or components of which the ceramic phase is only a part; another part is metallic or plastic. The useful properties of the combination are different from those of either phase alone. For example, porcelain enamel, a vitreous or crystal-plus-vitreous ceramic phase, is applied to metal structures to give oxidation and and corrosion resistance, smoothness, color, texture, and other desirable properties. The metal gives shape and strength to the article, and provides for easier forming and generally lighter weight than would be the case if it had been produced entirely as a ceramic (see Enamels).

Composites. As a rule of thumb, composites contain at least one phase which is macroscopic in at least one dimension; thus the minimum qualifying phase would have to have fibrous shape. Ceramic–metal composites, particularly those based on metallic honeycombs or webbings impregnated with a ceramic phase, have been employed for aerospace hardware, including reentry nose cones, rocket nozzles, ram-jet chambers, etc, under extreme conditions of thermal shock, high surface temperature, high-velocity erosion and/or ablation (see Ablative materials). Another important ceramic-based composite is glass–fiber reinforced plastic, which finds a variety of uses, including ablative nose cones in aerospace applications. Ceramic fiber-reinforced metals, metal-impregnated ceramics, and metal–ceramic laminates are receiving considerable research and development attention, particularly for high-temperature, high-strength materials capable of withstanding thermal shock or mechanical impact (see Composite materials).

Cermets. If the ceramic and metallic phases are randomly shaped and intimately dispersed one within the other on a microstructural scale, the resultant material is properly termed a cermet. A number of important cermets have been developed; by far the largest group of these are the metal-bonded hard carbides or cemented carbides (see Carbides).

Uses of Ceramics

Ceramics differ from other engineering materials (metals, plastics, wood products, textiles) in a number of individual properties, but perhaps the most distinctive difference to a designer or potential user of ceramic ware is the particularity of the individual ceramic piece. Ceramics are not readily shaped or worked after firing, except by very costly grinding; consequently, they normally must be used as is. Except for some simple tile, rod, and tube shapes of limited sizes, ceramics cannot be marketed by the foot or by the yard, nor cut to fit on the job.

All the useful properties, including shape, size, etc, must be provided in advance, beginning with the very early stages of ceramic processing, and not as an afterthought. The structural integrity of each piece must be preserved through a variety of thermal and mechanical stress exposures during processing and until it is finally installed and in service. If a ceramic should fail in service as a result of a variety of causes (brittle fracture on impact, thermal shock, dielectric breakdown, abrasion, melting slag corrosion, etc) it is not likely to be repairable, and usually must be replaced *in toto.*

Significant advancements have been made in fundamental understanding and technological control of the properties of ceramics, and of their utilization in many new, demanding, highly technical applications. The industry, in general, and the technical and electronic ceramic portions of it, in particular, have devised production and control techniques for mass producing complex shapes in bodies having carefully controlled electrical, magnetic, and/or mechanical properties, while maintaining dimensional tolerances that are good enough to permit relatively easy assembly with other components.

Many ceramics are produced in volume as standard items; refractory bricks and shapes, crucibles, muffles, furnace tubes, insulators, thermocouple protection tubes, capacitor dielectrics, hermetic seals, fiber boards, etc, are routinely stocked by a number of ceramic producers in a variety of compositions and sizes. It is usually quicker and cheaper to use stock items whenever possible. When stock items will not meet the need, most manufacturers are prepared to produce custom items to specifications. The more stringent the requirements for a given property of the ceramic, or the more restrictive the requirements for specific combinations of properties, sizes, and shapes, the more limited are the acceptable compositional, microstructural, and configurational parameters for the ceramic, and hence the greater the cost and difficulty of manufacture. Most ceramic manufacturers have experienced staff engineers and designers who are well qualified to work with potential customers on details of design of ceramic ware.

BIBLIOGRAPHY

"Properties and Applications of Ceramic Materials" under "Ceramics" in *ECT* 2nd ed., Vol. 4, pp. 793–832, by Hayne Palmour III, North Carolina State of The University of North Carolina.

1. W. D. Kingery, "Oxides for High Temperature Applications," *Proceedings of International Symposium on High Temperature Technology, Stanford Research Institute, Asilomar, Calif., Oct. 6–9, 1959,* McGraw-Hill Book Co., Inc., New York, 1960, pp. 76–89.
2. P. D. Johnson, *J. Am. Ceram. Soc.* **33**(5), 168 (1950).
3. G. Economos and W. D. Kingery, *J. Am. Ceram. Soc.* **36**(12), 403 (1953).
4. W. D. Kingery, H. K. Bowen, and D. R. Uhlmann, *Introduction to Ceramics,* 2nd ed., John Wiley & Sons, Inc., New York, 1976.
5. D. W. Lee and W. D. Kingery, *J. Am. Ceram. Soc.* **43,** 594 (1960); ref. 4, p. 675.
6. Ref. 4, p. 587.
7. G. G. Koerber, *Properties of Solids,* Prentice-Hall, Inc., Englewood Cliffs, N.J., 1962.
8. Ref. 4, p. 616.
9. Ref. 4, p. 643.
10. W. R. Buessem, "Internal Ruptures and Recombinations in Anisotropic Ceramic Materials" in W. W. Kriegel and H. Palmour, III, eds., *Mechanical Properties of Engineering Ceramics,* Interscience Publishers, New York, 1961, pp. 127–148.
11. T. K. Gupta, *J. Am. Ceram. Soc.* **55,** 249 (1972); ref. 4, p. 829.
12. D. P. H. Hasselman, "Thermal Stress Crack Stability and Propagation in Severe Thermal Environments" in W. W. Kriegel and Hayne Palmour, III, eds., *Ceramics in Severe Environments,* Plenum Press, New York, 1971.
13. J. B. Wachtman, Jr., W. E. Tefft, and D. G. Lam, Jr., "Young's Modulus of Single Crystal Corundum from 77 to 850 K" in ref. 10, pp. 221–283.
14. R. Chang, "The Elastic and Anelastic Properties of Refractory Materials for High Temperature Applications" in ref. 10, pp. 209–220.
15. J. K. Mackenzie, *Proc. Phys. Soc. (London) B* **63,** 2 (1950).
16. R. M. Sprigg, *J. Am. Ceram. Soc.* **44**(12), 628 (1961); D. P. H. Hasselman, *J. Am. Ceram. Soc.* **45**(9), 452 (1962).
17. R. Inglis, *Trans. Inst. Nav. Arch.* **55,** 219 (1913).

18. H. Orowan, *Z. Krist.* **A89,** 327 (1934).
19. A. A. Griffith, *Trans. R. Soc.* **A221,** 163 (1920).
20. R. W. Rice in R. K. MacCrone, ed., *Properties and Microstructure,* Academic Press, Inc., New York, 1978.
21. R. W. Rice and co-workers in R. C. Bradt, D. P. H. Hasselman, and F. F. Lange, eds., *Fracture Mechanics of Ceramics,* Vol. 4, Plenum Press, New York, 1978.
22. T. M. Hare and Hayne Palmour, III, in G. Y. Onoda and L. L. Hench, eds., *Processing of Ceramics Before Firing,* John Wiley & Sons, Inc., New York, 1978.
23. S. C. Carniglia, *J. Am. Ceram. Soc.* **55,** 243 (1972).
24. E. R. Parker, "Ductility of Magnesium Oxide" in ref. 10, pp. 65–87.
25. R. Von Mises, *Z. Agnew. Math. Mech.* **8,** 161, 1928.
26. R. C. Bradt, D. P. H. Hasselman, and F. F. Lange, eds., *Fracture Mechanics of Ceramics,* Vols. 3–4, Plenum Press, New York, 1978.
27. A. R. C. Westwood, *Effects of Environment on Fracture Behavior, Technical Report G2-15,* Research Institute for Advanced Study, Baltimore, Md., 1962.
28. G. R. Pulliam, *J. Am. Ceram. Soc.* **42**(10), 477 (1959).
29. S. M. Wiedhorn and co-workers, *J. Am. Ceram. Soc.* **57,** 336 (1974).
30. R. S. Tebble and D. J. Craik, *Magnetic Materials,* Wiley-Interscience, New York, 1969.

THOMAS M. HARE
North Carolina State University

CLAYS

SURVEY

Clays as they occur in nature are rocks that may be consolidated or unconsolidated. They are distinctive in at least two properties which render them technologically useful.

(*1*) Plasticity signifies the property of the clay when wetted that permits deformation by application of relatively slight pressure and retention of the deformed shape after release of the pressure. This property distinguishes clays from hard rocks.

(*2*) Clays are composed of extremely fine crystals or particles, often colloidal in size and usually platy in shape, of clay minerals with or without other rock or mineral particles. The clay minerals, mostly phyllosilicates, are hydrous silicates of Al, Mg, Fe, and other less abundant elements.

The very fine particles yield very large specific-surface areas that are physically sorptive and chemically surface-reactive. Many clay mineral crystals carry an excess negative electric charge owing to internal substitution by lower valent cations, and thereby increase internal reactivity in chemical combination and ion exchange. Clays may have served as substrates that selectively absorbed and catalyzed amino acids in the origin of life. They apparently catalyze petroleum formation in rocks (see Petroleum).

Because clays (rocks) usually contain more than one mineral and the various clay minerals differ in chemical and physical properties, the term clay may signify entirely different things to different technologists.

For geologists, clay refers to sediments or sedimentary rock particles with a diameter of 3.9 μm or less. Soil scientists define clays as disperse systems of the colloidal products of weathering in which secondary mineral particles of dimensions smaller than 2 μm predominate.

Ceramists, who probably process the greatest quantity of clay, usually emphasize aluminosilicate content and plasticity as in the following standard definition (1): Clay is " . . . an earthy or stony mineral aggregate consisting essentially of hydrous silicates and alumina, plastic when sufficiently pulverized and wetted, rigid when dry, and vitreous when fired at a sufficiently high temperature." Ceramists recognize that this definition is open to qualification for their definition (1) of flint fire clay as "practically devoid of natural plasticity," conflicts with their standard definition. The source of the plasticity is defined as (2): " . . . with more or less plasticity due to colloids of organic or mineral nature," but it may arise from other causes (3). Even the origin of the clay has been included in a definition (4): "clays are the weathered products of the silicate rocks . . .," but it has been recognized as a product of deep-seated alteration of silicate minerals by hydrothermal solutions rising from an igneous source as well as a product of surface weathering (5–6). Although clay minerals are usually considered breakdown products of silicates, largely by hydrolysis, they may be built up from hydrates of silica and alumina (7).

A clay deposit usually contains nonclaylike minerals as impurities, although these impurities may actually be essential in determining the unique and specially desired properties of the clay. Both crystalline and amorphous minerals and compounds may be present in a clay deposit (8).

The geologist views clay as a raw material for shale; the pedologist as a dynamic system to support plant life; the ceramist as a body to be processed in preparation for vitrification; the chemist and technologist as a catalyst, adsorbent, filler, coater, or source of aluminum or lithium compounds, etc. A broad definition includes the following properties of clays (rocks):

(*1*) The predominant content of clay minerals, which are hydrated silicates of aluminum, iron, or magnesium, both crystalline and amorphous. These range from kaolins, which are relatively uniform in chemical composition, to smectites, which vary widely in their base exchange properties, expanding crystal lattice, and proxying of elements. The illite group, typically a component of sediment (9), includes micas, although illite [*1273-60-3*] is considered a variation of muscovite [*1318-94-1*] because illite contains less potassium and more water. The chlorite clay minerals resemble metamorphic chlorite [*1318-59-8*] (10), but aluminous, or at least aluminum-rich, chlorites have also been found in soils (4,11). Vermiculite [*1318-00-9*], characterized by a highly expanding crystal lattice, and sepiolite [*15501-74-3*], and palygorskite [*12174-11-7*], which possess chain or fiber structures, must also be included. Moreover, clay minerals are excellent examples of mixed layering, both random and regular, in layer-structure silicates (Fig. 1).

(*2*) The possible content of hydrated alumina and iron. Hydrated alumina minerals like gibbsite [*14762-49-3*], boehmite [*1318-23-6*], and diaspore [*14457-84-2*] occur in bauxitic clays. Bauxites grade chemically into hydrated ferruginous and manganiferous laterites. Hence, finely divided M_2O_3, usually hydrated, may be a significant constituent of a clay. Hydrated colloidal silica may play a role in the slippery and sticky properties of certain clays.

(*3*) The extreme fineness of individual clay particles, which may be of colloidal size in at least one dimension. Clay minerals are usually platy in shape, and less often lathlike and tubular or scroll shaped (13). Because of their fineness they exhibit surface chemistry properties of colloids (14). Some clays possess relatively open crystal lattices and show internal surface colloidal effects. Other minerals and rock particles, which are not hydrous aluminosilicates but which also show colloidal dimensions and characteristics, may occur intimately intermixed with the clay minerals and play an essential role.

(*4*) The property of thixotropy in various degrees of complexity (3). This includes, in addition to technological applications, the loss of stability as shown by the sometimes catastrophic flow of quick clays, especially in Norway, Sweden, and Canada (15).

(*5*) The possible content of quartz [*14808-60-7*] sand and silt, feldspars, mica [*12001-26-2*], chlorite, opal, volcanic dust, fossil fragments, high-density so-called heavy minerals, sulfates, sulfides, carbonate minerals, zeolites, and many other rock and mineral particles ranging upward in size from colloids to pebbles. An extreme example is that of a clay from western Texas composed of 98.5% dolomite [*16389-88-1*], $CaMg(CO_3)_2$, and 1.5% iron oxide and alumina; it occurs in rhombic particles averaging 0.008 mm in diameter, and possesses "sufficient plasticity to be molded into bricklets" (16).

The synthesis of clay minerals has been extensively studied (17–21). Many experiments were performed at high temperature, some using synthetic chemicals, others

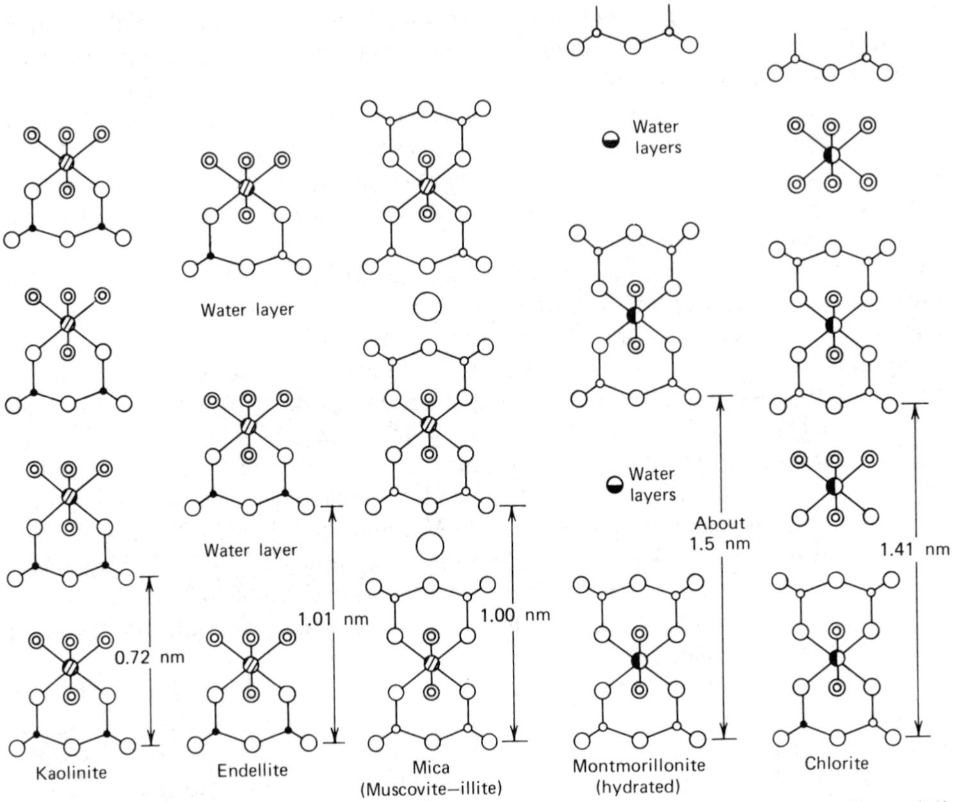

Figure 1. Diagrammatic representation of the succession of layers in some layer lattice silicates (12). O, oxygen; ◎, (OH); ●, silicon; ○, Si–Al; ⊘, aluminum; ◐, Al–Mg; ◯, potassium; ◒, Na–Ca.

using in part naturally occurring minerals. Organic compounds facilitated the synthesis of kaolinite at low temperature by condensing aluminum hydroxide into octahedrally coordinated sheets (18).

Geology and Occurrence

Clays may originate through several processes: (*1*) hydrolysis and hydration of a silicate, ie, alkali silicate + water → hydrated aluminosilicate clay + alkali hydroxide; (*2*) solution of a limestone or other soluble rock containing relatively insoluble clay impurities that are left behind; (*3*) slaking and weathering of shales (clay-rich sedimentary rocks); (*4*) replacement of a preexisting host rock by invading guest clay whose constituents are carried in part or wholly by solution; (*5*) deposition of clay in cavities or veins from solution; (*6*) bacterial and other organic activity, including the extraction of metal cations as nutrients by plants; (*7*) action of acid clays, humus, and inorganic acids on primary silicates; (*8*) alteration of parent material or diagenetic processes following sedimentation in marine and freshwater environments (22–23); and (*9*) resilication of high alumina minerals.

Every state in the United States has within its boundaries clays or shales that may be utilized in the manufacture of bricks, tiles, and other heavy clay products. Some blending of materials is often necessary to control shrinkage of the product, and the economics of manufacture are governed by the demands of fuel, labor, transportation costs, and the market.

Glacial clays, as unassorted glacial till or secondarily deposited melt water, are abundant in the United States north of the Missouri and Ohio Rivers. Quartz-rich sand, silt, or pebbles, especially limestone, may occur mixed with the clay.

Kaolins are plentiful in North Carolina, South Carolina, Georgia (24), Florida, and Vermont. Certain flint clays and other clays with a kaolinitic composition which can replace kaolins in some uses (25) occur in Missouri, Arkansas, Colorado, Texas, Ohio, Indiana, Oregon, and Pennsylvania.

Ball clays, ie, clays with high plasticity and strong bonding power, are obtained primarily from western Tennessee and Kentucky, though some are found in New Jersey. Other plastic clays, especially plastic fire clay, are extensively produced in Missouri, Illinois, Ohio, Kentucky, Mississippi, Alabama, and Arkansas.

Fire clays are those that resist fusion at a relatively high temperature, usually around 1600°C. Missouri, Pennsylvania, Ohio, Kentucky, Georgia, Colorado, New Jersey, Texas, Arkansas, Illinois, and Maryland are major producers of fire clays. Deposits also occur in South Carolina, Indiana, Alabama, California, West Virginia, and Oregon.

Loess is a quartz-rich, clayey silt, windblown in origin but, in some cases, reworked by water. It is prevalent along the Missouri, Mississippi, and Ohio Rivers and their tributaries. Loess has been used primarily for brick making.

Adobe is a calcareous, sandy to silty clay used extensively for making sundried brick for local use in the more arid southwestern states.

Slip clay for glazing pottery is produced near Albany, New York.

Bentonite [*1302-78-9*] is widely distributed geographically and geologically, and also varies widely in properties. Swelling bentonite occurs in Wyoming, South Dakota, Montana, Utah, Nevada, and California. Bentonite that swells little or not at all occurs in large quantities in Texas, Arkansas, Mississippi, Kentucky, and Tennessee.

Fuller's earth and bleaching clays, though found chiefly in Georgia and Florida, also occur in Arizona, Arkansas, California, Colorado, Illinois, Mississippi, Missouri, Oklahoma, Texas, and Utah.

High alumina clays refer in the ceramic industries to nodular clays, burley-flint clay, burley and diaspore, gibbsitic or bauxitic kaolins (clays), abrasives clays, and others. Since the depletion of diaspore varieties in Missouri and Pennsylvania, most bauxitic kaolin and clay is produced in Alabama and Arkansas.

Though each continent has clays of almost every type, certain deposits are outstanding (26–27). There are tremendous reserves of white kaolin on the Jari and Capim Rivers of Brazil, and deposits of bauxitic clay shared with countries across Brazil's northern borders. England has the famous kaolins of the Cornwall district. Refractory clays in Scotland provide the raw materials for a refractories industry. Deposits similar to those in Cornwall are found in Brittany and neighboring areas in France. Czechoslovakia, the Federal Republic of Germany, and the German Democratic Republic have large reserves of kaolin accompanied by quartz and mica. There are large deposits of bauxitic clays (with bauxite) in Hungary and Yugoslavia. Both flint clay and white kaolin are found in South Africa. Kaolins are found in the People's Republic of China. Japan has notable hydrothermal kaolins, sedimentary kaolin, smectite, and flint clay. Hydrothermal kaolins are also widely distributed in Mexico. Australia has large deposits of flint clay and kaolin clay naturally calcined by the burning of coal beds. New Zealand has hydrothermal clays. India has flint and lateritic clays. Sepiolite is found near Madrid, Spain. Localities producing bauxite almost always have a potential for producing associated high alumina clays.

The commercial value of a clay deposit depends upon market trends, competitive materials, transportation facilities, new machinery and processes, and labor and fuel costs.

Naturally exposed outcrops, geological area and structure maps, aerial photographs, hand and power auger drills, core drills, earth resistivity, and shallow seismic methods are used in exploration for clays (28). Clays are mined primarily by open-pit operation, including hydraulic extraction; however, underground mining is also practiced.

Specific information concerning the geology or occurrence of a particular deposit or variety of clay may be obtained from a state geological survey; general information may be obtained from the U.S. Geological Survey or the U.S. Bureau of Mines. Similar agencies may be contacted in foreign countries, or references may be consulted (8,29–31, and the general references).

Technological Classification

The technological classification of a clay (rock, deposit) should take into account the following factors: (*1*) The dominant clay mineral type including breakdown into its polymorphs, the sites and amount of charge on it, and shape of clay crystal and particle. (*2*) The clay minerals present in minor quantities, but perhaps coating the surface of the major constituent. (*3*) The particle-size distribution of the clay and other minerals. (*4*) Ion-exchange capacity (cation, anion) and neutral molecule sorption. (*5*) The type of exchangeable ions present on the clay and degree of saturation of exchange sites. (*6*) Hygroscopicity of the clay. (*7*) Reactivity of the clay with organic compounds. (*8*) Expansion potentialities of the clay mineral lattice. (*9*) Electrolytes and solutions in association with the clay deposit. (*10*) The accessory minerals, or mineral impurities, their sizes, homogeneity of mixture, and ion-exchange capacity. (*11*) Content of organic matter and especially its occurrence, size and discreteness of particles, its adsorption on and/or within the clay crystal units, and protective colloidal action. (*12*) Presence or absence of bacteria or other living organisms. The pH and other properties of a clay deposit may vary notably within a short time where bacteria are growing. (*13*) Content of hydrated alumina and/or silica, which are relatively soluble in ground water or in dilute acid or alkali. (*14*) The structure and texture of the clay deposit, such as lamination, orientation of mineral particles, and other gross features. (*15*) The rheological properties of both the natural and processed clay. (*16*) The engineering strength, and sensitivity to moisture, desalination, and shock, eg, the quick clays.

Mineralogy

The development of apparatus and techniques, particularly x-rays, contributed greatly to research on clay minerals. Today crystalline clay minerals are identified and classified (32) primarily on the basis of crystal structure and the amount and locations of charge (deficit or excess) with respect to the basic lattice. Amorphous (to x-ray) clay minerals are poorly organized analogues of crystalline counterparts.

Various techniques are used to study crystalline clay minerals including the polarizing microscope (33), chemical analysis and computation of the mineral formula taking into full account the substitution of atoms (34), staining (35–37), density, possible electrical double refraction (38), dehydration, base exchange (39–40), electron micrographs, transmission electron microscopy (tem) (41) and scanning electron microscopy (sem) (42–44), x-ray or electron powder diffraction patterns (32), differential thermal analysis and thermal balance, imbibition (45), infrared absorption (39,41), and field appearance, especially responses to weathering (46 and the general references) (see Analytical methods).

Clay minerals are divided into crystalline and paracrystalline groups and a group amorphous to x-rays (47).

Although the clays of the different groups are similar, they show vastly different mineralogical, physical, thermal, and technological properties (3). Chemical analysis alone may have limited value in revealing the clay's identity or usefulness. The mineral composition, which reveals the organization of the constituent elements is most important.

CRYSTALLINE AND PARACRYSTALLINE GROUPS

Kaolins. The kaolin minerals include kaolinite [*1318-74-7*], dickite [*1318-45-2*], and nacrite [*12279-65-1*] (all $Al_2O_3.2SiO_2.2H_2O$), and halloysite–endellite ($Al_2O_3.2SiO_2.2H_2O$ and $Al_2O_3.2SiO_2.4H_2O$, respectively) (47–50). The structural formulas for kaolinite and endellite [*12244-16-5*] are $Al_4Si_4O_{10}(OH)_8$ and $Al_4Si_4O_{10}(OH)_8.4H_2O$, respectively. The kaolinite lattice (12,51) consists of one sheet of tetrahedrally coordinated Si (with O) and one sheet of octahedrally coordinated Al (with O and OH), hence a 1:1 or two-layer structure. A layer of OH completes the charge requirements of the octahedral sheet. The so-called fire clay mineral is a *b*-axis disordered kaolinite (12); halloysite [*12244-16-5*] and endellite are disordered along both the *a* and *b* axes. Indeed, most variations in the kaolin group originate as structural polymorphs. Representative analyses (47–49) of the kaolin minerals are given in Table 1.

As can be seen from the optical constants given in Table 2, kaolinite and dickite are easily distinguished where they occur in recognizable crystals. Nacrite is relatively rare. Halloysite is usually exceedingly fine grained, showing a mean index of refraction of about 1.546. The index of refraction for endellite varies somewhat with the immersion liquid used; it ranges from 1.540 to 1.552 (48).

X-ray studies show that kaolin minerals have two-layer crystal lattices: a sheet of silica tetrahedra and an alumina–gibbsite sheet. Adjacent cells are spaced about 0.71 nm across the (001) plane. The interplanar spacings normal to the (001) cleavage are the most significant criteria used in x-ray differentiation between the clay mineral groups. Within the kaolin group other x-ray structural differences are used to distinguish the members (12).

Endellite, ie, the hydrated form of halloysite (48), expands to 1.0 nm along the *c* axis when solvated in ethylene glycol. Halloysite may be differentiated from kaolinite and dickite by treatment with potassium acetate and ethylene glycol (54). Halloysite–endellite minerals are usually tubular, or rolled or scroll-shaped in morphology; this was formerly interpreted as the result of a misfit between the octahedral and tetrahedral sheets (13). An alternative interpretation is that certain crystals have an inherent roundish, elongate morphology (55).

Table 1. Chemical Analyses of the Kaolin Minerals, %

Component	1[a]	2[b]	3[c]	4[d]	5[e]	6[f]
SiO_2	45.44	40.26	46.5	45.78	42.68	44.90
Al_2O_3	38.52	37.95	39.5	36.46	38.49	38.35
Fe_2O_3	0.80	0.30		0.28	1.55	0.43
FeO				1.08		
MgO	0.08			0.04	0.08	trace
CaO	0.08	0.22		0.50		trace
K_2O	0.14	0.74 (with Na_2O)		0.25 (with Na_2O)	0.49	0.28
Na_2O	0.66				0.28	0.14
TiO_2	0.16				2.90	1.80
H_2O removed						
at 105°C	0.60	4.45		2.05		
above 105°C	13.60	15.94	14.0	13.40	14.07	14.20
Total	*100.08*	*99.86*	*100.0*	*99.84*	*100.54*	*100.10*

[a] Kaolinite, Roseland, Va. (49).
[b] Halloysite, Huron Co., Ind. (47).
[c] Theoretical kaolinite.
[d] Washed kaolin, Webster, N. C. (6).
[e] Flint fire clay, near Owensville, Mo. (52).
[f] Typical sedimentary kaolin, S. C., Ga., Ala. Courtesy S. C. Lyons.

Table 2. Optical Constants of Kaolinite, Dickite, and Glauconite[a]

Optical constant	Kaolinite	Dickite	Glauconite
index of refraction			
α	1.561	1.561	1.597
β	1.565	1.563	1.618
γ	1.565	1.567	1.619
optical character	neg	pos	neg
extinction angle,°	3.5	ca 16	
dispersion	$r > v$		
axial angle, 2V	10–57°		20°
X	nearly perpendicular to (001)		dark bluish green[b]
Z, Y			yellow[b]

[a] Ref. 53.
[b] Pleochroic.

Staining kaolin minerals with aniline dyes produces varied artificial pleochroism which may be sufficiently selective to identify the mineral species (36). In differential thermal analysis, kaolinite shows a strong endothermic peak at about 620°C and a strong exothermic peak at about 980°C, which sharply differentiates it from the other clay mineral groups. Electron micrographs show kaolinite in roughly equidimensional, pseudohexagonal plates and halloysite in lath-shaped crystals. Kaolinite, the most abundant, and halloysite, the second most abundant of the kaolin group of minerals are those most used by industry. Problems of clay mineral differentiation within the kaolin group arise rarely; however, adequate means for identifying and distinguishing the various mineral species are available (30,36). The cationic base-exchange capacity of kaolinite is low, less than 10 meq/100 g of dry clay.

Kaolin most commonly originates by the alteration of feldspar or other aluminum silicate via an intermediate solution phase (56–57) usually by surface weathering (22,58) or by rising warm (hydrothermal) waters. A mica, or hydrated alumina solid phase may intervene between parent and kaolin minerals.

Large deposits of relatively pure kaolinite have developed from parent, feldspar-rich pegmatites, whereas others are secondarily deposited in sedimentary beds after transportation. Colloidal fractions of geologically ancient soils were presumably concentrated in old swamps and leached to develop kaolinitic clay deposits (6,25). Today kaolinite is formed by weathering in an oxidizing environment under acid conditions, and in a reducing environment where the bases such as calcium, magnesium, alkalies, and iron(II) are removed. Removal of the bases is essential in kaolin formation (59). With more intense leaching of silica from the clay an aluminous hydrate remains and the clay becomes bauxitic. Kaolinite may develop from the silication of gibbsite (7). Halloysite–endellite may be formed either by hydrothermal or weathering processes. Allophane [*12172-71-3*] may have led to halloysite and thence to kaolinite as crystallization became more highly ordered following weathering (60); however, this sequence does not always prevail. Some indications point toward a possible association of acid sulfate waters and mobile potassium with the origin of endellite–halloysite (58).

The textures of kaolin (rock) include varieties similar to examples observed in igneous and metamorphic as well as sedimentary rocks (57). Kaolin grains and crystals may be straight or curved, sheaves, flakes, face-to-face or edge-to-face floccules, interlocking crystals, tubes, scrolls, fibers, or spheres (57).

The kaolin group is transformed at high temperatures to a silica–alumina spinel structure (61) previously interpreted as gamma alumina, thence to mullite [*1302-93-8*] with or without accompanying cristobalite [*14464-46-1*] (62–63). The alkali metal, flux, and content of kaolin clay strongly influence the phases formed upon heating (62).

Serpentines. Substituting 3 Mg in the kaolin structure, $Al_2Si_2O_5(OH)_4$, results in the serpentine minerals, $Mg_3Si_2O_5(OH)_4$. In serpentines all three possible octahedral cation sites are filled, yielding a trioctahedral group carrying a charge of +6, whereas in kaolinite only two thirds of the sites are occupied by Al yielding a dioctahedral group also with a charge of +6. Most serpentine minerals are tubular to fibrous in structure presumably because of misfit, the reverse geometry of halloysite–endellite, between Mg octahedral and tetrahedral layers. Thus, structurally, serpentines are analogues of kaolin minerals, although they depart from aluminous clays in certain other properties.

Chrysotile [*12001-27-5*] (serpentine) occurs in both clino and ortho structures. Both one-layer ortho and clino, and six-layer ortho (as in nacrite) structures have been observed. Chrysotile transforms at high temperature to forsterite [*15118-03-3*] and silica.

Amesite [*12413-43-3*] approximates $(Mg_2Al)(SiAl)O_5(OH)_4$ in composition, cronstedite [*61104-63-0*], $(Fe_2^{2+}Fe^{3+})(SiFe^{3+})O_5(OH)_4$, and chamosite [*12173-07-2*], $(Fe^{2+},Mg)_{2.3}(Fe^{2+}Al)_{0.7}(Si_{1.14}Al_{0.86})O_5(OH)_4$. Garnierite [*12198-10-6*] is possibly a nickel serpentine; a cobalt serpentine has been synthesized (12).

Smectites (Montmorillonites). Smectites are the 2:1 clay minerals that carry a lattice charge and characteristically expand when solvated with water and alcohols, notably ethylene glycol and glycerol. In earlier literature, the term montmorillonite was used for both the group (now smectite) and the particular member of the group in which Mg is a significant substituent for Al in the octahedral layer. Typical formulas are shown in Table 3. Additional, less common smectites include volkhonskoite [*12286-87-2*] which contains Cr^{2+}; medmontite [*12419-74-8*], Cu^{2+}; and pimelite [*12420-74-5*], Ni^{2+} (12).

Smectites are derived structurally from pyrophyllite [*12269-78-2*], $Si_8Al_4O_{20}(OH)_4$, or talc [*14807-96-6*], $Si_8Mg_6O_{20}(OH)_4$, by substitutions mainly in the octahedral layers. Some substitution may occur for Si in the tetrahedral layer, and by F for OH in the structure. When substitutions occur between elements (ions) of unlike charge, deficit or excess charge develops on corresponding parts of the structure. Deficit charges in smectite are compensated by cations (usually Na, Ca, K) sorbed between the three-layer (two tetrahedral and one octahedral, hence 2:1) clay mineral sandwiches. These are held relatively loosely, although stoichiometrically, and give rise to major cation exchange properties of the smectite. Representative analyses of smectite minerals are given in Table 4.

The determination of a complete set of optical constants of the smectite group is usually not possible because the individual crystals are too small. However, suspensions of the clays evaporated on a glass slide usually deposit a pseudocrystal film from which α and γ component values may be determined. The clay films should be soaked in acetone to displace the air and then left to stand in the index immersion oil several hours before measuring indexes of refraction. Representative optical measurements are given in Table 5.

In the montmorillonite–nontronite series, as the iron(III) content increases from 0 to 28%, the α index ranges from 1.523 to 1.590, and the γ from 1.548 to 1.632.

X-ray diffraction patterns yield typical 1.2–1.2 nm basal spacings for smectite partially hydrated in an ordinary laboratory atmosphere. Solvating smectite in ethylene glycol expands the spacing to 1.7 nm, and heating to 550°C collapses it to 1.0 nm. Certain micaceous clay minerals from which part of the smectites' metallic interlayer cations has been stripped or degraded and replaced by H^+ or H_3O^+ expand similarly. Treatment with strong solutions of potassium salts may permit differentiation of these expanding clays (65).

Smectite [*12199-37-0*] from an oxidized outcrop is stained light blue by a dilute solution of benzidine hydrochloride. The color does not arise from smectite specifically, but from reaction with a high oxidation state of elements such as Fe^{3+} or Mn^{4+} (37).

Table 3. Typical Formulas of Smectite Minerals[a]

Mineral	CAS Registry No.	Formula[b]
montmorillonite	[*1318-93-01*]	$[Al_{1.67}Mg_{0.33}(Na_{0.33})]Si_4O_{10}(OH)_2$
beidellite	[*12172-85-9*]	$Al_{2.17}[Al_{0.33}(Na_{0.33})Si_{3.17}]O_{10}(OH)_2$
nontronite	[*12174-06-0*]	$Fe(III)\ [Al_{0.33}(Na_{0.33})Si_{3.67}]O_{10}(OH)_2$
hectorite	[*12173-47-6*]	$[Mg_{2.67}Li_{0.33}(Na_{0.33})]Si_4O_{10}(OH,F)_2$
saponite	[*1319-41-1*]	$Mg_{3.00}[Al_{0.33}(Na_{0.33})Si_{3.67}]O_{10}(OH)_2$
sauconite	[*12424-32-7*]	$[Zn_{1.48}Mg_{0.14}Al_{0.74}Fe(III)_{0.40}][Al_{0.99}Si_{3.01}]O_{10}(OH)_2X_{0.33}$

[a] Ref. 34.

[b] More substitution takes place than shown; $Na_{0.33}$ or $X_{0.33}$ refers to the exchangeable base (cation) of which 0.33 equivalent is a typical value.

Table 4. Chemical Analyses of the Smectite Minerals, %

Component	1[a]	2[b]	3[c]	4[d]	5[e]	6[f]
SiO_2	51.14	47.28	43.54	55.86	42.99	34.46
Al_2O_3	19.76	20.27	2.94	0.13	6.26	16.95
Fe_2O_3	0.83	8.68	28.62	0.03	1.83	6.21
FeO			0.99		2.57	
MnO	trace			none	0.11	
ZnO	0.10					23.10
MgO	3.22	0.70	0.05	25.03	22.96	1.11
CaO	1.62	2.75	2.22	trace	2.03	
K_2O	0.11	trace		0.10	trace	0.49
Na_2O	0.04	0.97		2.68	1.04	
Li_2O				1.05		
TiO_2	none			none		0.24
P_2O_6						
F				5.96		
H_2O removed						
at 150°C	14.81	19.72	14.05	9.90	13.65	6.72
above 150°C	7.99		6.62	2.24	6.85	10.67
Total	*99.75*	*100.37*	*100.02*	*102.98* *100.470–F*	*100.29*	*99.95*

[a] Montmorillonite, Montmorillon, France (34).
[b] Beidellite, Beidell, Colo. (34).
[c] Nontronite, Woody, Calif. (34).
[d] Hectorite, Hector, Calif. (34).
[e] Saponite, Ahmeek Mine, Mich. (64).
[f] Sauconite, Friedensville, Pa. (64).

Table 5. Optical Properties of the Smectite Minerals[a]

		Index of refraction			Optical	Axial	Physical
Mineral	Source	α	β	γ	character	angle, $2V$	form
beidellite	Beidell, Colo.	1.502		1.533			grains
nontronite	Woody, Calif.	1.560	1.585	1.585	negative	small	grains
hectorite	Hector, Calif.	1.485		1.516			film
saponite	Ahmeek Mine, Mich.	1.490	1.525	1.527	negative	moderate	grains
sauconite	Friedensville, Pa.	1.575		1.615	negative		film

[a] Refs. 34, 64.

Transmission electron micrographs show hectorite and nontronite as elongated, lath-shaped units, whereas the other smectite clays appear more nearly equidimensional. A broken surface of smectite clays typically shows a "corn flakes" or "oak leaf" surface texture (42).

The differential thermal analysis curves for smectite commonly show three endothermic peaks and one exothermic peak, which may fall within the ranges 150–320, 695–730, 870–920, and 925–1050°C, respectively. Beidellite exhibits endothermic peaks at about 200 and 575–590°C and an exothermic peak at 905–925°C.

High temperature minerals formed upon heating smectites vary considerably with the compositions of the clays. Spinels commonly appear at 800–1000°C, and dissolve at higher temperatures. Quartz, especially cristobalite, appears and mullite forms if the content of aluminum is adequate (51).

The cation-exchange capacity of smectite minerals is notably high, 80–90 meq or higher per 100 g of air-dried clay, and affords a diagnostic criterion of the group. The crystal lattice is obviously weakly bonded. Moreover, the lattice of smectite is expandable between the silicate layers so that when the clay is soaked in water it may swell to several times its dry volume (eg, bentonite clays). Soil colloids with high cation-exchange capacity facilitate the transfer of plant nutrients to absorbing plant rootlets.

The minerals of the smectite group have been formed by surface weathering, low temperature hydrothermal processes, alteration of volcanic dust in stratified beds (66), action of circulating water of uncertain source along fractures and in veins, and laboratory synthesis. The optimum weathering environment is one in which calcium, iron(II), and especially magnesium are present in significantly high concentrations. Potassium should be low or low in relation to magnesium, calcium, and ferrous iron. Organic matter that exerts reducing action is a usual concomitant, and a neutral to slightly alkaline medium generally prevails under conditions where the alkali and alkaline earth metals are not readily removed. The weathering environment for smectite is different from that in which kaolinite is formed (59). If the system permits effective leaching and H^+ ions become available in sufficient quantity to cause the metallic cations to be easily leached away, kaolinite tends to form. The reverse reaction rarely, if ever, takes place.

Bentonite is a rock rich in montmorillonite that has usually resulted from the alteration of volcanic dust (ash) of the intermediate (latitic) siliceous types. In general, relicts of partially unaltered feldspar, quartz, or volcanic glass shards offer evidence of the parent rock. Most adsorbent clays, bleaching clays, and many clay catalysts are smectites, although some are attapulgite [*1337-76-4*].

Illites or Micas. Illite is a general term for the clay mineral constituents of argillaceous sediments belonging to the mica group (9); it is not a single pure mineral (67). Mica minerals possess a 2:1 sheet structure similar to montmorillonite except that the maximum charge deficit in mica is typically in the tetrahedral layers and contains potassium held tenaciously in the interlayer space, which contributes to a 1.0 nm basal spacing. Because the micas in argillaceous sediments may be widely diverse in origin, considerable variation exists in the composition and polymorphism of the illite minerals.

The formula of illite can be expressed as $2K_2O.3MO.8R_2O_3.24SiO_2.12H_2O$ (9), and the crystal structure (68) by the formula $K_y(Al_4Fe_4.Mg_4Mg_6)(Si_{8-y})O_{20}(OH)_4$, where y refers to the K^+ ions that satisfy the excess charges resulting when about 15% of the Si^{4+} positions are replaced by Al^{3+}. For representative chemical analysis of illite (fine colloid fraction, Pennsylvania underclay, near Fithian, Vermilion County, Illinois), see Table 6.

Optical constants of illite minerals are difficult to obtain because of the small size of the available crystals. The highest (γ) index of refraction ranges from about 1.588 to 1.610, the birefringence is about 0.033, the optical character is negative, and the axial angle, $2V$, is small, on the order of 5°.

A 1.0 nm basal spacing exhibited in a diffractogram peak that is somewhat broad and diffuse and skewed toward wider spacings characterizes the x-ray diffraction pattern of illite. Polymorphs 1 *Md,* 1 *M*, 2 M_1, and 2 M_2 may be present (69); 1 *Md* and 1 *M* are most commonly reported. Muscovite derivatives are typically dioctahedral; phlogopite derivatives are trioctahedral.

Table 6. Chemical Analysis of Illite, Glauconite, and Attapulgite, %

Component	Illite	Glauconite	Attapulgite
SiO_2	51.22	48.66	55.03
Al_2O_3	25.91	8.46	10.24
Fe_2O_3	4.59	18.8	3.53
FeO	1.70	3.98	
MgO	2.84	3.56	10.49
CaO	0.16	0.62	
K_2O	6.09	8.31	0.47
Na_2O	0.17		
TiO_2	0.53		
H_2O removed			
at 110°C	7.49		
at 150°C			9.73
above 150°C			10.13
Total	*100.7*	*99.8*	*99.62*

Differential thermal and analysis curves of illite show three endothermic peaks in the ranges 100–150, 500–650, and at about 900°C, and an exothermic peak at about 940°C, or immediately following the highest endothermic peak. Minerals formed from illite at high temperature vary somewhat with the composition of the clay, but usually a spinel-structure mineral followed by mullite at still higher temperatures is observed (34).

The cation-exchange capacity of illite is 20–30 meq/100 g of dry clay. The interlayer potassium exerts a strong bond between adjacent clay structures. Illite that has lost part of its original potassium by weathering processes may be reconstituted with the sorption and incorporation of transient dissolved potassium (65).

Illite was defined as the most abundant clay mineral in Paleozoic shale and is widespread in many other sedimentary rocks; it is common in soils, slates, certain alteration products of igneous rocks, and recent sediments. Its origin has been attributed to alteration of silicate minerals by weathering and hydrothermal solutions, reconstitution, wetting and drying of soil clays, and diagenesis involving other three-layer minerals and potassium during geologic time and pressure under deep burial (23,70–71).

Glauconite. Glauconite [*1317-57-3*] (72–75) is a green, dioctahedral, micaceous clay rich in ferric iron and potassium. It has many characteristics common to illite, but much glauconite contains some randomly mixed expanding layers, interpreted as montmorillonitic. Glauconite occurs abundantly in sand size pellets or bigger, or in pellets within fossils, notably foraminifera, giving it an organic connotation (12). Occurrences as replacements, matrix, and flakes in sandstone and as a product of diagenesis (70,76) indicate other possible origins. Glauconite is typically formed in a marine environment (77), but glauconitic mica has been reported from nonmarine rocks (46,78).

The chemical analysis of glauconite (Bonneterre, Missouri), is given in Table 6, and the optical constants in Table 2. Powder x-ray diffraction patterns resemble those of illite in which intensities of even-numbered basal spacings are minimal.

The glauconitic green sands of New Jersey have been used in ion-exchange, water-softening installations (see Ion exchange), and as a source of slowly released potassium in soil amendments.

Chlorites and Vermiculites. Chlorite was identified as the mineral yielding a 1.4-nm basal spacing in clays some years after the clay mineral groups previously discussed were fairly well characterized. Chloritic clay minerals are therefore less well explored than the other clay minerals or the chlorite of igneous and metamorphic occurrences (10,78a). Chlorite is widespread in argillaceous sedimentary rocks and in certain soils. Structurally, it has been viewed as three-layer phyllosilicates separated by a brucite [*1317-43-7*], $Mg(OH)_2$, interlayer, or alternatively, as a four-layer group, or 242, of alternating silica tetrahedral and Mg octahedral sheets (79–81). Both 0.7- and 1.4-nm (basal spacing) polymorphs exist.

Aluminum chlorite [*25410-05-3*], in which gibbsite, $Al(OH)_3$, proxies in part for brucite, is being discovered in increasing occurrences and abundance (11,82). Chlorite-like structures have been synthesized by precipitating Mg and Al between montmorillonite sheets (83).

The interlayer sheet in chlorite is interpreted as brucitic, $Mg(OH)_2$, but in vermiculite it is thought to be octahedrally coordinated, 6 H_2O about Mg^{2+} (84–86). The basal spacing of vermiculite varies from 1.4 to 1.5 nm with the nature of the interlayer cation and its hydration. The cation-exchange capacity of vermiculite is relatively high and may exceed that of montmorillonite.

Regularly interstratified (1:1) chlorite and vermiculite has been attributed to the mineral corrensite (87). Corrensite [*12173-14-7*] is being discovered in greater abundance, and its distribution as sedimentary rocks, especially carbonate rocks, is being studied in greater detail (88–89).

Cookeite [*1302-92-7*], an aluminous chlorite containing lithium, has been found in high-alumina refractory clays and bauxites (90).

Attapulgite and Sepiolite. Attapulgite, named from its occurrence at Attapulgus, Georgia, and sepiolite (meerschaum), named from the Greek word for cuttle fish (whose bones are light and porous) possess chainlike structures, or combination chain–sheet structures (12). The attapulgite structure is similar to palygorskite minerals resembling cardboard, paper, leather, cork, or even fossil skin.

These clays have distinctive uses and properties not shown by platy clay minerals. The Georgia–Florida deposits originated from evaporating sea water (91). Attapulgite and other palygorskites sorb both cations and neutral molecules. Typical cation-exchange capacities are in the order of 20 meq/100 g dry clay. For chemical analysis of attapulgite see Table 6.

Sepiolite is used in drilling muds where resistance to flocculation in briny water is desired (see Petroleum). Sepiolite and attapulgite are best identified by their 110 reflections, 1.21 and 1.05 nm, respectively, in x-ray powder diffraction (92–93).

Mixed-Layer Clay Minerals. In addition to polymorphism due to the disordering and proxying of one element for another, clay minerals exhibit ordered and random intercalation of sandwiches with one another (12). For example, in mixed-layer clay minerals sheets of illite may be interspersed with montmorillonite, or chlorite with one of the others, either randomly or regularly. Corrensite [*12173-14-7*] has already been cited as an example of a 1:1, ie, regular, alternation of chlorite and vermiculite. Random mixing may consist of two types: (*1*) where there is little deviation from a mean ratio between two participants intercalated at random which yields a relatively sharply defined intermediate basal spacing between those of the two end members, and (*2*) a wide ratio between the two that are intercalated, yielding a wide band of spacings. Mixed layering originates, presumably, by either the degradation in random layers of one species, such as the random stripping of potassium from illite by H^+ ions, or by the random precipitation, reconstitution, or growth of layers of a different guest mineral within a layered host. Mixed layering is not restricted to clay minerals where it has been widely observed, but probably also occurs in numerous other mineral groups.

AMORPHOUS AND MISCELLANEOUS GROUPS

Allophane and Imogolite. Allophane is an amorphous clay that is essentially an amorphous solid solution of silica, alumina, and water (47). It may be associated with halloysite or it may occur as a homogeneous mixture with evansite, an amorphous solid solution of phosphorus, alumina, and water. Its composition, hydration, and properties vary. Chemical analyses of two allophanes are given in Table 7.

The index of refraction of allophane ranges from below 1.470 to over 1.510, with a modal value about 1.485. The lack of characteristic lines given by crystals in x-ray diffraction patterns and the gradual loss of water during heating confirm the amorphous character of allophane.

Allophane has been found most abundantly in soils and altered volcanic ash (60,94–95), but may be considerably more widespread than has been recognized. Quantitative estimation in mixtures depends upon a much higher rate of solubility than of its crystalline analogues. It usually occurs in spherical form but has been also observed in fibers.

Imogolite [*12263-43-3*] is an uncommon, thread-shaped paracrystalline clay mineral assigned a formula 1.1 $SiO_2.Al_2O_3.2.3–2.8H_2O$ (96). It can be classified as intermediate between allophane and kaolinite.

Table 7. Chemical Analyses of Allophane, %

Component	1[a]	2[b]
SiO_2	32.30	4.34
Al_2O_3	30.41	41.41
Fe_2O_3	0.23	0.86
MgO	0.29	0.22
CaO	0.02	0.20
$K_2O + Na_2O$	0.10	0.10
TiO_2	none	none
CuO	1.60	1.80
ZnO	4.06	4.30
CO_2	0.65	2.07
P_2O_5	0.02	9.23
SO_3	0.21	0.08
H_2O removed		
at 105°C	16.38	20.92
above 105°C	14.43	14.43
Total	*100.70*	*99.96*

[a] Allophane, Monte Vecchio, Sardinia (47).
[b] Allophane–evansite, Freienstein, Styria (47).

High-Alumina Clay Minerals. Several hydrated alumina minerals should be grouped with the clay minerals because the two types may occur so intimately associated as to be almost inseparable. Diaspore and boehmite, both $Al_2O_3.H_2O$ (Al_2O_3, 85%; H_2O, 15%) are the chief constituents of diaspore clay, which may contain over 75% Al_2O_3 on the raw basis (23). Gibbsite, $Al_2O_3.3H_2O$ (Al_2O_3, 65.4%; H_2O, 34.6%), and cliachite [*12197-64-7*], the so-called amorphous alumina hydrate (much cliachite is probably cryptocrystalline), as well as the monohydrates, occur in bauxite [*1318-16-7*] (29,31,97), bauxitic kaolin, and bauxitic clays (98–99).

The hydrated alumina minerals usually occur in oolitic structures (small spherical to ellipsoidal bodies the size of BB shot, about 2 mm in diameter) and also in larger and smaller structures. They impart harshness and resist fusion or fuse with difficulty in sodium carbonate, and may be suspected if the raw clay analyzes at more than 40% Al_2O_3. Their optical properties are radically different from those of common clay minerals, and their x-ray diffraction patterns and differential thermal analysis curves are distinctive.

High alumina minerals are found where intense weathering and leaching has dissolved the silica. It is generally believed that a very humid, subtropical climate is required for this (lateritic) stage of weathering.

BIBLIOGRAPHY

"General Survey" under "Clays" in *ECT* 1st ed., Vol. 4, pp. 24–38, by W. D. Keller, University of Missouri; "Survey" under "Clays" in *ECT* 2nd ed., Vol. 5, pp. 541–560, by W. D. Keller, University of Missouri.

1. ASTM Committee on Standards *J. Am. Ceram. Soc.* **11,** 347 (1928).
2. F. H. Norton, *Refractories,* 2nd ed., McGraw-Hill Book Co., Inc., New York, 1962.
3. R. E. Grim, *Applied Clay Mineralogy,* McGraw-Hill Book Co., Inc., New York, 1962.
4. H. Wilson, *Ceramics—Clay Technology,* McGraw-Hill Book Co., Inc., New York, 1927.
5. H. Ries, *Clays, Their Occurrences, Properties, and Uses,* John Wiley & Sons, Inc., New York, 1927, p. 1.
6. H. Ries, "Clay," in *Industrial Minerals and Rocks,* American Institute of Mining and Metallurgical Engineers, 1937, pp. 207–242.
7. M. Goldman and J. I. Tracey, Jr., *Econ. Geol.* **41,** 567 (1946).
8. T. Sudo, *Mineralogical Study on Clays of Japan,* Maruzen Co., Ltd., Tokyo, 1959.
9. R. E. Grim, R. H. Bray, and W. F. Bradley, *Am. Mineral.* **22,** 813 (1937).
10. M. D. Foster, *U.S. Geol. Surv. Prof. Pap.* **414-A,** (1962).
11. J. E. Brydon, J. S. Clark, and V. Osborne, *Can. Mineral.* **6,** 595 (1961).
12. G. Brown, ed., *The X-ray Identification and Crystal Structures of Clay Minerals,* Mineralogical Society, London, 1961.
13. T. F. Bates, F. A. Hildebrand, and A. Swineford, *Am. Mineral.* **35,** 463 (1959).
14. C. E. Marshall, *The Colloid Chemistry of the Silicate Minerals,* Academic Press, Inc., New York, 1949.
15. I. T. Rosenqvist, "Marine Clays and Quick Clay Slides in South and Central Norway," *Guide to Exc. No. C-13, 21st International Geological Congress, Oslo,* 1960.
16. H. Ries, *Am. J. Sci.* **44**(4), 316 (1917).
17. C. DeKimpe, M. C. Gastuche, and G. W. Brindley, *Am. Mineral.* **46,** 1370 (1961).
18. J. Linares and F. Huertas, *Science* **171,** 896 (1971).
19. R. Roy, *C.N.R.S. Groupe Fr. Argiles C.R. Reun. Etud.* **105,** 83 (1962).
20. B. Velde, *Clays and Clay Minerals in Natural and Synthetic Systems,* Elsevier, New York, 1977.
21. C. E. Weaver and L. D. Pollard, *The Chemistry of Clay Minerals,* Elsevier, New York, 1973.
22. W. D. Keller, *Principles of Chemical Weathering,* Lucas Bros., Columbia, Mo., 1957.
23. W. D. Keller, "Processes of Origin of the Clay Minerals," *Proceedings of the Soil Clay Mineral Institute,* Virginia Polytechnic Institute, Blacksburg, Va., 1962.

24. S. H. Patterson and B. F. Buie, "Field Conference on Kaolin and Fuller's Earth, Nov. 14–16, 1974" *Guidebook 14, Georgia Dept. Natl. Resources,* Atlanta, Ga., 1974.
25. W. D. Keller, J. F. Westcott, and A. O. Bledsoe, *Proceedings of the 2nd Conference of Clays and Clay Minerals, National Academy of Science-National Research Council Publication 327,* 1954, pp. 7–46.
26. S. H. Patterson and H. W. Murray, "Clays," *Industrial Minerals and Rocks,* 4th ed., AIME, 1975, pp. 519–585.
27. J. Vachtl, *Proc. XXIII Int. Geol. Cong.* **15,** 13 (1968).
28. M. Kuzvart and M. Bohmer, *Prospecting and Exploration of Mineral Deposits* Academia Press, Prague, Czechoslovakia, 1978.
29. Gy. Bardossy, *Acta Geol. Acad. Sci. Hung.* **6**(1–2), 1 (1959).
30. R. C. Mackenzie, ed., *The Differential Thermal Investigation of Clays,* Mineralogical Society, London, 1957.
31. S. H. Patterson, *U.S. Geol. Surv. Bull.* **1228** (1967).
32. C. M. Warshaw and R. Roy, *Bull. Geol. Soc. Am.* **72,** 1455 (1961).
33. T. R. P. Gibb, Jr., *Optical Methods of Chemical Analyses,* McGraw-Hill Book Co., Inc., New York, 1942, pp. 243–319.
34. C. S. Ross and S. B. Hendricks, *U.S. Geol. Surv. Prof. Pap.* **205-B,** 23 (1945).
35. G. T. Faust, *U.S. Bur. Mines Rep. Invest.* **3522,** (1942).
36. E. A. Hauser and M. B. Leggett, *J. Am. Chem. Soc.* **62,** 1811 (1940).
37. J. B. Page, *Soil Sci.* **51,** 133 (1941).
38. C. E. Marshall, *Z. Kristallogr. Mineral.* **90,** 8 (1935).
39. P. F. Kerr, ed., "Reference Clay Minerals," *American Petroleum Institute Research Project 49,* 1950.
40. C. S. Piper, *Soil and Plant Analysis,* Interscience Publishers, Inc., New York, 1944.
41. H. Beutelspacher and H. Van der Marel, *Atlas of Electron Microscopy of Clay Minerals and Their Admixtures,* Elsevier, New York, 1968.
42. R. L. Borst and W. D. Keller, *Proc. Int. Clay Conf. Tokyo* 871 (1969).
43. W. D. Keller and R. F. Hanson, *Clays Clay Miner.* **23,** 201 (1975).
44. W. D. Keller, *Clays Clay Miner.* **25,** 311 (1977).
45. J. Konta, *Am. Mineral.* **46,** 289 (1961).
46. W. D. Keller, *U.S. Geol. Surv. Bull.* **1150,** (1962).
47. C. S. Ross and P. F. Kerr, *U.S. Geol. Surv. Prof. Pap.* **185-G,** 135 (1934).
48. L. T. Alexander and co-workers, *Am. Mineral.* **28,** 1 (1943).
49. C. S. Ross and P. F. Kerr, *U.S. Geol. Surv. Prof. Pap.* **165-E,** 151 (1930).
50. G. T. Faust, *Am. Mineral.* **40,** 1110 (1955).
51. R. E. Grim, *Clay Mineralogy,* McGraw-Hill Book Co., Inc., New York, 1968.
52. M. H. Thornberry, *Mo. Univ. Sch. Mines Metall. Bull.* 8(2), 34 (1925).
53. C. S. Ross, *Proc. U.S. Nat. Mus.* **69,** 1 (1926).
54. W. D. Miller and W. D. Keller, "Differentiation between Endellite–Halloysite and Kaolinite by Treatment with Potassium Acetate and Ethylene Glycol," *Proceedings of the 10th National Conference of Clays and Clay Minerals,* 1961, 244–256.
55. F. V. Chukhrov, and B. B. Zvyagin, *Proc. Int. Clay Conf., Jerusalem,* **1,** 11 (1966).
56. L. B. Sand, *Am. Mineral.* **41,** 28 (1956).
57. W. D. Keller, *Clays Clay Miner.* **26,** 1 (1978).
58. W. D. Keller, *Tenth National Conference of Clays and Clay Minerals,* Pergamon Press, New York, 1963, pp. 333–343.
59. W. D. Keller, *Bull. Am. Assoc. Petrol. Geologists* **40,** 2689 (1956).
60. M. Fieldes, *N.Z. J. Sci. Technol.* **37,** 336 (1955).
61. G. W. Brindley and M. Nakahira, *J. Am. Ceram. Soc.* **42,** 311 (1959).
62. M. Slaughter and W. D. Keller, *Am. Ceram. Soc. Bull.* **38,** 703 (1959).
63. F. M. Wahl, R. E. Grim, and R. B. Graf, *Am. Mineral.* **46,** 1064 (1961).
64. C. S. Ross, *Am. Mineral.* **31,** 411 (1946).
65. C. E. Weaver, *Am. Mineral.* **43,** 839 (1958).
66. M. Slaughter and J. W. Earley, *Geol. Soc. Am. Spec. Pap.* **83,** (1965), 95 pp.
67. H. S. Yoder and H. P. Eugster, *Geochim. Cosmochim. Acta* **8,** 225 (1955).
68. R. E. Grim. *Bull. Am. Assoc. Petrol. Geologists* **31,** 1491 (1947).
69. A. A. Levinson, *Am. Mineral.* **40,** 41 (1955).

70. J. F. Burst, Jr., *Proceedings of the 6th National Conference of Clays and Clay Minerals,* Pergamon Press, Inc., New York, 1959, pp. 327–341.
71. W. D. Keller, Diagenesis of Clay Minerals—A Review," *Proceedings of the 11th National Conference of Clays and Clay Minerals, 1962,* Pergamon Press, Inc., New York, 1963.
72. J. F. Burst, Jr., *Bull. Am. Assoc. Petrol. Geologists* **42,** 310 (1958).
73. J. F. Burst, Jr., *Am. Mineral.* **43,** 481 (1958).
74. J. Hower, *Am. Mineral.* **46,** 313 (1961).
75. E. G. Wermund, *Bull. Am. Assoc. Petrol. Geologists* **45,** 1667 (1961).
76. P. M. Hurley and co-workers, *Bull. Am. Assoc. Petrol. Geologists* **44,** 1793 (1960).
77. P. E. Cloud, *Bull. Am. Assoc. Petrol. Geologists* **39,** 484 (1955).
78. W. D. Keller, *Fifth National Conference of Clays and Clay Minerals, National Academy of Science-National Research Council Publication 566,* 1958, pp. 120–129.
78a. A. L. Albee, *Am. Mineral.* **47,** 851 (1962).
79. S. W. Bailey and B. E. Brown, *Am. Mineral.* **47,** 819 (1962).
80. W. F. Bradley, *Second National Conference on Clays and Clay Minerals, National Academy of Science-National Research Council Publication 327,* 1954, pp. 324–334.
81. G. W. Brindley and F. H. Gillery, *Am. Mineral.* **41,** 169 (1956).
82. M. J. Shen and C. I. Rich, *Soil Sci. Soc. Am. Proc.* **26,** 33 (1962).
83. M. Slaughter and I. Milne, eds., *Proceedings of the 7th National Conference of Clays and Clay Minerals,* Pergamon Press, Inc., New York, 1960, pp. 114–124.
84. W. A. Bassett, *Am. Mineral.* **44,** 282 (1959).
85. A. M. Mathieson, *Am. Mineral.* **43,** 216 (1958).
86. G. F. Walker, *Clay Miner. Bull.* **3,** 154 (1957).
87. W. F. Bradley and C. E. Weaver, *Am. Mineral.* **41,** 497 (1956).
88. M. N. A. Peterson, *Am. Mineral.* **46,** 1245 (1961).
89. M. N. A. Peterson, *J. Geol.* **70,** 1 (1962).
90. H. A. Tourtelot and E. F. Brenner-Tourtelot, "Lithium in Flint Clay, Bauxite, Related High-Alumina Materials and Associated Sedimentary Rocks in the United States—a Preliminary Report," *U.S. Geol. Survey Open File Report 77-786,* 1977.
91. S. H. Patterson, *U.S. Geol. Surv. Prof. Pap.* **828,** (1974).
92. S. Caillere and S. Henin, "The X-ray Identification and Crystal Structures of Clay Minerals," *Mineralogical Society Great Britain Monograph, 325–342,* 1961, Chapt. VIII.
93. W. F. Bradley, *Am. Mineral.* **25,** 405 (1940).
94. K. S. Birrell and M. Fieldes, *N.Z. J. Soil Sci.* **3,** 156 (1952).
95. W. A. White, *Am. Mineral.* **38,** 634 (1953).
96. N. Yoshinga and S. Aomine, *Soil Sci. Plant Nutr. Tokyo* **8,** 22 (1962).
97. I. Valeton, *Bauxites,* Elsevier, 1972.
98. A. F. Frederickson, ed., "Problems of Clay and Laterite Genesis," *AIME Symposium Volume,* American Institute Mining and Mechanical Engineers, 1952.
99. M. Gordon, Jr., J. I. Tracey, Jr., and M. W. Ellis, *U. S. Geol. Surv. Prof. Pap.* **299,** (1958).

General References

References 3, 12, 14, 21–22, 24, 26, 30–31, 41, 51, 91, and 97–98 of the numbered bibliography may also be considered general reference works.

W. D. KELLER
University of Missouri, Columbia

USES

Clay materials are composed of extremely small particles of clay minerals. These minerals are generally crystalline, but in some cases their organization is so poor that diffraction indicates them to be amorphous. Clay minerals are essentially hydrous aluminum silicates, with iron or magnesium proxying wholly or in part for the aluminum in some, and with alkalies or alkaline earths present as essential constituents in others. Furthermore, clays may contain varying amounts of so-called nonclay minerals, such as quartz, calcite, feldspar, and pyrites.

The following factors control the properties of clay materials: the identity and relative abundance of the clay mineral components; the identity of the nonclay minerals and their shape, relative abundance, and particle-size distribution; the kind and amount of organic material; the kind and amount of exchangeable ions and soluble salts; and the texture, which refers to the particle-size distribution of the constituent particles, their shape, their orientation in space with respect to each other, and the forces binding them together.

Ceramic Products

Ceramic is defined as "relating to the art of making earthenware or to the manufacture of any or all products made from earth by the agency of fire, as glass, enamels, cements." To this list should be added brick, tile, heat-resisting refractory materials, porcelain, pottery, chinaware, and earthenware. In general, ceramic ware is produced by plasticizing the clay by the addition of water so that it may be shaped or formed by some means into the desired object. Ceramic products may also be formed by dispersing the clay in water to form a slip which is then cast into a plaster mold. In the case of porcelain enamel the slip is sprayed on a metal surface and then fired. After being shaped, the object is dried to increase its strength so that it may be handled, and is then fired at elevated temperatures (frequently in the range of 1090°C) until there has been some vitrification or fusion of the components to develop a glassy bond that makes the shape permanent and strong so that the object does not disintegrate in water (see Ceramics).

Properties. ***Plasticity.*** Plasticity may be defined as the property of a material that permits it to be deformed under stress without rupturing and to retain the shape produced after the stress is removed. When water is added to dry clay in successive increments, the clay tends to become workable, that is, readily shaped without rupturing. The workability and retention of shape develop within a very narrow moisture range.

Plasticity can be measured by determination of (*a*) the water of plasticity, ie, the amount of water necessary to develop optimum plasticity (judged subjectively by the operator) or the range of water content in which plasticity is demonstrated (Atterberg limits); (*b*) the amount of penetration of an object, frequently a needle or some type of plunger, into a plastic mass of clay under a given load or rate of loading and at varying moisture contents; and (*c*) the stress necessary to deform the clay and the maximum deformation the clay will undergo before rupture at different moisture contents and with varying rates of stress application.

In ceramics, plasticity is usually evaluated by means of the water of plasticity. Ranges of values for common clay minerals are given in Table 1.

Table 1. Ceramic Properties of Clay Minerals

Property	Kaolinite [1318-74-7]	Illite [12173-60-3]	Halloysite [12244-16-5]	Montmorillonite [1318-93-0]	Attapulgite [1337-76-4]	Allophane [12172-71-3]
water of plasticity[a], %	8.9–56.3	17–38.5	33–50	83–250	93	
green strength[b], kg/cm^2	0.34–3.2	3.2	5	5[c]		
dry strength[a]	69–4840	1490–7420	1965	1896–5723	4482	
linear shrinkage[a], %						
drying[d]	3–10	4–11	7–15	12–23	15	
firing	2–17	9–15	±20	±11	±23	+50

[a] Ref. 1.
[b] Ref. 2.
[c] Calcium montmorillonite.
[d] Values computed as % of dry length after drying test pieces of 6.45 cm^2 cross section in an oven at 105°C for 5 h.

Each clay mineral group can be expected to show a range of values, since particle size, exchangeable cation composition, and crystallinity of the clay mineral also exert an influence. Nonclay mineral component, soluble salts, organic compounds, and texture can also affect the water of plasticity.

In general, a relatively low value for water of plasticity is desired in ceramics and hence kaolinite, illite, and chlorite clays have better plasticity characteristics than attapulgite or montmorillonite. The plasticity values of the first group are changed only slightly by variations in the exchangeable cation composition. However, sodium gives lower values than calcium, magnesium, potassium, and hydrogen. In the case of montmorillonite, the water of plasticity varies considerably with the nature of the exchangeable cation, with sodium giving higher values than the other common exchangeable cations.

Clays composed only of clay minerals may have higher water of plasticity values than desired. Consequently, the presence of nonclay minerals in substantial amounts or the addition of such material serving to reduce the water of plasticity may improve the working characteristics of a clay.

Plasticity in clay–water systems is caused by a bonding force between the particles and water, a lubricant, which permits some movement between the particles under the application of a deforming force. The bonding force is in part a result of charges on the particles. It is now generally believed that the water immediately adsorbed on basal clay mineral surfaces is composed of water molecules in a definite orientation; that is, it is not fluid water (3). This water, composed of water molecules in a definite configuration, also serves as a bond between clay mineral particles. The orientation of the water would decrease outward from the clay mineral surface so that with little adsorbed water the bonding would be strong, whereas with much adsorbed water the bond would be weak or nonexistent. It has been suggested (4–5) that optimum plasticity develops when all the requirements of particle surfaces for oriented water are satisfied and there is a little additional water less rigidly fixed which can act as a lubricant when a deforming force is applied. Furthermore, influence of adsorbed cations on plasticity is to a considerable degree the result of their effect on the extent and perfection of the development of crystalline water, rather than directly on the bonding force between the particles.

The *green strength* of a clay body is the strength measured as transverse breaking strength that prevails while the plasticizing water is still present. As water is continuously added to a dry clay, the strength increases up to a maximum and then decreases. The strength at water of plasticity is, in general, lower than the maximum strength. Values for the common clay minerals are given in Table 1.

As in the case of plasticity, green strength values would be expected to vary with exchangeable cation composition to only a slight degree for kaolinite, illite, and chlorite clays, and to a considerable degree for montmorillonite clays. In the latter, sodium would be expected to provide higher maximum green strength than other common cations. Poorly crystallized varieties of kaolinite and illite yield higher green strength than well-crystallized varieties. The presence of considerable quantities of nonclay minerals reduces the green strength, whereas small amounts may actually increase the strength because they permit the development of a more uniform clay body. Green strength is also related to the particle size with smaller particle size providing the higher strength. If the clay mineral particles develop preferred orientation in certain directions during formation of the ware, the breaking strength will be somewhat greater in the directions transverse to such preferred orientation.

Drying Properties. *Drying shrinkage* is the reduction in size measured either in length or volume that takes place when a mass is dried in order to drive off the pore water and the adsorbed water. Values are given in Table 1.

In general, the drying shrinkage increases as the water of plasticity increases, and for a particular clay mineral it increases as the particle size decreases. In addition, drying shrinkage varies with the degree of crystallinity; thus ball clay (6), which contains relatively poorly ordered kaolinite, shows values at the high end of the range shown in Table 1. Similarly, mixed-layer clays would be expected to have relatively higher drying shrinkage than clays of similar composition, but with the clay mineral mixing being that of discrete particles. The nature of the adsorbed cation causes variations in the amount of drying shrinkage only as it affects the water of plasticity.

The presence of nonclay minerals tends to reduce drying shrinkage depending on their shape, particle-size distribution, and abundance. Granular particles with considerable particle-size range are most effective. The presence of nonclay minerals on the order of about 25% is generally desirable in ceramic bodies to improve their shrinking characteristic. Drying shrinkage is also related to texture; for example, if the clay mass shows parallel orientation of the basal plane surfaces of the clay minerals, shrinkage in the direction at right angles to the basal planes would be substantially greater than in the direction parallel to them (7).

In the initial drying phase of a clay body the volume shrinkage is about equal to the volume of water lost. Beyond a given moisture content there is either no further shrinkage or only a very small amount as the water is lost. The water lost during the shrinkage interval is called shrinkage water and seems to separate the component particles. The critical point at which shrinkage stops is reached when the moisture film around the particle becomes so thin that the particles touch one another sufficiently and shrinkage can go no farther. The water loss following the shrinkage period is called pore water.

In the production of ceramic ware the shape of the ware has to be retained after drying and should be free from cracks and other defects. Excessively slow drying and excessive control of humidity, airflow, etc, should be avoided. In general, clays containing moderate amounts of nonclay minerals are easier to dry than those composed wholly of clay minerals. Furthermore, clays composed of illite, chlorite, and kaolinite are relatively easier to dry than those composed of montmorillonite.

Dry strength is measured as the transverse breaking strength of a test piece after drying long enough, usually at 105°C, to remove almost all the pore and adsorbed water. Values are given in Table 1 and usually show a large range because of variations in particle-size distribution, perfection of crystallinity, and, especially for montmorillonite, the nature of the exchangeable ions.

Large amounts of nonclay mineral components, especially if the particles are well sorted, tend to reduce the dry strength. In general, the dry strength is higher when sodium is the adsorbed cation. The presence of organic material in some clays increases their dry strength and this appears to be partly the explanation for the high dry strength for some ball clays. A major factor in determining dry strength is the particle size of the clay mineral component; the maximum strength increases rapidly as the particle size decreases.

Firing Properties. Heating clay materials to successively higher temperatures results in the fusion of the material. In the 100–150°C range, the shrinkage and pore waters are lost with the attendant dimensional changes. In general, the rate of oxidation increases with increasing temperature. The oxidation of sulfides, which are present in many clays, frequently in the form of pyrite, FeS_2, begins between 400 and 500°C.

Beginning at about 500°C and in some cases continuing to 900°C, the hydroxyl water is driven from the clay minerals. The exact temperature, rate, and abruptness of loss of hydroxyls depend on the nature of the clay minerals and their particle size. Reduction of particle size, particularly if accompanied by poor crystallinity, tends to reduce the temperature interval. In general, kaolinite and halloysite minerals lose their hydroxyls abruptly at 450–600°C. The loss of hydroxyls from the three-layer clay minerals varies greatly with structure and composition, but is generally slower and more gradual than that for kaolinite and halloysite.

The loss of hydroxyls from the clay minerals is usually accompanied by a modification of the structure, but not by its complete destruction. In three-layer clay minerals it is not accompanied by shrinkage, whereas in kaolinite and halloysite loss of hydroxyl water is accompanied by shrinkage, which continues up into the vitrification range.

In the range of 800 to 950°C, the structure of the clay mineral is destroyed, and major firing shrinkage develops. Values for the firing shrinkage are given in Table 1.

The range in shrinkage values is due to variations in the size and shape of the clay mineral particles, the degree of crystallinity, and in the case of the three-layered clay minerals, variations in composition.

At temperatures above about 900°C new crystalline phases develop from all the clay minerals except those containing large amounts of iron, alkalies, or alkaline earths, in which case fusion may result after the loss of structure without any intervening crystalline phase. Frequently, there is a series of new high-temperature phases developing in an overlapping sequence as the mineral is heated to successively higher temperatures. This is followed by complete fusion of the mineral, which takes place in the case of kaolinite at 1650–1775°C. For the three-layered clay minerals, the fusion temperature varies from about 1000 to 1500°C, the lower values being found in materials relatively rich in iron, alkalies, and alkaline earths.

The initial high-temperature phases are frequently related to the structure of the original clay mineral, whereas the later phases developing at higher temperatures are related to its overall composition. In the development of high-temperature phases, nucleation of the new lattice configuration takes place first, followed by a slow gradual growth of the new structure and an increase in its perfection as the temperature is raised above that required for nucleation. Traces of various elements cause substantial changes in the temperature and rate of formation of high-temperature phases.

Miscellaneous. Other important properties are resistance to thermal shock, attack by slag, and thermal expansion in the case of refractories. For whiteware, translucency, acceptance of glazes, etc, may be extremely important. These properties depend on the clay mineral composition and the method of manufacture, such as the forming procedure and intensity of firing.

Raw Material. ***Brick.*** Almost any clay composition is satisfactory for the manufacture of brick unless it contains a large percentage of coarse stony material that cannot be eliminated or ground to adequate fineness. A high concentration of nonclay material in a silt-size range may cause difficulties by reducing greatly the green and firing strength of the ware. Montmorillonite should be absent or present only in very small amounts; otherwise the shrinkage may be excessive. Clays composed of mixtures of clay minerals with 20–50% of unsorted fine-grain nonclay materials are most satisfactory. Large amounts of iron, alkalies, and alkaline earths, either in the clay minerals or as other constituents, cause too much shrinkage and greatly reduce the vitrification range; thus, a clay with a substantial amount of calcareous material is not desirable. Face bricks, which are superior, can be made from similar material; it is even more desirable to avoid the detrimental components mentioned above. For a light buff or gray face brick, kaolinite clay is preferred.

Tile. Roofing and structural tiles are usually made from the same material as face brick. Drain tiles have a high porosity, which is frequently attained by firing at relatively low temperatures. Frequently, drain tiles are made from clays with about 75% of fine-grained nonclay mineral material, in addition to components that provide high green and dry strength and a low fusion point. Wall and floor tiles are frequently made of mixtures with talc and kaolin as major components.

Terracotta, Stoneware, Sewerpipe, Paving Brick. Clays composed of mixtures of clay minerals containing 25–50% fine-grained unsorted quartz are well suited for the manufacture of terracotta, stoneware, sewerpipe, and paving brick. A small amount of montmorillonite can be tolerated, but a large amount gives undesirable shrinkage and drying properties. In general, clays with low shrinkage, good plastic properties, and a long vitrification range should be used.

Whiteware. Porcelain and dinnerware are made up of about equal amounts of kaolin, ball clay, flint (ground quartz), feldspar, or some other white-burning fluxing material such as talc and nepheline. The kaolin clay is composed of well-crystallized particles of kaolinite. Ball clays are white-burning, highly plastic, and easily dispersible. They provide the plasticity necessary in the forming of the ware and adequate green and dry strength for handling. The chief component of most ball clays is extremely fine-grained and relatively poorly organized kaolinite. However, some ball clays are known, for example, in South Devonshire in Great Britain, that contain remarkably well-ordered kaolinite. Some ball clays also contain small amounts of illite and/or small amounts of montmorillonite, which may add to their desired properties. Also, many ball clays contain a small but appreciable amount of organic material that appears to enhance the desired properties. Small amounts of bentonite are also used in whiteware bodies as replacements of ball clay to increase dry strength.

Porcelain Enamel. The slurry used in enameling is commonly composed of ball clay, frits, and coloring pigments. The frits are finely ground particles of prepared glass with a low fusion point.

Refractories. Refractory products are prepared from a wide variety of naturally occurring materials such as chromite and magnesite or from clays predominantly composed of kaolinite. For many refractory uses a somewhat lower fusion point than that provided by pure kaolinite may be adequate, so that clay materials with a moderate amount of other components as, for example, illite, may be satisfactory. In some cases this may be desirable, as in the case of brick used in lining steel ladles where it seems to provide the necessary reheat expansion.

High-alumina clays are also used extensively for the manufacture of special types of refractories (see Refractories). Such clays may be bauxites, composed essentially of aluminum hydrate minerals (gibbsite, boehmite) and kaolinite, or they may be diaspore clays.

An interesting type of clay used widely in the manufacture of refractories is so-called flint clay. As its name suggests, this clay is very hard and has very slight plasticity even when finely ground. Flint clays are essentially pure, extremely fine-grained kaolinite clays. In some cases, the hardness appears to be due to the presence of a small amount of free silica acting as a cement, whereas in other cases it is the result of an intergrowth of extremely small kaolinite particles.

Molding Sands

Molding sands, which are composed essentially of sand and clay, are used extensively in the metallurgical industry for the shaping of metal by the casting process. Using a pattern, a cavity of the desired shape is formed in the sand, and into this molten metal is poured and then allowed to cool.

The molding sand may be a natural sand containing clay or a synthetically prepared mixture of clean quartz sand and clay. Synthetic sands are widely used because they can be prepared to meet property specifications and their properties are more easily controlled as they are used. Granular particles other than quartz sand, for example, calcined clay, olivine, zircon, or chromite, may be used in rare special instances. Ground bituminous coal (sea coal) and cereal binders may be added to the sand–clay mixture to develop certain properties.

A small amount of water (tempering water) must be added to the molding sand to impart plasticity to develop cohesive strength so that the sand can be molded around the pattern, and to give it sufficient strength to maintain the cavity after the pattern is removed and while the metal is poured into it. These properties vary greatly with the amount of tempering water as well as with the nature of the clay bond.

In foundry practice the same molding sand is used over and over again. The high temperature of the metal dehydrates and vitrifies some of the clay, and fresh clay must be added continuously as the sand is used.

The only adequate test for the satisfactory use of a clay in bonding molding sands is the result obtained by actual use in foundry practice. Laboratory tests only eliminate worthless clays.

The best model of the bonding action of the clays in molding sand is that of a wedge-and-block bond at the interface of the sand grains. The bonding action of clay and water is not that of a glue or adhesive causing the grains to adhere to each other.

The value of a clay for bonding molding sand is usually determined by the green and dry compression strengths of mixtures with varying amounts of the clay and to which varying amounts of tempering water have been added. Other properties such as the bulk density, flowability, permeability, and hot strength may be important. The American Foundrymen's Society has published standard procedures for determining the properties of bonding clays (8).

Raw Materials. The bentonites, composed essentially of montmorillonite and used extensively in bonding molding sands, are of two types. The type carrying sodium as a principal exchangeable cation is produced largely in Wyoming. The calcium-carrying type is produced in Mississippi and in many countries outside the United States, such as England, Germany, Switzerland, Italy, U.S.S.R., South Africa, India, and Japan. The natural calcium montmorillonite bentonites are occasionally treated with various sodium compounds so that their properties are similar to the properties of the natural sodium bentonites produced in Wyoming.

Plastic clays composed largely of poorly crystalline kaolinite but with small amounts of illite, and at times montmorillonite, are widely used in bonding molding sands especially in the United States. These clays are called fireclays because of the relatively high refractoriness. In the United States such clays are mined extensively in Illinois and Ohio from beds of carboniferous age, where they occur directly beneath beds of coal (underclays).

The third type of clay used in foundries is composed essentially of illite. Most illite clays have a bonding strength and plasticity too low for bonding use, but there are some varieties that have such properties approaching those of montmorillonite. The illite in such clays is fine-grained, poorly organized, and frequently associated with mixed-layer assemblages containing montmorillonite. Illite bonding clays are produced extensively in Illinois.

Properties. ***Compression Strengths.*** *Green compression strengths* in the range from about 35 to 75 kPa (5–11 psi) are desired in actual practice.

Figure 1 shows the maximum green strength for varying amounts of each type of clay up to 15%. This diagram refers to many uses of clays.

Calcium montmorillonite clay gives the highest green compression strength, whereas kaolinite and illite have about the same strength, which is considerably lower than that of either montmorillonite clay. In sands bonded with the calcium montmorillonite, the strength increases approximately in proportion to the amount of clay up to about 8%. Additional clay up to 10% causes only a very slight increase in strength, and clay in excess of this amount causes no further increase.

Sands bonded with the sodium montmorillonite show a similar relation between maximum green compression strength and the amount of clay, except that the reduction in strength per unit of added clay above about 8% is less than that for the calcium montmorillonite. The kaolinite- and illite-bonded sands show a continuing increase in maximum green compression strength up to 15% of clay, but less above 10% for kaolinite and 12% for illite.

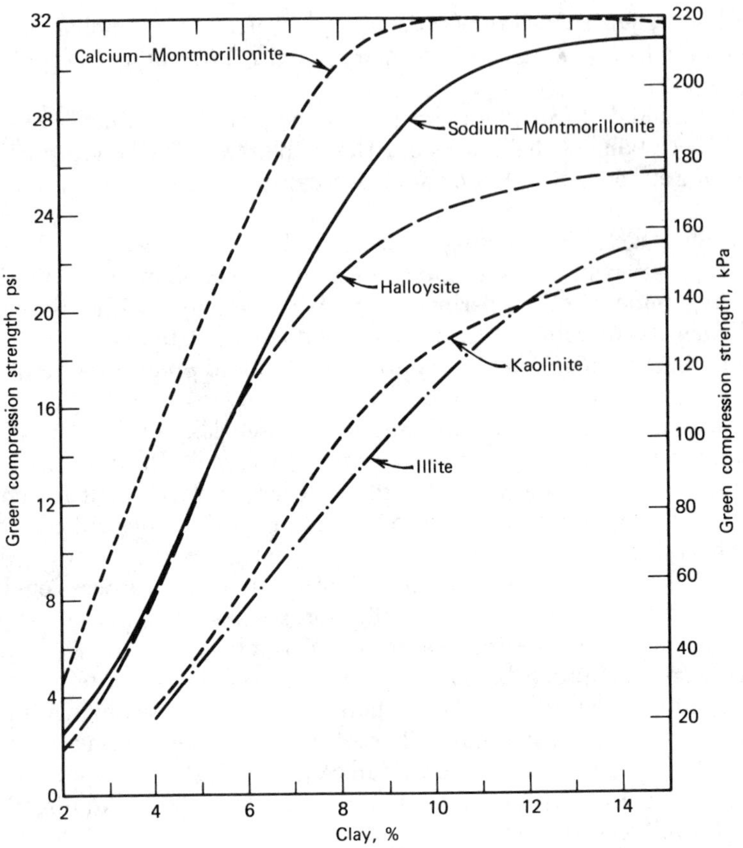

Figure 1. Maximum green compression strength developed by each type of clay in relation to the amount of clay in the sand clay mixture (9).

Considerably lower dry compression strength is developed by calcium montmorillonite (10) than by sodium montmorillonite clays. Furthermore, calcium montmorillonite clays require a certain minimum amount of tempering water to develop any dry strength which increases with the amount of tempering water up to a maximum value, which is reached abruptly. A striking feature of this clay is that maximum dry compression strength is about the same for all mixtures with more than about 6% clay.

Illite clays develop relatively high dry compression strength (10). Maximum strength is developed with relatively small amounts of tempering water.

The maximum dry compression strength obtained with kaolinite clays is reached abruptly and increases with the amount of clay in the mixture (10). In general, kaolinites yield less dry strength than illites and slightly more water is required to develop equivalent strength in a given mixture.

Drilling Fluids

In the drilling operation a fluid is pumped down through the hollow drill stem, emerging at the bottom of the hole through "eyes" in the bit. During actual drilling it circulates continuously to remove cuttings and rises to the surface in the annular space between the drill stem and the walls of the hole and then flows into a pit for removal of cuttings and entrained gas.

The removal of cuttings is readily facilitated by a fluid with a viscosity higher than that of water. Furthermore, it must be possible to pump the drilling mud. Its viscosity, on the order of 15 mPa·s(= cP), is controlled and it should be markedly thioxotropic; that is, should have gel strength. This is necessary so that the cuttings do not settle to the bottom of the hole and freeze the drill stem when pumping or agitation of the drill stem ceases temporarily.

The drilling fluid serves to confine the formation fluids. Normally such fluids are under pressure at least equal to that of a column of water of equal depth so that densities in excess of water alone are required. However, much higher densities frequently are required necessitating muds of high weight per unit volume which are attained by the addition of such materials as barites. The drilling mud must have sufficient viscosity and gel strength to maintain such weighting material in suspension. An impervious thin coating must be built up on the wall of the hole in order to impede the penetration of water from the drilling fluid into the formations. High impermeability is particularly important in drilling through formations that are permeable and contain fluids under relatively high pressure; it is also important when drilling water may soften formations and cause them to cave or heave into the hole. Argillaceous material added to the drilling fluid from the penetrated formations should be kept to a minimum because such material tends to increase viscosity.

Viscosity and gel strength of the drilling fluid should not be affected by variations in electrolyte concentration and should be subject to some control by chemical treatment. In some cases, high-lime muds with pH values in the range of 12–13 are used to maintain viscosity and gel strength in the presence of clay added from penetrated formations and of large amounts of electrolytes. These muds also contain a protective colloid that may be a tannin material, humic acid derivative, or gelatinized starch; otherwise the system would be completely flocculated.

The suitability of a clay for use in drilling mud is measured primarily by (*a*) the number of cubic meters (barrels) of mud with a given viscosity (usually 15 mPa·s or cP) obtained from one ton of clay in fresh water and in salt water; (*b*) the difference in gel strength taken immediately after agitation and 10 min later; (*c*) the wall-building properties, as measured by the water loss through a filter paper when a 15-mPa·s (= cP) clay is subjected to a pressure of 690 kPa (100 psi) for 30 min; and (*d*) the thickness of the filter cake produced in the standard water-loss test.

Clays and soils with more than a very small amount of nonclay minerals, particularly in silt and sand sizes, are not suitable for drilling muds because such materials dilute the desired properties and have abrasive action on pumps and other drilling equipment. Local soils and clays of low nonclay mineral content are frequently used to start a well, with special clays being added with increasing depth to control the properties of the drilling fluid.

The most widely used clay for drilling fluids is a bentonite from Wyoming composed of montmorillonite carrying sodium as the major exchangeable cation. This clay gives petroleum yields in excess of 15.9 m^3 per metric ton of clay (100 bbl/t) so that only about 5% clay is adequate to produce the desired viscosity. It also has very high gel strength and low permeability. This bentonite is used worldwide because of its valuable property of producing an impervious clay layer, even when very thin, on the wall of the drilling hole.

Recently, clays with yields up to 47.7 m^3/t of mud (300 bbl/t) have been developed by the addition of certain polymers to bentonites.

Muds containing clays composed of attapulgite and sepiolite show small variations in viscosity gel strength with large variations in electrolyte content, and considerable tonnage of these clays are consumed annually. In some deep drilling operations, relatively high temperatures, 370°C, are encountered to which sepiolite muds are especially resistant (11).

In oil-based muds, where oil rather than water is the fluid, clay mineral–organic complexes are used. In such complexes the clay materials are coated with various organic compounds that change them from hydrophilic to oleophilic (see Petroleum, drilling fluids).

Catalysis

Catalysts made from various clay minerals are extensively used (12) in the cracking of heavy petroleum fractions (see Catalysis; Petroleum). These catalysts are produced from halloysites, kaolinites, and bentonites composed of montmorillonite. In the United States, montmorillonite clays from Arizona and Mississippi, halloysite from Utah, and kaolinite clays from sedimentary formations in Georgia are used for this purpose. These clays must be low in iron and substantially free from various elements such as heavy metals which would favor either catalyst poisoning or an unfavorable product distribution.

Preparation. The process details are proprietary. In one process the clays are first treated with an acid solution, usually sulfuric, at moderately elevated temperatures. Subsequently, the clay is washed to remove alkalies and alkaline earths and partially reduce the alumina content. In another process, a mixture of kaolin and caustic soda is heated under controlled conditions to form a zeolite-type component. In both processes, the clay is prepared in the form of pellets or of "fluid" powder, and calcined at moderately elevated temperatures.

Oil Refining and Decolorization

Clay materials are used widely to decolorize, deodorize, dehydrate, and neutralize mineral, vegetable, and animal oils. Decolorization is generally the major objective of such processes. The oil is filtered through a granular product of 250–2000 μm (10–60 mesh) particles, or placed into contact with finely ground clay of approximately 741 μm (−200 mesh); it is then separated from the clay by a filter pressing operation. The percolation is essentially a low-temperature process, whereas the contact process takes place in the range of 150–300°C.

A wide range of clay materials have been used for decolorization, ranging from fine-grained silts to clays composed of almost pure clay minerals. The materials may be substantially crude clay, such as fuller's earth, or clay that has been prepared by chemical and physical treatment. The name fuller's earth comes from the use of these clays in cleaning or fulling wool with a water slurry of earth, whereby oil and dirt particles were removed from the wool. At the present time, this term is applied to any clay that has adequate decolorizing and purifying capacity to be used commercially in oil refining without chemical treatment. It does not indicate composition or origin.

Clays composed of attapulgite and some montmorillonites possess superior decolorizing powers. For preparation, these clays are dried at 200–315°C and ground to various sizes. The activity of the attapulgite clay is substantially enhanced by extrusion under high pressure at low moisture contents. Montmorillonite clays are not improved by such extrusion. Since only some montmorillonite clays possess substantial decolorizing properties, they have to be evaluated by field tests. Acid activation enhances the decolorizing power of some montmorillonite clays several-fold. Halloysite clays may have high decolorizing power (13–14), as does an unusual illitic clay from northern Illinois (15). Some sepiolite clays from Spain are excellent decolorizing materials (16). At the present time however, only attapulgite and montmorillonite clays are used for commercial decolorization.

Properties. A relatively small amount of clay must reduce color substantially. Furthermore, oil retention must be low; that is, only a small amount of oil is retained by the clay in the course of the decolorizing process. The latter property is particularly important if the oil cannot be reclaimed from the spent clay by a solvent or distillation. The clay must have good filtration characteristics; that is, the oil must pass through fairly rapidly, but the clay must not unduly bind the filters. In the case of edible oils, the earth must not impart an obnoxious odor or taste to the oil.

Paper

Paper is a thin uniform sheet of finely intermeshed and felted cellulose fibers. The cellulose for the highest-quality paper is obtained from cotton and linen, but for most purposes fibers obtained from a large variety of woods are satisfactory (see Paper).

A sheet of cellulose fibers is not well suited to high-fidelity printing because of transparency and irregularities of the surface. These deficiencies are corrected by the addition of binding agents such as starch and resin, and by the incorporation into the fiber stock of mineral fillers such as calcium carbonate or sulfate, and especially pure white clay. Ordinary filled paper still lacks the perfection of surface smoothness required for accurate production of the tiny dots in halftone printing. The quality of the filled sheet is often enhanced by coating its surface with a thin film of finely divided mineral pigment suspended in an adhesive mixture such as starch and casein. Depending on the desired effect, the coating pigment may be calcium carbonate, "satin white" (coprecipitated calcium sulfate and hydrated alumina), titanium dioxide, and especially relatively pure white clay with a specific particle-size distribution. The amount of coating ranges up to 25% of the weight of the paper, the amount of filler up to 35%.

Pure kaolinite clays are used for filling and coating paper, except when special properties are desired. The clays have to be easily dispersible in water and should have, in the crude state, a fairly wide particle-size distribution so that a range of products can be prepared commercially. The clay should be as nearly white as possible or easily bleached to a high degree of whiteness. For coating purposes, low viscosity in clay–water systems is essential; that is, a slurry of high clay concentration with the lowest possible viscosity must be obtained.

In general, the crude clays are washed to remove any grit that may be present, chemically bleached to increase their whiteness, fractionated by settling or centrifugation to develop products with a given particle-size distribution, and then dried.

The clay must be easily freed of grit content if there is any. The grit causes wear in the foundrinier wires of the papermaking machine, and in coating clays causes an abrasive surface on the paper which wears the press plates in printing. Many kaolins could be utilized for paper clay except that the grit cannot be removed economically. Ultraflotation and high-intensity magnetic separation are used in the kaolin industry to increase brightness by removing extremely fine pigment particles, generally iron and titanium minerals (see Flotation; Magnetic separation).

In some paper-coating operations, clays have been found useful that were calcined at temperatures in excess of that required for complete dehydration (540°C) and below that at which new high-temperature phases form when the clay begins to vitrify because calcination tends to increase brightness. After calcination the clay must be ground to an extreme fineness.

In recent years, the relation has been studied between variation in the properties of kaolins, eg, particle-size distribution, composition, particle shape, etc, and the properties of coated paper such as opacity, printability, gloss, etc (17,18). Kaolin producers now offer a variety of products with different compositions and therefore different coating properties. The paper industry now consumes about 75% of all kaolin production.

Attapulgite clay has been used in the manufacture of NCR (no carbon required) paper (19). The clay is used to coat the upper surface of a sheet of paper; the lower surface is coated with minute capsules of certain organic compounds. The lower surface of one sheet rests on the upper surface of the second sheet and the pressure of writing serves to break the capsules of the organic compound, which then penetrates the attapulgite clay causing a color reaction and the writing appears on the second sheet (see Microencapsulation). Recently other materials have largely replaced attapulgite in NCR paper.

At the present time, much research is underway to produce a chemically processed bentonite, kaolin, or sepiolite clay or their combinations for use in a variety of duplicating papers.

Bentonite clays composed of montmorillonite are used for several purposes in papermaking. Pitch, tars, waxes, and resinous material from the wood and from waste cuttings and mixed paper may tend to agglomerate and stick to screens and machine wires, brush rolls, etc, and cause defects in the paper. Addition of small amounts of bentonite prevents the agglomeration of such particles. The addition of 1–2% of such clay at the beater or pulper stage increases retention of pigments by the paper stock, and thus also the uniformity of their distribution throughout the paper. White, or at least very light, highly colloidal bentonites are preferred.

Miscellaneous

Clays are used as ingredients in a vast number of products, for example, asphalt, paints, and rubber. They are also used as agents in many processes, such as water purification, and as a source of raw material, eg, aluminum.

Adhesives. Clays, generally kaolinites, are used in a wide variety of adhesives, such as those prepared with lignins, silicate of soda, starch, latex, and asphalt. They are used extensively in adhesives for paper products and in cements for floor coverings, including linoleum, rubber, and asphalt tile.

In adhesives, clays are not merely inert diluents, but may improve the properties.

Thus, clays tend to reduce the amount of penetration of the adhesives into the members to be joined, cause a faster setting rate and superior bond strength, increase the solid content of the adhesive and the amount of bonded surface, and permit the control of the suspension and viscosity characteristics for more satisfactory application.

Aluminum Ore. Many clays contain as much as 30–40% alumina. The main sources of aluminum are bauxite ores, which are composed of aluminum hydrates, either gibbsite or boehmite (see Aluminum).

The United States has only limited bauxite deposits and imports its requirements from the West Indies, South America, and Australia. Recently, large new deposits have been found in Brazil, Australia, and West Africa.

During World War II, the bauxite sources for the aluminum industry in the United States were threatened and a large research effort was made toward extraction of alumina from clays. Several processes were investigated and several large pilot plants were put in operation. After the end of the war, it was more economical to import bauxite to the United States. However, the economic advantage of using bauxite over clay is relatively slight and improvement in the clay-extraction process or increases in the cost of imported ore may in the future make clay-extraction economical.

Alumina is extracted from clays (20–22) by (*a*) the acid process, where aluminum is dissolved (chiefly with sulfuric acid), usually after the clay has been heated to elevated temperatures (600°C) to increase the solubility of the aluminum; (*b*) the aluminum sulfate process, where the clay is baked with ammonium sulfate at approximately 400°C, followed by leaching with water to extract the aluminum sulfate and ammonium sulfate; (*d*) the lime sinter process, where the clay is treated with ground limestone and calcined at 1375°C to give a mixture of dicalcium silicate ($2CaO.SiO_2$) and pentacalcium trialuminate ($5CaO.3Al_2O_3$), followed by leaching with a dilute alkali carbonate solution to remove the aluminum; (*d*) the lime-soda sinter process, which is similar to the lime sinter process except that the clay is treated with calcium carbonate and sodium carbonate, calcined to give a mixture of dicalcium silicate and sodium aluminate ($NaAlO_2$), followed by dissolution of the aluminum in a dilute alkali carbonate solution; and (*e*) several high-temperature electrochemical processes in which there is a direct reaction of aluminum in the clay to form alumina or aluminum nitride, carbide, chloride, sulfide, etc; this reaction is carried out at very high temperatures, generally in an electric furnace. Kaolinite clays are used for these processes because of their high aluminum content and because they contain small amounts of iron, alkalies, and alkaline earths that may cause serious problems in the extraction processes.

In recent years particular attention has been given to the acid extraction process as being the most attractive financially and technically.

Asphalt Products. A large variety of asphalt products, eg, for use in highway construction, contain clay materials as fillers, extenders, and to control viscosity and penetration (see Asphalt).

Radioactive Waste Disposal. The use of clays has been suggested for the disposal of atomic wastes by adsorbing the active elements and then fixing them against leaching by calcination at temperatures adequate to vitrify the clay and thereby to bind the radioactive material in insoluble form (see Nuclear reactors). The disposal of the isotopes of cesium and strontium is particularly important because of their relatively long half-life, potency, and abundance. The problem for strontium can be solved fairly well by its fixation as a carbonate. Certain illite clays have unusual power for the fixation of cesium and chemical pretreatment or heating at moderately low temperatures may enhance the cesium fixation of some three-layer minerals.

Portland Cement. Portland cement, a mixture of calcium silicate and calcium aluminate minerals, is produced by the calcination of argillaceous limestones or mixtures of limestone and clay. Moderate amounts of iron, alkalies, and alkaline earths, except magnesia which should not be in excess of 5%, apparently are not detrimental to the properties of portland cement. Therefore, any of the clay minerals, with the exception of attapulgite, sepiolite, and some montmorillonites, could be used as an ingredient. However, kaolinite is preferred since it contributes only alumina and silica and is especially desirable in the manufacture of white portland cement and other special types requiring careful control of chemical composition (see Cement).

Pozzolanas. Pozzolanas are siliceous or siliceous and aluminous materials, natural or artificial, processed or unprocessed, which though not cementitious in themselves, contain constituents that combine with lime in the presence of water at ordinary temperatures to form compounds of low solubility with cementing properties (see Cement).

Pozzolanas are classified, depending upon the substances responsible for the pozzolanic action, in decreasing order of activity as follows: volcanic glass; opal; clay minerals (kaolinites, montmorillonites, illites, mixed vermiculite–chlorite clays, and palygorskite); zeolite; and hydrated aluminum oxide.

In the natural condition, the clay minerals are nonpozzolanic or only weakly so. With calcination, however, particularly in the range of 650–980°C, partial dehydration and structural changes result in significant reactivity for lime.

Medicinals and Cosmetics. Clays have been used for centuries in therapeutic intestinal adsorbent preparations against intestinal irritation. The clays are believed to function by adsorbing toxins and bacteria responsible for intestinal disorders and by coating the inflamed mucous membrane of the digestive tract (23). Kaolinite clays have been used widely. However, attapulgite clays that have been activated by moderate heating are five to eight times more active than kaolinite clays as adsorbents for alkaloids, bacteria, and toxins (24). By its highly effective adsorption of the aqueous part of inflammatory secretions, attapulgite aids in stool formation.

Clays of various kinds, but particularly those composed of montmorillonite, kaolinite, or attapulgite, have been used for a long time in the preparation of pastes, ointments, and lotions for external use; the best known of these uses is in the preparation of antiphlogistine for checking inflammation. Many cosmetic formulations contain montmorillonite, attapulgite, and kaolinite clays, taking advantage of the clay's softness, dispersion, gelling, emulsifying, adsorption, or other properties. No general specifications are reported for clays for these uses except that they be free from grit, very fine-grained, and dispersible in liquid (see Cosmetics).

Paint. Paint is essentially a fluid system in which solid bodies, usually identified as pigments, are suspended.

Various types of clays have been used for a long time as paint ingredients, at first as inert fillers. However, clays add valuable properties to paints and are essential components of some types of paint.

Kaolinite clays are used very widely in amounts varying from 2–5% in some enamels, to as much as 45% in some interior wall paints. Calcined kaolins are used as extenders for titania pigment in a variety of paints.

Kaolinite clays impart desirable and controllable surface characteristics, permit high pigment loading, add to the hiding power, and also have good oil-absorption and suspension properties (see Paint).

Bentonite clays, composed essentially of montmorillonite, are used extensively in both oil- and water-base paints. In the latter, the montmorillonite clays act as a suspending and thickening agent.

Montmorillonite is hydrophilic and thus may be difficult to disperse in some oil vehicles. Organic-clad montmorillonite clays, which are oleophilic, are now available that are tailor-made with a variety of organic compounds to meet the requirements of different vehicles, such as cellulose nitrate lacquers, epoxy resins, and vinyl resins.

Pelletizing Ores, Fluxes, Fuels. In the beneficiation of some ores, it is frequently necessary to pulverize the ore to accomplish separations and concentrations. The pulverized ore has to be pelletized or agglomerated into larger units. Bentonites of high dry strength are best suited for pelletizing. The largest market for bentonites is now pelletizing of iron ore. A process for pelletizing finely ground fluorspar flux at 90–200°C with about 1% of Wyoming bentonite and sulfite waste liquor has been patented (25), as has the pelletizing of powdered fuel (26) using about 5% bentonite and quebracho extract (see Pelletizing and briquetting).

Pesticides. Many pesticides are highly concentrated and are in a physical form requiring further preparation to permit effective and economical application. A number of diluents are available in solid or liquid form to bring the pesticide chemicals to field strength. Although carriers of diluents generally are considered to be inert, they have a vital bearing on the potency and efficiency of the pesticide dust or spray, since most dusts and sprays contain only 1–20% active ingredients and 80–99% carrier diluent. The physical properties of the carrier diluents are of great importance in the uniform dispersion, the retention of the pesticide by the plant, and in the preservation of the toxicity of the pesticide. The carrier must not, for example, serve as a catalyst for any reaction of the pesticide that would substantially alter its potency.

Clays composed of attapulgite, montmorillonite, and kaolinite are used for this purpose in finely pulverized as well as granular form. Granular formulations are reportedly less expensive, more easily handled, reduce loss due to wind drift, and produce a most effective coverage (see Insect control technology; Poisons, economic).

Plastics. Mineral filler used in reinforced polyester resins and plastics offers the following advantages (27): it produces a smooth surface finish; reduces cracking and shrinkage during curing; aids in obscuring the fiber pattern of the glass reinforcement; contributes to high dielectric strength, low water absorption, and high wet strength; aids resistance to chemical action and weathering; and controls flow properties. Clay fillers are used extensively in polyester resins and polyvinyl compounds in amounts as high as 60%. Examples of the latter usage are found in electrical insulations, phonograph records, and floor coverings. Clays may also be used effectively in rigid and flexible urethane foams to provide cost reduction and improve deflection values, uniformity, and cell size.

Kaolinite clays are used extensively in the filling of plastics because they are easily dispersible in the resin as compared with other mineral fillers (see Plastics). All fillers increase the viscosity of the mixture, but in the case of clays this property can be controlled by the selection of the proper particle-size distribution. The low specific gravity of kaolinite and its white color are desirable characteristics in plastic compounds. Kaolins clad with certain organic compounds have improved dispersion and flow properties in polyvinyl resin systems (27).

Rubber

Kaolins were first used in rubber as diluents (28), but they also provide desirable reinforcing and stiffening properties as well as increased resistance to abrasion. Kaolins are used with certain types of synthetic rubber for heels and soles, tubing, extruded stock of all types, wire insulation, gloves, adhesives, tire treads, and inner tubes. In general, kaolins of extremely small particle size are desired; thus, so-called hard kaolins with 90% of the particles finer than 2 μm are commonly required. Manganese content must be extremely low because of its deleterious catalytic action (see Rubber).

Montmorillonite clays are used as thickening and stabilizing additives to latex and as emulsion stabilizers in rubber-base paints and in rubber adhesives.

Organic-clad clays of varying types are also used in the rubber industry. Thus clays in which the inorganic cation has been replaced by a substituted organic onium base are reinforcing agents in natural and some synthetic rubbers (29). Coating the clay particles with a strongly adhering organic layer improves the properties as a rubber filler (30).

Water Clarification and Impedance

In the clarification of potable water, colloidal matter is removed by filtration and/or sedimentation processes. Alum is commonly used to flocculate the colloidal material to enhance its settling and filtration rate. A highly colloidal, easily dispersible clay added to the water before the alum is desirable. The alum flocculates the clay, which serves to gather up and collect the colloidal material that would otherwise not settle (31). Bentonites and attapulgite clays have been used for this purpose (see Water). Wyoming-type bentonites are used extensively to impede the movement of water through earthen structures and to retard or stop similar movement through cracks and fissures in rocks and in rock and concrete structures. For example, bentonite is used to stop the seepage of water from ponds and irrigation ditches and to waterproof the outside basement walls of homes (32).

The bentonite is used as a grout, in which a bentonite suspension is injected under pressure into a porous strata, a fissure or crack, or along the foundation of a building, or in a placement method in which the bentonite may be placed to form an unbroken blanket between the earth and the water or mixed with the surface soil of the structure to render the soil impervious (33). In the case of placing a bentonite blanket to seal a pond or lake, the pond need not always be drained as granular bentonite may be added to the water surface. It will settle to the bottom of the pond, forming an impervious blanket.

BIBLIOGRAPHY

"Clays" in *ECT* 1st ed., Vol. 4: "Ceramic Clays," pp. 38–49, by W. W. Kriegel, North Carolina State College; "Fuller's Earth," pp. 49–53, by W. A. Johnston, Attapulgus Clay Co.; "Activated Clays," pp. 53–57, by G. A. Mickelson and R. B. Secor, Filtrol Corporation; "Papermaking, Paint, and Filler Clays," pp. 57–71, by S. C. Lyons, Georgia Kaolin Company; "Rubbermaking Clays," pp. 71–80, by C. A. Carlton, J. M. Huber Corporation; and "Clays (Uses)" in *ECT* 2nd ed., Vol. 5, pp. 560–586, by Ralph E. Grim, University of Illinois.

1. W. A. White, *The Properties of Clays,* Master's Thesis, University of Illinois, Urbana, Illinois, 1947.
2. U. Hofmann, "Fullstoffe und keramische Rohmaterialien," *Rapport Europees Congrès Electronenmicroscopic,* Ghent, 1954, pp. 161–172.
3. P. F. Low, The Viscosity of Water in Clay Systems, *Proceedings of the 8th National Clay Conference,* Pergamon Press, New York, 1960, pp. 170–182.
4. R. E. Grim and F. L. Cuthbert, *J. Am. Ceram. Soc.* **28,** 90 (1945).
5. R. E. Grim, "Some Fundamental Properties Influencing the Properties of Soil Materials," *Proceedings of the International Conference of Soil Mech. Foundation Eng. 2nd, Rotterdam 1948,* **III,** pp. 8–12.
6. D. A. Holderidge, *Trans. Brit. Ceram. Soc.* **55,** 369 (1956).
7. W. O. Williamson, *Trans. Brit. Ceram. Soc.* **40,** 225 (1941).
8. *Testing and Grading of Foundry Clays,* 6th ed., Am. Foun. Soc., 1952.
9. R. E. Grim and F. L. Cuthbert, *Ill. State Geol. Surv. Rep. Invest. 102* (1945).
10. R. E. Grim and F. L. Cuthbert, *Ill. State Geol. Surv. Rep. Invest. 110* (1946).
11. L. L. Carney and R. L. Meyer, *Am. Inst. Min. Met. Eng. Soc. Petrol. Eng. Paper 6025* (1976).
12. T. H. Mulliken, A. G. Oblad, and G. A. Mills, *Calif. Dept. Nat. Resources Div. Mines Bull. 169* (1955).
13. V. Charrin, *Genie Civ.* **123,** 146 (1946).
14. F. J. Zvanut, *J. Am. Ceram. Soc.* **20,** 251 (1937).
15. R. E. Grim and W. F. Bradley, *J. Am. Ceram. Soc.* **22,** 157 (1939).
16. G. P. C. Chambers, *Silic. Ind.* **24,** 3 (April, 1959).
17. W. M. Bundy, W. D. Johns, and H. H. Murray, *Jour. Tech. Assoc., Pulp and Paper Ind.* **48,** 688 (1965).
18. W. M. Bundy and H. H. Murray, *Clays Clay Miner.* **21,** 295 (1973).
19. U.S. Pat. 2,550,469 (April 24, 1951), K. Barrett, R. Green, and R. W. Sandberg.
20. F. A. Peters, P. W. Johnson, and R. C. Kirby, *U.S. Bur. Mines Rept. Invest. No. 6229* (1963).
21. F. A. Peters, P. W. Johnson, and R. C. Kirby, *U.S. Bur. Mines Rept. Invest. No. 6133* (1963).
22. G. Thomas and T. R. Ingram, *Can. Dept. Mines Tech. Surv., Mines Branch Res. Rept. R45* (1959).
23. L. S. Goodman and A. Gilman, *The Pharmacological Basis of Therapeutics,* 2nd ed., The Macmillan Co., New York, 1955.
24. M. Barr, *J. Am. Pharm. Assoc., Pract. Pharm. Ed.* **19,** 85 (1958).
25. U.S. Pat. 2,220,385 (Mar. 5, 1940), F. C. Abbot and C. O. Anderson.
26. U.S. Pat. 2,217,994 (Oct. 15, 1940), G. G. Rick and C. E. Loetel.
27. J. R. Wilcox, *"Controlling Flow Properties with Fillers," Proc. Soc. Plastics Inc. 9, Sect. 27,* February, 1954.
28. C. C. Davis and J. T. Blake, "Chemistry and Technology of Rubber," *Am. Chem. Soc. Monograph Series No. 74,* American Chemical Society, Washington, D.C. (1937).
29. U.S. Pat. 2,531,396 (1950), L. W. Carter, J. C. Hendricks, and D. S. Bolley.
30. Brit. Pat. 630,418 (1949), N. O. Clark and T. W. Parker.
31. E. Nordell, *Water Treatment for Industrial and Other Uses,* Reinhold Publishing Corp., New York, 1951.
32. R. C. Mielenz and M. E. King, *Calif. Dept. Nat. Resources Div. Mines Bull.* **169,** 196 (1955).
33. C. D. Weaver, *Proc. Soil Sci. Soc. Am.* **11,** 196 (1946).

General Reference

R. E. Grim, *Applied Clay Mineralogy,* McGraw-Hill Book Co., New York, 1962.

RALPH E. GRIM
University of Illinois

COATED FABRICS

A coated fabric is a construction that combines the beneficial properties of a textile and a polymer. The textile (fabric) provides tensile strength, tear strength, and elongation control. The coating is chosen to provide protection against the environment in the intended use. A polyurethane might be chosen to protect against abrasion or a polychloroprene (Neoprene) to protect against oil (see Urethane polymers; Elastomers, synthetic).

Textile Component

The vast majority of textiles that are used for coating are purchased by the coating company. Several large textile manufacturers have divisions that specialize in industrial fabrics. These companies perform extensive development on substrates and can provide advice on choosing the correct substrate, hand samples, pilot yardage, and ultimately production requirements. A listing of suppliers of industrial fabrics can be found in *Davison's Textile Blue Book* (1).

Fiber. For many years cotton (qv) and wool (qv) were used as primary textile components, contributing the properties of strength, elongation control, and esthetics. Although the modern coated fabrics industry began by coating wool to make boots, cotton has been used more extensively. Cotton constructions, including sheetings, drills, sateens and knits, command a major share of the market. Cotton is easily dyed, absorbs moisture, withstands high temperature without damage, and is stronger wet than dry. This latter property renders cotton washable; it can also be drycleaned because of its resistance to solvents (see Drycleaning).

Polyester, by itself and in combination with cotton, is used extensively in coated fabrics. Polyester produces fibers that are smooth, crisp, and resilient. Since moisture does not penetrate polyester, it does not affect the size or shape of the fiber. Polyester resists chemical and biological attack. Because of its thermoplastic nature, the heat required for adhesion to this smooth fiber can also create shrinkage during coating (see Polyester fibers; Fibers, man-made and synthetic).

Nylon is the strongest of the commonly-used fibers. Since it is both elastic and resilient, articles made with nylon will return to their original shape. There is a degree of thermoplasticity so that articles can be shaped and then heated to retain that shape. Nylon fibers are smooth, very nonabsorbent, and will not soil easily. Nylon resists chemical and biological action. Nylon substrates are used in places where very high strength is required. Lightweight knits and taffetas are thinly coated with polyurethane or poly(vinyl chloride) and used extensively in apparel. In coating, PVC does not adhere well to nylon (see Polyamides).

Rayon and glass fibers are the least used because of their poor qualities. Rayon's strength approaches that of cotton but its smoother fibers make adhesion more difficult. Rayon has a tendency to shrink more than cotton, which makes processing more difficult. Glass fibers offer very low elongation, very high strength, and have a tendency to break under compression. Therefore, glass fabric is only used where support with low stretch is required and where the object is not likely to be flexed. For instance, glass might be used to support a lead-filled vinyl compound for sound dampening (see Glass; Insulation, acoustic).

Textile Construction. There are many choices in textile construction. The original, and still the most commonly used, is the woven fabric. Woven fabrics have three basic constructions: the plain weave, the satin weave, and the twill weave. The plain weave is by far the strongest because it has the tightest interlacing of fibers; it is used most often (2). Twill weaving produces distinct surface appearances and is used for styling effects. Because it is the weakest of the wovens, satin weave is used principally for styling. Woven nylon or heavy cotton are used for tarpaulin substrates. For shoe uppers, and other applications where strength is important, woven cotton fabrics are used.

Knitted fabrics are used where moderate strength and considerable elongation are required. Where cotton yarn formerly dominated the knit market, it has recently been replaced by polyester–cotton yarn, and polyester yarn and filament. Where high elongation is required, nylon is used. Knits are predominantly circular jersey; however, patterned knits are becoming more and more prevalent. When a polymeric coating is put on a knit fabric, the stretch properties are somewhat less than that of the fabric. Stretch and set properties are important for upholstering and forming. The main use of knit fabrics is in apparel, automotive and furniture upholstery, shoe liners, boot shanks—any place elongation is required.

Many types of nonwoven fabrics are used as substrates (3) (see Nonwoven textiles). The wet web process gives a nonwoven fabric with paperlike properties; low elongation, low strength, and poor drape. When these substrates are coated, the papery characteristics show through the coating, and fabric esthetics are not satisfactory. The nonwovens prepared by laying dry webs, compressing by needle punching, and then impregnating from 50–100% with a rubbery material, resemble in many respects split leather. These materials are used for shoe liners (see Leather-like materials). It is difficult to achieve uniformity of stretch and strength in two directions as well as a smooth surface; therefore, a high quality nonwoven of this type is very expensive. Spunbonded nonwovens are available in both polyester and nylon in a range of weights. The strength qualities are very high and elongation is low. Since they are quite stiff, these materials are used where strength and price are the major considerations. A lightly needled, low density nonwoven was marketed in 1970 and in the last three years has gained prominence in the coated fabrics industry. It is used in weights from 60–180 g and can be prepared from either polypropylene or polyester fibers. The light needling combined with careful orientation of the fibers and selection of the fiber length gives very good strength and more balanced stretch. Optionally, a thin layer of polyester-based polyurethane foam can be needled into the nonwoven to improve the surface coating properties. The furniture upholstery market was the first to accept this product. A 0.4 mm poly(vinyl chloride) skin on this nonwoven replaced expanded PVC on knit fabric at approximately half the cost. The finished product is softer, plumper, and may have better wear characteristics. A major automobile company has introduced this type of seating in one line of cars, and apparel manufacturers have exhibited interest in these types of construction.

Post Finishes of the Textile Component. The construction that results from either weaving or knitting is called a greige good. Other steps are required before the fabric can be coated: scouring to remove surface impurities; and heat setting to correct width and minimize shrinkage during coating.

Optional treatments include: dyeing if a colored substrate is required; napping of cotton and polyester–cotton blends for polyurethane coated fabrics; flame resistance treatments; bacteriostatic finishes for hygienic applications; and mildew treatments for applications in high humidity (see Textiles).

Polymer Component

Rubber and Synthetic Elastomers. For many years coated fabrics consisted of natural rubber (qv) on cotton cloth. Natural rubber is possibly the best all-purpose rubber but some characteristics such as poor resistance to oxygen and ozone attack, reversion and poor weathering, and low oil and heat resistance, limit its use in special application areas (see also Elastomers, synthetic).

Polychloroprene (Neoprene) introduced in 1933 rapidly gained prominence as a general purpose synthetic elastomer having oil, weather and flame resistance. The introduction of new elastomers in solid or latex form was accelerated by World War II. Currently, in addition to natural rubber and polychloroprene, other polymers in use include: styrene–butadiene (SBR), polyisoprene, polyisobutylene (Vistanex), isobutylene–isoprene copolymer (Butyl), polysulfides (Thiokol), polyacrylonitrile (Paracril), silicones, chlorosulfonated polyethylene (Hypalon), poly(vinyl butyral), acrylic polymers, polyurethanes, ethylene–propylene copolymer (Royalene), fluorocarbons (Viton), polybutadiene, polyolefins, and many more. Copolymerizations and physical blends make the number available staggering (see Copolymers; Olefin polymers; Polymers containing sulfur; Acrylic ester polymers; Vinyl polymers; Fluorine compounds; Acrylonitrile polymers; Silicon compounds; Urethane polymers).

In fact, the number of commercially available polymers in use is well over 1000. In each class there are several variations manifesting a wide range of properties. DuPont supplies about 24 types of polychloroprene. B. F. Goodrich supplies about 140 types of acrylonitrile elastomers (Hycar) and the same holds true for all the other types of coating polymers.

Most elastomers are vulcanizable; they are processed in the plastic state and cross-linked to provide elasticity after being put into final form. With the number of elastomer coatings available today almost any use requirement can be met. If there are limitations, they lie in the areas of processability and cost.

Elastomers are applied to the textile by either calendering or solution coating. Thin coatings are applied from solution and thicker coating by direct calendering.

A natural rubber-based formulation is shown in Table 1.

SBR (styrene–butadiene rubber) has replaced natural rubber in many applications because of price and availability. It has good aging properties, abrasion resistance and flexibility at low temperatures. A typical SBR-based formula is shown in Table 2.

Table 1. A Typical Natural Rubber Compound

Component	Parts
smoked sheet	100.00
stearic acid	1.00
ZnO	3.00
agerite white antioxidant	0.50
P-33 black	10.00
$CaCO_3$	75.00
clay	50.00
sulfur	0.75
methyl zimate } accelerators	0.25
telloy } accelerators	0.50
Total	*200.00*

Table 2. A Typical SBR Compound

Component	Parts
SBR	100.0
processing aid	5.0
stearic acid	2.5
ZnO	3.0
agerite white antioxidant	0.5
tackifier	20.0
$CaCO_3$	75.0
P-33 black	10.0
clay	75.0
sulfur	2.5
methyl zimate } accelerators tuex	0.5
Total	*294.0*

Neoprene offers resistance to oil, weathering, is inherently nonburning and is processable on either calenders or coaters. The cost of Neoprene and its reduced availability in recent years have led to the development of substitutes. Nitrile rubber–PVC blends and nitrile rubber–EPDM (ethylene–propylene–diene monomer) perform on an equivalent basis. The blends do not discolor like Neoprene, and light-colored decorative fabrics can be made.

A typical Neoprene-based formulation is shown in Table 3.

This mixture can be calendered or dissolved in toluene to 25–60% solids for coating.

Isobutylene–isoprene elastomer (Butyl) has high resistance to oxidation, resists chemical attack and is the elastomer most impervious to air. These properties suggest its use for protective garments, inflatables, and roofing.

Chlorosulfonated polyethylene (Hypalon) resists ozone, oxygen, and oxidizing agents. In addition it has nonchalking weathering properties and does not discolor, permitting pigmentation for decorative effects.

Nitrile elastomers (acrylonitrile–butadiene copolymers) have high resistance to oils at up to 120°C. If higher temperature protection is required, a polyacrylate elastomer can be employed up to 200°C.

Polyurethane. Polyurethanes have a number of important applications in coated fabrics. The most striking is footwear uppers because polyurethanes are lighter weight than vinyl polymers and have better abrasion resistance and strength. Polyure-

Table 3. A Typical Neoprene Formulation

Component	Parts
polychloroprene	100.0
stearic acid	1.5
antioxidant	2.0
MgO	4.0
clay	66.0
SRF black	22.0
circo oil	10.0
petrolatum	1.0
ZnO	5.0
ethylenethiourea	0.5
Total	*212.0*

thane-coated fabrics can be decorated to look like leather (qv). Earlier attempts to produce poromerics (coatings that transmit moisture much like leather) were not commercially successful because, although they approach leather in cost, they did not match it in comfort. However, poromerics are still available. Most of the urethane-coated fabrics are used in women's footwear where styling is important and lightweight is desirable. These products usually consist of 0.05 mm of polyurethane on a napped woven cotton fabric. The result is a lightweight product 0.88 mm thick that has good abrasion and scuff resistance (4). Urethane-coated fabrics have not been successful in either men's or children's shoes because greater toughness is required.

Low-weight coatings of polyurethane on very low-weight nylon fabric produce products suitable for apparel. This lightweight product, used for windbreakers and industrial clothing, resists water, provides thermal insulation, and has good drape. Coatings of urethane on heavier nylon structures are used for industrial tarpaulins to provide protection from the elements and extreme toughness. Polyurethane coatings have had limited application to furniture upholstery and practically none on automobile seating.

Poly(vinyl Chloride). By far the most important polymer used in coated fabrics is poly(vinyl chloride). This relatively inexpensive polymer resists aging processes readily, resists burning, and is very durable. It can be compounded readily to improve processing, aging, burning properties, softness, etc. In addition, it can be decorated to fit the required use. PVC-coated fabrics are used for window shades, book covers, furniture upholstery, automotive upholstery and trim, wall covering, apparel, conveyor belts, shoe liners, and shoe uppers. These few uses require millions of meters of coated fabrics each year and demonstrate the diverse properties of PVC coatings. Tables 4 and 5 show typical PVC formulations.

Table 4. A Typical Compound for Calendering PVC

Component	Parts
poly(vinyl chloride) resin (calender grade)	100.00
epoxy plasticizer	5.00
dioctyl phthalate	35.00
polymeric plasticizer	35.00
BaCdZn stabilizer	3.00
TiO_2 (pigment)	15.00
calcium carbonate (filler)	20.00
stearic acid (lubricant)	0.25
Total	*213.25*

Table 5. A Typical Plastisol PVC Formulation

Component	Parts
poly(vinyl chloride) resin (dispersion grade)	100.00
epoxy plasticizer	4.00
dioctyl phthalate	70.00
BaCdZn stabilizer	2.50
lampblack (pigment)	2.00
calcium carbonate (filler)	25.00
lecithin (wetting agent)	1.00
Total	*204.50*

Processing

Coated fabrics can be prepared by lamination, direct calendering, direct coating or transfer coating (see Coating processes). The basic problem in coating is to bring the polymer and the textile together without altering undesirably the properties of the textile. Almost any technique in applying polymers to a textile requires having the polymer in a fluid condition, which requires heat. Therefore, damage to the synthetic or thermoplastic fabric may occur.

Calendering. The polymer is combined in a Banbury mill with a filler, stabilizing agents, pigments, and plasticizers and brought to 150–170°C. The mixture ("compound") temperature is adjusted on warming mills and calendered directly onto a preheated fabric. The object is to get the required amount of adhesion without driving the compound into the fabric excessively, which would cause a clothy appearance and lower the stretch and tear properties of the coated fabric.

Coating. Coating operations require a much more fluid compound. Rubbers are dissolved in solvents. In the case of PVC, fluidity is achieved by adding plasticizers and making a plastisol. If lower viscosity is required, an organosol is made by adding solvent to the plasticized PVC. After the ingredients are mixed and brought to a coating head, the mixture is applied by either knife, knife-over-roll, or reverse roll coaters. Unless the fabric is very dense, or a high degree of penetration is desired, the coating cannot be placed directly on the textile. Transfer coating limits the penetration into the fabric. The mixture is coated directly on the release paper and penetration is limited either by the viscosity of the coating or partial solidification (gelling) of the coating prior to application of the textile. Most polyurethane coated fabrics are transfer coated. Expanded vinyl-coated fabrics consist of a wear layer, an expanded layer, and the textile substrate. The wear layer is coated on release paper (see Abherents) and gelled. A layer of vinyl-based compound containing a chemical blowing agent such as azodicarbonamide is applied. The fabric is placed on top of the second layer and sufficient heat is applied to decompose the blowing agent causing the expansion (5).

Lamination. In lamination a film is prepared by calendering or extrusion. It is adhered to the textile at a laminator either with an adhesive or by sufficient heat to melt the film (see Laminated and reinforced plastics).

Post Treatment. Coated fabrics can be decorated by printing with an ink. Usually the appearance of a textile or leather is the goal. The inks are applied as low-solid solutions by metal rotogravure rolls. Warm air drying is carried out in an oven. Because the ink dries rapidly, multiple print heads can be used (see Ink).

If a textured surface is desired, the coated fabric is heated to soften it and pressure is applied by an engraved embossing roll. Printing usually precedes embossing so that a flat surface is presented for printing. Special effects are obtained by embossing first and then printing or wiping the high points (see Printing processes).

The final layer is called the slip. Most coatings are tacky enough to stick to themselves (block) during stacking or rolling. The main purpose of the slip is to prevent blocking (see Abherents). Slips can be formulated as shown in Table 6 to improve abrasion resistance, seal the surface, adjust color and adjust gloss. Slips are low-solid solutions that are applied by metal rotogravure rolls. Air drying leaves about 200 g of solids per 100 m^2 of coated fabric.

Table 6. A Typical Slip Formulation for PVC-Coated Fabric

Component	Parts
vinyl chloride–vinyl acetate copolymer	100.00
polymethacrylate resin(s)	96.00
vinyl stabilizer	2.00
silica gel	18.00
methyl ethyl ketone	620.00
xylol	350.00
Total	*1186.00*

Economic Aspects

Poly(vinyl chloride) is the principal polymer employed (see Vinyl polymers). In the United States alone the consumption of PVC for calendered-coated fabrics was 28,000 metric tons in 1975 and 45,000 t in 1976. Coating of paper and textiles consumed 75,000 t in 1976.

The United States consumption of polyurethane for fabric coatings was 4500 t in 1975 and 5000 t in 1976 (6).

Health and Safety Factors

Some materials used in coating operations have been identified by the United States government as being hazardous to the workers' health.

Even when coatings are applied by extrusion and calendering, consideration should be given to handling the materials, evolution of gases during heating and post-finishes. For instance, there are strict regulations on exposure to vinyl chloride monomer. Emptying bags or bulk transfer must be monitored. The regulations do not apply to the handling or use of fabricated products made from poly(vinyl chloride).

When a coating machine is employed, attention must be given to exposure of the operator to the solvents.

In addition, particulate irritants such as asbestos (qv), pigments, and reactive chemicals are often involved.

Coating operation should not be initiated without consulting the *Federal Regulations on Occupational Safety and Health Standards, Subpart Z, Toxic and Hazardous Substances* (7).

Uses

Table 7 lists uses of coated fabrics and demonstrates how combinations of textiles and polymers can give significantly different products.

Table 7. Uses of Coated Fabrics[a]

Substrate	Coating	Use
nylon tricot nylon sheeting cotton sheeting	PVC, PU, SBR, Neoprene	clothing
polyester–cotton sheeting napped cotton drill nonwovens (high density)	PVC PU PU	shoe uppers
nonwovens (medium high density) cotton knits	PVC PVC	shoe liner-insoles
polyester nonwovens (light density) polypropylene nonwovens (light density) cotton knits polyester knits polyester–cotton knits napped cotton drills	PVC PVC PVC PVC PVC PU	furniture upholstery
nylon Helanca knits cotton single knits polyester–cotton pattern knits polyester nonwoven (light density on PU foam)	PVC PVC PVC PVC	auto upholstery
polyester nonwoven (light density) polyester stitched nonwoven polyester knit	PVC PVC PVC	landau tops
asbestos nonwoven	PVC	floor covering
polyester spunbonded	PVC PE	wallcovering
cotton sheeting glass scrim rayon scrim	PVC, SBR, Neoprene, silicone rubber, etc	tapes
polyester drill	PVC, SBR, Neoprene, natural rubber	hospital sheeting
absorbent cotton	PVC	Band-Aids
nylon scrim	PVC	window shades
cotton scrim	PVC	wallpaper
cotton sheeting polyester sheeting polyester–cotton sheeting	acrylic	lined drapes
glass scrim polyester scrim	lead-filled PVC, barytes-filled PVC, barytes-filled SBR	acoustical barriers
paperlike nonwovens	PVC	air and oil filters
dyed rayon drill	expanded PVC	soft-side luggage
rayon drill cotton drill polyester–cotton drill polyester drill nonwovens	PVC	luggage
nylon woven polyester woven	PVC, Neoprene, Hypalon, PU	tarpaulins
nylon woven cotton woven	PVC	awning
nylon scrim polyester scrim	PVC, EPDM, Hypalon, Butyl	pond and ditch liner
glass woven polyester woven	Neoprene, PVC, Hypalon	air supported structures

[a] PVC = poly(vinyl chloride); PU = polyurethane; EPDM = ethylene–propylene–diene-modified rubber; SBR = styrene–butadiene–rubber; PE = polyethylene.

BIBLIOGRAPHY

"Coated Fabrics" in *ECT* 1st ed., Vol. 4, pp. 134–144, H. B. Gausebeck, Armour Research Foundation of Illinois Institute of Technology; "Coated Fabrics" in *ECT* 2nd ed., pp. 679–690, by D. G. Higgins, Waldron-Hartig Division of Midland-Ross Corporation.

1. *Davison's Textile Blue Book,* Davison Publishing Co., Ridgewood, N.J., 1977.
2. N. J. Abbott, T. E. Lannefeld, and R. J. Brysson, *J. Coated Fibrous Mater.* **1,** 4 (July 1971).
3. S. P. Suskind, *J. Coated Fibrous Mater.* **2,** 187 (Apr. 1973).
4. H. L. Gee, *J. Coated Fabr.* **4,** 205 (Apr. 1975).
5. W. G. Joslyn, *Rubber Age* **106,** 49 (Feb. 1974).
6. *Mod. Plast.* **54,** 49 (Jan. 1977).
7. *Code of Federal Regulations, Title 29,* Chapter XVII, Section 1910.93 of Subpart G redesignated as 1910.1000 at 40 FR23072, U.S. Government Printing Office, Washington, D.C., May 28, 1975.

General References

F. J. Beaulieu and M. D. Troxler, "Substrates for Coated Apparel Applications," *J. Coated Fibrous Mater.* **2,** 214 (Apr. 1973).

"1977 Manmade Fiber Deskbook," *Mod. Text.* **2,** 16 (Mar. 1977).

"Generic Description of Major U.S. Manmade Fibers," *Mod. Text.* **58,** 17 (Mar. 1977).

"Names and Addresses of U.S. Manmade Fiber Producers," *Mod. Text.* **58,** 30 (Mar. 1977).

"77–78 Buyers Guide," *Text. World* **127,** (July 1977).

R. M. Murray and D. C. Thompson, *The Neoprenes,* E. I. du Pont de Nemours & Co., Inc., Wilmington, Del., 1963.

M. Morten, *Rubber Technology,* Van Nostrand Reinhold Co., New York, 1973.

J. Bunten, "Performance Requirements of Urethane Coated Fabrics," *J. Coated Fabr.* **5,** 35 (July 1975).

D. Popplewell and L. G. Hole, "Urethane Coated Fabrics," *J. of Coated Fabrics* **3,** 55 (July 1973).

H. A. Sarvetnick, *Polyvinyl Chloride,* Van Nostrand Reinhold Co., New York, 1969.

"Manufacturing Handbook and Buyer's Guide 1977/78," *Plast. Technol.* **23,** (Mid-May 1977).

Davison's Textile Blue Book, Davison Publishing Co., Ridgewood, N.J., 1977.

Rubber Red Book, Palmerton Publishing Co., New York, 1977.

FRED N. TEUMAC
Uniroyal, Inc.

COMPOSITE MATERIALS

Composites are combinations of two or more materials present as separate phases and combined to form desired structures so as to take advantage of certain desirable properties of each component. The constituents can be organic, inorganic, or metallic (synthetic or naturally occurring) in the form of particles, rods, fibers, plates, foams, etc. Compared with homogeneous materials these additional variables often provide greater latitude in optimizing, for a given application, such physically uncorrelated parameters as strength, density, electrical properties, and cost. Furthermore, a composite may be the only effective vehicle for exploiting the unique properties of certain special materials, eg, the high strength of graphite, boron, or aramid fibers (qv).

Some measure of coarseness of the homogeneous constituent structures is needed for a meaningful definition of composite material. The term as used here assumes that the average dimension of the largest single homogeneous geometric feature, in at least one direction, is small relative to the size of the total body in that direction; in addition, it assumes that the dimensions of the minor constituent phase are sufficiently large so that its characteristic properties are substantially the same as if it were present in bulk. Thus, laminated safety glass or copper-clad stainless steel, although composite *structures,* are not considered to be composite *materials* (see Laminated materials, glass; Laminated and reinforced metals). At the other extreme, gold ruby glass, which contains submicroscopic gold precipitate particles, is also not considered in this article. The above definition in terms of size serves as a guideline. This is occasionally violated as in the case of certain particulate-filled materials that contain submicroscopic silica gel, carbon black, etc (see Fillers).

Arbitrary control of the geometry, and often composition, of the constituent phases within wide limits is also usually implied in the term composite. Thus, a glass-epoxy composite could consist of glass fibers, glass beads, powdered glass, glass flake, or foamed glass impregnated with various epoxy formulations (see Epoxy resins). Wood, although consisting of cellulose fibers bonded together with lignin (qv) and other carbohydrate constituents, is not usually considered to be a composite, since it does not have a structure capable of arbitrary variation. On the other hand, wood particle-board or wood-flour filled resins would be classed as composite materials (see Laminated and reinforced wood). Certain other structures, such as the oriented eutectics in which rods and plate structures can be produced by controlled solidification with some latitude in size and composition, comprise an intermediate case. Certain such alloy and ceramic systems are showing promise for advanced applications. It is often useful to treat such naturally produced heterogeneous materials as composites. However, these constitute a rather special and somewhat restricted part of the disciplines and technologies common to composite materials.

Composite materials consist of a continuous matrix phase that surrounds the reinforcing-phase structures. Possible exceptions are (*a*) a laminated stacking of sheets in which the phases are kept separated, and (*b*) two continuous interpenetrating phases, such as an impregnated sponge structure, in which it is arbitrary as to which phase is designated as the matrix. The relative role of the matrix and reinforcement generally fall into the following categories:

(*1*) The reinforcement has high strength and stiffness, and the matrix serves to transfer stress from one fiber to the next and to produce a fully dense structure.

(*2*) The matrix has many desirable, intrinsic physical, chemical, or processing characteristics, and the reinforcement serves to improve certain other important engineering properties, such as tensile strength, creep resistance, or tear resistance.

(*3*) Emphasis is placed on enhancing the economic attractiveness of the matrix, eg, by mixing or diluting it with materials that will improve its appearance, processability, or cost advantage while maintaining adequate performance.

The first category constitutes the high performance composites. High strength fibers are used in high volume fractions, with orientations controlled and tailored for optimum performance. Considerations such as system performance benefits often determine the range of applications of this class of composites.

In the remaining two categories, cost is the more immediate consideration. Category (*2*) emphasizes improving engineering properties to extend the range of usefulness and marketability of a given matrix; moderate concentrations of fibers, often as discontinuous random fibers, and of flake and certain particulate reinforcements are used. The reinforced plastics fall in this class. In category (*3*) the emphasis is somewhat the inverse, ie, how to make an otherwise attractive material less costly *per se,* or how to process the material at lower cost without unacceptable degradation of properties through the use of particulate, flake, or fibrous fillers and colorants. This category largely consists of the filled polymers (see Colorants for plastics; Fillers; Laminated and reinforced plastics).

A composite material, as defined, although itself made up of other materials, can be considered to be a new material having characteristic properties which are derived from its constituents, from its processing, and from its microstructure.

Properties

Composites typically are made up of the continuous matrix phase in which are embedded: (*1*) a three-dimensional distribution of randomly oriented reinforcing elements, eg, a particulate-filled composite; (*2*) a two-dimensional distribution of randomly oriented elements, eg, a chopped fiber mat; (*3*) an ordered two-dimensional structure of high symmetry in the plane of the structure, eg, an impregnated cloth structure; or (*4*) a highly-aligned array of parallel fibers randomly distributed normal to the fiber directions, eg, a filament-wound structure, or a prepreg sheet consisting of parallel rows of fibers impregnated with a matrix. Except in case (*1*) the properties of the composite structure viewed as a homogeneous average material are more complex than are the more familiar isotropic materials which require two independent constants, such as the Young's modulus and the Poisson ratio (the ratio of the strain in the direction of the responsible applied principal stress to the strain produced in the transverse direction), to specify their elastic response. The other types of composites (*2, 3,* and *4*) require at least four independent constants, such as two Young's moduli and two Poisson ratios, for a comparable specification. These are needed to describe the dependence of the elastic response as affected by the orientation of the applied stress relative to that of the reinforcing fibers. These properties can be measured. Since composites in turn are often built up by laminating layers of composite sheets, these properties are needed to predict the overall response of the laminated structure. The fibers in each layer can be oriented differently from adjacent layers. If proper attention is not given to fiber symmetry, or the basic information needed for design is not available, peculiar effects can occur, such as a composite part twisting when a simple tensile load is applied. With an isotropic material, this would merely stretch the body. However, for purposes of designing optimum materials, it would be desirable to compute the properties of a composite considered as a homogeneous orthotropic material from the properties of the constituent matrix and reinforcements. The present state of analytical skills allows such predictions to be made with reasonable confidence in specialized cases. However, for many other situations, only upper and lower property bounds can be stated.

Micromechanics is the detailed study of the stresses and strains within a composite considered as a true heterogeneous system. This approach allows the effective average properties of the composite to be computed when the reinforcement has a simple geometric shape and is located in regular arrays. Such idealized models can be used to provide a semiquantitative framework for the behavior of real composite materials. Modeling of the properties of composites as a function of temperature, pressure, or other environments requires a corresponding knowledge of the behavior of the separate constituents plus that of their interactions, such as result from differences in thermal expansion. Much recent attention has been given to hygrothermal effects on the viscoelastic response in polymer matrix composites, ie, on the combined effect of temperature and moisture content on mechanical properties. Increasing water content decreases the stiffness of epoxies and other resins much as does increasing the temperature.

Prediction of Composite Properties. Certain properties, such as the colligative thermodynamic ones, can be accurately calculated from knowledge of the volume fractions and chemical composition of the constituent phases (see Thermodynamics). Other properties, such as thermal and electrical conductance and the elastic properties, can be calculated from idealized models which closely approximate real composite behavior. Other important properties, such as failure strength and fracture toughness, can only be approximated roughly.

Composite properties are often assumed to be representable by the rule of mixtures:

$$P = P_1V_1 + P_2V_2 + P_3V_3 + \ldots \tag{1}$$

in which P is the property value for the composite, and P_i and V_i are the property values and volume fractions of the ith phase. For a fully dense, two-component composite, the heat capacity and density are accurately given by the rule, where i has the values 1 and 2, $V_1 + V_2 = 1$, and P signifies either heat capacity or density. However, for the Young's modulus, E_{11}, of a continuous parallel fiber-reinforced composite in the direction of the fibers, one can only state rigorously that

$$E_{11} \geq V_fE_f + V_mE_m \tag{2}$$

although to a very good approximation the two sides can be taken to be equal. The subscripts f and m are now used to indicate fiber and matrix. However, if the fibers are parallel but are initially discontinuous or develop breaks, then V_f, the volume fraction of the fibers, must be replaced by βV_f where β is a constant less than 1 and reflects the fraction of the fiber rendered ineffective because of the loss of its ability to carry tensile load near a fiber end. The value of β depends on the fiber geometry, the elastic-deformation characteristics of the fiber and the matrix, and the interface. Thus within these limitations, equation 2 estimates one of the elastic constants needed. The Young's modulus transverse to the axis of the fibers is given by:

$$E_{22} \approx E_m(1 + \zeta\eta V_f)/(1 - \eta V_f) \tag{3}$$

where

$$\eta = (E_f - E_m)/(E_f + \zeta E_m), \tag{4}$$

and ζ is a constant determined by the fiber geometry. The major Poisson ratio, ie, the strain transverse to the direction of the fiber relative to the strain in the fiber direction when also stressed in the fiber direction, is approximated by:

$$\nu_{12} \approx V_f \nu_f + V_m \nu_m \tag{5}$$

and the minor Poisson ratio, ie, the ratio of strain in the fiber direction to the strain transverse when also stressed in that transverse direction, is given by:

$$\nu_{21} = \nu_{12} E_{22}/E_{11} \tag{6}$$

These equations permit the material parameters to be estimated from the fiber and matrix properties when the reinforcement consists of parallel fibers. For other reinforcement geometries, such as packing of spheres, these equations can be generalized using semiquantitative approaches in which ζ is allowed to take on geometrically dependent values that result from comparison with specialized micromechanical modeling or experiment.

Strength of Composites. The relatively high strength of composites on an equal weight or cost basis is often a major contributing factor to the importance accorded this class of materials. Strength, although relatively easy to measure, is even more difficult to predict in the case of composites than for homogeneous materials. As a first approximation, the rule of mixtures for strength, S, of a fiber-matrix composite in the direction of the fiber is often used

$$S_{\text{composite}} = V_f \overline{S}_{\text{fiber}} + V_m \sigma^* \tag{7}$$

where σ^* is the stress in the matrix at the failure strain, and $\overline{S}_{\text{fiber}}$ is the mean fiber strength. This is an intuitive equation that is often useful as a starting point to estimate strength in the absence of better information. However, it has no theoretical basis, even as an upper or lower bound.

The possibility of transferring load from one reinforcing element to another means that the effect of a fiber break can be localized. Thus the average strength of a composite can exceed that of its constituent reinforcement. As noted, composites are usually anisotropic. In aligned or continuous filament composites, the strength in the direction transverse to the fibers is much less than that parallel to the fibers. The high strength possible in fiber composites is, in most cases, attained at the expense of an absolute weakening in strength in other directions (relative to the strength that the bulk reinforcement material would have). Isotropic chopped-fiber composites are limited in strength because the geometric interferences between fibers limit their packing fraction. For these reasons composites must be carefully designed to be strong where needed and to ensure that the stresses remain low in the other directions.

Quantitative considerations of strength require a criterion for failure. In composites this can mean the stress level at which detectable loss of structural coherence occurs, the maximum apparent stress the structure can sustain, or in some cases the maximum apparent strain to cause separation into two parts. The failure process depends strongly on how the material is stressed. A stress applied nearly parallel to the reinforcement requires fiber breakage in order to propagate a crack. However, as the stress is applied increasingly off-axis, first shear failure and then transverse tensile failure of the matrix controls strength. Under compressive loading several failure modes are possible.

Much effort has been directed to finding failure criteria, such as a limit to the local energy of distortion, in analogy to the Von Mises criterion used to predict when metals will yield under complex stress conditions. Although such approaches provide a framework for presenting failure-stress information, there is as yet no generally valid, convenient model.

The discipline of fracture mechanics, developed to predict failure in homogeneous ductile materials, provides another approach. This assumes an ability to detect and measure the most dangerous strength-impairing flaws. Fundamentally, this method requires a crack to be a simple topological surface, a condition that often is violated in composites as a result of multiple splitting. This approach has been most successful when the stress and orientation variables are kept simple.

Experimentation and micromechanical modeling have provided valuable insight into the major factors that influence the various failure modes and other mechanical properties. Resistance to crack propagation transverse to fibers or sheets depends on the matrix-reinforcement bond strength. If the bond is too strong, the composite is brittle, and if it is too weak, the composite becomes excessively weak in the transverse direction. Voids distinctly reduce shear and compressive strengths. Fiber misalignments and matrix-rich regions can initiate buckling under compressive loading. Hence, for reasons of material characterization and quality control, tensile, shear, and compressive strengths are routinely determined but require special specimen configurations to avoid measurement artifacts because of the strong material anisotropy. Flexural testing avoids some of the complications. Relatively long composite bars are often used to measure tensile strength, and short, stubby bars are used to measure the shear strength, often termed short beam shear or interlaminar shear strength. The literature on strength and testing is extensive (see General bibliography).

Fabrication Methods

The methods used to make composite materials and structures depend, among other factors, on the type of reinforcement, the matrix, the required performance level, the shape of the article, the number to be made, and the rate of production. The orientation and positioning of continuous filaments are controllable, whereas short fiber, flakes, or particulates are apt to be more randomly distributed. However, varying degrees of preferred orientation can be achieved by appropriate shearing action, electrical fields, etc.

Large diameter, single-filament materials, such as boron, silicon carbide, or wires, are often fed in precisely controlled, parallel arrays to form tapes of sheet materials. Complex computer-controlled machines for laying such tapes in desired overlap angles over complex surfaces have been built for fabricating aircraft structures, rocket casings, pipes, etc, where the highest possible performance is required. The matrix in these cases is usually polymeric and is added with the filament. However metals can also be used as the matrix for making tapes and sheets. This requires that the filaments be positioned while the molten matrix is allowed to infiltrate and solidify around the fibers, or while consolidation by diffusion bonding is taking place.

In the case of finer filaments, such as fiberglass, carbon fiber, or boron nitride fiber, bundles (tows or roving) of hundreds to tens of thousands of loosely aggregated fibers are handled as an entity. When these fibers are to be incorporated into a polymer matrix composite, it is usually convenient to form a semiprocessed, shapable, intermediate ribbon or sheet product known as prepreg in which the fibers are infiltrated by the resin. This impregnated material is further processed so that it can be handled conveniently. Typically, the fiber bundles are laid down in arrays along with a desired resin system in a controlled way. The structure is then rolled, combed, or otherwise handled to spread out the fibers as evenly as possible and with a uniform thickness. The impregnated system is then partly cured (B-staged) to fix the geometry while allowing enough shape relaxation (drape) and adherence (tack) to permit complex shapes to be built up from sheets of this prepreg material. The fibers can also be woven into cloth, infiltrated with resin, and handled as a prepreg. A variety of special weaves is available for composite usage.

Another approach is to form dry structures first, such as wire armatures, which are then impregnated with the matrix material. When standard shapes of uniform cross section, such as bars, rods, I-beams, or channels, are required, a process known as pultrusion can be used. The resin (thermoplastics can be used) and the continuous fiber are formed to the desired shape while pulling on the product and fibers to produce a highly aligned fiber arrangement as the composite is formed in the orifice region.

All of these continuous fiber methods are capable of yielding high quality, nearly ideal composites. The final structures must be carefully consolidated using various combinations of pressure and vacuum to eliminate porosity, to ensure complete coalescence of the matrix structure, and to avoid matrix-rich pockets and fiber misalignments. Large, expensive equipment is often required in the case of big structures. Typical examples of composite structures that have been made using the continuous fiber methods include automotive springs and frames, tires, pressure vessels, helicopter blades, aircraft airfoils and fuselage structures, spacecraft, boats, chemical plant equipment, and such sporting goods as skis, golf shafts, tennis racquets, and vault jump poles.

For many classes of applications, the ultimate in mechanical performance is not required, but complex shape and appearance are important considerations. A broad range of methods is available for the discontinuous fiber, flake, and particulate composites which can be used in these cases. Prepreg sheets can be made using chopped or discontinuous fiber reinforcement. This has the advantage of being moldable into double-curved shapes without buckling. The matrix resins can be either thermoplastic or thermosetting. The reinforced thermoplastic sheets are particularly adapted for the rapid pressing into shapes needed for automobile body structures.

Another method useful where larger shapes or lower production volume is needed is the spray-up technique. Special spray guns are available into which are fed the continuous filament roving. The fibers are chopped and dispensed along with controlled amounts of a suitable matrix resin to build a composite material; personal protection equipment is required. The spray is directed against a carefully prepared mold surface that is treated to release the composite shape after curing (see Abherents). Large structures, such as boat hulls, furniture, bath fixtures, and tanks, can be made. It is possible to incorporate continuous fiber structures, such as stiffers or ribs, or other fittings. Good surface finishes are achieved by the use of gel coats sprayed on the mold surface.

Molding compounds useful for injection, transfer, compression, or other similar types of force-flow-forming can be compounded by use of chopped fiber or other types of fillers. These generally have a lesser reinforcement content because of rheological considerations (see Rheological measurements). This method is used where small to medium repetitive, often complex shapes are required. Both thermoplastic and thermoset resins can be used as the matrix in forming such products as power tool casings, gears, and washer agitators.

Regardless of the fabrication process, the inherent anisotropy and materials combinations require close attention to such factors as residual stresses and defects arising from volumetric changes on polymerization, differences in thermal expansion, and post-processing creep. Although the methods may be relatively straightforward, details such as pressure and temperature history, post-curing cycles, formulation of the matrix, and surface condition of the reinforcement can be crucial to the production of composites of high quality.

Reinforcements

If a reinforcement is to improve the strength of a given matrix, it must be both stronger and stiffer than the matrix, and it must significantly modify the failure mechanism in an advantageous way, or both. The requirement of high strength and high stiffness implies little or no ductility and, thus, relatively brittle behavior. Brittleness is, in fact, not unusual in reinforcement materials. Such materials are often used in the form of filaments because flaws markedly affect their strength, and a fiber geometry more effectively enhances the amount of material unaffected by flaws than do sheet or particulate shapes. Furthermore, a characteristic length-to-thickness (aspect) ratio must be maintained in order to transfer effectively from one reinforcement element to the next. Fibers are convenient for meeting this requirement, and thus, comprise the most important class of reinforcing materials. However, other classes of composites in which strength is not the most important requirement frequently use powders, flakes, short chopped fibers, or other forms of reinforcing materials.

Strong natural fibers such as plant fibers (see Fibers, vegetable) silk (qv), and asbestos (qv) have long been known to have been used to make some of the first commercial high performance composites: varnish and lacquer-impregnated fabrics were used to cover early aircraft; cotton reinforcement of rubber was first used in the making of pneumatic tires; and phenol–formaldehyde impregnated cloth was used for automotive timing gears during the same era. However, the events that led to the rapid development of high performance composites as a new class of engineering materials were: (*1*) the emergence of thermosetting- and thermoplastic-synthetic polymer technologies, (*2*) the commercial availability of high-strength glass fiber in the late 1930s, and (*3*) the accelerated development of composites for military uses during World War II. To a considerable extent, the success of composites is inextricably linked to the development and availability of cost-effective, strong, stiff fibers.

Glass fiber, the first of the synthetic fibers, is produced in nature from air-blown strands of molten volcanic glass (see Glass). Egyptian, Roman, and Venetian artisans used glass fibers for decorative effects. Continuous filament-drawing originated in the late seventeenth century. Its potential for high strength was noted by Griffith in 1920 (1). The need for temperature-resistant electrical insulation led to the development just prior to World War II of E-type glass for use as winding and cloth insulation. Reportedly, the discovery of glass fiber-reinforced plastics (GFRP) resulted from the accidental spillage of a polyester resin onto glass fiber cloth (see Laminated and reinforced plastics). The cured composite was recognized as having attractive mechanical properties. The E-glass fiber, although not developed as a structural material, has become the most widely used of the high performance reinforcements, because of its low cost and its reproducibly good properties. It has subsequently grown into transportation, furniture, construction, industrial, recreational and other important segments of the economy.

The tire industry has provided another major impetus for the development of strong, economical, high performance fiber reinforcements (see Tire cords). Competition and materials advances have resulted in the progression: cotton, rayon, tensioned rayon, nylon, polyester, glass, steel, and aramid fibers. A second major factor stimulating the development of high stiffness-to-weight and strength-to-weight composites has come from the aerospace requirements (see Ablative materials). Weight reductions have compounding beneficial effects, eg, decreasing the mass that needs to be carried aloft also decreases the fuel requirements, the size of the propulsion system, and the need for massive load supporting structures. Finally, substantial impetus resulted from the materials science activities in the 1950s and 1960s, which demonstrated that very high strength was achievable over a wider range of materials than had been previously supposed. Theory and experiment indicated that strengths approaching 0.1 and even 0.2 times Young's elastic modulus (the proportionality constant between stress and strain in elongated structures) could be achieved if the material is perfect, ie, free of mobile dislocations, stress-raising notches, steps, or inclusions. By contrast, the strength of ordinary (imperfect) materials is typically a hundred-fold smaller. Fibers offered the greatest probability of achieving structural perfection, and record strengths were reported in rapid succession in filamentary crystals (whiskers) and in glass fibers (see Refractory fibers). Fiber composites were identified as the way to take advantage of the enormous strengths of whiskers. However, the problems associated with their production, handling, and conversion into composites have not been economically solvable to date. Whiskers, even if available, continue to cost ca \$1–30/g, although projections of tenfold cost reductions have been made.

As a result, in spite of the great theoretical potential of whisker composites, attention since the mid-1960s has focused increasingly on continuous or mass-fabricated, high strength, high modulus fibers. These include glass, carbon and graphite, boron oxides, silicon carbide, and aramid filament.

The various types of fiber reinforcements listed in Table 1 provide a basis for comparisons. The data presented are approximate, ie, derived from various sources that may not use the same basis for measurement. Some fibers are available only as laboratory produced materials. The strengths cited often represent the maximum of what is practically possible, not necessarily what is typical of commercial material. The prices are those prevailing in 1977 or when the product was last available. In the case of metals, note that drawn wire is considerably more expensive than the bulk material. The noncontinuous or long-stranded types of reinforcements are presented in Table 2 according to their function. The cost of this class of materials in most cases is substantially less than in the case of the fiber materials. However, materials used in inorganic or metal composites, such as carbides, or refractory metal powders, can also be costly.

Nonstrand Reinforcements. A variety of reinforcements or fillers of plate-like or particulate geometries are frequently used in nonhigh-performance composites. These materials also function to extend the polymers, especially the more expensive engi-

Table 1. Characteristics of Candidate Reinforcing Fibers

Category	Material	E, GPa (10^6 psi)	S, GPa (10^3 psi)	Sp gr	Typical diameter, μm	\$/kg
glass	vitreous silica	72 (10.5)	5.9 (850)	2.19	10	12–100
	E glass	72 (10.5)	3.4 (500)	2.54	10	0.75–1.00
	S glass	85 (12.4)	4.5 (650)	2.49	10	4.50–11
carbonaceous	PAN[a] high strength	241 (35)	2.8 (400)	1.7–1.8	7	100–150
(material	PAN[a] high modulus	413 (60)	1.7 (250)	1.9–2.0	7	100–150
refers to	pitch	345 (50)	1.4 (200)		7	20–50
starting	rayon, very high modulus	689 (100)	3.5 (510)	2.0	7	[b]
process)	rayon, high modulus	517 (75)	2.6 (380)	1.8	7	1,000
polymer	aramid	103–152 (15–22)	2.8 (400)	1.44	12	20–50
	olefin	0.7 (0.1)	0.62 (90)	0.97	3–500	1.5–2.5
	nylon	3.4 (0.5)	0.86 (125)	1.14	3–500	1.5–3.0
	rayon	6.9 (1)	1.1 (155)	1.52	3–500	0.50–2.50
inorganic	alumina (monocrystal)	510 (74)	3.4 (500)	3.96	250	20,000–150,000[c]
	alumina (polycrystal)	379 (55)	1.0 (150)	3.96	3	20–60
	alumina (whisker)	510 (74)	21 (3000)	3.96	1–10	(35,000 in 1966)[b]
	alumina silicates	103–138 (15–20)	1.4 (200)	2.5–2.6	10–15	2–5
	asbestos	172 (25)	1.4 (200)	3.2	0.02	0.5–2
	boron (tungsten core)	3.79 (55)	2.8 (400)	2.63	140	300–500
	boron nitride	55–76 (8–11)	0.38 (55)	1.85	7	2500[c]
	silicon carbide (carbon core)	482 (70)	2.8 (400)	3.2	100	250–1,000[c]
	silicon carbide (polycrystal)	44 (64)	6.2 (900)	3.2	5–25	[b]
	silicon carbide (whisker)	482 (70)	21 (3000)	3.21	1–10	1,200[c]
	silicon nitride (whisker)	379 (55)	14 (2000)	3.18	1–10	[b]
	zirconia (polycrystal)	427 (62)	1.4 (200)	4.84	3	20–65
metal	beryllium	221 (32)	1.3 (185)	1.84	75	27,000
	molybdenum	358 (52)	2.2 (320)	10.2	25	4,500–5,000
	steel	200 (29)	4.1 (600)	7.2	75	40–50
	tungsten	407 (59)	4.0 (580)	19.4	25	2,000–4,000

[a] PAN is polyacrylonitrile.
[b] Current price unavailable.
[c] Price highly subject to change.

Table 2. Other Types of Reinforcements by Function

Polymer matrix
Extenders—clay, sand, ground glass, wood flour
Rheology control—mica, asbestos, silica gel
Color—titanium dioxide, carbon, pigments
Flame/heat resistance—minerals
Heat distortion resistance—fibers, mica
Shrinkage resistance—particulate minerals, beads
Toughness—fibers, carbon black, dispersed rubber phase
Metal/inorganic matrix
Hardness, wear resistance—metal, interstitial powders
Creep resistance—oxide dispersions

neering types (see Engineering plastics). Minerals, such as clay, talc (qv), sand, mica (qv), and asbestos (qv), are often used because they are inexpensive and impart desired characteristics to the matrix/reinforcement combination during processing and in its use properties or both. These properties include thixotropy, strength, ease of finishing, creep, and heat distortion temperature. Table 3 gives as an example the effect of various types of filler on nylon 66. Other fillers include short, chopped fibers, which are often milled into the resin to produce a molding compound, ie, a mixture having adequate strength in thin sections, ears, etc, and rheological characteristics that permit flow in a die cavity under pressure. For this purpose chopped textile, glass, and carbon fibers and wires are used. Flake materials are often used to provide strength in sheet structures. Other fillers include carbon black, silica gel, hydrated alumina, titanium dioxide, and inorganic pigments (see Fillers). On a weight percentage basis, fine particulates increase the effective viscosity more than coarse materials. This is one of the factors involved in the selection of fillers for specific applications. Thus, the different materials can affect properties in many different ways. For this reason specialized formulations are usually developed to match the application.

Glass Fibers. A wide variety of glasses can be used to make fibers by the old Modigliani process in which a thread is pulled away from a heated glass rod, much like pulling taffy (see Glass). However an enormous gain in productivity has been achieved in the more recent process in which a large number of fibers can be spun simultaneously by means of a heated platinum bushing having many small holes through which the glass can flow. The diameter of the holes, the temperature of the molten glass, and the rate of pulling determine the fiber diameter, which is ca 5–25 μm in commercial production. As the fibers are drawn, they are usually treated with sizing and coupling agents, ie, materials used to reduce strength degradation due to fiber–fiber abrasion, to transform a collection of loose parallel fibers into a coherent strand, and often to coat the fibers with agents that promote wetting and adherence to matrix resins. The properties of the glass fibers depend on the composition and to a lesser extent on processing history. Most (est 99%) of the glass fiber reinforcement is made from E glass. This glass is characterized by a range of compositions but has a very low alkali content, which results in a relative insensitivity to moisture, contributing to good electrical insulation and good strength retention over a wide range of conditions. The composition by weight of this type of glass is SiO_2 54 ± 2%, Al_2O_3 14 ± 2%, CaO + MgO 22

Table 3. Effect of Various Fillers/Reinforcements on Physical/Mechanical Properties[a] (As Exemplified by Nylon 66)

Property	Unrein-forced	Glass fiber	Carbon (graphite)	Mineral	Carbon/glass fiber[b]	Mineral/glass fiber[b]	Glass bead
reinforcement content, wt %	0	40	40	40	20C/20G	20M/16.5G	40
specific gravity	1.14	1.46	1.34	1.50	1.40	1.42	1.44
tensile strength, MPa	83	214	276	103	234	121	90
(psi × 10^3)	(12)	(31)	(40)	(15)	(34)	(17.5)	(13)
flexural modulus, GPa	2.8	11	23	7.6	19	65	5.5
(psi × 10^5)	(4.0)	(16)	(34)	(11)	(28)	(9.5)	(8)
impact strength, notched/unnotched, J/m	48/320	139/1014	85/694	37/427	96/854	53/694	53/294
(ft·lbf/in.)	(0.9/6)	(2.6/19)	(1.6/13)	(0.7/8)	(1.8/16)	(1/13)	(1/5.5)
heat deflection temperature, 1.82 MPa (264 psi), °C	66	260	260	227	260	243	88
thermal expansion, 10^{-5} m/(m·°C)	8.1	2.5	1.4	5.4	2.1	4.5	3.6
mold shrinkage, %	1.5	0.4		0.9		0.7	
water absorption, 24 h, %	1.6	0.6	0.4	0.45	0.5	0.5	0.65

[a] Courtesy of *Plastics World*.

[b] G = glass fiber; M = mineral.

± 2%, B_2O_3 10 ± 3%, Na_2O + K_2O less than 2%, plus other minor constituents. The general utility and low cost of this glass make it an important engineering material.

The strength of E glass, as can be seen in Table 1, ranks high among available fibers. However, for many applications greater stiffness is required. Accordingly, there has been a search for high modulus glasses. A number of experimental glass formulations have been found that are 50% or more stiffer than E glass. Many of these contain BeO as a constituent and are very difficult to process. The most successful of the (relatively) high modulus formulations is S glass having a nominal weight composition SiO_2 65%, Al_2O_3 25%, MgO 10%. This glass is about 20% stronger and stiffer than E glass and has good chemical stability. For certain applications these improved properties warrant the five- to tenfold increase in price over E glass.

Ranking lower in terms of volume usage, but outstanding in terms of its intrinsic strength, corrosion resistance, and temperature capability, is silica glass fiber. The strength of virgin fiber at 77 K (bp, N_2) has been measured to reach the theoretically estimated limit of 0.2 times the Young's modulus, ie, 14 GPa (2×10^6 psi). The high softening temperature of 1100°C necessitates more difficult drawing conditions than for the above glasses. Accordingly, the cost is even greater than for S glass. Very reproducible fibers are used in aluminum matrix composites and for high temperature applications.

Carbonaceous Fibers. Estimates of ultimate strength based on atomic bonding considerations and the observed strength of graphite whiskers indicate that graphite, in the direction of the basal plane, is among the strongest of known materials. Carbon fibers can be made by the controlled pyrolysis of organic fibers (see Carbon, carbon and artificial graphite). In general this process results in randomly oriented structure. However, in the early 1960s several approaches were recognized as leading to oriented fibers in which the basal plane of the graphitic structure is aligned with varying degrees of perfection parallel to the fiber axis.

The various carbonaceous fibers can be classified according to their starting materials, crystallinity, modulus, strength, density, etc. In all cases pyrolysis of an organic precursor is required. The requirement of removing the volatile decomposition products by diffusion limits the practical diameter of such fibers. Although fiber diameters somewhat in excess of 25 μm have been achieved, most fibers have diameters in the range 6–8 μm. For such small sizes filaments must be handled as bundles (tows), rather than as individual monofilaments. Commercially available tows contain ca 1000–60,000 fibers.

The first mass-produced, high performance carbon fibers were based on a rayon (qv) precursor, which was first pyrolyzed at relatively low temperatures and subsequently stress-graphitized. Commercial processes were developed by Union Carbide and by Hitco in the mid-1960s. In the pyrolyzed state the fiber has low strength and stiffness and appears to have a noncrystalline random structure using x-ray or electron microscopic techniques. By heating and stretching these filaments at 2200–2800°C, an oriented graphitic crystal texture develops. The strength and modulus values increase markedly with increasing degree of stretch, remaining approximately proportional to each other, ie, greater strength is accompanied by greater stiffness. Elastic moduli as great as 690 GPa (100×10^6 psi), corresponding to 70% of the value for graphite in the basal plane, have been achieved in experimental lots. The failure strain is ca 0.5%. Although much of the pioneering data on carbon fiber reinforced composites were generated using rayon-based fibers, this material has largely been supplanted commercially by the two grades described next.

Polyacrylonitrile (PAN) is a commonly available textile fiber and has been found to give high performance carbon and graphite fibers (see Acrylonitrile polymers). Both the process and the characteristics of these fibers differ from the cellulose-derived fibers, ie, stress graphitization is not required and the values of the modulus and strength of the resultant fibers are not simply related as a function of processing. The starting PAN polymer fibers are slowly oxidized at ca 150–300°C under conditions which stretch or, at least, restrain the fibers from shrinking. This transforms the polymer to give it a cross-linked ladder polymer structure. This material is no longer thermoplastic and can be heated to 1000°C to carbonize it or up to 3000°C to graphitize it, without requiring further tension in order to develop high strength and modulus. The stiffness increases monotonically with increasing final processing temperature. However, the strength passes through a maximum at about 1500°C. The absolute values depend on the starting fiber, initial degree of orientation, etc. This behavior allows fibers to be tailored to optimize various properties, such as strength, stiffness, or storable elastic energy. The latter property is related to impact toughness. The process for making these fibers is controllable and is convenient except for the long times of the order of many hours required to achieve the initial stretch oxidation. The basis for this technology can be traced in part to work by Shindo in Japan in 1960 (2) and to work at the Royal Aircraft Establishment (3) and at Rolls Royce in England (4) in the mid-1960s. PAN-derived fibers are being commercially produced by several manufacturers in the U.S., as well as the U.K., the Federal Republic of Germany, France, Japan, and the U.S.S.R.

Another approach uses pitch to produce fibers. These can be derived from a number of starting materials by pyrolysis carried out to the stage of yielding a liquid having a mesophase (planar, large fused-ring) structure. When this high-softening-temperature material is spun and then oxidized, a cross-linked, nonfusable fiber results that retains a high degree of molecular orientation. Such pitches have a high carbon content, hence, there is less problem in subsequently eliminating volatile decomposition products. Upon heating to carbonization or graphitization temperatures, an oriented, high performance fiber results. Stretch-graphitization can be used to further upgrade properties. This material appears to offer the greatest potential for achieving an economically low-cost product. Because the starting material is not a high volume fiber, to date the fiber is more variable than the rayon or PAN-derived fibers. Pitch-derived fibers are manufactured in the United States and Japan (see Tar and pitch).

Polymer Fibers. Almost any organic textile fiber can be incorporated into a composite structure. The carbon–carbon bond is very stiff, leading to a theoretical modulus of the order of 1000 GPa (ca 150×10^6 psi), based on the interatomic force constants for stretching. Against this as an upper bound, the stiffness of either natural or synthetic fibers is very low. This has been their major deficiency with respect to usage as reinforcements for high performance applications, even though failure strengths can be quite high. The cellulosic fibers exhibit the greatest stiffness and strength among the high volume, mass produced fibers (see Cellulose acetate and triacetate fibers). Because polymers typically exhibit complex time-dependent non-Hookean response, it is somewhat simplistic to use the concept of an elastic modulus, except as a limiting value at small strains for rapid loading conditions. In this sense, the highly oriented cellulose fibers, such as rayon and linen, have moduli that approach those of inorganic glass. Because of their relatively low stiffness, most of the organic fibers are used as reinforcements with even lower stiffness matrix materials, such as rubber and thermoplastics.

Recently, aramid fibers (qv) have been spun into high performance filaments that rival fiberglass in strength and have nearly twice the stiffness at about one-half of the density. These materials consist of highly aligned sheets of heterocyclic molecules. They have been considered to be intractable as solids. Hence, spinning from liquid crystal solution systems has been required (see Liquid crystals). These fibers have the unusual property of being strong in tension but relatively weak in compression. This is believed to be due to a microfibril rope-like structure of the individual filaments. These fibers are produced by the DuPont company under the trade name Kevlar and by Akzo as Arenka.

Inorganic Fibers. Whisker crystals have been made from a wide range of materials including Fe, Cu, Cr, Sn, Zn, Al_2O_3, Si, SiC, Si_3N_4, NaCl, among others. Typically whiskers exhibit outstandingly high strength and often have interesting magnetic, electrical, or other properties. They can be produced by various methods out of vapor or from condensed phases. A novel process for making low cost, very fine SiC whiskers involves the pyrolytic conversion of the silicon and the carbon contained in rice hulls. To be used in composites, they need to be collected, sorted, distributed into the desired configuration, etc. These practical considerations, their high cost, plus their noninfrequent degradation when processed into composites led to their displacement by continuous filament materials.

The most important of the continuous filaments is boron, which is produced from the thermal decomposition of BCl_3. This is deposited on a heated tungsten core having a typical diameter of 13 μm. Because of the high density and cost of the core, it is advantageous to build up boron layers to standard diameters of 100, 140, and even 200 μm. These fibers are stronger than carbon fibers, although on an equal weight basis, the strengths are about equal. Boron filaments can be made with uniform properties. They can be incorporated into both polymer and metal matrices, eg, aluminum. Their diameter, stiffness, and hardness require different fabrication techniques from those used with glass, carbon, or organic fibers. Individual filament positioning, the avoidance of sharp bends, and special cutting methods are used. The large diameter and perfection of the fiber stacking results in materials with high compressive strength. The principal drawback of this material is its relatively high cost, which is projected to be ca \$150/kg in large volume. However, because of its special properties it will remain an important engineering reinforcement material.

Silicon carbide can be produced in a manner very similar to that for boron. The resulting fiber has comparable properties to boron, but is more dense. The main advantage of this fiber is that it remains strong to a higher temperature than boron. It can be used to reinforce aluminum and titanium matrices. A new process for making SiC has been developed in Japan based on the pyrolytic decomposition of a spun filament of a polycarbosilane having a C:Si ratio of unity. This process is analogous to making carbon filaments. Excellent properties have been reported.

Other methods for producing mainly oxide filaments include spinning or extruding a mixture of very fine-grained oxide with an organic binder. The binder is subsequently burned off and the oxide sintered or otherwise consolidated. Crystallizable glass filaments can subsequently be heat-treated to produce micaceous or mullite-dispersed fibers. Boron nitride filaments, which have properties somewhat like carbon fibers but are much more oxidation resistant, are made by a somewhat analogous process.

Filaments of B_2O_3 glass are spun and subsequently converted to BN by the action of ammonia. One of the most ingenious methods has been the crystal growth of shaped continuous filaments of aluminum oxide, using a bushing that allows a great variety of sizes and cross-sectional shapes to be produced (5). A difficulty has been the inclusion of small voids in the structure.

To bring costs down to the level where the derivative composites can compete with alternative materials requires high volume production. Fibers such as polycrystalline, Al_2O_3 and ZrO_2, which can be used in quantity and, reasonably high performance for high temperature insulation (qv) best meet these requirements. Silicon carbide boron fibers, or both, may have enough demand for use as premium reinforcement to remain commercially useful. Other fibers will undoubtedly continue to be available as specialty materials (see Boron; Boron compounds; Carbides).

Metal Filaments. Metallic filaments are conventionally made by wire drawing, ie, the mechanical reduction of continuous strands of metal through successive dies, often with intermediate heat treatments. The process is quite expensive for small wire diameters. Methods have been developed for melt-casting filaments with glass capillaries as the envelopes. Processes also have been developed for producing discontinuous strands of fine filaments by a melt-spinning process. However, these processes apparently have not produced commercially competitive materials to date.

Steel wire used for tire reinforcement comprises the major, large-scale application for high-performance metal filaments. Typically such wire is drawn from special, high quality steel rods having a low inclusion content and a starting diameter of 5.5 mm to final standard sizes of 0.38 and 0.25 mm (see Steel). Three to five such wires, each having a tensile strength of ca 2.8 GPa (400×10^3 psi), are twisted into strands and woven or otherwise positioned into lamellar arrays for use in tire manufacture. The wires are brass plated to promote adherence to the rubber matrix. The cost of 0.5–1.5 kg of steel reinforcement used per tire is ca \$2/kg. Such wire is also used in enhancing the tensile strength of concrete (see Cement). Lower performance, coarser wire has long been used to produce a strong, flexible inner rim (bead) in tires (see Tire cords).

Matrices

Any solid that can be processed so as to embed and adherently grip a reinforcing phase is a potential matrix material. The polymers and metals have been the most successful in this role although cements, glasses, and ceramics have also been used. Thermosetting resins are particularly convenient because they can be applied in a fluid state, which facilitates penetration and wetting in the unpolymerized state, followed by hardening of the system at times and conditions largely controlled by the operator. Exothermicity, shrinkage, and the evolution of volatiles, if the polymerization is of the condensation type, are among the difficulties encountered using such resins.

Because of their cost, wide range of formulations, and generally good mechanical and electrical properties, the polyesters (qv) are an extensively used example of this class of resins. The epoxy systems are particularly useful because of their excellent adherence, low shrinkage, and freedom from gas evolution (see Epoxy resins). However, the epoxies are relatively expensive and are generally limited to service temperatures below 150°C. For many aerospace applications, higher use temperatures are required. Thermosetting systems having higher temperature capability tend to be resins that in their polymerization state contain nitrogen-heterocyclic moieties. Systems such as the imides, amide–imides, quinoxalines, imidazoles, etc, have been used. These are condensation–polymerization systems that are difficult to process; although stable at high temperature they have not succeeded in achieving mechanical properties in the composites that are as good as with the epoxy systems (see Embedding).

The reinforced thermoplastic resin systems are the most rapidly growing class of composites. In this class the focus is on improving the base properties of the resin to allow these materials to perform functionally in new applications or those previously requiring metals, such as die casting (see Engineering plastics). Thermoplastics may be of the crystalline or amorphous types. In the former type, the crystalline morphology may be significantly influenced by the reinforcement which can act as a nucleation catalyst. In both types there is a range of temperature over which creep of the resins increases to the point where it limits usage. The reinforcement in these systems can increase failure load as well as their creep resistance. Some shrinkage also occurs during processing, plus a tendency of a shape to remember its original form. The reinforcement can modify this response as well. Thermoplastic systems have advantages over thermosets in that no chemical reactions are involved causing release of gas products or exothermal heat. Processing is limited only by the time needed to heat, shape, and cool the structure, and the material can be salvaged or otherwise reworked. Solvent resistance, heat resistance, and absolute performance are not likely to be as good as with the thermosets.

Metal matrices offer even higher temperature capability, strength, and stiffness than is possible using polymers. Composite properties transverse to reinforcing fibers are superior, as is the fracture toughness. However, there is a penalty in terms of weight. More significantly, the fabrication of metal matrix composites is more difficult than in the case of polymers. Most metals react with fibers at elevated temperatures, especially in the molten state. Wetting is often uncertain. As a result it has been found necessary to coat fibers if melt infiltration is to be used, such as in the case of silica and boron filaments used with aluminum. Failure to do so causes marked degradation of strength of the fibers. Such reactions during processing can often be substantially reduced if the metal is retained in the solid state. This requires techniques such as diffusion bonding, roll bonding, and creep forming. These approaches depend on high temperature, pressure, and somewhat extended processing times. Conditions must be selected so as not to unduly damage the fibers mechanically.

Electrodeposition provides another technique for embedding reinforcement in a metal. In the case of metal wire-metal matrix composites, hot extrusion of a preform consisting of wires in the matrix can result in well-bonded material. To date aluminum, titanium, nickel, certain high temperature alloys, copper, and silver have been the most widely used. In moist, low temperature environments, galvanic couples between the reinforcement and the matrix can promote corrosion. At elevated temperatures reactions between the fiber and the matrix are difficult to avoid. Composites formed *in situ,* such as the oriented eutectics in which the fiber and the matrix are essentially in thermodynamic equilibrium at the time the total structure is formed by controlled solidification, offer a way of overcoming some of these problems.

Inorganic materials, such as glass, plaster, portland cement, carbon, and silicon, have been used as matrix materials with varying success. These materials remain elastic up to their point of failure and characteristically exhibit low failure strains under tensile loading but are strong under compression. Prestressed steel-reinforced concrete takes advantage of the latter characteristic.

The combined actions of various of the following factors in most cases have prevented the synthesis of high performance composites: (*1*) Chemical attack or dissolution of the reinforcement by the matrix. (*2*) Large thermal expansion mismatches. (*3*) The need for the modulus of the reinforcement to exceed that of the matrix. (*4*) Limited methods available for introducing and consolidating the matrix without introducing large, strength impairing, shrinkage effects.

In situ formation (and equilibration) of a ceramic matrix composite can be achieved in certain favorable cases using the solidification of eutectic compositions to achieve desired structures (see Ceramics). Another approach has been the melt infiltration of silicon into carbon fibers, which converts the carbon fibers into silicon carbide to yield a chemically stable system. Composites based on an inorganic matrix have received less attention than the other two matrix classes owing to the difficulties of fabrication and the relatively restricted range of promising systems identified to date. Reinforced concrete and fiberglass-reinforced plaster board represent the widest spread use of inorganic matrix composites.

Nomenclature

E	= Young's modulus
P	= property value
S	= strength
$\overline{S}$	= mean strength
V	= volume fraction
β	= efficiency factor determined by fiber geometry (fraction of ineffective fiber)
ζ	= a constant determined by fiber geometry
η	= a constant reflecting relative elastic behavior of fiber and matrix
ν_{12}	= major Poisson ratio
ν_{21}	= minor Poisson ratio
σ^*	= stress in matrix at failure strain

Subscripts

f	= fiber
m	= matrix
1, 2, 3, . . .	= fraction i
11	= parallel to fiber
12 or 21	= ratio of fiber direction to transverse
22	= transverse to axis of fiber

BIBLIOGRAPHY

1. A. A. Griffith, *Philos. Trans. Roy. Soc. A.* **221,** 163 (1920).
2. A. Shindo, *Rep. of Gov't Res. Inst. Osaka* **317** (Dec. 1961).
3. Brit. Pat. 1,110,791, W. Johnson, L. N. Phillips, and W. Watt; S. Allen, G. A. Cooper, and R. N. Mayer, *Nature* **224,** 684 (1969).
4. A. E. Standage and R. Prescott, *Nature* **211,** 169 (1966).
5. H. E. LaBelle and A. I. Mlavsky, *Nature* **216,** 574 (1974).

General References

J. E. Ashton, J. C. Halpin, and P. H. Petit, *Primer on Composite Materials: Analysis,* Vol. III, Technomic Publishing Co., Inc., Stamford, Conn., 1969.

L. J. Broutman and R. H. Krock, eds., *Modern Composite Materials,* Addison-Wesley Publishing Company, Reading, Mass., 1967.

L. J. Broutman and R. H. Krock, eds., *Composite Materials,* Vols. 1–6, Academic Press, New York, 1974:

A. G. Metcalfe, ed., *Interfaces in Metal Matrix Composites,* Vol. 1; G. P. Sendeckyj, ed., *Mechanics of Composite Materials,* Vol. 2; B. R. Noton, ed., *Engineering Applications of Composites,* Vol. 3; K. G. Kreider, ed., *Metallic Matrix Composites,* Vol. 4; L. J. Broutman, ed., *Fracture and Fatigue,* Vol. 5; E. P. Plueddemann, ed., *Polymer Matrix Composites,* Vol. 6.

S. W. Tsai, J. C. Halpin, and N. J. Pagano, eds., *Composite Materials Workshop,* Vol. I, Technomic Publishing Co., Inc., Stamford, Conn., 1968.

J. E. Gordon, *The New Science of Strong Materials,* Penguin Books Inc., Baltimore, Md., 1968; see especially Chapt. 8, "Composite Materials."

J. R. Vinson and T. W. Chou, *Composite Materials and Their Use in Structures,* Halsted Press, Division of John Wiley & Sons, Inc., New York, 1975.

W. J. Renton, ed., *Hybrid and Select Metal Matrix Composites: A State of the Art Review,* American Institute of Aeronautics and Astronautics, New York, 1977.

W. B. Hillig, *New Materials and Composites, Science* **191**(4228), 773 (1976).

Journals and Proceedings of Conferences

Fiber-Strengthened Metallic Composites, Symposium Proceedings, ASTM STP No. 427, American Society for Testing and Materials, Philadelphia, Pa., 1967.

Metal Matrix Composites, Symposium Proceedings, ASTM STP No. 438, American Society for Testing and Materials, Philadelphia, Pa., 1968.

Composite Materials: Testing and Design, Symposium Proceedings, ASTM, Philadelphia: *First Conference,* ASTM STP No. 438, 1968; *Second Conference,* ASTM STP No. 497, 1972; *Third Conference,* ASTM No. 546, 1974; and *Fourth Conference,* ASTM STP No. 617, 1977.

Fracture Mechanics of Composites, Symposium Proceedings, ASTM STP No. 593, American Society for Testing and Materials, Philadelphia, Pa., 1975.

E. Scala, E. Anderson, I. Toth, and B. R. Noton, eds., *Proceedings of the 1975 International Conference on Composite Materials,* Vol. 1 and 2, The Metallurgical Society of The American Institute of Mining, Metallurgical and Petroleum Engineers (AIME), New York, 1976.

R. T. Schwartz and H. S. Schwartz, eds., *Fundamental Aspects of Fiber Reinforced Plastic Composites,* Conference Proceedings, John Wiley & Sons, Inc., New York, 1968.

F. W. Wendt, H. Liebowitz, and N. Perrone, *Mechanics of Composite Materials* (*Proceedings of the Fifth Symposium on Naval Structural Mechanics*), Pergamon Press Inc., Elmsford, N.Y., 1970.

D. Johnson, ed., *Composites,* a journal published by IPC Science and Technology Press Limited, Surrey, England; Vol. 1, 1968, and continuing quarterly.

S. W. Tsai, ed., *Journal of Composite Materials,* Technomic Publishing Co., Inc., Westport, Conn.; Vol. 1 (1967) and continuing quarterly.

W. B. HILLIG
General Electric Co.

CORK

Cork [*61789-98-8*] is one of the few naturally grown closed-cell foams that has never been duplicated with synthetic material. It has been an item of commerce for at least 2500 years and was used by the early Greeks for shoes, floats, and stoppers.

As referred to in this article, cork is the outer bark of the cork oak, *Quercus Suber*, which grows mainly near the Mediterranean Sea on the Iberian Peninsula and along the shores of North Africa. The cork oak is unique in that the outer bark of cork, or the phellem, can be stripped from most of the tree trunk and major limbs without harming the tree. Furthermore, the cork bark undergoes rejuvenation and can be stripped repeatedly at specific intervals. The harvest starts when a new tree reaches maturity (20–25 yr) and the first, or virgin cork stripping is made. Successive strippings as a rule are made every nine years. The stripped cork bark is steamed, cleaned, graded, and baled for marketing. The yield per tree ranges from 15–100 kg. Mature trees of normal age between 50 and 200 years measure 1.5–4.0 m in trunk circumference and 10–20 m in height. The maintenance of trees and harvest of cork are now controlled to various degrees by the governments of the growing regions. Many countries have tagged and registered each cork oak and have established laws to protect the trees from being cut for firewood, lumber, or other use.

Structure and Physical Properties

In contrast to the bark of other trees, cork is nonfibrous. It is composed of tiny, closely packed cells that are tetrakaidecahedral, or 14-sided; six of the faces are quadrilateral and eight are hexagonal. This shape provides optimum packing of the cells without extra void space, which accounts for the excellent gasketing and flotation characteristics of cork. The cells are in the range of 0.025–0.050 mm in the longest dimension. The cell walls and resinous binder are highly resistant to water, most organic liquids, and all but strong acid and alkali solutions. Cork has a specific gravity of 0.1–0.3 which provides superior flotation and thermal-insulation properties. A fresh-cut surface of cork exhibits a high coefficient of friction even when subjected to water and oil.

Chemical Properties

Considerable work has been done over the past 200 years to study the chemical composition of cork. The various origins and growing conditions of natural cork cause variations in composition so that it is reasonable that a broad, general analysis must be used to represent the composition. A representative analysis is as follows:

Content	*wt %*
fatty acids	30
miscellaneous organics	17
lignin	16
other acids	13
ceroids	10
tannins	4
glycerol	4
cellulose	3
inorganic ash	3

The first two groups of chemicals are composed of the following:

Common name	*Systematic name*	*Formula*	*CAS Registry No.*
Fatty acids			
phellonic acid	22-hydroxydocosanoic acid	$C_{22}H_{44}O_3$	*[506-45-6]*
phellogenic acid	docosanedioic acid	$C_{22}H_{42}O_4$	*[505-56-6]*
phloionic acid	9,10-dihydroxyoctadecanedioic acid	$C_{18}H_{34}O_6$	*[23843-52-9]*
phloionolic acid	9,10,18-trihydroxyoctadecanoic acid	$C_{18}H_{36}O_5$	*[583-86-8]*
suberonic acid	a mixture		
suberolic acid	a mixture		
cordicinic acid	a mixture		
Miscellaneous organics			
resorcinol		$C_6H_6O_2$	*[108-46-3]*
hydroquinone		$C_6H_6O_2$	*[123-31-9]*
gallic acid		$C_7H_6O_5$	*[149-91-7]*
salicylic acid		$C_7H_6O_3$	*[69-72-7]*
glycerol		$C_3H_8O_3$	*[56-81-5]*
phloroglucinol		$C_6H_6O_3$	*[108-73-6]*
friedelin		$C_{30}H_{50}O$	*[559-74-0]*
oxalic acid		$C_2H_2O_4$	*[144-62-7]*
sterols		a mixture	

Processing

While enroute from the point of harvest to the manufacturing plant, the cork bark is visually classified by thickness, amount of noncork structure, color, and density. The better grades are used for cork-stopper production to be sold for bottling still wine, champagne, other alcoholic beverages, and perfumes. In spite of the development of plastic stoppers and closures for many of these applications, the volume of cork stoppers produced annually exceeds 5 billion. Natural stoppers are cut with the cylindrical axis normal to the bark thickness. Large, thin stoppers are cut parallel to the annual rings and laminates are made for large cork stoppers. In contrast to the old art of hand-cutting stoppers, most of the stopper production is now done by automatic sawing, punching, turning, and sorting equipment. The finished stoppers are cleaned, bleached, sterilized, and bulk-packaged for shipment from Spain, Portugal, and North Africa to all parts of the world. The 50% of the cork from stopper production that remains as waste is granulated to provide furnish for cork composition and corkboard thermal insulation (see Insulation, thermal).

Cork Composites

Today most of the cork harvested from trees ends up in cork compositions of many types. The cork-stopper waste and poorer grades of cork bark are ground, cleaned, and classified into various particle ranges. Binders, plasticizers, and cutting aids are mixed with the ground cork, and the composition is molded or extruded into blocks, bars, sheets, tubes, rods, and other shapes. The final products are cut from these stocks and finished for market. Compositions range from ground cork with 5% protein binder to ground cork with 70% rubber or plastic binder. Cork compositions are used in a wide variety of gasket and packing applications because of their inherent chemical resistance, true compressibility, and low specific gravity. Cork compositions are also used for such products as handles for fishing rods and sports rackets, coasters, table pads, bulletin boards, and wall covering. A unique use for composition cork was developed by the aerospace industry. Certain compositions have been used to provide ablative insulation covering for rocket-engine casings and fuel tanks (see Ablative materials). Other assemblies used cork compositions to protect Apollo vehicles on the way to and from the moon.

Economic Aspects

The evolution of cork from its position as a relatively low-cost raw material to its current position in economic competition with a variety of synthetic substitute materials has caused a gradual decline in world consumption of cork since 1960. Pressure from increasing labor and shipping costs and higher tariffs induced an average cork price escalation of 300% during the period of 1950–1976. This trend is reflected in the decline in the United States' total imports of both manufactured and unmanufactured cork through the past several decades, as shown:

Year	*U.S. imports, metric tons*
1941	154,000
1950	135,000
1961	53,000
1976	20,000

Currently, most of the baled cork imported to the United States is either scrap corkwood to be processed or granulated cork. The largest volume of imported manufactured cork is corkboard insulation in block and sheet form which is also used for such applications as decorative paneling, lamps, and furniture.

The major consumers of cork are now Western and Eastern European and Asiatic countries. Portugal and Spain supply most of the Western European and North and South American markets, and North Africa supplies the rest of the importing countries. Total world consumption of cork for 1976 was estimated at 300,000 metric tons valued in raw material cost at $130 million.

BIBLIOGRAPHY

"Cork" in *ECT* 1st ed., Vol. 4, pp. 480–487, by G. B. Cooke, "Cork" in *ECT* 2nd ed., Vol. 6, pp. 281–288, by G. B. Cooke, Essex Community College.

General References

E. Palmgren, *Cork Production and International Cork Trade,* International Institute of Agriculture, F.A.O., 1947.
J. Marcos de Lanuga, *The Cork of Quercus Suber, Inst. For. Invest. Exper. Madrid* **35**(82), 1964.
P. Pla Casadevall, *El Suro,* Universitat Politecnica de Barcelona, 1976.
Report FT 246 Annual 1976, U.S. Imports for Consumption and General Imports, 1976.

R. H. HUTCHINSON
Armstrong Cork Co.

DIATOMITE

Diatomite (diatomaceous earth, kieselguhr) is a sedimentary rock of marine or lacustrine deposition. It consists mainly of accumulated shells or frustules of hydrous silica secreted by diatoms, which are microscopic, one-celled, flowerless plants of the class *Bacillarieae*. Some are mobile, others stationary. From 60,000,000 years ago to as late as perhaps 100,000 years ago, the diatom plants thrived in the waters of the earth. They still exist, although in only a fraction of their prehistoric population.

Diatoms are single-celled photosynthetic plants consisting of two shells that fit together in the same manner as the two halves of a pill box (1). Reproduction is by division and at such a rate that it is estimated that one diatom could produce 10^{10} descendants in thirty days under the most favorable conditions. The prehistoric diatom plants extracted silica from the water and used this substance to form an encasing shell or exoskeleton. It is believed that during a period of many thousands of years intense volcanic activity caused both fresh and salt waters to develop a high silica content on which the diatoms flourished. At the end of a very brief life, the diatom settles to the bottom of the body of water and the organic matter decomposes, leaving the siliceous skeleton. These fossil skeletons (specifically frustules) are in the shape of the original diatom plant in designs as varied and intricate as snow flakes (2). Over 10,000 varieties have been classified (1). A few of these are shown in Figures 1 and 2.

Origin of Deposits

Diatoms still inhabit fresh, brackish, or sea waters, as undoubtedly was the case millions of years ago. Salinity changes produce different types of diatoms which may appear at different levels of the same deposit as changes in the salinity of the water occurred. Today's commercial deposits of diatomite are accumulations of the fossil skeletons in beds varying in thickness to as much as 900 meters (2) in some locations. Marine deposits must have been formed on the bottoms of bays, lagoons, or other bodies of quiet water, undisturbed by strong currents. Ocean currents, however, could have had considerable effect in carrying the diatoms from some distance and concentrating them in the quiet waters.

The main deposits of fresh water diatomite were laid down in large lakes. Possibly earthquakes, landslides, or lava flows formed dams in rivers to create lakes where the diatoms thrived. Several tens of square kilometers in Nevada, west of Tonopah, are covered with diatomite as are other large areas in that state.

The principal marine deposits were formed during the Tertiary era and more particularly the upper Miocene period. Major deposits of fresh water origin are more recent, dating from 1,000,000 years to as late as 100,000 years ago. During many thousands of years, owing both to earth upheavals and other causes, either the waters receded or the land was uplifted and the present day dry land deposits were formed. Most commercial deposits are at or comparatively near the surface.

Location of Deposits. Deposits of diatomite are known to exist in every continent, in nearly every country, in widely diversified and even unexpected environments. Over half of the states of the United States reportedly contain diatomaceous deposits. In some cases, deposits of marine and fresh water origin occur almost side by side, as do deposits of widely varying ages (2). Most of the deposits are not large enough or suf-

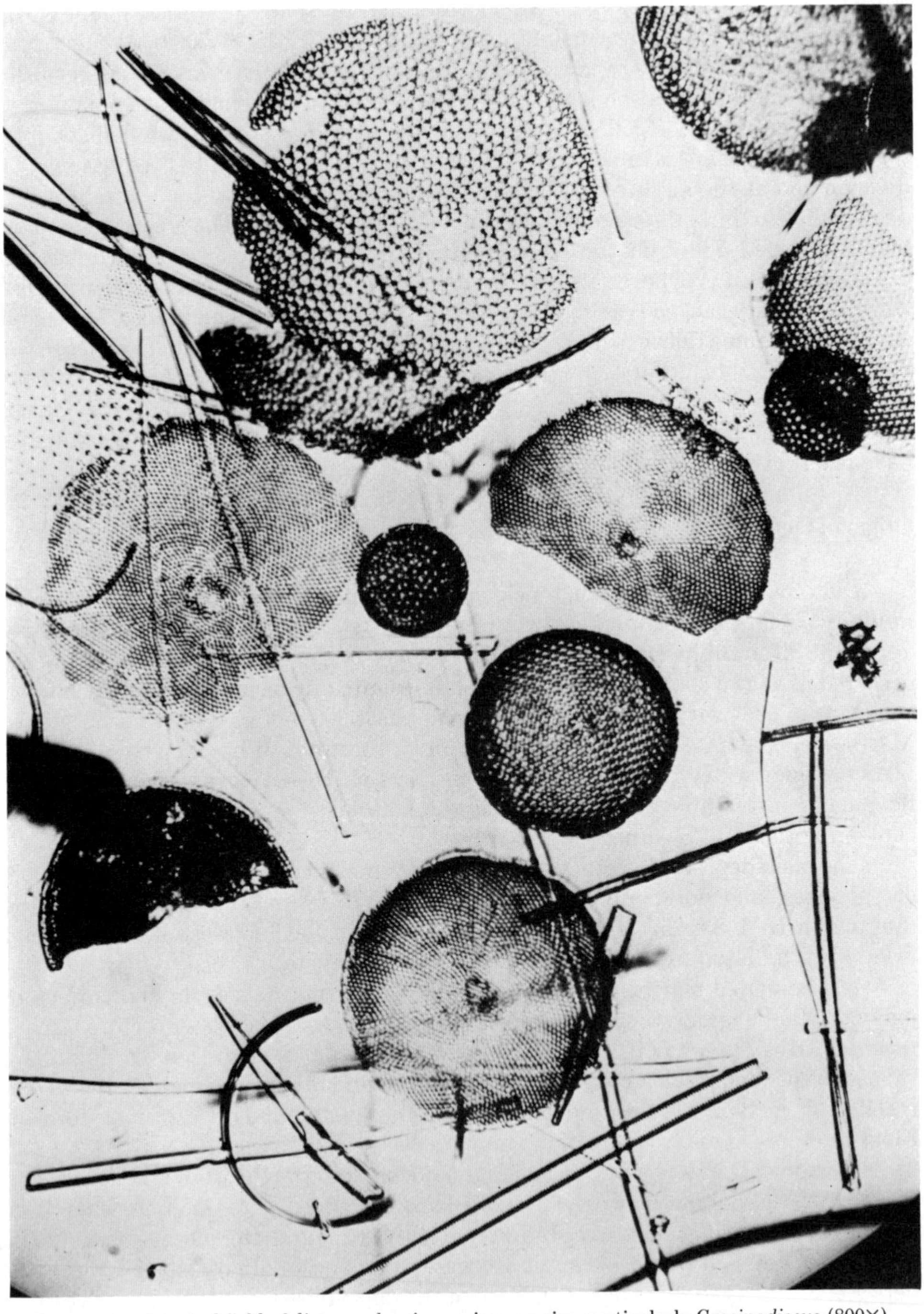

Figure 1. A typical field of diatoms showing various species, particularly *Coscinodiscus* (800×).

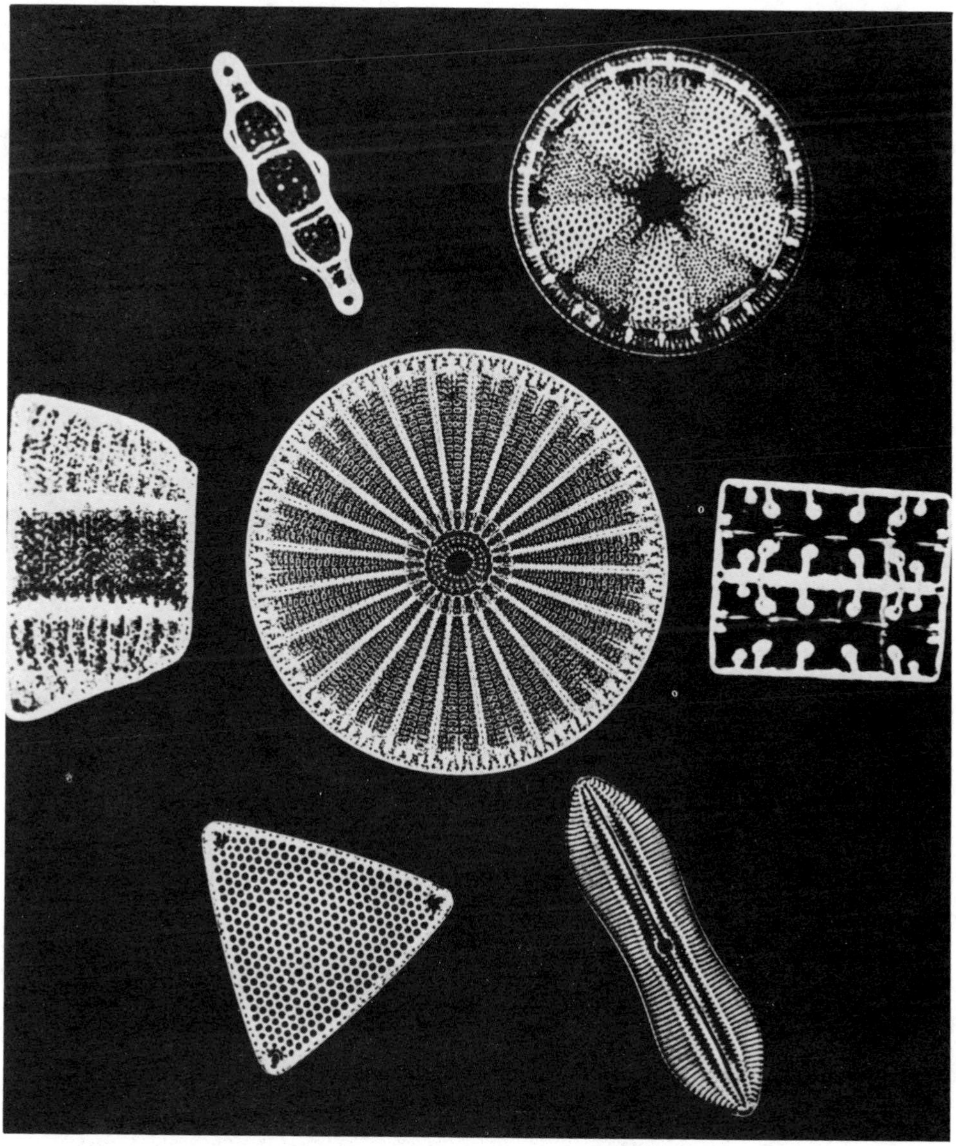

Figure 2. Example of different diatom shapes (500×).

ficiently pure to have commercial value; production figures show the location of the deposits that meet commercial standards in both respects.

The states in the United States having the most extensive commercial deposits are California, Nevada, Oregon, and Washington. The U.S. Bureau of Mines also reports the commercial operation of diatomite deposits in Arizona, Connecticut, Florida, Maryland, New Hampshire, New Mexico, New York, Virginia, and Idaho.

California contains the largest formation of diatomite in the United States; the Monterey formation extends from Point Arena in Mendocino County in the north to San Onofre in the south. The most extensive deposit in the area is near Lompoc, Santa

Barbara County, and it is of marine origin. Other important deposits (not now mined) are located in Monterey, Fresno, Shasta, Inyo, Kern, Orange, San Bernardino, San Joaquin, Sonoma, and San Luis Obispo counties. A large deposit in Los Angeles county was operated from 1930 to 1958. It had reached the point where the quantity and quality of crude diatomite available were overbalanced by the value of the land for real-estate development.

An extensive deposit in Oregon near Terrebonne was operated intensively from 1936 to 1961. Current Oregon production is from two deposits near Christmas Valley. Three different companies operate at least four deposits in Nevada. Of the several comparatively large deposits in Washington, only one, near Quincy, is being worked on a significant commercial scale (3).

Large fresh-water deposits are found in Canada, in British Columbia. Small deposits are located in Nova Scotia, New Brunswick, Quebec, and Ontario. Although deposits exist in other continents, the most important are in Europe, Africa, Japan, and the Union of Soviet Socialist Republics. Judging from reported output of finished diatomaceous earth products, the leading producers in Europe are France, the Federal Republic of Germany, Italy, and Denmark. Important deposits are located in Algeria and Kenya in Africa, with smaller operations in Mozambique, Rhodesia, and the Republic of South Africa. The Union of Soviet Socialist Republics is said to have very large deposits in the Caucasus Mountains.

Physical and Chemical Properties

Chemically, diatomite consists primarily of silicon dioxide, and is essentially inert. It is attacked by strong alkalies and by hydrofluoric acid, but it is practically unaffected by other acids. Because of the intricate structure of the diatom skeletons that form diatomite, the silicon dioxide has a very different physical structure from other forms in which it occurs (see Silicon compounds). The chemically combined water content varies 2–10%. Impurities are other aquatic fossils (sponge residues, *Radiolaria, Silico-flagellatae*), sand clay, volcanic ash, calcium carbonate, magnesium carbonate, soluble salts, and organic matter. The types and amounts of impurities are highly variable and depend upon the conditions of sedimentation at the time of diatom deposition. Variations exist among deposits as well as among parts of the same deposit. Typical chemical analyses of diatomite from different deposits are given in Table 1.

Table 1. Typical Spectrographic Analysis of Various Diatomites (Dry Basis)

	Deposit		
Constituent, %	Lompoc, Calif.	Basalt, Nev.	Sparks, Nev.
SiO_2	88.90	83.13	87.81
Al_2O_3	3.00	4.60	4.51
CaO	0.53	2.50	1.15
MgO	0.56	0.64	0.17
Fe_2O_3	1.69	2.00	1.49
Na_2O	1.44	1.60	0.77
V_2O_5	0.11	0.05	0.77
TiO_2	0.14	0.18	0.77
ignition loss	3.60	5.30	4.10

The color of pure diatomite is white, or near white, but impurities such as carbonaceous matter, clay, iron oxide, volcanic ash, etc, may darken it. The refractive index ranges from 1.41 to 1.48, almost that of opaline silica. Diatomite is isotropic.

The apparent density of powdered diatomite varies from 112 to 320 kg/m^3 and may reach 960 kg/m^3 for impure lump material. The true specific gravity of diatomite is 2.1–2.2, the same as for opaline silica, or opal (1).

Bed moisture may be over 65%, but in arid regions it often varies 15–25% (4).

The thermal conductivity is low but increases with the increase in percentage of impurities and the weight per unit volume. The fusion point depends on the purity, but averages about 1590°C for pure material, slightly less than the pure silica. The addition of certain chemical agents reduces the fusion point.

Diatomite has only weak adsorption powers, but shows excellent absorption (qv). Acids, liquid fertilizers, alcohol, water, and other fluids are absorbed by diatomite.

Mining and Processing

Diatomite deposits are usually discovered by observation of outcroppings and the value of the deposit is determined by geological prospecting and exploration. Usually samples are taken from the surface outcrops by digging or trenching; underground samples are secured from test holes, core-drill holes, or tunnels. Samples are examined microscopically, physically, and chemically to determine the suitability of the diatomite for various uses. Accessibility of the deposit to transportation is important. Estimates of tonnages available, based on the dry weight of the material per unit volume, are necessary to determine the potential profitability of operation as well as the life of the deposit.

Mining. Mining diatomite is relatively simple, especially in the larger deposits. In the past, some underground mining was practiced utilizing tunneling and room-and-pillar techniques. Today this method is rarely employed except in small-scale operations and in some nations where low-cost labor is available.

The most common method of mining diatomite is by quarrying or open-pit operation (5). The first step in opening a quarry area is removal of the overburden. This is highly mechanized: power shovels lift the crude material, which is then hauled to the processing plant. The diatomite is sometimes stockpiled, for air drying or for storage as blending crude, or for both reasons.

Processing. Manufacture of finished products from the crude diatomite has greatly advanced since the first serious research and development work in 1912. The principal reason for the large growth of the industry in the United States is the high degree of technological efficiency attained by the major producers. Every step from the selection of the crude material to the testing of the finished product before shipment is carefully monitored. The industry devotes much of its energy to the development of improved products and processes, including specially designed production equipment. The principal producers make every effort to supply custom-made materials for specialized applications. Figure 3 is a flow sheet showing the processing of diatomite at a plant at Lompoc, California.

Three general types are produced, with a range of grades in each. The term grade, as used here, refers not to the quality of the product but designates one of the series of products made for specific uses. Regardless of the ultimate use, the processing of each type is essentially the same. The crude diatomite, which may contain up to 60%

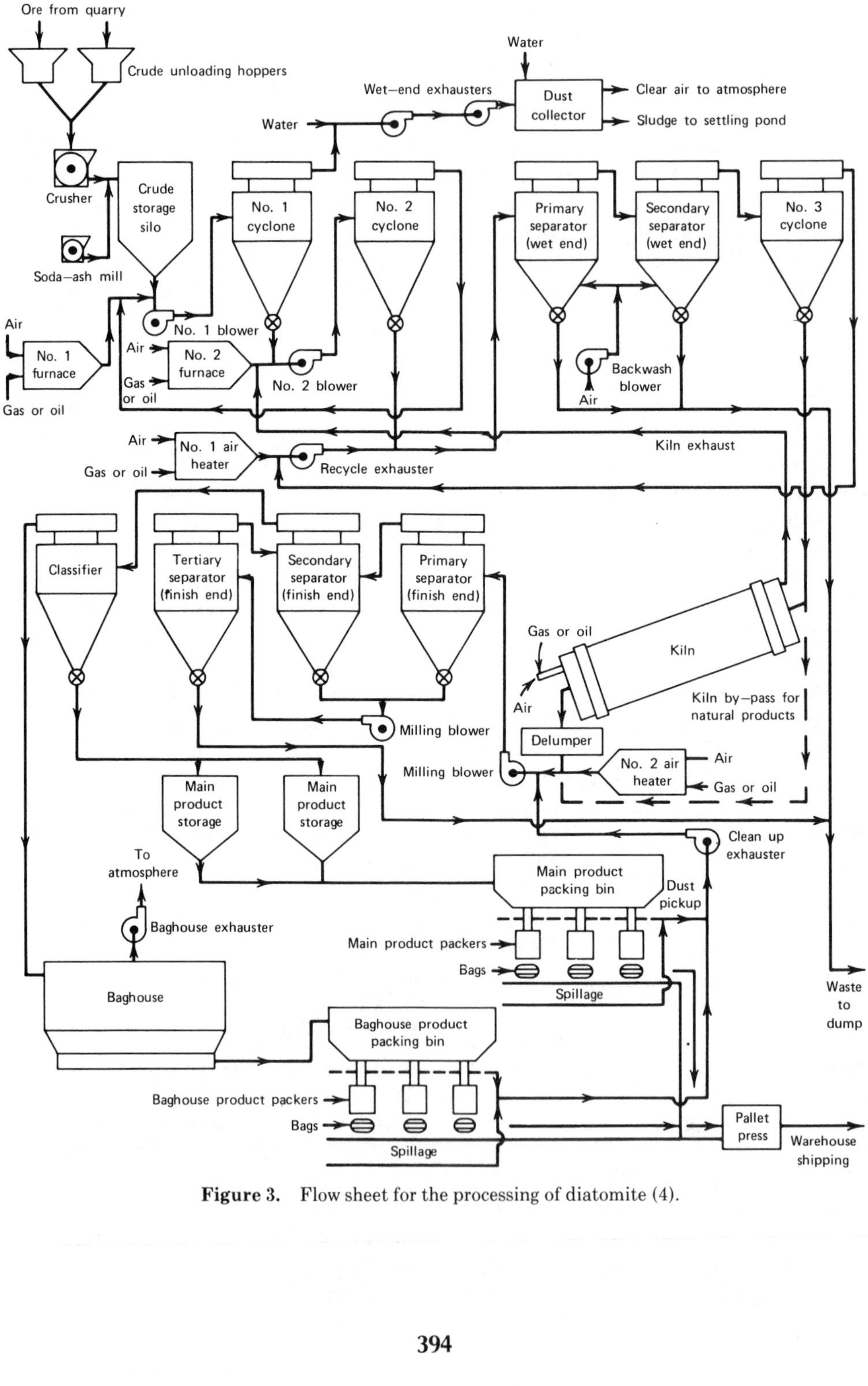

Figure 3. Flow sheet for the processing of diatomite (4).

moisture, is first broken up in hammer mills. This material is fed to dryers operating at relatively low temperatures, where virtually all of the moisture is removed. All manufacturing processes described with the exception of the calcination step, take place while the material is pneumatically conveyed. Coarse and gritty nondiatomaceous earth material is removed in separators and preliminary particle-size separation is made in cyclones. The resultant material is termed a natural product, and is the only type made by some of the smaller producers (6).

The second type is produced from natural diatomite, processed as above and also subjected to high temperature calcination in a rotary kiln at about 980°C. The calcined material is then again milled and classified to remove coarse agglomerates as well as extreme fines.

The third type of product is the so-called white or flux-calcinated diatomite material. This is obtained by calcination of the natural product in the presence of a flux, generally soda ash, although sodium chloride can also be employed. Such processing has the effect of reducing the surface area of the particles, changing the color from the natural buff cast to a true white, and rendering various impurities insoluble. Some of the diatomite is converted on calcination to cristobalite. The Threshold Limit Value (TLV) in mg/m^3 for calcined products is in the range of 0.094–0.24 and for flux-calcined diatomite is 0.074 (7).

In all processing of diatomite, the selection of the proper crude ore, the milling, calcination, and classification are of extreme importance. Each grade of material is manufactured to rigid specifications. The finished product must pass certain performance, chemical, and physical tests to ensure compliance with these specifications (see under Uses).

Economic Aspects

Owing to the light weight of diatomite, freight and trucking rates (on a weight basis) are high. The finished products are packed and shipped in laminated kraft-paper bags, usually containing 22.5 kg, for United States exports, or the product is shipped in bulk. Bagged products are shipped by truck or rail box car, normal box car loading being 27–36 metric tons. It may also be palletized and wrapped with a polyethylene shrink wrap. Bulk shipments are made in bulk pressure-differential trucks of 45 and 74 m^3 (1600 and 2600 ft^3). These carry 12–15 t and 16–21 t, respectively. Distribution by bulk truck is confined to the economical trucking distance from the producing plant.

For longer hauls, pressure-differential rail hopper cars are used. These hold between 36–45 t per car. In-plant storage is in conventional silos usually with a 60° cone bottom. The silo may be pressurized for discharge, but this is exceptional. Normally, diatomite is moved from the silo to an adjacent small pressure vessel for transfer to user sites. Or the material may be aerated and pumped using a modified diaphram pump. Bulk handling has the advantage of improved environmental conditions as well as lower handling cost in comparison to bagged material.

There are so many different products, and variations of these among manufacturers, that it is impossible to cite an average price for diatomite in the United States. The processed powders and aggregates range in price from $20.00–225.00/t for what might be called standard products in carload quantities. Materials for specialized applications, which require special processing, range up to $900.00/t in carload quantities. All these prices are FOB the diatomite plant.

Producers. Principal companies mining diatomite and processing it into finished products in the United States are Grefco, Inc., Lompoc, California, and Mina, Nevada; Johns-Manville Products Corporation, Lompoc; Eagle-Picher Industries, Inc., Sparks and Lovelock, Nevada; and Witco Chemical, Quincy, Washington. The remaining producers listed in *Minerals Yearbook* are Excel-Minerals Company, Taft, California; Airox Earth Resources, Inc., near Santa Maria, California; N L Industries, Inc., near Wallace, Kansas; Cyprus Mines, near Fernley, Nevada; and A. M. Matlock and American Fossil, Inc., both near Christmas Valley, Oregon.

Annual United States Production. The development of the diatomite industry in the United States is well illustrated by production figures for diatomite products as given in the yearbooks and special bulletins of the U.S. Bureau of Mines. In twenty-four years the tonnage has almost doubled, as shown in Table 2.

World Production. After the United States, the Union of Soviet Socialist Republics is the largest producer of diatomite products with an estimated 1974 production of 399,432 t. The ten highest producers among other countries, with annual production, are shown in Table 3.

Uses

It is reported that in 532 AD the Roman Emperor Justinian used diatomite bricks for lightweight construction of the dome when building the Church of St. Sophia in Constantinople (Istanbul). The names bergmehl, fossil flour, farine fossile, and mountain flour apparently originated when early poverty-driven people extended their

Table 2. Annual Production of Diatomite in the United States[a]

Years	Production, t	Year	Production, t
1954–1956	334,457[b]	1971	485,962
1957–1959	408,310[b]	1972	522,974
1960–1962	437,748[b]	1973	552,765
1963–1965	526,775[b]	1974	603,054
1966–1968	569,456[b]	1975	520,169
1969	543,302	1976	563,744
1970	542,962	1977	580,656

[a] Refs. 3, 8.
[b] Average annual production.

Table 3. Annual World Production of Diatomite[a]

Country	Production, t	Country	Production, t
U.S.S.R.	399,432	Iceland	22,513
France	208,794	Spain	20,879
Italy	59,007	Mexico	19,972
FRG	44,482	Denmark	19,972
Costa Rica	35,404	Argentina	19,972

[a] Estimated tonnage for all countries from Vol. 1 of *Minerals Yearbook*, U.S. Dept. of Interior, Bureau of Mines, U.S. Government Printing Office, Washington, D.C., 1974 (3); see also ref. 8.

meager supply of meal and flour by dilution with diatomaceous earth. Tripolite is a name given to diatomite formerly mined in Tripoli, North Africa. Kieselguhr is the name given to the diatomite first mined in Hanover, Germany, in 1836 or earlier, and is still used as a general name for all diatomite products in Europe. An incorrect name which persisted for many years was infusorial earth, incorrect because Infusoria comprises a group of the animal kingdom (1). Nobel developed the first important industrial use of diatomite as an absorbent for liquid nitroglycerin in the making of dynamite late in the nineteenth century (see Explosives and propellants).

The first substantial commercial shipment of diatomite in the United States was made in 1893, and consisted of material from a one-man quarry operation in the vast deposit near Lompoc, California. It went to San Francisco to be used for pipe insulation. Small-scale operation of parts of the Lompoc deposit continued from 1893 for about fifteen years until it was acquired by the Kieselguhr Company of America. This name was later changed to The Celite Company (9). The first significant research and development work on the production of diatomite materials for industrial use was started in 1912 (10). In the ca 70 years since, the industry has grown immensely and diatomite products are used in almost every country of the world.

Types of Products. Several hundred diatomite products are available, many of them processed for specific purposes. In general, these can be grouped according to use as follows: filteraids, fillers or extenders, thermal insulation, absorbents, catalyst carriers, insecticide carriers and diluents, fertilizer conditioners, and miscellaneous.

In physical form, powders make up by far the greatest proportion of diatomite products. Mean particle diameters range from 20 to 0.75 μm. Aggregates are available for special uses and range from 1.27-cm particles to fine powder. The other most common form is molded insulating brick. This does not take into account a number of types and shapes of thermal insulation that are composed of other ingredients mixed with diatomite.

Uses of Products. There are three principal ways in which finished diatomite products are utilized in manufacturing plants. One type of diatomite product, a filteraid, is used as an expendable processing aid. Another type, filler, becomes a component and remains as part of the manufactured product. A third type, insulation, is installed as a structural member in walls, bases, and other parts of heated equipment to prevent heat loss.

The value of diatomite for most uses depends on its unique physical structure, which is responsible for its high porosity, low density, and great surface area. The particles are quite irregular in shape. Regardless of seeming fragility, each fossil particle, because of its silica composition, is a rigid individual shape. These physical characteristics, in addition to chemical inertness, are of considerable importance to the uses of the material. Table 4 gives some properties of the products of typical diatomite products (11).

According to 1973 figures from the U.S. Bureau of Mines, the use of diatomite products is divided as follows: filtration, 61%; thermal insulation, 4%; miscellaneous, including fillers, 35%.

Filtration. The first use of diatomite as a positive filteraid was by Slater (9) for the filtration of sewage sludge (see Filtration; Water, sewage). The filteraid use developed, however, from the application in cane sugar refining (12). This is still one of the principal uses, including filtration of various liquors in refining cane, beet, and

Table 4. Properties of Typical Diatomite Products

Property	Filteraids			Fillers		
	Natural	Calcined	White	Natural	Calcined	White
relative filtration rates, based on natural with flow rate of unit value	1	1–3	3–25			
wet cake density, kg/m^3	2.4–3.5	2.4–3.7	2.6–3.4	2.4–4.5	2.4–4.8	2.4–6.4
(lb/ft^3)	(15–22)	(15–23)	(16–21)	(15–28)	(15–30)	(15–40)
sedimentation particle size distribution, %						
+40 μm	2–4	5–12	5–24	0–4	5–12	0–4
20–40 μm	8–12	5–12	7–34	0–10	5–12	1–7
10–20 μm	12–16	10–15	20–30	0–14	10–15	10–25
6–10 μm	12–18	15–20	8–33	1–32	15–20	15–35
2–6 μm	35–40	15–45	4–30	33–40	15–45	30–50
−2 μm	10–20	8–12	1–3	14–66	8–12	3–20
moisture, max, %	6.0	0.5	0.5	6.0	0.5	0.5
specific gravity	2.00	2.25	2.33	2.00	2.25	2.33
pH	6.0–8.0	6.0–8.0	8.0–10.0	6.0–8.0	6.0–8.0	8.0–10.0
refractive index	1.46	1.46	1.46	1.46	1.46	1.46
Gardner-Coleman oil absorption				150–170	130–160	95–160
ASTM rubout oil absorption				130–160	130–160	130–160
retained on 44 μm (325 mesh) screen, %	0–12	0–12	12–35	0–12	0–12	0–35
surface area by nitrogen adsorption, m^2/g	12–40	2–5	1–3	12–40	2–5	1–3

corn sugar and in clarification of syrups, molasses, etc (see Sugar). From this application, the material and technique was applied to a wide range of separation problems involving beer (qv), various chemicals, water, solvents, antibiotics, oils and fat, phosphoric acids, and many others. In recent years, the necessity of producing clean plant effluents has spurred the application of diatomite filtration for solids removal from waste streams from manufacturing plants (see Adsorptive separation).

Fillers. In a broad sense, diatomite mineral fillers (qv) are used primarily (*1*) where bulk is needed with minimum weight increase, (*2*) as an extender where economy of more expensive ingredients is a factor, or (*3*) where the structure of the particle is important. In other applications, diatomite can add strength, toughness, and resistance to abrasion, and in still others may act as a mild abrasive and polishing agent (see Abrasives).

The paint and paper industries are typical of those employing diatomite extensively as fillers and extender pigments. Also diatomite is extensively used as a polyethylene antiblock (see Abherents).

Insulation. Diatomite makes an efficient thermal insulator because of its high resistance to heat (fusion point about 1593°C) and its high porosity. Materials in the form of powders, aggregates, and bricks are most commonly used. At one time solid bricks were sawed directly from strata in a deposit, dried in a kiln, and then milled to size. In this natural form the diatomite would withstand direct service temperatures up to 870°C without undue shrinkage. This type is no longer available, although at one time it was very widely used as a brick course in walls, bases, and tops of heated equipment. Diatomite insulating bricks for all temperatures are now formed by adding a binder, molding the mixture to sizes and shapes as desired, and then firing in a kiln. Diatomite powders and aggregates are often installed loose over tops or in hollow wall-spaces of furnaces, kilns, ovens, etc. Calcined aggregates are supplied for mixture with water and portland cement to make insulating concrete for casting bases, doors, baffles, etc, for various types of heated equipment (see Insulation, thermal).

Calcium Silicate Insulation. This type of thermal insulation is produced by a number of manufacturers utilizing the reaction of lime and natural diatomite. It is usually precast in molds for pipe covering and other special shapes, in blocks for covering large areas, and in curved slabs for insulating tanks and other equipment. A fast reaction rate is essential for efficient production of calcium silicate insulation. Since the reaction time depends largely on a high surface area of diatomite used, special grades have been developed for this purpose. Most of this material is manufactured from a fresh-water diatomite which is characterized by its high surface area. The diatomite and lime are mixed with water in a slurry together with a catalyst and heated to ca 79°C. This heated slurry is poured into molds where an initial set takes place rapidly so that the desired shapes can be removed from the molds in a minimum of time. The material is then autoclaved and dried. This calcium silicate insulation is characterized by high temperature resistance, which permits its use at temperatures up to 677°C, whereas the well-known 85% magnesia insulation cannot be employed at temperatures exceeding 316°C.

Other Uses. Miscellaneous uses of diatomite are many, some being highly specialized and extensive. As a pozzolanic admixture in concrete mixes, it improves the workability of the mix, permitting easier chuting and placement in intricate forms (see Cement). Special grades are used as carriers for catalysts in petroleum refining, hydrogenation of oils, and in manufacture of certain acids. Diatomite powders are used as carriers for insecticides, and as a fluffing agent for heavier dusts (see Insect control technology). Certain diatomite powders, with nothing else added, act as a natural insecticide and are used to protect seeds and stored grain. This use is not extensive since outdoor application is very sensitive to low humidity and with stored grains, diatomite changes the density and grain value. The fertilizer industry uses large quantities of diatomite as an anticaking agent or conditioner, particularly for prilled ammonium nitrate. The diatomite greatly reduces absorption of moisture by the fertilizer, thus preventing caking and balling in the bag and making spreading easier (see Fertilizers).

BIBLIOGRAPHY

"Diatomite in *ECT,* 1st ed., Vol. 5, pp. 33–37, by H. Mulryan, Sierra Talc & Clay Company; "Diatomite" in *ECT* 2nd ed., Vol. 7, pp. 53–63, by E. L. Neu, Great Lakes Carbon Corporation.

1. R. Calvert, *Diatomaceous Earth,* American Chemical Society Monograph Series, Chemical Catalog Company, J. J. Little & Ives Co., New York, 1930.
2. W. W. Wornardt, Jr., *Miocene and Pliocene Marine Diatoms From California, Occasional Papers of the California Academy of Sciences, No. 63,* The California Academy of Sciences, Los Angeles, Calif., 1967.
3. *Minerals Yearbook,* Vol. 1, U.S. Department of Interior, Bureau of Mines, U.S. Government Printing Office, Washington, D.C., 1974, pp. 539–541.
4. W. G. Hull and co-workers, *Ind. Eng. Chem.* **45,** 256 (Feb. 1953).
5. C. V. O. Hughes, Jr., *Diatomaceous Earth Mining Engineering,* American Institute of Mining, Metal, and Petroleum Engineers, Littleton, Colorado, March 1953.

6. P. W. Leppla, *Diatomite,* Mineral Information Service, Department of Natural Resources, State of California, **6**(11), 1–5 (1953).
7. *Fed. Reg.* **36**(123), 242 (1971).
8. G. Coombs, *Diatomite Mining Engineering,* American Institute of Mining, Metal, and Petroleum Engineers, Littleton, Colorado, March 1977.
9. A. B. Cummins, *The Development of Diatomite Filter Aid Filtration, Filtration and Separation,* Uplands Press Ltd., Craydon, Eng., Jan–Feb 1973.
10. P. A. Boeck, *Chem. Met. Eng.* **12,** 109 (1914).
11. E. Brannigan, *Product Characteristics,* Internal Publication, Dicalite Division-Grefco, Inc., Los Angeles, Calif., 1977.
12. H. S. Thatcher, *Sugar Filtration Improved Methods Filtration,* Kieselguhr Co. of America, Lompoc, Calif., 1915.

E. L. NEU
A. F. ALCIATORE
Grefco, Inc.

EMBEDDING

The embedment of objects, the complete encasement of objects in a medium, practiced for centuries, has today become a science involving large numbers of scientists and engineers in all parts of the industrial world. The objectives of embedding may be either functional or decorative. Both are covered in this article with the greater emphasis on functional embedding. The most important functional area is the embedding of electrical and electronic circuitry. Embedding in electronics is a highly technical and specialized field, using very high performance materials and processes that are elaborate, precise, and tightly regulated, often by electronic controls.

Embedding technology, in its present form, began in the 1940s. Early embedding practices used waxes and bitumens, materials that are largely inadequate for modern needs because of their performance limitations in thermal stability, and in electrical and mechanical parameters. The rapid development in synthetic resins has produced today's high performance embedding materials. The early synthetic resins used for embedding had serious shortcomings. The first of these was a phenolic casting material developed by Baekeland in 1906. The use of phenolics has not become widespread because the acidic catalyst used in their manufacture is detrimental to electronic circuitry and components. The inherent brittleness in the phenolic resins leads also to an undesirable tendency to crack during temperature cycling.

Since functional embedding, mostly electrical and electronics, is by far the largest application of embedding, the presentation of embedding here provides information useful for that area. However, much of the information is equally important for nonfunctional or nonelectrical embedding. A later section of this article reviews some of the unique factors related to nonfunctional or nonelectrical applications.

Terminology for Embedding

The terminology covering the area of embedding is extensive, and unfortunately, there is little uniformity in the use of some of these terms. Nonspecialists may, for example, incorrectly use the terms "potting," "molding," or "encapsulating" to apply to the entire field. There are accurate and correct terms, however, each referring to the attainment of embedding by a different process. Thus embedding, which is the complete encasement of an object in a medium, can be achieved by several processes. The correct process-related terms are as follows:

Casting: this embedding process consists of pouring a catalyzed or hardenable liquid into a mold. The hardened cast part takes the shape of the mold and the mold is removed for reuse.

Potting: this embedding process is similar to casting except that the catalyzed or hardenable liquid is poured into a shell or housing which remains as an integral part of the unit.

Impregnating: this embedding process consists of completely immersing a part in a liquid so that the interstices are thoroughly soaked and wetted; the process is usually accelerated by vacuum or pressure, or both.

Encapsulating: this embedding process consists of coating (usually by dipping) a part with a curable or hardenable coating; coatings are relatively thick compared with varnish coatings.

Transfer Molding: this embedding process involves the transfer of a catalyzed or hardenable material, under pressure, from a pot or container into the mold which contains the part to be embedded.

Embedding Process Considerations

Often, a given embedded product can be made by two or more of the various embedding processes. Thus, an analysis of the important comparative advantages of the methods is required. This section presents some of the most important of these considerations. The comparison is summarized in Table 1 (see also Plastics technology).

Casting. For casting applications, the design of the mold and the design of the assembly should provide for minimum internal stresses during curing and for proper final dimensions after allowing for shrinkage.

In casting processes, the mold is cleaned thoroughly and a suitable release agent is applied, for many of the resins used normally adhere to the walls of the mold (see Abherents). The mold is then positioned around the part, and any points where leakage of the liquid resin from the mold might occur are sealed with a material such as cellulose acetate butyrate if they cannot be conveniently gasketed against leakage. Next, the resin and catalyst are mixed and are poured slowly into the mold so as to avoid air entrapment during the pouring.

The entire assembly is allowed to cure as required by the resin–catalyst system either at room temperature under its own exothermic heat or in an oven at some higher temperature. Finally, the part is separated from the mold. A typical cycle for producing embedded assemblies by the cast resin technique is shown in Figure 1.

Table 1. **Comparison Summary For Embedding Processes**[a]

Process	Advantages	Limitations	Material requirements	Applications
casting	requires a minimum of equipment and facilities; is ideal for short runs	for large-volume runs, molds, mold handling and maintenance; can be expensive; assemblies must be positioned so they do not touch the mold during casting; patching or correcting surface defects can be difficult	viscosity must be controlled so that the embedding material flows completely around all parts in the assembly at the processing temperature and pressure	most mechanical or electro-mechanical assemblies within certain size limitations can be cast
potting	excellent for large-volume runs; tooling is minimal; presence of a shell or housing assures no exposed components as can occur in casting	some materials do not adhere to shell or housing; electrical shorting to the housing can occur if the housing is metal	same material requirements as for casting except that materials that bond to the shells or housings are required	most mechanical or electro-mechanical assemblies, subject to certain size limitations and housing complexity limitations
impregnating	the most positive method for obtaining total embedding in deep or dense assembly sections such as transformer coils	requires vacuum or pressure equipment which can be costly; in curing, the impregnating material tends to run out of the assembly creating internal voids unless an encapsulating coating has first been applied to the outside of the assembly	low viscosity materials are required for the most efficient and most thorough impregnation	dense assemblies which must be thoroughly soaked; electrical coils are primary examples
encapsulating	requires a minimum of equipment and facilities	obtaining a uniform, drip-free coating is difficult; specialized equipment for applying encapsulating coatings by spray techniques overcomes this problem, however	must be both high viscosity and thixotropic so that they do not run off during the cure	parts requiring a thick outer coating, such as transformers

Table 1. (*continued*)

Process	Advantages	Limitations	Material requirements	Applications
transfer molding	economical for large-volume operations	initial facility and mold costs are high; requires care so that parts of assemblies are not exposed; some pressure is required and processing temperatures are often higher than for other embedding operations	should be moldable at the lowest possible pressure and temperature and should cure in the shortest possible time for lowest processing cost	for embedding small electronic assemblies in large-volume operations

[a] Refs. 1–2.

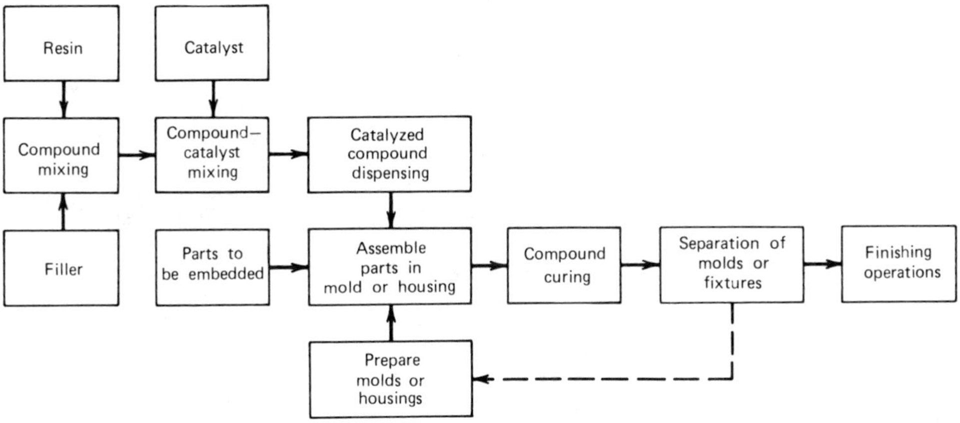

Figure 1. Typical process sequence for producing embedded electronic assemblies by casting (1–2).

The two types of molds in common use are metal and plastic, (Table 2). The relatively high thermal conductivity of metal molds is an advantage if oven curing of the resin is used. The heat is rapidly and uniformly transferred into the resin. The insulating properties of plastics may require that parts processed in plastic molds be preheated before the liquid resin is poured.

Potting. The potting process is similar to casting, except that the assembly to be potted is positioned in a can, shell, or other container. Since the shell or can will not be separated from the finished part, no mold-release treatment is necessary. However, if the container is metal and if the assembly is electrical, it is usually necessary to place a sheet of insulating material between the assembly and the can. The higher the applied voltage, the more stringent is this requirement.

Housings for potted assemblies are made of metal or plastic (Table 3). The same

Table 2. Considerations in Selection of Molds To Be Used in Casting[a]

Mold material and fabrications	Advantages	Disadvantages
machined steel	good dimensional control; can be made for complex shapes and insert patterns; good heat transfer; surfaces can be polished	assembly sometimes difficult; can corrode; usually requires mold release
machined aluminum	same as machined steel except more easily machined	same as machined steel, except for corrosion; easily damaged, because of softness of metal
cast aluminum	none over machined aluminum	same as machined aluminum; surface finish and tolerances usually not as good as machined aluminum; complex molds not as accurate as machined metal
sprayed metal[b]	none over machined metal; good surface possible	use usually limited to simple forms; not always easy to control mold quality; number of quality parts per mold limited; requires mold release
dip molded[b] (slush casting)	same as sprayed-metal molds	same as sprayed metal molds
cast epoxy	od dimensional control; surface can be polished; can be made for inserts and multiple-part molds; long life and low maintenance	dimensional control not quite as good as in machined-metal molds; requires mold release and cleaning; low thermal conductivity compared with metals
cast plastisols	parts easily removed from molds; molds are easy to make	short useful life; poor dimensional control
cast RTV silicone rubber	same as for plastisols; better life than plastisols	poor dimensional control, though better than plastisols
machined TFE fluorocarbon	no mold release required; convenient to make for short runs and simple shapes; withstands high-temperature cures	poor dimensional control
machined polyethylene and polypropylene	same as listed for TFE fluorocarbon except temperature capability and lower cost	poor dimensional control
molded polyethylene and polypropylene	same as listed for TFE fluorocarbon except temperature capability and lower cost	poor dimensional control

[a] Refs. 1–2.

[b] Although sprayed-metal molds and dip-molded molds are similar, differences in method of making these two types may give one an advantage over the other in specific cases.

Table 3. Considerations in Selection of Housing To Be Used in Potting[a]

Housing or container	Advantages	Disadvantages
steel	many standard sizes available; easily plated for solderability; good thermal conductivity; easily cleaned by vapor degreasing; good adhesive bond formed with most resins; easily painted; flame resistant	can corrode in salt spray and humidity; fitting of lids sometimes a problem; cut off of resin-filled can is sometimes difficult; possibility of electrical shorting
aluminum	same as for steel except plating ease; lightweight; corrosion resistant	same as for steel except aluminum is more corrosion-resistant; not easily soldered
molded thermosets (epoxy, alkyd, phenolic, diallyl phthalate, etc)	many standard sizes available; good insulator; corrosion resistant; color or identification can be molded in; terminals can sometimes be molded in; cut off of resin-filled shell easier than for metal cans; same type of material can be used for shell and filling resin, resulting in good compatibility	does not always adhere to resin, especially if silicone mold releases used to make shell; sealing of leakage joints can be difficult; physically weaker than steel, especially in thin sections; molding flash can cause fitting problems; cleaning of resin spillage can break shells
molded thermoplastics (nylon, polyethylene, polystyrene, etc)	same as listed for thermosets except last two items; often less prone to cracking than thermosetting shells, although this depends on resiliency of material	same as first three items listed in thermosets; adhesion can be poor due to excellent release characteristics of most thermoplastics; shell can distort from heat; cut off can be a problem, due to melting or softening of thermoplastics under mechanically generated heat

[a] Refs. 1–2.

comparisons of these materials apply to housings as to molds with respect to curing and thermal conductivity. Since the housing remains an integral part of the embedded unit, however, some additional considerations are in order. For instance, corrosion resistance in humidity and salt spray can be important, particularly if metal housings are used. Also, there may be standard metal containers available for certain types of products, such as transformers, which can reduce tooling costs. Thus, the choice between metals and plastics can be more important for potted units than for cast units, since the latter requires only a temporary mold. Comparisons of casting and potting techniques are shown in Table 4.

Impregnating. In the impregnating process, the liquid resin is forced into all the interstices of the component or assembly, after which the resin is cured or hardened. This can be an independent operation, or it can be used in conjunction with encapsulating, casting, or potting. Impregnation differs from encapsulation in that encapsulation produces only a coating, with little or no resin penetration into the assembly. Penetration is most important for certain electrical parts such as transformers.

Impregnation is sometimes accomplished by centrifugal means. The part is po-

Table 4. Comparison Considerations For Casting and Potting in Relation to Important Process and Product Parameters[a]

Parameter	Casting process	Potting process
skin thickness	difficult to control; components can become exposed in high-component-density packages	controlled minimum wall or skin thickness, due to thickness of shell or housing
surface appearance	cavities and surface blemishes often require reworking	established by surface appearance of shell or housing, though problems can arise if resin spillage not controlled
repairability	resin exposed for easy access	shell or housing must be removed and replaced
handling	handling and transfer of unhoused assembly can reduce yield	most handling of unembedded unit can be in housing
assembly	if molds are not well maintained, or it unit fits tightly into mold, handling can cause breakage of components	assembly is simplified since new shells or housings are always used, and wall thickness is controlled
manufacturing-cycle efficiency	production rate usually limited by quantity of molds	output not limited by tools
tool preparation and maintenance	relatively expensive	costs are minimal

[a] Refs. 1–2.

sitioned in a mold, the mold filled the resin, and the entire assembly spun at a high velocity.

If both encapsulation and impregnation are required, as in some transformer applications, the encapsulation dip coating is usually applied first. A hole is left in the coating so that the low-viscosity impregnating resin can be forced in after the shell created by encapsulation has hardened. This procedure provides a container, thus eliminating drain off of the impregnating material during its hardening or curing cycle.

Transfer Molding. Although most resin-embedded electronic packages were once produced by casting or potting techniques, the use of transfer molding is becoming widespread. Transfer molding offers advantages in economy and increased production rates for those assemblies that adapt to this technique and that are produced in large quantities. The advantages include a large reduction in the number of processing steps (over casting or potting), as shown in Figure 2, and a shorter curing time of the embedding compound. Transfer-molding materials cure in minutes; liquid casting and potting resins require hours.

Major limitations of the transfer-molding process are: (*1*) The assembly must be able to withstand pressures of 345–1724 kPa (50 to 250 psi). (*2*) The assembly must be able to take the curing temperature of 120–175°C. (*3*) Production volume must be large enough to justify equipment expenditures.

In transfer molding, a dry, solid molding compound, usually in powder or pellet form, is heated in a molding press to the point of becoming flowable or liquid, at which time it flows (is transferred) under pressure into a mold cavity containing the assembly to be embedded, as shown in Figure 3. The plastic remains in the heated mold for a short time until curing is completed.

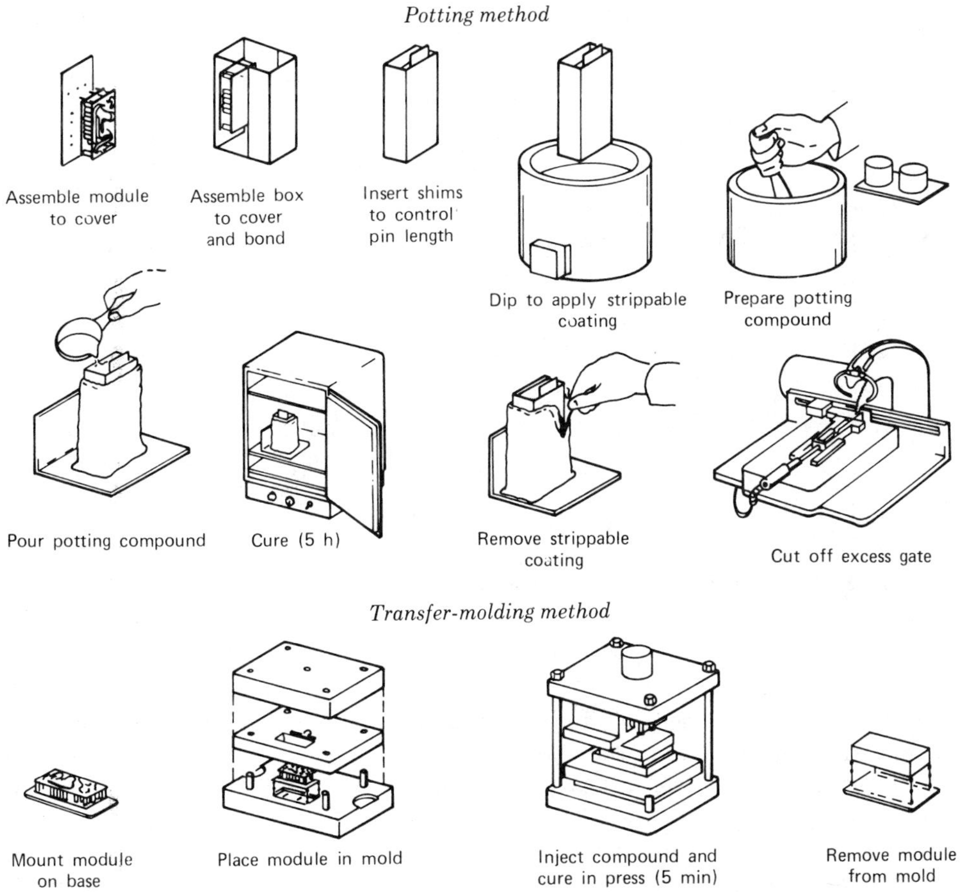

Figure 2. Comparison of processing steps for embedding electronic modules by potting and by transfer molding (1–2).

The transfer mold shown in Figure 3 embeds two similar parts, but a larger number of parts can be molded simultaneously. The fact that multicavity molds are common in transfer molding is one of the economies for large-volume runs. Many cavities can be filled as rapidly as a single cavity, thereby reducing the cost per part. Although mold cost increases as the number of cavities increases, it does not increase proportionately. Overall mold cost per part produced can be further reduced by incorporating cavities of different shapes into the same mold in proportion to the production volumes required, or by using mold inserts to vary the cavity configuration as required by changing production needs.

The small gate scar that remains on the finished part at the point where the molding compound goes out of the runners from the transfer cylinder into the cavity is usually unobjectionable.

Primary Embedding Materials. Although waxes and bitumens are still used occasionally, most materials used for embedding are plastics. Liquid or easily liquefied plastics are most commonly used. These plastic resins are most readily fabricated into complex structures, and they provide the proper physical characteristics required for embedding. As a class plastics are insulators. They therefore provide the necessary

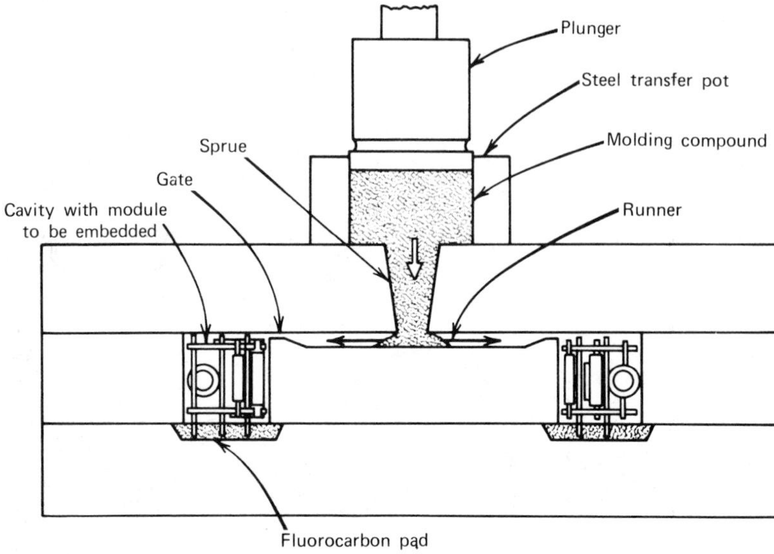

Figure 3. Typical transfer mold assembly showing flow of molding compound used for embedding (1–2).

insulation required in electronic and electrical applications. Useful plastics fall into two major classes, thermoplastics and thermosets.

Thermoplastics are characterized by their ability to flow upon the application of heat and substantial pressure. They become rigid again in a reversible manner upon subsequent cooling. Thermoplastics are polymers made up essentially of difunctional units. Simple chains formed of difunctional units can slip past each other to a limited extent upon heating or stressing; upon cooling or the relief of stress, the cohesive forces between molecules becomes predominant, and the plastic becomes rigid again. Typical examples of thermoplastics are polyethylene (see Olefin polymers) and polystyrene (see Styrene plastics).

Thermosets contain sizable portions of multifunctional units between which the cohesive forces are insufficient to prevent flow. These materials are liquids or solids at room temperature. Upon the addition of hardeners, ie, curing agents, and the application of heat cure, they harden into solids. Enough cross-links are formed so that flow is no longer possible. These changes are not reversible. Typical examples of thermosets are the phenolics (see Phenolic resins) and epoxies. Rubbery polymers are composed of chains that, because of the way the difunctional units are held together, tend to coil into helices that act as elastic springs, hence the name elastomers (see Elastomers, synthetic). In this case, the cross-linking reaction is commonly known as vulcanization. All degrees of rigidity and flexibility are available, since most of the thermosetting resins can be modified or obtained in flexibilities almost equal to those of elastomeric materials, and even softer than the low-melt thermoplastic materials.

As noted above, plastic resins most commonly used for embedding are the liquid resins and compounds, or those that can easily be liquefied by moderate heat or pressure, or both. There are many chemical types, each having different end properties.

Most liquid resins cure by heat or curing-agent influences, or both, give off heat during the curing process, and are thermosetting. Resin viscosity or fluidity, and the time-temperature curve (exotherm) for the exothermic reaction, which vary with each resin system, are key properties describing the nature of the individual resin through the curing cycle.

Viscosity. Flow properties of resins are important because of the need for flow and penetration at atmospheric or low pressures. If the viscosity is high, the formulation is difficult to pour and does not flow properly around inserts or components, thus allowing internal cavities to form. A high viscosity resin is usually too thick to allow evacuation of entrapped air, which also promotes cavity formations. High viscosity also makes mixing difficult. On the other hand, a resin whose viscosity is too low may leak through openings in the mold or container.

For most embedding applications, there is an optimum range of viscosity. For impregnating operations, extremely low viscosities, 100 mPa·s (= cP) or less, are desirable because complete impregnation of the parts under vacuum is required. In practice, however, impregnation is often achieved with viscosities considerably higher (up to 1000 mPa·s or higher). However, the higher the viscosity, the longer the cycling time or the higher the vacuum required for complete impregnation.

For embedment operations such as casting or potting, there is no lower limit on the viscosity, provided the mold or container is tight enough to prevent leakage. Usually, however, if impregnation is not required and if the components are not packed tightly, viscosities in the range of 1000–5000 mPa·s are satisfactory for casting and potting operations.

An encapsulation coating requires a thixotropic (nonflowing) material with an extremely high viscosity, because the part is dipped into the compound and the coated part cured without the use of a mold or container. The coating must not flow off during the curing operation.

Viscosity usually can be lowered by heating the resin, as shown in Figure 4, or by adding diluents, and it can be raised by adding fillers (qv). Not all resin viscosities show so strong a temperature dependence as that in Figure 4; silicones, for example, have a relatively flat viscosity curve (see Rheological measurements).

Exothermic Properties. Most polymeric resins used for embedment form by an exothermic reaction. It is essential that the exothermic properties of a particular system be known and that they be controlled. Too much heat may cause resin cracking during cure, or may adversely affect heat-sensitive components.

Three characteristics are commonly used for control measurements of these exothermic properties: gel time, peak temperature, and time-to-peak temperature. These quantities are measured from a single graphic plot of temperature versus time for a given resin-catalyst-curing agent system. A typical exothermic curve for a polyester resin is shown in Figure 5. Although the shapes of these curves vary widely from system to system, the curve for a given system should be closely reproducible. Exothermic curves also vary with mass of resin, as shown in Figure 6.

Gel time is the interval from the time the exothermic reaction reaches 65°C to the time when it is 5.5°C above the bath temperature (Fig. 5). The reason for starting the timing at 65°C rather than from the time at which the catalyst and resin are initially mixed is that it is not always practical to have the temperature of the ingredients precisely the same when the reaction starts, during and immediately after the initial mixing. The common base point is used to assure better reproducibility. Gelation

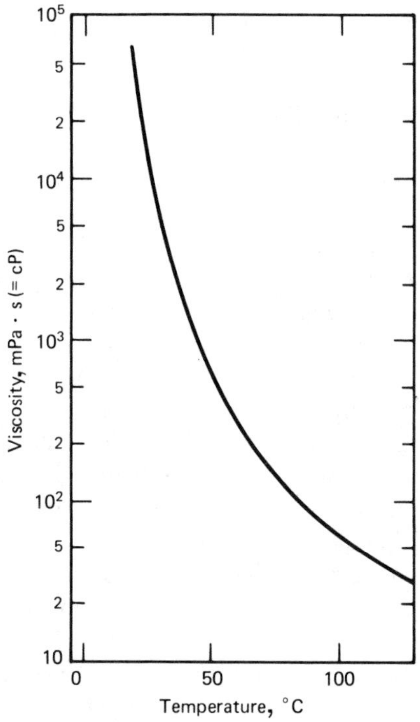

Figure 4. Viscosity–temperature relationship for a bisphenol epoxy resin (1–2).

usually occurs by the time the temperature has slightly exceeded the bath temperature, for this type of resin system. A practical indication of gel time is the time interval from first mixing to semisolid consistency.

The temperature rise is much greater after gelation has occurred than before gelation. This tendency is common for exothermic polymerizations. All changes in the resin, curing agent, or curing cycle vary this curve, which was nearly flat or very steep.

Transfer-Molding Resins. Although many plastic materials can be transfer-molded, most materials require transfer pressures too high to allow embedment of delicate electronic assemblies without damaging the components or distorting the position of the assembly in the mold. Development of plastic materials that can be transferred at low pressures has made possible the embedment of fragile assemblies by transfer molding. The most widely used materials are epoxies, although low-pressure molding materials have been developed from silicones, diallyl phthalates, phenolics, and alkyds (see Alkyd resins), the first two enjoying wide usage. Even though these materials are handled quite differently, their post-cure properties are similar to those of cured liquid resins of the same chemical type.

THERMOSETTING EMBEDDING RESINS AND FILLED COMPOUNDS

Many chemically distinct embedding resins are available, and there are many variations in each group. The most important of these are discussed in the following sections. Typical mechanical, physical, thermal, and electrical properties of several of these classes are shown in Tables 5 and 6.

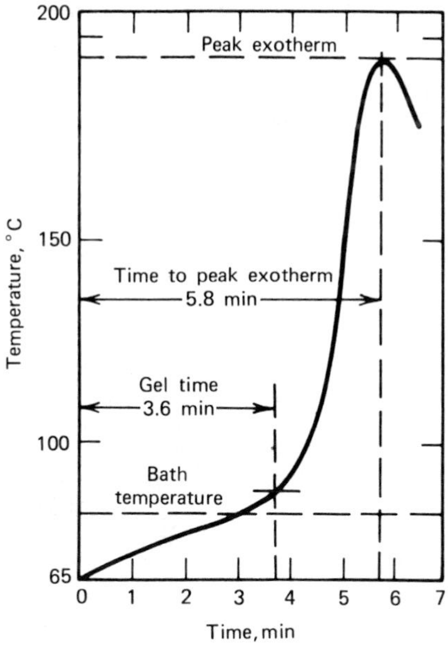

Figure 5. Exothermic reaction curve for a typical polyester resin, using benzoyl peroxide catalyst, and cured at 80°C (1–2).

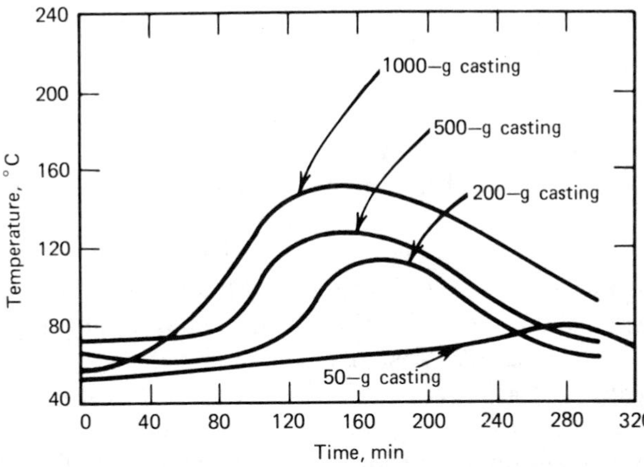

Figure 6. Exothermic curves, as a function of resin mass, for bisphenol epoxy and 5% piperidine curing agent (catalytic type) cured at 60°C (1–2).

Epoxies. The most used of the embedding resins are the epoxies, in many types and modifications (see Epoxy resins). All classes of epoxies have certain outstanding characteristics important in electronic assemblies. Chief among these properties are low shrinkage, excellent adhesion, excellent resistance to most environmental extremes, and ease of application for casting, potting, or encapsulation.

The original class of epoxies, the bisphenols, are the work-horses of the electronics industry. They are available as solids and as liquids over a wide viscosity range. The resins are syrupy; they are available in viscosities from 4000 mPa·s to semisolids at

Table 5. Typical Mechanical and Physical Properties of Several Common Embedding Resins[a]

Material	Tensile strength, kPa[b]	Elongation, %	Compr. strength,[b] kPa	Impact strength, Izod (J/cm of notch)[c]	Hardness	Linear shrinkage during cure, %	Water absorption, wt %
Epoxy							
rigid, unfilled	62,060	3	137,900	1.7220	Rockwell M 100	0.3	0.12
rigid, filled	68,950	2	172,400	1.3776	Rockwell M 110	0.1	0.07
flexible, unfilled	34,480	50	55,160	10.332	Shore D 50	0.9	0.38
flexible, filled	27,580	40	68,950	6.888	Shore D 65	0.6	0.32
Polyester							
rigid, unfilled	68,950	3	172,400	10.332	Rockwell M 100	2.2	0.35
flexible, unfilled	10,340	100		24.108	Shore A 90	3.0	1.5
silicone							
flexible, unfilled	3,447	175		no break	Shore A 40	0.4	0.12
urethane							
flexible, unfilled	3,447	300	137,900	no break	Shore A 70	2.0	0.65

[a] Refs. 1–2.
[b] To convert kPa to psi, multiply by 0.145.
[c] To convert J/cm to ft lb/in., divide by 3.44.

room temperature. Their viscosity can be reduced, of course, by heating. Other important classes of epoxies are cycloaliphatic diepoxides, novolac epoxies, and hydantoin epoxies.

Cured properties of bisphenol epoxies, as well as of other epoxies, are controlled by the type of curing agent used with the resin. Major types of curing agents are aliphatic amines, aromatic amines, catalytic curing agents, and acid anhydrides. These are discussed in Table 7 (see also Insulation, electric).

Although epoxy resins constitute the largest volume usage for embedding electronic packages, other resins are also important. These include silicones, urethanes, polyesters, thermosetting hydrocarbons, thermosetting acrylics, and polysulfides. Also, foams and low-density resins are widely used for electronic packaging for weight-reduction purposes.

Silicones. The silicone resins are convenient to use, they are available over a wide range of viscosities, and most of them can be cured either at room temperature or at low temperatures (see Silicon compounds, silicones). Silicones maintain their properties over a wide temperature range, generally from approximately −65 to 200°C and, in some cases, up to 300°C. Their excellent electrical properties, particularly low loss factors, and the fact that they generate little or no exothermic heat are additional advantages.

Three classes of silicones are used for embedding applications: RTV silicones and flexible resins; silicone gels; and rigid, solventless resins.

Table 6. Typical Thermal and Electrical Properties of Several Common Embedding Resins[a]

Material	Heat-distortion temp, °C	Thermal shock per MIL-I-16923	Coefficient of thermal expansion, ppm/°C	Thermal conductivity, W/(m·K)	Dissipation factor[b]	Dielectric constant[b]	Volume[c] resistivity, Ω-cm	Dielectric[d] strength, V/m	Arc resistance, s
epoxy									
rigid, unfilled	140	fails	55	16.7×10^{-2}	0.006	4.2	10^{15}	17.7×10^6	85
rigid, filled	140	marginal	30	62.8×10^{-2}	0.02	4.7	10^{15}	17.7×10^6	150
flexible, unfilled	<RT	passes	100	16.7×10^{-2}	0.03	3.9	10^{15}	13.8×10^6	120
flexible, filled	<RT	passes	70	50.2×10^{-2}	0.05	4.1	3×10^{15}	14.2×10^6	130
polyester									
rigid, unfilled	120	fails	75	16.7×10^{-2}	0.017	3.7	10^{14}	17.3×10^6	125
flexible, unfilled	<RT	passes	130	16.7×10^{-2}	0.10	6.0	5×10^{12}	12.8×10^6	135
silicone									
flexible, unfilled	<RT	passes	400	20.9×10^{-2}	0.001	4.0	2×10^{15}	21.7×10^6	120
urethane									
flexible, unfilled	<RT	passes	150	20.9×10^{-2}	0.016	5.2	2×10^{12}	15.8×10^6	15.8×10^6

[a] Refs. 1–2.
[b] Dissipation factor and dielectric constant are at 60 Hz and room temperature.
[c] Volume resistivity is at 500 V d-c.
[d] Dielectric strength is short time; to convert V/m to V/mil, multiply by 2.54×10^{-5}.

Table 7. Curing Agents For Epoxy Resins[a]

Curing agent type	Characteristics	Typical materials
aliphatic amines	Aliphatic amines allow curing epoxy resins at room temperature and thus are widely used. Resins cured with aliphatic amines, however, usually develop the highest temperatures during the curing reaction and therefore the mass of material that can be cured is limited. Epoxy resins cured with aliphatic amines have the greatest tendency toward degradation of electrical and physical properties at elevated temperatures.	diethylenetriamine (DETA), triethylenetetramine (TETA)
aromatic amines	Aromatic amine-cured epoxies usually have a longer working life than do aliphatic amine-cured epoxies. Aromatic amines usually require an elevated-temperature cure. Many of these curing agents are solid and must be melted into the epoxy, making them relatively difficult to use. The cured resin systems, however, can be used at temperatures considerably above those safe for aliphatic amine-cured resin systems.	*meta*-phenylenediamine (MPDA), methylenedianiline (MDA), diaminodiphenyl sulfone (DDS or DADS)
catalytic curing agents	Catalytic curing agents have a longer working life than aliphatic amine curing agents, and, like the aromatic amines, normally require curing at 95°C or above. In some cases, the exothermic reaction is critically affected by mass of resin mixture.	piperidine, boron trifluoride-ethylamine complex, benzyldimethylamine (BDMA)
acid anhydrides	The recent development of liquid acid anhydrides provides curing agents that are easy to work with, have lower toxicity than amines, and offer optimum high-temperature properties of the cured resin. These curing agents are becoming more and more widely used.	nadic methyl anhydride (NMA), dodecenylsuccinic anhydride (DDSA), hexahydrophthalic anhydride (HHPA), alkendic anhydride

[a] Refs. 1–2.

RTV Silicones and Flexible Resins. RTV silicones and flexible resins are by far the most widely used silicones. These flexible materials have excellent thermal-shock resistance and low internal curing stresses. Some can be cured at room temperatures and others at relatively low baking temperatures. Most are pigmented or colored, but several clear, flexible resins are available. These clear silicone materials are increasingly used because they have most of the good properties of the pigmented materials as well as optical clarity. The combination of flexibility and clarity facilitates cutting and repair when needed.

The cure of some clear, flexible resins is inhibited by contact with certain materials. Notable inhibiting materials are sulfur-vulcanized rubber and certain RTV silicone rubbers. This problem can usually be overcome with a coating of noninhibiting material on the component.

Silicone Gels. Silicone gels, as the name implies, exist in a gel state after being cured. Although these materials are very tough, they are usually used in a can or case. Silicone gels have the interesting advantage of allowing test probes to be inserted through the gel for electrical checking of circuits and components. After the test probes are withdrawn, the memory of the gel is sufficient to heal the portion that has been broken by the probes.

Rigid Solventless Silicones. Rigid solventless silicones are not used as widely as the other silicones because their resistance to thermal shock and cracking is less than that of the flexible materials, and because the rigid solventless resins are not as convenient to work with as the room-temperature curing materials. However, where the general properties of silicones are desired and rigidity is preferred to flexibility, the rigid solventless resins should be considered.

Polyesters. Polyester resins are widely used as embedding materials because of the wide range of flexibility and viscosity they offer (see Polyesters). These materials are popular because of their low cost and overall good electrical characteristics, especially low electrical losses. However, polyesters are inferior to epoxies with regard to adhesion, shrinkage, shock resistance, and humidity resistance.

Polyester encapsulating materials are formulated from: (*1*) a base resin, unsaturated polyester, (*2*) a monomer, and (*3*) a curing agent. A number of basic polyester resins are available. Depending on how they are prepared, these are generally referred to as general purpose, flexible, and fire-resistant resins, respectively. Some polyester resins incorporate bisphenol A or isophthalic anhydride in their structure for improved properties. Monomers are generally incorporated in polyester formulations to achieve low viscosity, low cost, and varied end properties. Properties of polyesters are affected not only by the type of monomer but also by its concentration in the formulation. Higher percentages of monomer generally cause greater shrinkage and more prevalent cracking. Curing systems do not greatly affect the end properties of the polyester, but they do affect storage life and processing conditions. Depending on the choice of catalyst and the reactivity of the polyester, cure cycles can vary from very rapid at room temperature to extended high temperature cures. The effect of fillers on polyesters is similar to that on epoxies, but more discretion must be used in their choice. Certain fillers are capable of prematurely gelling the polyester or of hindering its cure. Calcium carbonate is the most widely used filler for polyesters.

Polyurethanes. Cured polyurethanes are generally very tough and have outstanding abrasion resistance (see Urethane polymers). Formulations can be varied to produce soft elastomers or tough solids. Tear strength is high compared to that of other flexible materials. Polyurethanes offer excellent shock absorption because of their viscoelastic nature, and they produce very low internal stress. For these reasons they are often used on delicate electronic devices. Urethanes retain their good mechanical properties over a wide temperature range (cryogenic temperatures to 155°C for a short term, and 130°C continuous exposure). But they have limited life in high temperature, high humidity environments. Urethanes are also resistant to a wide range of solvents, offer excellent resistance to oxygen aging and have good electrical properties. Adhesion to most substrates is better than that of other encapsulation materials.

Cured polyurethanes are reaction products of an isocyanate, a polyol, generally polyether or polyester, and a curing agent. Polyester-based urethanes generally have better flexibility; but the use of polyethers offers better chemical resistance and hydrolytic stability. Curing agent selection influences curing characteristics as well as end properties. Diamines are the most common curing agents for polyurethanes.

Polysulfides. Liquid polysulfide resins are widely used in potting electrical connectors (qv). The cured polysulfide rubber is flexible and has excellent resistance to solvents, oxidation, ozone, and weathering (see Polymers containing sulfur, polysulfides). Gas permeability is low, and electrical insulation properties are good at temperatures between −53 and +150°C. At 25°C, cured polysulfide rubber has a volume resistivity of 10^9 Ω-cm and a 1-MHz dielectric constant of 7.5. Polysulfide rubber resins are the same chemical class as the polysulfide rubber resins used in modifying epoxy resins.

Polybutadienes. Polybutadienes are a recently developed group of thermosetting materials that have excellent electrical properties, high thermal stability, and outstanding resistance to water and aqueous solutions. Molding compounds are formulated from essentially all-hydrocarbon butadiene–styrene copolymers, polybutadienes, and blends of these two. They are high viscosity resins that are cured by peroxides, and exhibit high exotherms when formed. Disadvantages of these resins are brittleness and high shrinkage in curing. Currently, few manufacturers produce polybutadienes (see Elastomers, synthetic).

Low-Density Foams. Since most liquid resins can be made into low density foams by addition of selective foaming or blowing agents, most of the resins discussed previously are available in formulations that can be foamed in place. Epoxy and silicone foams are used in many embedding applications, but the polyurethanes are by far the most widely used foams. Urethane foams do not require blowing agents since gas is liberated during polymerization. These foams cure at room temperature or a low baking temperature and are relatively easy to work with, particularly the prepolymer foams.

Generally, foams have lower electrical losses, lower dielectric strength, lower thermal conductivity, and less mechanical strength than high density resins. Changes in these properties are usually almost directly proportional to foam density (see Foamed plastics).

Allylic Resins. Diallyl phthalate and diallyl isophthalate are the most widely used allylic resins. Although casting formulations are possible, transfer molding compounds comprise the largest embedding applications. These resins are noted for their excellent electrical properties and retention of those properties under environmental extremes. High insulation values are maintained in high humidity and up to 175°C for diallyl phthalate and 205°C for diallyl isophthalate (see Allyl monomers and polymers).

The allylic resins are generally filled to enhance their dimensional stability and mechanical properties. Diallyl phthalates offer very low after-shrinkage and good chemical resistance. Freedom from ionic and other corrosive impurities make them compatible with sensitive semiconductor elements. The electrical loss characteristics and dielectric strength do not vary as greatly with temperature as do those of epoxy and phenolic compounds. Allylic resins have high resistivity compared to other plastics.

Fillers. Fillers play a most important role in the use of resins for embedding. Fillers (qv) are additives, usually inert, capable of modifying nearly any basic resin properties in the direction desired. Fillers overcome many of the limitations of the basic resins. The proper use of fillers can produce major changes in properties such as thermal conductivity, coefficient of thermal expansion, shrinkage, thermal-shock resistance, density, reaction exotherms, viscosity, and cost. Although fillers can be used with all resins, they are used mainly with thermosetting resins, owing to the basic brittleness of unfilled thermosets, and the ease with which they can be mixed into liquid thermosets.

Costs and effects on resin properties of the more commonly used fillers are given in Table 8. Because of the large number of materials and suppliers available, this listing is not comprehensive.

Table 8. Costs and Effects on Resin Properties of Commonly Used Fillers[a]

		Property increase							Property decrease					
Type of filler[b]	Approx cost, ¢/kg	Thermal conductivity	Thermal-shock resistance	Impact resistance	Compressive strength	Arc resistance	Machinability	Electrical conductivity	Cost	Cracking	Exotherm	Coefficient of expansion	Density	Shrinkage
Bulk														
sand	2.204	X			X				*c*		X	X		X
silica	2.20–4.41	X			X				*c*		X	X		X
talc	2.20–8.82						*c*		X		X			X
clay	2.20–6.61		X		X		*c*		X	X	X	X		X
calcium carbonate	1.10–11.02						*c*		X		X			X
calcium sulfate (anhydrous)	4.41–8.82	X				X	*c*		X		X	X		X
Reinforcing														
mica	6.61–19.84		X	*c*					X	X				
asbestos	4.41–11.02		X	*c*						X				
wollastonite	4.41–6.61	X	X	*c*					X	X				
chopped glass	99–80		X	*c*						X				
wood flour	20.5		X	X			X	*c*	X				*c*	
sawdust			X	X			X	*c*	X				*c*	
Specialty														
quartz	4.41–11.02	X		X	X					X		*c*		X
aluminum	11.02–33.10	X	X		X						X	*c*		X
hydrate alumina	6.61–13.22					*c*								
Li-Al silicate		X			X					X	X	*c*		X
beryl		X			X				X	X	*c*		X	
graphite	13.22–66.12						X	*c*						
powder metals		X	X	X	X		X	X	X	X	X	X		X
low-density spheres	165.30–330.60												*c*	

[a] Refs. 1–2.

[b] Particle size of fillers are 74 μm (200-mesh) or finer, except for sand, hollow spheres, and reinforcing fillers that depend on particle configuration for the desired effect.

[c] Denotes most significant property of each filler listed.

Product Design With Thermosetting Embedding Materials. Table 9 matches some of the most important design objectives for embedded products with the best available embedding materials.

WAXES AND THERMOPLASTIC EMBEDDING MATERIALS

Owing to the low temperature stability of waxes and bitumens, these materials have only limited embedding application. The use of most thermoplastic resins is also restricted by the high molding pressures and temperatures required in their processing. The materials having some use are described briefly below.

Waxes. Wax is probably the original encapsulating and embedding material. It has excellent electrical properties and its use is extremely simple (see Waxes). Most waxes are fluid at 95°C for dipping or casting. Their low melting temperature limits their utility to approximately 52°C. Asphalts and tars have similar properties and are even less expensive. Cellulose esters (see Cellulose derivatives) are used in the same way as the waxes. Mechanically, they are stronger but their moisture absorption is high, and they are inferior to wax electrically.

Hydrocarbons. The hydrocarbon polymers, polyethylene, polystyrene, and polybutadiene, all have outstanding electrical properties, particularly in high-frequency applications where their low dielectric constants, 2.3–2.5, and outstanding low loss tangents, 0.0003–0.0005, make them the chosen materials. Polystyrene resins are not widely used for embedding applications because of certain practical limitations. They are generally useful only below 125°C; they have a long cure cycle and high shrinkage upon cure, causing a tendency to crack during the curing operation. Their normal expansion is high and curing is air-inhibited, causing the exposed surface to remain tacky.

Vinyls. Poly(vinyl chloride) and its copolymers are used in encapsulation techniques (see Vinyl polymers). Conformal coatings are possible using plastisols or organosols. Plastisol applications make use of a suspension of finely divided particles of polymeric liquid plasticizer. The component is dipped into the plastisol and the adhering layer is baked to form a solid impervious layer.

An organosol is a true solution of the plastic in solvent. It is usually applied by brushing or spraying. The solvent is subsequently removed in the baking operation. Organosol coatings must be much thinner than plastisol coatings owing to the danger of film rupture by solvent vaporization.

Decorative Embedding

The vast majority of embedding applications serve electrical or mechanical functional objectives, but decorative embedding also enjoys widespread use. Examples are primarily decorative items embedded in an unfilled resin having some degree of transparency, although nontransparent art objects are also cast using materials similar to those used in functional embedding. Also, scientific specimens are often embedded for cross-sectioning investigations.

Decorative embedding presents some unique limitations on selection of material and process. For use in decorative embedding other than nontransparent art objects, the transparency requirement for most decorative embedding poses several major limitations. Most embedding materials are not adequately transparent, especially in

Table 9. Selection of Embedding Materials For Important Product Design Objectives[a]

Design objective	Material candidates
adhesion of resin to assembly	Resins differ in their adhesive tendencies. Excellent bonding is usually obtained with the epoxies and urethanes. If adhesion of the base resin is a problem, primers can often be used to advantage. Cleanliness of parts is also important.
low electrical loss and/or low dielectric constant	Silicones and thermosetting hydrocarbons are outstanding among embedding resins in providing low electrical losses and low dielectric constants. Silicones are noted for their retention of good electrical properties even at elevated temperatures and high frequencies. Many other resins with low electrical values at room temperature improve as temperature is increased, particularly above 100°C.
thermal stability	Some of the most thermally stable resin systems are silicones, novolac epoxies, anhydride-cured epoxies, aromatic-amine-cured epoxies, and thermosetting acrylic resins. Most of the higher temperature resin formulations exhibit low weight loss in prolonged heat aging at temperatures to 200°C. There are many differences among these high-temperature resins, however, in retention of physical and electrical properties. Thus, it must be decided which of these high-temperature properties are most important. Thermal stability, both with respect to weight loss and retention of mechanical properties at elevated temperatures, can often be increased by addition of reinforcing fillers such as glass fibers.
cost	Polyesters are perhaps the best candidates where lowest material cost is required. Other possibilities however, are the low-density foams, which have a low cost per unit volume despite their higher basic resin cost, and epoxy resins heavily loaded with low-cost fillers.
room-temperature cure	Room-temperature curing formulations are available in silicone-rubbers and resins, urethanes, polyesters, epoxies, and polysulfides. Although room-temperature curing is mandatory in many applications, optimum resin properties are usually obtained by heat curing. In some instances, however, extended curing time at room temperature is equivalent to a heat cure.
low-temperature flexibility	Silicone rubbers and flexible silicone resins are the best materials for this requirement.
rigidity	Epoxy and polyester resins are prime candidates where rigidity is required. But some rigid formulations are either brittle or have cracking tendencies. Hence some compromise between hardness, toughness, brittleness, and crack resistance must be made.
flexibility	The most flexible materials are RTV silicone rubbers, urethane resins, and polysulfide rubbers. Flexible resins are produced from normally rigid resins by addition of various flexibilizers or modification of the base material, and resins such as epoxies and polyesters are available in flexible formulations. Resins are available in nearly any degree of flexibility or hardness desired.
clarity	Water-clear resins are available in the epoxy, polyester, and silicone materials. Clear epoxies are normally rigid; clear silicones are soft or flexible. Most other resins are not water clear but are amber or light colored so that parts can be seen through the cured, unfilled resins.
repairability	Repairability is easiest with the flexible materials because they can be cut easily. Especially repairable are the silicone gels. Rigid resins can be repaired by softening or dissolving in solvents or by heat softening.
low weight	Low-density foams, especially urethane foams, are most common, both in rigid and flexible formulations; many other resins, especially

Table 9. *(continued)*

Design objective	Material candidates
	epoxies and silicones, are also available in low-density foam formulations. Foams are normally available in densities from (0.9–9 kg/m^3). In addition to low-density foams, low-density resin systems (containing low-density fillers) are available in all embedding resins. Density of these resin systems is, of course, considerably higher than that of foams. However, physical and environmental properties are usually much better.
high thermal conductivity	Differences in thermal conductivities of base resins are slight. Hence high-thermal conductivity embedding materials are those incorporating large amounts of filler, especially large-particle fillers such as coarse sand, or aluminum oxide, magnesium oxide, or beryllium oxide. There is definitely a limit to the thermal conductivity obtainable in base resins, and thermal conductivity does not increase in proportion to the thermal conductivity of fillers used.

[a] Refs. 1–2.

thick sections. Most exhibit some light brown color. Second, even those that are transparent do not have a high transparency, so that cloudiness or a tinted appearance occurs in other than rather thin sections. Third, some that do have high transparency are used with monomers or solvents that can dissolve, attack, or discolor the object being embedded, especially biological or organic specimens. In this class are clear polyesters, which contain styrene or other monomers, and acrylics, which are used in liquid solvent carriers. A fourth basic limitation in decorative embedding is that, without close controls, the curing, baking, or polymerization process often involves heat which may darken either the embedding material or the object being embedded. It is usually desirable to use either a room temperature or low baking temperature embedding material, and an embedding material with lowest possible temperature rise during cure.

There are other problems in decorative embedding which demands use of embedding materials used at the lowest possible temperatures. These problems are (*1*) thermal expansion of the embedding material and embedded specimen, and (*2*) shrinkage of embedding material during hardening or cure. Both of these factors, especially in combination, readily result in cracking of the embedding material. The cracking problem is aggravated when high differentials of thermal expansion exist between embedding material and embedded specimen, and when the embedded specimen has sharp edges and corners, which result in stress point. Fillers are used in functional embedding to reduce this problem, but the transparency requirement prohibits use of fillers for decorative embedding.

Although critical limitations exist for decorative embedding, compatible combinations of embedding materials and embedded objects do exist. Through proper investigation and analysis, most requirements can be met. The problems, however, are very real, and must be both fully considered during investigation and fully controlled during manufacturing if quality products are to result.

Standards and Controls for Embedding

Embedding, like all industrial processes, requires a well-planned set of manufacturing standards. Frequently, the importance of the embedment process in the overall manufacturing procedure is overlooked because the material cost of the embedding resin may be insignificant relative to the cost of the electronic assembly. The importance of proper controls cannot be stressed too strongly from the standpoint of higher yield, lower cost, and increased efficiency, and from the standpoint of safety. Frequently, the potential hazard of handling embedding resins is not nearly so well recognized in an electronics plant as it would be in a chemical plant.

Quality control should cover both manufacturing and production control, including process specifications covering manufacturing instructions, in-process tests, safety procedures, well-organized record keeping, and any specialized manufacturing precautions and final product control, including tests designed to ensure that the product meets customer requirements and good monitoring and record keeping of final product properties to eliminate many processing and product problems before they arise. Several of the specifications and standard test methods most useful in assuring material and product control are presented in Table 10.

Finally, the key to reliable embedding techniques lies in material selection and maintaining the material quality by extensive testing. Applicable publications include MIL, FED, and ASTM specifications, as well as SPI (Society of the Plastics Industry), NEMA (National Electrical Manufacturers' Association), and U.S. Underwriters' Laboratories Standards. In most cases, material suppliers will provide a more exact definition of the quality control procedures for these materials and their applicable government and industrial specifications.

Table 10. Some Specifications and Standards for Embedding

U.S. Military Specifications
MIL-I-27,27A
MIL-I-6923D
MIL-I-17023C
MIL-T-5422, 5422B
MIL-STD-202, 202A
MIL-R-10509C
MIL-E-5272A
MIL-S-8516
U.S. Federal Specification
FED-L-P-406b
American Society For Testing and Materials Standard Methods
ASTM D 149, D 149-44, D 149-59
ASTM D 150-59T
ASTM D 257-57T
ASTM D 648
Institute of Electrical and Electronic Engineers Standards
IEEE No. 1
IEEE No. 50
National Association of Electrical Manufacturers' Classification on temperature classifications for electrical insulation

BIBLIOGRAPHY

"Embedding" in *ECT* 2nd ed., Vol. 8, pp. 102–116, by F. J. Modic and D. A. Barsness, General Electric Company.

1. C. A. Harper, *Handbook of Plastics and Elastomers,* McGraw-Hill Book Co., Inc., New York, 1975.
2. C. A. Harper, *Handbook of Materials and Processes for Electronics,* McGraw-Hill Book Co., Inc., New York, 1973.

General References

M. C. Volk, J. W. Lefforge, and R. Stetson, *Electrical Encapsulation,* Reinhold Publishing Corp., New York, 1962; see especially Chapt. 5, p. 54.

C. A. Harper, *Electronic Packaging with Resins,* McGraw-Hill Book Co., Inc., New York, 1961.

H. Lee and K. Neville, *Epoxy Resins,* McGraw-Hill Book Co., Inc., New York, 1957.

J. R. Lawrence, *Polyester Resins,* Reinhold Publishing Corp., New York, 1960.

M. B. Horn, *Acrylic Resins,* Reinhold Publishing Corp., New York, 1960.

L. E. Neilsen, *Mechanical Properties of Polymers,* Reinhold Publishing Corp., New York, 1962.

The Encyclopedia of Plastics Equipment, Reinhold Publishing Corp., New York, 1960.

Modern Plastics Encyclopedia, Modern Plastics, New York, annual publication.

J. Delmonte, *Plastics in Engineering,* 3rd ed., Penton Publishing Co., Cleveland, Ohio, 1949.

H. R. Simonds, *Source Book of the New Plastics,* Reinhold Publishing Corp., New York, annual publication.

C. A. Harper, *Plastics for Electronics,* Kiver Publications, Inc., Chicago, 1964.

C. G. Clark, "Potting, Embedment, and Encapsulation," *Space Aeronautics* (Dec. 1961).

J. Dexter, "Using Silicones to Meet Performance Demands in Electronic Equipment," *Mach. Des.* (May 24, 1962).

J. W. Hawkins, "Silicones—Coatings, Encapsulants, Potting, Embedding," *Electronic Design News* (July 1962).

F. L. Koved, "Encapsulating to Military Specifications," *Electron. Ind.* (July, 1963).

C. V. Lundberg, "A Guide to Potting and Encapsulating Materials," *Mater. Des. Eng.* (May, June 1960).

"Properties of Encapsulating Compounds," *Electron. Prod.* (April, 1963).

D. C. R. Miller, "High-Temperature Flexible Potting Resins Offer Unique Properties," *Electronics and Communications* (Oct. 1962).

J. M. Segarra, "A New Embedding Procedure for the Preservation of Pathological Specimens, Using Clear Silicone Potting Compounds," *Am. J. Clin. Pathol.* **40,** 655 (Dec. 1963).

H. L. Uglione, "Evaluation of Polyurethans, Polysulfides, and Epoxies, for Connector Potting and Molding Applications," *Insulation* (Apr. 1963).

"Potting and Encapsulation Technology Update," *Circuits Manufacturing* (Aug. 1977).

Journals

Electrical Design News (monthly); *Electronic Design* (weekly); *Electronics* (weekly); *Insulation* (monthly); *Materials in Design Engineering* (moithly); *Modern Packaging* (monthly); *Modern Plastics* (monthly); *Plastics World* (monthly); *Product Engineering* (weekly); *SPE Journal* (monthly); *Electronic Packaging and Production* (monthly); and *Circuits Manufacturing* (monthly).

CHARLES A. HARPER
Westinghouse Electric Corporation

ENAMELS, PORCELAIN OR VITREOUS

Historically, enamel has described decorative and protective glassy coatings on metal as well as glassy, decorative coatings on glass. Enamel has also implied certain organic coatings such as paints or lacquers; these are not discussed here (see Coatings, industrial; Paint). Glaze has most commonly referred to glassy coatings on ceramic bodies (see Ceramics).

In the United States, the term porcelain enamel designates the glassy coating on metal; however, in some other countries, vitreous enamel is the more common term for the same glassy coating. The ASTM defines porcelain enamel as a substantially vitreous or glassy inorganic coating bonded to metal by fusion at a temperature above 425°C (1).

Ceramic coatings, another term used for coatings on metal, connotes emphasis on the protective feature of the coating for the metal (see Refractory coatings). Ceramic coatings are often formulated and designed to contain mainly crystalline rather than glassy material (see also Colorants for ceramics; Glass).

Production processes for glass began about 2500–3000 BC; however, it is not clear when porcelain enamel originated. It is likely that metalsmiths explored the decorative technique of enameling through their use of colored glass inlays or by fabricating patterns of glass pieces on gold, silver, or copper articles. Porcelain-enameled metal art objects have been associated with early civilizations in the Middle East and in the Orient. It is believed that the techniques for the creation of decorative porcelain-enameled objects were passed from one Mediterranean group to another and these groups included Egyptians, Greeks, Romans, Spaniards, and Arabs. The Romans are thought to have introduced enameling to Great Britain. The art of porcelain enameling was traced from the fourth to the eleventh centuries in Byzantium (Istanbul) and was later introduced into Italy and parts of Western Europe (2).

Ancient and more recent enameled art objects evidence four enameling techniques: cloisonné, champlevé, basse-taille, and painting. *Cloisonné* enameling involves the preparation of a design of small wires or partitions (cloisons) which is soldered or fused onto the metal surface. Segments of the design are filled with different colorants of powdered glass and the entire ensemble is heated in order that the enamels are fused to the partitions and the metal substrate. After being ground and polished, this decorated surface displays a colored enameled design outlined by thin wires. *Champlevé* enameling is the technique of carving or gouging a design into the surface of a thick metal-base material such as copper, bronze, gold, or silver. The gouged areas of the design are filled with colored glass and the piece is fired. The polished, finished piece appears to consist of a design of enamel inlays in the metal surface. *Basse-taille* describes the technique of designing the metal surface in low relief and then of covering the entire surface with a transparent porcelain enamel. *Painting* involves enameling in layers. First the entire metal surface is covered with a dark enamel layer, then lighter-colored enamels are applied or painted on the first dark layer.

Ancient metal art objects were made from gold, silver, copper, or bronze. It was not until the early 1800s that porcelain enameling on cast iron was developed and practiced in Central Europe. As the sheet-steel manufacturing process developed about 1850, the industrial development of porcelain enameling of sheet steel naturally followed, first in Germany and Austria. About 1857 porcelain enameling was initiated in the United States by two manufacturers of porcelain-enameled kitchenware: the Grosjean Company of New York and the Vollrath Company of Sheboygan, Wisc.

Porcelain enameling protects against corrosion, decorates, and resists the attack of alkalies, acids, and other chemicals. This material is a nonporous sanitary coating imparting no odors or tastes. Since it is entirely inorganic, it does not serve as a feedstock for microorganisms. Because of its sanitary aspects, its protective and strengthening function, and its decorative character, porcelain enamel has been adopted as the most suitable material for bathtubs, laundry appliances, ranges, sinks, and refrigerator liners. The decorative and corrosion-resistant qualities of this coating have led to its many uses in architectural applications. Because of its extremely low porosity and resistance to chemical attack, porcelain enamel has found widespread use in the dairy, pharmaceutical, brewing, and chemical industries. Porcelain-enameled tanks and vessels made of heavy-gage sheet steel or cast iron are commonly used in the above industrial fields. Porcelain enamel on steel, cast iron, or aluminum is a most desirable composite system of materials. Sinks, stoves, ranges, refrigerators, clothes washers, dishwashers, and dryers represent major uses of these materials in the home-appliance industry. Cooking utensils, architectural panels, signs, silos, bathtubs, lavatories, brewing vessels, chemical storage tanks, gasoline service stations, roofing tile, guard rails, chalkboards, and many other products of commerce indicate the broad spectrum of home and industrial products finished with porcelain enamels.

Several common classifications of porcelain enamels are as follows:

Basis of classification	*Examples*
function	ground coats: single frit; multiple frit cover coat: directly applied to metal; applied to ground-coated metal
service	acid resistant; alkali resistant; hot water resistant; chemical resistant; abrasion resistant; electrically insulating; thermal-shock resistant; catalytic; pyrolytic
composition	alkali borosilicate; titania; lead-bearing; leadless enamels
metal coated	sheet steel, very low carbon (<0.0005% carbon), enameling iron (0.024% carbon or less), cold-rolled steel (0.06–0.10% carbon), hot-rolled steel (1010 or 1020 types; 0.10–0.20% carbon); cast iron; aluminum; copper; gold; silver; stainless steel
decorative character	clear; colored; white; stippled; matte; glossy; semimatte; beading
opacifying material	titania; zirconia; antimony oxide; molybdenum oxide
method of application	wet process: spraying, flow-coating, dipping, electrophoretic, electrostatic dry process: electrostatic; sifting
type of product	appliances; cooking utensils; sanitary ware; chemical equipment; jewelry; architectural panels; signs; hot water tanks; silos
firing temperature	540°C; low temperature, 595–760°C; normal, 790–870°C; high temperature, 870°C

The Enameling Process

The porcelain enameling process involves the re-fusing of powdered glass on the metal surface. The powdered glass is prepared by ball-milling a porcelain enamel glass engineered for specific properties. First, the glass is smelted from raw batch materials such as those listed for enamel glass compositions in Table 1. The enamel smelter is usually a box-shaped tank furnace. Continuous smelters, wherein the thoroughly mixed raw batch is fed in at one end and molten glass is flowing out at the other end, are common in commercial operations. Decomposition, gas evolution, and solution occur during smelting. After the molten glass has been smelted to a homogeneous liquid, it is poured in a thin stream into water or onto cooled metal rollers. This quenched glass, termed frit, is a friable material easily reduced to small particles by a ball-milling operation. Ball-milling the glass frit into small-sized particles can be carried out whether the frit is wet or dry (see Size reduction). Dry powders are used for dry-process cast-iron enameling and for electrostatic application on sheet steel (see Powder coating). Dry powders are also prepared and marketed for the subsequent preparation of slurries and slips used in the wet-process application techniques.

Process flow diagrams for sheet-steel and cast-iron enameling are shown in Figures 1 and 2.

The frit-making process involves all the technology from proper selection of raw materials through thorough mixing and smelting to uniform production of frit. Batch-type smelters, such as crucible furnaces, rotary smelters, or box-shaped smelters, are used for small production requirements or for research and development purposes (see Furnaces).

For conventional wet-process sheet-steel enameling, the porcelain enamel frit is ball-milled with clay, certain electrolytes, and water to form a stable suspension. This clay-supported slurry of small frit particles is called the slip and has a consistency similar to that of a thick coffee cream. The ingredients of the mill batch are carefully controlled. The amount and purity of all materials in the mill, including the clay and water, affect the rheological character of the slip as well as a number of the properties of the fired enamel such as acid resistance (increasing clay content causes lower acid resistance), reflectance, and gloss.

Within the past several years, the development of a one- or two-coat electrostatic application of dry frit particles has proceeded in various parts of the world. Dry ball-milled frit particles coated with a thin organic layer and agitated in a fluidized bed are delivered to spray guns that impart an electric charge to the sprayed particle. The sheet metal article to be coated is at ground potential. The silicone-based coating on the glass particle is applied during the grinding process. The glass particle without the coating has a resistivity of 10^{-6} Ω·cm; with the synthetic coating, the glass particle may achieve a resistivity as great as 10^{-18} Ω·cm. This greater resistivity is necessary so that the charged particle will adhere to the metal surface. An overall resistivity of 10^{-10} Ω·cm permits a coating as thick as 250 μm whereas the maximum thickness achieved with a particle having a 10^{-18} Ω·cm resistivity is about 25 μm. The coating thickness is self-limiting and thickness uniformity is excellent even on curved surfaces (±6 μm). This dry electrostatic process obviates the need for the drying and intermediate firing for two-coat articles and promises substantial savings in labor, material, and energy costs.

Table 1. Composition of Porcelain Enamel Frits, wt %[a]

	Sheet-steel ground coat	Titania cover enamel	Cast iron, high lead	Cast iron, low lead	Cast-iron ground coat	Cast-iron cover coat
Oxide composition						
KNaO	19.7	(14.0)	(9.4)	(18.9)	(11.4)	(20.8)
K_2O		3.5	2.3	4.6	7.1	4.4
Na_2O		10.5	7.1	14.3	4.3	16.4
B_2O_3	14.6	14.0	8.7	12.0	6.9	2.6
Al_2O_3	7.2		4.6	6.5	11.3	6.0
SiO_2	50.5	45.0	24.6	44.7	51.8	48.9
CaF_2	5.1					3.2
F_2		5.0	6.5	6.3		
CoO	0.6					
MnO	1.9					
NiO	0.2					
TiO_2		20.0				
P_2O_5		2.0				
PbO			33.6		18.4	
CaO			7.1	4.0		
ZnO			5.4			
BaO				4.0		
Sb_2O_5				3.8		8.7
AlF_3						5.0
NaF						7.5
Batch						
feldspar	30.3		17.3	22.4	35.0	22.6
borax (hydrous)	31.6		18.3	26.9	23.0	16.7
borax (anhydrous)		19.1				
quartz	20.0	42.0	10.6	22.3	15.0	26.0
soda ash	6.7			4.3		3.0
($NaNO_3$)	3.8	7.8	4.8	6.7	4.0	4.5
fluorspar	4.6		8.7	4.6		2.3
cobalt oxide	0.5					
manganese oxide	1.5					
nickel oxide	0.2					
monosodium phosphate		3.2				
zinc oxide			4.8			
red lead			30.7		23.0	
cryolite			2.9	5.5		10.7
boric acid			1.9			
titania		20.2				
Na_2SiF_6		1.2				
K_2SiF_6		7.8				
Sb_2O_5				3.1		
$BaCO_3$				4.2		
$NaSbO_3$						11.4
clay						2.4

[a] Ref. 2.

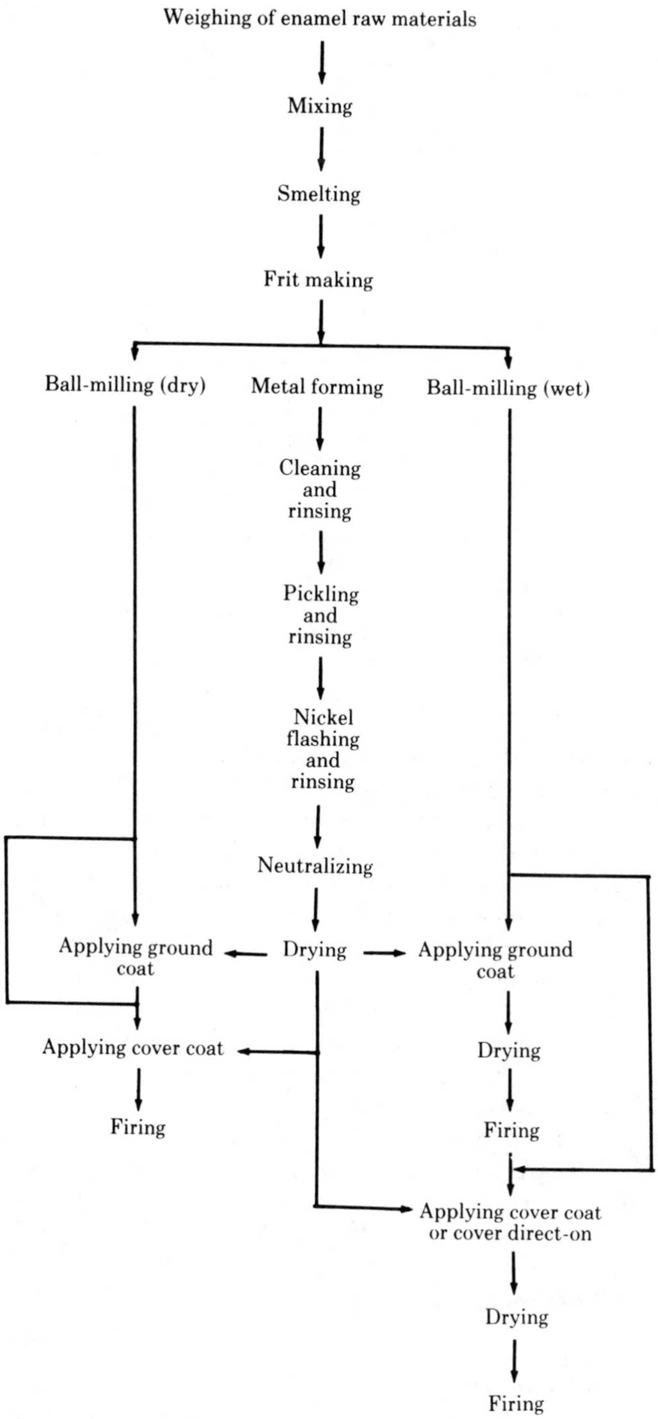

Figure 1. Process flow diagram of sheet-steel enameling.

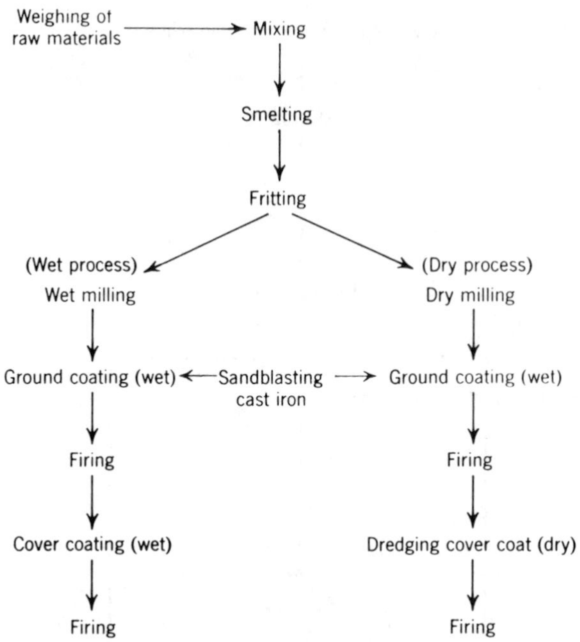

Figure 2. Flow diagram of cast-iron enameling.

The electrostatic powder processing of porcelain enamel is similar to the electrostatic technique used in preparing organic coatings (see Coating processes). Compared to the conventional wet-process porcelain enameling of sheet steel, the dry system eliminates the need for the clay suspending agents which represent refractory additions. As a consequence, the dry powders permit firing at a lower temperature as compared to that of the wet powders, thereby saving energy. At the lower firing temperatures, approaching the ferrite–austenite transition (ca 725°C), metal sagging and distortion are minimized, and additional materials savings can be achieved as higher carbon, nonpremium steels may then be used and less structural bracing of the steel article is required. The electrostatic dry process also has several additional advantages over the wet process. For example, much less waste of enamel occurs since the dry overspray is airborne and is recycled in a closed system. Productivity of the dry process is higher not only because of the single firing treatment, but also because fewer rejects are experienced. Compared to the wet-process, so-called direct-on cover-coat technique, which requires a decarburized steel and a special pickling treatment that promotes adherence, the dry electrostatic powder process with its two-coat, one-fire feature does not require use of premium steel and requires little or no pickling pretreatment. The problem of disposing of acid pickling wastes, which have a relatively high content of iron sulfates, has become less important in the absence of pickling pretreatment (see Metal surface treatments).

Prior to the recent development of the dry electrostatic process for sheet-steel enameling, the one-coat cover-coat direct-on process had been successfully developed and adapted by a large sector of industrial sheet-steel enamelers. This process requires the use of very low carbon steel (decarburized steel) so that carbon–oxide gas defects are eliminated. Also a special pickling process is required to assure good adherence of the enamel and this involves an accelerated etching of the metal using ferric sulfate additions in the sulfuric acid pickling operation, as well as subsequent heavy nickel flash deposits (see below). Cover enamels are applied by flow-coating, spraying, dip-draining, or electrophoresis.

The electrostatic spray application of enamel slip is a well-developed commercial process. The atomized enamel leaves the spray gun and passes through a high intensity d-c field of 100 kV. The negatively charged enamel particles are attracted to the surface of the metal article which is at ground potential. This method of coating, which is usually automated, produces a uniform coating thickness and good coverage around corners (wraparound).

Electrophoresis also is employed commercially and provides a dense uniform coating. In this process the sheet-metal article is positively charged and the negative electrodes are located in the slip. The metal article, which is immersed in the slip, attracts the charged frit particles which then pack into a tightly adhering coating layer. This process can be automated for uncomplicated shapes.

Ball-mill grinding is accomplished with porcelain or high-alumina balls 1.5–5.0 cm in diameter. Ball size, mill speed, mill charge, and ball charge are important parameters for the determination of the milling time required for optimum size distribution in prepared slip. Ground-coat enamels are ball-milled to a fineness of 95% of solids smaller than 74 μm (200 mesh). Cover coats are ground more finely, to 98% of solids finer than 44 μm (325 mesh).

Ground coats are applied to metal by spraying, dipping, or draining. Large articles may be coated only by spraying or flow-draining the slip, whereas small articles may be dipped into a tank of ground-coat slip. Cover coats may also be applied by similar methods as well as by electrostatic spraying or electrophoresis.

Dry-process cast-iron enameling involves the application of the cover coat by dusting or dredging the dry glass powder (using a long-handled vibrating sifter) onto the heated cast-iron article. The cast iron, which has been previously ground-coated, is removed from the furnace (870°C and higher) for the dusting operation and then is returned to the furnace. Generally, only two dusting and heating cycles are required for the development of a uniform coating.

Preparation of Metals. Sheet-metal parts are formed by the well-known processes of stamping, bending, and shearing. Many parts require welding, and it is important that this be carried out in a uniform, smooth manner so that the welded joint can be enameled without defects (see Welding). Cast-iron parts are formed by the usual cast-iron foundry methods; however, additional care is given to prevent contamination of the surface which causes defects in the enamel, particularly blisters or bubbles.

Enameling cannot be successful unless the metal is thoroughly cleaned and kept clean until the final coat is fired. Simply touching the surface by hand may cause defects. Cast-iron and thick steel parts are sandblasted without danger of excessive loss of metal and excessive warping. Sand, silicon carbide, and steel grit are satisfactory abrasives (qv). Products made from thin sheet materials are most satisfactorily and most economically cleaned by chemical methods which require alkali and soap solutions to remove grease and dirt and acid solutions to remove oxidized metal.

The commercial chemical treatment in the metal preparation process may be carried out with continuous cleaning and pickling equipment. The fabricated sheet-metal articles are supported on a special continuously moving rack which is first passed through the cleaning stage. The ware subsequently is subjected to pickling acids, nickel flashing, and neutralizing treatments. Metal articles also are cleaned and pickled in a batch process whereby baskets of metal articles are immersed in large tanks of various solutions and then water rinses.

The composition of the cleaning solution depends upon the type of oil, grease, and solid material to be removed, including the type and amount of drawing compounds used in the metal-forming operations. Vegetable oils may be saponified and removed by alkalies alone, but mineral oils must be removed with soap. A well-balanced cleaning compound contains an alkali, an alkali salt, which acts as a buffer material to maintain an approximately constant pH, and a soap (qv). Proprietary products that are especially adapted to the cleaning process are available and are usually used at a strength of ca 45 g/L (6 oz/gal) of water at boiling temperatures.

After being cleaned, the ware is immersed successively in one or more tanks of water at 80–95°C and then is transferred to the acid pickling solution. The pickling solution of 6–8% sulfuric acid is contained in lead-lined wooden or stainless steel tanks and is maintained at 60–65°C. The ware remains in the acid solution for as long as it is necessary in order to remove all oxide scale from the metal. Pickling inhibitors are rarely used as they have a tendency to cause enameling defects and prevent etching of the steel which is desirable. Rinse water which removes acid is maintained at a temperature of 80–90°C and rinsing time is usually 3 min. Plating galvanically or flashing a thin film of nickel on the iron after rinsing retards oxidation and enhances enamel bond. The ware is immersed in a solution containing 7.5 g/L (1 oz/gal) of nickel salts, such as nickel sulfate, $NiSO_4.6H_2O$, or the equivalent. The pH is maintained between 3.0 and 3.6 and the temperature at 75°C. The average time of immersion is 4–6 min and the deposit is usually 0.45–1.30 g/m^2 (0.04–0.12 g/ft^2). A wooden or stainless steel tank normally is employed as a container.

For the cover-coat direct-on process, a ferric sulfate etch is included in the metal pretreatment procedure. Hydrogen peroxide is added intermittently to a 1% ferric sulfate solution to reoxidize ferrous sulfate to ferric sulfate. This accelerated pickling procedure usually involves a sulfuric acid–ferric sulfate–sulfuric acid pickling sequence, and is designed to remove ca 20 g/m^2 (2 g/ft^2) of iron from the sheet-metal surface.

After being removed from the nickel bath, the ware is dipped into a hot or cold water rinse, quickly removed, and then transferred to a neutralizing bath where the last traces of acid are removed. Neutralizing with solutions of sodium carbonate and borax is common.

After being removed from the neutralizing solution, the ware is transferred to a dryer where it is maintained at a temperature of about 110–120°C and is provided with good air circulation, which ensures quick and complete drying without rusting of the metal. After being dried, the sheet-steel ware is ready for application of the enamel and after application of the coating, the ware is ready for the firing operation.

Firing. Firing may be carried out in intermittent box-type furnaces or continuous furnaces. The dryer and the furnace may form one continuous unit or separate units in the continuous firing process. In a continuous tunnel-type furnace, each coated item progresses through the furnace supported on firing racks especially designed to withstand long service at the repeated cycles of heating and cooling. Gradual heating and cooling of the ware are more characteristic of continuous firing than of intermittent firing. In the latter process, a batch of ware loaded on a large rack of special support tools is introduced into the box furnace.

Furnace temperature in the hot zone of continuous furnaces and box furnaces matures the coating in a matter of minutes. Ground-coats are fired in a box furnace for 3–6 min at 850–870°C. Cover-coat firing is generally carried out at shorter times and at slightly lower temperatures.

Enamel-firing temperature is related to the coating composition, metal thickness, and the type of metal used. Enamels for aluminum are fired at 510–540°C, whereas coatings for high temperature alloys (qv) may be fired at 930°C.

The industrial porcelain enamel process can be fully automated from the beginning of the metal pretreatment (pickling) procedure through the coating operations and the firing process. Flow-coating, electrostatic spraying, or electrophoretic deposition can be machine-programmed. Robot sprayers controlled by a computerized sensing mechanism can coat nonsimple shapes, and flow-coaters can also be programmed to uniformly coat some rather complicated shapes. Automated coating and firing of cast-iron bathtubs has also been carried out.

Energy Requirements. The relative energy intensiveness of preparing a porcelain enamel compared to two competitive materials, which were designed for a refrigerator liner, was studied by the Porcelain Enamel Institute (3). An acrylic enamel on steel, an acrylonitrile–butadiene–styrene plastic (ABS), and a direct-on white porcelain enamel on steel were compared, considering total energy needs for materials and processing. All three systems were within a 10% range about the arithmetic mean value, eg, 283.6 ± 1.42 MJ/m^2 (ca 25,000 ± 125 Btu/ft^2). Because plastic costs are directly related to oil and gas prices, whereas steel and glass costs are related to coal, it is projected that porcelain enamel will be much less energy cost-sensitive than its organic competitors. Furthermore, porcelain enamel manufacturers will probably widen the gap as the dry electrostatic powder process becomes the standard processing procedure for most mass-produced porcelain enameled sheet-steel articles.

Composition

Since porcelain enamels are substantially glassy coatings, the composition of this part of the enamel–metal materials system is based on glass-forming ingredients. The principal glass formers are B_2O_3, SiO_2, and P_2O_5 (see Glass). Other glass formers, such as GeO_2, BeF_2, and As_2O_3, are rarely used as the base for the glass because they are not economical and do not impart especially useful properties. Phosphate glasses, although low melting and commercially economical, are in general not sufficiently resistant to alkali or to hot water attack.

Table 1 lists the compositions in terms of wt % of the oxide components. Raw materials for the glass batch include minerals, such as the feldspars and quartz, since these are inexpensive sources of SiO_2 and Al_2O_3 (see Clays). The batch composition that is especially designed for cover coats is comprised primarily of manufactured chemicals of known, controlled levels of purity.

The composition of porcelain enamel glasses essentially is based on the alkali borosilicate glasses. Both B_2O_3 and SiO_2, the glass formers, are also called network formers, whereas the remainder of the ingredients are called network modifiers. It is considered that BO_3^{3-}, BO_4^{5-}, and SiO_4^{4-} structural units exist in the glass structure, and glass—as a rigid super-cooled liquid—has a short-range order of BO_3^{3-} triangles and BO_4^{5-} and SiO_4^{4-} tetrahedra, but long-range order of the units does not occur. These triangles and tetrahedra are joined at corners, ie, oxygen atoms at corners join two silicon or boron atoms and the continuous three-dimensional network of the glass structure is assured by this arrangement.

In the absence of network-modifier atoms that do not contribute to the continuity of the three-dimensional network of SiO_2 or B_2O_3 glass, these oxide glasses have a very high viscosity at their liquidus (melting temperatures). The viscosity at the liquidus temperature of glasses is in the range of 1–1000 Pa·s (10–10^4 P) whereas water nearing its freezing point has a viscosity of about 1 mPa·s (1 cP). This high viscosity attests to the strong interconnecting bonds of the network and to the high degree of association in the liquid glass. Modifier atoms, such as the alkalies, alkaline earths, or halides, cause the number of interconnecting bonds to decrease. Broken bonds resulting from a sodium atom bonding to an oxygen atom (Na—O—Si—O—Na instead of —Si—O—Si—O—Si—) or a fluoride atom bonded to a silicon atom (F—Si—O—Si—F instead of —O—Si—O—Si—O—) are associated with lower viscosity and lower firing temperatures of these modified glasses. Lead oxide, an ingredient in glass, plays the roles of both network former and network modifier. Lead oxide-bearing glasses have been widely used for cast-iron enamels.

Some generalities concerning the effect of specific ingredients of the porcelain enamel can be expressed as follows: an increase in SiO_2 content increases firing temperature and acid resistance, and lowers the expansion coefficient; an increase in alkali content decreases firing temperature and acid resistance, and raises the expansion coefficient; and an increase in the Al_2O_3 and ZrO_2 content increases alkali resistance.

Each porcelain enamel composition is designed to obtain specific performance characteristics of the enamel such as good adherence to the substrate, thermal expansion fit to the metal, desired chemical properties, such as acid resistance, catalytic effectiveness, hot water resistance, or alkali resistance, and desired physical properties, such as abrasion resistance, thermal-shock resistance, high gloss, high reflectance, and desirable color.

Fundamental Considerations in Porcelain Enamels

Adherence of Porcelain Metals. Cobalt-bearing materials incorporated in the frit composition enhance the adherence of the enamel to sheet steel. The mechanism by which glass adheres to metal is still not completely clear and the role of cobalt in promoting adherence is a continuing fundamental research subject. Adherence and wetting are associated: under oxidizing conditions the enamel glass wets the metal, and the oxide on the metal surface tends to dissolve into the enamel layer. Under reducing or neutral conditions in enamel firing, the oxide originally on the metal surface becomes completely dissolved and the glass then fails to wet the metal surface. There are additives or ingredients of the frit other than cobalt which seem to aid adherence. Molybdenum compounds incorporated in the enamel are considered adherence promoters.

The mechanism of adherence can be considered to be of two main types. One is physical adherence, which involves the physical gripping of the glass by a metal surface that has been roughened mechanically prior to enameling or roughened during enameling by the corrosive attack of glass on the metal and by dendritic attachments to the metal formed during the enamel-firing treatment. The other adherence mechanism involves the chemical bonding of the metal, metal oxide, and enamel glass.

Adherence in porcelain enamel terminology generally refers to the amount of glass remaining in an impacted, fractured area of the porcelain enamel system. If after impact the enamel surface is fractured to such an extent that it is stripped clean of the coating, it is said to have poor adherence; excellent adherence requires a large amount of fractured glass remaining on the impacted surface. A standard test has been devised (ASTM C 313-59) (1) using an adherence test apparatus to measure the amount of glass remaining on the base metal.

With respect to sheet-steel enamels, ca 0.25% of cobalt oxide is used to promote adherence of the ground coat to the sheet steel. In the case of cast-iron enameling which assures proper oxidation of the metal, no special adherence additives are needed.

The amount of force required to remove the enamel glass from the metal is not usually measured nor is the strength of the glass–metal bond generally evaluated.

Thermal Fit and Residual Stresses. Thermal expansion measurements, as determined by the interferometer test method (ASTM C 539-66), provide expansion (percent)–temperature data (see Fig. 3). The interferometer consists of optically flat plates of fused silica separated by three fragments of the specimen material. These plates are arranged horizontally and nearly parallel in an electrical furnace so that monochromatic light is reflected from the bottom and top plates, and interference fringes appear in the eyepiece. As temperature is increased, the distance of separation of these fused-silica plates is increased and a movement of interference fringes occurs across the field of view. The softening point refers to the temperature at which the glass specimens can no longer support the load of the top interferometer plate. This is noted by the reversal in the direction of fringe movement.

The expansion coefficient for the metal is constant over the entire temperature range, whereas the coefficient of linear expansion of the enamel glass increases with temperature. However, the glass expansion coefficient reverses above the softening point (Fig. 3**a**).

A porcelain enamel glass becomes less viscous as temperature is increased as in firing. At these elevated temperatures, the enamel is relatively fluid and it conforms to the metal surface.

As the porcelain enamel coating is cooled from firing temperatures (750–800°C) to the softening point, the fluid glass does not retain stress. However, as cooling proceeds below the softening point, the coefficient of expansion (or contraction) of the coating exceeds that of the steel, and tensile stresses begin to develop in the coating,

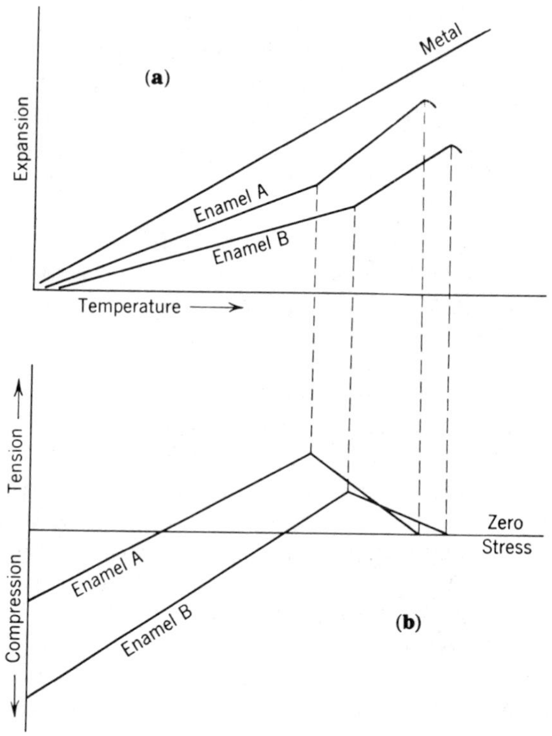

Figure 3. (**a**) Relative thermal expansion of porcelain enamel and sheet steel. (**b**) Stress development in enamel layer.

as shown in Figure 3**b**. On further cooling, the stress increases until the temperature at which the expansion coefficient of the glass equals that of the metal. With further cooling the coefficient of expansion of the glass becomes less than that of the metal, coating stresses decrease and compressive stresses develop.

Thermal expansion comparisons of coatings and metals have often been used to determine residual stresses in the coatings. The qualitative residual-stress analysis of Figure 3 shows how enamel A would develop high tensile stress and low residual compressive stress. Enamel B, with its higher softening point and lower expansion coefficient, develops less tensile stress and a much greater residual compressive stress.

Residual-stress analysis must take into account the cooling rates, the viscosity characteristics of the glass, the relative thickness of metal and coating, and the modulus of elasticity of coating and metal throughout the temperature range in which stress is developed. At high temperatures the glass behaves essentially as a viscous material, exhibiting viscous–elastic and then principally elastic properties as the temperature decreases.

Thermal expansion values can be calculated from measurements of thermal deflection of enamel–metal composites. Thermal expansion coefficient in the temperature range of 0–300°C can also be calculated using the additive formula:

$$P = AX_A + BX_B + CX_C + \text{etc}$$

where P is the property, such as linear expansion coefficient; A, B, C are the property factors for each ingredient in the composition; and X_A, X_B, X_C are ingredient compositions in wt %.

Factors for calculating the cubical thermal expansion coefficient of glasses have been determined and are listed in Table 2 (2). Glass compositions high in SiO_2 content have low coefficients of thermal expansion, whereas alkalies and alkaline earth materials raise the expansion coefficient (see Glass).

Residual compression in the coating is desirable since glass, as well as other ceramic materials, is much stronger in compression (about 2070 MPa or 300,000 psi) than in tension (about 69 MPa or 10,000 psi). If residual compression in the enamel layer is too great, the coating may fracture where the radius of curvature of the article is small. Since failure of the coating occurs because the tensile-stress limit has been exceeded, high residual compressive stresses in the coating increase the tensile load-bearing ability by the amount of the residual compressive stress that must be overcome before tension can be induced. If the tensile stress developed during the cooling or reheating of the enamel is too high, the coating may fracture. Crazing results when the damage occurs during cooling; a fine crack pattern called hairlining may be produced during reheating.

The linear expansion coefficient for common materials in the porcelain enamel system is as follows:

steel	11.7×10^{-6} cm/(cm·°C)
ground coat	$10–12.5 \times 10^{-6}$
cover coat	$8–11 \times 10^{-6}$
aluminum	23.5×10^{-6}
cast iron	10.5×10^{-6}

Composite Modulus of Elasticity. The modulus of elasticity of the enamel glass–steel composite system has been shown to lie between the modulus of the glass and that of the metal (5). The composite modulus can be calculated by the following expression:

$$E_c = (E_m - E_e)Q^3 + E_e$$

Table 2. Coefficient of Expansion Factors[a]

Oxide	Factor	Oxide	Factor	Oxide	Factor	Oxide	Factor
SiO_2	0.8	CaO	5.0	SnO_2	2.0	Cr_2O_3	5.1
Al_2O_3	5.0	MgO	0.1	TiO_2	4.1	CoO	4.4
B_2O_3	0.1	BaO	3.0	ZrO_2	2.1	CuO	2.2
Na_2O	10.0	As_2O_5	2.0	Na_3AlF_6	7.4	Fe_2O_3	4.0
K_2O	8.5	P_2O_5	2.0	AlF_3	4.4	NiO	4.0
PbO	4.2	Sb_2O_3	3.6	CaF_2	2.5	MnO	2.2
ZnO	2.1			NaF	7.4		

[a] Summation of (factors × weight percentages) × 10^{-7} = cubical expansion coefficient in vol/(vol·°C) (4). Courtesy of Garrard Press (2).

where E_c = modulus of elasticity of the composite; E_m = modulus of elasticity of the metal; E_e = modulus of elasticity of the enamel; and Q = thickness of the metal per total thickness of the composite.

Residual Compressive Stress. Residual compressive stress in commercial ground-coat enamels varies with enamel thickness as indicated below (6):

Ratio of enamel thickness to metal thickness	*Compressive stress, MPa (psi)*
0.8	69 (10,000)
0.6	110 (16,000)
0.4	138 (20,000)
0.2	221 (32,000)

Thinner coatings will, other factors remaining equal, yield higher compressive stresses. Higher residual compressive stress in the coating also can be obtained by using enamel glass with a lower expansion coefficient, or a metal with higher expansion coefficient and higher modulus of elasticity.

Maximum Strain. Strain in enamels that leads to failure is on the order of 0.002–0.003 cm/cm. Thinner enamels with their high residual compressive stresses are more flexible and can be strained to a greater degree. Some other physical properties of enamel glass are given in Table 3.

Appearance. Decorative porcelain enamels involve either a one-coat or multiple-coat system; in the latter, the dark cobalt-bearing ground coat is covered with the decorative second coat which affords the desired properties.

The most common cover or direct-on enamel has been white enamel. Whiteness or high diffuse reflectance is called opacity. The white, high-opacity enamel depends on crystalline opacifying agents, such as antimony oxide and zirconium oxide (before 1940) or titanium dioxide, which either remain well-dispersed in the glass during smelting and subsequent firing or are recrystallized from the enamel glass during the firing process. Opacifying pigments (qv) have an index of refraction much different from the 1.50–1.55 range of the glass matrix. The most effective opacifiers are given in Table 4 (7).

Commercial use of titanium dioxide as an opacifier for enamels did not begin until 1946 but recrystallizing white titania enamel is used today in practically all white sheet-steel enameling. Two polymorphic forms of titania, anatase and rutile, may be present in the enamel (see Titanium compounds, inorganic). Anatase is preferred since the anatase crystals are present in correct size range (0.1–0.2 μm) for maximum re-

Table 3. Some Physical Properties of Enamel Glass

Property	Value
density, g/mL	2.5–3.5
hardness (Mohs scale)	5–6
tensile strength, MPa[a]	34–103
compressive strength, MPa[a]	1380–2760
modulus of elasticity, GPa[b]	55–83
dielectric constant	5–10

[a] To convert MPa to psi, multiply by 145.
[b] To convert GPa to psi, multiply by 145,000.

Table 4. Most Effective Opacifiers in Porcelain Enamels

Opacifier	Index of refraction
NaF	1.336
CaF_2	1.434
Sb_2O_3	2.087–2.35
SnO_2	1.997–2.093
ZrO_2	2.13–2.20
TiO_2 (anatase)	2.493–2.554
TiO_2 (rutile)	2.616–2.903

flectance and, therefore, generate the most desirable bluish-white color. Smaller pigment particles provide too much light scattering, thereby giving a more undesirable bluish color; larger particles give a more cream-white color. In contrast to rutile, anatase crystals do not grow or change size with changes in firing temperature.

Spectrophotometer curves for bluish-white cover enamels opacified with oxides of antimony, zirconium, or titanium are shown in Figure 4. The titania enamel shows a characteristic absorption at the violet end of the spectrum.

Color stability of titania enamels can be obtained by adjusting the composition of the glass so that anatase recrystallizes predominantly over a wide temperature range. Figure 5 shows the effect of P_2O_5 content on the relative amounts of anatase and rutile recrystallizing from a titania enamel (7).

The color of cover-coat porcelain enamels can be produced in almost any hue, saturation, and brightness by using clear glass frits milled with colorant oxide pigments (see also Color). A typical mill batch for colored enamels is as follows:

Parts (wt)	*Component*
100	frit (clear alkali borosilicate-type glass)
4	clay (for producing stable suspension)
1/4	bentonite (for supporting suspension)
1/4	sodium nitrite (to disperse flocs)
3	oxide pigment colorant
45	water

Colored enamels may also be prepared by tinting the titania enamels during the smelting operation by the addition of colorant oxides.

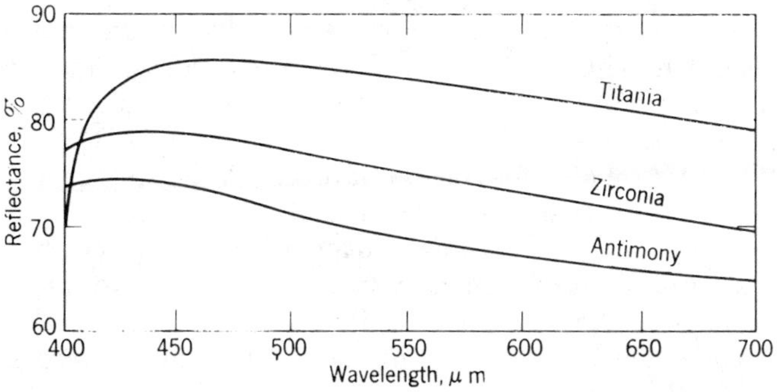

Figure 4. Spectrophotometer curves of titania, zirconia, and antimony enamels. Courtesy of Industrial and Engineering Chemistry (4).

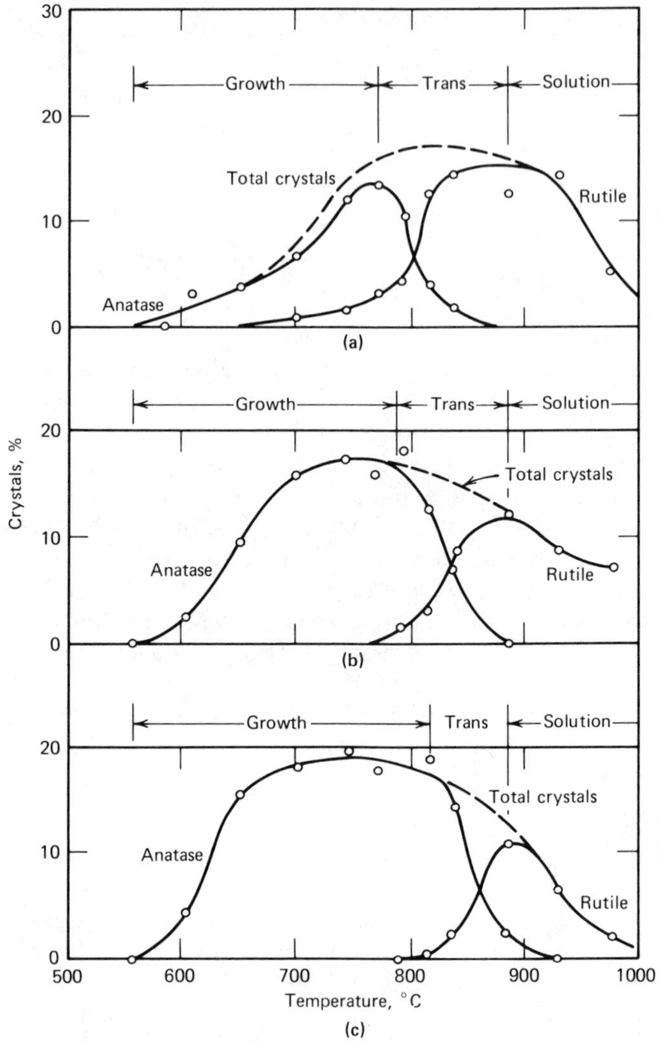

Figure 5. Crystallization of TiO_2 (**a**) in unstabilized enamel, (**b**) in phosphate-stabilized (2% P_2O_5) enamel, and (**c**) in phosphate-stabilized (4% P_2O_5) enamel. Courtesy of the University of Illinois.

Decorative one-coat finishes are also prepared by milling some cover-coat frit with ground-coat frit. This produces a fine speckled coating. Single-coat finishes of ground coat stippled with a white enamel prior to firing are also common.

Titania white enamels for sheet steel and antimony oxide-opacified enamels for cast iron exhibit reflectance values of 75% or higher. Although cast-iron ground coats are not colored, as are the sheet-steel ground coats which have dark cobalt-blue color, the thickness of cast-iron enamels is much greater than that of sheet-steel enamel coats. Dry-process cast iron requires generally two coating cycles to build a smooth and uniform coating and to achieve high reflectance.

Titania enamels have excellent hiding power. One thin coating of 0.05 or 0.08 mm completely masks the dark-colored ground coat and produce a white glossy coating with diffuse reflectance (green filter) of 80% or higher.

Decorating. In the process of commercial porcelain enameling, the decorating and painting steps involve coatings of various colors and textures. Common techniques for decorating include the following: stencil method of spraying an enamel on a stencil-covered surface; brushing the dried, unfired coating using a stencil or similar design; silk screen process (see Reprography); stamping a design using a rubber stamp with ceramic colors as inks, or dusting of a color pigment onto the area stamped with a gum or varnish; decalcomania using ceramic decals; graining, marbleizing, and other transfer processes whereby the designs impressed on an inked rubber cylinder from one flat designed plate are transferred by rolling the cylinder on the enameled surface (cylinders with the designed patterns are also used, after first inking with ceramic colors on a flat plate, then rolling onto the surface to be decorated); stippling or splattering droplets of the slip of a different colored enamel; and mottling to produce granite ware by adding cobalt or nickel sulfates to the ground-coat slip. These additions cause selective rusting and a mottled pattern in the fired finish.

Microstructure and Thickness. Sheet-steel enamels consisting of a ground coat and cover enamels have many bubbles of gas entrapped in the glass layer. Figure 6 is a cross-section photomicrograph of the enamel layer on sheet steel. The grains of ferrite in the cold-rolled steel can be identified in this acid-etched cross section. The ground coat contains more bubbles than the cover coat for three reasons: (*1*) the dissolved moisture in the ground-coat glass reacts with the metal at the firing temperature to produce hydrogen gas (some of this gas dissolves in the steel and some may produce bubbles in the molten glass); (*2*) carbon from the steel forms carbon oxide gases; and (*3*) there are more gas-producing, decomposable materials in the ground-coat mill batch such as carbonates and organic matter in the clay.

Total thickness of sheet-steel enamels consisting of a ground coat and a titania cover enamel is 0.12–0.20 mm. Cast-iron enamel coatings are much thicker, 0.50 mm or greater for dry-process coatings.

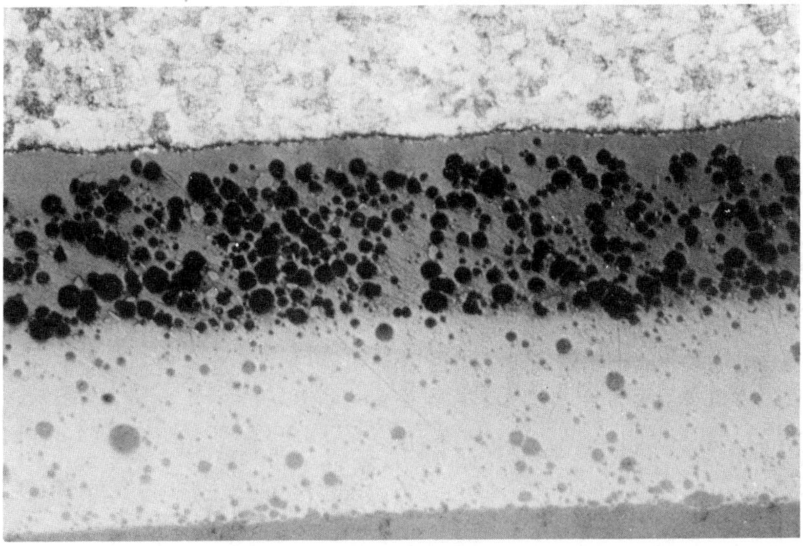

Figure 6. Cross-section photomicrograph of porcelain-enameled sheet steel. Titania cover coat over cobalt ground coat (150×).

One-coat titania enamels directly on steel are successful with certain special steels such as titanium-killed steels and extra-low-carbon steels. The thicknesses of such direct-on white enamels are 0.08–0.13 mm. Carbon or carbides at the surface of the metal form CO or CO_2 gases that produce a gas defect called primary boiling with ground-coat processing. For direct-white enameling, the carbon or carbides in these special steels must be eliminated or inactivated. As noted above, one-coat white enameling using wet or dry processes has been adopted by a substantial number of the industrial producers of sheet-steel porcelain enamels.

Enamel Testing

Abrasion Resistance. Porcelain enamel is the most scratch-resistant and hardest of commercial coatings (see Hardness). This property is used to distinguish between porcelain enamel and organic enamel or painted coatings. The rate of abrasive wear in surface abrasion increases with time, and the subsurface abrasion which follows exhibits a higher but constant rate of wear. Abrasion resistance can be evaluated by the loss of gloss or weight (ASTM C 448-64 Abrasion Resistance of Porcelain Enamels) (1).

Impact Resistance. Tests for impact resistance of porcelain enamels include falling weight tests such as a free-falling ball or a pendulum striking a rigidly held specimen. In such tests successively higher heights of fall are used until a visibly noticeable failure occurs. Factors affecting the impact resistance of porcelain enamels include the following: (*1*) lower modulus of elasticity of the metal, contributes to higher impact resistance; (*2*) the larger the radius of curvature of the article, the greater the impact resistance; (*3*) the thicker the metal and enamel, the greater the resistance to impact (because of the lower modulus of elasticity of the enamel, for equal thickness of the composite, the greater the ratio of enamel to metal, the greater the impact resistance); and (*4*) the physical structure of enamel, eg, excessive bubble content, large crystalline inclusions, and other discontinuities, often contribute to lower impact resistance. Weak bonding of the enamel to the metal also contributes to lower impact resistance.

The size of the fracture after impact failure has occurred is generally independent of the impact resistance. The size of the fracture is larger with a larger radius of curvature, a greater enamel thickness, and a poorer bond of enamel and metal.

Thermal Shock Resistance. Resistance of porcelain enamels to failure by thermal shock was developed for enamels for cooking utensils and other items subjected to high temperatures in service. Thermal shock is experienced by a heated enamel article on quenching with cold water. Thermal-shock tests involve repeated cycles of heating and quenching with water (heating for each successive cycle is carried out at progressively higher temperatures). A visible fracturing, as evidenced by spalling, constitutes a failure.

Water is an especially severe quenchant. Thermal-shock failures with water result from the water vapor entering the enamel layer through small, submicroscopic cracks formed at the instant of shock. The water vapor condenses in the crack and in the bubbles of the enamel near the cracks. On subsequent heating, the vapor from the entrapped water expands to cause spalling of the enamel layer. Other quenchant liquids, such as toluene, oils, and other organic liquids, also cause fine, almost invisible cracks but thermal-shock failures do not result with these quenchants on subsequent heating (ASTM C 385-58).

Thermal-shock resistance is a direct function of enamel thickness. The greater the residual compressive stress in the porcelain enamel, the greater is the resistance to thermal-shock failure. Thin coatings (with their greater residual compressive stress), such as one-coat enamels or the two-coat enamels with a low-expansion titania cover coat, provide excellent thermal-shock resistance.

Resistance to Chemical Attack. The resistance to alkali and acid attack is evaluated on the basis of loss in weight, loss in gloss, or cleanability of the surface.

A spot test with 10% citric acid is used at room temperature for acid resistance of glossy, light-colored enamels. In the spot test, loss of gloss and cleanability are determined in a qualitative manner (ASTM C 282-67). Resistance to boiling 6% citric acid is determined by the loss in weight (ASTM C 283-54). Lower alkali content of the glass yields higher acid resistance. Acid-resistant enamels for chemical service are compositions very high in SiO_2 and TiO_2 content. Alkali resistance is improved with increasing Al_2O_3 and ZrO_2 content. Resistance of enamels to water attack is also introduced in coatings for domestic water heaters (see also Coatings, resistant).

Other Properties and Tests. Physical and chemical properties of porcelain enamels can be evaluated by the following ASTM tests (1): C 282-67 Acid Resistance of Porcelain Enamels (Citric Acid Spot Test); C 313-59 Adherence of Porcelain Enamel and Ceramic Coatings to Sheet Metal; C 614-74 Alkali Resistance of Porcelain Enamels; C 282-67 Boiling Acid Resistance of Porcelain Enameled Utensils; C 538-67 Color Retention of Red, Orange, and Yellow Porcelain Enamels; C 743-73 Continuity of Porcelain Enamel Coatings; C 314-62 Flatness of Porcelain Enameled Panels; C 374-70 Fusion Flow of Porcelain Enamel Frits (Flow-Button Methods); C 346-59 Gloss of Ceramic Materials, 45-Deg Specular; C 540-67 Image Gloss of Porcelain Enamel Surfaces; C 539-66 Linear Thermal Expansion of Porcelain Enamel Frits by the Interferometric Method; C 632-69 Reboiling Tendency of Sheet Steel for Porcelain Enameling; C 347-57 Reflectivity and Coefficient of Scatter of White Porcelain Enamels; C 285-54 Sieve Analysis of Wet Milled and Dry Milled Porcelain Enamel; C 703-72 Spalling Resistance of Porcelain Enameled Aluminum; C 385-58 Thermal Shock Resistance of Porcelain-Enameled Utensils; C 409-60 Torsion Resistance of Laboratory Specimens of Porcelain Enameled Iron and Steel; and C 448-64 Abrasion Resistance of Porcelain Enamels.

Enamel Defects. Deviations from perfect continuity in porcelain enamel coatings and unusual departures from smoothness are described by special terms. For example, roughness is called orange peel; a severe case is called alligator hide. Blisters, pinholes, black specks, dimples, tool marks, and chipping are well-understood terms. Copperheads are defects of iron rust spots in the fired ground coat. Hairlines are defects of a strain pattern originating in the first part of the cover-coat firing and healed in the later stages of firing. Defects may result from accidents occurring at almost every stage of the enameling process. Porcelain enamel terminology is defined in "Standard Definitions of Terms Relating to Porcelain Enamel and Ceramic-Metal Systems," ASTM C 286-73 (1).

BIBLIOGRAPHY

"Enamels, Porcelain or Vitreous" in *ECT* 1st ed., Vol. 5, pp. 718–735, by R. M. King, The Ohio State University; "Enamels, Porcelain or Vitreous" in *ECT* 2nd ed., Vol. 8, pp. 155–173, by A. L. Friedberg, University of Illinois.

1. *1977 Book of ASTM Standards,* American Society for Testing and Materials, Philadelphia, Pa., 1977, Part 17.
2. A. I. Andrews, *Porcelain Enamels,* 2nd ed., Garrard Press, Champaign, Ill., 1961, p. 633.
3. *Porcelain Enamel and Energy,* Porcelain Enamel Institute, Inc., Arlington, Va., 1976.
4. P. Strong and S. Strong, *Ind. Eng. Chem.* **42**(2), 253 (1950).
5. P. S. Wolford and G. E. Selby, *J. Am. Ceram. Soc.* **29**(6), 162 (1946).
6. R. A. Jones and A. I. Andrews, *J. Am. Ceram. Soc.* **31**(10), 274 (1948).
7. R. D. Shannon and A. L. Friedberg, *University of Illinois Bull. Eng. Exp. Sta. Bull. No. 456* **57**(44), (Feb. 1960).

General References

A. E. Farr and G. Carini, "Evaluation of Dry Powder Porcelain Enamel Coatings," *Proc. Porcelain Enamel Institute Technical Forum* **37,** 45 (1975).

A. Lacchia, "Status and Trends of Electrostatic Enamel Powder-Coating Installations in Europe," *Proc. Porcelain·Enamel Institute Technical Forum* **38,** 78 (1976).

L. M. Dunning, "A Status Report on Porcelain Enamel Powder Coating Developments," *Proc. Porcelain Enamel Institute Technical Forum* **38,** 37 (1976).

D. R. Dickson and D. R. Larson, "Advances in Porcelain Enamel Powder Frit Materials," *Proc. Porcelain Enamel Institute Technical Forum* **39,** 102 (1977).

W. H. Steenland, "The PEI Energy Usage Survey," *Proc. Porcelain Enamel Institute Technical Forum* **39,** 79 (1977).

D. R. Sauder, "Use of Cold Rolled Steel in Enameling," *Proc. Porcelain Enamel Institute Technical Forum* **39,** 48 (1977).

A. L. FRIEDBERG
University of Illinois

ENGINEERING PLASTICS

Any discussion of engineering plastics falters at the outset on the matter of definition. The implication that there is a well-defined group of engineering plastics implies that there must also be a well-defined group of nonengineering plastics. Here the discussion breaks down, because there is virtually no plastic material that cannot, in some form, be termed an engineering plastic.

A practical definition must include not only property/performance criteria, but also market/pricing criteria that place certain resins in the engineering category to the exclusion of all others. Those resins that meet both sets of criteria are: nylon [*32131-17-2*], acetal [*9002-81-7*], thermoplastic polyester molding compounds, poly(phenylene oxide) [*25667-40-9*]-based resin (PPO), and polycarbonate [*25971-63-5*]. These five resin families have been the engineering plastics of the 1970s. The 1980s may see new members entering the group.

A definition based on property/performance criteria alone would include special grades and compounds of the commodity thermoplastics and a variety of polymer alloys and copolymers (qv). It would include all of the specialty thermoplastics that offer high strength along with high temperature performance, and it would include the thermoset resins that were the forerunners of the engineering thermoplastics.

The following discussion focuses on the five engineering resin families mentioned above, emphasizing their comparative advantages and how these have led to current usage patterns. To the extent that other resins compete against them, those resins are mentioned as well. Particular reference is made to acrylonitrile–butadiene–styrene (ABS), poly(phenylene sulfide), and polysulfone, which are recognized borderline engineering thermoplastics, as well as to phenolic molding compounds, and other thermosets, and to the fluoropolymers which offer outstanding but essentially different property advantages (see Acetal resins; Acrylonitrile polymers; Fluorine compounds, organic; Polyamides; Phenolic resins; Polyesters; Polycarbonates; Polyethers; Polymers containing sulfur).

In terms of properties, engineering plastics have a good balance of high tensile properties, stiffness, compressive and shear strength, as well as impact resistance, and they are easily moldable. Their high physical strength properties are reproducible and predictable, and they retain their physical and electrical properties over a wide range of environmental conditions (heat, cold, chemicals). They can resist mechanical stress for long periods of time. Flame retardance is not an essential requirement, but it has become an important added asset (see Flame retardants).

All of the individual physical properties of a resin can be quantified and compared through established testing procedures, but the balance of properties essential to a true engineering resin requires a broader view. A balance of properties exists when the achievement of one property does not demand a trade-off in another, ie, stiffness for low-temperature impact strength. Certain properties of commodity thermoplastics can be improved through the use of stabilizers, fibrous reinforcements, and particulate fillers (qv) to produce grades that compete directly with engineering plastics, but always with a corresponding reduction in other properties (1–2) (see Composite materials; Heat stabilizers; Uv absorbers).

The introduction of the term commodity leads to the market/pricing criteria that place the engineering plastics in a class of their own. Engineering plastics form a distinct group as compared to the high-volume/low-price commodity plastics and the low-volume/high-price specialty plastics (3).

The characteristics of the engineering plastics in these terms stand out in Table 1, which lists a sampling of thermoplastic resins along with average price, number of suppliers, and total consumption in 1978. The absolute figures have all changed since then, but the essential differences still hold.

The commodity plastics are characterized by average prices under \$1.10/kg, sales in the 500–3300 thousand metric ton range, and large numbers of suppliers. The engineering plastics are priced close together in the \$2.20–2.60/kg range on average, are sold in the 20–120 thousand metric ton range, and have few suppliers. The specialty plastics are priced at over \$4.40/kg, some of them well over, and annual consumption of all of them is roughly 10–15 thousand metric tons. Each of the specialty molding compounds has just one supplier.

These and other general characteristics of engineering plastics are cited in reference 4 as follows: predictable properties over a wide range of loadbearing conditions, patent protection on composition or process, or polymer and/or monomer, specialty chemical raw material bases, minimum number of competitors, high development costs and high cost of supplying markets, low volume as compared to commodities, high selling price, and high capital cost/output.

Nylon is an exception to some of these rules: its patents have expired, and its

Table 1. Thermoplastic Resins: Comparative Market Data[a]

Resin	Mid-1978 average list price, \$/kg	U.S. 1978 consumption, thousands of metric tons	Number of producers
Commodity			
low-density polyethylene	0.69	3248	13
high-density polyethylene	0.67	1893	13
polypropylene	0.66	1388	12
polystyrene	0.65	1759	15
poly(vinyl chloride)	0.60	2641	20
acrylonitrile–butadiene–styrene (ABS)	1.06	508	7
Engineering			
nylon	2.56	120	9[b]
acetal	2.20	45	2
thermoplastic polyester (TP) molding compounds	2.31	23	5
PPO-based resin[c]	2.51	64	1
polycarbonate	2.49	100	2
Specialty			
poly(phenylene sulfide), 40% glass-reinforced	4.52	2	1
polysulfone	6.50	5	1
poly(ether sulfone)	15.43	0.03	1
poly(amide–imide) molding compounds	22.92	0.2	1
aromatic polyester molding compounds	66.12	0.01	1

[a] Consumption figures for the commodity resins and nylon are from the SPI Committee on Resin Statistics as compiled by Ernst & Ernst, New York. All other consumption figures are Business Communications Co., Inc. estimates.

[b] Polymerizers only.

[c] PPO = poly(phenyl oxide).

economics are fiber-based. This is evidenced in Table 1, which shows that nylon has more suppliers than ABS, and still more if reprocessors of fiber scrap are counted in (see also Recycling). There are just five major suppliers of the virgin nylon-6,6 and -6 resins and compounds that account for most of the special grades of nylon used in engineering applications.

Properties

Since the 1950s when DuPont introduced nylon as the first engineering thermoplastic, it has been joined by four other resin families that have broadened the capabilities of plastics in replacement of metals, glass, and thermosetting resins. The relative capabilities of each of the major engineering thermoplastics are revealed by examination of key properties.

General-purpose nylon-6,6 has a heat deflection temperature (HDT) at 1.82 MPa (264 psi) of 104°C dry, as molded, but at equilibrium the HDT drops to 75°C (5). Its Thermal Index, as determined by Underwriters Laboratories (UL), is 75°C (6). Its flexural modulus is 2.8 GPa (410,000 psi) dry, but at equilibrium only 1.2 GPa (175,000 psi) for nylon-6,6, 965 MPa (140,000 psi) for nylon 6 (7). Notched Izod impact strength for nylon-6,6 is 1.12 J/cm (2.1 ft-lb/in.) of notch at equilibrium, but only 0.53 J/cm (1.0 ft-lb/in.) as molded (8). The Izod test, with its severe notch, is not necessarily the best method for comparing impact resistance, but it is frequently used in comparisons of engineering plastics.

Nylon's performance in a given application is dependent, therefore, on moisture content, and although impact resistance improves as equilibrium is reached, modulus drops sharply. Through modification, all of these properties have been improved, but these are the key properties on which engineering thermoplastics are based (9).

Acetals were introduced in 1960 (the DuPont homopolymer) and 1962 (the Celanese copolymer). Acetals brought somewhat higher heat deflection temperatures and higher U.L. Thermal Index values, but most important, their resistance to moisture pick-up made physical properties independent of moisture content. Flexural modulus of acetals as molded is the same (homopolymer) or slightly lower (copolymer) than nylon-6,6 as molded, but it is relatively constant after that. However, impact resistance does not improve with moisture pick-up as it does with nylon, but it is slightly higher to begin with.

Nylon and acetal, both crystalline resins, surpassed amorphous ABS in such properties as tensile strength, long-term heat resistance, and chemical resistance, but neither of them could approach ABS in notched impact resistance. In the 1960s, two amorphous resins were introduced that offered impact resistance equal to or better than ABS along with other important advantages.

With the introduction of polycarbonate in the early 1960s, a substantial gain was made in the U.L. Thermal Index rating. Polycarbonate can be used continuously at 115°C, 25 degrees higher than acetal, and 40 degrees higher than nylon. Polycarbonates also represented a major improvement in impact resistance, with a notched Izod value of 7.47 J/cm (14.0 ft-lb/in.) in 3.2 mm thickness. In addition, this resin was the first to offer transparency.

General Electric introduced its Noryl, a styrene-modified poly(phenylene oxide) resin, in 1966. Its heat deflection temperature, U.L. Thermal Index, and flexural modulus were not too different from acetal's, but with a notched Izod value of 2.67

J/cm (5.0 ft-lb/in.), impact resistance had been significantly improved. Noryl was between ABS and polycarbonate in price.

With the introduction of poly(butylene terephthalate) [*26062-94-2*] (PBT) thermoplastic polyester in 1970, new potential opened for thermoplastics in high-heat applications, but not among the unreinforced resins. Unreinforced PBT, as a crystalline resin, offers chemical resistance similar to nylon and acetal and certain common properties like lubricity, but without glass reinforcement its tensile strength and impact strength are relatively low, and its heat deflection temperature under 1.8 MPa (264 psi) load is very low (see Polyesters, thermoplastic).

An outstanding feature of PBT is the extent to which its heat resistance was improved by the addition of glass fibers (see Glass; Composite materials). With 30% glass reinforcement, PBT has a heat deflection temperature of 213°C, and a U.L. Thermal Index of 140°C. Glass-reinforced nylon has a higher HDT, but its U.L. rating is lower, and PBT has the important added advantages of lower moisture absorption and availability in flame-retardant grades without loss of properties.

All of these resins are now available in many different grades, and new grades that offer special price or property advantages are regularly introduced by their suppliers. Table 2 lists the key properties of the unfilled and glass-reinforced general purpose grades of these resins.

Table 2. Basic Property Comparison, Filled and Unfilled Engineering Resins[a]

	HDT, °C	U.L. Thermal Index, °C	Flex. modulus GPa[b]	Notched Izod, J/cm
Unfilled resins				
nylon-6,6 dry	104	75	2.8	0.53
nylon-6,6 (50% rh)	175	75	1.2	1.12
acetal copolymer	110	90	2.6	1.3
PPO-based resin[c]	129	90	2.5	2.67
polycarbonate	132	115	2.4	7.47
PBT[d]	54	120	2.3	0.53
Glass-reinforced (GR) resins				
nylon-6,6 dry 33% GR	249	105	8.9	1.07
nylon-6,6 (50% rh) 33% GR			6.2	
acetal copolymer 25% GR	163	95	7.6	0.85
PPO-based resin 20% GR	143	90	5.2	1.23
PPO-based resin 30% GR	149	90	7.6	1.23
polycarbonate 20% GR	146	120	5.5	1.34
PBT 30% GR	214	140	7.6	0.96

[a] Values are those for general purpose resins as they exist today, not necessarily as they existed at the time of their introduction. Data are from supplier literature, and are intended for rough comparison only.
[b] To convert GPa to psi, multiply by 145,000.
[c] PPO = poly(phenyl oxide).
[d] PBT = poly(butylene terephthalate).

As the table shows, the various resins are affected by glass reinforcement in some markedly different ways. Glass reinforcement raises flexural modulus for all of them. Heat deflection temperatures go up as well, as can be expected because glass fibers physically prevent deflection, but in the case of nylon and PBT, heat deflection temperatures soar. The 249°C HDT for nylon makes it eminently qualified for short-term high-heat exposure, and its long-term capability increases as well. In the case of PBT, glass reinforcement makes a true engineering resin out of a material that is not outstanding unfilled. Most important, its U.L. Thermal Index rises to 140°C, highest of all the resins in this group, which is particularly important with regard to its retention of electrical properties.

Glass reinforcement improves these properties in acetal compared to the neat resin, but in comparison to other glass-reinforced materials it is not outstanding. Glass reinforcement of PPO-based resin and polycarbonate results in decreased impact resistance, and in the case of polycarbonate, it also does away with transparency (see Laminated and reinforced plastics).

These brief comparisons do not accurately reflect the reasons for use of glass-reinforced resins because they ignore other significant properties (10). For example, in recent years, U.L.'s flammability ratings have become critical in the choice of one resin over another.

Along with the advantages offered by each resin, there are disadvantages. Overcoming the disadvantages is, of course, a major concern of resin suppliers. Ultimate goals for these resins would be: a moisture-resistant nylon without price penalty; a flame-retardant acetal; a PPO-based resin with better surface finish; polycarbonate without stress cracking; and an impact-resistant PBT without loss of modulus.

Processing

The melt-processability of thermoplastic resins is a characteristic that distinguishes them from thermosets. This pertains not only to the advantages of injection molding as compared to compression or transfer molding, but also to the variety of processing alternatives that extend the utility of the thermoplastics. There are thermosets that can be injection molded, but only the thermoplastics offer the options of extrusion into sheet, film and profiles, or blow molding (see Plastics technology).

The degree to which each of the engineering plastics is amenable to alternative processing methods varies, and the relative potential of each of them depends also on their potential in alternative processes, not just on their utility in injection molding.

In addition to its use in injection molding, nylon is extruded into monofilament and brush filament. Nylon-6 [*25038-54-4*] is used for sewing thread, fishing line, household/industrial brushes, and level-filament paint brushes. Nylon-6,6 [*9011-55-6*], stiffer than nylon-6, is used for sewing thread and household/industrial brushes. Nylon-6,12 [*24936-74-1*] dominates in personal-care brushes, and although the future may see some competition from poly(ethylene terephthalate) [*25038-59-9*] (PET), it is not evident as yet. In tapered-filament paint brushes, nylon-6,12's leading position has been taken over by PBT, a more expensive but more versatile filament.

Nylon is used as a wire coating, primarily as a protective abrasion-resistant coating over PVC-insulated wire (see Insulation, electric). Nylon film can be cast or blown, or extrusion-coated onto various substrates. Most nylon film is cast, and virtually all is sold to converters who add a sealant layer of low density polyethylene (LDPE), ethylene–vinyl acetate copolymer (EVA), or ionomer (see Film and sheeting materials). Its major market is vacuum packages of processed meats and cheese, usually combined with a PET sealant cover web. It is also used for fresh-meat packaging, and a new market has opened in medical-device packaging using techniques similar to those for formed-meat packaging. The most important properties in these applications are formability and heat resistance (see Packaging Materials).

Nylon strapping began replacing steel strapping in the early 1960s, even at higher cost, because of the general advantages of nonmetallic strapping. In recent years, nylon has met increasing competition in this market from polypropylene and PET.

Nylon is also extruded into rods, tubes, and shapes for machining, an important option for low-volume runs. In blow molding, nylon has been held back partly by cost, and partly because of the difficulties inherent in crystalline resins because of their sharp melting point. Nylon blow-molding resins have been developed with high melt strengths for parison forming, and the material is used to some extent for monolayer and coextruded bottles and for gas tanks in small equipment. Nylon-6 is also cast to produce very large bearings (see Bearing materials).

Nylon-11 [*25035-04-5*] is used for powder coatings and for flexible tubing. Nylon-12 [*24937-16-4*] is used for the same purposes, but to a greater extent in Europe than in the U.S. These resins have exceptional moisture resistance, but they are considerably less stiff than nylon-6 or -6,6. They are used to some extent in rotational molding.

In contrast to nylon, acetal offers few options outside of the injection molding category. An acetal terpolymer is available for injection blow molding, but apart from some carburetor floats, it has found little usage. Acetal is difficult to extrude, but it is extruded, as is nylon, into shapes for subsequent machining. Almost all acetal consumption is in injection molding, a factor that reduces its ultimate consumption.

As discussed below, the PET thermoplastic polyesters used for film and sheet and blow molding are not the same as those used for injection-molded engineering applications. PBT can be blow-molded, but rarely is. PBT is used almost entirely in injection molding, but there is some usage in tapered brush filaments and in extruded strip for small electrical parts.

Noryl's consumption in extrusion is relatively minor compared to injection molding, but it is used to some extent for stock shapes and to an increasing extent for sheet and profiles. Noryl sheet competes with flame-retardant ABS as it does in injection molding, and it can compete with less expensive resins like ABS and PVC where its properties permit the extrusion of thinner walls (11).

The transparency of polycarbonate, combined with its extrudability and impact resistance, made it a strong competitor for acrylic sheet in replacement of flat glass. Extruded sheet for glazing, lighting, and signs accounted for approximately 20% of polycarbonate's volume in 1978. Use in extruded profiles was minor, but polycarbonate has found a place in blow molding for 19-L (5-gal) water bottles, returnable milk bottles, baby nurser bottles, and miscellaneous packaging.

Economics and Marketing

Compared to the commodity plastics, investment costs per unit weight of engineering plastics are higher, capacity utilization rates are lower, and lead time before actual use is longer. One of the chief concerns for a producer of commodity plastics is whether the product can be made cheaply enough. The concerns of a producer of engineering plastics are whether the product can be made at all, and if so, whether it can be sold (3).

U.S. consumption of the five principal engineering plastics in 1978 is given in Table 3 by type of market.

Table 4 lists companies with their engineering resin products and shows the concentration of engineering resins among the various companies. The oil companies have only peripheral involvement with specialty plastics.

This article emphasizes the comparative aspects of engineering plastics in order to show where each has been successful. With this emphasis, a certain amount of perspective has been lost. There is a limit to the extent to which engineering plastics compete with one another in a given application because, by the end of the selection process, the property requirements usually match up with just one resin. Where more than one resin can do the job, price is the deciding factor.

The crystalline and amorphous resins rarely compete directly because of their basic property differences. Within each group, however, limited competition does exist. This article makes frequent reference, for example, to competition between nylon and acetal. In the context of total nylon consumption, however, there is no competition from acetal in markets that make up about 35% of nylon's volume. It is only in injection molding and some shapes for machining that competition might exist, but not in applications that call for low modulus, glass- and/or mineral reinforcement, special heat-stabilized high-impact grades, or special flammability characteristics.

Table 3. U.S. Consumption of Engineering Plastics, 1978 (Thousand Metric Tons)

	Nylon	Acetal	PBT	PPO-based resin	Polycarbonate
Extrusion					
monofilament, brush filament	9.5		1.8		
wire jacketing	5.0				
film, extrusion coating	10.0				2.3
strapping	2.7				
tubing, shapes	10.0	minor		minor	
sheet					22.7
Total extrusion	*37.2*		*1.8*		*25.0*
Injection molding					
automotive	39.0	13.6	9.0	10.5	10.0
electric/electronic	12.7	0.9	6.8	9.0	11.4
business machines				13.6	11.4
industrial/machinery	10.0	9.0	0.9		5.9
plumbing/hardware	6.4	9.5	1.4	2.7	1.4
appliances	5.0	5.5	1.8	22.7	18.2
consumer goods	5.0	6.4	0.9		3.6
miscellaneous	3.6	0.5	0.5	5.0	9.0
Total injection molding	*81.7*	*45.4*	*21.3*	*63.5*	*70.9*
Miscellaneous					
blow molding	<0.5	minor			4.1
stamped sheet	<0.5				
coatings	<0.5				
Total miscellaneous	*<1.5*				*4.1*
Total consumption	*120.0*	*45.4*	*23.1*	*63.5*	*100.0*

Table 4. Suppliers of Engineering Plastics

Company	Nylon	Acetal	TP[a]	PPO[b]	PC[c]	Other, related
DuPont	Zytel	Delrin	Rynite			fluoropolymers, polyimides, aramid
Celanese	[d]	Celcon	Celanex			
Monsanto	Vydyne					ABS[e]
Allied	Capron					fluoropolymers
American Hoechst	Fostalon					
General Electric			Valox	Noryl	Lexan	
Mobay Chemical			(new)		Merlon	PC[c]–ABS[e] alloys
GAF			Gafite			
Eastman			[d]			
Union Carbide						polysulfones, polyarylate amorphous nylon
Chevron	[d]					
Phillips						PPS[f]
Amoco						poly(amide–imide)
Gulf						polyimides
Arco						Dylark
Exxon						hi-temp film

[a] TP = thermoplastic polyester.
[b] PPO = poly(phenylene oxide)-based resin.
[c] PC = polycarbonate.
[d] There is no product with a trademark, but a resin is produced.
[e] ABS = acrylonitrile–butadiene–styrene terpolymer.
[f] PPS = poly(phenylene sulfide).

The applications that require only those properties that both nylon and acetal can offer represent only 15–20% of total nylon volume, and as long as acetal is more costly per unit volume, nylon will continue to be the resin of choice. In addition, as long as the only suppliers of acetal are also nylon suppliers, there is no reason to expect that the price balance will shift.

Lowering the price of acetal might be justified if it would open the door to nylon's $(110–140) \times 10^3$ metric ton annual volume, but it would only increase potential in sub-markets totaling about 20,000 t.

These considerations among other resins as well have determined the relative positions of the engineering plastics compared to the commodity and specialty plastics, and also the relative positions of the engineering plastics with respect to each other (3).

Engineering Plastics

Nylon. Nylons, also known as polyamides, carry numerical designations that refer to the number of carbon atoms in the amide links. Nylon-6,6 is the reaction product of hexamethylenediamine and adipic acid, each of which contains six carbon atoms. Nylon-6 results from the polymerization of caprolactam (see Polyamides). These two types of nylon account for over 90% of nylon resin consumption. Nylon-6,6 is the leader in the United States, nylon-6 leads in Europe. The relative market strength of the two nylons has more to do with historical development than with property differences, but differences do exist that are reflected in usage patterns.

Nylon-6,6 is stiffer, and predominates in all U.S. injection molding applications. Nylon-6 predominates in extrusion, and is somewhat more susceptible to moisture pick-up. However, the differences should not obscure the basic similarity between the two in comparison with other engineering plastics.

About 68% of nylon consumption in 1978 was in injection molding, and it is in these markets that nylon's capabilities as an engineering resin are most crucial.

In the auto industry, most of the nylon used is unfilled, and many of the applications are replacements for die-cast zinc. Glass- and/or mineral-filled nylons (12) play a major role as well. As in other markets, nylon competes with acetal in uses where wear resistance is required. Where strength and modulus are prime considerations, glass-reinforced nylon competes with polycarbonate. Mineral-nylon and mineral-PBT had been competing for exterior parts, but nylon has been most successful thus far. Both materials have exhibited product warpage problems. In some auto applications, polypropylene is a useful lower-cost alternative to nylon (13).

In industrial/machinery parts, nylon is widely used because of its natural lubricity, wear resistance, and chemical resistance. It is the oldest established resin for mechanical drive components such as bearings, gears, sprockets, pulleys, rollers, races, and chains. In these applications, acetal is nylon's closest competitor. Slightly higher in cost, acetal's resistance to moisture pick-up avoids problems with dimensional stability that require special design considerations with nylon.

The potential for plastics in friction-and-wear applications is continually being broadened by new technological developments in the field of tribology. Various fillers like MoS_2 and graphite improve the performance of a wide range of materials, but usually at a cost premium (14) (see Fillers). Nylon and acetal still dominate the field (see Bearing materials).

PBT has a better coefficient of friction against itself, and glass-reinforced PBT grades perform better than nylons and acetals after long hours at high temperatures (15) but, except in applications where flame retardance is required, unfilled PBT poses no major threats to the other crystalline materials in this market.

Also included in the industrial/machinery category are power tools and housings for equipment using gasoline engines. ABS is the favored resin for consumer electric hand tools, but glass- or mineral-filled nylons are favored for heavy-duty professional handles and housings. Because of their hydrocarbon resistance, nylon compounds perform well in lawn mowers, chain saws, and other gasoline-powered tools. Here again, PBT offers some competition, but mineral nylon has a good balance of properties and impact resistance of mineral grades is continually being improved.

In electric/electronic components, nylon is particularly important in wiring devices such as plugs and connector bodies, receptacles, and other parts, sometimes in replacement of phenolic molding compounds or rubbers (see Electrical connectors). The design freedom offered by thermoplastic nylon as opposed to thermosets has been a key factor, but the difficulty of flame-retarding nylon without loss of properties has been a hindrance in electrical applications, which are strictly tied to Underwriters Laboratories standards (see Flame retardants) (16).

Nylon drips when ignited, and because this removes the flame from the part, nylon's usage in a wide range of small-part electrical applications has been permitted. Glass reinforcement inhibits dripping, and GR parts have a greater tendency to burn with a smoky flame (17). A variety of flame retardants has been employed with nylons, but with limited success. Flame retardant compounds have had relatively low impact strength and poor heat stability in processing, which leads to the generation of toxic fumes, mold corrosion, and a variety of other problems. Except for acetal, which is not available in flame-retardant grades, nylon has the lowest percentage of sales (<5%) in flame-retardant grades among the engineering thermoplastics.

This is one reason why PBT is taking the lead in electric/electronic applications in general. In wiring devices, however, where nylon's properties (ie, elongation) are particularly valuable, its position is being improved by relaxation of U.L. requirements in some applications where the physical property advantages outweigh the lack of flame retardance. Volume losses to PBT in other areas, including electronic connectors, are likely to be regained only by suitable flame-retardant grades, and progress is being made in that direction (see also Insulation, electric).

In appliances, nylon plays a minor role compared to the amorphous resins. Moisture absorption and the lack of flame retardance are deterrents. Nylon use is also limited by its relatively low U.L. Thermal Index and low modulus. Most of nylon's usage in appliances is in moving parts and electrical components.

In the plumbing/hardware category, nylon's usage in plumbing is limited by its moisture pick-up, but in hardware it is used for miscellaneous parts including drapery slides and furniture supports. Lubricity is often a factor, and it competes with acetal and sometimes unfilled PBT. In consumer goods, nylon is used in a very diverse group of parts and products including spatulas, coffee urn spigots, garter grips, and brush backs.

DuPont is the major U.S. supplier of nylon by far, with its Zytel nylon-6,6 and a range of other nylons also under the Zytel name. The other major nylon-6,6 suppliers are Monsanto (Vydyne) and Celanese. Allied Chemical (Capron) and American Hoechst (Fostalon) are leading suppliers of nylon-6. Other nylon producers include Belding Chemical, Custom Resins, Inc., and Firestone Synthetic Fibers. Rilsan Corp. produces nylon-11 and 12. Several other companies supply nylons regenerated from fiber scrap, or monofilament, or special compounds based on resins from polymer producers.

Acetals. Applications for acetals are confined almost entirely to injection molding. They are difficult to extrude, but some stock shapes are made by extrusion and subsequent machining. Acetal homopolymers are made by the polymerization of formaldehyde; the high molecular weight polymers are stabilized by conversion of the end groups to esters. Acetal copolymers are made by the copolymerization of trioxane, a cyclic trimer of formaldehyde, with a small amount of comonomer.

As is the case with nylon-6,6 and 6, there are differences between the two types of acetal, but basic similarities exist between the two in terms of market position vs competing resins. The homopolymer is harder, more rigid, and has higher tensile and flexural strength. The copolymer is more stable in long-term high-temperature service, more resistant to hot water, resistant to strong bases, and has higher elongation.

The role of acetal among the engineering resins can best be described by departing from the standard list of markets (Table 3) which reveals little in the way of comparative data. Acetal is usually less expensive than nylon on a weight basis, but because its specific gravity (1.42) is higher than nylon's (1.13), acetal is more expensive on a ¢/cm^3 (¢/in.^{3}) basis. In applications where the two types of resin compete, therefore, nylon is chosen unless there is a need for special acetal properties, usually moisture resistance, as in gears. Dimensional stability is often critical in assuring a gap between the teeth for expansion due to heat build-up (18). Competition between nylon and acetal in bearings is very strong, but nylon has a slight edge. In all moving parts, acetal's outstanding fatigue resistance is a major asset. PBT has lower moisture absorption than either nylon or acetal below 65°C, but it also has lower unfilled strength.

In automotive applications, nylon and acetal compete in the same ways that they do in other markets, except that the property enhancement afforded to nylon through the addition of inorganic fillers and reinforcements gives it broader usage. In appliances, as with nylon, usage is mainly in moving parts. Acetal's use in electric/electronic applications is very minor. Because of its flammability, in particular, it is not found in wiring devices, connectors, and other U.L.-regulated parts.

Formaldehyde polymers pyrolyze to monomer when they burn. They burn without smoke, because there are no carbon-carbon chains in the structure, and they contribute little in the way of fuel to the fire because the carbon in polyformaldehyde is already partially oxidized, but they do burn. Acetals have the lowest oxygen index of all the plastic resins (see Flame retardants), and 15% oxygen in an O_2/N_2 mixture supports burning (17). Efforts to flame-retard acetals have not been successful owing to the nature of their pyrolysis, and lack of flame retardance has limited acetal's growth among the engineering resins.

Acetal is used in a wide variety of consumer products, including lighters, zippers, fishing reels, and writing pens, but its greatest strengths are in those consumer products that require chemical and water resistance, such as garden-chemical sprayers, household water softeners, paint-mixing paddles and canisters, and particularly plumbing applications where the copolymer exhibits superior resistance to continuous exposure to hot water. Usage of plastics in plumbing parts can require, depending on application, low coefficient of friction, abrasion resistance, strength under pressure, high-temperature chemical resistance, long-term property retention, fatigue resistance to retain burst strength, and other short- and long-term properties. Acetal offers a better balance of these properties than other engineering resins (19).

Celanese is the only U.S. supplier of acetal copolymer (Celcon), and DuPont the only supplier of acetal homopolymer (Delrin). The copolymer has about 55% of the acetal market.

Thermoplastic Polyester Molding Compounds. Thermoplastic polyester *molding compounds* are considered here rather than thermoplastic polyesters in general, because if all thermoplastic polyesters were included (see Table 1), total consumption would exceed that of any of the engineering resins and would be more in the neighborhood of the commodity resins. Poly(ethylene terephthalate) (PET) was used to produce over 136,000 metric tons of film in 1978, and consumption for beverage bottles is expected to be at least that high in the early 1980s (see Barrier polymers; Packaging materials).

This does not place the thermoplastic polyester engineering resins in the commodity category, however, because the resins used for the high-volume film and bottle markets are not the same as those used in engineering applications. The distinction was clear until 1978, because only more costly poly(butylene terephthalate) (PBT) was used for injection molding. Earlier attempts to mold PET had failed because of the slow crystallization rate of PET's shorter chain segments, but in 1978 DuPont introduced a new PET molding compound, and in 1979 Mobay entered the market with a family of thermoplastic polyester engineering resins, including two based on PET.

Post-1978 consumption data, therefore, includes both PBT and PET. This does not mean that engineering PETs are classed among the commodity resins, however, because PET resins for fibers, films, blow molding, and injection molding require separate production technologies. The thermoplastic polyester engineering resins used in 1978 were essentially PBT, and the designation PBT is used here because historical usage patterns have been developed with that resin.

Close to 70% of 1978 PBT volume went into automotive and electric/electronic applications, and much of the automotive usage was in electric/electronic parts. As noted earlier, the two characteristics that set PBT apart from the other crystalline resins are its high Thermal Index and its availability in U.L. 94 V-0 grades without a significant loss of properties.

Some of PBT's properties compared to nylon and acetal have been reviewed above. Generally speaking, PBT's relatively low impact resistance is a drawback, and the stiffness of glass-reinforced grades that makes it ideal for some applications is a detriment in others. PBT is not primarily a competitor for nylon and acetal, however. Its high temperature capability and V-0 rating at relatively low cost enable it to compete against phenolic molding compounds and other thermosets (20–21).

PBT has done well in the electric/electronic category because it brought temperature capability close enough to the thermoset range to allow conversion to more cost-effective thermoplastic processing. It has replaced thermosets in a wide range of electric/electronic components, although the heat resistance of PBT has sometimes been overrated. Although the Thermal Index of PBT is fairly close to the 150°C rating of phenolics, and the 215°C heat deflection temperature is much higher than most thermosets can offer, the actual performance of PBT at high temperatures is quite different. In one published test, thermoset bars retained their shape at 215°C but PBT bars sagged (22).

Phenolic molding compounds, sometimes called the oldest of the engineering resins, had become the standard insulating material in appliances, motors, wiring devices, circuit breakers, etc (23). The low-cost phenolics have been processed by compression and transfer molding in large volumes. In 1978 roughly one-third of the 136,000 t volume was used in appliances, and close to half went into industrial controls, circuit breakers, wiring devices, and other electric/electronic parts (see Phenolic resins).

Though inexpensive, phenolic resins are more costly in use because no effective method has yet been found for re-use of scrap (see Recycling). They are also deficient in impact resistance. More expensive thermosets such as alkyds and diallyl phthalate (DAP)-based products are used instead of phenolics for better arc resistance, higher surface resistivity, and better retention of electrical properties at high temperatures and in humid conditions, but the problems of scrap recovery and impact resistance remain (see Alkyd resins; Allyl monomers and polymers).

The impact resistance of PBT although not outstanding is better than that of the thermosets, and its combination of mechanical and electrical properties and long-term retention of these properties, combined with the advantages of thermoplastic processing, have made it a formidable contender for thermoset markets. PBT compounding has been directed at those markets, and the potential for the resin was enhanced with the development of a PBT with 180-s arc resistance and a U.L. 94 V-0 rating.

In addition to PBT various thermoset polyesters are competing as well in injection molding grades (24–26), and poly(phenylene sulfide) (see below) offers higher use temperature and the convenience of thermoplastic processing. Among the major engineering thermoplastics, however, PBT comes closest to the thermosets in balance of properties. Polycarbonate comes closest in dimensional stability, but PBT is the first of the major engineering thermoplastics to be considered for some of the most demanding thermoset applications.

Apart from electric/electronic usage, PBT is used in relatively small volumes in industrial machinery parts, where certain kinds of chemical resistance are required, or where glass-reinforced grades provide necessary heat resistance. It is used along with acetal in pump impellers and housings to an increasing extent. It is used in appliances to provide color where phenolics cannot be used, and for parts that need resistance to chemicals and fats. There is some usage in consumer products like zippers, where its self-coefficient of friction is better than nylon or acetal.

General Electric (Valox) is the leading supplier of thermoplastic polyester molding compounds, with Celanese (Celanex) a close second. GAF (Gafite) is a third supplier, and Eastman Chemical a fourth. Goodyear has offered PBT molding compounds, but its activities in this area are minor now. DuPont has entered the market in 1978 with its Rynite PET-based compounds, and Mobay announced entry in 1979.

Poly(phenylene Oxide)-Based Resin. The only PPO-based engineering thermoplastic commercially available in the United States is General Electric's Noryl, a blend of poly(phenylene oxide) and impact polystyrene (see Styrene plastics). The success of Noryl has been due to the development of a large family of formulations geared to specific markets. G.E. has patent protection on the basic resin and on the formulations introduced in the 1960s and 1970s. The major competitors for Noryl are high-heat grades of ABS and polycarbonate-ABS alloys.

The greater part of Noryl volume is in injection molding, along with some profile and sheet extrusion. An amorphous resin, it was introduced to fill the price/performance gap between ABS and polycarbonate. It competes most effectively against ABS where low-temperature impact strength, high-heat and moisture resistance and/or flame retardance are essential. Noryl has the lowest water absorption rate of the thermoplastic engineering resins, and it is completely resistant to hydrolysis. Also, flame retardance does not depend on the addition of halogens that would decrease impact strength.

The crystalline resins, in general, have relatively poor impact strength and good wear and chemical resistance. The amorphous resins, in general, have relatively good impact strength and relatively poor wear and chemical resistance. Because of these basic differences, the crystalline and amorphous resins seldom compete directly against each other.

Flame-retardant ABS applications have been a particular target for Noryl products because flame-retardant grades of ABS are very close to Noryl in price per unit weight, and when Noryl's lower specific gravity and shorter cycle times are considered, the latter material can often win out on cost. Metal substitution is also a target, as it is with all engineering resins, and this has been important in automotive applications and water-handling markets.

G.E. offers a range of standard Noryl non-flame-retardant grades with heat deflection temperatures from 113 to 150°C, and a range of flame retardant grades with HDTs from 88 to 150°C. Special grades are made for key markets like automotive and TV parts. Special-purpose formulations can be made by varying the relative content of PPO and polystyrene.

Flame retardance is imparted to Noryl by the incorporation of a proprietary flame retardant into the compounding procedure. Among the engineering resins, it has the highest percentage of volume in flame-retardant grades (about 50%). This feature has been important in appliances, business machines, and other electric/electronic applications. An important and growing market for Noryl is in structural-foam cabinetry for business machines, where it is the volume leader (27).

In the automotive market, Noryl is used for some exterior plated parts, but it has been a leading contender for interior parts that have to withstand greenhouse temperatures of 107–110°C. Appliances represent the largest single use category for Noryl, particularly if TV parts are included (Table 3).

Noryl is used for motor housings because of its heat resistance; for electrical enclosures because of its U.L. approval as sole support for current-carrying parts; for mixer housings, TV deflection yokes, and other parts where flame retardance is required; for dishwasher pump covers where hot-temperature moisture resistance is necessary; and for hydromassagers, steam curlers, and other units using moist heat. The greater percentage of Noryl used in appliances is in small kitchen or personal-care electrics, but it is used in some large appliances as well, particularly in control consoles. It is relatively important in environmental control appliances.

In the electric/electronic category, Noryl is used for outlet boxes, switch plates, connectors, compressor terminal covers, terminal blocks and strips, and other parts where competing materials might give problems with flash, warpage, inferior dimensional stability at high temperatures, or higher cost. In outlet and switch boxes, where demands on thermal and mechanical properties are low, commodity thermoplastics can sometimes do the job at lower cost. But for double-gang and ceiling boxes that have to meet the U.L. 514C standard, Noryl has the necessary strength and heat resistance. It is also used in lighting fixtures.

A major issue with respect to Noryl is whether it will be able to retain its markets in competition with newer grades of other styrenic resins. Where heat deflection temperatures of 105–110°C are required, along with flame retardance, almost no other materials compete with Noryl in impact resistance at the same price.

Where only the high heat-deflection temperature is required, there are several competitors including: high-heat ABS grades, a styrene copolymer (Arco's Dylark), an experimental styrene–maleic anhydride, and polycarbonate–ABS alloys.

For applications requiring flame retardance, but with HDTs below 93°C, ABS grades are available with additive flame retardants, or ABS–PVC alloys can be used. The alloys are ductile, tough materials, but they have disadvantages in processing stability because of the PVC content. The additive flame-retardant grades have better processing stability, but they are not as ductile and tend to be weak at low temperatures. Where both high heat-deflection temperatures and flame retardance are required, in a price range at the high end of the ABS scale, the closest competitor is a polycarbonate/ABS alloy with a V-0 rating (Mobay, Bayblend).

Polycarbonate. Polycarbonate is an amorphous polyester of carbonic acid, produced from dihydric or polyhydric phenols through a condensation reaction with a carbonate precursor. Bayer A.G., FRG, held the composition of matter patent that expired in 1979. G.E. produced the resin (Lexan) under a cross-licensing agreement with Bayer, and Mobay (a Bayer subsidiary) produces Merlon. Bayer and G.E. patents continue to protect certain processes and grades.

Because of its transparency and processability, polycarbonate is used in high-volume glass replacement applications not open to any of the other engineering thermoplastics. In window glazing and lighting it competes with acrylic resins and glass, in blow molding it is used for returnable milk bottles. In injection molding polycarbonate competes with the other engineering resins, and its transparency is often an asset.

Apart from transparency, however, polycarbonate has a unique balance of properties. It is relatively expensive compared to ABS, which can do the job in many applications that might otherwise go to polycarbonate. The cost-conscious auto industry used over 9000 t of polycarbonate in 1978, however, often in applications that required its outstanding dimensional stability.

Polycarbonate is very popular for small appliances, generally the first choice where heat or impact resistance, or both, is required. It has broad usage in newer types of hand-held appliances and in transparent sections of stationary appliances. It competes with Noryl and ABS in some uses like motor housings, and is often a replacement for die-cast zinc.

Polycarbonate is used in electric/electronic applications in a variety of uses that include covers for magnetic storage disks, switches, diode blocks, telephone dial rings and push buttons, and enclosures for control equipment. Its usage in connectors has been limited by its susceptibility to stress cracking.

In business machines, polycarbonate is used for structural-foam molded cabinetry, although Noryl can generally be used unless polycarbonate's superior strength or heat resistance is required. A special grade meets the requirements of U.L. Standard 748 for flammability in large complex assemblies like computer-room equipment.

In industrial machinery the greater part of polycarbonate's volume is in power tools, where it competes with ABS. The rest is in a variety of mechanical drive parts, protective equipment, etc. In power tools, polycarbonate has been meeting increasing competition from high-heat ABS grades.

Except in glazing, polycarbonate finds only relatively minor usage in construction. It is used in miscellaneous consumer and recreation products, as well as in institutional food service equipment where its transparency is an asset vs stainless steel. Where higher temperatures are encountered than polycarbonate can tolerate, polysulfone is used (see below). Polysulfone and TPX polymethylpentene (Mitsui Petrochemical, Japan) compete with polycarbonate in medical and laboratory equipment (see Hydrocarbon resins).

Polycarbonate must compete against many other resins in a variety of markets, including some not classified as engineering resins. Polycarbonate is irreplaceable where the full range of its properties are required, but other resins compete where only some of polycarbonate's advantages are called for. These include: acrylic resins for transparency and weather resistance; TPX for transparency, heat resistance, and broader chemical resistance; polysulfone for transparency, heat resistance, improved stress-crack resistance and mechanical properties (see Polymers containing sulfur); heat-resistant ABS for some opaque applications; mineral polypropylene for custom-formulated uses (see Olefin polymers); and DuPont's ST nylon for impact resistance combined with chemical resistance. Polyarylates (see below) began entering the market as new competitors in 1978.

Other Plastics Used in Engineering Applications. The five engineering plastics discussed in this article compete with other plastics in some markets, and to the extent that these other plastics do compete, they can also be called engineering plastics.

Competition from glass and/or mineral-filled polypropylene and flame-retardant ABS has been mentioned. Noryl also must compete against flame-retardant polystyrene in some electronic cabinetry where a relatively low Thermal Index is acceptable. Ultra-high-molecular weight polyethylene (UHMWPE) offers excellent wear resistance, but limited processability (see Olefin polymers).

Thermosets have long been available as insulators in electric/electronic applications, offering a wide range of capabilities in resistance to heat and environmental conditions. Thermoset molding compounds may be formulated to satisfy one or more important uses. Typical distinctive properties of thermosets include dimensional stability, low-to-zero creep, low water absorption, maximum physical strength, good electrical properties, high heat deflection temperatures, high heat resistance, minimal values of coefficient of thermal expansion, low heat transfer, and specific gravities in the 1.35–2.00 range (28–29).

Processing advantages have allowed thermoplastics to replace thermosets in many markets, but competition remains in some of the more demanding uses (30). Thermoplastics in general offer faster molding, lighter weight, possibility for thinner walls and more complex design, and greater impact resistance. Thermosets in general do not exhibit as much creep at elevated temperatures as thermoplastics, including the reinforced grades (see also Amino resins; Polyesters).

The engineering resins do not compete directly with epoxies, but some of the high-temperature resins are being used for that purpose (see Epoxy resins). The engineering resins rarely compete with polyurethanes or silicones (see Urethane polymers; Silicon compounds).

Fluoropolymers are often categorized along with the engineering plastics, but the two groups seldom compete. As a class, fluoropolymers do not offer the loadbearing capability of the engineering plastics, and loadbearing is generally one of the demands placed on plastics in engineering uses. In nonloadbearing uses, however, fluoropolymers have outstanding and unique properties, including resistance to very high and low temperatures, exceptional electrical properties, and low coefficient of friction (see Fluorine Compounds, organic).

Specialty Plastics

Specialty plastics include a mixed group of materials sold at relatively high prices compared to the engineering plastics and in relatively low volumes. The members of this group generally have high-temperature capability but this capability involves complex costly synthesis and usually difficulty in processing. In this group the polyimides (qv) can be used continuously at temperatures in the 260°C range.

There is no question that such materials can be used in engineering applications, but there are only two materials in this group that compete directly with the engineering plastics as defined here. These are poly(phenylene sulfide) (PPS) and polysulfone.

The relative heat resistance of plastic materials is generally measured either by ASTM D648, Deflection Temperature under Load, or U.L. 746B, Polymeric Materials—Long Term Property Evaluation (Relative Thermal Index). These two testing procedures have been referred to in connection with the main engineering plastics, but they become particularly relevant in attempting to define a high-temperature plastic. No comprehensive body of data exists to compare the mechanical, chemical, and electrical stress behavior of all materials under long-term elevated-temperature conditions (31–32).

The U.L. Thermal Index is a preferable indicator, if a choice must be made, because heat deflection temperature merely indicates the temperature at which a bar will first deflect. It provides no information about time to failure or mode of failure. Glass-reinforced nylon and polyester have very high heat-deflection temperatures, but since they are crystalline resins with sharp melting points there is an inherent risk of catastrophic failure.

The U.L. Thermal Index is a better indicator of continuous performance ability, and an acceptable dividing line between the major engineering thermoplastics and the high-temperature thermoplastics can be drawn at the U.L. 150°C Thermal Index. This is the generic rating for phenolic molding compounds, and most thermosets can be used continuously above that temperature. For thermosets, unlike thermoplastics, 150°C is usually taken for granted.

The significance of thermoplastics that can operate over 150°C does not lie in their role as replacements for thermosets in the low price range. With the exception of PPS they are too expensive. The can function in replacement of metals, glass, epoxies, fluoropolymers, and specialty thermosets in areas where thermoplastic processing advantages make the cost worthwhile (33).

Poly(phenylene sulfide) (Phillips Chemical Company, Ryton) is produced by the reaction of *p*-dichlorobenzene and sodium sulfide in a polar solvent. PPS, which is available for molding only in glass- and/or mineral-reinforced compounds, offers U.L. Thermal Index ratings of 240°C at a relatively low price. PPS also has outstanding chemical resistance, with no known solvents below 205°C. An additional asset is that PPS is inherently flame retardant. It is considered a thermosetting thermoplastic because optimal high-temperature properties can be obtained through annealing, in which some cross-linking takes place. PPS is also being used in high-temperature alloys (qv) (34).

Most PPS applications are in structural electric/electronic parts, where PPS competes against phenolic resins on the basis of cost savings through scrap re-use, and against other more expensive thermosets like diallyl phthalate (DAP) and diallyl isophthalate (DAIP) resins. Among the engineering thermoplastics, its closest competitor is PBT.

Polysulfone, Union Carbide's Udel, is a copolymer of 4,4-dichlorodiphenylsulfone and bisphenol A. This is a transparent amorphous resin with a U.L. Thermal Index of 150°C. Polysulfone is selected instead of polycarbonate where the higher use temperature is required, and sometimes for better stress-crack resistance at lower temperatures. Markets are often in glass and stainless steel replacement based on the advantages of transparency, heat resistance, hydrolysis resistance, suitability for food contact, and resistance to acids and alkalis.

Applications are in medical hardware, food processing and handling equipment, automotive, electric/electronic, and industrial parts, often for corrosion resistance. There is some usage of polysulfone as a substrate for circuit boards instead of the customary epoxy-glass composites (35). It is also used for process pipe, and as a matrix for carbon-fiber advanced composites for aircraft parts, in competition with carbon-epoxy composites at 110°C (see Ablative materials; Composites).

Plastics with still higher heat-resistance capabilities include some thermoplastics (ie, other polysulfones), some thermosets (ie, polyimides and aramids), and some that can be melt-processed but need cross-linking for optimal property development (ie, poly(amide–imide) resins). References 36–37 are exceptionally useful for comparisons of high-temperature thermoplastics (see Aramid fibers; Polyimides).

New Developments in Engineering Plastics

The cost of developing, introducing, and providing the marketing and technical back-up for engineering plastics is very high, and there is some opinion in the plastics industry that future engineering plastics will be chemical or physical modifications of current ones.

Desirable properties can be obtained by physically blending or alloying resins, or by incorporating inorganic or organic fillers and reinforcements. This is being done by primary resin suppliers, who use these methods to produce special grades, and by custom compounders, and to an increasing extent by end-users who tailor-make resins to fill their own specific needs (38–39).

The distinction between blends and alloys is not a clear one, but both terms are used for physical mixtures of two or more structurally different polymers (40–43) (see Copolymers). As compared to copolymers, in which the components are linked by strong chemical bonds, the components in alloys adhere primarily through van der Waal forces, dipole interactions, and/or hydrogen bonding (34).

Some blends are marketed as such, but others are proprietary formulations marketed only with a special trade name or grade designation. Mobay's Bayblend polycarbonate/ABS alloys are marketed as alloys; Uniroyal's Arylon-T poly(aryl ether) (sold to U.S.S. Chemical) is a polysulfone–ABS alloy but it is not marketed as such. Some grades of engineering plastics are generally presumed to be blends, but their composition is not revealed. Elastomers can be blended with engineering resins to increase impact strength (see Elastomers, synthetic).

Filled and reinforced compounds are readily identified, and the closing years of the 1970s have seen major advances in the use of glass fiber reinforcement (ie, stampable polypropylene and nylon sheet), carbon-fiber reinforcement (44), and the use of mineral reinforcements (see Carbon; Fillers; Laminated and reinforced plastics).

Copolymerization is another route to developing new resins. Polyarylates, eg, are polyesters produced from diphenyls and dicarboxylic acids. Carborundum's Ekkcel high-temperature (HDT 300°C) aromatic polyester is a polyarylate produced from 4,4-dihydroxydiphenyl and isophthalic acid. The first commercial polyarylate to be introduced into the U.S. engineering plastic market, based on bisphenol A and phthalic acids, was developed by Unitika in Japan and is marketed by Union Carbide as Ardel. At the close of the decade still another was introduced by Hooker Chemical, and more appeared to be in the developmental stage.

Although modification of existing resins may be the major route to new properties, what appear to be entirely new resins are still being introduced. Rohm & Haas announced a new engineering resin in mid-1979 of a polyimide type.

BIBLIOGRAPHY

1. R. H. Heinold, "Broadening the Capabilities of Polypropylene for Appliance Applications," *Proceedings, Soc. Plastics Engineers National Technical Conference,* Nov. 17–19, 1975, pp. 63–67.
2. J. Houston, "Automotive Market—Impetus to the Growth of Polyolefin Polymers," *Proceedings, Chemical Marketing Research Association meeting,* Feb. 7–10, 1978, pp. 142–168.

3. R. C. Wright, "Engineering Thermoplastics," *Proceedings, Fourth Annual Conference on Contingency Planning for Plastics,* PPC/Plastics Publishing Co., 1978, pp. 103–108.
4. B. Nathanson, "Outlook for Engineering Thermoplastics—United States and Western Europe," *Chemical Marketing Research Association meeting,* May 3–6, 1977.
5. *Deflection Temperature Under Load, ASTM Test Method D648,* American Society for Testing and Materials, annual.
6. *Underwriters Laboratories Recognized Components Index,* Underwriters Laboratories Inc., annual.
7. *Flexural Properties, ASTM Test Method D790,* American Society for Testing and Materials, annual.
8. *Impact Resistance of Plastics and Electrical Insulating Materials, Methods A and C, ASTM Test Method D256,* American Society for Testing and Materials, annual.
9. E. R. Rosenberg, "Plastics vs. Metals—to 1980 and beyond," *Proceedings, Second Annual Conference on Contingency Planning for Plastics,* PPC/Plastics Publishing Co., 1976, pp. 128–138.
10. J. E. Theberge, "Reinforced Thermoplastics" in *Modern Plastics Encyclopedia,* Vol. 55, McGraw-Hill Book Co., New York, 1978–1979, p. 140.
11. M. A. Rehm, "Reinforced Thermosets" in *Modern Plastics Encyclopedia,* Vol. 55, McGraw-Hill Book Co., New York, 1978–1979, p. 142.
12. V. Stayner "Non-Fibrous Property Enhancers" in *Modern Plastics Encyclopedia,* Vol. 55, McGraw-Hill Book Co., New York, 1978–1979, p. 186.
13. J. M. Smart, "Thermoplastics—Raw Materials for Replacing Diecasting Alloys," *paper presented at a Joint Conference of the Institute of Purchasing and Supply and the Diecasting Society,* Coventry, England, Oct. 12, 1977.
14. *Internally Lubricated Reinforced Thermoplastics,* Bulletin 254-278, LNP Corporation, 1978.
15. *What You Always Wanted to Know About Engineering Resins,* Celanese Plastics Company publication, 1976.
16. *Tests for Flammibility of Plastics Materials for Parts in Devices and Appliances, U.L. Standard 94,* 2nd ed., Underwriters Laboratories, Inc., 1978; 3rd imp., 1979.
17. "Fire Safety Aspects of Polymeric Materials" in *Materials: State of the Art,* Vol. 1, part of a 10 volume series sponsored by the National Materials Advisory Board, Technomic Publishing Co., 1977.
18. W. McKinlay and S. D. Pearson, *Plastics Gearing,* ABA/PGT Publishing Co., Manchester, Conn., 1976.
19. N. C. Baldwin, "The Use of Acetal Copolymer in Plumbing Applications," *Proc. 33rd Annual Technical Conference, Soc. of Plastics Engineers,* 1975, pp. 30–33.
20. A. M. Houston, *Mater. Eng.,* 42 (Feb. 1976).
21. S. Telofski, "Thermoplastic Forces the Obsolescence of Traditional Materials in Automotive Ignition Systems," *Soc. Plastics Engineers Divisional Technical Conference,* Sept. 27–28, 1977, p. 41.
22. *Plast. Technol.* **24,** 83 (Jan. 1978).
23. B. W. Perry, "Phenolics in the Engineering Plastics Arena," *Soc. Plastics Engineers, Connecticut Section, Regional Technical Conference,* Oct. 10–11, 1977, pp. 87–94.
24. *Plast. World* **33,** 57 (July 21, 1975).
25. D. Portman and J. C. Clark, "Glass-Reinforced Polyester for Circuit Breaker Applications," *Proceedings, 31st Annual Technical Conference,* Soc. of the Plastics Industry, Inc., Reinforced Plastics/Composites Inst., 1976, Section 10-F, p. 1.
26. R. D. Lake, J. T. Shreve, and R. L. Lovell, "Pelletized Thermoset Polyester Molding Compounds," *Proceedings, 31st Annual Technical Conference,* Soc. of the Plastics Industry, Inc., Reinforced Plastics/Composites Inst., 1976, Section 13-C, p. 1.
27. J. L. Throne, "Principles of Thermoplastic Structural Foam Processing: A Review," in N. P. Suh and N. H. Sung, eds., *Science and Technology of Polymer Processing,* MIT Press, 1979, pp. 77–131.
28. M. A. Rehm, "Reinforced Thermosets" in *Modern Plastics Encyclopedia,* McGraw-Hill Book Co., New York, 1978–1979.
29. *Plastics Design Forum* **2,** 58 (Jan./Feb. 1977).
30. *Plast. Eng.* **33,** 40 (April 1977).
31. A. M. Houston, *Mater. Eng.* **81,** 28 (June 1975).
32. *Plastics Design Forum* **2,** 18 (Nov./Dec. 1977).
33. *Plast. World* **34,** 28 (June 21, 1976).
34. R. T. Alvarez, S. B. Driscoll, and T. E. Nahill, "High Temperature Performance Polymeric Alloys," *Soc. of Plastics Engineers, Proceedings, Annual Technical Conference,* 1977, pp. 308–310.

35. *Mod. Plast.* **54,** 52 (June 1977).
36. J. E. Theberge, B. Arkles, and P. Cloud, *Mach. Des.* **47,** 73 (Feb. 6, 1975).
37. *Ibid.,* 79 (March 20, 1975).
38. *Plastics Compounding* **1,** 75 (May/June 1975).
39. P. J. Cloud and R. E. Schulz, *Plast. World* **33,** 36 (Sept. 22, 1975).
40. R. J. Jalbert and J. P. Smejkal, "Alloys" in *Modern Plastics Encyclopedia,* Vol. 53, McGraw-Hill Book Co., New York, 1976–1977, p. 108.
41. *Plast. World* **35,** 56 (Nov. 1977).
42. *Mod. Plast.* **54,** 42 (1977).
43. G. R. Forger, *Mater. Eng.* **86,** 44 (July 1977).
44. J. E. Theberge and R. Robinson, *Mach. Des.* **46,** 2 (Feb. 7, 1974).

General References

C. A. Harper, ed., *Handbook of Plastics and Elastomers.* McGraw-Hill Book Co., 1975.
Modern Plastics Encyclopedia, annual, McGraw-Hill Book Co. See in particular the "Design Guide," Vol. 55, Oct. 1978, pp. 463–497.
Mach. Des., Penton/IPC publication, annual materials reference issues.
R. D. Deanin and S. B. Driscoll, "Buying Properties," *Chemtech,* 209 (April 1978).
Plastic Applications in Automobiles and Trucks, annual listing of automotive applications for glass-reinforced plastics, Owens-Corning Fiberglas.
The International Plastics Selector, Inc., San Diego, Calif., *Desk-Top Data Bank Books, Plastics,* 3 volumes.

MARILYN BAKKER
Business Communications Co., Inc.

FELTS

Felt is a homogeneous fibrous structure created by interlocking fibers using heat, moisture, and pressure. Wool or wool combined with other fibers may be meshed using mechanical work and chemical action while the fibrous mass is kept warm and moist; no weaving is used to make this type of felt, which is often called a *pressed felt.* These felts may be combined with resins or chemicals or laminated with other materials, and many industrial applications result from the cutting, molding, or shaping of such felts.

Needled felts are strong, mechanically interlocked structures created by barbed needles penetrating and compressing synthetic fibers. Such felts may incorporate a woven base (foundation). Needled felts are used for products such as ink rollers or filters, or where chemical inertness or thermal stability may be required. Needled felts, with or without bases, are also used on paper machines for pressing or drying operations.

Woven felts include those felts made with a fabric of a special weave. Wool is usually one of the fibers of these felts, enabling the use of heat, moisture, and mechanical action to create a compact, interlaced structure that may have a deep nap or pile of surface fibers. Applications for these felts are similar to needled felts, although the conventional woven felt is now infrequently used on the paper machine.

Some useful properties of felts include: thermal and chemical stability, depending upon fiber content; high permeability and porosity; vibration and shock absorption; and wear resistance. These and other properties are manipulated by the engineer in fabricating mechanical parts and by the felt designer and paper manufacturer in providing the best felt for a particular paper and machine position. The following describes the manufacture and application of papermaking felts. In addition, a brief description of some other industrial uses of felt is included.

Function of Felts on the Paper Machine

In papermaking, felts of matted hair or wool were used as early as the eleventh century: a mold of woven cloth was dipped into a suspension of paper fibers so that a sheet could be formed; after most of the water had drained away, the mold was pressed against a felt and the sheet transferred or "couched" from the mold to the felt. The paper sheet was then allowed to air dry. By the eighteenth century, woven wool cloths were fulled or felted for the hand manufacture of paper. In 1799, the first paper machine was invented, and soon continuous felts were manufactured for pressing and drying operations (see Paper).

The two paper machine configurations most commonly used today, the Fourdrinier and multicylinder are shown in Figures 1–2. Papermaking generally follows the same principles for both machine types and most paper and board: (*1*) a web of paper is formed from an aqueous suspension of fibers (furnish) by gravity and suction through the screen or fabric; (*2*) the web is transferred to the pressing section where more water is removed by pressure and vacuum; (*3*) the sheet enters the dryer section, where steam-heated dryers and hot air complete the drying process; and (*4*) if desired, the sheet may be finished by coating or calendering before it is wound onto the reel.

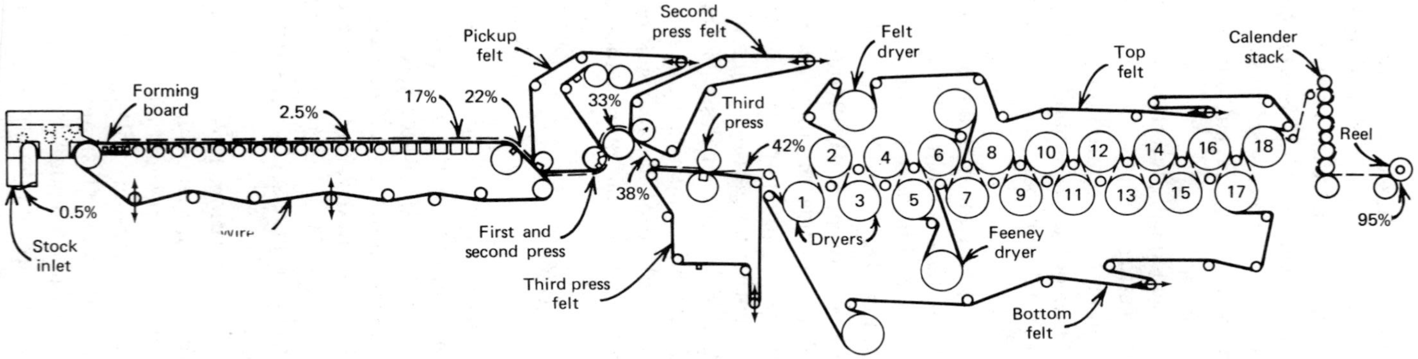

Figure 1. Fourdrinier paper machine. Consistency of paper web represented as percent solids.

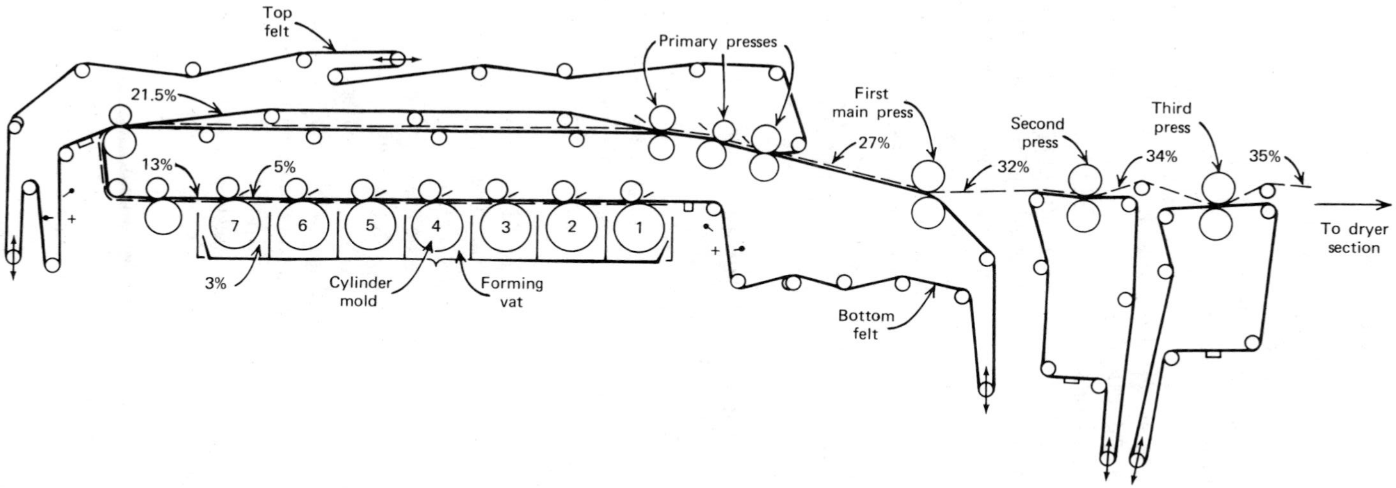

Figure 2. Multicylinder paper machine. Consistency of paper web represented as percent solids.

Manufacture of Felts for Papermaking

Design. Each felt is custom-made and carefully monitored by quality control throughout manufacture.

The following factors are considered in design:

Finish depends upon the competitive demands of the paper grade produced. The fineness of the felt, commensurate with machine type and production rates, is mainly determined by the market in which the paper is sold.

Drainage (for pressing) is the ability of the felt to handle water on the machine. The type of sheet, its filler, machine speed, press roll covers, and conditioning equipment are some variables that must be considered when analyzing drainage requirements. Drainage can be related to permeability, a standard measure of openness. For press felts, permeability gives an indication of expected drainage. For dryer felts and fabrics, it gives an indication of air passage and air pumping characteristics, both of which affect drying. Drainage is also considered with respect to void volume, which is defined in terms of fabric and fiber density; a press felt must have adequate void volume to absorb the water extracted from the sheet in the press nip. Moreover, felt compressibility should be low enough to maintain this adequate void volume.

Durability refers to the life of felts and relates to factors such as: strength, the ability of the felt to withstand operating tension and local strains such as wads going through the press; dimensional stability; and resistance to chemical degradation, heat, abrasion, and compaction.

Other factors such as the ability to run and guide without wrinkling and ease of cleaning must be considered during design.

Felt Classifications. In Tables 1–2, press felts and dryer felts are divided into groups based on constructional characteristics; each construction is divided into classes based upon the degree of fineness or coarseness; applications are also designated. Paper machine felt classifications vary internationally. The press felt categories are those currently recognized in the United States.

Materials. Depending upon the type of felt, the designer may vary the following: type of material; yarn form, weight, and count; weave; amount and fineness of batt (fibrous web); seam, if any; and chemical treatment, if any.

Both natural and synthetic fibers can be used. Table 3 summarizes some important fiber properties. In drying operations, for example, the high temperatures that could degrade the yarns must be accounted for. No one fiber is best for all applications. Basic fiber types and forms can be combined by blending or twisting operations during manufacturing. For example, a fine synthetic multifilament yarn can be twisted with a blended spun yarn for improved tensile strength.

One fiber type is frequently used for a machine direction yarn, and another is used in the cross-machine direction. In press felts, for example, use is made of polyester's excellent resistance to stretching and nylon's superior resistance to friction. Nylon could thus be selected for cross-machine yarns that are more subject to abrasive wear. Since machine direction yarns must withstand the tension on the paper machine, polyester is sometimes used to prevent excess stretching under high load conditions.

Manufacturing Processes. The five general manufacturing categories of yarn making, weaving, burling and joining, needling, and finishing are summarized in Figure 3. Not all felts go through all processes.

Table 1. Classification of Press Felts

Generic name	Constructional characteristics	Classes	Application
conventional press felt	Traditional felt woven of spun yarn in the machine and cross-machine directions and then mechanically felted into final form. Fiber content is predominantly wool with small percentages of synthetics.	coarse	For dissolving pulps, 9 point, roofing, flooring, sulfite, sulfate, or groundwood pulps.
		medium	For kraft, news, wrapping, groundwood specialties, Yankee tissue bottoms, cylinder tissue, Harper machines,[a] binders board, multicylinder tops, bottoms, chip and test liner.
		fine	For lightweight kraft, fine papers, and Fourdrinier board.
		superfine	For Yankee tissue and toweling pickup and various grades of fine paper requiring good sheet finish.
		extra superfine	For fine papers requiring extreme finish.
batt-on-base	Base fabric woven of spun yarns in the machine and cross-machine directions into which is needled a web of fiber batt. Fiber content up to 100% synthetic in batt and base.	coarse	Same as for conventional felts.
		asbestos cement	For top and bottom positions of special cylinder machines making asbestos cement products such as siding, shingles, interior and exterior wallboard and high and low pressure pipe.
		medium	Same as for conventional felts.
		fine	Same as for conventional felts.
		superfine	Same as for conventional felts.
knuckle-free or fillingless	Base fabric of spun machine direction yarns but without cross-machine direction yarns into which is needled a fiber batt. Fiber content up to 100% synthetic in batt and base.	medium	For a wide range of paper and board grades on all positions of both Fourdrinier and multi-ply board machines.
		fine	Same as above except where better sheet finish required.
		superfine	Same as above except for even better finish. Also for Yankee pickup.
		condenser	For condenser papers, this very fine fabric is specially finished to minimize conductivity.
batt-on-mesh	Three basic types: (*1*) Base fabric woven of fine synthetic multifilament yarns in the machine and cross-machine directions into which is needled a fiber batt. Fiber content is always 100% synthetic.	medium	Same as for conventional felts.
		fine	Same as for conventional felts.
		superfine	Same as for conventional felts.

Table 1 *(continued)*

Generic name	Constructional characteristics	Classes	Application
	(2) A very permeable, single-layer weave base fabric woven of synthetic multifilament machine direction yarns and synthetic monofilament cross-machine direction yarns into which is needled a fiber batt. Fiber content is always 100% synthetic.		
	(3) Base fabric woven of treated synthetic multifilament yarns in the machine and cross-machine directions into which is needled a fiber batt. The yarn resins when cured produce a relatively incompressible yarn structure. Fiber content is always 100% synthetic.		
combination	A two-layer weave, rigid base fabric of large void volume. Base composed mainly of synthetic monofilament with a portion of multifilaments into which is needled a fiber batt. Fiber content is usually 100% synthetic.		For all types of presses including plain presses for wide range of paper grades; best applied to latter presses or last press before dryer; for suction pickup as well as high speed Yankee pickup or first suction presses to eliminate shadow marking.
nonwoven	Construction without machine or cross-machine direction yarns. It is an all-needled fiber batt. Fiber content is usually 100% synthetic.		For suction, plain, grooved roll, and shrink sleeve presses primarily for fine paper and board machines where good finish is desired.
batt-on-base (with treated base yarns)	Base fabric of resin-treated yarns in the machine and cross-machine directions into which is needled a fiber batt. Fiber content is always 100% synthetic.	medium	Same as for conventional felts.
		fine	Same as for conventional felts.

[a] Paper machines.

Yarn Making. Synthetic multifilaments and monofilaments are usually prepared by melting the polymer and forcing it through fine extrusion dies. After extruding, the filaments are stretched to ensure orientation of the molecular chains. This builds high strengths at low elongation into the material. Spun yarns of cotton, asbestos, synthetics, or wool are prepared by standard textile techniques applied in the following order: blending to evenly mix the various components; carding to align the fibers somewhat parallel; and spinning to draft or stretch, twist, and wind the yarn onto bobbins.

Table 2. Classification of Dryer Clothing

Generic name	Constructional characteristics	Application
woolen dryer felt	Fabric woven of spun yarn in the machine and cross-machine directions and then heavily fulled (compacted) to produce a dense, smooth structure. Fiber content can be 100% wool[a] or wool and synthetics.[b]	for cigarette and condenser tissue and photographic paper
conven-tional dryer felt	Two- or three-layer weave of spun yarns in the machine and cross-machine directions. Fiber content can be cotton[a] and/or synthetics, or a combination of cotton, synthetics, and asbestos.[a]	for all grades of paper
needled batt-on-base dryer felt	Base fabric woven of spun yarns in the machine and cross-machine direction, into which is needled a fiber batt. Fiber content of batt is usually 100% synthetic; base may be cotton and/or synthetic.	for fine papers
needled batt-on-mesh dryer felt	Base fabric woven of synthetic multifilament yarns in the machine direction and monofilament yarns in the cross-machine direction, into which is needled a fiber batt. Fiber content is always 100% synthetic.	for fine papers
open mesh dryer felt	A two- or three-layer weave. Multifilament or spun yarns are woven in the machine direction; spun yarns or asbestos content yarns are woven in the cross-machine direction. Fiber content varies from high cotton-low synthetic to spun Nomex polyamide[c] machine direction yarns; cross-machine direction yarns are often combination asbestos wire.	for most grades of paper
dryer fabric	A two- or three-layer weave. Multifilaments and/or monofilaments are woven in the machine and cross-machine directions. Fiber content is always 100% synthetic.	for all grades of paper

[a] qv.
[b] See Fibers, chemical.
[c] See Aramid fibers; Polyamides.

Weaving. Yarns are woven into different patterns, chosen for dimensional stability, strength, finish characteristics, drainage ability, wear, and bulk. Felts can be woven into two forms, either as a piece of flat material that is later joined together, or in endless form without a seam. Before actual weaving, the warp yarns must be wound onto the warp beam (dressing). During drawing-in, warp ends are individually drawn through the eyes of the heddles in the loom harnesses that control the weave pattern.

Burling and Joining. Most felts are next inspected and burled, during which minor weaving and yarn imperfections are corrected. Felts that are not woven endless are joined by hand. The woven fabric then goes to needling (if it is to be a needled base) or to finishing and treating.

Needling. Mechanical bonding created by needling replaces fulling in conventional felts by creating controlled contraction and stabilization of the felt and by establishing final weight and size characteristics. A needled felt is composed of two parts: the base fabric and batt. The construction of the base fabric (weave, fiber content, etc) helps determine strength and stability. The batt, formed by carding fibers, is laid on the base just before passing through the needling machine. The number of batt

Table 3. General Fiber Properties

Fiber	Form available: Staple	Form available: Multi-filament	Form available: Mono-filament	Tensile strength	Abrasion resistance	Chemical resistance: Acids	Chemical resistance: Alkalies	Comments	Relevant article in *ECT*
dacron	X	X	X	excellent next to nylon	good to excellent next to nylon	good	fair to good	Good all-round fiber. Subject to moist heat hydrolysis. Do not use with strong alkalies.	Polyesters
orlon	X	X		good	above average	good	fair	Good combination of abrasion and heat resistance. Use where moist heat conditions prevail. Special heat finishing increases stability.	Acrylonitrile polymers
nylon	X	X	X	excellent	excellent	poor	good to excellent	Excellent abrasion resistance. Best of all fibers. Should be heat-set for high temperature uses. Do not use with strong acids.	Polyamides
nomex	X	X		good to excellent	good to excellent	fair	good	Good stability. Heat resistant.	Polyamides
polypropylene	X	X	X	excellent	excellent	excellent	excellent	Stretches under load even at low temperatures.	Olefin polymers
wool	X			fair	average	poor	poor	Can be felted.	Wool
cotton	X			average	average	poor	excellent		Cotton

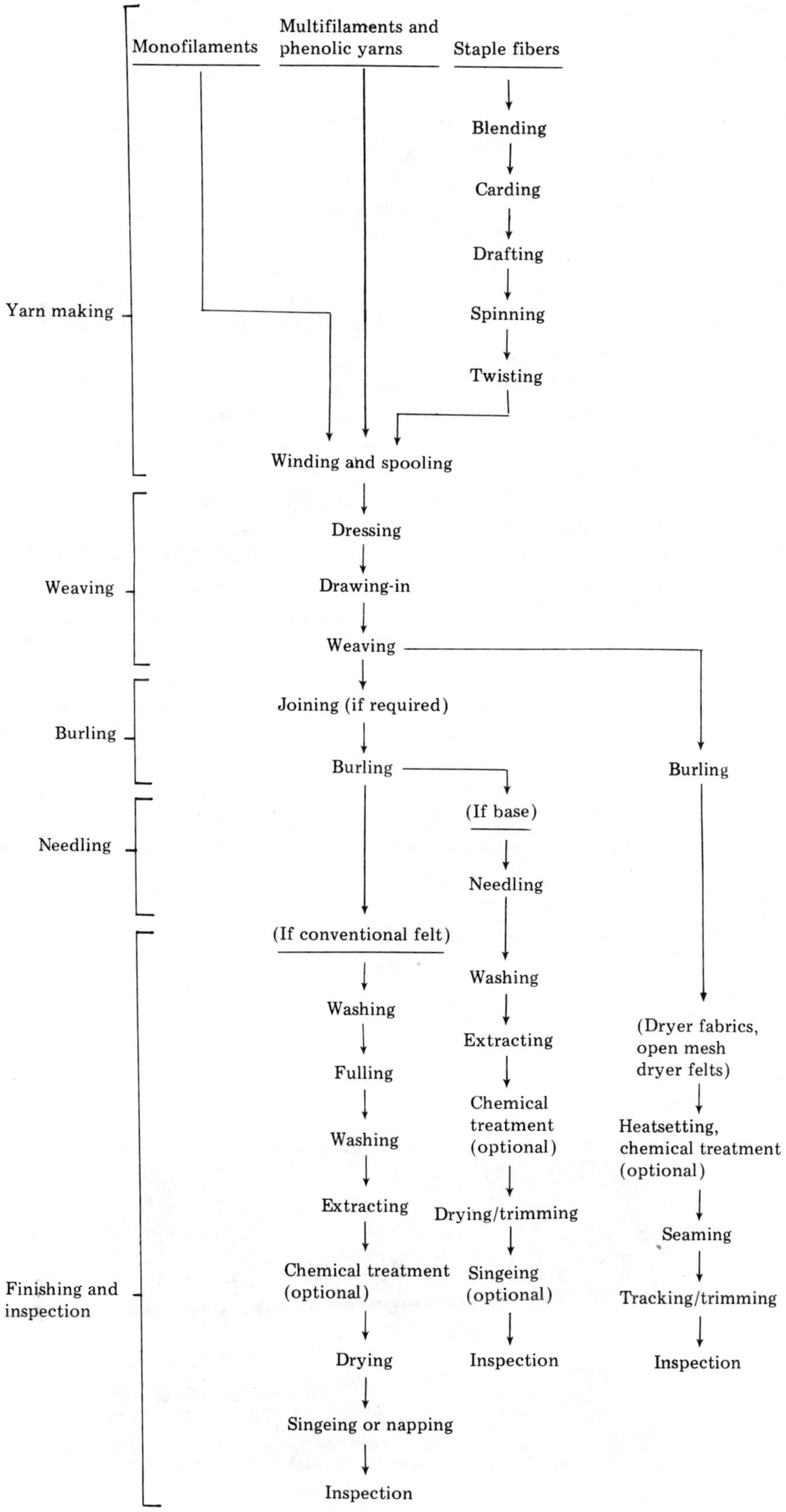

Figure 3. Common manufacturing processes.

layers may be varied and applied to either or both sides. Batt characteristics as well as the amount of penetration of the needles and the number of passes through the needle loom determine the finish, openness, and other characteristics of the felt.

The needles are equipped with tiny barbs facing towards the needle point and are fastened on a vertically reciprocating plate. As they descend, the barbs grasp a few batt fibers and force them into the base fabric (Fig. 4). The fibers are thus locked in the body of the felt and are partially oriented perpendicularly to the plane of the fabric. Each square centimeter of felt may be subjected to more than 400 needle penetrations.

Finishing and Treating. Conventional press felts are not needled but wet with a soap solution and then *fulled* or felted on a rotary fulling mill. This rotary consists of a set of driven rolls through which the felt runs continuously in a roped form. Just before the felt enters the nip of the rolls, it passes through a throat or vertical slot, which can be adjusted in width. The crowding and squeezing that occurs from this combined action causes the felt to shrink in width. As the felt emerges from the nip, it enters an open-ended box called a trap, which has a hinged vane at the top. This vane can be swung down to decrease the size of the opening at the back of the trap box. This results in a partial lengthwise restriction on the felt and is used to control the rate of length shrinkage. Shrinkage may be as great as 20% of the original length and 50%–60% of the starting width (Fig. 5).

This process of controlled felting determines final size and tension of the finished felt by mechanically bonding the individual wool fibers. A felted cloth of wool will change size for two main reasons: (*1*) wool fibers have scales pointing in one direction, thereby creating a directional friction effect that allows the fibers to move in one direction more easily than the other; (*2*) wool fibers possess the ability to stretch and recover from stretch when wet. As fulling progresses, the fibers become increasingly intertwined, changing the felt in length and width.

For both conventional and needled felts, the trademark line is applied next, and the felt is washed, extracted, and possibly treated with proprietary chemicals. Some treatments chemically modify the fibers and others coat the fibers and adhesively bond the yarns or fabric structure. Such treatments may be used to retard bacterial or chemical degradation, fiber shedding, abrasive wear, or compaction, or to enhance drainage, wet-up after installation, and stability.

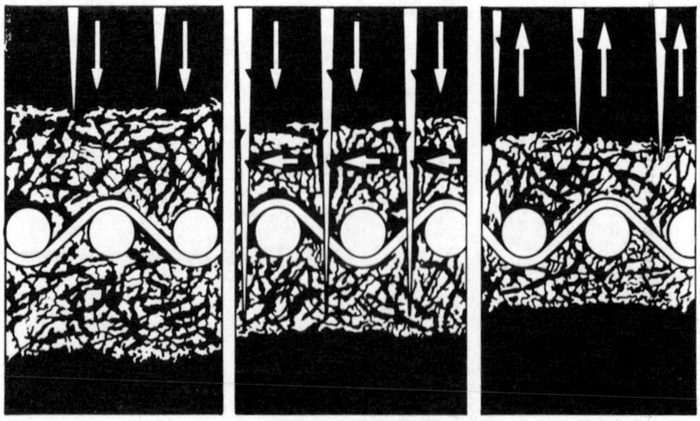

Figure 4. Needling.

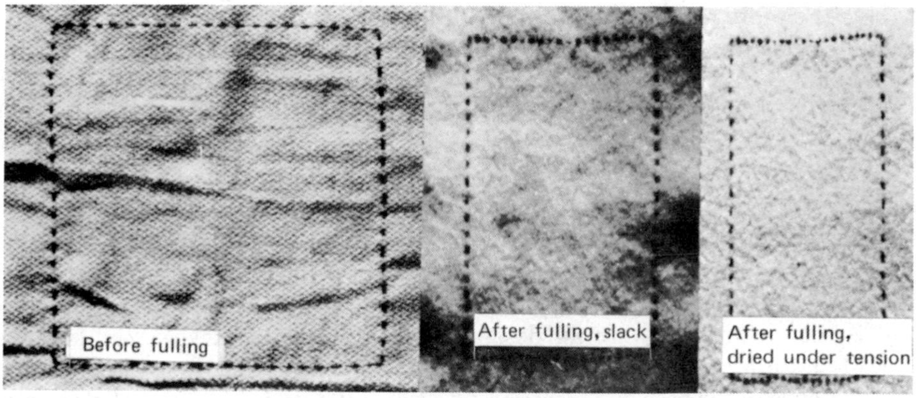

Figure 5. Conventional felt before and after fulling.

Felts are dried around a large steam-heated cylinder and trimmed if required; conventional felts may undergo napping, for extra cushioning or to increase water removal capacity. Singeing may be required to eliminate long surface fibers. Careful inspection follows to check manufacturing specifications, including proper length, width, and tension.

Other Finishing Processes. All-synthetic felts may be heat-set to size and tension. Dryer felts usually require a seam, since few are manufactured in an endless configuration; most are sold with a seam to join mechanically to the paper machine at installation.

Application of Press Felts

Evolution of Press Felt Constructions. In the past twenty years, press felt development has changed dramatically. Before 1960, most press felts were conventionally woven felts, and the plain press and the suction press were commonly found on paper machines. The economic value of the felts was measured by how many days the felt ran on the machine, and most felts, when removed, were worn out. Modern felts of highly synthetic constructions have grown from new manufacturing processes such as needling and from increased knowledge of press efficiency. These highly synthetic felts are seldom removed from the machine because they are worn out; they are generally removed because they are filled or compacted to the point where they no longer handle water uniformly or lack the drainage capabilities to maintain a high level of pressing efficiency.

As the concept of pressing has changed from the plain press to a suction press to the fabric press or shrink sleeve and to the grooved plain press or grooved suction press, changes in felt design have been made simultaneously. The conventional woven felt described in Table 1 is for the most part obsolete, although it remains used on some older machines that have not been updated. Its yarns and fibers are horizontally oriented.

The first significant departure from the conventional woven felt was the batt-on-base felt, a needled felt. The base of this felt resembles a conventional felt, with both machine and cross-machine direction spun yarns. Resistance to water flow within the base therefore remains high and pressing efficiency is limited by the buildup of hydraulic pressure. However, this needled felt is a substantial advance over conventional felts because the needling creates batt fibers that are more or less vertically oriented. Thus flow resistance in the vertical direction is much lower than in the conventional woven felt. Needled (batt-on-base) felts are widely used today; many are completely synthetic. If they can be kept open, their running time can be very long.

The knuckle-free or fillingless felt described in Table 1 has no cross-machine direction yarns, thereby lowering resistance to water flow in the machine direction and removing a major trap for paper fines and dirt. This was followed in its development by the various batt-on-mesh designs, which use an open mesh screen-type base fabric of multifilament and sometimes monofilament yarns. Such yarns reduce water flow resistance in the base and create a soft, pliable structure, the latter characteristic being desirable for ease of installation in certain machines. The batt-on-mesh felt with multifilaments in the machine direction and monofilaments in the cross-machine direction provides a very stable structure, a highly permeable base, and an excellent finish for fine paper applications on certain presses.

A further design development, the combination felt, has low flow resistance in the base and high water storage capability. These are the most effective press felts today to provide significant sheet dryness gains and improved moisture profiles on all types of presses including plain presses.

A recent development is the nonwoven felt, which eliminates the base weave to give the best possible openness and permeability; the absence of any type of organized yarn structure eliminates sheet marking and filling with dirt and paper fines, while providing the most uniform pressure distribution from felt to sheet in the press nip. The nonwoven felts are applicable to suction, plain, grooved roll, and shrink sleeve presses primarily for fine paper and board machines where good finish is desired (see Nonwoven textiles).

Press Felts for Papers and Board. Table 4 lists some felts used on typical machines; the example given are represented with large ranges in basis weights. Nomenclature for presses and felts can generally be seen from Figures 1–2. The following discussion mentions machines important for producing some selected paper grades, requirements for specific felt positions, and general characteristics of such felts.

Fine Paper. This category generally includes coated and uncoated publication, tablet, bond, litho, copying, business, cover stock, photographic, machine-glazed and modified-finish tissue, one-time carbon, and numerous other fine specialty grades.

Many two- and three-press open-draw Fourdriniers are still used for the production of fine paper. In the early 1950s, when the demand for fine paper grades exceeded the capability of open draw machines, the suction transfer came into use and provided a better transfer with fewer sheet breaks and, therefore, higher speeds. Next to be developed were the inverted first press and three-roll inclined press, the latter currently the most widely used configuration.

The suction-pickup felt for fine paper requires good openness, resistance to compaction, and enough surface fineness to promote a water-film bond between the sheet and felt; subsequent surface tension improves pickup at the wire and retards sheet loss. The pickup position of any fine paper machine configuration uses a relatively open base structure with a uniform layer of relatively permeable face-side batt. The batt is introduced to the inside surface to control an existing or potentially dangerous

Table 4. Felt Application Chart, Typical Felts on Their Representative Machines

Paper grade	Machine	Position	Machine speed, m/s (ft/min)	Press load, kN/m (pli[a])	Weight per unit area, kg/m^2 (oz/ft^2)
fine paper	three-roll inclined press	suction pickup	6.1–14 (1200–2800)	35–61 (200–350)	0.92–1.1 (3.0–3.7)
		second press	6.1–14 (1200–2800)	39.4–65.7 (225–375)	0.98–1.2 (3.2–3.8)
		third press	6.1–14 (1200–2800)	44–79 (250–450)	0.98–1.2 (3.2–4.0)
tissue	Yankee Fourdrinier (two-felt)	Yankee pickup	7.6–23 (1500–4500)	58–96 (330–550)	1.0–1.2 (3.4–3.9)
		Yankee bottom	7.6–23 (1500–4500)	30.6–44 (175–250)	0.79–.88 (2.6–2.9)
board	multicylinder	top	0.76–3.3 (150–650)	26–44 (150–250)	0.61–0.82 (2.0–2.7)
		bottom	0.76–3.3 (150–650)	26–44 (150–250)	0.76–0.92 (2.5–3.0)
		press	0.76–3.3 (150–650)	35–149 (200–850)	0.76–1.4 (2.5–4.5)
pulp	three-press Fourdrinier	third press (grooved)	1.3–2.5 (250–500)	175–263 (1000–1500)	1.2–1.3 (3.9–4.3)
asbestos cement	two-cylinder	bottom	0.56–0.81 (110–160)	18–44 (100–250)	1.4–1.5 (4.5–5.0)

[a] pli = pounds-force per lineal inch.

wear condition. The first press felt of open draw machines have characteristics similar to the pickup designs.

At the second and third presses, less water and fines are available in the sheet and the potential filling problem is reduced. The base fabrics can be closed up to present a smoother surface for uniform pressing. Because of the higher press loadings associated with the latter presses, the thickness of the batt layer is increased to eliminate the possibility of base or roll strike through. All constructions can be used on all machines, with emphasis on all-synthetic felts with batt-on-mesh, combination, or nonwoven constructions.

Newsprint and Offset. Newsprint is usually a lighter weight sheet than the average fine paper grade, and generally runs at higher speeds. Stock for newsprint is primarily composed of groundwood and is characterized by short fibers and a percentage of long fibered chemical pulp for added strength. Machines used for these grades are similar to those producing fine paper, with the three-roll inclined press and three-roll direct pickup press predominating.

Pickup felts for newsprint and offset should be permeable and not susceptible to filling: hence, batt-on-mesh designs are preferable. The three-roll direct pickup press does not require pickup quality to the same extent as pickup felts for other configurations. For other positions on all machine types, synthetic batt-on-base, batt-on-mesh, and combination felts are applicable. Compared to fine paper, these felts can be slightly heavier on the latter presses because of increased loading.

Tissue. Machines producing tissue are often referred to as Yankee Fourdriniers and produce grades of toilet, facial, napkin, and toweling tissues. The Yankee Fourdrinier with two felts has traditionally predominated, but there is a trend toward the single felt because it eliminates the bottom felt with its accompanying equipment and maintenance requirements. Pickup is generally not vacuum-assisted; thus characteristics are different from the suction pickup for fine paper. The Yankee pickup must be hard and dense with a superfine or very fine surface so that good pickup can be achieved with a uniform surface water film on the pickup. Batt-on-base Yankee pickup for a two-felt configuration has a synthetic content of 40–100%. Combination felts provide good pickup without requiring as much shower water to condition the felts, compared to other constructions.

The conventional woven Yankee bottom felt has been replaced mostly by the needled felt. Since openness, free drainage, and cushion are important, this felt is bulky and coarse compared to the Yankee pickup. Treated yarns in a batt-on-mesh or batt-on-base construction help maintain openness and drainage. Synthetic content is 50–100%.

Multi-ply Board. Grades included in this category are container board, folding boxboard, nonfolding boxboard, solid bleached board, and other board such as gypsum board. Many paperboard grades have a top liner, underliner, filler, and backliner, and are produced on machines having more than three cylinder vats. Most machines commonly have seven or eight cylinders.

The bottom felt on multicylinder machines, which picks up the web plies, must be open to allow for drainage, smooth enough to impart a desirable finish, strong and stable because of the higher press loadings and higher speeds, and resistant to the action of high pressure showers, which can wear the inside of the felt.

The top felt is very much like the bottom felt in construction, but is usually much lighter. Finish is more important, thus yarns are smaller. The top felt must also be open to prevent sheet-crushing. Tension requirements are not as severe as on the bottom felt. It is important that a balance in design of the top and bottom felts be achieved, so that both can handle the water adequately.

For the press felts, higher loaded presses, mainly grooved, are increasing in use. Therefore, heavier weight felts are needed to absorb more water from the sheet without crushing, and to impart good finish. New constructions such as batt-on-mesh, combination, an nonwoven felts are largely replacing batt-on-base.

Pulp. The pulp (qv) sheet is rarely used as a finished product and is generally converted to other grades. Pulp sheets used outside of the paper industry result in varied finished products including rayon (qv) yarn, cigarette filters (see Cellulose acetate), cellophane (see Film and sheeting materials), photographic and x-ray films (see Photography), plastic articles (see Plastics), lacquers, explosives (qv), and nonwovens (see also Cellulose). Most pulp machines are three-press Fourdriniers. To aid in drying, a short section of four to eight dryers sometimes precedes the third or hot press. High press loadings are also typical, particularly on the third press, which might have a loading as high as 263 kN/m (1500 pli). In some cases, up to 50% dryness out of the press section has been achieved.

It is particularly important that felts used on pulp machines provide good drainage; hence, such felts are coarse and open. As with most felts, long life is important. Pulp felts must also be bulky to withstand heavy press loadings. Finish is generally not a requirement. Needled felts have replaced conventional woven felts in most cases. Synthetic content is increasing, and all-synthetic batt-on-mesh and combination constructions are being used extensively.

Other Industrial Applications of Felt

Felt can be fabricated to produce parts to precise specifications for many diverse functions. Parts may be molded under heat and pressure, often resin-impregnated for stability and to provide special properties. For special shapes, felts may easily be cut without fraying and denser grades may be machined. Fiber content and felt construction can be varied for the application; impregnations and laminations are used to increase adaptability. Some varied functions of these fabricated felts are listed below.

The vibration and shock absorbing properties of wool felt suggest its application as a mounting under precision instruments; denser felts are used to accommodate heavier usage (see Wool). Felt can also be an effective sound absorber when applied as a surfacing, such as to line an air conditioner or a section of a typewriter or computer. Sound absorption depends on fiber content, felt construction, density, and air permeability (see Noise pollution; Insulation, acoustic).

The natural capillarity of felt enables it to be a liquid reservoir and medium for the transfer of fluids to lubricate bearings and other mechanical parts, particularly in closed systems. Felt seals may be used to retain grease or oil, while eliminating dust, moisture, and contaminants; other sealing applications include use in vacuum cleaners and weather stripping. From the wide variety of fibers and constructions, felts can be adapted to many types of wet and dry filtration (qv), such as respirators as well as filters for products ranging from vegetable oils (qv) to adhesives (qv). Another application is polishing felts; felts used in precision polishing of optical lenses provide a delicate abrasive surface that is long lasting and uniform.

Acknowledgement

Much of the text and all tables and figures are derived from *Paper Machine Felts and Fabrics,* Albany International Corp., Albany, New York, 1976; this is a fine reference for further reading.

BIBLIOGRAPHY

"Felt" in *ECT* 1st ed., Vol. 6, pp. 313–316, by Ivor Griffith, Philadelphia College of Pharmacy and Science.

General References

J. B. Casey, *Pulp and Paper; Chemistry and Chemical Technology,* 2nd ed., 3 Vols. Interscience Publishers, Inc., New York, 1961.

G. H. Ireland, *Paperboard on the Multi-Vat Cylinder Machine.* Chemical Publishing Company, Inc., New York, 1968.

Nuttall, G. H. *Theory and Operation of the Fourdrinier Paper Machine.* S. C. Phillips & Co., Ltd., London, 1968.

W. E. Becker, "Designing with Felt," *Mach. Des.,* 3 (June 26, 1969).

H. W. LoFaro, "Engineers Guide to Felt," *Mater. Eng.,* 34 (Nov. 1971).

F. M. Bolam, ed., *Design of the Wet End of the Papermachine,* Monographs in Paper and Board Making, Vol 3, Ernest Benn Limited for the Technical Section, The British Paper and Board Makers' Association, London, 1971.

J. D. Parker, *The Sheet-Forming Process,* Special Technical Association Publication, no. 9. Technical Association of the Pulp and Paper Industry, Atlanta, 1972.
J. F. Atkins, ed., *Paper Machine Wet Press Manual,* TAPPI Monograph Series, no. 34. Technical Association of the Pulp and Paper Industry, Atlanta, 1972.
Papermachine Clothing, Monographs in Paper and Board Making, Vol 5. Ernest Benn Limited for the Technical Section, The British Paper and Board Industry Federation, London, 1974.
J. Heller, *Papermaking.* Watson-Guptill Publications, New York, 1978.

MARY K. PORTER
Consultant, Troy, N. Y.
and the Albany International Corp.

FIBERS, CHEMICAL

The chemical fiber industry, a lineal descendant of prehistoric use of natural fibers in textiles, is now an important industry in its own right, a large segment of the chemical process industry, and heavily dependent on petrochemicals. Its history begins with the patent of an artificial silk by Count Hilaire de Chardonnet in France in 1885. Chardonnet silk was made by two of the three major spinning processes in use today. It was a regenerated cellulose fiber (now called rayon) made by converting a nitrated cellulose into fiber form and then chemically regenerating the cellulose to avoid the unacceptable flammability of cellulose nitrate. Initially, it was produced by extruding an alcohol solution of cellulose nitrate through small holes into water to effect fiber coagulation. In a later process it was generated by similarly extruding an alcohol—ether polymer solution into air, forming filaments that hardened as a result of solvent evaporation.

It is appropriate to introduce some definitions here. The word spinning as applied to chemical fibers refers to the process by which they are formed from polymeric substances. Chardonnet's first process is representative of wet spinning, in which a solution of a polymer is forced through one or more holes in a device called a jet or spinneret into a suitable coagulating bath. The individual thread-like strands are called filaments. Chardonnet's second process is called dry spinning, wherein a polymer solution is forced through the spinneret holes into a chamber where the solvent is evaporated from the filaments.

Chardonnet's fiber was derived from a naturally occurring long-chain polymeric material, cellulose. As the chemical fiber industry grew steadily from the 1880s to the 1930s, it continued to be based on use of naturally occurring polymers, primarily cellulose. Research in the 1920s and 1930s, particularly that of Staudinger and Carothers, laid the foundation for further advances, especially the development of the synthetic fibers. Staudinger, a 1953 Nobel prize winner, proved polymers were molecules of very high molecular weight. Carothers clearly defined the two major types of polymerization, condensation and addition; showed how condensation polymers of high molecular weight could be synthesized; and demonstrated that certain of these could be formed into filaments that exhibited the property of cold drawing and crystallization, resulting in oriented, strong fibers (1). Carothers' elegant research in DuPont's laboratories led in 1939 to two major innovations in the chemical fiber field. One was the commercialization of nylon fiber, the first truly synthetic fiber based on a polymer itself. The second was the technique of melt spinning. Instead of being dissolved as for wet or dry spinning, the nylon polymer was melted by heating, and the viscous liquid was forced through the spinneret holes to form filaments that hardened as they cooled.

The generally used term man-made, referred to here as chemical fibers, as applied to fibers differentiates between the natural fibers and those produced by chemical and physical means in manufacturing processes. The advent of nylon, and the subsequent proliferation of fibers from synthetic polymers, has brought about the subdivision of chemical fibers into two broad classes: those made from natural polymers such as the cellulosics (see Cellulose acetate and triacetate fibers; Rayon; Fibers, vegetable), and those derived from synthesized polymers. The latter, which are called synthetics (or sometimes noncellulosics), now represent the majority of chemical fibers. In this category fall the nylons, the polyesters, the acrylics, the polyolefins, and many others (see Acrylic and modacrylic fibers; Olefin fibers: Polyamides; Polyester fibers).

By 1977 world production of chemical fibers closely approached world production of natural fibers, and the production of synthetic fibers had grown to well over twice that of the cellulosics, with polyester taking the lead. Table 1 shows the changes in volume of various fibers between 1967 and 1977.

For polymers, whether naturally occurring or synthesized, to be suitable for fiber production they must meet several conditions. They must be capable of conversion into fibrous form and must be of sufficiently high molecular weight to permit development of adequate fiber properties. In order to develop such adequate fiber properties they must be capable of being oriented so that the long polymer molecules tend to lie parallel to the fiber axis.

The fiber is normally drawn (stretched) either during or subsequent to spinning to effect orientation. Drawing frequently imparts or enhances crystallinity resulting in enhanced fiber tensile strength. The choice of spinning process is primarily dictated by the nature of the polymer used and secondarily, in some cases, by the nature of the products desired. If the spinneret has a single hole, a monofilament is produced. Such monofilaments are desirable for certain uses, eg, fishing lines, sheer hosiery. Usually, however, the spinneret has many holes. For textile use multifilament yarns are produced, each thread line or "end" containing from fewer than 10 to 100 or more individual filaments depending on end-use, eg, apparel, upholstery, carpeting. Yarns of this type are usually referred to simply as filament yarns. Except for silk, however, natural fibers occur in relatively short lengths, usually measured in centimeters. These are called staple fibers, and yarns spun from them tend to have different aesthetic

Table 1. World Textile Fiber Production, 1967 and 1977[a], Thousands of Metric Tons

	1967	1977
Natural fibers		
raw cotton	10,180	14,138
raw wool	1,562	1,390
raw silk	32	49
Total	*11,774*	*15,577*
Chemical fibers		
cellulosics[b]	3,428	3,535
synthetics		
polyester	754	4,233
nylon	1,317	2,939
acrylic	542	1,765
olefin	150	949
other[c]	110	110
Sub total	*2,873*	*9,996*
Total	*6,301*	*13,531*

[a] Adapted from Textile Economics Bureau Data (2–3).
[b] Rayon and acetate, including acetate cigarette filter tow.
[c] Predominantly synthetic.

qualities from continuous-filament yarns. Chemical fibers are, in many cases, desired in staple form, either for use by themselves or for blending with other fibers. In the production of staple chemical fibers, polyester being preeminent, the spinneret contains hundreds or thousands of holes, and a very large bundle of tens of thousands of filaments, called a tow, is produced by combining output from several spinnerets. Such tows, after stretching, are crimped, ie, mechanically given a wavy form, and then cut into staple fiber of any length desired.

The physical, mechanical, and chemical properties vary widely among the various chemical fibers. They can also be varied to a degree in a given type of fiber by the method of spinning used and by the post-treatment, particularly extent of drawing. Key physical properties include the strength of the fiber, the degree of elongation or extensibility at the breaking point, and the modulus, or degree of stiffness. These are generally expressed relative to the fiber tex (denier times 0.1111), a measure of size defined as the weight in grams of 1000 (9000 for denier) meters of yarn or filament. The tex and the decitex, a unit ca 11% greater than denier, have come into use in Europe. Fibers of different densities thus have different cross-sectional areas for a given tex. Most of the chemical fibers in use today, except for glass and certain specialty products, have densities varying from slightly below 1.0 to ca 1.5 g/mL. The tenacity of a fiber or yarn is its breaking strength in grams divided by the dtex (or denier). Elongation at break is measured as a percentage of original length extended, and the modulus (or initial modulus), expressed in grams per dtex (or denier), is the force required to extend the fiber initially, divided by the degree of extension. As the properties of most fibers are affected by absorbed moisture and also by temperature, physical testing is done under controlled and defined conditions; ie, 21°C and 65% relative humidity (rh). Chemical fibers range widely in moisture regain from the dry state, from near zero for polypropylene and below 1% for polyester to ca 13% for rayon. Other properties significant to various uses include abrasion resistance, elasticity, resistance to drycleaning solvents and other chemicals, ease of dyeing and types of dyestuffs accepted, and dimensional stability, or resistance to shrinkage and wrinkling. Strength is important in industrial applications such as tire cord (qv). Abrasion resistance is significant in carpet-face yarns, but not particularly so in women's apparel. Though some of these properties influence what is called hand, the latter is a somewhat elusive, yet very important, property of a fabric. It denotes the tactile and aesthetic sensations experienced in feeling a fabric.

In addition to the several broad-based synthetic fibers introduced in the 1940s and the 1950s, there are now many high-performance specialty fibers introduced as a result of continuing research and development. For reference, Table 2 summarizes generic names and definitions established by the U.S. Federal Trade Commission for a number of chemical fibers. Several of them are discussed in more detail below.

Chemical Fibers Based on Natural Polymers

Rayon. Two other processes have long since superseded Chardonnet's method of regenerating cellulose, the cuprammonium process and the viscose process. They are based on two different methods of solubilizing the relatively intractable cellulose molecule. The cuprammonium, or Bemberg process, is currently of minor present significance. It is based on dissolving relatively purified cellulose, such as that from cotton linters or woodpulp, in an aqueous solution of ammonia, caustic soda, and basic

Table 2. Relevant F.T.C. Generic Names for Manufactured Textile Fibers

Generic name	Definition of fiber-forming substance[a,b]
acetate	cellulose acetate; triacetate where not less than 92% of the cellulose is acetylated
acrylic	at least 85% acrylonitrile units
aramid	polyamide in which at least 85% of the amide linkages are directly attached to two aromatic rings
azlon	regenerated naturally occurring proteins
glass	glass
modacrylic	less than 85% but at least 35% acrylonitrile units
novoloid	at least 85% cross-linked novolac
nylon	polyamide in which less than 85% of the amide linkages are directly attached to two aromatic rings
nytril	at least 85% long chain polymer of vinylidene dinitrile where the latter represents not less than every other unit in the chain
olefin	at least 85% ethylene, propylene, or other olefin units
polyester	at least 85% ester of a substituted aromatic carboxylic acid, including but not restricted to substituted terephthalate units and para-substituted hydroxybenzoate units
rayon	regenerated cellulose with less than 15% chemically combined substituents
saran	at least 80% vinylidene chloride
spandex	elastomer of at least 85% of a segmented polyurethane
vinal	at least 50% vinyl alcohol units and at least 85% total vinyl alcohol and acetal units
vinyon	at least 85% vinyl chloride units

[a] All percentages are by weight.
[b] Except for acetate, azlon, glass, novoloid, and rayon, the fiber-forming substance is described as a long chain synthetic polymer of the composition noted.

copper sulfate, followed by wet spinning into water. The regenerated cellulose fibers are stretched, extracted with sulfuric acid solution, washed, and dried. The process, first commercialized in Germany in 1899, was based on work by a Swiss chemist, Schweizer, in 1857 (see Cellulose; Rayon).

In 1891 the British workers Cross, Bevan, and Beadle found that cellulose is soluble in a mixture of caustic soda and carbon disulfide. The viscose process based on this finding was commercialized in 1903 in the United Kingdom (1910 in the United States) and is today predominant in rayon manufacture worldwide. It is a very complex process and its commercial utilization represented a triumph of ingenuity and technology. Basically, relatively purified wood-pulp cellulose, called dissolving pulp or chemical cellulose, is treated first with caustic soda, then with carbon disulfide, to form a xanthate which is, in turn, soluble in aqueous alkali solution. Control of conditions in the spinning solution, called viscose, and of the bath into which the viscose is extruded, importantly affect final fiber properties. The concentrations of sulfuric acid, sodium sulfate, zinc sulfate and other components of the spinning bath are especially important. The coagulated fiber formed in the spin bath is stretched, washed, and dried. For staple fiber production, the process sequence includes crimping and cutting. Hydraulic and coagulation circumstances limit the linear speed at which such wet spinning can be effected to not much over 100 m/min. Because of this, the process is best suited to the production of large total tex (denier) continuous filament yarns or of very large tows for production of staple fibers.

Rayon fibers, particularly staple fibers, serve in a wide variety of uses, alone and in combination with other fibers. High-tenacity, high total tex (denier) filament is used in tire cord and other industrial uses, although rayon's once-dominant tire cord position has long since yielded to nylon and polyester fibers.

Rayon staple is produced in a variety of types, including polynosic and high wet-modulus types, these retaining greater strength and dimensional stability when wet. Rayon staple is used in apparel, household goods, and various nonwoven fabrics (see Nonwoven textiles).

Despite the maturity of rayon, new product variants continue to be developed. Among the most recent developments are a highly crimped high wet-modulus item (4) closer to cotton in fabric hand and cover (opacity) either by itself or in blends with polyester staple, a higher absorptive capacity variant (5) for use in certain nonwoven products, and a hollow viscose rayon variant (6) that also provides more cotton-like hand and cover.

Cellulose Acetate. Unlike rayon, cellulose acetate [*9004-35-7*] is dry-spun, predominantly in continuous filament form. The fiber-forming base polymer is made by acetylating dissolving pulp with acetic anhydride usually in acetic acid solvent with sulfuric acid as catalyst. The fully acetylated cellulose initially produced is partially hydrolyzed to produce secondary acetate. Following catalyst removal, precipitation of the acetate flake, washing, and drying, the flake is dissolved in acetone containing a small amount of water. This spinning "dope" is subjected to extensive filtration and then dry-spun to evaporate the acetone solvent. The filament yarn so produced is wound at the bottom of the chamber on a bobbin or other suitable package for such further handling as end-uses dictate. Linear spinning speeds up to or exceeding 1000 m/min can be achieved.

Acetate fabrics have attractive appearance, pleasant hand, and excellent whiteness before dyeing. A wide variety of textile fabrics, both knit and woven, are produced. Women's apparel in which the relatively poor abrasion resistance of acetate fibers is not a particular problem, draperies, and upholstery are major use areas. Although acetate textile filament yarn is still an important chemical fiber, its production appears to have peaked ca 1970–1971. Since then acetate production has declined accompanied by closing of some plants and abandonment of acetate production by some manufacturers (7). A relatively small amount of acetate staple is also produced.

Cigarette filters represent a major and still growing use for acetate tow. The tow is a continuous-filament product containing several thousand filaments and made by combining the requisite number of multifilament yarns from several spinning chambers into a tow that is then crimped and baled for use in cigarette-making. Acetate filters predominate worldwide because of their combination of appearance, taste, and economics (see Cellulose acetate and triacetate fibers).

Cellulose Triacetate. Cellulose triacetate [*9012-09-3*] filament became a useful specialty fiber in the 1950s, initially in the United States and the United Kingdom. Its production is similar to that of acetate, differing in that the initial reaction product is not hydrolyzed to a product with less than 92% of the hydroxyl groups esterified, and the spinning solution for the triacetate flake is made with a mixture of methylene chloride and a small amount of methanol or ethanol. Its importance is that, unlike regular cellulose acetate, it can be heat-treated to induce a degree of crystallization, which produces dimensional stability and related "ease-of-care" characteristics permitting, for example, water washing. As a result heat-treated triacetate resembles such fully synthetic fibers as nylon or polyester.

Other Fibers Based on Natural Polymers. Natural polymers other than cellulose that are capable of producing fibrous structures in nature suggested themselves as bases for chemical fibers. Two types of such noncellulosic chemical fibers reached commercialization in the 1930s and 1940s. Although neither is any longer produced in the United States or of much commercial significance elsewhere, these fibers deserve mention.

Alginate Fibers. Both monofilament and multifilament fibers can be produced by wet spinning an aqueous sodium alginate solution based on alginic acid from seaweed (see Gums). The spin bath is acid to produce alginic acid fibers or a bath containing calcium ion is used to produce calcium alginate [*9005-35-0*] fibers. The latter was commercialized in the United Kingdom during World War II. These fibers are unsuitable for ordinary textile use, as they are very sensitive to weak soap or alkali solutions. Quite small amounts have been made for special uses in which the fiber is dissolved out of a textile structure. Production is understood never to have exceeded a hundred metric tons annually.

Protein Fibers. Just as cotton (qv) and flax (see Fibers, vegetable) are natural cellulose fibers so wool (qv) and silk (qv) are protein fibers in which polymeric amino acids provide the fibrous structure. These polypeptides are natural polyamides, analogous to the synthetic nylons, particularly nylon-6, though differing in that they are polymers of alpha amino acids (see Biopolymers).

Regenerated protein fibers have been made commercially since the 1930s. Several have been based on casein from milk, the fibers being produced by wet spinning (see Milk products). Most of these fibers are treated with formaldehyde to improve wet strength and reduce alkali sensitivity. The casein fibers have relatively high moisture absorption, similar to that of wool, and have been primarily used in staple form for blending with wool. A small amount is still made outside the United States.

The other regenerated protein fibers, based on wet-spinning proteins from peanuts, corn meal, and soybeans are chiefly of historical interest.

Synthetic Fibers

Polyamide Fibers. The importance of the development of nylon can hardly be overstated. It opened the way to a great number of synthetic fibers having a wide variety of chemical and physical properties. It freed the chemical fiber industry from the restrictions inherent in the use of natural polymers, permitting controlled tailoring of polymers.

Nylon-6,6. Nylon-6,6 [*9011-55-6*] is so designated because both the adipic acid and the diamine from which the polymer is synthesized contain 6 carbon atoms. Although coal tar intermediates were used, these starting materials are now derived from other intermediates based ultimately on petroleum hydrocarbons. The fiber-forming polymer is formed by heating nylon-6,6 salt under pressure to remove water as the adipic acid and diamine condense to form a long-chain, linear polymer in what is called a melt polymerization. When the desired average molecular weight is reached, pressure is reduced, and the polymer is usually extruded in spaghetti-like strands, cooled, and cut into chips. In some cases, the molten polymer is conveyed directly to melt-spinning equipment.

In spinning, the polymer chips are re-melted, either on a grid or in the melting zone of a screw extruder, metered via a gear pump, passed through a filter (usually sand) to remove solid impurities, and extruded vertically downwards through spinnerettes. The filaments are cooled and solidified, lubricated with a material designed to eliminate static and provide desired frictional characteristics, and wound on a rotating tube to form a "package." At this stage, the fiber is essentially unoriented, weak, and unsuitable for textile use.

Fiber strength is developed by subjecting the as-spun yarn to a drawing step, in which orientation and crystallization of the polymer molecules occur. This stretching is achieved by passing the yarn between two sets of rolls, the second of which operates at a surface speed of up to ca 5 times that of the feed roll, so that the yarn is drawn to several times the original length. In the draw zone, the yarn is usually passed over a pin to localize the draw point and enhance uniformity. Drawing is usually effected at ambient temperature, heat actually being generated in the individual filaments by the stretching process. Textile nylon is usually stretched and twisted simultaneously, the multifilament yarn being given a degree of twist to aid in subsequent processing. For certain uses, hot drawing (at temperatures, however, well below the polymer's crystalline melting point of about 263°C) or multiple drawing, or both, may be used to produce high tenacity with reduced extensibility.

Nylon-6,6 is a strong, tough, abrasion-resistant fiber that is relatively stable and relatively readily dyeable. It has good knot strength (ie, high percentage retention of tenacity when knotted) and good elasticity. By heat setting after being drawn, nylon fiber can be dimensionally stabilized at temperatures up to that of the annealing process, permitted the "boarding" of hosiery to preset shapes, pleat setting in garments, and importantly, providing ease-of-care characteristics.

Because of its broad range of characteristics, nylon-6,6 has a wide spectrum of uses. Textile nylon is used in hosiery, apparel, and home furnishings. Most of this fiber is multifilament, but monofilament nylon is used in sheer women's stockings. The high strength of nylon permits manufacture of lightweight and very sheer fabrics. Because of good abrasion resistance and resilience, nylon dominates the U.S. carpet market (8). A particular product for this area is BCF, or bulked continuous filament nylon. The bulking is accomplished by mechanically crimping multifilament yarn while it is heated, then cooling it in the bulky configuration. Nylon staple is also important in carpets.

Strong, durable nylon monofilaments are used in fishing lines and fishnets; high-strength nylon multifilament has a variety of automotive and industrial uses ranging from belting to filter fabrics. Tire cord (qv) is a major use, nylon having displaced high strength viscose rayon in this application.

Nylon-6. Although Carother's research in the late 1920s and 1930s touched on fiber-forming polymers from aminocaproic acid, commercialization of nylon-6 [*25038-54-4*] fibers, so named because of the six carbons in the caprolactam monomer, first occurred in Europe shortly after commercialization in the United States of nylon-6,6. Manufacture is based on the use of caprolactam, the internal cyclic amide of aminocaproic acid, derived petrochemically (see Polyamides).

Unlike nylon-6,6, caprolactam is polymerized by ring opening, followed by repetitive addition until the desired molecular weight is achieved. This reaction is also a melt polymerization in which an equilibrium content of monomer exists, the latter being removed after polymerization, usually by water washing. Nylon-6 polymer is also melt-spun, and the manufacture, fiber properties, and uses are very similar to those of nylon-6,6. Schlack at I.G. Farbenindustrie is credited with establishing appropriate conditions of converting caprolactam to a useful fiber-forming polymer.

Nylon-6 production has grown steadily, and has long since spread from Europe to the United States and elsewhere.

Qiana. In 1968 DuPont introduced a polyamide textile fiber different from either nylon-6 or 6,6. This fiber, named Qiana, is a condensation polymer containing alicyclic rings (9). As with other nylons, melt polymerization and melt spinning are used. The fiber is very silk-like in appearance and behavior, particularly with respect to fabric resilience after crushing. Qiana fiber is used mainly in apparel, particularly dresses, blouses, and shirts.

Aromatic Polyamides. Also introduced in the 1960s was a series of polyamides (10) in which the intermediates are wholly aromatic (see Aramid fibers). By virtue of the aromatic ring structures coupled with the relative regularity of the polymolecular chain, they are very high melting, very high in thermal stability, and possessed of generally high performance properties.

These polymers do not melt, for all practical purposes, other than at temperatures involving decomposition, and they are nearly insoluble. Consequently, conventional melt polymerization and melt spinning are not feasible. Polymers can be prepared from the dihalides of the dibasic acid by interfacial condensation with diamines at relatively low (<100°C) temperature or by reaction in suitable solvents (see Polymerization). In some cases sulfuric acid may be used as the solvent for these rather intractable polymers.

The intermediates, the polymerization, the fiber spinning and aftertreatment (drawing to orient, crystallizing and heat treatment) are all more costly than for the aliphatic nylons such as 6,6 or 6. The aramid fibers thus are higher in price, and find use in applications where exceptionally high strength, very high modulus (stiffness) or high resistance to heat, or both, are required.

Commercial aramid fibers (qv) are trademarked Nomex and Kevlar by DuPont. Nomex, based on meta-linked isophthalic acid and *m*-phenylenediamine, is used in specialty papers (11) for electrical insulation and aircraft structures, in protective garments, and in other applications requiring high thermal stability. Kevlar is an aromatic polyamide containing para-oriented linkages. It melts at over 500°C, is exceptionally high in strength with a tenacity more than twice that of high strength nylon or polyester, and a very high modulus. It finds use in heavy duty conveyor belts, and in composite structures with casting resins such as epoxies (see Embedding). It is contending against steel in radial tire reinforcement (12).

Polyester Fibers. Although Carothers had demonstrated that fiber-forming polyesters could be synthesized from glycols and dibasic acids, or from hydroxy acids, the laboratory work focused on polyamides for development. The original, and still growing, commercial success of polyester fibers is attributable to the work of Whinfield and Dickson in the United Kingdom (13). They demonstrated that the presence of aromatic rings in a linear polyester of sufficient molecular weight led to high melting points, good stability, and strong, tough fibers. Commercial production was initiated in the late 1940s and early 1950s (see Polyester fibers).

Polyester, now the largest-volume synthetic fiber in the world, surpassed nylon worldwide in 1972. The polymer, like nylon, is made by melt polymerization. Ethylene glycol reacts with either dimethyl terephthalate (DMT) or terephthalic acid (TPA) under heat and high vacuum. In the former case, methanol is eliminated in the condensation/transesterification reaction; in the latter, water is evolved. The intermediates are again petrochemicals-based. The polymer has a crystalline melting point of ca 250°C, similar to that of nylon-6,6.

Melt-spinning of polyester fibers is also similar to that of nylon. The as-spun fiber is amorphous and comparatively weak. Drawing, carried out hot because the glass transition temperature is higher than that of nylon, causes orientation of the polymer chains along with crystallization, and fiber strength is developed. The fiber may be heat-set (annealed) under tension to relax strains in the molecule and induce some additional crystallization. Like nylon, the extent and nature of the drawing can be used to control tenacity and elongation, these essentially being interchangeable. In more recent years, fiber production processes involving a combined spin-draw, or a combined drawing and texturing step, have proliferated.

Eastman Chemical Products, Inc. produces not only poly(ethylene terephthalate) [*25038-59-9*] under the trade name Kodel, but also a polyester based on condensing terephthalic acid with 1,4-dimethylolcyclohexane (14). This fiber is in many respects similar to the ethylene glycol-based polyester but melts higher and has a lower specific gravity and excellent recovery from stretch. It is designated Kodel II.

Polyester fibers are produced in both staple and continuous filament form. The former predominated in earlier years, but by 1977 polyester filament volume was 80% that of staple. The fiber is remarkably versatile, probably more so than any other chemical fiber. It is strong, abrasion resistant, relatively stable, higher in modulus than nylon, and of lower moisture regain. Its structure is such that polyester is more difficult to dye than most textile fibers, requiring somewhat specialized techniques.

Uses range widely over the apparel, home furnishings, automotive, and industrial fields. Enormous quantities of polyester staple are blended with cotton, rayon, or wool in spun yarns used for apparel. Good abrasion resistance also suits polyester to use in carpeting. Continuous filament yarn finds wide use in apparel; much of it is textured to produce bulk and opacity for this use.

In the automotive field, just as nylon tire cord (qv) to a great extent displaced rayon, high-tenacity polyester tire cord has made inroads against nylon, particularly in passenger car tires (15). Other industrial uses include belts, ropes, and filter fabrics.

Acrylic Fibers. Synthetic fibers based on acrylonitrile (qv) appeared in the United States in the 1940s (16). Subsequently, fibers of lower acrylonitrile unit content were introduced. These modacrylics present some use problems because they are lower in melting point and less resistant to the effects of heat than are acrylics, but they are relatively flame-resistant and useful in applications where this property is desirable (17) (see Flame retardant textiles; Acrylic and modacrylic fibers).

Acrylic and modacrylic fibers combined are now produced worldwide and represent the third largest in volume of the synthetic fibers produced. They are derived from petrochemical-based intermediates. Because acrylonitrile polymers cannot be melted without decomposition, melt spinning is not applicable. Highly polar solvents, such as dimethylacetamide, have been found to dissolve the acrylic polymers for either dry or wet spinning, both of which are used commercially (see Acetic acid). Virtually all acrylic fiber is produced in staple form. Earlier, filament acrylics were produced, but they proved very difficult to dye uniformly. A small amount of textile filament yarn is still produced in Japan, and acrylonitrile homopolymer filament is produced for carbonization in the manufacture of carbon fibers (see Carbon).

The fiber-forming acrylic polymers are high in molecular weight and are produced primarily in aqueous medium by free radical-initiated addition polymerization. The regular acrylics are usually copolymers with minor amounts of one or more comonomers, such as methyl acrylate, used to provide accessibility or sites for dyestuffs. Like other synthetics, acrylic fibers are stretched to develop orientation and fiber strength.

Acrylic fibers have good tenacity, although less than polyester or polyamides, excellent stability to sunlight, good dye acceptance as a result of the copolymer system used for this purpose, and a soft, pleasing hand of wool-like characteristics. Abrasion resistance, although well below that of nylon or polyester, is nevertheless good, and superior to that of wool. For these reasons, the acrylics find widespread use in both indoor and outdoor furnishings, including awnings and draperies, and in blankets, sweaters, and carpets (see Textiles). Blending low- and high-shrinkage yarns and shrinking the blend produces a bulky yarn. The levels of shrinkage are produced by different degrees of stretching in fiber production.

A variety of modacrylic fibers are produced in the United States and abroad. These are acrylonitrile copolymers or terpolymers with significant comonomer contents. Among the comonomers used are vinylidene chloride, vinyl chloride, and apparently acrylamides (qv). The fiber-forming modacrylic polymers are more soluble than the acrylics, some of them in acetone, for example, and are made commercially both by dry and wet spinning. The modacrylic fibers have various textile uses, including fake fur pile fabrics (see Furs, synthetic) and protective garments, among others. Modacrylics, because of their self-extinguishing or flame-resistant characteristics are also used in wigs (qv) and doll's hair.

Polypropylene. The discovery that catalyst systems could convert certain olefins, notably ethylene and propylene, to crystalline, ordered, or stereoregular polymers provided a basis for ordered polyolefin fibers of defined melting points (see Olefin fibers). Large-scale availability of low cost monomers from petrochemical operations, polymer specific gravities less than one, and melt-spinnability were key factors fostering the major development and commercialization of polyolefin fibers.

World polyolefin fiber production, most of it in the United States and Western Europe, is fourth in volume among the synthetics. Polypropylene [*9003-07-0*] is thought to constitute the majority of commercial polyolefin fiber. In stereoregular form, polypropylene has a crystalline melting point of about 165°C, compared to ca 125°C for low-pressure (regular) polyethylene [*9002-88-4*] and ca 110–120°C for high-pressure polyethylene (see Olefin polymers).

These fibers are melt-spun and produced in monofilament, multifilament, and staple forms. Both conventional melt spinning and a type of film extrusion are used. In the latter the film is mechanically slit, split, or fibrillated to produce coarser yarns where tex (denier) uniformity is not critical (18). In all cases the as-spun or extruded material is stretched to develop orientation and strength.

Polypropylene fiber is relatively low melting for a chemical fiber, limiting its uses. Offsetting this and other relative deficiencies is low cost, high strength, very great chemical inertness, and, because of the low density, high yardage of fiber of a given tex per kilogram (den/lb).

Because of the chemical inertness, dye acceptance is inadequate. For uses where color is requisite, either the base polymer is modified to provide dyeability or the fiber is spun-colored; that is, produced with pigments incorporated in the polymer and fiber during melting. Because of oxygen and light sensitivity, suitable stabilizers are also incorporated (19).

Textile uses include upholstery and carpeting, particularly indoor-outdoor carpets. In addition to face yarns for carpeting, polypropylene fiber, particularly that produced from film, is used for woven carpet backings. The low melting point, which prevents easy ironing, and the fabric hand have largely precluded polyolefins from application in apparel fabrics. Other important uses include rope and cordage, fishnets, and filter media. Negligible moisture absorption, resistance to decay by organisms, and low density, which causes the fiber to float make polyolefins particularly suited to such uses.

Vinyon and Other Vinyl Fibers. In addition to vinyl addition polymers based on acrylonitrile (ie, vinyl cyanide) others have been commercialized. Indeed, fibers known as Vinyon were commercialized in 1939 closely on the heels of the introduction of nylon-6,6, as a result of the joint efforts of the Union Carbide and American Viscose Corporations (now Avtex Fibers, Inc.) (see Vinyl polymers).

Poly(vinyl chloride) [*9002-86-2*] by itself tends to decompose on heating and is difficultly soluble in common solvents. However, a small amount of vinyl acetate copolymerized with it acts as an internal plasticizer, promotes solubility in solvents such as acetone, and yields a copolymer capable of producing useful fibers. Commercially the copolymers are about 85–90% vinyl chloride and 10–15% vinyl acetate units. Although textile filament was produced earlier, U.S. Vinyon production as of 1978 is entirely in staple form by Avtex. It is made by a dry-spinning process like that for cellulose acetate fiber. The fiber is temperature sensitive, starting to soften, shrink, and become tacky below the boiling point of water. It finds specialty use in applications requiring bonding and heat sealing, in conjunction with other fibers.

Other Vinyon fibers from vinyl chloride homopolymers, rather than copolymers, are produced abroad, notably in France, Italy, and Japan (20). They are produced primarily by spinning from solution, carbon disulfide being one of the solvents used. A variety of products are made but all share relatively limited thermal stability, tending to shrink at temperatures in the neighborhood of boiling water. They are, however, resistant to moisture and rotting, and inherently nonflammable, suiting them to specialty uses such as filter cloths, nonflammable garments, fishnetting, and felts (qv) for insulating purposes. Strength and residual shrinkage can be varied to fit the application.

Saran. Saran fibers based on vinylidene chloride polymers (21) are produced in several countries under various trade names. They are based on vinylidene chloride copolymerized with small amounts of vinyl chloride, and still smaller amounts of acrylonitrile.

Saran fibers are melt-spun and then stretched. They are characterized by a pale straw color and by resistance to water, fire, and light, and bacterial and insect attack. Their relatively low melting points require excessively low ironing temperature. They are difficult to dye, requiring pigmentation in the melt. The fibers find specialty use in certain types of upholstery, filter cloths, and fishnets. Both staple fibers and filament yarns are produced, the latter in both heavy tex (denier) monofilaments and in multifilament form. The relatively low cost of these fibers is one of their attractions (see Vinylidene polymers).

PeCe. PeCe fiber, a post-chlorinated vinyl chloride polymer, was developed in Germany and was of some significance there during World War II. The post-chlorination raises the chlorine content of the polymer from 57 to 64% and confers acetone solubility, permitting ready wet spinning. Like many other vinyl fibers, it is low melting and restricted to special applications.

Vinal. Vinal is the U.S. term applied to poly(vinyl alcohol) [*9002-89-5*] fibers. These fibers originated in Japan and considerable production continues there. Although vinyl alcohol does not exist as a monomer, its polymer can be obtained by the hydrolysis of the acetyl groups from poly(vinyl acetate). Vinyl acetate, a relatively abundant and modest-cost monomer, is polymerized and then saponified to poly(vinyl alcohol), which is wet-spun from hot water. In this form, the fiber is water soluble. To make it commercially useful, the fiber is treated with formalin and heat, rendering it water insoluble. As with other fibers, the relative tensile strength and extensibility can be varied by varying the degree of stretch used. The fiber has reasonable strength, a moderately low melting point (222°C), limited elastic recovery, good chemical resistance, and resistance to degradation by organisms. Both staple and filament forms, including monofilament, are produced. Vinal is used in bristles, filter cloths, sewing thread, fishnets, and apparel. Although other producers have closely studied this fiber, production has remained largely confined to Japan. Elsewhere, it would appear, the overall cost-performance ratio has not justified commercialization.

Polychlal. Polychlal fibers (22), unique to Japan, are closely related both to the Vinyon and Vinal types. They are produced from an emulsion of poly(vinyl chloride) and a matrix of poly(vinyl alcohol). This is essentially wet-spun into aqueous sodium sulfate and formaldehyde-treated. Production on a modest scale continues in Japan, and some of this fiber has been exported to the United States for use in flame-retardant apparel.

Other Synthetic Fibers. Synthetic fibers other than those already described have been commercialized. Although production is, in most cases, relatively small, these products are frequently important in specific applications.

Benzoate Fibers. A melt-spun polyester fiber is produced in Japan from the self-condensation polymer derived from p-(β-hydroxyethoxy)benzoic acid, the aim being to provide a silk-like fiber (22). Although limited production of this fiber continues, it does not appear likely to grow significantly.

Poly(hydroxyacetic Acid Ester) Fibers. Although most of the aliphatic polyesters, as distinct from the aromatic acid-based type, are low melting, poly(glycolic acid) [*26124-68-5, 26009-03-0*], also known as poly(hydroxyacetic acid) or polyglycolide, is an exception, melting at over 200°C (23). Fibers can be melt spun from homopolymers derived from glycolide or from copolymers derived from glycolide and a minor amount of lactide. The fibers are crystalline, and can be oriented by stretching. Because they are absorbable in the body, they have become important in surgical suture applications, where they replace catgut sutures in many such applications (see Sutures). Compared to the textile fibers, these are relatively high-priced, low-volume specialty items.

Spandex Fibers. Synthetic elastomeric fibers, developed to compete with or improve on natural rubber fiber, differ from ordinary textile fibers in having very high extensibility to break (500–600%) and high recovery from such stretching. The segments of the polyurethane are based on low molecular weight polyethers, such as poly(ethylene glycol), or polyesters, which then react with an aromatic diisocyanate to give a three-dimensional, elastomeric product. Though the chemistry is complex, basically the elastomeric products depend on the presence of hard and soft segments and cross-linking. The fibers produced are white and dyeable and are stronger and lighter than rubber, making them particularly suited for use in foundation garments, bathing suits, support hose, and other elastic garments (see Fibers, elastomeric; Urethane polymers).

Fluorine-Containing Fibers. Polymers based on polytetrafluorothylene [*9002-84-0*], DuPont's Teflon in the United States, have uniquely high chemical stability and inertness, no water absorption, low frictional characteristics, and high melting points with decomposition preceding and accompanying melting (see Fluorine compounds, organic). These properties clearly make conversion to fiber form difficult. Nevertheless, the problem was solved in the 1950s by emulsion spinning. This process appears basically to comprise sintering or fusing extremely small polymer particles that are themselves fibrous and have been aligned by passing a colloidal dispersion through a capillary. The fused fiber is subsequently drawn. Teflon fibers in both filament and staple forms are high priced and find, as would be expected, highly specialized applications, such as packings, special filtration fabrics, and other uses where corrosion resistance, lubricity, and temperature resistance are requisite.

Carbon Fibers. Carbon fibers (24), the subject of a great deal of research and development in the past decade, are characterized by extremely high strength and modulus along with high temperature resistance. These characteristics suggest themselves for particular applicability to high performance, reinforced composite structures where strength, stiffness, and lightness of weight are at a premium, eg, in special aerospace, industrial, and recreational applications. The composite structure may include other fibers, such as glass, along with a matrix resin, such as polyesters, epoxies, or polyimides.

Carbon fibers are made by controlled carbonization of an already formed fibrous structure based on an appropriate organic polymer. In particular carbonization has been applied to filament yarns of homopolymers of acrylonitrile, to high tenacity viscose rayon multifilament yarns, and to pitch for lower strength carbon fibers (see Ablative materials; Carbon; Composite materials).

Phenolic-Type Fibers. The Carborundum Company developed a fire-resistant fiber based on the technology of the well-known phenol–formaldehyde resin condensates. The latter, long used for moldings, laminates, and similar thermoset resin applications, are cross-linked, three-dimensional polymers made by curing a non-cross-linked precursor condensate of formaldehyde with phenolic compounds. Special processes were developed to produce fibers. The result, trade-named Kynol, is presently produced in Japan (Nippon Kynol Co., U.S. distributor, American Kynol Co.) and used for protective garments (see Novoloid fibers; Phenolic resins).

Polybenzimidazole Fibers. Poly(benzimidazole) [*26986-65-9*] fibers (25) have been produced semicommercially from condensation polymers based on the reaction of 3,3′-diaminobenzidine and diphenyl isophthalate. A process in which the fibers are dry-spun and then drawn was developed by the Celanese Corporation, partially supported by the government. The fibers are of the high performance type, of very high melting point, good strength and extensibility, nonflammable, and surprisingly for a synthetic fiber, of high moisture regain: 13%. These fibers are of interest for aerospace applications and potentially for industrial applications as well subject to incompletely defined economics (see Heat resistant polymers; Polyimides).

Glass. The world production of textile glass fiber listed at ca 800,000 metric tons for 1977 does not include substantial production of glass fiber batting materials. In many ways glass fiber is unique among the chemical fibers. It is an inorganic fiber, in contrast to the numerous organic, natural, and synthetic fibers, and the fibers are neither oriented nor crystalline. The fibers are strong, nonflammable, and rather heat-resistant, as well as highly resistant to chemicals, moisture, and attack by organisms. Not surprisingly, glass fibers are quite low in extensibility and higher in density than the organic fibers.

The fibers are produced by melt spinning. To offset their brittleness, they are spun into fine filaments. After lubrication, these are wound as a multifilament strand unless insulating batting is being produced; in the latter case the filaments are attenuated while still molten and are deposited in the form of a thick bat (see Insulation, thermal).

Glass fibers, in forms ranging from filament yarn to mats, woven rovings, and staple, find a variety of important uses. In addition to use as insulating bats, they are used in fabric form for draperies where the nonflammability, inertness, and resistance to the effects of sunlight are assets. Glass fiber cloth, as well as batting, is used in a variety of insulating applications and for filtration.

One of the most important uses for glass fiber is in reinforced plastics, particularly reinforced thermosetting polyester resins (26). Since the 1950s the construction of pleasure boats, both sail and power, has been shifted from hulls of wood to glass-reinforced polyester. Such composites are widely used in industrial and automotive applications. Glass-filled thermoplastic resins have also grown in volume. The glass fibers are used in nylon, polyacetal, and poly(butylene terephthalate) molding resins, for example (see Laminated and reinforced plastics). Glass fibers are also used in both radial and bias-ply automotive tire reinforcement (see Fiber optics; Glass).

Metal and Other Inorganic Fibers. Fine steel wire is, of course, well-known for radial tire reinforcement. However, a number of metallic and other inorganic fibers have been investigated for special applications. Metallic fibers have been used in small quantities with organic fiber carpet yarns to minimize static problems (see Antistatic agents). Fibers of boron, boron–tungsten, steel, beryllium, boron and silicon carbides, boron and silicon nitrides, alumina, zirconia, and other inorganics have received attention (27), particularly for high performance specialty uses. Most are quite costly, and the quantities employed quite small.

Fiber Modifications

It is axiomatic that the applicability of a given fiber is a function of the overall price-performance considerations for that fiber. Although certain fibers are barred from a particular use by some inherent limitation, there is much overlapping and inter-fiber competition. Moreover, as the chemical fiber industry has grown, modifications of individual fiber types have been developed to broaden uses.

These modifications may be chemical, physical, or a combination of both, to improve performance in a given use or to adapt it to uses for which it was not previously satisfactory. Since the preceding sections have only alluded briefly to such modifications, it is in order to review them in somewhat greater depth.

Physical Modifications. As noted, when chemical fibers are stretched, their tensile strength generally increases and their extensibility decreases. Use is made of this in producing highly drawn, high tenacity nylon and polyester for tire cord and industrial uses. By adjusting the draw ratio the fiber modulus can be altered, and polyester staple of a modulus suitable for blending with cotton is made this way.

By decreasing the filament tex, and therefore increasing the number of filaments in a yarn of given total tex, the suppleness of the yarn and the softness of hand in fabrics made from it are increased. This is the trend in polyester filament yarns for apparel (28).

Texturing or bulking of filament yarns is used to produce yarn with more of the characteristics of yarn spun from staple fibers while eliminating several steps involved in producing the latter. Texturing can also be used to produce stretch yarns, as from nylon. Most texturing of chemical fibers is based on the fact that if the yarn is distorted while heated to render it more plastic, then cooled, the yarn "remembers" its distorted structure and seeks to restore it. Several methods are briefly summarized in Table 3.

False-twist texturing has been used extensively particularly with polyester filament yarns. Basically, a running thread line is temporarily highly twisted between two points and heat-set. It is then untwisted, whereupon it kinks or bulks. Nylon of a high degree of stretch is thereby produced. Nonstretch bulked polyester yarn is also made this way; a second heater is used to set the bulked configuration. Initially, a spindle rotating at very high speeds and through which the yarn passes was used to insert the false twist. In recent years, the use of friction false-twist texturing has been growing. In this method, the running thread line proceeds over the edge of several disks in series, the twist being imparted by friction.

False-twist texturized polyester filament yarn has become a major item of commerce, with huge quantities being used in apparel. Originally, fully drawn filament yarn was textured. In recent years, it has been found advantageous in many cases for the fiber producer to make a partially drawn yarn with only some degree of orientation. This is known as POY (partially oriented or preoriented yarn). The texturer, who may also be the fiber producer, then simultaneously draws and textures this yarn.

Staple fiber bulk is enhanced by crimping before cutting into staple lengths and by using differential shrinkage fibers as described for acrylics. Rather elegant use of this principle is made in bicomponent acrylic staple fibers in which each fiber consists of two components fused together. These components are formulated to have different shrinkage potential. When subjected to hot water, a helical crimp develops.

Changing the cross section of the individual filaments in a yarn may affect fabric appearance and hand. In melt spinning the use of spinneret holes of special shapes

Table 3. Filament Yarn Texturing/Bulking Methods for Fibers

Means of yarn distortion	Comments
knitting, fabric heat-setting, de-knitting	called knit-de-knit; cumbersome, currently used little or not at all
passage over hot knife edge	Agilon process; spiral crimp; nylon stretch yarn an example
passage between gear teeth	gear-crimping process; zig-zag configuration; used for BCF carpet yarn
passage into tightly enclosed space	stuffer-box crimping; saw-toothed configuration; Ban-lon yarns; tows for crimped staple fiber
twisting, particularly false-twisting	stretch and bulky nonstretch yarns; textured polyester apparel yarn so produced
air turbulence	Taslan yarns; filaments looped randomly by air jet while over-feeding; not dependent on heat-setting; used for blends of nylon and polyester with acetates

can produce multilobal filament cross sections. Polyester yarn, for example, that is so spun produces a fabric without the sometimes objectionable shiny or plastic appearance. Alternatively, special cross sections can be used to provide "sparkle" yarns.

Chemical Modifications. Modified fibers can be produced by chemical alteration of the polymer, fiber, or fabric, or by the use of nonreacting additives. Improved dyeability is usually effected by use of a suitable comonomer. Polyester fibers ordinarily dyed with disperse dyestuffs are made cationically dyeable by use of a comonomer providing acidic sites in the base polymer.

Spun-colored fibers can be produced by incorporating a pigment before spinning. Acetate, triacetate, polyester, polyolefin, and acrylic fibers have used this approach for certain applications. In related processes, white pigments, particularly titanium dioxide, are used to dull the natural luster of chemical fibers, or optical brighteners may be added, such as to polyester, to increase the perceived whiteness of the fiber (see Brighteners, fluorescent).

Chemical approaches were of particular importance in seeking to meet U.S. Government regulations regarding flame-retardant garments for such uses as children's sleepwear. A halogenated phosphate compound known as Tris, tris(2,3-dibromopropyl) phosphate, used as an additive in acetate and triacetate spinning, and as a fabric additive with polyester, enabled fabrics from these fibers to meet federal requirements, but its use was dropped in light of work suggesting possible carcinogenicity. More recently, Hoechst Fiber Industries has introduced in the United States a flame-resistant polyester variant trademarked Trevira 271 based on copolymerization involving a phosphinic acid derivative (29) (see Flame retardants, phosphorus compounds).

The introduction of permanent-press treatment of polyester cotton staple fiber blends in fabric form was a major factor in the growth of polyester usage. Wash-and-wear characteristics can be imparted to fabrics by treatment with suitable thermosetting (cross-linking) resin systems, but an accompanying strength loss made this approach unsuitable for all-cellulosic fibers such as cotton or rayon. Blending the strong polyester fiber with the cellulosic overcomes this problem and yields a blend widely used in shirting and slacks fabrics.

Economic Aspects

Chemical fiber economics are complex, varying not only with the specific fiber and process, but also with the item (ie, staple or filament, tex, etc) of a given fiber, and with geography, scale, capacity utilization, and other factors. Certain generalities, however, provide guideposts. Of the noncapital-related costs, raw materials are the single largest item, followed by labor and energy. Both capital and production costs tend to be higher for filament yarn than for staple-fiber plants, and higher for low tex (denier) filament than for higher tex (denier) material. Because solvent and solution handling tend to increase energy consumption, labor, and environmental problems, both capital and operating costs for wet- and dry-spun fiber tend to be higher than those for melt-spun fibers.

High capital costs have for some years prevented construction of new rayon plants, except in special circumstances in certain Communist and developing countries. Rayon textile filament yarn has virtually disappeared because of high costs of wet-spinning fine tex items, despite the fact that cellulose is a renewable resource. Among the major synthetics, those with most favorable raw materials costs are polyester, polyolefins, and acrylics, with nylon-6 and 6,6 being significantly higher. Capital and energy costs, however, are less favorable for the acrylics. All these factors, plus versatility in use, have conspired to favor polyester in both staple and filament form.

Though highly profitable in much of their history, chemical fibers have been severely buffeted in recent years. In addition to the traditional cyclicality of the textile industry, during the 1970s overcapacity in both the fiber and petrochemicals industries and increases in energy and raw materials costs have significantly affected profitability for both U.S. and European fiber producers.

Table 4 lists typical market prices for several important fibers for 1970 and 1978, with 1973 and 1974 also included to reflect sharp increases in petroleum prices in that period. It must be emphasized that these prices reflect external factors as well as actual underlying production costs, that there are scores of other items at different prices, and that certain markets do not command these prices. All these economic pressures combine to place emphasis on the research and development of process improvements and cost reduction.

Present Status and Future Possibilities

Today there is one or more fibers to fit virtually any use, and there is an enormous range of properties in the broad spectrum of commercially available fibers. In addition, the direct conversion of synthetic polymers to nonwoven fabric form (30) has grown over the last few years. Polyethylene, polypropylene, and polyester resins, for example, can be converted to what are called spun-bonded fabrics in processes that combine melt-extrusion with a means of laying the filaments in a web and bonding the thermoplastic fibers at cross-over points to form fabrics useful in a variety of applications. Polypropylene carpet backing is just one example (see Nonwoven textiles).

Concurrent with these product developments has been a steady, continuing increase in uniformity of fiber products regarding properties, dyeability, and the like.

Scientific and technological possibilities are still plentiful, but in the future their exploitation is likely to be overshadowed by economic and political developments,

Table 4. Typical U.S. Market Prices Selected Chemical Fibers, 1970–1978[a], $/kg

Fiber and type	1970	1973	1974	1978
rayon staple, regular[b]	0.59	0.72	1.06	1.25
acetate filament, 8.3 tex (75 den)	1.61	1.47	2.02	2.42
16.7 tex (150 den)	1.23	1.10	1.63	2.11
polyester staple	0.90	0.84	0.97	1.19
polyester filament, 16.7 tex (150 den), drawn	2.99	1.94	1.87	1.43
nylon filament, 4.4 tex (40 den)	2.31	2.77	3.12	3.52
95.5 tex (860 den)	1.76	1.71	1.98	2.55
acrylic textile staple[c], low tex (denier)	1.96	1.71	1.76	1.63
polypropylene textile staple, natural[d]	1.01–1.08	0.79–0.86	1.03–1.10	1.14–1.21
177.8 tex (1600 den) BCF, colored filament	2.09	2.20	2.20	2.20

[a] Data from various industry sources, 1978.
[b] High wet modulus rayon staple ranges from $0.05–0.09 higher.
[c] Modacrylics tend to be higher, as do specialty items such as bicomponent acrylics.
[d] Natural refers to nonpigmented fiber.

particularly the increasing shortage of oil, increased cost of energy, environmental regulations, and other regulatory restrictions.

Overall, the chemical fiber industry can be expected to continue to grow, but at a rate lower than that of the 1950s and 1960s. Growth in the cellulosic sector is unlikely except for acetate cigarette filter tow. Growth will be in the synthetic fibers sector, and further inroads against the natural fibers are likely. Many chemical possibilities are in principle capable of exploitation. In the condensation polymers, eg, nylon or polyesters, the dibasic acid, the diamine, or the glycol can be varied widely to affect the properties of the fiber. For example, use of longer-chain glycols than ethylene glycol can impart greater resilience to polyester fibers based on terephthalic acid. Copolymerization in general, including block and graft copolymerization, offers seemingly endless possibilities for both addition and condensation polymers. Certain monomers not presently used commercially are known to produce interesting fibers; a notable case is pyrrolidone from which, by ring-opening polymerization, is obtainable nylon 4, a fiber of higher moisture regain than other synthetic fibers. Apparently, technical problems associated with polymer stability in melt-spinning and/or economic factors have prevented commercialization of nylon 4 fibers despite much work over several decades (see Polyamides).

Nevertheless, for the remainder of this century and perhaps indefinitely, it seems unlikely there will be any new, broad-based chemical fibers or that those relatively small-volume textile fibers already in existence will grow rapidly or become more widespread than at present. Thus, among the synthetics, the predominance of polyester, nylon, acrylic, and olefin fibers is likely to continue with continuing development of very specific modifications for specific uses.

BIBLIOGRAPHY

"Fibers" in *ECT* 1st ed., Vol. 6, pp. 453–467, by H. F. Mark, Polytechnic Institute of Brooklyn; "Fibers, Man-Made" in *ECT* 2nd ed., Vol. 9, pp. 151–170, by H. F. Mark, Polytechnic Institute of Brooklyn and S. M. Atlas, Bronx Community College.

1. H. Mark and G. Whitby, *Collected Papers of W. H. Carothers,* Interscience Publishers, Inc., New York, 1940.
2. *Textile Organon* **39**(6), 89 (June 1968).
3. *Ibid.,* **49**(6), 73 (June 1978).
4. *Mod. Text.* **52**(4), 19 (1976).
5. J. W. Schappel, F. R. Smith, and E. A. Zawistowski, *Textile Research Institute 48th Annual Research and Technology Conference, Atlanta Georgia, April 5–6, 1978.*
6. *Mod. Text.* **58**(6), 18 (1977).
7. *Ibid.,* **58**(3), 6 (1977).
8. *Chem. Eng. News.* **56,** 11 (Dec. 4, 1978).
9. G. Clayton and co-workers, *Text. Prog.* **8,** 1, 27 (1976).
10. *Ibid.,* pp. 98–115.
11. *Properties of Nomex Types E-54 and E-55 Aramid Paper,* DuPont Preliminary Information Memo 401, E. I. du Pont de Nemours & Co., Inc., April 1978.
12. R. E. Wilfong and J. Zimmerman, *J. Appl. Polym. Sci.* **31,** 1 (1977).
13. J. R. Whinfield, *Text. Res. J.* **23,** 289 (1953).
14. E. V. Martin, *Text. Res. J.* **32,** 619 (1962).
15. M. E. Denahm, *Mod. Text.* **59**(11), 13 (1978).
16. R. Meredith, *Text. Prog.* **7,** 21 (1975).
17. Ref. 9, pp. 61–69.
18. Ref. 9, p. 81.
19. Ref. 9, pp. 78–79.

20. R. W. Moncrieff, *Man-Made Fibers,* 6th ed., John Wiley & Sons, Inc., New York, 1975, pp. 522–532.
21. H. Mark, S. M. Atlas, and E. Cernia, *Man-Made Fibers Science and Technology,* Vol. 3, Wiley-Interscience, New York, 1968, pp. 303–326.
22. *Chemical Fibers of Japan 1970/1971,* Japan Chemical Fibers Association, Tokyo, 1970, p. 50.
23. Ref. 9, p. 50.
24. E. Fitzer and M. Heym, *Chem. Ind.* (16), 663 (1976).
25. R. H. Jackson, *Text. Res. J.* **48,** 314 (1978).
26. M. E. Dunham, *Mod. Text.* **59**(11), 16 (1978).
27. R. S. Goy and J. A. Jenkins, *Text. Prog.* **2,** 1, 31 (1970).
28. Ref. 8, p. 10.
29. W. S. Wagner in ref. 5.
30. R. A. A. Hentschel, *Chem. Technol.* **4,** 32 (1974).

General References

Refs. 9, 20, 21, and 27.
R. Hill, *Fibers from Synthetic Polymers,* Elsevier Scientific Publishing Company, Amsterdam, 1953.
J. J. Press, ed., *Man-Made Textile Encyclopedia,* Textile Book Publishers, New York, 1959.
E. M. Hicks and co-workers, "The Production of Synthetic Fiber," *Text. Prog.* **8,** 1 (1976).

WILLIAM J. ROBERTS
Consultant

FIBERS, VEGETABLE

Natural fibers of vegetable origin are constituted of cellulose (qv), a polymeric substance made from glucose molecules, bound to lignin (qv) and associated with varying amounts of other natural materials. A small number of a vast array of these vegetable fibers have industrial importance for use in textiles (qv), cordage, brushes and mats, and paper (qv) products. Cotton fiber makes up about three-quarters of the world vegetable-fiber tonnage (see Cotton).

Vegetable fibers are conveniently classified according to the part of the plant where they occur and from which they are extracted as shown in Table 1 (1–2): (*1*) leaf fibers are obtained from the leaves of monocotyledonous plants, mostly tropical, which are part of their fibro-vascular systems. Commercially important examples are abaca (Manila hemp), sisal, and henequen. These are the hard fibers used for cordage. The long, multicelled fibers do not readily split apart and are wire-like in texture. (*2*) Bast fibers are obtained from the bast tissue or bark of the plant stem. The group includes flax, hemp, jute, and ramie which are the so-called soft fibers that are converted into textiles, thread, yarn, and twine. The long, multicelled fibers in this case can be readily split into finer cells which are manufactured into textile and coarse yarns. (*3*) Seed-hair fibers, eg, cotton, kapok, and the flosses, are obtained from seeds, seedpods, and the inner walls of the fruit. These fibers are short and single-celled. (*4*) Miscellaneous

Table 1. Selected Vegetable Fibers of Commercial Interest [a]

Commercial name	Botanical name	Geographical source	Use
Leaf (hard) fibers			
abaca	*Musa textilis*	Borneo, Philippines, Sumatra	cordage
cantala	*Agave cantala*	Philippines, Indonesia	cordage
caroa	*Neoglaziovia variegata*	Brazil	cordage, coarse textiles
henequen	*Agave fourcroydes*	Australia, Cuba, Mexico	cordage, coarse textiles
istle (generic)	*Agave* (various species)	Mexico	cordage, coarse textiles
Mauritius	*Furcraea gigantea*	Brazil, Mauritius, Venezuela, tropics	cordage, coarse textiles
phormium	*Phormium tenax*	Argentina, Chile, New Zealand	cordage
bowstring hemp	*Sansevieria* (entire genus)	Africa, Asia, South America	cordage
sisal	*Agave sisalana*	Haiti, Java, Mexico, South Africa	cordage
Bast (soft) fibers			
China jute	*Abutilon theophrasti*	China	cordage, coarse textiles
flax	*Linum usitatissimum*	north and south temperate zones	textiles, threads
hemp	*Cannabis sativa*	*all temperate zones*	cordage, oakum
jute	*Corchorus capsularis; C. olitorius*	India	cordage, coarse textiles
kenaf	*Hibiscus cannabinus*	India, Iran, USSR, South America	coarse textiles
ramie	*Boehmeira nivea*	China, Japan, United States	textiles
roselle	*Hibiscus sabdarifa*	Brazil, Indonesia (Java)	cordage, coarse textiles
sunn	*Crotalaria juncea*	India	cordage, coarse textiles
cadillo	*Urena lobata*	Zaire, Brazil	cordage, coarse textiles
Seed-hair fibers			
cotton	*Gossypium* sp.	United States, Asia, Africa	all grades of textiles, cordage
kapok	*Ceiba pentranda*	tropics	stuffing
Miscellaneous fibers			
broom root (roots)	*Muhlenbergia macroura*	Mexico	brooms, brushes
coir (coconut husk fiber)	*Cocos nucifera*	tropics	cordage, brushes
crin vegetal (palm leaf segments)	*Chamaerops humilis*	North Africa	stuffing
piassava (palm leaf base fiber)	*Attalea funifera*	Brazil	cordage, brushes

[a] Ref. 2.

fibers are obtained from the sheathing leaf-stalks of palms leaves, stem segments, stems, and fibrous husks, eg, piassava and coir. These strawlike, woody, and coarse fibers are used for brush and broom bristles, matting, and stuffing.

Other classifications are applied according to different points of view such as commercial (hard and soft fibers), practical (textile, cordage, brushes and mats, stuffing and upholstery, papermaking and baskets, etc) (3), bast fiber crops for fine textiles, packaging, and soft cordage (4), morphological (hairs, bast, leaves, woody, and others) (3), and crop cultivation (cotton, jute, sisal, flax and hemp; and other fiber crops) (5) (see also Pulp; Bagasse; Sugar; Wood). Leaf fibers are also classed as structural fibers (6).

Cordage and fabrics were important to ancient man for fishing and trapping and applying motive power (7–8). The making of ropes and cords started in the paleolithic age (ca 20,000 BC); a mesolithic cave drawing (Spain) shows ropes partway down a cliff to recover honey (9). Evidence of predynastic Egyptian (ca 4000 BC) use of ropes and cords from reeds, grasses, and flax has been found (10). The vegetable fibers mostly used for rope were date palm fibers and papyrus. Matting was another important use of vegetable fibers in ancient Egypt, rushes and reeds, as well as papyrus grasses bound with flax string were used. Brushes from vegetable fibers were also used in ancient Egypt.

Flax, jute, ramie and sedges, rushes, and grasses have a long history of use in textiles and basketing.

Properties

Chemical Composition. Chemical analyses or compositions of various vegetable fibers are given in Tables 2 and 3. Cotton is at the high end of the range in cellulose content with >90%; jute and kenaf (bast fibers) and abaca (leaf fiber) have a relatively low cellulose content of 63–65%. The chemical compositions vary greatly between plants and within specific fibers depending on genetic characteristics, the part of the plant used and growth, harvesting, and preparation conditions.

Fiber Dimensions. Leaf fibers are multicelled, and not readily split into the component cells, whereas the bast fibers are readily broken down, permitting spinning. Strand and cell dimensions are given in Table 4. The cells themselves are composed of microfibrils which are comprised of groups of parallel cellulose chains.

Physical Properties. Mechanical properties of certain vegetable fibers are shown in Tables 5 and 6. The vegetable fibers are stronger but less extensible than cotton, ie, they have a higher breaking length and elasticity modulus with a lower strain (extensibility) and work modulus. The vegetable fibers approach glass in stiffness (resistance to deformation) and are considerably stiffer than man-made fibers, but have lower toughness (ability to absorb work) (see Fibers, chemical) (2). Kapok and other seed-hair fibers are relatively low in strength but have great buoyancy.

Vegetable fibers have spiral molecules that are highly parallel to one another. The spiral angles are low in flax, ramie, hemp, and other bast fibers, considerably less than cotton, accounting for the former fibers' low extensibility (12).

Table 2. Chemical Analysis of Vegetable Fibers[a], wt %

Fiber	Cellulose[b]	Moisture	Ash	Lignin[c] and pectins	Extractives
Leaf fibers					
abaca	63.72	11.83	1.02	21.83	1.6
bowstring hemp	69.7	9.7	0.7	13.7	6.2
caroa	60	10		12	18
cebu maguey (cantala)	75.8	5.5	1.4	14.1	3.2
henequen	77.6	4.6	1.1	13.1	3.6
phormium	63	11.61	0.63	23.07	1.69
piteira (Mauritius)	75.6	5.2	1.6	17.2	0.4
sisal	77.2	6.2	1.0	14.5	1.1
tula istle	73.48	5.6	1.65	17.37	1.9
Bast fibers					
Congo jute (cadillo)	75.3	7.7	1.8	13.5	1.4
hemp	77.07	8.76	0.82	9.31	4.04
jute	63.24	9.93	0.68	24.41	1.42
kenaf	65.7	9.8	1.0	21.6	1.9
ramie	91			0.65	
sunn	80.4	9.6	0.6	6.4	3.0
Seed and hair fibers					
cotton	90	8.0	1.0	0.5	0.5

[a] Ref. 2.
[b] See Cellulose.
[c] See Lignin.

Table 3. Chemical Composition of Various Vegetable Fibers[a], wt %

Type of fiber	Cellulose	Hemi-cellulose	Pectins	Lignin	Water-soluble compounds	Fats[b] and waxes[c]
cotton	91.8	6.4			1.1	0.7
flax (retted)	71.2	18.6	2.0	2.2	4.3	1.7
flax (nonretted)	62.8	17.1	4.2	2.8	11.6	1.5
Italian hemp	74.4	17.9	0.9	3.7	2.3	0.8
jute	71.5	13.4	0.2	13.1	1.2	0.6
Manila	70.2	21.8	0.6	5.7	1.5	0.2
New Zealand cotton	50.1	33.4	0.8	12.4	2.4	0.1
ramie	76.2	14.6	2.1	0.7	6.1	0.3
sisal	73.1	13.3	0.9	11.0	1.4	0.3

[a] Ref. 11.
[b] See Fats and fatty oils.
[c] See Waxes.

Plant Descriptions, Preparations, and Grades of Fibers

The vegetable fibers are grouped below partly according to their plant source and partly based on the importance of use within source groups (3–5,13).

Table 4. Dimensions of Fibers and Cells of Vegetable Fibers[a]

	Cells or ultimate fibers						Fibers or strands	
	Length, mm			Diameter, mm			Length, cm	Width, mm
Fiber	Min	Max	Av	Min	Max	Av	Range	Range
cotton	10	50	25	0.014	0.021	0.019	1.5–5.6	0.012–0.025
flax	8	69	32	0.008	0.031	0.019	20–140	0.04–0.62
hemp	5	55	25	0.013	0.041	0.025	100–300	
ramie	60	250	120	0.017	0.064	0.040	10–180	0.06–9.04
sunn	2	11	7	0.013	0.061	0.031		
jute	0.75	6	2.5	0.005	0.025	0.018	150–360	0.03–0.14
sisal	0.8	7.5	3	0.007	0.047	0.021	75–120	0.01–0.28
Manila (abaca)	2	12	6	0.010	0.032	0.024	180–340	0.01–0.28
phormium	2	11	6	0.005	0.025	0.013		
Mauritius	2	6		0.015	0.024			
kapok	15	30	19	0.010	0.030	0.018		
coir	0.3	1	0.7	0.010	0.024	0.020		
kenaf	2	11	3.3	0.013	0.034	0.023		
pineapple fiber	2	10	5.5	0.003	0.013	0.006		
nettle	4	70	38	0.020	0.070	0.042		
sansevieria	1	7	4	0.013	0.040	0.022		

[a] Ref 2.

Table 5. Mechanical Properties of Vegetable Fibers[a]

Fiber	Breaking length, km[b]	Ultimate strain at break, %[c]	Work of rupture modulus, kN·m/N	Initial linear elasticity modulus, N/tex[d]
kapok	16–30	1.2	0.1	1300
jute	27–53	1.5	0.3	1700–1800
hemp	38–62	2–4	0.6–0.9	180
flax	24–70	2–3	0.9	1800–2000
ramie	32–67	2–7	1.1	1400–1600
henequen	27–34	3.5–5		
manila hemp	32–69	2–4.5	0.6	
sisal	30–45	2–3	0.7–0.8	2500–2600
coconut	18	16	1.6	430

[a] Ref. 11.
[b] Tensile strength is reported on the basis of breaking length which measures strength per unit area.
[c] Elongation (stretch) at rupture.
[d] To convert N/tex to g/den, multiply by 11.33.

LEAF (HARD) FIBERS

Abaca. This fiber is extracted from the plant *Musa textilis*, native to the Republic of the Philippines. The fiber is also known as Manila fiber or Manila hemp. When mature, the individual abaca plant consists of 12 to 30 stalks radiating from a central root system. In appearance it is very similar to the banana plant (*Musa sapientum*). Each of the stalks is 2.7–6.7-m tall with a trunk 10–20-cm wide at the base. The sheaths forming the stalk expand into the overhanging leaf structure. The sheaths

Table 6. Coarseness and Breaking Length of Leaf Fibers [a]

Fiber	Coarseness, km/kg	Breaking length, km
Manila hemp		
mean	32	41.5
sisal	40	35.0
cantala	58	20.5
henequen	32	20.0
New Zeland flax	38	26.0
istle	34	27.0

[a] Ref 11.

before expanding are 2–4-m long, 13–20-wide, and ca 10-mm thick at the center. The fibers run lengthwise in the sheaths. The sheaths vary in length and width, and also in color. The mature stalks are cut off at the roots and at a point just below where they begin to expand. In the Philippines, the stripping method is generally used, ie, the fiber layer from the cut sheath is separated by inserting a knife just under the layer and a fiber strip or tuxy, 5–7-cm wide, is pulled off the sheath. Modern plantations have replaced this hand-pulling operation with power-driven pulling machines to accelerate production. After stripping, the fiber is air-dried; rain and sunlight need to be avoided to preserve color and luster.

A small proportion of Philippine abaca fiber is produced by mechanical decortication. The sheaths are fed into a decorticating machine in which the pulpy material is scraped from the fiber, then the fiber is washed and dried. The abaca fiber so produced is lower in quality than hand-cleaned fiber, ie, it has less sheen and is harsher in texture. Abaca fiber grown in Sumatra (Indonesia) and Central America is machine-decorticated exclusively.

The Philippines is the main source of abaca, contributing ca 90% of the total supply. The remainder is produced in Saba (North Borneo), Indonesia, and Central America. Philippine abaca is carefully graded by the government of the Philippines. The long cordage fiber is classified into four groups, determined by the fineness of the fiber and degree of separation of the individual fibers. Each group is further subdivided according to the range of color and associated fiber length. There are 18 grades, designated AB, CD, E, F, S2, S3, I, J1, G, H, H2, K, L1, L2, M1, M2, DL, and DM. These grades range in color from almost white to dark brown. Philippine abaca is longer and stronger than the corresponding grades produced elsewhere, and is generally lighter in color but bolder in texture.

Abaca is produced in two grades in Central America (introduced during World War II): clear and streaky. In Indonesia it is supplied in superior, good, fair, fair X, and B grades.

Sisal. This fiber is extracted from the leaves of the plant *Agave sisalana* which is widely cultivated in the Western Hemisphere, Africa, Asia, and Oceania. Other *Agave* sp. that furnish commercial vegetable fibers are henequen, cantala and istle (described below). However, only *A. sisalana* provides true sisal, which is further identified as African, Indonesian, Brazilian, and Bahama sisal.

The agaves have rosettes of fleshy leaves, usually long and narrow, which grow out from a central bud. As the leaves mature they gradually spread out horizontally and are 1–2-m long, 10–15-cm wide, and ca 6 mm thick at the center. The fibers are embedded longitudinally in the leaves, and are most abundant near the leaf surfaces. The leaves contain ca 90% moisture, but the fleshy pulp is very firm and the leaves are rigid. The fiber is removed when the leaves are cut because dry fibers adhere to the pulp. The fiber is removed by scraping away the pulpy material, generally by a mechanical decortication process.

In decortication, the leaves are fed through sets of fluted crushing rollers. The crushed leaves are held firmly at their centers and both ends are passed between pairs of metal drums on which blades are mounted to scrape away the pulp, and the centers are scraped in the same way. The fiber strands are washed and either air- or artificially dried. In Indonesia and Africa, for choice grades, the dried fiber is held against a revolving metal drum to remove remnants of dry adhering pulp.

Sisal fiber is graded according to the country and district of growth and further subgraded according to color, cleanliness, and length. There are eight Indonesian grades, designated, A, B, C, X, Y, X, D, and L; seven East African grades designated 1, 2, 3 long, 3 short, A, UG, and SCWF; five Haitian grades, designated A, B, X, Y, and S; eight West African grades, designated Extra, 1, 2, 2SL, 3 3L, A, and R; three Philippine grades, designated SR-1, SR-2, SR-3; and five grades of Brazilian sisal, designated Tipo 1, Tipo 3, Tipo 5, Tipo 7, and Tipo 9.

Henequen. This fiber is extracted from *Agave fourcryodes* and is also called Mexican henequen from its source, the state of Yucatan, Mexico. There are also El Salvador and Cuban henequens from *A. letonae*. These plants are harvested, decorticated, cleaned, and prepared for marketing by procedures similar to those used for sisal.

Mexican henequen is classified into seven grades according to color, cleanliness, and length. The grades are AA, A, B, B1, M, C, and M1.

Cantala. This variety of agave fiber is derived from *Agave cantala* and is also called Manila maguey from its Philippines source (also Indonesia). The fiber is prepared in the Philippines by retting the leaves in sea water followed by cleaning by hand or by decorticating as described for sisal. The Indonesian industry employs mechanical decortication without the retting step.

Catalpa is graded according to the method of decortication, retting, or machine decortication, into grades 1, 2, and 3.

Istle. This is the term for short, coarse fibers from the leaves of agaves and related plants growing wild in central and northern Mexico. There are three commercial varieties: tula istle from *A. lophanfu*, jaumave istle from *A. funhana*, and palma istle from *Samuela carnerosana*. The agave istle resembles small plants, whereas the palma istle looks like a small palm tree with leaves radiating from the top of the plant.

The istle fibers are extracted from the leaves by hand-scraping with subsequent sun-drying. The palma leaves are very gummy and are steamed before scraping.

Mauritius. This fiber is obtained from the giant cabuya plant, *Furcraea gigantea*, native to Brazil where it is called piteira or pita. It is cultivated commercially on the island of Mauritius and is known as Mauritius hemp or aloe. The cabuya plant resembles the agaves except that the leaves are heavier and larger.

The Mauritius fiber is extracted from the leaves by simple mechanical decortication by which the crushed leaves are scraped free from pulpy material, washed, steeped in a soapy solution and sun-dried.

There are five standard grades of Mauritius fiber: extra prime, prime, very good, good, and fair. Brazilian pita is graded according to standards for Brazilian sisal.

Phormium. This fiber is extracted from the plant *Phormium tenax*, native to New Zealand. The fiber is commonly called New Zealand flax or hemp, though it has no bast fiber characteristics. The plant is a perennial with a fan-shaped cluster of leaves 1.6–4.3-m long, and 6–10-cm wide. The leaves are green with a red midrib and red margins. The fibers are obtained by mechanical decortication of the cut leaves in a similar manner to sisal processing and then washed and sun-dried.

The phormium fibers are graded by New Zealand hard fiber standards according to color, strength and cleanliness and then subgraded into six classes: A, superior or superfine; B, fine; C, good; D, fair high-point, fair low-point; E, common; and F, rejections.

Sansevieria. This fiber is extracted from the perennial plant *Sansevieria* native to Africa, Arabia, India, and Ceylon. The fiber is also called bowstring hemp because of its primitive bowstring application. The plant grows wild or is cultivated in many countries. It is propagated with cultivated leaf cuttings. Extraction by mechanical decortication is similar to sisal processing.

Caroa. This leaf fiber is obtained from the plant *Neoglaziovia variegata*, which belongs to the pineapple family and is native to eastern and northern Brazil. The leaves for fiber production are collected from plants growing wild and are sword-shaped, 1–1.3-m long, and 2.5–5-cm wide. The fiber is extracted by hand-scraping after beating the leaves to break up the pulpy tissue, or after a retting process that partially ferments and softens the leaves. Caroa is graded by standards similar to those specified for Brazilian sisal. Its color and texture resemble sisal although it is apt to be incompletely cleaned.

Piassava. This fiber is extracted from *Attalea funifera*, a palm indigenous to Bahia, Brazil, where it grows wild and under cultivation. Bahia piassava and bass fiber are other terms. The plant is a tall feather palm with erect leaves 12 m in length. The fiber is taken from the sheathing leaf vases. The leaf sheaths generally need to be treated in water to free the fibers from the pulp by scraping. The mature lower leaves also may be cut at the base, the leaves crushed, and the fibers combed out with a knife. The fibers are sometimes hand-pulled from the leaf stem. The piassava fiber is classified into premium and second grades.

Broomroot. This fiber is extracted from the roots of the bunchgrass, *Muhlenbergia macroura*, a plant growing wild from Texas to Central America but produced commercially in Mexico. The plant grows 1–2-m tall, although the fiber is obtained from the roots by beating or rubbing the chopped roots to remove the bark-like covering. The fibers are bleached with sulfur fumes at the processing factory and graded into four qualities.

BAST (SOFT) FIBERS

Fine Textile Fibers

Flax. This fiber is extracted from the plant *Linum usitatissimum* L. which is grown chiefly in the USSR, Poland, France, Rumania, Czechoslovakia, Belgium, and Ireland. The plant is cultivated mostly for its oil-bearing seed (linseed), although it is also an important source of a vegetable fiber (see Vegetable oils). Flax is an annual plant with a slender, greyish-green stem growing to a height of 90–120 cm.

Several methods of harvesting are used and the correct time for harvesting for highest fiber quality and seed production is important. The plants are pulled by hand or machine for highest yield and quality, although mowing is practiced for some grades. After deseeding, the straw is retted, ie, the fibers are liberated through enzymatic action on pectinous binding material in the stem; dew- or water-retting or variations are employed. In the commonly used dew-retting, the straw is spread thinly on the ground and subjected to atmospheric precipitation with turning at intervals. Fungi enter the stem to cause the retting action. Water-retting is performed on bundles of dried straw which are immersed in rivers, pools, or ditches. A modern variation is a controlled warm-water procedure and, more recently, aerated retting is being used for reducing pollution of the effluents, water usage, and odors. The retted straw is dried before the scutching or fiber separation and cleaning step. This breaking of the straw was first done by hand-beating, and eventually in turbine-type machines consisting of multiple pairs of corrugated roll breakers and scutching wheels for cleansing and polishing. The fiber is then hackled for alignment and final cleaning. The short fibers from these operations (tow) and the longer fibers (line) are used for linen.

The qualities considered in grading flax are fineness, softness, strength, density, color, uniformity, luster, length, handle, and cleanliness. The line fiber from sources in different countries varies in grades and buyers have their own grading systems.

Hemp. This fiber is extracted from the annual plant *Cannabis sativa* which originated in Central Asia. The plant grows readily in temperate and tropical climates, but its commercial production for fiber is chiefly in the People's Republic of China and eastern Europe.

The hemp plant is grown for fiber from the stem, for oil from the seeds or for drugs from the flowers or leaves. Marijuana (see Hypnotics, sedatives, and anticonvulsants) is the narcotic alkaloid (cannabin) derived from a related hemp plant, which is not a commercial fiber-producing plant. The Marijuana Tax Act of 1937, nevertheless, includes *C. sativa.*

When mature, the stalks are 5–7-m tall and 6–16-mm thick. They are generally smooth and without branches or foliage except at the top. The stem structure is hollow. Thin-walled tissue adjoins the hollow center and outside of this is a layer of woody substance. The next layer consists of gummy tissue that cements the fiber layer to the woody layer. The fiber or bast layer is enveloped by a thin bark that constitutes the outside of the stalk.

The mature hemp stalk is harvested at the proper time to ensure highest quality and yield by hand-cutting and spreading or by a harvester-spreader for dew-retting. When water-retting is used, bundles of straw are dried in the field. Subsequently the leaves and seeds are removed by beating the tops on the ground, and they are then stacked before the water-retting. Water-retting is done in Italy, Spain, Hungary, Poland, and to a small extent in the USSR. Dew-retting is used mostly in the USSR. The retted and dried straw is further treated by either hand-breaking and hackling to remove the woody stem portion (hurd) or by mechanical breaking followed by turbine scutching. A wooden breaking device augments the hand-breaking.

The grading is by color, luster, density, spinning quality, cleanliness, and strength. Quality varies with country and regions within countries. The national classifications may include a number of grades and subgrades.

Ramie. This fiber is produced from the stems of *Boehmeria nivea*, a nettle native to central and western China but growing in regions varying from temperate to tropical, including the People's Republic of China (largest grower), the Philippines, Brazil,

Taiwan, Japan, and others. Another name for the raw fiber is China grass. The plant grows 1–2-m high or higher with stems 8–16 mm in diameter. It has perennial roots and yields three crops annually.

The ramie fiber is contained in the bark and is usually extracted by hand stripping and scraping. The plant is cut green and defoliated manually and bast ribbons are stripped from the woody stem. These are scraped in the field or stored in water for scraping in a central location. In Brazil, the Republic of China (Taiwan), and other countries the ramie ribbons are decorticated mechanically in small operations. Sometimes the ribbons are bleached by sulfur fumes.

Ramie fibers are graded according to their length, color, and cleanliness. Each country has its own grading system.

The raw ramie fibers contain 25–35% plant gums (xylan and L-arabinan) and small quantities of parenchyma cells which must be removed before the fibers can be spun. Degumming by boiling in aqueous alkaline solutions is sometimes done under pressure and at times with additives, followed by washing, bleaching with an oxidizing agent (eg, a hypochlorite), washing, neutralizing, and oiling to facilitate combing.

Packaging Fibers

Jute. This fiber is obtained from two species of the annual, herbaceous plant *Corchorus, C. capsularis* of Indo-Burma origin and *C. olitorius* from Africa. The major jute production area is India and Bangladesh and secondarily in other Asian countries and Brazil. The two sources of jute are differentiated by their seeds and seed pods. The *C. capsularis* is round-podded and is called white jute; the *C. olitorius* is long-podded and is known as tossa or daisee. The jute plants of both species have cylindrical stalks 2–4-m tall and 10–20 mm in diameter.

The plants are harvested by hand at an early seed stage using knives and the stems are left on the ground several days to promote defoliation. The defoliated stems are bundled and taken for wet-retting in canals, ditches, or ponds (slowly running water to carry off colored and acidic products) for periods of 10–20 d. The retting time for *C. insularis* is shorter than for *C. olitorius.* The retting can also be conducted on ribbons in concrete tanks. Stripping of the fibers follows immediately after stem-retting with sun-drying; the ribbon-retted material also is sun-dried.

The jute is made into crude bales for transport to a baling and grading center. Here, the fiber is designated as to species (white, tossa, or daisee), and source and grade according to the grades being used. The jute is then baled for shipment to domestic users or for export.

Kenaf and Roselle. These fibers are extracted from the stems of two closely related plants, *Hibiscus cannabinus* L. and *H. sabdariffa* L. var. *altissima*, known respectively as kenaf, Deccan hemp or mesta, and roselle. The two plants are native to Africa but are grown commercially in greatest quantity in the People's Republic of China, the USSR and Egypt for the kenaf, and India and Thailand for the roselle. The plants are herbaceous annuals growing in single stems to heights of 1–4 m for kenaf and 1–5 m for roselle with stem diameters of 10 to 20 mm. Kenaf is harvested as flowering begins whereas with roselle this takes place somewhat earlier. The kenaf can be hand-cut or mowed or pulled; the roselle may be hand-pulled. Both species may be field-dried for defoliation. Hand or mechanical ribboning followed by tank-retting is used on occasion, but the kenaf is often stem-retted followed by hand or mechanical stripping and washing. The washed fiber is finally dried on racks.

Kenaf is graded into three grades: A, good; B, medium; C, poor. The grading is based on color, uniformity, strength, and cleanliness. Roselle may be hackled to improve quality by removing dirt and increasing softness and luster. In Thailand roselle is classified into seven grades as to softness, color, and impurities.

Urena. This jutelike fiber is obtained from *Urena lobata* L., commonly called urena with many local names because of its widespread growth, although it is grown commercially mostly in Brazil and Zaire. This perennial shrub grows under cultivation as an annual with few branches and it grows to a height of 4–5 m with stems 12–18 mm in diameter. The stems are cut 20 cm above ground level because of the highly lignified base. The plants are defoliated green or after piling in the field; retting is similar to that for jute or kenaf and requires 8–10 d. The retted material is then stripped, washed, and sometimes hand-rubbed to remove impurities.

Urena fiber is officially graded in Brazil and Zaire into four or five grades according to color, uniformity of color, luster, strength, and cleanliness.

China Jute. This jute-type fiber is extracted from the stem of a mallow, *Abutilon theophrasti* Medic or *A. avecennae Gaetn.*, growing chiefly commercially in the People's Republic of China and the USSR, although also in Japan, Korea, and Argentina, and known as China jute or Tientsin jute, and by many other local names. The plant is a herbaceous annual growing to a height of 3–6 m with stems 10–18 mm in diameter. China jute is harvested by hand in the early flowering stage, the leaves are removed, and bundles are water-retted similar to jute operations. The harvesting, retting and fiber extraction may be mechanical in certain regions. In the hand operation the fiber is hackled, baled, and graded into two classes.

Soft Cordage Fiber

Sunn Hemp. This fiber is extracted from the stem of the legume, *Crotolaria juncea* L., indigenous to India and known as sunn hemp, brown hemp, Indian hemp, Benares hemp, and other national references. India is the largest producer followed by Bangladesh and Brazil. Sunn hemp is an erect herbaceous shrub growing to a height of 1–5 m with a thin, cylindrical, branch-free stem. The plant is harvested in the seed-pod stage by hand-cutting or pulling and left in the field for defoliation. The stems are water-retted, washed, stripped, and dried by methods similar to those for stem fibers described previously. For export, sunn-hemp fibers are hackled or dressed and graded by a system with many classifications according to length, strength, firmness, color, uniformity, and percentage of extraneous matter.

Seed and Fruit-Hair Fibers

Coir. This fiber is contained in the fruit or husk of the coconut palm *Cocos nucifera* L., which is positioned beneath the outer covering of the fruit and envelops the kernel or coconut. Sri Lanka(Ceylon) and India are important producers. The fruits are gathered just short of ripeness and the husks are broken by hand or by use of a bursting machine. The main supply of fiber is obtained as a residue for copra production. The extraction of the fiber involves retting at the edges of rivers and also in pits or, in modern operations, in concrete tanks. The retted husks are beaten with sticks to remove extraneous matter and the dried fiber is suitable for spinning. Rougher fibers require less retting and the fibers are extracted from the husks mechanically. The fibers may be washed before drying and may be hackled before direct use or baling for shipment. The fiber can also be removed from the husk by a decorticating machine in connection with copra production.

Kapok. This fiber is obtained from the seed pod of a tree, *Ceiba pentranda*. The tree grows to a height of ca 35 m, and is indigenous to Africa and Southeast Asia; it is also known as the silk-cotton tree. The fiber is exported from Thailand, India, and Indonesia. The pods are picked short of opening, hulled or broken open, and the floss is dried in an area enclosed with cloth or wire netting to contain the floss. The seeds are removed by hand or separated mechanically and further cleaned by an air separation method. The fiber is finally baled for shipment. The seeds contain up to 25% of a nondrying oil resembling cottonseed oil in properties and uses.

Miscellaneous. A number of seed-hair fibers have some minor uses, including East Indian balsa fiber (*Ochroma pyramdale*), Indian kumbi (*Cochlospermum gossypium*), American milkweed floss (*Asclepias* sp.) and cattail fiber (*Typha* sp.).

Production

The world production of jute and hard fibers is given in Table 7 for the period 1963 to 1976 (14).

The growth since 1970 has been at a zero or negative rate.

The distribution for the production of the major hard cordage fibers in 1976 is shown in Table 8.

U.S. imports of unmanufactured vegetable fibers have shown serious decreases in the period of 1963 to 1976 (Table 9).

The consumption of paper pulps made from vegetable fibers in the United States in 1972 is summarized in Table 10 with sources of pulp and types of paper products.

In addition, 216,000 metric tons of U.S. sugar cane bagasse pulp were consumed in making wall board and fine papers in 1972.

Vegetable Fiber Yields. The yields of prepared vegetable fibers are generally low, 1–6%, as shown in Table 11. Linen flax fiber and kapok are exceptions.

Table 7. World Production of Jute and Hard Fibers in Thousands of Metric Tons

	1963	1970	1976
jute	2421	2051	2034
abaca	119	93	67
sisal-henequen	826	763	628

Table 8. Percentage Distribution of Production of Major Hard Cordage, 1976[a]

	Thousands of metric tons	%
abaca	84.0	13
sisal	421.3	66
henequen	130.0	21
Total	*635.3*	*100*

[a] Ref. 15.

Table 9. U.S. Imports of Unmanufactured Vegetable Fibers[a]

	1963	1970	1975
	Metric tons		
flax (Belgium, Luxembourg)	2,246	1,581	411
hemp (Canada)	35	0	0
jute and jute butts (Thailand, Bangladesh)	78,811	20,625	20,953
kapok (Thailand)	12,157	11,703	7,499
abaca (Philippines, Ecuador)	28,174	21,492	22,239
sisal and henequen (Brazil, Tanzania, Kenya, Mozambique, Mexico)	92,448	59,677	14,465
istle or tampico (Mexico)	7,074	29	0
others[b]	4,753	7,551	12,655
Total	*225,698*	*122,658*	*78,222*

[a] Ref. 14.
[b] Crin vegetal, phormium, broom root, rice straw, coir, sunn.

Table 10. U.S. Consumption of Nonwood Paper Pulps in 1978[a]

Vegetable fiber	Source	Pulp consumption, t	Paper products
cotton linters	United States	100,000	rayon and fine papers
rags, textile residues	United States	70,000	absorbent felt and fine papers
seed flax straw	United States, Europe	60,000	cigarette and lightweight papers
abaca	United States (old ropes), Philippines	25,000	specialty papers, including tea bags
miscellaneous		30,000	
Total		*285,000*	

[a] Ref. 16.

Table 11. Yields of Prepared Vegetable Fibers[a]

Plant	Fiber form	Yield[b], %	Plant	Fiber form	Yield[b], %
Leaf fibers					
abaca	decorticated	2–3	hemp	dry-line fiber	3.5
sisal	decorticated	3[c]		dry tow	1.0
henequen	decorticated	2–3	ramie	decorticated	3.5
Seed-hair fiber			jute	dry-retted fiber	6.0
kapok		17[d]	kenaf	dry-retted fiber	4.8
Bast fibers			roselle	dry-retted fiber	4.4
flax	scutched line fiber	12–16	sunn hemp	dry-retted fiber	3.4
	scutched clean tow	4–8			

[a] Ref. 3–5.
[b] Green plant basis.
[c] Leaf basis.
[d] Pod basis (600 pods per tree).

Economic Aspects

Prices. The prices of vegetable fibers for cordage and noncordage uses have varied widely over the years depending on the supply and demand, world economic conditions, weather, and starting in the 1960s, competition from man-made fibers particularly polypropylene fibers (see Olefin fibers). In 1970, polypropylene had a price of ca \$400/t on the U.S. market and was projected to fall by 1980 to \$275/t, making vegetable fibers uncompetitive. In 1978, however, polypropylene had risen to \$1200/t because of the oil situation (17–18) (see also Olefin polymers).

The 1961–62 prices for unprocessed fibers imported into the United States for the hard fibers, abaca and sisal, were \$33.00 and \$17.82/100 kg, respectively. Recent prices for these fibers and a synthetic fiber are shown in Table 12.

Projections. Vegetable fibers, excluding cotton and flax, have been relatively unimportant in the total fiber supply and will become less important in relation to converted cellulosic and noncellulosic fibers as projected in Table 13.

Estimated world demands for hard fibers, as shown in Table 14, is expected to remain static (zero growth) for ropes, cables, nets and other cordage, paper and padding, and other uses.

Cordage goods from vegetable fibers accounted for 87% and synthetics for 13% of the total of 21,000 t estimated for the U.S. market in 1964–1966. The vegetable fiber

Table 12. Recent Prices for Various Fibers[a]

Fiber	Grade	Date	Price, U.S. \$/100 kg	Market
abaca	Davao I	av 1977	44.80	ex (export) ship, New York
		Dec. 1978	37.90	
sisal	British East Africa No. 1	av 1977	102.30	landed, New York
		Oct. 1978	102.80	
jute	raw Bangladesh White C	av 1977	46.90	CIF (cost, insurance, and freight),
		Oct. 1978	48.90	United Kingdom
flax	water-retted	av 1977	185.30	United Kingdom
		Sept. 1978	215.80	
polyester		av 1977	123.50	
fiber[b]		July 1978	116.90	wholesale, United States

[a] Ref. 18.
[b] See Polyester fibers.

Table 13. Percentage of Total Fiber Supply in the United States[a]

	Cotton	Wool	Rayon and acetate	Non-cellulosic	Other[b]
1950	64	6.7	25.3	3.5	0.2
1960	55.9	4.4	19.5	19.5	0.3
1980 (estd)	46	2.5	10.8	40.5	0.2
2000 (estd)	44	1.9	5.4	48.6	0.1

[a] Ref. 19.
[b] Including sisal, jute, kapok, abaca.

Table 14. Estimated World Demand for Hard Fibers, 1000 Metric Tons[a]

	1964–66	1968–70	1980
sisal-henequen	772	765	695
abaca	120	95	95
Total	*892*	*860*	*790*

[a] Ref. 20.

share slipped to 71% of a total of 927,000 t in 1968–1970 and is projected to drop to 60% of a total of 965,000 t in 1980. The influence of higher oil prices on the cost of the polypropylene resins used in the synthetic cordages may well change the percentages and trend shown above.

Although there was a zero or negative annual growth rate for nonwood plant fibers in U.S. papermaking in the 1970s, it is expected that these fibers will hold their small percentage of under 1% of the total U.S. pulp consumption through the year 2000 (21).

Uses

Information on the broad end uses of vegetable fibers is given in Table 1 and consumptions by general products are discussed subsequently. Uses by certain categories are described as follows (1–3,5,22).

Textiles and Woven Goods. Clothing, sacks and bags, canvas and sailcloth, and fabrics are made from the bast fibers flax, hemp, ramie, jute, kenaf, roselle, and nettle. Coarse sacks, coffee and sugar bags, floor coverings, and webbing are made from the leaf fibers sisal, henequen, abaca, and Mauritius hemp. Phormium, nettle, and cantala also are used in coarse bagging.

Cordage and Twines. Industrial ropes, hoisting and drilling cables, nets, and agricultural twines are made from abaca, sisal, and henequen. Abaca serves particularly for hawsers and ship cables. Cantala, phormium, and Mauritius hemp are also used for cordage. Hemp, flax, and jute are used for string and yarns, and ramie is used for thread. Hemp and kenaf are used for nets as well.

Brushes. Fibers suitable for bristles in scrubbing and scraping brushes and brooms include coir, piassava, istle and broomwort.

Stuffing and Upholstery Materials. Stuffing for mattresses and pillows and furniture is made from kapok, crin vegetal, sisal tow, and flosses. Kapok is also used in life preservers and coir for door mats.

Paper. In addition to the production of bond and fine writing papers from cotton and linen textile residues and roofing felt from low-grade rags, heavy-duty, multiwalled bags for industrial packaging of flour, cement, chemicals, fertilizers, and hardware are made containing salvaged vegetable fibers or residues from primary product manufacturing of sisal, jute, abaca hemp, henequen, phormium, caroa, and Mauritius hemp. Other specialty papers using these fibers are abrasive papers, and gaskets. Abaca fibers are used specifically in tea bag paper and flax tow and sunn in cigarette paper. Esparto (*Stipa tenacissima*) fibers are made into highest-grade printing papers. Kenaf is proposed for newsprint (see Paper).

Miscellaneous. Tie material, basketry and furniture are made from raffia (*Raphia raffia*), a palm leaf segment, and rattan (*Calamus* sp.), a stem fiber.

BIBLIOGRAPHY

"Fibers, Vegetable" in *ECT* 1st ed., Vol. 6, pp. 467–476, by David Himmelfarb, Boston Naval Shipyard; "Fibers, Vegetable" in *ECT* 2nd ed., Vol. 9, pp. 171–185 by David Himmelfarb, Boston Naval Shipyard.

1. J. Cook, *Handbook of Textile Fibers,* 4th ed., Morrow Publishing Co., Ltd. Watford, Hertz, England, 1968.
2. M. Harris, *Handbook of Textile Fibers,* Harris Research Laboratories, Inc., Washington, D.C., 1954.
3. R. H. Kirby, *Vegetable Fibers,* Interscience Publishers, Inc., New York, 1943.
4. J. M. Dempsey, *Fiber Crops,* The University Presses of Florida, Gainesville, 1975.
5. J. Berger, *The World's Major Fibre Crops: Their Cultivation and Manuring,* Centre d'Etude de l'Azote, Conzett and Huber, Zurich, 1969.
6. W. Von Bergen and W. Krauss, *Textile Fiber Atlas,* American Wool Handbook Company, New York, 1942.
7. K. R. Gilbert, "Rope Making" in C. Singer, ed., *History of Technology,* Vol. I, Oxford University Press at Clarendon, 1954.
8. J. Grant, "A Note on the Materials of Ancient Textiles and Baskets," in ref. 7.
9. G. E. Linton, *Natural and Man-made Textile Fibers,* Duel, Sloan and Pearce, New York, 1966.
10. J. H. Harris, *Ancient Egyptian Materials and Industry,* Edward Arnold Publishers Ltd., London, 1962.
11. T. Zylinski, *Fiber Science,* Office of Technical Services, U.S. Dept. of Commerce, Washington, D.C., 1964.
12. W. E. Morton and J. W. S. Hearle, *Physical Properties of Textile Fibers,* 2nd ed., John Wiley & Sons, Inc., New York, 1975.
13. "Wood, Paper, Textiles, Plastics and Photographic Materials" in *Chemical Technology: An Encyclopedic Treatment,* Vol. VI, Barnes and Noble Books, 1973.
14. *U.S.D.A. Agricultural Statistics,* U.S. Government Printing Office, Washington, D.C., 1977.
15. H. Jiler, ed., *1977 Commodity Yearbook, Hard Fibers,* Commodity Research Bureaus, Inc., New York, 1977.
16. J. E. Atchison, *An Update on Utilization of Nonwood Plant Fibers, Recent Developments and Some Observations on a Visit to China,* paper presented at the 1979 Pulp and Fiber Fall Seminar, American Paper Institute, Charleston, S.C., Nov. 4, 1979.
17. "General Outlook" in *FAO Agricultural Commodity Projections 1970–1980,* Vol. I, Rome, 1971, pp. 264–2822.
18. *FAO Monthly Bulletin of Agricultural Economics and Statistics* **1**(12), 45, 47, 49 (Dec. 1978).
19. G. W. Thomas, S. E. Curl, and W. F. Bennet, *Food and Fiber for a Changing World,* The Interstate Printers and Publishers, Inc., Danville, Ill. 1976.
20. R. Grilli, *The Future of Hard Fibers and Competition from Synthetics,* World Bank Staff, The John Hopkins University Press, Baltimore, 1975.
21. J. N. McGovern, *Non-Wood Plant Fiber Pulping,* TAPPI, CAR Report No. 67, 1976.
22. T. F. Clark in R. G. Macdonald, ed., *Pulp and Paper Manufacture,* Vol. II, McGraw-Hill Book Company, New York, 1969.

General References

J. E. Atchison, "Agricultural Residues and Other Nonwood Plant Fibers," *Science* **191,** 768 (1976).

D. Lapedes, ed., *Encyclopedia of Science and Technology,* Vol. 5, McGraw-Hill Book Company, New York, 1977.

G. E. Linton, *THE Modern Textile and Apparel Dictionary,* 4th ed., Textile Book Service, Plainfield, N.J., 1973.

JOHN N. MCGOVERN
The University of Wisconsin

FILLERS

A filler is a finely divided solid that is added to a liquid, semisolid, or solid composition to modify the composition's properties and reduce its cost. Fillers can constitute either a major or a minor part of a composition. The structure of filler particles can range from irregular masses to precise geometrical forms such as spheres, polyhedrons, or short fibers. Fillers are used for nondecorative purposes, although they may incidentally impart color or opacity to a composition. Although additives that supply bulk to drugs, cosmetics, and detergents are often referred to as fillers, they are more properly termed diluents, since their purpose is to adjust the dose or concentration of a composition, rather than to modify its properties or reduce its cost. Fibers and whiskers are not discussed here because they are generally regarded as reinforcements, not fillers (see Carbon; Composite materials; Fibers).

Although the use of simple diluents and adulterants almost certainly predates recorded history, the use of fillers to modify the properties of compositions can be traced at least as far back as early Roman times when artisans used ground marble in lime plaster, frescoes, and pozzolanic mortar. The use of fillers in paper and paper coatings made its appearance in the mid 19th century. Functional fillers (introducing new properties into a composition, rather than modifying preexisting properties) were commercially developed early in the 20th century when Goodrich added carbon black to rubber, and Baekeland formulated phenol–formaldehyde plastics with wood flour.

Fillers can be classified according to their source, function, composition, or morphology (see Table 1). None of these classification schemes are entirely adequate owing to overlap and ambiguity of their categories. Yet if fillers are to be systematically studied, some method of classification is necessary. Here the morphological scheme is used for the discussion of general filler properties, and the compositional scheme for the compilation of data on specific fillers (see Table 8, p. 211).

Properties

General Properties. ***Particle Morphology, Size, and Distribution.*** Many fillers, such as diatomaceous earth, pumice, and calcium carbonate, have morphological charac-

Table 1. Classification of Fillers

By particle morphology	By composition	
Crystalline	*Inorganic*	*Organic*
fibers	carbonates	celluloses
platelets	fluorides	fatty acids
polyhedrons	hydroxides	lignins
irregular masses	metals	polyalkenes
Amorphous	oxides	polyamides
fibers	silicates	polyamines
flakes	sulfates	polyaromatics
solid spheres	sulfides	polyesters
hollow spheres		proteins
irregular masses		

teristics that allow these materials to be identified microscopically with great accuracy, even in a single particle. Photomicrographs, descriptions, and other aids to particle identification can be found in ref. 1.

Filler particle morphology affects rheology and loading of filled compositions. When filled compositions flow, it becomes necessary for some adjacent particles to be lifted over, pushed under, or detoured around other particles. Thus, particles with fibrous, acicular (needlelike), or irregular morphology yield compositions more resistant to flow than compositions filled with spheres, polygons, or other fillers with regular morphologies. Similarly, fillers with regular shapes can be used in compositions at higher loadings than fillers with irregular shapes, all other properties being equal.

Because of the diversity of filler particle shapes, it is difficult to clearly express particle size values in terms of a particle dimension such as length or diameter. Therefore, the particle size of fillers is usually expressed as a theoretical dimension: the *equivalent spherical diameter,* the diameter of a sphere having the same volume as the particle. The degree to which a particle diverges from sphericity can be expressed as its *sphericity coefficient* ψ.

$$\psi = \frac{\text{surface area of sphere with same volume as particle}}{\text{surface area of particle}}$$

Devices for measuring the particle size of fillers usually assign particles to one of several size ranges. The values of these ranges are then collected and ordered to yield size distributions which are frequently plotted as cumulative weight percent vs diameter. However, collected as the number of particles in size ranges, the data can be converted to a weight distribution by the following equation:

$$W_i = \frac{N_i \bar{d}_i^3}{\sum_i N_i \bar{d}_i^3}$$

where W_i = weight fraction of particles in a size range; N_i = number of particles in this size range; and $\bar{d}_i$ = mean diameter for the size range.

Particle size distributions can be described in a concise manner by values that indicate properties or tendencies of the distribution, such as the mean, mode, or median diameter, and the spread, or standard deviation:

$$\sigma w = \left[\frac{\sum_i W_i d_i^2}{\sum_i W_i} - \bar{d}w^2 \right]^{1/2}$$

However, standard deviation only applies to normal distributions that seem to occur only for narrow size ranges of classified particles. Most particle size distributions are skewed toward their larger particle size classifications (2).

The size distribution of fillers whose particles are larger than ca 40 μm and which have at least moderate sphericity can be conveniently measured by sieving. For fillers whose particles range from 4 to 40 μm, the Coulter technique is preferred. Filler particle size distributions smaller than 4 μm are measured by microscopy, sedimentation, permeametry, or radiation-scattering methods (2–3) (see Size measurement of particles).

Surface Area and Energy. Surface area is the available area of fillers, be it on the surface or in cracks, crevices, and pores. The values obtained from different methods for measuring the surface area of a given filler may vary significantly. Despite these variations, surface area is a very important property. It helps determine the ease of dispersion, the rheology, and the optimum loading of filled compositions.

The external surface of the filler can be calculated from its equivalent spherical diameter and its sphericity coefficient, but this method provides little information on the actual area of the filler that influences physical and chemical processes. In practice, surface area is usually determined (4) from the quantity of a gas that adsorbs in a monolayer on the surface of the filler, ie, the monolayer capacity D_m. From this value the specific surface area can be determined (5), which is an area per unit weight and is usually expressed in m^2/g.

At the phase boundary between a filler particle and a liquid composition, there can be an energy barrier to the liquid's wetting the filler, or the filler's dispersing in liquid. This energy barrier, often termed the free surface energy of the system, results from each phase's greater affinity for molecules of its own substance rather than those of the other phase.

The free surface energy of fillers is a function of surface area and composition. It can be derived from the contact angle of liquids on filler surfaces, the heat of immersion of fillers and liquids, or by adsorption methods (6). Many commercial fillers are surface-coated or treated to modify their free surface energy.

Loading and Packing. The relative amounts of filler and matrix (binder) in a filled composition are usually termed the loading and are always expressed quantitatively; however, the quantitative indexes vary from industry to industry. Formulators in the plastics industry use phr = parts filler per 100 parts resin, % wt = weight percent filler, and % vol = volume percent filler. In the paint industry, pvc = volume percent pigment (filler) in the dry paint film, or pv = the volume ratio of pigment to binder are commonly used. In the paper industry, filler loading is usually expressed as ar = the weight of adhesives per 100 parts by weight of filler.

The optimum degree of filler loading in a composition is usually determined by balancing the functional physical properties of the overall composition against its cost over a range of loadings. This determination can be greatly aided by the use of predictive models based on the packing geometry of filler particles. Some experimental determinations are always necessary because of error introduced by simplifying assumptions in the models but, at the very least, the models can be used to establish minimum and maximum values for filler loading.

For large amounts of fillers, the maximum possible loading with known particle size distributions is estimated. This method (7) assumes that the interstices (voids) between particles are occupied by smaller particles, and that the voids between these particles are occupied by still smaller particles. Thus, a wide filler particle size distribution results in minimum void volume or maximum packing. To get from maximum packing to maximum loading, it is only necessary to express maximum loading in terms of the minimum amount of binder that fills the interstitial voids and becomes adsorbed on the surface of the filler. This adsorption value can be estimated from the surface area of the filler or determined empirically by an oil adsorption test (ASTM D 281 or equivalent).

For small amounts of filler, eg, in order to alter the rheology of the composition, the same methods used to estimate maximum packing and loading can be used to estimate minimum packing and loading. For instance, if maximum loading is obtained with a filler with a wide particle size distribution and a low surface area, then a narrow particle size distribution and a high surface area yields a filled system at minimum loading.

In modern industrial practice, compositions frequently contain pigments, reinforcements, and other materials in addition to fillers. Hence, more complex models are needed to predict the effect of filler loading. Optimum filler loading for paints and filled paper can be calculated (8). An excellent discussion of filler loading in reinforced plastics is given in ref. 9.

Functional Properties. The functional properties of fillers are those that fillers impart to compositions to enhance their performance or economic utility. Although the properties that are required by compositions vary from one application to another, a given physical or chemical property of the filler may or may not be functional and depends on the requirements of the application in which it is used. A quantification of functional properties per unit cost, therefore, provides a valid criterion for filler comparison and selection.

Specific Gravity. In applications where fillers are used in a liquid system and it is necessary to control the volume and/or the weight of the filled composition, the specific gravity (true density) of the filler is one of its functional properties. The specific gravity of fillers composed of relatively large, nonporous, spherical particles is usually determined with an air comparison pycnometer (ASTM D 2840). The specific gravity of finely divided, porous, or irregular fillers is determined with a liquid pycnometer (ASTM D 153).

Bulk Density. Bulk density (apparent density) refers to the total amount of space occupied by a given weight of a dry powder. It includes the volume of the particles plus the volume of the interparticle voids of the powder. Bulk density is a functional property of fillers in powdered compositions, and it also has to be considered in the storage and shipping of fillers.

A reliable method for determining bulk density, such as ASTM D 1895, distinguishes the loose bulk density from the tamped bulk density so that allowances can be made for the compaction of the powder during shipping or processing.

Optical Reflectance. In applications where esthetic appeal is important, fillers with controlled optical reflectance (color and brightness) preserve the color uniformity of the finished composition. Reflectance is usually determined by instrumental methods such as tristimulus green filter reflectivity (ASTM D 97) (see Color).

Refractive Index. The refractive index of a filler must be controlled in order to maximize or minimize its opacifying effect on the composition (10). Microscopic methods of determination are available (11).

Free Moisture. The free moisture of fillers is water present on the surface of the particles where it can contribute to interparticle bonding (reinforcing) or filler–matrix interaction (binder adsorption or catalysis). A standard method for the determination is given in ASTM D 280.

Thermal Stability. Fillers for high temperature applications must not decompose under conditions of use. Even in low temperature applications, thermal stability of fillers can be an important functional property if the composition is molded or otherwise subjected to heat during its processing. Determination of thermal stability varies from industry to industry depending on conditions. The paper industry uses ASTM D 776, the paint industry ASTM D 2485, and the adhesives industry ASTM C 792.

Thermal Expansion. In order to preserve the structural strength of filled compositions over a wide temperature range, all the components of the composition should have similar coefficients of thermal expansion. This prevents the stress-induced damage that occurs as components expand at different rates. A typical method for quantifying the expansion of solid fillers is ASTM D 176.

Applications

Elastomers. In the ANSI/ASTM standard D 1566, a filler is defined as a solid compounding material, usually in finely divided form, which may be added in relatively large proportions to a polymer for technical or economic reasons. However, at present in the rubber industry these materials are invariably termed pigments, even though they are almost never used solely as colorants. The use of the term is probably owing to the long record of the industry of compounding with pigment-grade carbon black (see Elastomers).

Most rubber technologists distinguish reinforcing pigments from nonreinforcing or inert pigments but would agree that these terms are loosely defined and often used only in a comparative context. The standard definition (ASTM D 1566) for reinforcing agents is a material, not basically involved in the vulcanization process, used in rubber to increase the resistance of the vulcanizate to mechanical forces. In practice, these materials impart abrasion resistance, tear resistance, tensile strength, and stiffness to an elastomer after vulcanization. Whether or not a particular pigment is reinforcing or nonreinforcing often depends on its composition, its physical form, and the type of elastomer in which it is compounded. For instance, carbon black in isoprene rubbers and pyrogenic silica in silicone rubbers are more active reinforcing agents in these respective elastomers than other pigments with equivalent particle size distributions. In general, however, most pigments become increasingly reinforcing as their particle size diminishes.

The nomenclature used here to identify grades of carbon black is in conformance with ASTM D 2516 (see Carbon, carbon black). In this system, carbon black grades are identified by four letters: the first one classifies the cure rate as slow (S) or normal (N), the second one is a digit classifying the typical particle size, and the last two characters are arbitrarily assigned digits. The particle size range indicated by the second digit of the identifying codes is given in Table 2. The ASTM abbreviations used for elastomers are given in Table 3, and Table 4 lists typical elastomer fillers and their uses.

Since most fabricated elastomer products contain 10–50 vol % of filler, their physical properties and processing characteristics depend to a great extent upon the

Table 2. Carbon Black Particle-Size Identification Numbers

Size number	Average particle size, nm
0	1–10
1	11–19
2	20–25
3	26–30
4	31–39
5	40–48
6	49–60
7	61–100
8	101–200
9	201–500

Table 3. Elastomer Abbreviations

Elastomer	ASTM abbreviation
acrylate–butadiene	ABR
butadiene	BR
chlorinated polyethylene	CM
chloroprene	CR
ethylene–propylene–diene	EPDM
ethylene–propylene	EPM
isobutylene–isoprene	IIR
synthetic isoprene	IR
nitrile–butadiene	NBR
nitrile–chloroprene	NCR
nitrile–isobutylene–isoprene	NIR
natural isoprene	NR
pyridine–butadiene	PBR
styrene–butadiene	SBR
styrene–chloroprene	SCR
styrene–isoprene	SIR

nature and quantity of the fillers. The rubber technologist has to manipulate the formula so as to optimize a large number of properties and keep costs down.

Latex Stock. *Activation.* In rubber technology, an activator increases the effectiveness of an accelerator, ie, a material that accelerates vulcanization of an elastomer. Zinc oxide is an effective activator in NR, IR, and other isoprene rubbers. It apparently reacts with a fatty acid in the rubber or with an added fatty acid to form a soluble zinc salt which in turn reacts with the accelerator to form a zinc–accelerator complex. This complex reacts during vulcanization to release nascent sulfur which rapidly vulcanizes the elastomer (12). Magnesium oxide is an effective activator in CM, CR, NCR, and other chlorinated elastomers. Magnesium oxide seems to control the pH during vulcanization and act as an acceptor for the chlorine which is liberated during vulcanization (13).

Recovery (Nerve). Uncured latex stock tends to recover its previous shape after being rolled or extruded during processing. This tendency can result in processing difficulties and reduce dimensional accuracy of finished products. The recovery of

Table 4. Typical Elastomer Fillers and Their Uses

Filler	Specific gravity	Typical compatible elastomers	Uses
alumina	2.7	NR, CR, SRs	hose, mats
asbestos	2.4	NR, SRs	mats, tile
barium sulfate	4.0–4.5	NR, CR, SRs	O-rings, belts
carbon blacks	1–2.3		
S212		IR, NR	tires
S315		IR, NR	cable jackets
N110		IR, NR	tires, pads
N219		IR, NR	tire carcasses
N220		NR, SBR	tire treads
N231		NR, IR, SBR	tire treads
N234		NR, SBR	tire treads
N242		NR, BR	tire treads
N293		NR, SRs	belts
N326		NR, IR	tire carcasses
N356		NR, SRs	tire retreads
N375		NR, SRs	tire treads
N472		NR, SRs	electrical goods
N539		NR, SRs	tire sidewalls
N550		NR, SRs	extruded goods
N660		BR, NR	tubes
N762		NR, SRs	foot wear
N774		NR, SRs	tire carcasses
N990		CR, EPDM	extruded goods
calcium carbonate	2.7–2.9	NR, SRs	footwear, mats
clay, hard	2.6	NR, SBR, EPM, EPDM	flooring, footwear
soft	2.6		molded goods
mica	2.82	NR, SR	molded goods
resins	1.1–1.3	NBR, CR, NR, SBR	footwear, coatings
silicas			
colloidal sol	1.3	NR, SBR	sponge
diatomaceous	2.0–2.35	NR, IRR	carpet backing
novaculite	2.65	NR, SRs	molded goods
wet process	1.93–2.2	IIR, CR, NBR, NR	hygienic goods
pyrogenic	2.2	silicone rubber	molds, electrical goods
surface treated		NR, SRs	specialty goods
talc	2.7–2.9	NR, CR, IIR, EPM	molded goods
natural materials[a]	1.0–1.1	NR, SBR, CR	tape, extruded goods
wood and shell flour	0.9–1.6	CR, NR, SRs	footwear

[a] Such as chlorinated or vulcanized vegetable oils, or rice hulls.

latex stock can be reduced by adding large-particle-size fillers, or fillers with a high degree of particle aggregation (agglomerated or structured fillers), such as carbon black or pyrogenic silica, or by increasing the loading of other fillers.

Tack causes two layers of material to adhere when they are pressed together. This property can be reduced by employing fillers with finer particle size distributions, or by dusting the stock with a laminar filler such as mica.

Retardation (*Scorch Resistance*). Scorch is the premature vulcanization of an elastomer during processing. The tendency of an elastomer to scorch can be reduced by compounding it with a filler with a lower pH or a larger particle size. The type of filler also has an effect on the tendency of a latex stock to scorch: highly reinforcing fillers, such as carbon black, silica, and zinc oxide, can promote scorching, whereas most clay fillers tend to retard scorching.

Vulcanized Elastomers. *Abrasion resistance* is the reduced tendency of a rubber particle to undergo surface attrition when subjected to a frictional force. The abrasion resistance of rubbers is a function of both filler type and form. The following fillers are listed in the order of their decreasing effectiveness in preventing abrasion: carbon black $\gg$ silica $\gg$ calcium silicate > zinc oxide > clay > calcium carbonate. Although, in general, abrasion resistance of fillers increases with increasing sphericity and decreasing particle size, fillers should be compared by evaluating their abrasion resistance using samples prepared at equal volume loadings and by a procedure such as ASTM methods D 1630 or D 2228.

Elongation is the extension produced by a tensile stress applied to an elastomer. It is reduced by reinforcing fillers and fibrous fillers such as wood flour, cotton flock, asbestos, or wollastonite. Regardless of what type filler is used, elongation decreases with increasing loading.

Hardness. The resistance of a fabricated rubber article to indentation, its hardness (qv), is influenced by the amount and shapes of its fillers. High loadings increase hardness. Fillers in the form of platelets or flakes, such as clays, impart greater hardness to elastomers than other particle shapes at equivalent loadings.

Modulus is the tensile force required to stretch a uniform cross section of an elastomer to a given elongation. Reinforcing fillers result in higher-modulus composites than nonreinforcing fillers at equivalent loadings. In general, modulus increases with decreasing particle size of fillers.

Permanent Set. When an elastomer is stretched and then allowed to relax, it will not completely recover its original dimensions. This divergence from its original form is called its permanent set. It is principally affected by the affinity of the elastomer for the filler surface and is, therefore, primarily a function of the surface energy of the filler.

Resilience is the ratio of energy output to energy input in a rapid full recovery of a deformed elastomer specimen. It is, therefore, a reliable index to the energy that is lost through internal friction of viscous flow. In general, fillers with the least effect on resilience are those that are least reinforcing. Zinc oxide is an exception because it has good resilience and also gives good reinforcement.

Tear Resistance. The resistance of an elastomer to tearing is affected by the particle size and the particle shape of the fillers it contains. It generally increases with decreasing particle size and increasing sphericity of fillers.

Tensile strength of a rubber compound is increased by fillers of small particle size and large surface area. For most fillers, tensile strength increases with loading to an optimum value after which it decreases with increased loading.

Paper. Paper is typically prepared by depositing cellulose pulp fibers on a continuous wire screen from a dilute water suspension (see Paper). Fillers or loading materials are finely divided solids that are incorporated into a paper sheet by adding them to the pulp slurry and then depositing them on the wire along with the pulp. Finely divided solids dispersed in water containing an adhesive and then coated on the paper after it is formed are usually termed pigments. Here the term filler is used for both loading materials and pigments. Table 5 lists typical paper fillers and their uses.

Table 5. Typical Paper Fillers and Their Uses

Filler	Specific gravity	Refractive index	Uses
alumina trihydrate	2.4	1.57	printing paper coating
barium sulfate	4.4	1.64	photographic paper
calcium carbonate	2.6	1.66	printing paper
calcium silicate	2.1	1.50	printing paper
clay	2.7	1.57	printing paper
polystyrene	1.0	1.59	printing paper coating
satin white[a]	1.5–1.8	1.4	art paper
silica, diatomite	2.0–2.4	1.4	paper board
precipitated	2.1	1.4	printing paper
sodium aluminosilicate	2.1	1.55	printing paper
talc	2.75	1.57	rotogravure paper
titanium dioxide[b]	3.9	2.55	bond, printing paper
zinc oxide	5.6	2.01	electrophotographic paper

[a] Calcium sulfoaluminates.
[b] Anatase.

Most paper used today in the graphic arts industry contains 1–40 wt % of filler. The optical and mechanical properties of filled paper are superior to those of unfilled paper.

Optical Properties. *Brightness* of paper can be defined as the reflectance of a layered stack of paper such that no change in reflectance occurs if the backing material is changed from black to white. The brightness of fillers is usually defined as the diffuse blue reflectance factor obtained by testing a pressed plaque of filler in accordance with TAPPI standard T 534 dm-76.

The ability of fillers to improve the paper brightness increases with their surface area and refractive indexes. It is maximum at an optimum filler particle size, about 0.25 μm in most cases, where the filler particle size is approximately one-half the wavelength of the light used for observation.

Gloss is the property of a surface to reflect light specularly. It is typically associated with such phenomena as shininess, highlights, and reflected images. The gloss of paper is usually quantified with a spectrophotometer and a standard procedure such as TAPPI T 442 su-72.

Paper gloss is influenced by the size and shape of filler particles at the surface. A roughness of the order of one fourth the wavelength of light can produce a perceptible reduction of specularly reflected light (14). Since each surface particle protrudes only partially above the surface, the average height of surface irregularities does not exceed one half of the particle diameter. Thus gloss reduction does not take place until filler particles exceed roughly 3 μm in diameter (15). Filler particles with a laminar, platelike shape increase gloss more effectively than spherical or irregular particles. This is because platelike particles tend to orient parallel to the surface of the sheet, thus reducing its roughness more than would be predicted for their particle size distribution.

Opacity of paper is a function of its light absorption and the amount of light scattering that occurs at the pigment–air, pigment–fiber, and fiber–air interfaces. Light scattering is the principal cause of opacity in paper containing white fillers, whereas in colored paper opacity depends on light absorption.

The filler properties that influence opacity are color, particle size, surface area, and refractive index. Dark pigments generally provide better opacity than light pigments because they absorb more light. Light scattering, like brightness, is a reflective phenomenon; the optimum particle size is again approximately one-half the wavelength of the light that is used for observation. The opacity of fillers increases with both their surface area and their refractive index.

Mechanical Properties. *Ink Absorbency.* Ink penetrates into paper during the printing process. The more the ink penetrates, the less glossy the print. The degree of ink absorption in paper can be controlled by the particle size and shape of the fillers. It is monitored by TAPPI method RC19.

Since the void volume of fillers decreases with decreasing particle size, ink absorbency can be reduced by using small particle fillers. Loading or coating paper with platelike fillers, such as clays, also reduces ink absorbency because these materials tend to overlap and thus reduce porosity (see Ink).

Retention. After filler is added to the cellulose pulp slurry in the course of paper making, it has to remain in the sheet when the pulp is deposited on the wire. If the filler is not well retained, the sheet has a two-sided character, ie, the wire side of the paper contains less filler than the top side. Poor filler retention increases costs, since the filtrate containing the lost filler (white water) may have to be processed to recycle the filler or upgrade the water quality for environmental reasons.

Retention is reduced by increasing filler solubility, specific gravity, and sphericity. It is increased by increasing the filler's particle size or by flocculating smaller particle size fillers.

Sheet strength of paper is measured as bursting strength (TAPPI T 403 os-76), internal tear resistance (TAPPI T 414 ts-65), tensile breaking properties (TAPPI T 494 os-70), internal bond strength (TAPPI T 506 su-68), and folding endurance (TAPPI T 511 su-69). Fillers generally reduce the paper sheet strength since they dilute the pulp and reduce the number of fiber-to-fiber interactions. However, this is not necessarily detrimental. It often significantly increases flexibility and stretchability and thus reduces the number of tears that occur when the paper is run on high-speed equipment (16).

Plastics. In the plastics industry, the term filler refers to particulate materials that are added to plastic resins in relatively large (over 5%) volume loadings (see Laminated and reinforced plastics).

At present, the performance specifications for most compounded plastic products are less strict than for other filled products, such as elastomers and paints, and cost reduction is the primary objective. Hence, plastic compounders tend to compound with the objective of optimizing properties at minimum cost rather than maximizing properties at optimum cost. Table 6 lists typical plastic fillers and their uses.

Resins. *Curing.* Fillers with high surface areas can retard the curing of plastics by adsorbing catalysts and/or promoters. Other fillers can accelerate curing if they contain active sites or trace quantities of catalytic materials. Low density fillers cause high temperatures during curing and subsequent long cool-down periods because of their insulating effect. High loadings of dense solid fillers can appreciably increase the curing time needed to fully cure thermosetting compounds.

Viscosity of resin–filler mixtures is affected by particle shape, loading, and degree of dispersion. It decreases with increasing sphericity and degree of dispersion but increases with increasing loading.

Table 6. Typical Plastic Fillers and Their Uses

Filler	Specific gravity	Typical compatible resins	Uses
alumina trihydrate	2.4	polyesters	fire resistant filler
carbon blacks	0.2–2.3	epoxies	electrical goods
calcium carbonate			
mineral	2.7	most resins	tile, molded goods
precipitated	2.6	most resins	pipe, putty
clays	2.6	most resins	flooring tile
feldspar	2.6	thermoplastics	plastisols
metals[a]	2.5–11.5	epoxies	radiation shields, solders
mica	2.82	most resins	sheet molded goods
polymers[a]			
solid spheres	1.1–1.3	most resins	molded goods
hollow spheres	0.2–0.5	thermosets	molded goods
silica			
diatomite	2.0–2.4	polyethylene	films
novaculite	2.65	thermosets	electrical goods
quartz flour	2.65	thermosets	molded goods
tripolite	2.65	thermosets	molded goods
wet process	1.9–2.2	thermoplastics	sheets, films
vitreous	2.18	epoxies	electrical goods
silicate glass			
solid spheres	2.5	most resins	molded goods
hollow spheres	0.22	thermosets	molded goods
flakes	2.01	thermosets	electrical goods
talc	2.7–2.9	PVC, polyalkenes	extruded and molded goods
wood and shell flour	0.19–1.6	most resins	molded goods

[a] See Table 8.

Cured Plastics. *Tensile strength* of filled plastic compositions is affected by filler particle shape and size, size distribution, surface area, and interfacial bonding. In general, it increases with the decreasing sphericity of its fillers. At equivalent volume loadings, small filler particle sizes and narrow size distributions give better tensile strengths than larger particle sizes and broad size distributions. Higher filler surface areas and, in general, stronger filler-to-matrix bonding also result in higher tensile strength compositions. Tensile strength of plastics is measured by ASTM D 638-76.

Compressive strength of filled composition is normally governed by the strength of the weakest component of the system, be it filler, matrix, or the bond between filler and matrix. Thus weak compressive fillers, such as celluloses, reduce the compressive strength of composites, whereas the reverse is true for strong, rigid fillers such as mineral oxides. Compressive strength of plastic composites is measured by ASTM D 621-64.

Fire Resistance. Many fillers, particularly inorganic oxides, are noncombustible and provide a measure of passive fire resistance to filled plastics since they reduce the volume of combustible matter in the filled composition and, depending on their density, may also serve as insulation (see Flame retardants). Fillers that contain combined water and carbon dioxide, such as alumina trihydrate or dawsonite

[*12011-76-6*], increase fire resistance by liberating noncombustible gases when they are heated. These gases withdraw heat from the plastic and can also reduce the oxygen concentration of the air surrounding the composition.

Electrical Resistance. Most fillers are mainly composed of nonconducting substances that should, therefore, provide electrical resistance properties comparable to the plastics in which they are used. However, some fillers contain adsorbed water or other conductive impurities that can greatly reduce electrical resistance. Some of the standard tests for the electrical resistance of filled plastics are dielectric strength (ASTM D 149), dielectric constant (ASTM D 150), arc resistance (ASTM D 495), and d-c resistance (ASTM D 257) (see Insulation, electric).

Paint. The liquid phase of paint formulations, usually termed the *vehicle,* typically contains volatile and nonvolatile fractions. When paint is applied, the volatile fraction of the vehicle evaporates and the nonvolatile fraction polymerizes to become a film matrix in which prime pigments and fillers are embedded (see Paint). Prime pigments are coloring and opacifying pigments. Fillers in paints are variously referred to as inerts, extender pigments, and supplemental pigments. Table 7 lists typical paint fillers and their uses. Compared to prime pigments, fillers have lower refractive indexes and tinting strength and cost less (see Pigments).

Fillers contribute to application, durability, protection, and decoration. The properties that they impart to paints are similar to the optical and mechanical properties they impart to paper.

Optical Properties. *Gloss.* As in paper technology, gloss is the spectrally reflected light typically associated with such phenomena as shininess, highlights, and reflected images. It is usually evaluated by comparing a paint sample with a highly polished, black glass surface (ASTM D 523-67).

The degree of gloss of the painted surface depends mainly on the smoothness of the reflecting surface. For this reason, filler size, shape, and surface energy influence the glossiness of filled paints. High gloss paints are formulated with fillers and pigments that have small particle dimensions. An average particle diameter of less than 0.3 μm is necessary to provide high gloss surfaces (17). Low-gloss paint films are obtained by using fillers with a large average particle size. As was the case in paper, laminar, platelike fillers increase the gloss of paints more effectively than spherical or irregularly shaped fillers of equivalent particle size. Pigments that are completely wetted by the vehicle permit higher gloss than pigments that are incompletely wetted (18).

Table 7. Typical Paint Fillers and Their Uses

Filler	Refractive index	Use
barium sulfate	1.6	sanding primers
calcium carbonate	1.6	most paints
clay	1.6	exterior paints
mica	1.6	exterior paints
silica, diatomite	1.5	traffic paints
precipitated	1.5	flatted paints
talc	1.6	exterior paints
zinc oxide	2.0	exterior paints

Hiding Power. Like opacity in paper, the hiding power of fillers in paint is determined by the difference between their index of refraction and the index of refraction of their surrounding medium. The indexes of refraction of common paint vehicles are ca 1.5. Air has a refractive index of 1.0. Since most fillers that are used in paint have refractive indexes of from 1.4–1.7, it is readily apparent that they contribute little to the hiding power of paint when they are completely embedded in the vehicle. However, if filler particles extend above the vehicle film into the air, they contribute hiding power. Hence, hiding power can be expressed by wet hiding power (ASTM D 2805-70) and dry hiding power (ASTM D 344). Dry hiding and high gloss are mutually exclusive.

Mechanical Properties. *Stain resistance* of paints is directly related to their porosity. Therefore, fillers that reduce porosity, ie, with low surface area, wide particle size distribution, and laminar particle shape, contribute to stain resistance. Stain resistance of paints is evaluated by ASTM D 3023.

Viscosity. The ease with which paint is applied and the uniformity of the paint film depends largely on the paint viscosity. If the viscosity is too high, paint does not level and, therefore, brush, roller, or spray marks result. On the other hand, if a paint has too little viscosity, it sags or runs. High surface area (high vehicle demand), a high degree of filler agglomeration, and low sphericity contribute to paint viscosity. The viscosity of filled paints is evaluated by ASTM D 1200 and ASTM D 562.

Weathering Resistance. The principal causes of paint weathering are moisture, abrasion, uv radiation, and microorganisms. Fillers that tend to yield porous coatings increase water resistance since water damage occurs most often when water becomes trapped between the substrate and the paint film and blistering results. Hard pigments such as silica and clay contribute abrasion resistance to paints. Since uv radiation oxidizes prime pigments and paint vehicles in the presence of oxygen and water vapor, fillers that absorb uv radiation and limit the paint porosity reduce this type of degradation. Alkaline fillers, such as zinc oxide and calcium carbonate, inhibit the growth of mildew and other microorganisms.

Economic Aspects

Sales of fillers in the United States was estimated at 380×10^6 dollars for 1977, an increase of 48% over 1972 (19). During 1978 the industry experienced a growth in excess of inflation of 3–4%. Mineral fillers represent over 80% of the filler market, primarily because of their lower cost. They have been upgraded by the introduction of new grinding, beneficiation, and surface-treatment technologies. Thus mineral fillers can meet the competition of the highly engineered synthetic fillers in many markets. On the other hand, synthetic fillers have grown during the 1970s at a rate far above the 7%/yr average growth rate for fillers in general (20).

Table 8 gives 1979 price information on specific fillers, including some physical properties and manufacturing processes.

Projections. The market for fillers has been projected to reach 3,000,000 metric tons at a value of 901×10^6 dollars by 1986 (20). This projection is based on the increased cost of petroleum-derived matrix materials which they will replace.

Elastomers. World markets for tires and nontire products are predicted to grow over 50% by 1987 (21). However, carbon black is forecast to grow only 2–3% annually during the same period. This lag is based on the fact that the feedstock for carbon black is decant oil, which is also the source of fuel oil.

Table 8. Specific Fillers

Fillers	CAS Registry Number	Manufacturing process	Specific gravity	Refractive index	Oil adsorption, wt %	Approximate price, 1979, $/kg
Inorganics						
carbonates						
barium carbonate	[*513-77-9*]	precipitated	4.43	1.6	14	0.30
calcium carbonate						
precipitated	[*471-34-1*]	precipitated	2.65–2.68	1.63	15–65	0.11
limestone	[*1317-65-3*]	mined	2.71	1.60	10.5–15	0.02–0.04
whiting	[*13397-26-7*]	wet ground	2.71	1.60	6.5–15.0	0.04–0.06
magnesium carbonate	[*546-93-0*]	precipitated	2.20	1.5–1.71	80	0.48
fluorides						
calcium fluoride	[*7789-75-5*]	mined	3.18	1.43		0.24
sodium aluminum fluoride	[*15096-52-3*]	precipitated	2.95	1.3		0.55
hydroxides						
aluminum hydroxide	[*21645-51-2*]	precipitated	2.4	1.58		0.15
metals						
aluminum	[*7429-90-5*]	atomized	2.55			3.08
bronze	[*12597-70-5*]	reduced	8.0			3.30
lead	[*7439-92-1*]	atomized	11.4			1.32
zinc	[*7440-66-6*]	atomized	7.0			0.99
oxides						
aluminum oxide	[*1344-28-1*]	furnace	3.8	1.6	13	0.29
magnesium oxide	[*1309-48-4*]	calcined	3.5		70	0.33
silicon dioxide	[*7631-86-9*]					
colloidal sol		ion exchange	1.3	1.4		
diatomaceous		mined	2.65	1.4	120	0.09
novaculite		mined	2.65	1.55	20	0.11
pyrogenic		oxidation	2.2	1.4	150	3.41
quartz flour		ground	2.65	1.55	32	0.04
tripolite		mined	2.65	1.55	31	0.06
vitreous		furnace	2.18	1.4	15	0.55
wet process		precipitated	1.9–2.2	1.4	160	0.35

titanium dioxide	*[1317-70-0]*	oxidation	3.9	2.55	24	0.99
zinc oxide	*[1314-13-2]*	oxidation	5.6	2.0	13	0.88
silicates						
asbestos	*[1343-90-4]*	air flotation	2.56	1.57	34	13
clay						
kaolin	*[1327-36-2]*	mined	2.60	1.56	36	3
calcined kaolin	*[1327-36-2]*	calcined	2.63	1.62	25	17
calcium silicate	*[1344-96-3]*	precipitated	2.33	1.4	375	15
feldspar		mined	2.6	1.53	19	6
glass						
ground glass		milled	2.5	1.5	15	11
flakes		milled film	2.5	1.5		176
hollow spheres		heat expanded	0.2	1.5	35	143
solid spheres		furnace	2.5	1.5	30	44
mica	*[1327-44-2]*					
muscovite		mined	2.75	1.59	47	26
phlogopite		mined	2.75	1.60	24	26
vermiculite		heat expanded	2.25	1.60		33
nepheline	*[14797-52-5]*	mined	2.6	1.6	20	6
perlite		heat expanded	0.17			33
pyrophyllite	*[12269-78-2]*	mined	2.85	1.59	36	6
talc	*[14807-96-6]*	mined	2.85	1.59	27	4
wollastonite	*[13983-17-0]*	mined	2.9	1.59	26	6

Table 8 (*continued*)

Fillers	CAS Registry Number	Manufacturing process	Specific gravity	Refractive index	Oil adsorption, wt %	Approximate price, 1979, $/kg
sulfates						
barium sulfate						
barytes	[*13462-86-7*]	mined	4.47	1.64	6	15
blanc fixe	[*7727-43-7*]	precipitated	4.35	1.64	14	75
calcium sulfate						
gypsum	[*7778-18-9*]	mined	2.35	1.53	21	4
anhydrite	[*14798-04-0*]	mined	2.95	1.59	25	6
precipitated	[*23296-15-3*]	precipitated	2.95	1.59	50	15
sulfides						
lithopone	[*8006-32-4*]	precipitated	4.2	1.8	14	68
zinc sulfide	[*12402-34-5*]	precipitated	4.0	2.37	13	66
Organics						
cellulose						
cork	[*61789-98-8*]	ground	0.25			1.21
corn cob		cut	1.2			0.22
flock		ground				0.55
shell flour		ground	1.4		15	0.13
wood flour		ground	0.65		25	0.11
fatty acids and esters						
sulfur-chlorinated vegetable oil		sulfur chlorination	1.04			1.54
vulcanized vegetable oil		vulcanized	1.0			1.54
polymers						
polystyrene	[*9003-53-6*]	FR catalysis	1.05	1.59		1.21
phenol–formaldehyde	[*9003-35-4*]	condensation	1.1			1.21
mineral rubber		petroleum residue	1.0			1.21

Paper. Throughout the 1970s writing, printing, and coated papers enjoyed an average annual growth rate of 4–5.2%. This growth rate is projected to decline through the 1980s to 2.5–3.0% (22). Fillers marketed to the paper industry are expected to grow faster than the industry itself because fillers enjoy an ever-increasing cost advantage over TiO_2, and at the same time improve brightness.

Plastics. Historically, sales of plastics have grown at about twice the growth rate of sales of fillers (20). The future growth of fillers for plastics, however, is expected to outstrip the plastics industry, at least on a volume basis because increased filler-to-resin ratio accompanies resin price increases owing to the increased cost of petrochemical feedstocks.

Paint. The sales of fillers to the paint market parallels industry trends. From the late 1960s to the late 1970s the industry grew at about 2%/yr, and it is forecast to grow at the rate of 2.8%/yr throughout the 1980s (23). Although the cost of paint vehicles will undoubtedly increase in the future, paint manufacturers will not respond by increasing their ratio of filler to vehicle because vehicle loading in paints is determined primarily by performance rather than economic considerations.

Health and Safety Factors

The major hazard involved in the handling and use of fillers is the inhalation of airborne particles in the respirable size range (less than 10 μm aerodynamic diameter). At the present time, filler dusts are classified into carcinogens, fibrogens, and nuisance particulates. Nuisance particulates are dusts that have a long history of little adverse affect on lungs and do not produce significant organic disease or toxic effect when exposures are kept under reasonable control.

The American Conference of Governmental Industrial Hygienists established TLVs for the airborne concentration of many fillers in workroom air (24). These range from a proposed 0.5 fiber/cm^3 for amosite asbestos (a carcinogen) through 10 mg/m^3 divided by the percent respirable quartz + 2 for dust containing quartz (a fibrogen) up to ca 1.1×10^9 particles/m^3 of air (3×10^7 particles/ft^3 air) for calcium carbonate (a nuisance dust). Suppliers usually have information on the safe handling of these products.

BIBLIOGRAPHY

1. W. C. McCrone, R. G. Draftz, and J. G. Delly, *Particle Atlas Two,* Ann Arbor Science Publishers, Ann Arbor, Mich., 1973.
2. T. Allen, *Particle Size Measurement,* Chapman and Hall, London, Eng., 1974, p. 93.
3. R. R. Irani and C. F. Callis, *Particle Size: Measurement Interpretation and Application,* John Wiley & Sons, Inc., New York, 1963.
4. G. J. Gregg and K. S. W. Sing, *Adsorption, Surface Area, and Porosity,* Academic Press, Inc., New York, 1967.
5. S. Brunauer, P. H. Emmett, and E. Teller, *J. Am. Chem. Soc.* **60,** 309 (1938).
6. A. W. Adamson, *Physical Chemistry of Surfaces,* 2nd ed., John Wiley & Sons, Inc., New York, 1960.
7. C. C. Furnas, *Ind. Eng. Chem.* **23,** 1052 (1931).
8. P. B. Mitton, "Opacity Hiding Power and Tinting Strength" in T. C. Patton, ed., *Pigment Handbook,* Wiley-Interscience, New York, 1973, p. 289.
9. J. V. Milewski, "Packing Concepts on the Utilization of Filler and Reinforcement Combinations" in H. S. Katz and J. V. Milewski, eds., *Handbook of Fillers and Reinforcements for Plastics,* Van Nostrand Reinhold, New York, 1978, p. 66.

10. A. N. Winchell and H. Winchell, *The Microscopical Characters of Artificial Inorganic Solid Substances*, Academic Press, Inc., New York, 1964.
11. N. H. Hartshorne and A. Stuart, *Crystals and the Polarizing Microscope*, 4th ed., Arnold, London, Eng., 1970.
12. R. B. Sucher, "Reinforcement" in *Basic Rubber Technology*, ACS Rubber Group Lecture, Philadelphia, Pa., 1955, p. 13.
13. *Ibid.*, p. 15.
14. R. S. Hunter, *ASTM Bull.* **186**, 48 (1952).
15. M. P. Morse, "Surface Appearance" in ref. 8, p. 289.
16. H. C. Schwalbe, *Paper Web Transactions of the Cambridge Symposium*, Vol. 2, British Paper and Board Makers Association, London, Eng., 1966, p. 692.
17. N. F. Miller, *Off. Dig. Fed. Soc. Paint Technol.* **34**, 465 (1962).
18. E. Singer, "Additives" in R. Myers and J. Long, eds., *Treatise on Coatings*, Vol. 3, Part I, Marcel Dekker, Inc., New York, 1974, p. 32.
19. C. H. Kline Co., *Am. Paint Coat. J.*, 12 (Aug. 29, 1977).
20. Business Communications Co., *Chem. Mark. Rep.*, 5 (June 20, 1977).
21. International Institute of Synthetic Rubber Producers, *Rubber World*, 34 (Jan. 1978).
22. K. E. Lowe, *Pulp Paper*, 60 (May 1976).
23. Predicasts Inc., *Chem. Week*, 53 (Nov. 15, 1978).
24. American Conference of Governmental Industrial Hygienists, *TLVS® Threshold Limit Values for Chemical Substances in the Workroom Environment with Intended Changes for 1978*, ACGIH, Cincinnati, Ohio, 1978.

JOHN G. BLUMBERG
JAMES S. FALCONE, JR.
LEONARD H. SMILEY
PQ Corporation

DAVID I. NETTING
ARCO Chemical Co.

FOAMED PLASTICS

Foamed plastics, otherwise known as cellular plastics or plastic foams, have been important to human life since primitive man began to use wood, a cellular form of the polymer cellulose. Cellulose (qv) is the most abundant of all naturally occurring organic compounds, comprising approximately one third of all vegetable matter in the world (1). Its name is derived from the Latin *cellula*, meaning very small cell or room, and most of the polymer does indeed exist in cellular form as in wood, straws, seed husks, etc. The high strength-to-weight ratio of wood, good insulating properties of cork and balsa, and cushioning properties of cork and straw have contributed both to the incentive to develop and to the background knowledge necessary for development of the broad range of cellular synthetic polymers in use today.

The first cellular synthetic plastic was an unwanted cellular phenol–formaldehyde resin produced by early workers in this field. The elimination of cell formation in these resins, as given by Baekeland in his 1909 heat and pressure patent (2), is generally considered the birth of the plastics industry. The first commercial cellular polymer was sponge rubber, introduced between 1910 and 1920 (3).

Many cellular plastics that have never reached significant commercial use have been introduced or their manufacture described in literature. Examples of such polymers are chlorinated or chlorosulfonated polyethylene (4), a copolymer of vinylidene fluoride and hexafluoropropylene [*9011-17-0*] (4), polyamides (4), polytetrafluoroethylene [*9002-84-0*] (5), styrene–acrylonitrile copolymers [*9003-54-7*] (6–7) and ethylene–propylene copolymers [*9010-79-1*] (8).

Cellular polymers have been commercially accepted in a wide variety of applications since the 1940s (9–18). The total usage of foamed plastics in the United States has risen from 245×10^3 metric tons in 1965 to over 1.4×10^6 metric tons in 1976, and has been projected to rise to about 6.6×10^6 t in 1990 (19). The markets and applications that make up this total usage are covered in detail in the section on applications.

Nomenclature

A cellular plastic has been defined as a plastic the apparent density of which is decreased substantially by the presence of numerous cells disposed throughout its mass (20). In this article the terms *cellular polymer, foamed plastic, expanded plastic,* and *plastic foam* are used interchangeably to denote all two-phase gas–solid systems in which the solid is continuous and composed of a synthetic polymer or rubber.

The gas phase in a cellular polymer is usually distributed in voids or pockets called cells. If these cells are interconnected in such a manner that gas can pass from one to another, the material is termed *open-celled.* If the cells are discrete and the gas phase of each is independent of that of the other cells, the material is termed *closed-celled.*

The nomenclature of cellular polymers is not standardized; classifications have been made according to the properties of the base polymer (21), the methods of manufacture, the cellular structure, or some combination of these. The most comprehensive classification of cellular plastics, proposed by Cooper in 1958 (22), has not been adopted and is not consistent with some of the currently common names for the more important commercial products.

A particular ASTM test procedure has suggested (23) that foamed plastics be classified as either rigid or flexible, a *flexible foam* being one that does not rupture when a $20 \times 2.5 \times 2.5$ cm piece is wrapped around a 2.5-cm mandrel at a uniform rate of 1 lap/5 s at 15–25°C. *Rigid foams* are those that do rupture under this test. This classification is used in this article.

In the case of cellular rubber, the ASTM uses several classifications based on the method of manufacture (24–25). These terms are used here. *Cellular rubber* is a general term covering all cellular materials that have an elastomer as the polymer phase. *Sponge rubber* and *expanded rubber* are cellular rubbers produced by expanding bulk rubber stocks and are open-celled and closed-celled, respectively. *Latex foam rubber*, also a cellular rubber, is produced by frothing a rubber latex or liquid rubber, gelling the frothed latex, and then vulcanizing it in the expanded state.

The term *structural foams* has not been exactly defined but are used here to refer to those rigid foams produced at greater than about 320 kg/m^3 density.

Theory of the Expansion Process

Foamed plastics can be prepared by a variety of methods. The most important process, by far, consists of expanding a fluid polymer phase to a low density cellular state and then preserving this state. This is the foaming, or expanding, process. Other methods of producing the cellular state include leaching out solid or liquid materials that have been dispersed in a polymer, sintering small particles, and dispersing small cellular particles in a polymer. The latter processes, however, are relatively straightforward processing techniques but are of minor importance.

The expansion process consists of three steps: creating small discontinuities or cells in a fluid or plastic phase; causing these cells to grow to a desired volume; and stabilizing this cellular structure by physical or chemical means.

Initiation and Growth of Cells. The initiation, or nucleation, of cells is the formation of cells of such size that they are capable of growth under the given conditions of foam expansion. The growth of a hole or cell in a fluid medium at equilibrium is controlled by the pressure difference ΔP between the inside and the outside of the cell, the surface tension of the fluid phase γ, and the radius r of the hole, according to equation 1:

$$\Delta P = 2\,\gamma/r \tag{1}$$

The pressure outside the cell is the pressure imposed on the fluid surface by its surroundings. The pressure inside the cell is the pressure generated by the blowing agent dispersed or dissolved in the fluid. If blowing pressures are low, the radii of initiating holes must be large. The hole that acts as an initiating site can be filled with either a gas or a solid which breaks the fluid surface and thus enables blowing agent to surround it (26–29).

During the time of cell growth in a foam, a number of properties of the system change greatly. Cell growth can, therefore, be treated only qualitatively. The following considerations are of primary importance: (*1*) the fluid viscosity is changing considerably, this tends to influence both the cell growth rate and the flow of polymer from cell walls to intersections leading to collapse; (*2*) the pressure of the blowing agent decreases, falling off less rapidly than an inverse volume relationship because new blowing agent diffuses into the cells as the pressure falls off according to equation 1; (*3*) the rate of growth of the cell depends on the viscoelastic nature of the polymer phase, the blowing-agent pressure, the external pressure on the foam, and the permeation rate of blowing agent through the polymer phase; and (*4*) as a consequence of equation 1, the pressure in the cell of small radius r_2 is greater than that in the cell of larger radius r_1. There will thus be a tendency to equalize these pressures either by breaking the wall separating the cells or by diffusion of the blowing agent from the small to the large cells. The pressure difference ΔP between cells of radii r_1 and r_2 is shown in equation 2:

$$\Delta P = 2\,\gamma\left(\frac{1}{r_2} - \frac{1}{r_1}\right) \tag{2}$$

Stabilization of the Cellular State. The increase in surface area corresponding to the formation of many cells in the plastic phase is accompanied by an increase in the free energy of the system; hence the foamed state is inherently unstable. Methods of stabilizing this foamed state can be generally classified as chemical, ie, the polymerization of a fluid resin into a three-dimensional thermoset polymer, or physical, ie, the cooling of an expanded thermoplastic polymer to a temperature below its second-order transition temperature or its crystalline melting point to prevent polymer flow.

The important considerations in the chemical method of dimensional stabilization are the following:

Chemistry of the System. This determines both the rate at which the polymer phase is formed and the rate at which it changes from a viscous fluid to a dimensionally stable thermoset phase. It also governs the rate at which the blowing agent is activated, whether it be owing to temperature rise or to insolubilization in the liquid phase.

Type and Amount of Blowing Agent. These factors govern the amount of gas generated, the rate of generation, and the pressure that can be developed to expand the polymer phase. They also determine the amount of gas lost from the system relative to the amount retained in the cells.

Additives to the Foaming System. Additives (cell-control agents) can greatly influence nucleation of foam cells, either through their effect on the surface tension of the system, or by acting as nucleating sites from which cells can grow. They can influence the mechanical stability of the foam structure considerably by changing the physical properties of the plastic phase and by creating discontinuities in the plastic phase that allow blowing agent to diffuse from the cells to the surroundings.

Environmental factors such as temperature, geometry of foam expansion, and pressure also influence the behavior of thermoset foaming systems.

In physically stabilized foaming systems the factors are essentially the same as for chemically stabilized systems but for somewhat different reasons:

Chemical Composition of the Polymer Phase. The type of polymer determines the temperature at which foam must be produced, the type of blowing agent required, and the type of cooling of the foam necessary for dimensional stabilization.

Blowing-Agent Composition and Concentration. These control the rate at which gas is released, the amount of gas released, the pressure generated by the gas, escape or retention of gas from the foam cells for a given polymer, and heat absorption or release owing to blowing-agent activation.

Additives. Additives have the same effect on thermoplastic foaming processes as upon thermoset foaming processes.

Environmental Conditions. Environmental conditions are important in this case because of the necessity of removing heat from the foamed structure in order to stabilize it. The dimensions and size of the foamed structure are important for the same reason.

Manufacturing Processes for Cellular Polymers

Cellular plastics and polymers have been prepared by a wide variety of processes involving many methods of cell initiation, cell growth, and cell stabilization. The most convenient method of classifying these methods appears to be based on the cell growth and stabilization processes. According to equation 1, the growth of the cell depends on the pressure difference between the inside of the cell and the surrounding medium. Such pressure differences may be generated by lowering the external pressure (decompression) or by increasing the internal pressure in the cells (pressure generation). Other methods of generating the cellular structure are by dispersing gas (or solid) in the fluid state and stabilizing this cellular state, or by sintering polymer particles in a structure that contains a gas phase.

Foamable compositions in which the pressure within the cells is increased relative to that of the surroundings have generally been called expandable formulations. Both chemical and physical processes are used to stabilize plastic foams from expandable formulations. There is no single name for the group of cellular plastics produced by the decompression processes. The various operations used to make cellular plastics by this principle are extrusion, injection molding, and compression molding. Either physical or chemical methods may be used to stabilize products of the decompression process.

EXPANDABLE FORMULATIONS

Physical Stabilization Process. Cellular polystyrene [*9003-53-6*], poly(vinyl chloride) [*9002-86-2*], copolymers of styrene and acrylonitrile, and polyethylene [*9002-88-4*] can be manufactured by this process.

Polystyrene. Polystyrene is the outstanding example of a material that can be fabricated into cellular form by this method. In general, there are two types of expandable polystyrene processes, depending on the applications—expandable polystyrene for molded articles, and expandable polystyrene for loose-fill packing materials (see Styrene plastics).

Expandable polystyrene for loose-fill packing materials is available in various sizes and shapes varying from round disks to S-shaped strands. These particles can These particles are prepared either by heating polymer particles in the presence of a blowing agent and allowing the blowing agent to penetrate the particle (30) or by polymerizing the styrene monomer in the presence of blowing agent (31) so that the blowing agent is entrapped in the polymerized bead. Typical blowing agents used in such processes are the various isomeric pentanes and hexanes, halocarbons, and mixtures of these materials (32). The same agents are used for styrene–acrylonitrile foams.

The fabrication of these expandable particles into a finished cellular-plastic article is generally carried out in two steps (33–36). In the first step the particles are expanded by means of steam, hot water, or hot air into low density replicas of the original material, called prefoamed or preexpanded beads. After proper aging enough of these prefoamed beads are placed in a mold to just fill it; the filled mold is then exposed to steam. This second expansion of the beads causes them to flow into the spaces between beads and fuse together, forming an integral molded piece. Stabilization of the cellular structure is accomplished by cooling the molded article while it is still in the mold. The density of the cellular article can be adjusted by varying the density of the prefoamed particles.

Expandable polystyrene for loose-fill packing materials is available in various sizes and shapes varying from round disks to S-shaped strands. These particles can be prepared either by deforming the polystyrene under heat and impregnating the resin with a blowing agent in an aqueous suspension (37) or by the extrusion method with various die orifice shapes (38).

The expansion of these particles into a product is usually carried out in two or three expansions by means of steam with at least one day of aging after each expansion (39). Stabilization is accomplished by cooling the polymer phase below its glass transition temperature during the expansion process.

Poly(vinyl Chloride). Cellular poly(vinyl chloride) can be produced from several expandable formulations as well as by decompression techniques. Rigid or flexible products can be made depending upon the amount and type of plasticizer used (40) (see Vinyl polymers).

Polyethylene. Polyethylene has a sharp melting point and the viscosity of polyethylene decreases very rapidly over a narrow temperature range above the melting point. This makes it very difficult to produce a low density polyethylene foam with nitrogen or chemical blowing agents because the foam collapses before it can be stabilized. This problem can be eliminated by cross-linking the resin before it is foamed, which slows the viscosity decrease above the melting point as shown in Figure 1 and allows the foam to be cooled without collapse of cell structure (see Olefin polymers).

Cross-linking of polyethylene can be accomplished either chemically or by high energy radiation. Radiation cross-linking is usually accomplished by x rays (42) or electrons (43–44). Chemical cross-linking of polyethylene is accomplished with dicumyl peroxide (45), di-*tert*-butyl peroxide (46), or other peroxides. Radiation cross-linking (41) is preferred for thin foams, and chemical cross-linking for the thicker foams.

Expandable polyethylene foam sheet can be made by a four-step process: (*1*) mixing of polyethylene, chemical blowing agent and cross-linking agent (in the case of chemical cross-linking) at low or medium temperature (examples of decomposable blowing agents used for expandable polyethylene are azodicarbonamide, 4,4′-oxybis(benzenesulfonyl hydrazide), and dinitrosopentamethylenetetramine (32)); (*2*) shaping at low or medium temperature; (*3*) chemical cross-linking at medium temperature or radiation cross-linking; and (*4*) heating and expanding at high temperature. Expansion of the cross-linked, expandable polyethylene sheet can be accomplished either by floating the sheet on the surface of a molten salt bath at 200–250°C and heating from above with ir heaters or circulating hot air or by expanding in the mold with a high pressure steam (see also Film and sheeting materials).

Chemical Stabilization Processes. This method has proved the most versatile in producing cellular plastics in that it has been used successfully for more materials than the physical stabilization process. It has generally been more adaptable for condensation polymers than for vinyl polymers because of the fast yet controllable curing reactions and the absence of atmospheric inhibition.

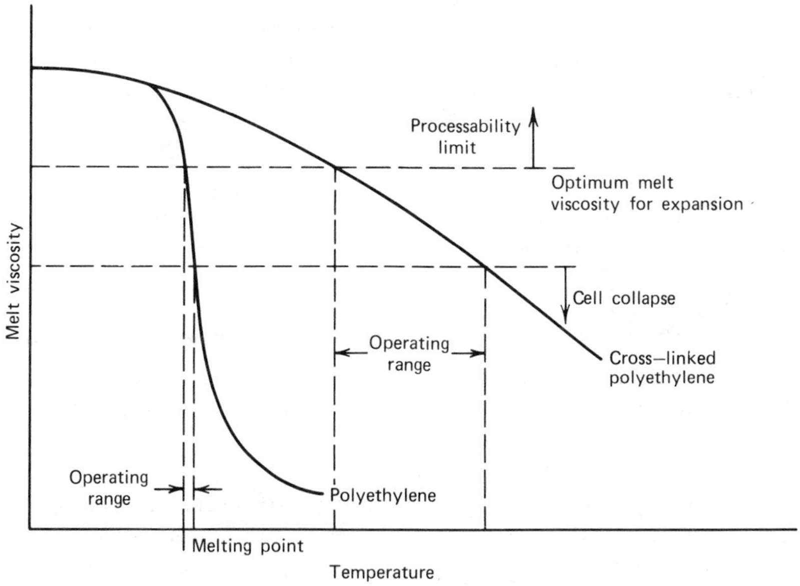

Figure 1. Effect of cross-linking on melt behavior of polyethylene (41).

Polyurethane Foams. The most important commercial example of the chemical stabilization process is the production of polyurethane [*27416-86-0*] foams. More literature has been published on the chemical and physical aspects of the polyurethane foam process than on all other plastic foam processes combined in spite of the relatively short commercial lifetime of polyurethane. An excellent summary of the chemistry and technology of these polymers has been published (12,47) (see Urethane polymers).

The chemical ingredients of a polyurethane foam system are a polyfunctional isocyanate (**1**) and a hydroxyl-containing polymer (**2**), along with the catalysts necessary to control the rate and type of reaction and other additives to control the surface chemistry of the process. A number of competing reactions can occur when (**1**) and (**2**) are brought together but the main product, shown in equation 3, is a polyurethane (**3**):

$$\underset{(1)}{OCN{-}R{-}NCO} + \underset{(2)}{HO{-}R'{-}OH} \longrightarrow \underset{(3)}{\sim\!\!\left[\overset{O}{\overset{\|}{C}}NH{-}R{-}NH\overset{O}{\overset{\|}{C}}{-}O{-}R'{-}O\right]_n\!\!\sim} \quad (3)$$

Originally carbon dioxide was generated *in situ* (by the reaction of isocyanate with water) as a blowing agent for both rigid and flexible polyurethane foams. It is still common practice to rely partly on this method of gas generation for flexible materials. Rigid cellular polyurethanes are now produced using volatile liquids, usually halocarbons, as expanding agents, producing gas as the foaming mixture heats up. These materials remain in the foam, and their presence as gas in the cells lowers the thermal conductivity of the foam considerably.

The general method of producing a cellular polyurethane is to mix the above ingredients and adjust the conditions and reactants such that the exotherm of the reaction causes expansion of the blowing agent and resultant foaming of the resins. The change of physical properties of the mixture is timed to coincide with the expansion of the foaming mixture so that dimensional stabilization by cross-linking of the polymer occurs at the time corresponding to minimum density.

The physical properties of the final cellular material can be varied broadly by controlling the degree of cross-linking in the final polymer as well as the structure of R and R′ in equation 3. The average molecular weight between cross-links is generally 400–700 for rigid polyurethane foams, 700–2,500 for semirigid foams, and 2,500–20,000 for flexible polyurethane foams (12). The variation of cross-link density is normally obtained by varying the structure of the hydroxyl-containing component (**2**), since the structure of the diisocyanate is limited to some 6 or 8 commercially available compounds (12,48). Because the variation between cross-links is controlled by the structure of the polyhydroxy resin component, it is common to use the equivalent weight per hydroxyl unit in the resin as a criterion for rigid, semirigid, and flexible foams. In these terms, the equivalent weights of resins used for rigid foams vary from 100 to 150, for semirigid between 200 and 1,000, and for flexible from 1,000 to 10,000. The equivalent weight in this case is defined as the ratio of molecular weight to hydroxyl functionality of the polyhydroxy resin.

Two general types of processes have been developed for producing cellular polyurethanes on a commercial scale (47). These two processes are commonly called the one-shot process and the prepolymer processes. In the one-shot process all the necessary ingredients for producing the foam are mixed and then discharged from the mixer onto a suitable surface. The reactions begin immediately and proceed at such a rate that expansion starts in about 10 s, the entire expansion is completed in 1–2 min. Curing may take several days.

In the prepolymer process the polyhydroxy component reacts with enough polyisocyanate to form a prepolymer with isocyanate end groups plus excess isocyanate:

$$OCN—R—NCO + HO—R'—OH \rightarrow OCN—R—NH—\overset{O}{\overset{\|}{C}}O—R'—O\overset{O}{\overset{\|}{C}}—\overset{H}{N}—R—NCO \quad (4)$$

prepolymer

R is typically a polyether structure but can also be a polyester. The prepolymer mixture then reacts with water to release carbon dioxide for expansion and to link the chains into a cross-linked matrix. This method is sometimes used for flexible foams.

In a semiprepolymer process, which has become more extensively used, a prepolymer containing excess isocyanate is mixed with more polyhydroxy resin and a separate blowing agent such as a halocarbon. In this case the prepolymer may contain only a few percent of the total polyhydroxy resin. This method is particularly useful for producing rigid foams.

After mixing, the resin can be dispensed by several different methods (47). Typical methods are (*1*) as an expandable liquid, (*2*) as a spray of small droplets of mixed resins which adhere to surfaces and foam on these surfaces, and (*3*) as a froth into which some gas has been mixed prior to exiting from the mixing head, which causes the liquid mixture to froth as its pressure is decreased to atmospheric pressure. A modification of method (*3*) allows the froth to be sprayed in chunks upon a surface with subsequent additional expansion during curing.

Polyisocyanurates. The isocyanurate ring formed by the trimerization of isocyanate has been known to possess high thermal and flammability resistance and low smoke generation (49–52). Because of these characteristics, aluminum foil-faced sheets of isocyanurate foam are now widely used as an insulation material. Cross-linking via the high functionality of isocyanurates causes an inherent friability of the foam. Chemical modifications of the foam with urethane groups is usually made to circumvent the friability of pure isocyanurate foam. When introducing the urethane group into the polymer backbone, the proper catalyst must be selected to favor or promote formation of desired compounds from the excess isocyanate during polymerization (53–54). Sodium or potassium coordination compounds with organic borate esters have been suggested as suitable catalysts for isocyanurate-modified polyurethane foams (55). Another isocyanurate catalyst consists of a mixture of trialkylaminohexahydrotriazine and a 1,2-epoxide (56). The manufacturing process for isocyanurate foam is similar to that for polyurethane foam (see Cyanuric and isocyanuric acids).

Polyphenols. Another increasingly important example of the chemical stabilization process is the production of phenolic foams (57–60). Phenolic foams are produced by cross-linking polyphenols (resoles and novolacs). The principal features of phenolic foams are low flammability, solvent resistance, and excellent dimensional stability over a wide temperature range (57). Because of these unique properties, phenolic foams are good thermal insulating materials (see Phenolic resins).

Most phenolic foams are produced from resoles and acid catalyst. Suitable water-soluble acid catalysts are mineral acids (such as hydrochloric acid or sulfuric acid) and aromatic sulfonic acids (61). Phenolic foams can be produced from novolacs with more difficulty than from resoles (57). Novolacs are thermoplastic and require a source of methylene group to permit cure. This is usually supplied by hexamethylenetetramine (62).

A typical phenolic foam system consists of liquid phenolic resin, blowing agent, catalyst, surface-active agent, and modifiers. Various formulations and composite systems (63–65) can be used to improve one or more properties of the foam in specific applications such as insulation properties (61,66–69), flammability (70–73), and open cell (74–76) (quality).

Several manufacturing processes can be used to produce phenolic foams (57,77): continuous production of free-rising foam for slabs and slab stock similar to that for polyurethane foam (59,78); foam-in-place batch process (59,79); sandwich paneling (61,80–81); and spraying (68,82).

Other Materials. Foams from epoxy resins (qv) (57–58,83–84), and silicone resins (see Silicon compounds) (29,58,85–86) can also be formed by chemical stabilization processes.

DECOMPRESSION EXPANSION PROCESSES

Physical Stabilization Process. Cellular polystyrene, cellulose acetate, polyolefins, and poly(vinyl chloride) can be manufactured by this process.

Polystyrene. The extrusion process for producing cellular polystyrene is probably the oldest process utilizing physical stabilization in a decompression expansion process (87). A solution of blowing agent in molten polymer is formed in an extruder under pressure. This solution is forced out through an orifice onto a moving belt at ambient temperature and pressure. The blowing agent then vaporizes and causes the polymer to expand. The polymer simultaneously expands and cools under such conditions that it has developed enough strength to maintain dimensional stability at the time corresponding to optimum expansion. In this case the stabilization is owing to cooling of the polymer phase to a temperature below its glass transition temperature. Cooling comes from the vaporization of the blowing agent, gas expansion, and heat loss to the environment. Polystyrene foams produced by the decompression process are commercially offered in the density range of 27–53 kg/m^3 (1.7–3.3 lb/ft^3) as well as at higher densities (88).

Several procedures can be used to produce polystyrene foam sheet for various applications. The extrusion of expandable polystyrene beads or pellets containing pentane blowing agent was originally used to produce low density foam sheet (89–90). The current method is to extrude polystyrene foam in a single-screw tandem line or twin-screw extruder and produce foam sheet by addition of pentane or fluorocarbon blowing agents into the extruder (91–92). Two general methods are used for the production of low density polystyrene foam sheet. For sheet thicknesses of less than 500 μm (20 mil), the blown-bubble method is normally used. This method involves blowing a tube from a round or annular die, collapsing the bubble, and then slitting the edges to obtain two flat sheets. For greater sheet thicknesses the sheet is pulled over a sizing mandrel and slit to obtain a flat sheet. Cooling of the expanded material by the external air is necessary to stabilize the foam sheet with a good skin quality.

Cellular polystyrene can also be produced by an injection-molding process. Polystyrene granules containing dissolved liquid or gaseous blowing agents are used as feed in a conventional injection-molding process (93). With close control of time and temperature in the mold and use of vented molds, high density cellular polystyrene moldings can be obtained (see also Styrene; Styrene polymers).

Cellulose Acetate. The extrusion process has also been used to produce cellular cellulose acetate [*9004-35-7*] (94) in the density range of 96–112 kg/m^3 (6–7 lb/ft^3). A hot mixture of polymer, blowing agent, and nucleating agent is forced through an orifice into the atmosphere. It expands, cools, and is carried away on a moving belt (see Cellulose acetate).

Polyolefins. Cellular polyethylene and polypropylene are prepared by both extrusion and molding processes. High-density polyolefin foams in the density range of 320–800 kg/m^3 are prepared by mixing a decomposable blowing agent with the polymer and feeding the mixture under pressure through an extruder at a temperature such that the blowing agent is partially decomposed before it emerges from an orifice into a lower pressure zone. Simultaneous expansion and cooling take place, resulting in a stable cellular structure owing to rapid crystallization of the polymer, which increases the modulus of the polymer enough to prevent collapse of cell structure (26,36,95). This process is widely used in wire coating and structural foam products. These products can also be produced by direct injection of inert gases into the extruder (96–97).

Low density polyethylene foam products (thin sheets, planks, rounds, tubes) in the range of 32–160 kg/m^3 (2–10 lb/ft^3) have been prepared by an extrusion technique using various gaseous fluorocarbon blowing agents (98–99). The techniques are similar to those described earlier for producing extruded polystyrene foam planks and foam sheets (see also Olefin polymers).

Structural Foams. Structural foams having an integral skin, cellular core, and a high strength-to-weight ratio are formed by means of injection molding, extrusion, or casting, depending on product requirements (100–101). Two processes are most widely used among the existing injection molding processes: the Union Carbide low pressure process (102) and the USM high pressure process (103) (see also Plastic building products; Plastics processing).

In the low pressure process, a short shot of a resin containing a blowing agent is forced into the mold where the expandable material is allowed to expand to fill the mold under pressures of 690–4140 kPa (100–600 psi). This process produces structural foam products with a characteristic surface swirl pattern produced by the collapse of cells on the surface of molded articles.

In the high pressure process, a resin melt containing a chemical blowing agent is injected into an expandable mold under high pressure. Foaming begins as the mold cavity expands. This process produces structural foam products with very smooth surfaces since the skin is formed before expansion takes place.

Extruded structural foams are produced with conventional extruders and a specially designed die. The die has an inner, fixed torpedo located at the center of its opening, which provides a hollow extrudate. The outer layer of the extrudate cools and solidifies to form solid skin; the remaining extrudate expands toward the interior of the profile. One of the most widely used commercial extrusion processes is the Celuka process developed by Ugine-Kuhlman (104).

Large structural foam products are produced by casting expandable plastic pellets containing a chemical blowing agent in aluminum molds on a chain conveyor. After closing and clamping the mold, it is conveyed through a heating zone where the pellets soften, expand and fuse together to form the cellular products. The mold is then passed through a cooling zone. This process produces structural foam products with uniform, closed-cell structure but no solid skin.

Polyurethane structural foam produced by reaction injection molding (RIM) is a rapidly growing product that provides industry with the design flexibility required for a wide range of applications. This process is more efficient than conventional methods in producing large area, thin wall, and load-bearing structural foam parts. In the RIM process, polyol and isocyanate liquid components are metered into a temperature controlled mold that is filled 20–60%, depending on the density of structural foam parts (105). When the reaction mixture then expands to fill the mold cavity, it forms a component part with an integral, solid skin and a microcellular core. The quality of the structural part depends on precise metering, mixing, and injection of the reaction chemicals into the mold.

Poly(vinyl Chloride). Cellular poly(vinyl chloride) is prepared by many methods (106), some of which utilize decompression processes. In all reported processes the stabilization process used for thermoplastics is to cool the cellular state to a temperature below its second-order transition temperature before the resin can flow and cause collapse of the foam.

A new type of physical stabilization process, unique for poly(vinyl chloride) resins, is the fusion of a dispersion of plastisol resin in a plasticizer. The viscosity of a resin–plasticizer dispersion shows a sharp increase at the fusion temperature. In such a system expansion can take place at a temperature corresponding to the low viscosity; the temperature can then be raised to increase viscosity and stabilize the expanded state.

Extrusion processes have been used to produce high and low density flexible cellular poly(vinyl chloride). A decomposable blowing agent is usually blended with the compound prior to extrusion. The compounded resin is then fed to an extruder where it is melted under pressure and forced out of an orifice into the atmosphere. After extrusion into the desired shape, the cellular material is cooled to stabilize it and is removed by a belt.

Another type of extrusion process involves the pressurization of a fluid plastisol at low temperatures with an inert gas. This mixture is subsequently extruded onto a belt or into molds, where it expands (107–108). The expanded dispersion is then heated to fuse it into a dimensionally stable form.

Injection molding of high density cellular poly(vinyl chloride) can be accomplished in a manner similar to extrusion except that the extrudate is fed for cooling into a mold rather than being maintained at the uniform extrusion cross section.

Chemical Stabilization Processes. Cellular rubber and ebonite are produced by chemical stabilization processes.

Cellular Rubber. The term cellular rubber refers to an expanded elastomer produced by expansion of a rubber stock, in contrast to latex foam rubber, which is produced from a latex. The following general procedure applies to production of cellular rubbers from a variety of types of rubber (109). A decomposable blowing agent, along with vulcanizing systems and other additives, is compounded with the uncured elastomer at a temperature below the decomposition temperature of the blowing agent. When the uncured elastomer is heated in a forming mold, it undergoes a viscosity change, as shown in Figure 2. The blowing agent and vulcanizing systems are chosen to yield open-celled or closed-celled cellular rubber. Although inert gases such as nitrogen have been pressurized into rubber and the rubber then expanded upon release of pressure, the current cellular rubbers are made almost entirely with decomposable blowing agents. The various types of such decomposable blowing agents used are ex-

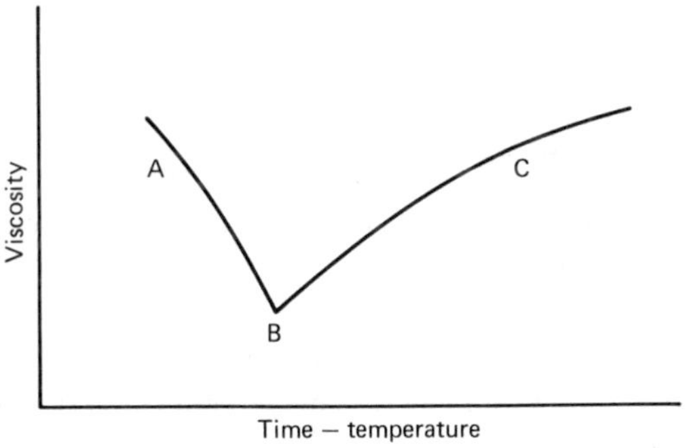

Figure 2. Viscosity of cellular rubber stock during production cycle (109).

emplified by sodium bicarbonate, 2,2′-azobisisobutyronitrile, azodicarbonamide, 4,4′-oxybis(benzenesulfonyl hydrazide), and dinitrosopentamethylenetetramine. Each of these blowing agents is typical of a class of compounds useful for expanding cellular rubber; the compound named is the most important commercial compound in its particular class (see Rubber; Elastomers, synthetic).

Open-Celled Cellular Rubber. To produce this type of rubber the blowing agent is decomposed just prior to point A in Figure 2, so that the gas is released at the point of minimum viscosity. As the polymer expands the cell walls become thin and rupture; however, the connecting struts have developed enough strength to support the foam. This process is ordinarily carried out in one step inside a mold under pressure.

Closed-Celled Cellular Rubber. The timing for blowing-agent decomposition is more critical in this case—it must occur soon enough after point A to cause expansion of the elastomer but far enough past point A to allow the cell walls to become strong enough not to rupture under the blowing stress. The expansion of closed-celled rubber is often carried out in two main steps: a partial cure is carried out in a mold that is a reduced-scale replica of the final mold; removed from this mold, it expands partly toward its final form. It is then placed in a mold of exactly the desired size and heated to complete the expansion and cure.

Most elastomers can be made into either open-celled or closed-celled materials. Natural rubber, SBR, nitrile rubber, polychloroprene, chlorosulfonated polyethylene, ethylene–propylene terpolymers, butyl rubbers, and polyacrylates have been successfully used (4,109–110).

A continuous extrusion process, as well as molding techniques, can be used as the thermoforming method. A more rapid rate of cure is then necessary to ensure the cure of the rubber before the cellular structure collapses. The stock is ordinarily extruded at a temperature high enough to produce some curing and expansion and then oven-heated to complete the expansion and cure.

A unique process for chemical stabilization of a cellular elastomer upon extrusion has been shown for ethylene–propylene rubber: the expanded rubber obtained by extrusion is exposed to high-energy radiation to cross-link or vulcanize the rubber and give dimensional stability (8).

Cellular Ebonite. Cellular ebonite [*12765-25-2*] is the oldest rigid cellular plastic. It was produced in the early 1920s by a process similar to the processes described for making cellular rubber. The formulation of rubber and vulcanizing agent is changed to produce an ebonite rather than rubber matrix (111).

DISPERSION PROCESSES

In several techniques for producing cellular polymers, the gas cells are produced by dispersion of a gas or solid in the polymer phase followed, when necessary, by stabilization of the dispersion and subsequent treatment of the stabilized dispersion. In frothing techniques a quantity of gas is mechanically dispersed in the fluid polymer phase and stabilized. In another method, solid particles are dispersed in a fluid polymer phase, the dispersion stabilized, and then the solid phase dissolved or leached, leaving the cellular polymer. Still another method relies on dispersing an already cellular solid phase in a fluid polymer and stabilizing this dispersion. This results directly in cellular polymers called syntactic foams.

Frothing. The frothing process for producing cellular polymers is an exact duplicate of the process used for making meringue topping for pies. A gas is dispersed in a fluid that has surface properties proper to produce a foam of transient stability. The foam is then permanently stabilized by chemical reaction. The fluid may be a homogeneous material, a solution, or a heterogeneous material.

Latex Foam Rubber. Latex foam rubber was the first cellular polymer to be produced by frothing. The basic steps in this process are as follows: (*1*) a gas is dispersed in a suitable latex; (*2*) the rubber latex particles are caused to coalesce and form a continuous rubber phase in the water phase; (*3*) the aqueous soap film breaks owing to deactivation of the surfactant in the water, breaking of the latex film and causing retraction into the connecting struts of the bubbles; and (*4*) the expanded matrix is cured and dried to stabilize it.

The earliest frothing process developed was the Dunlop process, which made use of chemical gelling agents (eg, sodium fluorosilicate) to coagulate the rubber particles and deactivate the soaps. The Talalay process, developed later, employs freeze–coagulation of the rubber followed by deactivation of the soaps with carbon dioxide. The basic processes and a multitude of improvements on each are discussed extensively in ref. 3. A discussion more oriented to current use of these processes is given in ref. 112.

Latex rubber foams are generally prepared in slab or molded forms in the density range 64–128 kg/m^3 (4–8 lb/ft^3). Synthetic SBR latexes have now replaced natural rubber latexes as the largest-volume raw material for latex foam rubber. Other elastomers used in significant quantities are polychloroprene [*9010-98-4*], nitrile rubbers, and synthetic *cis*-polyisoprene [*9003-31-0*] (112).

A recent method (113) of producing cellular polymers from a variety of latexes uses primarily latexes of carboxylated styrene–butadiene copolymers, although other elastomers such as acrylic elastomers, nitrile rubber, and vinyl polymers can be employed.

Urea–Formaldehyde Resins. Cellular urea–formaldehyde resins can be prepared in the following manner: An aqueous solution containing surfactant and catalyst is made into a low density, fine-celled foam by dispersing air into it mechanically. A second aqueous solution consisting of partially cured urea–formaldehyde resin is then mixed into the foam by mechanical agitation. The catalyst in the initial foam causes the dispersed resin to cure in the cellular state. The resultant hardened foam is dried at elevated temperatures. Densities as low as 8 kg/m^3 can be obtained by this method (114) (see Amino resins).

Syntactic Cellular Polymers. Syntactic cellular polymer is produced by dispersing rigid, foamed, microscopic particles in a fluid polymer and then stabilizing the system. The particles are generally spheres or microballoons of phenolic resin, urea–formaldehyde resin, glass, or silica, ranging 30–120 μm dia. Commercial microballoons have densities of approximately 144 kg/m^3 (9 lb/ft^3). The fluid polymers used are the usual coating resins, eg, epoxy resin, polyesters, and urea–formaldehyde resin.

The resin, catalyst, and microballoons are mixed to form a mortar which is then cast into the desirable shape and cured. Very specialized electrical and mechanical properties may be obtained by this method but at higher cost. This method of producing cellular polymers is quite applicable to small quantity, specialized applications because it requires very little special equipment.

In a variation on the usual methods for producing syntactic foams (115–116), expandable polystyrene or styrene–acrylonitrile copolymer particles (in either the unexpanded or prefoamed state) are mixed with a resin (or a resin containing a blowing agent) which has a large exotherm during curing. The mixture is then placed in a mold and the exotherm from the resin cure causes the expandable particles to foam and squeeze the resin or foamed matrix to the surface of the molding. A typical example is Dow Chemical Company's Voraspan, expandable polystyrene in flexible polyurethane foam matrix (117). These foams are finding acceptance in cushioning applications such as automobile crash-pad and molded furniture applications.

OTHER PROCESSES

Some plastics cannot be obtained in a low viscosity melt or solution that can be processed into a cellular state. Two other methods have been used to achieve the needed dispersion of gas in solid: sintering of solid plastic particles and leaching of soluble inclusions from the solid plastic phase.

Sintering. Sintering has been used to produce a porous polytetrafluoroethylene (15) (see Fluorine compounds, organic). Cellulose sponges are the most familiar cellular polymers produced by the leaching process (118). Sodium sulfate crystals are dispersed in the viscose syrup and subsequently leached out. Polyethylene (119) are poly(vinyl chloride) can also be produced in cellular form by the leaching process. The artificial leatherlike materials used for shoe uppers are rendered porous by extraction of salts (120) or by designing the polymers in such a way that they precipitate as a gel with many holes (121) (see Leatherlike materials).

Microporous polymer systems consisting of essentially spherical, interconnected voids, with a narrow range of pore and cell-size distribution have been produced from a variety of thermoplastic resins by the phase-separation technique (122). If a polyolefin or polystyrene is insoluble in a solvent at low temperature but soluble at high temperatures, the solvent can be used to prepare a microporous polymer. When the solutions, containing 10–70% polymer, are cooled to ambient temperatures, the polymer separates as a second phase. The remaining nonsolvent can then be extracted from the solid material with common organic solvents. These microporous polymers may be useful in microfiltrations or as controlled-release carriers for a variety of chemicals (see Microencapsulation; Pharmaceuticals, controlled release; Ultrafiltration).

A summary of the methods for commercially producing cellular polymers is presented in Table 1. This table includes only those methods thought to be commercially significant and certainly is not inclusive of all methods known to produce cellular products from polymers.

Table 1. Methods for Production of Cellular Polymers

Type of polymer	Extrusion	Expandable formulation	Spray	Froth foam	Compression mold	Injection mold	Sintering	Leaching
cellulose acetate	X							
epoxy resin		X	X	X				X
phenolic resin		X						
polyethylene	X	X			X	X	X	X
polystyrene	X	X				X	X	
silicones		X						
urea–formaldehyde resin				X				
urethane polymers		X	X	X		X		
latex foam rubber				X				
natural rubber	X	X			X			
synthetic elastomers	X	X			X			
poly(vinyl chloride)	X	X		X	X	X		X
ebonite					X			
polytetrafluoroethylene							X	

Properties of Cellular Polymers

Test Methods. Several countries have developed their own standard test methods for cellular plastics, and the International Organization for Standards (ISO) Technical Committee on Plastics T-61 has been making considerable progress on developing international standards (123). Information concerning the test methods for any particular country or the ISO procedures can be obtained in the United States from the American Standards Association (124). The most complete set of test procedures for cellular plastics is that developed by the ASTM. These procedures, the most used of any in the world, are published in new editions each year (125). There have been several reviews of those ASTM methods and others pertinent to cellular plastics (29,57,126–128).

Properties of Commercial Foamed Plastics. The properties of commercial rigid foamed plastics are presented in Table 2. The data shown are meant to demonstrate the broad ranges of properties of commercial products rather than a very accurate set of properties on a specific few materials. Specific producers of foamed plastics should be consulted for properties on a particular product (132,134, 139).

The properties of commercial flexible foamed plastics are presented in Table 3. The definition of a flexible foamed plastic is that recommended by the ASTM Subcommittee D 11-22. Again, these values are general in nature, and specific properties of specific foams should be obtained from the manufacturer of that product (132,139).

The properties that are achieved in commercial structural foams (density >0.3 g/cm^3) are shown in Table 4. Because these values depend upon several structural variables, they can be used only as general guidelines of mechanical properties from these products. Specific properties must be determined on the particular part to be produced because of the number of process and structural variables that influence these properties. A good guide recently has been published (142).

Structural Variables. The properties of a foamed plastic can be related to several variables of composition and geometry often referred to as structural variables. The particular set presented here was chosen (21) because it best meets the criteria of simplicity, completeness, and maximum compatibility with those structural variables reported in literature.

Polymer Composition. The properties of foamed plastics are influenced both by the foam structure and, to a greater extent, by the properties of the parent polymer. The polymer phase description must include the additives present in that phase as well.

Polymer State. The condition or state of the polymer phase (orientation, crystallinity, previous thermal history), as well as its chemical composition, determines the properties of that phase. The polymer state and cell geometry are intimately related because they are determined by common forces exerted during the expansion and stabilization of the foam.

Density. Density is the most important variable in determining mechanical properties of a foamed plastic of given composition. Its effect has been recognized since foamed plastics were first made and has been extensively studied.

Cell Structure. A complete knowledge of the cell structure of a cellular polymer requires a definition of its cell sizes, cell shapes, and location of each cell in the foam. Because such a specific definition is impractical, one must be content with measurable quantities that approach this absolute definition.

Cell Size. Cell size has been characterized by measurements of the cell diameter in one or more of the three mutually perpendicular directions (146) and as a measurement of average cell volume (147–148). Mechanical, optical, and thermal properties of a foam are all dependent upon the cell size.

Cell Geometry. The shape of the cells is governed predominantly by the final foam density and the external forces exerted on the cellular structure prior to its stabilization in the expanded state. In a foam prepared without such external forces, the cells tend to be spherical or ellipsoidal at gas volumes less than 70–80% of the total volume, and they tend toward the shape of packed regular dodecahedra at greater gas volumes. These shapes have been shown to be consistent with surface chemistry arguments (147,149–150). Photographs of actual foam cells (Fig. 3) show the general shapes to be consistent with the above argument but there is a broad range of variations in shape.

In the presence of external forces, plastic foams in which the cells are elongated or flattened in a particular direction may be formed. This cell orientation can have a marked influence on many properties. The results of a number of studies have been recently reviewed (57,151) (see also Foams).

Fraction Open Cells. An important character of the cell structure is the extent to which the gas phase of one cell is in communication with other cells. This is most commonly expressed as fraction open cells. When a large portion of cells are interconnected by gas phase, the foam has a large fraction of open cells, or is an open-celled foam. Conversely, a large proportion of noninterconnecting cells results in a small fraction of open cells, or a closed-cell foam.

The nature of the opening between cells determines how readily different gases and liquids can pass from one cell to another. Because of variation in flow of different liquids or gases through the cell-wall openings, a single measurement of fraction open cells does not fully characterize this structural variable, especially in a dynamic situation.

Table 2. Physical Properties of Commercial Rigid Foamed Plastics

Property	ASTM test	Cellulose acetate	Epoxy		Phenolic		Extruded plank	
density, kg/m^3 [a]		96–128	32–48	80–128	32–64	112–160	35	53
mechanical properties								
compressive strength, kPa [b] at 10%	D 1621	862	138–172	414–620	138–620		310	862
tensile strength, kPa [b]	D 1623	1,172		345–1,240	138–379		517	
flexural strength, kPa [b]	D 790	1,014		1,380–5,516	172–448		1,138	
shear strength, kPa [b]	C 273	965			103–207		241	
compression modulus, MPa [c]	D 1621	38–90	3.9	14.5–44.8			10.3	
flexural modulus, MPa [c]	C 790	38		17.2–41.4			41	
shear modulus, MPa [c]	C 273				2.8–4.8		10.3	
thermal properties								
thermal conductivity, W/(m·K) [c]	C 177	0.045–0.046	0.016–0.022	0.035–0.040	0.029–0.032	0.035–0.040	0.030	
coefficient of linear expansion, $10^{-5}/°C$	D 696		1.5	4.1	0.9	4.5	6.3	6.3
max service temperature, °C		177	205–260	205–260	132	205	74	
specific heat, kJ/(kg·K) [d]	C 351						1.1	
electrical properties								
dielectric constant	D 1673	1.12			1.19–1.20	1.19–1.20	<1.05	<1.05
dissipation factor		20			0.028–0.031	0.028–0.031	<0.0004	<0.0004
moisture resistance								
water absorption, vol %	C 272	4.5			13–51	10–15	0.02	0.05
moisture vapor transmission, g/(m·s·GPa) [e]	C 355		58				35	
references		21	129	129	130–131	130–131	21, 133	21

[a] To convert kg/m^3 to lb/ft^3, multiply by 0.0624.
[b] To convert kPa to psi, divide by 6.895.
[c] To convert MPa to psi, multiply by 145.
[d] To convert kJ/(kg·K) to Btu/(lb·°F), divide by 4.184.
[e] To convert GPa to psi, multiply by 145,000.

Table 3. Physical Properties of Commercial Flexible Foamed Plastics

Property	ASTM test	Expanded acrylonitrile–butadiene rubber	Expanded butyl rubber		Expanded natural rubber		Expanded neoprene		Expanded SBR	Latex	Foam	Rubber
density, kg/m^3 [a]		160–400	128–144	224–304	56	320	112	192	72	80	130	160
cell structure		closed	closed	closed	closed	closed	closed	closed	closed	open	open	open
compressive strength												
10% deflection, kPa [b]	D 1621											
25% deflection, kPa [b]	D 1621		11–13.8	25–38				52				
tensile strength, kPa [b]	D 1564	275				206		758	551	103		
tensile elongation, %	D 1564							500		310		
rebound resilience, %	D 1054		39–36	30–16						73		
tear strength, (N/m) [c] × 10^2	D 624											
max service temperature, °C		100			70	70	105	70	70			
thermal conductivity, W/(m·K)	C 177	0.036–0.043			0.036	0.043	0.040	0.065	0.030		0.050	
electrical properties refs.											3	3
cushioning properties ref.												
mechanical properties refs.		128	128		128		128		128	128	128	128

[a] To convert kg/m^3 to lb/ft^3, multiply by 0.0624.
[b] To convert kPa to psi, divide by 6.895.
[c] To convert N/m to lbf/in., divide by 1.75.

Table 2 (*continued*)

Polystyrene					Poly(vinyl chloride)		Polyurethane						Urea formaldehyde
Expanded plank			Extruded sheet				Polyether			Isocyanurate			
										Bun	Laminate		
16	32	80	96	160	32	64	32–48	64–128	144–192	304–400	32	32	13–19
90–124	207–276	586–896	290	469	345	1,035	138–344	482–1,896	2,000–3,800	8,270–13,800	117–206	97–282	34
145–193	310–379	1,020–1,186	2,070–3,450	4,137–6,900	551	1,207	138–482	620–2,000	1,585–3,100	4,800–8,960	248–290		
193–241	379–517				586	1,620	413–689	1,380–2,400		4,825–13,800			
	241				241	793	138–207	413–896		52,400	117		
	3.4–14				13.1	35	2.0–4.1	10.3–31					
	9.0–26				10.3	36	5.5–6.2	5.5–10.3					
	7.6–11.0				6.2	21	1.2–1.4	3.4–10.3			1.7		
0.037	0.035	0.035	0.035	0.035	0.023		0.016–0.025	0.022–0.030	0.027–0.036	0.049–0.060	0.019	0.020	0.026–0.030
5.4–7.2	5.4–7.2	5.4–7.2					5.4–7.2	7.2	7.2	7.2			
74–80	74–80	74–80	77–80	80			93–121	121–149	121–149	121–149	149	149	
							ca 0.9	ca 0.9	ca 0.9	ca 0.9			
1.02	1.02	1.02	1.27	1.28			1.05	1.1	1.2	1.4			
0.0007	0.0007	0.0007	0.00011	0.00014			13	18	32				
1–4	1–4	1–4											
<120	35–120	23–35	86	58	15		35–230	50–120	12		230		1,610–2,000
	131–132	134			135	135	136	136	136	136	137	137	138

Table 3 (*continued*)

Polyethylene extruded plank				Polyethylene sheet		Polypropylene			Polyurethane				Poly(vinyl chloride)			Silicone		
				Extruded	Cross-linked	Unmodified	Modified	Sheet	Supersoft		Polyether slab					Liquid	Sheet	Sheet
35	64	96	144	43	26–38	64–96	64–96	10	24	18	29	38	56	112	96	272	160	400–544
closed	closed	closed	closed	closed	closed	closed	closed		open	open	open	open	closed	closed	open	open	open	closed
48	75	124	360			550	206	4.8										
138	241	413	690	41		830	344		2.2	1.9	5.0	7	10.3	24	3.4	36 at 20%		
60	60	60	60	276	276–480	1100	1380	138–275	82	62	140	115			220	227	310	550–700
				50		25	75		190	220	258	163						
									31		49	55						
10.5	17.5	26	51	26							6	3						
82	82	82	82	82	79–93	135	135	121			125					350	260	232
0.053	0.058	0.058	0.058	0.040–0.049	0.036–0.040	0.039	0.039	0.039			0.040		0.035	0.040		0.078	0.086	
140	140	140	140	21, 140				131	141	43	43,	43	143	143		144	131	131
				140							142							
140	140	140	140	140	131	145	145	131	128	128	128	128	128	128	128	144	131	131

Table 4. Typical Physical Properties of Commercial Structural Foams

Property	ASTM test	ABS		Noryl	Nylon-6,6 glass-reinforced	Poly-carbonate	Thermo-plastic poly-ester	Poly-ethylene, high density	Poly-propylene		High impact polystyrene		Polyurethane			Poly-(vinyl chloride)
glass reinforced		no	yes	no	yes	no	30%	no	no	20%	no	20%	no	no	no	no
density, g/cm³		0.80	0.85	0.80	0.97	0.80	1.10	0.60	0.60	0.73	0.70	0.84	0.40	0.50	0.60	0.50
tensile strength, kPa[a]	D 1623	18,600	48,000	22,700	101,000	37,900	76,000	8,900	13,800	20,700	12,400	34,500	11,000	17,200	23,400	6,900
compression strength, kPa[a] at 10% compression	D 1621	6,900		34,500		51,700	76,000	8,900					5,500	12,400	19,300	
flexural strength, kPa[a]	D 790	25,500	82,700	41,400	172,000	68,900	137,900	18,800	22,000	41,400	31,000	58,600	22,000	31,700	41,400	
flexural modulus, GPa[b]	D 790	0.86	5.2	1.7	5.2	2.1	6.6	0.83	0.83	2.8	1.4	5.2	0.7	0.9	1.1	
max use temperature, °C		82		96	203	132	193	110	115							
ref.													141	141	141	

[a] To convert kPa to psi, divide by 6.895.
[b] To convert GPa to psi, multiply by 145,000.

Gas Composition. In closed-celled foams, the gas phase in the cells can contain some of the blowing agent (called captive blowing agent), gas components of air which have diffused in, or other gases generated during the foaming process. Such properties as thermal and electrical conductivity can be profoundly influenced by the cell gas composition. In open-celled foams the presence of air exerts only a minor influence on the static properties but does affect the dynamic properties such as cushioning.

Mechanical Properties. The mechanical properties of rigid foams vary considerably from those of flexible foams. The tests used to characterize these two classes of foams are, therefore, quite different, and the properties of interest from an application standpoint are also quite different. In this discussion the ASTM definition of rigid and flexible foams given earlier is used. A separate class of high density, rigid cellular polymers has grown rapidly in the last 5 to 7 years to become significant commercially. These are the structural foams with a density >0.3 g/cm^3. This class of polymer is treated as a separate class of rigid foams.

Rigid Cellular Polymers. Compressive strength and modulus are readily determined and are important in many applications. They have been widely used, therefore, as general criteria to characterize the mechanical properties of rigid plastic foams.

Rigid cellular polymers generally do not exhibit a definite yield point when compressed but instead show an increased deviation from Hooke's law as the compressive load is increased. Figure 4 demonstrates this behavior for several cellular polymers. For precision the compressive strength is usually reported at some definite deflection (commonly 5 or 10%). The compressive modulus is reported as extrapolated to 0% deflection unless otherwise stated. Structural variables that affect the compressive strength and modulus of a rigid plastic foam are, in order of decreasing importance: plastic-phase composition, density, cell structure, and plastic state. The effect of gas composition is minor, with a slight effect of gas pressure in some cases.

Figures 5 and 6 illustrate the large effect density and polymer composition have on the compressive strength and modulus. The dependence of compressive properties on cell size has been discussed (21). The cell shape or geometry has also been shown important in determining the compressive properties (21,57,151,158–159). In fact, the foam cell structure is controlled in some cases to optimize certain physical properties of rigid cellular polymers.

Strengths and moduli of most polymers increase as the temperature decreases (160). This behavior of the polymer phase carries over into the properties of polymer foams, and similar dependence of the compressive modulus of polyurethane foams on temperature has been shown (154).

Tensile strength and modulus of rigid foams have been shown to vary with density in much the same manner as the compressive strength and modulus. General reviews of the tensile properties of rigid foams are available (21,57,128,151,161).

Those structural variables most important to the tensile properties are polymer composition, density, and cell shape. Variation with use temperature has also been characterized (162).

Flexural strength and modulus of rigid foams both increase with increasing density in the same manner as the compressive and tensile properties. More specific data on particular foams are available from manufacturers literature and in references 21, 57–58, 128, and 161.

Shear strength and modulus of rigid foams depend on the polymer composition and state, density, and cell shape. The shear properties increase with increasing density

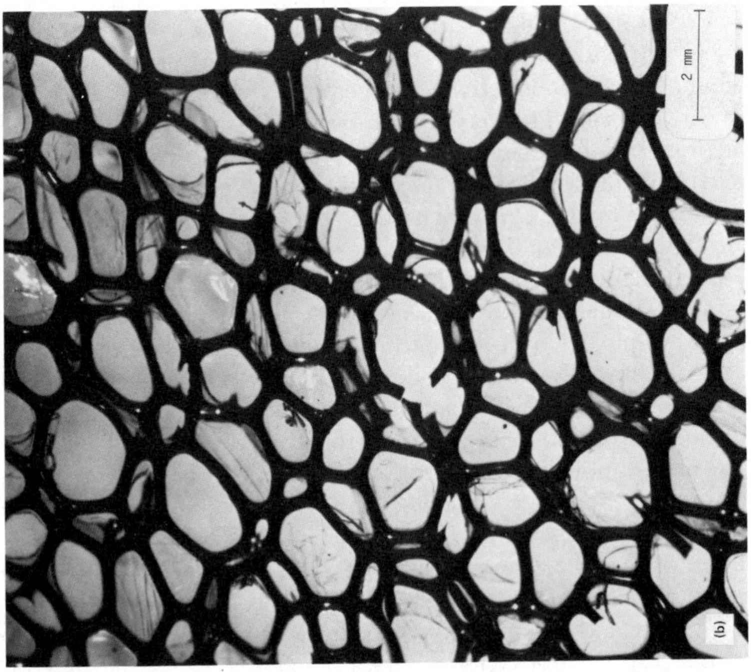
2 mm
(b)

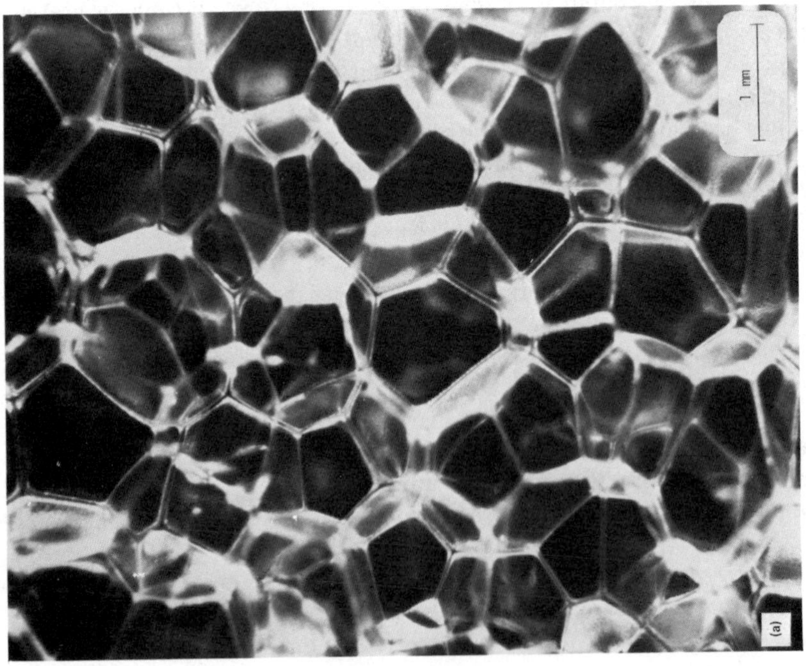
1 mm
(a)

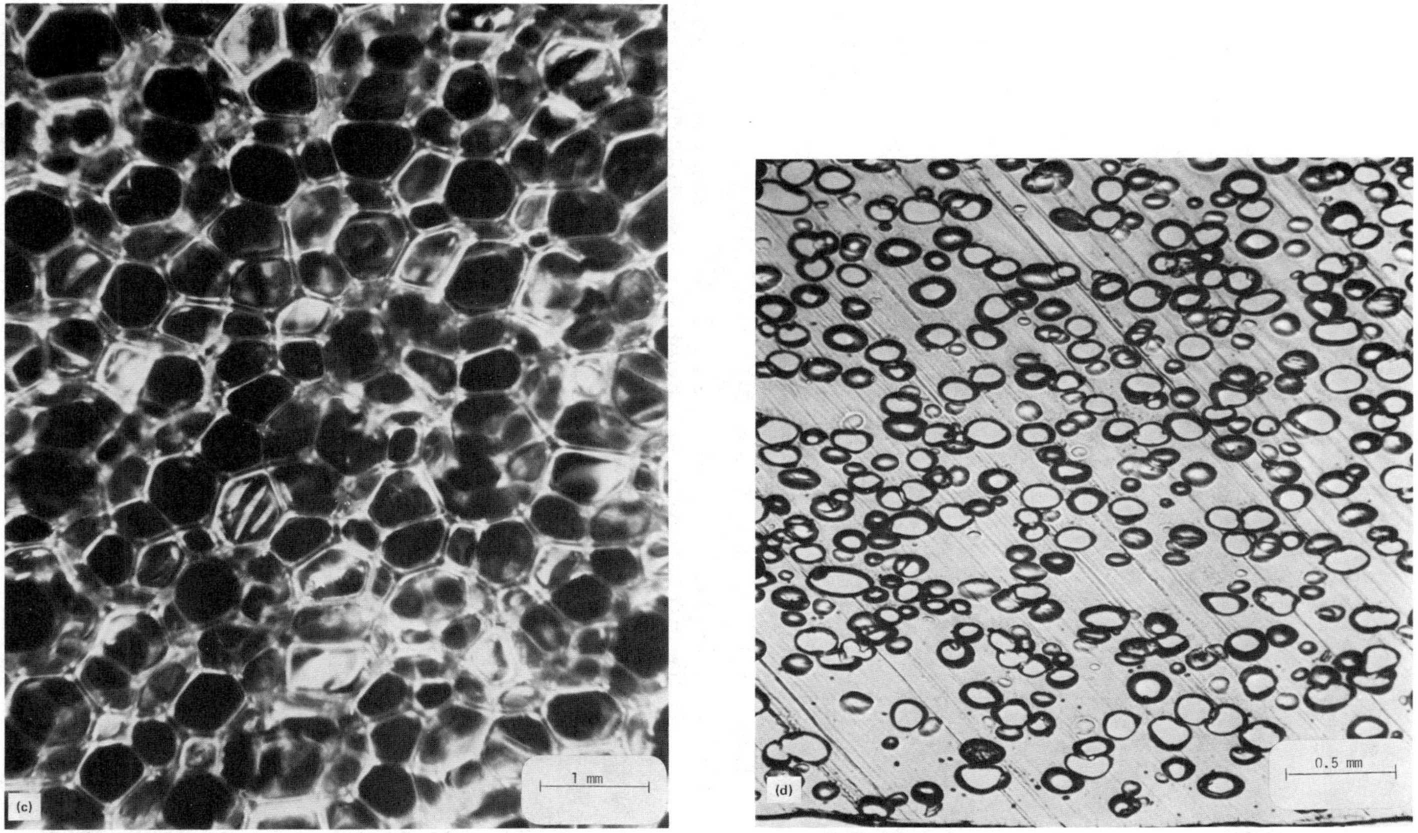

Figure 3. Photomicrographs of foam cell structure: (**a**) extruded polystyrene foam, reflected light, 26×; (**b**) polyurethane foam, reflected light, 12×; (**c**) polyurethane foam, transmitted light, 26×; (**d**) high density plastic foam, transmitted light, 50× (21). Courtesy of Van Nostrand Reinhold Publishing Corp.

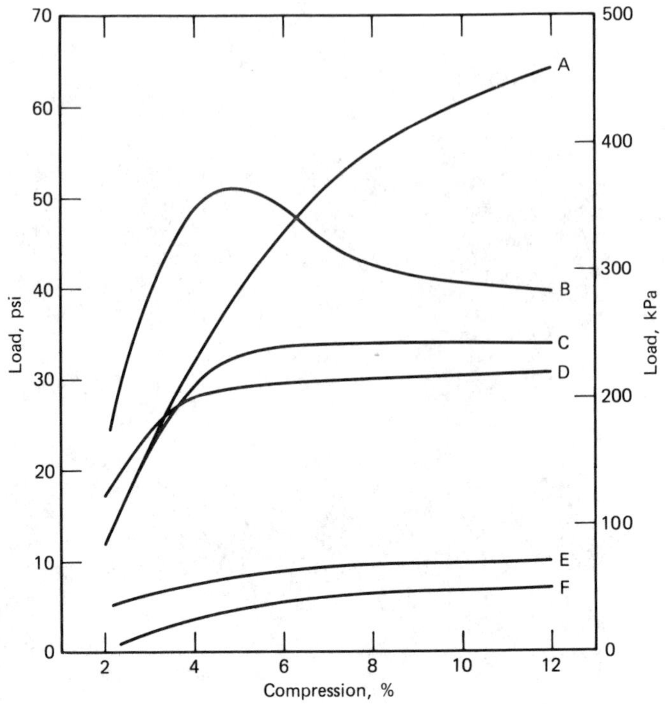

Figure 4. Compressive load vs deflection for rigid cellular polymers. A, polystyrene, 64 kg/m^3 (152); B, poly(vinyl chloride), 37 kg/m^3 (152); C, polyether–polyurethane, 30.5 kg/m^3 (152); D, polystyrene, 32 kg/m^3 (152); E, urea–formaldehyde, 29 kg/m^3 (152); F, polyethylene, 32 kg/m^3 (153). To convert kg/m^3 to lb/ft^3, multiply by 0.0624.

and with decreasing temperature (162). Specific values for a commercial rigid foam should be obtained from the manufacturer. Typical values are in Table 2.

Creep. The creep characteristic of plastic foams must be considered when they are used in structural applications. Creep is the change in dimensions of a material when it is maintained under a constant stress. The data of Brown (163) on the deformation of polystyrene foam under various static loads are shown in Figure 7. He states that there are two types of creep in this material: short-term and long-term. Short-term creep exists in foams at all stress levels, however, a threshold stress level exists below which there is no detectable long-term creep. Brown reports the minimum load required to cause long-term creep in molded polystyrene foam as shown in Table 5.

The successful application of time–temperature superposition (164) for polystyrene foam is particularly significant in that it allows prediction of long-term behavior from short-term measurements. This is of interest in building and construction applications where creep is a major consideration.

Structural Foams. Structural foams are usually produced as fabricated articles in injection molding or extrusion processes. The optimum product and process match differs for each fabricated article, so there are no standard commercial products for one to characterize. Rather there are a number of foams with varying properties. The properties of typical structural foams of different compositions are reported in Table 4.

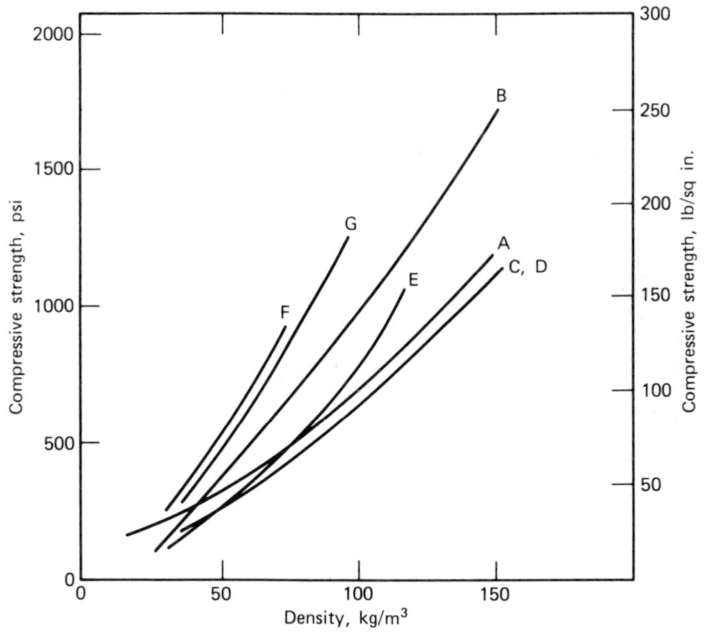

Figure 5. Effect of density on compressive strength of rigid cellular polymers. A, expanded polystyrene at 10% deflection (83); B, polyether polyurethane at yield and measured parallel to foam rise (154); C, phenol–formaldehyde foam at yield (56); D, expanded ebonite at 10% deflection (155); E, epoxy foam at yield (129); F, extruded polystyrene at 10% deflection (156); G, poly(vinyl chloride) (157). To convert kg/m^3 to lb/ft^3, multiply by 0.0624.

Table 5. Minimum Load to Cause Long-Term Creep in Molded Polystyrene[a]

Foam density, kg/m^3 (lb/ft^3)	Load, kPa (psi)
16 (1)	50 (7.3)
96 (6)	165 (24)
160 (10)	455 (66)

[a] Ref. 163.

The most important structural variables are again polymer composition, density, and cell size and shape. Since structural foams are generally not uniform in cell structure, they exhibit considerable variation in properties with particle geometry (142). Throne has extensively reviewed the mechanical properties of structural foams and their variation with polymer composition and density (142). Table 6 summarizes his conclusions as to the variation of structural foam mechanical properties with density as a function of polymer properties. The reader is cautioned to exercise care in the use of this information since in some cases the data are very limited. The design engineer should remember that the data are extracted from stress–strain curves and, owing to possible anisotropy of the foam, must be considered apparent values. These relations can provide valuable guidance toward arriving at an optimum structural foam, however.

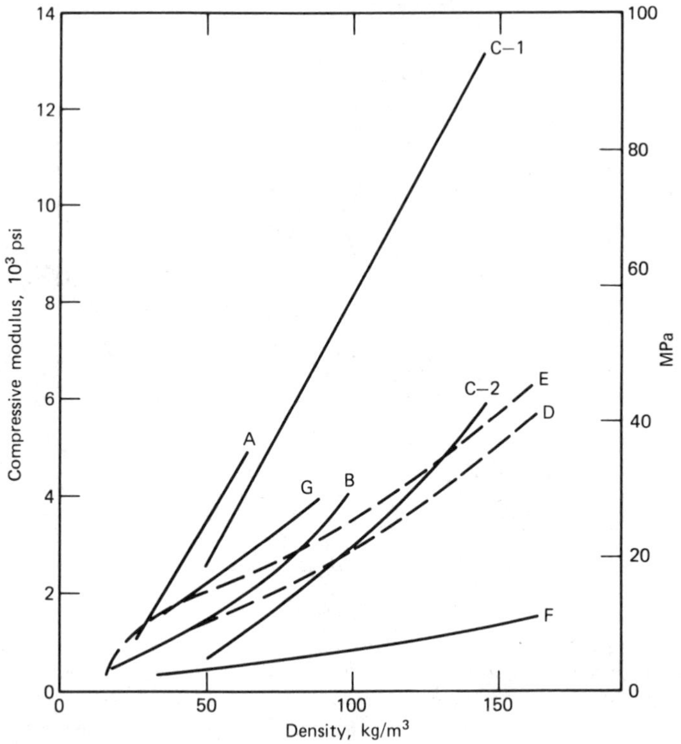

Figure 6. Effect of density on compressive modulus of rigid cellular polymers. A, extruded polystyrene (156); B, expanded polystyrene (155); C-1, polyether polyurethane (154); C-2, polyether polyurethane (154); D, phenol–formaldehyde (155); E, ebonite (155); F, urea–formaldehyde (155); G, poly(vinyl chloride) (157). To convert kg/m^3 to lb/ft^3, multiply by 0.0624.

Flexible Cellular Polymers. The application of flexible foams has been predominantly in comfort cushioning, packaging, and wearing apparel (165). This has resulted in emphasis on a different set of mechanical properties than for rigid foams. The compressive nature of flexible foams (both static and dynamic) is their most significant mechanical property for most uses. Other important properties are tensile strength and elongation, tear strength, and compression set. These properties can be related to the same set of structural variables as those for rigid foams.

Compressive Behavior. The most informative data in characterizing the compressive behavior of a flexible foam are derived from the entire load-deflection curve of 0–75% deflection and its return to 0% deflection at the speed experienced in the anticipated application. Various methods have been reported (3,165–169) for relating the properties of flexible foams to desired behavior in comfort cushioning. Other methods to characterize package cushioning have been reported (see Applications, Comfort Cushioning). The most important variables affecting compressive behavior are polymer composition, density, and cell structure and size.

That the polymer composition is the most important structural variable can be seen from Figure 8. Although the polystyrene and polyethylene foams are approximately the same density and the open-cell latex foam significantly more dense, all three show markedly different compressive strengths.

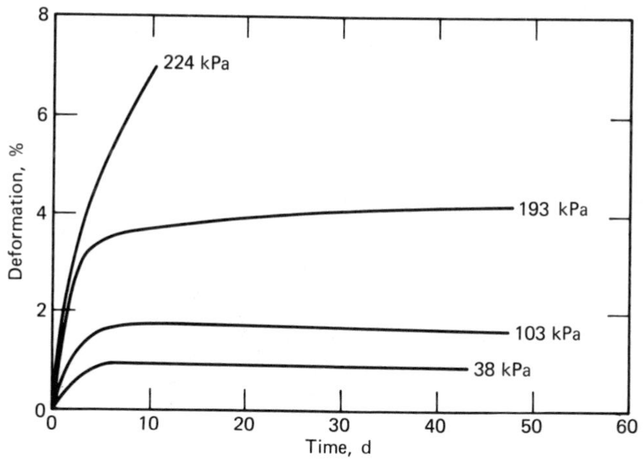

Figure 7. Deformation under various static loads for 48 kg/m³ (3 lb/ft³) molded polystyrene foams (163). To convert kPa to psi, multiply by 0.145.

Table 6. Design Criteria for Low Pressure Structural Foam[a]

Mechanical property	Equation[b]	Comments
tensile strength	$T_{foam} = T_o(\rho_{foam}/\rho_o)^2$	coefficient can increase with temperature
compressive modulus	$C_{foam} = C_o(\rho_{foam}/\rho_o)^2$	most data on short columns
shear modulus	$Sh_{foam} = Sh_o(\rho_{foam}/\rho_o)^2$	limited data
	$S_{foam} = S_o(\rho_{foam}/\rho_o)^{3/2}$	empirical, uncontrolled skin thickness, low foam density
flexural modulus	$S_{foam} = S_o(\rho_{foam}/\rho_o)$	law of mixtures, thick skins, high foam density
impact	$I_{foam} = I_o(\rho_{foam}/\rho_o)^4(t/t_o)^2$	tentative, few data from many test procedures
fatigue	$F_{foam} = F_o(\rho_{foam}/\rho_o)^2$	very tentative, rule-of-thumb
creep	$Cr_{foam} = Cr_o(\rho_{foam}/\rho_o)^2$	very tentative, rule-of-thumb

[a] Ref. 142.

[b] Where ρ_{foam} and ρ_o are the density of the foam and polymer, respectively, and t, t_o are the thickness of the foamed and unfoamed part.

The compressive behavior of latex rubber foams of various densities (3,170) is illustrated in Figure 9. Similar relationships undoubtedly hold for vinyl and flexible polyurethane foams as well. In the case of flexible polyurethane foams, there are many formulation variables that in addition to density can be changed to affect the compressive behavior of the foam (29,47,151). Figure 10 shows the compressive behavior of a good cushion latex foam rubber and of two polyurethane foams that have been formulated to match the latex foam performance (47).

Another variable that can exert considerable influence on the compressive behavior of flexible foams is the geometry of the sample. It has been demonstrated that various geometric coring patterns in polyurethanes (168,172) and in latex foam rubber (173) exert major influences on their compressive behavior.

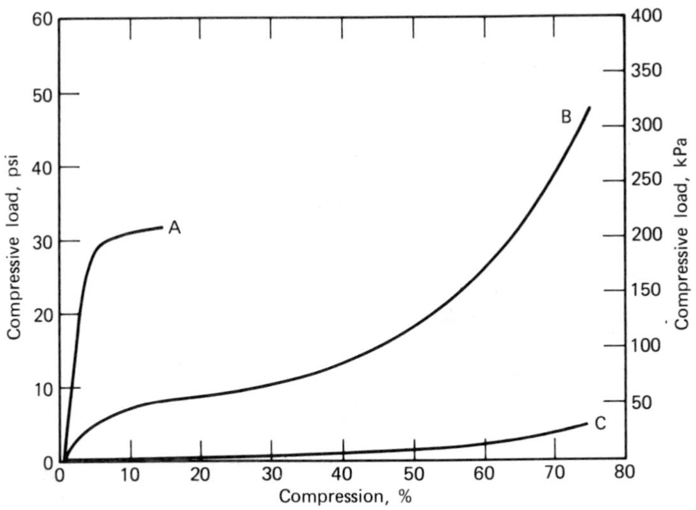

Figure 8. Load vs compression for plastic foams (153). A, polystyrene, 32 kg/m^3 (2 lb/ft^3); B, polyethylene, 32 kg/m^3; C, latex rubber foam. To convert kg/m^3 to lb/ft^3, multiply by 0.0624.

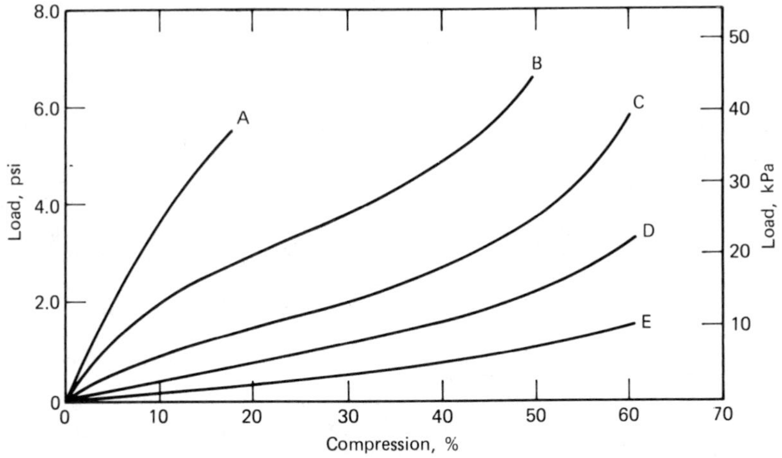

Figure 9. Effect of load on compression for latex foams of different densities (3,170): A, 304 kg/m^3; B, 208 kg/m^3; C, 179 kg/m^3; D, 139 kg/m^3; E, 99 kg/m^3. To convert kg/m^3 to lb/ft^3, multiply by 0.0624.

A good discussion of the effect of cell size and shape on the properties of flexible foams is contained in references (151,161). The effect of open-cell content is demonstrated in polyethylene foam (170).

Tensile Strength and Elongation. The tensile strengths of latex rubber foams has been shown to depend on the density of the foam (153,174) and on the tensile strength of the parent rubber (174–175). At low densities the tensile modulus approximates a linear relation with density but increases with a higher power of density at higher densities. Similar relations hold for polyurethane and other flexible foams (161,176–177).

The tensile elongation of solid latex rubber has been shown to correlate well with the elongation of foam from that latex (175). The elongation of flexible polyurethane has been related to cell structure (177–178).

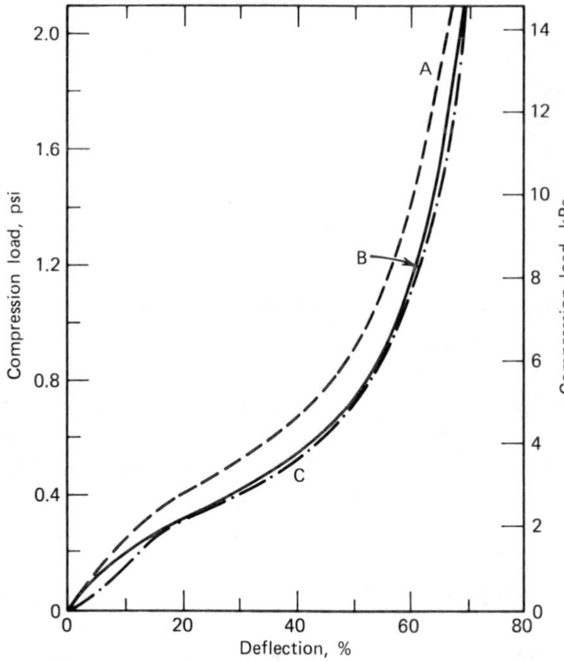

Figure 10. Compression–deflection characteristics of polyurethane and latex foams (171). A, polypropylene glycol urethane, 56 kg/m³ (3.5 lb/ft³); B, latex rubber, 101 kg/m³ (6.3 lb/ft³); C, polypropylene glycol urethane, 51 kg/m³ (3.4 lb/ft³).

Tear Strength. Gent and Thomas (174) have developed a relation for the tearing stress of flexible foams that predicts linear increase in the tearing energy with density; they also predict increased tearing energy with cell size. Both relationships are verified to a limited extent by experimental data.

Flex Fatigue. Considerable information on the measurement and cause of flex fatigue in flexible foams has been published (179–181). Changing compressive strength and volume upon repeated flexing over long periods of time is a major deterrent to the use of polyurethane foam in many cushioning applications. For polyurethane foams these changes have been correlated mainly with changes in chemical structure.

Compression Set. The compression set is an important property in cushioning applications. It has been studied for polyurethane foams (182–183), and is also discussed in recent reviews (29,58,161). Compression set has been described as flex fatigue and creep as well.

The thermal, electrical, acoustical, and chemical properties of all cellular polymers are of such a similar nature that the discussions of these properties are not separated into rigid and flexible groups.

Thermal Properties. ***Thermal Conductivity.*** More information is available relating thermal conductivity to structural variables of cellular polymers than for any other property. Several papers have discussed the relation of the thermal conductivity of heterogeneous materials in general (184–185) and of plastic foams in particular (129,146,154,186–188) with the characteristic structural variables of the systems. These references may be consulted for information more specific than the general treatment presented here. The manner in which each heat-transfer mechanism is applicable to cellular polymers is discussed, and the effect of each structural variable on these mechanisms is analyzed.

The following separation of the total heat transfer into its component parts, even if not completely rigorous, proves valuable to understanding the total thermal conductivity k of foams:

$$k = k_s + k_g + k_r + k_c \tag{5}$$

where k_s, k_g, k_r, and k_c are the components of thermal conductivity attributable to solid conduction, gaseous conduction, radiation, and convection, respectively.

As a good first approximation (184), the heat conduction of low density foams through the solid and gas phases can be expressed as the product of the thermal conductivity of each phase times its volume fraction. Most rigid polymers have thermal conductivities of 0.07–0.28 W/(m·K) and the corresponding conduction through the solid phase of a 32 kg/m^3 (2 lb/ft^3) foam (3 vol %) ranges 0.003–0.009 W/(m·K). In most cellular polymers this value is determined primarily by the density of the foam and the polymer-phase composition. Smaller variations can result from changes in cell structure.

Although conductivity through gases is much lower than that through solids, the amount of heat transferred through the gas phase in a foam is generally the largest contribution to the total heat transfer. This is because the gas phase is such a major part of the total value (ca 10 vol % in a 32 kg/m^3 foam). Table 7 lists values of the thermal conductivity for several gases that occur in the cells of cellular polymers.

As seen in Table 7, the thermal conductivities of the halocarbon gases are considerably less than those of oxygen and nitrogen. It has, therefore, proved advantageous to prepare cellular polymers using such gases that measurably lower the k of the polymer foam. Upon exposure to air the gas of low thermal conductivity in the cells can become mixed with air, and the k of the mixture of gases can be estimated by means of equation 6:

$$k_m = N_1k_1 + N_2k_2 \tag{6}$$

where k_m is the k of the gaseous mixture; N_1 and N_2 are the mole fractions of gases 1 and 2; and k_1 and k_2 are the thermal conductivity of pure gases 1 and 2.

Table 7. Thermal Conductivity at 20°C of Gases Used in Cellular Polymers

Compound	Thermal conductivity, W/(m·K)	Ref.
trichlorofluoromethane	0.0084	189
dichlorodifluoromethane	0.0098	189
trichlorotrifluoroethane	0.0072	189
dichlorotetrafluoroethane	0.0104	189
dichlorofluoromethane	0.0112	189
dichloromethane	0.0063	190
methyl chloride	0.0105	190
2-methylpropane	0.0161	190
carbon dioxide	0.0168	190
air	0.0259	190

Changes in total k calculated by equations 5 and 6 with change in gas composition agree well with experimental measurements (147,188,191–192).

There is ordinarily no measurable convection in cells of diameter less than about 4 mm (146). Theoretical arguments have been in general agreement with this work (154,188). Since most available cellular polymers have cell diameters smaller than 4 mm, convection heat transfer can be ignored with good justification. Studies of radiant heat transfer through cellular polymers have been made (146,154,188,193–194).

The variation in total thermal conductivity with density has the same general nature for all cellular polymers, as illustrated in Figure 11. The increase in k at low densities is owing to an increased radiant heat transfer; the rise at high densities to an increasing contribution of k_s.

The thermal conductivity of most materials decreases with temperature. When the foam structure and gas composition are not influenced by temperature, the k of the cellular material decreases with decreasing temperature. When the composition of the gas phase may change (ie, condensation of a vapor), then the relationship of k to temperature is much more complex (146,188,196).

It should be noted that the thermal conductivity of a cellular polymer can change upon aging under ambient conditions if the gas composition is influenced by such aging. Such a case is evidenced when oxygen or nitrogen diffuses into polyurethane foams that initially have only a fluorocarbon blowing agent in the cells (29,127,146,187–188,196–199).

Thermal conductivity of foamed plastics recently has been shown to vary with thickness (194). This has been attributed to the boundary effects of the radiant contribution to heat-transfer.

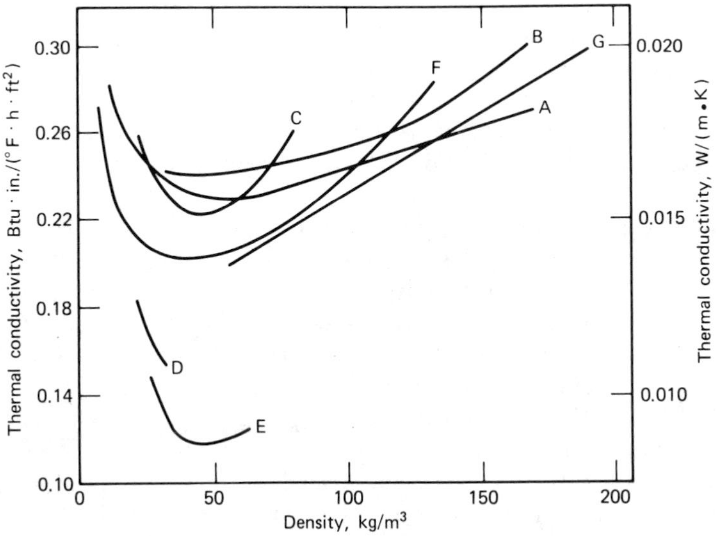

Figure 11. Effect of density on thermal conductivity of rigid cellular polymers. A, polystyrene (146); B, polystyrene (186); C, polyurethane–air (186); D, polyurethane–F-11 (CCl_3F) (195); E, polyurethane (186); F, phenol–formaldehyde (186); G, ebonite (186). To convert kg/m³ to lb/ft³, multiply by 0.0624.

Specific Heat. The specific heat of a cellular polymer is simply the sum of the specific heat of each of its components. The contribution of the gas is small and can be neglected in many cases.

Coefficient of Linear Thermal Expansion. The coefficients of linear thermal expansion of polymers are higher than those for most rigid materials at ambient temperatures because of the supercooled-liquid nature of the polymeric state. This large coefficient is carried over directly to the cellular state. A variation of this property with density and temperature has been reported for polystyrene foams (200) and for foams in general (21).

When cellular polymers are used as components of large structures, the coefficient of thermal expansion must be considered carefully because of its magnitude compared with those of most nonpolymeric structural materials. Many designs have been successful (201) when this fact has been considered.

Maximum Service Temperature. The maximum service temperature of cellular polymers cannot be defined precisely since the cellular materials, like their parent polymers (202), gradually decrease in modulus as the temperature rises rather than undergoing a sharp change in properties. The upper temperature limit of use for most cellular polymers is governed predominantly by the plastic phase. The act of fabricating the polymer into a cellular state normally builds some stress into the polymer phase; this may tend to relax at a temperature below the heat-distortion temperature of the unfoamed polymer. Of course, additives in the polymer phase or a plasticizing effect of the blowing agent on the polymer effect the behavior of the cellular material in the same way as the unfoamed polymer. Typical maximum service temperatures are given in Tables 2, 3, and 4.

Flammability. There has been an increased awareness on the part of industry, the government, and the consumer of the behavior of plastic foams in fire. The results of small-scale laboratory tests have been recognized as not predictive of the true behavior of plastic foams in other fire situations (203). Work now underway in industry and government is aimed at developing tests to evaluate the performance of plastic foams in actual fires. All plastic foams are combustible, some burning more readily than others when exposed to fire. Some additives (128,131), when added in small quantities to the polymer, markedly improve the behavior of the foam in the presence of small fire sources. All plastic foams should be used properly following the manufacturers recommendations and any applicable regulations (see also Flame retardants).

Moisture Resistance. Plastic foams have proven advantageous over other thermal insulations in several applications where they are exposed to moisture pickup. This is particularly true where they are subjected to a combination of thermal and moisture gradients in use. In some cases the foams are exposed to freeze–thaw cycles as well. The behavior of plastic foams has been studied under laboratory conditions simulating these use conditions as well as under the actual use conditions.

Hedlin (204) studied the moisture gain of foamed plastic roof insulations under controlled thermal gradients. His apparent permeability values are greater than those predicted by regular wet-cup permeability measurements. The moisture gains found in polyurethane are greater than those of bead polystyrene and much greater than those of extruded polystyrene.

Dechow and Epstein (205) reported moisture pickup and freeze-thaw resistance of various insulations and the effect of moisture on the thermal performance of these insulations. They showed that in protected membrane roofing applications the order of preference for minimizing moisture pickup is extruded polystyrene ≫ polyurethane > molded polystyrene. Water-pickup values for insulation in use for five years were extruded polystyrene 0.2 vol %, polyurethane without skins 5 vol %, molded polystyrene 8–30 vol %. These correspond to increases in K of 5–265%. For below-grade applications extruded polystyrene was better than molded polystyrene or polyurethane without skins in terms of moisture absorption resistance and retention of thermal resistance. Dechow and Epstein as well as a number of other investigators have related the increased water content with increased thermal conductivity of the insulations (206–210).

Electrical Properties. Cellular polymers have two important electrical applications. One takes advantage of the combination of inherent toughness and moisture resistance of polymers along with the decreased dielectric constant and dissipation factor of the foamed state to use cellular polymers as electrical-wire insulation (95). The other combines the low dissipation factor and the rigidity of plastic foams in the construction of radar domes. Polyurethane foams have been used as high voltage electrical insulation (211). See reference 21 for a compilation of the electrical properties of specific cellular polymers (see Insulation, electric).

Environmental Aging. Environmental aging of cellular polymers is of major importance to most applications of plastic foams. The response of cellular materials to the action of light and oxygen is governed almost entirely by the composition and state of the polymer phase (21). Expansion of a polymer into a cellular state increases the surface area; reactions of the foam with vapors and liquids are correspondingly faster than those of solid polymer.

Foams prepared from phenol–formaldehyde and urea–formaldehyde resins are the only commercial foams that are affected by water to any great extent (21). Polyurethane foams exhibit a deterioration of properties when subjected to a combination of light, moisture, and heat aging (47,197). All cellular polymers are subject to a deterioration of properties under the combined effects of light or heat and oxygen. A great deal of work has been done to develop additives that successfully eliminate this degradation (212). The best source of information on the use of specific additives is the individual manufacturer of the foam in question (see Plastics, environmentally degradable).

Miscellaneous Properties. The acoustical properties of polymers are altered considerably by their fabrication into a cellular structure. Sound transmission is altered only slightly because it depends predominantly upon the density of the barrier (in this case, the polymer phase). Cellular polymers by themselves are, therefore, very poor materials for reducing sound transmission. They are, however, quite effective in absorbing sound waves of certain frequencies (155). Materials with open cells on the surface are particularly effective in this respect. The combination of other advantageous physical properties with fair acoustical properties has led to the use of several different types of plastic foams in sound-absorbing constructions (213–214). The sound absorption of a number of cellular polymers has been reported (21,155,213,215) (see Insulation, acoustic).

The permeability of cellular polymers to gases and vapors depends upon the fraction of open cells as well as the polymer-phase composition and state. The presence of open cells in a foam allows gases and vapors to permeate the cell structure by effusion and convection flow, yielding very large permeation rates. In closed-celled foams the permeation of gases or vapors is governed by composition of the polymer phase, gas composition, density and cellular structure of the foam (191,197,213,216). Reference 217 contains a comprehensive treatment of the role of polymer and gas composition in permeation (see also Barrier polymers).

The penetration of visible light through foamed polystyrene has been shown to follow approximately the Beer-Lambert law of light absorption (21). This behavior presumably is characteristic of other cellular polymers as well.

Cellular polymers have been shown (155) to be of little or no food value for rodents but they do not act as a barrier to rodents; ie, rodents will eat through cellular polymers but they will not eat large quantities of the foam as a foodstuff. The resistance to rot, mildew, and fungus of cellular polymers can be related to the amount of moisture that can be taken up by the foam (155). Therefore, open-celled foams are much more likely to support growth than are closed-celled foams. Very high humidity and high temperature are necessary for the growth of microbes on any plastic foam.

Applications

Concern over energy conservation and consumer safety has provided a rapid growth in applications for insulation and cushioning in transport. A healthy and affluent economy is also expected to increase the consumer demand for comfort cushioning in furniture, bedding, and flooring, as well as for packaging such as drinking cups, and packaging inserts for consumer products (molded and loose-fill materials). The cost of finished wood articles has forced the industry to develop structural foams which are finding major applications as replacements for wood, metal, or solid plastics (see Engineering plastics).

The large growth in use of foamed plastics is based on economic, competitive, and technological factors. Because of the rapidly expanding and shifting market area served by the cellular polymers, applications of cellular polymers are discussed in general, and a minimum discussion is accorded to specific applications.

Comfort cushioning is the largest, single application of cellular polymers. As might be expected, the flexible foams are the major contributors to this field. Historically, cushioning in particular and flexible foams in general have constituted the greatest volume of cellular polymers. However, the growth rate of structural, packaging, and insulation applications is expected to bring their volume up to that of flexible foams within the next few years.

Good growth is expected for structural foams, which are projected to comprise more than one third of the total plastic foam consumption by 1990, up from 10% in 1976. On the other hand, cushioning, which now accounts for more than 40% of the total plastic foam consumption, will be expected to drop to less than 20% by 1990. Table 8 shows United States consumption of foamed plastics by resin, market, and function (19).

Comfort Cushioning. The properties of greatest significance in the cushioning applications of cellular polymers are compression–deflection behavior, resilience, compression set, tensile strength and elongation, and mechanical and environmental aging. Of these, compression–deflection behavior is undoubtedly the most important (see Mechanical Properties). The broad range of compressive behavior of various types of flexible foam is one of the strong points of cellular polymers, since the needs of almost any cushioning application can be met by changing either the chemical nature

Table 8. Market for Cellular Polymers, Thousands of Metric Tons[a]

Item	1963–1965	1976	1990[b]	Annual growth, % Historic	Annual growth, % Projected
Resin					
poly(vinyl chloride)	18	246	1871	24.3	15.6
polystyrene	68	370	1393	15.2	9.9
flexible polyurethane	120	542	974	13.4	4.3
rigid polyurethane	27	214	848	18.9	10.3
polyolefins		46	1077		25.3
ABS and other	11	46	479	12.3	18.2
Total	*244*	*1464*	*6642*	*16.1*	*11.4*
Market					
construction					
pipe		16	907		33.5
insulation	39	172	667	13.1	10.1
flooring	6	139	362	29.3	7.1
wood substitutes		56	612		18.6
wire		15	127		16.8
furniture					
structural–ornamental	2	48	306	31.4	14.1
cushioning and upholstery	80	302	547	11.7	4.3
packaging					
inserts	10	86	243	19.7	7.7
pallets		11	302		26.4
cups, trays, other	7	139	721	28.6	12.5
other					
transportation	37	234	470	16.6	5.1
appliances	7	37	293	14.6	15.9
consumer	5	40	306	19.8	15.8
textile	15	44	90	9.7	5.2
other	36	125	689	10.9	13.0
Total	*244*	*1464*	*6642*	*16.1*	*11.4*
Function					
cushioning	115	633	1269	15.3	5.1
insulation	52	230	920	13.2	10.4
extenders	2	153	2304	44.7	21.4
product protection	21	253	1291	22.9	12.4
other	54	195	858	11.2	11.2
Total	*244*	*1464*	*6642*	*16.1*	*11.4*

[a] Ref. 19.
[b] Projected.

or the physical structure of the foam. For example, flexible urethanes, vinyls, latex foam rubber, and olefins are used to make foamed plastic cushioning for automobile padding and seats, furniture, flooring, mattresses, and pillows. These materials compete with felt, fibers, innerspring, and other filling materials.

Flexible polyurethane foams are open-cell materials that allow free movement of air inside the foam when it is flexed. They are commercially available in densities of 13–80 kg/m^3 (0.8–5 lb/ft^3). Flexible foam for cushioning requires careful control of the load bearing characteristics, which depend primarily upon density. The indentation load deflection (ILD) test, ASTM D 1564-59T is often used as a criterion.

Fillers (qv) are often added to flexible urethane foams for improved cushioning and density control. The semiflexible polyurethane foams used extensively in automobile seats and interior safety applications have densities of up to 240 kg/m^3 (15 lb/ft^3) as compared to densities as low as 24 kg/m^3 (1.5 lb/ft^3) for the flexible foams used in household mattresses. The other flexible polyurethane foam markets include shoe liners, heating pads, furniture (117), and sponges.

Latex foam rubber was initially accepted as a desirable comfort-cushioning material because of its softness to the touch and its resilience (equal to that of a steel spring alone but with better damping qualities than the spring). It also has good endurance under repeated flexings and for extended periods of time. These advantages in physical properties and the ability to mold easily into a wide variety of shapes and sizes make latex foam rubber successful in the comfort cushioning field in spite of a higher price than competing noncellular cushioning materials.

Flexible cellular poly(vinyl chloride) was developed as a comfort cushioning material with compression–deflection behavior similar to latex rubber foam, and with the added feature of flame retardancy (40). It has a larger compression set than either latex rubber or polyurethane foams. The fact that the plasticizer in flexible vinyl foams can migrate to the surface restricts flexible vinyl foams in some applications. Furniture and motor vehicle upholstery is currently the largest market for flexible vinyl foams. Because of their better esthetics (leatherlike plastics), comfort, and favorable pricing, they are expected to show good growth in upholstery, carpet backing, resilient floor coverings, outerwear, footwear, luggage, and handbags.

Cellular rubber has been used extensively as shoe soles, where its combination of cushioning ability and wear resistance, coupled with desirable economics, has led to very wide acceptance. In this case the cushioning properties are of minor importance compared with the abrasion resistance and cost. Other major cushioning applications for cellular rubbers and latex foam rubbers are as carpet underlay and as cushion padding in athletic equipment (see Leatherlike materials).

Thermal Insulation. Thermal insulation is the second largest application of cellular polymers, and the largest application for the rigid materials. The properties of greatest importance in determining the applicability of rigid foams as thermal insulants are thermal conductivity, ease of application, cost, moisture absorption and transmission permeance, and mechanical properties (see Insulation, thermal).

The low thermal conductivity of low density cellular polymers has obviously been the main factor in their use as insulating materials. The thermal conductivities of other commercial insulating materials are compared with those of rigid cellular plastics in Table 9. The data show that those plastic foams containing a captive blowing agent have considerably lower thermal conductivities than other insulating materials, whereas other rigid cellular plastics are roughly comparable with the latter. Cellular plastics have been broadly accepted in various applications, not only because of their low thermal conductivities but because of other important considerations. For certain major applications these properties are outlined below.

Domestic Refrigeration. The very low thermal conductivity of polyurethanes plus the ease of application and structural properties of foamed-in-place materials gives refrigeration engineers considerable freedom of styling. This has resulted in an increasingly broad use of rigid polyurethane foams in home freezers and refrigerators which has displaced conventional rock wool and glass wool. Today, polyurethanes account for more than half the total insulation market for household refrigerators and

Table 9. Thermal Conductivity of Commercial Insulating Materials[a]

Material	Density, kg/m^3 [b]	Thermal conductivity at 20°C, W/(m·K)
sawdust	16	0.065
glass foam	144	0.058
vegetable fiberboard	208	0.045
wood fiberboard	368	0.058
corkboard	112	0.039
rock wool	32	0.048
perlite	104	0.053
vermiculite	80	0.064
cellulose	44	0.039
urea–formaldehyde	13	0.035
fiber glass	10	0.043
polystyrene foam (molded)	16	0.041
polystyrene foam (extruded)	32	0.029
polyurethane (or polyisocyanurate) foam	24	0.023

[a] Ref. 218.
[b] To convert kg/m^3 to lb/ft^3, multiply by 0.0624.

more than 85% of the total for household freezers (19). The remaining market is shared by other competitive materials such as polystyrene foam, glass fiber, and cork. Low moisture sensitivity and permanence are also very important (see Refrigeration).

Commercial Refrigeration. In this case again low thermal conductivity is important, as are styling and cost. Application methods and mechanical properties are of secondary importance because of design latitude available in this area. For example, large institutional chests, commercial refrigerators, freezers, and cold storage areas, including cryogenic equipment and large tanks for industrial gases, are insulated with polystyrene or polyurethane foams (see Cryogenics). Polystyrene foam is still popular where cost and moisture resistance are important; polyurethane is used where spray application is required. Polystyrene foam is also widely used in load-bearing sandwich panels in low temperature space applications.

Refrigeration in Transportation. Although styling is unimportant, the volume of insulation and a low thermal conductivity are of primary concern. Volume is not large, so application methods are not of prime importance. Low moisture sensitivity and permanence are again necessary. The mechanical properties of the insulant are quite important owing to the continued abuse the vehicle undergoes. Cost is of less concern here than in other applications (see Transportation).

Residential Construction. Owing to rising energy costs in recent years, the cost and low thermal conductivity are of prime importance in the wall and ceiling insulation of residential buildings. The cost of insulation to the customer per unit area of insulation and cost per R-value (thickness of insulation/K-factor) of insulation are measures of the cost–performance ratio of insulation materials. In the past, cellular plastics have not seriously penetrated this application, although their use has increased more than 200% over the past decade. For example, the combination of insulation efficiency, desirable structural properties, ease of application, ability to reduce air infiltration, and moisture resistance has led to use of extruded polystyrene foam in residential construction as sheathing, as perimeter and floor insulation under concrete and as a combined plaster base and insulation for walls.

The residential construction application may be divided into three major areas: insulation of new houses, retrofit insulation of old houses with inadequate insulation, and insulation of mobile homes. In residential sheathing insulation, fiberboard is still the most widely used product, although the use of extruded and molded polystyrene foam and of foil-faced isocyanurate foam is increasing depending on the cost, the amount of insulation required, and compatibility of insulation with other construction systems. In cavity-wall insulation, glass wool, polyurethane, urea-formaldehyde, and fiberglass are widely used, although fiberglass batt is the most economical insulation for stud-wall construction. In mobile and modular homes, cellular foams are used widely because of their light weight and more efficient insulation value.

Commercial Construction. The same attributes desirable on residential construction applications hold for commercial construction as well but insulation quality, permanence, moisture insensitivity, and resistance to freeze–thaw cycling in the presence of water are stressed more in commercial construction. For this reason cellular plastics have greater application here. Both polystyrene and polyurethane foams are highly desirable roof insulations in commercial as in residential construction.

Cellular polymers are also widely used for pipe and vessel insulation. Spray and pour-in-place techniques of application are particularly suitable, and polyurethane and epoxy foams are widely used. Ease of application, fire properties, and low thermal conductivity have been responsible for the acceptance of cellular rubber and cellular poly(vinyl chloride) as insulation for smaller pipes.

The insulating value and mechanical properties of rigid plastic foams have led to the development of several novel methods of building construction. The use of polyurethane foam panels as unit structural components presents interesting possibilities (219). Another development in building construction is the use of expanded polystyrene as a concrete base in thin-shell construction (220).

Packaging. Because of the extremely broad demands on the mechanical properties of packaging materials, the entire range of cellular polymers from rigid to flexible is used for packaging. The most important considerations in packaging are mechanical properties, cost, ease of application or fabrication, moisture susceptibility, thermal conductivity, and esthetic appeal.

The proper mechanical properties, particularly compressive properties, are the primary requirements for a cushioning foam (Fig. 12). The information necessary for proper design of the protective package should include: (*1*) the shock to which the item being packaged might be subjected, (*2*) an estimate of the type and degree of disturbances to which the entire package might be subjected, and (*3*) a definition of the exact capabilities of the cushioning material. Considerable progress has been made in some areas of this general problem. The reader is referred to the following sources for more specific information: package design (223); general vibration and shock isolation (224); protective package design (225); selection of cushioning material (221,226); and characterization of cellular polymers for cushioning applications (222,224–225).

Creep of a cushion packaging material when subjected to static stresses for long periods of storage or shipment is also an important consideration. Brown (163) showed that polystyrene foam does have considerable creep at high static loadings but that creep is insignificant under loadings in the static stress region of optimum package design (21) (see Fig. 7). The ability of polystyrene foam to withstand repeated impacts has also been studied (163,227).

The low density of most cellular plastics is important because of shipping costs

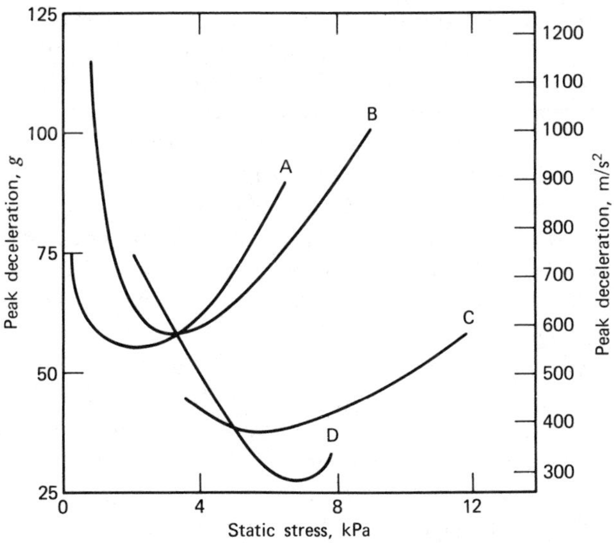

Figure 12. Dyanamic cushioning curves for cellular polymers at 76.2 cm drop height, 5.08 cm thickness. A, Polyurethane (221); B, polyethylene (222); C, expanded polystyrene (222); D, extruded polystyrene (222). To convert kPa to psi, multiply by 0.145.

for the cushioning in a package. Pelaspan-Pac (Dow), a low density, S-shaped, loose-fill, foamed polystyrene is a highly effective packing material (228). Other materials with a variety of shapes and sizes compete in this market.

The inherent moisture resistance of cellular plastics is of added benefit where packages may be subjected to high humidity or water. Many military applications require low moisture susceptibility. Foamed polystyrene is used as packaging inserts and as containers such as food trays, egg cartons, and drinking cups which require moisture resistance, rigidity, and shock resistance. Foamed polyurethane is also used as specialty packaging materials for expensive and delicate equipment.

The clean, durable, nondust-forming character of polyethylene foam has led to its increasing acceptance in packaging missile parts (229). Polyethylene foam sheet is also gradually displacing polystyrene foam sheet for packaging glass bottles and containers because of its greater resiliency and tear resistance (225) (see Film and sheeting materials).

Structural Components. In most applications structural foam parts are used as direct replacements for wood, metals or solid plastics. Structural foam parts find wide acceptance in appliances, automobiles, furniture, materials-handling equipment, and in construction. Use in the building and construction industry account for more than one half of the total volume of structural foam applications. Currently, high impact polystyrene is the most widely used structural foam, followed by polypropylene, high density polyethylene, and poly(vinyl chloride) (see Engineering plastics). The construction industry is also the area of application that exhibits the greatest growth potential for cellular plastics. Several of the more advanced applications which depend to a considerable extent on the excellent thermal properties of cellular plastics have been mentioned in the section on thermal insulation. These systems, such as the foam-cored panel (219) and thin-shell design utilizing cellular plastics (220), all make use of the structural properties of these materials. The design of entire structural systems that advantageously use such properties as the high bending stiffness-to-weight ratio of sandwich construction have been discussed recently (21,230). One obvious disadvantage of their use in such applications has been a lack of engineering design data on cellular plastics. As such data become available, the design engineer will be able to use cellular plastics as structural components, producing, no doubt, a large increase in the use of cellular plastics in construction applications.

For some structural applications an integral expandable sheet is prepared from high impact acrylonitrile–butadiene–styrene terpolymers [*9003-56-9*]. The sheet consists of solid ABS skins and an expandable core. Upon heating, the core expands and the entire composite can be formed into shapes. The resulting composite has the advantages both of chemical resistance of the plastic skins and the thermal insulation and buoyancy associated with the core, giving it the advantages of a sandwich construction as well as the immense latitude in design and use of a thermoformable material. Such items as automobile body parts, house trailers, and boats have been satisfactorily produced from this material (231). The sandwich-type structure of polyurethanes with a smooth integral skin produced by the reaction injection molding process provides a high degree of stiffness as well as excellent thermal and acoustical properties necessary for its use in housing and load-bearing structural components for the automotive, business machine, electrical, furniture, and materials-handling industry. A significant growth of this material is expected in many applications.

Buoyancy. The low density, closed-celled nature of many cellular polymers coupled with their moisture resistance and low cost resulted in their immediate acceptance for buoyancy in boats and floating structures such as docks and buoys. Since each cell in the foam is a separate flotation member, these materials cannot be destroyed by a single puncture.

The combination of structural strength and flotation has stimulated the design of pleasure boats using a foamed-in-place polyurethane between thin skins of high tensile strength (232). Other cellular polymers that have been used in considerable quantities for buoyancy applications are those produced from polyethylene, poly(vinyl chloride), and certain types of rubber. The susceptibility of polystyrene foams to attack by certain petroleum products that are likely to come in contact with boats led to the development of foams from copolymers of styrene and acrylonitrile which are resistant to these materials (6–7).

Electrical Insulation. The substitution of a gas for part of the solid polymer usually results in rather large changes in the electrical properties of the resulting material. The dielectric constant, dissipation factor, and dielectric strength are all generally lowered in amounts roughly proportional to the amount of gas in the foam (see Insulation, electric).

For low frequency electrical insulation applications, the dielectric constant of the insulation is ideally as low as possible. The lower the density of the cellular polymer, the lower the dielectric constant and the better the electrical insulation. The dielectric strength is also reduced by reducing the density; the insulation is, therefore, susceptible to breakdown from voltage surges from such sources as lightning and short circuits. The physical properties are also reduced proportionally to density, so that optimum density is determined by a compromise in properties. For many applications this compromise has been at an expansion of two or three volumes, mainly because the minimum physical properties required for fabrication and use are obtained at that point. Polyolefin foams have been most used as low frequency electrical insulation; poly(vinyl chloride) and polystyrene foams are used also. Ability to produce a completely homogeneous, closed-cell foam at lower densities in high speed wire-coating apparatus has been a problem at lower densities.

In high frequency applications, the dissipation factor is of greater importance. Coaxial cables using cellular polyolefins have been quite successfully used for frequencies in the megahertz range and above. Cellular plastics have also been used as structural materials in constructing very large radar-receiving domes (233). The very low dissipation factor of these materials makes them quite transparent to radar waves (see also Insulation, electric; Microwave technology).

Space Filling and Seals. Cellular polymers have become common for gasketing, sealing, and space filling. Cellular rubber, poly(vinyl chloride), silicone (144), and polyethylene are used extensively for gasketing and sealing of closures in the automotive and construction trade (109). Most cellular materials must be predominantly closed-celled in order to provide the necessary barrier properties. The combination of chemical inertness, excellent conformation to irregular surfaces, and ability to be compressed to greater than 50% with relatively small pressures and still function satisfactorily contribute to the acceptance of cellular polymers in these applications.

In the construction industry, cellular polymers are used as spacers and sealant strips in windows, doors, and closures of other types, as well as for backup strips for other sealants (qv).

Miscellaneous Applications. Cellular plastics have always been used for display and novelty pieces. Polystyrene foam combines ease of fabrication with lightweight, attractive appearance, and low cost to make it a favorite in these uses. Phenolic foam, on the other hand, enjoys its major use in floral displays. Its ability to hold large amounts of water for extended periods is used to preserve cut flowers. Cellular poly(vinyl chloride) is used in toys and athletic goods, where its toughness and ease of fabrication into intricate shapes have been valuable.

Cellular urea–formaldehyde and phenolic resin foams have been used to some extent in interior sound-absorbing panels and, in Europe, expanded polystyrene has been used in the design of sound-absorbing floors (234). In general, cost, flammability, and cleaning difficulties have prevented major penetration of the acoustical tile market. The low percent of reflection of sound waves from plastic foam surfaces has led to their use in anechoic chambers (214) (see Noise pollution).

Health and Safety

Plastic foams are organic in nature and, therefore, are combustible when exposed to open flames. They do vary in their response to small sources of ignition because of composition and/or additives (235). The products of combustion have been studied recently for a number of plastic foams (236). All plastic foams should be handled, transported, and used according to manufacturers' recommendations as well as applicable local and national codes and regulations.

The presence of additives or unreacted monomers in certain plastic foams can limit their use where food or human contact is anticipated. The manufacturers' recommendations or existing regulations again should be followed for such applications.

BIBLIOGRAPHY

"Foamed Plastics" in *ECT* 2nd ed., Vol. 9, pp. 847–884, by R. E. Skochdopole, The Dow Chemical Company.

1. E. Ott, H. M. Spulin, and M. W. Grafflin, eds., *Cellulose and Cellulose Derivatives,* 2nd ed., Part I, Interscience Publishers, Inc., New York, 1954, p. 9.
2. U.S. Pat. 942,699 (Dec. 7, 1909), L. H. Baekeland.
3. E. W. Madge, *Latex Foam Rubber,* John Wiley & Sons, Inc., New York, 1962.
4. A. Cooper, *Plast. Inst. Trans. J.* **29,** 39 (1961).
5. U.S. Pat. 3,058,166 (Oct. 16, 1962), R. T. Fields (to E. I. du Pont de Nemours & Co., Inc.).
6. A. R. Ingram, *J. Cell. Plast.* **1**(1), 69 (1965).
7. *Tyril Foam 80, Technical Data Sheet No. 2-1,* The Dow Chemical Co., Midland, Mich., Jan. 1964.
8. U.S. Pat. 3,062,729 (Nov. 6, 1962), R. E. Skochdopole, L. C. Rubens, and G. D. Jones (to The Dow Chemical Co.).
9. H. Junger, *Kunststoff-Rundschau* **9,** 437 (1962).
10. W. C. Goggin and O. R. McIntire, *Br. Plast.* **19**(223), 528 (1947).
11. A. F. Randolph, ed., *Plastics Engineering Handbook,* 3rd ed., Reinhold Publishing Corp., New York, 1960, pp. 137–138.
12. J. H. Saunders and K. C. Frisch, *Polyurethanes, Chemistry and Technology,* Vol. I, Interscience Publishers, a division of John Wiley & Sons, Inc., New York, 1963.
13. A. E. Lever, *Plastics (London)* **18**(193), 274 (1953).
14. *Plast. World,* 1 (May 1953).
15. *Ethafoam,* bulletin, The Dow Chemical Co., Midland, Mich., Apr. 1963.
16. U.S. Pat. 3,058,161 (Oct. 16, 1962), C. E. Beyer and R. B. Dahl (to The Dow Chemical Co.).
17. L. Nicholas and G. T. Gmitter, *J. Cell. Plast.* **1**(1), 85 (1965).
18. T. Wirtz, "Integral Skin Foam, A Progress in Urethane Molding," *paper presented at the 2nd SPI International Cellular Plastic Conference, New York, Nov. 8, 1968.*
19. G. P. Kratzschmer, *Plastic Trends, T44 Plastic Foams,* Predicasts Inc., Cleveland, Ohio, 1977.
20. *ASTM D 883-62T,* American Society for Testing and Materials, Philadelphia, Pa., 1962.
21. J. D. Griffin and R. E. Skochdopole, "Plastic Foams" in E. Baer, ed., *Engineering Design for Plastics,* Reinhold Publishing Corp., New York, 1964.
22. A. Cooper, *Plast. Inst. (London) Trans.* **26,** 299 (1958).
23. *ASTM D 1566-78,* Vol. 37, ASTM, Philadelphia, Pa., 1978.
24. *ASTM D 1056-62T,* ASTM, Philadelphia, Pa., 1962.
25. *ASTM D 1055-62T,* ASTM, Philadelphia, Pa., 1962.
26. R. H. Hansen, *SPE J.* **18,** 77 (1962).
27. A. R. Ingram and H. A. Wright, *Mod. Plast.* **41**(3), 152 (1963).
28. K. Hinselmann and J. Stabenow, *Ver. Dtsch. Ing.* **18,** 165 (1972).
29. K. C. Frisch and J. H. Saunders, *Plastic Foams,* Vol. 1, Part 1, Marcel Dekker, Inc., New York, 1972.
30. U.S. Pat. 2,681,321 (June 15, 1954), F. Stastny and R. Gaeth (to BASF).
31. U.S. Pat. 2,983,692 (May 9, 1961), G. F. D'Alelio (to Koppers Co.).
32. H. R. Lasman, *Mod. Plast.* **42**(1A), 314 (1964).
33. S. J. Skinner, S. Baxter, and P. J. Grey, *Trans. J. Plast. Inst.* **32,** 180 (1964).
34. *Ibid.,* p. 212.
35. S. J. Skinner and S. D. Eagleton, *Trans. J. Plast. Inst.* **32,** 321 (1964).
36. K. Goodier, *Br. Plast.* **35,** 349 (1962).
37. U.S. Pat. 3,697,454 (Oct. 10, 1972), R. L. Trimble (to Sinclair Koppers).
38. U.S. Pat. 3,066,382 (Dec. 4, 1962), M. L. Zweigle and W. E. Humbert (to The Dow Chemical Co.).
39. *Pelaspan-Pac,* bulletin, The Dow Chemical Co., Midland, Mich., 1974.
40. R. J. Meyer, *SPE J.* **18,** 678 (1962).
41. A. Osakada and M. Koyama, *Jpn. Chem. Q.* **5,** 55 (1969).
42. Jpn. Kokai 72 43,059 (June 22, 1972), J. Morita and co-workers (to Nitto Electric Industrial Co. Ltd.).
43. Jpn. Pat. 73 04,868 (Feb. 12, 1973), S. Nakada and co-workers (to Sekisui Chemical Co. Ltd.).
44. Br. Pat. 1,333,392 (Oct. 10, 1973), S. Minami and co-workers (to Toray Industries, Inc.).
45. U.S. Pat. 3,812,225 (May 21, 1974), K. Hosoda and co-workers, (to Furukawa Electric).
46. Fr. Pat. 1,446,187 (July 15, 1966), J. Zizlsperger and co-workers (to BASF).

47. J. H. Saunders and K. C. Frisch, *Polyurethanes, Chemistry and Technology,* Vol. II, Interscience Publishers, New York, 1964.
48. *Bayer Pocket Book for the Plastics Industry,* 3rd ed., Farbenfabriken Bayer, A.G., Leverkusen, FRG, Oct. 1963.
49. K. C. Frisch, K. J. Patel, and R. D. Marsh, *J. Cell. Plast.* **6**(5), 203 (1970).
50. R. Merten and co-workers, *J. Cell. Plast.* **4**(7), 262 (1968).
51. H. J. Papa, *Ind. Eng. Chem. Rod. Res. Dev.* **9,** 478 (1970).
52. R. H. Fish, *Proc. NASA Conf. Materials for Improved Fire Safety* **11,** 1 (1970).
53. D. E. Hipchen, "Modified Isocyanurates," paper presented at *Foams of the Future Seminar,* sponsored by Technomic Publishing Company, Inc., Pittsburgh, Pa., Mar. 17, 1976.
54. U.S. Pat. 3,799,896 (Mar. 26, 1974), E. K. Moss (to Celotex Corporation).
55. Ger. Pat. 2,034,166 (Jan. 21, 1971), G. M. Rambosek and co-workers (to 3M Co.).
56. Fr. Pat. 1,441,565 (June 10, 1966), L. Nicolas (to General Tire and Rubber Co.).
57. K. C. Frisch and J. H. Saunders, *Plastic Foams,* Vol. 1, Part 2, Marcel Dekker, Inc., New York, 1973.
58. C. J. Benning, *Plastic Foams,* Vol. 1, Wiley-Interscience, New York, 1969.
59. H. Weissenfeld, *Kunststoffe* **51,** 698 (1961).
60. Ref. 11, p. 149.
61. U.S. Pat. 3,821,337 (June 28, 1974), E. J. Bunclark and co-workers (to Esso Research Engineering).
62. *Japan Plastics Industry Annual,* 17th ed., Plastic Age, Tokyo, Japan, 1974.
63. U.S. Pat. 3,726,708 (Apr. 10, 1973), F. Weissenfels and co-workers (to Dynamit Nobel).
64. Fr. Pat. 2,185,488 (Feb. 8, 1974), J. Dugelay.
65. U.S. Pat. 3,830,894 (Aug. 20, 1974), H. Juenger and co-workers (to Dynamit Nobel).
66. U.S. Pat. 3,807,661 (Mar. 11, 1975), P. J. Crook and S. P. Riley (to Pilkington Brothers, Ltd.).
67. U.S. Pat. 3,907,723 (Sept. 23, 1975), M. Pretot (to Certain-teed Products Corp.).
68. Ger. Offen. 2,204,945 (Sept. 21, 1972), J. Tardy and co-workers.
69. Fr. Pat. 2,190,613 (Mar. 8, 1974), J. Tardy.
70. U.S. Pat. 3,766,100 (Oct. 16, 1973), H. A. Meyer-Stoll and co-workers (to Deutsche Texaco).
71. U.S. Pat. 3,741,920 (June 26, 1973), F. Weissenfels and co-workers (to Dynamit Nobel).
72. U.S. Pat. 3,694,387 (Sept. 26, 1972), H. Junger and co-workers (to Dynamit Nobel).
73. USSR Pat. 358,342 (Nov. 3, 1972), I. F. Ustinova and co-workers (to Kucherenko, V. A., Central Scientific Research Institute of Building Structures).
74. U.S. Pat. 3,101,242 (Aug. 20, 1963), J. M. Jackson, Jr. (to V. L. Smithers Manufacturing).
75. Jpn. Pat. 72 19,624 (June 5, 1972), M. Asaoka and co-workers (to Hitachi Chemical).
76. Br. Pat. 1,268,440 (Mar. 29, 1972), D. J. Rush and co-workers (to Midwest Research Institute).
77. K. Murai, *Plast. Age* **18**(6), 93 (1972).
78. V. D. Valgin and co-workers, *Plast. Massy* **1,** 28 (1974).
79. V. C. Valgin, *Europlast. Mon.* **46**(7), 57 (July 1973).
80. U.S. Pat. 3,400,183 (Sept. 3, 1968), P. I. Vidal (to Rocma Anstalt).
81. U.S. Pat. 3,214,793 (Nov. 2, 1965), P. I. Vidal (to Rocma Anstalt).
82. U.S. Pat. 3,122,326 (Feb. 25, 1964), D. P. Cook (to Union Carbide).
83. Ref. 11, p. 140.
84. *Epon Foam Spray 175, A Low Temperature Insulation,* bulletin, Shell Chemical Co., Houston, Tex., Nov. 1963.
85. Ref. 11, p. 164.
86. H. Vincent and K. R. Hoffman, *Mod. Plast.* **40**(1A), 418 (1962).
87. U.S. Pat. 2,515,250 (July 18, 1950), O. R. McIntire (to The Dow Chemical Co.).
88. V. L. Gliniecki, *Seventh Ann. Tech. Conf., Proc., SPI, New York, Apr. 24–25, 1963.*
89. *Mod. Plast.* **51**(1), 36, 40 (Jan. 1974).
90. T. P. Martens and co-workers, *Plast. Technol.* **12**(9), 46 (1966).
91. D. A. Knauss and F. H. Collins, *Plast. Eng.* **30**(2), 34 (1974).
92. F. H. Collins and co-workers, *Soc. Plast. Eng. Tech. Pap.* **19,** 643 (1973).
93. L. W. Meyer, *SPE J.* **18,** 1341 (1962).
94. Ref. 11, p. 139.
95. W. T. Higgins, *Mod. Plast.* **31**(7), 99 (1954).
96. U.S. Pat. 3,251,911 (May 17, 1966), R. H. Hansen (to Bell Telephone Lab.).
97. U.S. Pat. 3,268,636 (Aug. 23, 1966), R. G. Angell Jr. (to Union Carbide Corp.).
98. U.S. Pat. 3,065,190 (Nov. 20, 1962), D. S. Chisholm and co-workers (to The Dow Chemical Co.).

99. U.S. Pat. 3,067,147 (Dec. 4, 1962), L. C. Rubens and co-workers (to The Dow Chemical Co.).
100. J. L. Throne, *J. Cell. Plast.* **12**(5), 264 (1976).
101. R. W. Freund and co-workers, *Plast. Technol.,* 35 (Nov. 1973).
102. U.S. Pat. 3,436,446 (Apr. 1, 1969), R. G. Angell, Jr. (to Union Carbide Corp.).
103. *USM Foam Process, Technical Bulletin No. 653-A,* Farrel Company Division, Ansonia, Conn.
104. Fr. Pat. 1,498,620 (Sept. 11, 1967), P. Boutillier (to Ugine-Kuhlman).
105. J. J. Kolb in B. C. Wendle, ed., *Engineering Guide to Structural Foam,* Technomic Publishing Co., Inc., Westport, Conn., 1976, p. 161.
106. Ref. 11, p. 189.
107. U.S. Pat. 2,666,036 (Jan. 12, 1954), E. H. Schwencke (to Elastomer Chemical Co.).
108. U.S. Pat. 2,763,475 (Sept. 18, 1956), I. Dennis.
109. R. C. Bascom, *Rubber Age* **95,** 576 (1964).
110. L. Spenadel, *Rubber World* **150**(5), 69 (1964).
111. A. Cooper, *Plast. Inst. Trans.,* 51 (Apr. 1948).
112. T. H. Rogers, "Plastic Foams," paper presented at *Regional Tech. Conf., Palisades Section, Society of Plastics Engineers, New York, Nov. 1964.*
113. *Dow Latex Foam Process,* bulletin, The Dow Chemical Co., Midland, Mich., 1964.
114. *Iporka,* bulletin, Badische Anilin- und Soda-Fabrik A.G., Ludwigshafen am Rhein, FRG, July 1953.
115. *Dow Low Temperature Systems, 179-2086-77,* bulletin, The Dow Chemical Co., Midland, Mich., 1977.
116. U.S. Pat. 2,959,508 (Nov. 8, 1960), D. L. Graham and co-workers (to The Dow Chemical Co.).
117. J. B. Brooks and L. G. Rey, *J. Cell. Plast.* **9**(5), 232 (1973).
118. *Chem. Week* **19**(17), 43 (1962).
119. *Chem. Eng. News* **37**(36), 42 (1959).
120. U.S. Pat. 2,772,995 (Dec. 4, 1956), J. D. C. Wilson II (to E. I. du Pont de Nemours & Co., Inc.).
121. Can. Pat. 762,421 (July 4, 1967), M. E. Baguley (to Courtaulds Ltd.).
122. W. Worthy, *Chem. Eng. News,* 23 (Dec. 11, 1978).
123. *Mod. Plast.* **42**(4), 161 (1964).
124. W. E. Brown and F. C. Frost, in J. E. Schmitz, ed., *Testing of Polymers,* Wiley-Interscience, New York, 1965, p. 15.
125. *1978 Annual Book of ASTM Standards,* ASTM, Philadelphia, Pa., 1978.
126. W. H. Touhey, *J. Cell. Plast.* **4**(10), 395 (1968).
127. C. J. Hilado, *J. Cell. Plast.* **3**(11), 502 (1967).
128. R. E. Skochdopole, "Cellular Materials" in N. M. Bikales, ed., *Encyclopedia of Polymer Science and Technology,* Vol. 3, John Wiley & Sons, Inc., New York, 1965, p. 125.
129. R. P. Toohy, *Chem. Eng. Prog.* **57**(10), 60 (1961).
130. R. J. Bender, ed., *Handbook of Foamed Plastics,* Lake Publishing Co., Libertyville, Ill., 1965.
131. "Foam Plastics Chart," *Modern Plastics Encyclopedia* **54**(10A), 485 (1977–1978).
132. *Ibid.,* p. 776.
133. *Styrofoam SI Brand Plastic Foam,* bulletin, Dow Chemical U.S.A., Midland, Mich., 1978.
134. *Sweets Catalog File 7, Thermal and Moisture Protection,* Sweets Div., McGraw-Hill Information System Co., New York, 1978.
135. Y. Landler, *J. Cell. Plast.* **3**(9), 400 (1967).
136. R. K. Traeger, *J. Cell. Plast.* **3**(9), 405 (1967).
137. H. E. Reymove and co-workers, *J. Cell. Plast.* **11**(6), 328 (1975).
138. C. A. Schutz, *J. Cell. Plast.* **4**(2), 37, (1968).
139. *U.S. Foamed Plastics Market and Directory,* Technomic Publishing Co., Stamford, Conn., 1973.
140. *Ethafoam Brand Plastic Foam,* bulletin, Dow Chemical U.S.A., Functional Products and Systems Dept., Midland, Mich., 1976.
141. J. L. Eakin, *Plast. Eng.* **34**(7), 56 (1978).
142. J. L. Throne, B. C. Wendle, ed., *Engineering Guide to Structural Foams,* Technomic Publishing Co., Westport, Conn. 1976.
143. E. C. Van Buskirk and W. Pooley, paper presented at *American Chemical Society Meeting, Div. Paints, Plastics, Printing Ink, Chicago, Ill., Sept. 1958.*
144. C. E. Lee and co-workers, *J. Cell. Plast.* **13**(1), 62 (1977).
145. H. H. Lubitz, *J. Cell. Plast.* **5**(4), 221 (1969).
146. R. E. Skochdopole, *Chem. Eng. Prog.* **57**(10), 55 (1961).

147. R. H. Harding, *Mod. Plast.* **37**(10), 156 (1960).
148. D. M. Rice and L. J. Nunez, *SPE J.* **18,** 321 (1962).
149. *Br. Plast.* **35,** 18 (1962).
150. A. J. deVries, *Meded. Rubber Sticht. Delft* **326,** 11 (1957); *Rec. Trav. Chim.* **77,** 81, 209, 283, 383, 441 (1958).
151. C. J. Benning, *Plastic Foams,* Vol. I, Wiley-Interscience, New York, 1969.
152. T. L. Phillips and D. A. Lannon, *Br. Plast.* **34,** 236 (1961).
153. R. E. Skochdopole and L. C. Rubens, *J. Cell. Plast.* **1**(1), 91 (1965).
154. D. J. Doherty, R. Hurd, and G. R. Lester, *Chem. Ind. (London),* 1340 (1962).
155. A. Cooper, *Plast. Inst. Trans.* **26,** 299 (1958).
156. Ref. 128, pp. 103–104.
157. J. F. Hawden, *Rubber Plast. Age* **44,** 921 (1963).
158. R. H. Harding, *J. Cell. Plast.* **1**(3), 385 (1965).
159. R. H. Harding, *Resinography of Cellula Materials, ASTM Tech. Publ. 414,* ASTM, Philadelphia, Pa., 1967.
160. R. J. Corruccini, *Chem. Eng. Prog.* **53,** 397 (1957).
161. E. A. Meinecke and R. E. Clark, *Mechanical Properties of Polymeric Foams,* Technomic Publishing Co., Stamford, Conn., 1972.
162. R. M. McClintock, *Adv. Cryog. Eng.* **4,** 132 (1960).
163. W. B. Brown, *Plast. Prog.* **1959,** 149 (1960).
164. G. M. Hart, C. F. Balazs, and R. B. Clipper, *J. Cell. Plast.* **9**(3), 139 (1973).
165. G. H. Smith, *Rubber Plast. Age* **44**(2), 148 (1963).
166. G. J. Bibby, *Rubber Plast. Age* **45**(1), 52 (1964).
167. *Plast. Technol.* **8**(4), 26 (1962).
168. J. H. Saunders and co-workers, *J. Chem. Eng. Data* **3,** 153 (1958).
169. R. P. Marchant, *J. Cell. Plast.* **8**(2), 85 (1972).
170. T. H. Rogers and K. C. Hecker, *Rubber World* **139,** 387 (1958).
171. J. H. Saunders and co-workers, *J. Chem. Eng. Data* **3,** 153 (1958).
172. J. M. Buist and A. Lowe, *Trans. Plast. Inst.* **27,** 13 (1959).
173. J. Talaly, *Ind. Eng. Chem.* **46,** 1530 (1954).
174. A. N. Gent and A. G. Thomas, paper presented at *Proc. 7th Ann. Tech. Conf., Cellular Plastics Div., Soc. Plastics Ind., New York, Apr. 1963.*
175. L. Talalay and A. Talalay, *Ind. Eng. Chem.* **44,** 791 (1952).
176. M. A. Mendelsohn and co-workers, *J. Appl. Polym. Sci.* **10,** 443 (1966).
177. E. A. Blair, *Resinography of Cellular Materials, ASTM Tech. Publ. 414,* ASTM, Philadelphia, Pa., 1967, p. 84.
178. J. H. Saunders, *Rubber Chem. Technol.* **33,** 1293 (1960).
179. B. Beals, F. J. Dwyer, and M. A. Kaplan, *J. Cell. Plast.* **1**(1), 32 (1965).
180. R. P. Kane, *J. Cell. Plast.* **1**(1), 217 (1965).
181. *Plast. Technol.* **8**(4), 26 (1964).
182. S. M. Terry, *J. Cell. Plast.* **7**(5), 229 (1971).
183. *Ibid.,* **12**(3), 156 (1976).
184. R. L. Gorring and S. W. Churchill, *Chem. Eng. Prog.* **57**(7), 53 (1961).
185. M. E. Stephenson, Jr. and M. Mark, *ASHRAE J.* **3**(2), 75 (1961).
186. F. O. Guenther, *SPE Trans.* **2,** 243 (1962).
187. R. E. Knox, *ASHRAE J.* **4**(10), 3 (1962).
188. R. H. Harding, *Ind. Eng. Chem. Proc. Des. Dev.* **3,** 117 (1964).
189. *Freon Technical Bulletin #B-2,* E. I. du Pont de Nemours & Co., Inc., Wilmington, Del., 1975.
190. J. H. Perry, *Chemical Engineers Handbook,* 4th ed., McGraw-Hill Book Co., New York, 1963.
191. F. J. Norton, *J. Cell. Plast.* **3**(1), 23 (1967).
192. R. M. Lander, *Refrig. Eng.* **65**(4), 57 (1957).
193. B. K. Larkin and S. W. Churchill, *AIChE J.* **5,** 467 (1959).
194. B. Y. Lao and R. E. Skochdopole, paper presented at *4th SPI International Cellular Plastics Conference, Montreal, Can., Nov. 1976.*
195. L. E. LeBras, *SPE J.* **16,** 420 (1960).
196. G. A. Patten and R. E. Skochdopole, *Mod. Plast.* **39**(11), 149 (1962).
197. C. J. Hilado, *J. Cell. Plast.* **3**(4), 161 (1967).
198. R. R. Dixon, L. E. Edleman, and D. K. McLain, *J. Cell. Plast.* **6**(1), 44 (1970).

199. G. W. Ball, R. Hurd, and M. G. Walker, *J. Cell. Plast.* **6**(2), 66 (1970).
200. L. Vahl in ref. 115, p. 267.
201. C. H. Wheeler, "Foamed Plastics," *Proceedings of a Conference, Apr. 22–23, 1963, U.S. Army Natick Labs and Committee on Foamed Plastics, U.S. Dept. Comm. Office Tech. Serv. PB Rept. 181576,* p. 164.
202. G. A. Patten, *Mater. Des. Eng.* **55**(5), 117 (1962).
203. *FTC Consent Agreement, File #7323040,* The Dow Chemical Co., Midland, Mich., 1974.
204. C. P. Hedlin, *J. Cell. Plast.* **13**(5), 313 (1977).
205. F. J. Dechow and K. A. Epstein, *ASTM STP 660, Thermal Transmission Measurements of Insulation,* ASTM, Philadelphia, Pa., 1978, p. 234.
206. M. M. Levy, *J. Cell. Plast.* **2**(1), 37 (1966).
207. I. Paljak, *Mater. Constr.* (*Paris*) **6,** 31 (1973).
208. H. Mittasch, *Plaste Kautsch.* **16**(4), 268 (1969).
209. J. Achtziger, *Kunststoffe* **23,** 3 (1971).
210. C. W. Kaplar, *CRREL Internal Report No. 207,* U.S. Army CRREL, Hanover, N.H., 1969.
211. P. J. Palmer, *J. Cell. Plast.* **9**(4), 182 (1973).
212. N. Z. Searle and R. C. Hirt, *SPE Trans.* **2,** 32 (1962).
213. A. Cooper, *Plastics* **29**(321), 62 (1964).
214. *Mod. Plast.* **39**(8), 93 (1962).
215. G. L. Ball, III, M. Schwartz, and J. S. Long, *Off. Dig. Fed. Soc. Paint Technol.* **32,** 817 (1960).
216. E. F. Cuddihy and J. Moacanin, *J. Cell. Plast.* **3**(2), 73 (1967).
217. C. E. Rogers, "Permeability and Chemical Resistance" in E. Baer, ed., *Engineering Design for Plastics,* Reinhold Publishing Corp., New York, 1964.
218. *ASHRAE Handbook & Product Directory, 1977 Fundamentals,* ASHRAE, Inc., New York, 1977.
219. S. C. A. Paraskevopoulos, *J. Cell. Plast.* **1**(1), 132 (1965).
220. *Forming Thin Shells,* bulletin, The Dow Chemical Co., Midland, Mich., 1962.
221. R. K. Stern, *Mod. Packag.* **33**(4), 138 (1959).
222. R. G. Hanlon and W. E. Humbert, *Mod. Packag.* **35**(10), 158 (1962).
223. R. D. Mindlin, *Bell System Tech. J.* **24,** 353 (1945).
224. R. G. Hanlon and W. E. Humbert, *Package Eng.* **7**(4), 79 (1962).
225. K. Brown, *Package Design Engineering,* John Wiley & Sons, Inc., New York, 1959.
226. A. R. Gardner, *Prod. Eng.* **34**(25), 114 (1963).
227. C. Kienzle, "Plastic Foams," *paper presented at Regional Technical Conference, Buffalo, N.Y., Society of Plastics Engineers, Inc., Oct. 5, 1961,* p. 93.
228. P. W. Hemker, *J. Cell. Plast.* **5**(5), 295 (1969).
229. *Plast. World,* 15 (Mar. 1964).
230. *Styrofoam Sandwich Construction,* bulletin, The Dow Chemical Co., Midland, Mich., 1960.
231. *Plast. World,* 38 (Aug. 1963).
232. *Mod. Plast.* **42**(1A), 294 (1964).
233. E. B. Murphy and W. A. O'Neil, *SPE J.* **18,** 191 (1962).
234. F. Stastny, *Baugewerbe* **19,** 648 (Apr. 1957).
235. C. J. Hilado and R. W. Murphy, *ASTM Spec. Tech. Publ.* (*STP 685 Des. Build. Fire Safe*), *16-105* ASTM, Philadelphia, Pa., 1979.
236. C. J. Hilado, H. J. Cumming, and C. J. Casey, *J. Cell. Plast.* **15**(4), 205 (1979).

K. W. Suh
R. E. Skochdopole
The Dow Chemical Company

GLASS-CERAMICS

Glass-ceramics are polycrystalline solids produced by the controlled crystallization of glasses. They are primarily silicate-based materials that can be formed by highly automated glass-forming processes and converted to a ceramiclike product by the proper heat treatment. Glass-ceramics, which are also referred to in the technical literature as vitrocerams, devitrocerams, Pyrocerams (Corning Glass Works), sitalls, slagceramics and melt-formed ceramics, generally have a crystal content greater than 40–50% (see Ceramics). This excludes from the definition other types of glass products such as fluoride opals and copper–ruby glasses which contain minor amounts of a crystalline phase (see Gems, synthetic; Glass).

The liquidus temperature is a characteristic equilibrium temperature below which the liquid becomes thermodynamically unstable with respect to one or more crystalline phases. Whether crystals will actually form is dependent on kinetic factors such as the rate of cooling of the glass and the growth rate of the crystal species. In addition, there must be a nucleation site from which the crystallization can start. For most common inorganic glasses, crystallization begins at the interfacial boundaries, eg, glass–air, glass–refractory, etc, if they are exposed to temperatures just below the liquidus temperature for times generally on the order of minutes to hours. This relatively uncontrolled crystallization (commonly referred to as devitrification) results in columnar crystal growth perpendicular to the surface which generally affects the properties of the glass product in a detrimental manner. Since exposures to subliquidus temperatures always occur during the normal glassmaking process, the glassmaker has generally tried to make the glass formulations resistant to devitrification; this is done by adjusting the glass composition to minimize the crystal growth rate in the glass working range. Devitrification has been described as the chief factor which limits the composition range of practical glasses and is an everpresent danger in all glass manufacture and working (1).

One of the earliest references to an attempt to use the crystallization of glass to advantage dates to the early 1700s when deReaumur (2) in France attempted to crystallize soda-lime–silicate bottles by packing them in mixtures of sand and gypsum and heating them for a long time. The resulting bodies were weak and deformed since the crystal started to grow from the surface.

Little further work was done until the early 1900s when Tammann (3) started to investigate crystallization in more detail. This classic work was the basis for much of the present understanding of nucleation and crystallization. In the same period, a new industry arose in Europe based on the controlled crystallization of molten rocks, particularly basalts. Using many of the principles derived by Tammann, rocklike ceramics were produced and utilized in a range of applications. A very detailed summary is available (4) of the principles of melting and crystallization of fused basalts with photographs of production items.

In the 1930s Blau (5) illustrated the importance of controlled bulk crystallization. In the 1950s and 1960s there was an increased interest in controlled crystallization aided by nucleating agents since volume crystallization can enhance many of the properties; this is particularly true for the mechanical properties. Colloidal dispersions of metals were demonstrated to be good nucleants in some systems (6). The discovery of the effectiveness of certain refractory oxides as nucleation promoters (7–8) led to the commercialization of products based on the ability to closely control the crystallization process (see Crystallization).

The Glass-Ceramic Process

Figure 1 is a flow chart of a glass-ceramic manufacturing process. By definition, a glass-ceramic initially must be a glass. Thus, the first stages of the manufacturing process are similar to the processing of most commercial glasses; however, there are significant differences in the postforming stages. The environmentally related effects resemble those encountered in conventional glass processing. Pollution problems are generally minimal except for emissions during the melting of some compositions.

Melting. Glass compositions that are specially formulated for controllable crystallization are mixed like any other glass. The batch costs for the most common product (a low expansion cookware) is ca $0.33/kg. The range of batch costs is from ca $0.006/kg for glass-ceramics based on waste products to grêater than $14.00/kg for high purity products made for electronic applications (based on 1980 prices). In general, the glass-ceramic compositions are much more difficult to melt and homogenize than common soda-lime glass. Depending on the product requirements and the volume of the market, the commercial glass-ceramics may be made in continuous refractory-lined furnaces which produce glass 24 h/d; they may be melted in refractory-lined furnaces in an intermittent operation; or they may be melted in crucibles (either refractory- or platinum-lined) in an intermittent operation. Melting temperatures in excess of 1620°C are common. A large, continuous operation produces ca 90,000 kg/d; the intermittent operations vary from a few to 55,000 kg.

Forming. After melting, the resulting viscous liquid can be shaped by pressing, blowing, drawing, rolling, casting or some combination of these. Many of these processes are high speed, automatic operations which allow close control of dimensions. However, the viscosity–temperature relationship of most glass-ceramic compositions is such that they are generally much more difficult to form than the more common glasses. The most frequently used forming techniques are pressing and casting. In some applications the glass is powdered, mixed in a blender, and applied as a coating to

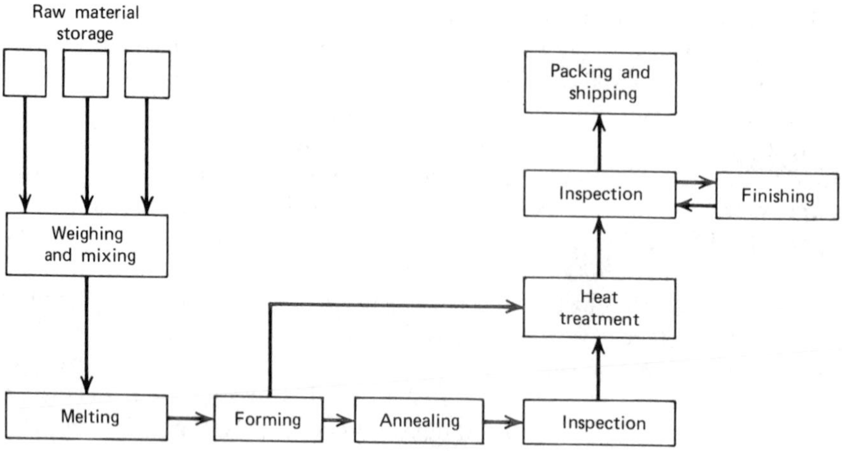

Figure 1. The glass-ceramic manufacturing process.

various metal and ceramic substrates. The coating is fused and crystallized during subsequent processing steps (see Powder coatings).

Postforming. The shaped body may be annealed (cooled slowly to avoid stresses) and inspected prior to heat treatment (Fig. 1), or the shaped object may be put directly into a heat-treatment furnace. The heat-treatment process may be continuous or intermittent.

During the heat-treatment process, the amorphous glass is transformed to a largely crystalline body. This complex transformation proceeds from distinct nucleation centers. The formation of crystals depends on the number of nuclei formed, the rate of crystal growth and the viscosity of the glass (3,5). The relationship between these factors and the heat-treatment process is shown in Figure 2. Figure 2 shows the nucleation formation rate and the crystal growth rate curves as a function of temperature for a condition where there is significant overlap of the nucleation and crystal growth rate curves (Fig. 2**a**) and a condition where the overlap is much less (Fig. 2**b**). T_L is the liquidus temperature. In Figure 2**a** (between T_2 and T_3) a crystal can grow as soon as a nucleus forms. The result is comparatively few crystals with a wide range of crystal sizes. In Figure 2**b**, a large number of nuclei can form (between T_1 and T_2) without appreciable crystal growth, which does not appear until the range T_2 to T_L. This is the rationale for the commonly used two-stage heat treatment which is shown in Figure 3.

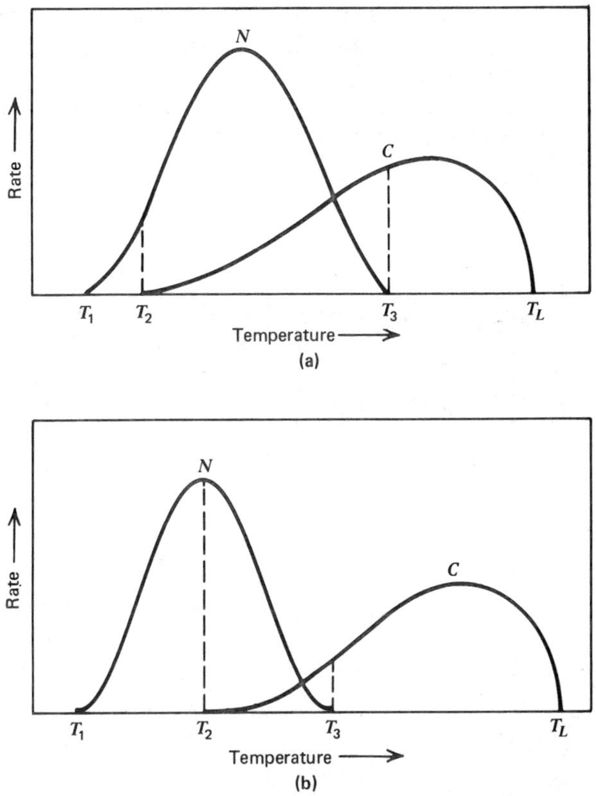

Figure 2. Number of nuclei N per unit of time, and the rate of crystallization C as a function of temperature (see text).

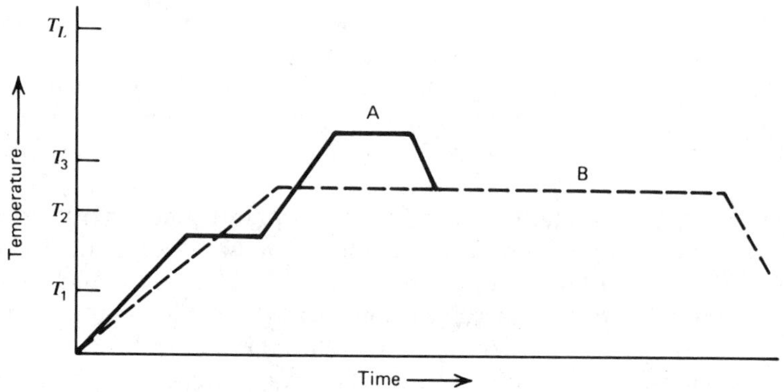

Figure 3. Heat-treatment cycle for controlled crystallization: A, two stage; B, isothermal.

One complication is the release of heat that accompanies crystallization. It can result in higher temperatures inside the body than on the surface and cause differential rates of crystallization throughout the piece, especially for articles having thick-walled sections. A solution is the use of a composition where the degree of overlap in the nucleation and crystal growth rate curves is such that a high rate of nucleation and a very slow crystal growth rate occur at the same temperature. A one-stage isothermal heat treatment can be used in which nuclei formation is followed by a slow rate of crystal growth such as shown in Figure 3. This slow rate of conversion allows the evolved heat to dissipate without producing unmanageable gradients. Figure 4 shows the change in the coefficient of expansion of a particular lithia aluminosilicate glass-ceramic composition at 760°C (9). The change in the expansion coefficient is roughly proportional to the degree of crystallinity. Crystallization starts at ca 20 h and the sample is 80% crystalline after 80–90 h.

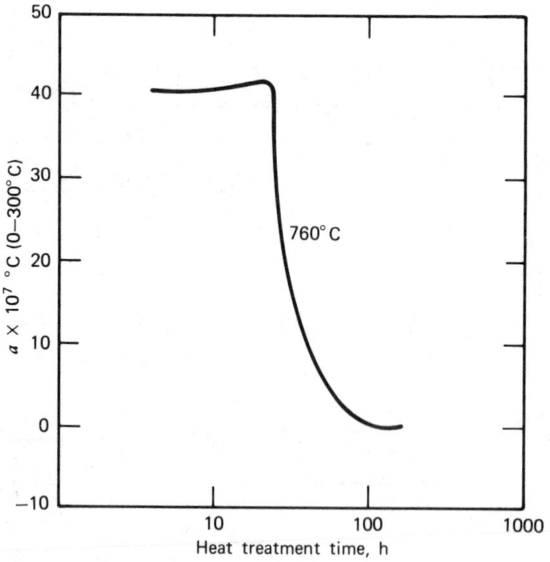

Figure 4. Thermal expansion vs heat treatment for a lithia aluminosilicate composition. α = coefficient of expansion.

Nucleation. One of the main factors that permits the formation of a conventional glass without crystallization is the very slow nucleation rate. Although some glass compositions are self-nucleating, most of the commercial glass-ceramic compositions rely on the presence of a nucleating agent to initiate internal crystallization. Although the mathematical treatment of homogeneous and heterogeneous nucleation is well developed (10), the exact role of the nucleating agent is not completely understood. The nucleant, which is usually a noble metal or a transition metal oxide, is incorporated in the batch and becomes an integral part of the glass structure during the melting. At heat-treatment temperatures of ca 30–100°C above the annealing point for times on the order of one hour or more, nucleating sites form. These nucleating sites are randomly distributed in concentrations of 10^9–10^{15} nuclei/cm^3 of glass and serve as sites for subsequent crystal growth.

Much of the information on nucleation is derived from detailed experimental observations on what occurs just prior to crystallization of the major phase (11). A number of different precrystalline events, such as enhanced amorphous phase separation and/or the appearance of trace nucleant phases, have been attributed to the nucleant but it is not always possible to identify a cause–effect relationship. Although there are probably several different nucleating mechanisms, the commonality of different nucleants is their introduction of sites of lower thermodynamic stability. The present understanding is not sufficient to allow *a priori* predictions of what is the most effective nucleant and what is the optimum nucleant concentration for a given base composition.

Crystallization. After nucleation, the material must generally be heated to higher temperatures in order for crystal growth to proceed; these temperatures are 750–1150°C, depending on the desired crystalline phase. The growth step is also complex (12). There may be metastable and equilibrium crystalline phases appearing simultaneously. Usually the composition of the crystals differ from the composition of the glass. In addition, the primary crystal phase (which is frequently a solid-solution phase) may start to transform by solid-state reaction to a new structural type. Solid solution refers to a family of compositions whose atomic arrangement is derived from some basic structural unit through the incorporation of selected foreign atoms. In silicate structures the Si^{4+} site is frequently occupied by an Al^{3+}, with the charge imbalance compensated by the addition of other cations. Some of these aspects can be illustrated by a transformation plot as Figure 5. This nonequilibrium diagram shows the phases as a function of the time and temperature of the heat treatment for a particular lithia aluminosilicate composition. It is only a first approximation for defining the heat-treatment limits since it does not contain all the information needed to optimize the crystallization process such as the phase concentrations.

The lower line represents the release of residual stress in the glass. Below this line the glass viscosity is such that the structural rearrangements required for the start of the glass-ceramic process do not occur in a practical time frame. The next higher temperature zone corresponds to the normal nucleation region. At higher temperatures the first major crystalline phase (a high-quartz solid solution) appears. With higher temperatures or longer times, the concentration of the high-quartz solid solution phase increases at the expense of the glassy phase (not shown in Fig. 5). With further heating or with longer times, or both, the high-quartz solid-solution phase transforms partially or completely to a new phase—a keatite solid solution (also commonly referred to as a β-spodumene solid solution); both of these solid solutions are based on phases of

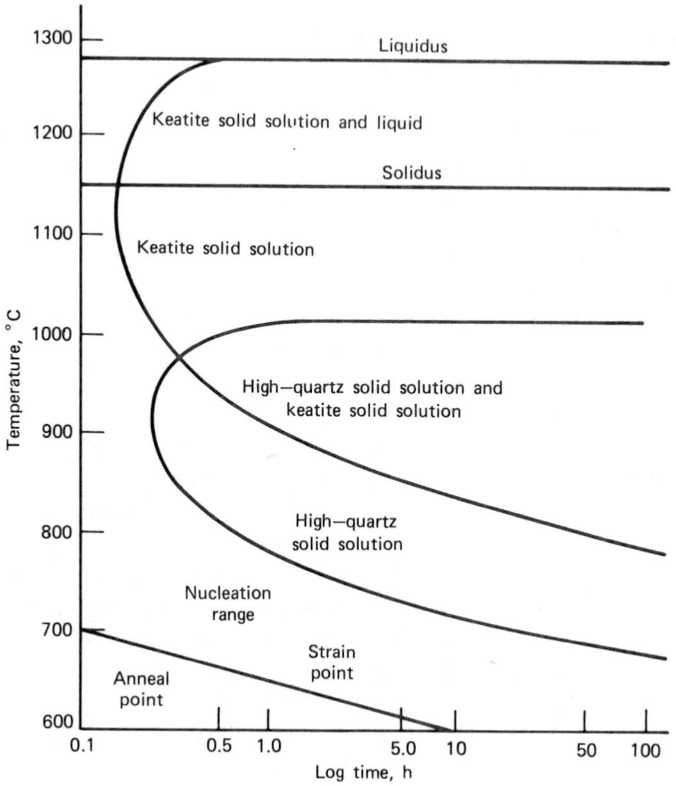

Figure 5. Schematic phase-transformation plot for a lithia aluminosilicate composition.

pure silica in which some of the silicon atoms have been replaced by aluminum atoms, with lithium atoms occupying interstitial sites to maintain charge balance. The keatite (or spodumene) solid solution is derived from the keatite structure—a high pressure form of silica (qv). As soon as the temperature is above the solidus, the amount of liquid starts to increase again.

After heat treatment, the material may undergo additional finishing steps such as cutting, grinding, polishing or decorating, or it may be packaged directly for shipment.

Properties

The properties of a glass-ceramic are dependent on such microstructural parameters (13) as: the amount and properties of the crystalline phase(s); the crystal sizes, shapes, and orientations; and the amount, composition, and distribution of the glassy phase (14). Some properties (eg, the thermal expansion coefficient) are additive functions determined by the properties and concentration of the individual phases. Other properties such as strength and maximum use temperature are not additive functions.

Figure 6 shows electron micrographs of two lithia aluminosilicate bodies with some of the microstructural features labeled. Although the crystalline phases are apparent,

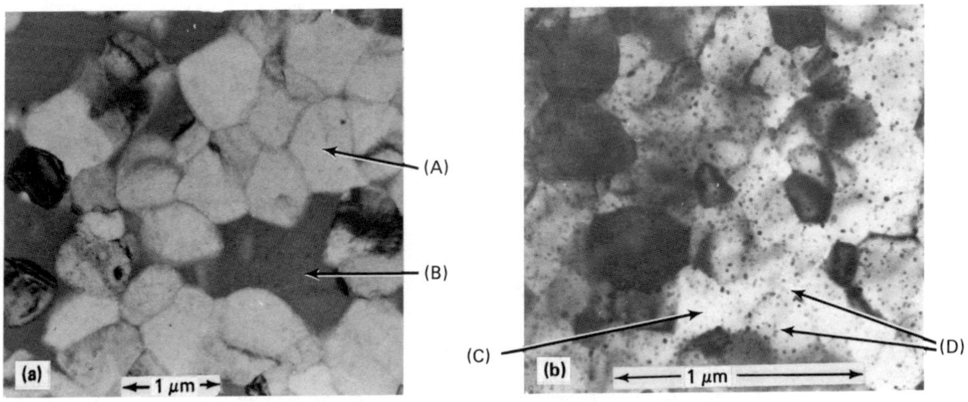

Figure 6. Transmission electron micrographs of two lithia aluminosilicate glass-ceramics, heat treated at a maximum temperature of (**a**) 800°C and (**b**) 1050°C. The phases present are: A, a high-quartz solid solution; B, a glassy phase; C, a keatite solid solution; D, an unidentified titania–zirconia solid solution (small grains are ca <0.05 μm).

the glass is not always readily visible (eg, Fig. 6**b**). A good deal of empiricism comes into play in achieving the desired properties since some properties are very sensitive to subtle microstructural changes.

The microstructure can be manipulated by changing the base composition, the nucleant system and the heat-treatment conditions. Commercial glass-ceramic materials generally contain at least seven or eight elementary species, each one added to fill a particular role. The constituents of the base composition determine the crystalline phases that occur, influence the melting and forming behavior, or influence the properties of the residual glass. A near-zero expansion body could not be produced if the chemical components of the formulation can only form high expansion phases. The nucleants not only initiate the crystallization but significantly affect the final-phase assemblage. By varying the nucleating agents and the heat treatment in a given base composition, it is possible to change the appearance and properties through changes in the type, amount, size, shape, and orientation of the crystal phases. Since it is unlikely that a glass-ceramic material can be developed with five or six specified properties all near the limits of observed values, the properties of a glass-ceramic material are invariably a compromise. Table 1 shows some of the properties of four commercial glass-ceramic materials. The characteristics of several base systems are discussed in the section Commercial Applications.

Thermal Properties. ***Thermal Expansion.*** Glass-ceramic materials have been reported with thermal expansion coefficients ranging from ca $(-60 \text{ to } 200) \times 10^{-7}/°C$. The thermal expansion properties, particularly the low expansions, are the primary properties that have been exploited commercially. The low expansion materials have been used for applications where thermal shock resistance is important such as cookware, stove tops, heat exchangers, etc. They have also been used in applications such as mirror substrates where minimal thermally induced distortions are important. Higher expansion materials have been used for sealing various metals and alloys. In some compositions it is possible to change the expansion coefficient over wide ranges by varying the heat-treatment conditions (eg, in some barium aluminosilicate compositions the expansion coefficient can be changed from $(55 \text{ to } 160) \times 10^{-7}/°C$).

Table 1. Properties of Four Commercial Glass-Ceramics[a]

Property	Example: 1 CER-VIT C-101	2 Corning 9608	3 Corning 9606	4 Corning 9658
Thermal				
coefficient of expansion, $10^{-7}/°C$				
at 0–38°C	0.0			
at 25–300°C		13	57	
at 25–400°C				94
thermal conductivity at 25°C, W/(m·K)	1.675	2.008	(mean at 3.398 20–800°C)	1.675
Mechanical				
Young's modulus at 20°C, GPa[b]	92.4	86.2	120	64.1
shear modulus at 20°C, GPa[b]	36.5		47.6	25.5
modulus of rupture at 20°C, MPa[c]	55	83	241	103
density at 20°C, g/cm^3	2.50	2.50	2.60	2.52
hardness, Knoop, kgf/mm^2 [d]				
100 g loading		593	657	
200 g loading	540			
500 g loading				250
Chemical				
corrosion rate, mg/cm^2, after 24 h in				
5% HCl at 95°C	0.3	0.12		87
5% NaOH at 85°C	3.5			8.5[e]
H_2O at 100°C	0.3			
Electrical				
dielectric constant at 100 Hz	11.7 (at 25°C)			
at 100 kHz				5.92 (at 25°C)
at 1 MHz	9.1 (at 25°C)	6.9 (at 20°C)	5.6 (at 20°C)	
at 1 MHz	14.2 (at 250°C)			
dissipation factors at 100 Hz	0.076 (at 25°C)			
at 10 kHz				0.003 (at 25°C)
at 1 MHz	0.023 (at 25°C)	0.023 (at 20°C)	0.017 (at 20°C)	
at 1 MHz	0.093 (at 250°C)			
$\log_{10}$ of volume resistivity, Ω·cm				
at 25°C	14	13.4	16.7	
at 250°C	8	8.1	10	10.8

[a] Data taken from property summaries provided by Owens-Illinois, Inc. and Corning Glass Works.
[b] To convert GPa to psi, multiply by 145,000.
[c] To convert MPa to psi, multiply by 145.
[d] See Hardness; 1 kgf/mm^2 = 9.8 MPa.
[e] After 6 h.

Thermal Stability. The maximum use temperature must be defined in terms of a time factor and the degree of loading. The crystalline phase(s) play an important role, but the maximum use temperature is limited by the residual glass. At elevated temperatures the residual glass may crystallize further or it may enhance deformation by viscous flow. The glass-ceramic materials are generally more refractory than the common glasses but less refractory than the common oxide refractories.

Since thermal shock resistance and thermal stability are a mutually desirable combination, most of the glass-ceramic work in this area has been on the low expansion lithia aluminosilicate compositions. The best of these compositions can survive >1000 h at 1000–1100°C without appreciable change in properties. More refractory glass-ceramic materials have not been commercialized owing to the difficulties of melting and forming them.

Thermal Conductivity. The thermal conductivity of glass-ceramics is slightly higher than that of conventional glasses and is generally higher than silicate ceramics such as porcelain. However, the thermal conductivity is much lower than that of the pure oxide ceramics such as alumina, magnesia, and beryllia. The thermal conductivity range is ca 1.7–5.4 W/(m·K).

Optical Properties. Transparent and opaque glass-ceramics can be made (see Fig. 7). The degree of transparency is a function of the optical properties of the crystalline species, the grain size, and the difference in the refractive indexes of the glass and the crystals. The transmission in the near infrared is typical of that of silicate glasses. The transparency has been utilized primarily for inspection purposes, but some compositions have been shown to display electro-optical activity (15).

Chemical Properties. The chemical durability can be varied over wide ranges. The rate of attack and the mechanism of attack are sensitive to pH. The crystalline phases and the residual glass generally are attacked at different rates. A product that exploits the solubility of lithium silicate phases is described in the section entitled Li_2O–SiO_2. The chemical durability of the common lithia aluminosilicate glass-ceramics is generally on a par with borosilicate glasses. A composition from the Li_2O–CeO_2–Al_2O_3–SiO_2 system is used commercially as a coating for chemical reactors because of its good chemical durability and mechanical toughness.

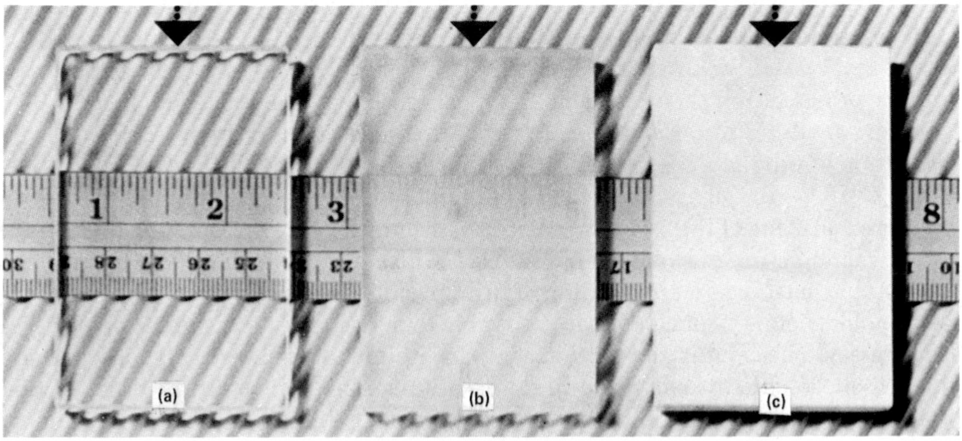

Figure 7. Glass-ceramic materials exhibiting a range of transparencies: (**a**) transparent; (**b**) translucent; (**c**) opaque.

Mechanical Properties. Glass-ceramics are brittle materials like ordinary glasses and ceramics, and exhibit many of the same properties, ie, there is a statistical distribution of strengths for any given set of samples and the results are a function of the testing conditions. The abraded modulus of rupture for most glass-ceramic materials is 55–250 MPa ((8–36) × 10^3 psi) compared with 40–70 MPa ((6–10) × 10^3 psi) for common glasses under identical abrasion and test conditions. High strength glass-ceramics (up to 1300 MPa or ca 190,000 psi) have also been reported (16). The high strengths are generally attributed to the formation of a surface-compressive layer, either through differential crystallization of a low expansion crystal phase, or by an external chemical ion exchange. The glass-ceramics are generally more abrasion resistant than the common glasses and require more energy for fracture because of the fine-grained microstructure. Glass-ceramics, based on slag and basalt (see $CaO–MgO–Fe_2O_3–Al_2O_3–SiO_2$ below) are notable for their abrasion resistance. Knoop hardness numbers as high as 900 have been measured (see Hardness). The modulus of elasticity for glass-ceramics generally falls in the range of 82–125 GPa ((12–18) × 10^6 psi). The elastic properties are an additive function of the concentration and elastic properties of the individual phases.

Electrical Properties. Most glass-ceramics are insulators. The conduction mechanism is primarily ionic and is more complicated than in a glass of the same composition because of the complex microstructure. The resistivity range is shown in Table 1. The dielectric properties also can be varied over wide ranges (15). The dielectric losses are primarily influenced by the mobility of the alkali ions and generally increase with increasing temperatures.

Magnetic Properties. Magnetic phases can be crystallized from a glass, but magnetic glass-ceramics are not commercially available. Most magnetic phases of interest do not contain atoms that make stable glasses; this limits the amount of magnetic phase that can be incorporated in the melt and subsequently crystallized (see Amorphous magnetic materials; Magnetic materials).

Commercial Applications

The size of the glass-ceramic market in the United States is not accurately known but 1979 sales were probably on the order of (\$100–200) × 10^6. The market outside the United States is probably comparable, with the market size strongly dependent on the level of USSR production of slag-based products. Examples of some glass-ceramic products and their principal chemical constituents are listed in Table 2. The following are descriptions of a few of the features that are important in the commercialization of products made from five of the more useful compositions.

$Li_2O–SiO_2$. In this system there is a large difference in chemical durability between the glass and some of the crystalline phases. Lithium silicate systems have been used for the manufacture of a photosensitive glass which can be chemically machined (6). By suitable masking and irradiation of lithium silicate glasses containing photosensitive noble metals as nucleants, it is possible to nucleate and crystalline only preselected areas. The crystallized portion contains lithium metasilicate crystals which are much more soluble in acid than the glass; these can be removed by etching, leaving the uncrystallized portion intact. It is possible to form intricate patterns in the glass (eg, form up to 1550 holes/cm^2 in a flat plate) for applications such as fluidic valves, printing-plate molds, and fine mesh screens.

Table 2. Commercial Applications

Basic chemical system	Unique properties	Product
$Li_2O–SiO_2$[a]	photosensitivity and differential solubility of phases leads to ability to form intricate patterns	substrates; fine-mesh screens; fluidic devices; printing-plate molds; spacers
$Li_2O–Al_2O_3–SiO_2$[b]	low expansion, thermal shock resistance	cookware; burner covers; preheaters; mirror substrates; low expansion housings; valve parts
$MgO–Al_2O_3–SiO_2$[a]	high strength, good dielectric properties	radomes; antenna windows
$K_2O–MgO–Al_2O_3–B_2O_3–SiO_2–F$	machinable	pacemaker parts; welder's fixture holder; welding nozzles; circuit bases
$CaO–MgO–Fe_2O_3–Al_2O_3–SiO_2$	abrasion resistance	industrial floor coverings; wall facings; abrasion resistant linings; high temperature insulators
$CaO–ZnO–Al_2O_3–SiO_2$	abrasion resistance	floor coverings; wall facings; counter tops
$B_2O_3–Al_2O_3–SiO_2$ $P_2O_5–Al_2O_3–SiO_2$	selective-ion evaporation at elevated temperatures	semiconductor doping sources
$Na_2O–Al_2O_3–SiO_2$	good strength	dinnerware
$Li_2O–CeO_2–Al_2O_3–SiO_2$[a]	chemical durability; mechanical toughness	coatings for chemical reactor accessories
$BaO–Al_2O_3–SiO_2$	specific expansion coefficient	alloy bonding
$Li_2O–BaO–SiO_2$	dielectric properties	insulators in devices

[a] 1979 United States sales probably over $1,000,000.
[b] 1979 United States sales probably over $50,000,000.

$Li_2O–Al_2O_3–SiO_2$. The lithia aluminosilicate system has the most commercial utility. Titania and titania–zirconia mixtures are the most commonly used nucleants. The properties of two commercial lithia aluminosilicate materials are listed in Table 1 (examples 1 and 2). The primary crystal phase in example 1 is a high-quartz solid solution; in example 2, it is a keatite solid solution. In addition to low expansion, some of the compositions are relatively stable up to 1000–1100°C; in other compositions, it is possible to make bodies that are either transparent or opaque. The commercial lithia aluminosilicate glass-ceramics are among the best glass-ceramic materials in terms of formability. The largest glass-ceramic body made to date is a 25-t casting used for an astronomical-mirror substrate.

$MgO–Al_2O_3–SiO_2$. Glass-ceramics from this system generally have good dielectric insulating characteristics at high frequency, high strength and moderate to high expansion (Table 1, example 3). Titania, zirconia, and mixtures of the two are the most commonly used nucleants and the principal crystalline phase is cordierite ($2MgO.2Al_2O_3.5SiO_2$). Since the magnesia aluminosilicates generally form relatively fluid melts, they are usually formed by casting. The combination of properties and formability has led to their use as radomes (see Microwave technology).

$K_2O–MgO–Al_2O_3–B_2O_3–SiO_2–F$. By proper compositional and heat-treatment control of silicates from the fluorine–mica family (13), it is possible to develop a body consisting of interlocking flakes of randomly oriented mica crystals. This unique microstructure limits crack growth and, as a result, these mica glass-ceramics are machinable to precise tolerances with conventional metal-working tools. In addition to machinability, they are strong, thermally shock-resistant and have good dielectric properties (Table 1, example 4). The commercial products are machined from pieces that are cut from cast slabs (see also Silica).

CaO–MgO–Fe_2O_3–Al_2O_3–SiO_2. Glass-ceramic materials called slagcerams have been produced by the crystallization of glasses made from metallurgical slags (17). The slag is generally modified by adding silica and/or alumina to make the slag glass more stable. The nucleating agents include transition metal oxide and sulfides as well as chrome and various forms of iron oxide. The nucleating effectiveness is dependent on the degree of reduction of the nucleating agents. The slag-based materials have excellent resistance to abrasion and wear, and have good chemical resistance to most common solvents. The slag-based melts are fluid, and generally are formed by a sheet or pressing operation. Products such as floor and wall coverings are attractive because of the low cost and the availability of the raw materials, and have been manufactured in the USSR and several other countries for many years.

Similar materials have been made from molten basalt—a plentiful, low cost rock (4,18). These basalt glass-ceramics are being produced in Europe and the USSR for use as abrasion-resistant pipelines, laboratory flooring, etc.

BIBLIOGRAPHY

"Slagceram" in *ECT* 2nd ed., Suppl. Vol., pp. 876–889, by S. Klemantaski, British Steel Corporation (Corporate Laboratories).

1. G. W. Morey, *The Properties of Glass,* 2nd ed., Reinhold Pub. Corp., New York, 1954.
2. M. deReaumur, *Mem. Acad. Sci.,* 370 (1739).
3. G. Tammann, *The States of Aggregation,* D. Van Nostrand Co., New York, 1925.
4. A. Portevin, *Mem. Soc. Ing. Civils Fr.,* 266 (1929); *Chem. Abstr.* **25,** 4207 (1931).
5. H. H. Blau, *Ind. Eng. Chem.* **25,** 848 (1933).
6. S. D. Stookey, *Ind. Eng. Chem.* **45**(1), 115 (1953).
7. S. D. Stookey and R. D. Mauer, "Catalyzed Crystallization of Glass—Theory and Practice" in *Progress in Ceramic Science,* Vol. 2, Pergamon Press, New York, 1962, pp. 77–101.
8. F. Albrecht, *Beispiele Angewandter Forschung, Frauenhofer* (*Society for the Promotion of Research, Munich*), 19 (1955).
9. J. E. Rapp, *Am. Ceram. Soc. Bull.* **52,** 499 (1973).
10. J. J. Hammel, "Nucleation in Glass—A Review" in *Advances in Nucleation and Crystallization in Glasses, Glass Division Symposium,* American Ceramic Society, Columbus, Ohio, 1971, pp. 1–9.
11. D. R. Stewart, "Concepts of Glass Ceramics" in *Introduction to Glass Science,* Plenum Pub. Corp., New York, 1972, pp. 237–271.
12. M. H. Lewis, J. Metcalf-Johansen, and P. S. Bell, *J. Am. Ceram. Soc.* **62,** 278 (1979).
13. G. H. Beall, *Glass Technol.* **19**(5), 109 (1978).
14. P. W. McMillan, *Glass Technol.* **15**(1), 5 (1974).
15. N. F. Borrelli and M. M. Layton, *J. Non-Cryst. Solids* **6,** 197 (1971).
16. D. A. Duke, J. F. MacDowell, and B. R. Karstetter, *J. Am. Ceram. Soc.* **50,** 67 (1967).
17. J. A. Topping, *J. Can. Ceram. Soc.* **45,** 63 (1976).
18. G. H. Beall and H. L. Rittler, *Am. Ceram. Soc. Bull.* **55,** 579 (1976).

General References

P. W. McMillan, *Glass-Ceramics,* Academic Press, Inc., New York, 1979; good introductory reference.

L. L. Hench and S. W. Freiman, eds., *Advances in Nucleation and Crystallization in Glasses, Glass Division Symposium,* American Ceramic Society, 1971; the proceedings of a symposium containing 26 papers relating to glass-ceramic materials.

DANIEL R. STEWART
Owens-Illinois

HIGH TEMPERATURE COMPOSITES

High temperature composites are a special class of composites (see Composite materials) in which the purpose is to produce a more desirable balance of properties over a range of elevated temperatures or at a given required maximum temperature than can be readily achieved with a homogeneous material. The motivation is often the reduction of weight as in the case of aircraft, space vehicles, or rotating machinery; the potential of raising the allowable temperature in a thermally limited application such as a gas turbine; and the need for better thermal protection. For example, because superalloys for turbines must simultaneously satisfy requirements relating to melting, oxidation, environmental corrosion, strength, stiffness, and creep, the addition of refractory metal fibers has been studied to improve performance with respect to the last three requirements (1).

As used here the term high temperature composites means synthetic multiphase systems of materials intended to be used at 200°C or above. This severely reduces the range of possible polymer matrix composites. On the other hand, composites that could function in this temperature range but would not normally do so are excluded. Thus, fiberglass reinforcement of concrete is not considered.

In addition to satisfying the principles applicable to composites in general, high temperature composites are subject to additional restrictions. Fibers, matrices, surface treatments, and fabrication procedures must be selected with great care because of stresses which can result from thermal expansion mismatch, and because chemical reaction, dissolution, or microstructural changes may degrade the materials. Environmentally caused degradation of the constituent materials during service must be considered also.

Most of the research and development over the last two decades has been in response to the enormous surge in aerospace activities. The peak effort occurred in 1969–1971 when about 600 papers and reports were published per year (2–3). In 1977 the rate was 200 per year. These numbers also include work not aimed at the high temperature performance of such materials.

High temperature composites are an evolving class of materials. There are various driving forces for making these materials increasingly important. The desire for more efficient and reliable thermal machines creates a continuing demand for improved high temperature materials. Energy costs for their extraction and the supply problems associated with conventional high temperature metals, such as chromium, cobalt, tantalum, and tungsten, stimulate a search for alternative materials. Monolithic ceramics, although often possessing excellent hot corrosion, oxidation, and high temperature–strength properties, are generally brittle materials, which lose strength dramatically as a result of flaws or mechanical damage. High temperature composites offer the potential of solving these needs and problems. To be successful the wide ranging problems of fabrication and high temperature reactions between the constituents of the composites must be overcome. The degrees of freedom and property control offered by a composite's approach provides the technical motivation for continuing to advance the science and technology of this class of high temperature materials and to overcome the barrier problems associated with the fabrication.

In view of the numerous possible reinforcing materials and matrix materials, the number of combinations comprising potential high temperature composite systems is enormous. However, analysis of the bibliographic references compiled in ref. 3 indicates that the technical activity has concentrated on relatively few fibers and matrices. This can be seen in Table 1. About 30% of the work has been devoted to each of boron and carbon fibers, followed in decreasing order by tungsten, steel, and silicon carbide. The remaining seven fiber types in aggregate account for 16%. In terms of matrix interest, the results show a similar selectivity: aluminum and magnesium 47%, carbon 2%, chromium and iron family metals 11%, titanium 9%, and the remaining four categories sharing 12%. This reflects the interest in lightweight, strong, moderately high temperature materials. High temperature resins are unquestionably another important class of matrix materials.

In this article the specific composites are grouped by matrix composition. This is partly because fabrication is an important consideration and the methods available depend more upon the matrix than upon the reinforcement.

Polymer Matrix

Polymers are particularly attractive as matrix materials because of their relatively easy processibility, low density, good mechanical properties, and often good dielectric properties (see Laminated and reinforced plastics). High temperature resins for use in composites are of particular interest to the high speed aircraft, rocket, space, and electronic fields. The temperature capability of a matrix resin is determined by its softening temperature, its oxidative resistance, or its intrinsic thermal breakdown. The more common linear alkyl-bonded polymers suffer from all of these various problems at relatively low temperature. Hence, there has been a substantial search for resin structures capable of sustaining temperatures of 300°C or even ≥500°C (4–6).

Several approaches have been followed. The C—H bond, since it is subject to oxidative attack, can often be replaced by the more resistant C—F bond. Although

Table 1. Publication Activity[a] in High Temperature Composite Systems, 1972–1978

Matrix fiber	Carbon	Al, Mg	Ti	Cr, Fe, Co, Ni	Other metals	Glass	Other ceramics	% of total
boron		335	43	1	1			31
carbon	251	102	5	12	10	8	11	32
SiC	1	23	23	11	1		8	5
carbides	2	1		16			1	2
borides–nitrides				2			3	<1
glass		6		1	3		4	1
Al_2O_3		11	9	10	3			3
other oxides		1		2	2	1	9	1
W		12	4	59	42			9
Mo			3	15	3			2
steel		55	1	3	7	1		5
other metals		19	27	4	11		35	8
% of total	21	46	9	11	7	1	6	

[a] Numbers refer to the publications in this interval as cited in ref. 3.

the fluorocarbon polymers, such as polytetrafluoroethylene, have good thermal stability, they are unsuitable as matrix materials because of their poor rigidity, propensity to flow, and poor fiber–matrix bonding (see Fluorine compounds, organic). Replacement of the C—C polymer backbone with siloxane —Si—O— has not adequately increased the thermal capability because of the tendency of silicones to depolymerize at elevated temperature. The most successful approaches have been to build thermally stable molecules based on aromatic and on heterocyclic ring structures. Among such polymer families that have been explored are those based on phenylene or on N-, O-, and S-containing aryl–heterocyclic repeat groups, such as quinoxaline, triazole, and imide structures. These polyring structures tend to restrict chain flexibility. Hence, thermally stable, linear linkages, such as —O—, —CO—, —NH—, —S—, —SO_2—, are introduced to promote rotational freedom. Nevertheless, in many cases the resins are refractory and require sintering to consolidate them into useful shapes.

Such resins are generally produced either by condensation polymerization directly or by addition polymerization followed by a condensation rearrangement reaction to form the heterocyclic entities. That is, H_2O is a reaction product in either case and creates inherent difficulties in producing void-free composites. Voids have a deleterious effect on shear strength, flexural strength, and dielectric properties, among others.

Three compromise approaches to achieve high performance and good processibility have been developed: (*1*) to use a fully prereacted thermoplastic as the matrix, (*2*) to allow the thermoset resin to react partially before consolidation into a composite structure, and (*3*) to carry out the polymerization *in situ*, ie, in the presence of the reinforcing material starting with the monomers or low molecular weight oligomers. In case (*1*) the softening point must be substantially above the intended use-temperature. Hence, a very high processing temperature is required which has the risk of causing pyrolytic degradation. Alternatively, a processing temperature near the softening point plus pressure are needed to achieve the required resin flow. In case (*2*) the resin is viscous and usually requires a volatile vehicle to achieve flow. As the vehicle is removed, the resin increases in viscosity. Thus, providing enough resin to fill space can be a problem. This problem is less severe in case (*3*), because the starting materials are more fluid owing to their low molecular weight. A class of *in-situ* polymerizable matrix resins was developed by TRW, Inc., under NASA sponsorship, in which the monomeric precursors are allowed to react and the H_2O removed prior to polymerization by imide cyclization (7). Addition polymerization is made possible by the use of norbornene end groups, which react without the further evolution of volatiles at 270–350°C.

A representative list of commercially available high temperature matrix resin systems is given in Table 2. These can be used to bond any inorganic reinforcing phase. However, they have been studied mainly in combination with carbon or glass fibers. Broadly speaking, these resins yield polymer-bonded composites that increase their upper-use temperature relative to epoxy resin systems by 100–150°C. These resins can also serve as useful high temperature adhesives.

With the best of the systems to date, prolonged exposure in air to 300°C causes the various measures of strength after 1000 h to be reduced to about half of their initial values. However, if only short thermal exposure is required, the range of useful temperatures can be extended considerably by selecting resins that char, rather than soften or depolymerize thermally (see Polyimides; Heat-resistant polymers; Aramid fibers; Novoloid fibers).

Table 2. Characteristics of High Temperature Matrix Resins

Class	Typical Solvents	Supplier, Designation	Type	Comments
thermoplastic (complete reaction of functional groups)	*N*-methyl-2-pyrrolidinone	DuPont, NR-150	polyimide	softens at 290–370°C
	m-cresol	Amoco, Torlon	polyamide-imide	softens at 275°C; used as molding resin
		Whittaker, PPQ 401	polyphenylquinoxaline	softens at 400°C
thermoplastic (partial reaction of functional groups)	*N*-methyl-2-pyrrolidinone	DuPont, Kapton	polyimide	
		Monsanto, Skybond	polyimide	
monomer reactants	dimethyformamide	TRW, P13N	polyimide	
		Rhodia, Kerimid–Kinel	polyimide	

Carbon Matrix

Carbon represents the ultimate high temperature end-member of polymer matrix materials. It has one of the highest temperature capabilities under nonoxidizing conditions among known materials (it melts or sublimes, depending on the pressure, at 3550°C). Additional considerations of chemical and thermal compatibility make it natural to use carbon and graphite fibers as the reinforcement material. The resultant carbon–carbon (C–C) composites (see Ablative materials) are especially desirable where extreme temperatures may be encountered, such as in rocket nozzles, ablative materials for reentry vehicles and disk brakes for aircraft. Other uses include bearing materials (qv) and hot-press die components.

The technology of producing C–C composites is complex because of the wide range of microstructures and densities possible in both the fibers and matrix. Fiber lay-up is done using techniques applicable to other types of composites including filament winding, cloth laminates, three-dimensional weaves, and discontinuous fiber methods such as injection molding, pressing, spraying, and slurry casting. The carbon matrix is produced by pyrolysis of resin or pitch infiltrants, or by chemical vapor deposition of carbon within the pores by the pyrolysis of methane, acetylene, benzene, etc, at reduced pressure (see Film deposition techniques). The carbon matrix can be transformed to graphite by heat-treating the composite to above ca 2500°C. The properties of the composite are a function of the starting fiber, fiber arrangement, fiber content, matrix material, density, and processing temperature (see Carbon and artificial graphite). The extensive literature (8–13) on C–C composites is devoted largely to details of processing descriptions, microstructure, articles produced, and some discussion of ablative behavior.

The strength properties of these composites are more variable than the case for similar fibers in a polymeric matrix. Upon pyrolysis the polymer or pitch infiltrants undergo a substantial weight loss of 10–60% and associated shrinkage in turn leads to cracking or porosity. Although the strength levels are less than if the same fibers were embedded in a polymeric matrix, impressive values can be obtained especially when considered above 1000°C and on an equal weight basis. Representative properties are given in Tables 3 and 4.

Metal Matrix

Metal matrix composites offer less pronounced anisotropy and greater temperature capability in oxidizing environments than do the polymeric and carbonaceous counterparts. Although most metals or alloys could serve as matrices, in practice the choices are sharply limited. The low density metals aluminum, magnesium, and titanium are particularly favorable for aircraft applications. In metal matrix composites, there is the intimate contact between constituents that are frequently reactive, and metals often allow relatively easy diffusion. Furthermore, the high temperatures experienced in fabrication and in service all contribute to the likelihood of interaction between the reinforcement and matrix phases. Chemical and physical stability are particularly important and pervasive concerns in metal matrix composites. Composite systems are classified according to their degree of reaction as follows (14): Class I, filament and matrix do not react and are mutually insoluble; Class II, filament and matrix do not react but exhibit some solubility; and Class III, filament and matrix react

Table 3. Properties of Composites Made with Low Strength Carbon Fibers[a]

Fiber and orientation	Matrix carbon, wt % fiber	Density, g/cm^3 [b]	Flexural strength, MPa[c]	Flexural modulus, GPa[d]
low strength fiber, graphite felt, Union Carbide grade WDF; random orientation	50–60	0.92–1.15	26–38	3.4–5.2
low strength fiber, graphite cloth, Union Carbide woven cloth grade WCB; unidirectional lay-up	60–70	1.17–1.26	66–93	3.4–5.2
high modulus carbon fibers, Modmor type I, chopped fibers; random orientation	70–75	1.58–1.62	43–100	3.1–3.8
carbon fiber–carbon composite low strength woven cloth; Carb-I-Tex grades 100, 500, 700	not known	1.38–1.44	76–124	9.0–19

[a] Adapted from ref. 9. Courtesy of the Propellants Explosives and Rocket Motor Establishment Procurement, Ministry of Defence, UK.
[b] To convert g/cm^3 to lb/in^3, divide by 27.68.
[c] To convert MPa to psi, multiply by 145.
[d] To convert GPa to psi, multiply by 1.45×10^5.

Table 4. Properties of Composites Made with High Strength Carbon Fibers[a]

Fiber and orientation	Composition, vol %			Density, g/cm^3 [b]	Flexural strength, MPa[c]	Flexural modulus, GPa[d]
	Fiber	Matrix carbon	Voids			
high modulus carbon fiber Modmor type I; unidirectional lay-up	55–65	15–27	10–25	1.63–1.69	345–524	138–172
high strength carbon fiber Grafil type II; unidirectional lay-up	62–65	20–24	13–18	1.47–1.49	1034–1241	152–172

[a] Adapted from ref. 9. Courtesy of the Propellants Explosives and Rocket Motor Establishment Procurement, Ministry of Defence, UK.
[b] To convert g/cm^3 to lb/in^3, divide by 27.68.
[c] To convert MPa to psi, multiply by 145.
[d] To convert GPa to psi, multiply by 1.45×10^5.

to form surface coating. Examples of each class are found in Table 5. Class I represents the ideal situation in which no degradation of properties occurs upon fabrication or in service. Composites, as long as they remain at sufficiently low temperature, can behave as if they belong to this class. Class II behavior can lead to geometric instability through coarsening and can result in secondary effects in the bulk of the fiber or matrix such as precipitation, recrystallization, or altered physical properties. Class III is the most common situation. If the reaction zone exceeds certain critical thickness, it can have substantial deleterious effect on mechanical properties (15). The enhanced kinetics of solid–liquid relative to solid–solid reactions restricts the useful fabrication techniques possible in such systems of practical interest as Al–B and Ti–B composites. Furthermore, reactivity considerations tend to favor the larger diameter fiber reinforcements such as B, SiC, and metal wires (see Laminated and reinforced metals; Metal fibers).

Table 5. Classification of Composite Systems[a]

Class I	Class II	Class III
copper–tungsten	copper(chromium)–tungsten	copper(titanium)–tungsten
copper–alumina	eutectics	aluminum–carbon (>700°C)
silver–alumina	columbium–tungsten	titanium–alumina
aluminum–BN-coated B	nickel–carbon	titanium–boron
magnesium–boron	nickel–tungsten[b]	titanium–silicon carbide
aluminum–boron[c]		aluminum–silica
aluminum–stainless steel[c]		
aluminum–SiC[c]		

[a] Ref. 14. Courtesy of A. G. Metcalfe and Academic Press, Inc.
[b] Becomes reactive at lower temperatures with formation of Ni_4W.
[c] Pseudo-Class I system.

Reference 16 enumerates six strategies to deal with the problem of fiber–matrix reactivity: (*1*) high speed (fabrication) processing to minimize reaction; (*2*) low temperature processing to minimize reaction; (*3*) development of low reactivity matrices; (*4*) development of (fiber) coatings to minimize reaction; (*5*) selection of systems with increased reaction tolerance; and (*6*) design to minimize effect of reduced strength. Although these strategies are presented with reference to titanium matrix composites, they are generally applicable to other systems. The processes available to consolidate the composite constituents either involve liquid infiltration or liquid-assisted sintering under as quick and mild conditions as possible. If liquid contact is not tolerable, then some variant of sintering or diffusion bonding is required. Such solid–state consolidation is relatively slow. Hence, pressure is often used to expedite the processes but it can also damage the reinforcement leading to property degradation. Pressure also increases the requirement for more complex equipment.

High temperature composites are subject to thermal cycling during the fabrication process and in service. Differences in the thermal expansivities of the constituent phases produce internal stresses and possible material degradation as a result of fatigue if the stress levels exceed the yield stress of the matrix. In systems where fiber-matrix bonding is weak, or where voids are present, thermal cycling can result in ratcheting, ie, in irreversible relative movement. The expansivities shown in Table 6 show that mismatch is unavoidable for most matrix-reinforcement combinations.

Table 6. Expansion Coefficients at RT

Matrix	Coefficient $10^6/°C$	Reinforcement	Coefficient $10^6/°C$
Si	2.5	C[a]	ca 0
Ti	8.5	SiO_2	ca 0
Fe, Ni, Co alloys	12–13	SiC	4
Cu	17	W	4.5
Ag	19	B	6
Al	24	Al_2O_3	8
Mg	26	Be	12

[a] Parallel to basal plane.

Aluminum and Magnesium Matrices. The melting points of aluminum and magnesium are 660 and 651°C, and their densities are 2.70 and 1.74 g/cm^3, respectively. Hence, they offer advantages over polymer-matrix composites relative to potential maximum use temperature and advantages over other metals relative to weight. Even though Mg has a substantially lower density than Al, the latter has been the more practically important matrix metal mainly because of corrosion and oxidation considerations. Therefore, unless Mg is specifically mentioned, the discussion in this section relates to Al-matrix composites.

The most widely studied and best developed metal-matrix composite system is that made using boron or Borsic SiC-coated boron fibers in combination with aluminum alloys (17). In part because carbon fibers are less expensive, there has been continuing interest in the use of carbon-fiber reinforcement (see Carbon and artificial graphite). Magnesium composites have also been explored based on these fibers. Other reinforcements that have been used include steel, Be, W, Al_2O_3, SiC, and vitreous silica.

Boron-Fiber Reinforcement. Aluminum and its alloys are used in aircraft because of their ease of fabrication and generally good properties. The use of boron fiber as a reinforcement adds substantially to the strength and stiffness with no penalty in density. Furthermore, the creep rupture resistance and tensile strength in the fiber direction are substantially improved as a function of temperature. The temperature dependence of the tensile strength of boron-fiber-reinforced composites containing 50 and 25 vol % are compared with the unreinforced matrix in Figure 1. The failure time as a function of stress and temperature is given in Figure 2 for Borsic-reinforced

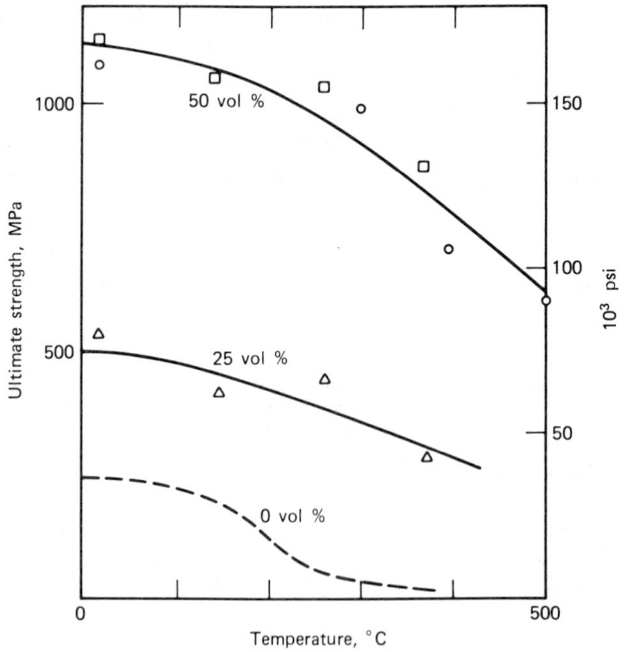

Figure 1. Failure stress of 6061 aluminum reinforced with the indicated amount of 100-μm-dia boron fiber as a function of temperature; △, □, Shaefer and Christian (17); ○, Kreider (17); ---, (18).

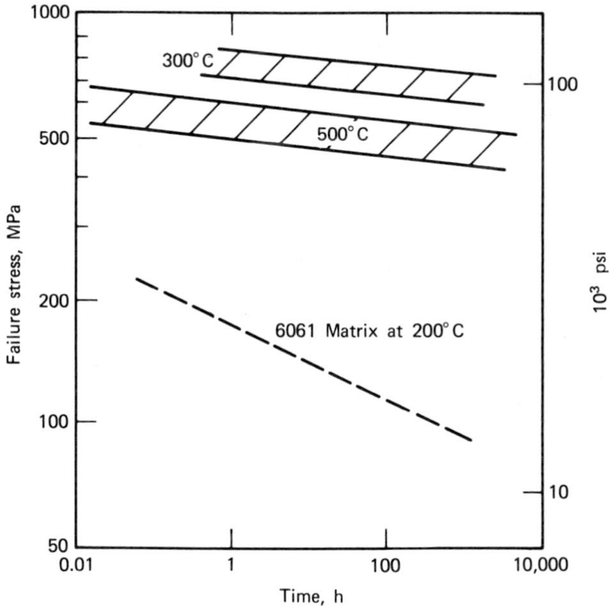

Figure 2. Time to rupture as a function of applied stress for 6061 aluminum matrix reinforced with 25 vol % of 100-μm-dia Borsic fiber. Adapted from ref. 17. Data on unreinforced alloy from ref. 18.

composites. As can be seen, useful strength can be maintained for long times even at 500°C. Uncoated boron reacts with the matrix at 400–500°C and cannot be used for sustained stress applications. In the transverse directions, the strength is governed by the matrix properties. Hence, cross-plying is usually required for high temperature applications.

The fabrication of the composite involves careful alignment and spacing of individual fibers. This is facilitated by the coarseness of the fibers which are available in 100 and 145 μm diameters. The fibers must be fixed in position, while consolidation of the structure is taking place. This frequently results in the desired final shape, such as a turbine fan blade. Standard shapes such as a Z-bar can also be produced. Such stock shapes normally require secondary fabrication for conversion to the final desired structure, eg, further cutting, shaping, bonding, or other means of joining. The most useful aluminum-matrix alloys are 2024 and 6061. The former is available as foil and powder and has good high temperature properties. The latter is tough, has good corrosion resistance, a relatively high melting point, and has been extensively investigated as a matrix material.

The methods of fabrication are determined in part by the degradation of the strength of the boron fiber if held for prolonged times at elevated temperature. Thus, contact with molten aluminum must not exceed 1 s, or in the case of solid-state pressure bonding, the maximum allowable time at 500°C is extendable to ca 1 h. A part can be made in a one-step process in which the boron fibers are positioned on an aluminum supporting structure and built-up to produce the final desired shape directly. Alternatively, the aluminum and boron can first be bonded to produce some standard shapes, such as ribbons, sheets (often called broad goods), or the other shapes described above. Ribbons can be made by quickly running an aligned array of fibers through molten aluminum. Broad goods can be made by precisely positioning a large number of parallel filaments on a foil substrate. These are then assembled and joined in a secondary fabrication process to achieve the desired article.

Collimation is achieved by winding from a single spool or multiple spools onto a drum using a screw feed to provide the necessary spacing. The filaments are held in place by use of a resin adhesive or in some cases by the use of aluminum-wire interwinding which serves as spacers. Upon curing of the resin, the foil-filament layer is slit and removed from the drum to form the sheets of broad goods. Required shapes to build the desired structure and filament orientations are cut from the broad-goods stock material. Sheets can also be made on a continuous basis, feeding as many as 600 filaments from individual spools onto the sheet in a carefully spaced array. The use of plasma-sprayed aluminum is used in some processes to replace the resin adhesive. This has the advantage of eliminating the need for resin removal during subsequent processing, as well as providing a product of greater mechanical stability, shelflife, and better matrix geometry.

To make the structure, the operator must stock the individual lamellae, place them into a mold, and subject them to a careful time–temperature pressure cycle to achieve the required material flow and diffusion bonding, without degrading the fiber strength. If the broad goods are made using the resin adhesive, the temperature cycle must provide for the complete removal of this material before the full pressure is applied to achieve the metallurgical bonding.

As with all filament-reinforced composites, boron–aluminum composites show strong anisotropy of mechanical and physical properties. The ratio of the strength transverse to the fibers relative to that in the parallel direction is ca 0.1 for 50 vol % of fiber at RT. This ratio is even less at higher temperatures. One way to reduce the anisotropy is to form ternary composites in which the boron–aluminum broad goods are interleaved with foils of a higher temperature material such as titanium. The benefit of using foil is shown in Table 7 (17). At relatively little penalty in density, a four-fold improvement in transverse properties is possible.

Components such as compression panels and stiffeners for the Atlas missile, the Space Shuttle Orbiter, and various supersonic aircraft have been built and have performed successfully in tests at ca 300°C. It is probable that this composite system will find increasing use in applications where long-term durability, high performance, and a wide range of environments, including temperatures as high as 300–400°C may be encountered.

Table 7. Comparison of Composites Showing the Benefit of Foil

Composition, vol %	Relative density	Strength parallel-to-fiber, MPa (10^3 psi)	Strength ratio, transverse to parallel
boron fiber, 50 6061 Al, 50	100	1138 (165)	0.12
boron fiber, 40 Ti–6, Al–4 V foil, 30 6061, Al, 30	120	1034 (150)	0.43

Carbon and Graphite-Fiber Reinforcement. Composites made from carbon or graphite fibers combined with aluminum or magnesium are outstanding in terms of their potential stiffness-to-density and strength-to-density ratios because of the low relative densities of both the metals and the filaments (19). The individual constituents are relatively inexpensive. This potential is tempered by the reactivity of the matrix metals towards carbon by the large surface-to-volume ratio of the carbonaceous filaments and by the galvanic corrosion possible if the composites are exposed to moist, conducting environments. Investigations and development efforts on these systems, although less extensive than those for the corresponding boron-fiber composites, have shown that good quality experimental composites can be produced practically.

The fineness of the carbon fiber (ca one fourth the diameter of a human hair) requires that the filaments be handled as bundles. Hence, the process for obtaining intimate contact between the individual filaments and the matrix metal requires achieving rapid and complete penetration without damaging the fibers. The typical procedure is to subject a strand to vapor- or liquid-phase treatments which may clean the fibers, promote adhesion and wetting, and/or provide a protective coating. The treated strand is then infiltrated by the molten metal or alloy to produce a kind of coarse wire. The wires are then handled much as is monofilament boron, ie, they can be individually positioned and oriented and subsequently hot-pressed into a coherent macrostructure.

Molten unalloyed aluminum and magnesium near their melting points do not wet untreated carbon fibers. Wetting, with the formation of Al_4C_3, occurs in the case of aluminum above 1000°C. This carbide decomposes in the presence of water vapor. Although alloying with certain elements, such as calcium, chromium, and nickel, reduces the temperature for wetting by ca 100°C, fiber coating is required to achieve wetting in the metal solidus-liquidus temperature regime. Carbides such as of titanium and niobium were unable to suppress Al_4C_3 formation. Other metals such as nickel form harmful brittle intermetallics at the surface. Among the metals, tantalum is particularly satisfactory in overcoming both of these problems. However, the Aerospace Corporation has developed a general coating procedure that is also applicable to magnesium as well as copper, silver, zirconium, lead, and other metals (20). This consists of codepositing boron and titanium on the graphite fibers by the vapor-phase reduction of BCl_3 and $TiCl_4$ with zinc vapor. This coating appears to interact with the matrix to form a titanium–boron-containing alloy near the fiber surface, and serves both to promote wetting and to act as a carbon diffusion barrier. This process is illustrated in Figure 3.

One of the earlier problems with carbon–aluminum composites was the relatively large scatter in strength. The improvement in coatings and use of aluminum alloys containing a large amount of silicon has been found to produce substantially more consistent properties. The values indicate that the full potential of the fibers can be achieved. A comparison of properties obtained using tensile bars prepared from various aluminum-matrix alloys is given in Table 8. Transverse tensile strength has generally not been as large as with boron–aluminum composites for comparable fiber loadings. This is apparently because of the greater relative transverse compliance of carbon fiber, to weaker fiber–matrix bonding, and/or to a weaker transverse strength of the fibers themselves.

To date, test-specimen shapes and representative prototype forms, such as sheets, plates, rods, stiffener sections, and cone–frustrum shapes, have been produced by

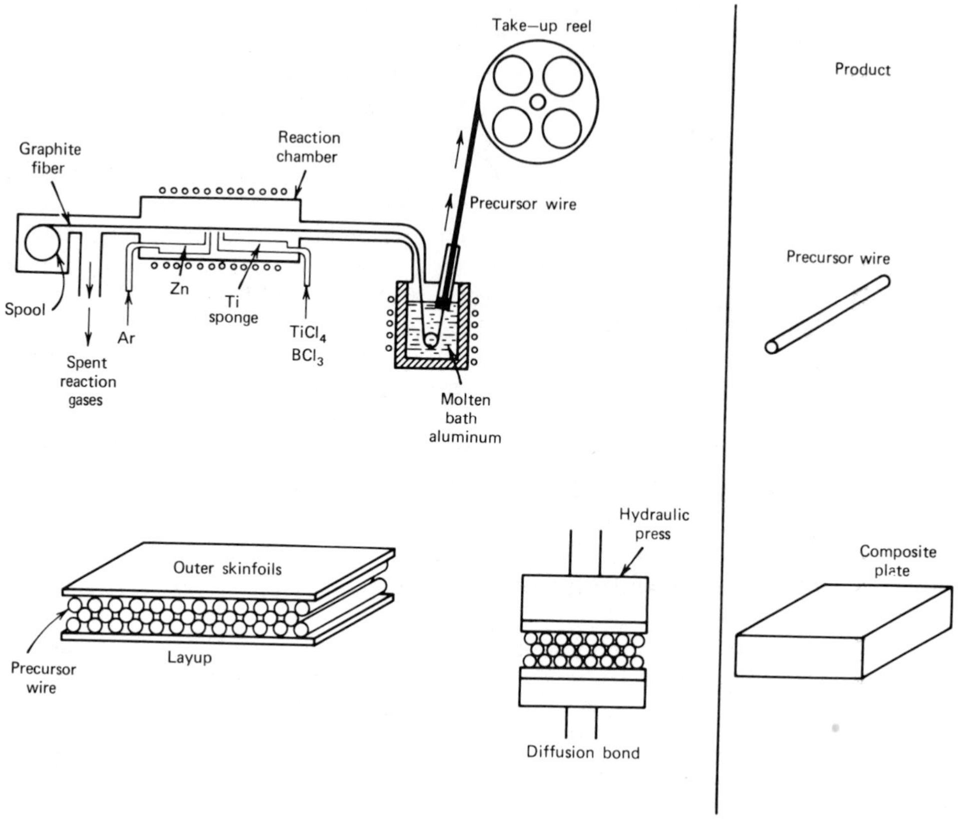

Figure 3. Schematic of composite fabrication. Courtesy of the Aerospace Corporation (20).

Table 8. Mechanical Properties of Aluminum Alloy–Thornel 50 Graphite Composite

Alloy	Fiber, vol %	Tensile strength, MPa[a]	Young's modulus, GPa[b]
201[c]	31	607	162
202[c]	32	690	175
6061[c]	29	607	154
Al–12 Si[d]	28	731	145

[a] To convert MPa to psi, multiply by 145.
[b] To convert GPa to psi, multiply by 1.45×10^5.
[c] Ref. 21.
[d] Ref. 19.

consolidating the infiltrated strands. Carrying out this consolidation at temperatures at which the alloys are partially molten appears to minimize mechanical damage during processing. Approximately 90% of the initial wire strength can be retained. Among the applications for aluminum–carbon composites are bearings. The coefficient of dry friction against aluminum is 0.15 with little wear when the fibers are oriented end-on against the sliding surface. Other uses include seals, sound insulation, mechanical damping material, electrical contacts, antennae, and various high temperature structures.

Silica-Fiber Reinforcement. Various other fibers have been used in composites with aluminum as the matrix including fused silica (qv), aluminum oxide, silicon carbide, steel, and beryllium (22). The fused silica–aluminum system was one of the first extensively studied and developed metal matrix composites (23). This work was undertaken at Rolls Royce with the intention of producing an improved material for use as aircraft-engine compressor blades. Fused silica can be drawn into fibers that retain high strength above the melting temperature of aluminum. However, when in intimate contact, SiO_2 and Al react causing degradation of the fiber strength. A key element in the development of this system was establishing a suitable combination of temperature and filament velocities which enabled the freshly drawn SiO_2 to be uniformly coated by aluminum as it passed through a bead of molten metal. The coated fibers were then hot-pressed into the desired shapes. Tensile strengths of 793 MPa (115,000 psi) at RT were reported, and at least half of this strength level was retained to 400°C. However, long-term property degradation occurred under the expected service conditions. The strength-degrading reaction between the fiber and matrix is first detectable at ca 300°C. Thus, this composite system was not introduced into its intended application. Nevertheless, up to 300°C this composite has attractive properties compared with the unreinforced alloys. However, no improvement in stiffness is possible because the elastic moduli of the metal and the fiber are nearly identical.

Aluminum Oxide-Fiber Reinforcement. The use of aluminum oxide fiber to reinforce aluminum has the obvious advantage of avoiding the problem of strength-degrading chemical attack. Such fibers have been available in the form of whiskers, polycrystalline filaments produced by DuPont under the designation Fiber FP and by International Chemical Industries as Saffil fiber. Continuous monofilaments by a kind of microscale continuous casting have been produced by the Tyco Laboratories and by A. D. Little Co. Fibers containing additional oxides are available commercially. Melt infiltration, although the preferred fabrication process, is difficult because of the problem of achieving good wetting to the fibers. Fiber coating can promote wetting. An alternative approach has been to alloy the metal (24–25). Lithium has been shown to be effective in this regard, probably owing to a small amount of reaction with the fiber. Other alloying constituents, such as magnesium and calcium, can also be used. Careful control is required of the time, temperature, and alloy composition to effect the optimum adhesion. For unidirectional composites containing 55 vol % of the DuPont fiber in an Al–2.5% Li alloy, the tensile strength in the parallel direction is essentially constant to 300°C at 552 MPa (80,000 psi). At 500°C the flexural strength decreases to 70% of its value at RT. Transverse properties have values and behaviors much like the matrix material. This system appears to offer potential for high temperature applications where the composition of the matrix has satisfactory environmental resistance. Various complex prototype shapes, such as rods, billets, beams, tubes, and tubing, have been produced by the melt-infiltration process.

In some ways magnesium is an even more attractive matrix in combination with aluminum oxide in that fiber wetting occurs without the need for special alloy additions. The mechanical properties on an equal weight basis are superior to the corresponding aluminum materials. Strength retention at elevated temperature is also better. Work on this system has been more limited.

Silicon Carbide-Fiber Reinforcement. Another reinforcement relatively available and only mildly reactive towards aluminum is silicon carbide. This is available in the form of whiskers, vapor-deposited monofilament similar to boron, and recently (26) as a solid-state pyrolysis product of a spun polydimethylsilane resin. Advantages of the SiC as a reinforcement are the high strength and possibly lower cost of this material (see Carbides). Infiltration of a bundle of vapor-deposited 100 μm fibers with molten aluminum to form wires has achieved average tensile strengths of 1170 MPa (170,000 psi) parallel to the fibers at a loading of 39 vol % (27). Similar strengths have been reported for composites produced from oriented β-SiC whiskers embedded in aluminum (28). To date there are few reports on this potentially important composite system.

Metal-Wire Reinforcement. Various metals including steel, tungsten, and beryllium have been used (22) to reinforce aluminum. These composites have been primarily used as model systems. In most cases fiber–matrix interactions would degrade properties at elevated temperatures. Substantial strengths have been achieved essentially in accordance with expectations of the rule of mixtures. An obvious advantage of metallic fibers is that they are less sensitive to surface damage. Thus, a wider range of fabrication processes is possible (see Laminated and reinforced metals).

Titanium Matrix. Because of their high temperature potential, there has been a considerable interest in titanium-matrix composites similar to the case for aluminum and magnesium. Although more dense than the latter metals, titanium is nevertheless a relatively low density material having a higher potential use temperature and good mechanical properties that, of course, diminish with increasing temperature. It is a very reactive metal, which is often used as a getter, and does not form a protective oxide. Hence, melt infiltration in most cases is not a useful fabrication method, and fiber–matrix interactions place greater restrictions on the possible materials combinations. The most important reinforcements have been boron, silicon carbide-coated boron, and silicon carbide.

Because of the high operating temperature of these composites, there has been a substantial amount of work done in trying to understand how this reactivity adversely affects the strength of the resultant composites, and how to overcome or minimize these effects (15–16).

Boron-Fiber Reinforcement. The large diameter of boron fibers combined with the good flow and high creep rates of titanium at 750–1000°C facilitates bonding of aligned filaments into a tape by creep-diffusion forming methods. One of the conditions required for good strength is that the titanium diboride layer formed at the interface not exceed 50 nm in thickness. Otherwise strength reductions of as much as 50% can result. As a result the time, temperature, and pressure conditions must be carefully controlled as is discussed in detail in ref. 16. The extent of reaction at the interface can be somewhat modified by the addition of alloying elements such as aluminum, molybdenum, vanadium, and zirconium. Coatings such as B_4C, SiC, and BN have relatively little affect on the reaction rate. Nevertheless, the use of Borsic silicon carbide-coated boron fiber does appear to be beneficial in terms of reducing the severity of mechanical property degradation when substantial interface reaction occurs. The properties of this composite given in Table 9, when compared with those of the matrix, suggest little improvement of absolute strength. The appropriate comparison for many applications is strength-to-weight for which the composites show benefit. Stiffness is clearly improved. These relative advantages improve with increasing temperature.

Table 9. Representative Properties of Titanium-Matrix Composites[a]

Fiber	Fiber, vol %	Matrix	Tensile strength, MPa[b] Parallel	Tensile strength, MPa[b] Transverse	Young's modulus, GPa[c] Parallel	Young's modulus, GPa[c] Transverse
B	25	Ti(75Al)	965	410	178	152
Borsic	50	Ti–6Al–4V	965	290	286	205
SiC	28	Ti–6Al–4V	980	650	250	195
Be	50	Ti–6Al–4V	700–1400		185–220	
	0	Ti–6Al–4V	1000	1000	110	110

[a] Ref. 16.
[b] To convert MPa to psi, multiply by 145.
[c] To convert GPa to psi, multiply by 1.45×10^5.

Silicon Carbide, Aluminum Oxide, and Metal-Fiber Reinforcements. The apparent benefit of silicon carbide coating on boron suggests the use of silicon carbide itself as the reinforcement for the highly reactive titanium matrix. Such composites have been made using techniques similar to those for the boron fibers (29). However, strength-degradation fabrication processing occurs. Strengths comparable to those for boron can be attained as shown in Table 9. However, increasing the volume fraction of the fiber seems to result in decreasing efficiency in the utilization of the fiber properties, and the resulting values have not been as consistent as for the case of boron.

Aluminum oxide has also been considered as a possible reinforcement even though it was expected to be reactive when in intimate contact with titanium. Complex interface reactions have been noted (30). Composites with properties not unlike those discussed above and showing good fiber utilization have been achieved. However, properties were degraded substantially on prolonged heating at 800°C.

Metals such as beryllium and molybdenum have been used to reinforce titanium because of the potential of using more conventional metal-processing methods (16). Some difficulties have been experienced in processing, owing in part to the loss of the high strength of highly worked beryllium when it is heated to relatively modest temperatures (500–750°C). Properties are shown in Table 9. Molybdenum, although capable of producing high performance composites, must be considered mainly as a model system. Improvements in mechanical properties are considerably offset by the increase in material density.

Superalloy Matrices. For even higher temperature usage, iron-, cobalt-, and nickel-based superalloys are commonly used as structural materials. Simultaneous unrelated or antagonistic requirements are often demanded of materials for use as gas-turbine blades, combustion systems, or rocket components, such as high strength and stiffness, combined with oxidation resistance, creep resistance, and hot corrosion resistance. The use of high strength filaments as reinforcements can reduce the problems introduced by these constraints.

Although the main alloy constituents of superalloys are chemically not as reactive as are the matrix metals discussed above, the high service and fabrication temperatures nevertheless mean that fiber–matrix interactions and the choice of fibers are major considerations. The reactivity or strength degradation of the fiber itself at 1000–1300°C precludes consideration of carbon, boron, silicon carbide, or glass fibers. Even such a stable oxide as alumina has been shown to react and degrade in contact with nickel-base refractory alloys under service conditions (31). However, metal-fiber reinforcement can be effective in this case (1,32). Wire alloys based variously on molybdenum, tungsten, tantalum and niobium have achieved tensile strengths at 1200°C as great as 530, 1940, 490, and 345 MPa (79,000, 281,000, 71,000, and 50,000 psi) and strength-to-density ratios of 53, 100, 29, and 34, respectively. On the basis of strength as well as reactivity considerations, tungsten has been investigated the most thoroughly particularly at the NASA-Lewis Research Center. The matrix alloy composition has a significant effect on the reactivity with the fiber. The best of several experimental matrix alloys had the composition Ni-25% W-15% Cr-2% Al-2% Ti. Other compositions lead to appreciable degradation of the fiber.

Shapes can be conveniently made by embedding the reinforcing fibers in powder metal in the desired configuration, sintering, and then hot isostatically pressing to achieve a fully dense structure. Such composites, even after correcting for the increased density, exhibit high temperature tensile strengths, stress rupture strengths, and impact strengths that outperform conventional superalloys several-fold at 1090°C. However, these composites are susceptible to oxidation and require protective coatings.

Although these composites satisfy many of the desired properties of high temperature structural materials, at this time they have not been introduced into service. This appears attributable in part to the increased complexity of fabrication over conventional alloys and the problems associated with introducing fine internal cooling passages, such as are required in modern turbines.

Glass Matrix

Glass (qv) is a convenient matrix candidate since it is an inorganic thermoplastic material. It is a relatively inert material and lends itself to some of the composite-processing methods applicable to polymers such as melt infiltration and compression molding. However, glass has an elastic modulus that is comparable to that of the reinforcing fibers in most cases and a failure strain that is less than that of the fibers. The reverse is more commonly true for the case of the polymers. Therefore, the strength of reinforcing fibers can only be utilized up to the failure strain of the matrix. In a well-bonded system, cracks originating in the matrix also propagate through the fibers; in others the cracks do not penetrate into the fibers, rather the fibers bridge the crack. However in the latter case, transverse strength is markedly poor and the fibers may be subject to environmental attack unless otherwise protected.

Carbon-Fiber Reinforcement. Glass has been used as a matrix material in combination with carbon, typically by infiltrating the fibers with a slurry of finely ground glass. After drying, hot pressing is used to consolidate the composite. Borosilicates, silica (qv), plus a wide range of other glass compositions, including crystallizable glasses, have been used or proposed (33–35). Fiber loadings as great as 70 vol % fiber have been reported. Using 60 vol % of Hercules HMS graphite fiber in combination with Corning 7740 borosilicate results in a composite having an elastic modulus of 234 GPa (34×10^6 psi), a parallel flexural strength of 966 MPa (140,000 psi) at RT. These properties are maintained up to 600°C, the softening temperature of the glass, where the strength increases over a narrow temperature range, and then decreases abruptly. At 100 h exposure to air at elevated temperature, the strength is 90% of its initial value at 450°C, and 70% at 540°C. Other work has shown a 50% loss of fiber strength in 40 min at 1000°C. Although the susceptibility to oxidation is substantially influenced by the degree of intimate contact achieved during the fabrication process, oxidation protection of composites will undoubtedly be required for extended service at elevated temperature.

The fracture toughness of glass–carbon composites is noteworthy. Work of fracture of as great as 23.5 kJ/m^2 (11.9 ft·lb/in^2) has been reported (36). This value is comparable with that of wood and of polymers, such as poly(methyl methacrylate), and is four thousand times greater than that of the glass itself.

Properties in the transverse direction of unidirectional composites have not been reported but can be expected to be low relative to those of metal-matrix composites or of the unreinforced matrix. However, cross-plying can be used to improve properties.

Oxide and Silicon Carbide-Fiber Reinforcement. Metals and wires although bondable to glass and themselves ductile are not especially effective reinforcements of brittle matrices. They can provide utility in preventing the broken composite from falling apart. However, high strength, high stiffness, and oxidation-resistant fibers, such as sapphire and silicon carbide, are potentially useful and are readily wetted by molten glass.

Preliminary studies of DuPont Fiber FP aluminum oxide filament with various glasses show an improvement in the strength of composites that is approximately linear with fiber content (37). At 30 vol % fiber in an aluminosilicate matrix, the average flexural strength is 276 MPa (40,000 psi). However, with fused silica as the matrix lower strengths resulted. These strengths are substantially below the rule of mixtures expectation. This is to be expected when matrix failure is strength-limiting.

In contrast, silicon carbide fibers in a crystallizable glass matrix have led to behaviors similar to those for carbon-fiber reinforcement. A composite consisting of about 50 vol % fibers in a glass having substantially the cordierite composition ($2MgO.2Al_2O_3.5SiO_2 + TiO_2$) resulted (38) in a material having a strength of 690 MPa (100,000 psi) and a fracture energy of 20 kJ/m^2 (1.8 Btu/ft^2). However, these properties required careful heat treatment which apparently resulted in substantial debonding between the fiber and the matrix. This combination of strength, fracture energy, and oxidation resistance is unique among ceramic materials.

Ceramic Matrix

These composites offer potential for overcoming some of the inherent flaw sensitivity of monolithic ceramics (qv), for improving the thermal shock resistance of otherwise very serviceable high temperature materials, and for facilitating fabrication in some cases (see also Glass ceramics). Properties can be expected to be very similar to the case of glass-matrix composites. The combination of matrix brittleness, shrinkage during consolidation if sintering is involved, reactivity, and the generation of internal stresses due to thermal expansion mismatch have proven to be formidable problems. Consequently, there has been much speculation, some preliminary investigation, and even less comprehensive work in this class of high temperature composites. A notable exception is the silicon–silicon carbide system discussed below.

Metal-Fiber Reinforcement. The primary benefit of metal-wire reinforcement of ceramic matrices to date has been to maintain macroscopic structural integrity and load-carrying capacity even though the matrix may have cracked. Thus, the work needed to actually separate a part into several pieces, as in a Charpy impact test, is increased. For silicon nitride, which is potentially technologically important as a turbine ceramic material, incorporation of 25 vol % of 0.64-mm-diameter tantalum has been found to increase the Charpy impact strength 36-fold at RT and 10-fold at 1300°C (39). Significantly, the acoustic damping provided by the metal also raised the threshold for detectable damage several fold. Similarly, the addition of tungsten and of molybdenum to several ceramic bodies including aluminum oxide and mullite substantially increases thermal shock resistance (40–41). Unfortunately, in most high temperature applications the metal fibers are susceptible to oxidative degradation whenever cracks or surfaces expose the metals (42). Carbon-fiber reinforcement is subject to the same problem.

Reinforcement in Oriented Eutectics. Lamellar or fiber-reinforced ceramic composites can be grown *in situ* in analogy to the oriented eutectic metallic structures (see High temperature alloys). The governing principles are generally applicable and are concerned with minimization of surface energy and the establishment of a stable thermodynamic potential gradient at the growth front (43). The generally higher melting temperatures, lower thermal conductivities, dependence of stoichiometry on redox conditions, and larger volumetric changes on solidification of ceramic systems affect the details of the process. Structures very much like those in the case of all-metallic systems have been achieved. More than thirty eutectic combinations with such metals as Cr, Nb, Ta, Mo, and W have been identified as providing reinforcement capabilities in conjunction with ceramic major phases such as $(Al,Cr)_2O_3$, CeO_2, Cr_2O_3, UO_2, and ZrO_2 (44). In addition, various all-ceramic systems, such as ZrO_2–MgO, NiO–CaO, and Al_2O_3–ZrO_2, have been shown to lead to similar structures (45). Fiber diameters of 1–5 μm can be achieved in this way. Among the benefits of such composites are improvements in fracture toughness, greater damage tolerance, and improved high temperature strength.

Silicon–Silicon Carbide Composite. This system is unusual in several respects. Elemental silicon melts at 1410°C and, like water, increases in volume when it freezes. It exhibits some highly strain-rate-sensitive ductility at elevated temperature. It is very reactive towards a wide range of materials but forms a protective oxide. In the liquid state, it reacts readily with carbon to form silicon carbide. Carbon fibers, when infiltrated with molten silicon, can be substantially transformed into silicon carbide, preserving the general geometry of the original carbon, and resulting in the Silcomp silicon–silicon carbide composite material (46). The grain size of the carbide is of the order of the typical diameter of 8 μm of available carbon (or graphite) fibers. Hence, upon conversion the fibers assume the appearance of a coarsely crystalline string which is embedded in a silicon matrix. By shaping and controlling the orientation and packing of the starting carbon fibers upon infiltration with silicon, a fully dense composite of controlled microstructure results. Density, strength, modulus, and degree of anisotropy of the resulting composite can be predetermined to a considerable degree (47). A wide range of shapes has been produced including bars, rings, tubes, sheets, and numerous prototype turbine shapes. This material has performed well as a gas-combustion chamber material since it is capable of sustaining temperatures up to ca 1300°C (48). Typical properties of various grades of this material are shown in Figures 4 and 5, in which the designations TH, CD, and F refer to unidirectional, woven, and omnidirectional materials containing ca 82, 70, and 25 vol % of silicon carbide, respectively. Those end-member compositions containing high volume fractions of silicon carbide

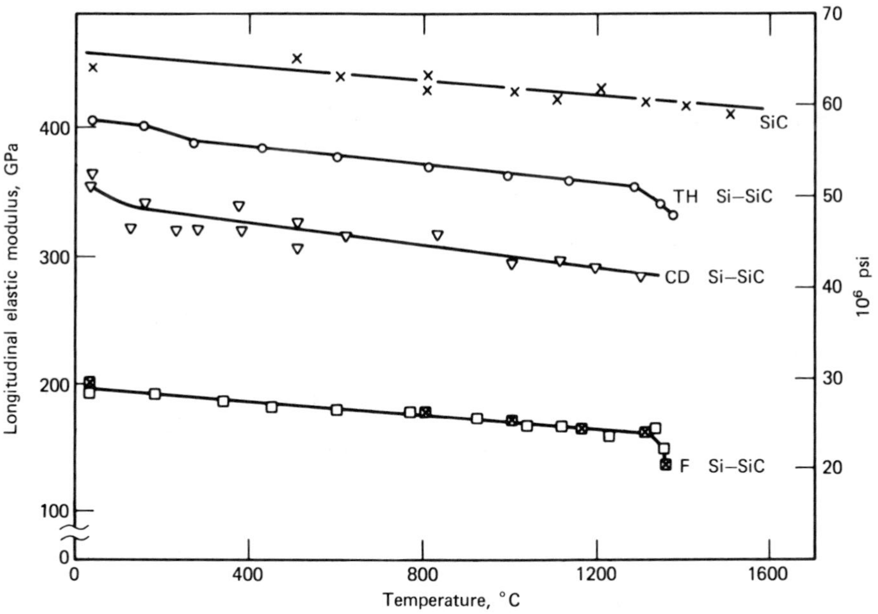

Figure 4. Modulus as a function of temperature for various grades of Silcomp Si–SiC. TH = unidirectional; CD = woven; and F = omnidirectional.

are similar in characteristics to reaction-bonded silicon carbide produced by other processes.

Ceramic and Carbon Fibers. The limited strain tolerance of ceramic fibers makes their use in combination with ceramic matrices difficult to achieve. Matrix shrinkage during consolidation is a particular problem, which can only be avoided in a few cases. For example, it is possible to take advantage of the process for producing reaction-bonded silicon nitride. This occurs by the nitridation of a porous compact of silicon in which the volume increase is accommodated by expansion into the interparticulate voids, without change of external shape or dimension. Thus, by mixing silicon carbide filaments with silicon powder and subsequent reaction with gaseous nitrogen, SiC-reinforced Si_3N_4 composites have been produced (49) with fiber contents of 5–25 vol %. Severalfold strength improvements over the unreinforced material were noted along with a very substantial increase of work of fracture, reaching a value of 0.9 kJ/m^2 (0.08 Btu/ft^2). This behavior is analogous to the toughening in a glass matrix (see above). However, this composite remains porous.

In other cases where a fully dense structure is desired, hot pressing of a fiber-containing starting material has been used, although some damage occurs to the fibers. By this technique dense composites have been achieved. In one study using silicon carbide and silicon nitride, both as matrix phases and as whisker reinforcement phases in all combinations, showed little benefit attributable to the reinforcement (50). A similar observation was made with 7-μm-dia SiC fiber hot-pressed in a Si_3N_4 matrix (51). However, when carbon fiber was included as the reinforcement, improvement of fracture toughness was noted. The same improvement was noted in independent studies including alumina (33,52) and magnesia (33) as matrices in combination with carbon fiber. However, when silicon carbide, or mullite fibers were incorporated into

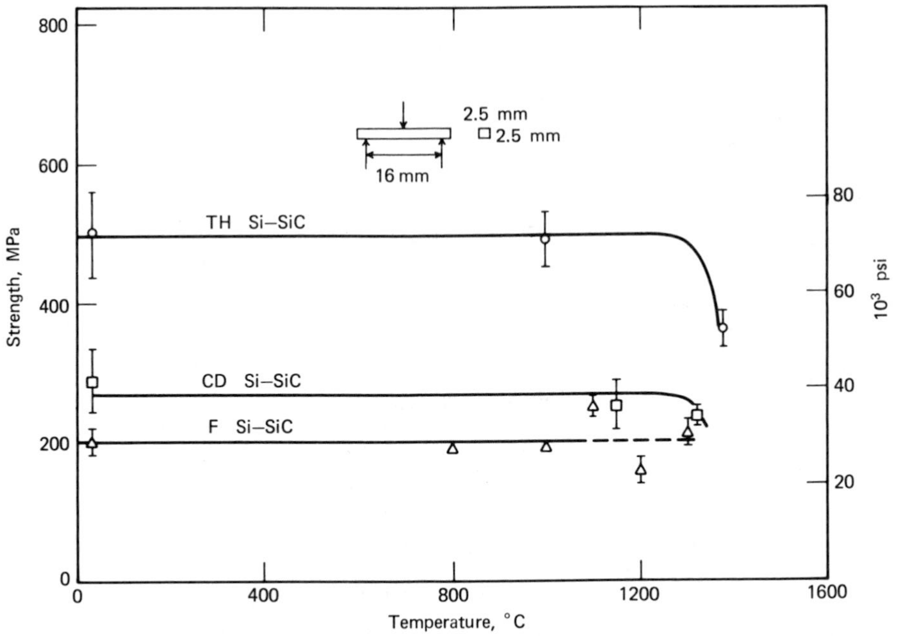

Figure 5. Strength as a function of temperature for three grades of Silcomp Si–SiC. TH = unidirectional; CD = woven; and F = omnidirectional.

alumina, no benefit (52) was noted. It appears that fiber decoupling from the matrix, as occurs in the case of carbon or graphite fibers, allows fiber pullout to occur on fracture, resulting in toughness. When the fiber-matrix bonding is strong, the cracks propagate through both the fiber and matrix relatively unimpeded.

Nonfiber Composites

Particulate and layer reinforcements or second-phase additions are also often used in high temperature materials. Laminated-metal structures, such as titanium-clad steel, are used for chemical-process reaction vessels. Other metal combinations are used as thermostat-sensor elements. These composite structures are described in detail in ref. 52.

Particulate composites are particularly convenient to fabricate by mixing, pressing, and sintering of two particulate materials. Cermets, which are a compacted mixture of metal and ceramic, usually oxide, constituents, typically do not exhibit a high degree of synergistic benefit. The ceramic phase reduces the ductility and creep characteristic of the metal phase but generally leads to brittleness and difficulty in secondary fabrication or shaping capability. The extremely fine-grained dispersions in alloys such as TD (thoria-dispersed) nickel and TD nichrome offer advantages over the metals themselves at the upper end of their use range.

Mixtures of ceramics or nonductile intermetallics are particularly interesting because in some cases they offer extreme temperature capability, eg, as rocket nozzle materials. In other cases, they hold some promise for overcoming to some degree the brittleness characteristic of monolithic ceramics. They have also been used as model systems to understand the factors affecting brittle failure.

Mixtures of zirconium diboride, silicon carbide, and graphite can be hot-pressed at a typical temperature of 2200°C to produce a fully dense material that exhibits good mechanical strength and thermal shock resistance (54). Another promising system is boron carbide plus carbon (51). Such materials are effective as high temperature materials in air because they react to form refractory oxides plus a glass phase, which volatilizes but also serves to form an impervious oxidation barrier layer. Other applications include hypersonic reentry-vehicle components such as leading edges and nose tips (see Ablative materials).

Some particulate composites have excellent thermal shock resistance (51). These include among others the experimental Al_2O_3–BN composites as well as commercially produced MgO–SiC, BeO–SiC, and BN–B_4C composites by the Ceradyne Corp. and BN–SiO_2 by the Carborundum Corp.

The presence of unstabilized zirconium oxide as a dispersed phase in a matrix of stabilized zirconium oxide (55), aluminum oxide (56), or thorium oxide (57), etc, results in a substantial enhancement of the fracture toughness of the matrix. This is owing to the local compression in the matrix and/or local microcracking at the crack–tip region when the high temperature tetragonal modification transforms abruptly (58) to the low temperature monoclinic form with a 1.4% linear expansion. The amount and particle distribution of the transformable ZrO_2 must be controlled carefully so that the resulting internal stress does not cause the material to fracture. Using this approach a doubling of fracture toughness has been achieved.

BIBLIOGRAPHY

" .amic Composite Armor" in *ECT* 2nd ed., Suppl. Vol., pp. 138–157, by George Rugger (Fabrication and plastics), Dept. of the Army and John R. Fenter (Ceramics and adhesives), Dept. of the Air Force.

1. R. A. Signorelli, "Metal Matrix Composites for Aircraft Propulsion Systems" in E. Scala and co-eds., *Proceedings of the 1975 International Conference on Composite Materials,* Vol. I, Metallurgical Society of AIME, New York, 1976, p. 411.
2. J. N. Fleck, *Bibliography on Fibers and Composite Materials—1969–1972, MCIC Report 72-09,* National Technical Information Service, Springfield, Va., 1972 (AD 746214).
3. D. M. Johnson, *Bibliography on Fibers and Composite Materials—1972–1978, MCIC Report 78-38,* Battelle's Columbus Laboratories, Columbus, Ohio, 1978.
4. I. Petker, *SAMPE Q.* **1,** 7 (1972).
5. J. I. Jones, *J. Macromol. Sci. Rev. Macromol. Chem.* **C2,** 2 (1968).
6. B. L. Lee and F. J. McGarry, *Study of Processing and Properties of Graphite Fiber/High Temperature Resin Composites,* Army Materials and Mechanics Research Center, Watertown, Mass., 1976.
7. T. T. Serafini, "Processible High Temperature Resistant Polymer Matrix Materials" in ref. 1, p. 202.
8. J. S. Evangelides, R. A. Meyer, and J. E. Zimmer, *Investigation of the Properties of Carbon-Carbon Composites and Their Relationship to Nondestructive Test Measurements, AFML-Tr-70-213,* Part II, Aug. 1971.
9. A. C. Parmee, Procurement Executive, Ministry of Defence, Rocket Propulsion Establishment, Westcott, UK, *paper presented at the Symposium on Materials and Space Technology IV, The British Interplanetary Society, London, Eng., Sept. 1971.*
10. "Carbon Composite Technology" in *Proceedings of the 10th Annual Symposium of the New Mexico Section of ASME and University of New Mexico College of Engineering, Albuquerque, N. Mex. Jan. 29–30, 1970.*
11. D. L. Schmidt, *SAMPE J.,* 6 (May/June 1972).
12. Ref. 3, pp. 47–50 for additional references.
13. Ref. 4, pp. 240–241 for additional references.

14. A. G. Metcalfe, ed., *Interfaces in Metal Matrix Composites* in L. J. Broutman and R. H. Krock, eds., *Composite Materials,* Vol. 1, Academic Press, Inc., New York, 1974, p. 1.
15. A. G. Metcalfe and M. J. Klein in ref. 11, p. 125.
16. A. G. Metcalfe in K. G. Kreider, ed., *Metallic Matrix Composites* in L. J. Broutman and R. H. Krock, eds., *Composite Materials,* Vol. 4, Academic Press, Inc., New York, 1974, p. 269.
17. K. G. Kreider and K. M. Prewo in ref. 16, p. 399.
18. *Metals Handbook,* 8th ed., Vol. I, American Society of Metals, Novelty, Ohio, 1961, p. 945.
19. E. G. Kendall in ref. 16, p. 319.
20. W. C. Harrigan, Jr., and R. H. Flowers, *Graphite-Metal Composites: Titanium/Boron Vapor-Deposit Method of Manufacture, ATR-78(8162)-1,* The Aerospace Corporation, El Segundo, Calif., Jan. 1978.
21. M. F. Amateau, W. C. Harrigan, Jr., and E. G. Kendall, *Mechanical Properties of Aluminum Alloy–Graphite Fiber Composites, SAMSO-TR-75-55,* The Aerospace Corporation, El Segundo, Calif., Feb. 1975.
22. K. G. Kreider in ref. 16, p. 1.
23. A. A. Baker and P. W. Jackson, *Glass Technol.* **9,** 36 (1968).
24. K. M. Prewo, *Fabrication and Evaluation of Low Cost Alumina Fiber Reinforced Metal Matrices, R77-912245-3,* United Technologies Research Center, East Hartford, Conn., May 1977.
25. A. R. Champion and co-workers, *paper presented at the Second International Conference on Composite Materials, Toronto, Canada, April 16–20, 1978.*
26. S. Yajima and co-workers, *Nature (London)* **261,** 683 (1976).
27. P. E. Gruber, *Proceedings of 22nd National Conference of the Society for the Advancement of Material and Process Engineering (SAMPE), San Diego, Calif., April 1977.*
28. F. Ordway, P. J. Lare, and R. A. Hermann, *Silicon Carbide Whisker–Metal Matrix Composites, AFML-TR-71-252,* March 1972.
29. T. C. Tsareff, Jr., G. R. Sippel, and M. Herman, *Proceedings of Symposium of the Metallurgical Society of AIME, Pittsburgh, Pa., May 12–13, 1969, Battelle Memorial Institute Report No. DMIC-243,* Columbus, Ohio, May 1969, pp. 38–42.
30. R. E. Tressler in ref. 14, p. 285.
31. R. L. Mehan and M. J. Noone in ref. 16, p. 159.
32. R. A. Signorelli in ref. 16, p. 229.
33. D. H. Bowen and co-workers, *Proceedings of Conference on Mechanical Properties of Ceramic Fibres and Composites, Materials and Test Group of the British Institute of Physics and the Basic Science Section of the British Ceramic Society, London, Eng., Dec. 15–16, 1971.*
34. U.S. Pat. 3,681,187 (Aug. 1, 1972), D. H. Bowen, R. A. Sambell, K. A. D. Lambe, and N. J. Mattingley (to United Kingdom Atomic Energy Authority).
35. K. M. Prewo, J. F. Bacon, and D. L. Dicus, *paper presented at the 24th National Meeting of the SAMPE, San Francisco, Calif., May 8–10, 1979.*
36. K. M. Prewo and J. F. Bacon, *Proceedings of the Second International Conference on Composite Materials, Toronto, Canada, April 1978.*
37. J. F. Bacon, K. M. Prewo, and R. D. Veltri, in ref. 36.
38. J. Aveston, *Proceedings of the Properties of Fiber Composites Conference, National Physical Laboratory, Teddington, Eng., Nov. 4, 1971.*
39. J. J. Brennan, "Increasing the Impact Strength of Si_3N_4 Through Fibre Reinforcement" in *Special Ceramics 6,* The British Ceramic Research Association, Stoke-on-Trent, UK, June 1975.
40. J. J. Swica and co-workers, "Metal Fiber Reinforced Ceramics" in *Wright Air Development Center (WADC) Technical Report 58-452,* Part II, Armed Services Technical Information Agency (ASTIA), Arlington Hall Station, Arlington, Va., Jan. 1960.
41. J. R. Tinklepaugh and co-workers in ref. 40, Part III, Nov. 1960.
42. D. G. Miller, R. H. Singleton, and A. V. Wallace, *Am. Ceram. Soc. Bull.* **45,** 513 (1966).
43. R. L. Ashbrook, *J. Am. Ceram. Soc.* **60,** 428 (1977).
44. J. Briggs and P. E. Hart, *J. Am. Ceram. Soc.* **59,** 530 (1976).
45. W. J. Minford, R. C. Bradt, and V. S. Stubican, *J. Am. Ceram. Soc.* **62,** 154 (1979).
46. W. B. Hillig and co-workers, *Am. Ceram. Soc. Bull.* **54,** 1054 (1975).
47. R. L. Mehan, *J. Mater. Sci.* **13,** 358 (1978).
48. W. B. Hillig, "The Application of Silcomp Composite Materials to Turbine Systems" in *Selective Application of Materials for Products and Energy,* Vol. 23, SAMPE, Azusa, Calif., 1978.
49. M. W. Lindley and D. J. Godfrey, *Nature (London)* **229,** 192 (1971).

50. W. H. Rhodes and R. M. Cannon, Jr., *High Temperature Compounds for Turbine Vanes, NASA CR-120966 (AVSD-0336-72-CR),* National Technical Information Service, Springfield, Va., Sept. 1972.
51. R. W. Rice and co-workers, *Proceedings of the Conference on Composites and Advanced Materials, American Ceramics Society, Cocoa Beach, Fla., Jan. 23–25, 1978.*
52. J. Barta, W. B. Shook, and G. A. Graves, *Proceedings of the Conference on Mechanical Properties of Ceramic Fibres and Composites, Materials and Test Group of the British Institute of Physics and the Basic Science Section of the British Ceramic Society, London, Eng., Dec. 15–16, 1971.*
53. E. S. Wright and A. P. Levitt in ref. 16, p. 38.
54. E. L. Strauss and T. F. Kiefer, *J. Am. Ceram. Soc.* **58,** 399 (1975); J. R. Fenter, *SAMPE Q.* **2**(4), 1 (1971).
55. R. C. Garvie and P. S. Nicholson, *J. Am. Ceram. Soc.* **55,** 152 (1972).
56. N. Claussen, *J. Am. Ceram. Soc.* **59,** 49 (1976).
57. R. M. Cannon and T. D. Ketcham, *Am. Ceram. Soc. Bull.* **58,** 338 (1979).
58. E. C. Subbarao, H. S. Maiti, and K. K. Srivastava, *Phys. Status Solidi A* **21,** 9 (1974).

WILLIAM B. HILLIG
General Electric Company

LAMINATED AND REINFORCED METALS

The desire to satisfy diverse and competing design requirements in a cost-effective manner has been coupled with the drive for energy conservation as a result of the energy shortage and increasing energy costs, and the need to develop cost-effective alternatives for critical materials.

Metallic laminates and fiber-reinforced metals are serious contenders for structural applications. In this article, these laminates are called metallic-matrix laminates (MMLs). MMLs are relatively expensive at this time (1980) because fibers, eg, boron fibers, are expensive and the manufacturing processes are time consuming and costly. They have been used mainly in aerospace structures. However, MMLs will find more extensive use as energy and other design considerations, including scarcity of critical materials, override the material costs.

A large body of information has been generated about MMLs over the last fifteen years. Significant developments of MMLs are reported in the proceedings of the Society for the Advancement of Materials and Processing Engineering (SAMPE). These proceedings include papers that are presented at the two SAMPE annual meetings (spring and fall). Three recently compiled bibliographies with abstracts (1–3) cover technical articles and government reports that have been published since about 1965. An extensive review of metal-matrix composites covering developments up to 1972 is in ref. 4. Boron-fiber reinforced aluminum (boron–aluminum composites) and graphite-fiber reinforced aluminum (graphite–aluminum composites) are reviewed in ref. 5. The present article is a selective review of the state-of-the-art of MMLs, it describes only material that is not directly defense-related. The emphasis of the review is on design–analysis procedures for structural components that have been made from MMLs (see Composite Materials).

Definitions of Selected Laminates

The types of MMLs that are reviewed herein include those made from fiber-reinforced metals, superhybrids, and those made from layers of different metals. Fiber-reinforced metals consist of unidirectional-fiber-composite (UFC) laminates, as depicted schematically in Figure 1**a**, and angleplied (APL) laminates, Figure 1**b.** In both UFC and APL laminates, metallic foils may be used between plies to enhance certain mechanical properties as is discussed later. Superhybrid composites (SHC) consist of outer metallic foils, boron–aluminum plies (B–Al), graphite-fiber–resin (UFC) inner or core plies, and adhesive film between these as shown in the photomicrograph in Figure 1**c.** Metallic laminates consist usually of alternate layers from two or more metals as depicted schematically in Figure 1**d.** The various procedures that are used to fabricate these laminates are described below. The combinations of materials that are used to make these laminates also are described below.

The basic unit used to study, design and fabricate UFC laminates is the single layer (ply, monolayer, lamina) which consists of stiff, strong fibers embedded in a metal matrix. The fibers and the matrix are generally called the constituent materials, or constituents, of the laminate (composite). Various constituents that have been used to make UFC are summarized alphabetically under the heading fiber-reinforced-metal laminates (first two columns, Table 1). A large number of constituent materials are used for both fibers and matrices. The materials for fibers range from alumina to whiskers. Those for matrices range from aluminum to superalloy. The constituents used thus far for SHC are those summarized under superhybrids in Table 1. The constituents that have been used for metallic laminates are summarized also in Table 1. An extensive list of constituent combinations for metallic laminates is tabulated in ref. 6, and constituent combinations for metallic laminates made by explosive bonding are tabulated in ref. 7.

Mechanical and physical properties of constituent fiber reinforcements for MMLs are summarized in Table 2. Corresponding properties for metal matrices and metallic constituents for MMLs are summarized in Table 3.

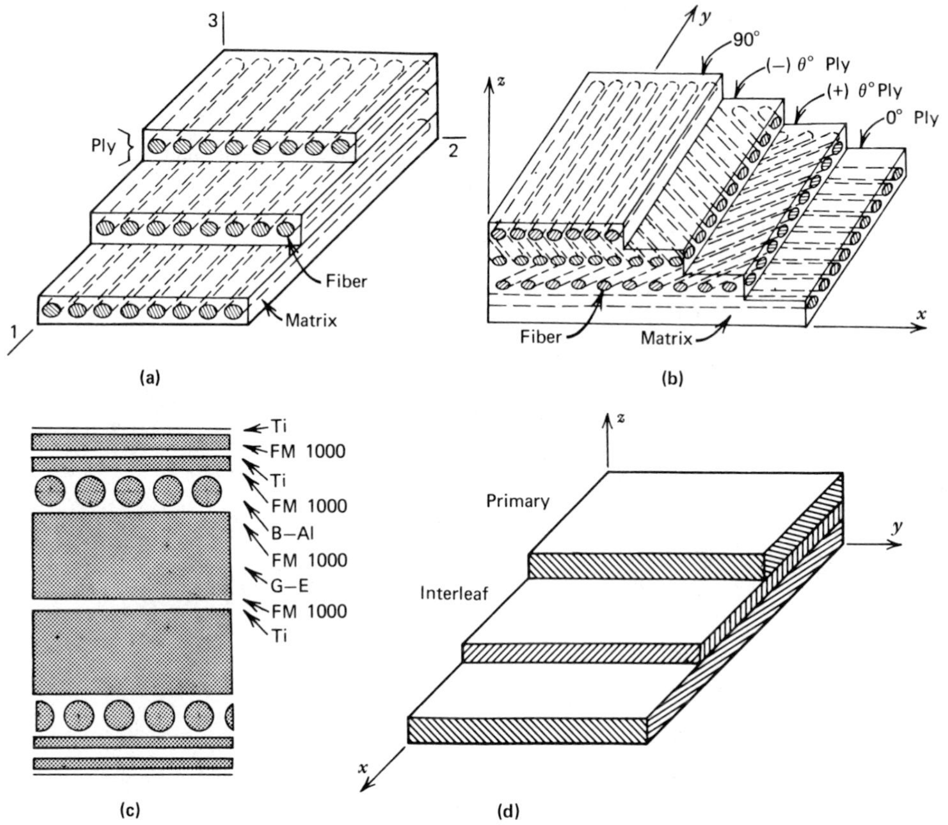

Figure 1. General types of metal-matrix laminates: (**a**) unidirectional fiber composite (UFC); (**b**) angleplied laminate (APL); (**c**) superhybrid composite (SCH); (**d**) metal–metal laminate (MML).

Hybrids

A general definition for a hybrid composite is a composite that combines two or more different types of fibers in the same matrix, or one fiber type in two different matrices or combinations of these (ICCM, International Conference on Composite Materials) (10). Superhybrids (Fig. 1**c**) are a generic class of composites that combine appropriate properties of fiber–metal-matrix composites, fiber–resin-matrix composites and/or metallic plies in a predetermined manner in order to meet competing and diverse design requirements (11–12). Tiber hybrids are trivial names for titanium–beryllium adhesively bonded metallic laminates (13).

Boron-fiber-reinforced 1100, 2024, 5052, or 6061 aluminum alloys have been investigated for use in fan blades for aircraft engines. Different diameter boron fibers (142 and 203 μm) may be included in the same hybrid. The high impact resistance of the 203-μm dia boron fiber in the 1100 aluminum alloy matrix (Fig. 2) is combined with the high transverse and shear properties of the 142- or 203-μm dia boron fiber in either a 2024, 5052, or 6061 aluminum alloy matrix. Fan blades made from some of these hybrids and subjected to the impact of a small bird (85 g) are shown in Figure 3. The advantages and disadvantages of these types of hybrids are described in ref. 14, and their use for fan blades is described in ref. 15.

Table 1. Summary of Constituent Materials for Metallic Laminates

Fiber-reinforced metal laminates

Fiber	Metal
alumina (FP)	aluminum
	lead
	magnesium
beryllium	titanium
boron	aluminum
	magnesium
	titanium
borsic	aluminum
	titanium
graphite	aluminum
	niobium
	copper
	lead
	magnesium
	nickel
	tin
	zinc
molybdenum	superalloy
silicon carbide	aluminum
	superalloy
	titanium
steel	aluminum
	nickel
tantalum	superalloy
tungsten	niobium
	superalloy

Superhybrids

Metal-matrix composite	
Fiber	Matrix
boron	aluminum
Resin-matrix composite	
Fiber	Matrix
graphite	epoxy
	polyimide
kevlar	epoxy
S-glass	epoxy
Metal foil	
titanium	

Metal–metal laminates

Primary	Interleaf
aluminum	aluminum
beryllium	titanium
steel	steel
titanium	aluminum
	titanium
tungsten	copper
	superalloy
tungsten	tantalum

Superhybrids have been developed primarily for use in fan blades of high-bypass-ratio turbojet engines. These types of superhybrids generally have: longitudinal strengths and stiffness comparable to advanced fiber composites; transverse flexural strength comparable to titanium; impact resistance comparable to aluminum; transverse and shear stiffness comparable to aluminum; and density comparable to glass-fiber–resin composites (11). In addition, superhybrids are notch insensitive and do not degrade when subjected to thermal fatigue (12). Impact-resistance data of superhybrids, other fiber composites, and some metals are summarized in Table 4. The high velocity-impact resistance of superhybrid wedge-type cantilever specimens relative to other composites and titanium is shown in Figure 4 (15). Large fan blades made by using superhybrids-shell–titanium-spar (either leading edge or center) have been described (16).

Experimental data generated at the NASA Lewis Research Center showed that tiber hybrids can be made which have: moduli equal to that of steel; tensile fracture stress comparable to the yield strength of titanium; flexural fracture stress comparable to the ultimate strength of titanium; and density comparable to aluminum (13). The relatively high stiffness of tiber hybrids and their relatively low density compared to

Table 2. Typical Properties of Constituent Fiber Reinforcements for Metal–Metal Laminates (Along the Fiber)[a]

Fiber	Density, g/cm^3 [b]	Mp, °C	Heat capacity, kJ/(kg·K)[c]	Thermal cond., W/(m·K)[d]	Coeff. of thermal exp., 10^{-6}/°C	Tensile strength, MPa[e]	Modulus, GPa[e]	Dia, μm	Remarks
boron on tungsten	2.49	2100	1.3	38	5.0	3,620	400	102–203	monofilament
borsic	2.71	2100	1.3	38	5.0	3,100	400	102–203	monofilament
boron on carbon	2.21	2100	1.3	38	5.0	3,450	360	102–203	monofilament
graphite									
PAN HM	1.86	3650	0.7	1003	−1.1	2,210	380	7	10,000 filaments per ton
PAN HTS (T300)	1.74	3650	0.7	1003	−1.1	2,340	210	8	3000 filaments per yarn
rayon (T50)	1.66	3650	0.7	1003	−1.1	2,170	390	6	1440 filaments per 2 ply yarn
Thornel 75 (T75)	1.83	3650	0.7	1003	−1.1	2,660	520	5	1440 filaments per 2 ply yarn
pitch (type P)	1.99	3650	0.7	1003	−1.1	1,380	340	5–10	2000 filaments per yarn
pitch UHM	2.05	3650	0.7	1003	−1.1	2,410	690	11	2000 filaments per yarn
silicon carbide on tungsten	3.32	2690	1.2	16	4.3	3,100	430	102–203	monofilament
silicon carbide on carbon	3.04	2690	1.2	16	4.3	3,450	400	102	monofilament
beryllium	1.86	1280	1.9	150	11.5	970	290	127	monofilament
alumina (FP)	3.96	2040			8.3	1,520	380	20	210 filaments per yarn
glass-S	2.49	840	0.7	13	5.0	4,140	80	9	1000 filament per strand
-E	2.49	840	0.7	13	5.0	2,700	70	9	1000 filaments per strand
molybdenum	1.02	2620	0.3	145	4.9	660	320	127	monofilament
steel	7.75	1400	0.5	29	13.3	2,070	210	127	monofilament
tantalum	16.88	3000	0.2	55	6.5	1,520	190	508	monofilament
tungsten	19.38	3400	0.1	168	4.5	3,170	390	381	monofilament
whisker ceramic[f] Al_2O_3	3.96	2040	0.6	24	7.7	42,760	450	10–25	
metallic (Fe)	7.75	1540	0.5	29	13.3	13,100	200	127	

[a] Adapted from ref. 8.
[b] To convert g/cm^3 to $lb/in.^3$, divide by 27.68.
[c] To convert kJ/(kg·K) to Btu/(lb·°F), divide by 4.184.
[d] To convert W/(m·K) to Btu·ft/(ft^2·h·°F), divide by 1.729.
[e] To convert MPa to psi, multiply by 145; to convert GPa to psi, multiply by 145,000.
[f] Ref. 9.

Table 3. Typical Properties of Metal Matrices and Metallic Constituents for Metal–Metal Laminates

Metal	Density, g/cm^3 [a]	Mp, °C	Heat capacity, kJ/(kg·K) [b]	Thermal cond., W/(m·K) [c]	Thermal exp. coeff., 10^{-6}/°C	Tensile strength, MPa [d]	Modulus, GPa [d]	Remarks
aluminum	2.8	580	0.96	171	23.4	310	70	6061 (T6)
beryllium	1.9	1280	1.88	150	11.5	620	290	annealed
copper	8.9	1080	0.38	391	17.6	340	120	oxygen-free hardened
lead	11.3	320	0.13	33	28.8	20	10	1% Sb
magnesium	1.7	570	1.00	76	25.2	280	40	AZ31B-H24
nickel	8.9	1440	0.46	62	13.3	760	210	nickel 200 hardened
niobium	8.6	2470	0.25	55	6.8	280	100	
steel	7.8	1460	0.46	29	13.3	2070	210	ultra-high strength (MOD.H-11)
superalloy	8.3	1390	0.42	19	16.7	1100	210	Inconel X-750
tantalum	16.6	2990	0.17	55	6.5	410	190	
tin	7.2	230	0.21	64	23.4	10	40	
titanium	4.4	1650	0.59	7	9.5	1170	110	Ti–6 Al–4 V
tungsten	19.4	3410	0.13	168	4.5	1520	410	
zinc	6.6	390	0.42	112	27.4	280	70	alloy agada

[a] To convert g/cm^3 to lb/in.^{3}, divide by 27.68.

[b] To convert kJ/(kg·K) to Btu/(lb·°F), divide by 4.184.

[c] To convert W/(m·K) to Btu·ft/(ft^2·h·°F), divide by 1.729.

[d] To convert MPa to psi, multiply by 145; to convert GPa to psi, multiply by 145,000.

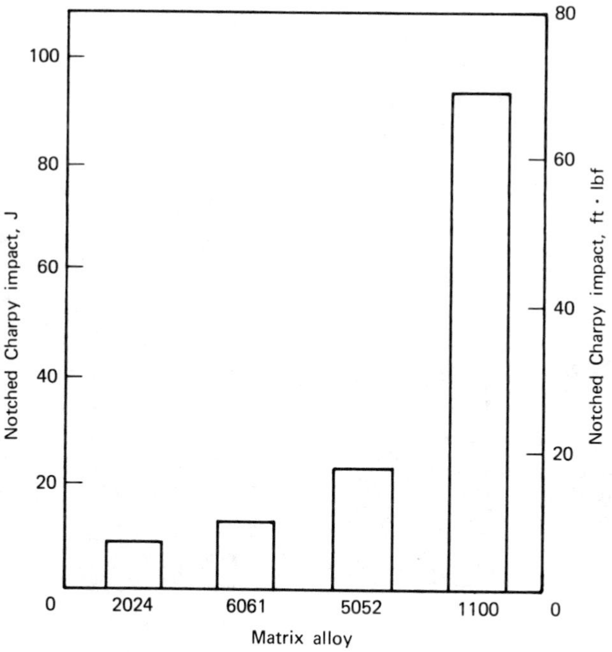

Figure 2. Impact resistance of boron–aluminum unidirectional metallic-matrix laminates (203-μm dia fiber, 0.50 fiber–volume ratio).

conventional metals make them good candidates for compression members in aircraft and space structures. Buckling stresses for plates and shells from tiber hybrids are compared with those from advanced unidirectional composites and from conventional metals in Table 5. Tiber hybrids have superior buckling resistance compared with either advanced composites or conventional metals. Another potential used for tiber hybrids is high-tip-speed fan blades (Fig. 5) for turbojet engines. Finite element analysis results showed that tiber-hybrid fan blades would have higher frequencies and lower tip distortions compared to those made from graphite fiber–resin composites (17). The comparisons for the first five frequencies are summarized in Table 6.

Properties

Physical, thermal, and mechanical properties of fibrous MMLs that have been made are summarized in Table 7. The last three entries in this table are superhybrid composites (Fig. 1**c**) where the core is made from graphite-fiber composite (see Carbon, carbon and artificial graphite), S-glass-fiber composite (see Glass), and Kevlar-fiber composite (see Aramid fibers).

Fibrous MMLs have relatively low transverse *T* strengths ranging from about 35–520 MPa (5,000–75,000 psi).

The transverse moduli range from 13.8 GPa (2×10^6 psi) for superhybrids to 207 GPa (30×10^6 psi) for SiC–Ti. The transverse properties for several fibrous MMLs are not available. These properties as well as those of metallic laminates (Fig. 1**d**), can be predicted approximately by the methods described below.

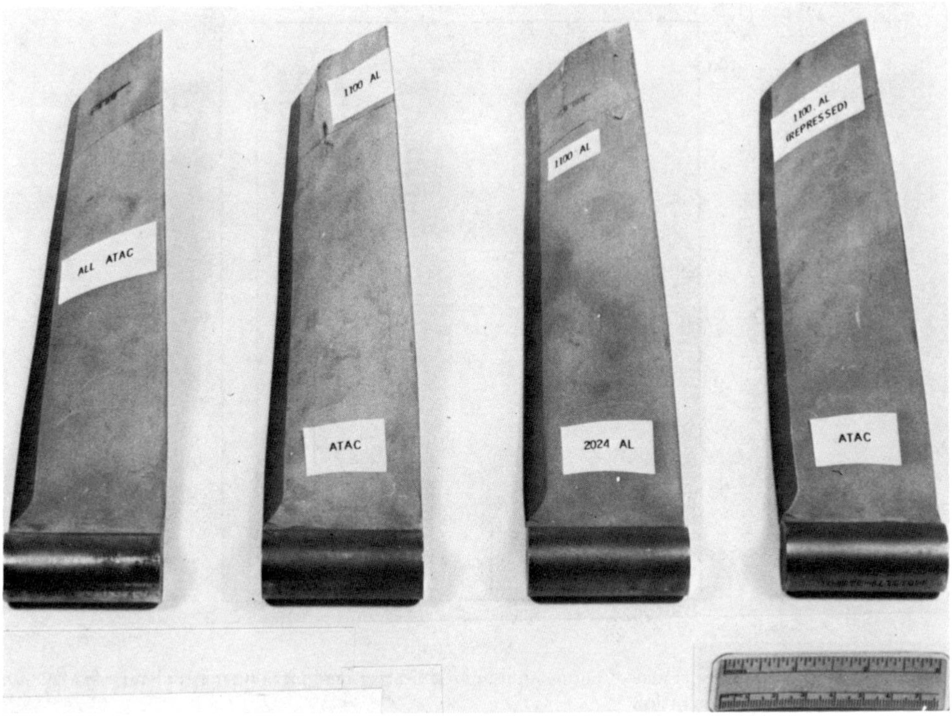

Figure 3. Boron–aluminum metallic-matrix fan blades after impact by a small bird (85 g).

Mechanical thermal and other physical properties are used in selecting fibrous MMLs for possible use in structural components during the preliminary design phases. These properties are then verified by selective testing and are subsequently used in the detailed analysis and the final design phases.

Fabrication Procedures

Laminates or composites from MMLs are fabricated using diffusion-bonding, roll-bonding, coextrusion, explosive bonding, and brazing in general.

Several other methods are used depending on the type of constituent metals to be used in the laminate. These methods include vacuum-infiltration casting, high energy forming, flow-molding, plasma-spraying, hot-pressing, continuous infiltration, power-metallurgy (qv) methods for discontinuous-fiber composites, explosive welding, and superplastic forming (4–5).

In diffusion-bonding, the filament or the interleaf layer (plies) are hot-pressed between layers of the matrix material. The pressure usually is 6.9–20.7 MPa (1000–3000 psi) and the temperature is 455–540°C. In roll-bonding the layers of metal–metal laminates are bonded by mill-rolling under specified heat and pressure. For fiber-reinforced metals, the ply (monolayer) is formed by diffusion-bonding, or any of the other methods, then the laminate of the specified number of plies is made by roll-bonding. In coextrusion, the constituents are assembled as in a billet and are extruded through a given die at specified temperatures and pressures depending on the con-

Table 4. Summary of Thin-Specimen Izod Impact-Strength Results and Comparisons with Some Other Materials (Specimen Nominal Dimensions: 1.27-cm Wide by 1.52-mm Thick)

Laminate type	Constituents	Test direction	Izod impact strength, kJ/m[a] Low	High	Number of specimens
I	Gr–Ep	longitudinal	1.45	1.59	4
		transverse	0.19	0.21	2
II	B–Al[b]	longitudinal	1.23	1.27	2
		transverse	1.01	1.10	2
III	B–Al[c]	longitudinal	0.68	0.96	2
		transverse	0.43	0.71	4
IV	Ti, B–Al	longitudinal	1.09	1.13	2
		transverse	0.79	1.00	4
V	superhybrid (Ti, B–Al, Gr–Ep)	longitudinal	2.82	3.20	2
		transverse	0.82	0.90	2
Other materials					
HT-S–PMR-PI[d]		longitudinal	0.91	0.92	2
		transverse	0.17	0.19	2
glass-fabric–epoxy			1.11	1.13	3
102 μm-dia B–6064 Al		longitudinal	1.13	1.21	2
aluminum 0.6061			3.36	4.07	2
titanium (6 Al–4 V)			11.23	11.38	2

[a] To convert kJ/m to ft·lbf/in., divide by 53.4×10^{-3} (see ASTM D 256).
[b] Diffusion-bonded.
[c] Adhesive-bonded.
[d] PMR = polymerization of monomeric reactants; PI = polyimide.

stituents used. The primary bonding mechanism in the coextrusion process is diffusion-bonding. Coextrusion is particularly suited for round and rectangular bar stocks. In explosive-bonding the constituent metal plies are bonded into a laminate by the high pressure generated through explosive means (see Metallic coatings, explosively clad metals). The amount of charge used is determined by the metallurgical bond required between the plies. This method is especially suitable for fabricating MMLs from metal plies with widely different melting temperatures. In brazing, bonding of the constituent metal plies into a laminate is accomplished by a third metal (brazing foil) which acts as a wetting liquid-metal phase and which has a lower melting temperature than either of the constituent metals (see Solders and brazing alloys). The plies to be bonded are stacked into a laminate with brazing foils between them. Then the temperature is raised between the melting temperature of the brazing foils and the constituent plies and appropriate pressure is applied. Upon solidification, the brazing foil bonds the adjacent constituent plies together into a laminate. Boron–aluminum plies are fabricated at about 600°C and <1.4 MPa (200 psi).

Superhybrids are fabricated by adhesively bonding titanium outer plies over boron–aluminum plies and graphite-fiber–epoxy inner plies (core) with a titanium ply at the center (11–12). The adhesive bond between the metallic plies and between the metallic and composite plies is provided by use of FM 1000 adhesive film approximately 25-μm thick. The bonding process is accomplished under specified pressure and temperature normally used for epoxy-matrix composites. This process consists of a 3-h cure at 150°C and <700 kPa (100 psi). The same bonding process (fabrication procedure) is used to fabricate the nongraphitic superhybrids, the tiber (titanium–beryllium) hybrids and the adhesively bonded metallic-ply laminates.

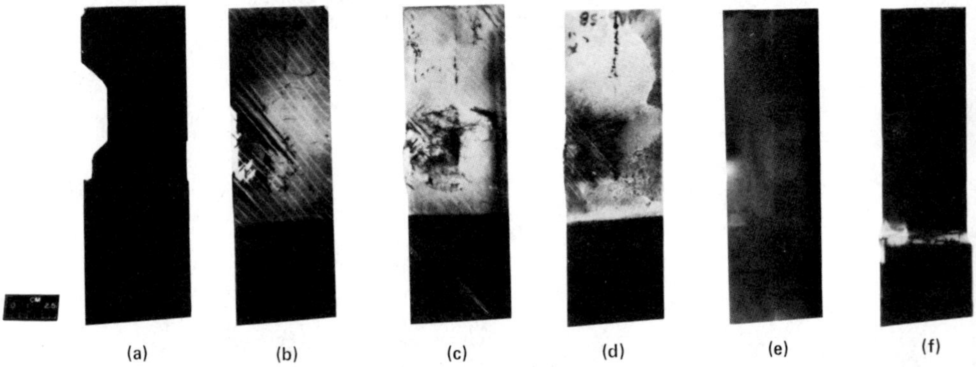

Figure 4. Relative high velocity impact assessment of superhybrids. Projectile was 2.5-cm dia gelatin sphere; specimens (**a**) through (**e**) oriented at 30° incidence angle, specimen (**f**) at 19° incidence angle. To convert kJ/m to ft·lbf/in., divide by 0.0534 (see ASTM D 256).

Specimen	Construction	Leading edge thickness, μm	Midchord thickness, mm	Projectile velocity, m/s	Slice, %	Kinetic energy per thickness, kJ/m
(**a**)	graphite–epoxy	737	3.81	252	55	43.45
(**b**)	graphite–glass–epoxy	711	3.73	284	50	55.94
(**c**)	boron–glass–polysulfone–graphite–epoxy	737	4.09	285	40	42.76
(**d**)	boron–glass–epoxy	762	4.14	269	40	37.04
(**e**)	titanium–boron–aluminum–graphite–epoxy superhybrid	635	3.96	281	50	63.63
(**f**)	solid titanium (6 Al–4 V)	356	3.89	222	50	23.65

Design–Analysis Procedures

Design requirements for structures may be: maximum strength with light weight, long life service with minimum strength degradation, notch or other defect insensitivity with high stiffness, impact resistance with high stiffness, damage tolerance with high stiffness and low cost. Many recent designs for aircraft, spacecraft, automobiles, complex machine parts, and electronic computers have design requirements that include several of the above as well as ease of fabrication.

Design of structures with metal-matrix laminates means cost-effective use of MMLs in structures and structural parts, ie, use of MMLs collectively satisfy all of the design requirements better than other materials for a given structure or structural part.

Alternative design concepts cannot be evaluated by trial and error in the building or fabricating of complex structures. Therefore, alternative design concepts for a specific case are evaluated on paper. The formal way for evaluating structural concepts with respect to given design requirements is the use of structural analysis. Structural analysis consists of a collection of mathematical models (equations). These equations describe the response of the structure when the structure is subjected to the anticipated loads which the structure will have to resist safely during its life span.

Table 5. Assessment of Tiber Hybrids for Possible Aerospace Structural Applications

Material	Specific buckling stress, $\sigma_{CR}/t_c\rho$ Plate	Cylindrical shell
tiber hybrids		
70% Ti–30% Be	402×10^3	4160
50% Ti–50% Be	532	6230
30% Ti–70% Be	818	9150
other composites		
B–Al (0.50 FVR)	883×10^3	5120
B–E (0.50 FVR)	464	1310
T75–E (0.60 FVR)	ca 238	1180
AS–E (0.60 FVR)	ca 233	1150
metals		
steel	331×10^3	2442
aluminum	307	2263
titanium	307	2263

Table 6. Comparisons of Frequencies of High-Tip-Speed Composite Blade

Mode, frequency	HTS–K 601 (±40°, ±20°)[a]	HTS–PMR (±40°, ±20°)[a]	Tiber hybrid (titanium–beryllium) (30%–70%)[b]
1, Hz	361	400	662
2, Hz	939	960	1608
3, Hz	1178	1418	2108
4, Hz	1485	1658	2333
5, Hz		2427	3253
density, g/cm³ [c]	ca 1.38	ca 1.52	ca 2.35

[a] 40°, 20° denotes angle measurement.
[b] Lamina thickness: titanium, 127 μm; beryllium, 254 μm.
[c] To convert g/cm³ to lb/in.³, divide by 27.68.

A general structural analysis model in equation form is given by:

$$M\ddot{u} + C\dot{u} + Ku = F \tag{1}$$

Equation 1 describes the structural response at any point in the structure in terms of acceleration $\ddot{u}$, velocity $\dot{u}$, and displacement u for a given mechanical and/or thermal load condition F. The structure geometric configuration and material are represented in equation 1 in terms of mass M, damping C, and stiffness K. Equation 1 can be applied to simple and complex structures, and to structures made from any material. In order to use equation 1 for a structure or structural part made from a given material, the values in M, C, and K for this material must be known.

Procedures for using equation 1 for the analysis and/or design of structures made from composite laminates are extensively discussed in refs. 18–19. Herein, the MML properties for M, C, and K are determined, as well as the strength and thermal properties which are needed to evaluate and/or select MMLs for specified design requirements.

If the MML behaves like a general orthotropic solid (20), then physical, thermal and mechanical properties that are needed for structural analysis of MMLs (Fig. 1**b**

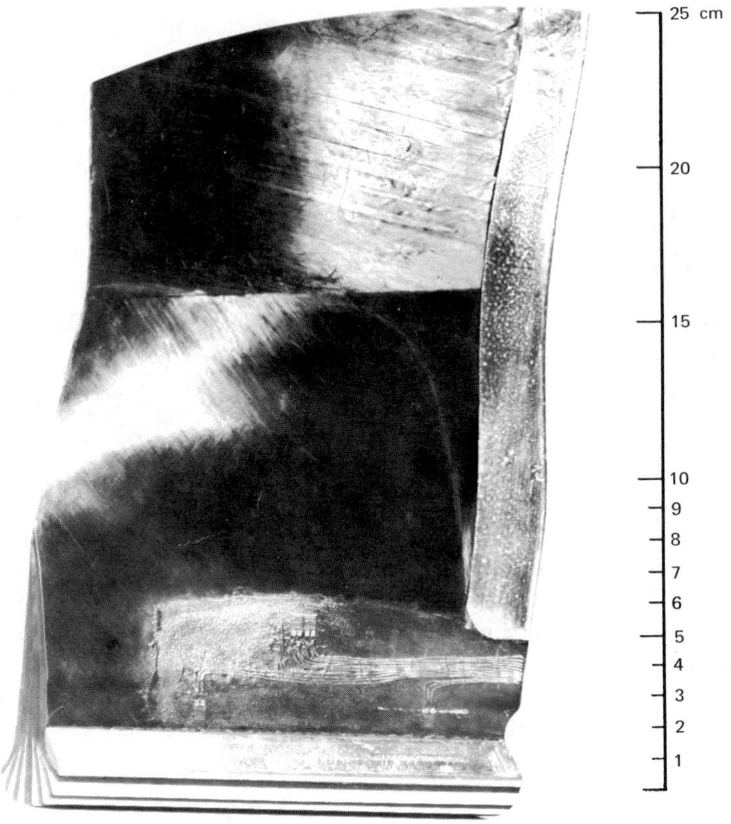

Figure 5. High-tip-speed composite fan blade.

and **d**) include: density ρ, heat capacity H_c, three thermal heat conductivities K, three thermal expansion coefficients α, three normal (Young's) moduli E, three shear moduli G, three Poisson's ratios ν, and nine strengths s. Except for ρ and H_c, the other properties are given with respect to three mutually orthogonal directions which are taken to coincide with the planes of elastic symmetry of the MML. These directions are either 1, 2, and 3 in Figure 1**a,** called the material axes of the single layer (ply), or x, y, and z in Figure 1**b** and **d,** called structural or load axes of the laminate (composite). It is customary to use subscripts to denote the directions along which the properties are given. The subscript l with combinations of 1, 2, or 3 subscripts are used to denote ply material-axis properties whereas the subscript c with combinations of x, y, or z subscripts are used to denote composite structural-axis properties. For example: E_{l11} denotes the modulus of elasticity (normal modulus) along the 1-direction and G_{l12} for the shear modulus in the 1–2 plane (Fig. 1**a**). The corresponding moduli along the structural axes of the MML for Figures 1**b** and **d** are E_{cxx} and G_{cxy}. These properties are summarized in symbolic form in Table 8 for MMLs made from plies with three types of symmetry. The material-axis properties are for the single layer (ply), and the structural-axis properties are for the laminate (composite).

Composite-mechanic theories based on constituent properties have been developed for predicting material-axes and/or structural-axis properties. Composite me-

Table 7. Typical Mechanical Properties of Metal-Matrix Composites

Fiber	Matrix	Reinforcement, vol %	Density, g/cm³ [a]	Longitudinal tensile strength, MPa [b]	Longitudinal modulus, GPa [b]	Transverse tensile strength, MPa [b]	Transverse modulus, GPa [b]
G T 50	201 Al	30	2.380	620	170	50	30
G T 50	201 Al	49		1120	160		
G GY 70	201 Al	34	2.380	660	210	30	30
G GY 70	201 A1	30	2.436	550	160	70	40
G HM pitch	6061 Al	41	2.436	620	320		
G HM pitch	AZ31 Mg	38	1.827	510	300		
B on W, 142 μm fiber	6061 Al	50	2.491	1380	230	140	160
borsic	Ti	45	3.681	1270	220	460	190
G T 75	Pb	41	7.474	720	200		
G T 75	Cu	39	6.090	290	240		
FP	201 Al	50	3.598	1170	210	(140)	140
SiC	6061 Al	50	2.934	1480	230	(140)	140
SiC	Ti	35	3.931	1210	260	520	210
SiC whisker	Al	20	2.796	340	100	340	100
B_4C on B	Ti	38	3.737	1480	230	>340	>140
G T 75	Mg	42	1.799	450	190		
G HM	Pb	35	7.750	500	120		
G T 75	Al–7% Zn	38	2.408	870	190		
G T 75	zinc	35	5.287	770	120		
G T 50	nickel	50	5.295	790	240		
G T 75	nickel	50	5.342	828	310	30	40
G (81.3 μm)	2024 Al	50	2.436	760	140		
G (142 μm)	2024 Al	60	2.436	1100	180		
superhybrid	grafitic	60	2.048	860	120	220	60
superhybrid	S-glass	60	2.159	740	80	190	30
superhybrid	Kevlar	60	1.799	700	80	190	10

[a] To convert g/cm³ to lb/in.³, divide by 27.68.
[b] To convert MPa to psi, multiply by 145; to convert GPa to psi, multiply by 145,000.

Table 8. Summary of Physical, Thermal, and Mechanical Properties Needed for Structural Analysis and Design for Metal-Matrix Laminates Made from Plies with Three Types of Symmetry in Symbolic Form[a]

	Composite with type of symmetry of:					
	Generally orthotropic		Transverse isotropic		Isotropic	
Property	Ply	Composite	Ply	Composite	Ply	Composite
density, g/cm^3	ρ_l	ρ_c	ρ_l	ρ_c	ρ_l	ρ_c
heat capacity	H_{cl}	H_{cc}	H_{cl}	H_{cc}	H_{cl}	H_{cc}
thermal heat conductivity	K_{l11}	K_{cxx}	K_{l11}	K_{cxx}	K_l	K_{cxx}
	K_{l22}	K_{cyy}	K_{l22}	K_{cyy}	K_l	$K_{cyy} = K_{cxx}$
	K_{l33}	K_{czz}	$K_{l33} = K_{l22}$	K_{czz}	K_l	K_{czz}
thermal expansion coefficient	α_{l11}	α_{cxx}	α_{l11}	α_{cxx}	α_l	α_{cxx}
	α_{l22}	α_{cyy}	α_{l22}	α_{cyy}	α_l	$\alpha_{cyy} = \alpha_{cxx}$
	α_{l33}	α_{czz}	$\alpha_{l33} = \alpha_{l22}$	α_{czz}	α_l	α_{czz}
elastic and shear moduli	E_{l11}	E_{cxx}	E_{l11}	E_{cxx}	E_l	E_{cxx}
	E_{l22}	E_{cyy}	E_{l22}	E_{cyy}	E_l	$E_{cyy} = E_{cxx}$
	E_{l33}	E_{czz}	$E_{l33} = E_{l22}$	E_{czz}	E_l	E_{czz}
	G_{l12}	G_{cxy}	G_{l12}	G_{cxy}	$G_l = \dfrac{E_l}{2(1+\nu_l)}$	$G_{cxy} = \dfrac{E_{cxx}}{2(1+\nu_{czy})}$
	G_{l23}	G_{cyz}	$G_{l23} = \dfrac{E_{l22}}{2(1+\nu_{l23})}$	G_{cyz}	$G_l = \dfrac{E_l}{2(1+\nu_l)}$	G_{cyz}
	G_{l13}	G_{cxz}	$G_{l13} = G_{l12}$	G_{cxz}	$G_l = \dfrac{E_l}{2(1+\nu_l)}$	$G_{cxz} = G_{cyz}$
Poisson ratios	ν_{l12}	ν_{cxy}	ν_{l11}	ν_{cxy}	ν_l	ν_{cxy}
	ν_{l23}	ν_{cyz}	ν_{l23}	ν_{cyz}	ν_l	ν_{cyz}
	ν_{l13}	ν_{cxz}	$\nu_{l13} = \nu_{l12}$	ν_{cxz}	ν_l	$\nu_{cxz} = \nu_{cyz}$
strengths	$S_{l11T,C}$	$S_{cxxT,C}$	$S_{l11T,C}$	$S_{cxxT,C}$	$S_{lT,C}$	$S_{cxxT,C}$
(T = tension;	$S_{l22T,C}$	$S_{cyyT,C}$	$S_{l22T,C}$	$S_{cyyT,C}$	$S_{lT,C}$	$S_{cyyT,C} = S_{cxxT,C}$
C = com-pression;	$S_{l33T,C}$	$S_{czzT,C}$	$S_{l33T,C} = S_{l22T,C}$	$S_{czzT,C}$	$S_{lT,C}$	$S_{czzT,C}$
	S_{l12S}	S_{cxyS}	S_{12S}	S_{cxy}	S_{lS}	S_{cxyS}
S = shear)	S_{l23S}	S_{cyzS}	S_{l23S}	S_{cyz}	S_{lS}	S_{cyzS}
	S_{l13S}	S_{cxzS}	$S_{l13S} = S_{l12S}$	S_{cxz}	S_{lS}	$S_{cxzS} = S_{cyzS}$

[a] Subscripts refer to directions shown in Figure 1.

chanics is subdivided into micromechanics, macromechanics, and laminate theory. Micromechanics embodies the various theories which are used to predict material-axis properties of unidirectional fiber composites (plies) using constituent fiber and matrix properties. Typical results predicted for boron–aluminum plies using composite micromechanics are shown in Figure 6. Macromechanics usually consist of transformation equations which are used to transform material axes properties along any other axes. Macromechanics also includes failure theories and criteria for plies subjected to combined stress states. Typical results predicted for boron–aluminum MMLs using composite macromechanics are shown in Figure 7 for thermal and elastic properties and in Figure 8 for strengths. Laminate theory embodies the equation and procedures that are used to predict the laminate properties using ply properties. Laminate theory is also used to generate the properties required to form M, C, and K (17,23–24). In addition, laminate theory is used to predict the residual stresses of the lamination in

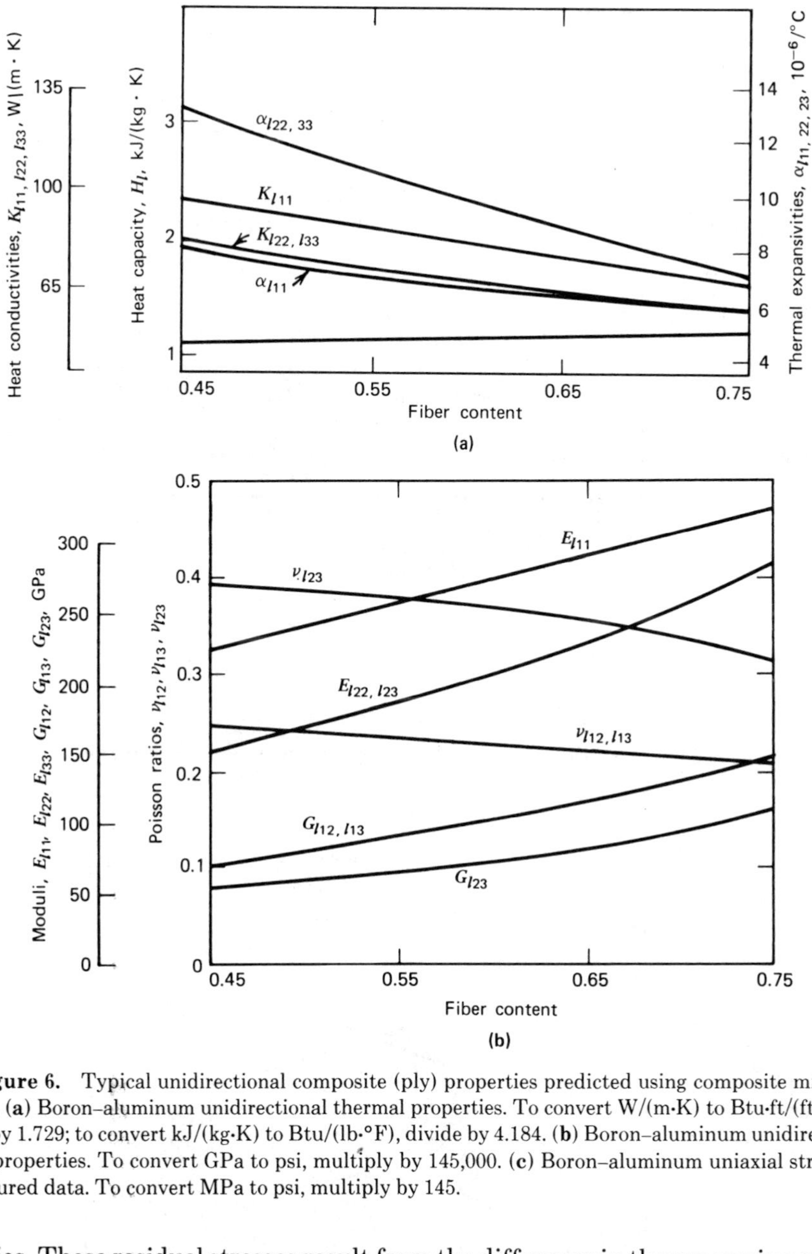

Figure 6. Typical unidirectional composite (ply) properties predicted using composite micromechanics. (**a**) Boron–aluminum unidirectional thermal properties. To convert W/(m·K) to Btu·ft/(ft²·h·°F), divide by 1.729; to convert kJ/(kg·K) to Btu/(lb·°F), divide by 4.184. (**b**) Boron–aluminum unidirectional elastic properties. To convert GPa to psi, multiply by 145,000. (**c**) Boron–aluminum uniaxial strengths; O measured data. To convert MPa to psi, multiply by 145.

the plies. These residual stresses result from the difference in the processing and use temperature as well as the difference in the thermal expansion coefficients (25–26). Typical results predicted for boron–aluminum MMLs using laminate theory are shown in Figure 9 for thermal expansion coefficients and elastic properties, and in Figure 10 for lamination residual stresses. Lamination residual stresses (strains) affect significantly the laminate mechanical behavior of boron–aluminum MMLs (26). Different types of heat treatment also affect the mechanical behavior of MMLs, especially the transverse properties (27).

MMLs are made from isotropic plies (Fig. 1**d**). The analysis is considerably simpler

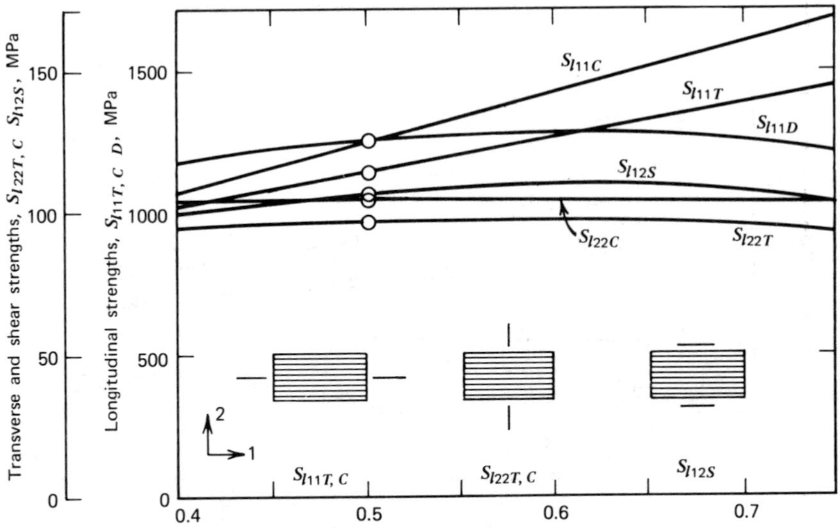

(c)

Figure 6. *(continued)*

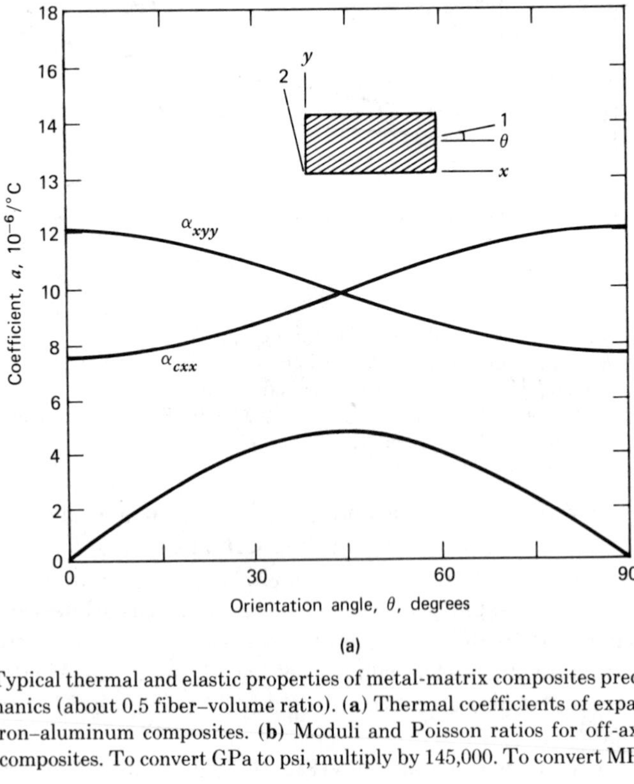

(a)

Figure 7. Typical thermal and elastic properties of metal-matrix composites predicted using composite macromechanics (about 0.5 fiber–volume ratio). (**a**) Thermal coefficients of expansion for off-axis unidirectional boron–aluminum composites. (**b**) Moduli and Poisson ratios for off-axis unidirectional boron–aluminum composites. To convert GPa to psi, multiply by 145,000. To convert MPa to psi, multiply by 145.

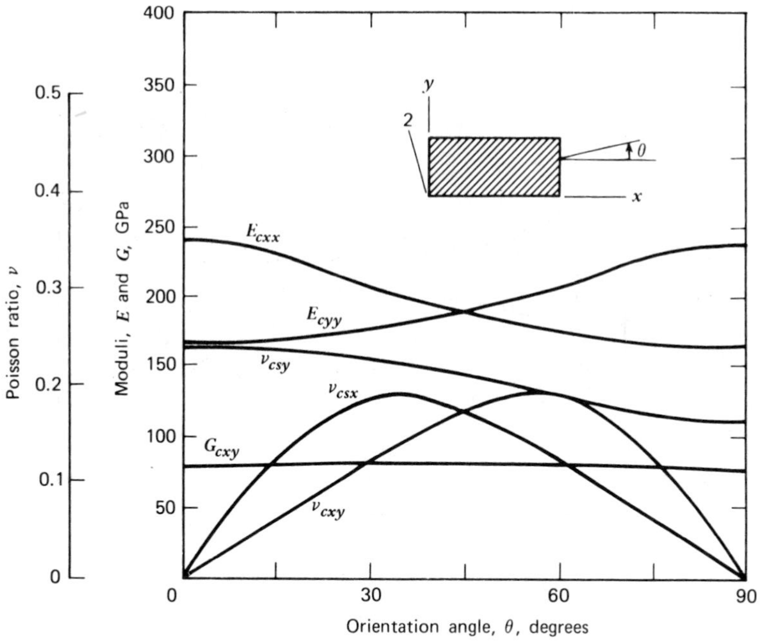

(b)

Figure 7. (*continued*)

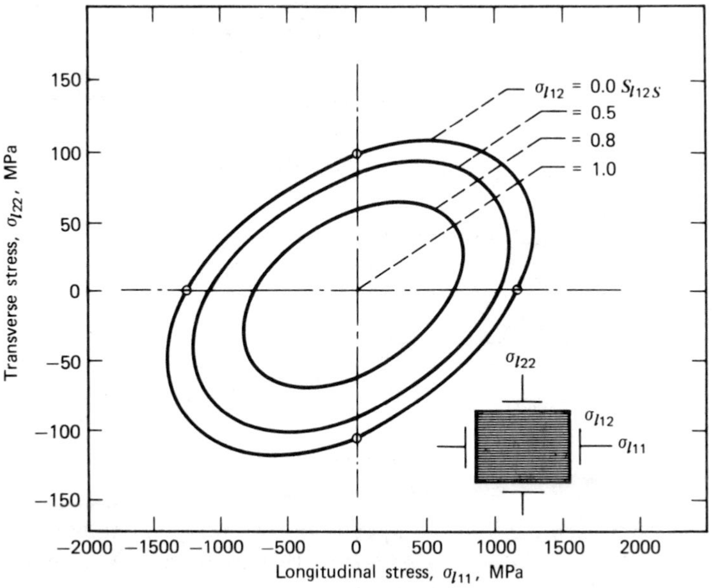

Figure 8. Typical strengths of metal matrix composites predicted using composite macromechanics (about 0.5 fiber–volume ration). (**a**) Failure envelopes under combined stress for boron–aluminum. ○ experimental data from ref. 22. (**b**) Boron–aluminum off-axis normal load failure envelopes. ○ experimental data. (**c**) Boron–aluminum off-axis shear load failure envelopes. ○ experimental data.

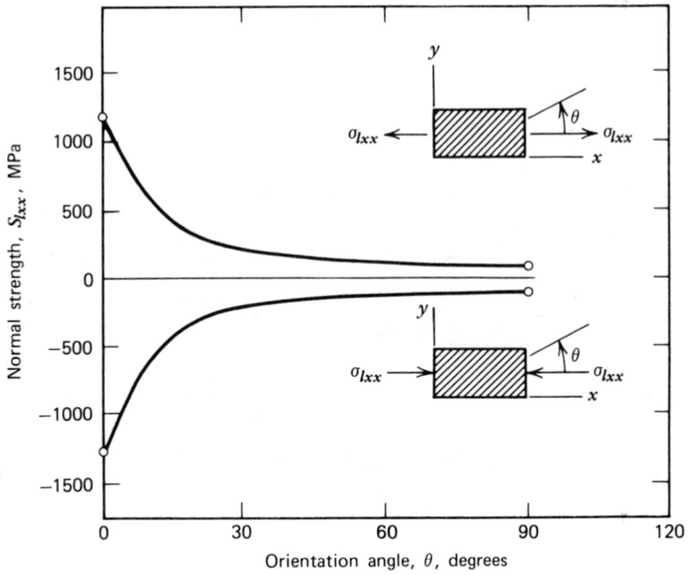

(b)

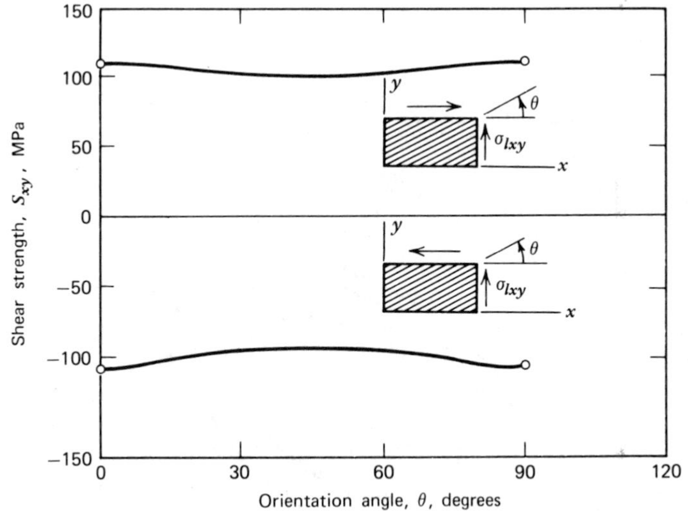

(c)

Figure 8. (*continued*)

as compared to that used for fiber-composite plies. One such analysis is described in detail in ref. 13. Typical results obtained using this analysis are summarized in Table 9 for titanium–beryllium (tiber) hybrid MMLs.

The thermal and mechanical properties of MMLs described above constitute only a minimum of those usually required to assess the suitability of a relatively new ma-

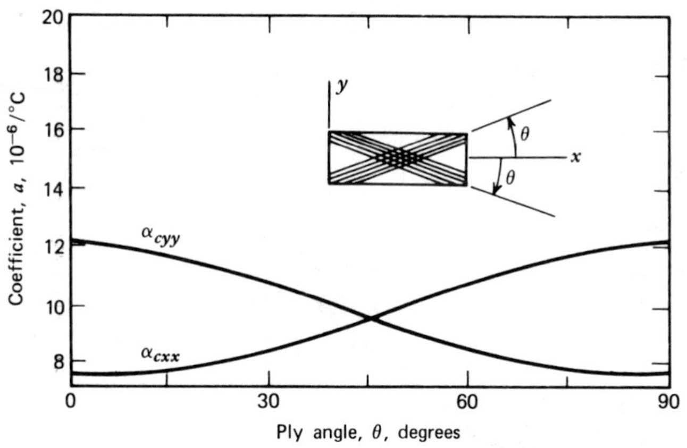

(a)

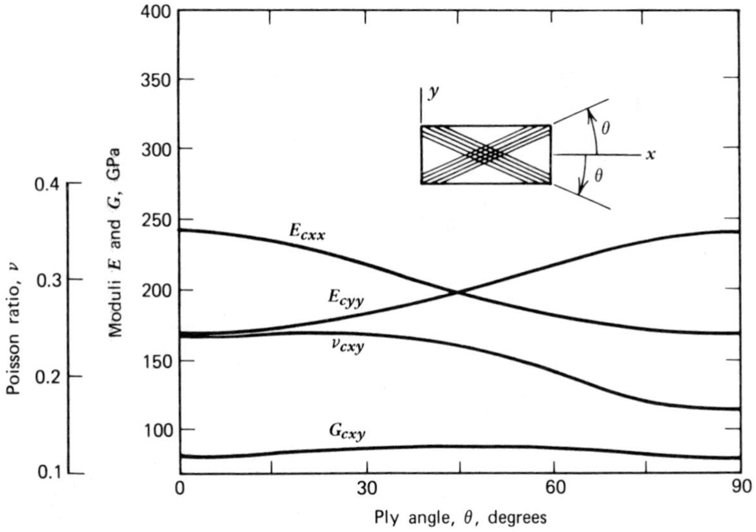

(b)

Figure 9. Typical thermal (**a**) and elastic (**b**) properties for metal matrix laminates (boron–aluminum angle-ply composites) predicted using laminate theory (about 0.5 fiber–volume ratio). To convert GPa to psi, multiply by 145,000.

terial at the preliminary design stages. Several other important factors need be considered simultaneously with the thermal and mechanical properties, eg, fatigue resistance, creep, impact resistance, erosion and corrosion resistance, service environment effects, notch insensitivity and fracture toughness, damage tolerance and repairability, fabrication and quality control, reliability and durability, inspectability and maintainability, design data development costs and reproducibility, design–analysis experience of the staff, and acceptance of the public agency which sets and administers structural integrity–safety requirements.

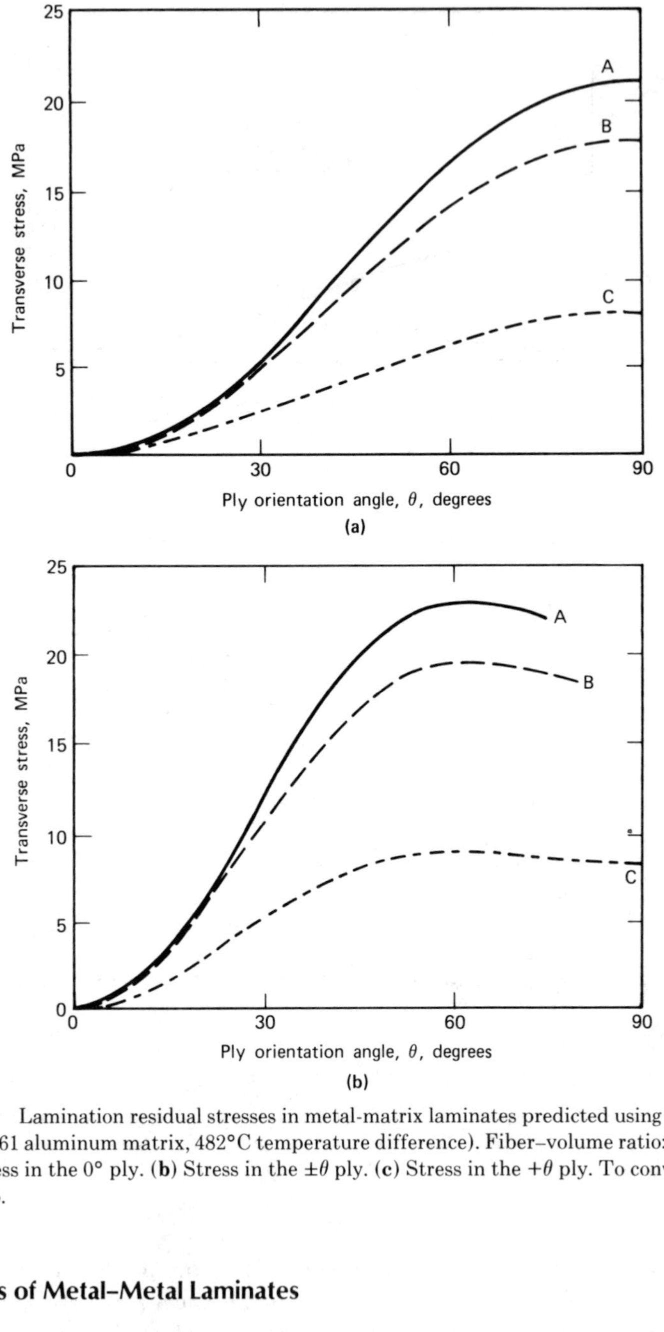

Figure 10. Lamination residual stresses in metal-matrix laminates predicted using laminate theory (boron fiber–6061 aluminum matrix, 482°C temperature difference). Fiber–volume ratio: A, 0.35; B, 0.55; C, 0.75. (**a**) Stress in the 0° ply. (**b**) Stress in the $\pm\theta$ ply. (**c**) Stress in the $+\theta$ ply. To convert MPa to psi, multiply by 145.

Special Types of Metal–Metal Laminates

The primary reason for making and investigating these types of MMLs is their potential for fracture control and damage tolerance. Fracture control is usually assessed by using a material property called fracture toughness. Special types of metal/metal MMLs that have been investigated (shown in Table 1) include different kinds of steels such as mild, high strength, and maraging (alloyed steel); aluminum–aluminum; titanium–titanium and titanium–aluminum; tungsten–superalloy and tungsten–tan-

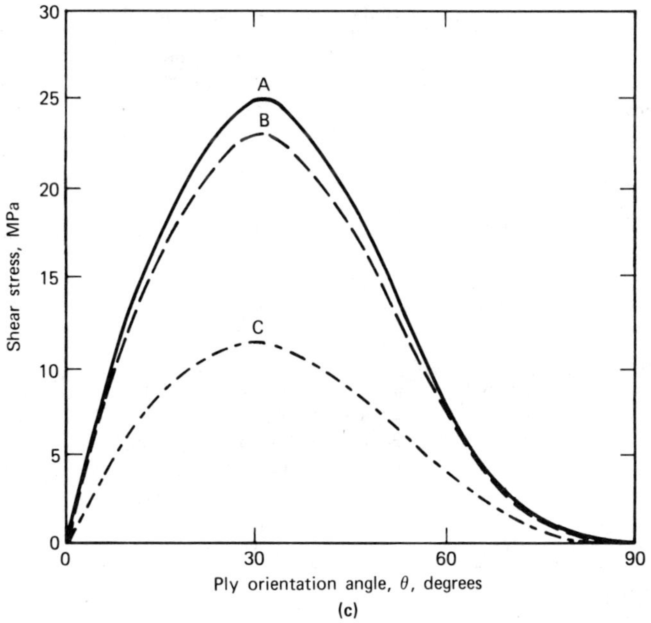

Figure 10. (*continued*)

talum; and titanium–beryllium. The fracture toughness of a plate-form material, with a cracklike defect, is the stress that this material can resist just prior to onset of rapid crack propagation. Fracture toughness varies for different materials. It also can vary for the same material but with different alloying elements, thickness and heat treatment. In addition fracture toughness depends on the use temperature. In principle, by interleafing materials with different fracture toughness, the fracture toughness of metal–metal MMLs can be altered.

For analysis–design purposes, fracture toughness is used to determine the level of stress that a structural member with an assumed crack length can safely support. This level of stress is determined using linear elastic fracture mechanics (LEFM) in general. Equation 2 is an elementary equation from LEFM:

$$\sigma = \frac{K_c}{\sqrt{a}} F \tag{2}$$

where σ is the average or gross stress (stress without the crack), K_c is the material fracture toughness parameter corresponding to primary loading conditions and anticipated crack propagation depicted schematically in Figure 11, a is the crack length, and F represents the stress state at the crack tip and depends on: material, geometry, and loading condition. Values for K_c for different materials are found in reports published by the Metals and Ceramics Information Center, Columbus, Ohio, as well as various handbooks dealing with aerospace structures, and pressure-vessel materials and design.

The designer can use MMLs to control fracture and, therefore, provide damage tolerance either by using plies of materials with different fracture toughness to divide the fracture-driving stress (crack divider), or by using plies with higher fracture

Table 9. Comparison of Measured and Predicted Properties of Tiber Laminates

	Tiber laminate and direction					
	I—40% Ti–58% Be		II—55% Ti–36% Be		III—63% Ti–31% Be	
Property identification	RD	TD	RD	TD	RD	TD
modulus, GPa[a]						
measured	203.5	206.9	165.5	165.5	175.9	169.0
predicted	211.0	212.4	163.5	165.5	157.9	160.0
difference, %	3.7	2.7	−1.2	0	−10.2	−5.1
Poisson ratio						
measured	0.20	0.25	0.26	0.27	0.26	0.28
predicted	0.27	0.27	0.28	0.29	0.29	0.29
difference, %	35.0	8.0	7.7	7.4	11.5	3.6
fracture stress, MPa[b]						
measured	646.9	466.2	723.5	713.8	787.0	716.6
predicted	677.3	642.8	713.8	694.5	767.0	749.0
difference, %	4.7	37.9	−1.4	−2.7	−2.5	−4.7
density, g/cm^3 [c]						
measured	2.85		3.24		3.40	
predicted	2.82		3.21		3.46	
difference, %	0.9		−0.8		1.6	

[a] To convert GPa to psi, multiply by 145,000.
[b] To convert MPa to psi, multiply by 145.
[c] To convert g/m^3 to $lb/in.^3$, divide by 27.68.

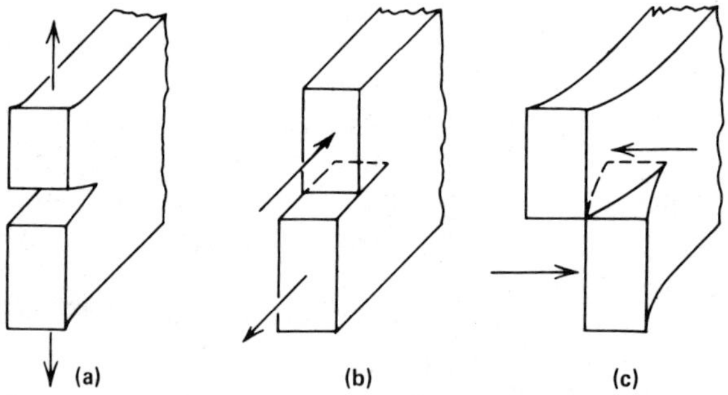

Figure 11. Primary load conditions and corresponding fracture modes used in fracture mechanics: (**a**) mode I loading; (**b**) mode II loading; (**c**) mode III loading.

toughness to arrest the fracture-driving stress (crack arrest). Both of these methods are illustrated schematically in Figure 12.

In order for either concept to work, the type of bond selected must meet three general criteria; it must be strong enough to constrain the laminate to respond structurally (with respect to displacement, buckling and frequency) like a homogenous material; weak enough to permit each ply to fracture independently of its neighbors; and brittle enough to fail by local delamination in the vicinity of the advancing crack front. Examples of the fracture toughness of aluminum–aluminum MMLs by diffusion, roll, or explosive bonding are described in ref. 28, and those made by adhesive bonding in ref. 29. Photomicrographs depicting arrested cracks in actual samples are shown in ref. 22; a concise treatment of fracture analysis for aerospace metals is given in ref. 21; and ref. 30 provides a comparable treatment for fatigue (see Ablative materials).

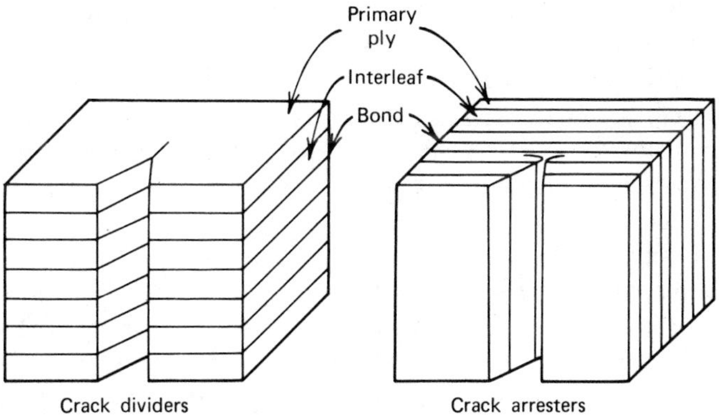

Figure 12. Metal laminate concepts for fracture control.

The root portion of helicopter blades is an example showing where MMLs are used for fracture control. This part of the blade may have cracklike defects because of the joint, and it is subjected to high cycle fatigue. Wings of military aircraft and helicopter booms are potential applications of MMLs in order to provide damage tolerance to projectile impact.

In addition to fracture control or damage tolerance, MMLs also are used in applications where the interleaf may be the stronger, stiffer material whereas the primary material provides erosion, corrosion, oxidation, or other service–environmental resistance. Examples of such an application are tantalum/tungsten MMLs which have been investigated for possible use in aircraft engine turbine blades. Tantalum coated with suitable coatings is used to resist the corrosive environment of the burning fuel, and tungsten is used for strength and stiffness in order to meet mechanical-design requirements.

Special Types of Fiber-Reinforced Metals

Boron-fiber–aluminum-matrix (B–Al) MMLs have been made and investigated more extensively than any other fiber-reinforced metal. These types of laminates combine several of the desirable features of aluminum and in addition provide about a threefold improvement in modulus and about a threefold improvement in strength over that of aluminum. The key disadvantage limiting its application is the high cost of the boron fiber; the main part of this boron-fiber cost is the tungsten substrate. In order to reduce the fiber cost, carbon fibers are used for the substrate, as well as boron fibers of larger diameter.

Boron-fiber–aluminum-matrix MMLs have excellent fatigue-, creep-, corrosion-, and erosion resistance. Galvanic action may degrade the interfacial bond, depending on the surface coating of the fiber. These MMLs have good temperature resistance up to about 150°C. They may be used with relatively small property-loss penalty up to 315°C in stiffness-controlled designs such as dimensional stability, and buckling and vibration frequencies. B–Al MMLs have improved fracture toughness compared to the aluminum matrix, and they are notch-insensitive. However, B–Al MMLs have about one half the impact resistance of aluminum. An extensive discussion on various mechanical properties of B–Al MMLs is given in ref. 31. The elevated temperature effects are discussed in ref. 32. The cryogenic temperature conditions have negligible effect on the tensile properties of B–Al MMLs from the limited available data (33).

Angleplied MMLs (Fig. 1**b**) from B–Al undergo inelastic deformations at relatively small load (about 10–20% of the fracture load) (34). In a cyclic-load condition, this inelastic deformation may progressively improve, degrade or have no effect on the mechanical properties of the laminate (35). Significant parameters affecting joints and joint design are discussed in ref. 36.

Boron–aluminum MMLs have been made for: aircraft fuselage skin and stringers, aircraft wing skins, aircraft-wing boxes, aircraft-engine fan and compressor blades, propeller shells, landing gear struts, thrust support structure for the space shuttle, shafts for torque transmission, and rocket-motor cases. An extensive discussion of the application of B–Al MMLs for aerospace structures is given in ref. 37. Because of the high cost [about $550/kg (1980 dollars)] B–Al MMLs have not been considered seriously for use outside the aerospace industry.

Graphite-fiber–aluminum-matrix (Gr–Al) MMLs are investigated mainly for B–Al MMLs replacement because of their low cost (about 25–50%) potential compared to B–Al MMLs. In addition, Gr–Al MMLs have excellent thermal-dimension stability. They also are good contenders in friction-wear applications because of the inherent lubricating properties of the graphite fibers. However, Gr–Al MMLs are susceptible to galvanic corrosion as are B–Al MMLs. Special surface treatment of the fibers is required to minimize this galvanic action. Gr–Al MMLs exhibit rule-of-mixtures properties along the fiber direction (Table 7). However, the transverse properties are relatively poor. Alternatives such as heat-treating and metal-foil-interleaving are used to improve the transverse properties. These alternatives are selected with considerable caution since they tend to reduce the longitudinal properties. Gr–Al MMLs have good fracture toughness and damage tolerance. They also have excellent mechanical- and thermal-fatigue resistance. Their corrosion and erosion resistance can be comparable to that of B–Al. An extensive discussion and a good review of several important aspects of Gr–Al MMLs up to 1976 is given in ref. 38.

Borsic-fibers–titanium and borsic–aluminum matrix MMLs have been investigated primarily for possible use in aircraft-engine fan blades. Borsic/titanium MMLs have about twice the stiffness and about 80% the density of titanium (39–40). The combination of these two properties is generally sufficient to eliminate the midspan shrouds which are presently used to meet vibration and flutter design requirements. This diminution of shrouds also is apparent with B–Al.

Tungsten-fiber–superalloy (TF–SA) MMLs are investigated for their potential use in aircraft-turbine blades. The excellent mechanical properties retention of the tungsten fibers at high temperatures (about 1100°C) is the key feature for investigating these types of laminates. However, TF–SA have two main disadvantages: high density, and thermal-fatigue degradation. The high density disadvantage may be circumvented to some extent by appropriate structural-design configurations, such as hollow blades. On the other hand, the thermal-fatigue degradation only can be minimized by metallurgical considerations. Most of the research to date for TF–SA MMLs was conducted at the NASA Lewis laboratory (41–42). Limited research recently has been initiated to make turbine blades from these laminates (43–44). Whisker-reinforced metals and ceramic laminates also are investigated for possible use in internal combustion engines and other high temperature applications in general. The major disadvantages of whisker-reinforced MMLs are their poor fracture toughness and poor impact resistance characteristics. These poor characteristics may be improved by designing the laminate to operate in preferential compression.

Some of the other fiber-reinforced MMLs listed in Table 7 are investigated for specific applications. For example, graphite-fiber–magnesium-matrix MMLs are investigated for space antennas because of their desirable thermal distortion and density properties, and graphite-fiber–lead-matrix MMLs are investigated for use in batteries where weight is an important design consideration. Some other MMLs listed in Table 7 are investigated for metallurgical considerations at the fiber–matrix interface (borsic–aluminum, silicon carbide–aluminum).

Nomenclature

a	= crack length
B–Al	= boron fiber–aluminum matrix
C	= damping
E	= modulus of elasticity
F	= stress state at crack tip
F	= mechanical and/or thermal load condition
G	= shear modulus
Gr–A	= graphite fiber–aluminum matrix
H_c	= heat capacity
ICCM	= International Conference on Composite Materials
K	= stiffness
$K_{subscript}$	= heat conductivity
K_c	= fracture toughness parameter
LEFM	= linear elastic fracture mechanics
M	= mass
MML	= metal matrix laminate
u	= displacement
$\dot{u}$	= velocity
$\ddot{u}$	= acceleration
SHC	= superhybrid composite
TF–SA	= tungsten-fiber–superalloy
UFC	= graphite-fiber–resin
α	= thermal expansion core
σ	= average or gross stress
ρ	= density

BIBLIOGRAPHY

"Ceramic Composite Armor" in *ECT* 2nd ed., Suppl. Vol., pp. 138–157, by George Rugger, Dept. of the Army, and John R. Fenter, Dept. of the Air Force.

1. M. F. Smith, *Metal Matrix Composites, NTIS/PS-78/0684, N79-10155,* Vol. 2, National Technical Information Service, Springfield, Va., 1978.
2. D. M. Cavagnaro, *Boron-Reinforced Composites, NTIS/PS-79/0476, N78-27189,* Vol. 2, NTIS, Springfield, Va., 1979.
3. D. M. Cavagnaro, *Boron-Reinforced Composites, NTIS/PS-78/0357, N78-27189,* Vol. 2, NTIS, Springfield, Va., 1978.
4. K. G. Kgeider, ed., *Composite Materials,* Vol. 4 of L. Broutman and R. Krock, eds., *Metallic-Matrix Composites,* Academic Press, Inc., New York, 1974.
5. W. J. Renton, ed., *Hybrid and Select Metal-Matrix Composites: A State-of-the-Art Review,* America Institute of Aeronautic and Astronautics, New York, 1977.
6. Ref. 4, p. 40.
7. Ref. 4, p. 49.
8. L. Rubin, *SAMPE J.* 4 (July–Aug. 1979).

9. L. R. McCreight, H. W. Rauch, and W. H. Sutton, *Ceramic and Graphite Fibers and Whiskers,* Academic Press, Inc., New York, 1965.
10. B. R. Noton and co-eds., *ICCM2, Proceedings of the 1978 International Conference on Composite Materials,* AIME, Warrendale, Pa., 1978, pp. 1337–1339.
11. C. C. Chamis, R. F. Lark, and T. L. Sullivan, *Boron/Aluminum-Graphite/Resin Advanced Composite Hybrids, NASA TND-7879,* NASA, 1975.
12. C. C. Chamis, R. F. Lark, and T. L. Sullivan, *Super-Hybrid Composites—an Emerging Structural Material, NASA TMX-71836,* NASA, 1975.
13. C. C. Chamis and R. F. Lark, *Titanium/Beryllium Laminates: Fabrication Mechanical Properties and Potential Aerospace Applications, NASA TM-73891,* NASA, 1978.
14. D. A. McDanels and R. A. Signorelli, *Effect of Fiber Diameter and Matrix Alloys on Impact Resistant Boron/Aluminum Composites, NASA TND-8204,* NASA, 1976.
15. J. W. Brantly and R. G. Stabrylla, *Fabrication of J79 Boron/Aluminum Blades, Final Report, NASA CR-159566,* NASA, 1979.
16. C. T. Salemmee and G. C. Murphy, *Metal Spar/Superhybrid Shell Composite Fan Blades, NASA CR-159594,* NASA, 1979.
17. C. C. Chamis, *J. Aircraft* **14**(7), 644 (1977).
18. C. C. Chamis, ed., *Structural Design and Analysis,* Part I, Vol. 7, Academic Press, Inc., New York, 1974.
19. C. C. Chamis, ed., *Structural Design and Analysis,* Part II, Vol. 8, Academic Press, Inc., New York, 1975.
20. Ref. 18, p. 8.
21. H. J. Oberson, Jr., *SAMPE J.* 4 (Nov.–Dec. 1977).
22. S. T. Mileiko and V. M. Anishenkov, *Sci. Adv. Mater. Process Eng. Ser.* **24**(Book 1), 799 (1979).
23. Ref. 19, p. 231.
24. C. C. Chamis, *Comput. Struct.* **3,** 467 (1973).
25. C. C. Chamis, *Lamination Residual Stresses in Multilayered Fiber Composites, NASA TND-6146,* NASA, 1971.
26. C. C. Chamis, *Residual Stresses in Angleplied Laminates and Their Effects on Laminate Behavior, NASA TM-78835,* NASA, 1978.
27. Ref. 5, p. 72.
28. R. D. Goolsby, "Fracture of Crack Divider Al/Al Laminates" in *Proceedings of the 1978 International Conference on Composite Materials, Metallurgical Society of AIME, Toronto, Canada, April 16–20, 1978,* pp. 941–960.
29. G. H. Koch, *SAMPE Q.* **11**(1), 7 (1979).
30. K. H. Miska, *Mater. Eng.,* 31 (June 1978).
31. Ref. 5, pp. 67–97.
32. P. G. Sullivan, *Elevated Temperature Properties of Boron/Aluminum Composites, NASA CR-159445,* NASA, 1978.
33. R. E. Schramm and M. B. Kasen, *Adv. Cryog. Eng.* **22,** 205 (1977).
34. C. C. Chamis and T. L. Sullivan, *A Computational Procedure to Analyze Metal Matrix Laminates with Nonlinear Lamination Residual Strains, NASA TMX-71543,* NASA, 1974.
35. C. C. Chamis and T. L. Sullivan, *Nonlinear Response of Boron/Aluminum Angleplied Laminates, Under Cyclic Tensile Loading: Contributing Mechanisms and Their Effects, NASA TMX-71490,* NASA, 1973.
36. S. Janes, *Application of Conventional Joining Techniques to Boron Fiber and Carbon Fiber Reinforced Aluminum, Report D2532 H1,* Battelle Institute, Frankfurt, FRG, 1976.
37. Ref. 5, pp. 99–157.
38. Ref. 5, pp. 159–255.
39. Ref. 4, pp. 269–318.
40. B. R. Collins, W. D. Brentnall, and I. J. Toth, *Properties and Fracture Modes of Borsic Titanium, AFML-TR-73-43,* AFB, Ohio, 1972.
41. R. A. Signorelli, *Review of Status and Potential of Tungsten-Wire–Superalloy Composites for Advanced Gas Turbine Engine Blades, NASA TMX-2599,* NASA, 1972.
42. Ref. 4, pp. 229–267.
43. W. D. Brentnall, *FRS Composites for Advanced Gas Turbine Engine Components, Final Report TRW-ER-7887-F, AD-AC50 59518ST,* Cleveland, Ohio, 1977.

44. D. W. Petrasek, E. A. Winsa, L. J. Westfall, and R. A. Signorelli, *Tungsten Fiber Reinforced Fe-CrAlY—A First Generation Composite Turbine Blade Material, NASA TM-79094,* NASA, 1979.

General References

Reference 4 is a general reference.

C. C. Chamis, "*Characterization and Design Mechanics for Fiber-Reinforced Metals, NASA TND-5784,* NASA, 1970.

K. M. Prewo, *Fabrication and Evaluation of Low Cost Alumina Fiber-Reinforced Metal Matrices, Report UTRC (United Technologies Research Center)/R77-912547-13, Contract F33615-76-C5199,* June 15, 1977.

A. K. Dhingra and W. H. Krueger, "Magnesium Castings—Reinforced with Du Pont Continuous Alumina Fiber FP, *paper presented at the 36th World Conference on Magnesium, Oslo, Norway, June 25–26, 1979.*

A. K. Dhingra, *Phil. Trans. R. Soc. London Ser. A* **294,** 559 (1980).

C. C. CHAMIS
National Aeronautics and Space Administration

LAMINATED AND REINFORCED PLASTICS

Reinforced plastics are combinations of fibers and polymeric binders or matrices that form composite materials (qv). The strongest geometry in which any solid can exist is as a wire, a fiber, or a whisker, ie, the strength of any solid is determined by the defects it contains—voids, cracks, discontinuities, etc—and the magnitude of their weakening influence depends upon their absolute size. Thus, if a fine-diameter fiber can be drawn or grown from a material, any defect it contains must be very small: the material will be stronger than in the bulk form (1). But fibers, even if they are strong when pulled, have very limited structural utility by themselves. They bend easily, especially if they are small in diameter, and when pushed axially, they buckle under very low forces. To remedy these inadequacies, a supporting medium is required, surrounding each fiber, separating it from its neighbors, and stabilizing it against bending and buckling. These are the functions of the matrix and they are best fulfilled when good adhesion exists between the two. If the fiber–matrix interaction is only a mechanical fit, without adhesion, the combined function is impaired and both components are underutilized.

Adhesion requires intimate contact, on the scale of a few hundred picometers (10^{-8} cm), over large surface areas. The best practical way to achieve this is by rendering one component a liquid, having it completely wet the surface of the other, and then solidifying it after the contact has been established (see Adhesives). Polymers readily make the liquid–solid phase change, and do not require much energy or elaborate processing to do so. Many thermoplastic polymers melt in the range of 150–250°C without large inputs of heat, and they readily solidify upon cooling. During the molten stage, even though their viscosity may be high, they are liquids and they possess the ability to wet most surfaces so their adhesion potential is readily available. The thermosetting polymers pass through a liquid phase just once during their life, while they are being polymerized and cross-linked into heat-infusible forms. During this liquid phase, the viscosity is very low, however, and they can easily infiltrate fiber bundles and fabrics, encapsulating every fiber; once this is done, solidification by catalysis or mild heating is quite straightforward.

The combination of strong fibers and synthetic polymers to form reinforced plastics and laminates derives from several basic considerations of materials science: the inherent strength of fine fibers, the wetting requirement for adhesion, and the ease of the liquid–solid phase change by synthetic polymers. These and other factors are responsible for the continuing rapid growth in the production of reinforced plastics, which has averaged about 10–12%/yr for many years and shows no signs of lessening, despite energy and raw materials availability problems.

At all levels of technological sophistication, the use of reinforced plastics continues to expand. Aircraft, marine, automobile and chemical markets hold additional opportunities, but one important fact must be recognized: reinforced plastics are not cheap, either by volume or by weight. At present, their cost ranges from $2.75 to as much as $165 per kilogram, in finished form. To compete with traditional materials—metals, wood, ceramics, concrete, and glass—their unique characteristics must be fully utilized in any application. The more this criterion can be fulfilled, the greater is the success of the application. Filament-wound pipe for corrosive service in oil fields is an almost perfect example (see Piping systems; Plastic building products). It is resistant to acids, water, oil, and decay. It is light, easy to transport, handle, and place. On-site joining with adhesives requires simple equipment, materials, and procedures; highly skilled labor is not necessary. Conventional tools and practices require only modest changes to be appropriate. Automated production promotes consistent, high quality units within an established context of industry standards, at tolerable costs. The cost in-place is absolutely competitive with metal pipe; the cost over the service life dramatically favors reinforced plastic.

Properties

The specific gravity of reinforced plastics is low, in the range of 1.5–2.25, compared to 3 for aluminum, 7.9 for steel, 2.5 for concrete, and 2.7 for natural granites and marbles. Only wood, at 0.5, is lower among the structural materials. This low density, and the ease of forming into intricately curved or corrugated forms, makes possible very high stiffness-to-weight structural elements and shapes. If strong, continuous fiber reinforcements are used, comparably high strength-to-weight ratios also can be obtained. These two factors, the high specific stiffness and specific strength as they are called, are the reasons why both commercial and military aircraft use fiber-reinforced plastics so widely. Weight can be kept to a minimum, extra fibers can be placed and oriented only where needed, section thickness changes easily are made, and joints, fasteners, and discontinuities are eliminated. Obviously, special design procedures and criteria are required for the materials, since their properties and behavior are much different from conventional, isotropic metals (2–4).

In contrast to most metals, polymers do not oxidize or corrode in normal moist air. Even in more severe environments—water, acids, bases, common solvents—specific polymers can be formulated to exhibit unusual resistance, often bordering on inertness (see also Coatings, resistant). Thus, reinforced plastics are used very widely in chemical applications, such as pipes, tubes, tanks, hoods, ducts, flues, fittings, and general hardware. The success of one of the most familiar products, boats, depends directly upon the resistance to water-based degradation, both biological and chemical, and the ease of curved, monocoque-shell construction without any joints.

The same general characteristics make reinforced plastics attractive to the automobile designer, now under such pressure to reduce vehicle weight, and this market is growing rapidly. Reinforced plastics offer another advantage: complex subassemblies involving welds, rivets, screws, and bolts executed in metals often can be molded in one piece, eliminating substantial labor costs even though the plastics may be more expensive in terms of materials costs. A classic example of this is the front-end grille assembly which also supports the many lights used in current automobiles. In metals, several dozen different parts require manufacture and assembly; in reinforced plastics, the whole unit is made in one molding operation, and it weighs 50% less. The cost in place is substantially reduced.

Most polymers are poor conductors of heat and electricity and if the fibers added to them have similar properties, strong, stiff insulating composites result. This was the motivation for the development of one of the earliest laminates, which is still very popular. As the supply of natural mica declined, its use to insulate motors, generators, and transformers expanded, so the need for a substitute became urgent. Thus, sheets of Kraft paper impregnated with phenol–formaldehyde polymer [*9003-35-4*] were squeezed together under heat and pressure to produce a replacement "for mica," a trade name familiar to all, although the current product is used more for decorative and protective purposes than for electrical insulation.

Fibers

Table 1 presents data about most of the fibers used in composites and laminates. The two predominant fibers are glass and cellulose.

Glass. Fibrous glass comprises well over 90% of the fibers used in reinforced plastics because it is inexpensive to produce and possesses high strength, high stiffness, low specific gravity, chemical resistance, and good insulating characteristics. Originally, the composition was developed for standoff insulators for electrical wiring, in which resistance to surface adsorption of water is important to reduce arcing. Only later was it discovered to be an excellent fiber-former: the operating window of temperature and rates within which fibers can be drawn from the melt is unusually large, and full-scale continuous production of such E-glass (Electrical grade) fibers became practical. Based on this ease of production, and the attractive properties of the fibers, uses were sought. Reinforced plastics and fireproof textiles have emerged as the principal markets.

In reinforced plastics, the glass is used in various forms. When chopped into short lengths (6–76 mm) and gathered into a felt or matte, it is easy to handle and low in cost. The discontinuities, however, penalize the strength properties of the final composite; as a result, a large variety of woven textile forms, or fabrics, also are available. These textiles use continuous fibers, gathered into yarns of low twist, and woven with essentially conventional textile processes. The fabrics cost more but offer better properties, although some sacrifices of formability into complex curved shapes may be incurred unless the design of the weave anticipates this. Contrary to what might be expected, sometimes the dry fabrics are not easily handled because wracking, unravelling, and other problems can arise as they are cut, tailored, and put into place in molds or on forms.

The best properties are achieved with nonwoven fabrics in which all the fibers are straight, continuous, and aligned parallel in a single direction (see Nonwoven textiles). Layers of these can be stacked atop each other, with each layer oriented in a specific direction, thus providing any option between high directionality, useful in a rod or a pole for example, to approximately nondirectional properties, referred to

Table 1. Fiber Properties

Material	Specific gravity	Tensile strength, MPa[a]	Tensile modulus, GPa[a]	Cost range, \$/kg
E-glass	2.6	3570	85	2–5
carbon[b]	1.6	2035	357	30–150
carbon[b]	1.9	1790	430	30–150
Kevlar 29[c,d]	1.44	2860	64	5–10
Kevlar 49[c,d]	1.44	3750	135	5–10
bulk spruce wood	0.46	104	10	0.60–1.00

[a] To convert MPa to psi, multiply by 145; to convert GPa to psi, multiply by 145,000.
[b] Properties depend upon carbon/graphite ratio.
[c] E. I. du Pont de Nemours & Co., Inc.
[d] See Aramid fibers.

as quasi-isotropic, which would be suitable for structural plates or sheets similar to metals. Obviously, such nonwoven fabrics are not easy to handle unless they are impregnated with a partially cured matrix (prepregs); if not, most of their use is in automated machine processes which operate continuously producing pipes, cylindrical tanks, rods, and other similar shapes generated by revolving forms or molds.

Although compounded to be resistant, E-glass is not impervious to water and, like all glasses, its surface also is sensitive to abrasion damage. To protect the surface, and to promote adhesion to the polymeric matrix, coupling agents are applied to the fibers as soon after forming as possible. These are complex molecules, one end of which is intended to react with the glass and the other with the polymer as the latter solidifies. Thus, a primary valence-bond bridge joins the reinforcement to the matrix. This straightforward idea continues to be the subject of much technical controversy and dispute but from an empirical viewpoint there is no question: a glass fiber-reinforced composite lacking a coupling agent will lose half its strength after a month under water, whereas one with a proper coupling agent shows no strength loss at all. Consequently, only rarely are coupling agents not used irrespective of the type of reinforcing fiber employed (5).

There is one other characteristic of glass that is not widely recognized. When subjected to tensile loads–pulls for prolonged periods of time, glass breaks at stress levels much below those measured in short-time (2–5 min) laboratory tests (6). This behavior, known as static fatigue, effectively reduces the useful strength of glass if it is intended to sustain such loads for months or years in service. The reduction can be as much as 70–80%, depending upon the load duration, temperature, moisture conditions, and other factors, and if fracture does occur, it gives little or no prior warning, since glass is a brittle material. Conservative design practices and periodic inspections are recommended preventive measures in such applications where failure could have critical consequences (see Glass).

Cellulose. Cellulose fibers come from many sources and are used in many forms. Cotton (qv), jute, hemp, sisal, and bagasse (qv) from sugar cane provide economical comparatively long fibers with attractive properties of stiffness, strength, low specific gravity, and resistance to handling damage (see Fibers, vegetable). Often these are spun into yarns, then woven into fabrics such as burlap or canvas (see Textiles). Disadvantages include water sensitivity, lack of flame resistance, and susceptibility to biodegradation. The common source of short cellulose fibers, of course, is wood and the conventional use form is paper. As mentioned above, Kraft paper is very commonly used in laminates, where it is impregnated with phenolic resin and then fused under high pressure and moderate temperature to form electrical insulation stock. A refined wood product, alpha cellulose, also is used as a reinforcing fiber in plastic molding compounds to reduce curing shrinkage and brittleness and to improve impact resistance (see Cellulose).

Graphite. The stiffest fibers known are composed of graphite, which theoretically can be almost five times more rigid than steel. Practically, those now produced commercially are 1.5–2 times stiffer and laboratory specimens a bit more than 3 times stiffer have been made. The fibers themselves are composites: only part of the carbon present has been converted to graphite, in tiny crystalline platelets specially oriented with respect to the fiber axis. The higher the graphite content, the stiffer becomes the fiber; unfortunately, stiffness and strength are inversely related (7) (see Carbon).

Despite much work over many years by numerous technical organizations, the cost of graphite fibers remains high: the cheapest and lowest quality still are 15–20 times more expensive than glass. Barring unusual breakthroughs, this disparity is likely to persist since the starting materials are expensive and the processes of carbonization and graphitization consume much time, energy, and materials, and require close control throughout. As a result, their use in composites is limited to applications that place a premium on saving weight: aircraft, missiles (see Ablative materials), some sports equipment, and certain specialized hardware. For a time it was thought that automotive applications would be forthcoming, but better use of conventional materials and public acceptance of smaller automobiles have delayed such developments. Further substantial cost reductions will be necessary to penetrate this market.

Aramids. Aramid fibers (qv) are relatively recent industrial products and can be considered as an evolutionary product following nylon (7). In the stiffness range between glass and steel, they are lighter than glass, comparably strong, and much tougher and absorb considerable energy before breaking, even under impact conditions. Aramid fibers are two to five times the cost of glass, and they are used in composites in sports and transportation equipment, and in protective systems where ballistic stopping exploits their superior impact resistance. These fibers are highly crystalline and directional in character. Their transverse strength is very low, so that scuff resistance is poor and fibrillation can become excessive when the fibers are abraded or worked mechanically.

Polymeric Binders

In principle, any polymeric resin that can be liquefied and thereby used to wet the reinforcing fibers can be used as a matrix for a composite material. In practice, there are examples of almost any resin in some kind of composite formulation. Realistically, however, the bulk of reinforced plastics produced is based on polyester, epoxy, or a few thermoplastic matrix materials (see Elastomers, synthetic—thermoplastic; Epoxy resins; Polyesters). Only these are discussed below.

Polyesters. A polyester is the reaction product of a diol and a dicarboxylic acid. About ten different diacids and seven or eight different diols are in common use, from which different polyesters are available. A polyester is a low viscosity, clear liquid, composed of linear molecules with the potential for further chemical combinations. The latter are effected by adding another low molecular weight liquid, a cross-linking agent, which forms bridges or cross-links between the polyester chains when catalyzed by chemical additives or by heat, irradiation, or other energy inputs. The mixture thickens, releases heat, solidifies, and shrinks. Again, seven or eight different monomers are commonly used, and a very wide variety of polyester matrix resins are commercially available; most are comparatively inexpensive and easy to apply. The flexibility and variety of composition also facilitates the tailoring of properties for specific applications. This art has reached a fairly high level of sophistication so that electrical, chemical, or structural grades (among many) have been developed for optimum performance in such service (8). The volumetric shrinkage upon hardening (curing) is approximately 8% and since any reinforcement present does not shrink, this can cause internal stresses and cracking as well as dimensional, inaccuracy, surface roughness, and instability. Much of the art of molding is concerned with minimizing and preventing such actions, which greatly affect the quality of the final composite.

Epoxy Resins. Epoxy resins shrink less than polyesters (4%), adhere to most surfaces better, are affected less by water and heat, cure in a more controllable manner, and are more expensive. The basic material is a diglycidyl ether, a low viscosity liquid of moderate molecular weight, formed by reaction of epichlorohydrin and Bisphenol A (see Chlorohydrins; Alkylphenols). This resin can be cross-linked or solidified by a number of different amines, anhydrides, and acids, and it is capable of reacting with many other polymers to form copolymer substances. Therefore, the epoxy family has composition and property versatility comparable to the polyesters (9). Materials used for the purpose include phenolics, melamines (see Amino resins), polyamides (qv), esters, and many elastomers. The result is a very large number of commercially available epoxies formulated to optimize certain characteristics sought in specific applications: metal coatings, electrical insulation, chemical resistance, structural strength, adhesives, etc. Despite their superior inherent adhesion, however, when epoxies are reinforced by fibers it is necessary to use coupling agents to retain the strength of the composite if exposed to water.

In practice, both polyester and epoxy resins contain a large number of additional substances. Since both are adversely affected by exposure to sunlight, uv absorbers (qv) often are added. Stabilizers are used to prolong storage life and to prevent gradual gelation. Viscosity modifiers, fire retardants, mold-release agents (see Abherents), additives to improve molded surface smoothness, impact improvers, chemical accelerators, and catalysts are among the materials commonly used in such resins. The result is a very complex system which is heterogeneous on a microscale.

Thermoplastic Materials. As mentioned above, a number of thermoplastic materials also are used as matrix resins. These include nylon, polystyrene, polyethylene, polypropylene, styrene/acrylonitrile, polycarbonate, and polysulfone (see Polyamides; Styrene plastics; Olefin polymers; Acrylonitrile polymers; Polycarbonates; Polymers containing sulfur). Usually the reinforcement is glass, although the use of graphite fibers is growing. All of these resins melt reversibly when heated; as a result, high production processes such as injection molding and extrusion are used to fabricate parts. Short ($\leqslant$6.4 mm) chopped fibers encapsulated in polymer enter such machines but the rigorous mixing and high shear flow patterns encountered cause much damage to them. In the finished part, the fiber length has been reduced by 10–100 times, so that its principal contribution is stiffening rather than strengthening. This can be especially important at elevated use temperatures where the fibers reduce creep and enhance dimensional stability; the burning characteristics also can be changed favorably by their presence. Impact resistance often is improved and, on balance, fiber reinforcement of many thermoplastics is attractive enough to ensure a continued, rapid growth of the field (10).

Fillers

Fillers differ from fibers in that they are small particles of very low cost materials; they are used extensively in reinforced plastics and laminates. Typical fillers include clay, silica, calcium carbonate, diatomaceous earth, alumina, calcium silicate, carbon black, and titanium dioxide. Sometimes they comprise as much as half the volume of a composite. They provide bulk at low cost, and they confer other valuable properties such as hardness, stiffness, abrasion resistance, color, reduced molding shrinkage, reduced thermal expansion, flame resistance, chemical resistance, and a sink for the heat evolved during curing. If coupling agents are used on their surfaces, many of them improve the impact strength of the composite. The original motive for using fillers was to replace part of the expensive resin phase, and this skill is important, but now their technology is becoming more complex and their use more versatile. Effects of particle size, size distribution, shape, surface treatment, and blends of particle types on the flow behavior and final properties of the composite are better understood, ensuring that their use will continue to expand (7) (see Fillers).

Products and Processes

Statistics in this area are imprecise, but as shown in Tables 2 and 3, a reasonable estimate of use categories is: marine (boats, decks, shields, etc)—20%; transportation (autos, trucks, trailers, body components)—24%; construction (corrugated sheet, space dividers, showers, tubs, lightweight control panels)—21%; chemical (pipes, ducts, hoods, tanks)—12%; electrical (printed circuits, insulation panels, switchgear)—9%; appliances, aircraft, recreational (housings, partitions, luggage racks, floor pan-

Table 2. Reinforced Plastics Markets[a]

	1979 (1000 metric tons)
Reinforced polyesters	
marine	159
transportation	191
construction, other (eg, farm equipment)	171
chemical	98
electrical	73
appliance, aerospace, consumer	106
Total	*798*
Epoxies	
electrical laminates	10
filament winding	9
other	5
Total	*24*

[a] Ref. 11.

Table 3. Reinforced Plastics Materials[a]

	1979 (1000 metric tons)
Reinforced thermoplastics	
nylon	20
polycarbonate	5
polyester	22
polypropylene	34
styrenics	15
other	9
Total	*105*
Decorative and industrial laminates	
Kraft phenolic	181 (est)
melamine phenolic	45 (est)

[a] Ref. 11.

els)—13%. In 1979, approximately one million (10^6) metric tons of reinforced plastics were used in the United States, the three principal resins being polyesters, thermoplastics, and epoxies. As mentioned above, over 90% of the fiber reinforcement was glass.

Hand Lay-Up/Spray-Up. The construction of boat hulls and comparably large articles is done by hand lay-up methods: one or two layers of dry fabric or matte are placed in molds or on forms and then liquid polyester is poured on them. The resin is distributed throughout the fibers by hand-roller action, and after several cycles, the desired thickness is achieved. Curing takes place by catalysis, either at room temperature or by means of heat lamps or warm-air ovens. Cycles are lengthy (8–24 h), production rates are low, molds are not expensive, and the quality of the product can vary widely. However, hulls as long as 49 m have been produced by this method. In the U.S., more than 90% of all boats less than 15 m are built by this procedure. (A variation of the technique, spray-up, uses a hand-held gun which propels chopped fiber and liquid resin against a mold surface until the desired thickness has been attained. This is used commonly for smaller-size boats).

Die-Molding. Most automotive parts and general hardware are made in heated matched metal dies or molds mounted in hydraulic presses operating semi-automatically. A charge of SMC (sheet-molding compound) or BMC (bulk-molding compound) is placed in the mold, then cured under moderate heat and pressure (120–175°C, 3.5–13.8 MPa (500–2000 psi)) for cycles ranging from 15–90 s. Typical recipes are 100 parts resin, 150 parts filler, and 30 parts chopped glass fiber. The molded piece is removed hot and allowed to air-cool before undergoing further operations such as trimming, assembly or painting. The quality of the product is fairly consistent, much subassembly work normal to metal practice can be avoided, and die costs are much lower than for sheet-metal stamping or metal die casting. Thus, shorter production runs are economical and more design flexibility and variety can be achieved.

Filament Winding. Pipes, tubes, and cylindrical tanks are made by a process known as filament winding. A mandrel or form is rotated about one axis as continuous yarn or roving which passes through a bath of liquid resin is wet-wound on it, usually in complex patterns controlled by automated machinery. Once the required wall thickness is reached, heat curing in autoclaves, ovens, or by heating the mandrel is begun. Because the reinforcement is continuous and its orientation subject to close control, very efficient structures can be produced consistently by this method, and in the case of smooth-bore pipes, the process can be continuous. Once cured, the pipe is removed from the metal mandrel and cut into proper lengths; mandrels for tanks and pressure bottles often are inflatable, soluble, or collapsible, so they can be removed through end ports. Pipe production for oil fields, the chemical processing industry, water distribution, and sewage disposal has reached very large volumes, in diameters from several centimeters to as much as 6 m; fittings also can be filament wound or die-molded. The attributes of light weight, ease of assembly (by field-applied adhesives), corrosion resistance, and economical production have combined to make this a very attractive business.

Flat or corrugated sheet stock normally is made in simple, multi-opening presses, often with prepreg (preimpregnated fabric or matte). Generally, quality and consistency are related directly to pressure and temperature, ie, high performance composites usually are made in this way. High fiber contents, good resin penetration, and full, controlled cures are easiest to achieve here and partially automated production methods keep costs moderate. Most of the electrical insulating board and printed circuit panels are high pressure laminates, using woven glass fabric or cellulose paper, with epoxy or phenolic resins.

Injection Molding. Injection-molded, fiber-reinforced thermoplastics compete directly with die-cast metals in many hardware and automotive parts. Higher production rates, lighter weight, and assembly simplification are their principal advantages. The fibers stiffen and allow higher operating temperatures. Often, however, the flow patterns of the molten material into the mold cavity cause significant orientation of the fibers and the localized anisotropy which results can adversely affect strength and thermal-expansion properties. Considerable skills are required of the mold designer and the machine operator. Also, the reuse of fiber-reinforced thermoplastic scrap material from runners, sprues, gates, and flash presents problems that can impair the overall efficiency of material utilization. Despite these difficulties, however, this is the fastest growing segment of the entire composites family (see Plastics processing).

Miscellaneous. The four processing techniques that have been mentioned are the dominant ones in use today: hand lay-up/spray-up; matched metal die molding (including flat-press laminating); filament winding; and injection molding. However, there are many others that may be variations of one of the above or are distinctive in themselves. For example, vacuum-bag techniques can be used to help remove air from a hand lay-up and to provide modest positive pressure on large, low cost molds too bulky for mechanical pressing. Better quality composites result from the reduction in void content and from the superior impregnation of fiber bundles by the liquid-matrix resin.

Pultrusion, in which continuous filaments are drawn through an orifice, which also meters out the encapsulating resin, can produce a variety of profiles containing high volume fractions (50–70%) of reinforcement. Very strong and stiff rods, tubes, and structural shapes result, but the properties are highly directional; transverse behavior depends almost entirely on the matrix which is at least an order of magnitude weaker and more flexible.

Transfer molding, borrowed from rubber technology, employs a piston to force compound to flow from a cool cylindrical reservoir into a hot mold cavity where it cures. Less fiber damage and fewer orientation effects are encountered, compared to injection molding, but cycles are longer and the process is discontinuous. This process has improved quality, permits more intricate detail and thicker sections, but it also has higher costs.

A variation of filament winding is centrifugal casting and some pipe is produced in this manner. Prewoven braids or sleeves of reinforcement can be impregnated quickly and effectively by the liquid resin and it is easy to maintain pressure on the assembly during curing. A very good quality product results.

Certain thermoplastic sheet materials such as polypropylene, nylon, polycarbonate, and others can be formed at temperatures well below their melting points. Even if these contain fibers or fabrics, some such materials still possess cold formability and they can be stamped, roll-formed, or bent into final shape simply by mechanical action. A substitute for sheet metal results that is light, stiff, strong, corrosion resistant, and has insulating properties.

Reaction injection molding (RIM) processes pump two or more liquid streams under high pressure into an impingement chamber where they are mixed intimately and then are forced into a mold cavity at much lower pressures. By heat, catalysis, and the mixing action, rapid polymerization occurs from reactions among the components in the cavity, which can tolerate no gaseous by-products. Urethanes are the principal polymers used in this process which emphasizes speed, large parts, and automation (see Urethane polymers). Reinforced RIM is being developed (RRIM) containing short chopped glass fibers or particulate fillers for stiffness, reduced shrinkage, and higher use temperatures. By this route, the "forgiving fender" for automobiles is predicted.

The area of fiber-reinforced plastics is complex and dynamic. A review of this type can describe only basic ideas and a few highlights, but comprehensive treatments are available (12–14) and several technical journals report new developments on a continuing basis (15–18).

Perhaps one of the most exciting developments in fibrous reinforcement is now emerging: molecular composites. Using special extrusion and drawing techniques, synthetic polymer fibers with ultrahigh orientations can be produced, and they exhibit strength and stiffness characteristics approaching theoretical limits, directly comparable to the strongest metals available (19). If the technology of these can be manipulated so they can be used in and reinforce matrices of the same or similar compositions, nearly ideal composite materials can be postulated (20).

BIBLIOGRAPHY

"Lamination and Laminated Products" in *ECT* 1st ed., Vol. 8, pp. 185–192, by H. W. Narigan and G. E. Vybiral, Panelyte Division, St. Regis Paper Co.; "Laminated and Reinforced Plastics" in *ECT* 2nd ed., Vol. 12, pp. 188–197, by C. S. Grove, Jr., Syracuse University, and D. V. Rosato, Consultant, Plastics World.

1. A. Kelly, *Strong Solids,* Clarendon Press, Oxford University Press, London, 1973.
2. R. M. Jones, *Mechanics of Composite Materials,* Scripta Book Co., Washington, D.C., 1975.
3. *Advanced Composites Design Guide,* 3rd ed., Vols. I–V, Air Force Materials Laboratory, Wright-Patterson Air Force Base, Oh., 1973 to date, with revisions.
4. L. R. Calcote, *Analysis of Laminated Composite Structures,* Van Nostrand Reinhold Co., New York, 1969.
5. "Composite Materials" in E. P. Plueddemann, ed., *Interfaces in Polymeric Matrix Composites,* Vol. 6, Academic Press, Inc., New York, 1974.
6. E. B. Shand, *Glass Engineering Handbook,* McGraw-Hill, New York, 1958.
7. H. S. Katz and J. V. Milewski, eds., *Handbook of Fillers and Reinforcements for Plastics,* Van Nostrand Reinhold Co., New York, 1978.
8. P. F. Bruins, ed., *Unsaturated Polyester Technology,* Gordon and Breach Science Publishers, New York, 1976.
9. H. Lee and K. Neville, *Handbook of Epoxy Resins,* McGraw-Hill, New York, 1967.
10. J. Agranoff, *Modern Plastics Encyclopedia, 1979–1980,* Vol. 56, No. 10A, McGraw-Hill, New York, 1979.
11. *Mod. Plast.* **57**(2), (Jan. 1980).
12. S. S. Oleesky and J. G. Mohr, *Handbook of Reinforced Plastics,* Reinhold Publishing Corp., New York, 1964.
13. G. Lubin, ed., *Handbook of Fiberglass and Advanced Plastics Composites,* Van Nostrand Reinhold Co., New York, 1969.
14. L. J. Broutman and R. H. Krock, eds., *Composite Materials,* Vols. 1–8, Academic Press, Inc., New York, 1974.
15. *Annual Proceedings—Reinforced Plastics/Composites Institute,* Society of the Plastics Industry, New York.
16. *J. Compos. Mater.,* Technomic Publishing Co., Westport, Conn.

17. *J. Sci. Technol. Reinforced Mater.*, I.P.C. Science and Technology Press Ltd., Bury St., Guilford, Surrey, England, GU25BH.
18. *Polymer Composites,* Society of Plastics Engineers, Inc., Brookfield Center, Conn.
19. A. Ciferri and I. M. Wards, eds., *Ultra-High Modulus Polymers,* Applied Science Publishers, London, 1979.
20. A. Ciferri, *Polym. Eng. Sci.* 15(3), (Mar. 1975).

FREDERICK J. MCGARRY
Massachusetts Institute of Technology

LAMINATED MATERIALS, GLASS

A laminate is an orderly layering and bonding of relatively thin materials. A commonly laminated material is glass. Usually, two pieces of float or sheet glass are bonded with poly(vinyl butyral) (PVB) (see Vinyl polymers, poly(vinyl acetals)) to produce a highly transparent safety glass, eg, an automotive windshield. This combining of transparent abrasion-resistant glass and resilient plastic achieves the durability and safety demanded of such products. Other materials that may be incorporated in laminated glass are colorants, electrically conducting films or wires, and rigid plastics. The value of the laminate is the utilization of the desirable properties from each of the constituents. In the case of laminated glass, the excellent weathering properties of the glass protect the impact-energy-absorbing plastic interlayer from deterioration, abrasion, and soiling.

Benedictus, a French chemist who accidentally broke a flask that contained dried-on cellulose nitrate, is credited with founding the laminated-glass industry (1). The first patent was issued in 1906 (2).

The growth of the laminated-glass market was slow until automobile numbers and automotive speeds increased to the point that glass-caused injury was of concern. By the late 1920s, laminated windshields were standard in United States automobiles. The most common construction was two pieces of plate glass bonded with cellulose nitrate. However, the plastic interlayer introduced problems of haze, discoloration, and loss of strength, and it was replaced by cellulose acetate in 1933. Cellulose acetate demonstrated improved stability to sunlight but lacked strength over a broad temperature range and produced haze. The advent of the poly(vinyl butyral) resins in 1933 permitted the development of the modern interlayers that are used to make the majority of laminated safety glass in use today; the resins were adopted for all automotive laminates by 1939.

Laminated glass is not a true composite material (see Composite materials). The glass needs the safety-net effect of the interlayer if impacted, and the interlayer needs the durability and rigidity of the glass for useful service other than during impacts. Exceptions where laminated glass more truly fits the definition of a composite are when it is used for noise attenuation (see Insulation, acoustic) or bullet resistance. In these applications, the alternate layering of rigid and soft materials achieves results beyond those produced by either alone.

Properties

Laminated materials frequently have limits on properties below those found in one of the components. Laminated glass, with a PVB interlayer, has a service temperature not exceeding 70°C (conservative), far below that of solid glass. The strength of laminated glass is dependent upon the number, thickness, and strength of the individual glass plies and upon the characteristics of the particular interlayers used. For the majority of laminates consisting of two plies of annealed glass and one PVB interlayer, the bending strength is about 0.6 of that for an equal thickness of solid glass.

Laminated glass becomes more rigid with a decrease in temperature and, below −7°C, it approaches the performance of solid glass. At temperatures above 38°C, it responds more nearly like two glass plies separated by a fluid. Some applications utilize heat-strengthened or tempered glass for additional strength. Figure 1 is an example of a wind-load chart for the combination of heat-strengthened and laminated glass (3). Wind-load information is used jointly by the architect, glazing contractor, and glass manufacturer to determine the permissible glazing area and glass thickness that is required to meet the design wind load.

Most laminated glass applications are concerned with impact strength, and minimum performance levels are required by specification. The impact strength of two plies of laminated annealed glass and various PVB thicknesses are reported in ref. 4. Aircraft laminates may utilize electrical resistance heating to improve laminate impact resistance (see Aircraft Windshields).

Automotive and architectural laminates of PVB develop maximum impact strength near 16°C, as shown in Figure 2. This balance is obtained by the plasticizer-to-resin ratio and the molecular weight of the resins. It has been adjusted to this optimum temperature based on environmental conditions and auto population at various ambient temperatures. The frequency and severity of vehicle occupant injuries vs temperature ranges at the accident location has been studied (5), and the results

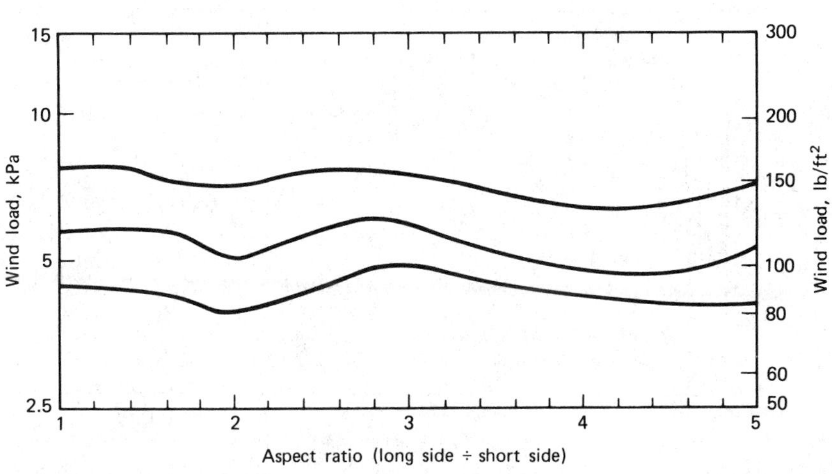

Figure 1. Wind-load data for heat-strengthened and laminated 3.2-mm glass. Architect's specified probability of breakage is 8/1000 laminates for a 1-min uniform wind-load duration. Four sides supported in weathertight rabbet. Courtesy of PPG Industries, Inc.

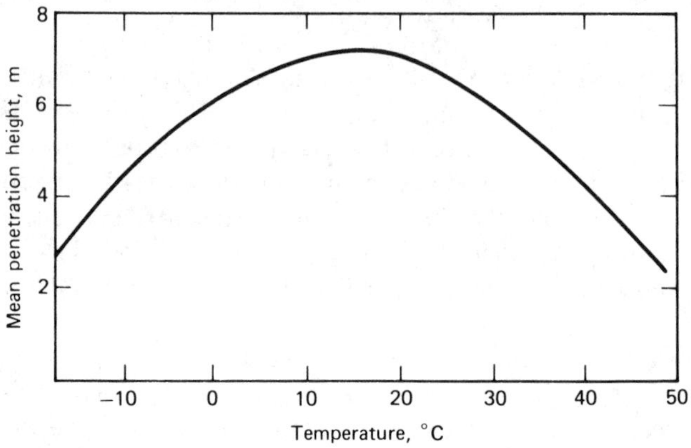

Figure 2. Typical penetration resistance vs temperature data from laboratory procedure (305 × 305 mm laminates, 0.76 mm PVB; 2.27-kg ball impact). Courtesy of Monsanto Co.

confirm the selection of the maximum performance temperature and the decreasing penetration resistance at temperature extremes.

The optical properties of laminated glass are required to be equal to solid glass, since most applications are in vision areas. Light scattering by the interlayer essentially is nonexistent if PVB is used. Clean-room practices can reduce the dust and lint that is attracted to the surfaces (see Sterile techniques). Visible-light transmittance of a typical automotive laminate (2.1 mm glass–0.76 mm PVB–2.1 mm glass) is nearly equal to solid glass (Fig. 3), and noticeable color usually is absent. Visible-light transmittance is about 88% for clear glass laminates and 78% for tinted laminates. All light is absorbed below 370 nm and several discrete absorption bands are in the infrared beyond 1100 nm. The uv absorption may be enhanced when additional protection of color dyes is required, eg, gradient shade bands in automotive windshields or merchandise in

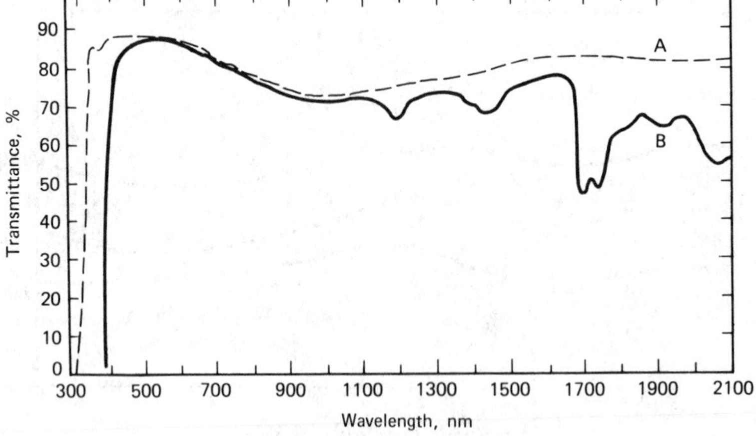

Figure 3. Visible-light transmittance of automotive laminate: A, 4.75-mm monolithic glass; B, 5-mm laminated glass. Courtesy of Ford Motor Co.

window displays (see Uv absorbers). Generally, the solar uv internal transmittance (neglecting losses resulting from surface reflections) is on the order of 30–35%, and the infrared is about 97% for 0.76-mm thick PVB.

The index of refraction of PVB at 1.48 is close enough to glass (which has a refractive index of 1.520) to couple the two glass plies, with a loss by reflectance of 0.02%:

$$R = \frac{(n_2 - n_1)^2}{(n_2 + n_1)^2} \tag{1}$$

where R is the reflectance at the interface and n_1 and n_2 are the different refractive indexes. The absorption coefficient for visible light (400–700 nm) generally is −0.25 to −0.45 for PVB. This produces a transmission loss within the PVB of 0.7–1.3%, which is mostly in the blue portion of the spectrum.

Subsequent to the lamination process, some defects may appear that were not visible previously. One of these phenomena is called a bull's eye when found in windshields. These are small depressions that are formed in the glass pair during bending by the presence of glass chips or other debris between the plies. Upon lamination, the pockets fill with PVB and become convex lenses. Conversely, shallow ridges on an internal glass surface may be absorbed in the PVB and the optics are improved. Various other optical distortions may be caused by nonparallel plies of glass or of PVB or nonuniform fabrication pressures.

Manufacture

Practically all laminated glass utilizes plasticized poly(vinyl butyral) (PVB) as the interlayer. Curved, laminated windshields are by far the principal product; silicone and cast-in-place urethane resins are used seldom. Laminators purchase PVB in rolls up to 500 m long, up to 270 cm wide, and from 0.38–1.52 mm thick. There are several plasticizers that are used and at different ratios of plasticizer-to-resin content, depending on the product being manufactured. Typically, Flexol 3 GH (bis(2-ethyl butanoic acid), triethyleneglycol ester) manufactured by Union Carbide, is utilized at about 44 parts per 100 parts of resin. Other plasticizers used are di-*n*-hexyl adipate and di-butyl sebacate. There are at least five companies offering these products with manufacturing facilities in the United States, Japan, Belgium, Germany, and Mexico. Since the plastic is an adhesive material, it is shipped either with a dusting of parting agent on the surface or is refrigerated so it does not cohere (see Abherents). The refrigerated material is clean, moisture adjusted, and ready to laminate. The dusted material requires washing and moisture conditioning. Removal of the parting agent by warm water, followed by a chilled water rinse, adds about 0.2% water content to the plastic which must be compensated either by overdrying before washing or drying after washing so as to achieve the desired 0.3–0.5% H_2O content. These steps are done more efficiently on the continuous roll before cutting.

The drying stage is carefully controlled to relax the sheeting of physical stresses as well as to adjust the moisture content. It consists of draping the sheeting over slat conveyors or driven rolls in a temperature- and humidity-controlled oven. The gradient-band sunshade that appears in many windshields is printed continuously on the interlayer roll at the PVB-manufacturing plant. To permit a more pleasing conformance to the curved glass, the interlayer may be preshaped which causes the extremities of the band to more nearly parallel the horizon in the installed windshield. The shaping of the interlayer may be carried out on the continuous roll using a vinyl expander prior to cutting the blanks (6). The radius of curvature is preset, depending on the particular windshield being manufactured. The interlayer then is cut to approximate laminate size and is accumulated in low stacks (150 mm) ready for assembly. Another method for shaping the interlayer involves warping in special ovens after the blanks are cut. The interlayer stacks must be stored in cooled, moisture-controlled rooms to prevent water absorption and blocking.

The glass for laminating may be annealed, heat-strengthened, tempered, flat or curved, clear or colored. Thicknesses of 1.5–12 mm or more are used. For flat laminates, the glass is cut to size, edged and treated, if specified, washed, and delivered to the clean room by conveyor. The washing process, in addition to cleaning, can affect the interlayer bond. Common water hardness residues at invisible levels can decrease the adhesion to the interlayer. The desired level is achieved by controlling the hardness of the final rinse water and by removing the water by air stripping as opposed to evaporative drying. The glass is cooled during drying to prevent premature grabbing when the interlayer is placed on the glass, thereby permitting easier positioning of the components.

In order to manufacture curved laminates, the glass is preshaped before laminating. This is achieved by simultaneously bending a pair of glass templates which are usually cut to the shape of the finished windshield and are separated by an inert powder to prevent fusing of the plies. The bending process occurs on a peripheral support metal fixture or mold; the pair slowly travels through a lehr so that the glass sags to shape by gravity. Glass temperatures of 600°C are required to achieve the shape, and the shaping is followed by annealing. Banded windshields usually are constructed with one or more pieces of tinted, heat-absorbing glass to enhance occupant comfort and to reduce air-conditioning load.

The clean room is operated at 18°C and 26% rh, which is an equilibrium condition for interlayer moisture control. The interlayer is placed on one piece of glass (Fig. 4) with the gradient band, if present, carefully positioned above the designated eye position. The adjacent piece is superimposed, excess interlayer is trimmed, and the sandwich is conveyed from the room through a series of heaters and rolls that press the assembly together while expelling air. Temperature is increased stepwise to 90°C and pressures of 170–480 kPa (25–70 psi) are applied. Solid rubber rolls usually are used with flat laminates, and curved glass requires segmented rolls on a swivel frame (Fig. 5) (7–8). For more complex shapes, peripheral gaskets may be applied and the assembly may be evacuated (9), or the entire assembly may be placed in a bag and evacuated. The bag may or may not be removed prior to autoclaving but, when using an oil autoclave where the oil would damage one of the components, an oil-resistant bag of poly(vinyl alcohol) is used (10). The tacked assembly is loaded onto racks for autoclaving, which may be either in an air or oil vessel that is capable of pressing the sandwiches at 1.38–1.72 MPa (200–250 psi) and 100–135°C for 30–45 min. Curved laminates and multiple laminates may require longer cycles.

In the autoclave cycle, the pressure is increased more rapidly than the temperature and then it is maintained towards the end of the cycle as the temperature is lowered, to reduce any chance of delamination. The exit temperature must be near 50°C to avoid thermal breakage. During the cycle, residual air is absorbed by the interlayer, and the embossed surface of the interlayer flows and wets the glass surface, thereby producing

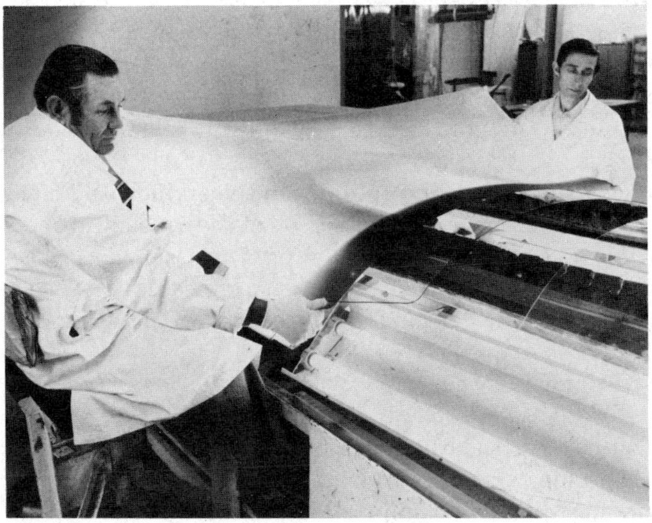

Figure 4. The vinyl interlayer is inserted into the matched set of bent glass in a cold room where the humidity and temperature are carefully controlled. Courtesy of Ford Motor Co.

Figure 5. The glass and vinyl sandwich is fed through a special deair machine to remove trapped air and increase the adhesion of the vinyl to the glass. Courtesy of Ford Motor Co.

the clear laminate. Occasional small residual bubbles of trapped air can be removed by an additional autoclaving cycle. The air–autoclave process is becoming more popular because it eliminates the subsequent washing of oily residues that are produced in oil autoclaving. The elimination of the process oil and subsequent wash-water waste products also are environmental improvements. In some cases, additional trimming of the interlayer or finishing of the glass edge is required. The protruding interlayer may be removed by wire brushing, but the glass edge is finished by abrasive belt-seaming or diamond-wheel edging (see Abrasives). The labeling of safety glass is done by grit-blasting through a mask or by enameling (see Specifications). A final step in windshield manufacture is the bonding of a small metal plate to the windshield. This plate, which is used to support the rearview mirror, is laminated to the glass using poly(vinyl butyral).

Radio antennas have been incorporated in some windshield models by inclusion of a very fine copper wire placed across the top of the windshield and vertically at the center. The wire is tacked to the interlayer prior to assembly and is embedded into the interlayer during autoclaving. An electrical connector that is soldered to the antenna wire is bonded to the bottom edge of the windshield for ease of connection.

Production and Shipment

Chemical attack, particularly from moisture and alkaline conditions, is prevented by use of acidic packing materials and open, ventilated packages. Good crate design and proper handling throughout shipment avoids mechanical damage. Glass-to-glass contact is never permitted. Long-term storage must be in well-ventilated areas, never in sealed containers.

Flat laminates are separated only by newsprint or plastic beads and are bound into a block to prevent movement between the laminates. Curved laminates are spaced to prevent abrasion and require supporting dunnage at several points to prevent breakage by excessive flexing during shipment. Banding and blocking are designed to add compressive forces only. Staining is not a problem in the uncovered areas, but the supporting members are specified to have an acidic content.

Laminated-glass products are considered noncombustible and are shipped without DOT hazardous-warning labels. Flat-laminate packs, because of their high density, do not fill the car or trailer and require sturdy bracing to prevent shifting. Glass products are shipped and stored in a vertical plane, and during transportation, they are placed so that each plate has an edge in the direction of travel.

Economic Aspects

The growth of laminated glass closely followed the growth of motor vehicles from the late 1920s to the 1960s. Windshields and side glasses of all domestic vehicles were laminated during this period. In the early 1960s, the flat laminated side glass was almost entirely replaced by curved tempered glass. The curved windshield laminated-glass market continued to follow automotive trends, but the flat-glass products redeveloped around architectural uses. Architectural products represent 5–10% of the laminated-glass volume. Increased consumer-safety awareness and security needs are expanding the flat-laminate market. Safety codes (eg, 16 CFR 1201 and ANSI Z97.1) specify laminated glass as one means of meeting their requirements (11–12).

Flat laminated glass and windshield-replacement production has been estimated to 1982 using trend lines of historical data (13). A similar technique has been used to determine original equipment manufacture (OEM) of windshields for worldwide use and the data are compiled in Table 1 (14). Laminated windshields, as opposed to tempered-glass windshields, are gaining in the market share outside of North America. From 37% of the non-North American market of 1976, they are estimated to have

Table 1. Laminated Glass and Windshield Production

Production	1976	1979	1982[a]
U.S. laminated flat glass, 10^6 m^2 [b]	2.7	3.5	4.6
wholesale values[a], 10^6 \$ U.S. (1979)	87	114	150
worldwide laminated windshield, 10^6 units			
U.S. and Canadian OEM[c]	13.1	14.7	17
U.S. replacements[b]	4.6	4.8	5.4
foreign	9.7	15.3	24
Total	*27.4*	*34.8*	*46.4*
wholesale value[a], 10^6 \$ U.S. (1979)	919	1137	1471

[a] Estimate.
[b] Courtesy of U.S. Glass Publications, Inc. (13).
[c] In part from ref. 14.

reached 55% by 1978 and are projected to obtain 75% by 1982. In addition to North America: Belgium, Italy, and the Scandinavian countries permit only laminated windshields, and other nations are increasing use by customer option.

Specifications

Practically all laminated glass is made, tested, and certified to comply with certain safety performance standards. In the United States, there are two types of standards: automotive and architectural. For the former, *ANSI Z26.1-1966* is used and is incorporated in the Federal Motor Vehicle Safety Standard 205 (15). It specifies safety performance, durability, and optical quality. Specific tests are required depending on the location in the vehicle where the glazing is to be used. Item 1, the most difficult to meet, may be used in any location in the vehicle. It requires, in addition to other tests, support of a 2.3-kg ball dropped from 3.7 m onto a 305-mm square of laminate. Item 2 (safety glazing for any location except windshields) may be met by laminated glass using thinner PVB because it does not require the 2.3-kg-ball test nor the distortion tests. Laminated glass also may be used in locations specifying item 3 (no visible transmittance requirement) or item 11A (bullet-resistant glass; also requires appropriate tests, eg, ballistic tests). Automotive safety glass is required to be labeled as to the manufacturer, code, item, and model number that will identify the type of construction.

Conformance to the standard is achieved by submitting samples to an approved laboratory for evaluation and submitting the laboratory report to the American Association of Motor Vehicle Administrators (AAMVA). The approved certificate is sent to the manufacturers with copies to the state and provincial jurisdictions for which the AAMVA serves as approvals agent (16).

Laminated glazing materials that are used in building locations specified by federal regulations are certified to comply to *Federal Standard 16 CFR 1201* by the Safety Glazing Certification Council (SGCC) (17). Other locations requiring safety glazing specified by state or local code may use *ANSI Z97.1-1975* (12). Glass complying to these standards is labeled permanently as to the standard (or standards) that it meets, including thickness, identification of the manufacturer, and plant. Also, it usually contains a date of manufacture. In situations where the large laminated sheets are cut into smaller pieces by the local distributor or installer, each piece is permanently labeled to indicate that it was cut from glass meeting the standard.

Certification to these standards is obtained by submitting a test report from an approved laboratory to the SGCC. Once certified, the product is assigned an SGCC certification number to identify it and the factory at which it was made. Subsequently, samples are selected randomly by the administrator at least twice a year to ensure continued adherence to the standard. Based on these reevaluation reports, SGCC authorizes continued use of the certification label and the product listing published in its directory. The building standards are concerned mainly with body impact, and they require testing by impact on the glazing with a 45-kg bag. Detailed testing procedures and interpretations are given in refs. 11–12.

Bullet-resistant glass products are tested according to *UL 752* (18). The test specifies that three shots are fired from 4.6 m and impacting within 100 mm of each other in a triangle, and that there is no penetration of the projectile nor any glass imbedded in the corrugated board. The level of approval is determined by the velocity and energy level of the bullet at the muzzle of the firearm. Additional tests required include impacts 38 mm apart and tests over temperature ranges of 13–35°C for indoor use and −31.7 to 49°C for outdoor use.

The above-mentioned codes contain requirements for accelerated durability tests. In addition, interlayer manufacturers and laminators expose test samples for several years under extreme weather conditions, eg, the Florida coast and Arizona desert. The laminated products weather extremely well, with no change in the plastic interlayer. Occasionally, clouding is noted around the edges when exposed to high humidity for long periods, but this is reversible.

Analytical and Test Methods

Interlayer moisture is one of the important controls for adhesion to glass. The moisture content equilibrates with the relative humidity to which the interlayer is exposed and, thus, is variable. Prior to lamination, interlayer moisture content is measured by one of three methods. The most rapid is by ir absorption using a spectrophotometric technique to determine a ratio of the 1.925 nm to the 1.705 nm wavelength peak (Fig. 3). A slower but less expensive method is weighing the interlayer before and after vacuum desiccation. The third and classical method is by Karl Fisher reagent. The infrared method, in addition to being the most rapid, permits measurement of the interlayer moisture content while the interlayer is in the laminate.

Interlayer bond strength is determined by either pummeling the laminate at −18°C to break away the glass and to determine the amount of adhering glass particles or by compressively shearing the laminate sample in a universal test machine. The optimum pummel range is three to six units on an arbitrary scale established by the industry. The relationship of pummel and compressive shear data to water content of the interlayer is given in Figure 6 and that of pummel to mean penetration height is given in Figure 7. These data are influenced also by residual hardness of the water used to wash the glass.

Subsequent to processing, an inspection is made for incomplete bonding, inside dirt, and glass quality. In the case of windshields, rigid optical standards are required, and these must be evaluated for the completed windshield. Extensive test requirements are described in the appropriate codes (11–12,15,18–24), and they include light stability, humidity, boil test, abrasion, and assorted impact tests.

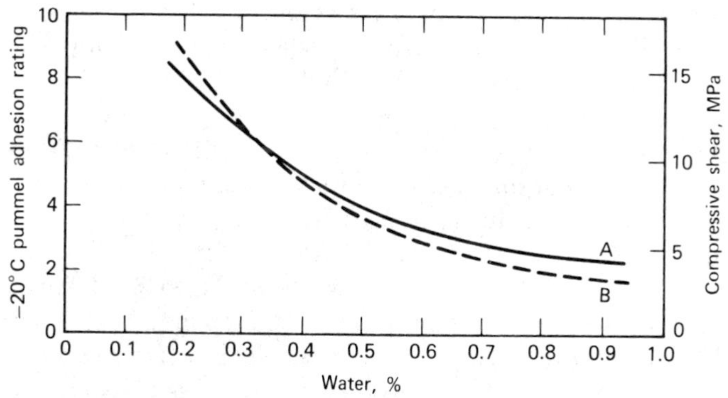

Figure 6. Typical effect of moisture on PVB adhesion. A, Pummel data from Monsanto Co.; B, compressive shear data from DuPont Co. To convert MPa to psi, multiply by 145.

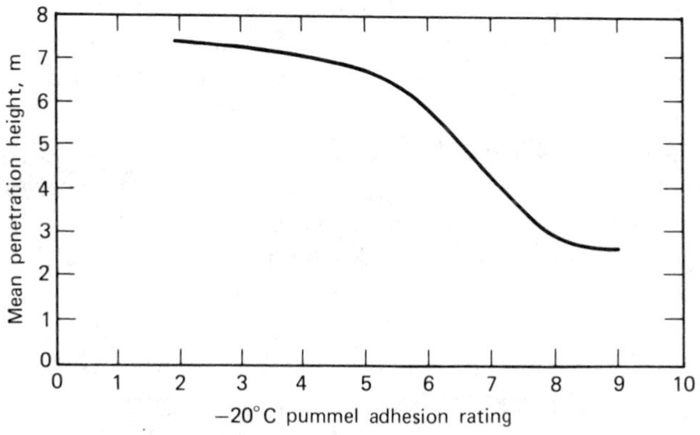

Figure 7. Typical variation of mean penetration height with adhesion (2.27-kg ball, impact at 20°C on 305 × 305-mm laminates; 0.40% water, 0.76 mm PVB). Courtesy of Monsanto Co.

Uses

Impact-Resistant Windshields. Performance difference between windshields manufactured in the FRG and the United States was reported in the early 1960s (25–26). Three variables contribute to the greater safety of the German windshields against impact: thinner glass (especially the inboard member), thicker plastic interlayer, and higher moisture content of the interlayer. The latter acts as a plasticizer and adhesion control (see Plasticizers). By reducing the adhesion of the interlayer to the glass, more interlayer area can be released and stretched during impact. Also, the thicker interlayer, in addition to having more inherent strength, causes more fracturing of the glass during impact than in the older model of windshield. This, in turn, increases the amount of released interlayer for impact-energy absorption. Upon review by the SAE Glazing Committee, it was agreed that the improvement was desirable, if it could be accomplished without taking the risk of increased water content. The U.S. PVB manufacturers subsequently developed controlled-adhesion interlayers without increasing moisture above the previous standard content, and the glass fabricators utilized this material to produce laminated windshields with more than twice the impact resistance of the pre-1966 windshields. This product was introduced in limited production in 1965 and was used in all United States car lines for the 1966 models (27).

The ASA (now ANSI) performance code for Safety Glazing Materials (*Z26.1*) was revised in 1966 to incorporate these improvements in windshield construction. The addition of test no. 26 requiring support of a 2.3-kg ball dropped from 3.7 m defined this level of improvement. It was based on a correlation established between 10-kg, instrumented, head-form impacts on windshields, on 0.6 × 0.9 m flat laminates and the standard 0.3 × 0.3 m laminate with the 2.3-kg ball (28). Crash cases involving the two windshield interlayer types were matched for car impact speeds and were compared (29). The improved design produced fewer, less extensive, and less severe facial lacerations than those produced in pre-1966 models.

Additional improvements have been incorporated since 1966 with the availability of thinner float glass. Glass thickness and interlayer thickness have been studied to optimize the product for occupant retention, occupant injury, and damage to the windshield from external sources (30–31). The thinner float-glass windshields are more resistant to stone impacts than plate-glass windshields. The majority of laminated windshields are made of two pieces of 2–2.5 mm of annealed glass and 0.76 mm of controlled-adhesion interlayer.

Special Laminated Windshields. Combinations of strengthened glass and interlayer offer advantages of lessened weight, higher impact resistance, lowered laceration potential, and resistance to bending stresses. These may be needed in high speed aircraft, helicopters, and motor vehicles. The additional strengthening is achieved by chemical or thermal processes. The chemical process utilizes a special glass of high alumina content, which, after bending to shape, is processed by ion exchange (qv) in molten potassium salts to produce a highly compressed skin and a center tension with a bending stress up to 276 MPa (40,000 psi). The product, offering reduced laceration potential, was used briefly on some United States-produced cars (32–33). A thermal process capable of inducing high stress into thin float glass and incorporating a precise forming capability is known as Triplex Ten-Twenty and is a patented process of the Triplex Safety Glass Company Limited (UK) (34–35). For automobile windshields, this process is used to induce a high stress into the inner glass whereas the outer glass is strengthened partially but to a level that is much less than that which would cause dicing by stone impact. Combined with 0.76-mm automotive PVB, these glass plies give a windshield that causes significantly less facial soft tissue damage upon impact than windshields made with annealed glass. This product is available commercially and is used in several European cars.

Another variation of special construction windshields is the Securiflex windshield made by St. Gobain (Fr.) (36). A fourth ply, a plastic film on the innermost surface, is formed and bonded onto a thin conventional windshield. It essentially prevents the occupant from scraping across the broken glass at impact.

A deicing–defogging windshield and back window were produced from 1974 to 1976 by Ford, and it is comprised of a Sierracin conductive film that is laminated within the interlayer rather than being coated onto the glass (37). This requires 1–2 kW for rapid deicing, necessitating a special alternator in the vehicle. The product has 70–75% visible transmittance and good ir reflectance which improves driving comfort as well as efficiently deicing the glazing.

Another type of deicing windshield and back window is made with resistance wires embedded in the plastic interlayer. Light diffraction from the wires and visible distortion from the heated interlayer have been recognized and modifications have been proposed (38–40) to undulate the wires randomly, and to vary the plasticizer content of the interlayer containing the wires (41). The process has been used in back windows and is in limited use in windshields (particularly for public service vehicles) in Europe.

Other automotive uses of laminated glass include colored glass and decorated glass. The privacy glass used in the side and rear glazing of vans frequently is laminated from a brown PVB and clear glass. Opera windows containing metallic ornaments and sufficient plastic interlayer to accomodate their thickness are popular. Laminated roof glazing usually is a combination of metallized glass and a colored PVB.

Aircraft Windshields. Aircraft windshields have extreme requirements in service temperature and pressurization and they must be resilient against high velocity bird impact. In addition, they must offer excellent visibility, both from optics and deicing capability, and an aerodynamic design. These highly specialized windshields are produced in low volume and are made by few companies (Sierracin, PPG Industries, Triplex Safety Glass Company). Construction varies with the need and service potential of the aircraft. Small planes of limited altitude and speed usually have acrylic monolithic windshields. Slower commercial aircraft use flat laminated glass made with aircraft-grade PVB (Monsanto Saflex PT). These aircraft require deicing capability which may be given by a conductive film that is pyrolytically or vacuum deposited on a glass surface or a conductive plastic film that is laminated in the sandwich. The third general class of aircraft windshield is for the modern, commercial, wide-body aircraft. These windshields become extremely complex, large in size, and expensive. A fourth type is for high speed, low flying military aircraft where birds, high skin temperature, and gunfire warrant extremely complex construction. The third and fourth types are multilayer constructions; typical examples are shown in Figures 8 and 9 (42–43).

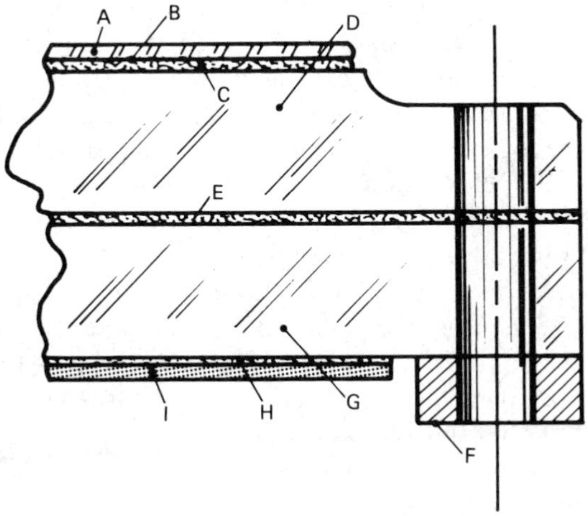

Figure 8. Cross section of Sierracin windshield used on Boeing 747 (42). A, 2.2-mm chemically strengthened glass; B, Sierracote 3 conductive coating; C, 1.9-mm PVB; D, 23-mm stretched acrylic; E, 1.3-mm PVB; F, laminated cloth spacer ring; G, 23-mm stretched acrylic; H, 0.6-mm PVB; I, 3.0-mm Sierracin 900.

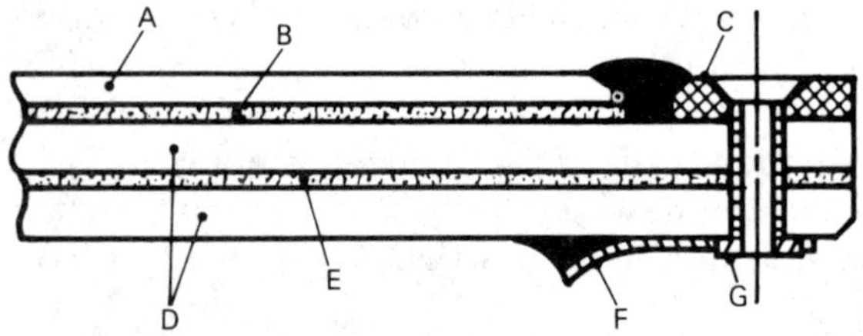

Figure 9. Sierracin lightweight, birdproof F-111 windshield cross section (43). A, 3.0-mm as-cast acrylic face ply; B, S-100 silicone interlayer; C, fiberglass retainer; D, 6.4-mm polycarbonate structural ply; E, S-120 polyurethane interlayer; F, stainless steel bearing strip; G, stainless steel bushing.

The Boeing 747 windshield (Fig. 8) is about 1 × 1.1 m and is curved to increase the pilot viewing area and to reduce air drag and air noise. Composed of seven plies, it weighs about 64 kg (42). The outer strengthened glass skin and the inner plastic shield may be replaced when damaged. Triplex Safety Glass Company also makes wide-body aircraft windshields, flat and curved, for Boeing and others. For the Boeing 747, two precurved, 12-mm plies of Ten-Twenty glass are laminated with PVB and covered with a 3-mm ply of Ten-Twenty glass bent to conform to the curved windshield. An electroconductive coating, Hyviz, is applied to the inner surface of the outer ply and then is laminated to the 12-mm Ten-Twenty ply (44).

The construction of the F-111 windshield shown in Figure 9 replaced a glass–silicone laminate previously used. The all-plastic windshield has improved impact resistance so that it is birdproof to 250 m/s (43). In this instance, the scratch resistance of glass was waived to obtain the impact performance at the allowed weight.

Architectural Products. Many specialized laminated glasses are made for architectural needs such as safety, sound attenuation, solar control, and security. These products may be further enhanced with colors and patterns for decorative effects. Safety glasses are specified in potentially higher risk breakage areas and overhead or sloped glazing (defined as more than 15° from vertical). Overhead glazing materials have varied in the past but more localities are accepting laminates. Sloped and overhead glazing frequently have heat-strengthened or tempered glass used in the construction of the laminate (45). Vertical passageway glazing usually is a 0.76-mm interlayer and sloped glazing is constructed with a 1.52-mm interlayer to accommodate the waviness of heat-treated glasses when they are used.

Noise attenuation is achieved effectively with laminated glass by the combination of the vibration damping effect of the plastic interlayer and usually an unbalanced glass thickness. Typical construction is 0.76–1.5-mm interlayer that is laminated with 3–10-mm glass. The type of glass or strength is not a factor in noise attenuation. A Sound Transmission Class Index of 34–41 is achieved with single laminate glazing and can be improved if combined with double glazing that has large air spaces. Mounting of the glass in an air-tight but flexible gasket reduces sound transmission (46). Airports, hotels, factory offices, and control rooms benefit from laminated acoustical glazing (47) (see Insulation, acoustic; Noise pollution).

Laminated glass is used for solar control, particularly where a highly reflective surface is not desired and where the laminate contributes other benefits. In these applications, a uniformly pigmented interlayer is obtained from the manufacturer and the laminate can be prepared by the conventional process. A broad range of colors and transmission levels is available with shading coefficients as low as 0.41. Pigmented interlayer is considered to be more color stable than dyed interlayer. Browns, blues, greens, pink, white, and clear plastic containing uv absorbers are readily available. Body-colored glasses may be used also, usually with clear interlayer. In these cases, the laminate is dependent only upon the solar properties of the glass.

All laminated glass increases the level of security to some extent. However, depending on the application, security glass is constructed of multiple layers of glass and PVB and, in some instances, additional rigid plastic is included. Laminated glass permits the same visual observation as normal glass but prevents or delays entry (or exit) until the attempt is detected. It complies with test UL 972 (48).

Bullet-resistant glass is constructed of many layers of glass and aircraft-type PVB depending on the level of resistance desired. Typical products are 38–50 mm thick and weigh 90–130 kg/m^2.

A third type of security glass is installed in modern penal institutions. This product is utilized for prisoner detention and obviates iron bars and their demeaning aspect. Typical construction utilizes three or more layers with at least one ply of thick PVB. Strengthened glass and electrically conductive circuits for alarms may be included. Large, heavy sections of similar construction have been used for underwater windows for boats, submarines, and aquariums. Four plies of fully tempered, 10-mm glass plus three plies of 1.9-mm PVB totaling 44.5 mm in thickness has a modulus of rupture of 172 MPa (25,000 psi) (49).

Glazing of laminated architectural glass requires additional care in the selection of sealants (qv) and drainage design. Sealants must be free of solvents (particularly aromatics) and mineral or vegetable oils (3) and must not provide pockets that would trap water at the glass–PVB edge. Similarly, the glazing detail must be designed with proper drainage (45). Generally, the practice is similar to that of glazing organically sealed insulating units (50–51).

BIBLIOGRAPHY

1. A. F. Randolph, *Mod. Plast.* **18**(10), 31, 98 (1941).
2. U.S. Pat. 830,398 (Sept. 4, 1906), J. C. Wood.
3. *PPG Glass Thickness Recommendations to Meet Architects' Specified 1-Minute Wind Load,* PPG Industries, Pittsburgh, Pa., 1979.
4. R. G. Rieser and G. E. Michaels, "Factors in the Development and Evaluation of Safer Glazing," *Proceedings of the Ninth Stapp Car Crash Conference,* University of Minnesota, 1965, pp. 181–203.
5. R. L. Morrison, "Influence of Ambient Temperature on Impact Performance of HPR Windshields," *paper presented at Fifteenth Stapp Car Crash Conference, SAE, 1971,* pp. 603–612.
6. U.S. Pat. 3,885,899 (May 27, 1975), D. J. Gurta and G. A. Koss (to Ford Motor Company).
7. U.S. Pat. 2,983,635 (May 9, 1961), R. E. Richardson (to Pittsburgh Plate Glass Company).
8. U.S. Pat. 3,009,850 (Nov. 21, 1961), J. P. Kopski and L. H. Schmidt (to Ford Motor Company).
9. U.S. Pat. 2,994,629 (Aug. 1, 1961), R. E. Richardson (to Pittsburgh Plate Glass Company).
10. U.S. Pat. 2,374,040 (Apr. 17, 1945), J. D. Ryan (to Libbey-Owens-Ford Glass Company).
11. *Standard 16 CFR 1201,* Consumer Products Safety Commission, Bethesda, Md.
12. *Safety Performance Specifications and Methods of Test for Safety Glazing Material Used in Buildings, ANSI Z97.1-1975,* American National Standards Institute, New York, 1975.
13. R. C. Cunningham, *U.S. Glass Metal and Glazing,* U.S. Glass Publications, Memphis, Tenn., Jan. 1979, p. 28.
14. *Ward's Automotive Yearbook,* 39th and 41st ed., Ward's Communications, Inc., Detroit, Mich., 1977 and 1979.
15. *Safety Code for Safety Glazing Materials for Glazing Motor Vehicles Operating on Land Highways, Z26.1-1966,* American National Standards Institute, New York.

16. *Manufacturer's Guide for Safety Equipment Services,* American Association of Motor Vehicle Administrators, Washington, D.C., 1979.
17. *CPSC Certified Products Directory,* Safety Glazing Certification Council, Hialeah, Fla., 1980.
18. *Standard for Bullet Resisting Equipment UL 752, ANSI SE 4.6-1973,* Underwriters' Laboratories, Inc., Melville, L.I., N.Y., 1973.
19. *ASS AS-R1-1968,* Standards Association of Australia, North Sydney, Australia, 1968.
20. *Brazilian Contran Resolution,* 483/74, Federal Official Gazette, Brazilia, Brazil, 1974.
21. *BS 5282-1975,* British Standards Institute, London, Eng., 1975.
22. *Specifications Relating to Safety Glass Requirements for Land Vehicles and Their Trailers,* Ministere De L'Equipement, Paris, Fr., 1975.
23. *Requirements on Safety Glass for Automotive Glazing,* Bundesministerim Ür Verkehr, Godesberg, FRG, 1973.
24. *A Tutte Gly Impettorati-Compartmentali Della Motorizzazione-Civille E Dei Transporti N Concessione E Sezioni,* Ministero Dei Transporti, Rome, Italy, Articles 218 and 297–302, 1959.
25. G. Rodloff, *Automobiltech. Z. (ATZ)* **64**(6), 1979 (1962); *English Translation 62-18916,* National Translation Center, Chicago, Ill.
26. G. Rodloff, *Automobiltech. Z. (ATZ)* **66**(12), 353 (1964); *English Translation 65-11982,* National Translation Center, Chicago, Ill.
27. J. C. Widman, *Recent Developments in Penetration Resistance of Windshield Glass, SAE 650474,* SAE, 1965.
28. E. R. Smith, "SAE Test Procedure for Quality Control of Windshields," *paper presented at Ninth Stapp Conference,* SAE, 1965, pp. 277–281.
29. D. F. Huelke, W. G. Grabb, and R. O. Dingman, *Automobile Occupant Injuries from Striking the Windshield, Report No. Bio-5,* Highway Safety Research Institute, Ann Arbor, Mich., 1967.
30. R. G. Rieser and J. Chabel, *Safety Performance of Laminated Glass Structures, SAE 700481,* SAE, 1970.
31. H. M. Alexander, P. T. Mattimoe, and J. J. Hofmann, *An Improved Windshield, SAE 700482,* SAE, 1970.
32. J. R. Blizard and J. S. Howitt, *Development of a Safer Nonlacerating Automobile Windshield, SAE 690474,* SAE, 1969.
33. L. M. Patrick, K. R. Trosien, and F. T. DuPont, *Safety Performance of a Chemically Strengthened Windshield, SAE 690485,* SAE, 1969.
34. S. E. Kay, J. Pickard, and L. M. Patrick, "Improved Laminated Windshield with Reduced Laceration Properties," *paper presented at Seventeenth Stapp Car Crash Conference,* SAE, 1973, pp. 127–169.
35. S. E. Kay, V. J. Osola, J. Pickard, and P. A. Brereton, *ATZ* **79**(9), 389 (1977).
36. U.S. Pat. 3,979,548 (Sept. 7, 1976), W. Schäfer and H. Raedisch (to St. Gobain Industries).
37. B. P. Levin, *Development of the Sierracin Electrically Heatable Safety Glass Interlayer, SAE 740156,* SAE, 1974.
38. U.S. Pat. 3,522,651 (Aug. 4, 1970), J. E. Powell and M. W. Lacey (to Triplex Safety Glass Company).
39. U.S. Pat. 3,895,433 (July 2, 1975), G. A. Gruss (to General Electric Company).
40. U.S. Pat. 3,954,547 (May 4, 1976), W. Genther (to St. Gobain Industries).
41. U.S. Pat. 3,903,396 (Sept. 2, 1975), P. T. Boaz and J. S. Maluchnik (to Ford Motor Company).
42. G. L. Wiser, "Sierracin® Glass/Plastic Composite Windshields," *paper presented at Conference on Transparent Materials for Aerospace Enclosures, U.S. Air Force and University of Dayton, June 25, 1969.*
43. J. B. Olson, "Design, Development and Testing of a Lightweight Bird-Proof Cockpit Enclosure for the F-111," *paper presented at the Conference on Aerospace Transparent Materials and Enclosures, Long Beach, Calif., Apr. 24–28, 1977.*
44. R. W. Wright, "High Strength Glass in Service—A Status Report," *paper presented at the Conference on Aerospace Transport Materials and Enclosures, Tech. Report AFML-TR-76-54, Atlanta, Ga., 1975.*
45. *Archit. Rec.* (6), 143 (1979).
46. *Architectural Saflex® for Sound Control, Tech. Bulletin No. 6295,* Monsanto Polymers and Petrochemicals, St. Louis, Missouri, 1972.
47. J. M. Clinch, *Study of Reduction of Glare, Reflection Heat and Noise Transfer in Air Traffic Control Tower Cab Glass, FAA-RD-72-65, AD747069,* NTIS, Springfield, Va., 1972.

48. *Burglary-Resisting Glazing UL 972,* Underwriters' Laboratories, Inc., Melville, L.I., N.Y., 1978.
49. *The New Look—Prisons Without Bars, Sierracin Field Report,* Sierracin Corporation, Sylmar, Calif., 1972.
50. *Alum. Curtain Walls* **6**, 24 (Sept. 1972) (Architectural Aluminum Manufacturers Association, Chicago, Ill.).
51. *FGMA Glazing Manual,* Flat Glass Marketing Association, Topeka, Kansas, 1974.

General References

R. N. Pierce and W. R. Blackstone, *Impact Capability of Safety Glazing Materials, PB195040,* Southwest Research Institute, San Antonio, Tex., 1970; contains detailed descriptions of test equipment, methods, and results for all types of glazings.
SAE Transactions (annual), *SAE Handbook* (annual), Society of Automotive Engineers, Warrendale, Pa.
Stapp Car Crash Conference series (annual, 1956 and continuing), Society of Automotive Engineers, Warrendale, Pa.; for safety and construction of automotive glass.

ROBERT M. SOWERS
Ford Motor Company

LAMINATED WOOD-BASED COMPOSITES

The bonding of wood (qv) with glue (qv) dates to the Pharaohs. Glued-wood products have become common, and development of a variety of adhesives (qv) and improvements in the gluing process have continued. During the last 35 years, the plywood industry, by far the largest user of wood adhesives, has become an extremely important factor in the construction field. The structural-laminating industry also has shown healthy growth as improved glues and design information have become available. Adhesives for furniture have shifted more and more from those based on the naturally occurring materials to synthetics. Bonding of wood with adhesives, which generally is far more efficient than with mechanical fasteners, has made possible a wide range of products and uses for which wood was not considered a few decades ago.

A requisite for good bonding is that the adhesive must wet the surfaces to be joined. Numerous factors (eg, pressure, temperature, and assembly time) play an important part during the formation of a glue bond. Wood is not a uniform substance but a complex material whose properties vary. The same bonding material and procedure rarely is suitable for all wood species.

Lumber is the well known product of the saw and the planing mill and, eg, in this discussion denotes structural members, the common 2-by-4s and 2-by-6s (ie, 5-by-10 cm and 5-by-15 cm) that are sawed from trees. Wood is used in a narrow and restricted sense—a cellular material with straight (or parallel) grain (fibers) that is obtained from trees and is free of strength-reducing characteristics such as knots, cracks, splits, or decay. Only limited quantities of wood in sizes larger than a small board are available from lumber but great quantities of small pieces are available. Lumber is graded within several species groups. Placement into a structural grade depends on the lumber's freedom from strength-reducing characteristics and the straightness of its grain.

Because grain in lumber is not always straight, the bending strength (MOR) of wood is about five times more than the bending strength of lumber. In the original tree, deviations from straight grain are useful. As cells form in the tree they align with the direction of the existing principal stress and naturally curve around limb and trunk junctions. The challenge is to convert trees into engineering materials of useful sizes and shapes with properties that approach, as closely as possible, those of wood.

Materials that are not the same in all directions are anisotropic (or orthotropic). Wood is highly anisotropic; its strength and stiffness parallel to the grain are about twenty times what they are perpendicular to the grain. Thus, wood has been structured for an optimum combination of strength and toughness in relation to stress parallel to the grain. Adapted as it is to the resistance of stress in one direction, wood's structural performance under stress along any other axis is markedly lower. For example, under stress perpendicular to grain, wood fibers separate easily and crush like a bundle of hollow tubes under compression. Anisotropy is advantageous to the tree because strength is oriented in the direction in which it is most needed; for the same reason anisotropy gives wood one of its natural merits as an engineering material. The largest stresses in beams and columns are parallel to the axes of the members; shear and bearing stresses generally are an order of magnitude smaller. Most engineering materials other than wood are isotropic (same properties in all directions) and, for efficient use, must be made artificially anisotropic in strength by the fabrication of ribs, corrugations, or the like.

Specific mechanical properties are convenient measures of a material's efficiency. Such properties are defined as the ratio of the property in question to density, eg:

$$\text{specific tensile strength} = \frac{\text{tensile strength}}{\text{density}}$$

Specific properties indicate how well a material utilizes the mass of its constituent molecules. For example, if one material is twice as dense as another material but both utilize their mass with equal efficiency, the tensile strength of the more massive material should be twice as great but the specific tensile strength should be the same. The material efficiencies of wood and lumber are compared to those of typical metal and masonry materials in Table 1. Wood is surprisingly efficient compared to materials that are microstructurally engineered. Most lumber that is supplied for structural uses is marketed as grade 2 or better, which is a medium-quality grade. Its properties can be considered typical of lumber properties in general. Some engineering properties of grade-2 southern pine lumber are compared with those of the wood it contains in Table 2.

The least amount of manufacture using trees results in lumber, a product of limited size and shape and decidedly lower in mechanical properties compared to wood. Hence, the first goal in manufacturing laminated wood-based composites is to produce from trees a structural material with better mechanical properties than those of lumber. A second goal is to produce useful sizes and shapes, including very large members, curved beams, and sheet and panel products. Laminating is a method that offers the following distinct advantages in the production of structural panel products or glued

Table 1. Structural Properties of Wood, Metals, and Masonries

Material	Specific tensile strength[a], km	Specific compressive strength, km	Specific modulus of elasticity (stiffness), Mm
wood	18.1[b]	9.93	2.41
no. 2 lumber	3.60[b]	5.92	2.04
oriented flakeboard[c]	9.49[b]	6.10	1.86
1020 steel	5.86	3.13[d]	2.69
4130 steel	11.2	8.98[d]	2.69
gray cast iron	2.97	12.0	1.43
2024-T3 aluminum	17.7	12.7	2.69
brass	6.14	4.93[d]	1.24
copper	3.88	3.53[d]	1.33
concrete	0.17[b]	4.07	0.89
class A brick	0.13[b]	0.89	0.62
granite	0.98[b]	6.25	1.56

[a] Specific tensile strength is the ratio of the property to density.
[b] Bending strength used instead of tensile strength.
[c] A particleboard research material made from Douglas-fir forest residues (see also Ref. 1.).
[d] Yield strength in tensile used instead of compressive strength.

Table 2. Comparison of No. 2 Lumber and Wood of Southern Pine

Material	Modulus of rupture (bending strength), MPa[a]	Modulus of elasticity (stiffness) in bending, GPa[b]	Compressive[c] strength, MPa[a]
wood[d]	97.5	13.0	53.5
no. 2 lumber[e]	19.4	11.0	31.9

[a] To convert MPa to psi, multiply by 145.
[b] To convert GPa to psi, multiply by 145,000.
[c] Parallel to grain.
[d] Ref. 2.
[e] These values are averages of test data; design values are smaller (see also Refs. 2–3).

structural members: ease of fabricating large structural elements from standard commercial sizes of lumber or veneers (thin layers or sheets of wood); elimination of checks or other drying defects associated with large, one-piece wood members because laminations are thin and can be dried readily before gluing; even dispersion of strength-reducing characteristics throughout the wood, which produces increased uniformity and higher values of most mechanical properties than those of corresponding properties in conventional lumber; the ability to fabricate larger panel products for covering applications; and the option of using lower quality raw material or different species for the individual laminations or components of the composite product.

Structural Panel Products

There are many wood-based composite panel products on the market and new products that are designed for specific purposes are being developed at an expanding rate. Some of the panel products provide strength and stiffness needed for a particular use; others provide required finish, sound reduction, insulation, covering, or other characteristics. These products are partly or entirely of wood-based material, ie, plywood, insulation board, hardboard, laminated paperboard, and particleboard (see Insulation, thermal). Manufacturing and finishing methods vary greatly to provide materials with specific desirable properties.

Although the resulting products vary widely in appearance and properties, all are manufactured in panel form and can cover large areas quickly and easily. Panels commonly are marketed in 1.22-by-2.44-m sheets and, generally, can be handled by one person. Some panels may be larger (eg, 1.22 × 3.05 or 1.22 × 3.66 m), and others are appreciably smaller (eg, decorative and acoustical ceiling tile). The most easily recognizable and widely used form is plywood, which is constructed from veneers that are unrolled from the tree (a log is rotated against a knife in a lathe). Adjacent layers of veneers are placed together with the grain at right angles and then they are glued to form the product. Production of most other panel products involves breaking the wood to small chips, flakes, or particles and then reassembling them into boards or panels.

The insulation board–hardboard–paperboard group is known as building fiberboard and includes such proprietary products as Celotex, Insulite, Masonite, Beaverboard, and Homasote.

Insulation board includes two categories: the semirigid type consists of low density products that are used as insulation and cushioning and the rigid type includes both interior board that is used for walls and ceilings and exterior board that is used for wall sheathing. Hardboard is a grainless, smooth, hard product that is used as prefinished wall paneling and exterior siding. Laminated paperboard serves as sheathing as well as in other covering applications but is not used as much as the other building fiberboards.

A special type of exterior particleboard sheathing panel that is 1.22-m wide has been approved by some building codes. It must be not less than 9.5-mm thick and must be applied to studs no more than 41 cm on center. A 9.5-mm-thick sheathing grade of laminated paperboard also is approved. Corner bracing is not required for either.

Roof sheathing provides the strength and stiffness needed for expected loads, racking resistance to keep components square, and a base for attaching roofing. Panels have several advantages over lumber, particularly in ease and speed of application and resistance to racking.

Softwood plywood is used extensively for roof sheathing, but other panel materials are satisfactory if they meet performance requirements. Special particleboards have been developed for such purposes and the Federal Housing Administration (FHA) has approved specific ones for some applications. Some insulation boards also meet the FHA requirements.

Use of adhesives for bonding wood has increased enormously over the past decades and glued products now vary in size from tiny wood jewelry to giant laminated timbers spanning >100 m. In general, the serviceability of a glued-wood assembly depends upon the kind of wood and its preparation for use, the type and quality of the adhesive, compatibility of the gluing process with the wood and adhesive used, type of joint or assembly, and moisture-excluding effectiveness of the finish or protective treatment that is applied to the glued products. Conditions in use naturally affect the performance of a glue bond. To give adequate performance, a glue joint should remain as strong as the wood under service conditions to which the glued product is exposed. If it does not, it becomes the weakest link in the assembly and the point at which failure might first occur.

Plywood. Plywood is a glued-wood panel that is composed of relatively thin layers, or plies, with the grain of adjacent layers at an angle to each other (usually 90°). The usual constructions have an odd number of plies to provide a balanced construction. If thick layers of wood are used as plies, often two corresponding layers with the grain directions parallel to each other are used; plywood that is so constructed often is called four ply or six ply. The outer pieces are faces or face and back plies, the inner plies are cores or centers, and the plies between the inner and outer plies are crossbands. The core may be veneer, lumber, or particleboard, with the total panel thickness typically not less than 1.6 mm or more than 7.6 cm.

Plywood has several advantages compared to solid wood, eg, it is relatively isotropic, has greater resistance to splitting, and has a form permitting many useful applications where large sheets are desirable. Because in some applications it is permissible to use plywood that is thinner than that normally available as sawed lumber, large areas may be covered with a minimum amount of wood fiber in the form of plywood.

The properties of plywood depend on the thickness and quality of the different layers of veneer, the order of layer placement in the panel, the glue used, and the control of gluing conditions in the gluing process. The grade of the panel depends on the quality of the veneers used, particularly of the face and back. The type of the panel depends on the glue joint, particularly its resistance to water. Generally, face veneers with figured (decorative) grain are used in panels where appearance is more important than strength (since this decorative pattern may significantly reduce strength and stiffness of the panels). On the other hand, face veneers and other plies also may contain certain sizes and distributions of knots, splits, or growth characteristics that have no undesirable effects on the strength properties for specific uses, eg, structural applications such as sheathing for walls, roofs, or floors.

Grades and Types. Both softwood and hardwood are used in the manufacture of plywood. (Softwood and hardwood generally include botanical groups of trees that have needlelike or scalelike leaves and broad leaves, respectively. These terms make no reference to the actual hardness of the wood itself.) In general, softwood plywood is intended for construction and industrial use and hardwood plywood is used where appearance is important. Originally, most softwood plywood was made of Douglas fir, but western hemlock, larch, white fir, ponderosa pine, redwood, southern pine, and other species are used as well (see Table 3).

A variety of plywood panels are made in which paper, plastic, and metal layers are combined with wood veneers (usually as the face layer) to provide special panel characteristics such as improved surface properties. Special resin-treated papers (overlays) can be bonded to plywood panels, either on one or both sides. These overlays are either of the high density or medium density types and are intended to provide improved resistance to abrasion or wearing, improved paint-holding properties, and

Table 3. Classification of Wood Species Used for Plywood[a]

Group 1	Group 2		Group 3	Group 4	Group 5
apitong[b,c]	cedar, Port	maple, black	alder, red	aspen	basswood
beech,	Orford	mengkulang[b]	birch, paper	bigtooth	fir, balsam
American	Douglas-fir[d]	meranti, red[b,e]	cedar, Alaska	quaking	poplar,
birch	fir	mersawa[b]	fir, subalpine	cativo	balsam
sweet	California red	pine	hemlock,	cedar	
yellow	grand	pond	eastern	incense	
Douglas fir[d]	noble	red	maple, bigleaf	western red	
kapur[b]	pacific silver	Virginia	pine	cottonwood	
keruing[b,c]	white	western white	jack	eastern	
larch, western	hemlock,	poplar, yellow	lodgepole	black (western	
maple, sugar	western	spruce	ponderosa	poplar)	
pine	lauan	red	spruce	pine	
Caribbean	almon	sitka	redwood	eastern white	
ocote	bagtikan	sweetgum	spruce	sugar	
pine, southern	mayapis	tamarack	black		
loblolly	red lauan		engelmann		
longleaf	tangile		white		
shortleaf	white lauan				
slash					
tanoak					

[a] Ref. 4.

[b] Each name represents a trade group of woods consisting of a number of closely related species.

[c] Species from the genus, *Dipterocarpus,* are marketed collectively: apitong if originating in the Philippines; keruing if originating in Malaysia or Indonesia.

[d] Douglas fir from trees grown in Washington, Oregon, California, Idaho, Montana, Wyoming, and the Canadian provinces of Alberta and British Columbia are classed Douglas fir no. 1. Douglas fir from trees grown in Nevada, Utah, Colorado, Arizona, and New Mexico are classed Douglas fir no. 2.

[e] Red meranti are limited here to species having a specific gravity of ≥ 0.41 based on green volume and oven dry weight.

better surfaces when appearance is important (as in concrete-form use). Some panels are overlaid, usually on the face only, with high density, paper-base decorative laminates, hardboards, or metal sheets. A backing sheet having the same properties (modulus of elasticity, dimensional stability, and vapor transmission rate) as the decorative face sheet must be used to provide a panel that does not warp as moisture changes occur. Overlays may be applied in the original lay-up or they may be applied to plywood after the panels have been surfaced. The two-step method permits a close thickness tolerance. Additional special products include embossed, grooved, and other textured panels. (Texturing may be achieved by wire brushing or sandblasting.) Such products are used primarily as interior paneling and exterior siding.

Prefinished plywood, particularly hardwood plywood, is available in a wide variety of forms. Finishes normally are applied in the plywood plant as clear or pigmented liquids. Various printed film patterns also are sometimes applied to plywood. Printed panels are processed using liquid finishing systems, and three-dimensional finishes can be achieved by passing the panels under an embossing roller. Application of these techniques improves the appearance for such uses as furniture and wall paneling. Clear, printed, and pigmented plastic films bonded to plywood also are used for the same purpose.

Plywood that has been treated for protection against fire or decay is available (see Flame retardants). Although it is technically feasible to treat veneers with chemical solutions and then glue them into plywood, a more general practice is to treat the plywood after gluing, either with fire-retardant or wood-preservative solutions, and then to redry the panels. To withstand such treatment, panels must be laminated with durable glues of the exterior type.

Large panels are made by end-jointing standard-size panels with scarf (sloping) joints or finger joints. End-jointing is used mainly with softwood panels for structural use as in boats or trailers. Requirements for joints are given in the product standards for conventional plywood (4–5).

Curved plywood sometimes is used as a specialty product, particularly in furniture. Much curved plywood is made by gluing the individual veneers to the desired shape and curvature in special jigs or presses. Flat plywood can be bent to simple curvatures after gluing, using techniques similar to those for bending solid wood such as steaming at atmospheric or low gauge pressure and soaking in boiling or nearly boiling water.

Dimensional Stability. *Arrangement of Plies.* Balanced construction largely eliminates the tendency of crossbanded products to warp as a result of uneven shrinking and swelling caused by moisture changes. Balanced construction plies are made by arranging them in similar pairs, a ply on each side of the core. Similar plies have the same thickness, kind of wood with particular reference to shrinkage and density properties, moisture content at the time of gluing, and grain direction. The importance of having the grain direction of similar plies parallel to each other cannot be overemphasized. Because the outer or face plies of a crossbanded construction are restrained by gluing on only one side, changes in moisture content induce relatively large stresses on the outermost glue joints. In general, the thinner, and thus, more flexible the face veneer, the less problem with face-checking.

Quality of Plies. All plies affect the shape and permanent form of thin plywood panels where dimensional stability usually is important. Thus, all plies need to be straight-grained, smoothly cut, and made of sound wood. In thick, five-ply, lumber-core panels, the crossbands particularly affect the quality and stability of the panel. In such panels, thin face and back plies of different species can be used without upsetting the stability of the panel. Imperfections in the crossbands, eg, marked differences in the texture of the wood or irregularities in the surface, are easily seen through thin surface veneers; the effect is called telegraphing. In wood that is cross-grained, the fibers deviate from a line that is parallel to the sides of the piece. Cross grain that runs sharply through the crossband veneer from one face to the other causes the panels to cup. Cross grain that runs diagonally across the face of the crossband veneer causes the panel to twist unless the two crossbands are laid with their grains parallel to one another. Failure to observe this simple precaution accounts for much warping in cross-band construction.

Appearance and dimensional stability are important in many hardwood plywood uses. The best woods for cores of high-grade hardwood panels are those of low density and low shrinkage, of slight contrast between earlywood and latewood, and of species that are easily glued.

In most species, a core that is made entirely of either quartersawed (edge-grained) or plainsawed (flat-grained) material remains more uniform in thickness through moisture content changes than one in which the two types of material are combined. Distinct distortion of surfaces is noted, particularly in softwoods, when the core boards are neither distinctly edge-grained nor flat-grained.

For many uses of softwood plywood (eg, sheathing), the strength characteristics of the panel are important, and appearance, moderate tendencies to warp, and small dimensional changes are minor concerns. Strength and stiffness in bending are particularly important in such panels for which veneers are selected mainly to provide strength properties. This selection often allows controlled amounts of knots, splits, and other irregularities that might be objectionable from an appearance standpoint.

Moisture Content. The tendency of plywood panels to warp is affected by changes in moisture content as a result of changes in atmospheric moisture conditions and by wetting of the surface by free water. Surface appearance also may be affected by warping. Whether plywood is made in hot or cold presses affects its moisture content. Most plywood is made in hot presses, from which the panels come out quite dry. The original moisture content of the veneer and the amount of water added by the glue must be kept low to avoid blister problems in hot-pressing the panels. Additional water is lost from the glue and the wood during heating. Cold-pressed panels have a fairly high moisture content when removed from the press; the values depend on original moisture content of the veneer, amount of water in the glue, and amount of glue spread. Such panels may lose considerable moisture while reaching equilibrium in service. Differences in the stability and appearance of plywood under service conditions may occur if panels that are produced by one process are mixed with those made by the other. However, either type of panel may be used satisfactorily if it is properly designed for the service condition.

Expansion or Contraction. The dimensional stability of plywood, which is associated with moisture and temperature changes, involves not only cupping, twisting, and bowing but includes expansion or contraction. The usual swelling and shrinking of the wood is effectively reduced because grain directions of adjacent plies are placed at right angles. The low dimensional change parallel to the grain in one ply restrains the normal swelling and shrinking across the grain in the ply glued to it. An additional restraint results because the modulus of elasticity parallel to the grain is about twenty times that across the grain. The total expansion or contraction of plywood is about equal to the parallel-to-grain expansion or contraction of the individual veneers. For all practical purposes, thickness changes in solid wood and in plywood are similar with corresponding changes in moisture and temperature.

Structural Design. Testing all of the many possible combinations of ply thickness, species, number of plies, and variety of structural components is impractical. However, mathematical formulas that are used to compute the stiffness and strength of these various combinations have been verified by tests (2). Depending upon the veneer or wood species property that is substituted, these formulas may be used, in general, for calculating the stiffness of plywood, stresses at proportional or ultimate limits, or for estimating working stresses. Plywood may be used under loading conditions that require the addition of stiffeners to prevent it from buckling. Plywood also may be used in the form of cylinders or curved plates.

A strip of plywood cannot be as strong in tension, compression, or bending as a strip of solid wood of the same size. Those plies having their grain direction oriented at 90° to the direction of stress can contribute only a fraction of the strength contributed by the corresponding areas of a solid strip with grain parallel to the stress. However, overall strength properties in the direction parallel and perpendicular to the face grain do tend to be equalized in plywood; in some interior plies, the grain direction is parallel to the face grain and, in others, it is perpendicular. As plywood of thin, crossbanded veneers is very resistant to splitting, nails and screws can be placed close together and close to the edge of plywood panels.

Highly efficient, rigid joints can be obtained by gluing plywood to itself or to heavier wood members as is needed in box beams and stressed-skin panels. (In stressed-skin construction, panels are separated from one another by spaced strips. The whole assembly is bonded to act as a unit when loaded.) Glued joints should not be designed primarily to transmit load in tension normal to the plane of the plywood sheet because of the rather low tensile strength of wood in a direction perpendicular to the grain. Rather, glued joints should be arranged to transmit loads through shear planes that are parallel to each other. Shear strength of plywood across the grain of wood (often called rolling shear strength because of the tendency to roll the wood fibers) is considerably less than shear strength parallel to the grain. Thus, sufficient area must be provided between plywood and flange members of box beams and plywood and stringers of stressed-skin construction to avoid shearing failure perpendicular to the grain in the face veneer, in the crossband veneer next to the face veneer, or in the wood member.

Manufacture. The processes of making plywood vary somewhat depending on the types and size of logs available and the type of product manufactured. Figure 1 presents a very general process for the production of hardwood plywood, and Figure

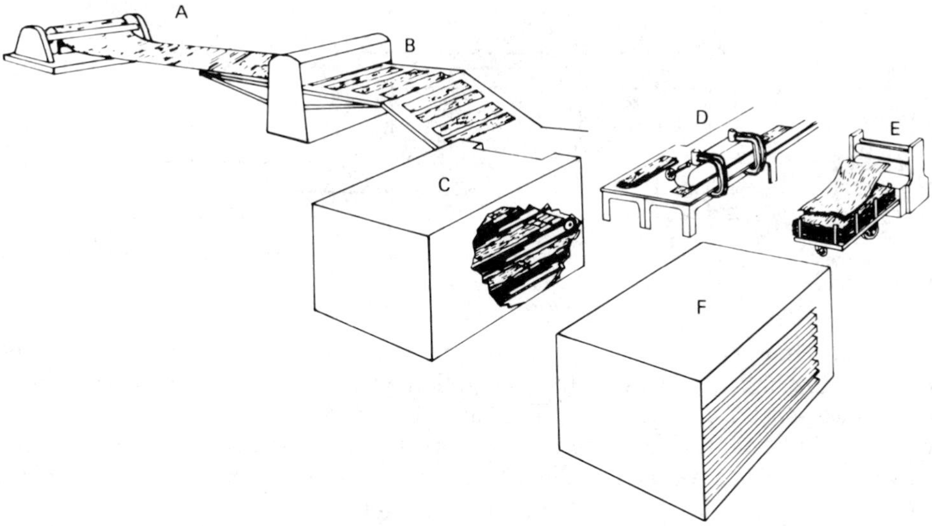

Figure 1. Manufacture of hardwood plywood. A, Veneer (80–90%) is cut by the rotary lathe method. As the lathe spindles move, the log is rotated against a knife. Speed with which knife and knife carriage move toward center of log regulates thickness of veneer. Before cutting, logs are steam heated to assure smooth texture and easier cutting. Another cutting method, slicing, is used primarily to cut face veneers from walnut, mahogany, cherry, and oak. Flitch is attached to log bed which moves up and down, cutting slice of veneer on each downward stroke. B, The clipper cuts veneer sheets into various widths. C, Dryers then remove moisture content to a level compatible with gluing. D, Veneer sheets of various sizes are clipped and spliced to make full-size sheets. E, Veneers then are coated with liquid glue, front and back, with a glue spreader. F, Heat and pressure that are applied in the hot press bonds the veneer into plywood. Panels are trimmed, sanded, and stacked for conditioning and inspection, after which they are ready for grading, strapping, and shipping. Courtesy of Hardwood Plywood Manufacturers Association.

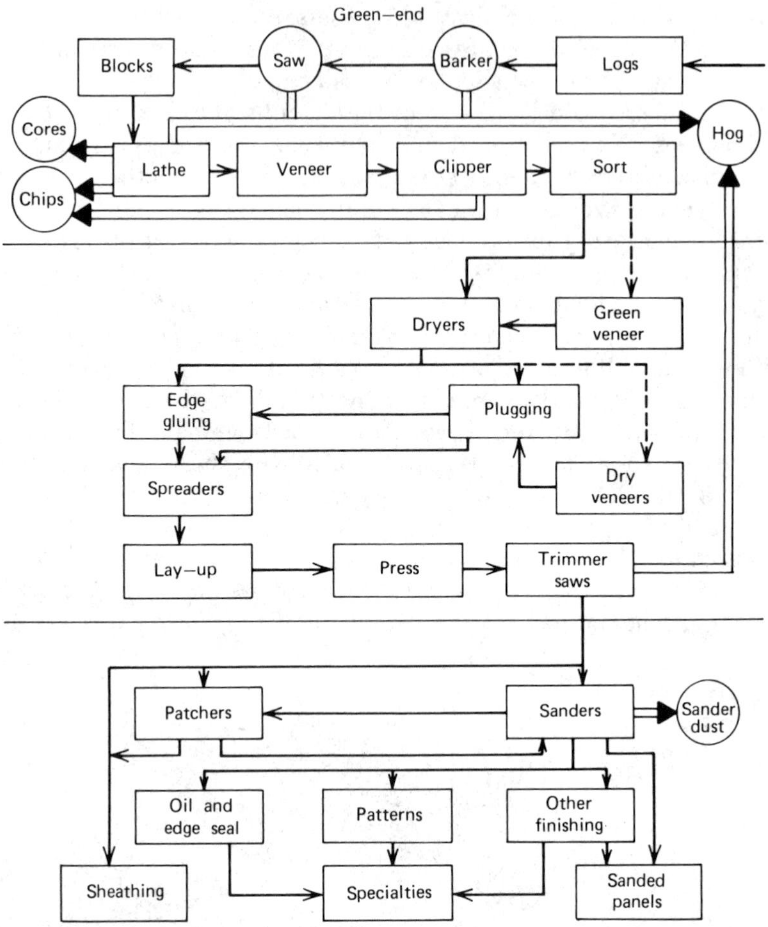

Figure 2. Softwood plywood manufacture: —, production flow; ═, waste and by-products; - - -, storage. Courtesy of the Plywood Research Foundation.

2 presents a more detailed manufacturing flow chart for the production of softwood plywood. In general, the logs are first sawed to the required length, usually long enough to produce panels 2.4 or 3.0 m long. The bark is removed from the log by methods which vary from hand debarking to the use of mechanical and hydraulic debarkers. The logs may be steamed or soaked depending on species, time of year, thickness of veneer being cut, etc.

Cutting. Cutting operations depend on the type of veneer to be produced. Rotary peeling is the most common procedure used in the production of veneer for assembly into construction and industrial use panels. During rotary peeling, a knife, equal in length to the width of the veneer, shaves a continuous sheet of veneer of uniform thickness from a log's surface.

In hardwood veneer cutting for decorative plywood grades (except for those mass-produced), manufacturing methods accentuate grain design and enhance the beauty of unusual patterns in knots, burls, and crotches. These effects are limited by the difficulty of producing plywood from the special cuts and, therefore, production is highly customized.

The great majority of all softwood species of plywood are made from rotary-peeled veneers. In rotary peeling, the thickness of the sheet usually is 0.56–5.1 mm. Thickness variations are controlled by the setting of the lathe. Speed of rotation depends upon the species of the wood being processed. The nonuniform character of wood does not permit the production of uniform quality veneer by any method of cutting. Large defects and unusable sections are clipped out but length is determined by the length of the bolt cut from the log and not by such clipping. Face and center plies generally are cut on one lathe and crossbands on another. Sequence matching, produced by knife slicing the veneer instead of peeling it, is used for some of the best custom panel work. By this technique, all of the veneer on one flitch is used in sequence just as it came from that log in slices. Fine decorative veneers often are book matched in making a panel: every other slice is turned over. If such veneer were rotary-peeled, the matching effect would be lost.

The veneer of hardwood species and that of Douglas fir generally is sorted to separate heartwood and sapwood. Sapwood of a species usually is lighter in color than the heartwood. If sapwood is included in a panel, separate staining treatment may be employed on the prefinishing line to make the appearance uniform.

Drying. The type of veneer frequently determines the method of drying (qv) and the equipment used. Highly figured decorative veneers may be air-dried or loft-dried in heated rooms and, later, plate-dried in special, quick-acting hot presses. Heavier veneers more commonly are processed in specially designed, long (7.6–91 m), tunnel-type dryers often having multiple tiers (up to 6 or 8). The veneer is moved continuously on rolls or belts along each tier and the drying time is controlled by the speed of the belts.

A main problem in veneer drying is moisture equalization; substantial amounts of veneer often are overdried to ensure that the wetter portions are adequately dried. Drying of Douglas fir may require as long as 10 min for 3.2-mm stock and 15 min for 4.2-mm material. The operating temperature is 165–182°C and generally is highest at the wet end, or entrance, of the dryers. In hardwood mills, drying temperatures are substantially lower, although still above the boiling point of water. The desirable moisture content after drying in hardwood plywood depends upon the adhesive used. In general, it is 3–6% for aqueous glues and 8–12% for film glues.

Sorting and Patching. After being dried, the veneers are sorted into various grades. Because veneer that is suitable for plywood faces is scarce and, consequently, valuable, every possible method is used to produce the maximum amount of this grade. Sheets smaller than the full panel size are taped or edge-glued together to produce the required size. Veneers with small defects are patched and used in faces in certain utilitarian types of plywood. Veneer that is not suitable for faces is graded for use as backs, crossbands, or centers. These grades are patched or unpatched, depending on the specification of the particular panel being produced. Patching removes a defective area by routing or punching and replaces it with a patch of sound wood of exactly the same size or with a synthetic-resin patch.

Assembling. Liquid adhesives are applied by machines (glue spreaders) each of which have a series of rubber-covered, grooved application rolls that apply the glue to the sheet of veneer. The number of grooves per linear meter (of the length of the roll), their depth, and the viscosity of the adhesive control the amount of material that is applied per unit area. In addition to roll spreaders, spray nozzle and curtain-coater adhesive-application systems are commonly used. In general practice, the adhesive is spread on two sides of one ply which is placed between two plies of unspread veneer in the panel assembly. The amount of adhesive used, called the spread rate, usually is expressed as the weight of liquid adhesive per 93 m^2 of glueline for hardwood species and per 93 m^2 of double glueline for softwood species. Veneers are assembled prior to plywood pressing. A five-ply construction is used in the following example of this procedure: A back, with its grain direction parallel to the long axis of the completed panel, is placed on the assembly table. The next veneer is a crossband placed so that its grain direction is perpendicular to that of the back. The crossband has adhesive on both surfaces. The center then is placed so its grain is perpendicular to that of the crossband and parallel to that of the back. Above the center, another crossband is positioned with adhesive spread on both sides. Lastly applied is the top or face which, like the back, has its grain direction parallel to the long axis of the completed plywood panel. The construction of such a panel may consist of a 3.2-mm center, two 3.2 mm crossbands, and a face and back each 2.5-mm thick. Whatever the measurements, the construction must be symmetrical about the center ply. Each layer, except the middle, must have a corresponding layer on the other side of the same thickness and same grain direction. The construction of a plywood panel affects its properties somewhat [although the effect of species predominates (see Table 4)]. The core material for most plywood panels is veneer; however, some lumber is used and, with the increasing scarcity of logs suitable for core-veneer production, the use of particleboard or flakeboard cores is increasing.

Press Curing. The assembly of veneers is transformed into plywood by bringing all faying (close-fitting) surfaces into intimate contact with each other (which is accomplished either by hot or cold pressing).

By far the largest production uses a hot press in which thermosetting resins are the adhesive basis. The hot presses are hydraulic and generally steam heated; minimum temperature is 113°C. Hot platen presses for hardwood plywood generally have 5–20 openings. For softwood plywood, presses with 30 or more openings are used. Temperatures for the manufacture of hardwood plywood vary from 113–143°C, with press times of 2–10 min or more. Furniture-plywood manufacture often involves temperatures up to 82°C with hot water-heated presses. The temperature used with softwood plywood generally is about 140°C and the press time is 5–7 min or more for greater thicknesses. Press pressures depend on the species and are chosen to bring about compaction (all plies brought into intimate contact) and 4% compression.

The practice of hot-stacking is common in softwood-plywood manufacture. Hot-stacking out of the press retains the heat which has been imparted to the panel; cure of the adhesive proceeds in the stack. Heat retention by stacking has a substantial effect on panels and appreciably shortens the required time in the hot press. Even in a well-designed plant, the hot press is the bottleneck of the process; thus, it is economical to keep the press periods as short as possible.

Cold-press curing of adhesives in the manufacture of plywood accounts for a small portion of production. Electric screw presses are used commonly, especially in the furniture-plywood industry. These cold presses (or baling presses) are used to compact the load of veneer assemblies, after which tie rods (which are adjustable by turnbuckles and I-beams) are arranged opposite to each other, top and bottom, and pressure is maintained on the bale until setting is accomplished. Depending on the adhesive, setting times may vary from 4 to 24 h but, generally, setting is an overnight operation.

Table 4. Example of Allowable Stresses for Plywood Conforming to *U.S. Product Standard PS 1,* for Softwood Plywood, Construction and Industrial, Dry Location, Normal Load Basis[a]

Type of stress	Species group[b]	Stresses, MPa[c]: Exterior A-A, A-C, C-C, and comparable grades of overlaid plywood, structural I A-A, A-C[d]	Exterior A-B, B-B, B-C C-C (plugged); plyform class I, plyform class II, and comparable grades of overlaid plywood structural I C-D;[d] structural II C-D;[e] standard sheathing[f] all interior grades[f]	All other grades of interior including standard sheathing
extreme fiber in bending, tension face-grain parallel or perpendicular to span (at 45° to face grain, use 1/6 value)	1	13.8	11.4	11.4
	2, 3	9.7	8.3	8.3
	4	8.3	6.9	6.9
compression parallel or perpendicular to face grain (at 45° to face grain, use 1/3 value)	1	11.4	10.7	10.7
	2, 3	8.3	7.6	7.6
	4	6.9	6.6	6.6
bearing (on face)	1	2.3	2.3	2.3
	2, 3	1.5	1.5	1.5
	4	1.1	1.1	1.1
shear in plane perpendicular to plies parallel or perpendicular to face grain (at 45°, increase value by 100%)	1	1.7	1.7	1.6
	2, 3	1.3	1.3	1.2
	4	1.2	1.2	1.1
shear, rolling, in plane of plies parallel or perpendicular to face grain (at 45°, increase value by 1/3)	all	0.4	0.4	0.3
modulus of elasticity in bending face grain parallel or perpendicular to span	1		12,400	
	2		10,300	
	3		8,300	
	4		6,200	

[a] Ref. 6.
[b] See Table 3 for lists of species in each group.
[c] To convert MPa to psi, multiply by 145.
[d] Use group 1 stresses.
[e] Use group 3 stresses.
[f] Exterior glue.

After being press-cured, the plywood is trimmed to proper size and is sanded to the desired thickness (except for the popular rough-textured panels used for siding or structural applications). Any defects discovered in inspection are repaired, and patching is done if necessary and the panels are inspected and graded.

Adhesives. By a large margin, the great bulk of adhesives used in the manufacture of plywood is based on either phenol–formaldehyde or urea–formaldehyde resins. A small amount of melamine–formaldehyde resin is used for severe exposure where the phenol–formaldehyde color is objectionable (see Amino resins; Phenolic resins). Resorcinol–formaldehyde and phenol–resorcinol–formaldehyde resins are used in patching and in scarfing to make panels of extra size. The most costly of the main resin types (except for resorcinol–formaldehyde) are melamine resins. Phenolics are less expensive and ureas are the least expensive. Phenolic–resin adhesives generally darken in color and may stain if they bleed through the face veneer; melamine and urea adhesives do not stain. Hybrid adhesives, known as melamine–urea–formaldehyde adhesives, are a compromise of quality and cost and sometimes are used in hardwood interior-grade plywood of the better grades. Urea–melamine adhesives also are used where high frequency curing techniques are used and where waterproof bonds are required. For hardwood interior plywood, the most common adhesive is urea–formaldehyde; both melamine–formaldehyde and phenol–formaldehyde are used under certain conditions, eg, hardwood doors for exterior openings, interior marine work (which is considered to be subject to severe exposure), and fire-resistant panels. Softwood exterior plywood is almost exclusively made with phenol–formaldehyde adhesives. Softwood interior plywood is made with phenol–formaldehyde adhesives, with blood albumen or soya protein, with mixtures of these two with and without phenol–formaldehyde adhesives and, occasionally, with urea–formaldehyde adhesives.

Finishing and Treating. Historically, the plywood industry has been concerned solely with the manufacture and distribution of plywood; if the use required it, other industries finished the material and still others treated the plywood for various special properties. Although this division of labor to some extent still occurs, increasing amounts of manufactured plywood, at least partially, are prefinished and often given special treatments. A great deal of hardwood plywood is completely prefinished as it leaves the plywood factory. Treatments vary from overlaying the plywood to impregnation. Overlays can be structural or decorative or may simply provide a means for improving the plywood surface to receive and retain paint. Overlays most commonly are a form of plastic or combination of plastic resin and paper. The three divisions are high density, medium density, and decorative, high pressure plastic overlays. A large volume of plywood that is covered with high density overlay is used for superior-grade, concrete-form boards. Such a construction yields smoother concrete surfaces and longer-lasting panels.

Some plywood is impregnated to achieve increased dimensional stability, decoration, fire or decay resistance, greater hardness, special color effects, and for forming finishes in place. Impregnation requires that the glue used in the plywood withstand the effect of impregnation. A very few species, which are not easily impregnated, also must be avoided. Methods for the impregnation of plywood closely follow those used for wood.

Impregnation to increase dimensional stabilization is almost exclusively based on the use of very low molecular weight, water-soluble, phenol–formaldehyde resins which have the peculiar property of being able to diffuse through the cell walls and react to reduce the natural hygroscopicity of wood. Certain water solutions of polyethylene glycols also give dimensional stabilization (see Glycols). Fire resistance is achieved by the use of a variety of inorganic salts, the most common ones being the ammonium phosphates and ammonium sulfate. Lesser portions of other inorganic salts are added, eg, borax, boric acid, zinc chloride, and potassium chromate for special effects such as reduction of hygroscopicity, increased resistance to leaching, and control of afterglow. Decay resistance is achieved by the use of certain toxic agents, eg, pentachlorophenol and the lower chlorinated phenols, copper compounds, metallic salts of naphthenic acids, and arsenic compounds (qv). Impregnation to achieve color effects is an old art which has never achieved a great deal of success; it does, however, hold some promise in the staining of sapwood to make it more uniform with the normal color of heartwood.

Production. Data for the United States production volume by type (ie, either interior or exterior) for softwood plywood produced in 1968–1977 is given in Table 5. Note that softwood plywood is defined as plywood having three or more wood plies with a face of softwood veneer. The back ply or inner plies may be any combination of hardwood or softwood veneer, lumber, particleboard, or hardboard.

Hardwood plywood quantities are reported by square meter, surface measure. The quantity of unfinished and prefinished hardwood plywood that was shipped and the total value of these shipments of 1968–1977 are listed in Table 6. Production, consumption, and export-import information for hardwood plywood for the years 1976 and 1977 are given in Table 7. Unlike softwood plywood, the quantity of shipped hardwood plywood does not accurately reflect the total amount of this material used

Table 5. Production of Softwood Plywood in the United States, 1968–1977 [a]

	Production, 10^6 m^2 [b,c]		
Year	Interior	Exterior	*Total*
1968	na	na	*1348.6*
1972	na	na	*1669.8*
1976	599.7	1070.9	*1670.6*
1977	630.6	1130.3	*1761.0*

[a] Ref. 7.
[b] All figures reported on 10^6 m^2, 9.5-mm basis. Thus, 93 m^2 of 6-mm thick plywood is equivalent to 61.9 m^2 on 10-mm basis, and 93 m^2 of 20-mm thick plywood is equivalent to 186 m^2 on 10-mm basis. To convert m^2 to ft^2, multiply by 10.76.
[c] na = not available.

Table 6. Quantity of United States Shipments of Hardwood and Prefinished Hardwood Plywood, 1968–1977, 10^6 m^2 Surface Measure [a]

	Hardwood plywood			
Year	Shipped as unfinished	Shipped as prefinished [b]	*Total*	Prefinished hardwood plywood
1968	80.2	85.3	*165.5*	131.3
1972	72.7	108.8	*181.5*	278.3
1976	68.2 [c]	48.0 [c]	*110.2*	194.3
1977	79.8	44.6	*124.4*	203.5

[a] Ref. 8. To convert m^2 to ft^2, multiply by 10.76.
[b] Made from purchased plywood.
[c] Shipped as prefinished reflects hardwood plywood produced and prefinished at the same location.

Table 7. Production, Shipments, Exports, Imports, and Apparent Consumption of Hardwood Plywood, 1977 and 1976 (Quantity in 10^3 m^2 surface measure)[a,b,c]

Product	Quantity	Manufacturers' shipment	Exports of domestic merchandise	Exports to manufacturers' shipments, %	Imports for consumption	Calculated import duty	Apparent consumption[d]	Imports to apparent consumption, %
		Quantity	Quantity	Quantity	Quantity		Quantity	Quantity
1977								
hardwood plywood, total[e]	137,359	12,4452	5,148	4.1	389,053	56,123	508,358	76.5
birch face ply[e]	34,316	33,312	na	na	25,282	4,673	na	na
Philippine mahogany face ply[d]	[b]	[b]	na	na	20,040	37,530	na	na
mahogany face ply[e]	833	468	na	na	764	212	na	na
walnut face ply[e]	6,950	5,550	na	na	579	93	na	na
other face ply[g]	57,427	73,494	na	na	142,387	13,617	na	na
1976								
hardwood plywood total[d]	127,119	116,198	5,547	4.5	406,033	51,510	519,997	78.5
birch face ply[e]	29,894	28,285	na	na	22,946	3,999	na	na
Philippine mahogany face ply[e]	[f]	[f]	na	na	275,905	38,811	na	na
mahogany face ply[e]	678	437	na	na	1,730	264	na	na
walnut face ply[e]	6,503	5,161	na	na	783	124	na	na
other face ply[g]	78,754	73,123	na	na	104,669	8,312	na	na

[a] To convert m^2 to ft^2, multiply by 10.76.

[b] Foreign trade data are classified by face species rather than by type.

[c] Ref. 8.

[d] Includes hardwood plywood unfinished or prefinished with a clear or transparent material which does not obscure the grain, texture, or makings of the face ply.

[e] Apparent consumption represents new domestic supply and is derived by subtracting exports from the total of manufacturers' shipment plus imports. Detail data only represents veneer core type II and board core therefore does not add to total.

[f] Includes all species of hardwood plywood, other than those reported above, which are prefinished with a clear or transparent material. Also includes all prefinished plywood not reported above, regardless of face species.

[g] Data for Philippine mahogany, face ply included with other face ply.

or consumed in a one-year period because a large amount of hardwood plywood is imported into the United States each year.

Specifications. The most commonly used specifications for plywood are the product standards established by the industry with the assistance of the U.S. Department of Commerce, NBS. Softwood plywood is covered by *U.S. Product Standard PS 1-74* (4); hardwood and decorative plywood is covered by *PS 51-71* (5). These specifications include requirements for species, grade, thickness of veneer and panels, glue bonds, moisture content, etc. They also provide standardized definitions of terminology, suitability, and performance test specifications, expected minimum test results, and explanations of the codes which normally are stamped on each panel. Some of the allowable stresses for softwood plywood that conform to *PS 1* are given in Table 4 (4).

Imported plywood generally is not produced in conformance with United States product specifications. However, some countries have their own specifications for plywood that is manufactured for export to the United States, and some of these follow the requirements of our applicable domestic product standards.

Uses. The American Plywood Association has published information on the projections of softwood-plywood-structural panel demand in the United States by market area for the years 1977, 1978, and 1979 (10). The estimated demand for softwood plywood by five broad market categories and information on the larger volume uses in each of these five categories is listed in Table 8. The bulk of softwood plywood is used where strength, stiffness, and construction convenience are more important than appearance. Some grades of softwood plywood are made with faces that are selected primarily for appearance and are used either with clear natural finishes or with pigmented finishes.

Hardwood plywood is used normally where appearance is more important than strength. Most of the production is intended for interior or protected uses, although a very small proportion is made with glues that are suitable for exterior service. A significant portion of all hardwood plywood is available completely finished. Typical uses of unfinished hardwood plywood include containers, curved or molded products, door skins, die boards, marine-grade materials, pin blocks, kitchen cabinets, furniture components, and exterior siding. Examples of prefinished hardwood-plywood uses include wall paneling or interior use surface-covering material, laminated hardwood block flooring, furniture components, door skins, kitchen and bathroom cabinets, and laminated door sides.

Wood-Based Fiber and Particle Panel Materials. Wood-based fiber and particle panel materials include insulation boards, medium-density fiberboards, hardboards, particleboards, and laminated paperboards. Various particleboards are known by the kind of particle used such as flakeboard, chipboard, chipcore, or shavings board. These panel materials are reconstituted wood (or some other lignocelluloselike bagasse) in that the wood is first reduced to small fractions of the original size and then is combined by special forms of manufacture into large and moderately thick panels. In final form, these materials retain some of the properties of the original wood but, because of the manufacturing methods, gain new and different properties than those of the wood. Because they are manufactured, they can be tailored to satisfy desired uses.

The wood-based panel materials are manufactured either by converting wood substance to fibers and then interfelting them into the panel material classed as building fiberboard, or by strictly mechanical means of cutting or breaking wood into

Table 8. Estimated United States Demand for Softwood Plywood by Market, 1977–1979 10^6 m^2 [a] (9.5-mm basis)

Market	Year 1977	1978	1979
new residential construction			
floor systems	310	280	300
roof decks	360	300	330
exterior siding	110	100	110
mobile homes	14	15	16
other	120	80	80
Total	*914*	*775*	*836*
distribution			
home repair and remodeling	210	220	230
other homeowner uses	130	140	140
miscellaneous other markets	70	80	90
Total	*410*	*440*	*460*
industrial			
material handling	50	60	70
transportation	50	50	50
other products made for sale	90	100	100
in-plant repair, maintenance, patterns, and jigs	80	80	80
Total	*270*	*290*	*300*
nonresidential construction			
building construction	80	90	90
concrete form	70	80	80
agricultural buildings	30	30	30
other	10	10	10
Total	*190*	*210*	*210*
international markets	30	30	60
Total market demand	*1814*	*1745*	*1866*
changes in mill and wholesale industry	−30	+20	+20
Total production	*1784*	*1765*	*1886*

[a] Ref. 10. To convert m^2 to ft^2, multiply by 10.76.

small discrete particles and then, with a synthetic-resin adhesive or other suitable binder, bonding them together in the presence of heat and pressure. These latter products are called particleboards (11–12).

Building fiberboards are made of fiberlike components of wood that are interfelted in the reconstitution and are characterized by a bond produced by the interfelting, and they are classified as fibrous-felted board products. Binding agents and other materials may be added during manufacture to increase strength, resistance to fire, moisture, or decay, or to improve some other property. Among the materials added are rosin, alum, asphalt (qv), paraffin, synthetic and natural resins, preservative and fire-resistant chemicals, and drying oils (qv). At certain densities and under controlled conditions of hot pressing, rebonding of the lignin effects an additional bond in the panel product.

Particleboards are manufactured from small components of wood that are glued together with a thermosetting synthetic resin or equivalent binder. Wax sizing is added to all commercially produced particleboard to improve water resistance. Other additives may be introduced during manufacture to improve some property or to provide added resistance to fire, insects (eg, termites), or decay. Particleboard is among the newest of the wood-based panel materials. It has become a successful and economical panel product because of the availability and economy of thermosetting synthetic resins, which permit consolidation of blends of wood particles and the synthetic resin and curing in a heated press. The thermosetting resins that are used are primarily urea–formaldehyde and phenol–formaldehyde. Urea–formaldehyde is lowest in cost and is the binder used in greatest quantity for particleboard that is intended for interior or other nonsevere exposures. Where moderate water or heat resistance is required, melamine–urea–formaldehyde resin blends are used. For severe exposures, eg, exteriors, or where some heat resistance is required, phenolics generally are used.

The kinds of wood particles that are used in the manufacture of particleboard range from specially cut flakes 2.5 cm or more in length (parallel to the grain of the wood) and only a few hundredths of a centimeter thick to fine particles approaching fibers or flour in size. The synthetic resin solids usually are 5–10 wt % of the dry wood. These resins are set by heat as the wood particle–resin blend is compressed either in flat-platen presses (similar to those used for hot-pressing hardboard and plywood) or in extrusion presses where the wood–resin mixture is squeezed through a long, wide, and thin die that is heated to provide the energy to set the resin. Particleboards produced by flat-platen presses are called mat-formed or platen-pressed particleboards and those produced in an extrusion press are called extruded particleboards.

Building fiberboards and particleboards are produced from small components of wood, hence the raw material need not be in log form. Many processes for manufacture of board materials start with wood in the form of pulp chips. Coarse residues from other primary forest products manufacture, therefore, are an important source of raw materials for both kinds of wood-based panel products. Particleboards and, to a lesser extent, building fiberboards, use fine residues as raw material (eg, planar shavings). Overall, about 70% of the raw material requirements for wood-base, fiber and particle panel materials are satisfied by residues.

Wood-based fiber and particle panel materials form an important part of the forest products industry in the United States. Not only are they valuable from the standpoint of integrated utilization, but the total production of more than 6.5×10^8 m^2 (7×10^9 ft^2) is important in terms of forest products consumed. Along with softwood plywood, wood-based fiber and particle panel materials are among the fastest growing components of the industry.

Fiber Panel. Broadly, the wood-based fiber panel materials (building fiberboards) are divided into two groups—insulation board (lower density products) and hardboard, which requires consolidation under heat and pressure as a separate step in manufacture. Insulation board and hardboard dimensional and quality requirements are given in refs. 13 and 14, respectively. The dividing point between an insulation board and a hardboard, on a density basis, is a specific gravity of 0.5 g/cm^3. Practically, because of the range of uses and specially developed products within the broad classification, further breakdowns are necessary to classify adequately the various products. Table 9 shows the building fiberboards identified by density.

Laminated paperboards require a special classification because the density of these products is slightly greater than the maximum for non-hot-pressed, fibrous-felted, wood-based panel materials. Also, because these products are made by laminating plies of paper about 2.5 mm thick, they have different properties along the direction of the plies than across the machine direction. The other fibrous-felted products have nearly equal properties along and across the panel.

Table 9. Densities of Building Fiberboards

Material	Density, g/cm³ [a]
insulation board	0.02–0.50
semirigid insulation board	0.02–0.16
rigid insulation board	0.16–0.50
hardboard	0.50–1.44
medium density hardboard	0.50–0.80
high density hardboard	0.80–1.20
special densified hardboard	1.35–1.45
laminated paperboard	0.50–0.59

[a] To convert g/cm³ to lb/ft³, multiply by 62.43.

Particle Panel. Mat-formed particleboards, because of differences in properties and uses, generally are classified by density into low (<0.59 g/cm³), medium (0.59–0.80 g/cm³), and high (>0.80 g/cm³) categories. All mat-formed particleboards are hot-pressed to cure the resins that are used as binders.

These mat-formed particleboards are homogeneous (the same kind, size, and quality of particle throughout the thickness), graduated (a gradation of particle size from coarsest in the center of the thickness to finest at each surface), or three layer (the material on and near each surface is different than that in the core). These boards also may be described by the predominant kind of particle, as shavings, flakes, slivers, or the combination in the instance of layered construction as flake-faced or fines-surfaced boards.

A generalized manufacturing flow diagram for a particleboard manufacturing process is presented in Figure 3. Particleboard production begins when the raw material

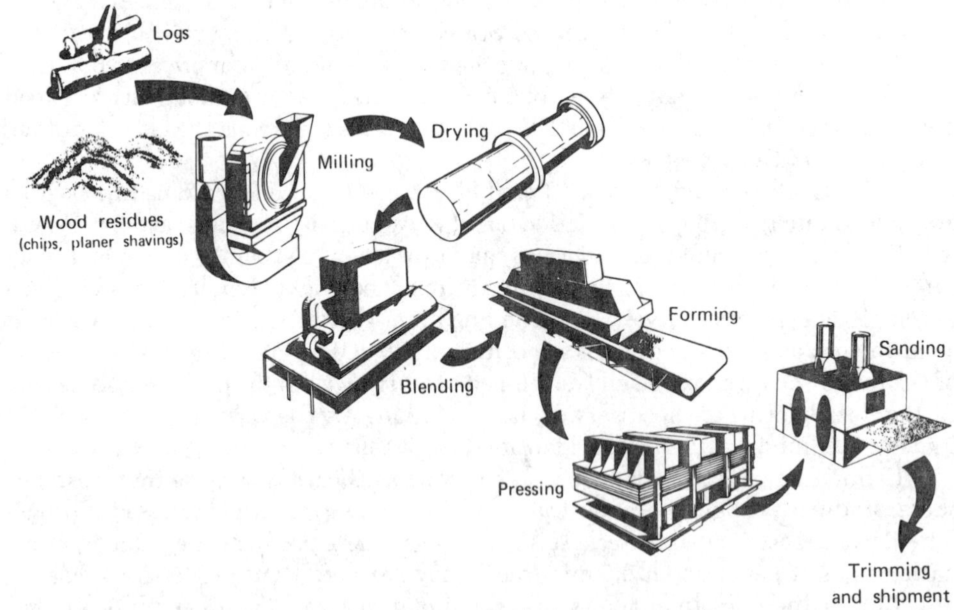

Figure 3. Manufacture of particleboard. Courtesy of the National Particleboard Association.

(wood chips, planer shavings, or logs) is reduced by flakers, hammermills, or other types of milling equipment to produce the desired types of tiny wood particles. Screens classify the particles into the proper mixture of sizes. Dryers remove excess moisture and uniformly control the moisture content to the desired level. Resin binders and other chemicals are sprayed onto the wood particles at a controlled rate in a blending operation. Forming machines deposit the treated particles onto belts or metal cauls, thereby forming mats. Particle mats are consolidated and the binders are cured in heated hydraulic presses with temperatures up to 200°C and pressures up to 6.9 MPa (1000 psi). After being pressed, boards are trimmed to the desired length and width. Sanding in high-speed belt sanders produces the smooth surfaces and accurate thickness tolerances characteristic of particleboards.

Extruded boards account for less than 5% of the total production of particleboard, and standards have not been developed for them to any appreciable extent. Most extruded particleboard is medium density because the compression applied to the wood particles during extruding does not increase the density beyond 0.80 g/cm^3. Particleboards thicker than about 16 mm can be extruded with hollow core sections similar to those molded in concrete blocks. Because of these hollow core sections, the equivalent density may fall below that of medium density, mat-formed particleboard. Extruded particleboards of that type, usually called fluted particleboards, are classified on the basis of weight per square meter for a specified thickness.

Waferboard and Other Structural Composite Panels. Waferboard is a structural board made of wood wafers that are cut to predetermined dimensions, randomly distributed, and bonded with a phenolic adhesive. The product generally is made from roundwood or forest residue rather than mill residue, and has been manufactured in Canada in considerable quantity since the late 1960s. A plant in the United States has had a waferboard plant in continuous production since the early 1970s, and in 1978 and 1979, seven waferboard plants have been announced, and two plant expansions are in the planning stages. Reasons for this strong interest in waferboard include a continuing strong demand for structural board products, a recognition of changing raw material supplies, and rapidly rising softwood plywood prices.

Another type of structural board, which has been manufactured in the Federal Republic of Germany and which soon will be manufactured in the United States in commercial quantities, uses flakes or strands that are narrower than those commonly used in waferboard. These flakes or strands are incorporated into three-layer panels that are composed entirely of oriented-strand-particleboard layers, which substitute for veneer. These panels, which represent one type of a family of panels commonly referred to as oriented-strand board (OSB) panels, initially will be used in floor and roof-sheathing markets where structural strength is important. It is anticipated that future markets for these types of products will expand from sheathing applications to uses in single-layer flooring, textured siding, and overlaid interior panels.

Another type of new structural panel combines an OSB core with veneer face and back material, each component produced separately, and laid-up on a conventional plywood line. The resultant panel looks like a plywood panel and frequently is used interchangeably with plywood. It has had very good market acceptance in the United States. One plant has been in production for more than 3 years and several others soon will begin production. A similar board with a nonaligned particle core has been introduced using an isocyanate adhesive as the binder. This is the first commercial use of this binder for a wood-board product in the United States. Markets include wall and roof sheathing and concrete forms.

Properties and Uses. *Semirigid Insulation Board.* Semirigid insulation board describes fiberboard products that are manufactured primarily for insulation and cushioning. These very low density fiberboards have about the same heat-flow characteristics as conventional blanket or batt insulation but have sufficient stiffness and strength to maintain their position and form without being attached to the structure. They may be bent around curves or corners and, when cemented, mechanically fastened, or placed between framing members, hold their shape and position even though subjected to considerable vibration.

Semirigid insulation boards are manufactured in sheets 13–38 mm thick. When greater thicknesses are required, two or more sheets are cemented together. Sheet sizes vary from 30 × 61 cm to 1.2 × 1.2 m. The thermal resistance factors R are 35–39 mW/(m·K) [0.24–0.27 (Btu·in.)/(h·ft^2·°F)] (see Insulation, thermal). Semirigid insulation boards are used for heat insulation in truck and bus bodies, automobiles, refrigerators, railway cars, on the outside of ductwork, and wherever vibrations are so severe that loose fill or batt insulation may pack or shift.

Rigid Insulation Board. Rigid insulation board is the oldest of the wood-based fiber and particle panel materials and is a structural insulation board. Structural insulating board is manufactured mainly for specific uses in construction although some is fabricated for special padding and blocking in packaging and a variety of other industrial uses. Interior-quality boards are used where high water resistance is not required but where a light-colored product is desired. Sheathing-quality boards are used where water resistance is required and are manufactured with added water-resistant materials (usually asphalt by impregnation and/or coating). Density is somewhat greater for sheathing-quality boards than for interior boards.

Strength and related properties of insulating board are included in Table 10 with those properties of other building fiberboards. The two basic insulating-board products, with only minor composition modifications in manufacture, are fabricated into a group of products designed to satisfy specific use requirements in construction. These requirements may call for structural strength and either high thermal insulation or good acoustical properties or both.

General-Purpose Board. There are two general-purpose structural insulating boards—building board and wallboard (the latter sometimes is called thin board because it is either 7.9- or 9.5-mm thick, whereas most other insulating board is 13-mm thick or thicker). Both general-purpose boards may be converted for a multiplicity of uses not specifically covered in the other products. The general-use boards usually are furnished with a factory-applied, flame-resistant finish. Building board is 13-mm thick and may be obtained in panels 1.2 × 2.4, 2.7, 3.0, or 3.7 m with square edges. Wallboard is furnished regularly 1.2 m wide in either 2.4- or 3.0-m lengths. Quality limits are set for these and other regular products in the standards.

Insulating Roof Deck. Insulating roof deck is a laminated structural insulating board product that is manufactured of several layers of sheathing-grade board and one layer of factory-finished interior board (either perforated or plain). It is used in exposed-beam ceiling constructions where the factory-finished interior board is applied face down. Insulating roof deck is made in 38-, 51-, and 76-mm nominal thicknesses in 0.6 × 2.4-m panels. The 38-mm-thick panel is made to span 61 cm, the 51-mm-thick panel to span 81 cm, and the 76-mm-thick material to span 1.2 m. Panel ends are square and sides are tongued and grooved.

Table 10. Strength and Mechanical Properties of Wood-Based Building Fiberboards[a]

Property	Structural insulating board	Medium density hardboard	High density hardboard	Tempered hardboard	Special densified hardboard
density, g/cm^3	0.16–0.42	0.53–0.80	0.80–1.28	0.93–1.28	1.36–1.44
modulus of elasticity, GPa[b]	0.17–0.86	2.24–4.83	2.76–5.52	4.48–7.59	8.62
modulus of rupture, MPa[b]	1.4–5.5	13.1–41.4	20.7–48.3	38.6–69.0	69.0–86.2
tensile strength parallel to surface, MPa[b]	1.4–3.4	6.9–27.6	20.7–41.4	24.8–53.8	53.8
tensile strength perpendicular to surface, MPa[b]	0.069–0.17	0.28–1.4	0.52–2.8	1.1–3.1	3.4
compressive strength parallel to surface, MPa[b]		6.9–24.1	12.4–41.4	25.5–41.4	183
shear strength (in plane of board), MPa[b]		0.69–3.3	2.1–4.1	3.0–5.9	
shear strength (across plane of board), MPa[b]		4.1–17.2	13.8–20.7	19.3–23.4	
24-h water absorption, vol %	1–10				
24-h water absorption, wt %		5–20	3–30	3–20	0.3–1.2
thickness swelling (24-h soaking), %		2–10	10–25	8–15	
linear expansion (50–90% rh), %[c]	0.2–0.5	0.2–0.4	0.15–0.45	0.15–0.45	
thermal conductivity (at 24°C), mW/(m·K)[d]	39–65	65–108	108–202	108–216	267

[a] The data presented are general, round-figure values, accumulated from numerous sources (see also ref. 2); for more exact figures on a specific product, individual manufacturers should be consulted or tests should be conducted. Values are for general laboratory conditions of temperature and humidity.
[b] To convert MPa to psi, multiply by 145; to convert GPa to psi, multiply by 145,000.
[c] Measurements made on material at equilibrium at each condition at room temperature.
[d] To convert mW/(m·K) to (Btu·in)/(h·ft^2·°F), divide by 144.1.

In climates where condensation occurs, insulating roof deck is furnished with a vapor-barrier membrane that is installed in the glueline between the layer of 13-mm-thick, interior-finish board and the first sheathing-quality layer. In this construction, the roof decking furnishes structural rigidity to support snow and water or wind loads, besides providing interior ceiling finish and thermal insulation. Thermal conductance factors for the various thicknesses are specified at 35, 26, and 17 mW/(m·K) [0.24, 0.18, and 0.12 (Btu·in.)/(h·ft^2·°F)] for the thickness of 38, 51, and 76 mm, respectively. For flat roofs, a built-up roof is applied directly to the top surface of the deck. If the pitch (slope) of the roof is sufficient, asphalt shingles may be attached to 51- and 76-mm-thick decking with special annular-grooved nails.

Roof Insulation. Above-deck, thermal insulation made of structural insulating board is manufactured in blocks 0.58 × 1.19 m or 0.61 × 1.2 m in 13-mm multiples of thickness between 13 and 78 mm. The blocks usually are multiple 13-mm thicknesses of insulation board and may be laminated or stapled together in the greater thicknesses. Insulation-board roof insulation is applied where the final roofing is of the built-up variety. It is secured in place by hot asphalt or roofing pitch or by mechanical fasteners, and has enough internal bond strength to resist uplift forces on the roof structure.

Ceiling Tile and Lay-in Panels. Ceiling tile, either plain or perforated, has a paint finish applied in the factory to provide resistance to flame spread. Interior-finish insulating board, when perforated or provided with special fissures or other sound traps, also will provide a substantial reduction in noise reflectance. Ceiling tiles usually are 30 × 30 cm and 30 × 61 cm in size, 13-mm thick, and have tongue and groove or butt and chamfered edges. They are applied to nailing strips with nails, staples, or special mechanical fastenings, or directly to a surface with adhesives.

A panel product similar to tile, but nominally 61 × 61 cm or 61 × 122 cm, is gaining popularity. These panels, called lay-in ceiling panels, are installed in metal tees and angles in suspended ceiling systems. They usually are 13-mm thick and are supported in place along all four edges. They frequently are used in combination with translucent plastic panels that conceal light fixtures. Finishes and perforation treatments for sound absorption are the same as for regular ceiling tile (see Insulation, acoustic).

Plank. Structural insulating-board plank is installed on side walls, often in remodeling, where it is used in conjunction with ceiling tile installations. Plank is 30-cm wide with matching long edges and is finished with a flame-resistant paint applied at the factory. Because of its low density, it is subject to abrasion when used for the lower part of walls where chairs or other furniture can bump it. It is used frequently in conjunction with wainscoting of wood paneling or one of the other wood-based panel materials like hardboard or particleboard.

Sheathing. Insulation board frequently is used to sheath houses in the United States. Sheathing is manufactured in three grades: regular density, intermediate density, and nail base. Regular-density sheathing is manufactured in both 13- and 20-mm thicknesses. Intermediate and nail base are made only 13 mm thick. Regular-density sheathing is furnished in two sizes, 0.61 × 2.4 m, or 1.2 × 2.4, 2.7, 3.0, or 3.7 m; the other two grades are furnished only in 1.2-m widths and 2.4- or 2.7-m lengths.

Regular-density sheathing usually is about 0.290 g/cm^3 in density and is sold with a thermal resistance R rating of 297 mW/(m·K) [2.06 (Btu·in.)/(h·ft^2·°F)] for 20-mm material and 190 mW/(m·K) [1.32 (Btu·in.)/(h·ft^2·°F)] for the 13-mm thickness. If the 0.6 × 2.4-m material is used as sheathing, it is applied with the long edges horizontal and adequate fastening (either nails or staples) around the perimeter and along intermediate framing, and requirements for racking resistance of the wall construction usually are satisfied. Horizontal applications with the 20-mm material require additional bracing in the wall system to meet code requirements for rigidity, as do some applications of the 13-mm-thick, regular-density sheathing applied with the long edges being vertical.

Costs of heating and requirements for air conditioning from summer heat may justify added thermal insulation over that required by minimum standards. When such added thermal insulation is used in construction of walls, intermediate and nail-base sheathing (with lower thermal resistance) is used. They are applied with their long edges vertical. With recommended fastening, such sheathing provides the racking rigidity and strength for the wall without added bracing.

The density of intermediate sheathing usually is about 0.350 g/cm^3 and that of nail base is about 0.40 g/cm^3. The insulation-board industry provides intermediate density and nail-base sheathing with rated thermal resistance R values; the R values are 176 and 164 mW/(m·K) [1.22 and 1.14 (Btu·in.)/(h·ft^2·°F)], respectively. Nail-base sheathing has adequate nail-holding strength so that asbestos (qv) and wood shingles for weather course (siding) can be attached directly to the nail-base sheathing with special annular-grooved nails. With the other grades of sheathing, siding materials must be nailed directly to framing members or to nailing strips attached through the sheathing to the framing. Because the method and amount of fastening are critical to racking resistance, local building codes should be consulted for requirements in different areas.

Sound-Deadening Board. Sound-deadening board is specially manufactured to provide a meaningful reduction in sound transmission through walls. Standard sizes are 13-mm thick, 1.2 m-wide, and 2.4- or 2.7-m long. In light-frame construction, sound-deadening board usually is applied to the wall framing; the final wall finish, such as gypsum board, is applied to the outside faces of the sound-deadening board. Acoustic efficiency of walls constructed with sound-deadening board depends on tight construction with no air leaks around the edges of panels, and close adherence to prescribed methods of installation.

Medium Density Hardboard. Medium density hardboard, formerly classified as medium density building fiberboard, is a relatively new wood-based panel product. Nearly all of the material being manufactured by the conventional methods that are used for other hardboard is being tailored for use as house siding. Medium density hardboard for house siding use is 9.5- and 11-mm thick and is fabricated for application as either panel or lap siding. Medium density hardboard sometimes is manufactured by a process that involves radio-frequency energy for curing thicker panels (usually about 19 mm although it is possible to make panels as thick as 76 mm) used mainly in furniture and cabinets as core stock or panel stock (see Radiation curing).

Panel siding is 1.2 m wide and commonly is furnished in 2.4-, 2.7-, or 3.0-m lengths. Surfaces may be grooved 51 mm or more on center parallel to the long dimension to simulate reversed board and batten or may be pressed with ridges simulating a raised batten. Lap siding usually is 30-cm wide with lengths to 4.9 m and is applied in the same way as conventional wood lap siding. Some manufacturers offer lap-siding products with special attachment systems that provide either concealed fastening or a wide shadow line at the bottom of the lap.

Most siding is furnished with some kind of a factory-applied finish, eg, the surface and edges are given a prime coat of paint and finishing is completed by application of one or more coats of paint. Two coats of additional paint, one of a second primer and one of topcoat, provide for a longer interval before repainting. However, there is a trend towards complete prefinishing of medium density hardboard siding. The complete prefinishing ranges from several coats of liquid finishes to cementing various films to the surfaces and edges of boards. Surfaces of medium density hardboard for house siding range from very smooth to textured; one of the newest simulates weathered wood with the latewood grain raised as though earlywood has been eroded away. Siding remains the most important use of medium density hardboard. This product is also marketed for industrial use under the name, Industrialite.

High Density Hardboard. Properties of high density hardboard are summarized in Table 10 but, in the trade, the various qualities are subdivided in smaller groups beyond those shown. An overlapping of properties is shown by the limiting values in the various standards for hardboard. Standard hardboard has a density of about 0.960–1.04 g/cm^3 which, usually, is unaltered except by humidification as it is produced by hot pressing. Tempered hardboard is a standard-quality hardboard treated with a blend of siccative resins (drying-oil blends or synthetics) after hot-pressing. The resins are stabilized by baking after the board has been impregnated. Usually about 5% solids are required to produce a hardboard of tempered quality. Tempering improves water resistance, hardness, and strength but embrittles the board making it less shock resistant.

Service-quality hardboard has a lower density than standard hardboard, usually 0.80–0.88 g/cm^3 and satisfies needs where the higher strength of standard quality is not required. Because of its lower density, service-quality hardboard has better dimensional stability than the denser products. It is used where water resistance is required but where the higher strength of regular treatment is not. Underlayment is service-quality hardboard, nominally 6.1 mm thick, that is sanded or planed on the back surface to provide a thickness of 5.461 ± 0.127 mm.

A substantial amount of high density hardboard is manufactured for special industrial use. Hardboard that is manufactured for concrete forms frequently is given a double tempering treatment. For some uses where high impact resistance is required (eg, backs of television cabinets), boards are formulated from specially prepared fiber and additives.

High density and medium density conventional hardboards are manufactured in several ways, and the result is reflected in the appearance of the final product. Hardboard is screen-backed or S-2-S (smooth two sides). When the mat from which the board is made is formed from a water slurry (wet-felted) and the wet mat is hot-pressed, a screen is required to permit steam to escape. In the final board, the reverse impression of the screen is apparent on the back of the board, hence the screen-back designation. A screen is required with mats formed from an air suspension (air-felted) when moisture contents going into the hot press are sufficiently high so as to require venting.

In some variations of hardboard manufacture, a wet-felted mat is dried before being hot-pressed which makes possible hot-pressing without using the screen, and an S-2-S board is produced. In air-felting hardboard manufacture, it also is possible to press without the screen, if the moisture content of the mats entering the hot press is low. In an adaptation of pressing hardboard mats, a caul with slots or small circular holes is used to vent steam; the board that is produced has a series of small ridges or circular nubbins which, when planed or sanded off, yield an S-2-S board.

Medium density hardboard that is manufactured using radio-frequency curing is produced from dry, fiber–resin blends. The mats are pressed between heated platens where the high frequency heat provides additional heat energy to cure the resin binder (usually urea–formaldehyde instead of the phenolics used with the more conventional hardboards); the product is S-2-S hardboard.

Commercial thicknesses of high density hardboard generally range 9.5–13 mm. Not all thicknesses are produced in all grades. The thicknesses of 2.5 and 2.1 mm are produced regularly only in the standard grade. Tempered hardboards are produced regularly in thicknesses of 3.2–7.9 mm. Service and tempered service are produced regularly in fewer thicknesses; however, none is less than 3.2 mm or is produced by all manufacturers or in screenback and S-2-S types. The appropriate standard specification or source of material should be consulted for specific thicknesses of each kind.

High density hardboards are produced in 1.2- and 1.5-m widths. Standard commercial lengths are 1.2, 1.8, 2.4, 3.7, and 4.9 m with a 5.5-m length being available in the 1.2-m width. Most manufacturers maintain cut-to-size departments for special orders. Retail lumberyards and warehouses commonly stock 2.4-m lengths; underlayment, however, is usually 1.2 m square.

About 15% of the hardboard used in the United States is imported. Imported board may or may not be manufactured to the same standards as domestically produced products. Before substituting an imported product in a use where specific properties are required, it should be determined that it has properties required for the use. Canadian products usually are produced according to the same standards as are United States products.

In addition to the standard smooth-surface hardboards, special products are made using patterned cauls so the surface is striated or produced with a relief to simulate ceramic tile, leather, basket weave, etched wood, or other texture. Hardboards are punched to provide holes for anchoring fittings for shelves and fixtures (perforated board) or with holes comprising 15% or more of the area (for installation in ceilings with sound-absorbent material behind it as an acoustical treatment or as air diffusers above plenums). High density hardboard is harder than most natural wood and, because of its grainless character, it has nearly equal properties in all directions in the plane of the board. It is not as stiff nor as strong as natural wood along the grain but it is substantially stronger and stiffer than wood across the grain. Specific properties in Table 10 can be compared with similar properties for wood, wood-base, and other materials.

Hardboard is used in construction as floor underlayment to provide a smooth undercourse under plastic or linoleum flooring, as a facing for concrete forms for architectural concrete, facings for flush doors, as insert panels and facings for garage doors, and material punched with holes for wall linings in storage walls and in built-ins where ventilation is desired. In furniture, furnishings, and cabinet work, conventional hardboard is used extensively for drawer bottoms, dust dividers, case goods and mirror backs, insert panels, television, radio, and stereo-cabinet sides, backs (die-cut openings for ventilation), and as crossbands and balancing sheets in laminated or overlay panels. Hardboard also is used in interior linings of automobiles, trucks, buses, and railway cars.

Densified Hardboard. Densified hardboard is manufactured mainly as diestock and electrical panel material. It has a density of 1.36–1.44 g/cm^3 and is produced in thicknesses of 3.2–51 mm in panel sizes of 0.91 × 1.2 m, 1.2 × 1.8 m, and 1.2 × 3.7 m. The 3.2-mm-thick board is specially manufactured for use as lofting board. As diestock, it finds use for stretch- and press-forming and spinning of metal parts, particularly when few of the manufactured items are required and where the cost of making the die is important in the choice of material. The electrical properties of the special densified hardboard meet many of the requirements set forth by the National Electrical Manufacturers Association (NEMA) for insulation resistance and dielectric capacity in electrical components; therefore, it is used extensively in electronic and communication equipment.

Laminated Paperboards. Laminated paperboards are made in two general qualities: interior and weather resistant. The main differences between the two qualities are in the kind of bond used to laminate the layers and in the amount of sizing used in the pulp stock from which the individual layers are made. For interior-quality boards, the laminating adhesives commonly originate from starch. Synthetic-resin adhesives are used for the weather-resistant board. Laminated paperboard regularly is manufactured in thicknesses of 4.8, 6.1, and 9.5 mm for construction uses although 3.2-mm thickness is common for industrial uses, eg, dust dividers in case goods, furniture, and automotive liners. Important properties are presented in Table 11.

Considerable amounts go into the prefabricated housing and mobile-home construction industry as interior wall and ceiling finish. In the more conventional building construction market, interior-quality boards are also used for wall and ceiling finish, often in remodeling to cover cracked plaster.

The common width of laminated paperboard is 1.2 m although 2.4-m widths are available in 3.7-, 4.3-, 4.9-m and longer lengths for building applications, eg, sheathing entire walls. Laminated paperboard intended for use where the surface is exposed has a surface that is ply-coated with a high quality pulp to improve surface appearance and performance. The surface finish may be smooth or textured.

Particleboard. Important properties for mat-formed particleboards are presented in Table 12. Similar values are not presented for extruded particleboards since they are never used without facings glued to them, and the facings influence the physical and strength properties. Extruded particleboards have a distinct zone of weakness across the length of the panel as extruded. They also have a strong tendency to swell in the lengthwise (extruded) direction because of the compression and orientation of particles from the extrusion pressures. Consequently, extruded particleboards always are used as corestock; mat-formed boards are used both as corestock and as panel stock where the only thing added to the surface is finish.

Table 11. Strength and Mechanical Properties of Laminated Paperboard[a]

Property	Value
density, g/cm^3	0.51–0.53
modulus of elasticity (compression), MPa[b]	
along the length of the panel[c]	2.1–2.7
across the length of the panel[c]	0.69–0.97
modulus of rupture, MPa[b]	
span parallel to length of panel[c]	9.7–13.1
span perpendicular to length of panel[c]	6.2–7.6
tensile strength parallel to surface, MPa[b]	
along the length of the panel[c]	11.7–14.5
across the length of the panel[c]	4.1–5.5
compressive strength parallel to surface, MPa[b]	
along the length of the panel[c]	4.8–6.2
across the length of the panel[c]	3.4–5.5
24-h water absorption, wt %	10–170
linear expansion of 50–90% rh, %[d]	
along the length of the panel[c]	0.2–0.3
across the length of the panel[c]	1.1–1.3
thermal conductivity (at 24°C), mW/(m·K)[e]	73

[a] The data presented are general round-figure values, accumulated from numerous sources; for more exact figures on a specific product, individual manufacturers should be consulted or actual tests made. Values are for general laboratory conditions of temperature and humidity. (See also ref. 2).

[b] To convert MPa to psi, multiply by 145.

[c] Because of directional properties, values are presented for two principal directions, along the usual length of the panel (machine direction) and across it.

[d] Measurements made on material at equilibrium at each condition at room temperature.

[e] To convert mW/(m·K) to (Btu·in.)/(h·ft^2·°F), divide by 144.1.

Table 12. Strength and Mechanical Properties of Mat-Formed (Platen-Pressed) Wood Particleboard[a]

Property	Low density particleboard	Medium density particleboard	High density particleboard
density, g/cm^3	0.40–0.59[b]	0.59–0.80	0.80–1.12
modulus of elasticity (bending), MPa[c]	1.0–1.7[b]	1.7–4.8	2.4–6.9
modulus of rupture, MPa[c]	5.5–9.7[d]	11.0–55.2	16.6–51.7
tensile strength, parallel to surface, MPa[c]		3.4–27.6	6.9–34.5
perpendicular to surface, MPa[c]	0.14–0.21[d]	0.28–1.4	0.86–3.1
compressive strength parallel to surface, MPa[c]		9.7–20.7	24.1–35.9
shear strength (in the plane of board), MPa[c]		0.60–3.1	1.4–5.5
(across the plane of the board), MPa[c]		1.4–12.4	
24-h water absorption, wt %		10–50	15–40
thickness swelling from 24-h soaking, %		5–50	15–40
linear expansion[e] (50–90% rh), %	0.30[f]	0.2–0.6	0.2–0.85
thermal conductivity (at 24°C), mW/(m·K)[g]	79–108	108–144	144–180

[a] The data presented are general round-figure values, accumulated from numerous sources; for more exact figures on a specific product, individual manufacturers should be consulted or actual tests made. Values are for general laboratory conditions of temperature and humidity. (See also ref. 2.)
[b] Lower limit is for boards as generally manufactured; lower density products with lower properties may be made.
[c] To convert MPa to psi, multiply by 145.
[d] Only limited production of low density particleboard so values presented are specification limits.
[e] Measurements made on material at equilibrium at each condition at room temperature.
[f] Maximum permitted by specification.
[g] To convert mW/(m·K) to (Btu·in.)/(h·ft^2·°F), divide by 144.1.

Quality criteria, testing, labeling, rating, and certification methods for mat-formed particleboard are established in *Commercial Standard CS 236-66* (15). For certain uses where special requirements must be satisfied, additional specifications outline the requirements for the particleboard. Particleboard is manufactured in both 1.2- and 1.5-m widths and wider, although for industrial sales, much is cut to size for the purchaser. In construction, as for other panel materials, the common size panel is 1.2 × 2.4 m.

The uses for particleboard are under development, and parallel those for lumber core in veneered or overlaid construction and for plywood. The two properties of particleboard that have the greatest influence on its selection for a use are that the panels have a uniform surface and stay flat as manufactured, particularly in applications where edges are not fastened to a rigid framework. For the majority of uses where exposures are interior or equivalent (furniture, cabinetwork, interior doors, and most floor underlayment), urea–formaldehyde resins are used. Boards with that kind of bond are classed as type 1 in the specifications (15). Where greater resistance to heat, moisture, or a combination of heat and moisture is required, type 2 boards that generally are bonded with phenolic resin are required.

In general, particleboards are manufactured in about the same thicknesses as softwood plywood; most manufacture is in thicknesses between 13 and 25 mm, although there are new developments for particleboard that are thinner than 13 mm or thicker than 25 mm. Much extruded particleboard is fluted in thicknesses that satisfy the need for cores for flush doors. Similarly, low density, mat-formed particleboard is manufactured for solid-core doors in thicknesses so that, when facings are applied, final door thicknesses are the standard 35 and 44 mm.

Recently, there has been a trend toward thinner particleboard products. Thicknesses of 6.1 and 9.5 mm are becoming common, both for special and general use, in the United States. In terms of volume, the main two uses for particleboard are as furniture and cabinet core, and floor underlayment. As corestock, particleboard has moved into a market formerly held by lumber core and, to a limited extent, by plywood. For example, certain grades of hardwood plywood now permit the use of particleboard as the core ply where, formerly, lumber core was specified.

Both three and five plies are employed in built-up constructions where particleboard is used as the core. Extruded particleboard nearly always requires five-ply construction because of the board's instability and low strength in one direction. A relatively thick crossband that has the grain direction parallel to the extruded direction, stiffens, strengthens, and stabilizes the core. Thinner face plies are laid with the grain at right angles to the crossband to provide the final finish.

The use of three- or five-ply construction with mat-formed particleboard corestock depends on the class and type of particleboard core (stiffness and strength), kind of facings being applied (plastic or veneer), and the requirements of the final construction. Balanced construction in lay-ups using particleboard is important; facings or crossbands with different properties can cause objectionable warping, cupping, or twisting in service. Edge-bonding of wood or of the facing material frequently is employed in panelized units that have particleboard as the corestock.

As a floor underlayment, particleboard provides leveling, the thickness of construction required to bring the final floor to elevation, and the indentation-resistant, smooth surface necessary as the base for resilient-finish floors of linoleum, rubber, vinyl, and vinyl–asbestos tile and sheet material. Particleboard for this use is produced in 1.2 × 2.4-m panels that commonly are 6.1-, 9.5-, or 16-mm thick. Separate use specifications cover particleboard floor underlayment.

Particleboard for siding, combined siding–sheathing, and as soffit linings and ceilings for carports, porches, and the like requires the durable adhesive, phenol–formaldehyde. For these uses, type 2, medium density board is required. Agencies, eg, the Federal Housing Administration, have established requirements for particleboard for such use. The satisfactory performance of particleboard in exterior exposure depends not only on the manufacture and kind of adhesive used, but on the protection afforded by the finish. Manufacturers recognize the importance of the combination by providing both paint-primed panels and those that are completely finished with liquid-paint systems or factory-applied plastic films.

Mobile-home manufacture and factory building of conventional housing are important and increasing users of forest products. Since particleboard is manufactured in hot presses as large as 2.4 × 12 m, panels are available in sizes larger than those generally used for conventional construction. With mechanical handling available in factories, large-sized panels can be attached effectively and economically. Two particleboard products have been developed to satisfy these uses. Mobile-home decking is used for combined subfloor and underlayment and has a type 1 bond but is protected from moisture in use by a bottom board, so that it generally provides a satisfactory service life. The United States industry has established separate standards for these products. They are marketed under a certified product-quality program.

Another use that commonly requires the more durable bond of a type 2 quality is a special corestock where laminated plastic sheets are formed on the face and back of a particleboard at the same time as they are bonded to it and, usually, a high density particleboard is used. The temperatures used in curing the resin-impregnated plastic sheets may reduce the strength of the bonds in boards made with urea–formaldehyde resins.

The properties of particleboard depend on the shape and quality of the particles used as well as on the kinds and amount of resin binder. Although most particleboard is produced using particles that yield a board of intermediate strength and stiffness, a substantial volume is comprised of flakes or other engineered particles for boards of higher strength and stiffness. Sometimes the boards of intermediate stiffness and bending strength are designated class 1 and those of the higher stiffness and bending strengths are designated class 2. Class 2 particleboards are more expensive than those in class 1, but they usually are justified for uses where the greater stiffness and strength are required. The same applies to those applications where a special surface such as a fine surface provides either a better base for finishing or less showthrough for an overlaid construction.

Production. The U.S. Department of Commerce has published production and market information for particleboard and medium density fiberboard for the year 1975 (16). A portion of the information presented in this Census Bureau document is given in Table 13. Reference 16 has complete production information by geographic area, and export–import breakdowns.

United States Panel Price Trends. Prices for every principal panel product group rose, albeit at varying rates depending on industry conditions, in the United States in the 1970s. Softwood plywood producer prices increased most, rising by 280% from 1970 to 1978 (Table 14). This rise reflected, in part, the strong demand for the product in the wake of the construction boom during this period. But the biggest factor was the escalation of the costs of materials and labor. Western softwood-plywood mills, which were dependent on a diminishing supply of high quality sawlogs, were forced to bid in a highly competitive log market with eager foreign and domestic buyers. The result was a substantial leap in log costs.

The price of no. 2 Douglas-fir sawlogs, which were used extensively for plywood sheathing, rose from \$34.70/1000 cm^3 (\$82 per thousand board feet (MBF)) in 1970 to about \$127/1000 cm^3 (\$300 MBF) in 1979. This fourfold jump in price was representative of the cost-push experienced by West Coast plywood producers. The supply of logs in southern United States, by contrast, was somewhat better, resulting in lower but still sizable cost increases for that segment of the industry. The cost and availability of resins also became a problem toward the end of the decade. The higher use of benzene in gasoline, caused by environmental laws, resulted in a rapid upward price spiral in benzene-derived resins. Plywood producers attempted to diminish the impact of this by using alternative glues and glue extenders, but nevertheless, were forced to absorb a large cost increase in this area as well.

Hardwood plywood prices exhibited a more moderate trend, rising by only 60% and can be attributed to the better supply of hardwood logs in the eastern United States where hardwood timber growth has exceeded removals consistently for the past three decades. Delivered oak prices in southern United States, eg, were estimated to have risen by 125%, a figure which contrasts sharply with the 265% increase in western United States log prices that are noted above.

Hardboard and insulating board do not require high-grade logs as input; thus, they were not subject to the same intense cost pressures experienced by softwood-plywood manufacturers. Hardboard prices rose by 60% over the period and insulat-

Table 13. United States Production and Market Information for Particleboard and Medium Density Fiberboard, 1975[a]

Product	United States total (19-mm basis), 10^6 m^2
particleboard, all types, total	235.9
platenboard production[b], total	232.6
floor underlayment, total	72.9
16-mm board	46.0
9.5-mm board	8.1
other thicknesses	18.8
mobile-home decking, total	21.0
19-mm thick	
16-mm thick	17.5
other platenboard (industrial board), total	149.3
19-mm board	53.7
16-mm board	28.2
13-mm board	19.8
9.5-mm board	6.2
6.1-mm board	0.4
all other	30.3
urea-formaldehyde resins	30.3
phenolic formaldehyde and other resins	10.7
extruded board production[c]	3.4
medium density fiberboard[d], total	20.0
19-mm thick	11.4
other thicknesses	8.6

[a] Ref. 16. To convert m^2 to ft^2, multiply by 10.76.

[b] Platenboard is an engineered product that is matformed of machined fiber particles (eg, granules, chips, slivers, flakes, and shavings) of a controlled moisture content and size and which are bonded with a synthetic resin or other added binder into panel form under controlled heat and pressure.

[c] Extruded board is produced by forcing machined fiber particles that are mixed with resin or other added binder through a long, heated die. The formation of the panel, the curing, and the pressing take place in one continuous operation.

[d] Medium density fiberboard is a panel product that is manufactured from lignocellulosic fibers that are combined with a synthetic resin or other suitable binder. The panels are manufactured to densities of 0.50–0.88 g/cm^3.

Table 14. Panel Board Prices in the United States, 1979, $/10³ m²[a]

Year	Softwood plywood sheathing[b]	Hardwood plywood[c]	Hardboard[c]	Insulating board[c]	Particleboard underlayment[b]
1970	861	11.08	11.09	11.95	463
1975	1453	12.93	12.70	15.50	721
1977	2271	13.78	15.39	19.81	1066
1978	2530	13.53	16.90	27.82	1539
1979	2411	17.76	17.76	21.31	990

[a] To convert m^2 to ft^2, multiply by 10.76.

[b] Ref. 17.

[c] Ref. 18.

ing-board prices rose by 78%. Particleboard prices rose somewhat faster at 115% with fairly sharp gains registered from 1977–1979. This may have resulted from the more restricted availability of planer shavings used for furnish raw material. With the increase in energy costs, many lumber and plywood mills began to use their planer shavings as fuel; this is reported to have restricted board output in several cases.

The broad trends outlined above (eg, inflation, raw material scarcities, and energy problems), look well entrenched for the foreseeable future and are likely to continue sending panel product prices higher. The upward trends may be temporarily interrupted by economic contractions, as happended in 1969–1970 and 1974–1975 but, once the economy returns to its normal growth track, the fundamental cost pressures are apt to reappear in higher prices for consumers.

Glued Structural Members

Glued-laminated or parallel-grain construction, as distinguished from plywood or other crossbanded construction, refers to two or more layers of wood that are joined with an adhesive so that the grain of all layers or laminations is approximately parallel. The size, shape, number, thickness of the laminations, and uses may vary greatly.

Although the properties of glued-laminated products are similar to those of solid wood of similar quality, manufacture by gluing permits production of long, wide, and thick items out of small and inexpensive material and often with less waste of wood than if solid wood were used alone. Curved members may be fabricated by simultaneously bending and gluing thin laminations to shapes that would be very difficult or impossible to produce from solid wood. The essentially parallel direction of grain of the wood to the longitudinal axis of these laminated products gives them strength that often is far superior to solid wood that is cut to the same size and shape.

The advantages of glued-laminated or parallel-grain construction include: ease of manufacture of large structural elements from standard commercial sizes of lumber or veneer; minimization of checking or other drying defects associated with large, one-piece wood members, in that the laminations are thin enough to dry readily before manufacture of members; the opportunity of designing on the basis of the strength of dry wood, for dry service conditions, inasmuch as the individual laminations can be dried to provide members that are thoroughly seasoned throughout; the opportunity to design structural elements that vary in cross section along their length in accordance with strength requirements; the possible use of lower grade material for less highly stressed laminations without adversely affecting the structural integrity of the member; and the manufacture of large laminated structural members from smaller pieces is increasingly adaptable to future timber economy, as more lumber comes in smaller sizes and in lower grades.

Certain factors involved in the production of laminated structural components are not encountered when solid sawed structural material is produced, ie, preparation of lumber for gluing and the gluing operation usually raise the cost of the final laminated product above that of solid sawn timbers in sizes that are reasonably available. Because the strength of a laminated product depends upon the quality of the glue joints, the laminating process requires special equipment, plant facilities, and manufacturing skills not needed to produce solid sawed timbers. These factors tend to emphasize the importance of the performance requirements of the intended use for the wood component or article. Generally, laminating provides the opportunity to produce engineered wood components for specific use applications.

Glued-Laminated Timbers. Structural glued-laminated timber (glulam) defines three or more layers of sawed lumber that are glued together with approximately parallel grain direction. The layers may vary as to species, number, size, shape, and thickness. Laminated wood was first used in the United States for furniture parts, cores of veneered panels, and sporting goods but now is widely used for structural timbers in building.

The first use of glued-laminated timbers was in Europe where, as early as 1893, laminated arches (probably glued with casein glue) were erected for an auditorium in Basel, Switzerland. Improvements in casein glue during World War I aroused further interest in the manufacture of glued-laminated structural members, at first for aircraft and later as framing members of buildings. In the United States, glued-laminated arches were installed in gymnasiums, churches, halls, factories, hangars, and barns. The development of very durable, synthetic-resin glues during World War II permitted the use of glued-laminated members in bridges, trucks, and marine construction where a high degree of resistance to severe service conditions is required. Glued-laminated construction forms an important segment of the woodworking industry.

Glued structural timbers may be straight or curved. Curved arches have been used to span more than 91.4 m. Straight members spanning up to 30.5 m are not uncommon, and some span as much as 39.6 m. Sections deeper than 2.1 m have been used. Straight beams can be designed and manufactured with horizontal laminations (lamination parallel to the neutral plane) or vertical laminations (laminations perpendicular to the neutral plane). The horizontally laminated timbers are the most widely used. Curved members are horizontally laminated to permit bending of laminations during gluing.

Internal Stresses. For best results in manufacturing glued-laminated timbers, it is important to avoid the development of appreciable internal stresses when the member is exposed to conditions that change its moisture content. Differences in shrinking and swelling are the fundamental causes of internal stresses; therefore, laminations should be of such character that they shrink or swell to a similar degree in the same direction. If laminations are of the same species or of species with similar shrinkage characteristics, if they are all flat-grained or all edge-grained material, and if they are of the same moisture content, the assembly will be reasonably free from internal stresses and have little tendency to change shape or to check. Laminations that have an abnormal tendency to shrink longitudinally because they have excessive cross-grain or compression wood should not be included.

Laminated timbers that are treated with preservatives after gluing and fabrication have given excellent service in bridges and similar installations. Laminations also can be treated and then glued if suitable precautions are observed. The treated laminations should be conditioned and must be resurfaced just before gluing. Not all preservative-treated wood can be glued with all adhesives, but if suitable adhesives and treatments are selected and the gluing is carefully done, laminated timbers can be produced that are entirely serviceable under moist, warm conditions that favor decay.

Species for Laminating. Softwoods, principally Douglas fir and southern pine, are most commonly used for laminated timbers. Other softwoods that are used include western hemlock, larch, and redwood. Boat timbers, on the other hand, often are made of white oak because the wood is moderately durable under wet conditions. Red oak that is treated with preservative also has been laminated for ship and boat use. Other species can be used, of course, when their mechanical and physical properties are suited for the purpose.

Quality of Glue Joints. Glue joint quality is one of the most critical manufacturing considerations. Using proven adhesives, it is possible to produce joints that are durable and essentially as strong as the wood itself. *U.S. Product Standard PS 56-73* (19) and *CSA Standard 0122-1969* (20) describe minimum production requirements for glulam.

For a large proportion of laminated timbers, because of their size, pieces of wood must be joined end-to-end to provide laminations of sufficient length. In most cases, the strength of timbers is reduced by the presence of end joints. The highest strength values are obtained with well-made plain scarf joints (21); the lowest values are with butt joints. Scarf joints with flat slopes have side-grain surfaces that can be well bonded and that develop high strength, whereas butt joints have end-grain surfaces that cannot be bonded effectively. Finger joints are a compromise between scarf and butt joints and strength varies with joint design.

Both plain scarf joints and finger joints can be manufactured with adequate strength for structural glued laminated timbers. The adequacy is determined by joint efficiency principles (21) or by physical testing procedures (19); the physical testing procedures are used more commonly.

The joint efficiency or joint factor is the joint strength as a percentage of the strength of clear, straight-grain material. The joint factor of scarf joints in the tension portion of bending members or in tension members is:

Scarf slope	*Joint factor, %*
1 in 12 or flatter	90
1 in 10	85
1 in 8	80
1 in 5	65

These factors apply to interior laminations as well as to the lamination on the tension face. In a beam of 40 equal laminations, eg, the outer lamination should be stressed no higher than 90% of the clear wood stress value if a scarf-joint sloping of 1 in 12 is used. The stress at the outer face of the second lamination is 95% of the outer-fiber stress, so that a scarf joint of about 1 in 10 would be satisfactory. Usually, however, laminators use the same joint throughout the members for ease in manufacture. Working stresses need not be reduced if laminations containing scarf-joint slopes of 1 in 5 or flatter are used in the compression portion of bending members or in compression members.

Finger-joint strengths vary depending upon the type and configuration of joint and the manufacturing process. High strength finger joints can be made when the design is such that the fingers have relatively flat slopes and sharp tips. Tips are a series of butt joints and, therefore, reduce the effective sloping area as well as act as possible sources of stress concentration. As a result, the strength of even the best finger joint developed to date has approached but not equaled the strength of a well-made plain scarf joint. If the joint factor for a specific finger joint is determined, allowable stresses may be calculated on the same basis as discussed for scarf joints.

The slope of scarf joints in compression members or in the compression portion of bending members should not be steeper than 1 in 5; a limitation of 1 in 10 is suggested for tension members and for the tension portion of bending members. Because there is some question as to the durability of steep scarf joints for exterior use or other severe exposure, scarf joints having slopes steeper than 1 in 8 should not be used under such conditions.

Joints should be well scattered in portions of structural glued-laminated timbers that are highly stressed in tension. Test results indicate that, in closely spaced end joints in adjacent laminations, failure progresses more or less instantaneously from the joint in the outer lamination through the others. Adequate longitudinal separation of end joints in areas of high stress, therefore, is desirable.

No data are available by which to substantiate any proposed spacing requirements. Spacing requirements depend on joint quality and stress level. No spacing requirement should be necessary for well-made, high strength joints or for joints that are stressed well below their strength. Suggested spacings of end joints are given in *Voluntary Product Standard PS 56* (19).

Allowable Design Stresses. The strength of laminated timbers principally is a function of two properties: the clear-wood strength (a species property) and the size and frequency of knots (a property of the lumber used in manufacture). In addition to knots, cross-grain and end-jointing efficiency also can influence strength. Because these three factors are not cumulative and because the lowest factor determines strength, current practice is to control cross grain and end joints such that the effect of each is less than that of the knots.

Clear-wood strength is obtained from data published for various species (21–22). A value of clear-wood strength is selected such that 95% of all sample data exceeds it. This value is reduced to account for a duration of load effect. A further reduction is applied to clear-wood strength, which includes a factor of safety. Current practice is to classify or grade lumber for laminating into different structural grades based, principally, on the size of knots and the slope of the grain. The effect of knots on strength properties depends on the stress under consideration.

Engineering formulas applicable to solid wood timbers generally are applicable to glulam timbers since properly glued joints have shear strength that is approximately equal to that of wood. However, because of the possible shapes and generally larger sizes of glulam, special design considerations are necessary. Allowable unit stresses for use in the design of glued laminated timber can be obtained from applicable building codes and other sources (20,23–26). These stresses must then be adjusted depending on the particular application.

Design. *In-Service Moisture Content.* Dry-use stresses for glulam members are published for an average moisture content of 12% and a maximum of 16%. When conditions of use permit in-service moisture contents to exceed 16%, as may occur in members exposed to precipitation or in covered locations of high humidity, wet-use stresses are applicable. Where in-use moisture contents are in the decay supporting range, special treatments should be considered. To obtain wet-use stresses when tabular information is not available, the following factors commonly are applied to the dry-use stresses:

Types of stress	*Wet-use factor*
bending and tension parallel to grain	0.80
compression parallel to grain	0.73

compression perpendicular to grain	0.67
shear	0.88
modulus of elasticity	0.83

Treatment. For pressure-impregnated wood that is treated for increased life, no reduction in allowable stresses is recommended. For wood that is pressure-impregnated with a fire-retardant treatment, a 10% reduction in the stresses is used (27). Difficult-to-treat species are sometimes incised prior to treatment in order to obtain higher preservative retention. Incising does significant damage to small localized regions of the exterior surface; however, the effect of incising upon overall strength normally is not considered a design factor.

Duration of Load. Duration of loading effects are the same as for lumber (27).

Curvature Factor. For sharply curved members, bending stresses should be calculated by methods related to curved beams (28). Practical limitations imposed by manufacturing generally prevents the divergence from normal bending theory from becoming large. For most designs where the normal bending theory is applied, the curvature factor includes an adequate reduction in the allowable bending stress to account for curvature. Bending stress that is induced in the individual laminations when they are bent to a form during manufacture results in residual stress which is only partially relieved by creep action. This residual stress must be considered in beam design. Common practice for curved members is to reduce the allowable design stress that is applicable to straight timbers (29).

Size Factor. The relative size of a bending member has effects on strength. Published stress data assume a depth of member of 30.4 cm and, in the United States, allowable unit stresses for glued-laminated timbers stressed in bending should be adjusted for any depth exceeding 30.4 cm. In addition, the span-to-depth ratio (L/d) and method of loading should be considered when adjusting bending stresses (29).

Lateral Stability. Unbraced bending members are subject to lateral buckling; consequently, bending stresses sometimes are reduced to permit longer, unbraced sections. In order to determine the allowable design stresses based on lateral stability considerations, the application of a slenderness factor in design is recommended (29).

Fastenings. Various types of fasteners are available to connect glued-laminated timber members. Common fasteners are timber connectors such as split rings or shear plates, bolts, and lag screws. Where lighter loads are involved, nails, spikes, and wood screws are considered (26). Glulam rivets, nonmetallic fasteners, spike grids, drift pins, and adhesives can be used for specific applications. Certain connection details have proven satisfactory, and these have been standardized (30).

Typical Members. Glulam members can be designed and fabricated to carry bending stresses, axial loads, or combined bending and axial loads. The design of each type of member is based on the application of specific design considerations.

Stresses and deflection of glued-laminated bending members are determined by utilizing basic formulas available in most design manuals. Deflection usually governs in members having long spans or in members subjected to light live loads (29). Camber is recommended for horizontal laminated beams in order to minimize ponding conditions as well as to avoid the possible impression of excessive deflection. The recommended magnitude of camber that is used is one and one-half times the dead-load deflection. Other amounts of camber should be considered depending on specific member configuration, loading condition, and desired appearance. Where horizontal members are not provided with adequate slope to allow for drainage, the possibility of ponding should be investigated (29). When a beam cross section is varied in depth along its length to meet specific requirements, consideration must be given to the combined effects of bending, compression, tension, shear, and radial stresses. The analysis involves an interaction formula in which stresses occur simultaneously to reduce beam capacity below that calculated by simple theory. Formulas also are used to calculate deflections of such beams (29,31). In cases such as purlins, where members are subjected to biaxial bending so the loading is not in the plane of the vertical axis of the member, the allowable stress in the member should be checked using conventional formulas (29).

Treatments. *Pressure Preservative.* Some use conditions require that members undergo pressure-preservative treatment to ensure long life. Design considerations, such as special connection details, flashing, and coatings, can provide some degree of protection; however, when the details of the design and construction are not sufficient to resist decay, fungi, insects, or marine borers, pressure preservative treatment is necessary. The effectiveness of the treatment depends on the chemicals used and their penetration and retention in the wood. Depending on the type of preservative used, the finished laminated timber can be treated, or the individual laminations may be treated prior to laminating. Creosote or oilborne chemicals may be used for treating after laminating. The use of waterborne salt treatments for treating glued laminated members after gluing is not recommended because of the degree of dimensional change and magnitude of checking which may occur in the laminated timber as the wood dries after treatment. The waterborne salts and oilborne chemicals in mineral spirits or volatile solvent carriers are the only preservatives recommended for treating individual laminations prior to gluing. Standards used in choosing and specifying preservative treatments are available (32–33); additional details related to preservative treatments are given in ref. 27.

Flame-Retardant Treatment. Laminated timber may be pressure-impregnated with mineral salts as flame-retardant treatments so as to reduce surface flamespread, smoke density, and fuel contribution from the laminated timber (34). When flame-retardant treatments are used, the reduction of strength of the member and its fastenings (27), the compatibility of treatment and adhesive, the use of special gluing procedures, and the effect on fabricating procedures should be investigated (33).

Coatings and Finishes. Glulam timbers are available with satins, sealers, or other finishes that are applied in the plant to satisfy an appearance requirement. If timbers will be subjected to low humidity conditions, end-grain coatings are recommended to minimize end-grain checking caused by rapid moisture loss through the end grain.

Uses. The American Institute of Timber Construction estimates that 590×10^6 cm^3 (250×10^6 board feet) of sawed, structurally graded dimension lumber was consumed by the glulam producers in 1978. Of this total, about 90% was used in making straight or slightly cambered members for use in roof-support structures. Curved members for arch-type roof supports accounted for the other 10% of the production.

In addition to fabricating glulam members in virtually any size or shape for simple bending members or axially loaded members, glulam fabrication is readily adaptable to other structural configurations. These include trusses, arches, domes, and other roof systems. Although many of the structural elements used can be applied in any given design situation, experience has shown that certain members are most applicable when used in specific span ranges, eg:

Member type	*Typical span range, m*
beams	6.1–39.6
bowstring truss	15.2–61.0
pitched truss	15.2–27.4
parallel chord truss	15.2–45.7
arches	9.1–91.4
domes	15.2–107

Wood–Plywood Glued Structural Members. Highly efficient structural components or members can be produced by combining wood and plywood through gluing. The plywood is utilized in load-carrying capacity and in filling large opening spaces. These components, including box beams, I-beams, stressed-skin panels, and folded plate roofs, are discussed in detail and completed designs are also available in the many technical publications of the American Plywood Association. These highly efficient designs, although adequate structurally, may suffer from lack of resistance to fire and decay unless treatment or protection is provided.

Beams with Plywood Webs. Box beams and I-beams with lumber or laminated flanges and glued plywood webs can be designed to provide desired stiffness, bending moment resistance, and shear resistance. The flanges resist bending moment, and the webs provide primary shear resistance. Either type of beam must not buckle laterally under design loads; thus, if lateral stability is a problem, the box-beam design should be chosen because it is stiffer in lateral bending and in torsion than the I-beam. On the other hand, the I-beam should be chosen if buckling of the plywood web is of concern because its single web, which has twice the thickness of that of a box beam, offers greater buckling resistance.

Stressed-Skin Panels. Constructions consisting of plywood skins glued to wood stringers are called stressed-skin panels. These panels offer efficient structural constructions for floor, wall, and roof components. They can be designed to provide desired stiffness, bending-moment resistance, and shear resistance. The plywood skins resist bending moment, and the wood stringers provide shear resistance.

Laminated Veneer Structural Members. Parallel-laminated veneer in thick sheets of any width or length is being examined as an alternative to solid sawed timber or glulam for structural-sized or specialty-type members. Research on this concept of improved resource utilization has been conducted for several years (35–39). Process variables that have been investigated include veneer thicknesses, adhesive types, drying methods, and laminating techniques.

Research in the last 10 years has shown that laminated veneer products offer advantages of increased product yield and improved product performance when compared to conventional lumber. Because of the dispersion of wood defects in the laminating process, lower limits of bending strength can be increased, and stiffness can be more uniform.

Laminated veneer structural members appear to have greatest market potential when used for specialty items (eg, truck decking and scaffold planking) or where high tensile strength is primary and stiffness is secondary (eg, in truss chords and flanges of built-up beams). Commercial quantities of material that is manufactured from

2.5–3.2-mm veneer are being produced and marketed (39). Research has shown that preservative treatment of difficult-to-treat species is improved considerably when laminations are peeled rather than sawed (40).

BIBLIOGRAPHY

"Plywood" in *ECT* 1st ed., Vol. 10, pp. 860–870, by J. G. Meiler, Coos Bay Lumber Company; "Plywood" in *ECT* 2nd ed., Vol. 15, pp. 896–907, by Charles B. Hemming, U. S. Plywood-Champion Papers, Inc.

1. R. L. Geimer, W. F. Lehmann, and J. D. McNatt, *For. Prod. J.* **25**(9), 72 (1975).
2. *Wood Handbook: Wood as an Engineering Material, U.S. Dept. Agric., Agric. Hand. 72, Rev.* U.S. Forest Products Laboratory, Madison, Wisc., 1974.
3. D. V. Doyle and L. J. Markwardt, *Properties of Southern Pine in Relation to the Strength Grading of Dimension Lumber, U.S. For. Serv. Res. Pap. FPL 64,* Forest Products Laboratory, Madison, Wisc., 1966.
4. *Construction and Industrial Plywood, Nat. Bur. Stand. Voluntary Product Standard PS 1-74,* U.S. Department of Commerce, Washington, D.C., 1974.
5. *Hardwood and Decorative Plywood, Nat. Bur. Stand. Voluntary Product Standard PS 51-71,* U.S. Department of Commerce, Washington, D.C.
6. *Plywood Design Spec. AIA File 19-F,* American Plywood Assoc., Tacoma, Wash., 1966.
7. *Bureau of the Census Current Ind. Rep. MA-24H(77)-1 on Softwood Plywood,* U.S. Dept. of Commerce, Washington, D.C., 1977.
8. *Bureau of the Census Current Ind. Rep. MA-24F(77)-1 on Hardwood Plywood,* U.S. Dept. of Commerce, Washington, D.C., 1977.
9. *Bureau of the Census Current Ind. Rep. MA-24F(78)-1,* U.S. Dept. of Commerce, Washington, D.C., 1978.
10. R. G. Anderson, *Plywood End-Use Marketing Profiles 1977–1979, American Plywood Assoc. Econ. Rep. E24,* Tacoma, Wash., 1978.
11. T. M. Maloney, *Modern Particleboard and Dry-Process Fiberboard Manufacturing,* Miller-Freeman Publications, San Francisco, Calif., 1977.
12. H. A. Miller, *Particle Board Manufacture,* Noyes Data Corporation, Park Ridge, N.J., 1977.
13. *Cellulosic Fiber Insulation Board, Nat. Bur. Stand. Voluntary Product Standard PS 57-73,* U.S. Department of Commerce, Washington, D.C., 1973.
14. *Basic Hardboard, Nat. Bur. Stand. Voluntary Product Standard PS 58-73,* U.S. Department of Commerce, Washington, D.C., 1973.
15. *Mat-Formed Wood Particleboard, Nat. Bur. Stand. Commercial Standard CS 236-66,* U.S. Department of Commerce, Washington, D.C., 1966.
16. *Bureau of Census Current Industrial Report MA-24L(75)-1,* U.S. Department of Commerce, Washington, D.C., 1975.
17. *Random Lengths* (annual), Lumber and Plywood Market Reporting Service, Eugene, Ore., 1979.
18. *Producer Prices and Price Indexes* (annual), U.S. Dept. of Labor, Bureau of Labor Statistics, Washington, D.C., 1979.
19. *Structural Glued Laminated Timber, Nat. Bur. Stand. Voluntary Product Standard PS 56-73,* U.S. Department of Commerce, Washington, D.C., 1973.
20. *Structural Glued Laminated Timber, Canadian Standard 0122,* Canadian Standards Association, Rexdale, Ontario, Canada, 1977.
21. A. D. Freas and M. L. Selbo, *U.S. Dept. Agric. Tech. Bull.* **1069,** U.S. Government Printing Office, Washington, D.C., 1954.
22. *ASTM Standard D 2555, Standard Methods for Establishing Clear Wood Strength Values, Annual Book of ASTM Standards,* Part 22, American Society for Testing and Materials, Philadelphia, Pa., 1977.
23. *Standard Specifications for Structural Glued Laminated Timber of Douglas Fir, Western Larch, Southern Pine, and California Redwood, AITC 117-74,* American Institute of Timber Construction, Englewood, Colo., 1974.
24. *Standard Specifications for Structural Glued Laminated Timber Using "E" Rated and Visually Graded Lumber of Douglas Fir, Southern Pine, Hem Fir, and Lodgepole Pine, AITC 120-71,* American Institute of Timber Construction, Englewood, Colo., 1971.
25. *Standard Specifications for Hardwood Glued Laminated Timber, AITC 119-71,* American Institute of Timber Construction, Englewood, Colo., 1971.

26. *National Design Specification for Stress-Grade Lumber and Its Fastenings,* National Forest Products Association, Washington, D.C., 1973.
27. *Design Guide and Commentary—Wood Structures,* American Society of Civil Engineers, New York, 1975.
28. R. O. Foschi, *Plane-Stress Problem in a Body with Cylindrical Anisotropy, with Special Reference to Curved Douglas Fir Beams, Publ. 1244, Dept. Forest. Rural Dev.,* Ottawa, Ontario, 1968.
29. *Timber Construction Manual,* 2nd ed., American Institute of Timber Construction, Englewood, Colo., 1974.
30. *Typical Construction Details, AITC 104-72,* American Institute of Timber Construction, Englewood, Colo., 1972.
31. A. C. Maki and E. W. Kuenzi, *Deflection and Stresses of Tapered Wood Beams, Forest Service Res. Pap. FPL 34,* Forest Products Lab., Madison, Wisc., 1965.
32. *AWPA Standards C1, C2, and C28,* American Wood-Preservers' Association, McLean, Va.
33. *Treating Standard for Structural Timber Framing, AITC 109-69,* American Institute of Timber Construction, Englewood, Colo., 1969.
34. *AWPA Standard C20,* American Wood-Preservers' Association, McLean, Va.
35. P. Koch, *For. Prod. J.* **23**(7), 17 (1973).
36. E. L. Schaffer, R. W. Jokerst, R. C. Moody, C. C. Peters, J. L. Tschernitz, and J. J. Zahn, *Press-Lam: Progress in Technical Development of Laminated Veneer Structural Products, USDA Forest Service Res. Pap. FPL 279,* Forest Products Lab., Madison, Wisc., 1977.
37. E. L. Schaffer, J. L. Tschernitz, C. C. Peters, R. C. Moody, R. W. Jokerst, and J. J. Zahn, *Feasibility of Producing a High Yield Laminated Structural Product, General Summary, USDA Forest Service Res. Pap. FPL 175,* Forest Products Lab., Madison, Wisc., 1972.
38. J. C. Bohlen, *For. Prod. J.* **22**(1), 18 (1972).
39. S. A. Nelson, *Structural Applications of Micro-Lam Lumber, American Society of Civil Engineers Meeting, Preprint 1714,* ASCE, New York, 1972.
40. J. L. Tschernitz, V. P. Miniutti, and E. L. Schaffer, *Am. Wood-Preserv. Assoc.* **70,** 189 (1974).

JOHN A. YOUNGQUIST
United States Department of Agriculture

LEATHER

Prehistoric man wore hides and skins which were preserved with juices that were extracted from tree bark. Leather provided footwear, garments, holsters for weapons, shelters, and many other products affording protection, comfort, durability, and esthetic properties in a combination that was not attainable from other materials. Despite efforts, particularly during the last 30 yr, to produce leatherlike materials (qv) from other fibers, sheet materials, and petrochemicals, and to substitute them in the marketplace for leather products, leather still is the product of choice by the consumer. The demand for leather far exceeds the supply, and the principal function of the substitutes has been to fill those needs for which the supply of leather is inadequate.

In the leather industry, supply bears no relationship to the demand for leather and leather goods. The hides and skins that are used for leather manufacture are by-products of the meat industry. As more meat has been consumed by a growing world population, more hides and skins have become available so that more leather is being manufactured than ever before. Yet the growth has been far below the growth of the consumer-goods industries and the demands for footwear, gloves, luggage, handbags, small personal leather goods, and garments. Synthetics have not displaced leather but have allowed it to be used where it is best suited and most needed. For example, as nonleather soling materials for shoes improve in quality, the shoe-upper leather manufacturer can buy more hides to make more uppers and more shoes can be produced to accommodate more people who have less need for the genuine comfort of an all-leather shoe. Leather-sole production and use stabilized at the point necessary for those people who could not tolerate synthetic soles because of excessive foot perspiration which could not be transmitted by the upper alone or because their feet were too tender to withstand the greater impact felt through synthetic soling materials. Substitutes for leather range from fabric (in garments) to rubber and synthetics (in shoes); rubber and synthetics account for nearly 90% of all shoe soles. Synthetic materials have been only moderately successful in displacing upper leather in shoes. Where comfort and durability are important, leather will continue to be the dominant material used. Similarly, leather upholstery has been replaced partially by synthetics, eg, vinyl-coated fabrics which provide alternative coatings for items such as cushioned chairs, sofas, and bar stools; but this field also has stabilized, and the use of leather again is growing as it meets needs for durability, long term flexibility, low heat conductivity, and resistance to tearing at stitches, none of which properties can be met by the synthetics; no synthetic material has ever duplicated all of the useful properties of leather. Leather has been used as industrial belting, where its properties of maintaining physical form, grease resistance, high strength, and extreme low temperature flexibility have been important.

Thus far, hides have been sold for leather manufacture at prices far exceeding those for any other potential use. If, for any reason, demand for leather should be less than supply, the price of hides could drop to about a hundredth of its present price before it would be used for markets not already saturated by the current trimmings and other losses from hides in process. However, because leather is so highly qualified for a variety of uses, the principal competitors for hides are other leather manufacturers for other end-use fields, and drastic price reductions of the type discussed most probably will not occur. (All aspects of the chemistry and technology of leather are reported in ref. 1.)

Physical Properties

Leather has unique properties which make it ideally suited for use in the manufacture of a variety of products, the most notable being footwear. Extremely large variations exist in leather properties because of differences in the types of skins employed, ranging from the loose open structure of a sheepskin to the hard shell-like area of a horse butt used to make the long-lasting, tough cordovan. Variations in processing induce large changes in physical properties even in the leather from a single animal type, eg, cattle. However, cattlehide leather, which is cut to different thicknesses, given different types of soaking, unhairing, pickling, tanning, lubrication, and finishing, can be manufactured with carefully specified and very different properties for varied uses.

Even if leather is considered for a single use, eg, shoe uppers, a large variation in properties is advisable to allow for different shoe types, eg, casual, military, work, or dress shoes. A range of values for various physical properties of cattlehide shoe upper leather, a representative value, and references from which these data were gathered are listed in Table 1. A general discussion of the factors involved in the tests for leather has been reported (2).

Table 1. Physical Properties of Shoe Upper Leather

Property	Range	Representative value	References
tensile strength, MPa[a]	15.26–37.48	27.6	3–5
elongation at break, %	29.5–73.0	40	3–5
stitch tear strength, N/cm[b]	1280–2275	1751	3, 5–7
tongue tear strength, N/cm[b]	226–961	525	3, 5
thickness, mm	1.5–2.4	1.8	4, 6
bursting strength, kN/cm[b]	1.10–24.5	17.5	3–4, 6
grain-cracking, N/cm[b]	525–1489	1051	3–4, 6
wet shrinkage temperature[c], °C	96–120	100	6, 8
apparent density, g/cm^3	0.6–0.9	0.75	2, 4
real density, g/cm^3	1.4–1.6	1.5	2, 4
flexibility (Flexometer)			
bending length, cm	6–9	7	4
flexural rigidity, mg/cm	10,000–50,000	20,000	4
bending modulus, MPa[a]	19.7–68.9	34.5	4
compression modulus, MPa[a]		0.345	4
cold resistance[d], K			
(without finish or fat liquor)			
workable without cracking at		92	9
heat resistance	shrinks depending on moisture content, anhydrous decomposition at 160–165°C		10
temper of chrome side leather (butt)			
flexibility factor, g		400	11
recovery, %		57	11
resilience, %		21	11

[a] To convert MPa to psi, multiply by 145.
[b] To convert N/cm to lbf/ft, divide by 14.6.
[c] Chrome-tanned leather values.
[d] Ordinarily limited by finish-crack or lubricant hardening properties.

One of the most important properties of leather and one for which there is a paucity of quantitative data, is the ability of leather to conform to the shape of the foot. Shoe sizes are quite standard, yet there is much variation in the shape of individuals' feet. However, these variously shaped feet fit into standard shoes because the leather gives at points of pressure. Yet, the leather maintains its shape and is not deformed beyond the point necessary to give foot comfort, so that it does not become baggy even though the distended part is subjected to repeated flexure. This may be ascribed to a favorable balance between plastic and elastic flow in leather. Plastic flow imparts the necessary give to provide foot comfort; elastic flow is responsible for the maintenance of shape by ensuring proper recovery during flexing action. Leather substitutes have not achieved this combination of properties.

Materials employed as leather substitutes have been able to function in uses where only protection and flexibility have been the chief requirements, eg, in upholstery, luggage, belts, and shoe soles. However, substitutes have not been able to conform to the shape of or to remove moisture from the foot. The amount of moisture from foot perspiration is considerable, although it varies greatly among individuals. For example, in an 8-h test of five subjects under conditions of moderate exercise at 23.9°C, the amount of perspiration emitted by one foot ranged 15–50 g with an average of 34 g (12).

It was believed that water vapor was transmitted through the free spaces within the leather (13) or by condensation in small pores of the leather on the more humid side and migration of the condensed liquid through capillary action toward the less humid side where it would evaporate (14–15). This belief has persisted even though it has been demonstrated that leather impregnation could drastically lower air permeability without significantly lowering water-vapor permeability (11).

Probably only a part of the transpiration of water through leather results from passage through leather interstices. The mechanism of the transmission of water vapor through leather has been demonstrated to be a function of the material rather than a linear diffusion process (16–17). The evidence indicates that water is sorbed at polar groups of the protein molecules and is conducted by an activated diffusion process along the fibers, even against an air-pressure head. Transmission of water vapor through leather is affected by filling the interstices of the leather with grease or other materials, but the range of air permeability changes ca 200 to 1 as contrasted with a change of ca 2 to 1 for water-vapor permeability. According to the mechanism of water-vapor transmission by active diffusion along polar groups of the internal protein surfaces of the leather, a certain amount of energy would be required to free sorbed or bound water from an active site before it can go on to the next site. This energy is the energy of activation for permeation and was measured at 15.0 kJ/mol (3.58 kcal/mol).

Manufacture

Raw Material. ***Chemical Composition.*** Fresh cattlehides, like most biological materials, contain 65–70% water, 30–35% dry substance, and less than 1% ash. The dry substance is largely made up of the fibrous proteins (qv), collagen, keratin, elastin, and reticulin. The main components of the ash, listed in decreasing concentration, are phosphorus, potassium, sodium, arsenic, magnesium, and calcium.

Collagen is the leathermaking protein of the hide. The tanning process consists of a number of collagen-purification steps prior to the actual tanning of the hide. If most of the minor proteins of the hide are not removed before the tanning, they seem to prevent the resulting leather from being soft and flexible. The corium is almost entirely collagen. This protein, which is ubiquitous in the animal kingdom, is unique in its amino acid composition and physical properties. Collagen is responsible for the strength and toughness of the raw hide and of the leather made from it. It is the principal component of connective tissue, and changes in collagen structure have been postulated to be important to aging. A great deal of research has been carried out regarding the structure and function of collagen, and a number of reviews have been published (18–21).

The hair is composed entirely of the protein, keratin, which is the second most common protein in the hide and, depending on the age of the animal and season of the year, may constitute 6–10% of the total protein. One of the first steps in the manufacture of leather is removal of keratin from the hide.

Smaller amounts of other proteins are associated with the hide. Myosin, a muscle protein, is found in the *erector pili* muscle in the grain layer. Various globulins, albumins, and mucoproteins also are present. They probably are residuals from blood as well as interstitial proteins. Elastin is the primary protein, other than collagen, that is present in the grain layer. This connective tissue protein is quite elastic and is resistant to temperature denaturation and proteolytic degradation. Very little elastin is found in the corium.

Lipids are the next most abundant chemicals in hides. The grain layer is ca 9% lipids on a dry weight basis; this fraction includes waxes, phospholipids, sterols, and fatty acids. Most of the lipids are found in the sebaceous glands around the hair follicles. The corium layer is quite variable in its lipid content and may contain from 1–11% lipids. These are, primarily, triglycerides which are randomly distributed in fat cells. Diet and age seem to correlate with the presence of fat cells in the corium.

A small amount of carbohydrates is present in the hide in the grain layer. These compounds largely are mucopolysaccharides which are believed to act as a lubricant for elastin.

Microstructure. The skin forms a tough, protective, thermal blanket on an animal and is vital to the animal's existence; it also provides an ideal raw material for leather. Microscopic examination reveals the physical structure that is responsible for skin's unique properties (Fig. 1). In general, all mammalian skins are composed of a dense, interwoven, fibrous mat in the center (dermis, including grain and corium); a thin, cellular outer layer (epidermis) with protruding hairs; and a loose, fatty, inner (subcutaneous) layer attached to underlying muscles. The looseness of the inner layer facilitates mechanical removal (flaying) of the skin in the slaughterhouse.

The dermis is composed of two layers with distinctly different structures. The upper portion, including the entire length of hair follicles in most species, is called the grain layer by tanners. An exception to this is pig skin, in which the hair follicles extend through the entire thickness of the dermis. The numerous components of the grain layer in cattle hide are illustrated in Figure 2 (enlarged from Fig. 1). Each hair follicle, or pocket, is associated with its surrounding epidermis and with an oil gland, a sweat gland, and an erector muscle. This cellular conglomerate serves protective, sensory, and excretory functions in the living animal as well as controls skin temperature by increasing or decreasing evaporative cooling at the surface. At the junction of epidermis and dermis, there is a coarse basement membrane which, after separation of the epidermis, becomes the grain surface (enamel) of leather. The grain layer also contains a network of small blood vessels and capillaries. Interwoven among these cellular components is a three-dimensional mat of thin collagen fibers which become the leather substance. Interspersed among these fibers, especially in the upper half of the layer, are thinner, elastic tissue fibers that are arranged mostly parallel to the surface. When the cellular structures have been decomposed by chemical processing, the resultant fibrous mat has many voids and reduced density.

The corium is a much simpler structure than the grain layer and is composed mostly of thick bundles of collagen fibers. These bundles are interwoven in a fairly random, three-dimensional pattern, but the angle of weave in a vertical direction tends to vary with animal species and, especially, with location on the body. The excellent strength and stretch properties of leather depend largely on corium fibers and their prevailing orientation. There also are some scattered blood vessels and elastic tissue fibers in the corium layer and variable fat deposits in animals that were intensively fed. In fresh skin, the dermal structures are embedded in a semifluid gel or ground substance. Chemical processing opens fine spaces between fibers but otherwise does not alter the gross physical structure of the corium as it does the grain layer.

At the molecular level, collagen is synthesized in the dermis by elongated cells called fibroblasts and then aggregates into unit fibrils and fibers. Because of their

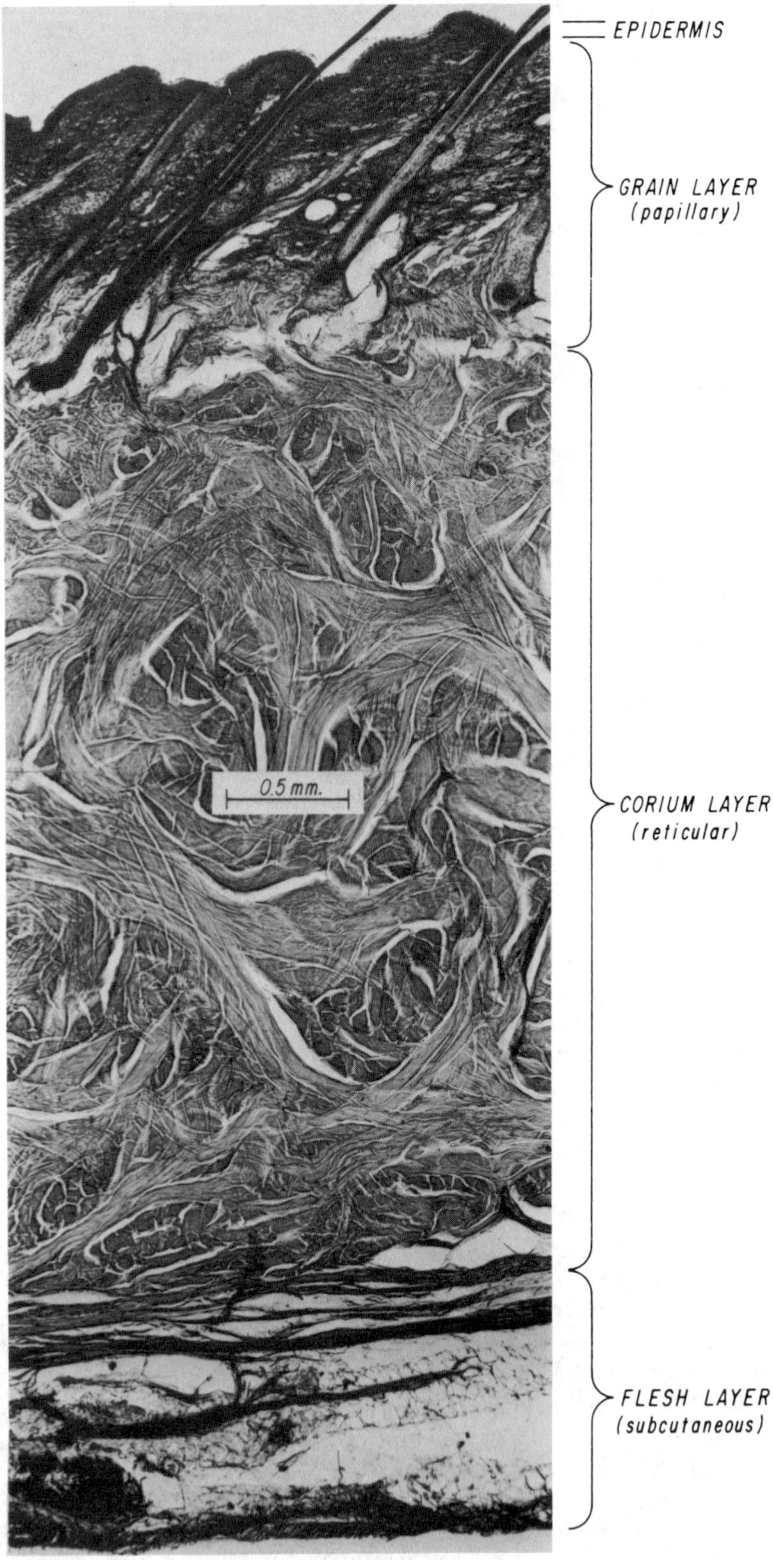

Figure 1. Cross section of cowhide showing subdivision into four principal layers. The dermis, which includes the grain and corium layers, becomes the fibrous substance of leather.

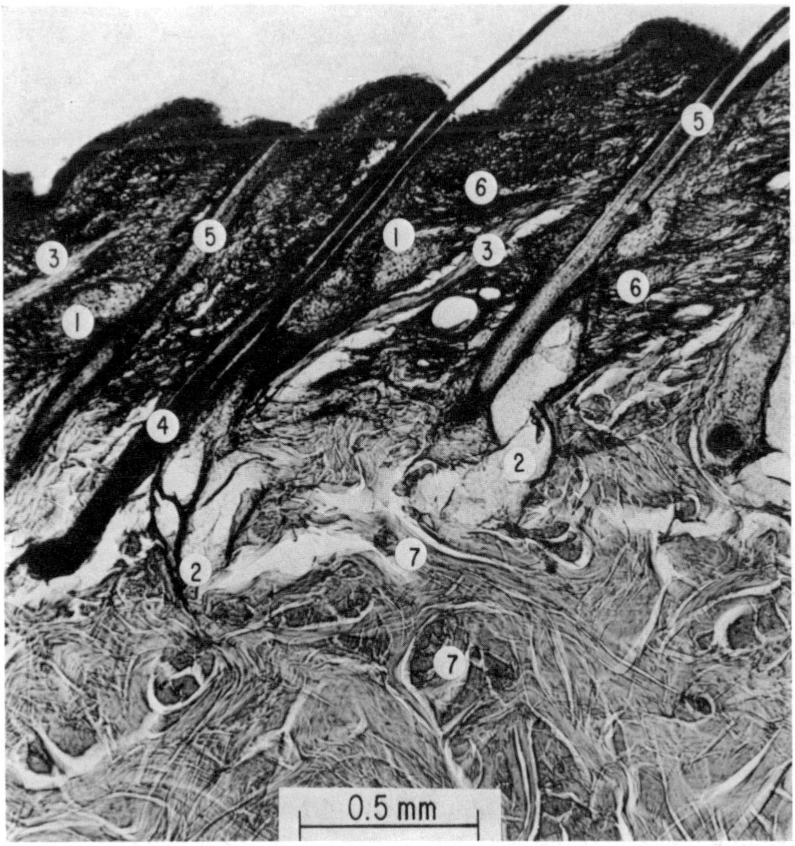

Figure 2. Enlargement of grain layer and upper corium showing components: (1) oil or sebaceous gland; (2) sweat or sudoriferous gland; (3) hair muscle or *erector pili;* (4) growing hair; (5) degenerating hair; (6) fine fibers of elastic tissue; (7) collagen fiber bundles shown in longitudinal and transverse section.

extremely small size, the unit fibrils must be visualized with an electron microscope. Figure 3 (22) shows a thin section of cowhide at high magnification (ca ×17,000). Lengthwise fibrils, with their characteristic cross striations, can be seen in one corner; elsewhere, the fibrils are seen in cross section as they are organized into two small fibers. These fibers are arranged with others in parallel alignment to form larger fibers and, finally, fiber bundles. Ordinary light microscopy may be used to examine the interrelationships of all structures at lower, more practical magnifications.

Although mammalian skins generally are similar in most respects, the species that are used for leather display a number of minor variations which are often important. For example, skins from wool-bearing sheep contain an unusually large number of hair follicles per unit area and a characteristic fatty layer at the base of the follicles which affect processing procedures as well as final properties. Anatomical differences must be known in order to make the best use of each type of material. These differences also add to the esthetic appeal and variety of leather, since modifications of the surface grain pattern are formed by follicle openings. Inspection of full-grain leathers with a hand lens reveals distinctive patterns which, in most cases, can be used

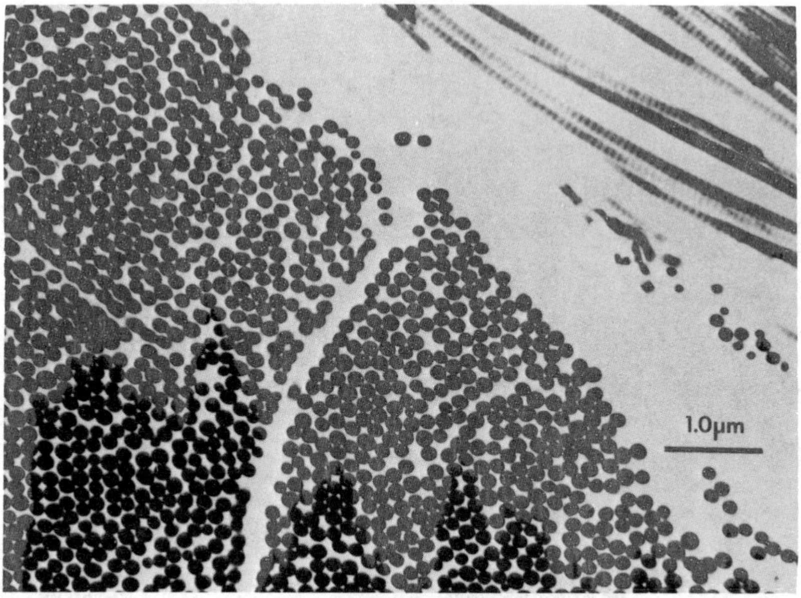

Figure 3. Electron micrograph of hide cross section showing, at upper right, longitudinal collagen fibrils with typical striations and, elsewhere, a cross section of fibrils comprising two small fibers (22).

to fingerprint the species of origin, unless an embossing technique has been used to imitate a different species. This inspection may be augmented by a similar inspection of a freshly cut edge to expose the internal structure and, thus, add to the reliability of identification (11,23–27).

Defects. Significant defects which seriously mar the surface appearance or impair the physical properties of leather can occur anywhere in the manufacturing process, ie, from the living animal to the finished product. Some of these may result from poor flaying technique in the slaughterhouse, inadequate preservation and handling methods, or improper use of chemicals and heat in the tannery. Wherever they occur, defects are costly to some phase of the industry and, ultimately, to the consumer. Although much progress has been made in minimizing this wastage, much remains to be done.

A comprehensive review illustrating the nature and cause of most skin and hide defects has been published (28). Many animals suffer scratches or punctures from barbed-wire fences, prods, or other sharp objects which leave their marks as visible scars on the skin. A more serious type of scar, which is man-made and, thus, preventable, is the brand produced by application of a hot iron for identification purposes. Hot branding is a traditional practice in the U.S. and is responsible for major economic losses. A less damaging system of freeze branding was developed (29) but, for various reasons, it was not found suitable by the beef-cattle industry, although it is commonly used for horses and dairy cattle. A better, nondestructive identification system is needed. Potentially useful ideas include the insertion of an electronic transponder under the skin of the animal to transmit a coded number to a portable interrogator/receiver or the use of a laser device to bleach the hair pigment in a given pattern (see Lasers).

Among the various skin diseases that affect domestic animals, those that are caused by parasites probably affect leather quality the most. For many years the damage from grub or warble holes in cattlehides and associated losses of meat caused by migrating larvae (*Hypoderma* spp.) have been costly to both tanners and meat packers despite the development of preventive treatments. Another problem is the grain damage to cattlehides and goat skins caused by demodectic mange mites (*Demodex* spp.); although this damage (30) is less severe but still significant, more research effort is needed to discover an effective treatment. The cockle defect of sheepskin seriously damages suedes, shearlings, and other types of leather and has been attributed (31) to a parasite, the common sheep ked (*Melophagus ovinus*), and good preventive treatments are known. However, because it does not seriously threaten the life of the animal, the condition is not routinely treated, and it continues to damage leather and reduce the values of meat and wool (qv).

Another parasite problem, which continues despite the availability of good treatment methods, is the damage from many species of ticks (30). Ticks also can transmit serious diseases, but the insects are most prevalent in hot climates and are less important to maintaining hides in the U.S. A number of skin diseases caused by bacteria, fungi, or viruses are most prevalent in warmer climates, and few of them seriously threaten U.S. livestock.

An important biological defect that is most prevalent in a major breed of U.S. beef hides, but has received only limited attention, is the vertical-fiber defect (32) known as pulpy butt. This consists of an inherent, abnormally vertical orientation of the corium fibers which leads to extremely weak leather and many shoe failures. The condition is localized in the choice area of certain hides, is difficult to detect in raw material, and is thought to be hereditary. More research might resolve the remaining genetic uncertainties and, thus, justify a breeding approach to control it.

Animal growers play a more significant role in the production of good leather than they realize. Full use of available technology could improve the quality of the product. Among the defects mentioned, only brands and grubs are obvious enough to be detected in the classification of raw material, and their presence lowers the purchase price. The other defects remain hidden until processing. The economic losses are most apparent in the case of brands. The price differentials between branded and unbranded hides amounted to ca $\$50 \times 10^6$/yr (33), and this loss was passed to the cattle growers. Another $\$50 \times 10^6$ was lost by the tanning industry in wasted costs for processing unusable parts of branded hides. The total loss of $\$100 \times 10^6$ has risen because of inflation.

Chrome Tanning. ***Practice.*** The primary function for a tanning agent is to stabilize the collagen fibers so that they are no longer biodegradable. Over 95% of all the leather manufactured in the United States is chrome tanned. The manufacture of leather can be divided into three separate phases. Historically, the first took place in the beamhouse where the skin was prepared for tanning. The next phase was done in the tanyard. The last phase is finishing. The beamhouse was so named because the steps carried out in this part of the process originally were carried out by hand, with the skin laid over a log or beam. Tanning was done in an outdoor tanyard in pits or holes in the ground containing the tanning solutions. Finishing was developed much later and protects the grain surface of the hide and enhances its appearance. Although methods have changed greatly, the beamhouse and tanyard still exist in modern tanneries. Discussions of this subject are available (1,27). Some of the proteins in the raw skin must be removed in order to make satisfactory leather. In addition, the collagen fiber structure must be modified to allow full and uniform penetration of the tanning chemicals. All of the following factors have an effect on the product of any step of the leathermaking process: temperature, time, concentration of reactants, and the amount of mechanical action.

The hides generally are received by the tanner in a salt- or brine-cured condition. The curing dehydrates the hide, which may contain as much as 14% of its weight as salt. A fresh hide contains ca 64% water; after brine curing it contains ca 45% water and 41% protein, only ⅔ of which is leathermaking collagen. The first step taken by the tanner is to remove the salt and to rehydrate the fibers by soaking. A detergent usually is added to speed up the hydration. The soaking procedure also removes water-soluble proteins and washes the hide free of manure and dirt. The period of soaking generally is from 12–24 h.

Next, the hair must be removed. The keratin has a large content of sulfur-containing amino acids, mainly cystine (see Amino acids). This amino acid's disulfide cross-links stabilize the protein molecules, resulting in a stable hair fiber. Hair removal can be accomplished with a saturated solution of calcium hydroxide (lime) alone or in combination with a sharpening agent, eg, arsenic or cyanide (both of which are no longer used) or sodium sulfide or sodium sulfhydrate. Lime by itself does not dissolve the hair but only loosens it in the base of the hair follicle for easy removal by an unhairing machine. This labor-intensive apparatus scrapes the loosened hair from the surface of the skin and is termed a hair-save process. Lime (qv), by itself, requires from 5–7 d to loosen the hair.

Because of the small market for hair and the importance of time, a hair-burn process more commonly is used. Although sulfide at a pH greater than 11.5 can dissolve the hair in as little as 30–40 min, the usual sulfide unhairing process takes from 4–6 h. Although it is rapid and requires less manpower than the hair-save process, it contributes heavily to the pollution load (BOD) in the tannery effluent.

In many tanneries, the relatively brief unhairing step is followed by a longer (4–16 h) liming step. The spent unhairing liquors with the dissolved hair are drained from the hides and a fresh saturated lime solution is added. The action of lime not only loosens the hair but opens up the collagen fiber structure. Collagen swells outside of its isoelectric point in either acid or base in 8–48 h. This swelling leads to subsequent fiber separation and allows rapid penetration of tanning chemicals. Additional proteins also are removed in the liming stage, and some hydrolysis of amide side chains of the collagen to acid side chains takes place and aids the tanning reaction, since the acid groups are the primary source of binding for chromium tanning agents.

The liming step, when complete, is followed by deliming and bating. The hide is washed to remove soluble lime and hair particles. At this point, the stock is at a pH of 12.5. The most widely used deliming salt is ammonium sulfate, which lowers the pH to 8–9, a range in which the enzymes in the bate can act properly. The bate is a preparation of pancreatic enzymes that usually are absorbed on sawdust. The effective enzymes are proteases which break down additional miscellaneous proteins (see Enzymes, industrial). Bating action usually is short, ie, ca 1–4 h. Immediately after the bating, the hides are pickled with sulfuric acid to lower the hide pH to less than 3. Sodium chloride is added to prevent acid swelling. A full pickle requires at least two hours. Because the addition of a strong acid to water generates a great deal of heat, care must be taken to prevent denaturation of the collagen. Once the hide is in the acid condition, it is prepared for the tanning operation.

At a pH of 2.8, chrome sulfate (usually ca 33% basicity) is soluble. After the tanning solution has been allowed to fully penetrate the hide, the pH is raised slowly with sodium bicarbonate. By the time a pH of 3.4–3.6 is obtained, the chrome has reacted with the collagen to produce a fully preserved, tanned hide. At this point, the hide is said to be in the blue. Although specific procedures are followed in the beam house for a particular type of finished leather, the blue stage allows many options in the next phases of leather manufacture.

The tanned hides generally are stacked overnight and the chrome further fixes onto the collagen. They then are put through a hide wringer so that they are almost dry to the touch and then are sorted for quality and thickness. Each hide is selected to be made into a particular leather product line, eg, to be heavy- or lightweight shoe leather, to be naked grain, or, because of poor grain condition, to be heavily finished. Once the choice has been made, the hide is split to the desired thickness. The split, which is from the flesh side of the hide, is either sold to a split tanner or is processed further in the same tannery to make split leather.

Theory. Chrome tanning generally is carried out by adding the acidified hide to an aqueous solution of trivalent chromium sulfate of 30–50% basicity. By using a combination of gel-permeation chromatography, gel electrophoresis, ion-exchange chromatography, and spectroscopic techniques, it was established that there are at least 10 ionic and neutral complexes in a 33% basic chrome sulfate solution (34) (see Chromium compounds). The structures of eight of these were determined along with the relative amounts of each. The six compounds present in highest concentration are shown in Figure 4. The simplest ion, Cr^{3+}, with six coordinated water molecules, was present as 9% of the total chrome. The routes of formation of the other complexes resulting from replacement of water with sulfate ion (left side of the figure) or by hydroxide ion (right side of the figure) are indicated. The salient features of the route

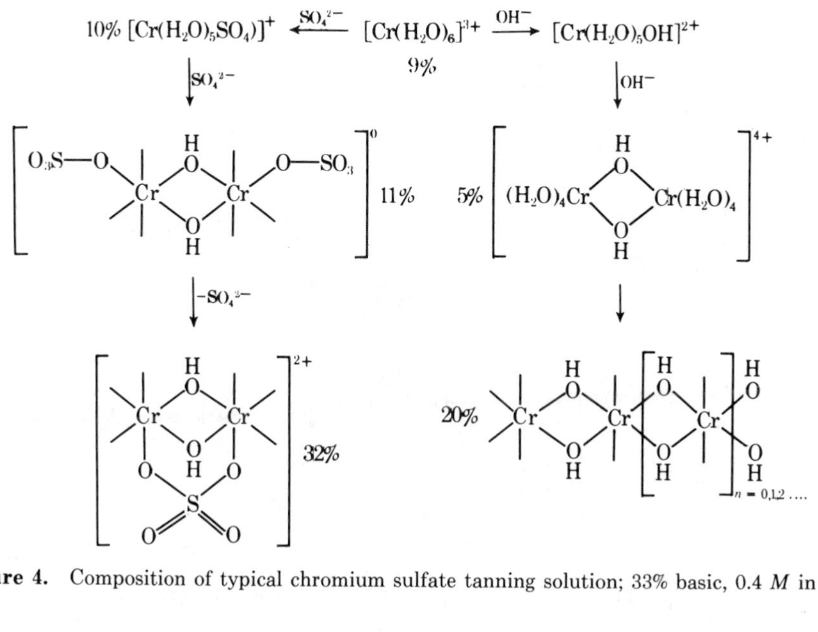

Figure 4. Composition of typical chromium sulfate tanning solution; 33% basic, 0.4 *M* in Cr^{3+} (34).

of formation of the most abundant species, ie, the 2+-charged, binuclear complex with a bidentate sulfate bridge at the lower left of Figure 4, are the replacement of H_2O by SO_4^{2-} in $[Cr(H_2O)_6]^{3+}$ to give the 1+ monosulfate ion, the formation of the third complex in a series of steps in which OH^- replaces H_2O in a pair of ions, and condensation to form the olate-bridged binuclear structure with monodentate sulfate groups. One sulfate group then rotates into the plane of the other, displacing it from the coordination complex to form the bidentate sulfate bridge. This purely descriptive and oversimplified mechanism gives an idea of the type of reaction possible, consistent with experimental evidence and the thermodynamics and stereochemistry of chrome complexes as recently reviewed (35).

The reactions of these basic chromium sulfate tanning solutions with hide collagen have been studied (36). It is firmly established that cross-linking is accomplished by bonding of the various chromium species shown in Figure 4 with free carboxyl groups in the collagen side chains. The liming of hides is effective in providing additional carboxyl groups by chemical hydrolysis of amide side chains.

The reactions that can take place as the carboxylate ions, which are attached to the collagen, enter one of these complexes are shown in Figure 5. The carboxylate group can displace water from the $[Cr(H_2O)_6]^{3+}$ ion (9% abundance) to form monodentate bonds, as shown in the upper equation, or bidentate coordinate bonds with binuclear complexes, as the one shown in the lower equation.

Two mechanisms by which cross-linking can occur (Fig. 6) are straightforward entry of two carboxylate ions into the same chrome complex and olation which involves elimination of water and formation of a linkage between two complexes. The olation reaction is favored by an increase in the alkalinity of the reaction mixture. As the reactions proceed and multinuclear complexes form with multiple olate bridges, hydronium ions are released and highly stable oxalate bridges are formed, as is shown on the right side of Figure 6. There is evidence that bidentate sulfate groups remain in the final complex after curing and drying. Apparently they play a role in improving stability of the complexes.

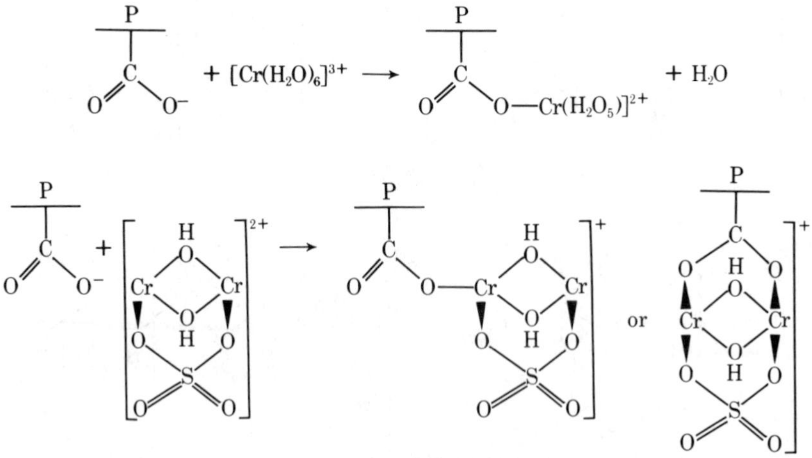

Figure 5. Complex formation with protein (P) carboxyl groups.

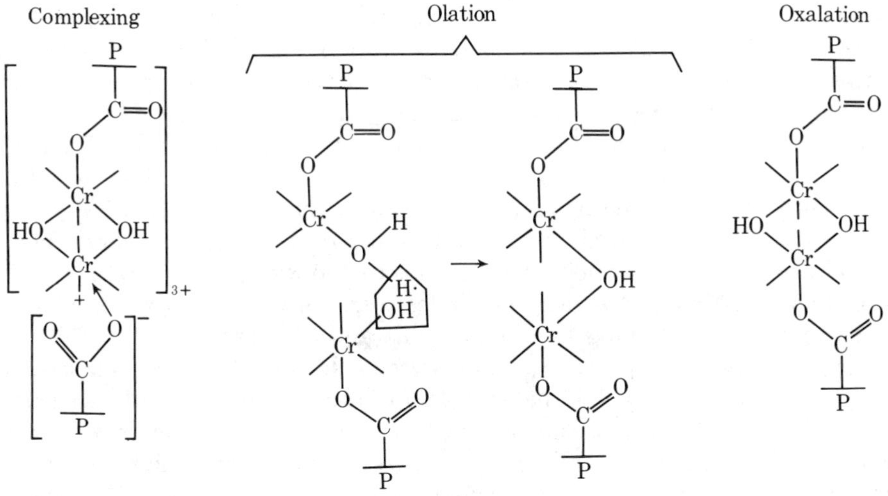

Figure 6. Cross-link formation from chromium complexes (P = protein).

Similar coordination complexes are involved in other mineral tannages. All of these complexes can be reversed or modified by acids, salts, strong bases, and chelating agents. Chromium complexes, although more difficult to form, have the advantage over other complexing cations of reacting much more slowly in these ligand replacement reactions and, therefore, producing leather that is more stable and serviceable in use. Chromium 3+ also is unique in its resistance to oxidation.

Vegetable Tanning. ***Practice.*** Vegetable tanning has decreased considerably in importance in the United States; there are only five tanneries producing vegetable-tanned leathers. Leathers made with a full vegetable tannage are used for shoe soles, belts, saddles, upholstery, lining, and luggage. Vegetable tanning produces a fullness and resiliency characteristic of only this type of tannage. It has certain molding characteristics so that, in sole leather, a shoe is produced that adapts to the shape of the individual foot. Vegetable-tanned leathers also have good strength and dimensional stability and, thus, find use in power-transmission belts. Their hydrophilic character is a great aid in shoe linings for the removal of perspiration from the foot.

Vegetable tannins are the water-soluble extracts of various parts of plant materials, including the wood, bark, leaves, fruits, pods, and roots. Some sources contain up to 20% tannin. The extraction process yields a mixture of tannins and nontannins; the higher the proportion of tannins to nontannins, the more valuable the extract. The tannin content is analyzed by an empirical method involving the reaction of the extract with a specially prepared hide powder under specified conditions. In general, the same steps are carried out in preparing the hide for vegetable tannage as are carried out in preparing the hide for chrome tannage, eg, the hair is removed by the same chemicals. However, a much slower process is used and, frequently, involves as long as five days of soaking in pits. The hides are exposed first to saturated solutions of lime and then are moved into other pits in which small amounts of sodium sulfide are present. Initial exposure of the hair keratin to the alkali in the absence of sulfide produces a reaction known as immunization. The main chemical reaction occurring in this process is a conversion of the cystinyl residues in the keratin to lanthionyl residues; the latter provides a much more stable cross-link to the protein than the former. The resulting hair fiber is much more resistant to chemical attack than is the nonimmunized fiber. At the end of the five days, the epidermis and hair have been loosened in the follicles by the chemicals and are mechanically scraped from the hide in unhairing machines. This lengthy liming and dehairing step is necessary to open up the fibers of the hide in such a way as to prepare them for penetration by the vegetable tannins.

The hides then are mechanically fleshed and returned to vats where they are chemically delimed with ammonium sulfate and bated with pancreatic enzymes. The pH is adjusted ultimately to ca 5. The hides then are placed on frames which are lowered into vats containing tannin solutions and are gently agitated. They are subjected over a number of days to a series of vegetable tannins of increasing strengths. Slow, thorough penetration of the hide is accomplished without case hardening (surface accumulation of tannins). This may be done by moving the frame from pit to pit where tanning solutions of increasing strengths are present or by pumping solutions of tannins from pit to pit over a period of time while the hides remain in one pit. The entire process takes ca three weeks. The tanning liquors normally contain phenolic syntans (synthetic tannins) for color control and naphthalene syntans for sludge dispersancy.

Recently, a more rapid, minimum-effluent vegetable tanning system, known as the Liritan process (37), has been developed. The limed and bated hides are treated for 24 h in a pit with 5% sodium hexametaphosphate (Calgon) solution and sufficient sulfuric acid to achieve a pH of 2.8 at the end of that time. This part of the process has become known as the Calgon pickle. The solution is reused daily, being regenerated with additional Calgon and sulfuric acid, and is discarded only once a year. The treatment presumably prepares the hides for a more rapid vegetable tanning process, and the recommended one with varied concentrations of wattle (mimosa) takes 11 d. Again, the tannin liquors are recirculated and reused. Further finishing of leathers that have been prepared by the Liritan process is the same as for those prepared by conventional processes. First introduced in 1960, the Liritan process has spread throughout the world and is used by the major vegetable-leather tanners in the U.S.

The tanned hides are further processed in order to clean the surface of the hides of excessive amounts of tannins (usually unbound tannin) and then are wrung free of excess water and are oiled. Oiling is carried out in a drum with oils added to lubricate the leather fibers plus a variety of materials that may be drummed into the leather for the purpose of achieving specific properties in the final product. The recipes that are used are largely the inventions of the individual tanner and may include powdered lignin preparations, naphthalene syntan, Epsom salt, corn sugar, salts of organic acids, bicarbonates, and borax. Most of the solids are added as such, the sugar is usually added in the molten form, and the oil, of course, as the liquid. The leather then is hung so as to dry slowly for one week. At the start, air at ca 90–95% rh and no more than 37.8°C is slowly circulated through the drying loft. The relative humidity is decreased slowly so that, at the end of the week, the moisture content of the leather is ca 10–12%.

The grain surface then is sponged with a dilute oil preparation and is rolled repeatedly under considerable pressure with a highly polished metal (usually brass) cylinder on a large pendulum-type machine. The leather is moved back and forth under the pendulum arm by an operator as the arm strikes down on the leather. The pressure also is regulated by the operator. This operation packs the fibers of the leather and imparts a characteristic gloss to the grain. The leather is allowed to dry, is dip-washed in a solution containing a small amount of wax, is redried, and is given a final dry rolling.

Theory. Functionally, the vegetable tannins are polyphenolic compounds. They are empirically divided into two groups: the hydrolyzable tannins and the condensed tannins. The hydrolyzable tannins are derivatives of pyrogallol. The distinguishing characteristic is that, as a group, they produce solutions when boiled in dilute mineral acids. Chestnut and myrobalan are typical hydrolyzable tannins. The condensed tannins are derivatives of catechol and commonly undergo condensation reactions when heated with mineral acids, thereby producing precipitates (see Hydroquinone, resorcinol, and catechol). Hemlock and wattle are of this type.

It is highly likely that no cross-linking of the protein, other than by the formation of hydrogen bonds, takes place as a result of vegetable tanning. This results from a displacement of hydrogen-bonded water molecules by the phenolic groups of the tannins with the formation of hydrogen bonding between these groups and the peptide bonds of the protein chains. The large size of these molecules and the large amounts of tannin that are used produce a coating on the fibers as well as fill the voids of the leather. In some cases, as much as 50 wt % of tannin is incorporated into the hide.

Sole leather is the predominant leather manufactured by vegetable tanning. Processes for other types of leathers (belts, harnesses) made by vegetable tanning are very similar. Cattlehides are the raw material used for this purpose, with cowhides producing the lighter (thinner) leathers, and steerhides the heavier (thicker) leathers. Historically, sole-leather weight (thickness) has been measured in irons (one iron = 0.53 mm). Cowhides yield leathers ranging from six to nine irons and are used primarily for lightweight shoes, eg, women's shoes; steerhides yield leathers ranging from nine to twelve irons and are used in heavier shoes, eg, men's shoes.

Sole-leather tanning begins with the whole cattlehide; however, parts are removed at different phases of the process and, ultimately, only ca 50% of the hide is sold as sole leather. This 50% is referred to as a double bend or as two bends resulting from cutting the double bend in two down the backbone (see Fig. 7). The bellies, which are ca 25% of the hide, are removed from the pickled hide and are sold for conversion into lining or work-glove leather. The shoulders, which comprise ca 20% of the hide, are removed after tanning and are used for welting leather.

Other Tannages. Chrome and vegetable tannages have greater commercial importance than those that are considered below. Most of these tannages are pretannages or post-tannages for chrome or vegetable processes and are employed to give such qualities as filling, lighter shade for dyeing, and reduction of tannery effluent. Only occasionally are they used alone.

Mineral Tannages. The principal mineral tannage other than chromium is zirconium tannage, and it has been extensively reviewed (38). Zirconium tannage has been compared to chrome tannage but seems to form complexes faster. Stronger-acidity pickles usually are employed to slow the reaction, and the pH is raised more slowly. Extra salt is used to repress swelling at the lower pH. Pretannage with aldehyde sometimes is employed. Zirconium tannage leads to fairly firm, full white leathers, which unlike chrome, are white throughout the cross section. The mechanism of zirconium tannage is unresolved but there seems to be an initial uptake of anionic zir-

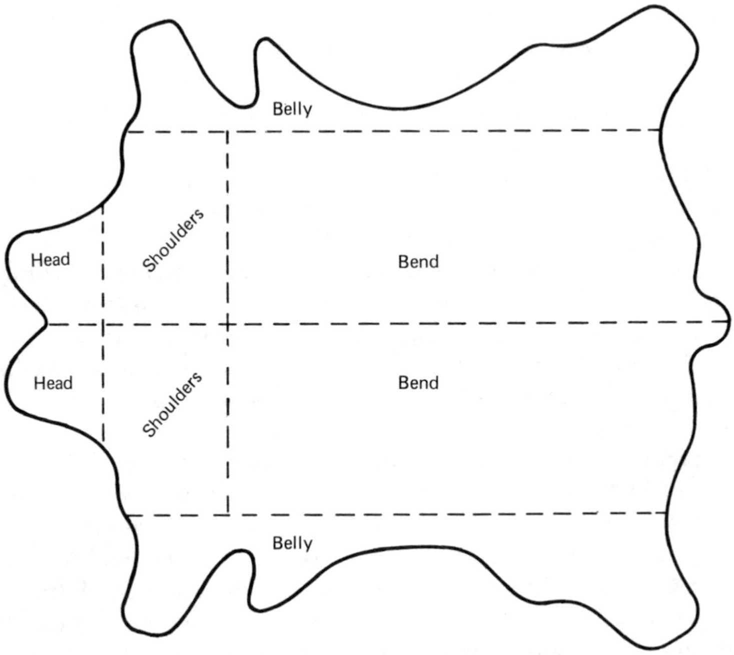

Figure 7. Segmentation of hides for vegetable-tanned leather manufacture.

conium salts at low pH (41). However, for satisfactory leather quality, it is necessary to have subsequent neutralization, which converts the zirconium complex to a cationic state at which time some polymerization may occur.

Aluminum tannage requires very low initial pH and large quantities of neutral salt to repress both swelling and rapid flocculation of aluminum salts as the pH is raised. This type of tannage leads to leathers that tend to dry out and to become hard and horny unless much salt is used; even then, the leather tends to be flat and papery in feel. Aluminum tannage is used in combination with other tannages. Renewed interest in such combination tannages is being generated in view of increasing regulation against chrome tannery wastes. The following recommendations have been made: a mimosa vegetable pretannage followed by aluminum tannage (39), a combined tannage with chromium, alum, and glutaraldehyde with inclusion of acrylic acid polymers to increase aluminum fixation (40–41), and a combination chromium and aluminum tannage to reduce chrome waste (42).

Iron tannage also tends to give flat, papery leather. Initial work indicated poor resistance to aging (43). However, it seems probable that the early leathers received insufficient tannage (44) and two reports indicate better aging resistance than originally was supposed (45–46). Iron-tanned leathers have very little commercial importance, possibly in part because of the intense black coloration produced by iron reaction with phenolics, which are the principal components of the vegetable tannis and syntans so widely employed as retans in tanneries.

Polyphosphates are excellent pretannages for vegetable tanning (47–50). Optimum molecular weights of the polyphosphates are from 1500–2500 (51–53). The Liritan polyphosphate–vegetable combination tannage process (37), as a no-effluent rapid tannage for sole leather (see Vegetable Tanning), is used by sole-leather tanneries throughout the world.

Silica tannage is effected by the formation of silicic acid and its penetration into the skin in the form of sols of monomeric silicic acid or its low molecular weight polymers. However, for tannage to occur, polymerization of the sols must take place within the skin (44). The silicic acid is fixed to the protein by coordinate bonds and, at the same time, is polymerized in long chains. Difficulties in effecting this reaction reproducibly and leather weakness have hindered the use of silica tannage, although it has been used in some instances to obtain fuller leather in combination with more stable tannages.

Polyphenolic Syntans. Polyphenolic syntans (54–55) are low molecular weight condensation polymers of aromatic phenols, usually with formaldehyde (see Phenolic resins). They usually are sulfonated for increased water solubility. Initially, work on these materials was carried out in order to find a replacement for the natural vegetable tannins that would be consistent in action and easily purified. The variety, availability, and cheapness of the natural tannins, in general, have precluded this approach, except for syntans that were developed as strategic materials when supplies might be cut off in time of war. During World War II, Orotan was developed by the Rohm and Haas Company; it was claimed that Orotan could completely substitute the natural vegetable tannins as the tanning material in vegetable-tanned leathers (56). However, during peacetime and because of their greater cost, the syntans are used extensively only in combination tannages, usually following chrome tannage. They are used to control leather fullness, area yield, color, and the electrical charge of the leather. The latter affects dyeing and fatliquoring operations.

Resin Tannages. The principal resin tannages are those performed with the aminoplast resins. These are low molecular weight, polyfunctional, organic compounds whose reactive groups are *N*-methylol or *N*-alkoxymethyl groups. They usually are made from polyamines, eg, melamine or urea or dicyandiamide, by reaction with formaldehyde and water to form *N*-methylol groups or by reaction with formaldehyde and an alcohol to form *N*-alkoxymethyl groups (see Amino resins). A simple monomeric member of the group is hexa(methoxymethyl)melamine. Alkoxylation need not be carried so far as with hexamethoxymethylmelamine. Residual amino groups can be left. Monosubstitution rather than disubstitution also can be done. The materials are capable of self-condensation to release alcohol, or water for *N*-methylol types, or formaldehyde and they react strongly with amines or alcohols.

The materials used for tanning usually are of low molecular weight, the degree of polymerization being from 2–3, and are water soluble. The reactivity of the *N*-methylol types is greater than that of the *N*-alkoxymethyl types, but the latter are more stable to handling. The low molecular weight materials are incorporated into the leather by drumming. Then the pH is dropped and, under acid catalysis, both self-condensation and reaction with amino and hydroxyl groups of the protein can occur. A review of some tanning procedures has been given (57) and includes discussion of tannage with styrene–maleic anhydride copolymer, tannage with isocyanates, and tannage with furfuryl alcohol, which is condensed *in situ* within the hides by acid catalysis. As yet, none of these is of significant commercial importance.

Acrylic resin syntans containing multiple carboxylic acid groups have gained increased commercial attention (58–59). Such polycarboxylic acid polymers can be expected to increase fixation of mineral tannages (40–41) and can eliminate the possible toxic hazards from unreacted phenols or formaldehyde which might be present in some of the other types of syntans.

Oil Tannage. Oil tanning produces leathers with unique characteristics (60). Oil-tanned chamois leather is very soft and stretchy, absorbent of water which can be readily squeezed out again and, because it also absorbs grease, is an ideal material for cleaning such items as auto windows and spectacles. The grease is removed from the chamois by alkaline washes with soaps, and the chamois can be reused many times. Because of the softness and suppleness imparted by oil tannage, it also has been used for tanning furs. Here, however, to avoid excessive stretchiness, it usually is employed as a combination tannage with alum, chrome alum, and/or aldehyde tannage.

Chamois leather originally was made from the skin of the chamois, a goatlike antelope found in the French Pyrenées. Now, however, it is made principally from sheepskins with the grain layer split off, although other loose-textured skins are employed. The tanning process consists of stuffing the beamed skins with as much oil as possible (principally unsaturated marine oils) and then subjecting the oiled skins to conditions favoring autoxidation of the oils. Theories of tannage involve not only polymerization of the oil but projections of tannage with aldehydes formed from oxidative chain scission of the oil, epoxy formation, and acrolein formation. However, in view of the necessity for free-radical formation prior to the above processes, the ease of formation of these free radicals on the methylene group alpha to two unsaturated double bonds, and subsequent formation of peroxide free radicals from these (61), it seems reasonable to believe that direct free-radical attack on the collagen occurs. This would form strong covalent bonds that are able to withstand caustic washing. This concept should certainly be a part of consideration of the mechanism of oil tannage. It is analogous to the concept of graft polymerization, a process much studied for leather modification in recent years, both in the United States (62–77) and abroad (78–81). Detailed studies of the autoxidation of oils (82) would seem to give credence to a strong possibility of oil tannage being a form of graft polymerization when such reactions are carried on in the presence of collagenous hide tissue.

Sulfonyl Chloride Tannage. Sulfonyl chloride tannage for chamois-type leather was developed in Germany in World War II with oils from a coal-gas liquefaction process. The fraction that was used boiled from 220–320°C, and it was hydrogenated and then sulfochlorinated to yield a mixture of aliphatic sulfonyl and disulfonyl chlorides, which was termed Immergan. A pretannage with formaldehyde was advised for using these sulfonyl chlorides to make chamois-type leather (60,83–85). However, some researchers (86) found that pretannage with formaldehyde, though helpful, was not necessary. In addition, they noted that the leathers that were produced had excellent cold-temperature properties which were highly beneficial for arctic or other cold-weather usage.

Aldehyde Tannage. Only two aldehydes have commercial use as tanning agents: formaldehyde and glutaraldehyde. With one exception, they are never used as the only tannage but most commonly are used in conjunction with some other tanning agent, usually chromium. The one exception is in the tanning of light-colored, including white, glove leathers (27,87) (see Uses). The functional group in proteins that reacts with aldehydes to form cross-links is the primary amino group of protein side chains of lysinyl residues. Details about the reactions which take place between the aldehydes and the protein amino groups are not known. However, some of the potential reactions which should be considered are shown in Figure 8 (88). Formation of a methylol or substituted methylol derivative certainly is the first step; however, following this formation, several alternatives are possible. Direct substitution of an amino group for the hydroxyl is one. This leads directly to a hydrolyzable cross-link in a protein. Addition of more aldehyde can take place and, for those aldehydes having hydrogens on the α carbon, Schiff's base formation also can take place. In the latter case, further steps in the reaction sequence can lead to condensation reactions of the type shown. These have been proposed for glutaraldehyde cross-linking. Glutaraldehyde is a difunctional reagent and both ends of the 5-carbon chain can take part in these reactions.

One important difference between the tannages achieved with these two aldehydes is that formaldehyde forms cross-links that are disrupted easily by simple hydrolysis, whereas glutaraldehyde forms cross-links that are completely stable even to hydrolysis with strong acid. Thus, there is a considerable difference in the relative stabilities of the two tannages. Another difference is that formaldehyde tannage yields a white leather, whereas glutaraldehyde tannage produces a light-tan leather. When these aldehydes are used with other tanning agents, the color imparted by the additional tanning agent changes the results. One combination that has been used successfully to produce a white leather is that of formaldehyde and aluminum.

The principal raw materials for the production of high quality, light-colored glove leathers are kid skins and hair sheepskins. These skins are processed with conventional chemicals through soaking and unhairing, fleshing and degreasing, and pickling. They then are tanned with from 6–12% formalin, based on the weight of the pickled skins. The starting pH for this tannage usually is ca 2 or 3 in ambient temperature conditions. During the tanning reaction, which usually takes ca 16–18 h, the pH gradually is raised and the temperature is increased. The final temperature of ca 35°C and a pH of ca 7 is common. Further processing includes the use of retannages and fatliquors to maintain the white or light color of the leather. For darker-colored glove leathers, a combination of chromium and formaldehyde tanning commonly is used. Aluminum sometimes is used to increase the stability of the white or lighter-colored leathers, and this combination also is frequently used for fur tannages.

The retannage of vegetable-tanned leather with formaldehyde, which reacts under

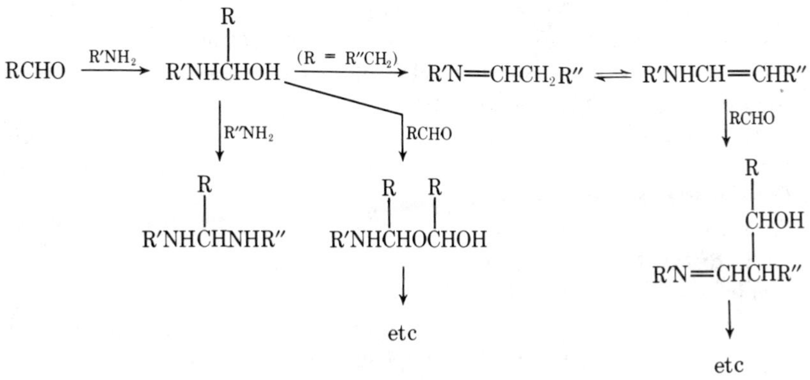

Figure 8. Aldehyde–amine reactions.

acidic conditions and increased temperatures with the polyphenolic vegetable tannins, provides a more stable tannage. Formaldehyde retannage of vegetable-tanned insole leather increases the resistance of the leather to perspiration.

Glutaraldehyde is most frequently used as a retannage for chrome-tanned leather. Glutaraldehyde tanning results in the formation of extremely stable cross-links in the protein. Although numerous investigations of the nature of these cross-links have been made, their identity is unknown. The stability of this tannage to hydrolytic conditions has led to its use in the production of leathers with improved resistance to hot soap solutions, to perspiration, and to alkalis. Thus, leather for use in gloves, garments, work shoes, and nurses' shoes frequently is made by retanning chrome-tanned leather with glutaraldehyde. Other improvements resulting from the use of glutaraldehyde retanning include a more level coloring and a better mellowness of the leather.

Although the most frequent use of glutaraldehyde is as a retanning agent, following chrome tanning, it can be used quite effectively with vegetable tanning either as a pretanning agent or as a retanning agent. When used in the former manner, the subsequent penetration of the vegetable tannin extract is very rapid and the properties (eg, shrinkage temperature, water solubles, and perspiration resistance) are much improved. When used in the latter manner, it leads to improvements in these properties in addition to an increase in thickness of the leather.

Post-Tanning. ***Retanning, Dyeing, and Fatliquoring.*** During the early aging period, when the leather is stacked, there is a gradual decline in the pH of the stock as increased chrome fixation, olation, and oxolation occur. Consequently, after transfer to a drum, the stock is neutralized to a desired pH, frequently is retanned, and then is dyed and fatliquored in rapid succession in the same drum.

The acidic, chrome-tanned leather is highly cationic in nature, since it is well below the isoelectric point of collagen which, after the liming process, has an isoelectric point of 5.4. Neutralization pushes leather closer to the isoelectric point and reduces the surface charge on the protein. This is important because many of the dyes that are employed are anionic. If the leather is too cationic, the precipitation of the anionic dye at the surface occurs rapidly, the dye is confined too much to superficial layers, and the dye is considered too astringent. Neutralization of the surface layer slows the cationic/anionic interaction and allows deeper penetration of the dyestuff. Retannage also affects the surface charge on the leather. Many of the syntans employed for retannage are salts of naphthalene sulfonates or other organic sulfonates having an anionic charge. They occupy many of the cationic sites on the leather and, thus, change the net charge and affect dye penetration and combination. The natural vegetable tannins and polyphenolic syntans also are anionic in character and affect the dyeing process. If highly vegetable-retanned, the leather composite becomes anionic and usually requires the utilization of basic dyes. Hence, vegetable retans are used not only to impart vegetable-tanned leather characteristics, eg, the dimensional stability to changes in relative humidity, increased fullness and temper, and ability to be tooled, but for their effects on leather dyeing and appearance.

Fatliquoring is the application of oil-in-water emulsions to the leather. It usually succeeds the dyeing process, although some tanneries apply them simultaneously rather than in rapid succession. The oil emulsions usually are drummed in at elevated temperatures. Commercially used oils usually are of animal or vegetable origin and of relatively low iodine number, although cod and other marine oils are sometimes included but usually in minor amounts. Fatliquoring emulsions may be anionic, nonionic, or cationic. Blends of nonionic with one of the other classes frequently are employed to assist oil penetration; the nonionic surfactant prolongs the stability of the oil after its anionic or cationic charge has been neutralized. Protective colloids, eg, gums, starches, and proteins, also can be used to prolong emulsion stability, and emulsions prepared in their presence frequently are of finer particle size than otherwise is obtained. Anionic fatliquors usually are prepared from mixtures of either sulfated or sulfonated oils with raw oils. Cationic fatliquors usually are blends of alkylated long-chain amines together with raw oils. However, synthetic surfactants of the desired type may be added to both. The polar components tend to be bound to the protein, probably mostly by ion-dipole interaction. The raw oils have more freedom to migrate but tend to stay in the area of the bound oil, probably by a process of mutual solubility. The principal function of the fatliquoring oils is to reduce the amount of fiber cohesion during the drying process. Lubrication of the leather is an important but secondary effect. Because oil distribution is much affected by the distribution of ionic charges in various strata of the dyed and undyed leathers, a variety of tempers, flexibilities, and stretch characteristics can be attained. Dyeing and fatliquoring are reviewed in detail in refs. 89–92.

Drying. Once all of the wet operations have been completed, the leather can be dried which not only involves removal of excess water, bringing the moisture content close to that of the finished leather, but completes the reactions of some of the materials (eg, tanning agents, fatliquors, and dyes) with which the hide has been treated (see Drying) (27,93–94). Leather drying occurs by removal of water first from the surface of the leather and then by a diffusion-controlled process from within the leather. If drying at the surface occurs too rapidly before the diffusion can occur, the surface becomes dehydrated and, eventually, hard and the interior remains moist. If dehydration proceeds too far, the surface can never be rewetted completely. Thus, the drying operation must be carefully controlled. Properly tanned leather can be dried more rapidly and easily than untanned collagen. One of the criteria of a tannage is that the leather is soft when dry. The extent to which this is the case depends on the type of tannage and other materials (fatliquors especially) present. Even so, drying is accomplished under carefully controlled conditions of temperature, humidity, and air circulation.

The most fundamental and oldest method involves simple air drying of hides and skins that are hung over supports. Currently this is done, but the drying chambers are programmed in some manner to control the conditions. Since the hides and skins are not held in any shape, ie, restrained, shrinkage occurs with a resulting area loss. To a limited extent, this problem has been overcome by tacking the hides and skins flat to boards. A variation of the above is toggling. In this practice, the hides and skins are stretched out over screens or perforated plates, called toggle frames, and are clamped in place. The frames that contain the wet hides and skins are placed in drying units. Large units are available with several frames mounted on tracks so that the frames can be easily slid into and out of the units.

Another method involves pasting the hides or skins, grain-side down, to plates of glass, porcelain, or metal. These plates are mounted on tracks that move into and out of large drying units. The leather dries from the flesh side only and a smoother, flatter grain is achieved on the leather.

The most recently developed method is vacuum drying. The hides and skins are spread out, grain down, on a smooth, highly polished, heated, stainless steel plate in which the temperature is carefully controlled. A cover is lowered and sealed over the plate and the space between is evacuated. This method permits rapid drying with few problems with the product quality because the conditions can be controlled much more accurately than by other methods.

Coatings. In the process of making leather from hides or skins, great pains are taken to make a uniform product. A manufacturer of a particular type of leather will use a particular animal species. Yet because of variations in individual animals and their nutritional and environmental histories even within that species, manufacture starts with a very variable substrate. The skins vary in thickness, in the angle of weave (95) of the fibers, and in natural defects, not only from animal to animal, but from area to area of the skin from the same animal. The leather manufacturer minimizes these differences by using processes as uniformly as possible, by splitting and shaving the leathers to uniform thicknesses, by tanning and filling to occupy the voids in the looser-structured flanks, and by using mordants to obtain as level dyeing as possible. However, it is the finishing process with its application of natural or synthetic polymers and of colorants, within and on the surface of the leather, which produces the uniformity, appearance characteristics, and resistance to scuffing and abrasion which are required for a commercial product.

Coatings (qv) for leather function as decoration and protection (96). The application is difficult in so far as providing uniformity because of the decidedly nonuniform substrate. Leather coatings also have a requirement for much greater flexibility and extensibility than other coatings. For example, the required resistance to flexural fatigue is very high in the vamp of a shoe where the leather is folded upon itself and straightened with each step for hundreds of thousands of flexes. Also, because of its fibrous network, leather can be stretched greatly multidirectionally and, thus, exert extreme local extensions and stresses in the coatings. This happens particularly in garment leather where the loose fiber weave of goatskins or sheepskins, particularly in the flanks, stretch up to 300% of its original length. The leather finish must undergo such stretching even at subzero temperatures without breaking or checking under the strain.

The ability to meet these demanding requirements for uniformity, flexibility, and extensibility depends to a large extent on primer coatings which are similar in formulation to those of latex house paints (see Paint), but which are made with specially developed emulsion polymers, mostly of the polyacrylate type although butadiene copolymers also are employed. The latex polymers for leather basecoats have to be of exceptionally high molecular weight and exceeding low glass-transition temperature T_g (97).

Unfortunately, latex coatings of the type described offer little protection against abrasion and/or scuffing and higher modulus polymers are required. These are provided in upper coats applied from solvent solutions. The natures of some of the polymer systems have been reviewed (98). Mostly nitrocellulose or vinyl chloride copolymer lacquers are employed. Polyurethanes are used increasingly, but their principal application for leather has been in the specialized production of patent-leather coatings where thermoplastic, followed by moisture-cured, polyurethanes have completely replaced the bodied linseed-oil varnishes that used to be employed (see Urethane polymers).

If the high modulus polymer topcoats are applied directly over the extremely low modulus basecoats, the difference in modulus leads to intercoat failure during flexing.

It is necessary to provide multiple coatings with a gradually increasing modulus in each coat from bottom to top. This prevents a stress build-up within any particular coating. Usually no less than three and, more often, four or five coats are employed. The development of a class of vinylidene chloride–acrylate latex copolymers (99), which are partially crystallizable and, although of low glass-transition temperature, are less thermally sensitive and of somewhat higher modulus than equally low T_g polyacrylates, has been found to aid the transition.

An additional benefit of using multiple coatings is the achievement of special fashion effects; the following example is illustrative:

Semi-aniline shoe-upper leather coating

1. Light-colored pigmented base coat (seasoning machine application): 38.1 μm coating with low T_g polyacrylate.

2. Same color pigmented latex spray coat (spray machine): 25.4 μm coating with slightly higher T_g vinylidene chloride–acrylate copolymer.

Smooth and fuse to the leather at 10.3 MPa (1500 psi) in a heated press at 94°C.

3. Spray with light-colored, highly plasticized nitrocellulose solution coating (ca 17.8 μm) containing organic pigment that is similar in color but livelier in appearance than the pigmented base coats. Level by plating at 104°C, 13.8 MPa (2000 psi).

4. Spray with a slightly less plasticized nitrocellulose coating (ca 7.6 μm) containing a transparent organic dyestuff of darker color than the color formerly employed. Spray at an angle to get uneven dye application for special effect desired. This will give an uneven transparent look with the bright, solid, lively color underneath showing through.

5. Spray with a clear nitrocellulose top (ca 12.7 μm) that is low in plasticizer to obtain abrasion resistance. Use a smooth plate in the press at high temperature (ca 110°C). Use just enough pressure to provide smoothness and slip to the surface. The leather resulting from this process will be lively, deep-toned, and rich in appearance.

To preserve the long-term flexibility achieved in the base coats by high molecular weight emulsion polymers, the polymers used in the upper coats also have been of as high a molecular weight as was practically feasible but with limited solubility. Almost all upper coats for leather have been at 10% or less solids. Hence, 90% of the coatings' weight has entered the environment as volatile organic compounds (VOC). The EPA has recommended limiting VOC of surface coatings to much lower solvent emissions for a series of industries (100) and is expected to do so for leather coatings also. Consequently, emphasis is being placed on potential utilization of water-based upper coats, which include not only the vinylidene chloride–acrylate copolymer latex topcoats (97) but water-based polyurethanes (101–102) and tougher water-based acrylics (103).

These water systems do not provide the liveliness, clarity, gloss, and fashion effects that are obtainable with the current solvent system upper coats. Early reports of radiation-curable, 100% active leather-coating systems may permit avoidance of this difficulty (104–105). Such radiation-curable systems consist of coatings in which the only solvents employed are acrylic or other vinyl monomers. The leather and its wet coating are passed under a beam of electrons or strong ultraviolet light; polymerization occurs in seconds, and the solvents, instead of evaporating, become part of the coating. The coatings have good gloss and clarity, solvent is not emitted to the environment, and energy and space requirements are reputed to be far less than in conventional ovens (see Radiation curing).

A special treatment was introduced into leather finishing during the 1960s which has assumed considerable commercial importance, particularly for corrected grain leathers. This was the development of specially penetrating polymeric precoating systems that migrate rapidly into the thermostatic or grain layer of the leather to fill this upper layer with reinforcing polymers which reduce scuffing of the leather and improve its handling in the shoe factories. In addition, the break or fine folds in the leather surface are greatly improved in appearance so that the coarser grain surface appearance of shoe-upper leathers made from steer and cowhides could rival or even surpass the appearance of fine calfskin leathers after repeated flexing of both types of leather. Specially developed polymers were of acrylic (106–108), polyurethane (109) and acrylic–urethane copolymer (110) types. Because of greater ease in handling, lower cost, and better break improvement, acrylics have been favored. The modes of action and general requirements for the polymers involved have been described (111). In recent years, the process of making leather/polymer composites in conjunction with the normal wet-processing operations prior to finishing has been receiving considerable attention both in the U.S. Department of Agriculture's Eastern Regional Research Center (62–77) and in the Central Leather Research Institute at Madras, India (79–80). Evidence of grafting monomers onto the leather protein structure to form new chemical entities has been adduced. The degree of grafting has since been questioned (112–113) but whatever the mechanism, the modification of leather is sufficiently great to introduce new sets of properties. One startling example is the development of shearling and garment leathers whose polymer, fatliquor, and dyes are unextractable with dry cleaning solvents so that they have been successfully cleansed in consumer, coin-operated, cleaning machines (114).

Energy Consumption. On the basis of quantity of energy consumed per unit of product produced, the leather-manufacturing industry would be categorized with the aluminum, paper, steel, cement, and petroleum-manufacturing industries as a gross consumer of energy (115). Since the leather manufacturing industry does not produce the quantities that these other industries produce, its total energy consumption is considerably less. Studies made during the petroleum shortage in the early 1970s indicated that the leather-manufacturing industry was highly wasteful in its energy usage. Estimates at that time indicated that only ca 30% of the energy consumed was required on a theoretical basis. With the inefficiencies that exist in energy conversion and usage, it certainly would not be possible to eliminate all of the waste. However, an energy wastage in the amount of 70% of energy input offers considerable opportunity for improvements.

About 40% of the energy used by the tanning industry is consumed in drying the leather, but only about one quarter of this was needed. The remaining three-quarters, or 30%, of the industry usage was wasted by inefficiencies in the drying equipment and as exhaust air. Since that time, improvements have been made in the drying operation, but there are no estimates as to how great these improvements are. Obviously, the potential exists for substantial savings in this operation.

Finishing of the leather consumes another 37% of the total energy consumption. By the use of new technology for applying and curing finishes, substantial savings could be realized. In particular, gravure coating of leather is a practical, energy-efficient method of finish application (116) and switching from thermally-cured finishes to radiation-cured finishes would be extremely beneficial (104).

The beaming and tanning of the hides account for only ca 13% of the total energy consumption, with the possibility for a proportionately smaller savings of energy. The final 10% of the energy consumption is what has been labeled comfort heat, which is apparently well used, with the potential of extremely small savings.

A 15% reduction in energy consumption could be realized with minimum investment and up to 30% could be achieved after substantial investment in auxiliary equipment. Beyond this point, significant changes in currently used processes would be required. Some of the modifications to current processing technology that are made in order to reduce pollution also reduce energy consumption. Process modifications that result in lower water usage reduce not only the total effluent discharge from the tannery but reduce the energy required to bring the water to process temperatures.

Economic Aspects

Hides of bovine cattle are the primary raw material used in the United States leather-tanning industry. In 1978, 84% of the leather produced in the United States came from bovine hides. The leading animal skin alternative is sheepskin; however, the number of sheep in the U.S. has been declining steadily. Alternative animal skin sources include pigskins, goatskins, horsehides, and deer and elk hides. Pigskins probably will be the best alternative for cattlehides. Current pigskin production for leather manufacture is estimated at ca 3×10^6 skins annually, with four pigskins being equivalent to one cattlehide in area. However, recent technological developments in skinning swine carcasses offer some potential for increased availability (117).

World production of hides and skins in 1977 and 1978 was ca 4.5×10^6 metric tons (118). The United States is the largest leather-producing country and accounts for just over one fourth of the world production. World trade in hides and skins was ca 1.6×10^6 t in 1977 and 1978, and the United States accounted for ca 44% of this trade which amounts to ca 60% of United States production. In the U.S., the bulk of the hides (almost 70%) are produced in the midwest and southwest parts of the country.

The marketing of hides, worldwide, has been undergoing changes. Other principal producers of cattle have limited their supplies of cattlehides to the rest of the world in order to encourage their own tanning industries and to improve their balance of payments. As a result, the United States has become almost the sole source of cattlehides in international trade. Also, the bottom of a cattle population cycle was reached, and the resulting scarcity of hides caused a doubling of hide prices within 4–6 mo. The demand for leather worldwide far exceeded the supply of raw material, and over 60% of the domestically produced hides in the U.S. were exported. As a direct result of these exports, wettings, ie, hides entering processing, decreased from a high of ca 20×10^6 hides in 1976 to ca 15×10^6 hides in 1979. There is considerable pressure to slow or stop exportation of raw hides and, instead, to export partially processed hides. The latter practice is increasing and may provide an important export market commodity. In addition, alternative leathermaking raw materials are being sought; there is considerable potential for pigskins to meet this need. Pigskin leather may become a major lightweight shoe and garment material in the 1980s.

When a hide is removed from the carcass, the first concern is to preserve or cure that hide so that it is protected until it can be transported to a tannery for manufacture into leather. When time and labor were less expensive, this was done by salting the hides in a solid pack. Hides were laid out flat and covered with a minimum of 1 kg salt/kg hide. The salt dissolved in the water in the hide, formed brine, and dehydrated the hide. Dehydration inhibits autolysis and the growth of bacteria. After the hides had drained for 30 d, they were sorted and bundled for delivery to the various tanners. Today the vast majority of hides are cured in a saturated salt solution in a raceway that can hold several hundred hides. They are continually tumbled in this salt solution, which is held at saturation and, thus, removes water from the hide to prevent microbial growth. A bactericide also is added to assist in preserving the hides. Full salt penetration and dehydration require from 18–24 h. The cured hides generally are fleshed to remove fatty tissue adhering to the underside of the hide and they are sorted, bundled, and tied for shipment. This whole process can be carried out within 2–3 d, and large inventories are no longer necessary.

Hides are sold in a variety of categories, the main types being steerhides and cowhides. They may be native, butt-branded, or Colorados. Native means that there are no brands; a butt-branded animal is branded within 30 cm forward of the tail, and Colorados are branded on the side beyond 30 cm from the tail. Each type is sold in several weight classes. From this variety of selections the tanner chooses the best starting material for the particular leather to be manufactured. One of the unique aspects of this industry is that its raw material is bought by weight, and the product is sold by area. Naturally, the tanner buys the material which optimizes output in area of leather per unit weight of starting raw material. The exception to this is sole leather which is sold by weight.

In 1978, there were an estimated 380 plants involved in tanning and finishing leather in the U.S. (118). These plants belonged to 187 companies and were concentrated in seven states: California, Maine, Massachusetts, New Jersey, New York, Pennsylvania, and Wisconsin, and employed ca 21,500 workers. They sold over half of their leather to the footwear industry with the balance going into leather garments, gloves, handbags, belts, and miscellaneous products.

Health and Safety Factors

In 1973, the U.S. EPA published (119) a set of proposed regulations designed to control effluent quality in the leather manufacturing industry. These standards were rejected by the courts, and a second proposal was issued to the industry in 1979 (120). Whether or not these also are changed by the courts, it is clear that the industry must clean its effluents; however, some of the smaller tanners will not have the capital to invest in the necessary equipment. Clean air and water laws also are having an affect. In 1970, there were 400 tanneries in the U.S.; in 1979, there were only 208.

The effluent streams which must be treated can be separated into two areas of the tanning operation, the beamhouse and the tanyard. The main pollutants found in beamhouse effluent are sulfide and BOD, and the pH is ca 12. It typically contains sulfide and is saturated with $Ca(OH)_2$. Dissolved sulfide under neutral or acid conditions forms the poisonous gas, hydrogen sulfide. Accidental release of this gas has been responsible for the death of a number of tannery workers over the last few years. Some type of oxidation, eg, catalytic (manganous salts) or chemical (peroxide or permanganate), may be used to convert sulfide to sulfate. The second major component of the beamhouse waste treatment is BOD from the hair dissolved in the unhairing step. Keratin can be removed by acid precipitation or by biological treatment in a conventional activated-sludge treatment plant. A physical-chemical approach, which was developed by the USDA, provides for simultaneous recovery of the sulfide for reuse in the tannery and most of the protein as a solid cake, which may be used as a feed supplement or disposed of as a solid waste (see Pet and other livestock feeds).

The second, and even more difficult, stream to treat is that from the tanning operation. This effluent contains chromium as well as conventional pollutants, eg, BOD and suspended solids. All wastes containing chromium are considered by EPA to be hazardous. This is a controversial classification of real importance to the tanning industry. Chromium ions in the 3+ valence state have not been demonstrated to be harmful to either plants or animals; there is evidence that chromium is a micronutrient necessary for proper growth. On the other hand, chromium in the 6+ state in certain salt combinations has been shown to be toxic and carcinogenic. The EPA regulations proposed for 1984 set a limit of 3 ppm total chromium for indirect dischargers [effluent goes to publicly owned treatment plant (POTW)] and 0.5 ppm for direct discharges into a river or stream. With current technology, these will be difficult limits to meet. It is expected that recycling of chrome tanning solutions or recovery and reuse by precipitation of the chrome is needed to achieve these limits. Meeting the pretreatment limitations for the conventional pollutants can be achieved with conventional biological treatment. Direct discharges need some type of tertiary treatment, possibly followed by some polishing steps.

Once effluent quality is achieved, however, the tanner faces another problem: solid waste containing chromium is considered hazardous and must be disposed of in a sanitary landfill. This presents an additional expense over disposal in an unsecured landfill and the number of available sites for sanitary landfills is decreasing rapidly. One solution is to find a by-product use for the solid wastes that contain chromium. A second possible route is the incineration of chrome wastes and recovery of chrome for reuse (see also Recycling, metals).

The finishes that conventionally are applied to leather are contained in a variety of solvents that are removed during drying and enter the atmosphere. The concentration and quantity of VOC are being regulated in some states. The solution is either energy-consuming combustion processes to remove the VOC or a change in the finishing technology.

Uses

In 1978 the United States leather manufacturing industry had estimated shipments of $\$1.5 \times 10^{12}$ of leather (99), and the principal domestic markets for this leather were in leather footwear ($\$4 \times 10^{9}$), luggage ($\$660 \times 10^{6}$), handbags ($\$570 \times 10^{6}$), small personal leather goods ($\$410 \times 10^{6}$), leather apparel ($\$280 \times 10^{6}$), and gloves ($\$180 \times 10^{6}$). These industries, however, lack developing technology and, consequently, have low productivity and they continue to be pushed out of the market by importers of leather and leather goods. In 1978, leather and leather goods imports into the United States were valued at $\$3.2 \times 10^{9}$ and exports of hides and skins, leather, and leather goods amounted to $\$700 \times 10^{6}$, resulting in a net deficit in the balance of (leather) trade of $\$2.5 \times 10^{9}$. It is encouraging to note that United States tanners are turning to overseas markets for their leather with some degree of success. Another interesting trend is the construction by Western European shoe companies of shoe manufacturing facilities in the United States, which could provide additional markets for United States leather.

Hide Uses. Various components of the hides not made into leather constitute by-products, some of which have value. Hide collagen, when it can be separated from the hide in a native state, has use in food products, cosmetics (qv) pet treats, and biomedical products. Hide collagen that has been tanned or treated with biocides and other hide materials has some uses but, in many cases, is a waste product. The greater use that is made of these hide components, the greater the return is to the tanner and the lower the cost of manufacture of the principal product, leather. Compositionally, collagen is low in several essential amino acids and totally lacks tryptophan; hence, it is not a nutritionally complete protein. Its use in foods depends on other, mainly physical, properties which it possesses; the same is true concerning its use in cosmetics and biomedical products. Because collagen is a nonspecific protein, it can be used in cosmetics (eg, in moisturizing creams, lotions, and hair sprays) and biomedical products (eg, burn dressings, implant coverings, and as a hemostat), since it does not cause an allergic response when used on or in the body. Its use as a pet treat depends on the fact that it has a texture that dogs prefer for chewing and can easily be flavored.

The collagen used in these applications is obtained mainly from limed splits. These are provided by tanners who split the hides to the desired thickness for grain leather manufacture following the unhairing and reliming steps rather than after tanning. The split has been thoroughly cleaned of extraneous material, is relatively free of bacterial contamination (pH 12.5), and is stable for a short period of time. For some applications, only uncured hides have been used for production of the limed splits to avoid any possible residual contamination with the biocides commonly used in hide curing. This practice, however, requires that the hides be processed within a relatively short time after removal from the animal. Aside from gelatin manufacture, the largest food use is in sausage casings; however, its use in other food products is being studied. It is expected that the nonleather uses for hide collagen will grow.

By-Products. The weight of finished grain leather generally is from 20–30% of the weight of the hide, depending on the thickness of the leather and the rawhide selection (121–122). The remaining 70% consists of either by-products that can be sold or wastes that must be disposed of. Over the years, as markets have changed, some wastes have become by-products and various by-products have become wastes. The largest by-product by weight is the split. It may account for from 30–50% of the incoming hide. If the hide is split right after the liming operation, the option to use the split for collagen casings is available. If it is split in the blue, it can be sold to a split tanner to be made into split leather. Although this market fluctuates, the split is almost always a salable by-product.

Another 15% of the raw hide appears in the effluent as dissolved and suspended organic matter which consists largely of the protein that was dissolved in the beamhouse processes. Some potential exists for obtaining this protein in a form that could be used for animal feed. However, this generally is not done and the protein is considered a waste requiring disposal. One reason that the protein is not recovered is that there is not enough in the effluent of a single tannery to make recovery economically worthwhile. This is a general problem with by-products of the tanning industry.

Twenty percent of the raw hide shows up as solids that contain chromium. In the form of shavings, blue trim, buffing dust, and crust trim, these solids offer potential for some type of leather fiberboard product, although little if any is produced right now, and most tanners must pay to have the chromium by-products hauled away. Once again, the individual tanner frequently does not handle enough raw hide to produce leather fibers economically.

About 5% of the hide coming into the tannery ends up as raw material for a renderer. The value of fleshings is lowered if they are from salted hides. The remaining 7% of the hide is composed of solids not containing chrome, primarily the trim which sometimes can be sold along with the fleshings to the renderer; otherwise, it is considered a solid waste.

Cattle-hair markets have almost disappeared because, with the advent of hair burning or dissolving processes, which rapidly remove the hair from the hide, no hair is left. This once salable by-product has become an effluent problem.

BIBLIOGRAPHY

"Leather and Tanning" in *ECT* 1st ed., Vol. 6, pp. 289–309, by Fred O'Flaherty, University of Cincinnati; "Tanning Materials" (in three parts) in *ECT* 1st ed., Vol. 13, pp. 578–599 by F. L. Hilbert, U.S. Process Corp., and R. L. Stubbings, Lehigh University; "Leather" in *ECT* 2nd ed., Vol. 12, pp. 303–343, by F. O'Flaherty, University of Cincinnati, R. L. Stubbings, Institute of Leather Technology.

1. F. O'Flaherty, W. T. Roddy, and R. M. Lollar, eds., *The Chemistry and Technology of Leather,* Vol. I–IV, Reinhold Publishing Corp., New York, 1956–1965.
2. J. R. Kanagy in ref. 1, Vol. IV, 1965, pp. 369–416.
3. W. T. Roddy, J. Jacobs, and J. Jansing, *J. Am. Leather Chem. Assoc.* **44,** 308 (1949).
4. G. O. Conabere and R. H. Hall in *Progress in Leather Science, 1920–1945,* British Leather Manufacturers' Research Association, London, 1948, pp. 265–279.
5. W. T. Roddy, R. Echerlin, and J. Jensing, *Quartermaster General's Research Reports on Leather Technology,* April 15, 1948.
6. *United States Military Specification MIL-L-3122.*
7. J. A. Wilson, in *International Critical Tables of Numerical Data, Physics, Chemistry, and Technology,* Vol. II, McGraw-Hill Book Co., Inc., New York, 1927, pp. 250–254.
8. M. P. Balfe and F. E. Humphreys in Ref. 4, p. 417.
9. R. A. Vickers, *Leather Manuf.* (Sept. 1950).
10. S. S. Kremen and R. M. Lollar, *J. Am. Leather Chem. Assoc.* **46,** 34 (1951).
11. J. A. Wilson, *Modern Practice in Leather Manufacture,* Reinhold Publishing Corp., New York, 1941, pp. 629–704.
12. S. J. Kennedy, Quartermaster General Research Laboratory, Natick, Massachusetts, personal communication.
13. R. G. Mitton, *J. Soc. Leather Trades Chem.* **31,** 44 (1947).
14. J. A. Wilson and G. O. Lines, *Ind. Eng. Chem.* **17,** 570 (1925).
15. R. S. Edwards, *J. Int. Soc. Leather Trades Chem.* **16,** 439 (1932).
16. J. R. Kanagy and R. A. Vickers, III, *J. Am. Leather Chem. Assoc.* **45,** 211 (1950).
17. J. R. Kanagy and R. A. Vickers, *J. Res. Natl. Bur. Stand.* **44,** 347 (1950).
18. G. N. Ramachandran, ed., *Treatise on Collagen,* Vol. 1, Academic Press, New York, 1967.
19. B. S. Gould, ed., *Treatise on Collagen,* Vol. 2, Academic Press, New York, 1968, Parts A and B.
20. J. H. Highberger in ref. 1, Vol. I, 1956, pp. 65–193.
21. R. M. Koppenhoefer in ref. 1, Vol. I, 1956, pp. 41–64.
22. A. L. Everett and R. J. Carroll, *Leather Manuf.* **82**(8), 29 (1965).
23. W. T. Roddy in ref. 1, Vol. I, 1956, pp. 4–40.
24. M. Dempsey in ref. 4, 1948, pp. 3–27.
25. *Hides, Skins, and Leather Under the Microscope,* British Leather Manufacturers' Research Assoc., Surrey, Eng., 1957.
26. F. Karl and R. M. Schwartzman, *Veterinary and Comparative Dermatology,* J. B. Lippincott Co., Philadelphia, Pa., 1964.
27. T. C. Thorstensen, *Practical Leather Technology,* Robert E. Krieger Publishing Co., Inc., Huntington, New York, 1975.
28. J. J. Tancous, W. T. Roddy, and F. O'Flaherty, *Skin, Hide, and Leather Defects,* Tanners' Council Research Laboratory, Cincinnati, Ohio, 1959.
29. A. L. Everett, N. W. Hooven, Jr., J. Naghski, and R. G. Koeppen, *J. Am. Leather Chem. Assoc.* **70,** 188 (1975).

30. A. L. Everett, R. W. Miller, W. J. Gladney, and M. V. Hannigan, *J. Am. Leather Chem. Assoc.* **72,** 6 (1977).
31. A. L. Everett, I. H. Roberts, and J. Naghski, *J. Am. Leather Chem. Assoc.* **66,** 118 (1971).
32. A. L. Everett and M. V. Hannigan, *J. Am. Leather Chem. Assoc.* **73,** 458 (1978).
33. *Tanners' Council News,* (Feb. 3, 1976).
34. N. P. Slabbert, *Proc. XIV Congr. Int. Union Leather Chem. Technol. Socs.* **I,** 240 (1975).
35. H. M. N. H. Irving, *J. Soc. Leather Technol. Chem.* **58,** 51 (1974).
36. S. G. Shuttleworth in ref. 1, Vol. II, 1958, pp. 281–322.
37. S. G. Shuttleworth, *J. Soc. Leather Trades Chem.* **47,** 143 (1963).
38. I. C. Somerville in ref. 1, Vol. II, 1958, pp. 323–348.
39. N. P. Slabbert, *Proc. XVI Congr. Int. Union Leather Chem. Technol. Socs.* **II,** 88 (1979).
40. A. Zissel in ref. 39, p. 52.
41. C. W. Beebe, W. F. Happich, F. P. Luvisi, and M. V. Hannigan, *J. Am. Leather Chem. Assoc.* **52,** 560 (1957).
42. C. Krawiecki in ref. 39, p. 71.
43. H. R. Procter, *The Principles of Leather Manufacture,* E. & F. N. Spon, Ltd., London, 1922.
44. P. Chambard in ref. 1, Vol. II, 1958, pp. 349–387.
45. J. R. Kanagy, *J. Am. Leather Chem. Assoc.* **33,** 565 (1938).
46. S. T. Tu, *J. Am. Leather Chem. Assoc.* **43,** 181 (1948).
47. J. A. Wilson, *J. Am. Leather Chem. Assoc.* **32,** 113, 494 (1937).
48. U.S. Pats. 2,140,041 (Dec. 13, 1939); 2,140,042 (Dec. 13, 1939); 2,172,233 (Sept. 5, 1940), J. A. Wilson (to Hall Laboratories, Inc.).
49. Ger. Pats. 671,019 (Jan. 30, 1939); 671,712 (Feb. 14, 1939); 672,747 (March 9, 1939), K. Lindner (to Chemische Fabrik J.A. Benckiser G.m.b.H.).
50. K. Lindner, *Collegium* **816,** 146 (1938).
51. R. Lasserre, *Bull. Assoc. Fr. Chim. Ind. Cuir Doc. Sci. Tech. Ind. Cuir.* **18,** 9 (1956).
52. R. S. Meldrum, *J. Soc. Leather Trades Chem.* **37,** 278 (1953).
53. K. H. Gustavson and A. Larsson, *Acta Chem. Scand.* **5,** 1221 (1951).
54. H. G. Turley in ref. 1, Vol. II, 1958, pp. 388–407.
55. Austria Pat. 58,405 (1911), E. Stiasny.
56. H. G. Turley, *J. Am. Leather Chem. Assoc.* **40,** 58 (1945).
57. P. S. Chen in ref. 1, Vol. II, 1958, pp. 408–425.
58. W. C. Prentiss and C. R. Sigafoos, *J. Am. Leather Chem. Assoc.* **70,** 481 (1975).
59. W. C. Prentiss and J. J. Hodder, *IV Encontro National dos Quimicos Tecnicos da Industria do Couro,* Rio de Janeiro, 1979.
60. A. Kuntzel in ref. 1, Vol. II, 1958, pp. 426–454.
61. E. H. Farmer, *Trans. Faraday Soc.* **38,** 340, 356 (1942); **42,** 228 (1946).
62. A. H. Korn, S. H. Feairheller, and E. M. Filachione, *J. Am. Leather Chem. Assoc.* **67,** 111 (1972).
63. S. H. Feairheller, M. M. Taylor, and A. H. Korn, *Polym. Prepr. Am. Chem. Soc. Div. Polym. Chem.* **13,** 359 (1972).
64. S. H. Feairheller, E. H. Harris, Jr., A. H. Korn, M. M. Taylor, and E. M. Filachione, *Polym. Prepr. Am. Chem. Soc. Div. Polym. Chem.* **13,** 736 (1972).
65. A. H. Korn, M. M. Taylor, and S. H. Feairheller, *J. Am. Leather Chem. Assoc.* **68,** 224 (1973).
66. E. H. Harris, M. M. Taylor, and S. H. Feairheller, *J. Am. Leather Chem. Assoc.* **69,** 182 (1974).
67. U.S. Pat. 3,843,320 (Oct. 22, 1974), S. H. Feairheller, A. H. Korn, E. H. Harris, Jr., E. M. Filachione, and M. M. Taylor (to U.S.D.A.).
68. S. H. Feairheller, H. A. Gruber, M. M. Taylor, and E. H. Harris, Jr., *Polym. Prepr. Am. Chem. Soc. Div. Polym. Chem.* **17,** 814 (1976).
69. E. H. Harris, Jr. and S. H. Feairheller, *Polym. Eng. Sci.* **17,** 287 (1977); *AIChE Sym. Ser.* **74,** 131 (1977).
70. A. H. Korn, E. H. Harris, and S. H. Feairheller, *J. Am. Leather Chem. Assoc.* **72,** 196, abstr. of presentation (1977).
71. H. A. Gruber, E. H. Harris, Jr., and S. H. Feairheller, *J. Appl. Polym. Sci.* **21,** 3465 (1977).
72. M. M. Taylor, E. H. Harris, and S. H. Feairheller, *J. Am. Leather Chem. Assoc.* **72,** 294 (1977).
73. M. M. Taylor, E. H. Harris, and S. H. Feairheller, *Polym. Prepr. Am. Chem. Soc. Div. Polym. Chem.* **19,** 618 (1978).
74. H. A. Gruber, E. H. Harris, and S. H. Feairheller, *J. Am. Leather Chem. Soc.* **73,** 410 (1978).

75. H. A. Gruber, M. M. Taylor, E. H. Harris, and S. H. Feairheller, *J. Am. Leather Chem. Assoc.* **73,** 530 (1978).
76. M. M. Taylor, M. V. Hannigan, and E. H. Harris, *J. Am. Leather Chem. Assoc.* **74,** 167, abstr. of presentation (1979).
77. E. H. Harris, H. A. Gruber, and M. M. Taylor, *J. Am. Leather Chem. Assoc.* **75,** 6 (1980).
78. J. Kudaba, E. Ciziunaite, and D. Jonutiene, *Liet. TSR Aukst. Mokyklu Mokslo Darb. Chem. Chem. Technol.* **10,** 147 (1969); *Chem. Abstr.* **73,** 4982b (1970).
79. K. P. Rao, K. T. Joseph, and Y. Nayudama, *J. Polym. Sci. A-1* **9,** 3199 (1971).
80. K. P. Rao, K. T. Joseph, and Y. Nayudama, *J. Appl. Polym. Sci.* **16,** 975 (1972).
81. K. Studniarski and A. Alabrudzinska, *Leder* **30,** 49 (1979).
82. W. O. Lundberg, ed., *Autoxidation and Antioxidants,* Vol. I, Interscience Publishers, a division of John Wiley & Sons, Inc., New York, 1961.
83. *Austauschfettstoffe für Leder,* booklet by I. G. Farbenindustrie, Frankfort am Main, FRG.
84. G. W. Schultz and A. Schubert, *An Investigation of the Leather Industry,* Technical Report by Office of Quartermaster General, APO 757, 1945.
85. C. F. Payan, *Some Aspects of the German Leather Industry,* Technical Report by British Intelligence Objectives Subcommittee BIOS Final Report No. 150, Item 22, Sept.–Oct. 1945.
86. J. B. Brown, M. F. White, W. T. Roddy, and F. O'Flaherty, *J. Am. Leather Chem. Assoc.* **42,** 625 (1947).
87. E. F. Mellon in ref. 1, Vol. II, 1958, pp. 66–97.
88. J. W. Harlan and S. H. Feairheller in M. Friedman, ed., *Protein Crosslinking Nutrition and Medical Consequences,* Vol. 86A, Plenum Publishing Corp., New York, 1977, pp. 425–440.
89. J. S. Kirk in ref. 1, Vol. III, 1962, pp. 1–15.
90. G. Otto in ref. 1, Vol. III, 1962, pp. 16–60.
91. C. E. Retzsch in ref. 1, Vol. III, 1962, pp. 61–72.
92. M. H. Battles in ref. 1, Vol. III, 1962, pp. 73–107.
93. C. F. Dudley in ref. 1, Vol. III, 1962, pp. 108–126.
94. L. Buck in ref. 1, Vol. III, 1962, pp. 127–183.
95. M. Dempsey in ref. 4, pp. 319–414.
96. P. R. Buechler in N. M. Bikales, ed., *Encyclopedia of Polymer Science and Technology,* Vol. 13, Interscience Publishers, a division of John Wiley & Sons, Inc., New York, 1970, pp. 486–505.
97. P. R. Buechler, *J. Am. Leather Chem. Assoc.* **72,** 79 (1977).
98. R. Shaw in ref. 1, Vol. IV, 1965, pp. 194–222.
99. U.S. Pat. 3,048,496 (Aug. 7, 1962), P. R. Buechler and B. B. Kine (to Rohm and Haas Co.).
100. *Control of Volatile Emissions from Existing Stationary Sources,* U.S. Environmental Protection Agency, Vols. I–VIII, 1976–1978.
101. V. E. Müller, *Angew. Makromol. Chem.* **14**(203), 75 (1970).
102. G. J. Katz, *Leather Manuf.* **96**(7), 34 (1979).
103. C. E. Cluthe, F. A. Desiderio, and W. C. Prentiss, *J. Am. Leather Chem. Assoc.* **73,** 22 (1978).
104. P. R. Buechler, *J. Am. Leather Chem. Assoc.* **73,** 56 (1978).
105. M. A. Knight and A. G. Marriott, *J. Soc. Leather Technol. Chem.* **62,** 14 (1978).
106. U.S. Pat. 3,103,447 (Sept. 10, 1963), J. A. Lowell, H. L. Hatton, F. J. Glavis, and P. R. Buechler (to Rohm and Haas Co.).
107. U.S. Pat. 3,231,420 (Jan. 25, 1966), J. A. Lowell, E. H. Kroeker, and P. R. Buechler (to Rohm and Haas Co.).
108. Belg. Pat. 637,052, I. V. Mattei (to Rohm and Haas Co.).
109. U.S. Pat. 3,066,997 (Dec. 4, 1962), M. B. Neher and V. G. Vely (to Titekote Corp.).
110. U.S. Pat. 3,441,365 (Apr. 29, 1969), J. A. Lowell and P. R. Buechler (to Rohm and Haas Co.).
111. P. R. Buechler, *Leather Manuf.* **78**(2), 19 (1961).
112. E. F. Jordan, Jr., B. Artymyshyn, A. L. Everett, M. V. Hanningan, and S. H. Feairheller, *J. Appl. Polym. Sci.,* in press.
113. E. F. Jordan, Jr. and S. H. Feairheller, *J. Appl. Polym. Sci.,* in press.
114. S. J. Viola and S. H. Feairheller, *J. Am. Leather Chem. Assoc.* **70,** 227, abstr. of presentation (1975).
115. M. S. Maire and P. A. Sundgren, *Weekly Bull. Leather Shoe News,* Boston, Mass., May 25, 1974.
116. W. C. Prentiss and C. E. Cluthe, *J. Am. Leather Chem. Assoc.* **73,** 471 (1978).
117. S. H. Feairheller, D. G. Bailey, and M. S. Maire, *Leather Manuf.* **97**(2), 10 (1980).
118. *U.S. Industrial Outlook—1979,* U.S. Dept. Commerce, 1979, Chapt. 36, pp. 383–402.

119. U.S. Environmental Protection Agency, *Fed. Regist.* **39,** 12958 (Apr. 9, 1974).
120. U.S. Environmental Protection Agency, *Fed. Regist.* **44,** 38747 (July 2, 1979).
121. R. Donovan, *Leather Manuf.* **95**(7), 16 (1978).
122. M. S. Maire, *Leather Manuf.* **93**(9), 12 (1976).

DAVID G. BAILEY
PETER R. BUECHLER
ALFRED L. EVERETT
STEPHEN H. FEAIRHELLER
United States Department of Agriculture

LEATHERLIKE MATERIALS

Leatherlike material normally refers to synthetic materials that are used as substitutes for leather in those applications where the latter traditionally has been used. In footwear applications, these products generally are referred to as man mades, synthetics, or poromerics. In seating and upholstery, they usually are called coated fabrics or vinyls (see Coated fabrics). In apparel and accessory applications, they usually are given a name that is suggestive of the currently popular styling, eg, the wet look in the early 1970s, supersuede in the mid 1970s, and totes or disco in the late 1970s.

Leather is the material of choice; synthetics are materials of economic necessity. The labor shortage during World War II caused the substitution of neoprene for leather in shoe soles and insoles (see Elastomers, synthetic). The profit squeeze in the U.S. shoe industry during the 1950s led to the adoption of vinyl-coated fabrics for women's shoes. Periodic shortages caused by the cyclical nature of leather supplies cause continued inroads by new products into traditional leather markets. These include the use of urethane-coated fabrics in shoes and apparel and the development of poromerics for men's shoes (see Urethane polymers). Each of these products is designed to simulate certain properties of leather that are relevant to a specific application (see Leather). The suitability of synthetics as leather substitutes depends upon their being able to duplicate certain desirable characteristics of leather. These characteristics arise from the physico-chemical structure of the leather (1). The high permeability for air and moisture vapor is one of leather's most desirable properties in its main use area, ie, footwear.

It has been shown that high moisture absorption is necessary for a feeling of coolness and comfort in the uppers of closed shoes. The modulus of leather is low for the first 10–15% of elongation, after which point it increases abruptly. This type of modulus characterizes leather as highly adaptable to the lasting process in the manufacture of shoes.

Most tanned hides are not used in their original thicknesses but are split into two or more layers. Leather can be split easily to yield whatever thickness is needed for a given application. The various splits usually are sold as weight grades (kg/m^2 or oz/ft^2)

rather than by gauge (mm or mils). Leather also is readily skived to allow smooth folds in certain applications. However, it is difficult to incorporate the ability to be split or skived into many types of synthetics. Splitting is eliminated by manufacturing various weight grades, but skiving often remains a problem.

Any artificial, leatherlike material ideally should have mechanical properties that are similar to those of leather, as well as breathability and moisture vapor permeability. However, these ideal properties often are compromised in favor of economic considerations, and many products that have been marketed successfully merely look like leather but have very few of its physical properties. Therefore, it can be argued that the most important characteristics of leatherlike synthetics are esthetic ones.

Types

Leatherlike materials have one common characteristic; they look like leather in the goods in which they are used. For purposes of a technological description, however, they may be divided into two general categories: coated fabrics and poromerics. Poromerics are manufactured so as to resemble leather closely in both its barrier and transition properties, eg, for heat or moisture, and its workability and machinability. The barrier or permeability properties normally are obtained by manufacturing a controlled microporous structure.

The modulus characteristics of synthetics should match those of leather at 80°C rather than at room temperature. Although the latter temperature commonly is used as the standard in the industry, the former more closely matches the lasting conditions.

Coated Fabrics. All commercial nonporomerics are coated fabrics. Several types of fabrics have been used and the coating may be either vinyl or urethane, depending upon the desired balance between physical properties and economic considerations.

The earliest important coated fabric was solid vinyl on cotton sateen. Solid vinyl has the economic advantage of being relatively inexpensive. It can be embossed readily with a leatherlike grain and it fulfills, at least to a first approximation, the visual esthetic requirements for a leather substitute. However, it does not acquire a leatherlike crease or break with usage. This break, which can be observed in any leather shoe, garment, or chair is the formation of tiny permanent wrinkles in areas that are subjected to repeated flexing. Solid vinyl also feels cold and hard when touched, and it does not allow the transmission of air or moisture vapor. The latter deficiency is particularly disadvantageous in footwear applications.

A second type of coated fabric is expanded vinyl, which is of more recent origin than solid vinyl and gives a distinct improvement over the latter in tactile esthetics. Expanded vinyl is manufactured so as to have internal air bubbles which give the material enhanced flexibility and cause it to feel warmer and softer to the touch than solid vinyl. Figure 1 is a photomicrograph of a solid-vinyl coated fabric, and Figures 2 and 3 are examples of expanded vinyls; Figure 2 is a footwear-grade product and Figure 3 is an upholstery-grade product. The former is constructed of a woven fabric, and the latter has a knit backing and a thinner skin layer. The bubble structure, which is a closed-cell type of foam, does not confer any appreciable degree of permeability on the material. Expanded vinyl has the same shortcomings in the area of moisture-vapor transmission as does the solid product (see also Foamed plastics).

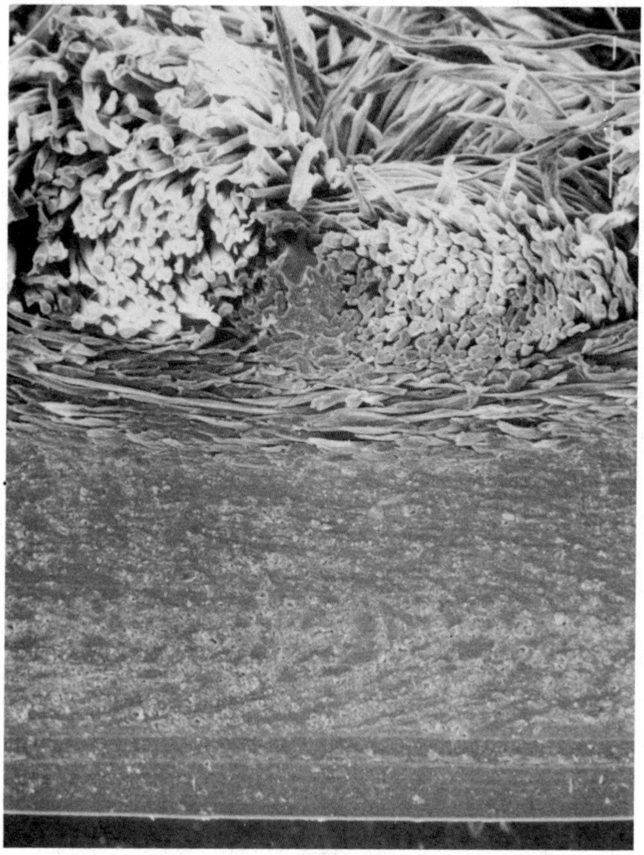

Figure 1. Cross-sectional view of solid vinyl (100×).

The most recent development in coated fabrics is the use of urethane coating. Urethane-coated fabrics can be constructed so as to give natural break characteristics and, thus, have a leatherlike appearance. Since urethane is a hydrophilic polymer, urethane-coated fabrics show a certain amount of moisture permeability. Values from 5–18 g/(m^2·h) have been reported (4). Although this represents some improvement over the total impermeability of vinyl, it is far below that of the better poromerics. Studies indicate that a minimum of 40 g/(m^2·h) and, preferably, 60–80 g/(m^2·h) is needed for comfort in a closed shoe. On the other hand, 15–20 g/(m^2·h) sometimes is sufficient permeability for semiopen shoes. Figure 4 is a photomicrograph of a typical urethane-coated fabric.

Poromerics. Poromerics have been developed in order to match the moisture vapor permeability, skivability, and nonfraying characteristics of leather; the products have microscopic, open-celled structures. The structure of a poromeric is more homogeneous than that of a coated fabric, since it is either all polymer or polymer reinforced with nonwoven fibers. The size and configuration of the cells is such that the passage of moisture vapor but not liquid water can occur. Although the vapor transmission rates of commercial poromerics vary considerably, the better ones can equal that of leather.

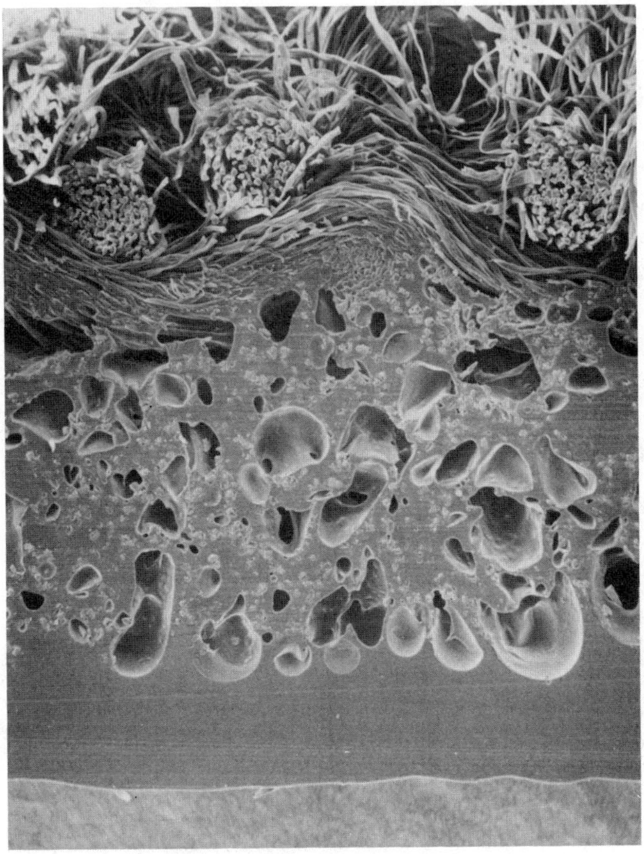

Figure 2. Cross-sectional view of expanded vinyl (50×).

The attempt to obtain a skivable material generally involves using a nonwoven fibrous structure. One product, Porvair, is skivable because it has a homogeneous, unreinforced, all-urethane structure. Skivable products that are based on a woven/nonwoven combination structure have been introduced. The nonwoven approach has not been popular in the United States, but it has been very successful in Japan (eg, Clarino). The homogeneous product has been quite successful in Europe, and the woven/nonwoven approach is preferred in the United States (see Nonwoven textiles).

Figure 5 is a photomicrograph of Porvair. An unreinforced structure has the advantage of relative simplicity and flexibility in manufacture. Its principal drawback is that its modulus characteristics are the least leatherlike of any of the commercial synthetics. The nonreinforced poromeric has a rubberlike modulus curve. The pronounced modulus difference from leather requires considerable modification of lasting machinery in shoe applications. This disadvantage, however, must be balanced against the ability of a homogeneous product to be split or skived in any manner. Splitability allows simple adjustment of gauge for a wide variety of applications, and skivability allows application in areas requiring tight folding of the material.

A woven, reinforced product, eg, DuPont's Corfam, gives a more leatherlike

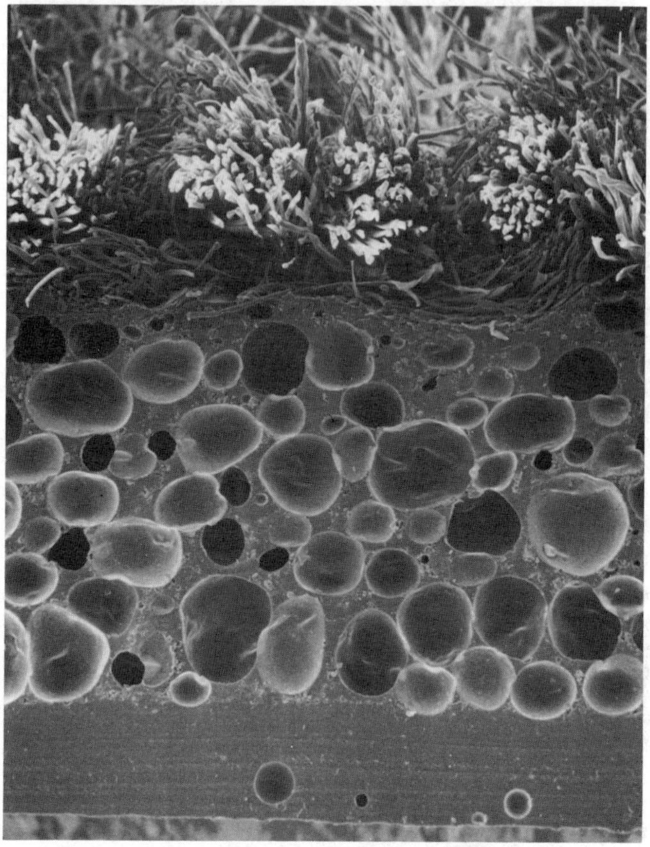

Figure 3. Cross-sectional view of expanded vinyl (50×).

modulus curve than Porvair. However, the Corfam type of poromeric cannot be skived. One highly successful reinforced poromeric that can be skived is Clarino. This product is based on a loosely meshed nonwoven, about which a poromeric urethane is formed so that the fibers are not solidly bonded to the urethane but lie within the micropores. The different structures are distinguished in Figures 6, 7, and 8 of Aztran, Clarino, and Corfam, respectively.

Manufacture and Processing

Coating of Fabrics. The type of fabric used as a coating substrate is determined by the intended application of the product. Woven fabrics are used for most footwear products, whereas knits are used for upholstery, apparel, or accessories where greater extensibility is required. The first vinyl-coated fabrics had cotton as the substrate. Recently, however, polyester and polyester/cotton-blend textiles also have been used; the choice depends upon a balance of factors, eg, cost, availability, and physical properties (see Polyesters). Fabric coating is reviewed in ref. 5.

The weight of solid vinyl coatings varies from 200–270 g/m^2 (6 to 8 oz/yd^2) and the thickness of the vinyl from 0.30–0.75 mm (12–30 mils), depending upon the specific

Figure 4. Cross-sectional view of polyurethane fabric (70×).

use. The vinyl usually is a plasticized vinyl chloride homopolymer that is compounded with various stabilizers and pigments as desired. The ester-type plasticizer normally is 30–40 wt % of the vinyl, depending upon the required degree of softness and flexibility. The formulated vinyl is applied to the continuously moving textile web by coating, calendering, or extrusion (see Coating processes).

The form of vinyl polymer that is used depends upon the coating method. The first form is general-purpose resin, which is made using suspension polymerization and is processed as a melt. The second form is vinyl plastisol which is a suspension of fine vinyl particles in a plasticizer in which the vinyl is not readily soluble at room temperature. Additional nonsolvent may be added to reduce the viscosity in which case the mixture is called an organosol. The fine particles usually are obtained by spray drying emulsion-polymerized vinyl. This procedure yields particles from 0.5–5.0 μm in diameter. Plastisols are easier to handle, formulate, and pigment than are general purpose resins; however, the former are more expensive. Vinyl is used as plastisol in casting which may be done by either a direct or a transfer method. In the former, the plastisol is knife-coated directly onto the fabric which then is passed through an oven for fusion of the coating. Often, more than one coat is needed to obtain a smooth coating of the desired weight. In such cases, the multiple coats usually are applied sequentially

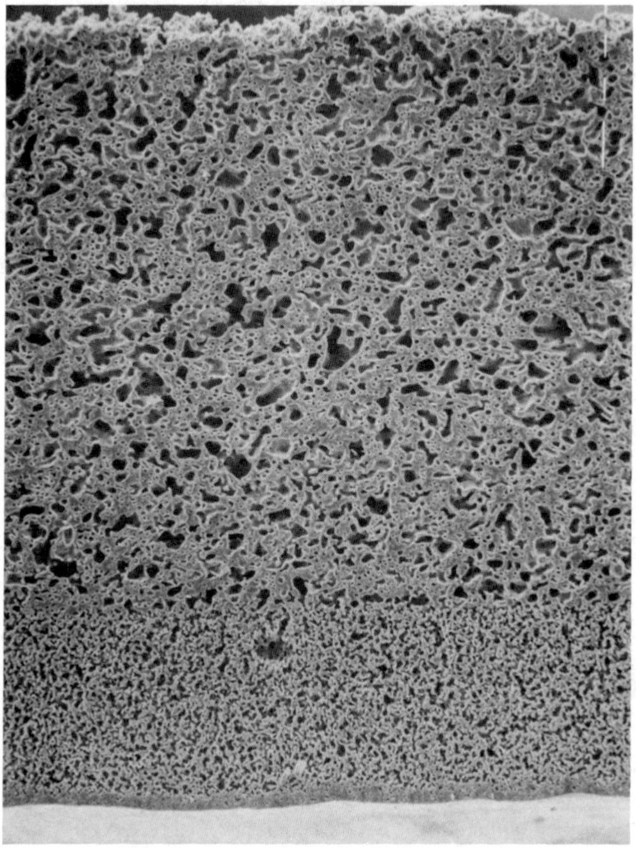

Figure 5. Cross-sectional view of Porvair (65×).

on a single line. Although the direct-coating method is simple, it tends to produce a product that is relatively stiff as a result of the filling of the interstices in the fabric by the vinyl.

The loss of textile flexibility can be avoided by using the transfer-coating technique. The plastisol is deposited, usually by reverse-roll coating, on a web of release paper. An adhesive layer is applied to the plastisol after it has been passed through an oven. The fabric backing is laminated against the adhesive layer, and the sandwich is passed through another oven for fusion of the adhesive layer. Finally, the web is cooled, and the release paper is removed. As the adhesive layer does not penetrate very deeply into the fabric, a fairly flexible product can be obtained. Transfer casting is applicable to a wider variety of plastisol formulations than is direct casting.

In calender coating of solid vinyl, compounded vinyl first is formed into a thin uniform sheet on a calender and then is combined with the fabric by one of two methods. The two may be combined by fusing the vinyl to the fabric with heat and pressure, or they may be laminated with an adhesive layer. Calender coating lacks the processing flexibility of cast coating and generally is used for long runs of a single product.

Extrusion coating is similar to calender coating, except that the molten vinyl

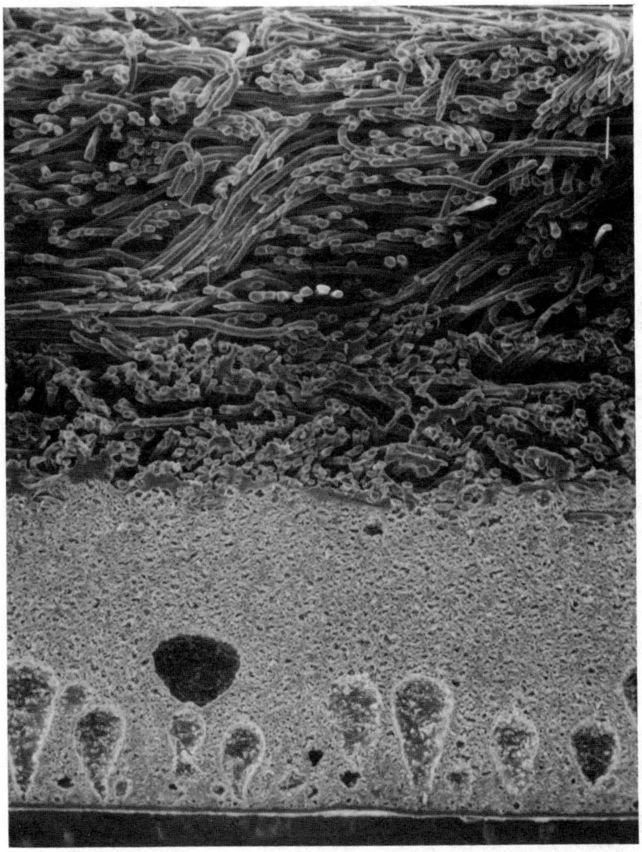

Figure 6. Cross-sectional view of Aztran (55×).

compound is extruded through a sheet die directly into a nip where it is fused to the fabric. The products resemble those produced by heat-laminated calender coating. Like calender coating, extrusion coating is only suitable for long runs of a single product. Both of these techniques, however, have the advantage over cast coating of involving less-expensive, general-purpose grades of vinyl.

Expanded vinyl coatings are prepared by either the transfer coating or calender coating technique (6). In the case of transfer coating, at least two layers are necessary, ie, a skin layer and a foam layer. In U.S. manufacture, the foam layer acts as the adhesive layer. It is partially gelled, but not yet foamed, prior to its contact with the fabric. The usual practice in Europe is to use a three-layer coating, which includes a separate adhesive layer in addition to the skin and foam layers. The latter process requires a production line with three ovens, whereas the U.S. method involves only two ovens.

Foaming or blowing the interlayers is accomplished by including a blowing agent in the formulation of the vinyl. Upon being heated, the blowing agent decomposes to yield a gas, eg, N_2 or CO_2. The gas, in turn, produces a closed-cell foam structure. When expanded vinyl is made by calender coating, two layers of vinyl must be applied, ie, an expandable middle layer and a solid skin layer; the skin layer may be applied either before or after the foam layer is blown.

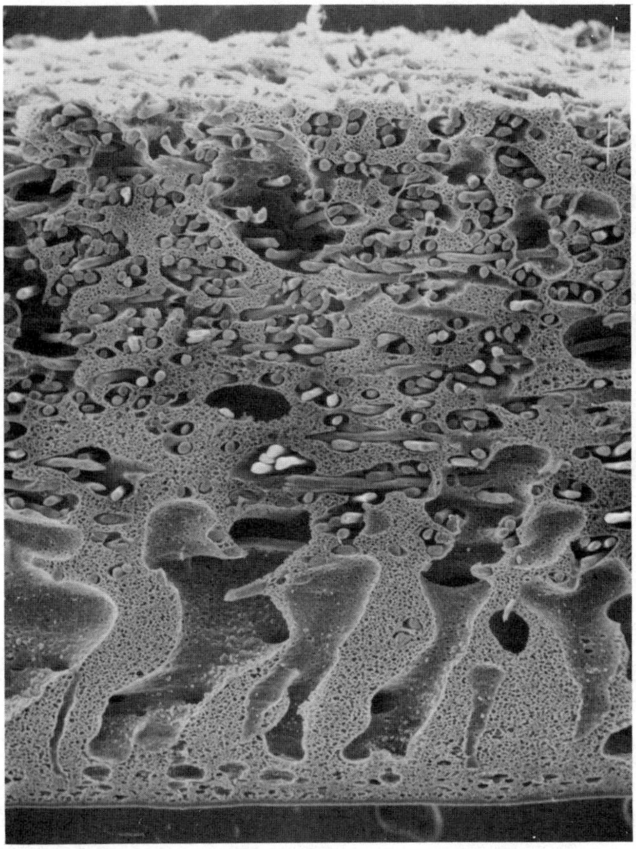

Figure 7. Cross-sectional view of Clarino (70×).

The Oakes process, a mechanical alternative to chemical blowing, has been adopted by a number of European manufacturers. By this process, a foam is produced by beating air into the vinyl coating which then is cast onto the fabric. Problems of odor and bleeding have been caused by the surfactants used to produce the foams; however, unlike other vinyl products, these coatings are somewhat permeable. They are used in shoe linings and counters, whereas blown foams predominate in applications not requiring permeability, eg, open shoe uppers, handbags, and upholstery.

Urethane-coated fabrics first were made by direct coating of two-component systems. The components, a polyol and a polyisocyanate, were mixed as liquids and they reacted on the fabric to form the solid urethane (7) (see Isocyanates, organic). Because of problems regarding adhesion to the fabric and controlling softness, production changed to cast coating of thermoplastic urethane solutions. Urethanes have an advantage over vinyl in that their molecular structure renders them inherently flexible. The commercial products are block, or segmented, ester–urethane or ether–urethane copolymers. Whereas vinyl must be softened by the addition of a plasticizer, the urethanes are softened by decreasing the urethane segment lengths relative to the ester or ether segments (see Plasticizers). This segmented structure gives urethanes the widest span between glass transition temperature T_g and melting point T_m of any

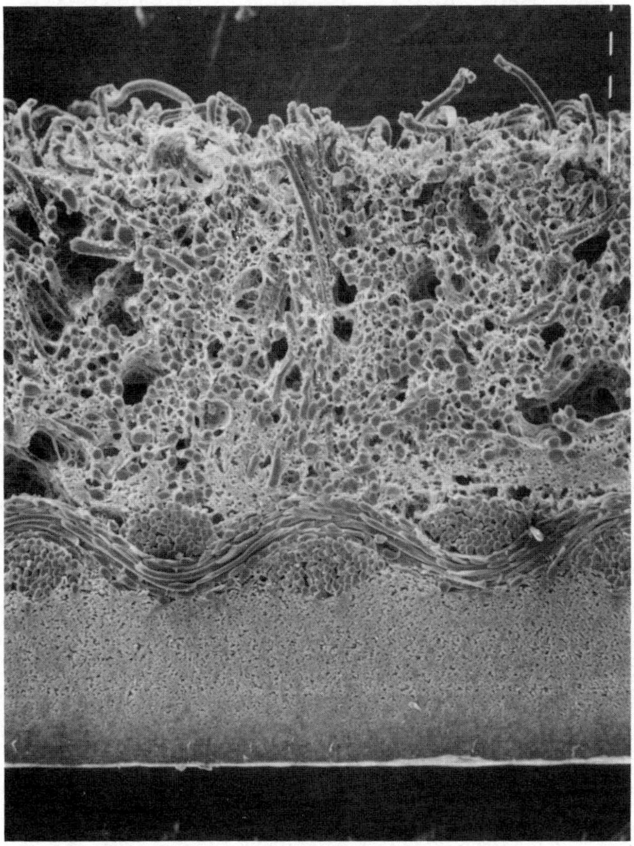

Figure 8. Cross-sectional view of Corfam (70×).

commercial polymer. The temperature spread is wide enough to produce a product that can be subjected to very high temperatures in shoe lasting (eg, when used with hot-melt adhesives) but remain sufficiently flexible for use in outdoor winter temperatures.

Urethanes also have tensile strengths that are as much as ten times greater than those of vinyls. Because of their strength and their higher cost, urethanes are coated at much lower thicknesses than vinyl. Dry coating weights from 50–100 g/m^2 (1.5–3.0 oz/yd^2) commonly are used. These low coat weights give the product a low heat capacity which gives the coated fabric a warm or cool hand under temperature conditions where vinyl-coated fabrics feel hot or cold; this heat capacity effect is of particular importance in upholstery applications. The thin coating also enhances the break characteristics of the product. To make the break more leatherlike, urethanes usually are coated onto fabrics that are napped on the coating side. The nap provides volume into which the coating can buckle to produce the creases of the break.

Urethanes do not require the stabilization against thermal degradation that vinyl does, but they do require stabilization against hydrolytic decomposition. The level of stabilization depends upon the type of urethane, with polyether-based types being considerably more stable to hydrolysis than polyester-based ones.

One of the major disadvantages of urethanes, with respect to vinyls, is that urethane solutions require strong solvents, eg, dimethylformamide (DMF) or tetrahydrofuran, both of which are expensive and toxic. Solvent recovery systems are available, but a large-volume line is needed to justify their cost.

Formation of Poromeric Structures. The manufacture of poromeric structures, particularly those with sufficient permeability for footwear applications, has presented the largest technological problem in the leather-substitute field. In the mid 1960s, over 20 chemical companies in the United States had projects directed toward this objective. The failure of DuPont's Corfam and Goodrich's Aztran were symptomatic of the problems in this area.

The simplest poromeric products are those that contain no reinforcing fiber, eg, Porvair (8). The most direct method of forming such a homogeneous microporous web is by polymer coagulation which involves extruding a web of urethane solution onto a suitable support. The web is immersed in a nonsolvent (for the polymer) which is miscible with the solvent used. The usual solvent/nonsolvent combination is DMF/water. When the urethane web is immersed in the water, DMF migrates to the aqueous phase and water begins to migrate into the web; simultaneously, the polymer precipitates from solution. The net effect of this precipitation/migration is the formation of a solid urethane that contains microscopic channels or pores.

Figure 9 is a flow diagram for the production of a typical poromeric, and Figure 10 is a diagram for the production of Porvair. Close control of the pore size is necessary in order to produce a product that has high permeability without having pores so large that they are visible to the naked eye. Pore size can be controlled by a number of factors, eg, urethane composition and solids content or time and temperature of the dunking cycle. Other process variations include adding small, measured amounts of nonsolvent to the urethane solution before extrusion and blowing steam onto the extruded web just before dunking.

One method of controlling pore size involves the use of a finely ground inert filler which is removed later in the process (see Fillers). For example, finely ground sodium

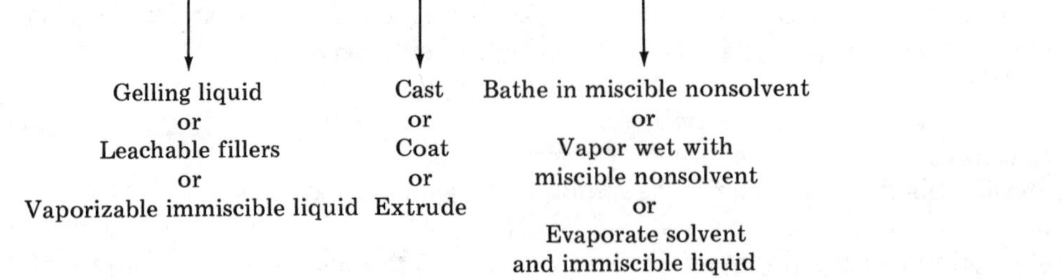

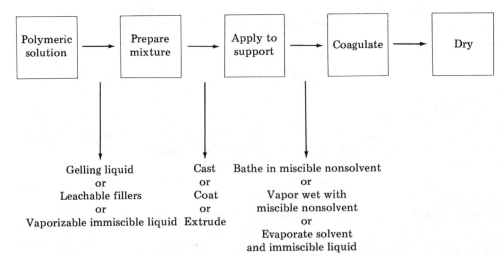

Figure 9. Flow diagram for the production of a typical poromeric.

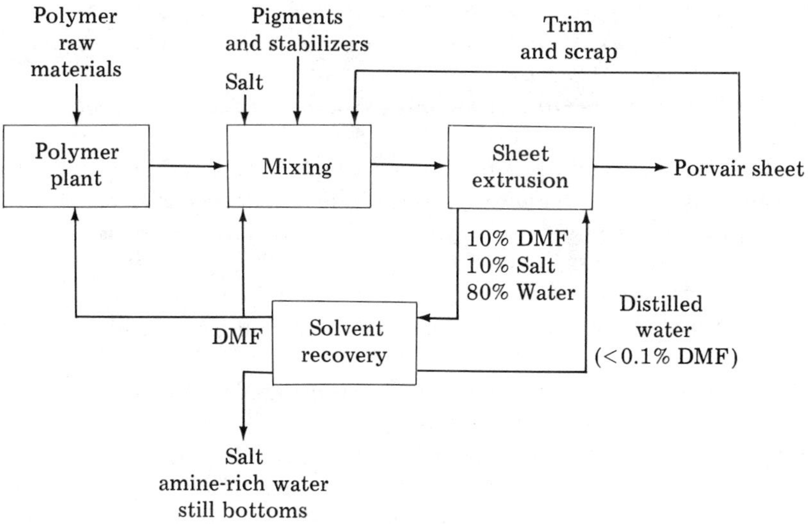

Figure 10. Diagram for the production of Porvair.

chloride or ammonium sulfate is mixed with the urethane/DMF solution. When the solution is coagulated by dunking in water, the salt slowly leaches into the aqueous phase leaving cells in the web. The final solid structure from this process consists of large cells of ca 13 μm in diameter that are interconnected with smaller channels ca 1 μm in diameter.

The various techniques for producing fiber-reinforced poromerics result in a large variety of structures (9). The major variations include the type of fiber structure used and the degree to which it is impregnated with polymer. Although some poromerics, eg, Corfam, include a woven-fabric reinforcement, recent practice has tended toward using only nonwoven materials; the nonwoven-based products tend to match leather more closely in modulus. On the other hand, poromerics that are based upon woven or knitted fabrics generally have better drape characteristics. The nonwoven fibers usually are polyester or nylon; both confer strength and resiliency. A proportion of rayon (qv) commonly is used to give a degree of moisture absorption, and a small amount of polypropylene also is used (see Olefin polymers).

In general, these fibers are formed into a web by air-laying them to form a batt which is needle-punched to give the web density and integrity; needle punching is the most critical step in determining the web properties. As for general-purpose nonwoven textiles, a punching density of 60–90 punches per square centimeter is sufficient to yield a web that has planar integrity. However, needle punching results in a web of relatively low cohesive strength and of a uniform density on a macro scale but not on a micro scale. To achieve microscale uniformity, a punching density of 900–1500 punches per square centimeter would be needed. This would give a fabric with the same surface uniformity as a 90 × 60- or 100 × 100-count woven fabric. Such high punching densities are impractical by traditional needling methods. DuPont tried to avoid this problem first by using shrinkable fibers and then by putting a fine percale woven fabric between the nonwoven and the urethane layer (Corfam). This solution had drawbacks both in the high cost of the percale and in its total lack of extensibility.

The Japanese have developed an elegant nonwoven approach in the production of Clarino by producing fibers that are close in properties to the collagen fibers in leather; the method that was developed yields an islands-in-the-sea fiber. Two immiscible molten polymers are extruded to give a fiber with a cross-section, as shown in Figure 11. By subsequent selective solvent extraction of either the island or the sea phase, a hollow supple compressible fiber or a bundle of very fine fibers is obtained (10). Clarino is based upon a nonwoven web of the former type of fiber made of nylon. The latter type of fiber is used in Ultrasuede. The use of fibers of complex morphology or of bundles of fine fibers allows the formation of a high integrity microscopically uniform, nonwoven substrate at a relatively low needle-punching density. In addition to improving the process economics, this leads to a product of considerably enhanced suppleness.

After needle punching, the nonwoven is impregnated with a reinforcing resin and is coated with a urethane solution. The coating step is followed immediately by a coagulation step similar to that described for nonreinforced poromerics. Various products differ in the degree of impregnation of the fabric. The degree of impregnation has a pronounced effect on moisture absorption and permeability. Some products may be only lightly treated with a polymer, usually as a latex, to bind the fibers at their points of intersection (see Latex technology). This binding gives an open structure which, together with the coagulated coating layer, gives high permeability. In other cases, the nonwoven web may be heavily saturated with polymer solution, which subsequently is coagulated with the coating layer; this gives a structure which is microporous throughout. An extreme case of a polymer-saturated web is Clarino; the heavy impregnation, unique fiber structure, and carefully controlled coagulation conditions yields a structure that contains micropores which contain individual unbonded fibers.

A technique that has gained considerable commercial importance is coagulation coating (11) by which urethane/DMF solutions are cast onto a fabric similarly as for normal coated fabrics. However, instead of being dried by evaporation, they are coagulated in a water bath. This procedure produces a structure that has a poromeric layer on top of a fabric. These products are intermediate in properties between reinforced poromerics and urethane-coated fabrics and have the air permeability of the former and the physical properties of the latter.

Composites are manufactured by laminating a fabric backer, whether woven or nonwoven, to a previously prepared, unreinforced poromeric (see Composite materials).

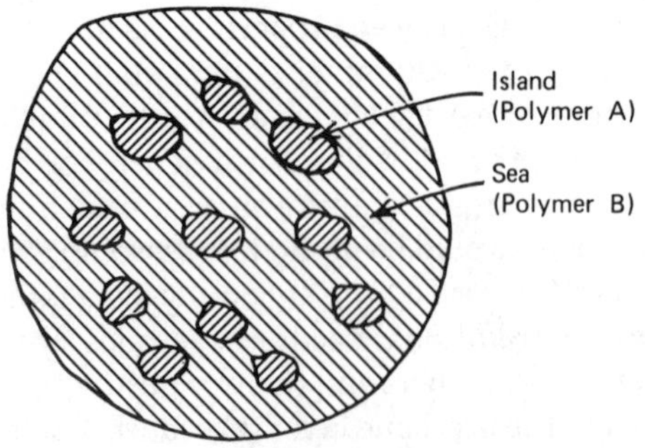

Figure 11. Cross-sectional view of islands-in-the-sea fiber.

The difficulty with this manufacturing technique is in designing a bonding method that does not greatly reduce the permeability of the final structure or unduly increase its stiffness; however, this problem can be overcome by applying a discontinuous adhesive layer. The percentage of adhesive coverage must be closely controlled so that a good balance between permeability and adhesion is achieved. The microporous poromeric layer contributes elastic recovery and barrier properties, the fabric contributes leatherlike modulus, and the combination yields the product's tear strength. Composites represent one of the few instances where the addition of a plastic coating to a fibrous layer increases, rather than decreases, the tear strength of the fibrous substrate. However, a problem that is encountered with composites is that of achieving adhesion that is resistant to the moist, high temperature conditions that are encountered in shoe lasting. Traditionally, leather shoe uppers are steamed to soften them before lasting, and shoe manufacturers tend to steam synthetics, even though this is not necessarily needed.

The best overall properties in laminated structures have been obtained with a combination woven/nonwoven structure. The structure is prepared by heavily napping a sateen fabric, shearing the nap, and then impregnating the fabric with a reinforcing resin. The process results in a backing that is skivable and that has a leatherlike hand.

Finishing. In addition to having suitable physical properties, it is necessary for a synthetic leather to have a leatherlike appearance. The appearance of a piece of leather results principally from two factors: color and texture. The finished surface of the material must have some resistance to abrasion and appropriate tactile properties. Three types of surfaces are commercially important: grained, patent, and suede, and each requires different finishing techniques. The method used for producing grain on a surface affects the manufacturing process for the basic material, eg, with a transfer-coated fabric, grain can be formed during the casting of the coating. This is done by casting onto a release paper that has had a negative version of the grain embossed into it (see Abherents). When the coating is laminated to the fabric, a product with the desired grain is obtained. With most poromerics, grain is obtained by embossing the complete structure. Embossing usually is preceded by one or more printing, spraying, or coating steps. Coating may be in a very thin layer that is just sufficient to give the product the desired color. Printing may be used where it is desired to have a color variation that matches the grain.

Embossing is a process of thermoforming, ie, heating the material, deforming it, and cooling to set it in its new shape. The web is passed over heated rolls that are engraved with a negative version of the desired grain. The grain can vary from very light (eg, calf) to very heavy (eg, reptile). In order that the grain be permanently impressed upon the elastomeric urethane, the embossing temperature must be very close to the material's melting point. However, at such high temperatures there is a risk that the pore structure may collapse (12). Even a very thin, impermeable layer on the surface can drastically reduce the overall moisture vapor transmission rate (MVT) of poromerics. It can be shown that for a structure of n layers,

$$K2_T = \left(\sum_{i=1}^{n} \frac{1}{K_i} \right)^{-1},$$

where K_T is the MVT of the total structure and K_i is the MVT of the ith layer in g/(m^2·h). Even one low value of K_i can markedly reduce K_T.

Maintaining permeability is not possible in a patent finish. However, as patent leather also has negligible permeability, this is not really a problem. Producing a patent effect requires a highly smooth surface which is coated with a clear layer (13); the latter gives the patent surface its appearance of depth. Producing a smooth, dust-free surface for coating is the critical step. One technique is to heat-fuse the surface in contact with a Mylar sheet by passing the Mylar and product web over a heated drum. Subsequent stripping of the Mylar then leaves a smooth, glossy surface. The lacquer used in the coating step may be urethane, vinyl, or acrylic. The urethanes may be either thermoplastic or two-component reactive systems; the reactive systems, although more difficult to handle, give a smoother, glossier surface than the thermoplastics. The viscosity increase resulting from the reaction occurs at a slower rate than that resulting from solvent evaporation. This slower rate allows the surface to become more level before it solidifies. Knife, spray, and roller-coating techniques have been used to apply patent lacquers.

A drawback of patent leather has been its tendency to crack during lasting; however, recent studies (14) indicate that synthetic patent leathers are not as prone to this problem. The difference is attributed to the greater extensibility at lasting temperature of the coatings used on the synthetics.

Inexpensive coated fabrics can be given a suedelike surface by floccing (15) which involves the bonding of short fibers more or less perpendicularly to the surface. The appearance mimics that of suede, which is made by napping leather. Suedes also have been produced from the type of heavily napped fabric described above (16). Rather than being laminated to poromeric layers, these fibers are sanded to give them a suedelike surface. Impregnation is controlled so as to provide the material with a high level of porosity.

A process that has been developed for vinyls involves coating an expanded vinyl on a fabric substrate and then tearing the cellular layer while it is hot. The tearing leaves a rough, irregular cell edge structure that has a fibrous appearance (17). A similar edge structure can be produced by sanding the surface of an expanded vinyl (18). The aesthetic qualities of the sanded surface can be improved if butadiene–acrylonitrile copolymer is included in the vinyl formulation (19). These products, which are based upon expanded vinyls, continue to hold a large share of the U.S. synthetic suede market.

Sueded, urethane-coated fabrics usually are produced by molding or hot embossing. A silicone rubber mold having cavities that closely resemble suede filaments is made; normally, the master mold is a flocked material. The thermoplastic urethane film is pressed into the mold, heated until it flows into the cavities, cooled, and removed. A semicontinuous version of this process was practiced in the U.S. in the early 1970s using flat-bed heated presses (20) which allowed the production of sheets only as large as the press size, ie, 127 × 137 cm. A continuous version of the process has been reported (21).

Sueded poromerics normally are produced from columnar cellular structures. The coagulation of the urethane solution is achieved under compositional and temperature conditions such that cylindrical, rather than spherical, cells are produced. The surface is abraided with rotary sanding machines to produce a suedelike effect (22). The columnar technique also has been applied to expanded vinyls (23).

None of the above methods can truly match the tactile qualities of suede, which result from the extremely fine diameter of the collagen fibers in leather. Conventional synthetic-fiber spinning techniques cannot produce fibers of sufficiently low tex (denier). Fine fibers have been made, however, using the islands-in-the-sea technique developed for Clarino (10). Commercial products, eg, Ultrasuede, have been produced by forming the islands-in-the-sea fibers into a nonwoven fabric and then dissolving the sea phase. This leaves a very highly entangled, nonwoven web of very fine fibers which then are impregnated with urethane. The unique nature of the nonwoven yields excellent drape characteristics, making it an exception to the general rule that woven-based products have better drape. The very low tex of the fibers also gives the product a feel that is almost indistinguishable from that of genuine suede.

Economic Aspects

In almost all cases, synthetic materials are considered to be less expensive substitutes for leather. Therefore, the economics of leatherlike synthetics are dominated by the state of the leather market. The leather market, in turn, is largely determined by the market for meat and dairy products, as hides are by-products of the cattle industry. As this latter business tends to be cyclical, so do hide prices. Because leather prices are only a fraction of the cost of finished leather products (10–20% in shoes), a rather large upward swing in leather prices is required to produce a significant advantage by switching to synthetics.

Substitution of synthetics for leather can be explained by the cyclical shortage of hides rather than by their cost. Hides must be processed within a short time after their production to avoid spoilage. The availability of hides is controlled by the slaughter rate which is controlled by the supply of and demand for meat. It has been observed that the demand side behaves normally, ie, it increases with increasing population and standard of living. On the other hand, the supply side behaves quite erratically, with a severe dip occurring approximately every seven years (eg, 1965, 1972, and 1979). This dip seems to be tied to the life cycle of cattle. Increasing future cattle herds requires an increase in calf production which requires that some heifers be bred as cows, thus removing them from the pool of slaughterable cattle, ie, current production must be sacrificed for the benefit of future increases. It has been observed that synthetics have an increased share of the footwear market during the dips in leather availability. However, once a synthetic does gain use in a market area, its use continues even when leather prices enter a subsequent down cycle. This phenomenon can be attributed to the properties of the synthetics. As they are uniform in composition and mostly thermoplastic in nature, they are well suited to quicker and more efficient manufacturing techniques. Cutting can be achieved in multiple layers from rolls; setting of uppers can be carried out rapidly by dry heat rather than by long steaming as with leather; and soling can be done by direct injection molding instead of the time-consuming sewing or cementing processes used with leather.

The future competitive position of synthetics may be seriously hampered by rising petroleum costs. With the single exception of cotton substrates for coated fabrics, all of the components of synthetics are derived from petrochemical feedstocks (qv). For example, poly(vinyl chloride) (PVC) is produced by way of ethylene from naphtha. Urethanes are based on aromatic isocyanates which are derived from benzene or toluene. Prices of these petrochemicals have increased dramatically since 1973 and probably will continue to do so. Eventually, these materials could be made from coal; however, although this would assure supply, it would do little to reduce cost. It should

Table 1. Worldwide Production of Coated Fabrics in 1978, 10^6 m^2 [a,b]

	Urethane	PVC
Far East	80	400
Western Europe	150	550
Eastern Europe	30	300
North America	55	450
South America	15	100
Africa	10	25
Total	*340*	*1825*

[a] Ref. 24.
[b] To convert m^2 to ft^2, multiply by 10.76.

Table 2. Estimated Consumption of Poromerics in 1978, 10^6 m^2 [a,b]

	All uses	Footwear only
Western Europe	2.6	1.9
Eastern Europe	3.7	2.3
North America	1.4	0.9
Far East	3.1	1.3
Total	*10.8*	*6.4*

[a] Ref. 24.
[b] To convert m^2 to ft^2, multiply by 10.76.

be noted that leather prices are not entirely independent of the petroleum market. Cattle ranching, particularly in the U.S. where cattle are grain fed rather than range fed, is an energy-intensive industry. Steeply increased energy costs could discourage the production and consumption of beef and, thereby, decrease the hide supply. Prices (1979) for synthetics range from ca \$4.30/m^2 (\$0.40/ft^2) for vinyls to ca \$13.45/m^2

Table 3. Worldwide Production of Poromerics in 1978, 10^6 m^2 [a,b]

Company	Product	Total capacity	Estimated production
Far East			
Kuraray	Clarino	8.8	6.9
Kanebo	Patora		
Toray	Alcantara		
Teijin	Cordley		
Eastern Europe			
Pronit	Polcorfam (Poland)	3.3	1.9
Exico	Barex (Czechoslovakia)		
Graboplast	Graboxan (Hungary)		
VEB Vogtlandische	Ekraled (GDR)		
Western Europe			
Porvair	Porvair and Vantel	2.3	2.0
Yagi	Patora		
Societe Belge	Kabipor		
North America			
Uniroyal	Capilair	na	na
Scott-Chatham	Tanera		

[a] Ref. 24.
[b] To convert m^2 to ft^2, multiply by 10.76.

(\$1.25/ft^2) for poromerics, as compared to leather prices which range from \$13.45/m^2 (\$1.25/ft^2) for side leather to \$75.00/m^2 (\$7.00/ft^2) or more for high-grade, calf leather.

The worldwide production of coated fabrics in 1978 is given in Table 1. Note that vinyls are produced in considerably larger quantity than are urethanes, as might be expected from their lower cost. The worldwide consumption pattern by market for urethane-coated fabrics in 1978 was clothing, 50%; footwear, 15%; bags, 25%; upholstery, 5%; and industrial, 5% (24). Clothing and accessories are the principal markets for these products; whereas shoes comprise the principal market for poromerics, as shown in Table 2. This is to be expected, as poromerics originally were developed specifically for footwear applications. The poromerics that were produced commercially in 1978 are listed in Table 3.

The 1978 U.S. consumption of various upper materials for footwear is listed in Table 4. These figures exclude the 250×10^6 pairs of rubber and fabric shoes produced that year. The trend of U.S. footwear consumption and manufacturing is shown in Table 5. The large increase in imports is approaching 50% of domestic consumption. The shift to foreign production, especially to developing nations, results from the labor-intensive nature of shoe manufacturing. Producing a pair of shoes requires 0.5–1.0 h of labor, depending upon style, and ca 0.15 m^2 (1.6 ft^2) of upper material. Another trend shown in Table 4 is the relatively flat consumption rate, except for a dip during the 1974–1975 recession, which can be explained by a change in consumer tastes. It is estimated that consumption of rubber and fabric shoes (eg, sneakers) increased from $(70–250) \times 10^6$ pairs from 1965–1978. Rubber and fabric shoes are manufactured using nonleatherlike products: the soles normally are of rubber, eg, plasticized PVC, urethane, neoprene, or SBR (styrene–butadiene rubber); and the uppers vary from injection-molded PVC in ski boots to nylon fabric in tennis shoes. Thus, these shoes can offer serious competition to those made of leatherlike synthetics in some low priced markets.

Table 4. Footwear Upper Materials Consumed in the U.S. in 1978, 10^6 pairs[a]

	Leather	Vinyl	Other	Total
production	235	104	80	419
imported	174	146	54	374
Total	*409*	*250*	*134*	*793*

[a] Data from American Footwear Industries Association.

Table 5. The U.S. Market for Upper Materials in Shoes, 10^6 pairs[a]

	1965	1970	1975	1978
U.S. production	626	562	413	419
imported	88	242	286	374
Total	*714*	*804*	*699*	*793*

[a] Data from American Footwear Industries Association.

BIBLIOGRAPHY

"Poromeric Materials" in *ECT* 2nd ed., Vol. 16, pp. 345–360, by J. L. Hollowell, E. I. du Pont de Nemours & Co., Inc.

1. F. O'Flaherty, W. T. Roddy, and R. M. Loolar, eds., *The Chemistry and Technology of Leather, ACS Monograph Series, No. 134,* Reinhold, New York, 1965.
2. W. Riess, *Proceedings SATRA North American Int. Conf., Ontario,* 1974, pp. 56–68.
3. H. Herfeld, *Index 78 Programme,* Session 6, paper 4.
4. L. C. Hole and B. Keech, *SATRA Bull.,* 37 (March 1968); *Proc. Xth Congr. International Union Leather Chem. Soc. London, 1969.*
5. H. R. Lasman in ref. 2, pp. 80–89.
6. U.S. Pat. 2,964,799 (Dec. 20, 1960), P. E. Roggi and R. A. Chartier (to U.S. Rubber).
7. H. J. Koch, *Melliand Textilbe.* **51,** 1313 (1970); R. R. Grant, *Urethane Plast. Prod.* **1,** 1 (1971).
8. U.S. Pat. 3,696,180 (Oct. 3, 1972), V. R. Cunningham and T. S. Dodson (to Porous Plastics Ltd.); U.S. Pat. 3,729,536 (April 24, 1973), E. A. Warwicker (to Porvair); U.S. Pat. 3,860,680 (Jan. 14, 1975), E. A. Warwicker and R. Hogkinson (to Porvair); U.S. Pat. 3,968,292 (July 6, 1976), A. W. Pearman and S. J. Wright (to Porvair); U.S. Pat. 4,157,424 (June 5, 1979), D. L. Bontle (to Porvair).
9. T. Hayaski, *Chem. Tech.,* 28 (Jan. 1975); L. C. Hole, *Rubber J.,* 152, 72 (1970); R. C. Hole and R. E. Whittaker, *J. Mater. Sci.* **6,** 1 (1971).
10. U.S. Pat. 3,865,678 (Feb. 11, 1975), M. Okamoto and co-workers (to Toray); U.S. Pat. 3,873,406 (March 25, 1975), K. Okazaki, A. Higuchi, and N. Imaeda (to Toray); U.S. Pat. 3,908,060 (Sept. 23, 1975), K. Okazaki, A. Higuchi, and N. Imaeda (to Toray); U.S. Pat. 4,103,054 (July 25, 1978), M. Okamoto and S. Yoshida (to Toray); U.S. Pat. 4,136,221 (Jan. 23, 1979), M. Okamoto and Y. Yoshida (to Toray).
11. *Mod. Plant. Int.* 12 (Dec. 1975).
12. U.S. Pat. 3,764,363 (Oct. 9, 1973), F. P. Civardi and H. G. Kuenstler (to Inmont); U.S. Pat. 3,931,437 (Jan. 6, 1976), F. P. Civardi and H. G. Kuenstler (to Inmont).
13. U.S. Pat. 2,801,949 (Aug. 6, 1957), A. W. Bateman (to E. I. du Pont de Nemours & Co., Inc.).
14. G. Hole and D. Hill, *SATRA Bull.* **18**(9), 85 (1978).
15. U.S. Pat. 3,222,208 (Dec. 7, 1965), N. J. Bertollo (to Interchemical Corp.).
16. U.S. Pat. 3,998,488 (Oct. 26, 1976), F. P. Civardi (to Inmont); U.S. Pat. 4,055,693 (Oct. 25, 1977), F. P. Civardi (to Inmont); U.S. Pat. 4,122,223 (Oct. 25, 1978), F. P. Civardi and F. C. Loew (to Inmont).
17. U.S. Pat. 3,709,752 (Jan. 9, 1973), R. Wisotzky and R. E. Patersen (to Pandel-Bradford); U.S. Pat. 4,048,269 (Sept. 13, 1977), R. Wisotzky and J. C. Bolger (to Pandel-Bradford); U.S. Pat. 4,052,236 (Oct. 4, 1977), V. C. Kapasi and co-workers (to Pandel-Bradford).
18. U.S. Pat. 3,041,193 (June 26, 1962), E. Hamway and B. Edwards (to General Tire and Rubber Co.).
19. U.S. Pat. 3,949,123 (April 6, 1976), R. N. Steel (to Uniroyal).
20. U.S. Pat. 3,533,895 (Oct. 13, 1970), K. Norcross (to Nairn-Williamson); U.S. Pat. 3,632,727 (Jan. 4, 1972), K. Norcross (to Nairn-Williamson).
21. U.S. Pat. 4,044,183 (Aug. 23, 1977), N. Forrest; U.S. Pat. 4,124,428 (Nov. 7, 1978), N. Forrest.
22. U.S. Pat. 3,284,274 (Nov. 8, 1966), D. G. Hulsander and W. F. Manwaring (to E. I. du Pont de Nemours & Co., Inc.); U.S. Pat. 3,912,834 (Oct. 14, 1975), S. Imai, T. Eguchi, and M. Shimokawa (to Kanegafuchi).
23. U.S. Pat. 3,776,790 (Dec. 4, 1973), G. N. Harrington and F. P. Civardi (to Inmont).
24. S. D. Cleaver, Senior Information Officer, SATRA, private communication, March 7, 1980.

F. P. Civardi
G. Frederick Hutter
Inmont Corporation

MATERIALS STANDARDS AND SPECIFICATIONS

Standards have been a part of technology since building began, both at a scale that exceeded the capabilities of an individual, and for a market other than the immediate family. Standardization minimizes disadvantageous diversity, assures acceptability of products, and facilitates technical communication. There are many attributes of materials that are subject to standardization, eg, composition, physical properties, dimensions, finish, and processing. Implicit to the realization of standards is the availability of test methods and appropriate calibration techniques. Apart from physical or artifactual standards, written or paper standards also must be examined, ie, their generation, promulgation, and interrelationships.

A standard is a document, definition, or reference artifact intended for general use by as large a body as possible; whereas a specification, although involving similar technical content and similar format, usually is limited in its intended applicability and its users. The International Organization for Standardization (ISO) defines a standard as the result of the standardization process, "the process of formulating and applying rules for an orderly approach to a specific activity for the benefit and with the cooperation of all concerned and in particular for the promotion of optimum overall economy taking due account of functional conditions and safety requirements" (1). Standardization involves concepts of units of measurement, terminology and symbolic representation, and attributes of the physical artifact, ie, quality, variety, and interchangeability. A specification, however, is defined as "a document intended primarily for use in procurement which clearly and accurately describes the essential technical requirements for items, materials, or services including the procedures by which it will be determined that the requirements have been met" (2). The corresponding ISO definition is: "A specification is a concise statement of a set of requirements to be satisfied by a product, a material or a process indicating, whenever appropriate, the procedure by means of which it may be determined whether the requirements given are satisfied. Notes—(*1*) A specification may be a standard, a part of a standard, or independent of a standard. (*2*) As far as practicable, it is desirable that the requirements are expressed numerically in terms of appropriate units, together with their limits." A specification is the technical aspects of the legal contract between the purchaser of the material, product, or service and the vendor of the same and defines what each may expect of the other.

There are psychological barriers to making intelligent selections and applications of standards. For some, a standard has attributes of superiority; it is considered an unattainable ideal which only can be approached but never met. For others it represents the opposite, ie, a lowest common denominator to which all must be reduced. For most, however, the terms impart a sense of a condition from which all creativity and individuality have been removed.

Standards

Objectives and Types. The objectives of standardization are economy of production by way of economies of scale in output, optimization of varieties in input material, and improved managerial control; assurance of quality; improvement of interchangeability; facilitation of technical communication; enhancement of innovation and technological progress; and promotion of the safety of persons, goods, and the environment. The likely consequences of choosing a material that is not standard, other than in very exceptional circumstances, are that the selected special would be unusually costly; require an elaborate new specification; be available from few sources; be lacking documentation for many ancillary properties other than that for which it was chosen; be unfamiliar to others, eg, purchasers, vendors, production workers, maintenance personnel, etc; and contribute to the proliferation of stocked varieties and, thus, exacerbate problems of recycling, mistaken identity, increased purchasing costs, etc.

Physical or artifactual standards are used for comparison, calibration, etc, eg, the national standards of mass, length, and time maintained by the NBS or the Standard Reference Materials collected and distributed by NBS. Choice of the standard is determined by the property it is supposed to define, its ease of measurement, its stability with time, and other factors (see also Fine chemicals).

Paper or documentary standards are written articulations of the goals, quality levels, dimensions, or other parameter levels that the standards-setting body seeks to establish.

Value standards are a subset of paper standards and usually relate directly to society and include social, legal, political, and to a lesser extent, economic and technical factors. Such standards usually result from federal, state, or local legislation.

Regulatory standards most frequently derive from value standards but also may arise on an *ad hoc* or consensus basis. They include industry regulations or codes that are self-imposed; consensus regulatory standards that are produced by voluntary organizations in response to an expressed governmental need, especially where well-defined engineering practices or highly technical issues are involved; and mandatory regulatory standards that are developed entirely by government agencies. Examples of regulatory standards from the materials field include safety regulations, eg, those of the OSHA; clean air and water laws of the EPA; or rulings related to exposure to radioactive substances. Regulatory standards deliberately may be set in advance of the state of the art in the relevant technology, in contrast to the other types of standards (see Regulatory agencies).

Voluntary standards are especially prevalent in the United States and are generated by various consortia of government and industry, producers and consumers, technical societies and trade associations, general interest groups, academia, and individuals. These standards are voluntary in their manner of generation and in that they are intended for voluntary use. Nonetheless, some standards of voluntary origin have been adopted by governmental bodies and are mandatory in certain contexts. Voluntary standards include those which are recommended but which may be subject to some interpretation and those conventions as to units, definitions, etc, that are established by custom.

Product standards may stipulate performance characteristics, dimensions, quality factors, methods of measurement, and tolerances; and safety, health, and environmental protection specifications. They are introduced principally to provide for interchangeability and reduction of variety. The latter procedure is referred to as rationalization of the product offering, ie, designation of sizes, ratings, etc, for the attribute range covered and the steps within the range. The designated steps may follow a modular format or a preferred number sequence.

Public and private standards also may be distinguished. Public standards include those produced by government bodies and those published by other organizations but promoted for general use, eg, the ASTM standards. Private standards are issued by a private company for its own interests and, generally, are not available to parties other than its vendors, customers, and subcontractors.

Consensus standards are the key to the voluntary standards system since acceptance and use of such standards follow directly from the need for them and from the involvement in their development of all those who share that need. Consensus standards must be produced by a body selected, organized and conducted in accordance with "due process" procedures. All parties or stakeholders are involved in the development of the standard and substantial agreement is reached according to the judgment of a properly constituted review board. Other aspects of "due process" involve proper issuance of notices, record keeping, balloting, and attention to minority opinion.

Generation, Administration, and Implementation. The development of a good standard is a lengthy and reiteratively involved process, whether it be for a private organization, a nation, or a international body. The generic aspects of the development of a standard are shown in Figure 1. Once it has been determined that a need for a standard exists, information relevant to the subject must be gathered from many sources, eg, libraries and specialists' knowledge, field surveys, and laboratory results. Multidisciplinary and multifunctional teams must digest this information, array and analyze options, achieve an effective compromise in the balanced best interests of all concerned, and participate in the resolution of issues and criticisms arising from the reviews and appeals process. Among its functions, the administrative arm of the pertinent standards organization sets policy, allocates resources, establishes priorities, supervises reviews and appeals procedures, and interacts with organizations external to itself. The affected entities usually comprise a very large, diverse, and overlapping group of interests, ie, economic sectors: industry, government, business, construction, chemicals, energy, etc; functions: planning, development, design, production, maintenance, etc; and organizations and groups of individuals: manufacturers, consumers, unions, investors, distributors, etc. No standard can be fully effective in meeting its objectives unless attention is paid to the implementation function which includes promulgation, education, enforcement of compliance, and technical assistance. Usually in the choice of a standard for a given purpose, the more encompassing the population to which the standard applies, ie, from the private level to the trade association or professional society to the national level or to the international level, the more effective and the less costly the application of the standard is. Finally it must be recognized that standardization is a highly dynamic process. It cannot function without the continued feedback from all affected parties and it must provide for constant review and adaptation to changing circumstances, improved knowledge and control.

Standard Reference Materials. An important development in the United States, relative to standardization in the chemical field, is the establishment by the NBS of standard reference materials (SRMs), originally called standard samples (4). The objective of this program is to provide materials that may be used to calibrate measurement systems and to provide a central basis for uniformity and accuracy of measurement. SRMs are well-characterized, homogeneous, stable materials or simple artifacts with specific properties that have been measured and certified by NBS. Their use with standardized, well-characterized test methods enables the transfer, accuracy, and establishment of measurement traceability throughout large, multilaboratory measurement networks. More than 1000 materials are now included, eg, metals and

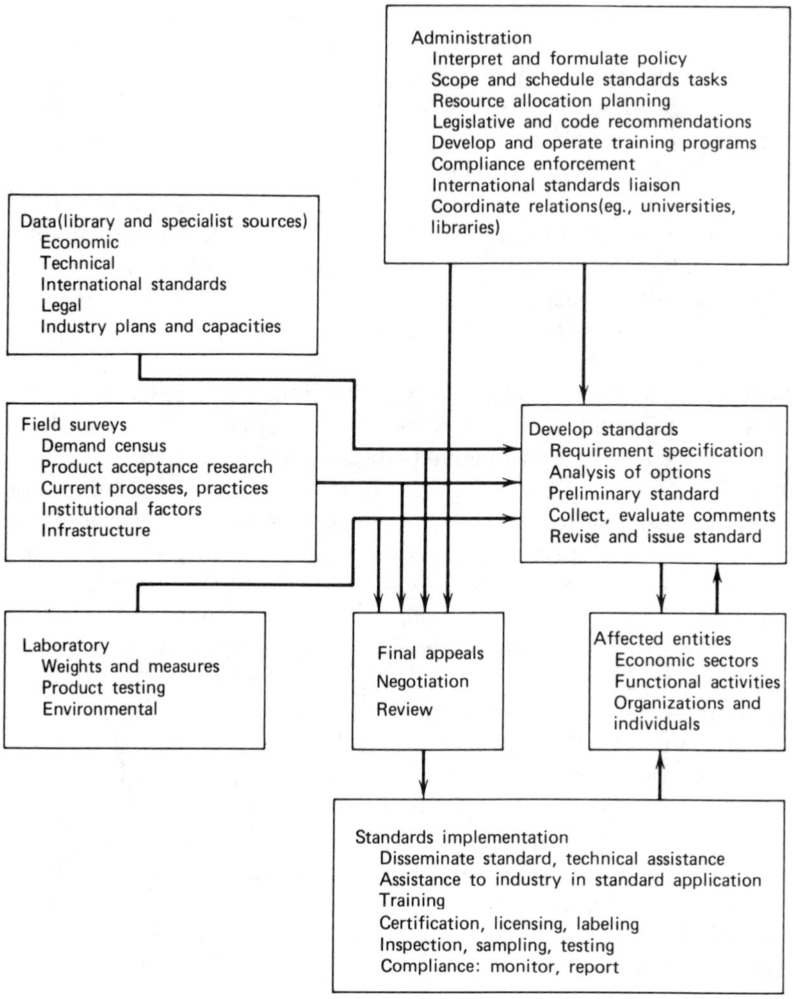

Figure 1. Flow chart of the standardization process (3).

alloys, ores, cements, phosphors, organics, biological materials, glasses, liquids, gases, radioactive substances, and specialty materials. The standards are classified as standards of certified chemical composition, standards of certified physical properties, and engineering-type standards. Although most of these are provided with certified numerical characterizations of the compositions or physical properties for which they were established, some others are included even where provision of numerical data is not feasible or certification is not useful. These latter materials do, however, provide assurance of identity among all samples of the designation and permit standardization of test procedures and referral of physical or chemical data on unknown materials to a known or common basis.

Major shifts in the nature of the materials included in the SRM inventory are expected. In the compositional SRMs, increased attention is anticipated to trace-organic analysis for environmentally, clinically, or nutritionally important substances;

to trace-element analysis in new high technology materials, eg, alloys, plastics, and semiconductors; and for bulk analyses in the field of recycled, nuclear and fibrous materials (see also Trace and residue analysis). It is expected that there will be concern not just for certification of the total concentration of individual elements but for the levels of various chemical states of those species. With regard to the development of physical property SRMs, density standards, dimensional standards at the micrometer and submicrometer level, and materials relative to standardization of optical properties will be among the more active areas. Major developments in SRMs for engineering properties will include materials suitable for nondestructive testing (qv), evaluation of durability, standardizing computer and electronic components, and workplace hazard monitoring.

Standard reference materials provide a necessary but insufficient means for achieving accuracy and measurement compatibility on a national or international scale. Good test methods, good laboratory practices, well-qualified personnel, and proper intralaboratory and interlaboratory quality assurance procedures are equally important. A systems approach to measurement compatibility is illustrated in Figure 2. The function of each level, I–VI, is to transfer accuracy to the level below and to help provide traceability to the level above. Thus, traversing the hierarchy from bottom to top increases accuracy at the expense of measurement efficiency.

Analytical standards imply the existence of a reference material and a recommended test method. This subject with reference to analytical, electronic-grade, and reagent chemicals has been discussed (see Fine chemicals). Analytical standards other than for fine chemicals and for the NBS series of SRMs have been reviewed (5). Another sphere of activity in analytical standards is the geochemical reference standards maintained by the U.S. Geological Survey and by analogous groups in France, Canada, Japan, South Africa, and the GDR (6).

Chronological standards are needed for an extremely diverse range of fields, eg, astrophysics, anthropology, archaeology, geology, oceanography, and art. The techniques employed for dating materials include dendrochronology, thermoluminescence, obsidian hydration, varve deposition, paleo-magnetic reversal, fission tracks, racemization of amino acids, and a variety of techniques related to the presence or decay of radioactive species, eg, ^{14}C, ^{10}Be, ^{18}O, and various daughter products of the U and Th series. Since the time periods of interest range from decades to millions of years and the available materials may be limited, no one technique presents a general solution. Although some progress has been made on age standardization and calibration through the efforts of the Sub-Commission on Geochronology of the International Union of Geological Sciences, much remains to be done (7). A great benefit could be extended to a broad range of scientific and cultural communities by the establishment of a physical bank of chronological standards that are analogous to those standards set by the NBS and the U.S. Geological Survey for compositional and physical properties.

Standard Reference Data. In addition to standard reference materials, the materials scientist or engineer frequently requires access to standard reference data. Such information helps to identify an unknown material, describe a structure, calibrate an apparatus, test a theory, or draft a new standard specification. Data are defined as that subset of scientific or technical information that can be represented by numbers, graphs, models, or symbols. The term, standard reference data, implies a data set or collection that has passed some screening and evaluation by a competent body and

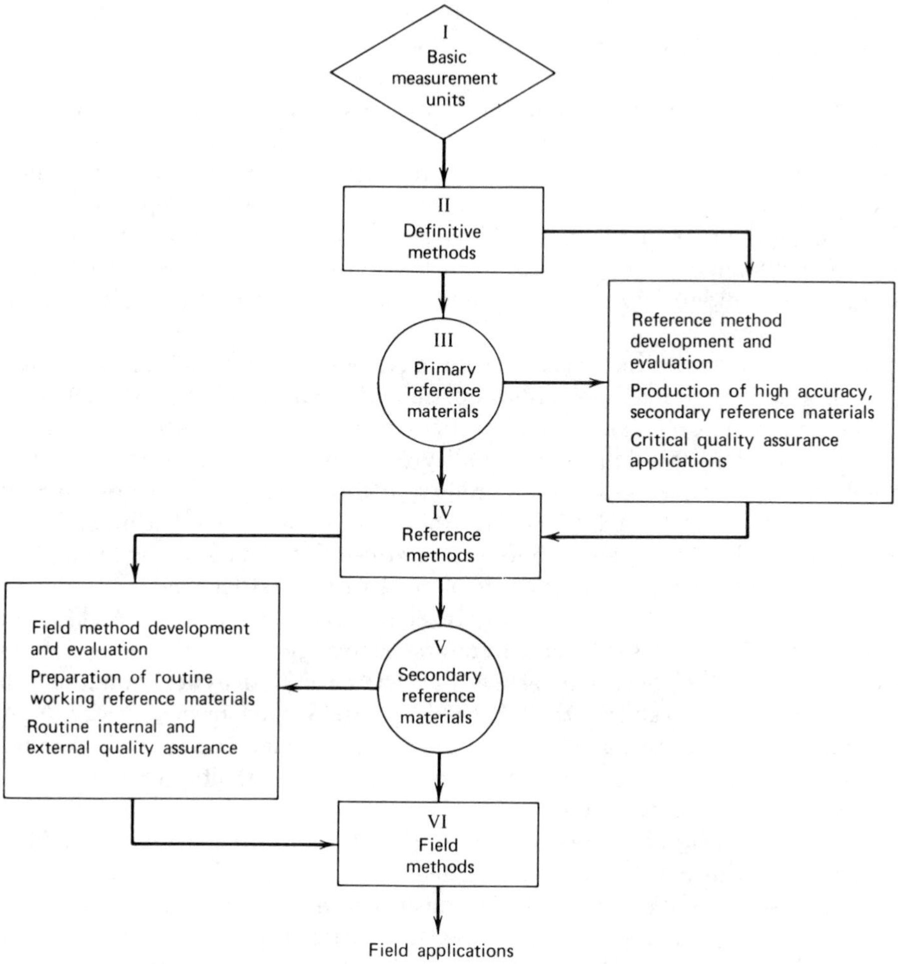

Figure 2. Measurements standards hierarchy (4).

warrants their imprimatur and promotion. Such a data set may be generated expressly for this purpose by especially careful measurements made on a standard reference material or other well-characterized material, eg, the series of standard x-ray diffraction patterns generated by the NBS. In some cases, a reference data set may not represent any specific set of real experimental observations but a recommended, consistent set of stated reliability that is synthesized from limited, fragmentary, and conflicting literature data by review, analysis, adjustment, and interpolation (8). A convenient reference summarizing and classifying data sources relevant to materials has been published (9).

Standards for Nondestructive Evaluation (NDE). Nondestructive evaluation standards are important in materials engineering in evaluating the structure, properties, and integrity of materials and fabricated products. Such standards apply to test methods, artifactual standards for test calibration, and comparative graphical or pictorial references. These standards may be used as inspection guides, to define terms describing defects, to describe and recommend test methods, for qualification and certification of individuals and laboratories working in the NDE field, and to specify materials and apparatus used in NDE testing. NDE standards have been reviewed with regard to what standards are available, what are satisfactory, what are lacking, and what need improvement (10). Other references include useful compilations of standards and specifications in NDE (11–12) (see Nondestructive testing).

Traceability. Measurements are traceable to designated standards if scientifically rigorous evidence is produced on a continuing basis to show that the measurement process is producing data for which the total measurement uncertainty is quantified relative to national or other designated standards through an unbroken chain of comparisons (13). The intent of traceability is to assure an accuracy level sufficient for the need of the product or service. Although calibration is an important factor, measurement traceability also requires consideration of measurement uncertainty that arises from random error, ie, variability within the laboratory, and from systematic error of that laboratory relative to the reference standard. Although the ultimate metrological standards are those of mass, length and time, maintained at the NBS and related to those defined by international standardizing bodies (see below), there are literally thousands of derived units, for only a fraction of which primary reference standards are maintained.

Traceability also is used by materials engineers for the identification of the origin of a material. This attribution often is necessary where knowledge of composition, structure, or processing history is inadequate to assure the properties required in service. Thus, critical components of industrial equipment may have to be related to the particular heat of the steel that is used in the equipment apart from having to meet the specification. The geographical derivation of certain ores and minerals often is specified where analytical measurements are inadequate, and much recycled materials can be used only if traceability to the original form can be established.

Basic Standards for Chemical Technology. There are many numerical values that are standards in chemical technology; a brief review of a few basic and general ones is given below. Numerical data and definitions quoted in this section are taken from refs. 14–16 (see also Units) and are expressed in the International System of Units (SI).

Atomic Weight. The present definition of atomic weights (1961) is based on ^{12}C, which is the most abundant isotope of carbon and whose atomic weight is defined as exactly 12 (see Isotopes).

Temperature. Temperature is the measurement of the average kinetic energy, resulting from heat agitation, of the molecules of a body. The most widely used scale, ie, Celsius, uses the freezing and boiling points of water as defining points. The ice point is the temperature at which macroscopic ice crystals are in equilibrium with pure liquid water under air that is saturated with moisture at standard atmospheric pressure (101.325 kPa). One degree on the Celsius scale is 1.0% of the range between the melting and boiling points of water. The unit of thermodynamic temperature is the Kelvin, defined as 1/273.16 of the thermodynamic temperature of the triple point of water. The relation of the Kelvin and Celsius scales is defined by the International Practical Temperature Scale of 1968 (17) which is related to the temperature standard reference points published by ASTM (16) and others. The international temperature scale between −190 and 660°C is based on a standard platinum resistance thermometer and the following formula for the resistance R as a function of temperature t

$$\text{below } 0°\text{C} \quad R_t = R_o\,[1 + At + Bt^2 + C(t - 100)t^3]$$

$$\text{above } 0°\text{C} \quad R_t = R_o\,[1 + At + Bt^2]$$

where A, B, and C are arbitrary constants (see Temperature measurement).

Pressure. Standard atmospheric pressure is defined to be the force exerted by a column of mercury 760 mm high at 0°C. This corresponds to 0.101325 MPa or 14.695 psi. Reference or fixed points for pressure calibration exist and are analogous to the temperature standards cited above (18). These are based on phase changes or resistance jumps in selected materials. For the highest pressures, the most reliable technique is the correlation of the wavelength shift $\Delta\lambda$ with pressure of the ruby R_1 fluorescence line and is determined by simultaneous specific-volume measurements on cubic metals correlated with isothermal equations of state which are derived from shock-wave measurements (19). This calibration extends from 6–100 GPa (0.06–1 Mbar) and may be represented

$$\mathrm{P(Mbar)} = \frac{(19.04)}{5}\left\{\left[\frac{\lambda_o + \Delta\lambda}{\lambda_o}\right]^5 - 1\right\}$$

where λ_o is the wavelength measured at 100 kPa (1 bar) (see Pressure measurement).

Length. One meter is defined as exactly 1,650,763.73 wavelengths of the radiation in vacuum corresponding to the unperturbed transition between the levels $2p_{10}$ and $5d_5$ of krypton-86 (the orange-red line).

Mass. The unit of mass is the kilogram and is the mass of a particular cylinder of Pt–Ir alloy which is preserved in France by the International Bureau of Weights and Measures.

Time. The unit of time in the International System of units is the second: "The second is the duration of 9,197,631,770 periods of the radiation corresponding to the transition between the two hyperfine levels of the fundamental state of the atom of cesium-133" (20). This definition is experimentally indistinguishable from the ephemeris-second which is based on the earth's motion.

Standard Cell Potential. A very large class of chemical reactions are characterized by the transfer of protons or electrons. Substances losing electrons in a reaction are said to be oxidized, those gaining electrons are said to be reduced. Many such reactions can be carried out in a galvanic cell which forms a natural basis for the concept of the half-cell, ie, the overall cell is conceptually the sum of two half-cells, one corresponding to each electrode. The half-cell potential measures the tendency of one reaction, eg, oxidation, to proceed at its electrode; the other half-cell of the pair measures the corresponding tendency for reduction to proceed at the other electrode. Measurable cell potentials are the sum of the two half-cell potentials. Standard cell potentials refer to the tendency of reactants in their standard state to form products in their standard states. The standard conditions are 1 M concentration for solutions, 101.325 kPa (1 atm) for gases, and for solids, their most stable form at 25°C. Since half-cell potentials cannot be measured directly, numerical values are obtained by assigning the hydrogen gas–hydrogen ion half reaction the half-cell potential of zero V. Thus, by a series of comparisons referred directly or indirectly to the standard hydrogen electrode, values for the strength of a number of oxidants or reductants can be obtained (21), and standard reduction potentials can be calculated from established values (see also Electrochemical processing; Batteries).

Standard cell potentials are meaningful only when they are calibrated against an emf scale. To achieve an absolute value of emf, electrical quantities must be referred to the basic metric system of mechanical units. If the current unit A and the resistance unit Ω can be defined, then the volt may be defined by Ohm's law as the voltage drop across a resistor of one standard ohm (Ω) when passing one standard ampere (A) of current. In the ohm measurement, a resistance is compared to the reactance of an inductor or capacitor at a known frequency. This reactance is calculated from the measured dimensions and can be expressed in terms of the meter and second. The ampere determination measures the force between two interacting coils while they carry the test current. The force between the coils is opposed by the force of gravity acting on a known mass; hence, the ampere can be defined in terms of the meter, kilogram, and second. Such a means of establishing a reference voltage is inconvenient for frequent use and reference is made to a previously calibrated standard cell.

Ideally, a standard cell is constructed simply and is characterized by a high constancy of emf, a low temperature coefficient of emf, and an emf close to one volt. The Weston cell, which uses a standard cadmium sulfate electrolyte and electrodes of cadmium amalgam and a paste of mercury and mercurous sulfate, essentially meets these conditions. The voltage of the cell is 1.0183 V at 20°C. The a-c Josephson effect, which relates the frequency of a superconducting oscillator to the potential difference between two superconducting components, is used by the NBS to maintain the unit of emf, but the definition of the volt remains the Ω/A derivation described above (see Superconducting materials).

Concentration. The basic unit of concentration in chemistry is the mole which is the amount of substance that contains as many entities, eg, atoms, molecules, ions, electrons, protons, etc, as there are atoms in 12 g of ^{12}C, ie, Avogadro's number $N_A = 6.022045 \times 10^{23}$. Solution concentrations are expressed on either a weight or volume basis. Molality is the concentration of a solution in terms of the number of moles of solute per kilogram of solvent. Molarity is the concentration of a solution in terms of the number of moles of solute per liter of solution.

A particular concentration measure of acidity of aqueous solutions is pH which, usually, is regarded as the common logarithm of the reciprocal of the hydrogen-ion concentration (qv). More precisely, the potential difference of the hydrogen electrode in normal acid and in normal alkali solution (−0.828 V at 25°C) is divided into 14 equal parts or pH units; each pH unit is 0.0591 V. Operationally, pH is defined by pH = pH (soln) + E/K, where E is the emf of the cell:

$$H_2|\text{solution of unknown pH}\|\text{saturated KCl}\|\text{solution of known pH}|H_2$$

and $K = 2.303\ RT/F$, where R is the gas constant, 8.314 J/(mol/K) (1.987 cal/(mol·K)), T is the absolute temperature, and F is the value of the Faraday, 9.64845×10^4 C/mol. pH usually is equated to the negative logarithm of the hydrogen-ion activity, although there are differences between these two quantities outside the pH range 4.0–9.2:

$$-\log q_{H^+}m_{H^+} = \text{pH} + 0.014\ (\text{pH} - 9.2) \text{ for pH} > 9.2$$

$$-\log q_{H^+}m_{H^+} = \text{pH} + 0.009\ (4.0 - \text{pH}) \text{ for pH} < 4.0$$

Energy. The SI unit of energy is the joule which is the work done when the point of application of a force of one newton is displaced a distance of one meter in the direction of the force. The newton is that force which, when applied to a body having a mass of one kilogram, accelerates a body one meter per second squared. The calorie is the quantity of heat absorption of water per gram per degree Celsius at 15°C and it is equal to 4.184 J (see Units).

Specifications

Objectives and Types. A specification establishes assurance of the fitness of a material, product, process, or service for use. Such fitness usually encompasses safety and efficiency in use as well as technical performance. Material specifications may be classified as to whether they are applied to the material, the process by which it is made, or the performance or use that is expected of it. Product or design specifications are not relevant to materials. Within a company, the specification is the means by which the engineering function conveys to the purchasing function what requirements it has for the material to be supplied to manufacturing. It has its greatest utility prior to and at the time of purchase. Yet a properly written and dated specification with accompanying certificates of test, heat or lot numbers, vendor identification, and other details pertinent to the actual material procurement constitutes an important archival document. Material specification records provide information regarding a proven successful material that can be used in a new product. Such records also are useful in the rebuilding of components and as defense evidence in a liability suit.

Content. Although formats of materials specifications may vary according to the need, the principal elements are title, statement of scope, requirements, quality assurance provisions, applicable reference documents, preparations for delivery, notes, and definitions. The scope statement comprises a brief description of the material, possibly its intended area of application, and categorization of the material by type, subclass, and quality grade. Requirements may include chemical composition, physical properties, processing history, dimensions and tolerances, and/or finish. Quality assurance factors are test methods and equipment including their precision; accuracy and repeatability; sampling procedures; inspection procedures, ie, acceptance and rejection criteria; and test certification. Reference documents may include citation of well-established specifications, codes and standards, definitions and abbreviations, drawings, tolerance tables, and test methods. Preparations for delivery are the instructions for packing, marking, shipping mode, and unit quantity of material in the shipment. The notes section is intended for explanations, safety precautions, and other details not covered elsewhere. Definitions are specifications of terms in the document which differ from common usage.

Strategy and Implementation. Great reliance used to be placed on compositional specifications for materials, and improvements in materials control were sought by increasing the number of elements specified and by decreasing the allowable latitude, eg, maximum, minimum, or range, in their concentration. However, the approach is fallible: the purchaser assumes enough knowledge about materials behavior to completely and unerringly associate the needed properties with composition; analyses must be made by purchaser, vendor, or both for each element specified; and the purchaser bears responsibility for materials failures when compositional requirements have been met. Property requirements alone or in combination with a less exacting compositional specification usually is a more effective approach. For example, the engineer may specify a certain class of low alloy steel and call for a particular hardenability but leave the vendor considerable latitude in determining composition to achieve the desired result.

The most effective specification is that which accomplishes the desired result with the fewest requirements. Properties and performance should be emphasized rather than how the objectives are to be achieved. Excessive demonstration of erudition on the part of the writer or failure to recognize the usually considerable processing expertise held by the vendor results in a lengthy and overly detailed document that generally is counterproductive. Redundancy may lead to technical inconsistency. A requirement that cannot be assessed by a prescribed test method or quantitative inspection technique never should be included in the specifications. Wherever possible, tests should be easy to perform and highly correlatable with service performance. Tests that indicate service life are especially useful. Standard test references, eg, ASTM methods, are the most desirable, and those that are needed should be selected carefully and the numbers of such references should be minimized. To eliminate unnecessary review activity by the would-be complier, the description of a standard test should not be paraphrased or condensed unless the original test is referenced.

Effective specification control often can be established other than through requirements placed on the end-use material, ie, the specification may bear upon the raw materials, the process used to produce the material, or ancillary materials used in its processing. Related but supplementary techniques are approved vendor lists, accredited testing laboratories, and preproduction acceptance tests.

Economic Aspects

The costs and benefits associated with standardization are determined by direct and indirect effects. A proper assessment depends on having suitable base-line data with which to make a comparison. Several surveys have shown typical dollar returns for the investment in standardization in the range of 5:1–8:1 with occasional claims made for a ratio as high as 50:1.

Savings include reduced costs of materials and parts procurement; savings in production and drafting practice; reduction in engineering time, eg, design, testing, quality control, and documentation; and reduction in maintenance, field service, and in-warranty repairs. It is curious fact that in most companies a very small number of individuals are authorized to write checks on the company's funds, but a large number of people are permitted to specify materials, parts, processes, services, etc, which just as definitely commit company resources. Furthermore, actions involving specification and standard setting frequently lack adequate control and may not be monitored regularly. Thus, awareness, appreciation, and involvement of top management in any industrial standardization program are essential to its success.

The DOD estimates conservatively that materials and process specifications represent almost 1% of the total hardware-acquisition costs. The operation of a single ASTM committee dealing with engine coolants has been estimated at over $150,000/yr (2). Costs of generation of a single company specification range from a few hundred to several thousand dollars. The total U.S. cost for material and process specifications is greater than $\$3 \times 10^8$. Since these costs can be so great it is imperative to ensure that monies are not spent unwisely in the specification and standardization field. Although there are justifiable instances for "specials" or documents intended to fill the needs of an individual company or other institution, savings will usually be realized by adopting a standard already established by an organization at a higher hierarchical level—trade association, national, international.

The ideal specification regards only those properties required to assure satisfactory performance in the intended application and properties that are quantitative and measurable in a defined test. Excessively stringent requirements not only involve direct costs for compliance and test verification but constitute indirect costs by restricting the sources of the material. Reducing the margin between the specification and the production target increases the risk that an acceptable product is unjustly rejected because the test procedure gives results that vary from laboratory to laboratory. A particularly effective approach is to recognize within a specification or related set of specifications the different levels of quality or reliability required in different applications. Thus, *Military Handbook V* recognizes class A, class B, and class S design allowables where, on the A basis $\geqslant$99% of values are above the designated level with a 95% confidence; on the B basis, $\geqslant$90% are above with a 95% confidence; and, on the S basis, a value is expected which exceeds the specified minimum (22).

From the customer's point of view, there is an optimal level of standardization. Increased standardization lowers costs but restricts choice. Furthermore, if a single minimal-performance product standard is rigorously invoked in an industry, competition in a free market ultimately may lead the manufacturer of a superior product to save costs by lowering his product quality to the level of the standard, thus denying other values to the customer. Again, excessive standardization, especially as applied to design or how the product performance is to be achieved, effectively can limit technological innovation.

Legal Aspects

The increasing incidence of class action suits over faulty performance, the trend toward personal accountability and liability, and the increasing role of consumerism have affected standardization. Improvement in the technical quality of standards, the involvement of all of the possible stakeholders in their creation, and their endorsement by larger standardizing bodies help to minimize the legal exposure of the individual engineer or of his company. A particular embodiment of these attitudes is the certification label, ie, a symbol or mark on the product indicating that it has been produced according to the standards of a particular organization. The Underwriters Laboratories seal on electrical equipment is a familiar indication that the safety features of the product in question have met the exacting standards of that group. Similarly, the symbol of the International Wool Secretariat on a fabric attests to the fiber content and quality of that material, and the API monogram on piping, fittings, chain, motor-oil cans, and other products carries analogous significance.

Antitrust laws sometimes have been invoked in opposition to the collaborative activities of individual companies or private associations, eg, ASTM, in the development of specifications and standards. Although such activities should not constitute restraint of trade, they must be conducted so that the charge can be refuted. Therefore, all features of due process proceedings must be observed. Actions aimed at strengthening the voluntary standards system have begun (23). A recommended national standards policy has been generated by an advisory committee that was initiated by, but is independent of, the ANSI (24). The Federal Office of Management and Budget issued a circular establishing a uniform policy for federal participation and the use of voluntary standards (25). In general, the circular calls for federal agency participation in the development, production, and coordination of voluntary standards and encourages the use, whenever possible, of applicable voluntary standards in Federal procurement.

Education

Since few universities or engineering schools incorporate much explicit treatment of the subject of specifications and standards in their curricula, resort must be made to seminars, workshops, and short courses sponsored by professional societies and trade associations. Far too little is being done in this area. The needed training encompasses sources of information in the field, how to prepare specifications and standards, how to tailor requirements for cost effectiveness, and the cross-referencing and correlation of specifications and standards.

Outlook

International trade is increasing rapidly in volume and in its significance to individual national economies and, thus, will require more extensive adoption of international standards and, in the immediate future, more extensive cross-referencing of equivalent national specifications. International trade also is increasing in complexity as well as volume. No longer is international trade comprised principally of raw materials sold by undeveloped countries to industrialized countries in exchange for manufactured products. Consider for example, the U.S.-designed car that is equipped with a German engine and French tires, built in part from Japanese steel and Dutch plastics, and the needs that the composite implies for materials standardization. Further, there is always the hazard of the intentional or unintentional use of standards as technical barriers to trade. The recent international treaty, the General Agreement on Tariffs and Trade (GATT Code) is intended primarily to prevent just such a possibility. The implications of this Code for the predominantly voluntary standards programs used in the United States are reviewed in ref. 26.

New technology, eg, nuclear energy, introduces demands for new and better standards and specifications. Extension of temperature and pressure capabilities in the laboratory and the factory demand new accepted standards of calibration. Pressure equipment in the GPa range and tokamak nuclear fusion apparatus with 10^6 °C operating temperatures are being built but the state of the art of standards in these fields is far behind such values (see Fusion energy). Microminiaturization of the active components of electronic equipment requires updated standards for smoothness and dimensional and compositional measurement and control within tens of nanometers. Standards and specifications work may soon affect the biomaterials field, eg, regarding laboratory-created microorganisms (see Genetic engineering). The current and future impact of the computer on standardization is realized through enormously increased automation with computer-aided design and computer-aided manufacturing (CAD/CAM) and the integration of testing with automated manufacturing.

New environments, a selection of which is presented in Table 1, and in which all the usual engineering functions must be performed, also pose problems and opportunities for material standards. Sensors must measure the attributes of the new environment, construction materials must withstand new exposure regimes, and new performance criteria must be specified.

As the SI system is adopted increasingly worldwide, the elimination of confusion from contemporary but distinctly different French, German, Italian, etc, metric units will be realized. Only Brunei, Burma, Yemen, and the United States have not formally

Table 1. Characteristics of Some New Environments[a]

Space	Ocean	Human body	Nuclear reactor	Laboratory
extreme vacuum	high pressure	moist	high temperature	high temperature
radiation	nearly constant	complex and	neutron flux	high pressure
nonpenetrating	temperature	diverse electro-	reactive coolants	high magnetic fields
penetrating	saline water	chemistry of	radioactive	plasmas
temperature	silt and colloidal	various body	sources	
ascent	suspensions	fluids	high thermal flux	
reentry	marine life	complex flexural	inaccessibility	
ambient	mechanical	behavior		
lack of normal	instability	multicomponent		
gravitational	waves	composite,		
field	tides	highly damped		
micrometeorites	currents	in the		
		mechanical		
long-term missions	opaque to EM[b]	and electrical		
inaccessibility	radiation	sense		
	inaccessibility	multielement		
		constitution of		
		body fluids		
		gases		
		wastes		
		nutrients		
		antibodies		
		hormones		
		enzymes		
		reactive to foreign		
		materials		
		inaccessibility		

[a] Ref. 27.
[b] EM = electromagnetic.

adopted the SI as the predominant national system of metrology. The participation of the United States in the metrication movement has been favored by the passage of the Metric Acts of 1866 and 1975 and the subsequent establishment of the American National Metric Council (private) and the U.S. Metric Board (public) to plan, coordinate, monitor, and encourage the conversion process (16) (see Units and conversion factors).

Changing environments of civilization introduces the increased requirement for quality and reliability in all products and especially in those of high dollar value and in components of highly integrated technological systems. Increasing concern over the environment and safety issues has led to new standards for exposure of organisms to certain materials or to noise or electromagnetic radiation (see Industrial hygiene and toxicology; Noise pollution). The decreasing availability of natural resources forces industry to make use of leaner ores and more frugal applications of materials in short supply; both will result in new analytical standards and compositional specifications. The use of specifications in coping with problems of residual and additive elements in both virgin and recycled materials has been reviewed (28) (see Recycling).

Sources

There are many hundreds of standards-making bodies in the United States. These comprise branches of state and Federal government, trade associations, professional and technical societies, consumer groups, and institutions in the safety and insurance fields. The products of their efforts are heterogeneous, reflecting parochial concerns and different ways of standards development. However, by evolution, blending, and accreditation by higher level bodies, many standards originally developed for private purposes eventually become *de facto*, if not official, national standards. Individuals seeking access to standards and specifications are referred to the directories listed in refs. 29–30. Selected references, principally from these two sources, that are especially relevant to chemistry and chemical technology, are listed below (see also Information retrieval).

Equipment and Instrumentation Standards

Instrument Society of America
400 Stanwix Street
Pittsburgh, Pa. 15222

Standards and Practices for Instrumentation 5th ed., 1977. Instrumentation standards and recommended practices abstracted from those of 19 societies, the U.S. Government, the Canadian Standards Association, and the British Standards Institute. Covers control instruments, including rotameters, annunciators, transducers, thermocouples, flowmeters, and pneumatic systems.

American Institute of Chemical Engineers
345 East 47th Street
New York, N.Y. 10017

Standard testing procedures for plate distillation columns, evaporators, solids mixing equipment, mixing equipment, centrifugal pumps, dryers, absorbers, heat exchangers, etc.

Scientific Apparatus Makers Association
1140 Connecticut Ave., NW
Washington, D.C. 20036

Standards for analytical instruments, laboratory apparatus, measurement and test instruments, nuclear instruments, optical instruments, process measurement and control, and scientific laboratory furniture and equipment.

General Sources

American National Standards Institute (ANSI)
1430 Broadway
New York, N.Y. 10018

ANSI, previously the American Standards Association and the United States of America Standards Institute, is the coordinator of the U.S. federated national standards system and acts by: assisting participants in the voluntary system to reach agreement on standards needs and priorities, arranging for competent organizations to undertake standards development work, providing fair and effective procedures for standards development, and resolving conflicts and preventing duplication of effort.

Most of the standards-writing organizations in the United States are members of ANSI and submit the standards that they develop to the Institute for verification of evidence of consensus and approval as American National Standards. There are ca 11,000 ANSI-approved standards, and they cover all types of materials from abrasives to zirconium as well as virtually every other field and discipline. Presently, ANSI adopts the standard number of the developing organization, eg, ASTM, unless they (ANSI) have sponsored the standards coordinating activity. ANSI also manages and coordinates participation of the U.S. voluntary-standards community in the work of nongovernmental international standards organizations and serves as a clearinghouse and information center for American National Standards and international standards (31).

MTS Systems Corp.
Box 24012
Minneapolis, Minn. 55424

Standards Cross-Reference List, 2nd ed., 1977 includes standards issued by an agency but adopted and renumbered or redesignated by another one; compiled and cross-referenced to aid in their location and identification.

The American Society for Testing and Materials (ASTM)
1916 Race Street
Philadelphia, Pa. 19103

The ASTM *1980 Annual Book of ASTM Standards* comprises over 48,000 pages and contains all currently formally approved (ca 6000) ASTM standard specifications, test methods, classifications, definitions, practices, and related materials, eg, proposals. These are arranged in 48 parts as follows:

Part 1—Steel piping, tubing, and fittings.

Part 2—Ferrous castings, ferroalloys.

Part 3—Steel: plate, sheet, strip, and wire, metallic coated products, fences.

Part 4—Steel, structural, reinforcing, pressure vessel, railway, fasteners.

Part 5—Steel-bars, forgings, bearings, chain, springs.

Part 6—Copper and copper alloys.

Part 7—Die-cast metals: aluminum and magnesium alloys.

Part 8—Nonferrous metals: nickel, lead, tin alloys; precious, primary, and reactive metals.

Part 9—Electrodeposited coatings, metal powders, sintered P/M structural parts.

Part 10—Metals: physical, mechanical, corrosion testing.

Part 11—Metallography, nondestructive testing.

Part 12—Chemical analysis of metals and metal-bearing ores.

Part 13—Cement, lime, ceilings and walls, manual of cement testing.

Part 14—Concrete and mineral aggregates, manual of concrete testing.

Part 15—Road: paving bituminous materials, traveled surface characteristics.

Part 16—Chemical-resistant materials: vitrified clay and concrete, asbestos, cement products, mortars, masonry.

Part 17—Refractories, glass, ceramic materials, carbon and graphite products.

Part 18—Thermal insulation, building seals and sealants, fire tests, building constructions, environmental acoustics.

Part 19—Soil and rock: building stones, peats.

Part 20—Paper: packaging, business-copy products.

Part 21—Cellulose, leather, flexible barrier materials.

Part 22—Wood, adhesives.

Part 23—Petroleum products and lubricants (I) D 56–D 1660.

Part 24—Petroleum products and lubricants (II) D 1661–D 2896.

Part 25—Petroleum products and lubricants (III) D 2891—latest; aerospace materials.

Part 26—Gaseous fuels, coal and coke, atmospheric analysis.

Part 27—Paint: tests for formulated products and applied coatings.

Part 28—Paint: pigments, resins, and polymers.

Part 29—Paint: fatty oils and acids, solvents, miscellaneous, aromatic hydrocarbons, naval stores.

Part 30—Soap, coolants, polishes, halogenated organic solvents, activated carbon, industrial chemicals.

Part 31—Water.

Part 32—Textiles: yarns, fabrics, and general test methods.

Part 33—Textiles: fibers, zippers.

Part 34—Plastic pipe and building products.

Part 35—Plastics: general test methods, nomenclature.

Part 36—Plastics: materials, film, reinforced and cellular plastics, high modulus fibers and composites.

Part 37—Rubber: natural and synthetic, general test methods, carbon black.

Part 38—Rubber products: industrial specifications and related test methods, gaskets, tires.

Part 39—Electrical insulation: test methods, solids, and solidifying fluids.

Part 40—Electrical insulation: specifications, test methods for solids, liquids, and gases, protective equipment for liquids and gases.

Part 41—General test methods, nonmetal: laboratory apparatus, statistical methods, space simulation, durability of nonmetallic materials.

Part 42—Analytical methods: spectroscopy, chromatography, computerized systems.

Part 43—Electronics.

Part 44—Magnetic properties; metallic materials for thermostats, electrical resistance heating, and contacts; temperature measurement; illuminating standards.

Part 45—Nuclear standards.

Part 46—End-use and consumer products.

Part 47—Test methods for rating motor, diesel, and aviation fuels.

Part 48—Index: subject index, numeric list.

General Services Administration
Federal Supply Service
18th and F Streets
Washington, D.C. 20406
Publishes *Index of Federal Specifications and Standards, 41CFR 101-29.1.*

Defense Supply Agency

Publishes *Department of Defense Index of Specifications and Standards*, a monthly with annual accumulations; available from Superintendent of Documents, GPO, Washington, D.C. 20402.

Global Engineering Documentation Services, Inc.
3301 W. MacArthur Boulevard
Santa Ana, Calif. 92704
An information broker, not an issuer of standards. The world's largest library of government, industry, and technical society specifications and standards, including obsolete documents dating from 1946. Publishes an annual *Directory of Engineering Documentation Sources*.

National Bureau of Standards (NBS)
Standards Information & Analysis Section
Standards Information Service (SIS)
Bldg. 225, Room B162
Washington, D.C. 20234
Maintains a reference collection on standardization, engineering standards, specifications, test methods, recommended practices, and codes obtained from U.S., foreign, and international standards organizations. Publishes various indexes and directories, including: (*1*) *An Index of U.S. Voluntary Engineering Standards*, NBS Special Publication 329, 1971, Supplement 1, 1972, and Supplement 2, 1975; (*2*) *World Index of Plastics Standards*, NBS Special Publication 352, 1971; (*3*) *Tabulation of Voluntary Standards and Certification Programs for Consumer Products*, NBS Technical Note 948, 1977; (*4*) *An Index of State Specifications and Standards*, NBS Special Publication 375, 1973; (*5*) *Directory of United States Standardization Activities*, NBS Special Publication 417, 1975; (*6*) *Index of International Standards*, NBS Special Publication 390, 1974; (*7*) *Index of U.S. Nuclear Standards*, 1977; and (*8*) *Directory of Standards Laboratories*, 1971.

National Standards Association, Inc.
5161 River Road
Washington, D.C. 20016

Visual Search Microfilm Files (VSMF)
Information Handling Services
15 Inverness Way East
Engelwood, Col. 80150
VSMF carries government specifications, ASTM, AMS, and many other specifications and standards. Copies of these may be obtained on an individual basis or broad categories of this service may be obtained on a subscription basis.

Journals

ANSI Reporter and Standards Action
American National Standards Institute
The biweekly *ANSI Reporter* provides news of policy-level actions on standardization taken by ANSI, the international organizations to which it belongs, and

the government. *Standards Action*, also biweekly, lists for public review and comment standards proposed for ANSI approval. It also reports on final approval actions on standards, newly published American National Standards, and proposed actions on national and international technical work. These two publications replace *The Magazine of Standards* which ANSI, formerly The American Standards Association, discontinued in 1971.

ASTM Standardization News (formerly *Materials Research and Standards* and, earlier, *ASTM Bulletin*)
American Society for Testing and Materials
A monthly bulletin which covers ASTM projects, national and international activities affecting ASTM, reports of new relevant technology, and ASTM letter ballots on proposed standards.

Journal of the American Society of Safety Engineers
American Society of Safety Engineers
850 Busse Hwy.
Park Ridge, Ill. 60068
A monthly that reviews safety standards.

Journal of Research of National Bureau of Standards
The journal is published in four parts: (*1*) physics and chemistry, (*2*) mathematics and mathematical physics, (*3*) engineering and instrumentation, and (*4*) radio science.

Journal of Physical and Chemical Reference Data
American Chemical Society
1155 16th St., NW
Washington, D.C. 20036: quarterly

Journal of Testing and Evaluation
American Society for Testing and Materials
A bimonthly in which data derived from the testing and evaluation of materials, products, systems, and services of interest to the practicing engineer are presented. New techniques, new information on existing methods, and new data are emphasized. It aims to provide the basis for new and improved standard methods and to stimulate new ideas in testing.

Metrologia
International Committee of Weights and Measures (CIPM)
Pavillon de Breteuil
Parc de St. Cloud, France
Includes articles on scientific metrology worldwide, improvements in measuring techniques and standards, definitions of units, and the activities of various bodies created by the International Metric Convention.

Standards Engineering
Standards Engineering Society
6700 Penn Avenue South
Minneapolis, Minn. 55423

A bimonthly in which general news and technical articles dealing with all aspects of standards and U.S. and foreign articles on standard materials and calibration and measurement standards are presented.

Materials

Biochemical Compounds

National Research Council
Committee on Biological Chemistry
National Academy of Science
Washington, D.C. 20418
Specifications and Criteria for Biochemical Compounds, 2nd ed., 1967.

Carbides

Cemented Carbide Producers Association
712 Lakewood Center North
Cleveland, Ohio 44107
Standards developed by Cemented Carbide Producers Association, ie, standard shapes, sizes, grades, and designation and defect classification.

Castings

Investment Casting Institute
8521 Clover Meadow
Dallas, Texas 75243

American Die Casting Institute
2340 Des Plaines Ave.
Des Plaines, Ill. 60060

Steel Founders Society of America
20611 Center Ridge Rd.
Cast Metals Federation Bldg.
Rocky River, Ohio 44116

Cement

Cement Statistical and Technical Association
Malmo, Sweden
Review of the Portland Cement Standards of the World (1961). Describes standards in the 42 countries that have issued their own national specifications. Chemical, physical, and strength requirements of each country are reviewed.

Chemicals

Chemical Manufacturers' Association
1825 Connecticut Ave., NW
Washington, D.C. 20009
Manual of Standard and Recommended Practice for chemicals, containers, tank car unloading and related procedures.

Chemical Specialties Manufacturers Association
1001 Connecticut Ave., NW
Washington, D.C. 20036
Standard Reference Testing Materials for insecticides, cleaning products, sanitizers, brake fluids, corrosion inhibitors, antifreezes, polishes, and floor waxes.

Color Association of the United States
24 East 39th Street
New York, N.Y. 10016
Color standards for fabrics, paints, wallpaper, plastics, floor coverings, automotive and aeronautical materials, china, chemicals, dyestuffs, cosmetics, etc.

Friction Materials
Friction Materials Standards Institute
E210, Rte. 4
Paramus, N.J. 07652

Leather
Tanners' Council of America
411 Fifth Avenue
New York, N.Y. 10016
Standards for color, hide trim, leather weight or thickness.

American Leather Chemists' Association
c/o University of Cincinnati
Cincinnati, Ohio 45221
Chemical and physical test methods for leather.

Metals and Alloys
Aluminum Association
818 Connecticut Ave., NW
Washington, D.C. 20006
Standards for wrought and cast aluminum and aluminum alloy products, including composition, temper designation, dimensional tolerance, etc.

Society of Automotive Engineers (SAE)
400 Commonwealth Drive
Warrendale, Pa. 15096
SAE Handbook. An annual compilation of more than 500 SAE standards, recommended practices, and information reports on ferrous and nonferrous metals, nonmetallic materials, threads, fasteners, common parts, electrical equipment and lighting for motor vehicles and farm equipment, power-plant components and accessories, passenger cars, trucks, buses, tractor and earth-moving equipment, and marine equipment.

AMS Index. A listing of more than 1000 SAE Aerospace Material Specifications (AMS) on tolerances; quality control and process; nonmetallics; aluminum, magnesium, copper, titanium, and miscellaneous nonferrous alloys; wrought carbon steels; special purpose ferrous alloys; wrought low alloy steels; corrosion- and heat-resistant steels and alloys; cast iron and low alloy steels; accessories, fabricated parts, and assemblies; special property materials; refractory and reactive materials.

Copper Development Association
405 Lexington Avenue
New York, N.Y. 10017
Standards for wrought and cast copper and copper alloy products; a standards handbook is published with tolerances, alloy data, terminology, engineering data, processing characteristics, sources and specifications cross-index for 6 coppers and 87 copper-based alloys that are recognized as standards.

Tin Research Institute
483 West 6th Avenue
Columbus, Ohio 43201

Zinc Institute
292 Madison Avenue
New York, N.Y. 10017

Lead Industries Association
292 Madison Avenue
New York, N.Y. 10017

American Iron & Steel Institute
1000 16th Street, NW
Washington, D.C. 20036
Standards for steel compositions, steel products manufacturing tolerances, inspection methods, etc.

Ferroalloys Association
1612 K Street, NW
Washington, D.C. 20006

Metal Powder Industries Federation
PO Box 2054
Princeton, N.J. 08540

Silver Institute
1001 Connecticut Ave., NW
Washington, D.C. 20036

International Magnesium Association
c/o Bell Publicom
1406 Third National Bldg.
Dayton, Ohio 45402

Paper

Technical Association of the Pulp and Paper Industry
One Dunwoody Park
Atlanta, Ga. 30338

Tappi Standards and *Yearbook* cover all aspects of pulp and paper testing and associated standards.

American Paper Institute
260 Madison Avenue
New York, N.Y. 10016
Physical standards, sizes, gauges, definitions of paper and paperboard.

Petroleum Products

Institute of Petroleum
61 New Cavendish Street
London, W1, England
Annually publishes *Institute of Petroleum Standards for Petroleum and its Products*.

American Petroleum Institute
1801 K Street, NW
Washington, D.C. 20006
Fosters development of standards, codes, and safe practices in petroleum industries and publishes the same in its journals and reference publications.

Steam

International Association for Properties of Steam
c/o Dr. H. White
National Bureau of Standards
Washington, D.C. 20234

Treating and Finishing

Metal Treating Materials
1300 Executive Center, Suite 115
Tallahassee, Fla. 32301

National Association of Metal Finishers
111 E. Wacker Drive
Chicago, Ill. 60601

Welding

American Welding Society
2501 N.W. Seventh Street
Miami, Fla. 33125
Codes, Standards, Specifications. A complete set of codes, standards, and specifications is published by the Society and is continuously updated. Covers fundamentals, training, inspection and control, and process and industrial applications.

Wood

American Lumber Standards Committee
20010 Century Blvd.
Germantown, Md. 20767

American Wood Preservers Bureau
2740 S. Randolph Street
Arlington, Va. 22206

National Hardwood Lumber Association
332 S. Michigan Ave.
Chicago, Ill. 60604

National Standards, Worldwide. Most countries have a national standards organization that both leads the standardization activities in that country and acts within its own country as sales agent and information center for the other national standardizing bodies. In the United States, the ANSI performs that function. The organizations for the leading industrial countries of the world are as follows:

Australia: SAA, AS, Standards Association of Australia, 80-86 Arthur Street, North Sidney NSW 2060.

Austria: ON, ONORM, Oesterreichlisches Normungsinstitut, Leopoldsgasse 4, A-1021 Wien 2.

Belgium: IBN, Institut Belge de Normalisation, 29 Avenue de la Brabanconne B-1040 Bruxelles 4.

Brazil: ABNT; NB, EB, Associacao Brasileira de Normas Tecnicas, Caixa Postal 1680, Rio de Janeiro.

Canada: CSA, Standards Council of Canada, 2000 Argentia Road, Suite 2-401, Mississauga, Ontario.

China: China Association for Standardization, PO Box 820, Beijing, People's Republic of China.

Czechoslovakia: Urad pro normalizaci a mereni, Vaclavske namesti 19, 113 47 Praha 1, Czechoslovakia.

Denmark: DS, Dansk Standardiseringsraad, Aurehjvej 12, DK-29000, Hellerup.

Finland: SFS, Suomen Standardisoimisliitto, Box 205 SF-00121 Helsinki 12.

France: AFNOR, NF, Assoc. Francaise de Normalisation, Tour Europe, Cedex 7, 92080 Paris-La Defense.

FRG: DNA, DIN, Deutsches Institut fur Normung, 4-10 Burggrafenstrasse, D-1000 Berlin 30.

India: ISI, IS, Indian Standards Institution, Manak Bhavan, 9 Bahadur Shah Zafar Marg, New Delhi 110002.

Iran: Institute of Standards and Industrial Research of Iran, Ministry of Industries and Mines, PO Box 2937, Tehran.

Ireland: IIRS, I.S., Institute for Industrial Research and Standards, Blasnevin House, Ballymun Road, Dublin-9.

Israel: SII, Standards Institution of Israel, 42 University St., Tel Aviv 69977.

Italy: UNI, Ente Nazionale Italiano de Unificazione, Piazza Armando Diaz 2, 120123 Milano.

Japan: JISC, JIS, Japanese Industrial Standards Committee, Agency of Industrial Science and Technology, Ministry of International Trade and Industry, 1-3-1 Kusumigaseki Chiyoda-Ku, Tokyo 100.

Mexico: DGN, Diraccion General de Normas, Tuxpan No. 2, Mexico 7, D.F.

Netherlands: NNI, Nederlands Normalisatie-instituut, Polakweg, 5 Rijswijk (ZH)-2280.

Poland: Polski Komitet Normalizacji, Miar i Jakosci, Ul. Elektoraina 2, 00-139 Warszawa.

Romania: Institutul Roman de Standardizare, Casuta Postala 63-87, Bucarest 1.

Spain: Instituto Nacional de Racionalizacion y Normalizacion, Aurbano 46, Madrid 10.

Sweden: SIS, Standardiseringskommission i Sverige, Tegnergatan 11, Box 3295, Stockholm S 10366.

United Kingdom: BSI, BS, British Standards Institution, 2 Park Street, London W1 A 2BS, England.

USSR: GOST, Gosudarstvennyj Komitet Standartov, Soveta Ministrov S.S.S.R., Leninsky Prospekt 9b, Moskva 11 7049.

Nuclear Standards

American Nuclear Society (ANS)
555 N. Kensington Avenue
La Grange Park, Ill. 60525
Information center on nuclear standards.

American National Standards Institute (1974)
1430 Broadway
New York, N.Y. 10018
Catalog of Nuclear Industry Standards

National Bureau of Standards
Index of U.S. Nuclear Standards
W. I. Slattery, ed.
National Bureau of Standards, Special Pub. 483 (1977)
Washington, D.C.

Safety Standards

The American Society of Mechanical Engineers (ASME)
United Engineering Center
345 East 47th Street
New York, N.Y. 10017

The ASME Boiler and Pressure Vessel Code, under the cognizance of the ASME Policy Board, Codes and Standards, considers the interdependence of design procedures, material selection, fabrication procedures, inspection, and test methods that affect the safety of boilers, pressure vessels, and nuclear-plant components, whose failures could endanger the operators or the public (see Nuclear reactors). It does not cover other aspects of these topics that affect operation, maintenance, or nonhazardous deterioration.

American Insurance Association (AIA)
85 John Street
New York, N.Y. 10038

Handbook of Industrial Safety Standards, Association of Casualty and Surety Companies, New York, 1962. Compilation of industrial safety requirements based on codes and recommendations of the ANSI, the National Fire Protection Association (now part of AIA), the ASME, and several government agencies.

National Fire Protection Association
470 Atlantic Avenue
Boston, Mass. 02210
National Fire Codes. 1980 ed., issued in 16 volumes: Volume 13 is devoted exclusively to hazardous chemicals, but most other volumes have some coverage of material hazards, use of materials in fire prevention or extinguishing, hazards in chemical processing, etc. More than 200 standards are described.

American Public Health Association
1015 18th St., NW
Washington, D.C. 20036
Standard Methods for Examination of Water and Wastewater, 14th ed., 1975; *Methods of Air Sampling and Analysis*, 2nd ed., 1977.

National Safety Council
444 N. Michigan Avenue
Chicago, Ill. 60611
Industrial safety data sheets on materials and materials handling and safe operation of equipment and processes.

Underwriters Laboratories
207 East Ohio Street
Chicago, Ill. 60611
Standards for Safety is a list of more than 200 standards that provide specifications and requirements for construction and performance under test and in actual use of a broad range of electrical apparatus and equipment, including household appliances; fire-extinguishing and fire protection devices and equipment; and many other nongenerally classifiable items, eg, ladders, sweeping compounds, waste cans, and roof jacks for trailer coaches.

Safety Standards
U.S. Department of Labor
GPO
Washington, D.C. 20402
Industrial safety hazards.

American Conference of Governmental Industrial Hygienists
P.O. Box 1937
Cincinnati, Ohio 45201
Practices, analytical methods, guides to codes and/or regulations, threshold limit values.

Factory Mutual Engineering Corporation
1151 Boston-Providence Turnpike
Norwood, Mass. 02062

Standards for safety equipment, safeguards for flammable liquids, gases, dusts, industrial ovens, dryers, and for protection of buildings from wind and other natural hazards.

Code of Federal Regulations
Title 49, Transportation, Parts 100 to 199
Superintendent of Documents
GPO
Washington, D.C. 20402
Safety regulations related to transportation of hazardous materials and pipeline safety.

Code of Federal Regulations
Title 29, Occupational Safety and Health
Superintendent of Documents
GPO
Washington, D.C. 20402
Safety regulations and standards issued by OSHA.

Code of Federal Regulations
Title 40, Environmental Protection Administration
Superintendent of Documents
GPO
Washington, D.C. 20402
Safety regulations and standards issued by the EPA.

Code of Federal Regulations
Title 21, Radiological Health
Superintendent of Documents
GPO
Washington, D.C. 20402

Weights and Measures

National Conference on Weights and Measures
c/o National Bureau of Standards
Washington, D.C. 20234

National Conference on Standards Laboratories
c/o National Bureau of Standards
Boulder, Col. 80303

U.S. Metric Association
Sugarloaf Star Rte.
Boulder, Col. 80302

U.S. Metric Board
1815 N. Lynn Street
Arlington, Va. 22209

American National Metric Council
1625 Massachusetts Ave.
Washington, D.C. 20036

Metrology and Fundamental Constants
A. F. Milone and P. Giacomo, eds.
North Holland Publishing Co.
Amsterdam, 1980

Proceedings of an international course that was organized to review comprehensively metrology and to illustrate links between metrology and the fundamental constants. Status of research is presented and future work and priorities are outlined.

International Bureau of Weights and Measures
Pavillion de Breteuil
F-92310
Sevres, France

International Standardization. International standardization began formally in 1904 with the formation of the International Electrotechnical Commission (IEC) and involves the national committees of more than 40 member nations who represent their countries' interests in electrical engineering, electronics, and nuclear energy. In 1947, the International Organization for Standardization (ISO) was formed to review standardization activities in fields other than electrical; it is comprised of more than 80 member countries. Both organizations are autonomous but they maintain a coordinating committee to answer jurisdictional questions and both occupy the same building in Geneva. With the recent rapid growth in world trade, the activities of the IEC and ISO have increased many fold. The United States is represented in ISO by ANSI and in IEC by a U.S. National Committee that is a part of ANSI. Although the two organizations are dominant in drafting documentary standards, the influence and activities of other international organizations are substantial. Among these are IUPAC; NATO; the European Economic Community (EEC); COPANT, the Pan American Standards Commission; and CODATA, a subsidiary of the International Council of Scientific Unions. In contrast to the other groups, CODATA concentrates its attention on the evaluation of data and the methodology of compilation, presentation, manipulation, and dissemination of data in all fields of science and technology. Much of its work consists of appraisal of standard data and standards for presentation of data (32). The role of the U.S. government in international standardization activities has been examined by a special ASTM task force (33). The addresses of these organizations or their subsidiary standards groups are as follows:

International Bureau of Weights and Measures
Pavillon de Brateuil
F-92310, Sevres, France

International Electrotechnical Commission (IEC)
1, rue de Varembe, 1211 Geneve 20 Switzerland/Suisse

International Organization for Standardization (ISO)
1, rue de Varembe, CH 1211 Geneve 20 Switzerland/Suisse

North Atlantic Treaty Organization (NATO)
Military Committee, Conference of National Armament Directors
1110 Brussels, Belgium

European Economic Community (EEC)
200 rue de la Loi
1049 Brussels, Belgium

COPANT
Av. Pte, Roque Soenz Pena 501
7 Piso
OF 716, Buenos Aires, Argentina

CODATA
51, Boulevard de Montmorency
75016 Paris, France

International Union for Pure and Applied Chemistry (IUPAC)
Bank Court Way, Cowley Centre
Oxford OX4 3YF, England

Among its publications in the standards field are *Manual of Symbols and Terminology for Physico-chemical Quantities and Units*, Buttersworths, London, 1970; and *Nomenclature of Inorganic Chemistry*, Buttersworths, London, 1970.

BIBLIOGRAPHY

1. *ISO Standardization Vocabulary*, Geneva, 1977 (available from ANSI).
2. N. E. Promisel and co-workers, *Materials and Process Specifications and Standards*, NMAB Report-33, Washington, D.C., 1977.
3. D. Lebel and K. Schultz, *Technos*, 4 (Apr.–June 1975).
4. G. A. Uriano, *ASTM Standardization News* **7,** 8 (Sept. 1979).
5. G. W. Latimer, Jr. in C. T. Lynch, ed., *Handbook of Materials Science*, Vol. I, CRC Press, 1974, p. 667.
6. F. J. Flanagan, *Geochim. Cosmochim. Acta* **37,** 1189 (1973).
7. R. H. Steiger and E. Jaeger, *Planet. Sci. Lett.* **36,** 359 (1977).
8. C.-Y. Ho and Y. S. Touloukian, *Proceedings 5th Biennial International CODATA Conference*, Boulder, Col., 1977, pp. 615–627.
9. J. H. Westbrook and J. D. Desai, *Ann. Rev. Mater. Sci.* **8,** 359 (1978).
10. H. Berger, "Nondestructive Testing Standards—A Review," *Symposium Report No. STP624,* ASTM, Philadelphia, Pa., 1977.
11. *Handbook for Standardization of Nondestructive Testing Methods*, Vols. I and II, MIL HDBK-33, 1974.
12. R. E. Englehardt, "Bibliography of Standards, Specifications and Recommended Practices" in *Nondestructive Testing Information Analysis Center Handbook*, Mar. 1979.
13. B. C. Belanger, *ASTM Standardization News*, 8 (Sept. 1979).
14. E. R. Cohen and co-workers, *CODATA Bull. 11*, (Dec. 1973).
15. "Quantities, Units, Symbols, Conversion Factors, and Conversion Tables" *ISO Reference 31*, 15 sections, Geneva, 1973–1979.
16. "Standard for Metric Practice," *ASTM E 380-79*, Philadelphia, Pa., 1979.

17. *Metrologia* **12,** 7 (1976).
18. F. P. Bundy and co-workers in B. D. Timmerhaus and M. S. Barber, eds., *High Pressure Science and Technology*, Vol. 1, Plenum Press, New York, 1979, pp. 773, 805.
19. H. K. Mao, P. M. Bell, J. W. Shaver, and D. J. Steinberg, *J. Appl. Phys.* **49,** 3276 (1978).
20. *Proceedings 1967 General Conference of Weights and Measures*, International Bureau of Weights and Measures, BIPM, Parc de Saint-Cloud, France.
21. J. F. Hunsburger, "Electrochemical Series" in R. C. Weast, ed., *Handbook of Chemistry and Physics*, CRC Press, Boca Raton, Fla., 1980–1981.
22. "Metallic Materials and Elements for Aerospace Vehicle Structures," *Military Handbook V*, Department of Defense, Dec. 15, 1976.
23. *The Voluntary Standards System of the United States of America—An Appraisal by the American Society for Testing and Materials*, ASTM, Philadelphia, Pa., 1975.
24. *ASTM Standardization News* **16,** 8 (May 1978); *Fed. Regist.*, 7 (Dec. 1978).
25. *ASTM Standardization News* **8,** 21 (Mar. 1980); *ANSI Rep.* **14**(2), (1980).
26. D. L. Peyton, *Implementing the GATT Standards Code*, ANSI, New York, 1979.
27. J. H. Westbrook in A. B. Bronwell, ed., *Science and Technology in the World of the Future*, John Wiley & Sons, Inc., New York, 1970, p. 329.
28. J. H. Westbrook, *Phil. Trans. Roy. Soc. London* **A295,** 25 (1980).
29. S. J. Chumas, "Directory of United States Standardization Activities," NBS SP 417, Washington, D.C., 1975.
30. E. J. Struglia, *Standards and Specifications—Information Sources*, Gale Research, Detroit, Mich., 1965.
31. *ANSI Progress Report*, PR 35, New York, 1980.
32. S. A. Rossmassler and D. G. Watson, eds., *Data Handling for Science and Technology*, North Holland Publ. Co., Amsterdam, 1980.
33. *ASTM Standardization News* **8,** 16 (Apr. 1980).

J. H. WESTBROOK
General Electric Company

METAL FIBERS

A wide range of products, eg, textile products, paper, floor covering, insulation products, and many composites, depend on fiber manipulation technology for their economical manufacture and on the inherent fiber characteristics for their end-use properties. However, not many products have embodied the use of metal fibers. Until fairly recently, metals and alloys generally have been available at reasonable cost only in the form of wire of diameters significantly greater than natural fibers. The large diameter and the inherently high elastic modulus of metals and alloys has limited the use of many well-developed and economical fiber-manipulation processes; exceptions are woven wire products for which special looms and knitting equipment have been developed. Also, processes for producing metals and alloys with acceptable properties and dimensions have been developed slowly and are incomparable in scale and volume to those used in the synthetic organic fiber or the glass-fiber industry (see Fibers, chemical; Glass). The most significant reason for the slow development of economical processes is the difficulty of forming metal filament directly from the liquid phase. This arises from the unusually high ratio of surface tension to viscosity, which results in liquid-jet instability, and from the inability to attenuate significantly the stream outside the spinnerette as is the usual practice in the synthetic-fiber industry. Despite these shortcomings, a metal-fiber industry is emerging and will have a large impact on product and process design and performance.

Properties

Fiber. Fiber properties result from a combination of the material properties, the effect of processing the material into fiber form and, in some cases, the geometry of the final fiber. The mechanical, physical, and chemical characteristics of metals and alloys are high modulus and high strength, high density, high hardness, good electrical and thermal conductivity, can be magnetic, high temperature stability, good oxidation resistance, and corrosion resistance to varied chemical groups.

It may be that only one of the characteristics determines the choice of a particular metal or alloy for a specific application. For example, where a high specific modulus, ie, modulus-to-weight ratio, is desirable, beryllium with its very low density but high modulus may be the preferred material. In the generic material group, a metal or alloy usually can be formed and have properties that are functionally and economically acceptable for a particular application. However, for metal fibers, the choice of material is restricted since only a small number of metals and alloys are available in fiber form, particularly those with diameters <50 μm.

Most of the metal-fiber industry is directed to markets where one or more of the following properties predominate: strength–stiffness, corrosion resistance, high temperature oxidation resistance, and electrical conduction. These markets generally are satisfied by the following materials: carbon and low alloy steels, stainless steels, iron, nickel, and cobalt-based superalloys.

Property modifications can be made. For example, commercial stainless steels made according to standard melting practice contain small quantities of impurities which manifest themselves in the solidified ingot as small, nonmetallic inclusions. These inclusions are relatively unimportant in the majority of applications of the material because of their small size. However, in the case of stainless steel fiber, the diameter of the fiber can be of the same order as that of the inclusion and, consequently, the inclusion can significantly affect the strength of the fiber. The apparent strength of the fiber tends to decrease and the probability of finding an inclusion in the test sample increases with an increase in the length of the test fixture. The breaking strength of 12-μm fibers processed from five different heats of steel are plotted in Figure 1 as a function of the length of the fiber in the test machine. The gauge length effect is significantly more pronounced for heats G, H, and I as compared to heats E and F. The inclusion count for the five steels is presented in Table 1 and, as the data indicate, steels with low inclusion counts show the least sensitivity to fiber length.

Table 1. Inclusion-Count Data for Stainless Steel Heats

Mean inclusion diameters, μm	Number of inclusions for various heats				
	E	F	G	H	I
1	5	16	170	224	219
2		3	21	41	39
3		1	11	22	22
4			2	9	14
5			3	2	11
6				1	10
7					5
8				1	
9					
10					2

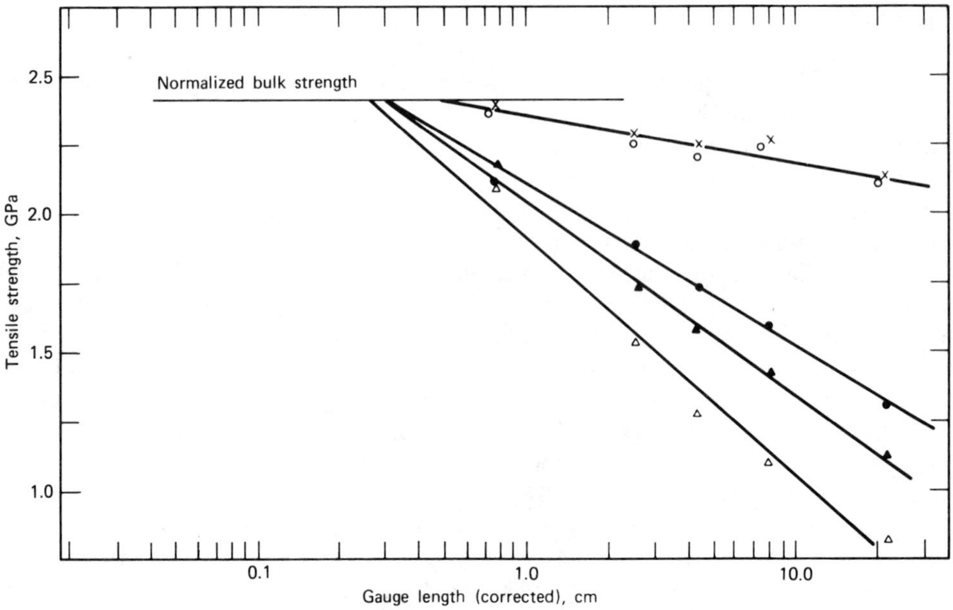

Figure 1. Gauge length dependence of the tensile strength of stainless steel filaments (diameter = 12 μm): X, heat E; ○, heat F; ●, heat G; ▲, heat H; △, heat I. To convert GPa to psi, multiply by 145,000.

The normalized bulk strength value identified in Figure 1 corresponds to the intrinsic value for the specific stainless steel as measured in sufficient diameters, ie, >50 μm, to be uninfluenced by inclusion defects. The normalization is to a base-line chemistry since each heat shows small composition variations within the type specification. Even with the ultraclean, ie, low inclusion-count steels, breaking-load discrepancies within a fiber may exist resulting from small variations in fiber cross-section area. These cross-section variations, which are inherent to the bundle-drawing process, typically may result in a coefficient of variation in breaking load of about 6–8% for 12-μm diameter fiber, but only 3–4% for 25-μm fiber. Thus, fine-filament manufacturers using, eg, the bundle-drawing process must not only carefully control the process parameters but must establish strict specifications for the starting material in order to provide a quality product. Similar requirements are necessary for other fiber-forming techniques.

In a few cases, smallness improves material properties, eg, the mechanical properties of metal whiskers (1). The preparation of metal whiskers by vapor deposition or decomposition of a gaseous compound results in a slow buildup of material at high temperature. The resulting fiber is essentially free from dislocations and, consequently, exhibits mechanical properties which approach theoretical, ie, values related to the actual breaking of metal-to-metal bonds rather than the sliding of atoms or atom planes past each other as is the usual deformation mechanism.

Physical and chemical properties of metal fibers also may tend to be modified or exaggerated because of the small fiber diameter which results in high surface-to-volume ratios. For example, in electrical conduction, direct current is carried uniformly through a fiber cross section, whereas at high frequencies, electricity is carried close to the conductor (fiber) surface; the surface current density increasing with increasing frequency. Thus, the ratio of alternating-current to direct-current resistance of a small-diameter fiber is close to unity to significantly high frequency ranges; consequently, the fine fiber is an efficient high frequency conductor compared to large diameter wire.

In terms of chemical characteristics, the high surface-to-volume ratio is advantageous where the fiber serves as a catalyst but is disadvantageous where it is desired to minimize reaction of the fiber with the environment. The latter situation becomes important in many applications of fibers, eg, filtration (qv), seals, and acoustic treatment (see Insulation, acoustic). Fibers are chosen carefully to provide long life for the product.

Assembly. The properties of any assemblage of fibers often are determined by the particular arrangement of the fibers in that structure. For example, the mechanical tensile strength of a number of fibers or filaments in a yarn depends on the degree of twist imparted to the group of fibers during the spinning of the yarn. Too low a twist results in the fibers being loaded nonuniformly with application of tension and in progressive fiber failure which produces a low yarn strength. At optimum twist, the load applied to the yarn is equally distributed on all fibers and the yarn exhibits strength which is the sum of the breaking loads of all of the fibers.

The mechanical properties of sintered-fiber, randomly oriented structures, as exemplified by elastic modulus and tensile strength, have been reported by a number of investigators and there is considerable variation even for the same alloy fiber. A composite of tensile data obtained from a variety of samples of 304-type stainless steel fiber mat is shown in Figure 2. The samples include porous structures prepared by press and sinter techniques using 12-, 25-, and 125-μm fibers with aspect ratios of 25, 62, and 187, as well as sintered air-laid web of 12-μm fiber calendered to a density of about 20% of theoretical. The curve in Figure 2 signifies the probable maximum at-

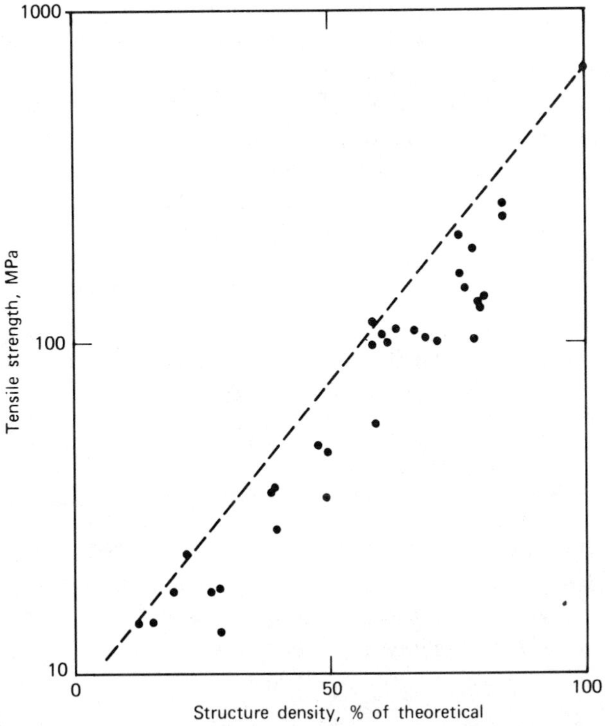

Figure 2. Tensile strength of porous, stainless steel structures as a function of structure density. To convert MPa to psi, multiply by 145.

tainable strength for 304-type stainless steel sintered-fiber structures as a function of structure density for the particular fiber types employed, ie, chopped conventionally drawn wire and bundle-drawn fiber.

Measurement of the elastic modulus of sintered-fiber porous structures, again of 304-type stainless steel, have been reported over a density range of about 40–85% (2). The data was generated from samples composed of 50- and 100-μm diameter fiber, 4 mm in length. The data is reasonably consistent with the theoretical relation between elastic modulus and structure density that was established for porous materials (3):

$$E_m = E_0\,(1 - 1.9\,P + 0.9\,P^2)$$

where E_0 = modulus of elasticity (in tension) of 100% dense material, E_m = derived structure modulus (in tension) of the porous structure, and P = fractional porosity. The compressive modulus of elasticity is approximately one order of magnitude lower than the tension data. When compressed the fibers tend to buckle at low stress levels resulting in low structure modulus.

Measurement of the thermal conductivity of various types of porous materials has been reported (4). The samples range from sintered spherical powder of OFHC copper and 304L stainless steel in the 10–30% porosity range to fibrous stainless steel mats composed of 12-μm diameter filament with a porosity of 78%. The empirical relationship that best fits the data (5) is

$$\frac{\lambda}{\lambda_0} = \frac{1 - \epsilon}{1 + 11}\,\epsilon^2$$

where λ = sample thermal conductivity, λ_0 = solid material thermal conductivity, and ϵ = fractional porosity. The relationship is reliable for porosities up to about 80%. An attempt to derive thermal conductivity values from the measurement of electrical conductivity using a modified Wiedemann-Franz relationship is not successful in the case of stainless steels of >40% porosity (6).

The complications arising from changes in flow velocity; media geometry, including pore size and tortuosity; and fluid properties often require empirical approaches in order to satisfactorily describe observed results. All media present a finite resistance to this flow of fluids. In many cases, it is the magnitude of this resistance that is a determining economic factor in the choice of a particular porous material to perform a specific function, whether it be a filtration application or a lubricating device.

The resistance to flow is expressed in terms of the pressure drop across the medium per unit of length and the flow rate per unit area. The simplest relationship involving only viscous flow is Darcy's Law:

$$\frac{\Delta p}{L} = \frac{\mu}{K}\frac{Q}{A}$$

where Δp = pressure drop (101.3 kPa = 1 atm); L = media thickness, cm; μ = viscosity, mPa·s (= cP); Q = flow rate, cm^3/s; A = cross-section area, cm^2; K = permeability, darcys. The relationship defines the common permeability unit, the darcy (units of cm^2). When fluid flow involves inertial energy losses and viscous drag, the expression is modified to account for the increased energy loss:

$$\frac{\Delta p}{L} = \frac{\mu}{K}\cdot\frac{Q}{A} + \frac{\rho}{K^1}\cdot\frac{Q^2}{A^2}$$

where ρ = fluid density, and K^1 = permeability factor associated with kinetic energy loss. Compressible fluids require additional energy-loss terms to account for the work performed in fluid compression.

Thus for each porous structure and each specific fluid, there is a unique relationship defining the permeability of the medium under a particular set of conditions; however, in many cases the relationship cannot be derived from first principles. For the same medium type but with differing porosity values, the permeability is different. A relationship for laminar-flow conditions has been derived and satisfactorily predicts the change of permeability with change in porosity for the same fiber and structure geometry (fiber orientation):

$$K \propto \frac{\epsilon^3}{(1-\epsilon)^2}$$

where ϵ = fractional porosity. The permeability of randomly oriented fiber structures can be very high when compared to other porous materials primarily because low density structures of good mechanical integrity can be fabricated. The use of fine fibers reduces viscous drag and, thus, compliments the special structure features.

The high permeability properties of randomly oriented fiber structures are particularly attractive in filtration applications. Comparative data for three nominal 20-μm rated filter media composed of sintered powder metal, woven wire cloth, and a randomly oriented fiber structure illustrate the advantage of the latter: the measured nitrogen permeability, normalized to the fiber structure data, are 0.12 cm, 0.42 cm, and 1.00 cm, respectively.

In addition to the high permeability values, the randomly oriented fiber structure exhibits another desirable property with regard to filtration, ie, the high dirt-holding capacity or on-stream life. This property can be quantified by the time taken to develop a particular cut-off Δp and can be as much as a factor of two compared to sintered powder media and a factor of 1.5 compared to the woven-wire structure.

Manufacture and Processing

Often there are two distinct elements in the fabrication of fiber products, ie, the formation of the fiber and the assembly of the fibers into a useful structure or form. The majority of commercial applications involve a large degree of secondary processing by various fiber-manipulation techniques.

Certain fiber-forming processes yield free fiber and others tend to produce a primitive fiber assembly, eg, a tow or mechanically interlocked bundles comparable to the bale of natural fiber. Thus, the commercially available form depends on the type of forming process employed. It also may depend on the business strategy of the manufacturer, who may limit the availability of the primitive form in order to attain the benefits of the value added by further in-house processing.

Fiber dimensions, as defined by the natural- and synthetic-fiber industry, tend to be restricted to diameters or equivalent diameters for noncircular cross sections of less than 250 μm. This would correspond to normal limitation for further processing by common textile-fiber manipulation techniques. This limitation also applies to metal fibers.

Fiber Forming. The principal methods that have been developed for metal-fiber forming relate to the basic starting material form. Mechanical processing incorporates processes that rely on plastic deformation to produce a fiber from a solid precursor.

In liquid-metal processing or casting, the fiber is formed directly from the liquid phase.

Mechanical Processing. The mechanical processes involve material attenuation by gross deformation or they involve the parting of material from a source, eg, a strip or rod. The first group encompasses wire-drawing techniques and solid-state extrusion, and the second group consists of cutting or scraping-type operations, ie, slitting, broaching, shaving, and grinding.

The various processes are identified in Figure 3 in relation to their source material; both current and potential commercial processes are given. For processes, eg, conventional wire drawing, the manufacturing cost is highly dependent on the diameter of the final product. Material attenuation must take place in a series of small, usually equal steps of ca 20% area reduction per step. Since drawing speed cannot be continually increased to compensate for the lower through-put per step, succeeding reducing steps involve lower and lower efficiency, ie, quantity produced per unit time. The modified wire-drawing process, ie, bundle drawing, circumvents this strong diameter dependency by drawing many wires or filaments simultaneously through a reduction die. Consequently, the mass flow (kg/h) of the wire-drawing machines is increased dramatically and results in a large reduction in manufacturing cost. Some of this ad-

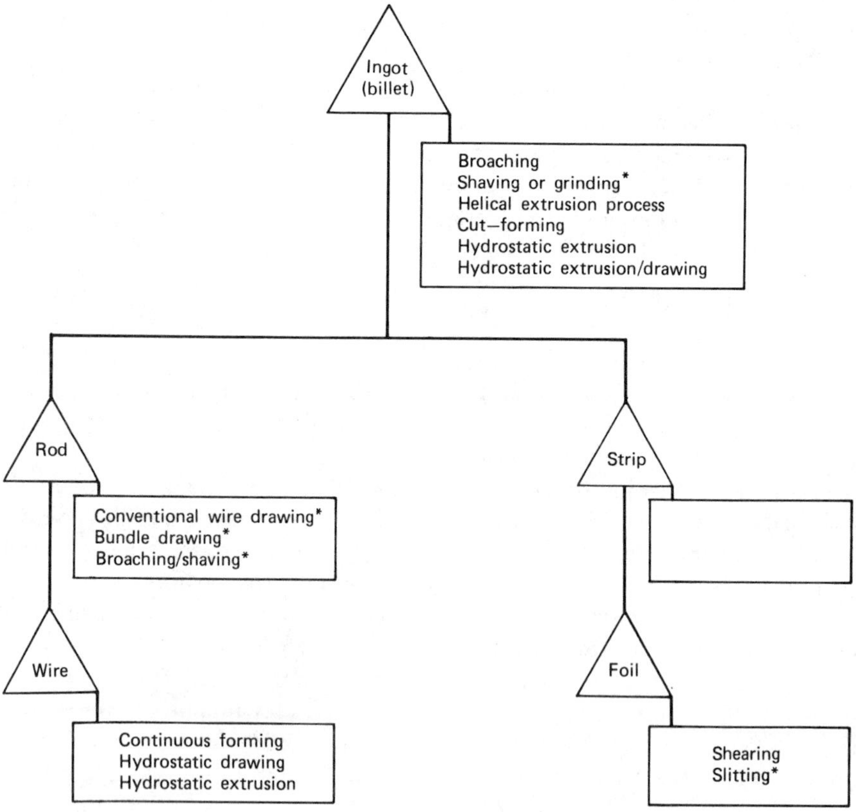

Figure 3. Mechanical attenuation and material separation processes for metal filament forming: △, stock material form; □, fiber-forming processes; *, commercially available product.

vantage, however, is offset by the equipment and processing that are necessary to form the initial bundle of wires and that are required to separate the fibers at the conclusion of the process. On balance, the overall economy improvement factor is highly favorable and can be as much as 40 or 50 to one for fibers of ca 12 μm in diameter. The bundle-drawing process (BDP) is an important source of quality fiber. A schematic process flow diagram comparing conventional wire drawing is shown in Figure 4.

The wire drawing, modified wire drawing, and extrusion processes produce continuous filaments of basically circular cross section and of unique properties that are partly a consequence of the processing history. Attenuation or constraining the material in the reducing die produces changes in the internal structure or morphology of the metal, eg, reduction in grain size, development of preferred orientation and,

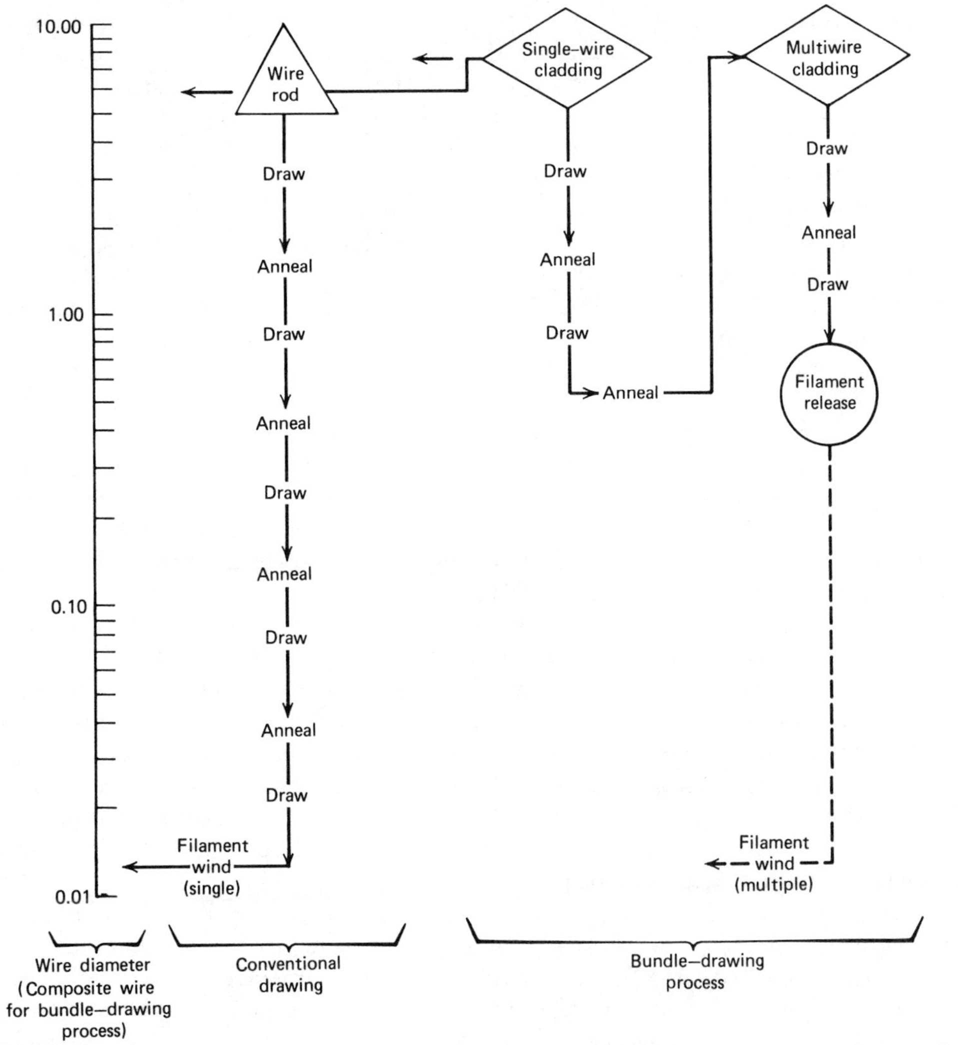

Figure 4. Flow diagram comparing conventional and bundle-drawing processes: △, starting material; ◇, ○, special operation.

in some cases, induced phase transformation. In multiple-step reduction processes, the effect of such changes in material properties causes succeeding steps to become increasingly more difficult and usually it is necessary to eradicate the work-hardening effect by annealing the material. The drawing–annealing schedule can be arranged to provide efficient processing and desirable mechanical properties in the final fiber product.

The slitting, broaching, shaving, and grinding machines and devices designed to convert standard mill forms of metal to fiber have been described (7–9). The wire-shaving technique produces a fiber product at low cost and with wide application (10). Modern machines operate automatically on a multiplicity of wires and can produce metal wool with an equivalent diameter of <25 μm. The process involves the shaving of wire, which may be ca 12 mm in diameter, in a series of steps by serrated, chisel-type tools; the size and spacing of the serrations determining the cross-section dimensions of the individual fibers. The industry standards define fiber-wool grades as grade 3 (largest) through 0 to the very fine 0000 grade, with the extremes corresponding to a mean fiber width of 178–241 μm and 15–25 μm, respectively (11). The cross section of the shaved fiber generally is triangular with the apices defining sharp edges along the fiber length. The coarser grades contain very long fiber and, even in the finer grades, the fiber is of sufficient length to cling together without unraveling in the rolls or pads, which are the usual form for shipping. Material compositions are mainly ferrous alloys, either low carbon steel, or 400 series ferrite stainless steel.

Many variations of the wire-shaving process have been developed. A group of devices or machines have been designed for the shearing or shaving of stacks of thin metal foil, although no such commercial fiber product is on the market (8). The product produced by these processes usually is rectangular or square in cross section and the length depends on the process employed: for edge-shearing or slitting the fiber length can be equal to the foil length, whereas for end-shearing the width of the foil is the limiting factor. From the material standpoint, the versatility of mechanical processing is limited only by the required availability of the foil form. The slitting of metal foil in single thicknesses to provide continuous filament usually is unattractive because of the associated low production rate. For some specialized applications, however, eg, for decorative textile thread, the cost may be justified especially if the foil costs are reduced by substitution of metal coated plastics.

Other mechanical processing techniques identified in Figure 3 either are not used for fiber production or are emerging technology. Hydrostatic extrusion and extrusion-drawing techniques offer opportunities for fiber-forming materials which are difficult to work or deform (12). These processes also allow for comparatively high reduction ratios, ie, the ratio of the diameter of the metal wire or rod stock to the diameter of the final wire or filament. Helical extrusion and continuous forming offer potential for a low cost product but it is too early to predict the lower limit of product dimensions consistent with economic production (13–14). A 1 mm diameter copper wire can be produced directly from a 150 mm diameter and 1 m long billet by helical extrusion and a 1 mm diameter aluminum wire can be converted directly from 10 mm diameter wire stock by continuous forming; the cut-form process is a combination of the two processes (15). The first step is the formation of a metal chip from a billet; the second step involves feeding the chip into a forming die. The advantages of this process are the attractive economics of step one and the improved product quality associated with the second step. The process is in the early developmental stage.

The fiber characteristics and relative economics of the commercially significant fiber-forming techniques by mechanical processing are summarized in Table 2. Scanning electron micrographs of representative fiber products are presented in Figure 5.

Liquid-Metal or Casting Processes. Melt spinning of glass and certain polymers is an established technique for mass production of fine filaments or fibers. The liquid material is forced through a carefully designed orifice or spinerette and solidifies in a cooled environment, usually after considerable attenuation, before being wound on a spool. However, this process cannot be adapted easily to metals. The development of liquid-metal fiber-forming processes has revolved around overcoming the inherently low viscosity of molten metals. The low viscosity and the high surface tension of liquid metals make it extremely difficult to establish free-liquid jet stability over a length sufficient to allow freezing of the metal into a fiber before the jet separates into droplets. The problem has been solved with varying degrees of success by one of the following approaches: altering the surface of the liquid jet by chemical reaction; promoting jet stabilization by indirect physical means, eg, an electrostatic field; or accelerating the removal of heat from the jet to promote solidification before breakup occurs. An alternative approach which has not been commercially exploited is that of placing a glass envelope around the molten metal to control the formability and thus circumvent the problem of jet stabilization (16–17). However, the final application of the product may require the removal of the glass envelope, which would add considerably to the product cost since it may account for 75 vol % of the material produced. The various liquid-metal fiber-forming processes of historical and commercial interest are identified in Figure 6.

Melt spinning involving a free-liquid jet is used for a variety of low-melting-point metals or alloys including Pb, Sn, Zn, and Al. Various techniques are involved in the cooling and quenching process including co-current gas flow, mists, and liquid media. The processes permit production of fiber of 25–250 μm in diameter and of continuous length. Melt-spin processes applied to higher-melting-point metals, particularly the

Table 2. Mechanical Fiber-Forming Processes and Related Fiber Characteristics

Process	Typical fiber diameter, μm	Typical length	Materials	Cross-section shape	Economics
conventional wire drawing	$\geq$12	continuous	all ductile metals and alloys	round (other sections possible)	304 stainless steel ca \$3.00/kg at 250 μm; ca \$3000/kg at 12 μm
bundle drawing	$\geq$4; typically 8 or 12	continuous	ductile metals and alloys	rough surface	304 stainless steel ca \$100/kg at 12 μm
broaching or shaving, eg, wire rod, and billet	$\geq$8	short to continuous	most ductile metals and alloys	generally triangular	low carbon steel \$2–4/kg depending on grade
slitting and shaving, eg, foil and sheet	ca 25 and greater	short (0.0004–4 cm) or continuous	most ductile metals and alloys	ductile square or rectangle	shaving fiber <\$10/kg (not available currently)

Figure 5. Scanning electron micrographs of fibers produced by mechanical fiber-forming processes: (**a**) 25 μm stainless steel, conventional wire drawing; (**b**) 19 μm 304-type stainless steel, bundle-drawing process; (**c**) 0000-grade steel wool, shaving process; (**d**) transverse shaving marks of 0000-grade steel wool, shaving process; (**e**) felt metal (FM), 1100 series, 347 stainless steel; (**f**) sheared, low carbon, steel fiber edge.

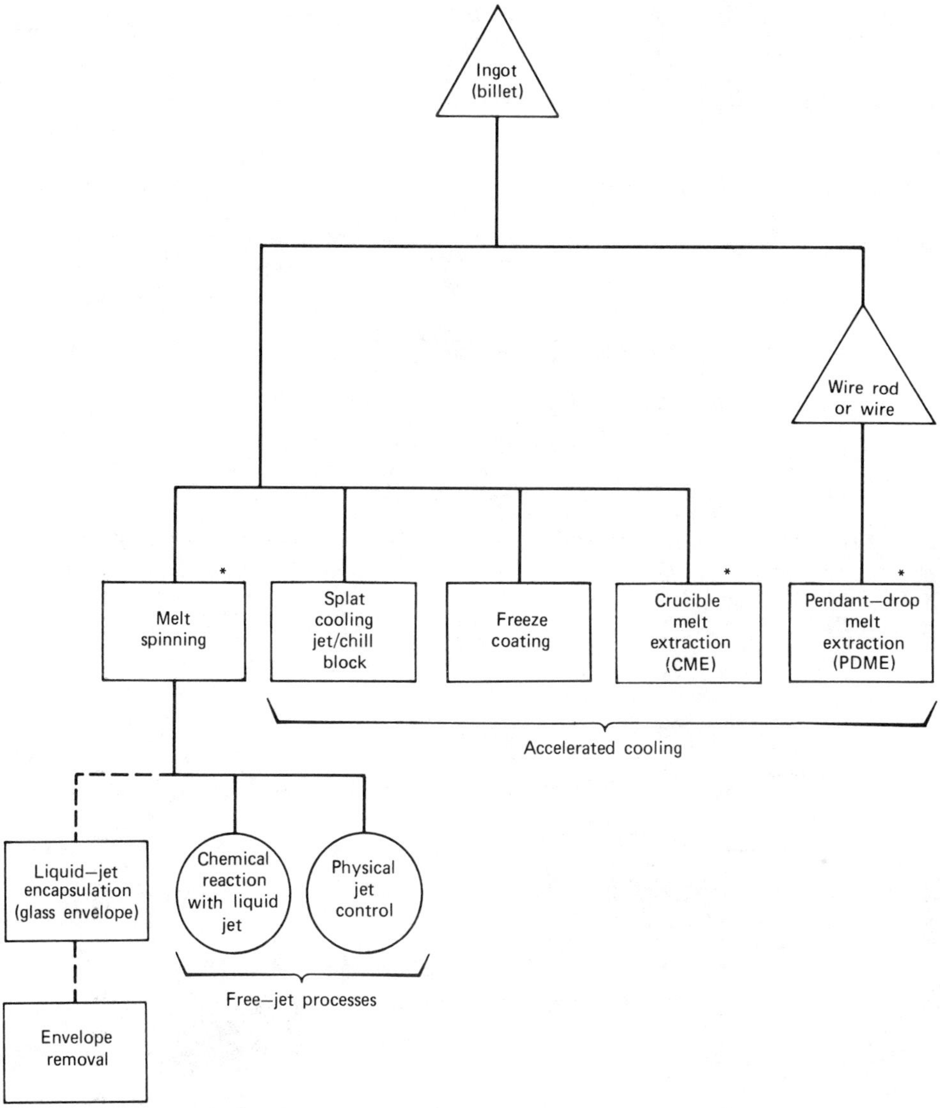

Figure 6. Liquid-metal, fiber-forming processes: △, stock material form; □, basic fiber process; ○, process modifications; *, commercially available fiber.

ferrous alloys, requires the removal of very large quantities of heat from the liquid-metal jet in a very short time. Generally, those techniques that are successful for the low melting alloys cannot be used for the high melting alloys. An alternative approach involves the reaction of the liquid jet and the cooling medium, or some addition to that medium which results in the formation of a case or envelope of sufficient strength to prevent breakup of the liquid-metal jet.

In some cases, the chemical composition of the melt must be adjusted prior to

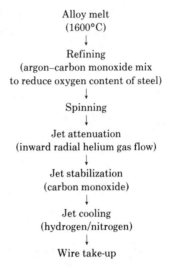

Alloy melt
(1600°C)
↓
Refining
(argon–carbon monoxide mix
to reduce oxygen content of steel)
↓
Spinning
↓
Jet attenuation
(inward radial helium gas flow)
↓
Jet stabilization
(carbon monoxide)
↓
Jet cooling
(hydrogen/nitrogen)
↓
Wire take-up

Figure 7. Flow diagram for silicon steel and aluminum steel melt spinning with jet stabilization.

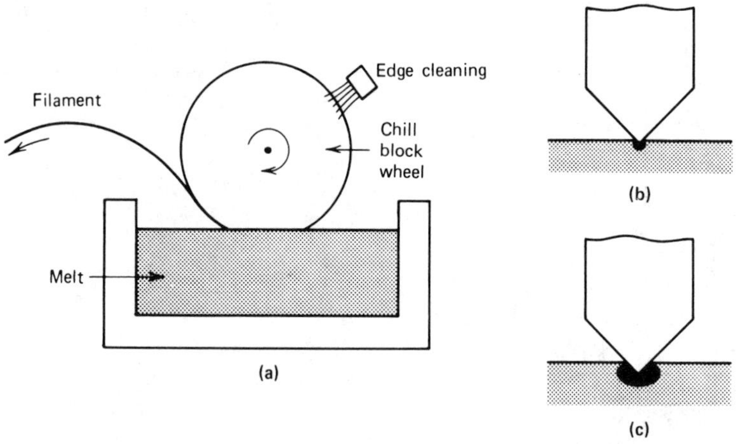

Figure 8. (**a**) Crucible melt-extraction process (CME); (**b**) and (**c**) crescent shaped, larger filaments.

spinning in order to prevent dissolution of the stabilizing film in the molten jet. A typical process flow diagram for the melt spinning of steel wire is shown in Figure 7 (18). The process is designed for two steel compositions, an aluminum steel (1.00 wt % Al, 0.36 wt % C) and a silicon steel (2.40 wt % Si, 0.32 wt % C). Carbon monoxide is the stabilizing medium and surface films of aluminum oxide or silicon oxide are formed. A continuous filament of 75–200 μm in diameter is spun at 240–120 m/min with continuous operation to 100 h. A useful degree of attenuation of the molten-metal jet outside the orifice is achieved by using a radially directed helium gas flow. Consequently, the final filament diameter can be controlled independently of the spinning orifice or spinerette geometry.

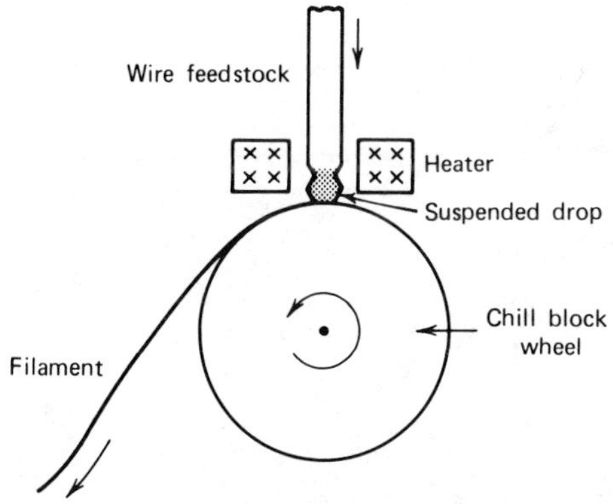

Figure 9. Pendant-drop, melt-extraction process (PDME).

Table 3. Liquid-Metal Fiber-Forming Processes and Related Fiber Characteristics

Process	Typical fiber diameter	Typical length	Materials	Cross-section shape	Economics
melt spin	≥25 μm	continuous	Pb low melting point metals and alloys	circular	plumbers wool (lead) <$2/kg
melt spin with jet stabilization (chemical)	≥75 μm	continuous	most metals and alloys (special compositions)	circular	
crucible melt extraction (CME)	≥25 μm	continuous or controlled length	most metals and alloys	small diameter, circular; large diameter, crescent shaped	stainless steel $2–7/kg
pendant-drop melt extraction (PDME)	≥25 μm	continuous or controlled length	most metals and alloys	small diameter, circular; large diameter, crescent shaped	

The alternative approach to fiber forming from the melt is by greatly accelerated heat transfer (19–20). The processes essentially are based on the quenching of the molten-metal jet on a chill plate positioned close to the orifice or at a point prior to the onset of jet instability and breakup. A number of different configurations for the chill plate have been proposed, ie, from rotating drums to concave disks. The quench rate of the molten jet on the chill plate is as high as 10^6 °C/s. Quench rates of this magnitude can produce nonequilibrium conditions in certain alloys and a characteristic grain structure associated with the impressed unidirectional heat flow and the kinetics of the nucleation–growth process. The fibers produced by these chill-plate processes generally are ribbonlike but can be continuous.

The crucible melt extraction (CME) process is an extension of the chill-plate processes and is illustrated in Figure **8a** (21). There is no spin orifice and, thus, no need for forming a liquid stream. The chill surface, the shaped rim of a revolving disk, is dipped into the crucible so that the shaped edge just contacts the liquid-metal surface.

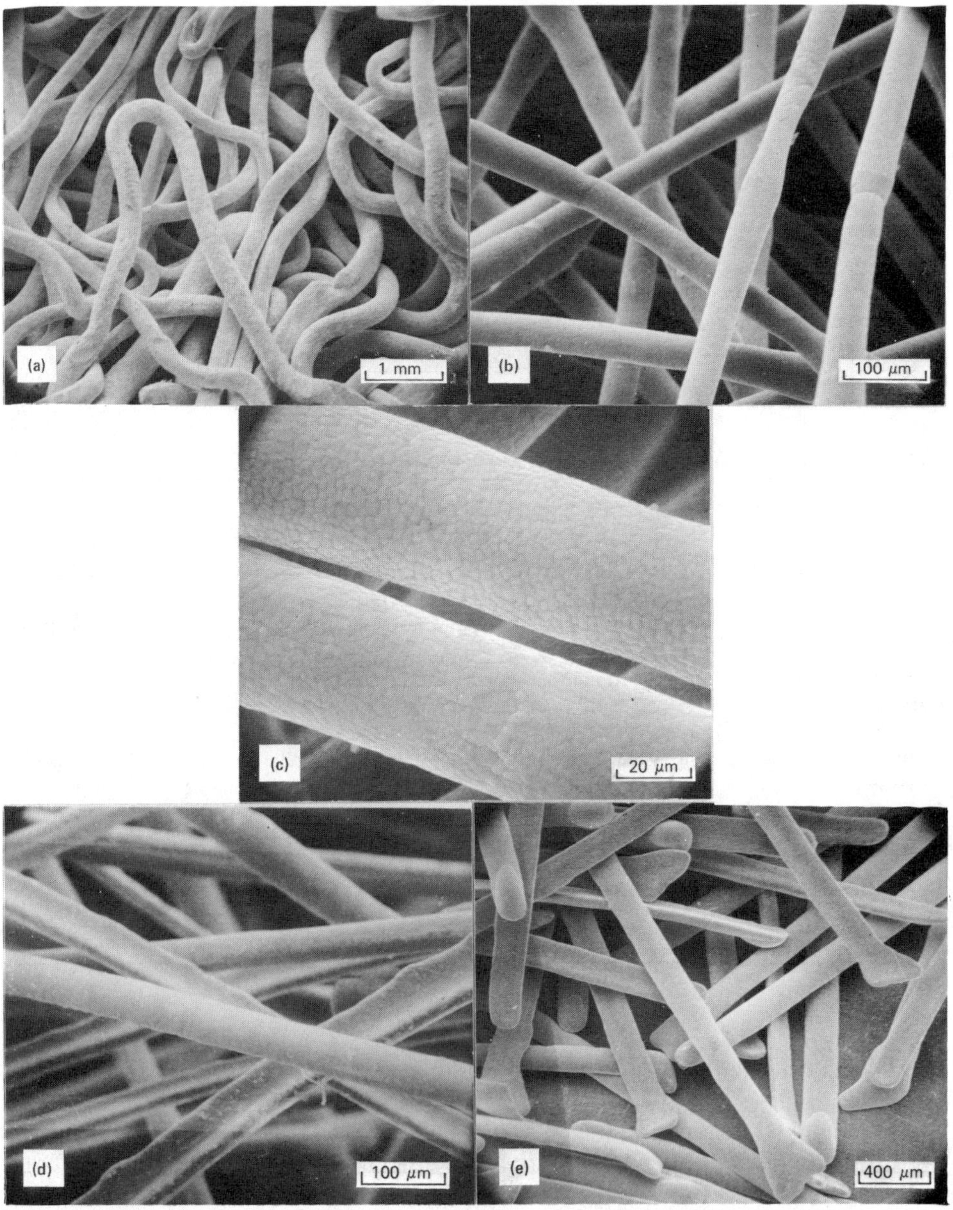

Figure 10. Scanning electron micrographs of fibers produced by liquid-metal or casting processes: (**a**) 250-μm lead filament, melt spinning; (**b**) 40-μm aluminum filament, melt spinning; (**c**) 35-μm melt spin aluminum filament showing surface structure; (**d**) 50 × 100-μm stainless steel filament, PDME; (**e**) 150-μm stainless steel filament (3 mm long), PDME.

Table 4. Miscellaneous Metal-Fiber Forming Processes

Filament material source	Process	Product	Status
metal powder	slurry forming, extrusion with binder followed by sintering	potentially most metals and alloys; specifically W–Ni (250 μm in dia)	experimental
any metal form	vapor deposition on glass fiber or plastic strip which is subsequently slit[a]	aluminum-coated product for conductive or decorative applications	commercially available
	freeze coating on glass or other fiber	aluminum on glass	experimental
inorganic chemical or mineral	electroplating on a helical mandrel[b]	50 × 12-μm Ni continuous filament	experimental
	chemical decomposition in an organic fiber precursor[c]	Ni–Cr alloy fiber with dimensions equal to precursor	experimental
	halide reduction by hydrogen[d]	whiskers of many metals, eg, Cu, Ag, Fe, Ni, and Co; length, 3–20 mm	mostly experimental

[a] Ref. 22.
[b] Ref. 26.
[c] Ref. 25.
[d] Ref. 23–24.

The cool disk edge immediately causes solidification of a small volume of liquid metal, which is carried out of the crucible and is ejected from the wheel by centrifugal action. The shape of the fiber cross section is dependent on the wheel-edge geometry and the depth of immersion, but it can be made circular for small diameter filaments (25–75 μm); larger filaments tend to be crescent shaped, as indicated in Figures 8**b**–8**c**.

A further evolution in the chill-plate concept is the pendant-drop melt-extraction process (PDME) (21). In this process, the orifice for jet forming is eliminated and the crucible is replaced by a suspended drop held by surface tension to the feedstock, which is a wire or wire rod. The PDME process conveniently circumvents jet stability problems and solves the basic material compatibility problems associated with the molten-metal-orifice and the molten-metal-crucible interfaces. The basic elements of this process are schematically represented in Figure 9. The process can be applied to a wide range of metals and alloys and can produce fibers with equivalent diameters as small as ca 25 μm.

The PDME and CME processes have been varied to provide discrete fiber lengths as opposed to continuous filament. The introduction of discontinuities on the chill-wheel rim at fixed intervals results in the casting of short fibers with the discontinuity spacing defining the fiber length. The short fibers tend to have a dog-bone configuration which, for certain applications, is advantageous. A summary of the liquid-metal fiber processes in terms of fiber characteristics and comparative economics is presented in Table 3. Scanning electron micrographs of representative fibers produced by the various casting techniques are reproduced in Figure 10.

Table 5. Special Metal-Fiber Characteristics and Processability

Fiber characteristics	Process parameter	Remedy
surface roughness	friction–wear	surface treatment:
high hardness	wear	lubrication; change material of rubbing surfaces in equipment
irregular shape, high density	fluid dynamics	modify process media conditions, eg, flow velocity, density, viscosity
high modulus	general processability	use fiber with smaller cross-section (lower section modulus)
high yield	resistance to take permanent set (spring back)	anneal to reduce yield strength

Table 6. Fiber-Structure Fabrication Processes

Technology	Preferred fiber form	Process	Product structure
papermaking	free fiber; length >12 mm for fibers 12–250 μm	wet slurry casting on screen	green random-fiber mat
textile processing	continuous filament	twisting	continuous filament yarn
		breaking	sliver–roving
	sliver–roving	drafting–spinning	staple yarn (blends)
		carding	random web
		air laying	random web
	yarn (staple or continuous filament)	weaving	fabric sheet
		knitting	fabric sheet or tube
		braiding	ribbon or tube
		tufting	carpetlike structures
	random web	needle punching	feltlike products
miscellaneous	chopped fiber; length >4 mm	flocking	short-pile structure on substrate
	continuous filament yarn	filament winding	composite structures
	random web, woven or knitted structures	vacuum or gravity infiltration of matrix material	composite structures

Miscellaneous Processes. Production of low cost metal filaments or products with unique properties has been reviewed (7,22–26). A sampling of significant technology other than mechanical and liquid-metal processes is presented in Table 4. Many of these processes could, with additional development and the use of recent advances in material science and related technology, lead to attractive processes.

Assembly. The various assembly processes of metal fibers into structures are similar to those developed for natural and synthetic fibers. The main fiber characteristics and the process parameters most likely to be affected by those characteristics,

Table 7. Secondary or Special Processing Techniques

Process	Typical application or particular advantage
annealing	softens fiber, increases fiber ductility, assists in further processing
sintering	fuses fibers at contact points, provides increased mechanical strength in fibrous structures
fiber alignment	produces anisotropic characteristics in web structures or composite materials
plating	provides protection from corrosive environments
coatings	provides interfiber lubrication
calendering (rolling)	precise pore size and density control, particularly in nonwovens
pressing	precise pore size and density control, particularly in nonwovens
crimping	geometrical elasticity in fiber, yarn, or nonwovens

and suggested solutions or remedies to alleviate processing problems are given in Table 5. Applications of metal-fiber products may require a combination of fiber types, eg, metal fibers and organic fibers. Thus, processes must be adapted to accomodate blends of fibers of vastly differing properties and processing characteristics.

The basic fiber-processing techniques in relation to the preferred fiber form for each process and the resulting products are summarized in Table 6. In textile processing particularly, the product of one process often becomes the preferred fiber form for a second process; thus, sequential processing is common for the more advanced products. In addition to the principal fiber-structure processing techniques enumerated in Table 5, a variety of secondary or special processes are important in modifying or optimizing the fiber-structure properties; these are listed in Table 7 with their typical applications.

Uses

Applications for metal fibers are of two types. One is the substitution for other fiber types or for metal powder in the case of porous metals to improve performance or to provide a cost benefit, eg, high temperature oxidation resistance or improved permeability in porous structures. The other application is in the development of new products that are based on the unique fiber properties or property combinations of metals, eg, in high gradient magnetic separation (see Magnetic separations).

A representative listing of applications, with brief descriptions and main advantages associated with the use of metal fibers, is presented in Table 8. The fiber structure applications are exploitations of the structure and the inherent fiber properties. The composite applications benefit from both the fiber and fiber structure characteristics and their interaction with the matrix material (see Composite materials; Laminated and reinforced metals).

Table 8. Metal-Fiber Applications

Application	Description	Status[a]	Special advantages and principal fiber function
Textile products and porous structures			
filters			
surface	screen or wire-mesh products in disk or cartridge configuration used for low contamination level fluids, eg, hydraulic fluids	C	mechanical strength, nonmigrating, corrosion resistant
depth	nonwoven or random web structures in disk or cartridge configuration; general industrial applications	C	high permeability, high dirt-holding capacity
electrostatic	particle-capture augmentation of charged particles using low density web structures	R	improved filtration efficiency, electrical conduction
	filter-cake density control using designed electric-field configuration in bag-house filters	D	improved cake permeability
magnetic (HGMS)	magnetic attraction of ferromagnetic or paramagnetic particles through field distortion associated with ferromagnetic fibers in a uniform magnetic field; used for decontamination of kaolin	C	high field distortion associated with fine filaments
seals			
abradable	zero-clearance seal formed by turbine blade tips mating with deformable, porous-mat engine-housing lining	C	improved engine performance (efficiency)
gaskets	rope-type structures and nonwoven mats which conform to surface irregularities under compression; general industrial high temperature applications	C	high temperature, resilient structures
abrasion	metal-wool abrasive products for material removal and polishing	C	sharp cutting edges, shaved fibers
antistatic			
textiles	blended yarn (stainless steel–nylon fiber) up to 10 wt % steel woven into fabric for clothing applications	C	electrical conduction and flexibility
carpets	blended yarn introduced into tufting operation to provide <1% steel fiber in face yarns	C	electrical conduction and flexibility
filter bags	needle-punched blend of organic fiber and steel to provide static control in bag-house filtration applications	C	electrical conduction and flexibility
brush	steel-fiber brush for static control of paper in photocopying devices	C	electrical conduction and flexibility
insulation	low density web (<5%) with suitable attachments used for aerospace applications, eg, rocket-engine nozzle insulation	C	high temperature stability, nonfriable
acoustics	acoustic impedance control for tuned resonator-duct liners in jet engines, etc	C	fine fiberweb structures provide improved linearity in absorption characteristics
catalysts	worn wire-mesh stacks for nitric acid production (75 μm diameter, Pt–Rh wire)	C	optimum surface-to-volume ratio

Table 8 (*continued*)

Application	Description	Status[a]	Special advantages and principal fiber function
electromagnetic interference control and field production	Faraday suits of stainless steel/nylon fabric for high voltage transmission-line maintenance: heat shrinkable cable-termination shields embodying air-laid web	C	electrical conduction, flexibility
fluid flow	a family of products including, flow restrictors, snubbers, silencers, vents, demisters, homogenizer plugs, etc, using porous fiber structures	C	mechanical integrity, controlled permeability
heated fabrics	built-in flexible electrical conductors for heating, clothing, etc	C	electrical conduction, flexibility
electrodes	battery plaques, current collectors for high energy batteries, fuel cells; electrodes for electrolytic capacitors	D	high surface area, distributed conductor network
bearings	liquid lubricant or air bearings utilizing porous fiber structures	D	controlled porosity, mechanical integrity
wicks	capillary structure for liquid-phase transport in heat pipes	D	high temperature stability, controlled size, interconnected pores
shock mounts	friction damping in shock and vibration isolation mounts	C	high fiber-to-fiber friction, resilience
flame trap	protection of flammable fluids using sintered fiber-metal porous plugs	C	high thermal capacity and heat conductivity
transpirational cooling	surface cooling by controlled fluid flow through component structure, eg, turbine blades	D	structural integrity with designed permeability
fluidizer plate	support and diffuser plate for fluidized bed, eg, grain cars		porous-plate structure with load-bearing capabilities
ceramic attachment	compliant layer of fibers for ceramic attachment to metal surfaces	C	resilient fiber structure accomodating differential thermal expansion coefficients of substrate and ceramic
yarn and cable products	yarn and cable applications demanding flexibility with high strength, eg, in instrument control cables and medical sutures	C	good hand, high knot strength, bio-compatibility, high flexibility
Composite applications			
tire cord	steel tire cord containing 0.15–0.25-mm, brass-plated filament in designed cord configurations	C	strength, dimensional stability, flexibility, heat conduction, bonding to rubber
tire tread	tread impregnation with chopped wire for off-the-road vehicle and aircraft tires	C	abrasion and cut resistance
timing belts	reinforced, nonstretching belts	?	dimensional stability, strength
brake lining	high friction composite material with dispersed short metal fiber for heat conduction	C	improved operating performance by temperature control using conductive fibers

Table 8 (*continued*)

Application	Description	Status[a]	Special advantages and principal fiber function
refractory bricks (linings)	furnace-lining reinforcement with castable refractories embodying chopped fiber or melt extraction process fiber to reduce friability	C	mechanical integrity, crack arrestor function
concrete	reinforcement with short fibers to increase load-bearing capability	C	crack arrestor
conductive plastics	plastic housings for electrical equipment to control electromagnetic interference (EMI)	C	electrical conduction
	fabric containing short metal fiber for shielding–reflecting microwave radiation	C	efficient coupling
superconductors	continuous filament, small-diameter superconductors embedded in high thermal conductivity matrix	C	more efficient use of superconductivity material and protection during conduction state transition

[a] C = commercial, D = development, and R = research.

BIBLIOGRAPHY

1. R. V. Coleman, *Metall. Rev.* **9**(35), 261 (1964).
2. P. Ducheyne, E. Aernoudt, and P. De Meester, *J. Mater. Sci.* **13,** 2650 (1978).
3. J. K. Mackenzie, *Proc. Phys. Soc.* (*London*) **B63,** 2 (1950).
4. R. P. Tye, *A.S.M.E. Publication No. 73-HT-47*, American Society of Mechanical Engineers, New York, 1973.
5. J. Y. C. Koh and A. Fortini, *NASA Publication No. CR-120854*, 1972.
6. R. W. Powell, *Iron and Steel Institute Special Report No. 43*, p. 315.
7. C. Z. Carroll-Porczynski, *Advanced Materials*, Chemical Publishing Co., Inc., New York, 1969.
8. U.S. Pat. 3,122,038 (Feb. 25, 1964), J. Juras.
9. W. M. Stocker, Jr., *Am. Mach.* (May 15, 1950).
10. L. E. Browne, *Steel* (Feb. 25, 1946).
11. *Federal Specification FF-S-740a*, U.S. Government Printing Office, Washington, D.C., Oct. 1965.
12. H. Ll. D. Pugh and A. H. Low, *J. Inst. Met.* **93,** 201 (1964–1965).
13. D. Green, *J. Inst. Met.* **99,** 76 (1971).
14. *Ibid.*, **100,** 295 (1972).
15. T. Hoshi and M. C. Shaw, *J. Eng. Ind. Trans. ASME*, 225 (Feb. 1977).
16. G. F. Taylor, *Phys. Rev.* **23,** 655 (1924).
17. U.S. Pat. 1,793,529 (Feb. 24, 1931), G. F. Taylor.
18. *AIChE Symp. Ser.* **74**(180), (1978).
19. U.S. Pat. 2,879,566 (1959), R. B. Pond.
20. U.S. Pat. 2,976,590 (1961), R. B. Pond.
21. R. E. Maringer and C. E. Mobley, *J. Vac. Sci. Technol.* **11,** 1067 (1974).
22. J. F. C. Morden, *Met. Ind.*, 495 (June 17, 1960).
23. S. S. Brenner, *Science*, **128,** 569 (1958).
24. S. S. Brenner, *Acta Met.* **4,** 62 (1956).
25. W. H. Dresher, *Technical Report AFML-TR-67-382*, WPAFB, Dec. 1967.
26. E. H. Newton and D. E. Johnson, *Technical Report AFML-TR-65-124*, WPAFB, April 1965.

JOHN A. ROBERTS
Arco Ventures Co.

METALLIC COATINGS

SURVEY

Metallic coatings provide a basis material with the surface properties of the metal being applied as coating. The functional composite so produced has an appearance or utility not achieved by either component singly and in fact becomes a new material (see also Composite materials). The base material almost always provides the load-bearing function and the coating metal serves as a corrosion- or wear-resistant protective layer (see Corrosion and corrosion inhibitors). The base material most often is another metal, but it can be a ceramic, paper, or a synthetic fiber (1). The bulk of all metallic coatings provide a protective function in one of five principal ways: they are anodic to iron and can protect it by cathodic protection, eg, Al [*7429-90-5*], Mg [*7439-95-4*], Zn [*7440-66-6*], and Cd [*7440-67-7*]; they form highly protective, passive films in aqueous media, eg, Cr [*7440-47-3*], Ni [*7440-02-0*], Ti [*7440-32-6*], Ta [*7440-25-7*], and Zr [*7440-67-7*]; their oxides are slow growing and adherent and, therefore, protect at high temperatures, eg, Al, Cr, and Si [*7440-21-3*]; they are noble metals and corrode little or not at all and function as barriers to corrosive agents, eg, Au [*7440-57-5*], Ag [*7440-22-4*], Cu [*7440-50-8*], Pt [*7440-06-4*], Rh [*7440-16-6*], Pd [*7440-05-3*], etc; and their compounds, which are formed by uniting with the basis metal or one of its constituents, are very hard and provide wear resistance, eg, B_4C [*12069-32-8*], SiC [*409-21-2*], TiC [*56780-56-4*], WC [*12070-12-1*], Cr_2O_3 [*1308-38-9*], etc. In certain coating processes, eg, chemical vapor deposition (CVD), the hard-phase compounds are grown directly on the substrate.

The protective function is primary. Decorative and reflective metallic coatings must be highly corrosion resistant, at least in ambient environment, to maintain a visually satisfying appearance and to provide satisfactory service over the anticipated useful life of the composite. The corrosion resistance of individual metallic elements and their principal alloys is discussed in refs. 2–5. Types are reviewed in refs. 6–9. Coating technology is discussed in refs. 10–14 and a discussion of high temperature oxidation- and sulfidation-resistant coatings primarily for aerospace applications is given in refs. 15–17 (see also Electroplating).

Metallic coatings have been used since ancient times. One of the earliest known applications involved the cementation of copper or bronze with arsenic to produce a silvery coating of Cu_3As [*12005-75-3*] on art objects (18). Between 1 and 600 AD, Andeans plated copper objects by galvanic displacement by which 0.5–2 μm thick films of gold or silver are deposited (19).

Metallic coatings provide functionality at low cost. Sudden changes in availability of materials have led to a critical shortage of and, hence, a large increase in the cost of cobalt. This increase in cost has strongly affected the technology of hard facing alloys and has promoted the development of several new substitutes. Strict government regulations on pollution control have delayed some UK plating firms' efforts to establish effective treatment measures and, in the United States, has led to the closing of some firms. However, the new regulations have had much less affect on some new coating technologies, which have grown rapidly. Strict regulations have caused a general stagnation in the metal-finishing industry for the last five years (1975–1980). However, continued growth in the electroplating of plastics and printed circuit boards is expected. Thus, it is the continued appearance of new applications for coatings coupled with emerging coating technologies that provide improved quality and cost effectiveness. Many of the new techniques will have enormous technological impact in the coming years because of growth in supporting industries, eg, electron-beam dissociation has been developed at IBM for depositing conducting metal coatings only 0.05-μm thick to form the elements of the Josephson-effect transistor (see Integrated circuits).

Diffusion

In the various processes for diffusion coating, the basis metal is contacted with the coating metal, which is in the liquid or solid state or is brought to the surface by vapor transport. The two materials are held at an elevated temperature for a sufficient time to allow lattice interdiffusion of the two materials. A solid-solution alloy forms and may be accompanied by the formation of one or more intermetallic compounds that provide a compositional transition zone within the coating. The growth of the coating often is limited by diffusion of one species through one of the intermetallic layers, resulting in a parabolic rate of coating thickness increase with time. Whether the coating metal is brought into contact as a liquid, or is carried by another solvent, or is transported by a vapor-phase mechanism, surface contamination and oxide films must be removed by a suitable fluxing process either prior to or during the contact period and oxidation-free conditions must be maintained during diffusion. Three basic techniques are used in diffusion coating: hot dipping, cementation, and use of liquid carriers. In hot dipping, an excess of liquid-coating metal or alloy usually is carried on the product which may be finished with steam or air (20) or it is wiped on rollers as it moves out of the bath to control the total thickness of the final coating. In many cases, the coating either is thicker on one side than the other or the coating is on one side only. In cementation and liquid-bath diffusion processes, the coating consists only of intermetallic compound and solid-solution regions, the excess coating metal having been removed by a postcoating process. Postcoating heat treatments often are used to equilibrate these intermetallics.

Hot-Dipped Coatings. ***Aluminum.*** Hot-dipped aluminum coatings on steel strips have been produced for 35 yr and, in 1976, 300,000–500,000 metric tons was produced in the United States (21). Aluminum forms a very protective oxide film with outstanding high temperature oxidation resistance and it provides the attractive appearance of pure aluminum. It also provides galvanic protection for steel (qv) but must be about twice the thickness of zinc if it is to impart acceptable corrosion resistance. The high melting point of aluminum, relative to zinc, causes recrystallization and, therefore, produces softening of the cold-worked steel during hot dipping. Nevertheless, hot dipping is the most widely used method of coating steel, primarily plain carbon, low alloy, and stainless steel, with aluminum.

During the process, the aluminum is maintained at 680–720°C and the steel strip is immersed for ca 5–15 s. Commercial practice is to utilize the Sendzimir line which first uniformly oxidizes the iron and then passes it to a reducing atmosphere in which a layer of pure iron quickly forms on the surface (22). On immersion, the iron layer reacts quickly with the aluminum. After dipping, the material is finished by jet finishing, rolling, and quenching. For high temperature applications, the bulk of the coating should consist of an outer layer of Fe–Al and Fe_3Al [*12004-62-5*] at the steel–coating interface with an average surface concentration of <50% aluminum and, ideally, 12%. Generally, this is accomplished by heat treating the strip at 820–930°C after dipping. For low temperature applications and to promote formability, the thickness of the intermetallic layer should be minimized. Silicon, although it detracts from the coating's corrosion resistance, impedes the growth of the intermetallic region. Aluminum hot-dip processing is difficult to control because the activation energy for the exponential growth of the intermetallic layer is 170–180 kJ/mol (41–43 kcal/mol). Also, simultaneous growth and spalling is possible if dipping times exceed 60 s (23). If the above activation energy values are used at a bath temperature of 680°C and an immersion time of 3.5 s, a 10-μm intermetallic layer develops. At 700°C, the immersion time must be less than 2 s. Cleanliness of the strip prior to immersion is essential to avoid barrier films of Al_2O_3. Commercial practice involves hydrogen treating, fluxing, and precoating treatments using a thin layer of copper or a film of ethylene glycol.

An aluminized material, Aluma-Ti (Inland Steel Co.), was designed to have heat-resistant properties equivalent to type 409 stainless steel for applications up to 815°C, eg, in motor vehicle exhaust systems. The material is basically a type 1 aluminized coating, eg, aluminum with a small addition of silicon, but the steel basis metal is an aluminum-killed steel with sufficient titanium to tie up residual carbon and nitrogen and to provide ca 0.3% titanium in solution. Aluma-Ti resists formation of a porous intermetallic layer and a subsurface iron oxide layer which tends to promote spalling of the intermetallic layer. Loss of the intermetallic layer and the presence of a continuous subsurface oxide stops the continued diffusion of aluminum into the basis steel, which is necessary to provide long term protection. Aluma-Ti in tests has shown superior spalling resistance to both type 1 aluminized and 409 stainless steel. Mechanical properties of Aluma-Ti compare well with those of type 409.

Armco Steel Corp. has produced a steel alloy which, after being aluminized, also is superior to 409 stainless. Its composition includes 2% Cr, 2% Al, 1% Si, and 0.5% Ti. In practice, the steel surface is sufficiently oxidized under controlled conditions in the oxidizing section of the Sendzimir line so that a region near the surface and including the iron is oxidized. When the oxidized strip passes to the reducing section, the iron oxide is reduced to iron which contains a fine dispersion of the stable oxides, ie, of Ti, Al, and Si. Subsequent aluminizing is satisfactory as long as sufficient oxidation of iron has taken place so that a continuous oxide film of the more stable oxides cannot form.

Aluminum hot-dipped steel sheet products are particularly useful in appliances for their heat-reflective qualities and in exhaust system components where temperatures are as high as 538°C. Because of their high surface quality, sheet products are used in buildings and other applications as a replacement for more expensive alloy steels.

Lead, Terne, and Tin. Hot dipping of pure lead [*7439-92-1*] is not used extensively despite its low cost which is less than tin and zinc. Lead does not alloy with steel and must be alloyed with other metals, eg, tin, antimony, zinc, or silver which improve fluidity and facilitate bonding with iron. Terne metal [*39428-85-8, 54938-78-2*] is more commonly used and contains 15–50% tin. It is applied where atmospheric corrosion resistance in the absence of abrasion is required at low cost. It is made predominantly in continuous electroplating lines up to 1067 mm wide with a coating weight of 150 g/m^2 and a nominal one-side thickness of 15 μm. British Steel Corporation produces 18,000 t/yr (24).

Hot-dipping of tin also has been superseded largely by electrolytic coating techniques. However, changes in effluent standards for electroplate wastes may revive hot dipping to some extent. Hot dipping of tin coatings usually is done in a two-compartment cell or pot that is partitioned in the upper region only, so that the plate enters through a flux layer, continues into the molten tin on one side of the partition, and leaves the tin bath emerging through a palm oil layer at the other end of the cell. Rollers are immersed in the hot oil layer and they control the thickness of the finished plate. Most of the oil is removed after the plated steel leaves the cell; the small amount that remains protects against in-storage discoloration and acts as a lubricant in subsequent forming operations. The coating is produced by the formation of a layer of $FeSn_2$ [*12023-01-7*]; its growth is limited by diffusion to practical thicknesses of ca 0.5 μm. The thickness of the outer coating of pure tin depends on the speed of the plate leaving the pot and the pressure of the rollers. Products, eg, wire, are tinned in a similar fashion.

A new terne-forming process is being implemented by Broderick Structures, Ltd. The material is a composition formed by cold-roll bonding at high pressure. Prior to the roll bonding, the sheet is treated resulting in the formation of a thin terne plate which, on contact with lead sheet during the rolling operation, becomes integrally and permanently bonded without deformation of the underlying steel. This process is the most economical solution to meet the need for structural lead sheet. The process should broaden the applications of lead, particularly for architectural uses.

Terne is used in the auto industry for gasoline tanks; in roofing eg, flashing; in plumbing, eg, for laboratories; and as a gasket material. More than 90% of tin plate is used as tin cans for food packaging. Tin-dipped wire resists corrosion by sulfur from rubber insulation layers.

Zinc. Over 40% of all zinc that is produced is used to protect steel products; the corrosion rate of iron is 25 times that of zinc in the atmosphere and in water. Zinc is anodic to its normal metal impurities and to steel, has a high hydrogen overvoltage, and forms insoluble basic salts. World production of galvanized steel is ca 1.4×10^7 metric tons per year and involves over 100 galvanizing lines of which ca 65 are Sendzimir lines (25). About 5×10^6 t/yr of galvanized sheet are produced in the United States.

The most important advance in hot-dip galvanizing is Sendzimir's process by which the surface is preoxidized at 650°C and then hydrogen-reduced at 850–950°C. The temperature is lowered to 400°C with the strip still protected in hydrogen until it enters the zinc bath. In this way, flux at the entrance to the bath is avoided and small amounts of aluminum are used to inhibit formation of zinc–iron intermetallic intermediate layers. The bath temperature is maintained at 450–460°C by the sensible heat of the incoming strip.

The important intermetallic region consists of three successive layers on the steel, namely, Fe_3Zn_{10} [*12182-98-8*], $FeZn_7$ [*12023-07-3*], and $FeZn_{13}$ [*12140-55-5*], followed by a thicker layer of pure zinc. The intermetallic region is one tenth of the total coating region. The most modern lines involve jet finishing rather than rolls because line speeds can be increased from 76–92 m/min to 185 m/min which increases the economic benefits of the process (20). As the strip rises vertically out of the zinc bath, it carries an entrained viscous layer of molten zinc. A row of horizontal jets of air are impinged perpendicularly to the strip with one on each side, as shown in Figure 1, and cause a return flow of liquid metal into the bath. Sensors above the row of air jets meter the thickness of the coating and adjust the velocity of air flow by electronic feedback circuits so that the desired thickness on each side can be maintained continuously throughout the run.

Armco Steel Corp. adds 0.01–0.10% magnesium in the galvanizing coating with 0.2–0.4% aluminum or 0.3% chromium which provide significant improvement in the atmospheric and marine corrosion resistance of galvanized steel (see also Coatings, marine). In a test involving 200 wet–dry cycles in seawater at pH 8.4, a threefold reduction in weight loss corrosion is observed for Zn–0.29 wt % Cr–0.4 wt % Mg which compares well with standard galvanizing (0.17 wt % Al–0.2% wt Pb).

The use of standard two-sided galvanized steel, eg, in automotive bodies, causes some difficulties in welding and the adherence and brightness of paint coatings is not as good as on bare steel. Consequently, many producers make a one-side, hot-dipped, galvanized product or electrolytically strip the thinner side of a differentially coated product. Paint adherence properties of a galvanized product have been improved by galvannealing which consists of heat treating the coating in-line to reduce the surface zinc to an iron–zinc intermetallic layer so that it accepts paint more readily than normal galvanized surfaces. Inland Steel produces sheet of which one side is galvannealed and the other side is hot-dipped zinc.

With regard to paint adherence on a standard galvanized product, research has shown that paint retention depends on the orientation of the zinc crystals in the spangle

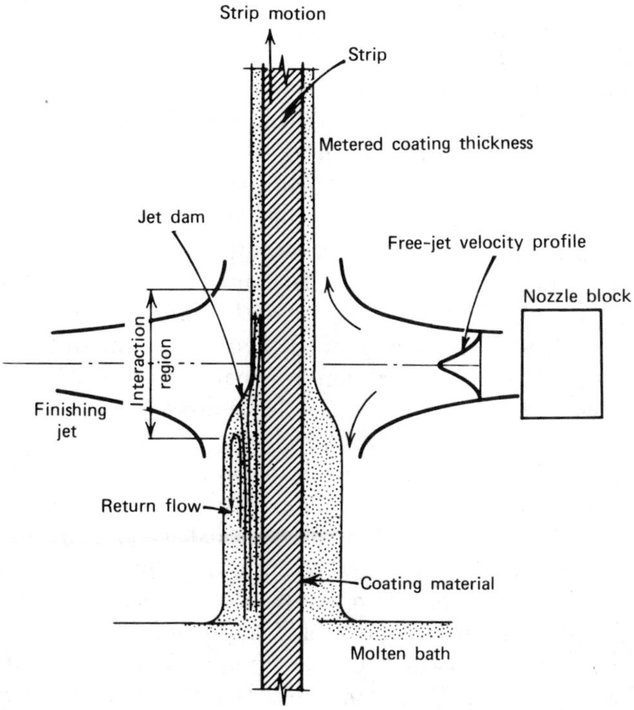

Figure 1. Schematic representation of jet-finishing process for hot-dip coating of strip.

(26). The crystal orientation affects the amount of carbonaceous residue which causes poor paint adherence. The sulfur-containing organic matter present on the surface is more readily removed by alcohol washing when the basal plane of the crystals is parallel to the steel substrate. Therefore, processing of a galvanized product to produce such a spangle orientation is preferred and undoubtedly will become accepted practice where the product use requires painting.

The characteristic spangle of galvanized sheet results from the rate of crystallization of the molten zinc, which depends on the condition of the starting steel and the presence of minor additions to the melt. The latter lower the melting point of the zinc and, thereby, lower the cooling rate of the molten layer. Large producers, eg, National Steel, offer regular spangle, minimized spangle, and flat bright; the latter is recommended for painted parts. Commercial grades of standard galvanized sheet, eg, G90, has 0.275 kg/m^2 of zinc amounting to a 0.19-μm thick layer on each side. The heaviest grade, G235, has ca 3 times as much zinc as the G90 material.

The various methods of applying zinc to steel surfaces have been compared with regard to capital costs and maintenance, and it was concluded that hot-dip galvanizing was the most advantageous method of application (25). A recent report included analyses of emissions from 17 hot-dip galvanizing lines; total particulate emissions from hot-dip galvanizing operations in the United States is estimated to be ca 1600 t/yr (27). Waste disposal from one galvanizing wire-coating operation involves two 36,000-L sulfuric acid pickling tanks and one dip-type rinse tank in an ammonium-fluxed, zinc-coating line. After being collected, neutralized, and aerated, the rinse water is discharged into the city sewer (28). Oxidized, solid sludge also is acceptable for discharge, eliminating the need for clarification equipment. Based on these limited examples, it can be tentatively concluded that no significant environmental hazards are associated with hot-dip zinc galvanizing.

Aluminum–Zinc. Within the past several years a hot-dip-processed coating has been introduced by Bethlehem Steel Corporation. The product is a 55% Al–Zn coating on steel; it is called Galvalume in the United States and Zincalume in Australia (29–30). The coating offers the resistance of aluminum with the galvanic protection of zinc to protect cut edges. The coating is applied by dipping at 593°C. The coating thickness on each side of the sheet normally is 20 μm and the thickness is controlled by automated jet finishing. The coating consists of a multiphase outer region containing 80 vol % α-Al, 22 wt % Al–Zn eutectoid, and a silicon-rich minor constituent. A layer of a quaternary intermetallic Fe–Al–Si–Zn (10% of the total thickness) is bonded to the steel at the steel interface. These regions are depicted in the micrograph of the coating cross section shown in Figure 2. Corrosion tests conducted over 14 yr show that the equivalent thickness of 55 wt % Al–Zn offers 2 to 6 times longer life than regular galvanized steel in a variety of atmospheres, including marine exposure (31). In moist condensate, 55 wt % Al–Zn outlasts galvanized samples after 4 yr. At temperatures up to 700°C, the 55 wt % Al–Zn is equivalent to type 1 Al-coated samples. Current production rates of 350,000 t/yr will increase to 750,000 t/yr in the early 1980s. The principal uses are for building roofing and siding and for high temperature parts used in appliances and automotive parts. In many products, it has replaced both aluminum-coated and galvanized sheet because of its superior performance and lower cost. Its salt-corrosion resistance makes it suitable for automotive body parts and its high temperature resistance allows its use in mufflers and exhaust-system components. Because it can be offered in high strength grades and has exceptional corrosion re-

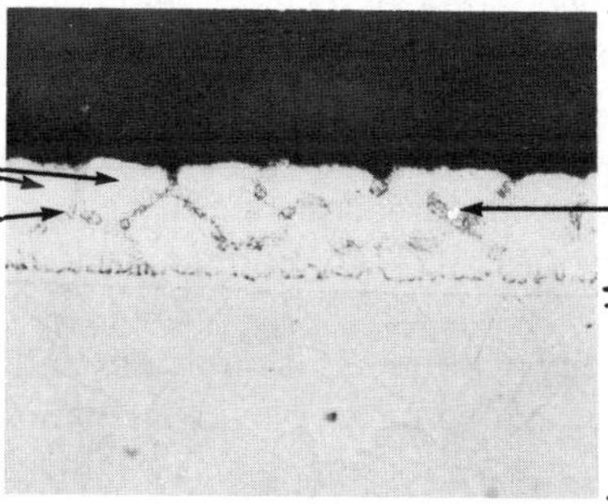

Figure 2. Random cross section of a 55 wt % Al–Zn coating (×500, Amyl-Nital etch). The steel substrate is bonded by a thin intermetallic layer which is overlaid with a solidified layer containing aluminum-rich dendrites, a zinc-rich interdendritic region, and silicon-rich particles.

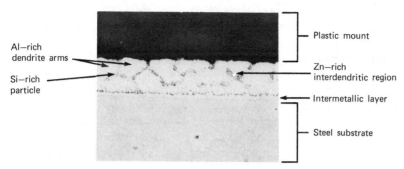

sistance, it is expected that the number of uses will continue to grow as more information on its performance becomes available.

Cementation Coatings. The cementation process is conducted in a mixture of inert diluent particles, eg, alumina or sand; the coating metal in powder form; and a halide activator, which is poured or packed into a metal container with the part to be coated. The container usually is sealed against air entrainment but is arranged so that volatilization of the activator can drive the air from the pack mix. The pack is heated to 800–1100°C and held for ca 1–24 h, depending on the thickness of the desired coating. The coating metal is transferred to the basis material by the formation of a volatile metal halide which is transferred through the pack mix by volume diffusion. Decomposition of the halide at the part surface provides a coating metal which diffuses into the basis metal, thus forming compounds as dictated by the equilibrium phase diagram but limited by the activity of the coating metal. This activity often is controlled by using alloys or intermetallic compounds of the coating metal. The size of the part that can be coated in this manner is somewhat limited by the time required to heat a large pack vessel and the ability to heat the contents uniformly. Small parts, eg, turbine blades, screws, and nuts, are handled commercially. However, pipe and tubing as long as 14 m and up to 2 m in diameter that is intended for refinery and chemical-plant service have been coated by pack aluminizing (32). The pack cementation process has been applied primarily to Al, Cr, B [*7440-42-8*], and Zn (Sherardizing), although some work has been done with Si, Ti, and Mo [*7439-98-7*] and many other elements on an experimental basis. Details regarding many of the diffusion coating techniques and possible elements that can be utilized have been reported (33). Two new pack cementation processes have been described; one involves magnesium in the development of an adherent, diffused, sacrificial coating anodic to the substrate (34) and another involves manganese-rich layers which work-harden rapidly on running-in, to form a wear-resistant, adherent coating on carbon steel (35).

Aluminizing. Pack-aluminized coatings on superalloy turbine blades and vanes increase the parts' resistance to oxidation and sulfidation at high temperatures. This high temperature resistance to gaseous corrosion results from the formation of an adherent Al_2O_3 protective oxide film which grows very slowly at high temperatures and acts as a barrier to further metal loss.

There are two very different coating structures that can be obtained for diffusion aluminum coatings, depending on the activity of the aluminum in the pack mix: inward and outward-diffusing types. When the aluminum activity in the pack is low, nickel is the predominant diffusing constituent ($D_{Ni}/D_{Al} = 3$ where D is the diffusion coefficient. Units are cm^2/s.), whereas at the stoichiometric composition, this ratio changes abruptly and aluminum becomes the predominant diffusing species in high aluminum, NiAl [*12003-78-0*] ($D_{Ni}/D_{Al} = 0.1$) (36–38). Similar studies on Al–Fe interdiffusion have been carried out (39).

One of the problems with pack cementation processing is the cost of removal of the pack constituent from the coated parts, particularly those parts having intricate passages or reentrant angles. A pack process that has good throwing power for complex parts with internal passages can be obtained with parts that are held over the pack mixture rather than being immersed in it; this also alleviates the problem of pack-mix removal. A pack process has been developed which involves either NaCl:$AlCl_3$ at the stoichiometric weight ratio of 3:7 ($NaAlCl_4$) or Na_3AlF_6 as the activator (40). Gas-phase deposition of aluminum in the pack process increases with decreasing partial pressure of the transporting agent, and the rate of deposition is highest for the most stable aluminum alkaline-earth halide. This process undoubtedly will be commercially important for the coating of advanced turbine hardware with intricate air-cooling passages.

Diffusion aluminizing of steels for petrochemical applications, eg, the protection of ethylene pyrolysis tubes and downstream heat exchangers from coke formation and carburization is being considered by Alon Processing, Inc. Applications are expected to increase for coal-gasification and -liquefaction systems to provide resistance to sulfidation, carburization, and abrasion.

A new application of the pack aluminizing process has been developed by the Alloy Surfaces Co., which is using the process to prepare catalytic surfaces called BD catalysts. These materials are made by first subjecting a nickel metal or nickel-alloy substrate, eg, a gauze, sheet, screen, or other desirable configuration, to the pack aluminizing treatment to form an intermetallic compound, eg, NiAl. A leaching operation partially removes the aluminum to form a fragmented, high surface area coating. The catalyst, which is similar to Raney nickel, can be used in methanation; it imparts high durability and the opportunity for exceptional space velocity. Other basis metals and diffusing elements also have been used to form catalytic surfaces on metals, eg, platinum, silver, rhodium, iron, palladium, stainless steel, and titanium. Because aluminum diffuses readily in Ni, Cu, Fe, Ti, Pt, and Pd and readily forms intermetallic compounds with them, a significant emerging source of novel catalytic materials will be available which is based on the pack cementation process.

Chromizing. Chromizing is a very popular and economical process by which the corrosion and wear resistance of low cost steels can be improved. The pack chromizing process, like aluminizing, is a relatively simple method of providing a diffused surface region on a steel part or on a nickel-base alloy. The chromizing process is carried out in a pack containing 30–60% chromium; a few percent of an activator, eg, NH_4Cl or, preferably, NH_4I; and inert diluent, eg, Al_2O_3. Heating the pack for 20–24 h at 950–1100°C produces a diffused coating layer ca 150–200 μm thick. During the initial period in the pack, the process is controlled by decomposition of the gas but, during the remainder of the treatment, the growth of the diffused layer is governed by the diffusion of chromium into the base metal. The chromizing of Armco iron has been studied (41). At the pack temperature, the initial diffusion process begins in the gamma phase (fcc) but, when the chromium content reaches ca 12%, a transformation to alpha occurs. This results in a moving front of alpha phase that has a higher diffusivity for chromium than for the gamma phase (at 1273 K, Cr diffuses 3.5 times faster in alpha than in gamma) so that a sharp discontinuity in chromium concentration occurs at the phase boundary. Chromium also diffuses more rapidly in the grain boundaries and tends to precipitate a $Cr_{23}C_6$ carbide phase and, at the same time and because of the affinity of chromium for carbon, the region below the coating is decarburized. These changes in composition affect the mechanical properties of the coated object, eg, the fatigue resistance; depending on the steel, the chromizing process does not provide any increase in fatigue life and may substantially deteriorate it. Nevertheless, sheet metal components of low carbon steel can undergo extensive forming operations after chromizing without cracking, spalling, or peeling even if bent 180°. If the steel to be chromized contains >0.3 wt % carbon, a chromium carbide forms on the surface of the steel as a result of the diffusion of carbon within the steel toward the chromium-rich surface; the resultant coating is not as corrosion resistant. If corrosion resistance is desired on these steels, they can be decarburized prior to chromizing, or it is possible to use steels with small percentages of the carbide stabilizing elements Nb, Ti, or Zr, eg, the high strength, low alloy steels, to achieve the required strength.

The hard chromium carbide surface coating may be desirable for wear resistance. The carbide grows outwardly in contrast to the inward diffusion of chromium in low carbon alloys, and the thickness of carbide depends on the rate of carbon diffusion through the carbide. The rate of diffusion of Cr in steel, like that of aluminum in NiAl, also depends on the stoichiometry since lowering the chromium activity in the pack increases the extent of diffusion. The type of carbide which forms depends on the chromium and carbon content in the steel. High Cr, low C steels form 23:6 carbide. High Cr, high C steels form a 7:3 carbide with a possible 23:6 outer layer. Low Cr, high C steels may form 23:6 and 7:3 carbides but will form a Fe_3C-based cementite between the 7:3 carbide and the substrate. Low Cr, low C steels may form either 23:6 or 7:3 carbides in equilibrium with the gamma phase.

A growing market for chromized steel is for automotive hardware, eg, mufflers. Pump components, which have been made from cast iron or stainless steel, can be made from chromized gray cast iron which increases component life by a factor of 10. Output of tube-drawing dies has been improved by chromized–carbide layers. Small components, eg, nuts, bolts, washers, etc, can be chromized by utilizing a rotary furnace. New and service-worn gas-turbine blades of both nickel and cobalt-base alloys are conditioned for service by chromizing.

Chrom-Aluminizing. Although dry corrosion resistance of either aluminized or chromized basis metals and alloys is excellent, further improvements are effected using a two-step process of chromizing followed by aluminizing. Above ca 900°C, chromized materials begin to rediffuse which results in a reduction in the surface chromium concentration. Aluminizing following chromizing can change the nature of the protective oxide from Cr_2O_3 to Al_2O_3; the latter is more stable and is less volatile at high temperatures. Thus, improved scaling resistance is possible above 1000°C and performance is superior to 18-8 stainless steels and Inconel (42). Two-step chrom-aluminizing treatments have been optimized for a number of nickel- and cobalt-based alloys used in high performance turbine engines. However, these duplex pack cementation coatings are more expensive and, therefore, are not widely used for common applications.

Boronizing. Boronizing is a diffusion process by which a boron-rich layer, which is similar to a carbided or nitrided surface layer, can be formed on steel or on another substrate. Extremely hard surface layers result, provided borides are formed. The pack cementation process is one of several techniques for boronizing and is carried out in closed retorts into which the parts to be boronized are packed with a mixture of 50% boron powder, 49% alumina and 1% $NH_4F(HF)$. Ammonium bifluoride acts as an activator to carry boron to the surface of the pack. After being sealed, the retort is heated at 800–900°C for 6–24 h. The steel is heat-treated for mechanical strength after boronizing and the nature of the coating is not affected because its hardness is a result of the formation of boride intermetallics. There are two iron borides, FeB [*12006-84-7*] and Fe_2B [*12006-85-8*], at 16.25 and 8.84 wt % boron, respectively; the former compound is the harder of the two but generally is not used because of its increased brittleness. The diffusional thickness of the boron compound follows a parabolic relation with time and is greatest for low alloy steels. It is difficult to form boronized layers exceeding 10 μm on alloy steels with Cr or refractory elements, eg, a high speed steel. On plain carbon steels, a 150-μm layer can be formed in ca 6 h. Boron has limited terminal solubility in the allotropic iron structures; thus, Fe_2B is the only constituent of the coating regardless of the application of postcoating treatments designed to modify it. If the activity of boron in the pack is high, FeB forms in the saturated Fe_2B. The integrity of the coating depends on the nature of the interface between the basis metal and Fe_2B. If there is a jagged saw-toothed interface, as occurs with plain carbon steels, the interfacial strain is accommodated more effectively and the coating is more adherent and less prone to microcracking. On the other hand, the smooth interface resulting from boronization of alloy steels leads to less desirable adhesion, and spallation and microcracking are likely, particularly at corners and edges. Thus, the corrosion resistance of boronized plain carbon and alloy steels differs widely.

The outstanding feature of the boronized coating is its hardness, relative to nitriding and carburizing layers, on steel. The Vickers microhardness of boronized steel typically is between 1600 and 2000, whereas a nitride layer is ca 600–900 and a carburized surface is ca 700–800. Heating to 1000°C affects the hardness of the latter coatings but not that of the boronized layer. The great affinity of boron for oxygen results in a thin oxide layer on the boride; the layer appears to provide an antiwelding surface which reduces the interaction of other metals with it. Therefore, boronizing is a particularly effective measure to improve the wear characteristics of steel surfaces (43). In recent tests of the relative cost effectiveness of various materials exposed to a number of abrasives, it has been demonstrated that boronized low alloy steel is as much as three times better than plasma Ni–Co–Cr carbide and significantly better than manual arc-coated WC (44). One company in the UK (Ronson Products Ltd.) uses boronized mild-steel jigs to carry parts through an abrasive polishing process.

Boronized parts, eg, Borofused wire dies, also show chemical corrosion resistance to HCl, HF, and/or H_2SO_4. However, there is no evidence of protection against corrosion by HNO_3 nor is there significant improvement in regard to conventional rusting of ferrous alloys, probably because of the tendency to microcracking which results from the brittle borides. Molten zinc is very corrosive to mild steel parts; one solution in galvanizing is the substitution of boronized mild-steel fixtures for the previously used more expensive titanium.

Siliconizing. In siliconizing from the pack, the source of silicon can be elementary silicon, ferrosilicon, or silicon carbide. The inert diluent material usually is Al_2O_3 and the activator generally is NH_4Cl at ca 2–5% of the weight of the pack mixture. Siliconizing of Armco iron or low carbon steel is not an efficient process; eg, 10 h of siliconizing at 1100–1200°C is required and there is considerable consumption of the pack source. Iron articles are limited in the silicon content of the coating layer to 5–12% silicon at the expense of a considerable increase in grain size. Although the silicon-rich surface improves the corrosion resistance of the material in weakly corrosive media, the high temperature corrosion resistance above ca 700°C of silicon-containing coatings that are applied in this way is not satisfactory, the improvement over untreated material being less than a factor of two. However, siliconizing followed by chromizing results in thicker diffusion layers, because silicon aids in stabilizing the alpha phase in which chromium diffuses more rapidly. Siliconizing is more effective as a deposition process for coating the refractory metals, eg, Ti, Nb, Ta, Cr, Mo, and W, where silicides can form. The pack silicide coating for Mo is one of the oldest and best-studied coating systems. The coating depends on the formation of $MoSi_2$ [*1317-33-5*] and the oxide responsible for protection is SiO_2 [*7631-86-9*]; the latter effectively protects molybdenum in air up to 1700°C. Small $MoSi_2$-coated molybdenum rocket engines were used very successfully in the Apollo program for attitude control of the command module and Lunar Excursion Module (LEM) (45). Where very high temperature oxidation resistance is required, the silicide-coated refractory metals offer the best performance.

Sherardizing. Sherardizing or dry galvanizing is the oldest and the most widely used pack cementation-type diffusion process (46). Parts to be coated are cleaned and packed in zinc dust in a metal container which, after sealing, is rotated slowly and heated to 350–375°C for 3–12 h. The diluent phase is comprised of zinc oxide and other impurities which are obtained by recovering used dust and by innoculating the new charge. Iron and iron oxide are undesirable contaminants and must be removed periodically from the charging inventory.

The coating consists of a zinc-rich intermetallic, $FeZn_7$ [*12023-07-3*], which can be accompanied by $FeZn_3$ [*60383-43-9*] if the process is carried out at >375°C or, if the part is subsequently heat treated. Because it is a diffused coating, the surface structure of the part is replicated in the coating and uniformly applied over the surface. As with other intermetallics formed by diffusion, the coating tends to be microcracked. However, the $FeZn_7$ is anodic to the steel and protects it sacrificially.

Because of its uniformity of coverage, the Sherardizing process is unexcelled in providing protection on steel hardware, eg, nuts, bolts, washers, and other close-fitting parts. It is likely that a steady market will persist for this coating. However, with the advent of mechanical plating and ion plating as techniques for fastener improvement, the range of materials coated by Sherardizing will depend on the specific application and the relative economies of these processes.

Liquid-Carrier Diffusion Coatings. Diffusion coatings also can be made by immersion of the basis metal in a liquid bath containing the dissolved coating metal. The bath can be composed of fused salt mixtures or liquid metals, eg, Ca or Pb, which can dissolve small amounts of the coating metal but in which the basis metal is not dissolved. The coating takes place because of the difference in activity of the coating metal in the bath and that of the basis metal. Generally, because of limitations in the amount of material that can be dissolved in such baths, the surface composition of the coating metal in the basis metal is lower than for other coating processes. Other limitations are related to the physical problems of containing and operating high temperature baths which may corrode containers and sometimes damage the basis metal.

Fused Salt. Transfer of metal can be accomplished in molten salt mixtures. One such bath for chromizing has a composition of 40 mol% NaCl, 40 mol % KCl, and 20 mol % $CrCl_2$ (47). The $CrCl_2$ is formed by adding $CrCl_3$ and sufficient Cr metal to reduce the trichloride *in situ* to $CrCl_2$. Carbon steels at 1000°C form chromium carbides in accordance with the Fe–Cr–C-phase diagram. In 30 h of exposure, 11 μm of $M_{23}C_6$-type carbide is formed at the outer surface over a 24-μm thick layer of M_7C_3. The carbides grow mainly by diffusion of carbon from the interior of the steel through the carbide layer to the outer surface where it meets chromium and forms the carbide. One difficulty with the operation of fused-salt baths is the necessity to scrupulously avoid contamination by oxygen. This is usually accomplished by purging with purified argon.

Toyota Central Research and Development Laboratories has announced a carbide-coating process which involves the immersion of parts into a molten-borax bath at 800–1200°C in ambient atmosphere (48). Metals are placed in the bath in the form of powders. Carbides, eg, Cr_7C_3, VC [*12070-10-9*], and NbC [*12069-94-2*], are formed on the surface of carbon-containing parts. Articles, eg, metal-working dies, knives, and machine components, have been coated with carbide layers that are characterized by excellent wear, abrasion, and corrosion resistance. Applications for this technique are expected to grow worldwide.

Boriding of low carbon steels can be done in fused mixtures of boric acid and potassium borate and diffused coatings containing titanium carbide and boron carbide can be formed on carbon steels from fused mixtures of boric acid, TiO_2, and potassium fluoride at 1020–1170°C (49). Diffusion layers 200–250 μm thick can be formed in 1 h or less.

Liquid Calcium. One liquid-metal diffusion-coating technique involves the use of molten calcium as the transfer agent (50). A solubility of only 0.1% for the coating element is required in the calcium bath for rapid transfer of the solute to the part. An argon-protected calcium bath with ca 10 wt % of the desired coating element is prepared, heated to 1100°C, and agitated, and the part to be coated is immersed for ca 1 h. In this manner, a 50-μm thick coating of Cr is obtained on iron in which the surface concentration of Cr is 45%. On removal from the bath, the part can be cooled in air or quenched in oil and, once it is cool, excess calcium may be removed in hot water or dilute HCl. Multiple diffusions can be made simultaneously by addition of the appropriate concentrations of the required elemental constituents. In this way, Ti, V, Cr, Mn, Co, and Ni have been simultaneously diffused into iron. However, with iron substrates, it usually is difficult to obtain coatings with less than 50% iron at the surface. Outward diffusion of carbon in iron also can occur and, if the bath contains chromium, carbides form in the bath until the chromium is saturated in carbon; then, chromium carbide ($Cr_{23}C_6$) forms on the surface of the iron. It is possible to select chromium or chromium carbide as the coating for substrates with high carbon or cast iron because calcium also is an effective decarburizing agent. Liquid metals are efficient media for transfer of metallic and nonmetallic elements between two metals and diffusion can be as rapid as when the two metals are in direct contact.

The surface of austenitic stainless steel can be transformed to a ferritic layer by removing the Ni into the bath and increasing the chromium level by treatment in a calcium–chromium bath. Such a ferritic layer can improve the chloride stress-cracking resistance of the austenitic steel. High carbon steel and cast iron can be decarburized at the surface and coated with a ductile layer of corrosion-resistant chromium. Oxidation-resistant coatings containing Al, Cr, Si, and Ni can be applied to steel from calcium baths. Refractory metals, eg, Mo, can be coated readily with Al, Si, or Cr. An example of a multielement coating composition on steel is 45 wt % Cr–52 wt % Fe–2 wt % Ni–1 wt % Al; this coating has excellent resistance to the CASS (copper accelerated salt spray) test (3). The practical exploitation of this process can be expected to be retarded by the problems of fire safety and control of alkali fumes associated with the finishing process. It is more likely that less hazardous liquid-metal baths, eg, lead, will provide a more economical approach.

Lead Bath. Materials Sciences Corp. has developed a process called Dilex 101 to chromize steel using molten lead baths in ca 1100°C for 4 h (51). The optimum chromium content of the bath is 0.85%, based on the weight of lead. The alloy bath may involve other diffusing elements, eg, Co, Ni, Y, Mo, Ti, Nb, V, Ta, W, Si, and Mn. Multiple-element, diffusion coatings, eg, Cr–Al coatings for heat-exchanger materials, have been applied successfully by this method.

It is not likely that this process can be used for parts that are required to operate at high temperatures because of the possibility of contaminating the part with lead which can deteriorate the creep-rupture properties of nickel-based alloys and high alloy steels. However, many other applications are being investigated, eg, coating powder metallurgy (qv) parts, wire, tubing, valves, fittings, etc.

Spraying

Sprayed coatings generally are applied to structures or parts which either are not conveniently coated by other means because of their size and shape or are susceptible to damage by the heating requirements of other coating techniques. Slurry coatings and electrostatic powder coatings require heating to the fusion temperature either by massive heating of the part or by localized heating, eg, by induction, electron-beam, or laser techniques and generally in a protective atmosphere.

Flame-spraying and arc-spraying techniques are used in a large variety of industrial applications, eg, in both shop and field situations because equipment usually is portable and can be taken to the work site. However, laser, electrostatic, and slurry coatings must be formed in the shop.

Flame Spraying. ***Oxyacetylene.*** Flame spraying is the simplest of the thermal spray techniques; it is used where heating the substrate above 315°C would cause undesirable tempering, recrystallization, oxidation, or warping. Flame spraying uses oxyacetylene or oxypropane flames with flame temperatures at ca 2750°C which is adequate to spray most ferrous and nonferrous coatings and oxides, eg, alumina and zirconia, but only to densities from 85–95%. Either wire or powder is fed into the flame. The heat of the flame melts the coating material and accelerates it toward the workpiece where particles fuse as interlocking laminates, each layer fused to the previous one. The flame oscillates over the part surface, giving a uniform coating. More than one pass can be made with several materials as required; thus, worn parts can be salvaged by building up material in the work area and machining back to the original dimension (52). One utility has found that rebuilding their transit-department vehicles has been consistently less expensive than the cost of replacement (53). Worn shafts, axles, packing sleeves, and journals are sprayed with a nickel–aluminum bond coat followed by aluminum bronze to a 3.2-mm thickness before remachining. Flame-sprayed materials, usually Zn or Al, also have been applied to large storage tanks and to at least one bridge for corrosion protection (54–55).

Detonation Gun. The detonation gun, invented by Union Carbide, overcomes many of the limitations of the flame-spray process. The detonation gun is a cannonlike device that detonates metered mixtures of oxygen and acetylene in a combustion chamber. Powder particles of the coating material, which are carried in a nitrogen stream, also are metered into the chamber prior to spark ignition. The shock wave leaving the barrel at ca 2770 m/s accelerates the powder particles to ca 770 m/s. Particles also are heated by the 3000°C combustion gases at which temperature most coating materials melt. The high temperature of the particles at this characteristic velocity produces a coating having exceptionally high bond strengths, ie, >98 MPa (>14,200 psi), and porosities <1%. The detonations are repeated 4–8 times each second and are accompanied by short nitrogen purges to clean the barrel.

The majority of coating materials applied by the detonation gun (D-gun) are oxides and carbide mixtures with various bond metals, eg, NiCr and Co. The coatings provide exceptional resistance to wear (56). A disadvantage of the process is the supersonic velocities that are produced and that require double-walled, soundproof cubicles and remote-control operation. The process also is line-of-sight, which limits the type of product that can be produced. The D-gun process has been estimated to be the most expensive of the spray techniques, depending on part size, fixturing, masking, and use (57). However, initial cost is compensated by the superior life of these coatings, which often outlast conventional metal spraying and weld surfacing by 8 to 1.

Arc Spraying. ***Wire-Arc.*** In wire-arc spraying, two wires are fed to a gun through two electrical conduits which bring the wires together at a 30° angle. On contact, an arc is struck and melts the wire ends. Compressed air drives the liquid metal forward to the work. As the arc is broken, the wires are advanced to repeat the process. The arc temperature (ca 3800°C) causes deposition of molten droplets ca 3–8 times faster and with more fluidity than oxyacetylene flame-spray units. At an arc current of 250 A, 7, 12, or 42 kg/h of Al, stainless steel, or zinc, respectively, can be sprayed. The bond quality is better than that for flame-sprayed material. The equipment is light and portable and is simple to use with any coating material which can be made into wire form. One arc-spraying facility includes an enclosure 7 × 13 × 3-m high which is purged and filled with purified argon which also is the gas used to drive the molten material from the arc-spray unit (58). Operators are provided with space suits equipped with independent breathing apparatus. Large components up to 2.75 m in diameter have been coated with 1-mm thick coatings of titanium. One experimental heat-exchanger tube sheet which was coated in this protective chamber has been running for more than 5000 h in a seawater desalination plant with no problems. The high quality of the coating product facilitates machining, welding, and forming operations. Moreover, freedom from internal porosity is obtained. These coatings have been produced without the substrate heating requirements of weld surfacing yet they are characterized by good bonding and high density.

A definite advantage of the arc-wire spray is the simplicity of operation and the absence of the degree of noise which is characteristic of plasma torches. Inability to deposit material that is not fabricated into wire is perhaps a disadvantage but more of the material reaching the substrate surface is molten leading to, in many cases, much better bonding and excellent high density.

Plasma. The plasma-spraying process utilizes the available energy in a controlled electric arc to heat gases to ≥8000°C. The low voltage arc is ignited between a water-cooled tungsten cathode and a cylindrical water-cooled copper anode. Argon, nitrogen, or hydrogen or suitable mixtures of these gases are heated in the annulus and are expelled at high velocity and temperature into a characteristic flame. Powder material, either metallic or nonmetallic, is fed into the flame just downstream of the anode. Particles of the powder are melted and accelerated toward the work to be coated. Since it is hotter than the oxygen–gas flames or arc-wire systems, the plasma device can deposit W, Mo, tungsten carbide, and numerous ceramic materials. Usually the particle velocity is 125–300 m/s but, with plasma guns working in subatmospheric chambers, velocities of up to 460 m/s and extremely fine, dense, and wear-resistant coatings can be obtained. Metco, Inc. has supplied a significant portion of plasma-spray equipment, powdered material, and applications technology. The structure differences and mechanical behavior of plasma- and detonation-gun coatings have been reviewed (59). Plasma-sprayed metallic coatings have characteristic porosities of 5–15%. A great deal of effort has been made to increase the particle velocity and thereby reduce the porosity to improve the corrosion resistance of these coatings, particularly at high temperatures. An improved plasma process called Gator-gard is characterized by a particle velocity of 1230 m/s, which provides coating densities >99%. Deposit density and efficiency also is related to the uniformity and purity of the powder materials. A narrow distribution produces the best results because large particles may pass unmelted through the flame, whereas small particles are vaporized and lost. Many companies can provide spherical, uniform, high quality powder material by various techniques. If a metallic envelope can be provided over the sprayed material then hot isostatic pressing can be used to heal internal defects and, thereby, improve the finished density of plasma-sprayed coatings (60).

The plasma-spray coating technique has been used to deposit molybdenum and Cr on piston rings, cobalt alloys on jet-engine combustion chambers, tungsten carbide on blades of electric knives, wear coatings for computer parts, etc. The technique is used to increase component life and to reduce machinery down time (61–62) (see Plasma technology).

Laser Coating. Laser power is applied to produce sprays of powdered material for coating purposes (see Lasers). Powder particles can be accelerated in the laser beam and melted before striking the substrate material where rapid solidification takes place (63). Power from a 25-kW CO_2 laser is directed into an evacuated chamber and focused on the substrate surface. A carrier gas, eg, helium, is used to transport powdered material through a gold-plated water-cooled nozzle which projects powder into the laser beam. The accelerated molten particles impact on the substrate and, depending on the power density at the substrate, either coat the substrate with a quenched structure or are incorporated into a region of the substrate which has been melted by the beam. Solid particles, eg, carbides, in the latter process can be incorporated into the molten matrix with little dissolution, thereby producing a modified surface region that is up to 1 mm thick and impregnated with hard particles for wear resistance. A similar process, called laser alloying, uses high power laser energy to melt a thin layer of material at the surface of a part, to which alloying elements are added, to produce a chemically modified region on subsequent solidification (64). In one example, a paint that contains an alloy with 30% chromium is placed on the outer edge of an engine valve. A ring-shaped laser beam scans the painted area and produces a chromium-rich region in the valve. A cost analysis of laser alloying versus conventional hard-facing techniques indicates that an 80% savings can be achieved by the laser process (65). In laser cladding, a prepositioned coating material is melted so that it is bonded to the substrate surface but not alloyed with it, much as in braze surfacing, conventional hard facing by plasma techniques, or flame spraying (66).

For example, a Stellite Alloy No. 1 is prepared as a hard surface material by the oxyacetylene flame technique and has a structure with massive carbide particles with high (Rockwell) hardness, ie, HRC 70, imbedded in a soft matrix of HRC 45. The average hardness is HRC 51. However, the laser-clad Stellite Alloy No. 1 produces a uniquely homogeneous structure with extremely fine carbide particles, resulting in an average hardness of HRC 60. The cost/benefit of laser-clad coatings results not only from the unique microstructures produced but from the use of lower cost powder material, reduced amounts of necessary coating, rapid processing rate leading to high production rates, and lack of postcladding machining or clean-up operations. The projected in-production coating rate for the laser-cladding operation on a valve seat is 5 s compared with 15 s for other processes. The laser beam can be time-shared at many work stations to prorate the high cost of the laser equipment. In addition, laser beams can be directed into small holes that are inaccessible by other spray-coating techniques. However, the cost of the laser and optical system is a present limitation.

The laser will be used increasingly in the generation of amorphous coatings, eg, by the Laserglaze technique (67). Combinations of laser processing with other techniques may be useful in certain instances. Laser fusing of flame-sprayed Metco 15F (Ni–17 wt % Cr–4 wt % Fe–4 wt % Si–3.5 wt % B–1 wt % C) produces high quality coatings up to 0.25 mm thick on mild steel 10 times faster than by torch methods (68). Evaporation of high melting metals, eg, Pt, which, because of their high thermal conductivity, are difficult to evaporate by conventional electron-beam techniques can be accomplished with lasers. Some success with platinum laser-vapor deposition has been reported for the fabrication of beam-lead integrated circuits using a 20 W YAG laser (69).

Electrostatic Powder Coating. A relatively new application of an old technique is the electrostatic deposition of powders, eg, Al, Cr, Ni, and Cu. British Steel Corp. has been developing a line for an electrostatic aluminum-deposition process (Elphal) with an ultimate capacity of 75,000 t/yr of sheet steel (70). Cleaned strip or sheet is electrostatically coated with metal powder, cold-rolled to compact the coating, and sintered by heating to develop a bond to the steel. The principal advantage is the protection provided at 500°C in service, eg, heating appliances, automotive exhaust systems, and heat exchangers. USSR work on this type of process also is underway. Aluminum alloy powder that is sprayed onto aqueous alkali silicate-coated steel has been deposited to 150 g/m^2 at 7 kV/cm. The strip, which is dried at 350°C, is heat-treated at 500°C for 1 h to develop optimum corrosion resistance (71). The chief advantages of the electrostatic method are the wide variations in coating composition available through prealloying of the powder, the absence of any hot-metal baths, and the conservative use of coating material resulting from the self-leveling and low over-spray characteristics of the process. However, because of the relatively limited markets for specialty alloy coatings on rolled sheet, the role of this type of process in sheet and strip coating probably will not be as large as for hot-dipped or electrocoated material.

A related process involves electrophoretic deposition from a liquid mixture of powdered material (72). Usually, 0.04-mm (325-mesh) powders of the metallic coating material are added to a stirred solution of isopropanol and nitromethane at between 2 and 10%. Addition of ionic agents stabilizes the dispersion by electrostatic repulsion. A d-c potential of ca 100 V is applied between two stainless steel electrodes and the item to be coated. Deposition rates are high, eg, 0.005–0.05 mm/min. Particles, not ions, are plated out by electrophoretic migration. Uniformity of such coatings are excellent as a result of the self-leveling effect on the initial deposit which attaches at points of high field strength. The insulating effect of the coating reduces the rate of deposition in these regions in favor of points with lower field strength. Once applied, the coatings are densified by hydrostatic pressing followed by sintering or sintering alone if coatings are not to be handled in the green state. The sintering temperatures can be as high as 1100°C, depending on the coating and basis material. Among the advantages cited for this method are the ambient temperature of application, uniformity, versatility regarding the material used in the coating, ease of automation, and low cost which is one half that of pack cementation coatings and one tenth that of vapor deposited coatings that offer similar protection (see Powder coating).

Slurry Coatings. Metallic coatings can be applied simply by applying powders of the desired metal or alloy in a paint medium and by brushing, dipping, or spraying it onto the basis material. The coating is cured and then fired, during which time, the organic portion vaporizes and the metallic particles fuse to form a dense, metallurgically bonded coating. The coatings are versatile, both as to their composition and the shapes of the basis materials that can be coated. Slurry coatings were developed for commercial use in the 1960s for gas-turbine hardware and for coatings on niobium alloys. A Si–20 wt % Cr–20 wt % Fe, fused slurry coating is used commercially on niobium-alloy afterburner components for the F-100 jet engine and on niobium-alloy rocket nozzles for the space shuttle where the coatings are exposed to temperatures of 1370–1650°C. Similar coatings were developed for NASA by Lockheed for tantalum–tungsten alloy components (73). A family of nickel–chromium slurry coatings, Nicrocoat, was developed for high temperature, and corrosion and abrasion resistance by Wall Colmonoy Corp. for furnace fixtures, heater tubes, thermocouple protection tubes and automotive parts (74). More recently, Lockheed Palo Alto Laboratories has developed a slurry coating of 33 wt % Cr–63 wt % Al–4 wt % Hf for IN 800 alloy which is exceptionally resistant to sulfidation (75). Homogenization after coating by laser-fusion treatment has increased the corrosion and abrasion resistance of these coatings.

Combinations of slurry-fusion and pack cementation techniques have been developed for experimental evaluation by the Solar Division of International Harvester Company for advanced turbine-engine applications designed specifically for operation in marine environments where protection against sulfidation attack is critical. Slurries of Ni–Co–Cr have been applied to parts and allowed to react in cementation packs to deposit aluminum. The coatings that are produced protect against hot corrosion and oxidation as well as contemporary, vapor-deposited coatings. The low cost and versatility of combination methods of this kind is promising for other technologies where hot corrosion problems are recognized as limitations, eg, in high sulfur crude processing and coal-gasification units for synthetic fuel production (see Fuels, synthetic).

Mechanical and Liquid-Metal Cladding

Metallic coatings also can be applied by mechanical methods where the coating material is forced into intimate contact with the basis metal such that the forces at the interface disrupt and disperse the boundary oxide films existing on each of the constituent metals. Formation of a metallurgical bond may be augmented by mechanical attachment and thermal interdiffusion. Metallic coatings also can be melted into place by a weld-surfacing or casting operation. These techniques generally are applied to large, heavy basis metals in the form of plates or large forgings and usually produce quite thick coatings in comparison to other types of processes.

The selection criterion for a given method of producing a composite plate is the required thickness of the desired protective material and the thickness of the basis material to which it is applied. If 1 cm is adequate, the solid coating (cladded) alloy often is used alone. For a basis material greater than 1 cm and up to 6 cm thick, roll-bonding techniques are used. Explosion bonding is most often used when the basis metal thickness is ca 6–8 cm (see Metallic coatings, explosively clad metals). Beyond a 10-cm basis-metal thickness, only weld-overlay techniques or electroslag-casting techniques are practical.

Roll Bonding. The main sources of roll-bonded, clad plate are the Lukens Steel Company and the Phoenix Steel Corporation. In roll bonding heavy plate, a four-ply sandwich is made in which two basis-metal backer plates enclose two of the clad metal plates which are nickel plated on the sides exposed to the backer, and a parting agent is between the clad so that, when the rolling is complete, the sandwich can be opened to produce two clad plates (see Abherents). The nickel electroplate metallurgically bonds the clad to the backer. The sandwich is sealed around the edges by strips that are welded to the backers before hot rolling.

One of the most important applications of roll-bonded heavy sheet is in the fabrication of clad vessels. One of the largest and most experienced fabricators in the United States is the Nooter Corporation of St. Louis, Mo. Most of the vessels are used in chemical plants, eg, for acid-gas removal from process streams. A growing market is clad vessels for coal-gasification plants. Another new market is in shipbuilding. The 62-m ship Copper Mariner has been built with roll-bonded 90 wt % Cu–10 wt % Ni cupronickel [*11114-42-4*] on steel plate, with the clad occupying 10% of the composite thickness. Fuel savings and lower maintenance costs resulting from lack of fouling with marine organisms is between $5,000 and $10,000/yr. In terms of the compensated tonnage coefficient benefits, LNG carriers should benefit the most from clad construction. A minimum practical thickness for normal shipyard handling is 0.15 cm clad which is fabricated ideally by roll bonding. For certain specialty service in chemical-process equipment, tantalum-lined vessels are being produced with 0.25–0.38 mm thick elastomer-bonded tantalum [*7440-25-7*] sheet on steel plate.

Skive inlaying is another roll-bonding process which usually involves a precious metal bonded to a freshly skived or machined surface by pressure rolling, as shown in Figure 3 (76). The freshly skived surface is atomically bonded to the precious metal surface which contacts it for a short time in comparison to that required to develop an oxide film. The process provides savings of 50–90% in comparison to electroplating because the precious metal that is used is restricted to the contact area. Many inlays may be atomically bonded without heat, including solder, Cu–Ni–Si alloy, copper, silver, and gold. These inlays can be bonded into steel, stainless steel, aluminum, brass,

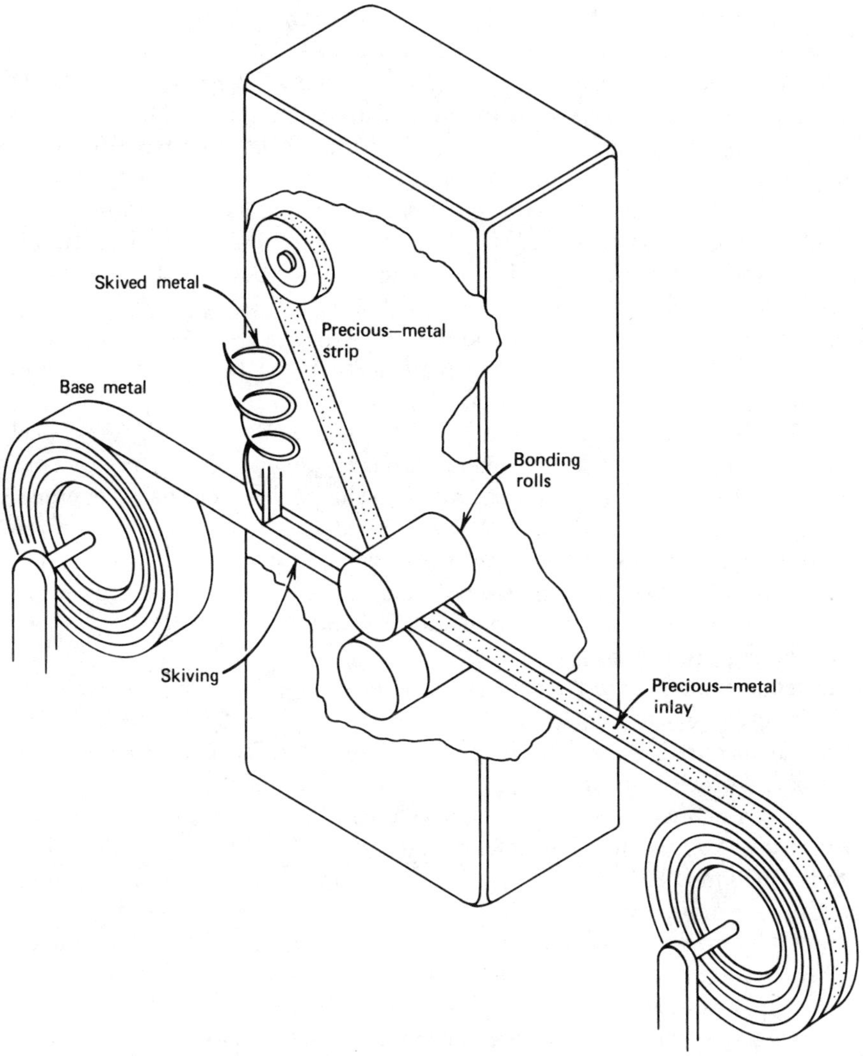

Figure 3. Pictorial view of skive-inlaying technique. A base-metal strip is skived by a blade to reveal atomically clean metal. A precious-metal strip is roll-bonded immediately into the skived groove to produce the inlay. No adhesives or other bonding is required.

bronze, and beryllium–copper. Coils up to 3050 m long and 1.6–102 mm wide have been made. Configurations to suit almost any design requirement can be produced, including edgelays, multiple stripe, and full-width cladding (76). Some of the applications for this new technology include contact springs for data-processing equipment and bonding pads for automobile voltage regulators.

Powder Rolling. Clad strip also can be produced from two kinds of powder by introducing two different powders on opposite sides of a baffle plate that extends longitudinally between two rollers. As the powders fall into the rolls, they are pressure-bonded into a strip with a green density that is dependent on roll speed and roll gap; thin sections have higher green density than thick sections. Sintering at 800–

1000°C follows. Copper and iron have been produced in a 0.1-mm composite strip and characterized by an ultimate strength of 275 MPa (40,000 psi) and an elongation of 25% when sintered to 900°C (77). Chromized steel strip, a development of Bethlehem Steel, has been produced by passing the steel strip, which is lightly coated with tridecyl alcohol, over a fluidized pumping bed of ferrochrome powder. The bed contains brushes which mechanically impinge loose powder up onto the sheet from the freeboard of the fluid bed. The powdered sheet is roll-bonded after being reversed so that the opposite side of the sheet can be coated over a second fluidized bed (78). The strip must be annealed before rolling so that it is soft enough to be penetrated by the powder on contact with the rolls. After rolling, the open-wound coils are diffused at 885°C in hydrogen for 28 h. The coil then is wound tightly to complete the process. The chromium content after heat treatment is 25% at the surface and tapers to 15% at a depth of 0.05 mm.

Roll bonding of Al–Zn alloys is accomplished in a similar way by passing steel sheet that is treated lightly with tridecyl alcohol through a powder bed of a mixture of 26-μm Al and Zn powders. About 325 g/m^2 attaches to the work which then is compacted by work rolls. The strip is heated to 399°C in an 18% H_2–N_2 atmosphere for 5 min to bond the coating. An ambient-temperature skin pass finishes the sheet by providing a smooth attractive surface. The coating is ductile, adherent, and metallurgically bonded (79). New powder compositions also include silicon additions which improve the interface alloying properties.

Mechanical Impingement Coatings. The mechanical plating process developed by 3M Company is the cold welding of a ductile metal onto the surface of a metal substrate by mechanical energy. The process is designed for small parts, eg, fasteners, and is a barrel process in which parts, water, glass shot, and proprietary additives are tumbled with fine metal powder. The glass shot and powder sizes are 0.2–5.5 mm and ca 4 μm, respectively. The plate thickness usually is 0.00245–0.0177 mm but can be as high as 0.076 mm. Alloys can be plated by mixing powders in the barrel. Generally, the coatings are limited to Cd, Zn, Sn, Pb, In, Ag, Cu, brass, and tin/lead solder (80–81). The main advantages of mechanical plating are the absence of hydrogen embrittlement; the coatings can be made with alloys; there is a simple, time-dependent control of coating thickness; the energy requirements are low; it is a room temperature, nonfuming process; and it requires little predisposal waste treatment of used solutions. However, it is limited to small parts and cannot produce a cosmetic coating with fine surface finish; nor is it possible to deposit nonductile powders, eg, Cr or Ni, or active metals, eg, Al.

Mechanical plating is used to coat ca 3×10^6 critical parts per month in one plant and provides corrosion protection without hydrogen embrittlement failures; costs are about the same as for electroplating with afterbake (82).

Mechanical vapor plating, which was developed by Alloying Surfaces Co., involves modulated pulsations which expose powder particles of material to as much as 4425°C at the particle–basis metal interface. Materials, eg, WC, TiC, Mo, Ni, Cr, and Ni borides, are deposited to a 0.45-μm finish at 6 cm^2/min (83). Applications for this process include wear-resistant surfaces and special tooling, eg, metal-forming dies and cutting tools.

Explosive Bonding. ***Weld Surfacing.*** Weld-deposited overlay coatings for small area requirements usually are achieved using manual stick electrodes or metallic inert-gas-shielded arc (MIG) or tungsten inert-gas-shielded arc (TIG) techniques (see

Welding). For larger areas, single or multiple submerged arc, strip cladding, and automated plasma-transferred arc spraying of powders or wire generally are used. The choice of cladding material and the method of application is determined largely by the intended purpose; the product or process to be contained or the general surface property requirement; metallurgical compatibility; relative coefficients of expansion, possible unfavorable circumstances required of the site for the cladding operation, eg, access to the surface; and the inspection requirements for the operation.

Plasma-Transferred Arc Cladding. Plasma and arc techniques can be used for clad overlay deposition. One technique uses a plasma torch in the transferred-arc mode to control basis-metal melting, whereas powders which are introduced in the normal plasma-spray method provide the overlay material. Greater efficiency and deposition rate over oxyacetylene and gas tungsten arc methods are obtained. Deposit rates are 4.5–5.5 kg/h. In gas–metal–plasma arc-weld cladding, the plasma arc is transferred to the work but by an independently controlled power supply (84). A second supply source powers an arc between the two advancing wires and the plasma to melt filler metal. Figure 4 is a schematic diagram of the plasma unit, the current path from power source no. 1 which controls basis-metal melting, the arc-wire circuit powered by source no. 2, and the mechanism for advancing the overlay metal in wire form. In this process,

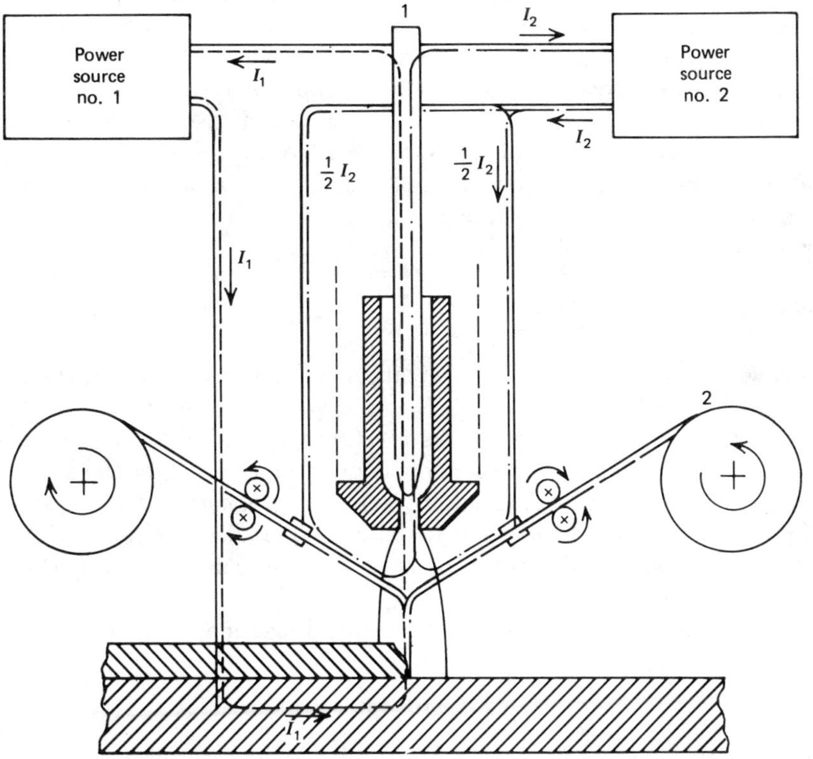

Figure 4. Sketch of gas–metal–plasma arc-weld surfacing technique. Power source no. 1 controls the heat from the plasma into the basis metal. Power source no. 2 controls the heat input from the plasma to the advancing filler wires, thereby independently providing control of melt rate and degree of interface mixing. 1, plasma torch; 2, wire feed. I_1 = current of power source no. 1. I_2 = current of power source no. 2.

which does not utilize a slag cover but depends on the plasma gas and an outer plenum gas as a pool shield, a mechanical oscillator moves the weld head perpendicular to the welding direction to produce a wide weld bead, which can be observed through a window in the protective gas-shield cap. The melting process and metal transfer to the weld pool are smooth and stable even at production deposition rates up to 80 kg/h. The unique control that is possible with independent supplies should make this process widely applicable to situations permitting wire-fed overlay material in a mechanized manner.

Electroslag Cladding. A technique similar to the submerged-arc process, but which works with more massive equipment and produces much thicker coatings, is the electroslag weld-overlay coating process. The basic process was developed in the USSR's Paton Electric Welding Institute in Kiev (85). The heat energy necessary to melt the basis metal and the filler alloy is generated by electrical current in a molten, electrically conducting slag which also purifies and protects the filler metal as it advances through the slag layer into the molten pool. The molten slag and metal pool are held against the basis metal by water-cooled copper retainers, which must conform to the contour of the object to be clad. One application for this process is in the refurbishing of mill rolls which can be performed on ESR (electroslag remelted) ingot-production units with only minor change (86). In operations, eg, cladding of steel arbors, deposition rates of 16 kg/h per electrode are possible with up to 45 electrodes used in unison and positioned appropriately around the circumference of the roll. Rolls up to 5.5 m long and 0.6 m in diameter have been produced.

A comparison of costs for submerged-arc and electroslag processes for similar cladding situations has been made (87). Electroslag is ca 25% more efficient than submerged arc, largely because of the lower labor requirements of the former process. An additional advantage of electroslag processing is cleanliness and reduced slag inclusions which contribute to a reduced coating spallation rate that often results in longer service life of the roll cladding.

Chemical Coatings

Chemical Vapor Deposition. Chemical vapor deposition (CVD) is the gas-phase analogue of electroless plating (qv): CVD is catalytic, occurs on surfaces, and involves a chemical reduction of a species to a metallic or compound material which forms the coating; the reactions are temperature dependent but occur at much higher temperatures (10–13,88–93). A CVD process involving a metal carrier compound, which is reduced by a gaseous reducing agent to deposit a metallic coating, is distinguished from the purely thermal decomposition of an unstable compound into its parts, one of which is a metal.

Chemical Deposition. The types of chemical reactions utilized in CVD are reduction reactions, eg,

$$WF_6 + 3\ H_2 \rightarrow W + 6\ HF$$

displacement reactions, eg,

$$SiCl_4 + CH_4 \rightarrow SiC + 4\ HCl$$

and disproportionation reactions, eg,

$$2\ GeI_2 \rightarrow Ge + GeI_4$$

These reactions generally require temperatures from 500 to 1200°C and often as high as 1500°C. Moreover, the structure of a given deposit may be different at different reaction temperatures. Much of the development work has centered on reactions that allow the deposition to proceed at lower temperatures. For example, a W_3C [*12012-18-9*] coating on steel can be produced at 300°C (94). Because few basis-metal substrates can tolerate high temperatures, the key to wider application of CVD lies in lowering the deposition temperature to make the process applicable to common structural materials.

Chemical vapor deposition coatings tend to be purer than non-CVD-produced coatings. Control of chemical composition is a matter of controlling the gaseous reactants entering the reactor; graded coatings and mixed (sequential) coatings are possible through selection of appropriate gases. Because the reactants are gaseous, the throwing power of the process is excellent. However, the kinetics of the deposition involves hydrodynamic flow and diffusion processes which, if not taken into account, can affect the quality of the deposit.

The largest bulk industrial application of CVD coatings is in wear-resistant overlays, eg, TiC [*12070-08-5*] and TiN [*25583-20-4*] on cemented-carbide cutting tools, but they must be applied at 800°C. A layer of TiC that is 4–8 μm thick can increase the tool life fivefold; α-alumina coatings deposited by CVD on cutting tools increases the wear life still further. However, the most technically sophisticated application of CVD techniques is in the production of electronic materials, where the extreme purity of CVD deposits and the variety of elements and compounds that can be deposited are a distinct advantage. The types of possible electronic materials produced by CVD include semiconductors (qv), insulators, conductors, magnetic materials (qv), and superconductors (95–96). However, the lack of relevant experimental data to characterize the various operational parameters is a major obstacle to the acquisition of a fundamental insight into CVD processes that must develop in order to broaden the utilization of this technique for electronic purposes. Notwithstanding, the applications of CVD to a variety of technologically important applications in the industrial sector continues. Tantalum coatings 20–30 μm thick which impart adequate acid corrosion resistance have been deposited on the inner surface of long carbon steel pipes by hydrogen reduction of tantalum pentachloride (97). A CVD silicon coating that is 50 μm thick has been developed for a nickel-based superalloy (Nimonic 105) by hydrogen reduction of $SiCl_4$ at 1090°C (98). The coating is ductile and corrosion resistant at high temperatures. The need for wear-resistant coatings in components for coal-gasification pilot plants has lead to the development of CVD techniques for deposition of TiN on low carbon steel at and below 1000°C. Ball seat valves up to 0.3 m in diameter have been coated (99). Chemical vapor deposition may be utilized in the production of solar absorber stacks because continuous multicomponent fabrication techniques, which can be achieved by changing the fractional composition of the reactant gas stream, can produce a deposit sequence, ie, silicon, silicon nitride, silicon oxynitride, silicon dioxide (100) (see Solar energy). Such graded refractive-index profiles provide antireflection properties over a large incidence angle and, thereby, provide superior performance in thermal-collector designs involving large optical acceptance angles. By substitution of electron kinetic energies for thermal energy, plasma discharges promote CVD at low temperatures (101). Lasers also have been used as a heat source in CVD because of the localized nature of the heat source and because of the avoidance of excessive substrate heating. Films of TiO_2 from mixtures of $TiCl_4$, H_2, and CO_2 have been prepared in this way (102) (see Film deposition techniques).

Thermal Decomposition. A CVD-related process is the thermally induced decomposition of a compound into a metal and a gaseous by-product. It differs from CVD in that no reducing agent is required; also the reagent need not be a gas. For example, when solid films of previously evaporated silver chloride are contacted in vacuum by an electron beam, they can be made to dissociate to form metallized silver. In this way circuit paths for integrated microelectronic devices that are 0.43 μm wide and 0.7 μm wide are made (103).

The decomposition technique is particularly attractive for *in situ* deposition on powder assemblages prior to pressure densification. Tungsten-coated Eu_2O_3 powders have been produced by decomposition of $W(CO)_6$. In a slightly different way, tantalum-coated powder is made by first blending the Eu_2O_3 powder with a slurry of $TiH_{0.5}$ in amyl acetate which acts as a binder. Thermal decomposition of the hydride produces tantalum-coated oxide particles which can be pressed and sintered to form a cermet (104) (see also Glassy metals). Vapor-formed nickel deposits on plastic injection-molding dies have been prepared commercially (105). The carbonyl is fed from storage containers to a vaporizer–mixer chamber where it is diluted with a carrier gas and fed into the coating chamber at atmospheric pressure. Nickel is deposited on the mold and is characterized by excellent throwing power, and brightness and its cost is one third that of producing comparable machined steel molding dies. A similar process involving the decomposition of nickel carbonyl has been used to produce 20-μm thick dendritic nickel coatings for selective photothermal energy absorbers at \$0.76–1.79/m^2 (1980) (106). Because of the low temperatures of these types of decomposition reactions in comparison with other CVD reactions, new applications for this coating method should continue.

Vacuum Coatings

In vacuum deposition, the desired coating metal is transferred to the vapor state by a thermal or ballistic process at low pressure. The vapor is expanded into the vacuum toward the surface of the precleaned basis metal. Diffusion-limited transport and gas-phase prenucleation of the coating material is avoided by processing entirely in a vacuum that is sufficiently low to ensure that most of the evaporated atoms arrive at the basis metal without significant collisions with background gas. This usually requires a background pressure of 0.665–66.5 mPa (0.005–0.5 μm Hg). At the basis metal, the arriving atoms of coating metal are condensed to a solid phase. The condensation process involves surface migration, nucleation of crystals, growth of crystals to impingement, and often renucleation. Thermal sources based on resistance (I^2R) heating, induction heating, electron-beam heating, and laser irradiation have been used to vaporize the coating material. These processes are physical vapor deposition techniques, ie, thermal energy raises the material to its melting point or above, whereupon it is vaporized by evaporation and adiabatic expansion. In high rate depositions, the reaction forces of atoms leaving the surface form a depression in the surface of the liquid. In these processes, the energy of the physically evaporated atom usually is fractions of an electron volt, depending on the physical properties of the coating material, ie, melting point. If the evaporated coating metal is made to intercept atoms or ions of a special background gas with which it may react to form a compound and then strikes the basis metal, the process is reactive evaporation.

The coating material also can be maintained in a solid form and then suitably bombarded by positive ions of rare gas generated by a glow discharge or other ion source. The coating material or target generally is negatively biased by several hundred to a few thousand volts. The high velocity ions that impinge on the coating material dislodge surface atoms by sputtering. Sputtered atoms are ejected from the coating metal surface with energies of between one and ten electron volts. In the sputtering process, the high energy ions impart a fraction of their energy in a collision cascade within the target that reaches back toward the coating metal surface. About five percent of the energy in the collision cascade reaches the surface and results in sputtered atoms; the rest of the energy is dissipated in the target as heat. Sputtering usually is carried out in an inert gas at 0.13–1.3 Pa (0.001–0.01 mm Hg) so that a glow discharge can be supported by electron impact which provides a source of ions to maintain a steady-state process. If the sputtered atoms are made to intercept a reactive gas species on their way to the substrate basis metal or other object so that a compound is formed which subsequently condenses on the substrate, the process is reactive sputtering. Alternatively, the coating material first can be evaporated or sputtered to vapor and then partially converted to the ionized state by electron bombardment. In ion plating, the ionized portion of the vapor cloud can be accelerated to high energies, eg, a few thousand electron volts, by maintaining a negative bias on the substrate so that ions sputter on arrival at the basis-metal substrate.

If the accelerating voltages are high, eg, 80–100 kV, the ions become permanently embedded as atoms in the near-surface region of the crystal lattice of the basis metal and the process is ion implantation (qv). At these energies, the sputtering process is minimal because the collision cascade does not impart sufficient energy at the surface to eject many surface atoms. Having been neutralized by conduction electrons, the ions become part of the surface of the target. If the ions are not of a rare gas but ions of metal atoms, they chemically alter the alloy's near-surface region. The properties of the near surface may be drastically changed by this internal alloying by the implantation process and, in certain circumstances, the near-surface region of the basis metal can be made amorphous in the affected region.

An enormous variety of elemental metallic films, compounds, semiconductors, insulators, and amorphous coatings have been made by these processes. Metallic coatings have been applied by vacuum-deposition techniques to a host of substrate materials, including paper, cloth, metal strip, metal parts of various sorts, plastic, glass, and semimetals. These techniques have been used most widely in the complex multistep coating processes involved in generating electronic microcircuits and memory elements (see Integrated circuits; Vacuum technology).

Physical Vapor Deposition. ***Evaporation.*** Electron-beam vapor sources are used extensively for metallic coating because they can be fed continuously with solid evaporant material either as wire or rod (107–108). Evaporation of alloys requires that a constant rate and inventory of liquid is maintained. When melting begins, the more volatile elements are gradually depleted in the inventory of liquid. With time, the liquid becomes enriched in the less volatile elements and the vapor composition becomes constant and identical to the composition of the feed (109). In this way, many alloy compositions have been successfully evaporated to form alloy coatings (110). The process is used to coat high temperature, corrosion-resistant coatings, eg, CoCrAlY [*59299-14-8*], for gas-turbine blades and vanes, and coating rates of up to 30 μm/min are achieved. Dense coatings with minimal defects are produced by glass-bead peening and heat treating to at least 950°C (111). However, two disadvantages occur with evaporated coatings: the atoms arrive essentially on line-of-sight from a virtual point source, necessitating that the objects to be coated are suitably rotated if uniform coatings are desired; and the temperature gradients in the source cause chemical and density gradients in the vapor cloud so that, depending on the position of the substrate, the composition of the deposit may vary. For microelectronic-device manufacturing where electron-beam sources are used, elaborate part-rotation operations involving planetary devices ensures uniform coating thickness and chemistry on the silicon chips.

A continuous electron-beam evaporation system with a total evaporation power of 150 kW has been developed to coat industrial 3 × 3.6-m plate glass for architectural applications (112). The cost of decorative evaporated coating, depending upon the part design, is as cost effective as electroplating and, in some circumstances, less expensive (113).

Vacuum metallizing is potentially effective on ABS (acrylonitrile–butadiene–styrene), nylon, polycarbonate, or Noryl that has been base-coated with polyester or urethane using a Cr–15 wt % Fe–5 wt % Ti metallization coating followed by an acrylic or urethane top coat (114). A cost of ca $\$0.25/m^2$ of coated surface for such coatings or about one half that for electroplating can be achieved. Cost advantages of vacuum metallizing are chiefly in coating large parts (115).

Reactive Evaporation. In reactive evaporation, a compound deposit is produced by the reaction of the evaporant and a chemically active background gas (116). In activated reactive evaporation, an electrical discharge from a probe above the evaporation source, as shown in Figure 5, increases the reaction cross section, thereby increasing the probability of compound-producing collisions (117). Compounds, particularly carbides, eg, TiC, can be formed at high deposition rates with adjustable stoichiometry, depending on the preselected ratio of hydrocarbon gas and evaporated titanium atoms. A boron alloy evaporated in the presence of ammonia has produced a coating containing 20% of cubic boron nitride, a form of hard nitride previously formed only at very high pressures (118). The properties of coatings produced by reactive evaporation and related processes are given in refs. 12–14.

Ion Plating. Ion plating is conducted with either evaporation or sputtering to provide a vapor deposit of material on a basis metal or substrate which is maintained at a negative potential (119–121). The background gas (eg, argon) pressure in evaporative ion plating is increased to 0.1–0.15 Pa (0.75–1.15 μm Hg) in order to create a negative glow around the object to be coated. Argon ions bombard the substrate and remove undesirable oxides and other contamination from the surface. When evaporation or sputtering begins, most of the atoms from the vapor source enter the dark space, ie, the ion-accelerating region near the surface, around the part; and become partially (1%) ionized and, therefore, accelerated in the dark space toward the substrate or part surface. The energy over the thermal energy acquired by the coating atoms provides substrate heating and high surface mobility of the coating atoms. The energy is sufficient to imbed some of the atoms 0.1–0.2 nm beneath the surface. The material in the near surface region is highly disrupted and defects are introduced, which promotes coating–substrate interdiffusion. The result of this surface activity is greatly improved coating adherence, uniformity, and the ability to deposit a substantial fraction on the dark side of the vapor source, eg, on the surfaces not in line-of-sight with the source. The introduction of high rate, electron-beam evaporation sources and

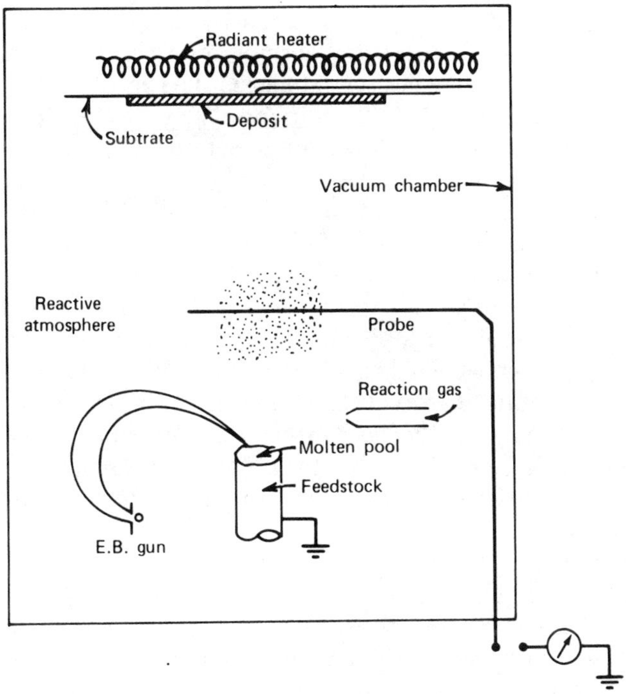

Figure 5. Schematic illustration of activated reactive-evaporation process. An electron-beam vapor source produces an expanding vapor cloud. The reaction gas and metal vapor are partially ionized by glow discharge, which is maintained by current from the probe, to facilitate compound formation. The compound metal coating is condensed on the heated substrate.

induction-heated vapor sources have improved the technology (122–123). A major commercial application, called Ivadizer, is the production coating of aircraft fasteners with aluminum by ion plating (124). Cost of the process is competitive with many other conventional processes, and there are no toxic wastes or fumes associated with it. Also, no hydrogen embrittlement is possible as with electrochemical techniques.

Sputtering and Ion Implantation. When the source material is maintained in the solid state and is bombarded with ions from a gas discharge, surface atoms are dislodged with high energy. If the discharge is created between the substrate and the coating metal or target, the process is diode sputtering. If a hot filament and anode circuit provide a separate low voltage electron-beam discharge, the process is triode sputtering. Alternatively, application of r-f power to the target results in r-f diode sputtering. Dual targets allow a-c sputtering. If the process is carried out in a strong transverse magnetic field it is magnetron sputtering. Many other variations and combinations exist, but the fundamental process of atom ejection remains the same. Addition of substrate biasing capability provides precoating etching and sputter–ion-plating possibilities. Background gas dopants that are added to the discharge gas provide reactive sputtering. All of these variants have been utilized to provide better deposit properties, ie, the ability to sputter dielectric materials; the achievement of a high deposition rate; and the production of compounds, highly pure materials, or epitaxially grown deposits. Glow-discharge sputtering has been reviewed recently (125).

Magnetron Sputtering. A magnetron discharge is used to achieve high rates of deposition by sputtering. Recently, an enormous growth in the applications of the process has taken place. The applications of magnetron sputtering have been reviewed for cylindrical magnetron sources (126) and for planar magnetron sources by Waits (127). Deposition power as high as 80 kW into sputtering targets as large as 0.35 × 1.8 m can be achieved (128). Because radiation damage by fast electrons and neutrals is so much less in magnetron sputtering than with previously used sputtering techniques, the former process is being considered for thermal-sensitive semiconductor device fabrication as a supplement to evaporated coatings. Magnetron sputtering also is used to produce experimental coatings on laser-fusion target microspheres that are 100–500 μm in diameter (see Fusion energy). Sputtering rate increases from 0.75–12.6 kA/min for platinum are expected to open up many new applications for magnetron sputtering of platinum-group metals. Selective solar-absorber coatings that are deposited by magnetron sputtering appear to be inexpensive (129). Automotive applications, chiefly for exterior trim, are expected to increase to $1 × 10^7 in 1980 largely because of an 80% reduction in energy costs over electroplated coatings and the very low environmental effects of the technique.

The use of magnetron sputtering for coating gas-turbine components has been studied (130). A special advantage of sputtering for these components is the ability to coat very complex alloy systems that contain elemental constituents with both very high and very low vapor pressures; these alloys are difficult to apply by other techniques. The structural metallurgy associated with high rate sputtered deposits for coatings applications has been discussed (131). Substrate temperature, smoothness of the interface, and rate of deposition are the principal factors which influence coating structure. Future advantages for the process are the suppression of undesirable phase constituents and the development of dispersoid distributions by the incorporation of rapid solidification processes into the magnetron-sputtering technique.

Ion Implantation. When coating atoms are ionized and achieve sufficient energy, they can be driven into the surface of the basis metal to form a near-surface region where the chemistry has been altered by the addition of the ions, which are neutralized on entry into the host lattice (132). Because this process requires sophisticated ion-beam accelerators which function at 80–150 kV, the cost is very high, ie, $1–5/cm^2 (133). Ion-beam doping of GaAs, GaP, and other compound semiconductor materials is common. A 30-kV, high throughput ion-implantation system for B, P, As, and Sb doping of 7.6 cm thick wafers through a mask has been operating since 1974 and involves the use of beam currents of 5 mA; 450 wafers per hour are produced with doses up to 10^{15} ions per square centimeter (134). Application of ion implantation to the formation of metastable phases in metals has been suggested (135). Because the lattice temperature that is produced by an incoming energetic ion may be 500–1000 K, it has been calculated that a cooling rate of 10^{12}–10^{15} K is probable and is far in excess of the 10^6 K/s rate that is characteristic for liquid quenching of metals. Many atoms have been implanted in a copper host lattice as substitutional solid solutions, eg, Ag, Sb, I, Xe, W, Pt, Au, Hg, Tl, Pb, and Bi (136). If the dose rate of the ion is increased, for example Ta in copper, the surface structure becomes completely amorphous. In another example, phosphorus that is implanted into 316 stainless steel induces a surface amorphous state (137).

The applications of ion implantation to tribology and corrosion science have been reviewed (138). Wear rates of a nitrogen-implanted nitriding steel are an order of magnitude lower than the unimplanted material. Nitrogen that is implanted into cutting knives for paper and high speed steel taps for phenolic plastics increase the useful lives of the parts two- to fivefold. The fatigue life of AISI 1018 steel is increased twofold by nitrogen implantation. The oxidation of a chromium-rich stainless steel at high temperature is reduced by implantation of Y ions which cause a reduction in the rate of oxide spallation (139). Ion-implanted Cr in steel produces the same corrosion resistance as an Fe–Cr alloy (140). A commercial Fe–18 wt % Ni–8 wt % Co maraging steel can be implanted with Cr to improve corrosion resistance without the problems associated with prolonged heating of the alloy structure during chromium cementation coating (141). A number of other studies have been made to demonstrate the effect of ion implantation on the corrosion resistance of stainless steels (142), aluminum, and titanium. For example, titanium implanted with 10^{16} Pd atoms per square centimeter at 90 kV produces a 5-at %, Pd-enriched, subsurface region. The corrosion potential of the Pd-implanted sample in boiling, 1 *M* H_2SO_4 is ca 1000 mV more noble than the pure Ti and quite close to that of pure Pd; a dramatic reduction of the corrosion rate also is observed (143). A possible but untried combination for near-surface materials modification is the incorporation of laser annealing, laser fusion, and ion-implantation techniques (144). Because of its cost, ion implantation likely will remain largely a microelectronics fabrication method. It has already become an important part of magnetic bubble memory technology in that it is used to control bubble states, flux cap the bubbles, induce desirable increases in the local lattice parameter, and increase the etching rate (145). Because it is a clean, well-controlled process, ion implantation may be used eventually to mitigate corrosion, particularly where very costly implantable materials can be made to function as protective coatings (see also High temperature alloys).

Health and Safety Factors

All coating techniques are based on technologies that have inherent hazards, eg, high temperature, the use of liquid-metal or molten salt baths, high voltages, and often, the use of toxic chemicals. However, the coatings industry and the suppliers of coating equipment have provided carefully engineered systems, which if handled according to recommended practice, are relatively benign with respect to operator and user personnel hazards.

Abrasive cleaning can create oxide and metallic particulates which must be collected and carefully disposed of. Each abrasive process must be considered separately along with the basis-metal requirements for the established air-loadings limits for personnel exposure. Acid-etching baths and organic solvent washes must be constructed, ventilated, and controlled to avoid operator hazards. Information regarding established exposure levels to the types of materials discussed below are reviewed in ref. 146.

Diffusion coating can involve liquid-metal baths which must be isolated from operator contact but, more importantly, must be clear of water sources which could provide an explosion hazard. Corrosion of metal containers for metal baths also must be considered and accepted practice must be followed to avoid unexpected spills of liquid metal. In coating processes involving lead, vapors, excess coating material, and cleaning effluents must be properly controlled and disposed of. Tin, which has a low melting point and low vapor pressure and is nontoxic, is the most hazard-free of the hot-dipped coatings. Zinc, aluminum, and aluminum–zinc hot-dipped coatings require some care because of their higher melting points, greater reactivities and the nature of the fluxes required in their related processes.

Cementation coating is safe because of the use of closed retorts. Care must be exercised in some pack-dismantling operations regarding the use of fluoride activators. Proper inventory of pack materials, adequate ventilation, and dust-avoiding work stations are required and should be monitored to ensure compliance with published requirements. Similarly, care must be exercised with liquid-metal carrier baths and fused-salt baths to avoid moisture and to prevent undesirable release of either the bath materials, their fluxing agents, or their coating constituents.

Sprayed coatings are less hazardous than hot-dipped ones since the volume of liquid metal produced at any instant is quite low. However, sprayed coatings are largely hand operations involving torches. Aerosols are of concern mainly with plasma spraying and sometimes arc-wire spraying; water walls or properly vented enclosures are necessary. The concerns in this type of coating operation are not unlike those in welding operations and the same general rules for safe practice apply. NIOSH has concluded that dust and fumes from metallizing by flame spraying at Shell Oil Company's refinery at Wood River, Illinois, does not constitute a health hazard. Measures have been taken to avoid personnel contact with materials which, in some instances, had caused dermatitis (147). Arc and plasma methods have the additional hazards of radiation burns resulting from exposure to the uv radiation that is released during operation; welding practices for operator safety must be followed. Laser-coating operations are similar, in that direct contact with the radiation must be avoided and care is required to avoid difficult-to-detect stray reflections. Laser-beam processing in conjunction with established standards can be considered a safe operation. In those operations, eg, flame spraying, plasma spraying, electrostatic powder coating, D-gun and slurry coating, powder rolling, etc, which involve the use of submicrometer powders of the coating material, care is required in the handling, storage, and use of these powders to prevent dust and possible dust explosions and, in many cases, reactions with water or other reagents that might otherwise be nonreactive for bulk forms of these materials. Specialists in the coatings technology area should always be consulted to determine the precautions required for each powder material.

In CVD processes, the coating-metal carrier often is a volatile, reactive liquid or gas, often a fluoride or other halide which is easily hydrolyzed to acid halides. However, these reagents have established storage and delivery requirements and manufacturers can supply details regarding the proper utilization of these reagents and the measures required to safeguard against operator exposure.

Vacuum coatings, of necessity, are produced in well-controlled environmental chambers where air, moisture, and radiation are isolated by the walls of the device. Often high voltages are used as in electron-beam evaporation and sputtering and in the latter, r-f power at high voltage is common. Proper control of electron-beam evaporation and sputtering power sources and their distribution and entry points into equipment is necessary; properly engineered and maintained equipment is required and reputable manufacturers of such equipment should be consulted when necessary to ensure continued immunity against the possible operator hazards involved.

BIBLIOGRAPHY

"Metallic Coatings" in *ECT* 1st ed., Vol. 8, pp. 898–922, by W. W. Bradley, Bell Telephone Laboratories, Inc.; "Metallic Coatings" in *ECT* 2nd ed., Vol. 13, pp. 249–284, by William B. Harding, The Bendix Corporation.

1. A. Pinto, *Mod. Packag.* **52,** 25 (1979).
2. "Properties and Selection of Nonferrous Alloys and Pure Metals" in *Metals Handbook*, 9th ed., Vol. 2, American Society for Metals, Metals Park, Ohio, 1979.
3. H. H. Uhlig, *Corrosion Handbook*, John Wiley & Sons, Inc., New York, 1948.
4. F. L. LaQue and H. R. Copson, *Corrosion Resistance of Metals and Alloys*, 2nd ed., Reinhold Publishing Corp., New York, 1963.
5. "Metal/Environment Reactions" in L. L. Shrier, ed., *Corrosion*, 2nd ed., Vol. 1; "Corrosion Control" in L. L. Shrier, ed., *Corrosion*, 2nd ed., Vol. 2, Butterworths, London, Eng. available from American Society for Metals, Metals Park, Ohio; H. H. Uhlig, *Corrosion and Corrosion Control*, 2nd ed., John Wiley & Sons, Inc., New York, 1971.
6. R. D. Gabe, *Principles of Metal Surface Treatment and Protection*, 2nd ed., Pergamon Press, Inc., Elmsford, N.Y., 1978.
7. V. E. Carter, *Metallic Coatings for Corrosion Control* (*Prof. Conf.*), Newnes-Butterworths, London, Eng., 1977.
8. *Materials and Coatings to Resist High Temperature Corrosion* (*Proc. Conf.*), Verein Deutscher Eisenhuttenleute Dusseldorf, FRG, New York, May 1977, Applied Science Publishers, Ltd., 1978.
9. B. Chapman and J. C. Anderson, *Science and Technology of Surface Coatings*, Academic Press, New York, 1974.
10. R. C. Krutenat, ed., *Proc. of Conference on Structure/Property Relationships in Thick Films and Bulk Coatings*, San Francisco, Calif., Jan. 28–30, 1974, American Institute of Physics, 1974, LC 74-82950; *J. Vac. Sci. Technol.* **11,** 633 (1974).
11. R. E. Reed, ed., *Proceedings, 2nd Conference on Structure/Property Relationships in Thick Films and Bulk Coatings*, American Institute of Physics, New York, 1975 (LC 75-18563); *J. Vac. Sci. Technol.* **12,** 741 (1975).
12. R. F. Bunshah, ed., *Metallurgical Coatings 1976, Proceedings of the International Conference*, Elsevier Sequoia S. A. Lausanne, 1977; *Thin Solid Films* **39,** 1 (1976); *Thin Solid Films* **40,** 1 (1977).
13. R. F. Bunshah, ed., *Metallurgical Coatings, 1977, Proceedings of the International Conference*, Elsevier Sequoia S.A. Lausanne, 1977; *Thin Solid Films* **45,** 1 (1977).
14. R. F. Bunshah, ed., *Metallurgical Coatings, 1978, Proceedings of the International Conference*, Elsevier Sequoia S.A. Lausanne, 1978; *Thin Solid Films* **53,** 1 (1978); *Thin Solid Films* **54,** 1 (1978).
15. *High Temperature Oxidation-Resistant Coatings*, National Academy of Sciences, Washington, D.C., 1970 (LC 78-606278).
16. H. N. Hausner, ed., *Coatings of High-Temperature Materials*, Plenum Press, New York, 1966.
17. J. Huminik, Jr., ed., *High Temperature Inorganic Coatings*, Reinhold, New York, 1963.
18. C. S. Smith in W. J. Young, ed., *Application of Science in Examination of Works of Art*, Museum of Fine Arts, Boston, Mass., 1973, p. 96.
19. H. Lechtman, *J. Met.* **31**(12), 154 (1979).
20. J. A. Thornton and H. F. Graff, *Met. Trans. B* **7,** 607 (1976).
21. J. C. Zoccola and co-workers, "Atmospheric Corrosion Behavior of Al-Zn Alloy Coated Steel," *ASTM Symposium on Atmospheric Corrosion*, May 1976.
22. S. G. Denner and co-workers, *Iron Steel Int.* **48,** 241 (1974).
23. S. G. Denner and R. D. Jones, *Met. Tech.* **4**(3), 167 (1977).
24. K. Gale, *Steel Times* **206,** 896 (1978).
25. L. Pugazhenthy, *Corros. Maint.* **1**(2), 153 (1978).
26. D. Kim and H. Leidheiser, Jr., *Surface Technol.* **5,** 379 (1977).
27. P. J. Drivas, *Contract EPA-68-01-3156, Pacific Environmental Services, Inc.*, Santa Monica, Calif., Mar. 1976.
28. J. K. Fuller, *Wire J.* **10**(1), 60 (1977).
29. J. B. Horton, A. R. Borzillo, N. Kuhn and G. J. Harvey, *12th International Conference on Hot-Dip Galvanizing*, Paris, Fr., May 20–23, 1979.
30. G. J. Harvey, *Met. Australia* **8**(8), 176 (1976).
31. J. B. Horton, A. R. Borzillo, N. Kuhn, and G. J. Harvey, *12th International Conference on Hot-Dip Galvanizing*, Paris, Fr., May 20–23, 1979; R. D. Jones, *Iron Steel Inst.* **51**(3), 149 (1978); *Plat. Surf. Finish.* **65**(9), 30 (1978).
32. W. A. McGill and M. J. Weinbaum, *Met. Prog.* **116**(2), 26 (1979).
33. N. S. Gorbunov, *Diffuse Coatings on Iron and Steel*, Academy of Sciences, Moscow, USSR, 1958, translated by S. Friedman, A. Artman and Y. Halprin, OTS 60-21148.
34. U.S. Pat. 4,125,646 (Nov. 14, 1978), M. F. Dean and R. L. Blize (to Chromalloy American Corp. Off. Gaz.).
35. *Prod. Finish.* **30**(2), 10 (1977).

36. A. K. Sarkhel and L. L. Seigle, *Met. Trans.* **7A,** 899 (1976).
37. B. K. Gupta, A. K. Sarkhel, and L. L. Seigle, *Thin Solid Films* **39,** 313 (1976).
38. S. Shankar and L. L. Seigle, *Met. Trans.* **9A,** 1467 (1978).
39. R. W. Heckel, M. Yamada, C. Duchi, and A. J. Hickel, *Thin Solid Films* **45,** 367 (1977).
40. R. S. Parzuchowski, *Thin Solid Films* **45,** 349 (1977).
41. B. Weiss and M. R. Meyerson, *Trans. Met. Soc. AIME* **245,** 1633 (1969).
42. A. N. Mukherji and P. Prathakaram, *Anticorr. Methods Mater.* **25**(1), 5, 12 (1978).
43. T. S. Eyre, *Wear* **34,** 383 (1975).
44. R. H. Biddulph, *Thin Solid Films* **45,** 341 (1977).
45. Ref. 15, p. 126.
46. Brit. Pat. 5,647 (1900), S. Cowper-Coles; S. Trood, *Trans. Am. Inst. Met.* **9,** 161 (1919).
47. L. Zancheva and co-workers, *Met. Trans.* **9A,** 909 (1978).
48. *Met. Prog.* **117**(1), 89 (1980).
49. V. A. Parenov and F. K. Aleinikov, *Liet. TSR Mokslu Akad. Darb.* [*B*] **4**(107), 93 (1978).
50. G. F. Carter, *Met. Prog.* **93**(6), 124 (1968).
51. B. K. Granat, *Metalworking News*, 32 (Oct. 11, 1976).
52. *Weld. J. Miami Fl.* **55,** 675 (1976).
53. *Weld. J. Miami Fl.* **56**(8), 41 (1977).
54. "Corrosion Prevention with Thermal Sprayed Zn and Al Coatings," *Third International Congress on Marine Corrosion and Fouling*, Oct. 2–6, 1972, Sponsored by AWS Committee on Thermal Spraying, Special NBS Publication.
55. *Bull. Routier* **4**(3), (Nov.–Dec. 1978).
56. E. P. Cashon, *Tribology* **8**(3), 111 (1975).
57. R. R. Irving, B. D. Wakefield, and T. C. Dumond, *Iron Age* **211**(15), 65 (1973).
58. H. Kayser, *Thin Solid Films* **39,** 243 (1976); *Proc. International Conf. on Metallurgical Coatings*, Elsevier-Sequoia S.A. Lausanne, 1976.
59. R. C. Tucker, Jr., *J. Vac. Sci. Technol.* **11,** 725 (1974).
60. U.S. Pat. 4,182,223 (July 13, 1977), F. J. Wallace, N. S. Bornstein, and M. A. DeCrescente (to United Technologies Corporation).
61. G. J. Robson, *Australian Conf. on Manufacturing Engineering*, Aug. 17–19, 1977, pp. 211–216.
62. F. J. Wallace, *Plat. Surf. Finish.* **62,** 559 (1975).
63. R. J. Schaefer, J. D. Ayers, and T. R. Tucker, *NRL Report #3953*, Naval Research Laboratory, Washington, D.C., Mar. 30, 1979.
64. U.S. Pat. 4,015,100 (March 29, 1977), D. S. Gnanamuthu and E. V. Locke (to Avco Everett).
65. D. Belforte, *High Power Laser Surface Treatment*, SME Technical Paper IQ-77-373, Soc. Mfg. Eng., Dearborn, Mich.
66. U.S. Pat. 3,952,180 (April 20, 1976), D. S. Gnanamuthu (to Avco Everett).
67. D. A. Van Cleave, *Iron Age* **219**(5), 25, 30 (1977).
68. G. C. Irons, *Weld. J. Miami Fl.* **57**(12), 29 (1978).
69. M. S. Hess and J. F. Milkosky, *J. Appl. Phys.* **43,** 4680 (1972).
70. K. Gale, *Engineer* **234**(6064), 34 (1972).
71. V. A. Paramonov and co-workers, *Korroz. Zashch.* (8), 24 (1973).
72. *Iron Age*, 80 (May 7, 1964).
73. C. M. Packer and R. A. Perkins, *J. Less Common Met.* **37,** 361 (1974).
74. F. M. Miller and N. T. Bredzs, *Met. Prog.* **103**(3), 80, 82 (1973).
75. P. R. Clark, C. M. Packer and R. A. Perkins, *Quarterly Report, DOE Contract No. EF-77-C-012592*, Lockheed Palo Alto Research Laboratory, Palo Alto, Calif., Apr. 1–June 30, 1979, p. 9.
76. R. J. Russell, *Mater. Eng.* **81**(3), 48 (1975).
77. K. Tamura and N. Miyamoto, *J. Jpn. Soc. Powder Metall.* **20**(1), 10 (1973).
78. E. H. Mayer and R. M. Willison, *presentation, 75th General Meeting American Iron and Steel Institute*, May 25, 1967, pp. 1–12.
79. V. Siran and co-workers, *Met. Finish.* **76**(9), 48 (1978).
80. E. A. Davis in *Coatings for Corrosion Prevention, 1978 Symposium ASM Materials and Processing Congress*, Philadelphia, Pa., Nov. 9, 1978, American Society for Metals, Metals Park, Oh., 1979, p. 35, LC-79-17292.
81. *Prod. Finish.* (*London*) **31**(7), 22 (1978).
82. *Prod. Finish* (*Cincinnati*) **40**(1), 65 (1975).
83. R. A. Serlin, *Cutting Tool Eng.* **30**(7–8), 6 (1978).

84. E. Smars and G. Backstrom, *Proc. International Conf. on Exploiting Welding in Production Technology*, Welding Institute of Cambridge, Eng., 1975, pp. 179–187.
85. P. Blas'kovic, *Schweisstechnik* (9), (1975); J. S'Krianiar, *Metals Technology Conference B*, Australian Institute of Metallurgy, 1976, paper 16-3-1.
86. C. Kubisch, P. Pressler, P. Machner, and O. Kleinhagauer in G. K. Bhat, ed., *Fifth International Symposium on Electroslag and Other Special Melting Technology, 1974*, Pittsburgh, Pa., Carnegie-Mellon Institute, 1975.
87. W. R. Foley and W. R. Huber, *Iron Steel Inst.* **51**(4), 72 (1971).
88. W. A. Bryant, *J. Mater. Sci.* **12,** 1285 (1977).
89. K. K. Yee, *Int. Met. Rev.* **23**(1), 19 (1978).
90. A. C. Schaffhauser, ed., *Proc. of Conf. on Chemical Vapor Deposition of Refractory Metals, Alloys and Compounds*, American Nuclear Society, Hinsdale, Ill., 1967.
91. J. M. Blocher, Jr., and J. C. Withers, eds., *Proc. of Second Int. Conf. on Chemical Vapor Deposition*, Electrochemical Society, New York, 1970.
92. F. A. Glaski, ed., *Proc. of Third Int. Conf. on Chemical Vapor Deposition*, American Nuclear Society, Hinsdale, Ill., 1972.
93. G. F. Wakefield and J. M. Blocher, Jr., eds., *Proc. of Fourth Int. Conf. on Chemical Vapor Deposition*, Electrochemical Society, Princeton, N.J., 1973.
94. N. J. Archer and K. K. Yee, *Wear* **48,** 237 (1978).
95. T. L. Chu and R. K. Smeltzer, *J. Vac. Sci. Technol.* **10**(1), 1 (1973).
96. J. J. Tietjen, *Ann. Rev. Mat. Sci.* **3,** 317 (1973).
97. C. Beguin, E. Horrath, and A. J. Perry, *Thin Solid Films* **46,** 209 (1977).
98. P. Felix and Erdos, *Werkst. Korr.* **23,** 627 (1972).
99. J. B. Stephenson, D. M. Soboroff, and H. O. McDonald, *Thin Solid Films* **40,** 73 (1977).
100. B. O. Seraphin, *J. Vac. Sci. Technol.* **16**(2), 193 (1979).
101. M. J. Rand, *J. Vac. Sci. Technol.* **16,** 410 (1979).
102. S. D. Allen and M. Bass, *J. Vac. Sci. Technol.* **16**(2), 431 (1979).
103. J. P. Ballantyne and W. C. Nixon, *J. Vac. Sci. Technol.* **10,** 1094 (1973).
104. C. S. Morgan, *Thin Solid Films* **39,** 305 (1976).
105. R. O. Betz, *Iron Age* **220**(1), 47 (1977).
106. D. P. Grimmer, K. C. Herr, and W. J. McCreary, *J. Vac. Sci. Technol.* **15**(1), 59 (1978).
107. T. Santala and C. M. Adams, Jr., *J. Vac. Sci. Technol.* **7,** 522 (1970).
108. R. F. Bunshab and R. S. Juntz, *Trans. Vac. Met. Conf., American Vacuum Society*, New York, 1965, p. 200.
109. R. Nimmagadda, A. C. Raghuram, and R. F. Bunshah, *J. Vac. Sci. Technol.* **9,** 1406 (1972).
110. H. R. Smith, Jr. and C. D'A. Hunt, *Trans. Vac. Met. Conf.*, American Vacuum Society, 1964, p. 227.
111. S. Shen, D. Lee, and D. H. Boone, *Thin Solid Films* **53,** 233 (1978).
112. A. D. Grubb, *J. Vac. Sci. Technol.* **10**(1), 53 (1973).
113. C. C. Storms, *Proc. Soc. of Vac. Coaters, 19th Annual Conference*, 1976, pp. 1–3; A. Mock, *Mater. Eng.* **87**(4), 51 (1978).
114. D. M. Lindsey, *Prod. Finish. (Cincinnati)* **43**(10), 34 (1979).
115. S. Kut, *Finish. Ind.* **1**(7), 17 (1977).
116. G. K. Wehner, *Phys. Rev.* **102,** 690 (1956); *Phys. Rev.* **114,** 1270 (1959); *J. Appl. Phys.* **30,** 1762 (1959); G. S. Anderson and G. K. Wehner, *J. Appl. Phys.* **31,** 2305 (1960).
117. R. F. Bunshah and A. C. Raghuram, *J. Vac. Sci. Technol.* **9**(6), 1385 (1972).
118. *Battelle Today* (13), 1 (June 1979).
119. D. M. Mattox, *Electrochem. Tech.* **2**(9–10), 295 (1964).
120. D. M. Mattox, *J. Vac. Sci. Technol.* **10**(1), 47 (1973).
121. D. M. Mattox, *IPAT 79, Proc. Inter. Conf. Ion Plating and Allied Techniques*, London, July 1979, CEP Consultants, Ltd., Edinburgh, UK, p. 1.
122. C. T. Wan, D. L. Chambers, and D. C. Carmichael, *Proc. of Fourth Inter. Conf. on Vac. Met.*, The Iron and Steel Institute of Japan, Tokyo, 1974, p. 231.
123. G. W. White, *SAE paper 730546*, Detroit, Mich., May 14–18, 1973.
124. *Finish. Ind.* **1**(7), 19 (1977).
125. W. D. Westwood, *Prog. Surf. Sci.* **7**(2), 71 (1976).
126. J. A. Thorton, *J. Vac. Sci. Technol.* **15**(2), 171 (1978).

127. R. K. Waits, *J. Vac. Sci. Technol.* **15**(2), 179 (1978).
128. J. L. Hughes, *J. Vac. Sci. Technol.* **15**(4), 1572 (1978).
129. J. A. Thornton, *Proceedings, AES Coatings for Solar Collectors*, Atlanta, Ga., Nov. 9–10, 1976, American Electroplaters Society, Inc., Winter Park, Fla., 1976, pp. 63–77.
130. R. J. Hecht, *Air Force Materials Laboratory Contract No. F33615-78-C-5070*, Project No. 212-8, Wright Patterson AFB, Ohio, Pratt and Whitney Aircraft Group, West Palm Beach, Florida, Quarterly Progress Reports, 1978–1979, FR-11081, FR-11649, FR-11906, FR-12170 and FR-12557.
131. J. A. Thornton, *Ann. Rev. Mater. Sci.* **7,** 239 (1977).
132. W. L. Brown, ed., *Ion-Implantation—New Prospects for Materials Modification*, IBM Laboratories, Yorktown Heights, New York, sponsored by American Vacuum Society; *J. Vac. Sci. Technol.* **15,** 1629 (1978).
133. T. C. Wells, *Surf. J.* (*London*) **9**(4), (1978).
134. J. G. McCallum, G. I. Robertson, A. F. Rodde, B. Weissman, and N. Williams, *J. Vac. Sci. Technol.* **15,** 1067 (1978).
135. J. A. Borders, *Ann. Rev. Mater. Sci.* **9,** 313 (1979).
136. J. A. Borders and J. M. Poate, *Phys. Rev. B* **13,** 969 (1976).
137. W. A. Grant, *J. Vac. Sci. Technol.* **15,** 1644 (1978).
138. J. K. Hirronen, *J. Vac. Sci. Technol.* **15,** 1662 (1978).
139. J. E. Antill and co-workers in B. L. Crowder, ed., *Proc. Third Inter. Conf. on Ion Implantation in Semiconductors and Other Materials*, IBM, Yorktown Heights, 1972, Plenum Press, New York, 1973.
140. B. D. Sartwell, A. B. Campbell, and P. B. Needham in F. Chernow, J. A. Borders, and D. Brice, ed., *Proc. Fifth Inter. Conf. on Ion Implantation in Semiconductors and Other Materials*, Boulder, Co., Aug. 1976, Plenum Press, New York, 1977.
141. B. S. Corino, Jr., P. B. Needham, Jr., and G. R. Connor, *J. Electrochem. Soc.* **125,** 370 (1978).
142. S. B. Agarwal, Y. F. Wang, C. R. Clayton, and H. Herman, *Thin Solid Films* **63,** 19 (1979).
143. J. K. Hirronen, *J. Vac. Sci. Technol.* **15,** 1667 (1978).
144. H. S. Rupprecht, *J. Vac. Sci. Technol.* **15,** 1674 (1978).
145. J. C. North, R. Wolfe, and T. J. Nelson, *J. Vac. Sci. Technol.* **15,** 1675 (1978).
146. N. I. Sax, *Dangerous Properties of Industrial Materials*, 4th ed., Van Nostrand Reinhold Co., New York, 1975.
147. R. S. Kramkowski and E. Shmunes, *Final Report No. NIOSH-TR-058-74; HHE-72-87-58*, NIOSH, Cincinnati, Oh., June 1973.

R. C. KRUTENAT
Exxon Research and Engineering Company

EXPLOSIVELY CLAD METALS

Explosives (qv) were used increasingly in the 1950s in metal-working operations because the explosives provided an inexpensive source of energy and precluded need for expensive capital equipment (1–2). Research in explosively clad metals began during the same period (3–7).

Explosive cladding, or explosion bonding and explosion welding, is a method wherein the controlled energy of a detonating explosive is used to create a metallurgical bond between two or more similar or dissimilar metals. No intermediate filler metal, eg, a brazing compound or soldering alloy, is needed to promote bonding and no external heat is applied. Diffusion does not occur during bonding.

In 1962, the first method for welding metals in spots along a linear path by explosive detonation was patented (8); however, the method is not used industrially (see Welding). In 1963, a theory that explained how and why cladding occurs was published (9). Research efforts resulted in 27 U.S. process patents which standardized industrial explosion cladding. Several of the patents describe the use of variables involved in parallel cladding which is the most popular form of explosion cladding (10–13).

During the 1960s and early 1970s, research and development work on explosive cladding was conducted throughout the world. In the United States, much of the research occurred at Battelle Memorial Institute (14), Drexel University and Frankford Arsenal (15), DuPont (16–24), Stanford University (25–27), and the University of Denver (28). Research in other countries includes that conducted in Japan (29), Ireland (30–31), and the USSR (32–33). Several excellent reviews on metal cladding have been published (34–36). From 1964 to 1980, ca 200,000 metric tons of clad metals have been produced in noncommunist nations.

Advantages and Limitations

The explosive-cladding process provides the following advantages over other metal-bonding processes:

(*1*) A metallurgical, high quality bond can be formed between similar metals and between dissimilar metals that are incompatible for fusion or diffusion joining. Brittle, intermetallic compounds, which form in an undesirable continuous layer at the interface during bonding by conventional methods, are minimized, isolated, and surrounded by ductile metal in explosion cladding. Examples of these systems are titanium–steel, tantalum–steel, aluminum–steel, titanium–aluminum, and copper–aluminum. Immiscible metal combinations, eg, tantalum–copper, also can be clad.

(*2*) Explosive cladding can be achieved over areas that are limited only by the size of the available cladding plate and by the magnitude of the explosion that can be tolerated. Areas as small as 1.3 cm^2 (37) and as large as 27.9 m^2 (18) have been bonded.

(*3*) Metals with tenacious surface films that make roll bonding difficult, eg, stainless steel/Cr–Mo steels, can be explosion clad.

(*4*) Metals having widely differing melting points, eg, aluminum (660°C) and tantalum (2996°C), can be clad.

(*5*) Metals with widely different properties, eg, copper/maraging steel, can be bonded readily.

(*6*) Large clad-to-backer ratio limits can be achieved by explosion cladding. Stainless steel-clad components as thin as 0.025 mm and as thick as 3.2 cm have been explosion clad.

(*7*) The thickness of the stationary or backing plate in explosion cladding is essentially unlimited. Backers over >0.5 m thick and weighing 50 t have been clad commercially.

(*8*) High quality, wrought metals are clad without altering their chemical composition.

(*9*) Different types of backers can be clad; clads can be bonded to forged members, as well as to rolled plate.

(*10*) Clads can be bonded to rolled plate that is strand-cast, annealed, normalized, or quench-tempered.

(*11*) Multiple-layered composite sheets and plates can be bonded in a single explosion, and cladding of both sides of a backing metal can be achieved simultaneously. When two sides are clad, the two prime or clad metals need not be of the same thickness nor of the same metal or alloy.

(*12*) Nonplanar metal objects can be clad, eg, the inside of a cylindrical nozzle can be clad with a corrosion-resistant liner.

(*13*) The majority of explosion-clad metals are less expensive than the solid metals that could be used instead of the clad systems.

Limitations of the explosive bonding process are listed below.

(*1*) There are inherent hazards in storing and handling explosives and undesirable noise and blast effects from the explosion.

(*2*) Obtaining explosives with the proper energy, form, and detonation velocity is difficult.

(*3*) Metals to be explosively bonded must be somewhat ductile and resistant to impact. Alloys with as little as 5% tensile elongation in a 5.1-cm gauge length, and backing steels with as little as 13.6 J (10 ft·lbf) Charpy V-notch impact resistance can be bonded. Brittle metals and metal alloys fracture during bonding.

(*4*) For metal systems in which one or more of the metals to be explosively clad has a high initial yield strength or a high strain-hardening rate, a high quality bonded interface may be difficult to achieve. Metal alloys of high strength, ie, >690 MPa (10^5 psi) yield strength, are difficult to bond. This problem increases when there is a large density difference between the metals. Such combinations often are improved by using a thin interlayer between the metals.

(*5*) Geometries that are suited to explosive bonding promote straight-line egression of the high velocity jet emanating from between the metals during bonding, eg, for the bonding of flat and cylindrical surfaces.

(*6*) Thin backers must be supported, thus adding to manufacturing cost.

(*7*) The preparation and assembly of clads is not amenable to automated production techniques, and each assembly requires considerable manual labor.

Theory and Principles

To obtain a metallurgical bond between two metals, the atoms of each metal must be brought sufficiently close so that their normal forces of interatomic attraction produce a bond. The surfaces of metals and alloys must not be covered with films of oxides, nitrides, or adsorbed gases. When such films are present, metal surfaces do not bond satisfactorily.

In fusion welding, the surfaces of the two metals are melted by heating and the contaminated surface layers are brought to the surface of the melt pool. However, pressure-welding processes do not involve melting for removal of surface contaminants. Instead, the work pieces are plastically deformed, often after they have been heated to an appropriate temperature, which breaks the surface films and creates fresh uncontaminated areas where bonding can occur. Cold pressure welding, inertia welding, and ultrasonic welding are variations of this process.

Explosive bonding is a cold pressure-welding process in which the contaminant surface films are plastically jetted from the parent metals as a result of the high pressure collision of the two metals. A jet is formed between the metal plates, if the collision angle and the collision velocity are in the range required for bonding. The contaminant surface films that are detrimental to the establishment of a metallurgical bond are swept away in the jet. The metal plates, which are cleaned of any surface films by the jet action, are joined at an internal point by the high pressure that is obtained near the collision point.

Parallel and Angle Cladding. The arrangements shown in Figures 1 and 2 illustrate the operating principles of explosion cladding. Figure 1 illustrates angle cladding that is limited to cladding for relatively small pieces (38–39). Clad plates with large areas cannot be made using this arrangement because the collision of long plates at high stand-offs, ie, the distance between the plates, on long runs is so violent that metal cracking, spalling, and fracture occur. The arrangement shown in Figure 2 is by far the simplest and most widely used (10).

Jetting. A layer of explosive is placed in contact with one surface of the prime metal plate which is maintained at a constant distance from and parallel to the backer plate, as shown in Figure 2**a**. The explosive is detonated and, as the detonation front moves across the plate, the prime metal is deflected and accelerated to plate velocity V_P; thus, an angle is established between the two plates. The ensuing collision region progresses across the plate at a velocity equal to the detonation velocity D. When the

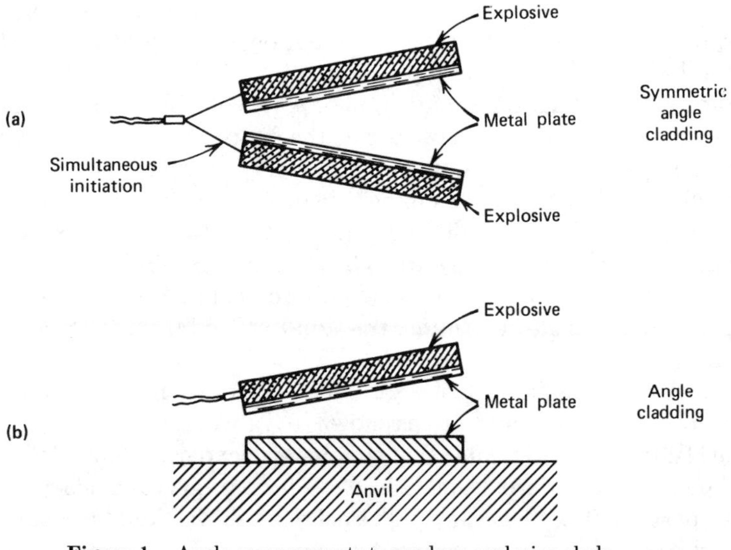

Figure 1. Angle arrangements to produce explosion clads.

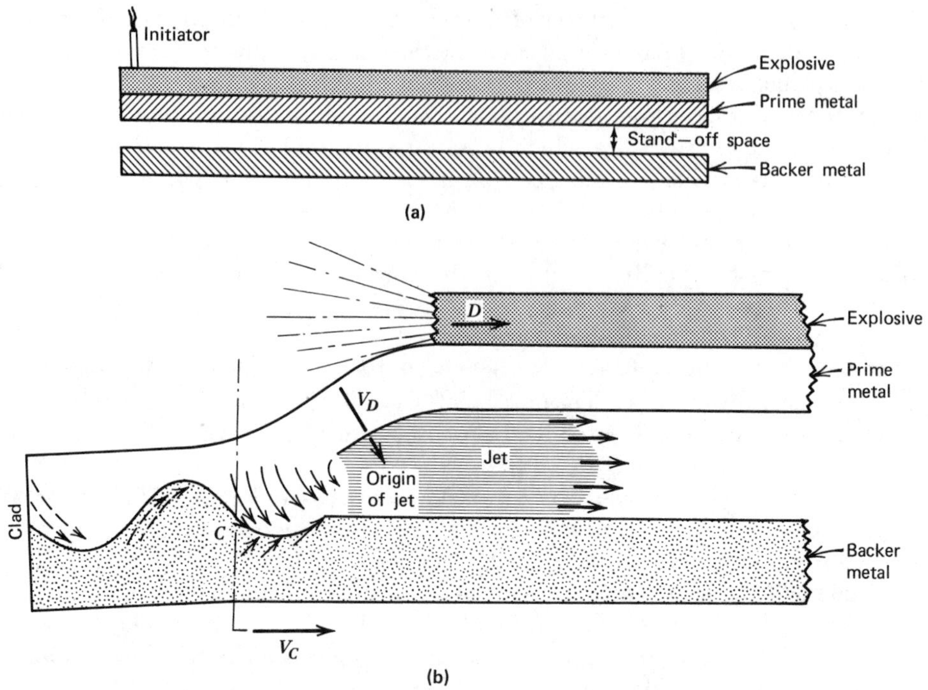

Figure 2. Parallel arrangement for explosion cladding and subsequent collision between the prime and backer metals that leads to jetting and formation of wavy bond zone.

collision velocity V_c and the angle are controlled within certain limits, high pressure gradients ahead of the collision region in each plate cause the metal surfaces to flow hydrodynamically as a spray of metal from the apex of the angled collision. Jetting is the flow process and expulsion of the metal surface (9). Photographic evidence of jetting during an explosion-bonding experiment is given in Figures 3**a**, **b**, and **c** (19). The jet, which moves in the direction of detonation, is observed between the deflected prime metal and the backer metal.

Typically, jet formation is a function of plate collision angle, collision-point velocity, cladding-plate velocity, pressure at the collision point, and the physical and mechanical properties of the plates being bonded. For jetting and subsequent cladding to occur, the collision velocity has to be substantially below the sonic velocity of the cladding plates, usually ca 4000–5000 m/s (16). There also is a minimum collision angle below which no jetting occurs regardless of the collision velocity. In the parallel-plate arrangement shown in Figure 2, this angle is determined by the stand-off. In angle cladding, the preset angle determines the stand-off and the attendant collision angle (see Figure 1).

The amplitude and frequency of the bond-zone wave structure varies as a function of the explosive and the stand-off, as shown in Tables 1 and 2 (12).

Bond Nature. It is preferable that commercial, explosively bonded metals exhibit a wavy, bond-zone interface. Bond-zone wave formation is analogous to fluid flowing around an obstacle (9). When the fluid velocity is low, the fluid flows smoothly around the obstacle but, above a certain fluid velocity, the flow pattern becomes turbulent,

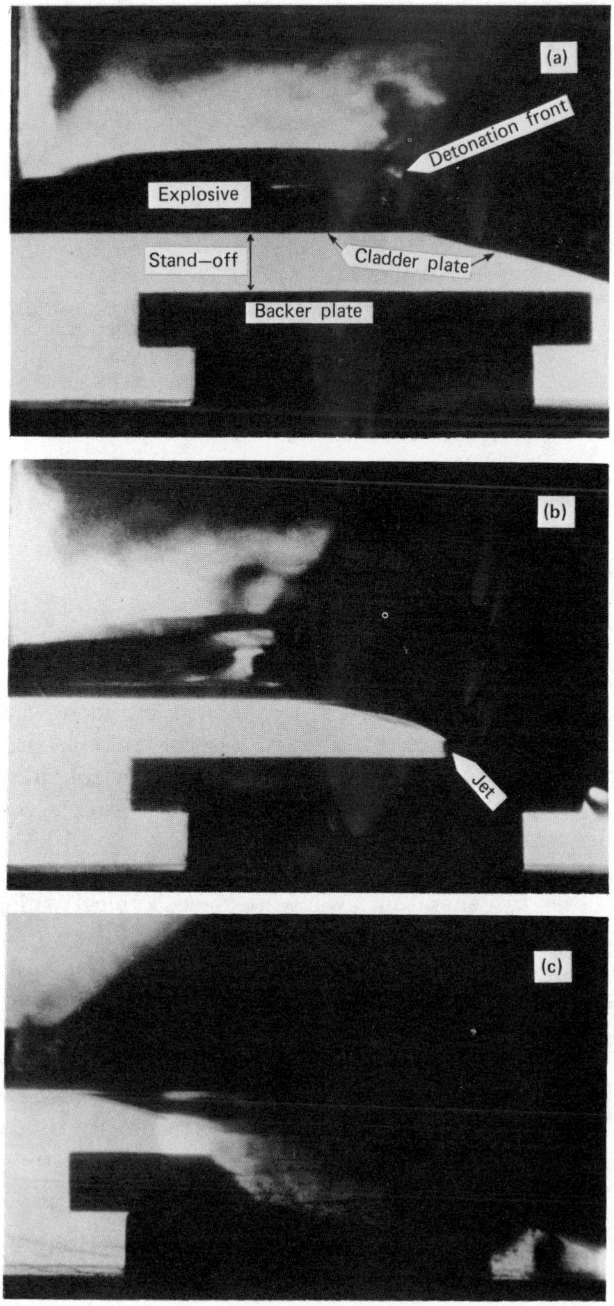

Figure 3. (**a–c**) Cladding of aluminum to aluminum showing jet formation (19).

Table 1. Measured Explosion-Cladding Parameters and Bond-Zone Characteristics for Cladding 3.2-mm Grade A Nickel to 12.7-mm AISI 1008 Carbon Steel[a]

				Bond-zone characteristics			
Collision velocity, m/s	Parallel stand-off, mm	Flyer-plate velocity, m/s	Flyer-plate angle, deg	Type	Wavelength, μm	Amplitude, μm	Equivalent melt thickness, μm
1650	1.14	215	7.4	straight and wavy	112	10	<1
2000	1.14	250	7.0	wavy	103	11	<1
2500	1.14	270	6.6	wavy	236	39	1.6
3600	1.14	410	6.5	wavy	254	38	5.1
2000	2.16	310	8.7	wavy	318	41	<1
2500	2.16	337	8.2	wavy	425	76	3.9
3600	2.16	510	8.25	wavy	590	96	9.8
1650	3.96	325	11.2	wavy	520	52	<1
2000	3.96	372	10.5	wavy	567	88	<1
2500	3.96	407	9.7	wavy	671	121	6.0
3600	3.96	625	9.95	wavy	739	146	28.8
2000	6.35	420	11.8	wavy	790	132	<1
2500	6.35	462	10.8	wavy	895	171	9.0
3600	6.35	700	11.2	wavy	965	162	24.0
1650	10.54	425	14.8	wavy	1018	169	<1
2000	10.54	460	13.0	wavy	623	97	<1
3600	10.54	775	12.5	wavy	1333	284	59.2
1650	17.78	465	16.5	straight			not detectable

[a] Ref. 12.

as illustrated in Figure 4. In explosion bonding, the obstacle is the point of highest pressure in the collision region. Because the pressures in this region are many times higher than the dynamic yield strength of the metals, they flow plastically, as evidenced by the microstructure of the metals at the bond zone. Electron microprobe analysis across such plastically deformed areas shows that no diffusion occurs because there is extremely rapid self-quenching of the metals (16).

Under optimum conditions, the metal flow around the collision point is unstable and it oscillates, thereby generating a wavy interface. Typical explosion-bonded interfaces between nickel plates made at different collision velocities are illustrated in Figure 4 (24). A typical explosion-bonded interface between titanium and steel is shown in Figure 5. Small pockets of solidified melt form under the curl of the waves; some of the kinetic energy of the driven plate is locally converted into heat as the system comes to rest. These discrete regions are completely encapsulated by the ductile prime and base metals. The direct metal-to-metal bonding between the isolated pockets provides the ductility necessary to support stresses during routine fabrication.

The quality of bonding is related directly to the size and distribution of solidified melt pockets along the interface, especially for dissimilar metal systems that form intermetallic compounds. The pockets of solidified melt are brittle and contain localized defects which do not affect the composite properties. Explosion-bonding parameters for dissimilar metal systems normally are chosen to minimize the pockets of melt associated with the interface.

Table 2. Measured Explosion-Cladding Parameters and Bond-Zone Characteristics for Cladding 3.2-mm Grade 1 Titanium to 12.7-mm AISI 1008 Carbon Steel[a]

				Bond-zone characteristics				
Collision velocity, m/s	Parallel stand-off, mm	Flyer-plate velocity, m/s	Flyer-plate impact angle, deg	Type	Wavelength, μm	Amplitude, μm	Equivalent melt thickness, μm	Tensile elongation, %
2000	1.14	330	9.9	wavy	103	8	<1	32
2500	1.14	400	9.7	wavy	254	19	1.2	32
3600	1.14	580	9.3	[b]	250	23	14.0	
2000	2.16	420	12.2	[b]	215	17	<1	34
2500	2.16	465	11.0	[b]	482	47	3.0	29
3600	2.16	710	11.3	[b]	468	59	11.6	28
2000	3.96	495	14.2	[b]	373	31	<1	31
2500	3.96	520	12.1	[b]	768	89	3.1	29
3600	3.96	845	13.4	[b]	868	122	9.2	
2000	6.35	530	15.2	[b]	610	53	<1	32
2500	6.35	565	13.0	[b]	1009	130	8.2	29
3600	6.35	945	15.0	[b]	1228	189	18.5	
2000	10.54	560	15.5	[b]	1013	96	<1	32
2500	10.54	600	14.0	[b]	1300	167	3.6	27
3600	10.54	1040	16.5	[b]	1360	230	21.8	23

[a] Ref. 12.
[b] Melted layer and waves.

When cladding conditions are such that the metallic jet is trapped between the prime metal and the backer, the energy of the jet causes surface melting between the colliding plates. In this type of clad, alloying through melting is responsible for the metallurgical bond. As shown in Figure 6, solidification defects can occur and, for this reason, this type of bond is not desirable.

The industrially useful combinations of explosively clad metals that are available in commercial sizes are listed in Figure 7. The list does not include triclads or combinations that corrosion or materials engineers or equipment designers may yet envision. The combinations that explosion cladding can provide are virtually limitless (14).

Processing

Explosives. The pressure P generated by the detonating explosive that propels the prime plate is directly proportional to its density ρ and the square of the detonation velocity, V_D^2 (40):

$$P = \frac{1}{4} \rho V_D^2$$

The detonation velocity is controlled by adjusting the packing density or the amount of added inert material (41).

The types of explosives that have been used include the following (14,41) (see

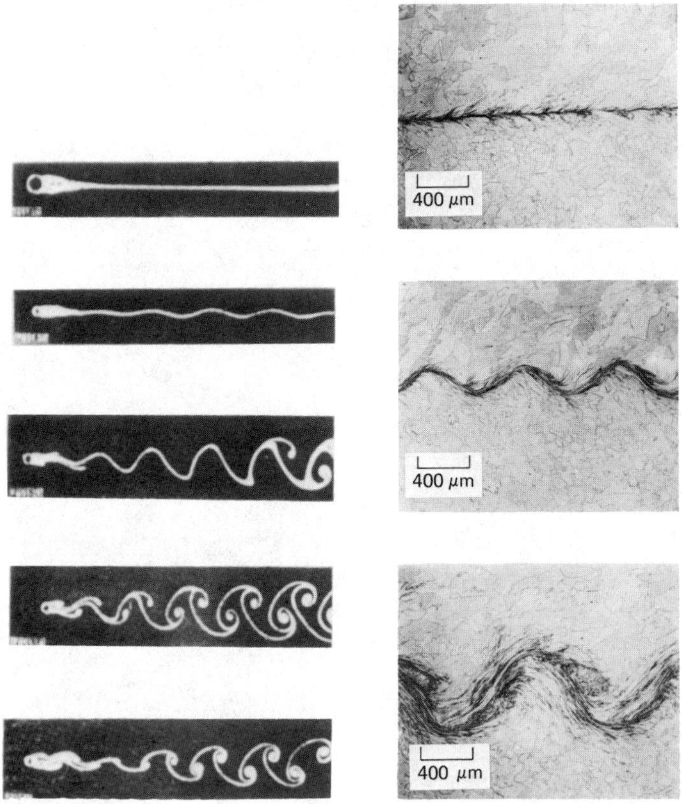

Figure 4. At left are photographs of fluid flow behind cylinders at increasing flow velocities top to bottom. At right are photomicrographs of nickel–nickel bond zones made at increasing collision velocities; top, about 1600 m/s; middle, about 1900 m/s; bottom, about 2500 m/s (24).

Explosives):

High velocity (4500–7600 m/s)	*Low–medium velocity (1500–4500 m/s)*
trinitrotoluene (TNT)	ammonium nitrate
cyclotrimethylenetrinitramine (RDX)	ammonium nitrate prills sensitized with fuel oil
pentaerythritol tetranitrate (PETN)	ammonium perchlorate
composition B	amatol
composition C_4	amatol and sodatol diluted with rock salt to 30–35%
plasticized PETN-based-rolled sheet and extruded cord	dynamites
primacord	nitroguanidine
	diluted PETN

Metal Preparation. Preparation of the metal surfaces to be bonded usually is required because most metals contain surface imperfections or contaminants that undesirably affect bond properties. The cladding faces usually are surface-ground using an abrasive machine and then are degreased with a solvent to ensure consistent bond

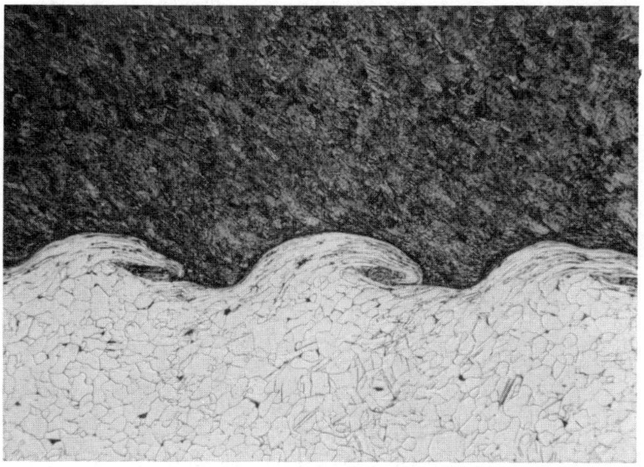

Figure 5. Photomicrograph of titanium, top, to carbon steel, bottom, explosion clad (100×).

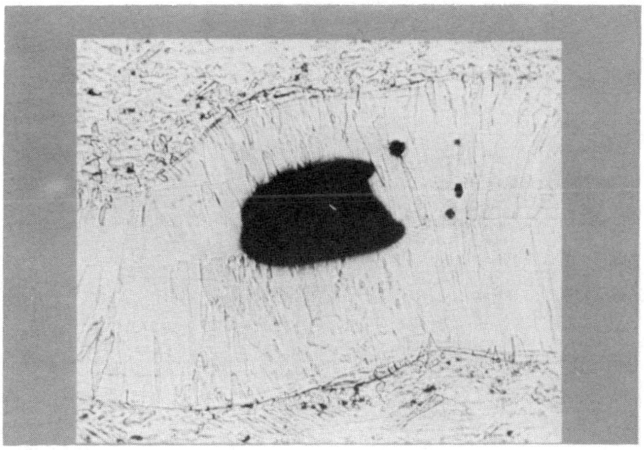

Figure 6. Solidification defects in the copper–copper explosion clad evidence the occurrence of melting at the interface (9) (100×).

strength (41). In general, a surface finish that is ≥3.8 μm is needed to produce consistent, high quality bonds.

Fabrication techniques must take into account the metallurgical properties of the metals to be joined and the possibility of undesirable diffusion at the interface during hot forming, heat treating, and welding. Compatible alloys, ie, those that do not form intermetallic compounds upon alloying, eg, nickel and nickel alloys, copper and copper alloys, and stainless steel alloys clad to steel, may be treated by the traditional techniques developed for clads produced by other processes. On the other hand, incompatible combinations, eg, titanium, zirconium, or aluminum to steel, require special techniques designed to limit the production at the interface of undesirable intermetallics which would jeopardize bond ductility.

Assembly, Stand-off. The air gap present in parallel explosion cladding can be maintained by metallic supports that are tack-welded to the prime and backer plates or by metallic inserts that are placed between the prime and backer (41–43). The inserts

	Zirconium	Magnesium	Stellite 6B	Platinum	Gold	Silver	Niobium	Tantalum	Hastelloy	Titanium	Nickel alloys	Copper alloys	Aluminum	Stainless steels	Alloy steels	Carbon steels
Carbon steels	•	•			•	•	•	•	•	•	•	•	•	•	•	•
Alloy steels	•	•	•					•	•	•	•	•	•	•	•	
Stainless steels			•		•	•	•	•		•	•	•	•	•		
Aluminum [*7429-90-5*]		•				•	•	•		•	•	•	•			
Copper alloys						•	•	•		•	•	•				
Nickel alloys		•		•	•			•	•	•	•					
Titanium [*7440-30-6*]	•	•				•	•	•		•						
Hastelloy									•							
Tantalum [*7440-25-7*]					•		•	•								
Niobium [*7440-03-1*]				•			•									
Silver [*7440-22-4*]						•										
Gold [*7440-57-5*]																
Platinum [*7440-06-4*]				•												
Stellite 6B																
Magnesium [*7439-95-4*]		•														
Zirconium [*7440-67-7*]	•															

Figure 7. Explosion-clad metal combinations that are commercially available.

usually are made of a metal that is compatible with one of the cladding metals. If the prime metal is so thin that it sags when supported by its edges, other materials, eg, rigid foam, can be placed between the edges to provide additional support; the rigid foam is consumed by the hot egressing jet during bonding (41–44) (see Foamed plastics). A moderating layer or buffer, eg, polyethylene sheet, water, rubber, paints, and pressure-sensitive tapes, may be placed between the explosive and prime metal surface to attenuate the explosive pressure or to protect the metal surface from explosion effects (14).

Facilities. The preset, assembled composite is placed on an anvil of appropriate thickness to minimize distortion of the clad product. For thick composites, a bed of sand usually is a satisfactory anvil. Thin composites may require a support made of steel, wood, or other appropriate materials. The problems of noise, air blast, and air pollution (qv) are inherent in explosion cladding, and clad-composite size is restricted by these problems (see Noise pollution). Thus, the cladding facilities should be in areas that are remote from population centers. Using barricades and burying the explosives and components under water or sand lessens the effects of noise and air pollution (14).

An attractive method for making small-area clads using light explosive loads employs a low vacuum, noiseless chamber (14). Underground missile silos and mines also have been used as cladding chambers (see also Insulation, acoustic).

Analytical and Test Methods

When the explosion-bonding process distorts the composite so that its flatness does not meet standard flatness specifications, it is reflattened on a press or roller leveller (see ASME-SA-20). However, press-flattened plates sometimes contain localized irregularities which do not exceed the specified limits but which, generally, do not occur in roll-flattened products.

Nondestructive. Nondestructive inspection of an explosion-welded composite is almost totally restricted to ultrasonic and visual inspection. Radiographic inspection is applicable only to special types of composites consisting of two metals having a significant mismatch in density and a large wave pattern in the bond interface (see Nondestructive testing).

Ultrasonic. The most widely used nondestructive test method for explosion-welded composites is ultrasonic inspection. Pulse-echo procedures (ASTM A 435) are applicable for inspection of explosion-welded composites used in pressure applications (see Ultrasonics).

The acceptable amount of nonbond depends upon the application. In clad plates for heat exchangers, >98% bond usually is required (see Heat-exchange technology). Other applications may require only 95% of the total area to be bonded. Configurations of a nonbond sometimes are specified, eg, in heat exchangers where a nonbond area may not be >19.4 cm^2 or 7.6 cm long. The number of areas of nonbond generally is specified. Ultrasonic testing can be used on seam welds, tubular transition joints, clad pipe and tubing, and in structural and special applications.

Radiographic. Radiography is an excellent nondestructive test (NDT) method for evaluating the bond of Al–steel electrical and Al–Al–steel structural transition joints. It provides the capability of precisely and accurately defining all nonbond and flat-bond areas of the Al–steel interface, regardless of their size or location.

The clad plate is x-rayed perpendicular from the steel side and the film contacts the aluminum. Radiography reveals the wavy interface of explosion-welded, aluminum-clad steel as uniformly spaced, light and dark lines with a frequency of one to three lines per centimeter. The waves characterize a strong and ductile transition joint and represent the acceptable condition. The clad is interpreted to be nonbonded when the x ray shows complete loss of the wavy interface (see X-ray technology).

Destructive. Destructive testing is used to determine the strength of the weld and the effect of the explosion-welding process on the parent metals. Standard testing techniques can be utilized on many composites; however, nonstandard or specially designed tests often are required to provide meaningful test data for specific applications.

Pressure-Vessel Standards. Explosion-clad plates for pressure vessels are tested according to the applicable ASME Boiler and Pressure Vessel Code Specifications. Unfired pressure vessels using clads are covered by ASTM A 263, A 264, and A 265; these include tensile, bend, and shear tests.

Tensile tests of a composite plate having a thickness of <3.8 cm require testing of the joined base metal and clad. Strengthening does occur during cladding and tensile strengths generally are greater than for the original materials. Some typical shear-strength values obtained for explosion-clad composites covered by ASTM A 263, A 264, A 265, which specify 138 MPa (20,000 psi) minimum, and B 432, which specifies 83 MPa (12,000 psi) minimum, are listed in Table 3 (see High pressure technology).

Chisel. Chisel testing is a quick, qualitative technique that is widely used to determine the soundness of explosion-welded metal interfaces. A chisel is driven into and along the weld interface, and the ability of the interface to resist the separating force of the chisel provides an excellent qualitative measure of weld ductility and strength.

Ram Tensile. A ram tensile test has been developed to evaluate the bond-zone tensile strength of explosion-bonded composites. As shown in Figure 8 the specimen is designed to subject the bonded interface to a pure tensile load. The cross-section area of the specimen is the area of the annulus between the OD and ID of the specimen.

Table 3. Typical Shear Strengths of Explosively Clad Metals

Cladding metal on carbon-steel backers	Shear strength[a], MPa[b]
stainless steels	448
nickel and nickel alloys	379
Hastelloy alloys	391
zirconium	269
titanium[c]	241
cupronickel	251
copper	152
aluminum (1100-Ah4)	96

[a] See ASTM A 263, A 264, A 265, and B 432.
[b] To convert MPa to psi, multiply by 145.
[c] Stress-relief annealed at 621°C.

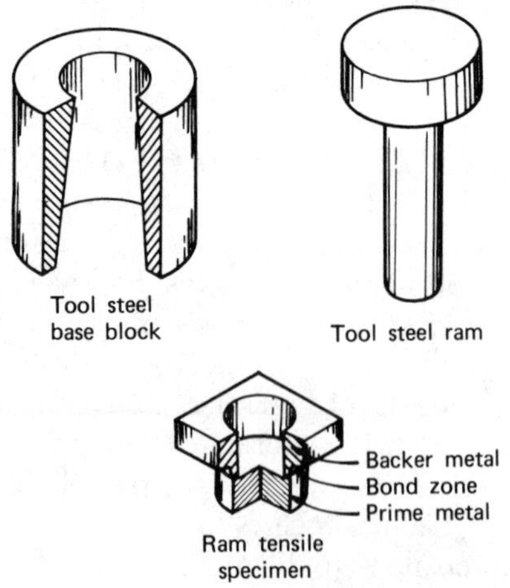

Figure 8. Machined explosion-clad tests sample and fixture for ram tensile test of bond zone.

The specimen typically has a very short tensile gauge length and is constructed so as to cause failure at the bonded interface. The ultimate tensile strength and relative ductility of the explosion-bonded interface can be obtained by this technique.

Mechanical Fatigue. Some mechanical fatigue tests have been conducted on explosion-clad composites where the plane of maximum tensile stress is placed near the bond zone (20).

Thermal Fatigue and Stability. Explosion-welded plates have performed satisfactorally in several types of thermal tests (18). In thermal-fatigue tests, samples from bonded plate are alternately heated to 454–538°C at the surface and are quenched in cold water to less than 38°C. The three-minute cycles consist of 168 s of heating and 12 s of cooling. Weld-shear tests are performed on samples before and after thermal cycling. Stainless steel clads have survived 2000 such thermal cycles without significant loss in strength (18). Similarly welded and tested Grade 1 titanium–carbon steel samples performed in similar satisfactory fashion.

Metallographic. The interface is inspected on a plane parallel to the detonation front and normal to the surface. A well-formed wave pattern without porosity generally is indicative of a good bond. The amplitude of the wave pattern for a good weld can vary from small to large without a large influence on the strength, and small pockets of melt can exist without being detrimental to the quality of the bond. However, a continuous layer of melted material indicates that welding parameters were incorrect and should be adjusted. A line-type interface with few waves indicates that the collision velocity of the plate was not great enough and/or that the collision angle was too high for jetting to occur. A well-defined wave pattern in which the crest of the wave is bent over to form a large melt pocket with a void in the swirl is indicative of a poor bond. In this case, the plate velocity is too high as is the collision angle.

In some materials, eg, titanium and martensitic steels, shear bands are adjacent to the weld interface if the cladding variables are excessive. This is the result of thermal adiabatic shear developed from excessive overshooting energy, and a heat treatment is required to eliminate the hardened-band effect. When the cladding variables and the system energy are optimum, thermal shear bands are minimized or eliminated and heat treatment after cladding is not required. Several types of metal composites require heat treatment after cladding to relieve stress, but intermetallic compounds can form as a result of the treatment. A metallographic examination indicates if the heat treatment of the explosion-bonded composite has resulted in the formation of intermetallic compounds.

Hardness, Impact Strength. Microhardness profiles on sections from explosion-bonded materials show the effect of strain hardening on the metals in the composite (see Hardness). Figure 9 illustrates the effect of cladding a strain-hardening austenitic stainless steel to a carbon steel. The austenitic stainless steel is hardened adjacent to the weld interface by explosion welding, whereas the carbon steel is not hardened to a great extent. Similarly, aluminum does not strain harden significantly.

Impact strengths also can be reduced by the presence of the hardened zone at the interface. A low temperature stress-relief anneal decreases the hardness and restores impact strength (20). Alloys that are sensitive to low temperature heat treatments also show differences in hardness traverses that are related to the explosion-welding parameters, as illustrated in Figure 10 (36). Low welding-impact velocities do not develop as much adiabatic heating as higher impact velocities. The effect of the adiabatic heating is to anneal and further age the alloys. Hardness traverses in-

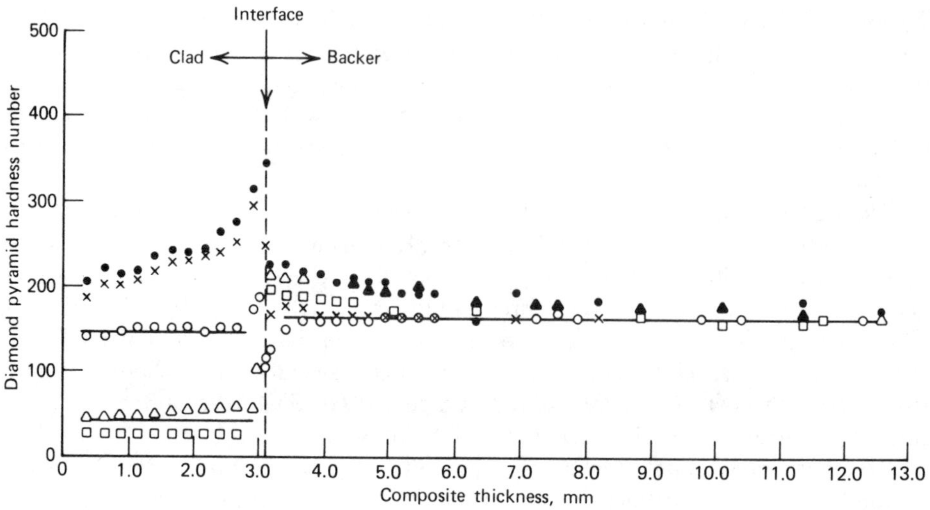

Figure 9. Microhardness profile across interfaces of two types of explosion clads that show widely divergent response resulting from their inherent cold-work hardening characteristics (17). 3.2 mm type 304L stainless/28.6 mm, A 516-70: — control (before cladding). ● = clad + flat. × = clad + stress-relief annealed at 621°C + flat. ○ = clad + normalize at 954°C. 3.2 mm 1100-H14 aluminum/25.4 mm, A 516-70: — control (before cladding). △ = clad + flat. □ = clad + flat + stress-relief annealed at 593°C (17).

dicate the degree of hardening during welding and what, if any, subsequent heat treatment is required after explosion bonding. Explosion-bonding parameters also can be adjusted to prevent softening at the interface, as shown in Figure 10.

Safety Aspects

All explosive materials should be handled and used following approved safety procedures either by or under the direction of competent, experienced persons and in accordance with all applicable federal, state, and local laws, regulations, and ordinances. The Bureau of Alcohol, Tobacco, and Firearms (BATF), the Hazardous Materials Regulation Board (HMRB) of the Department of Transportation (DOT), the Occupational Safety and Health Agency (OSHA), and the Environmental Protection Agency (EPA), Washington, D.C., have federal jurisdiction on the sale, transport, storage, and use of explosives. Many states and local counties have special explosive requirements. The Institute of Makers of Explosives (IME), New York, provides educational publications to promote the safe handling, storage, and use of explosives. The National Fire Protective Association (NFPA), Boston, Mass., similarly provides recommendations for safe explosives manufacture, storage, handling, and use.

Uses

Cladding and backing metals are purchased in the appropriately heat-treated condition because corrosion resistance is retained through bonding. It is customary to supply the composites in the as-bonded condition because hardening usually does not affect the engineering properties. Occasionally, a postbonding heat treatment is used to achieve properties required for specific combinations.

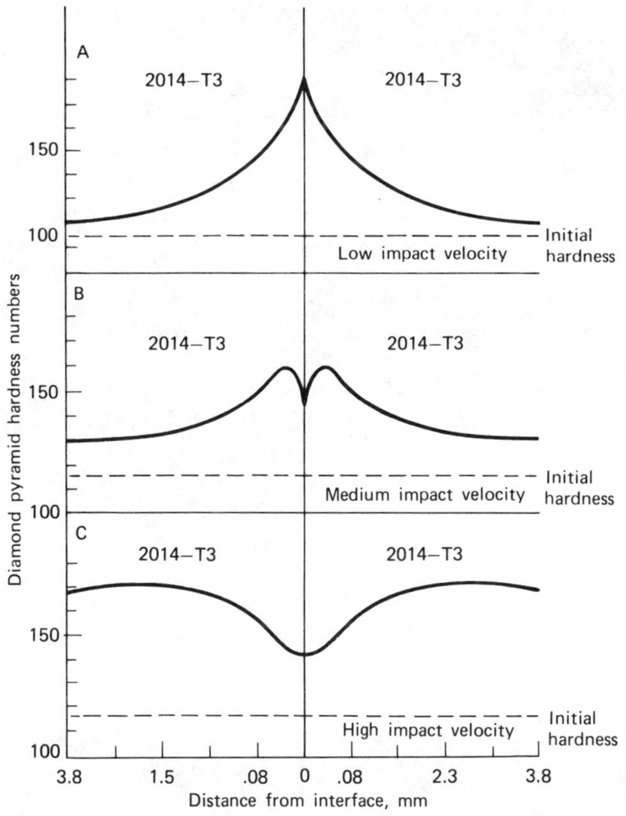

Figure 10. Microhardness profiles across interface of explosion-clad age-hardenable aluminum alloys (36).

Vessel heads can be made from explosion-bonded clads, either by conventional cold- or hot-forming techniques. The latter involves thermal exposure and is equivalent in effect to a heat treatment. The backing metal properties, bond continuity, and bond strength are guaranteed to the same specifications as the composite from which the head is formed.

Applications such as chemical-process vessels and transition joints represent ca 90% of the industrial use of explosion cladding.

Chemical-Process Vessels. Explosion-bonded products are used in the manufacture of process equipment for the chemical, petrochemical, and petroleum industries where the corrosion resistance of an expensive metal is combined with the strength and economy of another metal. Applications include explosion cladding of titanium tubesheet to Monel (Fig. 11), hot fabrication of an explosion clad to form an elbow for pipes in nuclear power plants, and explosion cladding titanium and steel for use in a vessel intended for terephthalic acid manufacture (Fig. 12).

Precautions must be taken when welding incompatible clad systems, eg, hot forming of titanium-clad steel plates must be conducted at 788°C or less. The preferred technique for butt welding involves a batten-strap technique using a silver, copper, or steel underlay (see Fig. 13). Precautions must be taken to avoid iron contamination

Figure 11. Tubesheet of titanium explosively clad to Monel. Tubes were titanium and shell was Monel 400. Courtesy Nooter Corporation.

Figure 12. Titanium–carbon steel vessel made from explosively clad plate for use in terephthalic acid manufacture. Courtesy of Nooter Corporation.

of the weld either from the backer steel or from outside sources. Stress relieving is achieved at normal steel stress relieving temperatures, and special welding techniques must be used in joining tantalum–copper–steel clads (45–46).

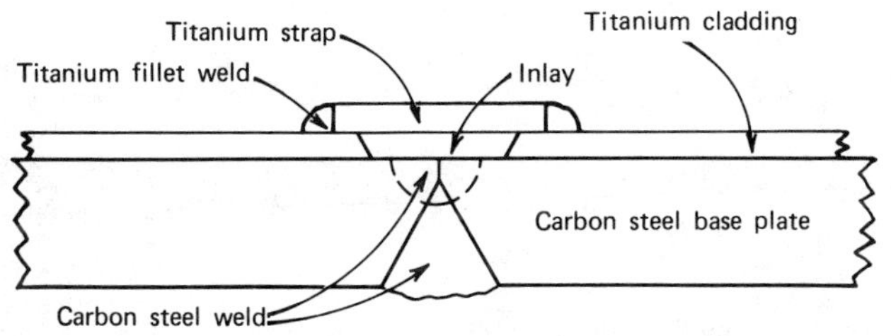

Figure 13. Double-v inlay, batten-strap technique for fusion welding of an explosion-clad plate containing titanium and zirconium.

Conversion-Rolling Billets. Much clad plate and strip have been made by hot and cold rolling of explosion-bonded slabs and billets. Explosion bonding is economically attractive for conversion rolling, because the capital investment for plating and welding equipment needed for conventional bonding methods is avoided. Highly alloyed stainless steels and some copper alloys, which are difficult to clad by roll bonding, are used for plates made by converting explosion-bonded slabs and billets. Conventional hot-rolling and heat-treatment practices are used when stainless steels, nickel, and copper alloys are converted. Hot rolling of explosion-bonded titanium, however, must be performed below ca 843°C to avoid diffusion and the attendant formation of undesirable intermetallic compounds at the bond interface. Hot-rolling titanium also requires a stiff rolling mill because of the large separation forces required for reduction.

Perhaps the most extensive application for conversion-rolled, explosion-bonded clads was for U.S. coinage in the 1960s (47). Over 15,900 metric tons of explosion-clad strip that was supplied to the U.S. Mint helped alleviate the national silver-coin shortage. The triclad composites consist of 70–30 cupronickel/Cu/70–30 cupronickel.

Transition Joints. Use of explosion-clad transition joints avoids the limitations involved in joining two incompatible materials by bolting or riveting. Many transition joints can be cut from a single large-area flat-plate clad and delivered to limit the temperature at the bond interface so as to avoid undesirable diffusion. Conventional welding practices may be used for both similar metal welds.

Electrical. Aluminum, copper, and steel are the most common metals used in high current–low voltage conductor systems. Use of these metals in dissimilar metal systems often maximizes the effects of the special properties of each material. However, junctions between these incompatible metals must be electrically efficient to minimize power losses. Mechanical connections involving aluminum offer high resistance because of the presence of the self-healing oxide skin on the aluminum member. Because this oxide layer is removed by the jet, the interface of an explosion clad essentially offers no resistance to the current. Thus, welded transition joints, which are cut from thick composite plates of aluminum–carbon steel, permit highly efficient electrical conduction between dissimilar metal conductors. Sections can be added by conventional welding. This concept is routinely employed by the primary aluminum-reduction industry in anode-rod fabrication. The connection is free of the aging effects that are characteristic of mechanical connections and requires no maintenance. The mechanical properties of the explosion weld, ie, shear, tensile, and impact strength, exceed those of the parent-type 1100 aluminum alloy.

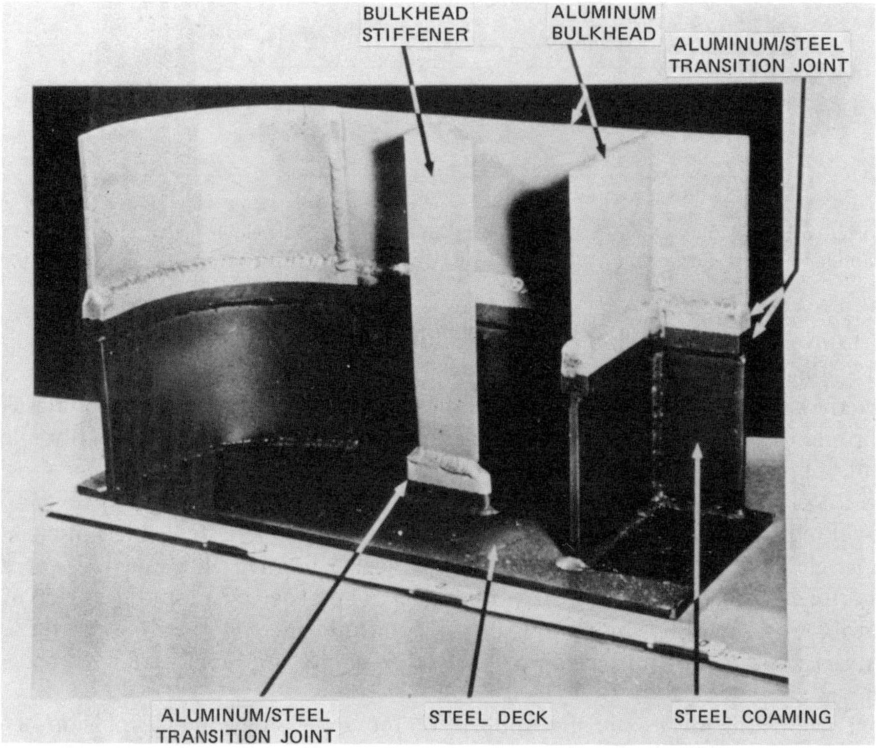

Figure 14. Sample showing typical aluminum superstructure and deck connection made possible by use of explosion clad aluminum–carbon steel transition joint.

Usually, copper surfaces are mated when joints must be periodically disconnected because copper offers low resistance and good wear. Junctions between copper and aluminum bus bars are improved by using a copper–aluminum transition joint that is welded to the aluminum member. Deterioration of aluminum shunt connections by arcing is eliminated when a transition joint is welded to both the primary bar and the shunting bar.

The same intermetallic compounds that prevent conventional welding between aluminum and copper or steel can be developed in an explosion clad by heat treatment at elevated temperature. Diffusion can be avoided if the long-term service temperature is kept below 260°C for aluminum–steel and 177°C for copper–aluminum combinations. Under short-term conditions, as during welding, peak temperatures of 316°C and 232°C, respectively, are permissible. Bond ductility is maintained, although there is a reduction in bond strength as the aluminum is annealed. Bond strength, however, never falls below that of the parent aluminum; therefore, nominal handbook values for type 1100 alloy aluminum may be used in design considerations. The bond is unaffected by thermal cycling within the recommended temperature range.

Marine. In the presence of an electrolyte, eg, seawater, aluminum and steel form a galvanic cell and corrosion takes place at the interface. Because the aluminum superstructure is bolted to the steel bulkhead in a lap joint, crevice corrosion is masked and may remain unnoticed until replacement is required. By using transition-joint

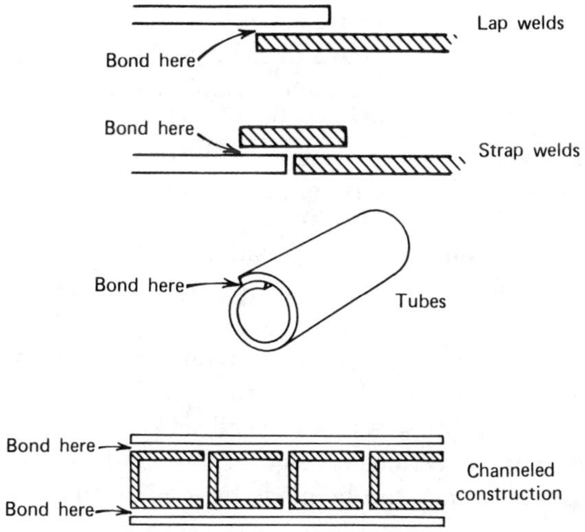

Figure 15. Explosion-clad welding applications (7).

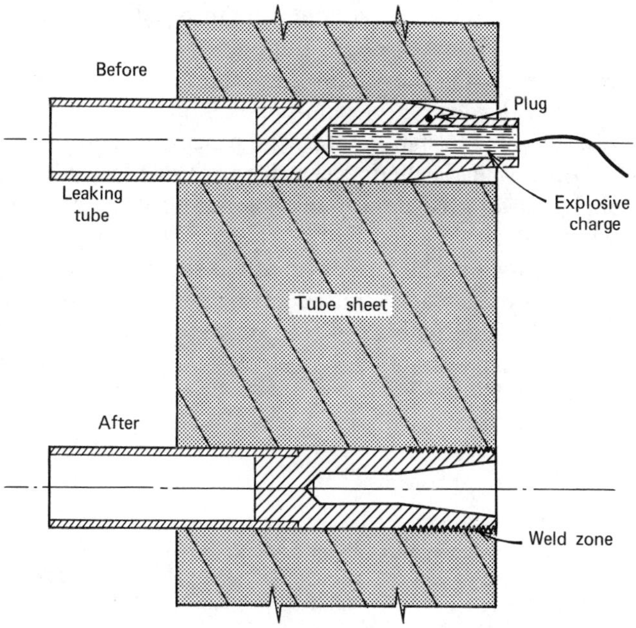

Figure 16. Tube-to-tubesheet plugging (48).

strips cut from explosion-welded clads, the corrosion problem can be eliminated. Because the transition is metallurgically bonded, there is no crevice in which the electrolyte can act and galvanic action cannot take place. Steel corrosion is confined to external surfaces where it can be detected easily and corrected by simple wire brushing and painting.

Explosion-welded construction has equivalent or better properties than the more complicated riveted systems. Peripheral benefits include weight savings and perfect electrical grounding. In addition to lower initial installation costs, the welded system requires little or no maintenance and, therefore, minimizes life-cycle costs. Applications of structural transition joints include aluminum superstructures that are welded to decks of naval vessels and commercial ships as illustrated in Figure 14.

Tubular. Explosion welding is a practical method for providing the means to join dissimilar metal pipes, eg, aluminum, titanium, or zirconium to steel or stainless steel, using standard welding equipment and techniques. The process provides a strong metallurgical bond which assures that the transition joints provide maintenance-free service throughout years of thermal and pressure/vacuum cycling. Explosion-welded tubular transition joints are being used in many diverse applications in aerospace, nuclear, and cryogenic industries. They operate reliably through the full range of temperatures, pressures, and stresses that normally are encountered in piping systems. Tubular transition joints in various configurations can be cut and machined from explosion-welded plate, or they can be made by joining tubes by overlap cladding. Standard welding practices are used to make the final joints.

Nonplanar Specialty Products. The inside walls of hollow forgings that are used for connections to heavy-walled pressure vessels have been metallurgically bonded with stainless steel. These bonded forgings, or nozzles, range from 50 to 610 mm ID and are up to 1 m long. Large-clad cylinders and internally clad, heavy-walled tubes have been extruded using conventional equipment. Other welding applications have been demonstrated, including those shown in Figure 15.

Tube Welding and Plugging. Explosion-bonding principles are used to bond tubes and tube plugs to tube sheets. The commercial process resembles the cladding of internal surfaces of thick-walled cylinders or prressure vessel nozzles, as shown in Figure 16; angle cladding is used (48). Countersink machining at the tube entrance provides the angled surface of 10–20° at a depth of 1.3–1.6 cm. The exploding detonator propels the tube or tube plug against the face of the tube- sheet to form the proper collision angle which, in turn, provides the required jetting and attendant metallurgical bond. Tubes may be welded individually or in groups. Metal combinations that are welded commercially include carbon steel–carbon steel, titanium–stainless steel, and 90–10 cupronickel/carbon steel.

BIBLIOGRAPHY

1. J. Pearson, *J. Met.* **12,** 673 (1960).
2. R. S. Rinehart and J. Pearson, *Explosive Working of Metals*, MacMillan, New York, 1963.
3. J. J. Douglass, *New England Regional Conference of AIME*, Boston, Mass., May 26, 1960.
4. *Ryan Reporter*, Vol. 21, No. 3, Ryan Aeronautical Company, San Diego, Calif., 1960, pp. 6–8.
5. C. P. Williams, *J. Met.*, 33 (1960).
6. "High Energy Rate Forming" in *Product Engineering and American Machinist/Metalworking Manufacturing*, McGraw Hill, New York, 1961 and 1962.
7. A. H. Holtzman and C. G. Rudershansen, *Sheet Met. Ind.* **39,** 401 (1961).
8. U.S. Pat. 3,024,526 (Mar. 13, 1962), V. Philipchuk and F. Le Roy Bois (to Atlantic Research Corp.).
9. G. R. Cowan and A. H. Holtzman, *J. Appl. Phys.* **34**(Pt. 1), 928 (1962).
10. U.S. Pat. 3,137,937 (Jun 23, 1964), G. R. Cowan, J. J. Douglass, and A. H. Holtzman (to E. I. du Pont de Nemours & Co., Inc.).
11. U.S. Pat. 3,233,312 (Feb. 8, 1966), G. R. Cowan and A. H. Holtzman (to E. I. du Pont de Nemours & Co., Inc.).
12. U.S. Pat. 3,397,444 (Aug. 20, 1968), O. R. Bergmann, G. R. Cowan, and A. H. Holtzman (to E. I. du Pont de Nemours & Co., Inc.).

13. U.S. Pat. 3,493,353 (Feb. 3, 1970), O. R. Bergmann, G. R. Cowan, and A. H. Holtzman (to E. I. du Pont de Nemours & Co., Inc.).
14. V. D. Linse, R. H. Wittman, and R. J. Carlson, *Defense Metals Information Center, Memo 225*, Columbus, Ohio, Sept. 1967.
15. J. F. Kowalick and D. R. Hay, *Met. Trans.* **2,** 1953 (1971).
16. A. H. Holtzman and G. R. Cowan, *Weld. Res. Counc. Bull. No. 104*, Engineering Foundation, New York, Apr. 1965.
17. A. Pocalyko and C. P. Williams, *Weld. J.* **43,** 854(1964).
18. A. Pocalyko, *Mater. Prot.* **4**(6), 10 (1965).
19. O. R. Bergmann, G. R. Cowan, and A. H. Holtzman, *Trans. Met. Soc. AIME* **236,** 646 (1966).
20. J. L. DeMaris and A. Pocalyko, *Am. Soc. Tool and Mfg. Engrs. Paper AD66-113*, Dearborn, Mich., 1966.
21. O. R. Bergmann, *ASM Met. Eng. Quart.*, 60 (1966).
22. T. J. Enright, W. F. Sharp, and O. R. Bergmann, *Met. Prog.*, 107 (1970).
23. C. R. McKenney and J. G. Banker, *Mar. Technol.*, 285(1971).
24. G. R. Cowan, O. R. Bergmann, and A. H. Holtzman, *Met. Trans.* **2,** 3145 (1971).
25. D. E. Davenport and G. E. Duvall, *American Soc. Tool & Mfg. Engrs. Tech. Paper SP60-161*, 1960–1961.
26. G. R. Abrahamson, *J. Appl. Mech.* **28,** 519 (1961).
27. D. E. Davenport, *Am. Soc. Tool & Mfg. Engrs. Tech. Paper SP62-77*, Dearborn, Mich., 1961–62.
28. H. E. Otto and S. H. Carpenter, *Weld. J.* **51,** 467 (1972).
29. T. Onzawa and Y. Tshii, *Trans. Jpn. Weld. Soc.* **6,** 28 (1975).
30. A. S. Bahrani, T. J. Black, and B. Crossland, *Proc. Roy. Soc.* **296A,** 123 (1967).
31. B. Crossland and J. D. Williams, *Met. Rev.* **15,** 80 (1970).
32. A. A. Deribas, V. M. Kudinov, and F. I. Matveenkov, *Fiz. Goreniya Vzryva* **3,** 561 (1967).
33. S. K. Godunov, A. A. Deribas, A. V. Zabrodin, and N. S. Kozin, *J. Comp. Phys.* **5,** 517 (1970).
34. B. Crossland and A. S. Bahrani, *Proc. First Int. Conf. Center High Energy Forming*, University of Denver, Denver, Co., 1967.
35. A. A. Ezra, *Principles and Practices of Explosives Metal Working*, Industrial Newspapers, Ltd., London, 1973.
36. S. H. Carpenter and R. H. Wittman, *Ann. Rev. Mater. Sci.* **5,** 177 (1975).
37. J. L. Edwards, B. H. Cranston, and G. Krauss, *Met. Effects at High Strain Rates*, Plenum Press, New York, 1973.
38. U.S. Pat. 3,264,731 (Aug. 9, 1966), B. Chudzik (to E. I. du Pont de Nemours & Co., Inc.).
39. U.S. Pat. 3,263,324 (Aug. 2, 1966), A. A. Popoff (to E. I. du Pont de Nemours & Co., Inc.).
40. M. A. Cook, *The Science of High Explosives*, Reinhold Publishing Corp., New York, 1966, p. 274.
41. A. A. Popoff, *Mech. Eng.* **100**(5), 28 (1978).
42. U.S. Pat. 3,140,539 (July 14, 1964), A. H. Holtzman (to E. I. du Pont de Nemours & Co., Inc.).
43. U.S. Pat. 3,205,574 (Sept. 14, 1965), H. M. Brennecke (to E.I. du Pont de Nemours & Co., Inc.).
44. U.S. Pat. 3,360,848 (Jan. 2, 1968), J. J. Saia (to E. I. du Pont de Nemours & Co., Inc.).
45. U.S. Pat. 3,464,802 (Sept. 2, 1969), J. J. Meyer (to Nooter Corporation).
46. U.S. Pat. 4,073,427 (Feb. 14, 1978), H. G. Keifert and E. R. Jenstrom (to Fansteel, Inc.).
47. J. M. Stone, *Paper presented at a Select conference on Explosive Welding*, Hove, Eng., Sept. 1968, pp. 29–34.
48. R. Hardwick, *Weld. J.* **54**(4), 238 (1975).

General References

The Joining of Dissimilar Metals, DMIC Report S-16, Battelle Memorial Institute, Columbus, Ohio, Jan. 1968.

S. H. Carpenter, *Nat. Tech. Info. Ser. Rept. No. AMMRC CTR74-69*, Dec. 1978.

C. Birkhoff, D. P. MacDougall, E. M. Pugh, and G. Taylor, *J. Appl. Phys.* **19,** 563 (1948).

L. Zernow, I. Lieberman, and W. L. Kincheloe, *Am. Soc. Tool. & Mfg. Engr., Tech. Paper SP60-141*, Dearborn, Mich., 1961.

R. H. Wittman, *Metallurgical Effects at High Strain Rates*, Plenum Press, New York, 1973.

R. H. Wittman, *Amer. Soc. Tool & Mfg. Engrs. Tech. Paper AD-67-177*, Dearborn, Mich.

G. Bechtold, I. Michael, and R. Prummer, *Gold Bull. Chamber Mines S. Afr.* **10**(2), 34 (1977).

B. H. Cranston, D. A. Machusak, and M. E. Skinkle, *West. Electr. Eng.*, 26 (Oct. 1978).

A. A. Popoff, *Amer. Soc. of Mfg. Engrs., Tech. Paper AD77-236*, Dearborn, Mich., 1977.
T. Z. Blazynski, *International Conference on Welding and Fabrication of Non-Ferrous Metals*, Eastbourne, May 2 and May 3, 1972, Cambridge, Eng., The Welding Institute, 1972.
J. F. Kowalick and D. R. Hay, *Second International Conference of the Center For High Energy Forming*, Estes Park, Co., June 23–27, 1969.
J. Ramesam, S. R. Sahay, P. C. Angelo, and R. V. Tamhankar, *Weld Res. Suppl.*, 23s (1972).
L. F. Trueb, *Trans. Met. Soc. AIME* **2,** 147 (1971).

ANDREW POCALYKO
E. I. du Pont de Nemours & Co., Inc.

MICROENCAPSULATION

"Small is better" would be an appropriate motto for the many people studying microencapsulation, a process in which tiny particles or droplets are surrounded by a coating to give small capsules with many useful properties. In its simplest form, a microcapsule is a small sphere with a uniform wall around it. The material inside the microcapsule is referred to as the core, internal phase, or fill, whereas the wall is sometimes called a shell, coating, or membrane. Most microcapsules have diameters between a few micrometers and a few millimeters, as illustrated in Figure 1.

Many microcapsules, however, bear little resemblance to these simple spheres. The core may be a crystal, a jagged adsorbent particle, an emulsion, a suspension of solids, or a suspension of smaller microcapsules. The microcapsule even may have multiple walls.

In the formation of some microcapsules, a finely divided solid such as a drug is suspended in a polymer solution. Droplets then are formed and solidified. In this case, the wall actually has become the continuous phase of the particle. The solid is now said to be held in a matrix of polymer.

Figure 1. Microcapsules come in a wide size range. Courtesy of Eurand America, Inc.

The reasons for microencapsulation are countless. In some cases, the core must be isolated from its surroundings, as in isolating vitamins from the deteriorating effects of oxygen, retarding evaporation of a volatile core, improving the handling properties of a sticky material, or isolating a reactive core from chemical attack. In other cases, the objective is not to isolate the core completely but to control the rate at which it leaves the microcapsule, as in the controlled release of drugs or pesticides (see Pharmaceuticals, controlled release). The problem may be as simple as masking the taste or odor of the core, or as complex as increasing the selectivity of an adsorption or extraction process.

Properties Important in Choice of Process

In choosing among processes for a particular application, the following physical properties must be carefully considered.

Core Wettability. Coacervation is the formation of a second polymer-rich liquid phase from a polymer solution, eg, by addition of a nonsolvent. In coacervation coating, the crucial property is the wettability of the core by the coacervate. As long as solid particles are properly wetted, they are frequently easier to coat than liquid cores. If a liquid core material is highly insoluble in the coacervate-forming solution, proper wetting may be difficult. To the dismay of workers in this field, both solids and liquids sometimes present a wettability problem.

In principle, the wettability of a solid in a particular coacervation system is determined easily. However, in practice, a surface of the proper configuration rarely is available for accurate measurement of a wetting angle or spreading coefficient. The wettability usually is determined directly during the microencapsulation process by examining the ability of the coacervate particles to coat the core particles adequately.

Core Solubility. In a coacervation system, it is critical that the core not be soluble in the polymer solvent and that the polymer not partition strongly into a liquid core. In interfacial polymerization systems, determination of the solubilities of the reactants in the phases permits a choice of solvents and polymers.

In some cases, cores can be employed that have some solubility in the external polymer solution. In spray-coating, it is possible to coat a water-soluble solid with an aqueous polymer solution because the water evaporates so rapidly after the spray droplets hit the core that there is little penetration or dissolution of the core.

Wall Permeability and Elasticity. The polymer permeability indicates whether a core can be isolated or a drug released at the required rate (1) (see Membrane technology). The microcapsules must be able to tolerate handling but may be required to break above a threshold pressure. The wall polymer, capsule size, and wall thickness determine elasticity and friability.

Other Properties. Other important variables concern the range of concentrations and temperatures over which the wall polymer is sticky or tacky, causing clumping. Stickiness of the wall solution during drying may be the most difficult problem to overcome. Similarly, spray-coating is impeded by stringy polymer solutions.

Melting points, glass-transition temperatures, degree of crystallinity, wall-degradation rate, and many other properties have to be considered.

Techniques

A great many microencapsulation techniques are available and new ones are being developed each year (see General References). Most procedures have seemingly endless variations, depending on wall-polymer solubility, core solubility, particle size, wall permeability, surface free energies, desired release pattern, physical properties, etc.

Formation. The pan coating process, widely used in the pharmaceutical industry, is among the oldest industrial procedures for forming small coated particles or tablets. The particles are tumbled in a pan or other device while the coating material is applied slowly. The classical example is the sugar-coating of medicinals. The expense of this process and the need to encapsulate fine particles and liquid cores stimulated the development of the new microencapsulation processes (see Coating processes; Pharmaceuticals).

Air-Suspension Coating. Air-suspension coating of particles by solutions or melts gives better control and flexibility (2). The particles are coated while suspended in an upward-moving air stream, as shown in Figure 2. They are supported by a perforated plate having different patterns of holes inside and outside a cylindrical insert. Just sufficient air is permitted to rise through the outer annular space to fluidize the settling particles (see also Fluidization). Most of the rising air (usually heated) flows inside the cylinder, causing the particles to rise rapidly. At the top, as the air stream diverges and slows, they settle back onto the outer bed and move downward to repeat the cycle. The particles pass through the inner cylinder many times in a few minutes.

As the particles start upward they encounter a fine spray of the coating solution. Only a small amount of solution is applied in each pass. Hence, the solvent is driven off and the particles are nearly dry by the time they fall back onto the outer bed. Particles as large as tablets or as small as 150 μm can be coated easily. Since many thin layers of coating are sprayed onto all surfaces of the randomly oriented particles, a uniform coating is applied, even on crystals or irregular particles.

A great variety of coating materials have been used with this process, including waxes, cellulosic compounds and water-soluble polymers. Development and coating are done by Coating Place, Inc., Verona, Wisc. Equipment of several sizes is sold worldwide by Glatt Industries. The largest unit is 117 cm in diameter and requires

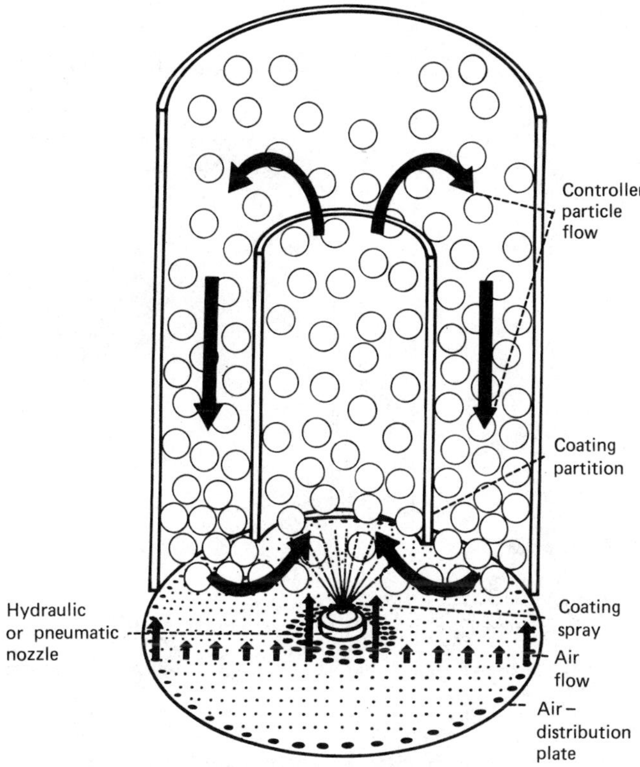

Figure 2. Wurster coating chamber. Perforations in air distribution plate are designed to control direction of particle movement. Courtesy of Coating Place, Inc.

over 283 m^3/min (10^4 ft^3/min) of air in a typical operation. Up to 400 kg can be handled in one batch, and cycle time can be as short as 15 min, although it is typically 1–2 h.

Centrifugal Extrusion. Several processes have been patented by the Southwest Research Institute, in which liquids are encapsulated using a rotating extrusion head containing concentric nozzles (3) (Fig. 3). In this process, a jet of core liquid is surrounded by a sheath of wall solution or melt. As the jet moves through the air it breaks, owing to Rayleigh instability, into droplets of core, each coated with the wall solution. While the droplets are in flight, a molten wall may be hardened or a solvent may be evaporated from the wall solution. Since most of the droplets are within ±10% of the mean diameter, they land in a narrow ring around the spray nozzle. Hence, if needed, the capsules can be hardened after formation by catching them in a ring-shaped hardening bath.

This process is excellent for forming particles 400–2000 μm in diameter. Since the drops are formed by the breakup of a liquid jet, the process is only suitable for liquids or slurries. A high production rate can be achieved, ie, up to 22.5 kg of microcapsules can be produced per nozzle per hour; heads containing 16 nozzles are available.

This is an inexpensive process for producing large drops having meltable coatings

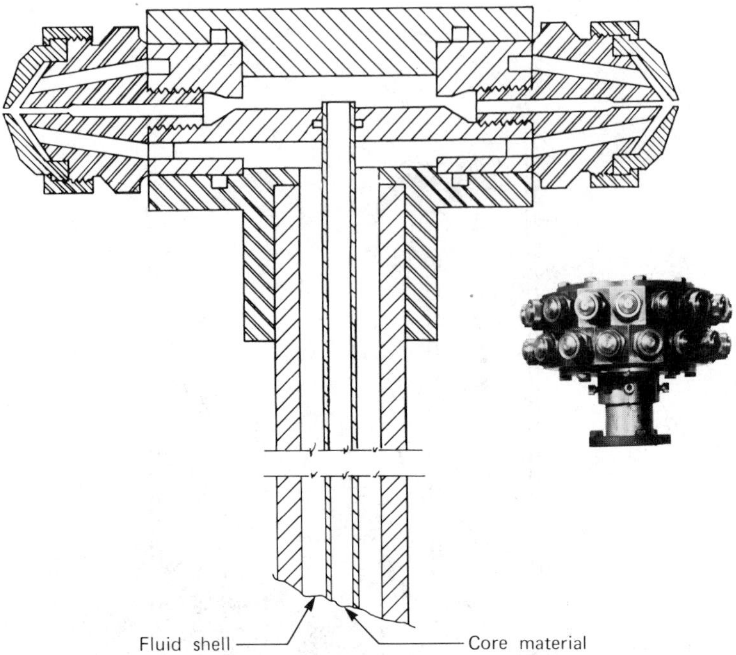

Figure 3. Rotating head to form biliquid jets. Courtesy of Southwest Research Institute.

such as fats or waxes. However, it also has been used extensively for coating liquids with synthetic polymers and gelatin and to form calcium alginate shells. The only wall materials that are unsuitable are those that are stringy and do not permit the clean breakup of the jet. With this process it is possible to encapsulate aqueous solutions in waxy wall materials, a difficult problem for some processes.

The 3M Company also has a process in which a biliquid column is formed, with subsequent breakup of the jet into coated droplets (4). In this case also, melt systems work well and it is easiest to form 800–4000-μm droplets. The fill phase in this process should have a viscosity of at least 20 mPa·s (= cP).

Vacuum Metallizing. The National Research Corporation has developed vacuum-deposition techniques to coat particles with a wide variety of metals and some nonmetals (5). The particles can be as small as 10 μm. The entire operation is carried out under vacuum, with the particles conveyed slowly down a refrigerated vibrating table while being exposed to a metal-vapor beam from a heated furnace. The beam condenses on the cool small particles, eventually coating them evenly. The process can handle only solids and usually forms a coating that is not completely impervious to gases or liquids. The solidification of the vapors occurs so rapidly that the particles do not tend to agglomerate unless the vaporized material is waxy or sticky. The process forms an even coating on rough or irregular surfaces, since all surfaces of the vibrating particles are exposed equally to the vapor beam (see Film deposition techniques).

Since the coating material must be evaporated and the substrate particles must be kept in vacuum and under refrigeration, the process may be expensive compared to other ways of depositing organic coatings. However, it is unique for depositing metals on fine particles.

Liquid-Wall Microencapsulation. A liquid as the wall of a microcapsule offers several advantages (6). Since the wall can be broken at will, the wall material can be recovered and reused. The core also can be recovered if desired. In addition, the permeabilities of many molecules through a liquid wall are better controlled and predicted than through a solid wall, where morphology can vary greatly.

The liquid core can be a single droplet, an emulsion, or a suspension. It can consist of a variety of chemical reactants, which suggests a wide spectrum of possible uses. The liquid wall is stabilized by addition of a surfactant, such as saponin or Igepal, and through the addition of strengthening agents. It is possible to form thin, aqueous membranes around organic droplets, such as toluene, or oil-based membranes around aqueous solutions. These microcapsules are so stable that they remain suspended with the mildest agitation and coalesce only slowly when stagnant. Droplets of one hydrocarbon, surrounded by a stable aqueous membrane, can be made to pass upward through another hydrocarbon without intermixing. Extraction can be carried out based on the selective permeability of the aqueous membrane. Because the process is inexpensive and the membrane material can be recycled, liquid-membrane microcapsules are attracting attention for a variety of industrial uses such as removal of contaminants from wastewater.

For many applications, the droplets containing reactant (eg, an aqueous solution) can be emulsified into the encapsulating immiscible liquid. This highly stable emulsion can then be gently dispersed into the liquid to be treated. Hydrocarbon droplets containing emulsified aqueous microdroplets are shown in Figure 4.

Spray-Drying. Spray-drying serves as a microencapsulation technique when an active material is dissolved or suspended in a melt or polymer solution and becomes trapped in the dried particle. The main advantage is the ability to handle labile materials because of the short contact time in the dryer; in addition, the operation is economical. In modern spray dryers the viscosity of the solutions to be sprayed can be as high as 300 mPa·s (= cP), meaning that less water must be removed from these concentrated solutions.

Recent examples of advances in spray-drying for controlled release are the

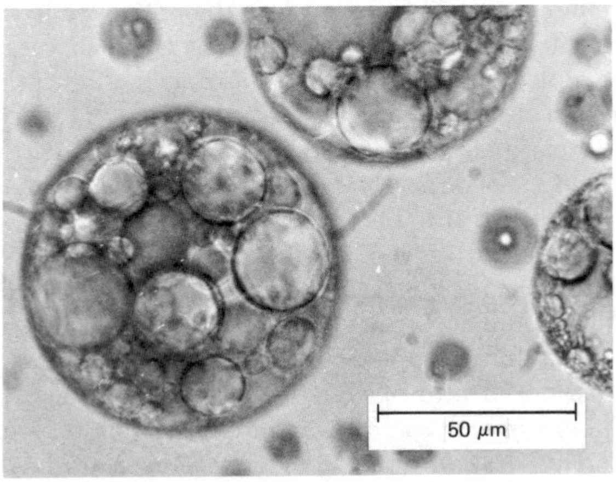

Figure 4. Stable hydrocarbon emulsion drops containing an encapsulated aqueous emulsion. Courtesy of Exxon Research and Engineering Company.

products in which flavor oils are emulsified into a solution of a polymer, such as gum arabic, and spray-dried to make fine particles. Particles of 250 μm (60-mesh) are ideal for inclusion in such products as bath powders. However, for use in an aerosol product, all particles must be below 74 μm (200 mesh) (see Aerosols). With careful gel formulation, the amount of fragrant oil on the surface of the final product may be less than 1% (see Flavors and spices; Perfumes).

When the core material is incorporated into a meltable fat or wax, the process is called spray-chilling, since the emulsion or suspension has only to be chilled below its melting point to form particles. This has been found practical for microencapsulation of such products as citric acid (qv), vitamin C (see Vitamins), and ferrous sulfate.

In a variety of other microencapsulation processes spray-drying could be a final step, since the capsules are in an aqueous dispersion from which they could be isolated economically by spray-drying (see Drying).

Hardened Emulsions. Since the aim of microencapsulation is the formation of many tiny particles, the first step in a number of processes is the formation of an emulsion or suspension of the core material in a solution of the matrix material (see Emulsions). This emulsion can be emulsified in another liquid, and the droplets hardened.

Emulsions and suspensions can be converted directly to hardened microcapsules by several methods, eg, aspirin is encapsulated by first suspending it in a solution of ethyl cellulose (7). This suspension is dispersed in saturated ammonium sulfate solution and the solvent driven off by heating. This general process also has been used for enzymes (8) and many other core materials. Polystyrene, ethyl cellulose, and silicones are reported to work well as matrix materials.

In another process, a water solution is dispersed in melted waxes or fats. This emulsion is then dispersed in an aqueous solution held at a temperature above the melting point of the wax or fat. The aqueous solution is then cooled to solidify the wax particles (9).

Processes based on emulsion hardening are particularly useful when a drug has high solubility in the polymer solution, preventing formation of a separate polymer wall around the drug by other methods. Initially, the solution of polymer and drug is emulsified in an immiscible liquid. Solvent is then removed by application of heat and/or vacuum. As the organic solvent evaporates, the drug crystallizes inside the polymer solution droplet. Control of the rate of drug crystallization at this point in the process is critical in obtaining particles with good long-term release characteristics. Emulsion-hardening methods are being applied to the encapsulation of various drugs in biodegradable poly(*d*,*l*-lactic acid) to obtain injectable particles. Active in this area are Southern Research Institute, Battelle Memorial Research Institute and Washington University (St. Louis).

Liposomes and Surfactant Vesicles. The incorporation of drugs into liposomes (phospholipid vesicles) and surfactant vesicles has attracted recent attention (10). Liposomes, such as lecithin (qv) and phosphatidylinositol, are smectic mesophases of phospholipids organized into bilayers (see Liquid crystals). Liposomes can be prepared in the laboratory by rotary evaporation of a chloroform solution of the phospholipid and cholesterol, with the subsequent removal of the thin lipid film from the wall of the flask by shaking with aqueous buffer. In order to incorporate a hydrophobic drug into these liposomes, the drug is dissolved in the chloroform solution before evaporation. A water-soluble drug can be incorporated by dissolving it in the buffer solution used to form the liposome. The untrapped drug is removed by gel filtration.

The liposomes formed are typically in the 0.1–0.15 μm range but can be prepared larger. With sonication during the formation step, the liposomes can be as small as 25 nm (see Ultrasonics). The amount of drug that can be entrapped depends on the ratio of the lipid to the trapped aqueous phase and on the solubility of the drug. Some drugs are trapped to the extent of a few percent, but up to 60% bleomycin, eg, can be encapsulated. The liposomes are typically stored below the phase-transition temperature of the solid wall but generally are used above this temperature.

Release rates from the liposomes can be small, in spite of the large area–volume ratio. The release of 8-azaguanine is reported to be around 10% in 20 h. Liposomes can be targeted to specific cells by associating them with immunoglobulins raised against the target cells. This has been accomplished with several drugs. Surfactant vesicles can be formed also from dioctadecyldimethylammonium chloride by sonicating it in a concentrated drug solution and gel-filtering the suspension (11). The amount of drug trapped is many times that trapped in liposomes.

Phase Separation. In several microencapsulation processes, the core material first is suspended in a solution of the wall material. The wall polymer then is induced to separate as a liquid phase, eg, by adding a nonsolvent for the polymer, decreasing the temperature, or adding a phase inducer, another polymer that has higher solubility in the solvent. In the last case, incompatibility between the two polymers causes the first polymer to separate as another phase. When the wall polymer separates as a polymer-rich liquid phase, this phase is called a coacervate and the process is called coacervation.

As the coacervate forms, it must wet the suspended core particles or droplets and coalesce into a continuous coating. The final step is the hardening and isolation of the microcapsules, usually the most difficult step in the process. An understanding of the complex physical chemistry of phase-separation microencapsulation is helpful in designing the process (12).

Complex Coacervation. The first commercially valuable microencapsulation process, developed by the National Cash Register (NCR) Company in Dayton, Ohio, was based on coacervation. The coacervate was formed from the reaction product complex between gelatin and gum arabic (13). The first application was in No-Carbon-Required carbonless copy paper. Carbonless copy systems are still by far the largest market for microencapsulated products. Applications of this technology in the paper industry are now handled by Appleton Papers, Inc. Applications in other industries are handled by Eurand America, Inc., Dayton, Ohio.

In this process, gelatin having a high isoelectric point and gum arabic, which contains many carboxyl groups, are added to a core-containing suspension at pH 2–5 above 35°C to ensure that the coacervate is in the liquid phase. As the gelatin and gum arabic react, viscous liquid microdroplets of polymer coacervate separate. If the core particles are easily wetted by these microdroplets, a wall of the liquid coacervate forms on the core particles. This wall can be hardened by several means; eg, by the addition of formaldehyde. Finally the mixture is cooled to 10°C, the pH adjusted to 9, and the microcapsule suspension filtered. The capsules shown in Figure 1 have walls produced by complex coacervation.

In a modification of this process, two gelatins of differing isoelectric points are used as the reacting species. In this case, the coacervation is reversible before hardening, and distorted capsules can be heated to restore their spherical shape. Such microcapsules can be hardened by the addition of glutaraldehyde.

The complex coacervation process works well in the microencapsulation of solids and oily materials. The walls are polar and have low permeability to nonpolar molecules. Toluene and carbon tetrachloride can be contained so well that only a small fraction of the contents is lost from a thin layer of 1-mm capsules over a period of two years at 21°C at 70% humidity.

Thermal Coacervation. The effect of temperature on solubility is sometimes strong enough to cause a dissolved polymer to form a coacervate upon cooling. However, the temperature effect is nearly always used in conjunction with another method of coacervate formation, as illustrated in the following section.

Polymer–Polymer Incompatibility. This broadly applicable technique, also developed at NCR (14–16), works well for encapsulating many solids and some liquids. A coacervate of the wall material is induced to form from solution by the presence or addition of a phase-inducer. The general phenomenon of polymer–polymer incompatibility is based on the fact that the free energy of mixing for polymers is positive owing to a positive enthalpy change and negligible entropy change.

An example of this process is the microencapsulation of activated carbon in ethyl cellulose using cyclohexane as the solvent. The carbon is slurried in a 2% suspension of ethyl cellulose in cyclohexane, followed by the addition of 2% of low molecular weight polyethylene as a phase-inducer. The entire system is heated to 80°C, where both polymers dissolve. The system then is cooled slowly to room temperature. The change in temperature and the presence of the polyethylene cause the formation of a coacervate containing the ethyl cellulose which coats or wraps around the activated carbon particles. At room temperature, the particles can be separated and dried. Careful attention must be paid to rates of cooling, agitation, etc, to obtain satisfactory, free-flowing microcapsules.

A system also has been developed by NCR to microencapsulate some polar liquids and aqueous solutions (17). In this process, the wall polymer is partially hydrolyzed ethylene–vinyl acetate copolymer which coacervates upon the addition of polyisobutylene. Smooth coherent walls can be formed around glycerol and aqueous solutions of methylene blue or ferrous ammonium sulfate. The walls are nearly impenetrable by ions and all but the smallest polar molecules. The walls can, however, be penetrated at reasonable rates by uncharged organic molecules. A few of these microcapsules containing ferrous ammonium sulfate are shown in Figure 5.

Another NCR variation of this process can be carried out at constant temperature, an advantage in encapsulating temperature-sensitive materials. After the core material is dispersed in the solution of a wall-forming polymer, poly(dimethyl siloxane) is added as the phase-inducer. A possible disadvantage of this process is that a small amount of the silicone is trapped in the polymer wall, making dispersion difficult for some uses.

Nonsolvent Coacervation. A number of patents have issued in which microcapsules are formed or hardened by the addition of nonsolvents for the wall polymer. In a particularly successful example of such a technique (18), the core, typically an aqueous solution, is dispersed in a methylene chloride solution of cellulose acetate butyrate (CAB). Toluene, a nonsolvent for the polymer, is then added slowly to cause coacervation of CAB and the wrapping or coating of the core with the polymer as it

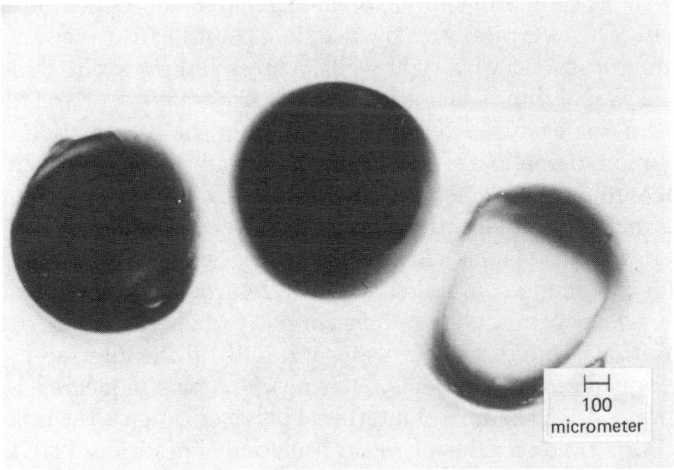

Figure 5. Clear capsule: ferrous ammonium sulfate encapsulated in partially hydrolyzed ethylene–vinyl acetate copolymer. Dark capsules: same after contact with sodium salicylate, showing formation of Fe(III) salicylate complex inside the capsule.

separates from solution. The wall is hardened by the addition of petroleum ether to extract solvent from the walls. Typically, spherical microcapsules are obtained with thin elastic walls. This process has been used to encapsulate some organic compounds by emulsifying them in aqueous solution before coacervation.

Interfacial Coacervation. In another technique, a polymer coacervate forms directly at an interface without the addition of a phase-inducer (19). For example, tris-buffered hemolyzate is emulsified in ice-cooled ethyl ether, saturated with water, which contains 1% Span 85. A solution of collodion is added. Weak microcapsules develop within 45 min. They contain hemolyzate and have a collodion coacervate wall. The wall is hardened with *n*-butyl benzoate, after which the microcapsules are separated by centrifugation, washed, and stored in Tween 20.

In another process, an aminoplast wall is induced to coat the surface of droplets (20). A water-insoluble fill material is emulsified into an aqueous solution containing a preformed urea–formaldehyde polymer of sufficiently low molecular weight to retain water solubility (see Amino resins). After emulsification, a water-soluble acid such as formic acid is added to lower the pH to 1.5–3, causing the formation of the urea–formaldehyde polymer shell, microencapsulating the emulsion droplets. The microcapsules usually range 10–50 μm and are used in 3M Microfragrance scents for advertising.

Another method of forming microcapsule walls at an interface is based on the observations that some cellulosic polymers are soluble in cold water but not in hot water (21), eg, hydroxypropyl cellulose and ethyl hydroxyethyl cellulose. These materials precipitate from solution when the temperature exceeds a critical value of 40–70°C. To carry out microencapsulation based on this phenomenon, the oil to be encapsulated is emulsified in an aqueous solution of the cellulosic material, at a low temperature where it is still soluble, in the presence of a surfactant. The emulsion is then heated until the cellulosic polymer gels or precipitates from solution, wetting the oil droplets. While the suspension is hot, the cellulosic wall is solidified by reaction with a water-soluble agent such as dimethylolurea, urea–formaldehyde resin, or methoxymethylmelamine resin. This is typically a slow reaction, resulting in somewhat porous walls. To produce microcapsules with tight walls, a cross-linking agent is included before emulsification, eg, polyfunctional acyl chlorides, isocyanates, or anhydrides. Water-soluble cross-linking agents also can be included in the aqueous solution. Capsules from 1 to 50 μm in diameter are made easily and wall thickness is well-controlled.

Chemical Methods. ***Interfacial Polymerization.*** In interfacial polymerization, the two reactants in a polycondensation meet at an interface and react rapidly (22). The basis of this method is the classical Schotten-Baumann reaction between an acid chloride and a compound containing an active hydrogen atom, such as an amine or alcohol. Polyesters, polyurea, polyurethane, or polycarbonates may be obtained. Under the right conditions, thin flexible walls form rapidly at the interface.

Microencapsulation of proteins, enzymes, etc, was pioneered by Chang (see General References). An example of interfacial polymerization is the process developed by Pennwalt Corporation for the microencapsulation of pesticides (23). In this process, a solution of the pesticide and a diacid chloride (eg, in toluene) are emulsified in water and an aqueous solution containing an amine and a polyfunctional isocyanate is added. Base is present to neutralize the acid formed during the reaction. Condensed polymer walls form instantaneously at the interface of the emulsion droplets. The isocyanate is not required for wall formation but acts as a cross-linker, giving a tougher, less-permeable wall. Surfactants and suspending agents are used in the process to control particle size. The first commercial product, Penncap-M, contained methyl parathion in microcapsules having a mean diameter of 30 μm. Figure 6 is a photograph of a microcapsule of another insecticide, Knox-out 2FM, trapped on the leg of a cockroach.

Other methods for carrying out similar chemical reactions at the interface have been patented by Moore Business Forms to produce microcapsules for carbonless copy paper (24). A hydrophobic marking fluid plus an oil-soluble reactant are emulsified into water, followed by addition of the other reactant which immediately causes the formation of the condensation polymer at the drop interface.

The processes developed by Pennwalt and Moore have the advantages that they are capable of being run on a large industrial scale and that the reactions are very fast. If the final product can be used in aqueous suspension, the drying step can be eliminated. Hence, the cost is relatively low. Variations in cross-linkers, reactant concentrations, solvent composition, and processing steps permit optimization of the process to obtain different capsule properties.

The importance of carbonless copy paper has stimulated microencapsulation studies by many paper companies. Another application of interfacial polymerization is that of the Mead Corporation (25). Rather than place the reactants of a polycondensation system in each phase, they form an emulsion in which the organic phase contains a monomer and the aqueous phase an initiator that causes polymerization at the droplet surface. This process can be used with monomers such as styrene, methyl methacrylate, vinyl acetate, and acrylonitrile. Initiators (qv) include hydrogen peroxide, acetyl peroxide, zinc peroxide, and alkali metal hydroxides.

Surface polymerization processes to make opacifiers, pigments and dyes for paper coating also have been developed (26). A toluene solution of a reactant, such as toluene diisocyanate, is emulsified in an aqueous solution containing a polymeric film-forming emulsifying agent such as methyl cellulose, starch, or poly(vinyl alcohol). The emul-

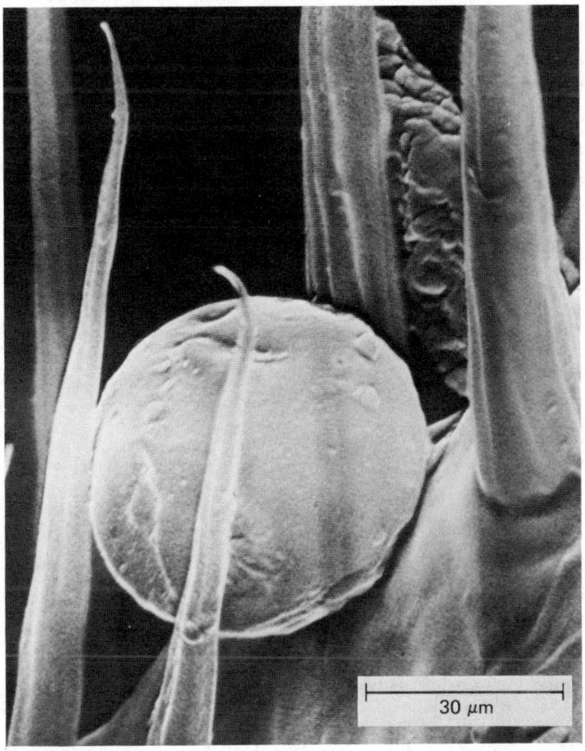

Figure 6. A Knox-out 2FM microcapsule trapped on the leg of a cockroach. Courtesy of Pennwalt Corporation.

sifying agent, which is present at the surface of the emulsion droplets, is then cross-linked and insolubilized by reaction with the organic reactant. If the microcapsules are to be used as opacifiers, the internal solvent is driven out and the microcapsules should have diameters <1 μm. Even with typical wall thickness of 5–60 nm, these small capsules tolerate calendering at 3.4 MPa (500 psi).

A different method of microcapsule formation is based on the fact that alginic acid anion forms an insoluble precipitate instantly upon contact with calcium ions. Hence, if droplets of an aqueous solution of sodium alginate are placed in a calcium chloride solution, a membrane quickly forms around the droplets (see Gums). If the initial droplets contain enough alginate and are allowed to remain in the calcium-containing solution long enough, they gel completely.

A method for forming extremely small microcapsules was developed at the Swiss Federal Institute of Technology in Zurich (27). The particles are so small that the process is called nanoencapsulation. Micelles are formed in a water-in-oil emulsion, using polymerizable surface-active agents such as acrylamide (qv). The surfactant is polymerized in various ways such as by x-ray or gamma radiation. Particle sizes range from 80 to 250 nm. It has been possible to suspend up to 40 vol % of water in hexane in this manner. The particles are sufficiently small that they cannot be observed with a microscope but can be seen with the Tyndall effect. Measured pore diameters in the walls appear to be 2–5 nm. The wall is 3–5 nm thick or more. Hence, the particles can be sturdier than liposomes. The final particles, containing a drug, are sustained-release vehicles which are small enough for parenteral use.

Microencapsulation techniques based on surface polymerization at elevated temperatures have been developed (28). In these processes, two film-forming reactive materials are dissolved together with an oily liquid, at low concentration in a volatile solvent, at a temperature sufficiently low to effectively prevent reaction. This oil phase is emulsified in an aqueous phase and the temperature increased, eg, to 90°C, where the solvent evaporates from the droplets and the film formers increase in concentration at the surface of the droplets, reacting to form a wall. A wide variety of polymer walls can be formed in such a process, such as polyols, polyisocyanates, polythiols, and polyamines.

An all-aqueous system for interfacial polymerization has been patented (29). This technique is based on the reaction between polar reactive materials, such as hydroxyethyl cellulose, in one phase and inorganic silicate, such as lithium hectorite clay, suspended in the other phase. When the material to be dispersed, such as a pigment suspension containing the organic material, is mixed with the clay suspension, the reaction is essentially instantaneous, causing the formation of particles of one aqueous suspension in another. This technique is used, eg, for the formation of suspended particles of colored water-dispersible pigment in a latex paint formulation.

In-Situ Polymerization. In a few microencapsulation processes, the direct polymerization of a single monomer is carried out on the particle surface. In one process, eg, cellulose fibers are encapsulated in polyethylene while immersed in dry toluene (30). In the first step, a Ziegler-type catalyst is deposited on the surface of the fiber at 20–30°C by the addition of $TiCl_4$ followed by triethylaluminum. When the catalyst is formed, the surface appears dark brown. The addition of ethylene, propylene, or styrene results in immediate polymerization directly on the surface. The polymerization is typically carried out to a final weight ratio of 50:50 polyethylene–cellulose. This coating has a molecular weight of up to 2×10^6 and a melting point of 56–58°C, vs the theoretical maximum of 61°C. This process also has been applied to coating glass. The coating, typically translucent or opaque, can be clarified by melting after formation (see Olefin polymers).

Unique coatings can be obtained with the Union Carbide Parylene process (31). Di-*p*-xylylene or a derivative is pyrolyzed at 550°C where it dissociates yielding free radicals in the vapor state. When the vapor is cooled below 50°C, the free radicals polymerize to a high molecular weight polymer:

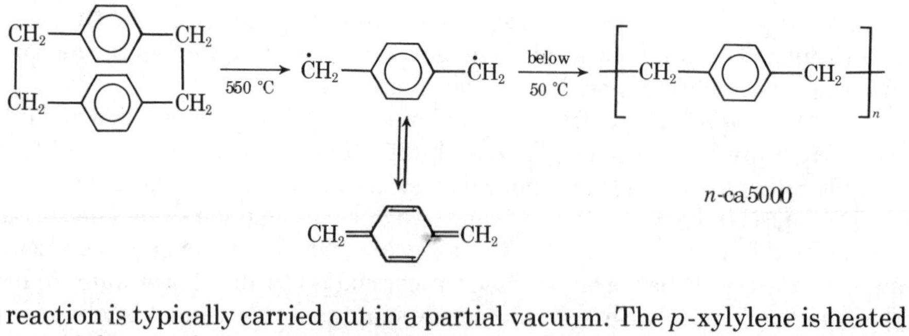

n-ca 5000

The reaction is typically carried out in a partial vacuum. The *p*-xylylene is heated in one part of the vessel whereas the material to be coated is cooled in another section. The product is a linear thermoplastic, insoluble in organic solvents up to 150°C, which has many attractive properties. It can be deposited on any solid surface, and has very low permeability to gases and moisture, although it is somewhat permeable to aromatic compounds. The coating retains some ductility at cryogenic temperature, has excellent electrical resistance, and a softening range of 290–400°C (depending on the derivative). Usual deposition rates are about 0.5 μm/min. Coating thickness ranges 0.2–75 μm. The coating is uniform, even over sharp projections, as shown by the coating over the stylus point shown in Figure 7 (see Film deposition techniques).

Matrix Polymerization. In a number of processes, a core material is imbedded in a polymeric matrix during formation of the particles. A simple method of this type is spray-drying, in which the particle is formed by evaporation of the solvent from the matrix material. However, the solidification of the matrix also can be caused by a chemical change.

Matrix particles can be formed by suspending the core in gelatin solution, emulsifying the gelatin solution, and then hardening the droplets with formaldehyde or glutaraldehyde. Solid particles can be obtained in some of the processes mentioned previously if the wall-forming chemicals are present in sufficient quantity and if the reaction time is prolonged.

In silastic microspheres containing a suspended drug in silicone oil, addition of a catalyst such as stannous octanoate polymerizes the silicone prepolymer (32).

An inexpensive process for encapsulating water-insoluble oxidation-resistant liquids, in the form of emulsified droplets in starch xanthate, has been developed by the U.S. Department of Agriculture (33). First, the xanthate derivative of starch is made by reaction with a base and CS_2 (see Starch). A concentration of about 10%

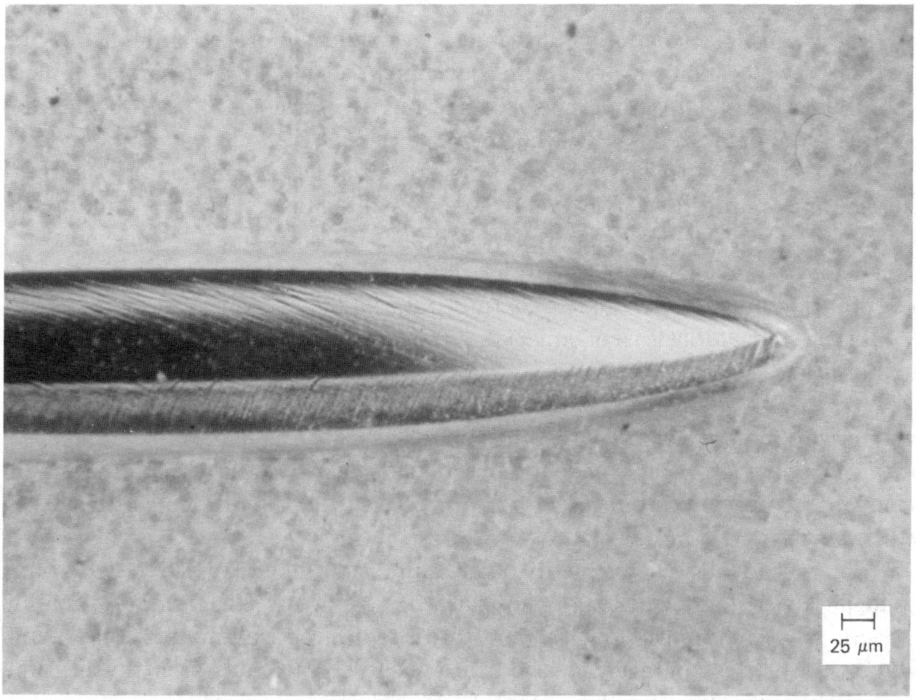

Figure 7. Parylene coating over stylus point. Courtesy of Union Carbide Corporation.

xanthate has the proper consistency for the subsequent steps. A core material such as a pesticide is added with gentle stirring, giving a stable suspension. If the core material is sensitive to base, the pH is reduced from near 11 to 6–7 with acetic acid. After the dispersion is formed, the xanthate is solidified by oxidation and cross-linking with sodium nitrate or hydrogen peroxide. With agitation, solid granules form in a few minutes, and 80% of the water in the granules can be squeezed out physically. The resulting crumbs, containing approximately 50% moisture at this stage, can be ground to the desired particle size, then dried to, eg, 5% moisture. No active ingredient is removed with the water or with the grinding steps. The material appears to be self-sealing. Drying temperature can be as high as 121°C (if the core does not degrade). After the material has been dried, it can no longer be ground without losing active ingredient.

The granules may contain up to 40% active ingredient, although a few percent is more typical for pesticides. A large number of water-insoluble compounds have been encapsulated by this technique. They are extracted with difficulty into nonpolar organic solvents, indicating that such solvents do not penetrate the wall readily. On the other hand, alcohol and water increase the release rate. In the soil, the core is probably released by the action of moisture and bacteria.

The xanthate matrix material can be cross-linked with formaldehyde or epichlorohydrin to decrease its permeability even further.

Process Selection

The range of applicability of the microencapsulation processes described above is extremely wide, and many applications overlap. A slight change in surface properties or solubilities may determine the success or failure of the operation. New materials require experimentation and testing.

There are, nevertheless, some factors that exclude certain processes and favor others. For example, if the material to be encapsulated is a solid and cannot be used as a slurry, liquid-jet methods are not suitable. Phase-separation methods or air-suspension coating may work well if the solid particles are large. If they are very fine, incorporation into a matrix of the coating material may be appropriate.

If the core material is an organic liquid, air-suspension coating cannot be employed but liquid-jet methods may work well for particles larger than a few hundred micrometers. If smaller particles are needed, coacervation techniques or interfacial polymerization may be applicable. If the particles must be hard or have thick walls, interfacial polymerization may be too slow and expensive. If the material forms stable emulsions in a polymer solution or a meltable medium, solidified particles of these emulsions might be the answer.

Microencapsulation of a highly polar liquid is, in general, more difficult but the liquid-jet method, interfacial polymerization, or some of the NCR phase-separation methods may be applicable. If the core and the desired wall material have similar solubilities, phase-separation methods are excluded but matrix polymerization or emulsion solidification may be suitable.

Release Methods and Patterns

Even when the aim of a microencapsulation application is the isolation of the core from its surroundings, the wall must be ruptured at the time of use. Many walls are ruptured easily by pressure or shear stress, as in the case of breaking dye particles during writing to form a copy. Capsule contents may be released by melting the wall, or dissolving it under particular conditions, as in the case of an enteric drug coating. In other systems, the wall is broken by solvent action, enzyme attack, chemical reaction, hydrolysis, or slow disintegration.

Microencapsulation can be used to slow the release of a drug into the body. This may permit one controlled-release dose to substitute for several doses of nonencapsulated drug and also may decrease toxic side effects for some drugs by preventing high initial concentrations in the blood. There is usually a certain desired release pattern. In some cases, it is zero-order, ie, the release rate is constant. In this case, the microcapsules deliver a fixed amount of drug per minute or hour during the period of their effectiveness. This can occur as long as a solid reservoir of dissolving drug is maintained in the microcapsule.

A more typical release pattern is first-order in which the rate decreases exponentially with time until the drug source is exhausted. In this situation, a fixed amount of drug is in solution inside the microcapsule. The concentration difference between the inside and the outside of the capsule decreases continually as the drug diffuses.

In most processes, a small amount of drug is not encapsulated properly. In use, these accessible drug particles release rapidly into the surrounding fluid, causing a burst effect. Sometimes this is desirable, and functions as an initial loading dose to increase the blood concentration toward the therapeutic level. However, if undesirable, the poorly encapsulated particles have to be removed before use.

Typical release patterns for drugs are shown in Figure 8.

An excellent reference on the mathematics of controlled-release systems (34) gives solutions for rates and amounts released as a function of time for the typical cases of constant release rate, exponential release rate, and release from matrices. Many examples and an extensive reference list are provided.

Economic Aspects

Wall polymers may cost \$2–40/kg. Solvent cost is substantial, and may represent over half the cost of the entire operation. Hence, installation of solvent-recovery

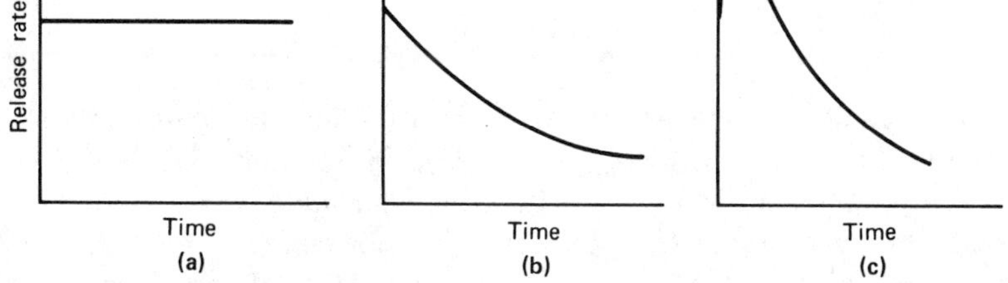

Figure 8. (**a**) Constant release; (**b**) first-order release; (**c**) first-order release with initial burst.

equipment offers both environmental and economic advantages. The most economical wall materials are waxes and fats, which can be melt-coated. This eliminates all solvent costs.

Wall thickness, particle size, and particle density may have a strong affect on the coating cost of finished product. For small particles, the area to be covered with polymer is high for a given weight, and the rate at which the particles can be produced is limited for processes such as spray-coating, vapor deposition, or jet methods. On the other hand, these considerations are not important in coacervation or interfacial polymerization systems.

As with any manufacturing process, the costs of microencapsulation decrease significantly with increasing volume of production. Melt-coating solid particles at high volume in a trouble-free system may give a product at \$0.50/kg, whereas expensive wall materials or time-consuming operations can increase costs to \$10–20/kg.

Characterization and Evaluation

Size and Size Distribution. Determination of the size distribution of the capsules, as produced, is an important measurement. It also is critical in studying the effect of the process variables on the final product. A wide range of particle-size-measuring techniques is available (see Size measurement of particles).

Loading Fraction. The amounts of polymer coating and core can be measured directly by isolating the microcapsules, dissolving or crushing the wall, and making a mass balance on the core and polymer. The fraction of the core in the microcapsules or matrix is frequently a function of particle size, and the deviation of the actual percent loading from the design loading is sometimes a function of the loading itself.

Release Properties. In many applications, such as carbonless copy paper, microcapsules are ruptured by pressure. Rupture pressure is measured under controlled conditions similar to those of the application. Most companies have developed their own test apparatus for purposes of quality control and process development.

In a more complex situation, the release of a core drug or the adsorption of an external solute is time-dependent. The kinetics of these processes is usually determined by contacting the microcapsules with a fixed volume of fluid at constant temperature and measuring the change in solution concentration. To avoid the effect of a stagnant fluid layer adjacent to the microcapsules, the bottles are placed in a shaker bath or mounted on a rotating-arm apparatus. Similar results are obtained in both tests. Since solute interactions can have strong affects on distribution coefficients and, hence, on diffusional driving forces, the test fluid should resemble the fluid used in the application. A case in point is drug release into blood or intestinal fluid, where tests in buffers can be highly misleading.

Health and Safety Factors

In any application of microencapsulation in foods, pharmaceuticals or veterinary products, only wall materials approved by the FDA should be used. The GRAS list (generally recognized as safe) can be consulted. Some materials, such as starches, cellulosic compounds, and gums, have been used for many years as food additives or as excipients in drug formulation. However, any process or material not previously approved must be submitted to the FDA. The approval requirement extends to methods of solvent removal, conditions of storage, methods of handling, etc.

Uses

Carbonless Copy Paper. The most significant application of microcapsules is in business forms permitting copies to be made without the need for carbon paper. Industry-wide production of carbonless-copy business forms is approximately 500,000 t/yr. In carbonless business forms, a dye intermediate, such as crystal violet lactone, is microencapsulated to form particles less than 20 μm in diameter. These are deposited in a thin layer on the underside of the top sheet of paper. The receiving sheet or copy is coated with another reactant such as acidic clay. The microcapsules are designed to resist breakage under normal conditions of storage and handling, but to break under the high local pressure of a pen or pencil point. Upon breaking of the capsules, the two chemicals react, producing a dye copy of the original. Figure 9 is a photomicrograph of NCR microcapsules on the paper produced by Appleton Papers Inc.

Flavors and Essences. Since many of the flavors and essences in candies, foods, and perfumes are volatile or are affected by oxidation, microencapsulation of these materials offers advantages (see Flavors and spices). Spray-drying has proved to be an excellent, economical way to form particles containing suspended flavors and scents in a polymer matrix. The short contact time of this high-volume process minimizes loss of volatile cores.

Eurand America, Inc. has used coacervation and 3M has used its aminoplast-wall process for new approaches in advertising by microencapsulating essences and coating them onto paper to be used in magazines and promotional literature. The capsules are broken by scratching, releasing the scent of products such as soaps or perfumes. The materials that have been successfully microencapsulated range from lemon oil and essence of roses to the aroma of dill pickles or chocolate mint cookies.

There is much interest in microencapsulating insect pheromones such as *cis*-2-decyl-3(5-methylhexyl)oxirane (disparlur), the scent that guides the male gypsy moth to the female. Disparlur could be used to lure males to traps or, if it was widely dis-

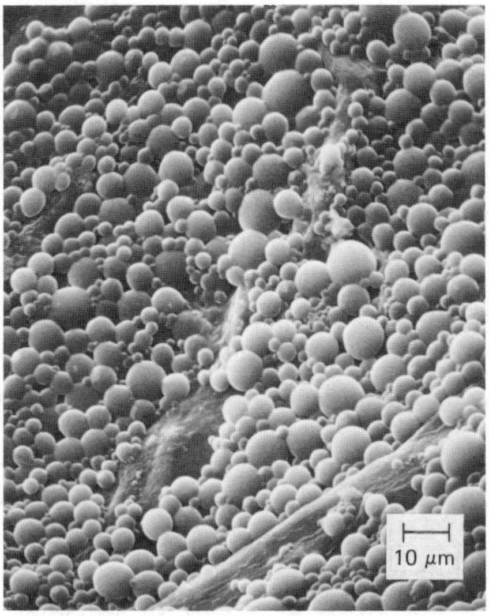

Figure 9. Microcapsules on carbonless copy paper. Courtesy of Appleton Papers Inc.

persed, to confuse them in their attempt to find females. Microencapsulation greatly increases the time over which the volatile scent is active. Pennwalt Corporation and Zoecon, Inc. are among the companies active in this area.

Pesticides and Herbicides. Some volatile pesticides and herbicides are highly toxic substances. Microencapsulation decreases their toxicity upon contact and controls their release rate (see Herbicides; Insect control technology). Pennwalt Corporation markets several microencapsulated pesticides, such as Penncap M, which is microencapsulated methyl parathion. The microencapsulation permits early safe entry into a sprayed field, and increases the time of effectiveness from 1–2 d to 5–7 d. Since the tiny capsules (about 30 μm) are in the form of an aqueous suspension, spills can be cleaned easily and the capsules can be flushed easily from the skin. In tests on rabbits, encapsulation lowers the dermal toxicity by a factor of 12. These microcapsules are made by interfacial polymerization using bifunctional acid chlorides and bifunctional amines, cross-linked with a polyfunctional isocyanate. The walls of the capsules are about 1 μm thick.

Pharmaceuticals. By choice of polymers, particle-size distribution, wall thickness, and processing conditions, the release profile of a drug can be controlled (see Pharmaceuticals, controlled-release).

Probably the largest-selling drug product based on controlled release is Contac, the cold-relief remedy. Its relatively large particles are pan-coated. The coatings dissolve at different times. Another commercial controlled-release drug form is timed-release aspirin, marketed eg by, Sterling Drug Company in the United States and by Rhone-Poulenc S.A. in Europe. Tests show that microencapsulation increases the time of effective action at least by a factor of two. Furthermore, the polymeric coating combined with slower release decreases gastric irritation.

A large number of controlled-release pharmaceuticals have been developed by Eurand S.p.A., the European licensor for National Cash Register technology. The three basic forms (35) are Diffucap, which consists of particles about 0.5 mm in diameter from which the drug is released by diffusion; Chronodrug, in which the membrane disintegrates; and Microcap, which consists of very fine particles that release by both diffusion and disintegration. Among their many controlled-release drugs are ampicillin, indomethacin, potassium chloride, amytryptyline, and phenylbutazone.

An ingenious method for attaining a constant release rate is the Oros osmotic pump (Alza Corporation) shown in Figure 10. In this device, the drug to be released is enclosed, along with a soluble material, in a wall that has a known permeability to water. Upon immersion of the device, water enters slowly by osmosis, causing the internal pressure to increase and dissolving part of the contents. The internal pressure forces the saturated solution through a small hole that has been drilled through the wall. The diameter of the hole is calculated to give a known amount of hydraulic resistance to the flowing drug solution. When the device reaches a steady rate of osmotic water influx, the solution flows at a steady rate, giving a constant rate of drug delivery until the remaining solid contents are dissolved. Then the drug release decreases exponentially. Since the osmotic flux of water depends on the difference in water concentration, there is little affect of external solute concentration or stirring on the drug-release rate, which closely approximates the design release rate.

At present, pharmaceutical uses are the focus of much effort in microencapsulation, and a variety of new controlled-release products should appear in the next decade.

Figure 10. Oros osmotic pump discharging dyed stream. Courtesy of Alza Corporation.

Medical Applications. A number of medical problems might be solved by microcapsulation to achieve isolation or controlled release. In the late 1950s, Chang studied the preparation of artificial red blood cells. A procedure was developed for encapsulating aqueous solutions, such as red blood hemolyzate, in a very thin membrane of nylon through the use of interfacial polymerization. Although the original aim of artificial blood cells was put aside, the general technique has been used for microencapsulation of enzyme solutions as a remedy for enzyme deficiency, such as phenylketonuria or acatalasemia. A further study examined the administration of L-asparaginase microcapsules to remove L-asparagine in treating asparagine-dependent tumors, and demonstrated temporary remission of such tumors in mice. The enzyme microcapsules were given to the mice intraperitoneally, where the capsule wall prevented the usual rapid destruction of the enzyme.

The microencapsulation of adsorbents such as activated carbon has been studied for the removal of drugs or waste metabolites from patients. In the case of detoxification by direct perfusion of blood through a bed of activated carbon or other sorbent, a coating is frequently used to increase blood compatibility, giving some control over protein adsorption and platelet interaction (36). Several companies market products in this field, including Smith and Nephew, Ltd. in the U.K., Gambro in Sweden, Asahi in Japan, and Becton-Dickinson, Inc. and Extracorporeal Medical Specialties, Inc. in the United States.

In the intestinal removal of toxins, there is severe competition for sites on the carbon surface by the hydrophobic molecules in the intestinal tract. This competition can be overcome partially by selectively permeable polymer coatings on the carbon. Capacities for creatinine in the fluid from the intestinal tract of the pig were increased nearly fourfold in this manner (37).

Injectable microcapsules can control the long-term release of drugs given intramuscularly or subcutaneously. The capsules must be below 100 μm in diameter to pass through a typical hypodermic needle after suspension in an injection medium. To avoid leaving debris in the body, the microcapsules are made with poly(*d*,*l*-lactic acid), which slowly degrades in the body without harmful effect (see Biopolymers). This system can also slowly release narcotic antagonists for eventual use in the treatment of drug addiction (38). A sustained-release form of astiban has been studied as an approach to the treatment of schistosomiasis mansoni, a tropical parasite disease (39). A further intriguing application is the injection of progesterone-loaded microcapsules, to obtain systemic release of the drug for contraception (40).

Droplets of fluids containing islets of Langerhans have been encapsulated in calcium alginate (41). (Islets are the pancreas cells responsible for secreting insulin into the blood in response to increased glucose.) This may be another step toward the use of mammalian islets cells from one species to treat the effects of diabetes in another species. The microcapsule wall would prevent direct contact of the foreign tissue with the lymphocytes and antibodies of the host, slowing or eliminating immunological rejection. Another approach is the direct deposition of a polymeric wall onto the surface of living cells (42) (see Insulin and other antidiabetic agents).

Workers at the University of Chicago School of Medicine incorporated a drug, together with magnetite particles (10–20 nm), in albumin solution (43). The suspension was emulsified in cotton seed oil, sonicated, and the albumin set with aldehydes or heat. The final particles are less than 1.5 μm in diameter, and can pass into blood capillaries. The microspheres can be localized by an external magnetic field to obtain local drug release. Experiments in rats show that the local effects obtained with the microspheres match those obtained with one hundred times as much drug given intravenously. The same workers have coupled protein A (a surface protein from *Staphylococcus aureas* which binds selectively with immunoglobulin G) onto the iron in the microspheres to permit separation of B-lymphocytes from T-lymphocytes. T-cells contain no surface IgG, but B-cells do and are bound to the microspheres.

Among the large number of applications of the Exxon liquid-microcapsule technique are several of medical importance. In one, blood oxygenation is carried out by passing small bubbles of air or oxygen through blood but with the normally deleterious blood-gas interface eliminated by encapsulating the gas in a thin fluorocarbon wall. In another application, aimed at use in kidney failure, urea is removed from the intestinal tract by using oil droplets containing an emulsion of concentrated citric acid solution (44). Uncharged NH_3, generated by enzymatic hydrolysis of urea to ammonia and carbon dioxide, passes through the oil phase into the citric acid where it becomes ionized and can no longer pass through the oil. Hence, the selective permeability of the continuous oil phase allows ammonia to be trapped inside the citric acid droplets.

Veterinary Applications. A number of drug-release and feed-additive problems in the veterinary field are attracting the attention of workers in microencapsulation (see Veterinary drugs). The U.S. Department of Agriculture has developed the following method for replacing some saturated fat in milk with unsaturated fat, should this be of benefit in the diet (45). Safflower oil is emulsified in 10% sodium caseinate solution and the emulsion is spray-dried. The protein is cross-linked with formaldehyde. When this material is added to the feed of dairy cows, the coating and the unsaturated oil pass through the rumen without the usual enzymatic hydrogenation. In the abomasum, the microcapsules break at the low pH and the unsaturated oil is then incorporated into the milk by the cow. In tests, up to 24% of the milk fat contained linoleic acid. There is no change in the taste of the fresh milk containing the unsaturated fat, but there is sufficient oxidation in three days to affect the taste adversely (see Milk and milk products).

Adhesives. A unique use of microcapsules is the formation of adhesive mixtures which only set when the air concentration is extremely low (see Adhesives). These are referred to as anaerobic adhesives. They are manufactured primarily by Loctite Corporation and by Omni-technik in Munich, FRG. These microencapsulated adhesives can be coated onto bolts or other fasteners. When a nut is screwed onto a coated bolt, the microcapsules break, releasing the adhesive into the void spaces of the fastener. The oxygen concentration is low and the adhesive sets. Such adhesives can be made to set over a controllable time, and to cure at room temperature. The cured assembly has very high breakaway strength and performs better than a toothed lockwasher in vibration tests. A typical formula is based on the polymerization of tetraethylene glycol dimethacrylate, to which has been added cumene hydroperoxide, *N*,*N*-dimethyl-*p*-toluidine, benzoic sulfimide, and *p*-benzoquinone (46). The first three produce free radicals, but air and the benzoquinone prevent the material from turning solid until the oxygen concentration is low. The solidification is irreversible.

Southwest Research Institute (SWRI) is producing 1.5–1.8 mm microcapsules filled with water to be mixed with a fast-setting gypsum in a novel system for shoring up the ceilings of mines. As mines are dug, support plates, held by rods, must be placed on 1.2–1.5 m centers. The support rods are typically 1.8 m long and 2.2 cm in diameter. They can be set in polymers but these frequently are expensive, difficult to apply, and may require appreciable setting time. In the SWRI system, the water-containing microcapsules are mixed with fast-setting gypsum in a bag the shape of a sausage casing. The bag is inserted into a drilled hole (2.5 cm dia) and the rod rammed into place. This ruptures the capsules, releasing the water. Within a few minutes the bolt withstands a pull of 89 kN (20,000 lbf). The capsules contain 65% water and have a wall of wax, resin, and polymer of low water permeability to give long shelf life. The system is inexpensive and nonflammable.

Visual Indicators. Some liquid-crystal materials change color over a narrow temperature range because of changes in internal structure. Such materials tend to be oily liquids and are hard to handle and keep in place. They are stabilized by microencapsulation. Among the first uses were thermometers on which only the temperatures near 20°C were visible. Several companies are now developing such systems for medical use (see Chromogenic materials; Liquid crystals).

Appleton Paper Company has recently described the Magne-Rite system in which encapsulated metal flakes are suspended in oil. The application of a field of 0.05–0.1 T (500–1000 G), eg, with a magnetic stylus, causes the flakes to align, forming an image within 1–3 ms. A display system has also been developed by Fuji Photo Film Company in Japan.

Outlook. The range of possible applications of microencapsulation is breathtaking. As with all new techniques, microencapsulation will find those situations in the market place where it provides unique products. Cost will limit large-scale industrial uses for all but a few processes but uses in pharmaceuticals and medicine should grow and the markets in specialty products should continue to expand.

BIBLIOGRAPHY

"Microencapsulation" in *ECT* 2nd ed., Vol. 13, pp. 436–456, by James A. Herbig, The National Cash Register Co.

1. J. Crank and G. S. Park, *Diffusion in Polymers*, Academic Press, Inc., New York, 1968.
2. U.S. Pat. 2,648,609 (Aug. 11, 1953); 2,799,241 (July 16, 1957), D. E. Wurster (to Wisconsin Alumni Research Foundation).
3. J. T. Goodwin and G. R. Somerville, *Chem. Tech.* **4,** 623 (1974); J. T. Goodwin and G. R. Somerville in J. E. Vandegaer, ed., *Microencapsulation: Process and Applications*, Plenum Press, New York, 1974, pp. 155–163.
4. U.S. Pat. 3,423,489 (Jan. 21, 1969), R. P. Arens and N. P. Sweeney (to 3M Company).
5. U.S. Pat. 2,846,971 (Aug. 12, 1958), C. W. Baer and R. W. Steeves (to National Research Corporation).
6. N. N. Li, *AIChE J.* **17,** 459 (1971).
7. U.S. Pat. 3,703,576 (Nov. 21, 1972), M. Kitajima, Y. Tsuneoka and A. Kondo (to Fuji Photo Film Co., Ltd.).
8. U.S. Pat. 3,691,090 (Sept. 12, 1972), M. Kitajima, T. Yamaguchi, A. Kondo, and N. Muroya (to Fuji Photo Film Co., Ltd.).
9. U.S. Pat. 3,726,805 (April 10, 1973), Y. Maekawa, S. Miyano and A. Kondo (to Fuji Photo Film Co., Ltd.).
10. J. H. Fendler and A. Romero, *Life Sci.* **20,** 1109 (1977); D. Papahadjopoulos, *Liposomes and Their Use in Biology and Medicine*, New York Academy of Sciences, New York, 1978.
11. A. Romero, C. D. Tran., P. L. Klahn and J. H. Fendler, *Life Sci.* **22,** 1447 (1978).
12. C. Thies, *Polym. Plast. Technol. Eng.* **5,** 1 (1975).
13. U.S. Pat. 2,800,457 (July 23, 1957), B. K. Green and L. S. Schleicher (to The National Cash Register Co.).
14. U.S. Pat. 3,155,590 (Nov. 3, 1964), R. E. Miller and J. L. Anderson (to The National Cash Register Co.).
15. Br. Pat. 1,099,066 (Jan. 10, 1968), (to The National Cash Register Co.).
16. U.S. Pat. 3,415,758 (Dec. 10, 1968), T. C. Powell, M. E. Steinle and R. A. Yoncoskie (to The National Cash Register Co.).
17. U.S. Pat. 3,674,704 (1972), R. G. Bayless, C. P. Shank, R. A. Botham, and D. Werkmeister (to The National Cash Register Co.).
18. D. L. Gardner, R. D. Falb, B. C. Kim, and D. C. Emmerling, *Trans. ASAIO* **17,** 239 (1971).
19. T. M. S. Chang, *Artificial Cells*, Charles C Thomas, Springfield, Ill., 1972, pp. 16–18.
20. U.S. Pat. 3,516,846 (June 30, 1970); 3,516,941 (June 23, 1970), G. W. Matson (to Minnesota Mining and Manufacturing Company).
21. U.S. Pat. 4,025,455 (May 24, 1977), D. R. Shackle (to Mead Corp.).
22. P. W. Morgan, *Condensation Polymers by Interfacial and Solution Methods*, Wiley-Interscience New York, 1965.
23. E. E. Ivy, *J. Econom. Entomol.* **65,** 473 (1972).
24. U.S. Pat. 3,429,829 (Feb. 25, 1969), H. Ruus (to Moore Business Forms, Inc.).
25. M. Gutcho, *Capsule Technology and Microcapsules*, Noyes Data Corp., Parkridge, N.J., 1972, p. 152.
26. U.S. Pat. 3,779,941 (Dec. 18, 1973), M. P. Powell (to Champion International Corporation).
27. P. Speiser in J. R. Nixon, ed., *Microencapsulation*, Marcel Dekker, Inc., New York, 1976, pp. 1–12.
28. U.S. Pat. 3,726,804 (April 10, 1973), H. Matsukawa and M. Kiritani (to Fuji Photo Film Co., Ltd.).
29. U.S. Pat. 3,852,076 (Dec. 3, 1974), S. C. Grasko (to Jack W. Ryan).
30. Fr. Pat. 2,648,609 (April 27, 1960), J. A. Orsino and co-workers (to National Lead Company).
31. W. F. Gorham, *J. Polym. Sci. Part A-1* **4,** 3027 (1966).
32. C. Deng, B. B. Thompson, and L. A. Luzzi in T. Kondo, ed., *Microencapsulation*, Techno Inc., Tokyo, Japan, 1979, p. 149.
33. W. M. Doane, B. S. Shasha, and C. R. Russell, *Am. Chem. Soc. Symp. Ser.* **53,** 74 (1977).
34. R. W. Baker and H. K. Lonsdale in A. C. Tanquary and R. E. Lacey, eds., *Controlled Release of Biologically Active Agents*, Plenum Press, New York, 1974, pp. 15–71.
35. M. Calanchi in ref. 27, pp. 93–102.
36. H. Yatzidis, *Proc. Eur. Dial. Transplant Assoc.* **1,** 83 (1964); T. M. S. Chang, *Can. J. Physiol. Pharmacol.* **47,** 1043 (1969).

37. K. K. Goldenhersh, W. Huang, N. S. Mason, and R. E. Sparks, *Kidney Int.* **10,** S251 (1976).
38. C. Thies in ref. 30, p. 143.
39. P. Gopalratnam, M.S. Thesis, Washington University, St. Louis, Mo., 1978.
40. L. R. Beck, V. Z. Pope, D. R. Cowsay, D. H. Lewis, and T. R. Tice, *Contracept. Del. Syst.* **1,** 79 (1980).
41. F. Lim and A. M. Sun, *Science* **210,** 908 (1980).
42. P. Aegerter, M. S. Thesis, Washington University, St. Louis, Mo., 1980.
43. J. Widder, A. E. Senyei, and D. F. Ranney, *Adv. Pharm. Chemother.* **16,** 213 (1979).
44. W. J. Asher, K. C. Bovèe, T. C. Vogler, R. W. Hamilton, and P. G. Holtzapple, *Clin. Nephrol.* **2,** 92 (1979).
45. J. Bitman, T. R. Wrenn, and L. F. Edmondson in ref. 32, p. 195.
46. U.S. Pat. 3,218,305 (Nov. 16, 1965), V. K. Krieble (to Loctite Corporation).

General References

J. R. Nixon, ed., *Microencapsulation*, Marcel Dekker, Inc., New York, 1976.

T. Kondo, ed., *Microencapsulation*, Techno Inc., Tokyo, Japan, 1979.

J. E. Vandegaer, *Microencapsulation: Processes and Applications*, Plenum Press, New York, 1974.

T. M. S. Chang, ed., *Artificial Cells*, Charles C Thomas, Springfield, Ill., 1972.

M. H. Gutcho, *Microcapsules and Microencapsulation Techniques*, Noyes Data Corp., Park Ridge, N.J., 1976.

A. C. Tanquary and R. E. Lacey, eds., *Controlled Release of Biologically Active Agents*, Plenum Press, New York, 1974.

A. F. Kydonieus, ed., *Controlled Release Technologies: Methods, Theories, and Applications*, CRC Press, Cleveland, Ohio, 1980.

D. R. Paul and F. W. Harris, eds., *Controlled Release Polymer Formulations, ACS Symposium Series No. 33*, American Chemical Society, Washington, D.C., 1976.

W. Sliwka, "Microencapsulation," *Angew. Chem. Int. Ed.* **14,** 539 (1975).

H. B. Scher, ed., *Controlled Release Pesticides, ACS Symposium Series No. 53*, American Chemical Society, Washington, D.C., 1977.

ROBERT E. SPARKS
Washington University

NONWOVEN TEXTILE FABRICS

SPUNBONDED

Spunbondeds are a significant and growing area of the nonwovens industry and, in 1979, amounted to about one third by weight of the North American nonwovens market (1). Staple-fiber nonwovens result from shortening the textile chain in the preparation of textiles from fibers; spunbondeds result from the preparation of synthetic-fiber nonwovens to achieve chemical-to-fabric routes. The introduction of fabrics prepared from continuous filaments by such routes dates from the mid-1960s in Europe and North America (2–4). However, chemical-to-fabric routes had been described, patented, or commercialized earlier (5–7).

The development of spunbondeds has been confined to the Western world and Japan and their development to profitability has been reviewed (8). Spunbonded lines have a high unit cost, ca \$(1.2–1.5) $\times 10^7$ in 1977, but can produce vast quantities of product (5). Spunbondeds are no longer identified as textile substitutes but as materials in their own right. However, although limited penetration of the domestic textile field has been achieved, spunbondeds are deficient in esthetic properties, eg, drape, conformity, and textile appeal. These deficiencies result at least in part from their bonded structure which is of rigid fiber-to-fiber links. The major area for staple-fiber, nonwoven fabric sales, ie, cover stock, could represent a major growth area for spunbondeds. The nonwovens market, in general, is expected to grow rapidly but the direction of growth is unclear (9–14). However, it is likely that, in the short term, spunbondeds will grow in areas where they have a foothold, eg, in civil engineering and in the carpet, automotive, furniture, durable-paper, disposable-apparel, and coated-fabrics (qv) industries.

Properties

Appearance. Spunbonded nonwoven textiles generally are manufactured as roll goods in widths up to ca 5.5 m with basis weights of 5–500 g/m^2 or greater. In some cases, shaped articles such as filters may be produced as unit items but these are the exception. The basis weight of the majority of fabrics are 50–180 g/m^2. Generally, they are white, although gray and colored versions are produced for deliberate application benefits and occasionally as a result of recovered or scrap raw material forming part of the feedstock.

The fabrics generally have a random fibrous texture but they can have an orderly arrangement of fibers or be filmlike or netlike. Thicknesses generally are 0.1–5 mm; the majority of fabrics are 0.1–2.00 mm thick. The fibers comprising the fabrics generally are continuous and circular but, depending on the particular fabric, they may be in short lengths or in a mixture of lengths and they may have relatively uniform, noncircular cross sections or nonuniform, circular or noncircular cross sections. Filament diameters are 0.1–50 μm, with the preferred range being 15–35 μm, except for microdenier spunbondeds which are 0.1–1.0 μm in diameter. Spunbonded nonwovens have a characteristic stiff or feltlike handle which is the result of their rigid structure in which the filaments lack shear and, therefore, have a restricted relative movement.

The fabrics may have a glazed, plasticlike finish or a fibrous look. In some cases, the surface texture is discontinuous and may consist of a pattern of regular semiglazed dots which occupy 10–20% of the surface area, with the rest of the material having a more fibrous appearance. Such fabrics are point-bonded fabrics and generally are less stiff than fabrics which are bonded throughout. Thick, feltlike fabrics often are characterized by the presence of holes up to 1.0 mm in diameter from surface to surface; this is the result of needle punching, which is a common bonding method. In thinner fabrics, the pattern in which filaments have been laid down may sometimes be visible, as may the existence of heavier and lighter areas, thus indicating the limitations of the filament distributing apparatus.

Physical and Chemical. Physical and chemical properties, eg, specific gravity, moisture absorption, electrical behavior, chemical and solvent resistance, temperature resistance, dyeability, resistance to microorganisms and insects, and light stability of spunbonded fabrics, are determined by the material composition. Reference to the equivalent textile or polymer composition generally gives a satisfactory indication of these properties. Light stability, where it is important in relation to the fabric application, invariably must be approved by the customer in a test relevant to the use situation. In general, spunbonded textiles are engineered to be of adequate stability for their design applications, and the availability of excellent stabilizers enables virtually any desired degree of stability to be achieved. Illustrative tensile strength-loss curves are shown in Figure 1.

Mechanical, Hydraulic, and Textile Properties. Properties of spunbonded textiles are related to the mechanical properties and physical form of their fiber constituents and to their geometrical construction and binding. These properties are described by a wide range of tests (see Analytical and Test Methods) (15). Important areas for fabric characterization are thickness, basis weight, opacity, compressibility, isotropy, uniformity, tensile and modulus properties, recovery, burst properties, tear properties, permeability, porosity, abrasion resistance, creep, flex life, dimensional stability, static properties, absorbency, fraying properties, wrinkle and crease resistance, seam strength, and drape/handle characterization. Some of these properties are listed in Table 1. The basis weight per unit area is often the most important fabric parameter to the customer, since it is related to the yield and, thus, the area cost. The fabric thickness may be of importance in some applications and is measured by simple techniques which take into account the fabric uniformity, compressibility, etc (15).

Properties which are common to conventional textiles and nonwovens often are compared when spunbondeds are used in textile applications. These include dimensional stability to washing or to dry heating, eg, for coating substrates, and typically, values as low as 1.0% area shrinkage from machine washing may be achieved by the appropriate selection of the raw material, eg, polyester, and of the process. The abrasion resistance of spunbondeds and their resistance to pilling, snagging, filamentation, or

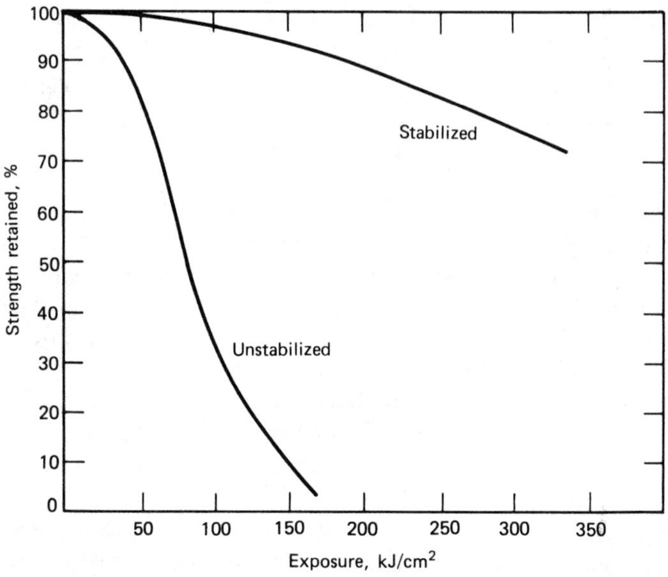

Figure 1. The effect of outdoor Florida weathering on the tensile strength of polyethylene/polypropylene, bicomponent, spunbonded fabrics. (Fabrics facing due south at 45°.) To convert kJ/cm^2 to langleys, multiply by 239.

cobwebbing also may be important and well-documented test procedures that are based on conventional textile testing have been described (15) (see Textiles, testing). Generally, the properties of spunbondeds are within a wide range of values. Spunbondeds excel in fray and crease resistance. The moisture absorption of most spunbondeds is very low, as is common with most synthetic-fiber textiles. However, the water holding of the fabrics can be quite high because of the high void content, especially with needled fabrics, where water may be retained by capillary action. The esthetic characteristics of spunbondeds are rarely quantified, except in the research environment.

Properties for objective comparison with conventional textiles include shear, drape, and nonrecoverable extension; numerous tests have been devised to assess fabric softness and hand (16). In some instances, shear, or the resistance of the fabric to in-plane shearing stress, may be the most important single factor determining fabric performance (17). The resistance of a woven fabric to in-plane shearing stresses is low because the yarns of which it is composed are free to move relative to each other in the fabric plane since they are interlaced. In a bonded nonwoven fabric, the filaments are not only locked together by rigid bonds, but the usually random arrangement of filaments produces the equivalent of triangulation in an engineering structure, thus presenting great resistance to in-plane movement of the filaments. Since the ability of a textile to accept shear deformation is a necessary condition for conformable fitting to a three-dimensional surface, spunbonded structures are excluded from many textile market sectors if they lack shear. Spunbondeds differ from most conventional textiles in their out-of-plane deformation behavior. This is usually quantified by some test, eg, drape or bending length (16).

Manufacture and Processing

A variety of raw materials and routes are used in the manufacture of spunbonded nonwovens. Virtually all of the spunbonded fabrics that are marketed are prepared from thermoplastic polymers (see Polymers, thermoplastic). A number of patents dealing with the manufacture of spunbonded fabrics based on dry-spun acetate fibers have been published (18–19); however, such work has not led to the marketing of acetate spunbonded fabrics. Trade names and manufacturers of spunbonded fabrics are listed alphabetically according to trade name in Table 2.

The basic polymers that are used are those that are common in fiber- and film-forming operations, ie, polyamide (nylon-6 and nylon-6,6), polyester, isotactic polypropylene, and polyethylene. Nylon-6,6 probably is used more commonly than nylon-6, although the latter is incorporated in a number of spunbonded fabrics. In some fabrics, both types of nylon are used, which allows thermal bonding based on the difference in melting points of the two materials (20). Until recently, certain versions of Terram and Mirafi were based on a nylon-6:6,6 copolymer which could be bonded thermally at a low temperature in the presence of saturated steam (21) (see Copolymers). Other nylon copolymers may be employed in the manufacture of fusible interlinings where a low melting point is needed. Nylons have a number of advantages, such as, easy bonding and dyeability, in nonwovens for certain applications. Their major disadvantage is their high energy content, which has led to their increasing cost and decreasing economic attractiveness as compared to polyesters and polyolefins. Generally, there is little need for precise end-group control for spunbondeds, and only medium tenacity and normal shrinkage filaments are required.

Poly(ethylene terephthalate), because of its desirable blend of useful properties for textiles, has become the major synthetic textile fiber. Its use in spunbonded nonwovens has not mirrored its penetration of the textile market, probably for economic reasons; although its high processing temperature and the necessity for moisture-free handling prior to extrusion lessens its attractiveness. However, it is used in a number of fabrics at normal textile molecular weights and mainly where its good dimensional stability at elevated temperatures is desirable (2–3). For example, poly(ethylene terephthalate) is used in carpet backings which undergo hot wet treatments under high tensions during carpet manufacture. As with other fibers, its properties and, hence, those of the fabrics, may be tailored, within limits, during the fiber-production process. For thermal-bonding processes, the melting point of a proportion of the fiber may be reduced either by leaving it undrawn or by employing a copolymer (22). The most usual technique is the replacement of 10–30 mol % of the terephthalic acid by isophthalic acid, since the latter compound is available relatively cheaply as a raw material for unsaturated polyester resin manufacture (see Polyesters, unsaturated).

In the polyolefin field, both isotactic polypropylene and polyethylene are used for spunbonded fabric manufacture (2); the factors dictating the choice are both economic and technical (see Olefin polymers). In recent years, polypropylene has been

Table 1. Typical Properties of Spunbonded Materials

Property	Cerex[a]	Corovin PP-S[a]	Fibertex 200[a]	Lillionette 506[a]	Lutradur H7210[a]	Net 909 P520[a]	Novaweb AB-17[a]	Terram 1000[a]	Trevira 30/150[a]	Typar 3351[a]	Tyvek 1073-B[a]
composition	nylon-6	polypropylene	polypropylene	nylon	polyester	high density polyethylene, polypropylene, or blend	polyethylene/polypropylene	low density polyethylene/polypropylene	polyester	polypropylene	high density polyethylene
bonding	autogenous area bonded	thermally point-bonded	needled	area resin bonded	area thermally bonded	fibrillated film	foamed film	area thermally bonded bicomponent	area resin bonded	area thermally bonded using undrawn segments	area thermally bonded (calendered)
basis weight, g/m^2	10–68	20–100		30–100	50–250	9–54	10–30	70–300	85–170	54–142	39–110
typical basis weight, g/m^2	34[b]	75[c]		60[b]	100[c]	27[d]	16–18[d]	101/170	142[c]	119[d]	75[d]
thickness, mm	0.12[e]		1.27[e]		0.38[f]	0.12[d]		0.7[d]	0.35[f]	0.33[d]	0.2[d]
breaking strength, N[l]	182/116[g]	130[h]	578[g]	103/98[g]	200/180[h]	9.8/7.8[d,i]	15/1.4[d,j]	850[g,k]	62/55[h]	512[g]	79/93[d,i]
breaking elongation, %		45[h]	125[g]	89/77[g]	40/40[h]		30/80[d]	80[g,k]	40/40[h]		26/33[d]
tear strength, N[l]	36/30[m]	15[n]	200[m]	34/32[o]	50/50[p]			250[q]		255[r]	4.5/4.5[s]
burst strength, kPa[t]	276[u]		1724[u]	245[v]				1100[w]			1180[x]

opacity	intermediate	high	high	intermediate	intermediate	intermediate	high	high		88%[y]
flex life, cycles		106[z]								>10[aa]
air permeability, L/(dm²·min)	1094[u]	70[bb]	3040[cc]	287[cc,dd]					304[ee]	
water permeability, L/(dm²·min)			30[ff]					24[gg]		641[hh]
pore size, μm			80–100[ii]		300[d]			100[jj]		

[a] For name of manufacturer, see Table 2.
[b] ASTM D 1910.
[c] DIN 53854.
[d] Test method not given.
[e] ASTM D 1777.
[f] DIN 53855.
[g] ASTM D 1117-1682 (grab test).
[h] DIN 53857 (strip test).
[i] N/10 mm.
[j] N/50 mm.
[k] Sample width 200 mm.
[l] To convert to dyne, multiply by 10^5.
[m] ASTM D 2263.
[n] DIN 53356.
[o] ASTM D 2261.
[p] DIN 53859.
[q] BS (British Standard test) 4303-1968.
[r] Trapezoid test.
[s] Elmendorf test.
[t] To convert kPa to psi, multiply by 0.145.
[u] ASTM D 231.
[v] ASTM D 774-67.
[w] BS 4768-1972.
[x] Mullen test.
[y] Eddy opacity test.
[z] BS 3424; Method 11B.
[aa] MIT Flex test.
[bb] DIN 53887.
[cc] ASTM D 737.
[dd] L/min.
[ee] Frazier Air Porosity.
[ff] U.S. Corp of Engineers, unit cm/s.
[gg] Manufacturer's test method.
[hh] Water vapor ($g/m^2/24$ h).
[ii] Equivalent opening sieve.
[jj] Pore size 0_{90} μm.

Table 2. Trade Names and Manufacturers of Spunbonded Fabrics

Admel	MPC-Spunbond Inc., Japan
Asahi Spunbond	Asahi Chemical Industry Co., Japan
Axtar	Toray Industries, Inc., Japan
Bem-Liese	Asahi Chemical Industry Co., Japan
Bidim	Rhone-Poulenc Textile, France, South Africa, Brazil Monsanto Textile Co., U.S.
Cerex	Monsanto Textiles Co., U.S.
Colback	Colbond bv, Netherlands
Conwed	Conwed Corporation Inc., U.S.
Colbond	Colbond bv, Netherlands
Corovin	J. H. Benecke GmbH, FRG
Celestra Crowntex	Crown Zellerbach Corporation, U.S.
Delnet Delweve	Hercules, U.S.
Dipryl	Sodoca S.a.r.l., France
Duon	Phillips Fibers Corporation, U.S.
Dynac	Toyobo Spun Bond, Japan
Enkamat	Enka Glanzstoff AG, FRG
Evolution/Evolution II	Kimberly Clark Corporation, U.S.
Fibretex	Crown Zellerbach Corporation, U.S.
Kimcloth	Kimberly Clark Corporation, U.S.
Kridee	VEB Textile Verpackungsmittel, GDR
Kyrel	J. P. Stevens, U.S. (imported from Asahi)
Lillionette	Snia Viscosa SpA, Italy
Lutrabond	Lutravil Spinnvlies, FRG
Lutradur	Lutravil Spinnvlies, FRG
Lutrasil	Lutravil Spinnvlies, FRG
Marix	Unitika Ltd., Japan
Mirafi	Fiber Industries Inc., U.S.
Novaweb	Bonded Fibre Fabrics, Ltd., UK
Netlon	Netlon Ltd., UK
Net 909	Smith and Nephew Ltd., UK
Petex	Juta, Czechoslovakia
Petromat	Phillips Fibers Corporation, U.S.
Polyfelt	Chemie Linz AG, Austria
Polyweb	Riegel Products Corporation, U.S.
Reemay	E. I. du Pont de Nemours & Co., Inc., U.S.
Sharnet	Inmont, U.S.
Silheim	Toray Ltd., Japan
Sodospun	Sodoca S.a.r.l., France
Sualen	VEB Chemiefaserkombinat Wilhelm Pieck, GDR
Supac	Phillips Fibers Corporation, U.S.
Syntex	MPC-Spunbond Inc., Japan
Tapyrus	Tonen Petrochemical Co. Ltd., Japan
Tafnel	MPC-Spunbond Inc., Japan
TCF	Mitsubishi Rayon Co., Ltd., Japan
Terram	ICI Fibres, UK
Texizol	Chemosvit np., Czechoslovakia
Thinsulate	Minnesota Mining & Manufacturing Co. Inc., U.S.
Toyobo Spun-Bond	Toyobo Co. Ltd., Japan

Table 2 (*continued*)

Trevira-Spunbond	Hoechst AG, FRG and U.S. (1980)
Typar	E. I. du Pont de Nemours & Co., Inc., U.S. and Luxembourg
Tyvek	E. I. du Pont de Nemours & Co., Inc., U.S.
Unisel	Teijin Ltd., Japan
Viledon-M	Carl Freudenberg, FRG
Vivelle	ICI Mond, UK, FRG

the least expensive fiber-forming polymer available and it is less dense than all of the polymers, particularly polyester whose fabric volume yield is less than 70% of that of polypropylene. Thus, unless major technical factors inhibit the choice of polypropylene, it is the favored raw material for spunbonded fabrics. Because of its low density, polypropylene fabrics float on water, which can be an advantage, for example, in the installation phase with civil-engineering fabrics. Standard predegraded fiber- or film-forming grades normally are used in this application, together with whatever additive masterbatches are dictated by processing or fabric application considerations. Conventional spinning and drawing technology usually is employed to yield medium tenacity filaments. Generally, it is not necessary to employ special techniques to yield low shrinkage filaments unless the fabrics are to be used in some critical high temperature application, eg, a coating substrate. Where thermal bonding techniques are used, the fabric is necessarily subjected to a heat treatment which tends to stabilize it against further shrinkage. In contrast to nylon and polyester, delustering pigments, eg, titania, normally are not included in polypropylene that is destined for fiber manufacture, so that polypropylene spunbonded fabrics have a fairly lustrous appearance. Colored pigments sometimes are employed where colored fabrics are required, since polypropylene cannot be dyed easily.

Polypropylene filaments have a further advantage in spunbondeds manufacture in that, under the influence of heat and pressure, quite strong bonds can be formed, even with drawn filaments, without destroying them. This is particularly useful in point bonding where high pressures are available (23). Bonding also may be achieved by employing undrawn and, hence, lower softening temperature filaments or by the use of a variety of binders (24). A further advantage of polypropylene, which also applies to polyethylene, is that scrap fabric or spinning waste can be reused and does not require elaborate recovery techniques. This is especially important in spunbondeds manufacture where the effectiveness of the operation may depend upon achieving a good conversion efficiency and, inevitably, up to 5% of the fabric has to be trimmed as edge waste.

Polyethylene is more expensive than polypropylene, it has a slightly lower fabric yield, and it cannot be made to produce filaments of the high strengths of polypropylene. However, its lower melting point permits less expensive and easier processing, and it is used in a number of materials where its lower service temperature is not a disadvantage. It is used in the manufacture of extruded nets, tack-spun items, film-fibril spunbondeds, and civil-engineering fabrics. The use of polypropylene and polyethylene in some types of spunbonded fabrics also results from their ability to fibrillate easily. As with polypropylene, fiber or film grades of polyethylene are used because their purity and melt viscosity are suitable for precision extrusion. Both low and high density polyethylenes are used in this area; the new linear, low pressure polyethylenes offer potential advantage of higher strength and lower cost. Polyethylene can be used in a self-bonding mode, either while still tacky after extrusion, or under the influence of heat and pressure (25). It also can be used as the binder with other fibers mentioned above (26). Several combinations of the four fiber-forming polymers are possible and a number are employed in various spunbondeds.

Spunbonded fabrics are produced by many routes which are combinations of alternative process steps (2). From these combinations, a small number of major processes can be identified by which the majority of spunbonded fabrics are produced. The predominant technology consists of continuous filament extrusion, followed by drawing, web formation by the use of some type of ejector, and bonding of the web.

Basic Spun-Filament Route. Spunbonded fabrics based on the extrusion of continuous filaments were introduced commercially as Reemay by DuPont in 1965 (4). The continuous-filament process may best be illustrated by reference to Figure 2 which shows diagrammatically the basic elements of typical mainstream, spunbonded fabric processes (2–3).

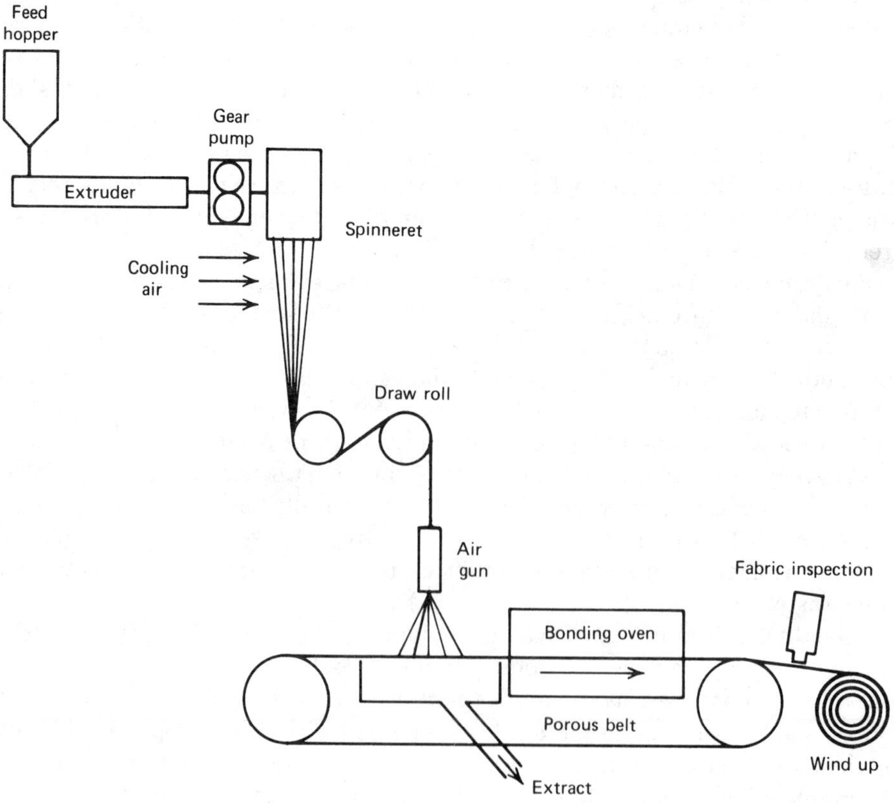

Figure 2. Flow sheet for typical spunbonded-fabric manufacture.

Typically, the polymer is melt spun in an identical manner to that used for the manufacture of continuous-filament textile or industrial yarns (27). In the case of polyester filaments, a homogenous granular polyester feedstock of suitable intrinsic viscosity is dried to a low moisture content and is fed to a suitable melter, either a screw pressure melter or a screw extruder, where it forms a pressurized (650–7000 kPa or 80–1000 psig), heated (265–280°C) stream from which it is fed to a metering pump which both meters the polymer stream and boosts its pressure to 7–35 MPa (1000–5000 psig). The molten polyester is extruded from a stainless steel spinneret containing from a few to several hundred holes ca 0.2 mm in diameter by way of a sand or other type of filter. A flow of cooling air solidifies the filaments below the extrusion point. If a polypropylene spunbonded fabric is to be prepared, similar melt-spinning principles apply, although the details differ.

In the early days of melt spinning, maximum wind-up speeds were limited, for mechanical reasons, to ca 1000–1500 m/min. Thus, in any combined spin-draw process, because of the effect of draw ratio, the spun yarn haul-off speed or the spinning speed could be as low as 200 m/min. This severely limited the throughput of the melt-spinning process, which, uncombined, could be run easily at the upper mechanical limit to the haul-off speed of 1000–1500 m/min which was more favorable economically. The development of improved winders capable of operating at up to 4000 m/min or higher and the knowledge that, as the spun yarn haul-off speeds increase, the required draw ratio is reduced to 2:1 or less, allowed the development of partially oriented yarns (POYs) (28). The more natural combination of drawing and texturing could be used for POYs, since false-twist texturing is limited in speed to below ca 1000 m/min because of the high levels of false twist which have to be inserted into the yarns.

This same limitation in haul-off speeds applies to spunbonded fabric manufacturing routes, and a number of techniques have been employed to circumvent it. However, since no wind-up unit is involved and, hence, no traverse, the mechanical limitations on speeds are not so low. If separate spinning and drawing processes are used, the attractions of an integrated process disappear; alternatively, if spin-draw is used, the melt-spinning process may not operate at maximum efficiency. This has been overcome either by hauling off the spun filaments at the highest possible speeds and accepting the reduced filament properties which result from the lack of a drawing stage or by tolerating the reduced extrusion rates employed with the spin-draw route (3,27). In the former route, haul-off is achieved by the use of a pneumatic jet or gun, which is capable of forwarding the filaments at speeds up to those for POYs, ie, ca 3000 m/min (27). This has led to filament tenacities that are ca 50–70% of those which would have been obtained by conventional drawing and to extensions that are considerably higher than those of textile filaments.

In the integrated spunbonded fabric process, the overall line speed may be limited by factors other than spinning-machine output. For example, the throughput of filaments per gun, the gun-scanning speed, and the web-bonding speed all may be controlling elements for the overall manufacturing rate of the fabric. The gun is one of the key elements in the manufacture of spunbonded fabrics.

The alternative roll spin-draw route has much to recommend it, since it permits closer matching of the properties of the filaments and, therefore, of the fabrics to the requirements of the application. A diagrammatic version of this route is shown in

Figure 3; the diagram illustrates a number of features of a more sophisticated spunbonding process. Extruded filaments are supplied from two spinnerets: those from one spinneret pass directly to the gun without drawing and those from the second spinneret pass over a guide roll 1 to a heated feed roll 2 and over a slot heater 3 to a draw roll 4 before entering the gun. Roll 4 is arranged to run faster than roll 2; thus, the filaments are elongated by an appropriate amount. The gun 5 traverses, from side to side above the moving porous belt 6, on which filaments are deposited in a layered arrangement. From belt 6, the mat of filaments passes to an oven which is maintained at a suitable temperature so that the undrawn filaments are preferentially melted and bond the drawn filaments from spinneret 7.

The electrode 8 and the grounded plate 9 improve the filament separation and, therefore, the cover and uniformity of the fabric. If an electrostatic charge is induced onto the filaments and as soon as they are released from the tension of the gun, they repel and are attracted to grounded surfaces. Thus, electrode 8 causes charging of the filaments either by direct charging from a high voltage source or by triboelectric charging, ie, the charge is generated by the friction of the filaments against a grounded, conductive surface (27). The mutually repelled, charged filaments are attracted to plate 9 and lose their charge after being precipitated in web form. In the absence of a grounded target, the charged filaments continue to repel each other after hitting

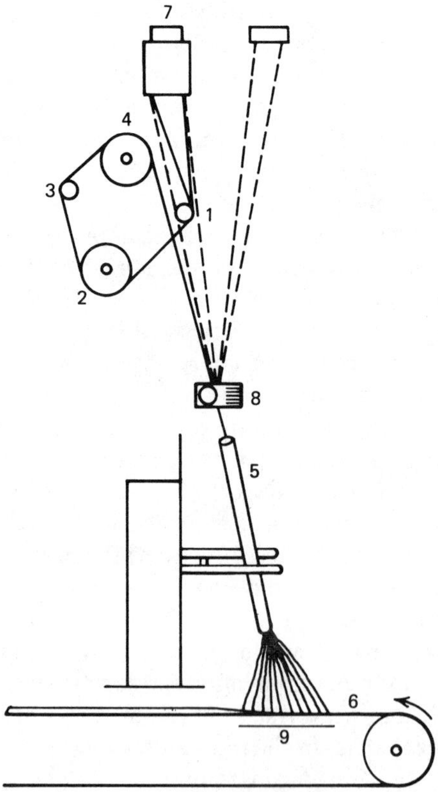

Figure 3. The spin-draw route to spunbonded fabrics showing the optional use of a separate feed of bonding filaments (27).

the porous belt and form an unmanageable mass. For simplicity, the porous belt may form the target if, for example, it is made of wire mesh and is earthed. Static eliminators or high humidities may be employed to assist in the removal of charge, because it may take several seconds to leak to the grounded target since the filaments are nonconducting. If the filaments are charged triboelectrically, and since it is often convenient for the grounded surface to act as a yarn guide, ceramics such as chromium oxide often are used to resist abrasion. It is possible to modify the polymer composition or to select suitable polymer/guide combinations to enhance the triboelectric charging effect.

Because spin finishes contain an antistatic agent, the use of such a filament treatment would vitiate the charging process and would seriously hinder the formation of uniform filament webs. Since it is very difficult to wind packages of filament yarn without a spin finish because of static problems, split, spunbonded-fabric processes are not common. In principle, the winding of packages and the subsequent separate formation of the yarns into webs offers some advantages, eg, a better match between the spinning and the web-making processes (29); however, these are outweighed by process difficulties. The formation of webs from continuous filament tow has been described (30).

An alternative to charging the filaments to obtain controlled laydown in the form of a mat is to suck the air from below the porous belt (3). An overall downward air flow pins down the filaments and reduces disturbance by the air from the guns. This technique is used almost universally in mainstream spunbonded-fabric manufacture, although it is not without problems.

Another technique which is employed to prevent web disturbance after laying is spontaneous bonding or tacking. In the case of certain nylons or nylon copolymers, freshly spun filaments are tacky for a few seconds until they crystallize (3). Thus, webs of these filaments exhibit a significant coherence which can eliminate the need for suction. This process may be enhanced by treating the filament mats with water vapor or steam to increase the crystallization rate (31). This treatment also may cause spontaneous elongation of the filaments which results in some filament interlocking as well as tacking. The process can yield webs which are stable enough to wind and unwind prior to bonding, thus allowing a process break if web formation and bonding are mismatched in speed. Polypropylene and polyester do not exhibit spontaneous tackiness but, the webs may be temporarily tacked by slight heat, eg, 120°C, and pressure to give handleable webs prior to further bonding (32–33).

After web formation, either with or without temporary consolidation, the webs must be bonded to achieve their final strength. Bonding usually is achieved integrally with the web-formation process and at the web-formation speed of between a few and tens of meters per minute. Generally, for thermal area-bonding, some type of drum oven is used (34). From the oven, the web passes to a winder where it is rolled either for direct packaging or for later rewinding and breakdown into rolls. For thermal point-bonding, a heated textile calender is used and consists of two heated steel rolls; one is plain and the other bears a pattern of raised points (35). Control of filament orientation is necessary to concentrate the strength of a polypropylene carpet-backing fabric in the warp and weft directions to provide greater resistance to stretching and neckdown during carpet processing (4). This is achieved by arranging for the guns to scan at high speeds, some in the machine direction and some in the cross-machine direction (36). Since the scanning speeds are comparable to the filament-forwarding velocities, great control of the filament orientation is obtained on the porous belt.

The filament spray is scanned by deflecting it just as it emerges from the gun exit by means of oscillating air jets. This technique for achieving a measure of deliberate filament orientation gives a combination of randomly oriented filaments and a proportion of filaments of the desired orientation. Recently, for purposes of manufacturing spunbonded nonwovens for textile applications, a technique has been proposed for achieving 100% orientation of filaments in any chosen direction (37). The technique makes use of scanning jets that involve a very narrow filament spray and that confine the filament spray between pairs of plates extending from the gun exit to the porous belt. Thus, webs can be prepared for spunbonded fabrics in which the filament orientation is controlled as closely and as uniformly as in woven fabrics. The introduction of warp orientations depends on traversing a complete gun/plate assembly at right angles to the direction of the belt movement.

As illustrated in Figure 4, a gun for forwarding filaments consists essentially of a means for conducting the bundle of filaments to a point at the high velocity, low pressure zone of a venturi. Air is entrained with the filaments as a result of the low pressure generated in the jet. Once the filaments have reached the low pressure zone, the filament bundle and the entrained air move with the expanding supply air to form a cone or a fan of separated filaments; the exact shape depends on the detailed design of the gun (3,38–39). The pressure of the supply air may be from hundreds to thousands of kilopascals (several to tens of atmospheres) and air velocities in the low pressure

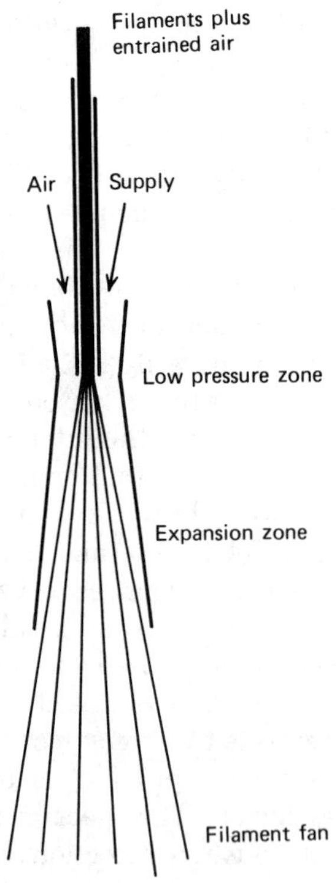

Figure 4. The principal features of an air gun used to forward filaments in spunbonded-fabric manufacture.

zone may be several times supersonic. The air velocity must always be greater than the filament velocity in order to generate tension in the filament bundle. The filaments may be electrostatically charged prior to entering the gun to assist in separating them by mutual repulsion once they enter the expansion zone; tension generated in the low pressure zone holds the bundle together up to this point. Air flows of free air may be tens or hundreds of liters per minute and tensions of tens of grams are generated in the threadline.

The objective of good gun design is achieving the best separation of the filament bundle into individual filaments that are suitably distributed over the desired spray profile. If good filament separation is not achieved, then the fabric will have a rope appearance and consist of ropes of filaments and areas of low filament cover. The greater the quantity of supply air per filament, the better is the separation. However, equally important considerations in gun design are achieving maximum filament throughput at minimum air pressure and consumption, and the tension generated by the gun and the spray width. The greater the spray width, the fewer the number of guns required to cover a given width of fabric.

The tension requirements of the gun depend on the process configuration. If a roll-drawing route is employed, the gun needs only to generate sufficient tension to receive the filaments from the last godet, ca 0.003–0.01 N/tex (0.03–0.1 gf/den). If only gun drawdown is used, higher tensions of ca 0.01–0.1 N/tex (0.1–1.1 gf/den) are required at higher velocities (higher velocities because sufficient drawing tension must be generated by air drag below the spinneret). Thus, much more powerful guns with high supply pressures are required in order to maintain a favorable excess of supply air velocity relative to filament velocities of $\geqslant$2300 m/min (27,38). It is not possible to achieve the highest filament tenacities by gun drawdown because of the limited tensions which can be achieved at the highest gun-supply pressures, particularly with filament diameters at the upper end, ie, 35 μm, of the working region where surface/volume ratios for air friction purposes are reducing rapidly. For this reason, gun-drawdown processes are less in favor than roll-drawing processes.

The filaments emerging from the gun form a spray pattern which may be circular or elliptical; an elliptical profile, which approximates a line of filaments, is preferred. In some types of spunbonding operations, the spinnerets are arranged in line, and this linear arrangement is continued through the guns and to the porous belt (3). This is difficult to arrange if the filaments are to be roll-drawn and, in general, throughput and technical considerations rule out this arrangement. Usually, it is necessary to generate uniform sheets of web from a number of fans of filaments of elliptical plan with the longer axis of a few to tens of centimeters in length.

The weight distribution of deposited fiber depends on the gun design but tends to be approximately triangular on both ellipse axes. Thus, if a single gun is held above the porous belt with the larger ellipse axis at right angles to the direction of belt travel, a strip of web of approximately constant density will be deposited along any line parallel to the direction of belt movement. A uniform web over a porous belt several or many times wider than the gun spray width can be formed by combining a number of such strips so that gaps and overlaps are not a problem and enough coherence is obtained between strips to ensure adequate cross-machine-direction fabric strength (40–42). The web can be achieved by arranging for several layers of strips to be laid from multiple guns which are overlapped like shingles. If the gun is traversed from side to side of the porous belt, the path traced out by the web is as shown in Figure 5.

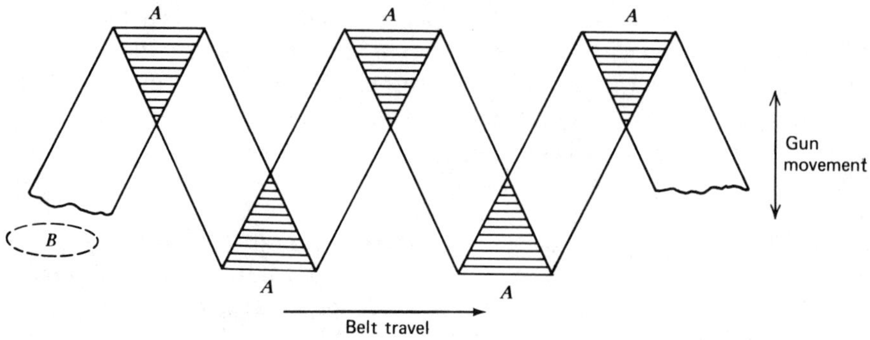

Figure 5. Path traced on a moving porous belt by a traversing spray (B = spray pattern in plan view).

In the deeply hatched triangular areas A, the web thickness is twice that in the remainder of the strip. The size of areas A depend on the relative rates of movement of the gun and belt and on the spray width. If the belt advances by one spray width for every cycle of gun movement, then adjacent areas A touch; at this stage all of the porous belt becomes covered by a double web thickness. If the belt advances by less than one spray width per cycle of the gun, then the leading and trailing edges of the strip of web overlap, as do adjacent areas A on each edge, giving high density areas in the web.

The effect of moving guns also may be achieved by using stationary or swivelling deflector plates from which the spray is bounced onto the belt (43–46). Scans may be arranged along and at right angles to the direction of belt movement to achieve the desired orientation of the filaments in the web (36). Normally, in spunbonded-fabric manufacture, the filaments crumple as they hit the porous belt. Crumpling must occur unless the apparent relative velocities of the filaments and the belt are zero. Thus, if crumpling is to be avoided and if the filaments arrive at the belt at 2000 m/min, then the laydown speed also must be 2000 m/min. If a total scan width of one meter is desired, a scan rate of about 17 Hz is required to lay the filaments without crumpling. Such a rate is possible for small excursions obtained by scanning but not for the large excursions used with traversing guns. However, it is possible to superimpose a scanning motion onto a traversing motion.

Bonding. As with staple-fiber nonwoven textiles, there are three major bonding techniques, ie, thermal, chemical/adhesive, and needling, that are used for spunbondeds. Except with needling, a choice may be made of point- or area-bonding. Point-bonding consists of cohering the filaments in small, discrete, and closely spaced areas of the fabric. Area-bonding involves using all available bond sites in the fabric; however, every filament contact is not necessarily bonded, since not every contact necessarily is capable of forming a bond (47). The three basic techniques of bonding do not lead to three completely different varieties of fabric physical properties. Thermal and chemical/adhesive bonding produce fabrics which are not radically different.

Needling is the less versatile technique since the only bonding variants are punch density, depth, and needle design; the products are thick fabrics of good flexibility, low initial modulus, high ultimate strength, high tear strength, and low fatigue life (2). Needling is not suitable for lightweight fabrics, ie, <100 g/m^2, because the needling action tends to concentrate the filaments in the already denser areas and, thus, destroys the fabric uniformity. Modern needle looms operate at high speed over high widths and, thus, the technique is perfectly applicable to spunbonded lines (48). It is less expensive to operate and requires much less sophisticated control than thermal bonding does. Many Docan process licensees use it and it tends to be popular for civil-engineering fabrics which are generally medium to heavyweight, ie, 120–500 g/m^2 (49).

Thermal and chemical/adhesive bonding offer the choice of point or area-bonding; the choice of either is dictated mainly by the ultimate fabric application. Because the bonded area usually is only 5–25% of the total fabric area, point-bonded fabrics tend to be more flexible and textilelike than area-bonded fabrics. The essential requirement for thermal bonding is the presence in the web of some material which holds together crossing or adjacent filaments and which is activated by heat. Usually the material is a polymer and often is the same material as that from which the web is manufactured. In the simplest form, thermal bonding makes use of the sticky nature of the extruded filaments emerging from the spinning machine (25). In the majority of thermally bonded fabrics, secondary heating is employed; the web is formed from fibers which have cooled to below their softening point and additional heat must be supplied to cause fusion.

In the simplest arrangement, the heat may be supplied by a bank of ir heaters, which are particularly applicable in the case of materials which have a wide softening range, eg, copolymers for fusible interlinings. Generally, fairly accurate temperature control is necessary. In the case of drawn polypropylene filaments, the useful bonding range is only 0.5°C, outside of which either no bonds are formed or the filament properties deteriorate (50). The use of hot-air drum ovens also is common (see Fig. 6). In hot-air drum ovens, accurate temperature control is achieved by high air flows. The porous pinning belt provides a restraining pressure to prevent movement of the filaments during bonding and it helps to prevent fabric shrinkage. If the belt pressure

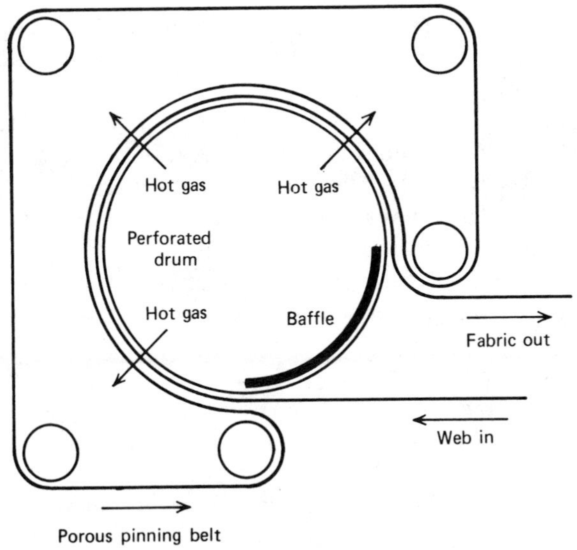

Figure 6. Typical bonding oven used in the manufacture of heat-bonded nonwovens.

is insufficient to give the desired degree of web compression, then additional nip rolls may be employed at the fabric exit while the filaments are still soft. Highly accurate temperature control also may be obtained by the use of pressurized steam ovens. In the steam ovens, the web, which is trapped between two porous belts, is passed through suitable seals into a pressurized steam box where the heat of condensation plus specific heat of the steam almost instantaneously heat the belts and web to the steam-box temperature. After the fabric leaves the exit seal, which may provide substantial nip pressure on the web, any residual moisture is evaporated by the heat that is retained in the belts and web.

Since the temperature of the steam box is very sensitive to the steam pressure, which can be easily and accurately controlled, very good temperature control is possible. Operating temperatures are limited to about the softening point of polypropylene, because temperatures above this require very high steam pressures which are difficult to seal. A further advantage of steam-heated bonding is that discoloration of materials that are sensitive to oxidation, eg, nylon-6,6, does not occur.

For lightweight fabrics, where heat transfer through the fabric is rapid, thermal area-bonding may be achieved by hot calendering the web through plain rolls; this gives the highest pressures if maximum fabric strength is required. If controlled web compression is needed, then fixed-gap, hot calendering is possible. In principle, other types of heating, eg, dielectric or ultrasonic, could be used for thermal area-bonding.

The introduction of bonding sites for activation by thermal bonding may be accomplished in a variety of ways, many or all of which also are applicable to staple-fiber nonwovens. The simplest technique but the most difficult to control is softening the filaments of the web until they stick at the crossover points. A common technique is to combine filaments of the same material but of different softening points; an example is the use of undrawn polypropylene filaments as a binder for webs of drawn polypropylene (50). A softening point difference of 3°C is generated. The most sophisticated way of achieving this, eg, in Typar, is the introduction of undrawn segments at intervals along the polypropylene filaments of which the remainder is drawn (50). Undrawn polyester also may be used as a bonding agent (27). A similar technique is to include a different material of a lower softening point in the web filaments. However, this is complicated, since a separate extrusion system is required, for spunbonded fabrics. The different material may be introduced as a powder which is distributed throughout the web. With the powder technique, it is difficult to achieve uniform bonding throughout the width, length, and depth of the fabric.

Another sophisticated technique for achieving a uniform distribution of a polymeric binding agent is to use bicomponent filaments in which the binding agent forms a thin sheath around each web filament or is present as an integral sector of the filament cross section; such bicomponent filaments are called, respectively, core/sheath and side/side heterofilaments (see Polyamides, polyamide fibers). Combinations in spunbonded fabrics have been polyethylene/polypropylene (51), nylon-6/nylon-6,6 (21), nylon-6/polyester (52), and nylon-6:6,6 copolymer/polypropylene (21). All of these are of the core/sheath configuration; the sheath material is given first in each case.

Although, in principle, the manufacture of heterofilaments does introduce significant complications to the extrusion process, heterofilaments can be economically manufactured to the same precision as single-component filaments (53). The advantage of the heterofilament route to thermal bonding is the high degree of control that it confers on the properties of the spunbonded fabric. By variation of the characteristics of the sheath material, the core-to-sheath ratio and the proportion of heterofilaments present in the web, in combination with variations in the characteristics of the core material, the drawing conditions and the filament diameter, the balance of tensiles, and tear and other properties of the textile can be varied over a wide range to tailor the properties closely to the application.

Adhesive, or resin bonding also is common to staple-fiber nonwovens. The process usually consists of applying an acrylic or similar resin that is dispersed in water to the web, removing the surplus by mangling or other means, and passing the web over drying cans or rolls which evaporate the water and polymerize the resin. The resin accumulates at the filament contact points prior to water removal and polymerization as a result of surface tension. The web is bonded and, if a thinner product is required, also may be calendered which flattens the filaments and the bonds. Generally, resin bonding is not used in the thermoplastic spunbonded fabric area except for special applications, because it is a wet process which offers no particular advantages against a choice of alternative dry-bonding techniques (54).

Solvent bonding also has been disclosed for certain specialized applications (55), but chemical bonding is more important commercially. The latter is a system of bonding which is applicable to a number of nylon polymers (56). When the filaments are in close contact, the presence of anhydrous hydrogen chloride, or any of a number of other gaseous substances, initiates the formation of interfilament bonds, possibly by breaking interchain hydrogen bonds by forming an HCl complex with the amide group. When the HCl is desorbed, the hydrogen bonds immediately reform between molecular chains from different filaments. The bonding agent may be present in the gaseous phase or it may be dissolved in an inert organic carrier.

The relationship of fabric properties to the manufacturing variables is complex. Broadly, the tensile properties of the fabric are determined by the tensile properties of the filaments. High tenacity, low extension filaments lead to the formation of high tensile strength, low extensibility fabrics. However, the bonding medium properties also exercise an influence on the tensile properties. A high modulus binder tends to yield a high initial fabric modulus and, usually, the binder modulus is related to the binder tensile strength. Since needled fabrics have no binder, this corresponds to a very low modulus binder, and, hence, low initial-modulus fabrics. A high concentration of binder, which tends to lock the filaments together tightly, tends to effect high tensile strength. However, as the proportion of binder increases, the fabric becomes more filmlike and, therefore, less resistant to dynamic tearing. It also becomes stiffer, less flexible, and less permeable to gases and liquids. Conversely, with reduced binder proportions, fabrics are more feltlike or fibrous, less abrasion resistant, more flexible, more resistant to dynamic tearing, and more permeable. Again, needled fabrics are equivalent to zero binder levels. The influence of binder at levels of up to 30% or so, where it does not form a major proportion of the fabric, is summarized in Figure 7. The introduction of crimp into the filaments tends to increase the fabric flexibility. However, since the flexibility is a function of the average free filament length between bonds, a similar effect on fabric flexibility may be obtained by point-bonding, which results in a much greater average free filament length.

The most common form of point-bonding is thermal point-bonding, which is effected by the use of a heated calendar using virtually any of the techniques available

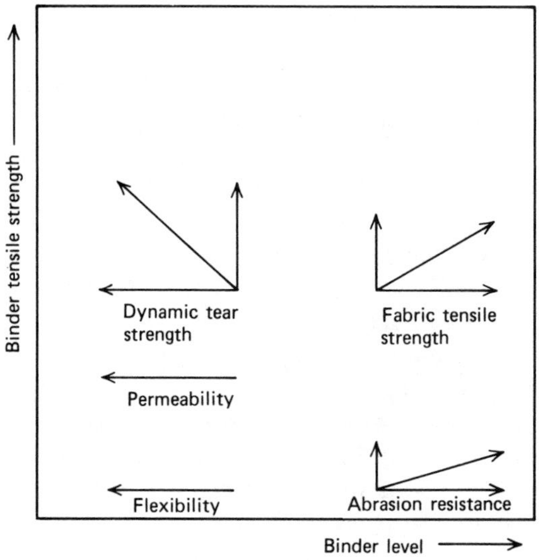

Figure 7. The qualitative influence of binder parameters on spunbonded textile properties. The arrows indicate increasing levels of the indicated property.

for area bonding, ie, bicomponent filaments, filament blends, undrawn filaments, or direct bonding of the drawn filaments. Since bonding is by both heat and pressure, the bonding of drawn filaments is easier to control than it is in area-bonding where temperature is the only variable. Based on this relative ease of bonding of single-component webs, point-bonded, drawn polypropylene filament fabrics are common because of the cost advantage of this polymer and its well-established filament-spinning route (32). The chemical and adhesive bonding routes can also be adapted for point-bonding, although in the case of adhesive bonding which is used for applying the adhesive spots, the process is termed print bonding. As far as is known, it is not used at all with spunbonded fabrics; one of the disadvantages is the inevitable relative coarseness of the print patterns since control over the bond size, adhesive penetration, etc, is imprecise. Chemical point-bonding is applicable to nylon fabrics and is based on the same technique as chemical bonding (57).

In thermal point-bonding, a heated calendar is used. Point-bonding calenders may be characterized by a 1–3-m working width and consist of heated steel work rolls between which the web is passed for point-bonding. The pattern may be cut on one or both rolls, the diameters of which are related to the working width of the calender for stiffness purposes. For a 3-m working width, rolls of 0.5-m diameter are not unusual. Of critical importance is the uniformity of pressure across the working width of the machine. Since the hydraulic or pneumatic pressure which is applied at the ends of the rolls to create the working nip pressure is several tons, it is impossible to prevent some bowing of the work rolls no matter how stiff the roll geometry is. Thus, some expedient must be adopted to obtain adequate pressure uniformity. Possibilities are backing the work rolls with crowned supporting rolls, crowning the work rolls, or setting the work rolls at a slight angle to each other. The most common solution is to employ crowned support rolls, since crowning the work rolls creates complications in pattern cutting.

Crowning consists merely in symmetrically reducing the roll diameter progressively from the center to the ends of the rolls; the total reduction is only a fraction of a millimeter. The pressure on the ends of the support rolls causes a balancing pressure to the center of the work rolls as a consequence of the crowning. One or both of the work rolls may be heated. Since accurate temperature control, typically ±1°C at up to 225°C, is essential, oil heating is common although smaller calendar rolls may be electrically heated. With oil heating, the heated rolls are hollow and contain suitable oil inlet and outlet rotating seals and inner distribution channels to ensure high and uniform heat transfer. Calendering speeds depend on the fabric basis weight, the pattern design, the web material, and other factors. However, a range of 5–100 m/min probably is available; the higher speeds correspond to the lower basis weights.

Duplex bonding reduces the presence of area bonding by allowing the heat required to form the point bonds to penetrate from both sides of the web (53). In duplex bonding, part of the pattern is cut onto each roll and both rolls are heated to the bonding temperature. For example, opposing helical lands on the rolls generate a pattern of lozenge-shaped point bonds (58). Since the bonding heat has only half as far to penetrate into the web, the roll temperatures may be lower and the residence time in the nip zone may be shorter; thus, there is a reduction in the degree of adventitious area-bonding. Further, if the rolls are identical in land width and spacing, the fabric is identical on both faces. Any part of the web in contact with a heated surface undergoes bonding of one sort or another if the surface is at or above the bonding temperature. If the web which is in contact with the heated surface is under pressure, then primary bonds, ie, the lozenges, are formed; if the contact is without the bonding pressure, then secondary bonds result. In the case of the helical lands, each side of the fabric contains diagonal lines of secondary bonds interrupted by lozenge-shaped primary bonds which pass through the thickness of the fabric and which occur at the intersection of the secondary bonds. Only the primary bonds are common to both sides of the fabric; the helical lands of one work roll can create lines of secondary bonds only on one side of the fabric. The presence of secondary bond lines affects the fabric's flexibility. A fold can occur easily when the fold line is along a bond line. Folding at right angles to the bond line is less easy. Thus, fabrics with bond lines at right angles to each other have hindered flexibility.

The continuous secondary-bond lines reduce the fabric flexibility and, therefore, a further development of duplex bonding has evolved to overcome this problem (59–60). In principle, the use of identical point patterns on both rolls should eliminate all secondary bonding. However, mechanical considerations prevent the use of this solution, ie, such patterns can never be held in perfect register. Thus, the solution has been to use on the rolls point patterns of different pitch which can never mesh and damage the rolls but which, when interacting, generate a point-bond pattern of the required technical and esthetic characteristics. An example is shown in Figure 8 and illustrates the formation of a bark-weave design by the superimposition of two checkerboard designs. No continuous lines of secondary bonding can be formed from either of the two checkerboard rolls; therefore, fabric stiffness is minimized. However, isolated, secondary point bonds are associated with each primary point bond of the bark-weave pattern, and these isolated secondary point bonds are useful in improving the abrasion resistance of the fabric. An intermediate effect may be obtained by the combination of nonmeshing points and lands. Further esthetic improvements may be achieved by the duplex combination of nonmeshing basic point or line-point patterns with specialized half-tone point or line patterns (61).

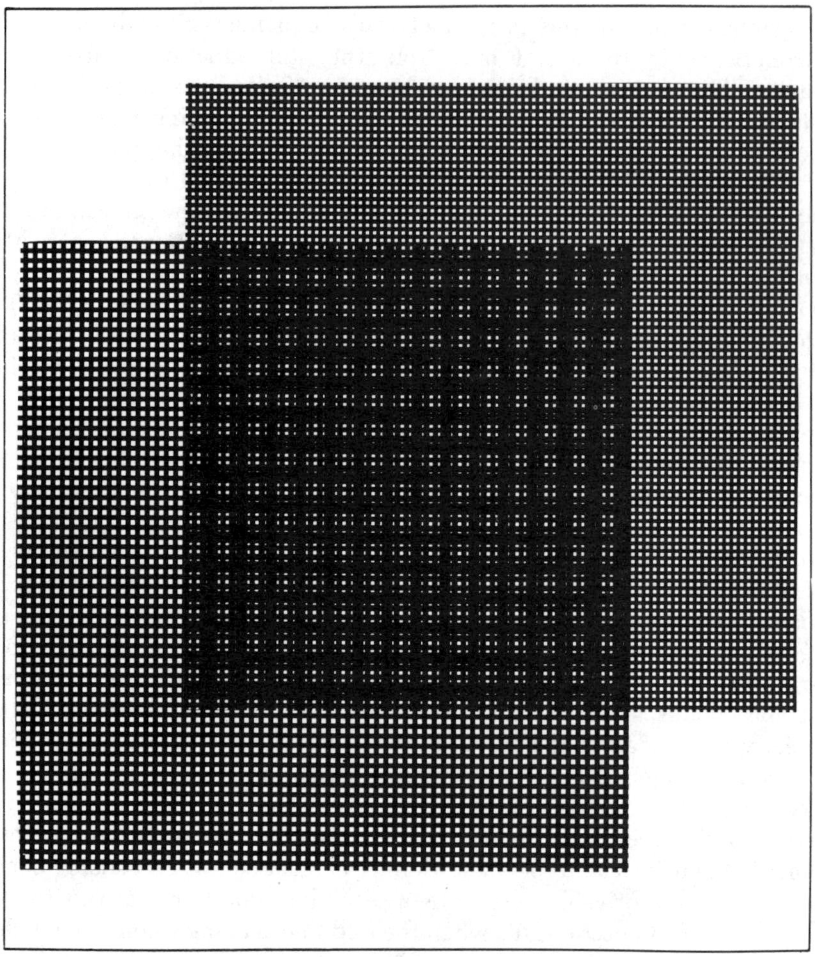

Figure 8. The formation of a duplex bark-weave point-bond pattern by the superposition of two checkerboard designs (53).

The properties of point-bonded fabrics are closely related to those of area-bonded fabrics of the same composition. Point-bonding variables are total bond area, bond spacing, and bond shape; the bond size is determined if the three variables are fixed. The total primary bond area usually is 7–12%. In a duplex system, a primary bond area of 9% implies that the point area on each roll amounts to 30% of the roll area if the design is balanced, ie, equal proportions of secondary bonding form on each side of the fabric. In such a case, 21% of each side of the fabric area is secondary bonded. A primary bond area of 9% is obtained by the interaction of point areas of 20% and 45%. However, in the latter case, one side of the fabric has 11% of its area secondary bonded and the other side, 36%. Increasing the percentage primary bond area moves the fabric properties toward those of area-bonded fabrics, whereas changing the point-spacing or shape has a complex effect on the overall property balance.

Melt-Blown Textiles. Historically, the oldest chemical-to-fabric route is the melt-blown route. Work beginning in the late 1930s provided the basis for processes that are licensed to a number of companies (5–6,62). In this process, the fibers differ from those of the basic process both in diameter, ie, they are microdenier fibers with diameters of 0.1–1 μm, and in continuity, ie, they are staple fibers up to several centimeters long. Products include battery separators, hospital–medical products, ultrafine filters, insulating batting, and semidurable tablecloths.

The manufacturing technique for melt-blown textiles, which is inherently more expensive than that for conventional spunbondeds, consists of extruding the fiber-forming polymer through a single-extrusion orifice directly into a high velocity heated air stream (62–64). A diagram of the typical nozzle arrangement is shown in Figure 9. Typically, air at 300°C and 450 kPa (50 psig) is used. Virtually all of the fiber-forming thermoplastic polymers are suitable for melt blowing, but polypropylene and polyester are used commercially. The fiber diameter depends inversely on the air pressure but the use of higher pressures increases the cost.

After fiber formation, a web is collected by the interposition of a suitable screen conveyor. No binder is used; the fibers are held together by a combination of fiber interlacing and thermal bonding resulting from the residual heat of the extrusion and blowing process. Since the fibers are not drawn in the conventional textile sense, they are unoriented and have a low tensile strength. For some applications, they may be calendered to give thin, microporous sheets. They also may be embossed, vacuum- or thermo-formed, or laminated to other materials. Conventional or net-type spunbondeds may form the conveyor material to provide a high strength, low porosity laminate (62).

Flash-Spun Textiles. Flash-spinning is an alternative technique for the conversion of fiber-forming polymer into fibers. The fibers that are produced are in the form of a three-dimensional network of thin, continuous interconnected ribbons that are <4 μm thick that are termed film-fibrils or plexifilaments (65). Plexifilaments are produced by extruding the fiber-forming polymer through a single orifice as a high temperature, high pressure solution in an inert solvent with a boiling point 25°C or more below the melting point of the polymer. Thus, a yarn bundle is formed without the need for individual spinneret holes. The solvent must not act as a solvent for the

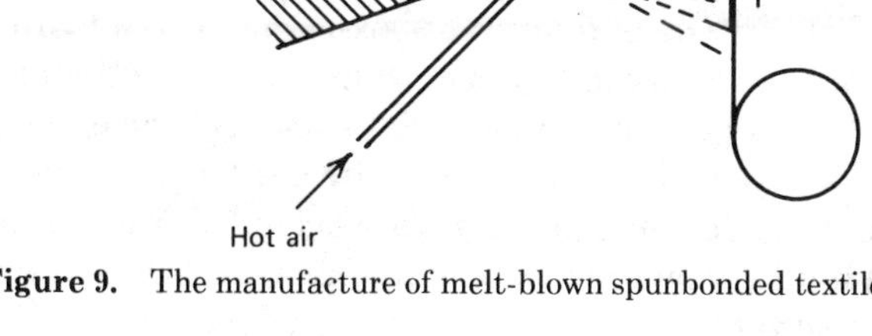

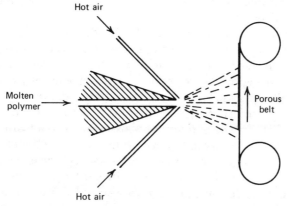

Figure 9. The manufacture of melt-blown spunbonded textiles.

polymer at temperatures below its normal boiling point, unlike solution spinning where solvents are used in which the polymers are stable at temperatures well below the solvent boiling point. Typically, high density polyethylene is used as the fiber-forming polymer and methylene chloride is used as a solvent. A preferred polymer extrusion temperature can be obtained which is within 45°C of the critical temperature of the solvent, ie, 193°C, with polymer concentrations being 2–20%. The presence of a dissolved inert gas, eg, CO_2, is used to increase the degree of fibrillation.

It is necessary to use a polymer or polymer mixture which is capable of crystallizing rapidly to a high degree of crystallization, if high strengths are to be obtained. Since the plexifilaments may be extruded at high speeds, up to nearly 16 km/min, they can have a high orientation and, therefore, a high strength, up to 0.37 N/tex (0.033 gf/den), in the as-spun state. The plexifilament networks may take several forms, depending on the conditions of manufacture. For the manufacture of spunbonded fabrics, the preferred form is one in which the fibrillated strands may be spread transversely into a lacey sheet, when they are hot calendered, to give a strong, paperlike product (4–5,8–9,66–67). Both area- and point-bonding may be used; the latter gives a more flexible product. As in solvent-spinning processes, the recovery of solvent influences the economics. A typical product which is based on this technique is Tyvek, which is used for book covers, maps, banners, signs, flags, tags, labels, packaging materials, and single-use garments for industrial, medical, and consumer applications.

Foam Spinning. In the continuing search for the simplest possible chemical-to-fabric process, several techniques have been proposed for converting extruded film into fabrics by slitting or fibrillating them (2). A technique which is similar to the flash-spinning process but which is related to film extrusion has been patented (68–69). In this process, fiber-forming polymers are extruded through a radial die. However, the molten polymer, typically polypropylene, contains a foaming agent, eg, azodicarbonamide, which evolves gas at the extrusion temperature (see Foamed plastics; Initiators). As the film is cooled immediately after extrusion by a blast of cold air, a foamed film forms and is reheated to above its glass-transition temperature and is stretched. The biaxial stretching or drawing process converts the foamed structure into a balanced fibrous tubular network which may be either collapsed or slit and wound up as textile roll goods. A diagrammatic indication of the equipment used in this process is given in Figure 10. The overall draw ratio, as given by the ratio of the diameters of the die and the pullring, may be 1.5–8; the upper limit is reached when fibers start to break out of the fiber network.

Integral Fibrillated Nets. A technique for the direct conversion of film to fabric or to a textilelike material has been commercialized as the Net 909 process (70–71). Extruded film is embossed while molten by being passed through a pair of heated nip rolls, one of which is engraved so as to form a pattern of usually hexagonal bosses on the sheet. When the embossed film is biaxially stretched, preferential drawing of the thin portions takes place, thereby causing splits to occur in the film and resulting in a sheet material that consists of a series of bosses connected by fibrillar strands.

Direct Extrusion. The possibility of forming useful textilelike materials by the direct extrusion of thick polymeric monofilaments has been exploited in such applications as land reclamation. The principle of direct extrusion is that filaments 0.2–1.5 mm in diameter of fiber-forming polymers are extruded from suitable spinneret orifices, usually in a multicurtain arrangement, and are allowed to drop ca 10 to 20 cm and to collect on a suitable moving collector, eg, a spiked roller (25). The technique

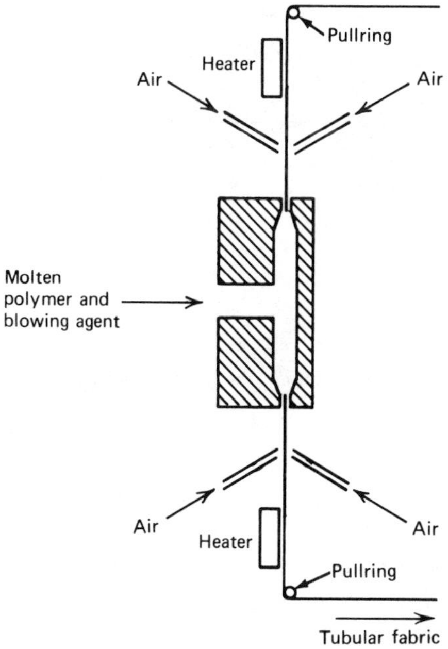

Figure 10. Die and drawing equipment used in the foam-spinning process for the manufacture of nonwoven textiles.

results in the formation of a coarse mat of thick, looped, undrawn monofilaments of any desired mat thickness from 5 mm to 20 mm or more which is suitable for upholstery material, for filtration (qv), as vertical and horizontal drainage, in soil stabilization and antierosion applications, as underwater mats in hydraulic engineering, and for numerous other purposes.

Integral Extruded Nets. The Netlon process for the manufacture of integral extruded plastic nets dates from 1955 (72). This process has been licensed worldwide and the products are familiar as polyethylene or polypropylene net packaging of fruit, agricultural nets, bird netting, plastic fencing, etc. Since the basic patent appeared, many similar but alternative techniques have appeared leading to a wide variety of integral net products (72–73). The basic process involves a spinneret composed of two rotating die members. The spinneret holes are situated at the annular intersection of the faces of the two members such that a proportion of each hole is in each member. As the die members rotate in opposite directions, strands emerging from the die orifices are full thickness only at the instant at which the portions of each hole are coincident (see Fig. 11). The tubular net that is produced is cooled, usually in a water bath, and is stretched slightly over a mandrel before being slit either lengthwise or on the bias to give substantially unoriented roll goods with either diamond or square mesh.

If full orientation of the strands is required, then monoaxial or biaxial stretching, usually with heating, is carried out. With suitable extrusion systems, the extruded product has the appearance of a perforated plastic sheet with 180 or more holes per square centimeter at the extrusion stage and 120 or more after the sheet is passed over the mandrel, or the strands may merge into a nonporous structure. When the material is biaxially oriented, the net structure is regenerated with well-defined strands (13)

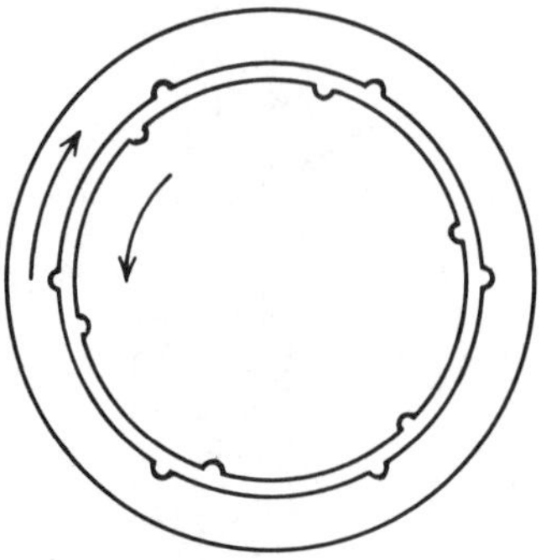

Figure 11. Split die used in the manufacture of integral extruded nets by the Netlon process.

and a relatively high number of holes per square centimeter, perhaps 8 or more, depending on the draw ratio.

Tack-Spun Textiles. The tack-spinning process for the preparation of pile-surfaced products is sometimes used with spunbonded textile substrates (74–75). Fiber-forming polymer that is in the form of an extruded film is fed with a carrier layer, which may be paper (qv), nonwoven textile, etc, into the nip of a pair of rolls which are heated to above the melting point of the film but well below the melting point, if any, of the carrier. When the film contacts the hot roll, it melts and is carried through the nip with the carrier material. As the heated roll and carrier surfaces part, fibers are drawn out of the melt and are immediately cooled by an air blast (76). The pile density and height may be controlled by varying the operating parameters of the process. The fibers that are formed are basically unoriented. Typically, either natural or pigmented polyethylene is used as the fiber-forming polymer and paper is used as the carrier. The product may be used for decorative packaging, hospital disposable applications, etc.

Fabrics From Yarns. In principle, it is possible to prepare nonwoven fabrics from continuous filament yarns or tows, and such split processes could have some advantages although they do not have chemical-to-fabric appeal. The advantages of the process accrue from potentially better matching of the fiber and fabric production rates and improved overall utilization of the plant. A number of papers and patents have appeared for such processes, but no commercial products are available based on such techniques, with the possible exception of a Japanese spunbonded textile (29–30).

Rayon Textiles. The technical knowledge for the manufacture of spunbonded rayon textiles has been available at least since 1972 in the UK and Japan (37,77–80) (see Rayon). However, the relatively low spinning speed (ca 125 m/min) and the increasingly expensive energy requirements relative to moisture removal in a wet process such as for spunbonded rayons lead to unfavorable economics, especially at a time when manufacturers of competitive dry-laid staple nonwovens are seeking binderless or thermal bonding processes to increase process speeds and reduce energy costs.

Cellulose Esters. Like spunbonded rayon fabrics, a process for the manufacture of spunbonded nonwoven textiles from cellulose esters (qv) is available, but is not in major commercial use (18–19). The route was developed for the improved manufacture of cigarette filter rods by obviating the use of additional plasticizers or a curing stage. The production of cellulose ester nonwovens involves the dry-spinning process.

Economic Aspects

In 1979, spunbonded fabrics represented about one third by weight of all nonwovens production in the United States (1). In Europe in 1978, spunbondeds comprised ca 29% of all nonwovens. Their development and manufacture has occurred primarily in the United States and Western Europe. Only small tonnages are produced in Japan and virtually none is exported (81). Some spunbonded fabrics are produced in Eastern Europe but no figures are available and production is thought to be small. Because spunbondeds and nonwovens are new materials, data are not easy to locate nor necessarily reliable, especially as definitions for the various types of nonwovens are not universally agreed upon and cash volumes may be related to roll goods or finished consumer products. However, estimates of world spunbonded fabric production in recent years are summarized in Table 3. The weight of fabric is composed of a number of different basis weights. In the United States in 1977 the average weight was ca 60 g/m^2. Thus, the U.S. market in 1977 was ca 10^9 m^2 of fabric out of a total of approximately 6.0×10^9 m^2 of all types of nonwovens (82). In 1979, over one half of the U.S. spunbonded capacity was based on polypropylene, about one fifth each on polyester and polyethylene, and under a tenth on polyamide (1). Spunbondeds are expected to continue their growth in the United States at a true rate in value terms of 12%/yr during 1979–1982 (1).

Analytical and Test Methods

The simplest test for characterizing the strength of spunbonded textiles is the strip tensile test in which a strip of the material, which usually is 25 or 50 mm wide, is clamped in jaws of the same width in a conventional machine for textile testing. A gauge length of 200 mm is used and the jaws are separated every minute to provide a stress-strain curve of the material (84). An alternative tensile test is the grab tensile test in which a specimen that is considerably wider than the jaws is used (15). Special sample widths may be necessary to achieve reproducible results (51). Another type of tensile test is the plane-strain test (51,85). Grab tensile test stress-strain curves for a number of spunbonded fabrics are illustrated in Figure 12, and the isotropy of a

Table 3. World Production of Spunbonded Fabrics (1973–1979), thousand metric tons

Year	U.S.	Europe[a]	Japan	*Total*
1973		11		
1974	42[b]	17	1.8[e]	*60.8*
1975		20	1.7[e]	
1976		28	2.3[e]	
1977	57[d]	39	3.3[e]	*99.3*
1978	64[c]	44	3.7[f]	*111.7*
1979	84[b]			

[a] European Disposables and Nonwovens Association (EDANA).
[b] Ref. 82.
[c] Ref. 83.
[d] International Nonwovens and Disposables Association (INDA).
[e] Ref. 81.
[f] Ref. 79.

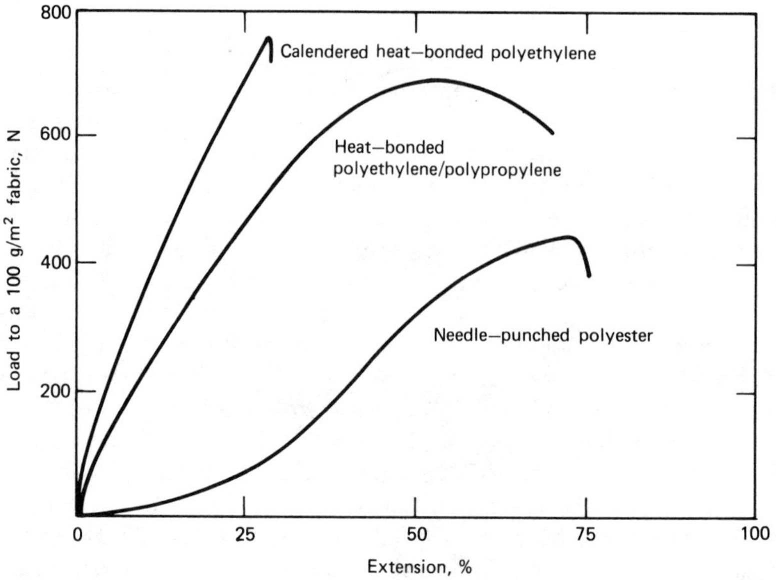

Figure 12. Typical stress-strain curves for spunbonded textiles of various types (ASTM D 1682; 200-mm sample width used).

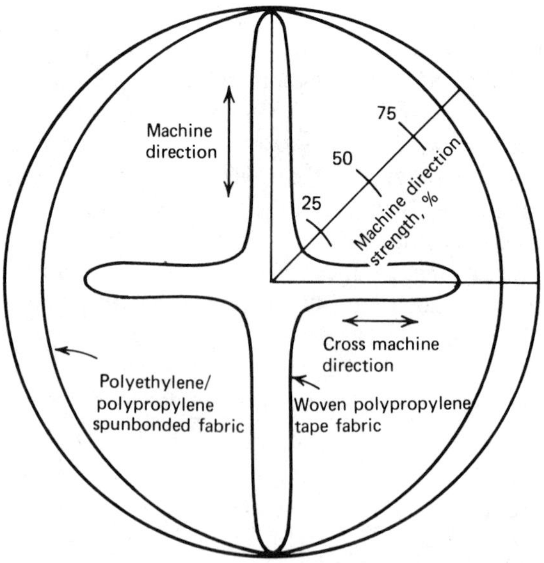

Figure 13. Polar tensile-strength distribution of a typical spunbonded fabric compared with that of a woven fabric (DIN 53857).

typical spunbonded fabric as compared with a woven fabric is illustrated in Figure 13. The work to rupture and the modulus of the fabrics under different types of test situations may be relevant to certain applications. A multidirectional tensile test, eg, the French cylinder test, sometimes may be used when considering the overall fabric resistance to multidirectional stress (86).

Typical measures of tear strength are the trapezoid and wing tear tests; both are carried out at low rates of tearing (15). A test which is commonly used for high tear rates is the falling pendulum or Elmendorf tear test. Although this test is noted as being generally unsuitable for nonwovens, it often is used as the only well-established dynamic tear test (87). Less well known but becoming increasingly adopted in Europe in the civil-engineering field is the cone-drop test, in which the diameter of the puncture that is produced by a heavy pointed cone falling from a specific height is recorded (88).

The burst strengths of spunbondeds may be measured by the Mullen test (89). In civil engineering, a plunger test, which is known as the CBR test because of its relationship to the California Bearing Ratio test and which gives both the strength and deformation to failure, is being used increasingly (86).

Hydraulic tests include measurement of the permeability of the fabric by air and water perpendicular to and in its plane and measurement of the pore size distribution, often by a graded sieving technique (15,51). The void ratio of the fabric, or its porosity, may be calculated from the thickness and the basis weight (51).

Health and Safety Factors

Considerations relevant to textiles of similar composition, which generally are inert, nontoxic, and nonallergenic, probably are broadly applicable. Flammability, allergic reactivity, food contact considerations, linting in surgical applications, etc, should be considered for health and safety. Specialized information usually is available from manufacturers and from the literature.

Uses

The applications for nonwoven textiles often are categorized as disposable or durable; disposable may be equated with dry or wet-laid staple-fiber nonwovens and durable applications often involve spunbonded fabrics. In the United States, disposable applications accounted for ca 4.4×10^9 m^2 of fabric and durable applications accounted for 1.6×10^9 m^2 of fabric in 1977 (12,82). About 60% of the total volume of disposable textiles was used in the cover-stock field for items, eg, diapers, sanitary napkins, tampons, underpads, etc. Until recently, very little spunbonded fabric was used as cover stock although, because of its dominance in volume terms, it is a market that is attractive to the spunbonded producer (90). The lightweight (15–25 g/m^2) fabrics for this market have well-established, demanding specifications and are highly price competitive and sell at only a few cents per square meter. At a total market value of $\$1.55 \times 10^8$, the average price in 1977 was ca $\$0.05/m^2$ compared to an average price for fabrics for all durable applications of ca $\$0.27/m^2$. In Europe, sanitary and medical applications accounted for approximately 16% by weight of total nonwovens fiber consumption compared to ca 32% in the United States (91).

A detailed breakdown of nonwoven-fabric applications is given by market sector in Table 4. In order of importance, the spunbonded market sectors are carpet, bedding/home furnishing, disposable apparel and durable paper, coated fabrics, furniture components, interlinings, and civil-engineering fabrics. Spunbondeds also are marketed in smaller proportions in a number of other sectors and in many miscellaneous applications.

Table 4. Estimated U.S. Consumption of Nonwoven Textiles by Market in 1977[a]

Application	Consumption				Spunbonded content
	10^6 m²	10^3 t	10^6 $	%	
coverstock	2600	64	155	17.7	very small
surgical packs	460	34	90	10.3	small
interlinings	195	24	72	8.2	small
coated fabrics	155	22	62	7.1	intermediate
wipes	376	22	59	6.8	very small
filters	343	18	55	6.3	intermediate
carpet components	343	23	55	6.3	large
bedding/furnishings	272	19	52	5.9	intermediate
disposable apparel	125	8	21	2.4	large
durable papers	134	7	20	2.3	large
furniture	46	5.4	14	1.6	intermediate
fabric softeners	176	3.6	13	1.5	very small
civil engineering	13	2.3	7	0.8	large
other disposable	269	16	42	4.8	intermediate
other durable	403[b]	42	157	18.0	intermediate
Total	*5910*[b]	*310.3*	*874*	*100.0*	

[a] Ref. 12.
[b] Ref. 82.

The first spunbonded fabrics were developed for use in the carpet sector to replace jute which was used as a primary carpet backing; however, a number of factors prevented the market penetration that had been planned (62). Spunbonded primary backing entered segments of the carpet market where it offered unbeatable technical advantages, eg, a backing for printed carpets because of its superior dimensional stability and for fine gauge tufting for which its finer filament structure is superior to woven tape fabrics. Spunbondeds also are used in the carpet field as secondary backings, where the esthetics of melt-pigmented spunbonded fabrics are advantageous, and as an underlay carrier, where strength and lightness, hence, cost competitiveness, are beneficial. The carpet sector annual growth rate is leveling and 5% growth in weight of product consumed was forecast for the period 1977–1980 (82).

An area of much greater potential growth is in furnishings and automotive components. The former, which is termed hidden furniture fabrics, was expected to show a growth of 80% in value from 1977 to 1980 (12); another report forecast an annual 18% volume increase in coated and laminated fabrics for the latter outlets from 1977 to 1981 (82). Furniture applications are basically structural and include spring pocketings, dust covers, cushion backings, bottoming cloths, skirt liners, and pull strips (92). In the automotive area, there are eight main applications: paneling, visors, seat backing, seat listing, insulators, substrates for coated or laminated upholstery, landau tops, and trunk linings. Spunbonded fabrics comprise 36% of the nonwovens market in furniture applications and 43% of the nonwovens automotive market. The penetration of nonwovens into the automotive market is expected to be accelerated by the energy crisis since weight reductions may be obtained by their use.

Many durable nonwoven components are used in apparel which were made from woven materials 10–15 yr ago (93). These include metal fly-clasp reinforcements, belt-loop linings, pocket stays, waistbands, shirt collars, center placket and cuff interlinings. Probably only a small proportion of these nonwovens is spunbonded. One use for spunbondeds is as a fusible binder. Some spunbondeds are used in shoes as internal components and in luggage and accessories (92) (see Leatherlike materials). The advantages of nonwovens in these areas are improved stability, ease of handling, isotropy and uniformity, and suitability to automation with consequent savings in labor process steps and energy. Growth in these areas is not expected to be dramatic (82). Very significant growth is anticipated, however, in the disposable apparel area which is based on spunbonded fabrics and which is expected to double in value by 1981 compared to the 1977 level (12). The advantages of disposable garments are appreciated in the nuclear field, hospitals, laboratories, and many fields where smooth-surfaced overalls, smocks, laboratory coats, etc, resist contamination and, when contaminated, may be inexpensively and safely disposed of.

A 26 vol % annual growth rate in the bedding and home-furnishing area has been forecast in the United States (82). A significant proportion of nonwoven fabric used in this field is spunbonded and much of this may be point bonded. Typical uses are mattress pads, quilts, bedspreads, draperies and headers, and other accessory fabrics (12).

Civil-engineering fabrics form the final and major growth market for spunbonded fabrics; as a proportion of the total U.S. market for nonwovens, this sector is still quite small and was thought at ca 2.3×10^7 m^2 to be about half the size of the European civil-engineering fabric market in 1978 (94). However, it is expected to grow to ca $(1.0–1.5) \times 10^8$ m^2 in North America by 1985 when it will equal the European market. Use of nonwovens in civil engineering in the rest of the world is expected to equal that in Europe and North America, producing a total world market of $(3–4) \times 10^8$ m^2 in 1985. Virtually all of the fabrics used in this application are spunbonded. The application considerations for nonwoven fabrics in civil-engineering uses have been described (51). The fabric performs three functions when installed in a civil engineering or a geotechnical application, ie, separation, filtration, and reinforcement. Specifically, these functions provide embankment stabilization, drainage, and improved soil/fabric composite strength.

Spunbonded civil-engineering fabrics have expanded into a multitude of related civil-engineering and agricultural uses, eg, erosion control, revetment protection, railroads, canal and reservoir lining protection, highway and airfield blacktop cracking prevention, etc. National and international specifications are being agreed upon and established (10). Largely as a result of cooperation in producing fabrics of the appropriate specification with the civil-engineering industry by leading spunbondeds producers, geotextiles are being included in civil-engineering projects to an increasing extent at the project-design stage (10). The particular properties of spunbondeds which are responsible for this revolution are chemical and physical stability, high strength/cost ratio, and their unique and highly controllable structure which can be tailored to provide the desired characteristics, especially with respect to water and soil permeabilities. Spunbonded fabrics also are used for agricultural shading and insulation and other agricultural uses, wall coverings, medical and surgical disposable reinforcements, decorative packaging, capillary mattings, reinforced plastics, textile linings, electrical and building applications, wipes, scrims, blinds, battery separators, sound absorbents, gaskets, abrasives, composites, etc (see Insulation, acoustic).

Outlook

There is no clear agreement on the detailed direction of the growth of the nonwovens industry. Disposables may outgrow durables from 1978–1988 (83), although durables may lead disposables in short-term (1977–1981) growth. All authorities agree that nonwovens will grow at a significant rate in the next few years and that a real term sales-value growth figure of ca 8% will apply to which will be added a price inflation of ca 7%/yr (1,10–12,82–83,95–96). If durables do lead, it is likely that the newer techniques, eg, spunbondeds, will spearhead the growth and it is considered that this technology and spunlaid fabrics will grow at 19%/yr from 1978 to 1982 (1,12). Spunbonded lines are expensive and, as the output may be (1–4) $\times 10^7$ m^2/yr per line, very firm and profitable outlets must be anticipated for investment to be made. Thus, both technical and economic factors are retarding the technical progress of spunbondeds, and market growth in the near future is likely to occur by expansion of the applications for fabrics produced by existing technology.

BIBLIOGRAPHY

1. J. R. Starr, *European Disposables and Nonwovens Association*, AGM Paper, Munich, June 12, 1980.
2. A. Newton and J. E. Ford, *Text. Prog.* **5**(3), 1 (1973).
3. L. Hartman, *Text. Manuf.* **101,** 26 (Sept. 1974).
4. O. L. Shealey, *Text. Inst. Ind.* **9,** 10 (Jan. 1971).
5. F. F. Hand, Sr., *Text. Ind.* **143,** 86 (July 1979).
6. R. G. Mansfield, *Text. World* **129,** 83 (Feb. 1979).
7. Brit. Pat. 836,555 (Nov. 9, 1955), F. B. Mercer (to Plastic Textile Accessories Limited).
8. R. G. Mansfield, *Text. World* **127,** 81 (Sept. 1977).
9. W. W. Powell, *Mod. Text.* **LVII,** 39 (July 1977).
10. *Text. Manuf.* **79**(2), 19 and **79**(3), 26 (1979).
11. *Text. World* **129,** 57 (Aug. 1979).
12. J. R. Starr, *Am. Dyest. Rep.* **68,** 45 (March 1979).
13. M. E. Denham, *Mod. Text.* **LX,** 8 (June 1979).
14. *Pulp and Pap. Int.*, 59 (Mar. 1978).
15. *INDA Standard Test Methods*, International Nonwovens and Disposables Association, New York.
16. *Softness—Hand Research Study*, International Nonwovens and Disposables Association, New York, 1974.
17. J. Skelton, *Text. Res. J.* **46,** 862 (1976).
18. U.S. Pat. 3,148,101 (Sept. 8, 1964), W. T. Allman and co-workers (to Celanese Corporation).
19. U.S. Pat. 3,669,788 (June 13, 1972), W. T. Allman and co-workers (to Celanese Corporation).
20. A. E. Pedder and S. J. D. Hay, *Nonwovens Year Book 1976*, Texpress, Stockport, UK, pp. 35, 62.
21. Brit. Pat. 1,157,437 (July 9, 1969), B. L. Davies (to Imperial Chemical Industries Limited).
22. U.S. Pat. 2,836,576 (May 27, 1958), J. A. Piccard and F. K. Signaigo (to E. I. du Pont de Nemours & Co., Inc.).
23. U.S. Pat. 3,855,045 (Dec. 17, 1974), R. J. Brock (to Kimberley Clark Corporation).
24. Brit. Pat. 993,920 (June 2, 1965), P. J. Couzens (to Imperial Chemical Industries Limited).
25. Neth. Pat. 7,607,664 (Jan. 11, 1977), A. Rasen and co-workers (to Akzo NV).
26. Ger. Pat. 2,812,429 (Sept. 14, 1978), H. A. Booker, B. L. Davies, A. J. Hughes, and C. Shimalla (to Fiber Industries Inc.).
27. U.S. Pat. 3,338,992 (Aug. 29, 1967), G. A. Kinney (to E. I. du Pont de Nemours & Co., Inc.).
28. U.S. Pat. 3,771,307 (Nov. 13, 1973), D. G. Petrille (to E. I. du Pont de Nemours & Co., Inc.).
29. R. Krĉma, *Textiltechnik* **28,** 775 (1978).
30. Jpn. Pat. 48 056,967 (Aug. 10, 1973), (to Teijin Ltd.).
31. Brit. Pat. 1,288,802 (Sept. 13, 1972), W. G. Parr (to Imperial Chemical Industries Limited).
32. U.S. Pat. 3,855,046 (Dec. 17, 1974), P. B. Hansen and B. Pennings (to Kimberly Clark Corporation).

33. U.S. Pat. 3,989,788 (Nov. 2, 1978), L. L. Estes, Jr., A. F. Fridrichsen, and V. S. Koshkin (to E. I. du Pont de Nemours & Co., Inc.).
34. *Nonwovens Rep. Int.*, 1 (Oct. 1979).
35. V. D. Freedland, *Nonwovens Ind.* **10,** 33 (Aug. 1979).
36. U.S. Pat. 3,991,244 (Nov. 9, 1976), S. C. Debbas (to E. I. du Pont de Nemours & Co., Inc.).
37. Brit. Pat. 2,006,844 (Oct. 26, 1977), P. M. Ellis and R. D. Gibb (to Imperial Chemical Industries Limited).
38. Ger. Pat. 2,785,158 (July 6, 1972), H. O. Dorschner and co-workers (to Metallgesellschaft AG).
39. Brit. Pat. 1,436,545 (May 19, 1976), J. Brock (to Imperial Chemical Industries Limited).
40. Ger. Pat. 2,706,976 (Dec. 7, 1978), V. Semjonow (to Hoechst AG).
41. Ger. Pat. 1,560,790 (Mar. 27, 1975), L. Hartmann (to Lutravil Spinnvlies GmbH and Co.).
42. Brit. Pat. 1,231,066 (May 5, 1971), C. H. Weightman (to Imperial Chemical Industries Limited).
43. U.S. Pat. 4,163,305 (Aug. 7, 1979), V. Semjonov and J. Foedrowitz (to Hoechst AG).
44. Ger. Pat. 2,421,401 (Mar. 22, 1979), K. Mente, W. Raddatz, and G. Knitsch (to J. H. Benecke GmbH).
45. U.S. Pat. 4,017,580 (Apr. 12, 1977), J. Barbey (to Rhone-Poulenc Textile).
46. Neth. Pat. 7,710,470 (Mar. 28, 1979), D. J. Viezee, P. J. M. Mekkelholt, and J. A. Juijn (to Akzo NV).
47. K. West, *Chem. Br.* **7,** 333 (Aug. 1971).
48. D. Ward, *Nonwovens Yearbook 1979*, Texpress, Stockport, UK, pp. 19–26.
49. Ger. Pat. 1,966,031 (May 19, 1971), H. O. Dorschner, F. J. Carduck, and C. Storkebaum (to Metallgesellschaft AG).
50. U.S. Pat. 3,322,607 (May 30, 1967), S. L. Jung (to E. I. du Pont de Nemours & Co., Inc.).
51. *Designing with TERRAM*, ICI Fibres, 'Terram Group,' Pontypool, Gwent, UK, Aug. 1978.
52. *Nonwovens Rep. Int.*, 1 (Jan. 1980).
53. K. Porter, *Phys. Technol.* **8,** 204 (Sept. 1977).
54. U.S. Pat. 4,125,663 (Nov. 14, 1978), P. Eckhardt (to Hoechst AG).
55. U.S. Pat. 4,181,640 (Jan. 1, 1980), G. P. Morie, C. H. Sloan, W. J. Jackson, Jr., and H. F. Kuhfuss (to Eastman Kodak Company).
56. U.S. Pat. 3,542,615 (June 16, 1967), E. J. Dobo, D. W. Kim, and W. C. Mallonee (to Monsanto Company).
57. U.S. Pat. 4,075,383 (Feb. 21, 1978), R. M. Anderson and co-workers (to Monsanto Company).
58. L. G. Kasebo, *International Nonwoven and Disposable Association, Technical Symposium Paper*, Washington, Mar. 1974.
59. Brit. Pat. 1,474,101 (May 18, 1977), D. C. Cumbers and D. Williams (to Imperial Chemical Industries Limited).
60. Brit. Pat. 1,474,102 (May 18, 1977), D. C. Cumbers and D. Williams (to Imperial Chemical Industries Limited).
61. Brit. Pat. 1,499,178 (Jan. 25, 1978), K. Porter (to Imperial Chemical Industries Limited).
62. *Nonwovens Ind.* **10,** 10 (Sept. 1979).
63. U.S. Pat. 3,972,759 (Aug. 3, 1976), R. R. Buntin (to Exxon Research and Engineering Co.).
64. U.S. Pat. 4,168,138 (Sept. 18, 1979), D. McNally (to Celanese Corporation).
65. U.S. Pat. 3,081,519 (Mar. 19, 1963), H. Blades and J. R. White (to E. I. du Pont de Nemours & Co., Inc.).
66. U.S. Pat. 3,442,740 (May 6, 1969), J. C. David (to E. I. du Pont de Nemours & Co., Inc.).
67. U.S. Pat. 4,069,078 (Jan. 17, 1978), M. D. Marder, R. Osmalov, and G. P. Pfeiffer (to E. I. du Pont de Nemours & Co., Inc.).
68. U.S. Pat. 4,085,175 (Apr. 18, 1978), H. W. Keuchel (to PNC Corporation).
69. Jpn. Pat. 53 078,374 (Dec. 20, 1976), (to Teijin Ltd.).
70. Brit. Pat. 914,489 (July 19, 1960), D. E. Seymour and D. J. Ketteridge (to Smith and Nephew Research Limited).
71. Brit. Pat. 1,548,865 (July 18, 1979), A. G. Patchell, W. O. Murphy, and R. Lloyd (to Smith and Nephew Research Limited).
72. U.S. Pat. 3,959,057 (May 25, 1976), J. J. Smith.
73. U.S. Pat. 4,123,491 (Oct. 31, 1978), R. L. Larsen (to Conwed Corporation).
74. Ger. Pat. 1,902,880 (Aug. 20, 1970), K. Seiffert.
75. Brit. Pat. 1,503,669 (Mar. 15, 1978), A. A. A. Giovanelli and E. W. Schmidt (to Imperial Chemical Industries Limited).

76. Brit. Pat. 1,378,638 (Dec. 27, 1974), D. M. Fisher and co-workers (to Imperial Chemical Industries Limited).
77. M. J. Welch and J. A. McCombes, *Nonwovens Ind.* **11,** 10 (May 1980).
78. *Nonwovens Rep. Int.*, 5 (Sept. 1978).
79. *Nonwovens Rep. Int.*, 3 (Jan. 1980).
80. *Textile World* **129,** 101 (Nov. 1979).
81. S. Tauchibayashi, *Nonwovens Ind.*, 14 (Mar. 1980).
82. L. E. Seidel, *Textile Ind.* **142,** 41 (Jan. 1978).
83. *Mod. Text.* **LX,** 7 (Oct. 1979).
84. *DIN 53857, B1.2*, DIN Deutsches Institut für Normung eV., Berlin, FRG.
85. C. R. Sissons, *C.R. Coll. Int. Sols. Text.* **II,** 287 (1977).
86. W.Wilmers, *Str. Autobahn* **31**(2), 69 (1980).
87. *ASTM D 1424-63* (*Reapproved 1975*), Part 32, ASTM Standards, American Society for Testing and Materials, Philadelphia, Pa., 1979, p. 352.
88. S. L. Alfheim and A. Sorlie, *C.R. Coll. Int. Sols. Text.* **II,** 333 (1977).
89. *ASTM D231-62* (*Reapproved 1975*), Part 32, ASTM Standards, American Society for Testing and Materials, Philadelphia, Pa., 1979, p. 119.
90. D. K. George, *Nonwovens Ind.* **11,** 22, 43, 49 (Mar. 1980).
91. *Nonwovens Rep. Int.*, 4 (Apr. 1979).
92. *Text. Ind.* **143,** 66 (Feb. 1979).
93. F. F. Hand, Sr., *Text. Ind.* **144,** 92 (Feb. 1980).
94. R. G. Mansfield, *Text. World* **128,** 52 (Dec. 1978).
95. *Mod. Text.* **LX,** 11 (May 1979).
96. *Mod. Text.* **LX,** 27 (July 1979).

K. PORTER
ICI Fibres

STAPLE FIBERS

Nonwoven textile fabrics are porous, textilelike materials, which usually are in sheet form, are composed primarily of fibers, and are manufactured by processes other than spinning, weaving, knitting, or knotting (see Textiles). These materials also have been called bonded fabrics, formed fabrics, or engineered fabrics. The thickness of the sheets may vary from 25 μm to several centimeters. The weight may vary from ca 5 g/m^2 to ca 1 kg/m^2. The texture or feel may range from soft to harsh. The sheet may be compact and crisp as paper, supple and drapable as a fine conventional textile, or high loft like natural down; also, it may be highly resilient or limp. Its tensile properties may range from barely self-sustaining to impossible to tear, abrade, or damage by hand. The fiber components may be natural or synthetic, from 1–3 mm to essentially endless. The total composition may consist only of one type or a mixture of fibers within which frictional forces provide the basis for the product's tensile properties. However, the finished product may include a lesser or greater proportion of film-forming polymeric additive functioning, which is an adhesive binder, to impart these properties. All or some of the fibers may be welded chemically, ie, with a solvent, or physically, ie, with heat. A scrim, gauze, netting, yarn, or other conventional sheet material may be added to one or both faces or embedded within the nonwoven as reinforcement.

The above description does not exhaust the possible variations in composition and in physical properties which are or can be made. A wide range of products exist, with paperlike materials at one extreme to woven fabrics at the other.

There have been many attempts to formulate a definition of nonwoven fabrics as well as a more affirmative name (1–3). Felted fabrics from animal hairs, eg, wool (qv), are excluded from the class, even though their structure fulfills descriptions of nonwoven textile fabrics.

Conventional textile fabrics are based on yarns and, in special cases, monofilaments. Yarns are composed of fibers which have been parallelized and twisted by a process called spinning to form cohesive and strong one-dimensional elements. In making textile fabrics, the yarns (or the monofilaments) are interlaced, looped or knotted together in a highly regular repetitive design in any of many well-known ways to form a fabric (see Fig. 1). The fabric strength and other physical properties are derived from friction of individual fibers against each other in each yarn, and friction between adjacent yarns.

The present nonwoven technology started ca 1938. Nonwovens are derived from conventional textile technology where long, ie, staple, fibers are used (see Fig. 2), and conventional paper (qv) technology where short fibers are used. In contrast to conventional textiles (4–5), the base structure of all nonwovens is a fibrous web; thus, the basic element is the single fiber. The individual fibers are arranged randomly. Tensile and other, eg, stress–strain and tactile, properties are imparted to the web by adhesive or other chemical and physical bonding of fiber to fiber; fiber-to-fiber friction which, primarily, is created by entanglement; and/or reinforcement of the web by added structures, eg, woven and knitted fabrics, yarns, scrims, nettings, films, and foams.

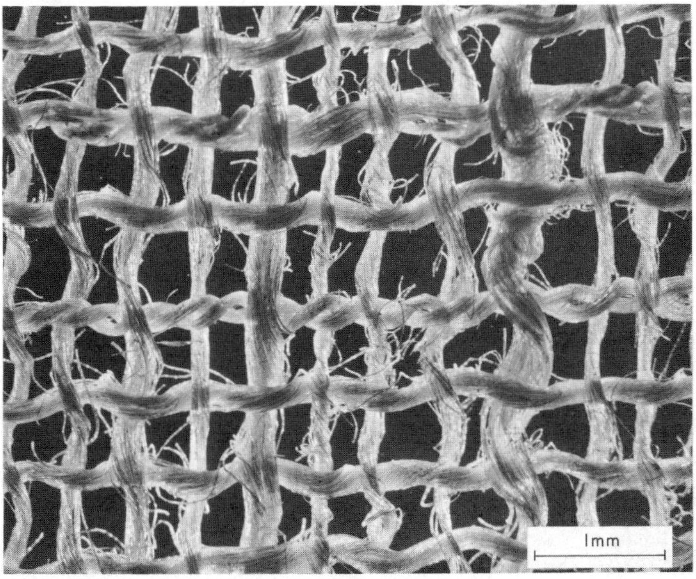

Figure 1. An ancient Egyptian woven linen fabric which closely resembles a modern woven gauze, except for the improved uniformity of weight and twist in modern yarns.

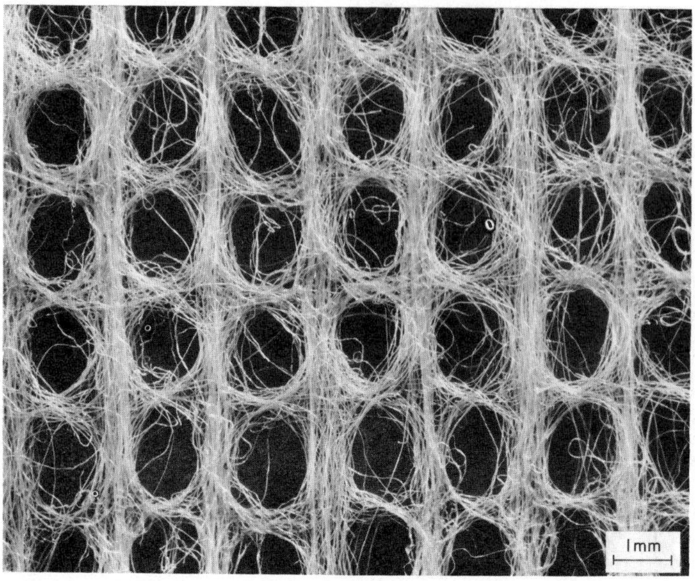

Figure 2. A rearranged card web of rayon fibers also resembles a woven gauze. Billions (10^9) of meters of nonwovens, such as shown here, have been produced.

The modern history of staple nonwoven textile fabrics began with the publication of patents representing two different approaches for the manufacture of flexible bonded sheets. The first, issued in 1936, describes an intermittently bonded nonwoven fabric made by printing an adhesive in a predetermined repetitive pattern onto an oriented fiber web (6). The bonded areas provide strength and the unbonded areas, where the fibers are free to move relative to each other, ensure a degree of softness (see Fig. 3). The second approach was patented in 1942 and describes a method for making a flexible bonded sheet using a random mixture of potentially fusible and nonfusible fibers (7). A sheet which is held together by a system of random intermittent adhesive bonds is produced when the web is subjected to physical or chemical conditions to selectively soften the fusible fibers while appropriate pressure is applied. Nonwovens also were made by saturating webs with water-based latexes which then were heat-dried.

There are a few basic elements which must be varied and controlled to produce the great range of nonwoven fabric types which are available. These include: the fibers, including chemical types and physical variations; the web and the average geometric arrangement of its fibers as predetermined by its method of forming and subsequent processing; the bonding of the fibers within the web, including the properties and geometric disposition of the adhesive binder or, alternatively, the frictional forces between fibers, primarily as created by close fiber contact or entanglement; and reinforcements, eg, yarns, scrims, films, and nettings. In practice, each element can be varied and, thus, can exert a powerful influence, alone and in combination, on the final fabric properties.

Staple fibers originally were defined as the approximately longest or functionally the most important length of natural fibers, eg, cotton (qv) or wool (8). In the present context, as applied to regenerated cellulose and synthetic fibers, staple fibers are of

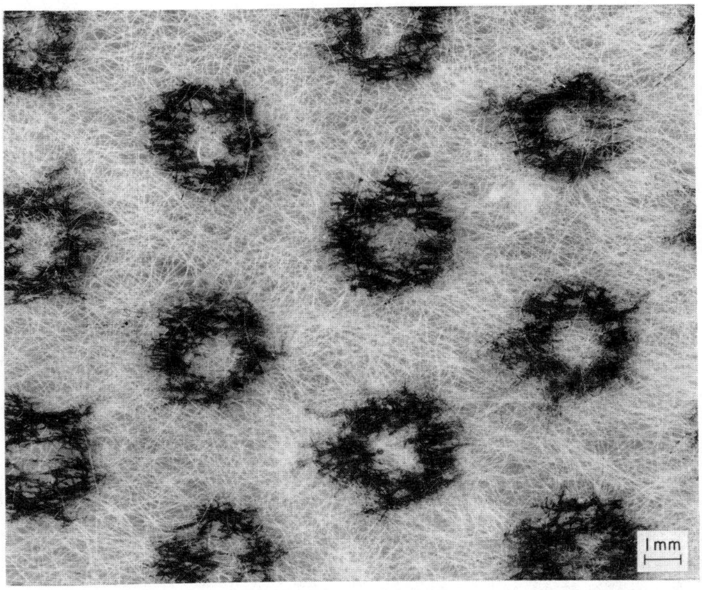

Figure 3. This is a typical print-bonded nonwoven fabric, made from card web and bonded by the rotogravure process with poly(vinyl acetate) latex. The binder, in a regular pattern of annular circles, is stained differentially by a solution of I_2.KI.

relatively uniform length, ca 1.3–10.2 cm, and can be processed on conventional textile machinery. Since regenerated and other extruded fibers essentially are endless as formed, they are cut during the manufacturing process to a specific length to meet a processing or market need. Extruded fibers also are produced as continuous filaments. Woven or nonwoven fabrics that are made from staple fibers appear fuzzier and feel softer and fuller than fabrics that are made from endless filaments. The processes for forming webs from staple fibers are entirely different than those in use for continuous filaments. Staple and filament fiber webs lead to products which differ substantially in their properties.

Components and Processes

Fibers. The fibers, as defined by their chemical composition and as a result of their physical–mechanical properties, determine the ultimate fabric properties. Other determinants, such as web structure and bonding maximize inherent fiber characteristics, eg, strength, resilience, abrasion resistance, chemical properties, and absorbency or hydrophobicity.

Any of the natural fibers which are available in commercial quantities can be used. In practice, wood pulp, which is far shorter than staple length, is the only one used in large amounts because of its high water absorbency, bulk, and low cost. Occasionally it is used in mixtures with staple-length rayon (qv) or polyester and, less often, with nylon (see Polyamides). Cotton and manila hemp also are used but in small amounts. Cotton has excellent inherent properties for use in nonwovens (9). Recent simplifications in cotton-fiber cleaning and bleaching should increase its commercial usage (10). Fibers such as wool, silk (qv), and linen are far too costly for serious consideration.

Regenerated fibers include viscose rayon and cellulose acetate (see Cellulose acetate and triacetate fibers). Viscose rayon was the overwhelmingly predominant fiber in use until recent years, when its cost rapidly escalated and new uses were developed for which rayon is not suitable. It is an easy fiber to process into webs and is bonded readily by several different kinds of adhesives, notably latexes, and by mechanical entanglement. It is available as regular and high wet modulus types; both, but primarily regular modulus, are used. However, it is not a strong fiber and this deficiency is evident in products made from it. Rayon is almost ideal for surgical and sanitary products, filters for food and industrial solvents, wiping cloths, certain lightweight facings, hand towels, and hospital garments. Common denominators for these uses are biodegradability, compatibility with body tissues, inertness to common solvents, good water absorbency, and moderate cost; therefore, they are well-suited for disposability. In recent years, viscose-rayon manufacturing plants in the United States and Western Europe have been closing because of increasing cost of the fiber. Considerable research is underway to replace the viscose process and to improve rayon fiber properties (11–12). Products comprised of wood pulp are gaining in importance because of the latter's low cost and equal purity–absorbency properties. However, the short fiber length, 1–3 mm of common commercial grades, creates problems in bonding, if product softness is needed. Polyester fibers (qv) are now (1981) the fibers used in the greatest quantity in staple form.

All synthetic fibers are potentially available for use. However, only polyester poly(ethylene terephthalate), nylon (types 6,6 and 6), vinyon, polypropylene (see Olefin polymers), and poly(vinyl alcohol) are significant commercially (see Vinyl polymers); polyester fibers represent by far the largest volume usage. Polyester cost is slightly lower than viscose rayon, and its strength and resilience are superior. Polyester fibers are hydrophobic which is desirable in lightweight facing fabrics, eg, in disposable diapers. They impart an easily perceptible dry feel to the facing even when the inner absorbent wood pulp is saturated. As practical methods are developed to process and bond polyester fibers, they are displacing rayon fiber facings in the marketplace.

Nylon fibers have been used in nonwovens, eg, in garment interlinings, because of their strength and resilience. However, their cost is substantially higher than polyester and they do not have sufficient compensating properties for many uses; therefore, they are used in staple form in moderate amounts.

Polypropylene fibers are strong, low in cost, and highly hydrophobic. Because of their low density, their yields, ie, high polymer volume, large diameter, is better than other fibers. Even though they are derived ultimately from petroleum, their projected future cost is favorable (see Petroleum, products). Polypropylene is difficult to bond with the widely used latexes, but effective binders are being developed. When the bonding problem is solved, staple polypropylene fibers probably will be used in large amounts (see Fibers, chemical; Olefin fibers).

Poly(vinyl alcohol) fibers, which are cross-linked using formaldehyde, are manufactured in substantial commercial quantities and only in Japan. They are strong, water insoluble, and hydrophilic. They are comparable to rayon in their response to latex binders and produce nonwovens which are slightly stiffer, slightly less absorbent but substantially stronger (13). The comparison of original fiber strengths of rayon with poly(vinyl alcohol) and the tensile properties of the respective finished fabrics illustrate the retention of fiber properties in the finished nonwoven. Poly(vinyl alcohol) fibers are not widely used because of unfavorable availability and cost.

Cellulose acetate fibers, although moderately inexpensive, are weak and difficult to bond. Hence, these fibers are not used in significant amounts.

In the early stages of the nonwoven industry, the staple fibers which were used were, with minor exceptions, those available for the conventional textile industry. The exceptions were the small amounts of thermoplastic fibers made primarily as nonwoven binders in the 1940s and 1950s, ie, plasticized cellulose acetate for fusible–nonfusible fiber mixtures (7) and vinyon HH fibers of poly(vinyl chloride)/acetate for wet-formed nonwovens used for tea bags. In recent years, the fiber industry has provided polypropylene fibers and undrawn or only slightly drawn polyester fibers as more effective thermoplastic-fiber binders (14).

Manufacturers in England (ICI), Japan (Chisso), and the U.S. (Enka) have developed quite effective binder fibers especially for the nonwoven industry. These are bi-component fibers characterized by a low melting skin and a high melting core. They appear to be, by far, the best thermoplastic binder fibers made to date. When the binder fibers are heated, the skin melts or softens to bond to adjacent fibers but the core is not affected. The fibers can be used alone or blended with conventional fibers. Bonding can be effected by even heat and pressure or in a pattern.

Several manufacturers have been developing modified cellulose fibers that are made with a microcrimp superimposed on a conventional fiber crimp (11), that are produced from cellulose solvent processes other than viscose (11,15), that have a round or irregular crosssection with a correspondingly shaped lumen to increase absorbency (16), and that are alloyed with sodium polyacrylate to increase absorbency (American Enka, Avtex) (11) (see Acrylic and modacrylic fibers; Acrylic acid).

In addition to the substantial numbers and variety of natural, regenerated, and synthetic fibers which are readily available, mechanical differences are introduced intentionally in extruded fibers to affect processing conditions and web and final product properties. The variations which may be produced include fiber length, diameter, crimp, cross-section, spin finish, draw ratio, and inclusion of delustering agent.

Length. Synthetic and regenerated fibers are made by extrusion; therefore, they can be prepared in any length by appropriate cutting operations. Extruded fibers are furnished in precision cut, uniform lengths of ca ±0.3 cm. Staple extruded fibers are provided in lengths of ca 2–6 cm. Fibers that are intended for wet-processing into webs are cut much shorter.

Diameter. Regenerated and synthetic fibers are produced in a wide range of diameters for conventional textile uses; these diameters also are accepted and used in making nonwovens. Fiber diameter is described by tex, which is the weight of a 1000 meters of filament. The practical metric unit is decitex (= 0.9 den), ie, the weight of 10 km of filament. Fiber diameter is directly proportional to the square root of its tex and industry proportional to the specific gravity of the fiber polymer.

The fiber staple length usually is chosen for convenience in the nonwoven manufacturing process; when longer than ca 25 mm, length has a small effect on product properties. Decitex, by contrast, has a great effect on properties of the finished fabric and is carefully selected. Fiber decitex, hence diameter, affects the hand of the fabric; bonding properties, whether by adhesive or by friction; fabric cover or opacity, especially in light weights; fabric resilience and web thickness; fabric liquid or gas filtering properties; and machine processing requirements.

The diameter of typical elliptically shaped fine United States cotton fibers is ca

8 μm (minor axis) and 20 μm (major axis), which is comparable to viscose rayon fibers of 1.67 dtex (1.5 den) (17). Such a decitex yields satisfactory products and is convenient for processing. However, the proliferation of end uses has required the use of higher decitex (den) fibers for many products; thus, there is no standard decitex (den).

Crimp. Crimp refers to a fine, periodic, three-dimensional sine-wave or saw-tooth shape along its length. The scale of this wave is crimps per centimeter. Natural fibers, eg, cotton or wool, have an inherent, irregular crimp. Synthetic staple fibers virtually always are crimped deliberately during manufacture by a mechanical stuffing or gear crimping process. Rayon staple fibers also are crimped either mechanically or chemically.

In mechanical processing, the crimp is necessary so that the teeth of the cards and the licker-in cylinder can grip and separate the fibers. The crimp enhances interfiber friction which is necessary for the web so that it can sustain itself during the manufacturing process. In the finished nonwoven, the presence of residual crimp in the fibers gives the product a third dimension and, therefore, a degree of resilience and a degree of textile appearance and feel. Fibers which are intended to be made into webs by wet or paper-making processes customarily are not crimped, since this would increase uncontrolled fiber entanglement in the water dispersion.

Cross Section. Conventional viscose rayon fibers are roughly circular with characteristic crenulations. High wet-modulus viscose rayon fibers do not show crenulations and are either round or bilobal. Special, highly water-absorbent rayon fibers have been made with a round or cross section and a central lumen (16). Cotton fibers have a distinctive, irregular cross-sectional shape which is like a kidney bean but with a central, frequently flattened lumen.

The common synthetic fibers, ie, polyester, nylon (6 or 6,6), and polypropylene, normally are circular in cross section, although they have been extruded in a variety of shapes. It has been proposed that specific, noncircular shapes are better for bonding efficiency and fabric softness; however, customarily, round cross-section fibers are used (18).

The bi-component fibers, which were developed specifically for the nonwoven industry as combination structural and binder fibers, have an approximately spherical cross section. There are two important varieties: in one, a constituent is distributed around a higher melting core; in the other, the two components are present more or less side by side.

Spin Finish. Spin finish refers to the lubricant which is added in small amounts (ca 0.2 wt %) to the surface of extruded fibers to enhance their processability in the mechanical operations, up to and including spinning, by controlling lubrication and friction and subduing static electricity. A fiber with no finish cannot be made into yarn. The effectiveness of the spin-finish is critical for the commercial acceptance of a fiber and, as a result, its chemical identity frequently is proprietary.

Finishes permit dry processing at high speeds and, for synthetic fibers, frequently are hydrophilic to enhance the wettability of the fibers by water-based latex binders. Where the fibers are intended to be dispersed in water or wet-laid for the primary web-forming step, rayon fibers have been provided with no spin-finish called hard finish, whereas the normally hydrophobic synthetic fibers, eg, polyester, have a hydrophilic finish. The proper finish on a synthetic fiber helps dispersability in the wet web-forming process so that longer fibers can be used.

Delustered Fibers. Most of the research effort to improve rayon is to achieve the properties of cotton (11). Delustering is accomplished by dispersing a small amount (ca 1–2 wt %) of TiO_2 pigment in the viscose dope prior to extrusion. The use of delustered fibers increases the opacity and, therefore, the covering power of lightweight nonwovens and gives these, as well as heavier-weight products, an attractive white matte appearance. A special, highly delustered rayon fiber with a higher TiO_2 content has been made for the nonwoven industry. The incorporation of delustering agents does not have a significant effect on the tensile properties of the fiber.

Delustering agents also are used in synthetic fibers. Colored pigments can be dispersed in extruded fibers, and the resultant dope-dyed fibers have some minor use in the nonwoven industry.

Draw Ratio. Synthetic fibers for nonwoven use are drawn in the same manner as for general textile use, except in special cases, eg, in making bonding fibers. Poly(ethylene terephthalate) fibers, either undrawn or drawn much less than standard, have a substantially lower softening temperature and are used widely as bonding fibers; they are activated by heat and pressure.

Fiber Webs. A web is the common constituent of all nonwoven fabrics. Polymer nettings, foams, and scrims which are made from cross-laid yarns incorrectly have been claimed to be nonwoven fabrics. The characteristic properties of the base web are determined by fiber geometry, largely as determined by mode of formation; fiber characteristics, both chemical and mechanical; web weight; and further processing, including compression, fiber rearrangement, and fiber entanglement. The mode of web formation is a primary determinant of fiber morphology, including fiber orientation and frictional engagement. Therefore, it has a strong influence on the properties of the finished product. Among the important aspects of fiber morphology are the average directional fiber orientation, whether isotropic or anisotropic; the geometric shape of the fibers, predominantly stretched-out, hooked, or curled; interfiber engagement or entanglement; and residual crimp. Of these, fiber orientation properties are the most important.

There are only a few ways to form fiber webs and most are derived from classical textile or paper-making processes. The methods are dry formation, including carding and air-laying (19); and wet-laying in large volumes of water by modified paper-making methods and in high viscosity, low volume foam. Development of fundamental new methods for forming webs or major modifications of old methods is difficult.

Dry Formation. *Carding.* Carding is a mechanical process whereby clumps of staple fibers are separated into individual fibers and simultaneously made into a coherent web (20). The operation is carried out on a machine which utilizes opposed moving beds of fine, angled, closely spaced needles or their equivalent to pull and tease the clumps apart. Card clothing is the structure comprised of needles, wires, or fine metallic teeth embedded in a heavy cloth or in a metal foundation. The carding machine may be considered as a convenient frame to support the clothing so that it may operate at maximum efficiency (20). The opposing, moving beds of needles are wrapped on a large cast-iron main cylinder, which is the heart of the card, and a large number of narrow, cast-iron flats, which are held on an endless belt that is placed over the top of the main cylinder.

The needles of the two opposing surfaces must be inclined in opposite directions and must be moving at different speeds relative to each other. The main cylinder moves at a higher surface speed than the flats. The clumps between the two beds of needles are separated into fibers and are statistically aligned in the machine direction as each fiber is theoretically held at each end by individual needles from the two beds. The individualized fibers engage each other randomly, and, with the help of their crimp, form a coherent web at and below the surface of the needles of the clothing on the main cylinder.

All of the mechanisms are built around the main cylinder and, since the various settings must be precise, the cylinder is massive. In addition to the flats, the carding machine includes means to carry a crude lap or batting onto the cylinder where the carding action takes place. It contains other mechanical means to strip or doff the web off the cylinder after completion of the process. The doffed web is deposited gently onto a moving belt where it is plied with other webs and carried to the next step which, presumably, is rearranging or bonding. Most nonwovens, including the most lightweight products, are made of webs combined from several cards which are aligned in tandem in order to average nonuniformities in the webs.

The orientation ratio of webs at the doffer of a conventional cotton-system card is ca 3:1 in the machine direction. This web is highly sensitive to stretching or drafting. Drafting as little as ca 5% greatly increases the web anisotropy ultimately to 10:1 or more. The physical properties of nonwoven fabrics are highly dependent on fiber orientation; even saturation with binder cannot overcome the effects of fiber orientation on tensile properties. Consequently, nonwovens that are made from webs from conventional cards characteristically have high machine direction and low cross-machine direction strengths. Fabrics made from card webs can be torn readily along the machine direction, ie, parallel to fiber orientation direction, and can hardly be torn across the predominant fiber orientation axis.

In the preparation of fiber assemblies for spinning into yarns, it is desirable that the fibers are oriented in a parallel manner; in making nonwoven fabrics, it usually is desirable that fibers are not highly oriented. Cards are available which are very wide, ie, as much as 355 cm; can be run at high speeds, eg, as much as 137 m/min; produce webs with low long-to-cross orientation ratios; and are less susceptible to becoming highly oriented by drafting than conventional textile cards.

A sizable portion of the industry has solved the low cross-tensile strength dilemma by producing webs which have been cross-laid. Cross-laying is a process by which oriented webs are laid down at or near alternating 45° angles on another oriented web on a moving belt. Cross-laying has not been successful with lightweight webs because they cannot be placed accurately enough to avoid unsightly edge lines; the process is quite successful for heavy weights. A major use is for preparing webs for subsequent needle-bonding. Garnett cards, or modified cotton cards, can be used in this process.

Air-Laying. The earliest and most successful means to overcome the severe anisotropy of webs that are made by carding is the formation of webs by capturing fibers on a screen from an air stream. The fibers in the air stream are individualized and totally randomized so that there should be no preferential orientation when they are collected on the screen. Actually, this ideal situation is not fully attained, and there is a slight orientation in the machine direction and imparted to the web by the air stream and the forward motion of the screen. Fibers that are deposited from an air stream appear to be more curled, thereby having a shorter effective length than fibers in a card web. Drafting the web in handling also orients the fibers to a degree but less so than in card webs.

The length of fibers used in air-laying varies from 1.9 to 6.4 cm (21). The shorter lengths are used more commonly because they allow higher production speeds with better web uniformity. Long fibers require a higher air volume, ie, a lower fiber density in the air stream, to avoid tangling. Use of high production speeds requires the use of high air velocity which promotes nonuniform air flow which, in turn, has a disastrous effect on web quality.

Air-laying cannot be carried out at speeds as high as those used in carding; this problem is especially true for lightweight webs. The quality (primarily, the uniformity) of air-laid webs is enhanced if preformed webs, eg, card webs, are fed into the air-laying equipment; however, this adds to the cost of the process. The adjustment and maintenance of the equipment must be rigorously optimized for best results in speed and web quality. Nevertheless, the advantages of an isotropic or nearly isotropic base web for nonwovens are so considerable that air-laying is an important commercial practice.

A typical example of a specialized variation of web forming is the dual rotor process and apparatus (22). This is an air-laying process which produces a web containing two different fibers, eg, wood pulp and staple rayon or polyester. Supplies of fibers are fed to oppositely rotating licker-ins that are rotated at speeds which are optimum for the fibers being individualized by the licker-ins. The fibers from each supply are entrained in their respective air streams which are impelled at high speed toward each other. The degree of mixing of the dissimilar fibers can be controlled to a considerable degree. Another significant innovation is the development of a card which has a short air-scrambling process at the doffer, thereby producing a card web with a nearly isotropic fiber web (23).

The dry mechanical processes commonly have metal needles or teeth which open clumps and individualize fibers by strong, tearing forces. The tearing action unavoidably breaks fibers, so that the true fiber lengths span a range and, on average, are shorter than the starting fibers. Fiber damage varies from process to process, fiber to fiber and, perhaps, day to day on a single production line. Analysis of fiber-length distribution originally was tedious but is greatly simplified and can be used to quantify fiber breakage, which may be a source of product variation that has not been adequately recognized (24–25).

Wet Formation. Wet formation is a variation of the methods by which paper is made. Fibers are dispersed in water. The water is filtered through a moving screen, leaving a fiber web on the screen. The web is transferred to appropriate belts and/or felts and is dried. Wood pulp fibers disperse readily in water, but longer fibers present many practical difficulties.

High decitex (den), ie, high diameter, high wet-modulus fibers, are dispersible at lengths up to ≥1.9 cm or more. Long, fine diameter fibers, ie, 1.7–3.3 dtex (1.5–3 den), have been dispersed successfully on laboratory equipment and in Japanese handmade paper processes by using very high molecular weight hydrocolloids as dispersing aids (26) (see Dispersants). Without dispersing aids, the long fibers entangle and form a nonuniform web. It appears that 0.95 cm is the upper feasible length and 0.64 cm is the upper practical commercial length for fine decitex hydrophilic fibers, eg, 1.67 dtex (1.5 den) viscose rayon. Fine decitex fibers, ie, 1.67 dtex (1.5 den), of staple length, 1.3 cm or longer, are not in commercial use as ingredients of wet-formed webs.

An unconventional wet-forming process for the dispersion of relatively long, 1.67

dtex (1.5 den), 1 cm fibers has been developed. It is based on the dispersion of the fibers in a low volume viscous foam (27).

It appears that wet formation should produce isotropic webs. At the moderate process speeds of nonwoven manufacture, web orientation is slightly anisotropic. However, high water volumes are needed, and this is adequately handled by the Rotoformer or by an inclined wire (28–29) (see also Paper).

Bonding. Bonding implies the use of an adhesive ingredient or heat, including embossing, or chemical treatment, but bonding also can be effected by interfiber friction from mechanical entanglement. The adhesive methods are numerous and varied. Binder can be applied in many different practical ways, with a choice of numerous kinds of adhesives; greater or lesser amounts; and in uniform, nonuniform, random, or patterned distribution in the web. In choosing and applying binder, the developer of a nonwoven fabric has the largest scope in selection and application of any of the principal elements, ie, fiber, web, binder, and reinforcement. The selected binder may be added in the form of a synthetic water-based emulsion or dispersion, which commonly is called a latex; a solution; a foam; thermoplastic particles, including fibers among others; monomer or oligomer for polymerization in place; plastisols; or polymers that react with the fibers. Any film-forming polymeric substance, which can be dissolved or dispersed and/or partially or fully fused while in contact with the fibers of a web, may be used as an additive binder.

In addition, methods have been devised where most or all of the base fibers of the web have been chemically or thermally softened or fused to give them adhesive properties. The activated fibers subsequently are resolidified to connect fibers to fibers and, therefore, function as a binder. This method, in which the fibers of the web provide the bonding substance, is autogenic bonding.

The most common commercial bonding system is based on synthetic, aqueous latexes. Latexes have many practical advantages, including availability of numerous varieties with different chemical and physical properties, ease of handling, and low cost. They can be modified easily or tailor-made for special needs. The use of latexes is so widespread that they comprise a substantial industry in their own right (30). Present latex or water-dispersed binders, with rare exceptions, are vinyl polymers which are emulsion polymerized by free-radical mechanisms (31).

In additive adhesive bonding, the web is formed first, usually is compressed, and the binder substance is added to it. Thermoplastic binder fibers are incorporated in the web as it is formed. In its final functional form, the binder is a film-forming polymer. For the formation of a continuous chain of bridging throughout the web, each fiber must be bonded at least two times (preferably more) to different fibers. Fibers which escape any bonding are likely to fall out of the web during use, causing undesirable dusting. Fibers which are bonded only once do not contribute to tensile properties but may enhance bulk, softness, and absorbency.

The mechanism of adhesive bonding is not definitively understood. The customary explanation is that bonding is directly related to adhesion, ie, shear or peel adhesion or a combination of the two. The alternative to adhesion is the mechanical entrapment of microfibers within encompassing bonds. At the Chicopee laboratory, corona-discharge treatments were applied to polymer films in a reactive environment; the result was greatly enhanced peel adhesion of latex to the film. The same treatment, as applied to fiber webs of nominally the same chemical composition as the film and bonded by the same latex, produced nonwoven fabrics with very little improved strength (32). The peel adhesion properties of a latex on nylon-6,6 film were improved five-fold by modifying the zeta potential or electrophoretic mobility of the latex. Applying this modified latex to a web of nylon-6,6 fibers, however, did not produce a measurably stronger nonwoven fabric (33).

Experiments were carried out at North Carolina State University where polyester film was treated with diacetyl, 2,3-butanedione, an adhesion assistant, and coated with an acrylic monomer which was polymerized by electron-beam radiation. The diacetyl pretreatment greatly improved the peel adhesion of the polymer to the polyester film. A parallel experiment was carried out using a web of poly(ethylene terephthalate) fibers in place of the poly(ethylene terephthalate) film. They were pretreated with diacetyl, and bonded with monomer that was polymerized in place by electron-beam radiation. The resultant fabric showed no improvement in tensile properties over a fabric that had not been pretreated with the adhesion enhancer (34).

However, there are contrary data which indicate that specific adhesion can be important. An example is a commercial, lightweight nonwoven fabric composed of viscose rayon fibers bonded with ca 2 wt % cellulose, which is applied as a viscose solution and is regenerated. Microscopic examination of fabric cross sections show no interface between viscose binder and viscose fibers, indicating fusion of the two. Viscose rayon fiber fabrics require 10–15 wt % added latex binders to achieve comparable tensile properties. If adhesion is very high, it can be a controlling factor in bonding. In addition, where the binder content is so low that fiber embedment is not possible, eg, in lightly bonded wood-pulp webs, adhesion is important (see Pulp).

Experiments have been carried out to investigate bonding by chemical reaction, ie, formation of covalent bonds between binder and fibers, based on the chemistry of fiber-reactive dyes and ionic bonds (35–36) (see Dyes, reactive). Some latex bonding of rayon, as carried out in large commercial volume, is enhanced by the reaction of excess latex cross-linking agent, eg, *N*-methylolacrylamide, with the polymer and active sites on the fiber (37).

Latex. Synthetic latexes have the most extensive history as binders and are the most widely used. The estimated total usage of latexes in 1979 in nonwoven fabrics was 54,000 metric tons (dry basis) (30). Although it is possible to make a reasonably complete listing of the major categories of commercial latexes, the number of minor variations in use is large. The main types include polyacrylates, eg, poly(ethyl acrylate) and copolymers; vinyl acetate–acrylic acid copolymers; poly(vinyl acetate)s; poly(vinyl chloride)s; ethylene–vinyl acetates; styrene–butadiene carboxylates; and polyacrylonitriles (see Latex technology).

Most of the latex binders include a small amount of a self-cross-linking agent, typically *N*-methylolacrylamide (see Acrylamide). Other, nonformaldehyde cross-linking systems are being developed. The advent of the self-cross-linking latexes at the end of the 1950s made durable, washable, wet-abrasion resistant nonwovens practical and, possibly, was the most important factor in the subsequent rapid growth of nonwovens made from staple fibers. The self-cross-linking monomer does not increase the modulus of the cured binders to a significant degree but greatly improves water and solvent resistance.

All of the above latexes are effective binders for rayon webs. Except for a few obvious latex properties, eg, film-forming ability, modulus of the cast film, solvent or water resistance, and wettability, the relationship of latex properties to final fabric properties is not reliably predictable, although a substantial effort has been expended to establish such relationships (38–40). However, manufacturers of latexes and of fabrics, have developed rapid screening procedures and can readily search for latexes which satisfy commercial needs.

The latexes which are, for the most part, universally good binders for rayon and other hydrophilic fibers are not successful, ie, they do not lead to the formation of strong fabrics with reasonable amounts of added binder, when used on synthetic fibers, in particular, polyester and polypropylene. Even though the synthetic fibers have tenacity values (N/tex or gf/den) two to four times greater than that of rayon, the synthetic fiber webs which are bonded with conventional latexes are much weaker, ie, typically, one half to one fourth as strong.

Latex binders can be efficient impregnants on 100% polyester webs if the following parameters are optimized: a hydrophilic finish on the binder and a binder formulation which wets the fibers readily (37). Presumably, the binder effectiveness is dependent on its micromorphology in the web and on the fibers, as well as on its specific adhesion. In addition, effective binders for polyester fibers have been developed from water-dispersable polyester polymers (31).

Needs of the Manufacturing Process. Latex binders may be applied to dry-formed webs by overall saturation, application in pattern by the rotogravure process or a modified silk-screen process, spraying on one or both web faces, and application as a foam. However, these methods subject the latex to severe shear stress. Therefore, a high degree of mechanical stability at either full or diluted concentration is required. The latex binder must be able to accept miscellaneous additives without becoming unstable. It should be stable in storage for many months but, once applied, it should dry and cross-link under reasonable time and temperature conditions. Consecutive production batches must not vary in composition, viscosity, and other rheological properties; surface tension; latex particle size; and size distribution. It should be a good film former, but it should be readily dispersable for clean-up. This list can be expanded by processors with specific needs.

Needs of the Nonwoven User. Each commercial product has a complex of properties which are directly dependent on the binder. These may include softness, strength, water absorbency or repellency, wet–dry abrasion resistance, and water or solvent resistance. The binder must have no toxic, allergenic, or odorous ingredients. Its film properties, including water-white color, strength, and modulus, must be stable for a reasonable period of time. In addition, each user may have specific needs.

Binder Formulation. It is not customary for a latex to be used exactly as received from the supplier. The latexes usually are manufactured at ca 50 wt % solids; the user often dilutes the latex with water. An acid or latent acid catalyst, a defoaming agent, a viscosity modifier, usually a thickener, and/or a surfactant or a water repellent may be added (see Defoamers; Surfactants and detersive systems). Less often a pigment, either white or colored, a rust-inhibitor, a fire-retardant salt but only if the latex is salt-tolerant, and a binder-migration control agent are added.

The amount of binder which is added to the web varies from ca 5 wt % to ca 50 wt %, depending upon desired fabric properties, commonly 10–35 wt %. The higher ranges of binder content tend to obscure fiber properties; one measure of binder and process efficiency is the minimum binder content required for needed tensile and other physical properties.

Application of Latex Binder. *Saturation.* Saturation or impregnation involves immersing the fiber web in the latex and removing the excess latex. The web can be fed directly into the bath under an idler roll and exit between the mangle rolls. An unbonded web is not strong enough for this simple process and requires support on a carrier or, more likely, is carried between two open-mesh screens. There are many variations in web handling and threading practices.

As the web enters the nip of the mangle, it is exposed to the excess liquid which simultaneously is moving in the reverse direction. The liquid can severely damage the fragile web which, in practice, means that the speed at which this kind of saturation process can be run is limited. The problem is especially severe for lightweight webs. One approach to solving the problem is the use of mechanical means to divert the exiting stream from contact with the web. However, the customary method for impregnation is to circumvent the saturation process by metering latex onto the web using an engraved roll and a very fine, close pattern. The close pattern allows the latex to diffuse, and the web is effectively saturated with a controlled binder content.

In concept, if not execution, saturation bonding with latex is the simplest way to make a nonwoven fabric. Variations in properties and amount of binder added, the method of drying and curing, and the degree of compacting the web enable the formation of a great diversity in products.

Print Bonding. One of the frequently desired attributes in a nonwoven fabric is a textilelike flexibility, ie, hand or drape. This can be accomplished by saturating a web with a relatively small amount of soft binder, ie, one that has a low film modulus or low second-order transition temperature (38–40). One deficiency of such a saturated staple fiber web is that all fibers and fiber ends are tied down, giving the product a thin, flat dimension and a slick surface. Another method which is used to optimize fabric hand is intermittent bonding: binder is distributed intermittently so that the resultant product has alternating areas with and without binder. The bonded areas provide the strength and the unbonded areas provide a degree of bulk, flexibility, and surface texture.

Random intermittent bonding has been achieved by use of a mixture of fusible and nonfusible fibers; the former are activated by heat and pressure. Another approach was the printing of latex binders on the web in a pattern using an engraved roll. Billions (10^9) of meters of nonwovens have been made by the second method, which is particularly suitable for products as lightweight as 15 g/m^2 but which can be efficiently applied to webs as heavy as ca 100 g/m^2 (ca 4 oz/yd^2). Print bonding is applicable readily to card webs of any hydrophilic fiber, eg, rayon, both standard and high wet modulus; cotton; poly(vinyl alcohol); and their blends (13). Research is being carried out to optimize print bonding of hydrophobic fibers, including polyesters and polyolefins. Print bonding is a widely used and mature technology.

Commercial equipment has been developed to apply latex binder by a patterned-screen method. Binder formulations are modified to greatly increase viscosity so that, according to machine design, the latex is forced through the screen pattern into the web. Line conditions and binder compositions are different from those customarily used in rotogravure printing. Screen printing characteristically deposits the binder more precisely and with less sidewise diffusion or migration than its older counterpart. Rotogravure printing is most conveniently applied to webs that are composed of long fibers, are prewet with water, and are produced by carding so that fibers are oriented preferably in the machine direction. All of the above conditions are not required in screen printing.

Print-Pattern Design. The earliest print-pattern design is a series of parallel lines running transverse to the machine direction and is used when the fibers are oriented predominantly in the machine direction (6). The rationale for this pattern was that it maximizes cross-strength; as if each transverse binder line is continuous and acts as a filling yarn in a woven fabric. A cursory examination using a low power microscope shows that the line is discontinuous and that the binder is deposited preferentially at fiber crossovers and other high surface energy locations. This simple line pattern is fairly effective and a slight variation, ie, a pattern of horizontal, parallel wavy lines, is in wide use. It was soon recognized that significant improvements in fabric properties, other than simple tensile strength, can be effected by the design of the print pattern.

Other designs were developed to increase fabric cross-directional toughness, ie, the area under the stress–strain tensile curve; resilience; and wet-abrasion resistance. Cross-directional toughness is improved substantially by patterns based on rectangular or oval units (41–42). Cross-resilience is improved by diagonal line patterns, and wet-abrasion resistance is improved by a miniaturization of the pattern (43). Since the latex binders frequently are pigmented, the pattern design becomes a decorative design. Consequently, the design of a new pattern becomes a combination of technology and esthetics.

In the rotogravure system, the latex binder usually is printed onto a wet web. The water-based latex penetrates through the web and simultaneously diffuses sidewise; this phenomenon is migration. Excess migration blurs the carefully designed pattern, sometimes unpredictably, and generally is undesirable. The degree of migration can be controlled by coagulating the latex in place as it contacts the web by using physicochemical methods (44) or, when the web reaches the heat of the dryers, by using heat-unstable surfactants or other heat-reactive additives in the latex. Heat-unstable additives also are used in saturated webs to prevent migration of binders to the fabric faces during drying (46–47). The ability of patterns to substantially improve cross-tensile properties is limited. Fiber orientation has a far greater effect; eg, no print-pattern or other bonding method can upgrade cross-tensile strength nearly as much as changing the long–cross ratio of fiber orientation (see also Printing processes).

Spray Bonding. Binders, particularly latexes, can be applied by direct spraying onto a web. Several requirements must be fulfilled for this process to be practical, including low viscosity binders which can tolerate the extreme shear conditions at the spray nozzle, nonplugging nozzles, and an airless spray to avoid disrupting the web. The spray usually is applied to a dry web which may be composed of short and/or long fibers, including wood pulp alone and mixtures with staple fibers. The spray tends to penetrate only partly into the web and is used primarily where high loft properties are desired in the final product. The spray can be applied to one or both faces of the web. In heavy applications, the sprayed binder penetrates fully.

Foam Bonding. Binder may be applied to the web as a water–air foam which implies that less water is used and increased energy savings are achieved in the drying process. Foam characteristically is deposited on the web surface. Binder that is applied as foam which collapses during drying is called froth. It is also possible to apply foam using formulations permitting the foam structure to survive the drying process.

Thermoplastic Bonding. A thermoplastic bond is a physical or adhesive bond made from an added thermoplastic polymer which has been softened or fused as it is in close contact with the fibers. Upon solidification and while still in close contact, the polymer binds the structural fibers. Thermoplastic bonding is one of the oldest and most effective types of bonding. The thermoplastic material may be in the form of fibers, rods (short-cut, high denier filaments), powder, granules, netting, film, or particles in irregular shapes (7,47–48). Each system has economical or functional merits for an intended use. Bonding with thermoplastic particles is a versatile technique because of the great number of variations which may be practiced, including choice of polymer for strength, melting temperature and rheological properties of the melts, and adhesive and chemical properties; amount of polymer added; shape and size of activable particles; and heat and/or solvent vapor, pressure, and time conditions of activation.

Fibers are the most widely used form. Thermoplastic fibers have the added advantage of simplicity of mixing with base-web fibers. The fibers customarily are intermixed at random and activated by heat and pressure or an intermittently patterned heat and pressure. Until recently, polypropylene fibers were the most effective and practical type. Recently, ICI (UK), Chisso (Japan), and Enka (U.S.) have introduced bimodal, staple fibers which have promising properties.

Autogenic Bonding. Autogenic bonding is the bonding of a web by partially solubilizing the base-web fibers in close contact using heat, pressure, or solvents and resolidification to form polymer bonds. The concept has been popular but not often commercially successful. Typical examples include: immersion of a compressed web of cellulose fibers in concentrated H_2SO_4 or other cellulose solvent followed by quenching and washing in water; immersion of a compressed viscose rayon fiber web in 8–10 wt % NaOH, within which concentration range viscose rayon becomes gelatinous, followed by quenching and washing with water; even or patterned exposure of a 100% thermoplastic polyolefin fiber web to heat and pressure, followed by cooling; exposure of nylon fiber-web to HCl gas, followed by washing to remove HCl; and addition of aqueous solutions of certain salts, ie, LiCl, KCNS, and others, to nylon, acrylic, or polyester fiber webs, removal of water by heat to locate the resultant concentrated salt solution at fiber crossovers (thereby dissolving fiber polymer), followed by washing to solidify the bond and to remove salt (49).

Although autogenic bonding appears attractive, it has flaws in practice. Since the bond substance comes from the fibers, the latter are subject to damage. Furthermore, experience has shown that the range of useful bonding is limited; there is a strong tendency toward only a narrow range of practical process variability. Too little treatment produces underbonding; too much treatment damages fibers and embrittles the resultant product.

Solution Bonding. Bonding with organic solutions of polymers is an efficient process by the criterion of strength attained binder added, especially at low binder content (50). The practical problems of solvent hazards and cost and the difficulties of handling and applying viscous but low solids solutions have negated the widespread use of solution bonding. However, water solutions of binders have specialized uses. Poly(vinyl alcohol), which usually is cross-linked with a formaldehyde donor, is practical as a binder and has a commercial history (52). Another interesting water solution system is viscose, ie, cellulose xanthate dissolved in aqueous NaOH. This has been used on a small but successful commercial scale for many years on rayon and cotton-fiber webs. The product is effectively bonded with much lower binder added than is customary and is a classic example of and evidences excellent binder efficiency when specific adhesion is very high.

Polymerization of Monomer. Only recently, with the advent of electron-beam polymerization, has polymerization of added monomer or oligomer *in situ* in a fiber web been feasible (52–53). The process is notable for its low power requirements. Nonwoven fabrics have been bonded experimentally by this process, but results have not been good enough to justify the high cost of the special monomers required and difficulties in application (see Radiation curing).

Reactive Bonding. In one experiment, a binder is applied to a web of cellulose fibers on which it reacts to form covalent bonds (54). The chemical that is used is based on fiber-reactive dye systems using cyanuric chloride, which reacts with the hydroxyl groups of cellulose fibers and polymeric diamines of the binder. This method should be equal or superior to those of systems with high specific adhesion.

An inexorable problem associated with adhesive bonding is that maximum strength and softness are mutually irreconcilable. Strength may be increased only by the sacrifice of suppleness and vice versa:

$$\text{strength} + \text{softness} = K$$

where K is an arbitrary constant. Exceptionally effective adhesion, softer and stronger binders, or optimization of binder morphology may increase the value of K, but only to a limited degree.

Bonding by Fiber-to-Fiber Friction. Needle-punching describes a process which is in large-scale use and by which a base web is reinforced by passing it under banks of rapidly reciprocating barbed needles (55). The barbs entrap bundles of fibers of the web, forcing them downward through the web, thereby forming numerous small loci of fiber entanglement. The process is relatively slow, is most useful for heavy webs, and is capable of making very strong, supple products.

Laboratories supported by the governments of the GDR and of Czechoslovakia have developed a family of nonwoven fabrics that are made by extensive overstitching or knitting of fiber webs with yarns. The resultant products are unique, although they have a superficial resemblance to soft woven or knitted textiles. Tradenames associated with these nonwovens include Arachne, Malivlies, Schusspol, Maliwatt, Malimo, Malipol, and Voltex (56). Similar products are also available in the United States. These products can be used as manufactured or as a base for plastic coating.

An ingenious method for making a soft, highly absorbent all-cotton pad consists of exposing a web of unbonded cotton fibers to a mercerizing-strength NaOH solution. A card web of relatively short cotton fibers is carefully floated without tension on the cold, aqueous NaOH solution. Mercerization without tension occurs and is accompanied by vigorous curling and tangling of the fibers. After thorough washing and drying, the resultant material is bulky, resilient, and strong enough for uses as a specialty absorbent and as a carrier of reagents (57).

When a web of staple fibers is exposed to fluid forces, eg, multiple water jets, the fibers can be rearranged to form yarnlike bundles of fibers, thereby altering the web appearance and function. A web which is textureless in appearance can be made to resemble woven, knitted, or embroidered fabrics or can be given an entirely novel appearance by this process. From the inception of fiber rearrangement, it was recognized that some fiber entanglement takes place, and the patents describe these products as self-sustaining (58–60). However, they do not have adequate strength, without additional bonding, for broad usage. Webs with fibers that are rearranged as described and bonded with latexes have been manufactured in large amounts since the late 1950s.

Spunlace Bonding. Fibers in a web can be tangled to a substantial degree by the application of fine, columnar water jets at very high pressures (61). As the jets penetrate the web and deflect from a wire screen or other permeable backing, some of the fluid splashes back into the web with considerable force. Fiber segments are carried by the turbulent fluid and become entangled on a semimicro scale. These products represent a new species of fabric where there is an irregular interlocking, hence, entanglement on a fiber-to-fiber scale. Manufacture of spunlace products presents difficult engineering problems and is highly energy-intensive. The products have excellent strength properties, except in very light weights. A major fault is their lack of stretch recovery. In addition, they tend to have low abrasion resistance, unless they are extensively processed (62). They have a fundamental advantage over chemically bonded counterparts in that strength and suppleness are not mutually exclusive. Both can be maximized at the same time. Spunlace products have not become the universal nonwovens but they occupy an important and expanding section in the product spectrum.

Reinforcements. Fiber webs can be reinforced by woven fabrics, plastic nettings, cross-laid yarn scrims, foams, and polymer films. The combination of webs with a reinforcement usually is enhanced by adhesive bonding, although mechanical entanglement also is feasible. However, it is desirable that adhesives are effective for both components and that the stress–strain characteristics of both are similar so that they efficiently complement each other.

Manufacture

The manufacture of nonwoven fabric roll goods is achieved by combining the raw materials, ie, fibers, reinforcing structures, binders, or fibers alone, by two or more of the processes of web forming, adhesive bonding, and fiber entanglement. Although there are few processes and components, each exists in so many variations that there are numerous possible methods.

In all methods, the web is formed first. The chemical type of fiber and its decitex, length, etc, are chosen for ultimate product properties. If staple fibers are used, the web-forming process is always dry, either involving carding or air-forming. If the final product is intended to be bonded by thermoplastic fibers, they are included in the original fiber blend that composes the web. Other thermoplastic particles are likely to be added as the web is formed or webs are plied. After the web is formed, it may be wetted for ease of handling, especially if it is to be bonded with a water-based adhesive, eg, a latex or an aqueous solution. The latex may be added evenly, ie, by saturation, by printing in a pattern, by spraying, or by application as a foam. The type, amount, and geometric placement of the binder can be varied, depending on the resultant desired properties. If the fibers are rearranged to give the product a characteristic appearance, rearrangement is carried out at the web-forming–bonding step.

After the binder is applied when using, it is activated by heat thermoplastic adhesive; dried; and self-cross-linked using latex, foam, or solution; or exposed to specific solvent or heat conditions by autogenic bonding. During the bonding, the web customarily is compressed to a desired degree. A reinforcing structure may be incorporated in the web during its formation, between plies or on one or both faces after formation and before application of binder. The reinforcing web may be saturated with binder

Table 1. United States Consumption of Nonwoven Fabrics [a]

	Roll-stock value, $\$10^6$ 1978	Roll-stock value, $\$10^6$ 1983 [b]	Growth, %/yr
Type			
disposable	455	930	15
durable	405	780	14
Process			
dry (carding, air-forming, needle-punching)	490	950	14
wet (modified paper process)	90	160	12
composite	70	105	9
spunbonded (on-line polymer to nonwoven)	190	435	18
spunlaced (entangled-fiber bonding)	20	60	24
End Use			
diaper-cover stock	120	250	16
surgical packs and gowns	100	200	15
interlinings and interfacings	80	130	10
coated and laminated fabrics base	75	140	13
home furnishings			
carpets and bedding	135	280	16
wipes and towels	70	135	14
sanitary napkins, tampons, and underpads	40	70	12
filtration media	65	135	16
all others	175	370	16

[a] Ref. 64.
[b] Estimated.

Table 2. Estimated Worldwide Consumption of Nonwoven Fabrics [a], 1978, 10^3 Metric Tons

North America	
dry-formed	195
wet-formed	36
composite	32
spunbonded	54
spunlaced	6.8
Far East	
dry-formed	23
spunbonded	4.5
wet-formed	5.4
Western Europe	
dry-formed	79
wet-formed	20
spunbonded	27
Other	
dry-formed	29
wet-formed	2.3
spunbonded	6.8

[a] Ref. 64.

and placed between plies of the web; in this way, it represents a specialized means to combine the bonding and reinforcing steps.

Where the bonding is by mechanical entanglement rather than by an adhesive or another physicochemical method, the web is transported to the entanglement zone.

There it is either passed under a bed of reciprocating, barbed needles, overstitched with yarns, mercerized, or subjected to fiber-scale entanglement by fine high pressure jets of water.

Economic Aspects

There is no ideal or prototype nonwoven fabric. There are approximately 150 important end uses and the products are made on more than 1000 production lines worldwide. In 1979, the value of nonwoven fabrics that were produced as roll goods in the United States was ca $\$10^9$ and the converted value was ca $\$3 \times 10^9$. In that year, the amount of nonwovens manufactured approached 10% of the entire textile industry, whereas weaving accounted for ca 65%, and knitting comprised ca 25% (63).

A breakdown of consumption of nonwoven fabrics in the United States and worldwide is given in Tables 1 and 2.

BIBLIOGRAPHY

1. P. Coppin and co-workers, "The Definition of Nonwovens Discussed," *European Disposables and Nonwovens Association Symposium, Gothenberg, Sweden, June 6–7, 1974.*
2. *Guide to Nonwoven Fabrics*, INDA, The Association of the Nonwoven Fabrics Industry, New York, 1978.
3. *ASTM D 123-70*, ASTM, Philadelphia, Pa., 1970.
4. W. D. Freeston, Jr., and M. M. Platt, *Text. Res. J.* **35,** 48 (1965).
5. S. Backer and D. R. Petterson, *Text. Res. J.* **30,** 704 (1960).
6. U.S. Pat. 2,039,312 (May 5, 1936), J. H. Goldman.
7. U.S. Pat. 2,277,049 (March 24, 1942), R. Reed (to The Kendall Co.).
8. E. R. Schwarz, K. R. Fox, and N. V. Wiley in H. R. Mauersberger, ed., *Matthews Textile Fibers*, 5th ed., John Wiley & Sons, Inc., 1947, Chapt. XXIV.
9. A. R. Winch, *Eighth INDA Technical Symposium*, Orlando, Fla., March 19–21, 1980, p. 224.
10. A. R. Winch, *Text. Res. J.* **50,** 64 (1980).
11. R. Remirez, *Chem. Eng.* **86**(7), 113 (March 26, 1979).
12. *Chem. Week* **124**(23), 27 (June 6, 1979).
13. U.S. Pat. 3,930,086 (Dec. 30, 1974), C. Harmon (to Johnson & Johnson).
14. U.S. Pat. 3,507,943 (April 21, 1970), J. J. Such and A. R. Olson (to The Kendall Co.).
15. A. F. Turbak and co-workers, *ACS Symp. Ser.* **58,** (1977).
16. M. J. Welch and J. A. McCombes in ref. 9, p. 3.
17. R. F. Nickerson in ref. 8, Chapt. V, p. 178.
18. G. G. Allen and L. A. Smith, *Cellul. Chem. Technol.* **2,** 80 (1968).
19. F. M. Buresh, *Nonwoven Fabrics*, Reinhold Publishing Company, New York, 1962.
20. G. R. Merrill and co-workers, *American Cotton Handbook*, 2nd ed., Textile Book Publishers Inc., New York, 1949, Chapt. 7.
21. G. B. Harvey, *Formed Fabric Ind.* **6**(9), 10 (1975).
22. U.S. Pat. 3,740,797 (June 26, 1973), A. P. Farrington (to Johnson & Johnson).
23. Manufactured by E. Fehrer, *Textilmaschinenfabrik und Stahlbau*, Linz, Austria.
24. W. L. Balls, *A Method of Measuring the Length of Cotton Hairs*, London, Eng., 1921.
25. *Evaluation of the Motion Control Text System*, USDA Agricultural Marketing Service, Memphis, Tenn., Nov. 1974.
26. U.S. Pat. 3,794,557 (Feb. 26, 1974), C. Harmon (to Johnson & Johnson).
27. B. Radvan and A. P. J. Gatward, *Tappi* **55,** 748 (1972).
28. U.S. Pat. 2,781,699 (Feb. 19, 1957), W. J. Joslyn (to The Sandy Hill Iron and Brass Works).
29. U.S. Pats. 2,045,095; 2,045,096 (June 23, 1936), F. Osborne (to C. H. Dexter Co.).
30. C. H. Kline & Co., *INDA Newsletter* **79,** 3 (Dec. 1979).
31. K. R. Barton, *Nonwovens Ind.* **10**(5), 28 (1979).
32. U.S. Pat. 3,661,735 (May 9, 1972), A. Drelich (to Johnson & Johnson).

33. U.S. Pat. 3,639,327 (Feb. 1, 1972), A. Drelich and P. Condon (to Johnson & Johnson).
34. W. K. Walsh, "Radiation Processing of Textiles," *Symposium, North Carolina State University, Raleigh, May 26–27, 1976.*
35. G. G. Allen, G. Bullick, and A. N. Neogi, *J. Polym. Sci.* **11,** 1759 (1973).
36. M. L. Miller, "Ionic Bonding in Rayon Nonwovens," Ph.D. dissertation, University of Washington, 1972.
37. W. F. Schlauch, "Recent Developments in Nonwoven Binder Technology," *Nonwoven Fabrics Forum, Clemson University, Clemson, S.C., 1979.*
38. J. W. S. Hearle and P. J. Stevenson, *Text. Res. J.* **34,** 181 (1964).
39. J. W. S. Hearle, R. I. C. Michie, and P. J. Stevenson, *Tex. Res. J.* **34,** 275 (1964).
40. R. I. C. Michie, *Text. Res. J.* **36,** 501 (1966).
41. U.S. Pat. 2,705,687 (April 5, 1955), D. R. Petterson and I. S. Ness (to Chicopee Manufacturing Corp.).
42. U.S. Pat. 3,009,823 (Nov. 21, 1961), A. Drelich and V. T. Kao (to Chicopee Manufacturing Corp.).
43. U.S. Pat. 2,880,111 (March 31, 1959), A. Drelich and H. W. Griswold (to Chicopee Manufacturing Corp.).
44. U.S. Pat. 4,084,033 (April 11, 1978), A. Drelich (to Johnson & Johnson).
45. U.S. Pat. 3,944,690 (April 6, 1976), D. Distler and co-workers (to Badische Anilin- und Soda-Fabrik Aktiengesellschaft).
46. U.S. Pat. 4,119,600 (Oct. 10, 1978), R. Bakule (to Rohm and Haas Co.).
47. U.S. Pat. 2,880,112 (March 31, 1959), A. Drelich (to Chicopee Manufacturing Corp.).
48. U.S. Pat. 2,880,113 (March 31, 1959), A. Drelich (to Chicopee Manufacturing Corp.).
49. U.S. Pat. 3,542,615 (Nov. 24, 1970), E. Y. Dabo and co-workers (to Monsanto Co.).
50. A. Drelich and P. N. Britton, "Physical Combinations of Cellulose Fibers and Polymers," *149th Meeting of the American Chemical Society, Detroit, Mich., April, 1965.*
51. U.S. Pat. 3,253,715 (May 31, 1966), E. V. Painter and co-workers (to Johnson & Johnson).
52. C. Houng, E. Bittencourt, J. Ennia, and W. K. Walsh, *Tappi* **59,** 98 (1976).
53. U.S. Pat. 4,146,417 (March 27, 1979), A. Drelich and D. Oney (to Johnson & Johnson).
54. G. G. Allan and T. Mattila, *Tappi* **53,** 1458 (1970).
55. P. Lennox-Kerr, ed., *Needle Felted Fabrics*, The Textile Trade Press, Manchester, Eng., 1972.
56. J. D. Singelyn, "Principles of Stitch Through Technology," *Nonwoven Fabrics Forum, Clemson University, Clemson, S.C., 1978.*
57. U.S. Pat. 2,625,733 (Jan. 20, 1953), H. Secrist (to The Kendall Co.).
58. U.S. Pat. 3,081,514 (March 19, 1963), H. Griswold (to Johnson & Johnson).
59. U.S. Pat. 2,862,251 (Dec. 2, 1958), F. Kalwaites (to Chicopee Manufacturing Corp.).
60. U.S. Pat. 3,033,721 (May 8, 1962), F. Kalwaites (to Chicopee Manufacturing Corp.).
61. U.S. Pat. 3,485,706 (Dec. 23, 1969), F. J. Evans (to E. I. du Pont de Nemours & Co., Inc.).
62. M. M. Johns and L. A. Auspos, "The Measurement of the Resistance to Distanglement of Spunlaced Fabrics," *Seventh INDA Technical Symposium, New Orleans, La., March 1979.*
63. E. Vaughan, Clemson University, and D. K. Smith, Chicopee, private communication, Sept. 1979.
64. J. R. Starr, consultant, private communication, Boston, Mass., 1980.

General References

F. M. Buresh, *Nonwoven Fabrics*, Reinhold Publishing Company, New York, 1962.
R. Krcma, *Manual of Nonwovens*, Textile Trade Press, Manchester, Eng., 1971.
E. M. Passot, "Computer Evaluation of Fiber Crimp and Spot-Bonding for Drapable Nonwovens," Ph.D. dissertation, University of Washington, 1974.
M. S. Caspar, *Nonwoven Textiles*, Noyes Data Corporation, Park Ridge, N.J., 1975.
INDA (*Association of the Nonwoven Fabrics Industry, New York*) *Technical Symposia*, 1973 to present.
Annual Symposium Papers, Clemson Nonwoven Fabrics Forum, Clemson University, Clemson, S.C., 1969 to present.

ARTHUR DRELICH
Chicopee Division
Johnson and Johnson

NOVOLOID FIBERS

Novoloid fibers are cross-linked phenolic–aldehyde fibers typically prepared by acid-catalyzed cross-linking of a melt–spun novolac resin with formaldehyde (see Phenolic resins). Such fibers are generally infusible and insoluble, and possess physical and chemical properties that clearly distinguish them from other synthetic and natural fibers (see Fibers, chemical). Novoloid fibers are used in flame- and chemical-resistant textiles and papers, in a broad range of composite materials, and as precursors for carbon and activated-carbon fibers, textiles, and composites.

The generic term novoloid was recognized in 1974 by the United States Federal Trade Commission (FTC) as designating a manufactured fiber containing at least 85 wt % of a cross-linked novolac (1). In granting its official approval, the FTC noted that its criteria for new generics required that a candidate fiber have a chemical composition radically different from other fibers resulting in distinctive physical properties of significance to the general public.

The basic novoloid patents were issued to the Carborundum Company, Niagara Falls, New York, in the early 1970s (2). At present, Nippon Kynol, Inc. (Japan), and American Kynol, Inc., as exclusive licensees of Carborundum, manufacture and sell novoloid fibers under the trademark Kynol.

Fiber Properties

Except where specifically noted, the properties described here are those of commercial Kynol novoloid fibers produced by melt-spinning and aqueous curing. Fiber properties are strongly affected by the chemical and physical characteristics of the novolac resin, the fiber diameter, and the method and degree of curing (cross-linking). These variables may be adjusted to produce the balance of properties required for a particular application; the data given here represent typical values or ranges.

Novoloid fibers are generally elliptical in cross-section, with a ratio of diameters of approximately 5:4. They are light gold in color and darken gradually to deeper shades with age and exposure to heat or light, although there is no significant concomitant change in other fiber properties. The fibers have a very soft touch or hand, and are generally without appreciable crimp, although crimp may be imparted by thermomechanical means to facilitate spinning.

Thermal Properties. Novoloid fibers are highly flame resistant, but are not high temperature fibers in the usual sense of the term. A 290 g/m^2 woven fabric withstands an oxyacetylene flame at 2500°C for 12 s or more without breakthrough. However, the practical temperature limits for long-term application are 150°C in air and 200–250°C in the absence of oxygen. This apparent paradox must be understood in terms of the chemical structure of the fiber (see also Heat resistant polymers).

Flame Behavior and Resistance: Combustion. A measure of the flame resistance of a given material is the limiting oxygen index (LOI), ie, the percent concentration of atmospheric oxygen required for self-supporting combustion (see Flame retardants). The LOI of novoloid fiber materials varies with the configuration of the fibers (as yarn, felt, fabric, etc), but is generally in the range of 30 to 34, ie, higher than that of natural organic textile fibers and of all but the most exotic synthetic organic fibers. (Measured values of LOI are significantly influenced by the specific test method and apparatus employed, and, therefore, even higher values have been reported in the literature.) Moreover, when exposed to flame, novoloid fibers do not melt but gradually char until completely carbonized without losing their initial fiber form and configuration. This behavior is attributable to the cross-linked, amorphous, infusible structure of the fiber and to its high (76%) carbon content.

Combustion of an organic polymer substance generally involves not only oxidative attack at the surface but, more important, decomposition and volatilization of the interior material leading to oxidation in the gaseous phase. The process may be accelerated by softening or melting, which hastens decomposition and permits escape of flammable volatiles. Some insight into flame behavior may be gained, therefore, by consideration of behavior on heating alone, in the absence of the final, oxidative combustion step.

Thermogravimetric and differential thermal analytic data (tga and dta, respectively) on novoloid fibers indicate that above 250°C in the absence of oxygen novoloid fibers undergo gradual weight loss until, close to 700°C, the fiber is fully carbonized, with a carbon yield of 55–60% (3–4). Thus the high initial carbon content results in limited production of volatiles (weight loss 40–45%), only a portion of which is, in fact, flammable. Melting does not occur, and shrinkage is small, suggesting that the cross-linked structure of the material promotes gradual coalescence of the aromatic units into a stable, amorphous char; the relatively low thermal conductivity of the material aids in the moderation of this process.

The foregoing observations are helpful in understanding the high flame resistance of novoloid fibers and fabrics, particularly since oxygen generally is depleted in the actual zone of combustion. Exposure to flame results in formation of an amorphous surface char that serves both to radiate heat from the fiber and to retard the evolution of flammable volatiles; the latter is limited by the high carbon content. The amorphous (glassy) nature of the char presents a minimum reactive surface to the flame. Moreover, in a textile-fabric structure the charred fibers at the fabric face provide a protective insulating barrier retarding penetration of both heat and oxygen into the interior of the fabric; minimal shrinkage enhances the mechanical stability of this barrier. Finally, the formation of H_2O and CO_2 as products of decomposition and combustion provides an ablative type of cooling effect (see Ablative materials).

Since novoloid fibers are composed only of carbon, hydrogen, and oxygen, the products of combustion are principally water vapor, carbon dioxide, and carbon char. Moderate amounts of carbon monoxide may be produced under certain conditions; but the HCN, HCl, and other toxic by-products typical of combustion of many flame-resistant organic fibers are absent. The toxicity of the combustion products is thus very low or negligible (5). Smoke emission is also minimal, less than that of virtually any other organic fiber.

Oxidative Degradation. The mechanism of oxidative degradation of novoloids is similar to that of phenolic resins in general (6), and appears to involve mainly oxidative attack on the methylene linkages between aromatic units, with initial formation of a peroxide (see Hydrocarbon oxidation). This peroxide is subject to further decomposition to a carbonyl which, in conjugation with the phenolic hydroxyl, apparently forms the chromophoric keto–enol group which is thought to contribute to the characteristic golden color of the fiber (4).

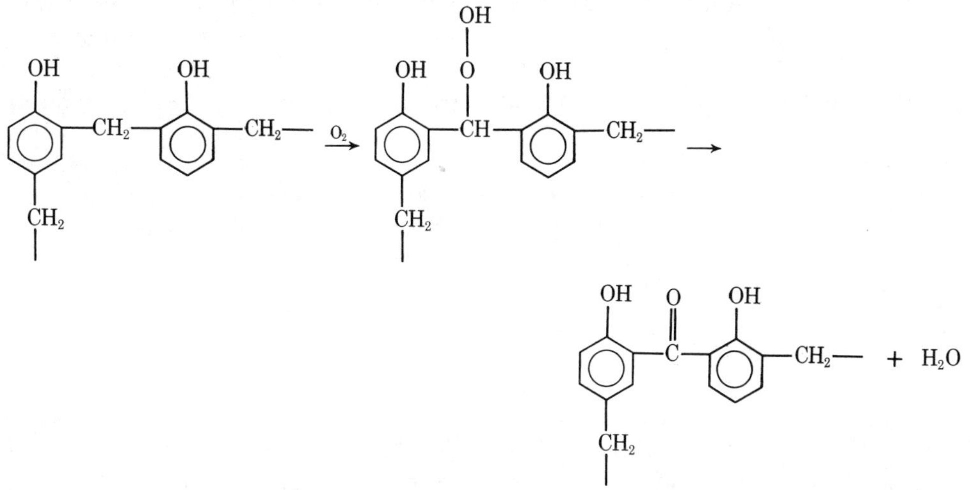

The subsequent, complex mechanism of oxidative degradation involves principally the same carbonyl group and leads to chain scission. Formation of quinoids and related compounds is probably responsible for further darkening.

Oxidative degradation is a comparatively slow process which, in the severe environment of a flame, is far less significant in determining fiber behavior than is the nonoxidative char formation described above. At ordinary temperatures it leads only to darkening of the fiber surface, with no significant effect on properties. However, oxidation is accelerated by increasing temperatures, leading to significant loss of weight and strength as temperatures approach 200°C. Therefore, the practical temperature limit for long-term applications in air is about 150°C.

Punking. Decomposition of the peroxide formed at the methylene linkage is an exothermic reaction. As a result this reaction may, under certain conditions, become self-sustaining, leading to the phenomenon known as punking.

Punking may occur in materials containing methylene bonds subject to peroxide formation, provided a large surface area is available for oxidation and the rate of heat removal is low. These conditions are met by large masses of tightly packed novoloid fibers, which are excellent heat insulators. Given sufficient oxygen for peroxide formation, sufficient fiber bulk for efficient insulation, and an initial temperature high enough for peroxide decomposition to generate more heat than is removed by ventilation or other means, internal temperatures may increase to 570–580°C, at which point ignition of the fiber may occur. The result is a smoldering combustion that continues until the fiber mass is consumed.

Punking may easily be prevented by appropriate control of storage and use. Thus, novoloid materials should not be stored in bulk at high temperatures or subjected to lengthy heat treatment over 120°C without adequate ventilation, and should be cooled below 60°C after heat treatment or high temperature use. Applications involving prolonged exposure to high temperatures with limited opportunity for heat escape—such as steam-pipe insulation—should be avoided. Punking does not occur in the absence of oxygen, eg, when the fibers are encapsulated by a matrix material (see also Peroxides, organic).

Chemical Resistance. Novoloid fibers display excellent chemical and solvent resistance. They are attacked by concentrated or hot sulfuric and nitric acids and strong bases, but are virtually unaffected by nonoxidizing acids, including hydrofluoric and phosphoric acids; dilute bases; and organic solvents.

Other Properties. Novoloid fiber materials display excellent thermal insulating characteristics. A nonwoven batting at a density of 10 kg/m^3 showed a thermal conductivity of 0.04 W/(m·K) at 20°C and less than 0.03 W/(m·K) at −40°C (7) (see Insulation, thermal). Retention of properties at very low temperatures is excellent. Efficacy of sound absorption is high (see Insulation, acoustic). Ultraviolet radiation, although leading to deepening of color, has minimal effect on fiber properties; resistance to γ-radiation is also high.

Textile and Paper Properties

The textile properties of novoloid fibers are summarized in Table 1. The ranges shown for strength, elongation, and modulus reflect the dependence of these properties on fiber diameter, with the higher values corresponding to finer diameters.

Novoloid fibers are processed by suitably modified conventional textile techniques. The moderate tensile strength (comparable to that of cellulose acetate) and lack of inherent crimp require that suitable precautions be taken, particularly in carding staple fibers, to prevent excessive fiber breakage. Thorough opening is required, and carding speed is lower than that for most conventional fibers. A modified woolen system is probably best for spinning. Blending with other fibers such as aramids improves processing speeds and increases yarn tensile strength (see Aramid fibers).

Table 1. Textile Properties of Novoloid Fibers

Property	Value
diameter, μm (tex[a])	14–33 (0.22–1.1)
specific gravity	1.27
tensile strength, GPa[b] (N/tex[c])	0.16–0.20 (0.12–0.16)
elongation, %	30–60
modulus, GPa[b] (N/tex[c])	3.4–4.5 (2.6–3.5)
loop strength, GPa[b] (N/tex[c])	0.24–0.35 (0.19–0.27)
knot strength, GPa[b] (N/tex[c])	0.12–0.17 (0.10–0.13)
elastic recovery[d], %	92–96
work-to-break, J/g (= mN/tex[e])	26–53
moisture regain at 20°C, 65% rh, %	6

[a] To convert tex to den, multiply by 9.
[b] To convert GPa to psi, multiply by 145,000.
[c] To convert N/tex to gf/den, multiply by 11.33; N/tex = GPa ÷ density.
[d] At 3% elongation.
[e] To convert mN/tex to gf/den, multiply by 0.01133.

Heavy yarns (300 tex) of 100% novoloid fiber are readily spun. With suitably modified equipment 100% novoloid yarns as fine as 30 tex and blended yarns of 20 tex are routinely produced.

Spun novoloid yarns are used in the production of woven fabrics in weights from 100 to >550 g/m^2, as well as in knitted products such as gloves. Blending with other fibers improves the low tensile strength and abrasion resistance of 100% novoloid materials. Novoloid fibers may be dyed with cationic or disperse dyes; however, color range and stability are limited by the inherent gold color of the fiber and its tendency to darken with exposure to heat and light.

Dry-formed webs and felts are produced by needle-punching and other felting techniques. Novoloid papers are produced by wet-laid paper methods. For papermaking, the physical characteristics of the fiber generally require admixture of a binder, such as PVC, polyesters, or epoxies. For 100% novoloid paper, a suitable proportion of partially cured fiber may be added. Hot-pressing or calendering of the web expresses the uncured novolac resin and promotes binding by cross-linking to the fiber surface (see Paper; Pulp, synthetic).

Manufacture

Novoloid fibers are infusible, insoluble, intractable three-dimensional network polymers. To produce fibers from such a material, fiberization must precede cross-linking; ie, preformed precursor fibers must undergo chemical modification.

Melt-Spinning. Novoloid fibers are typically produced by first melt-spinning a novolac resin. Novolac resins are manufactured from phenols and aldehydes using an acid catalyst and excess phenol. A suitable novolac for fiber production is that made from phenol and formaldehyde. Such a material may be melt-spun to produce an extremely friable, whitish fiber that is both readily fusible and soluble in organic solvents. Some molecular orientation parallel to the fiber axis may be observed at this stage.

Aqueous Cure. The melt-spun fibers are then cured by immersion in an acidic aqueous solution of formaldehyde, followed by gradual heating to promote the acid-catalyzed cross-linking reaction. Curing commences with the formation of a skin on the fiber surface, and proceeds inward as the acid (usually HCl) and formaldehyde diffuse into the material.

This process transforms the novolac resin into a three-dimensionally cross-linked network polymer through the formation of methylene and dimethyl ether linkages. The molecular orientation disappears, and the material becomes unoriented and

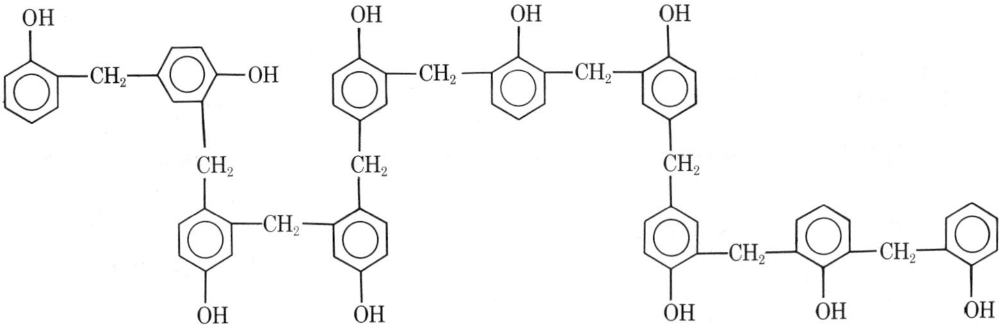

novolac resin

amorphous. As curing proceeds, the proportion of dimethyl ether linkages declines, and the cross-link density may approach that of cured phenolic resins. Generally, however, the density of cross-linking remains comparatively low. Although tensile strength increases by a factor of roughly ten, the cured fibers remain flexible and elastic.

The extent of cross-linking is limited by steric factors and by adjustment of curing conditions. As a result, methylol groups formed on sites where cross-links are not completed typically constitute 5–6 wt % of the cured fiber, and are available for further reaction with matrix materials at the fiber surface. These groups play a key role in the formation of novoloid-fiber composites. The proportion of these groups may be reduced by heating the cured fiber to 180°C.

The combination of melt-spinning followed by aqueous curing permits the large-scale production of textile-grade fibers with closely controlled and highly reproducible properties in diameters from <14 to >33 μm (0.2–1 tex or 2–10 den) and lengths from 100 μm to continuous filament. Such fibers are currently available for ca 10 $/kg (staple fiber). Price reductions of 30–40% are expected as production increases.

Partial Cure. Interruption of the curing process at an early stage gives partially cured fibers containing uncured novolac resin within an envelope of cross-linked material. The uncured core remains fusible and soluble, and may be extracted by organic solvents such as methanol to leave hollow novoloid fibers. Alternatively, the uncured material may be left in place to be removed later by thermomechanical means when it may serve, for instance, as a self-contained binder for novoloid-fiber papers.

Blown Fibers. For finer fibers or when precise control of diameter and length are not required, other manufacturing methods may be used. Very fine precursor fibers can be produced by air attenuation of molten resin injected into a stream of hot air. When collected and cured as above, these blown fibers have a fuzzy or woolly appearance, random lengths, and random diameters in the range of 2–5 μm. Alternatively, fiberization may be carried out by centrifugal spinning in a cotton-candy-type machine.

Self-Curing Fibers. Self-curing novoloid fibers also may be produced. Curing agents such as hexamethylenetetramine or paraformaldehyde are routinely blended into novolacs to provide bulk-molding resins which cure upon heating. A similar method has been demonstrated for the production of novoloid fibers (8). For successful fiber formation, the blend of resin and curing agent must be melted and fiberized quickly

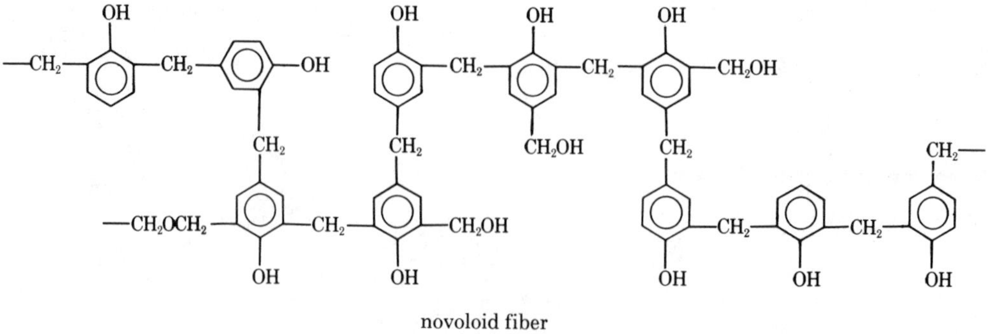

novoloid fiber

without curing, and then cured without remelting. Rapid fiberization is accomplished by hot air impingement, centrifugal spinning, or other suitable means, followed by curing in an acidic gas such as hydrogen chloride or boron trifluoride. This method permits very rapid curing, with cross-linking completed in minutes. Commercial feasibility of this method has not yet been established.

Acetylation. Novoloid fibers produced by the above methods are typically golden, the coloration deepening with age and exposure to heat or light (reverse fading). Color development may be prevented by heating the cured fibers with acetic anhydride to produce bleached or white novoloids.

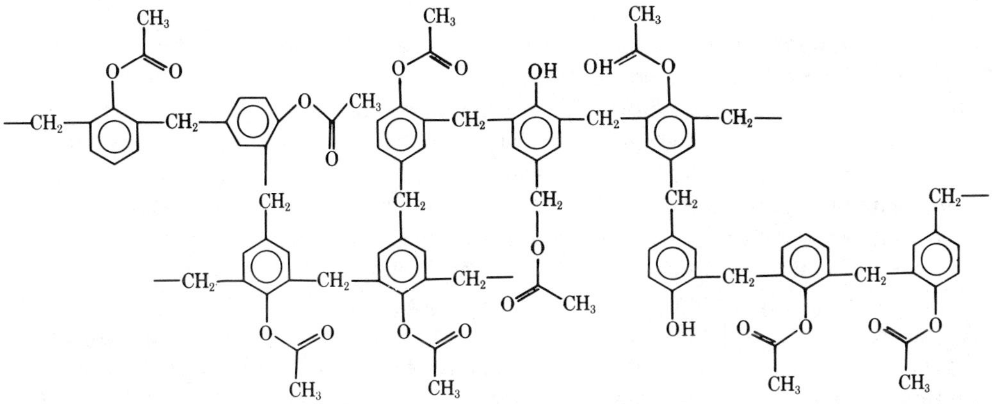

acetylated novoloid fiber

Acetylation of the phenolic hydroxyl groups inhibits peroxide formation at the methylene linkage, through both steric and inductive effects. Thus oxidative degradation and formation of chromophoric groups is prevented, and acetylated novoloid fibers are white and not subject to reverse fading. The limitation in dyeability is overcome and a broad spectrum of colors becomes available. Thermal stability is enhanced, and the possibility of punking is diminished. However, tensile strength and flame resistance are reduced.

Acetylation significantly increases fiber costs.

Analysis

Table 2 shows typical elemental analyses and approximate empirical formulas for novolac resin, cured novoloid fibers, and acetylated novoloid fibers.

Table 2. Empirical Formulas and Elemental Analyses

Material	Empirical formula (approximate)	Elemental analysis, wt % C	H	O
novolac resin	$C_{69}H_{59}O_{10}$	78.3	5.6	15.4
cured novoloid fiber	$C_{63}H_{55}O_{11}$	75.8	5.5	17.9
acetylated novoloid fiber	$C_{61}H_{56}O_{14}$	72.0	5.5	22.5

Health and Safety Factors

A toxicity evaluation of Kynol novoloid fibers in accordance with regulations under the *Federal Hazardous Substances Act* demonstrated that the fiber is neither toxic nor highly toxic under the oral and skin contact categories; is not an eye or primary skin irritant; and has a low order of acute inhalation hazard (9). The low toxicity of combustion products and limited smoke evolution are discussed above, as is the need for appropriate precautions in storage and use to forestall punking.

Uses

The combination of infusibility, excellent flame and chemical resistance, minimal evolution of smoke or toxic gases, and low specific gravity permit novoloid fibers (either alone or in combination with other materials) to replace asbestos in many uses (see Asbestos). Furthermore, the specific physical and chemical properties of these fibers lead to numerous other applications, particularly in new composite materials (see Composite materials). Novoloid fibers also are precursors of low modulus carbon and activated carbon fibers (see Carbon and artificial graphite).

Textile and Related Applications. Woven, knitted, and felted novoloid fibers, either alone or in combination with other materials, are used in protective apparel including fire suits, insulating gloves, and clothing for foundry workers (see Flame-retardant textiles). Fabrics may be aluminized or elastomer-coated for added protection and strength. Moderate tensile strength and abrasion resistance, and restricted dyeability, limit these applications.

Heavy woven novoloid fabrics, with or without aluminum or elastomeric coating, can be substituted for asbestos textiles in applications requiring resistance to flame and metal splash; they are lighter in weight and easier to manipulate. Because of its high chemical and solvent resistance, novoloid fiber also is utilized in gaskets and braided packings.

Novoloid spun yarns and papers are employed as filler yarns and wrapping tapes in high performance communications cables for installation where maximum circuit integrity must be maintained for safety reasons; eg, in nuclear-energy plants, highway tunnels, high-rise buildings, and underground concourses.

Composites. Novoloid fibers are used in combination with a wide variety of matrix materials such as thermosetting and thermoplastic resins, elastomers, and ceramics. Owing to a direct cross-linking reaction between fiber and matrix, in which the methylol groups formed during curing play the key role, novoloid composites frequently display unexpected synergistic improvements in properties. Typical examples are the composites of novoloid fiber with resole resin and chlorinated polyethylene (CPE).

As the fiber component of organic matrix composites, novoloid fibers are characterized by easy and uniform dispersion and generally excellent fiber-to-matrix wetting and adhesion. They improve properties such as heat resistance, impermeability, compressive strength, shock resistance, dimensional stability, and hardness. However, because of the comparatively low strength of the fiber itself, tensile strength generally is not markedly increased. The low specific gravity of novoloid fibers often leads to a reduction in weight of the composite material.

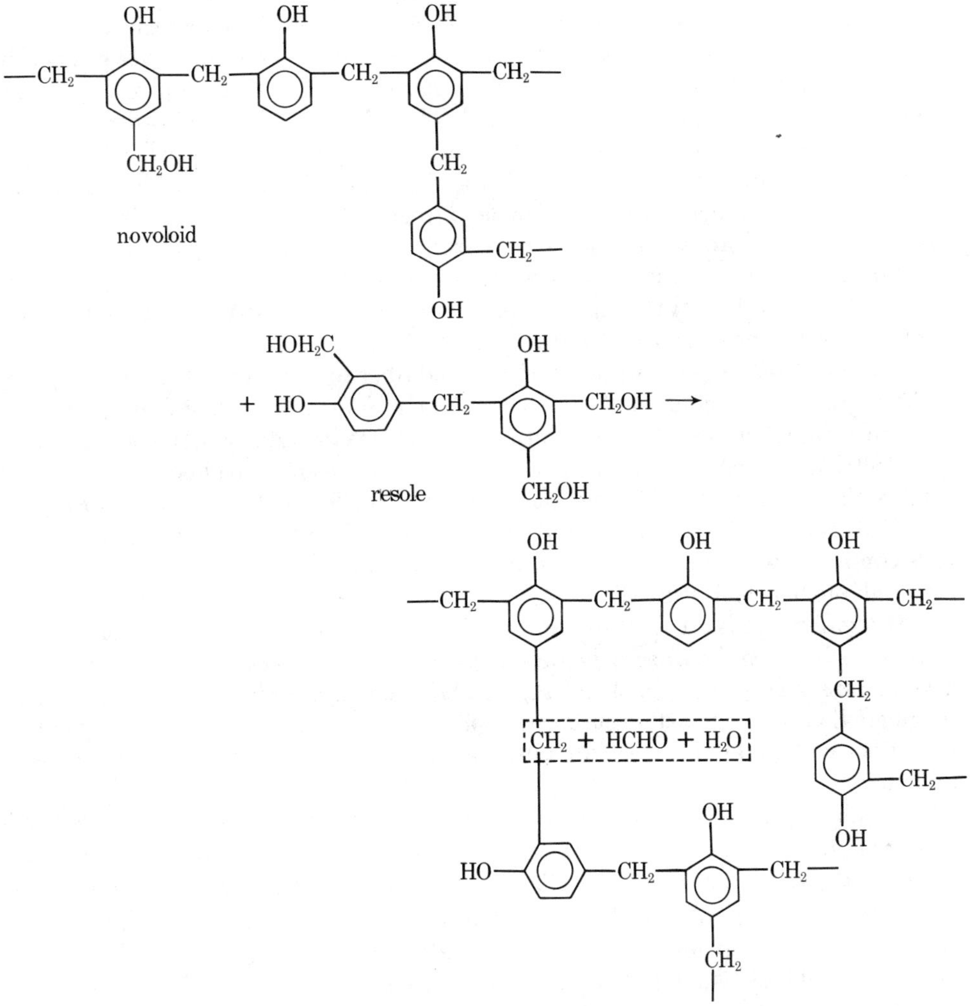

novoloid + resole →

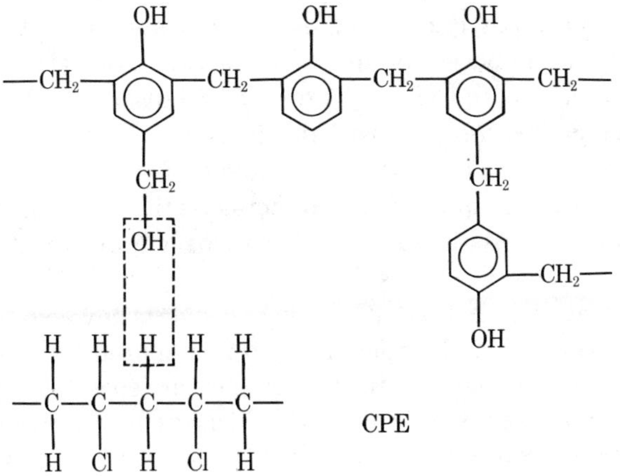

novoloid

CPE

Thermosetting Resins. Owing to the large degree of similarity in chemical structure, novoloid fibers are particularly suitable as fillers for phenolic resins. As much as 70 wt % fiber may be incorporated into a resole matrix. The composite has better thermal stability (up to 230°C), thermal insulation, electrical resistance (especially after boiling in water), high temperature shock resistance, chemical resistance, and machinability than resins filled with other fibers. Further evidence of the high compatibility and synergism of novoloid fiber and resole resin is the fact that, although the carbon yield on carbonization of novoloid fibers is 55–57% and that of resole alone 50–52%, a 50/50 composite of the two materials yields 65–70% carbon (see Fillers).

Similar results are obtained with epoxy and other reactive resins. Applications for such fiber-reinforced thermoset plastics include heat-resistant automotive and electrical parts, and chemical-resistant materials.

Thermoplastic Resins. Addition of novoloid fiber to PVC raises the melting point of the resin; at a fiber content of ca 15 wt %, the material does not melt. Similar results are seen with polyamides (qv). Potential applications include printed circuit boards and electrical panels. In combination with polyester resins, novoloid fibers provide a lightweight, readily machinable material with excellent cold resistance for piping and similar applications. Furthermore, woven novoloid fabrics and thin-paper surface veils combined with polyester resins provide long-lasting tank linings for corrosive materials such as HCl and HF.

Elastomers. It is difficult to reinforce cured CPE, a flame-retardant elastomer, with fibers because the fiber is degraded by HCl released on curing. Novoloid fibers, however, are unaffected by HCl; in fact, the fiber–matrix cross-linking reaction permits the fibers themselves to act as the curing agent for pure (unformulated) CPE, yielding a product with good hardness, dimensional stability, and resistance to flame, heat, and chemicals.

Similar results are achieved with numerous other elastomers including ethylene–propylene, acrylic, and nitrile rubbers, chlorosulfonated polyethylene, and polytetrafluoroethylene (PTFE). High ratios of uniform fiber incorporation, better than 50 wt %, may be obtained with conventional rubber-processing equipment. Because of the relatively low resistance of novoloid fibers to sulfuric acid, sulfur-curing systems should be avoided or applied with caution.

Applications for novoloid-filled elastomers include cushion boards and rollers; chemical, solvent, and fuel hoses; ducts; gaskets; packings; and the coating or saturation of novoloid fabrics for use in hoses and ducts and as protective sheets and curtains. A 550 g/m^2 novoloid fabric coated with 650 g/m^2 of formulated CPE, eg, has a tensile strength of 4–5 GPa (580,000–725,000 psi) and sheds molten steel.

Ceramics. Addition of small amounts of novoloid fibers to unfired ceramics stabilizes the material during rapid firing and increases the strength and integrity of the final product. Applications include daubable furnace linings and sprayable fire-protective cable coatings (see Ceramics).

Carbon Fibers. Heating in an inert atmosphere converts novoloid fibers gradually to pure carbon at yields of 55–57%. Neither pretreatment nor tension is required during carbonization. The rate at which carbonization can be carried out is governed mainly by oven configuration and the rate at which evolved gases are removed. Shrinkage of the fiber on carbonization is predictable and low, on the order of 20%.

As a consequence of this simplicity and stability of processing, novoloid fibers may readily be carbonized either as a continuous tow, or in a final product configuration (felt, fabric) pre-formed from the readily processable novoloid precursor fibers.

Novoloid-based carbon fibers are amorphous. Even treatment at the usual graphitization temperatures above 2000°C does not result in the formation of the typical well-ordered graphite molecular structure. Novoloid-based carbon is thus a low modulus, low strength material, at least in comparison to the high modulus fibers prepared from rayon and polyacrylonitrile (PAN) precursors.

Novoloid-based carbon fibers are obtainable in regular and high purity grades. Treatment of the precursor by controlled increase of temperature to about 800°C in an inert atmosphere gives highly stable carbon fibers with a carbon content of about 95%. Further heating to 2000°C yields fibers of 99.8% or greater carbon content.

Table 3 gives the typical properties of novoloid-based carbon fibers compared to PAN- (high modulus) and pitch- (low modulus) based carbon fibers. The decrease in specific gravity of novoloid-based carbon as temperature exceeds 1000°C is remarkable and has not been adequately explained. It does not appear related to any observable development of microporosity or voids. Despite the apparent decrease in interatomic spacing d_{002}, the x-ray peak remains broad and the material is essentially amorphous in nature. Of further interest is the fact that although fibers carbonized at 600°C retain the elliptical cross section of the novoloid precursor, those treated at 2000°C tend to become round.

Novoloid-based carbon fiber is soft and pliable, produces little dust or fly on processing, and has good lubricity. It retains much of the good dispersability and affinity for matrix materials of its novoloid precursor. In addition, its heat, chemical, and electrical characteristics are comparable to those of other carbon fibers. Thus,

Table 3. Typical Properties of Carbon Fibers

Property	Precursor: Novoloid		Pitch		Polyacrylonitrile	
type	low modulus		low modulus		high modulus	
treatment temperature, °C	800	2000	1000	2000	1500	2000
specific gravity, g/cm^3	1.55	1.37	1.63	1.55	1.8–1.9	1.9–2.0
carbon content, wt %	95	99.8+	95	99.5+	93	99.5+
x-ray diffraction profile, 002, 2θ, degrees	23.0[a]	25.0[a]	24.0[a]		25.0[b]	26.1[c]
interlayer spacing, d_{002}, pm	395	351				336
tensile strength, MPa[d]	500–700	400–600	500–1000		1500–3000	
elongation, %	2.0–3.0	1.5–2.5	1.5–2.5		1.0–1.5	
modulus, GPa[e]	20–30	15–20	30–50		150–300	
heat resistance, °C						
tga	436	541	416		519	
air	350	380	350		350	
specific resistivity, mΩ·cm	10–30	5–10	10–30		1–10	
affinity with PTFE, CPE, epoxies[f]	good		fair		poor	

[a] Broad.
[b] Medium.
[c] Sharp.
[d] To convert MPa to psi, multiply by 145.
[e] To convert GPa to psi, multiply by 145,000.
[f] PTFE = polytetrafluoroethylene; CPE = chlorinated polyethylene.

it is well suited for production of braided packings, either impregnated or unimpregnated (no coating is required for braiding even of the high purity filament yarn); bearings and other low friction materials; carbon-fiber-containing conductive elastomers with uniform, predictable, and high conductivity; and highly heat- and chemical-resistant gaskets and sheets. Paper made from chopped carbon fiber is suitable for resin impregnation to produce sheet electrodes for electrostatic precipitators (see Air pollution control methods). Yarns, fabrics, and felts of novoloid-based carbon are used in protective curtains, vacuum-furnace insulation, and static-elimination brushes.

Activated Carbon. Novoloid-based activated carbon is formed by a one-step process combining both carbonization and activation, in an oxygen-free atmosphere containing steam and/or CO_2, at ca 900°C. The material can be activated either as a continuous tow or after the formation of fabrics or felts. The degree of activation, as measured by the BET method, is controlled by time rather than temperature, and can approach 3000 m^2/g. Table 4 gives typical properties of novoloid-based activated-carbon fibers; data for PAN- and rayon-based activated-carbon fibers and for granular activated carbon are presented for comparison.

The pores of novoloid-based activated carbon are generally uniform in size and

Table 4. Typical Properties of Activated Carbon Fibers

Property	Precursor				
	Novoloid	Novoloid	PAN[a]	Rayon[a]	Granules
yield, wt %	33	22	15	7	
diameter, μm	9.2	8.5	ca 5	ca 10	
pH	7.3	7.5	8.0		6.3
pore capacity					
specific surface area[b], m^2/g	1500	2000	870, max	1450, max	910, typical
pore volume[c], cm^3/g	0.63	0.75	0.28	0.53	0.42
mechanical properties					
tensile strength, MPa[d]	400	350	300	70	
elongation, %	2.8	2.7			
modulus, GPa[e]	14	12	50+		
gas adsorption capacity at 20°C[f], wt %					
benzene	53	67	34	46	33
toluene	57	80	32	47	35
trichloroethylene	83	104	53		61
moisture content at 20°C, 65% rh, wt %	37	20	30		27
methylene blue adsorption capacity[g], mL/g	200	300	40		100

[a] Owing to their partially crystalline structure, these examples represent the practical limits of activation for these fibers. PAN = polyacrylonitrile.
[b] BET method (Brunauer, Emmett, Teller).
[c] Steam-adsorption method.
[d] To convert MPa to psi, multiply by 145.
[e] To convert GPa to psi, multiply by 145,000.
[f] JIS K-1474 Japan industry standard.
[g] JIS K-1470 Japan industry standard.

straight, rather than branched as in granules. Pore radius distribution shows a single peak at about 1.5–1.8 nm; radius and volume increase with increasing activation, and thus may be controlled for selective adsorption of polypeptides and similar large molecules. Pore configuration and the high surface-to-volume ratio of the fibers, compared to granular activated carbon, permit extremely rapid adsorption and desorption. A further advantage over granules is the reported potential for rapid and convenient reactivation of novoloid-based activated-carbon felts and fabrics by direct-resistance heating *in situ* with a moderate electrical current (10).

Owing to the strength and flexibility of novoloid-based activated carbon fibers, wet-laid paper webs can be formed from activated-carbon fiber and kraft pulp without any binder. Effective surface area of the finished sheet is therefore close to that calculated from the activated-carbon fiber content.

Applications for activated-carbon paper include medical supplies such as bandages for malodorous wounds; quick-acting air and liquid-purification filters; ozone eliminators for electrostatic copiers; and protective wrappings for electronic sensors. Activated-carbon fabrics hold promise for use in compact dialysis equipment.

Glassy Carbon. When a novoloid–resole composite is baked at high temperatures over an extended period of time, it completely carbonizes to form a uniform, amorphous, glassy or vitreous carbon in which the portions formerly comprising fiber and matrix are virtually indistinguishable. This material is harder than glass, has extremely low gas permeability, and is hardly affected by 5 h immersion in 50:50 sulfuric–nitric acid at 100°C. Applications are foreseen in extremely rigorous chemical, electrical, mechanical, and heat environments. Production cost is expected to be significantly lower than that of hitherto available glassy carbon products.

BIBLIOGRAPHY

"Phenolic Fibers" in *ECT* 2nd ed., Suppl. Vol., pp. 667–673, by James Economy and Luis C. Wohrer, The Carborundum Company.

1. *Fed. Reg.* **39,** 1833 (1974).
2. U.S. Pats. 3,650,102 (March 21, 1972); 3,723,588 (March 27, 1973), J. Economy and R. A. Clark (to the Carborundum Company).
3. J. Economy and L. Wohrer, "Phenolic Fibers" in N. M. Bikales, ed., *Encyclopedia of Polymer Science and Technology*, Vol. 15, Wiley-Interscience, New York, 1971, pp. 370–373.
4. J. Economy, L. C. Wohrer, F. J. Frechette, and G. Y. Lei, *Appl. Polym. Symp.* **21,** 81 (1973).
5. J. Economy, "Phenolic Fibers" in M. Lewin, S. M. Atlas, and E. M. Pearce, eds., *Flame-Retardant Polymer Materials*, Vol. 2, Plenum Press, New York, 1978, pp. 210–219.
6. R. T. Conley and D. F. Quinn, "Retardation of Combustion of Phenolic, Urea–Formaldehyde, Epoxy, and Related Resin Systems" in ref. 5, Vol. 1, 1975, pp. 339–344.
7. Dynatech R/D Company, report to American Kynol, Inc., April 26, 1976.
8. U.S. Pat. 4,076,692 (Feb. 28, 1978), H. D. Batha and G. J. Hazelet (to American Kynol, Inc.).
9. Industrial Health Foundation, Inc., report to the Carborundum Company, Sept. 1972.
10. J. Economy and R. Y. Lin, *Appl. Polym. Symp.* **29,** 199 (1976).

General References

References 3–5, and 10 are general references.

J. Economy, L. C. Wohrer, and F. J. Frechette, "Non-Flammable Phenolic Fibers," *J. Fire Flammability* **3,** 114 (April 1972).

R. Y. Lin and J. Economy, "Preparation and Properties of Activated Carbon Fibers Derived from Phenolic Precursors," *Appl. Polym. Symp.* **21,** 143 (1973).

J. Economy, "Kynol Novoloid Fibers," *Polym. News* **II** (7/8), 13 (1975).

H. D. Batha, "Resistance of Kynol Fabrics to [Molten] Metals" in V. M. Bhatnagar, ed., *Fire Retardants: Proceedings, 1976 International Symposium on Flammability and Fire Retardants*, Technomic, Westport, Conn., 1977, pp. 81–87.

K. Ashida, M. Ohtani, T. Yokoyama, K. Kosai, and S. Ohkubo, "Full Scale Investigation of the Fire Performance of Urethane Foam Cushions Using Novoloid Fiber Products as Interlayer," *J. Cellular Plast.*, 311 (Nov.–Dec. 1978).

J. Economy, "Now that's an interesting way to make a fiber," *Chemtech* **10**(4), 240 (April 1980).

JOSEPH S. HAYES, JR.
American Kynol, Inc.

PAPER

Paper consists of sheet materials that are comprised of bonded small discrete fibers. The fibers usually are cellulosic in nature and are held together by secondary bonds which, most probably, are hydrogen bonds (see Cellulose). The fibers are formed into a sheet on a fine screen from a dilute water suspension. The word paper is derived from papyrus, a sheet made in ancient times by pressing together very thin strips of an Egyptian reed (*Cyperus papyrus*) (1).

The early use of paper was as a writing medium which replaced clay tablets, stone, parchment, and papyrus sheets. Papyrus sheets are not considered to be paper because the individual vegetable fibers are not separated and then reformed into the sheet. Paper apparently originated in China in 105 AD and was made from flax and hemp or bark fibers of certain trees (2). The manufacture of paper from bark and bamboo spread from China to Japan where its manufacture began in ca 610. Papermaking from flax and hemp spread through Central Asia, the Middle East, and eventually into Europe. Spain and Italy were the early European papermaking centers. The first European paper was made in Spain in 1150, in France by 1189, in Germany by 1320, and in England by 1494. Papermaking was introduced in the United States in ca 1700 by William Rittenhouse in Philadelphia; a history of U.S. papermaking is given in ref. 3.

Paper is made in a wide variety of types and grades to serve many functions. Writing and printing papers constitute ca 30% of the total production. The balance, except for tissue and toweling, is used primarily for packaging. Paperboard differs from paper in that it generally is thicker, heavier, and less flexible than conventional paper.

Fibrous Raw Materials

The main components used in the manufacture of paper products are listed in Table 1. More than 95% of the base material is fibrous and more than 90% originates from wood (qv). Many varieties of wood, eg, hardwood and softwood, are used to produce pulp. In addition to the large number of wood types, there are many different manufacturing processes involved in the conversion of wood to pulp. These range from mechanical processes, by which only mechanical energy is used to separate the fiber from the wood matrix, to chemical processes, by which the bonding material, ie, lignin (qv), is removed chemically. Many combinations of mechanical and chemical methods also are employed (see Pulp). Pulp properties are determined by the raw material and manufacturing process, and must be matched to the needs of the final paper product.

In mechanical pulps, the fibers are separated by mechanical energy. Because there is no chemical removal of wood components, the process results in a high pulp yield (95%). The chemical composition of mechanical pulps is similar to those of native wood, ie, they contain significant amounts of lignin and hemicellulose in addition to the basic cellulose component of the fiber. In the stone groundwood process, fibers are separated from the wood by grinding logs against revolving stone wheels. The resultant fibers are fractured and much debris is generated. These pulps are used where opacity and good printability are needed. The presence of large amounts of lignin, however, results in poor light stability, permanence, and strength. Such pulps are bleached with sequences that maintain a high pulp yield and do not remove lignin. Typical bleaching chemicals are either alkaline hydrogen peroxide or sodium hydrosulfite (dithionite). Brightness levels of 80% usually can be achieved.

Newer methods of mechanical pulp production involve disk refiners to produce pulp from wood chips. Chips are passed between closely spaced revolving disks and the fiber is broken free from the wood material. The pulps generally contain less debris and longer fibers than stone groundwood pulp. Refiner mechanical pulp (RMP) is produced at atmospheric pressure from a disk refiner; it is the oldest of the refiner processes. In the thermomechanical pulp (TMP) process, chips are steamed at 120°C prior to fiberization in a pressurized disk refiner. Compared to stone groundwood pulp, TMP is comprised of longer, less damaged fibers and less debris, which results in improved strength but some opacity loss. Thermomechanical pulps can be used to reduce or eliminate chemical pulps in many blends. Softwood is the preferred raw material to produce optimum strength in the product.

Table 1. Main Components of Paper and Paperboard (in 1980), wt %[a]

	World	North America
mechanical/semichemical wood pulp	20.8	21.5
unbleached kraft chemical wood pulp	18.5	25.2
white chemical wood pulp[b]	26.6	28.6
waste fiber	25.3	20.3
nonwood fibers	4.2	0.9
fillers/pigments	4.6	3.5

[a] Ref. 4.
[b] Includes unbleached sulfite.

Certain chemical treatments can be employed during the TMP process to achieve improved strength. Sodium sulfite and hydrogen peroxide have been used either for chip pretreatment or posttreatment of the TMP pulp; such pulp is chemithermomechanical pulp (CTMP). The strength improvements, which may be 50%, are obtained at some sacrifice to yield and opacity.

The yields of mechanical pulps are 90–95%; the lower yields are associated with chemical treatment. No major commercial pulps are produced in the next lower yield range, ie, 80–90%. The next major class of pulps, semichemical pulps, generally is characterized by a yield of less than 80%; significant amounts of material are removed by chemical action.

Semichemical pulps are produced by mild chemical digestion of chips prior to reduction to pulp in a disk refiner. Yields are 70–80%, and the pulp contains a lower lignin and hemicellulose content than the wood from which it was derived. The main use for this product is in corrugated media, in which the stiffness resulting from the lignin and hemicellulose components is a product advantage. Hardwoods usually are the base woods for semichemical pulps. Use of the neutral sulfite process is declining because of environmental problems resulting from the lack of a suitable recovery system. Newer processes, which are based on sodium carbonate/sodium sulfide, ie, the green liquor process, or sodium hydroxide/sodium carbonate, are replacing the neutral sulfite sequence. Compared to the neutral sulfite pulps, these alkaline semichemical pulps are darker in color and have slightly higher lignin contents.

Chemical pulps have greatly reduced lignin and hemicellulose contents compared to the native wood, as these components are dissolved during chemical digestion. Because the lignin is removed, much less mechanical energy is needed to separate the fibers from the wood matrix, and the resulting pulp fibers are undamaged and strong. Chemical pulps are used principally for strength and performance in a variety of paper and paperboard products.

The principal process for producing chemical pulps is the kraft process: mixtures of sodium sulfide and sodium hydroxide are the pulping chemicals, and yields are 46–56%. The higher yield pulps contain about 10% lignin. They are used in bags or linerboard where strength is important. The lower yield pulps typically are bleached to remove virtually all lignin and to produce high brightness (90%+). These pulps are used where permanence and whiteness are needed in addition to strength. Bleaching technology is multistage and is based on sequential use of chlorine, hypochlorite, and chlorine dioxide. The use of oxygen compounds, eg, gaseous oxygen, ozone (qv), and hydrogen peroxide (qv), is becoming more important as environmental problems become more severe. Variations in the pulping and bleaching sequences produce varying brightness, yield, and strength.

Manufacture of small amounts of chemical pulp still is based on an acid sulfite process. A shift from calcium-base to soluble-base acid sulfite systems is occurring in response to environmental constraints on the disposal of waste pulping liquor. Sulfite pulps are used primarily to provide high purity and brightness.

Reclaimed fiber accounts for ca 25% of the total fiber used in the United States. A variety of sequences are used to disperse and clean the waste fiber. Emphasis is on mechanical screening and cleaning and only limited chemical treatment. The properties of these pulps depend largely on the input raw material and generally are lower in strength and brightness than a comparable virgin pulp.

Nonwood fibers are used in relatively small volumes. Examples of nonwoody pulps and products include cotton linters for writing paper and filters, bagasse (qv) for corrugated media, esparto for filter paper, or Manila hemp for tea bags. Synthetic pulps which are based on such materials as glass (qv) and polyolefins also are used (see Pulp, synthetic). These pulps are relatively expensive. They usually are used in blends with wood pulps where they contribute a property which is needed to meet a specific product requirement.

Physical Properties

Most properties of paper depend upon direction. For example, strength is greater if measured in the machine direction, ie, the direction of manufacture, than in the cross-machine direction. For paper made on a Fourdrinier paper machine, the ratio of the two values varies from about 1.5 to 2.5. An even greater anisotropy is observed if either of the in-plane values is compared to the out-of-plane strength. Paper is quite weak in the thickness direction.

Paper may be considered an orthotropic material, ie, one possessing three mutually perpendicular symmetry planes (5). The three principal directions are the machine, cross-machine, and thickness directions. There are several reasons for this anisotropy. Wood pulp fibers are long, slender, and usually ribbonlike, rather than circular, in cross section. During the deposition of the slurry onto the wire, the fibers tend to line in the direction of the moving wire, and the extent to which they do this depends largely on the ratio of the jet and wire speeds. The sheet tends to become stronger in the machine direction as more fibers line up in the same direction. Another important factor is the tendency for paper to become stronger if it is dried under restraint, ie, prevented from shrinking during drying. Machine-made paper is dried by being passed through an array of drum dryers. The paper is under tension in the machine direction during this process.

Because the fibers generally are flat or ribbonlike, they tend to be deposited on the wire in layers. There is very little tendency for fibers to be oriented in an out-of-plane direction, except for small undulations where one fiber crosses or passes beneath another. The layered structure results in the different properties measured in the thickness direction as compared to those measured in the in-plane direction. The orthotropic behavior of paper is observed in most paper properties and especially in the electrical and mechanical properties.

The basis weight W, as commonly expressed in the United States and determined in accordance with T 410 of the Technical Association of the Pulp and Paper Industry (TAPPI), is the mass in grams per square meter. It can also be expressed as pounds of a ream of 500 sheets of a given size, but the sheet sizes are not the same for all kinds of paper. Typical sizes are 43.2 × 55.9 cm for fine papers, 61.0 × 91.4 cm for newsprint, and 63.5 × 96.5 cm for several book papers. The most common designation is pounds per 3000 square feet (1.62 g/m^2) for paper. The basis weight of board usually is expressed as pounds per thousand square feet (kg/205 m^2). For example, the material comprising a 69-pound (31-kg) linerboard weighs 69 lb/1000 ft^2 (337 g/m^2). Typical basis weights are tissue and toweling, 16–57 g/m^2; newsprint, 49 g/m^2; grocery bag, 49–98 g/m^2; fine papers, 60–150 g/m^2; kraft linerboard, 127–439 g/m^2; and folding boxboard, 195–586 g/m^2.

The thickness or caliper is the thickness of a single sheet measured under specified conditions (see TAPPI T 411). It usually is expressed in micrometers. Calipers for a number of common paper and board grades are capacitor tissue, 7.6 μm; facial tissue, 65 μm; newsprint, 85 μm; offset bond, 100 μm; linerboard, 230–640 μm; book cover, 770–7600 μm.

The tensile strength is the force per unit width parallel to the plane of the sheet that is required to produce failure in a specimen of specified width and length under specified conditions of loading (see TAPPI T 404). The strength of paper also is expressed in terms of a breaking length, ie, the length of paper that can be supported by one end without breaking. Breaking lengths for typical papers are from ca 2 km for newsprint to 12 km for linerboards. The values for the stronger papers compare favorably with other engineering materials. For example, breaking lengths for aluminum are ca 20–25 km.

Stretch is the extension or strain resulting from the application of a tensile load applied under specified conditions (see TAPPI T 457). The numerical result usually is expressed as a percentage of elongation per original length and includes the elastic and the inelastic extensibility of the paper. Stretch is greatest in the cross-machine direction except for creped grades. It is becoming more common to evaluate the elongation as a continuous function of the applied load. The initial slope of the load–elongation curve, ie, load/width versus strain, defines the modulus of elasticity E in the machine or cross-machine directions. By this definition, E describes units of load per width.

The bursting strength is the hydrostatic pressure required to rupture a specimen when it is tested in a specified instrument under specified conditions. It is the pressure required to produce rupture of a circular area of the paper (30.5 mm dia) when the pressure is applied at a controlled rate (see TAPPI T 403). It is related to tensile strength and extensibility and is used extensively throughout the industry for packaging and container grades.

Tearing strength, or the internal tearing resistance, is the average force required to tear a single sheet of paper under standardized conditions by which the specimen is cut prior to tearing (see TAPPI T 414). Internal tearing resistance should be distinguished from initial or edge-tearing resistance.

Stiffness is related to bending resistance. It is most commonly measured by determining the force required to produce a given deflection or by measuring the deflection produced by a given load when the paper specimen is supported rigidly at one end and the deflecting force is applied at the free end. A fundamental measure of stiffness is the flexural rigidity of the sample. Flexural rigidity is the product of the modulus of elasticity and the second moment of the cross section I. If evaluated in accordance with TAPPI T 451, stiffness values are proportional to EI/W. All other factors being the same, the stiffness of paper varies with the cube of the thickness and directly with the modulus of elasticity.

Folding endurance refers to the number of folds a paper can withstand before failure when tested according to TAPPI T 423.

Moisture content is determined by drying the sample at 100–105°C, until no change in weight is observed (see TAPPI T 412). Unless otherwise specified, the difference in the before and after drying weights is expressed as a percentage of the original weight of the sample.

Water resistance refers to that property of a sheet that resists passage of liquid water into or through the sheet. The tests usually are designed to simulate use conditions; consequently, there are several dozen different test methods. The dry indicator method is employed quite extensively (see TAPPI T 433).

Water-vapor permeability refers to a specific permeability of the paper to water vapor. Two common gravimetric methods for evaluating this property are given in TAPPI T 448 and T 464; the tests differ in the temperature and vapor pressure differences which cause the permeation. The permeability usually is reported in grams of vapor permeating one square meter of paper per twenty-four hours. Because of the unusually high affinity of cellulose for water and water vapor, water-vapor permeability generally does not correlate with permeability to other vapors and gases.

The common optical properties of paper are brightness, color, opacity, transparency, and gloss. Brightness is the reflectivity of a sheet of pulp or paper for blue light, ie, ca 457 nm (see TAPPI T 452). The reflectivity is the reflectance for an infinitely thick sample. Color is measured by evaluating the spectral reflectivity, as given in TAPPI T 442. Opacity relates to that property of a sheet which prevents dark objects in contact with the back side of the sheet from being seen. It usually is evaluated by contrast ratio, which is the ratio of the diffuse reflectance of the sheet when backed by a black body to that of the sheet when backed by a white body of given absolute reflectance value (see TAPPI T 425). Transparency is that property of a paper by which it transmits light, so that objects can be seen through the paper. Transparency ratio is a measure of transparency as judged when a space separates the specimen and the object being viewed. Gloss is the ability of the surface to reflect light specularly. There are numerous definitions of gloss as it relates to appearance criteria for paper, eg, specular gloss (see TAPPI T 480) and low angle gloss, which often is used as a smoothness test for linerboard (see TAPPI UM (useful method) 558).

Chemical Properties

The chemical composition of paper is determined by the types of fibers used and by any nonfibrous substances incorporated in or applied to the paper during the papermaking or subsequent converting operations. Paper usually is made from cellulose fibers obtained from the pulping of wood. Occasionally, synthetic fibers and cellulose fibers from other plant sources are used. Paper properties that are affected directly by the fibers' chemical composition include color, opacity, strength, permanence, and electrical properties. Development of interfiber bonding during papermaking also is strongly influenced by the composition of the fibers. Because residual lignin in the fibers inhibits bonding, groundwood pulp is used in newsprint and in some book and absorbent papers which do not require a highly bonded structure. Hemicelluloses in chemical pulps contribute to bonding; therefore, pulps containing hemicelluloses are used for wrapping papers and other grades which require bonding for strength, and in glassine, which requires bonding for transparency.

In most papers, the chemical composition largely reflects those nonfibrous materials that were added to the paper to achieve the desired physical, optical, or electrical properties. Examples of chemicals and resultant properties are dyes and optical brighteners to enhance appearance, resins to impart wet strength, rosin or starch size to reduce penetration of aqueous liquids, pigment coatings to provide a smooth surface for printing, mineral fillers to increase opacity, polymers applied by saturation or extrusion to impart mechanical or barrier properties, and cationic polyelectrolytes and resistive polymers used in the interior and on the surface, respectively, of papers for dielectric recording.

The performance of a limited number of papers depends upon chemical reactions of noncellulosic additives. Specialized paper coatings in which chemical reactions occur at the time of use are essential in photographic, thermal, and carbonless copy papers (see Photography; Microencapsulation; Reprography). Strips of paper saturated with color-forming reagents permit rapid, inexpensive urinalyses. Phosphates or halogenated compounds are incorporated in papers to promote flame retardancy (see Flame retardants).

For acceptable performance, some grades of paper should not contain certain chemical species. Papers that are used for electrical insulation must be free of electrolytes, and papers that are used for permanent documents must be low in acidity. Reducible sulfur compounds, ie, sulfide, elemental sulfur, and thiosulfate, should not be present in papers that receive metallic coatings or in antitarnish papers that are used for wrapping polished silver or steel items. Chemical substances that unintentionally could become a component of food must not be included in papers for food-contact applications unless the chemicals are approved by the FDA as indirect food additives.

Manufacture and Processing

Stock Preparation. Stock preparation denotes the several operations that must be undertaken in order to prepare the furnish from which paper is made. Detailed discussions are given in refs. 6–9 and in the literature. During the stock-preparation steps, papermaking pulps are most conveniently handled as aqueous slurries, so that they can be conveyed, measured, subjected to desired mechanical treatments, and mixed with nonfibrous additives before being delivered to the paper machine. In the case of adjacent pulping and papermaking operations, pulps usually are delivered to the paper mill in slush form directly from the pulping operation. Purchased pulps and waste paper are received as dry sheets or laps and must be slushed before use. The objectives of slushing is to separate the fibers and to disperse them in water with a minimum of mechanical work so as not to alter the fiber properties. Slushing is accomplished in several types of apparatus, eg, the Hydrapulper or the Sydrapulper which are illustrated in Figures 1 and 2, respectively.

Beating and Refining. Virtually all pulps are subjected to certain mechanical actions before being formed into a paper sheet. Such treatments are used to improve the strength and other physical properties of the finished sheet and to influence the behavior of the system during the sheet-forming and drying steps. Beating and refining may be considered synonymous.

During refining, the cellulose fibers are swollen, cut, macerated, and fibrillated. The most desirable action is the development of internal fibrillation, which makes the wet fiber more compliant or flexible. This flexibility enhances the number of interfiber contacts during formation of the paper and bonding during subsequent pressing and drying operations. Although refining markedly alters the physical properties, no major chemical changes occur. A sheet that is formed from an unbeaten pulp has a low density, is rather soft, and is weak. If the same pulp is well-beaten, however, the resultant paper is much more dense, hard, and strong. If taken to the extreme, beating produces very dense, translucent, glassine-type sheets.

Refining greatly increases the wet specific surface of pulp fibers, the swollen specific volume, and the fiber flexibility. Although hydration in the chemical sense does not occur, the affinity for water is enhanced. Because of the unique cellulose–water

Figure 1. The Hydrapulper. Courtesy of Black Clawson Co.

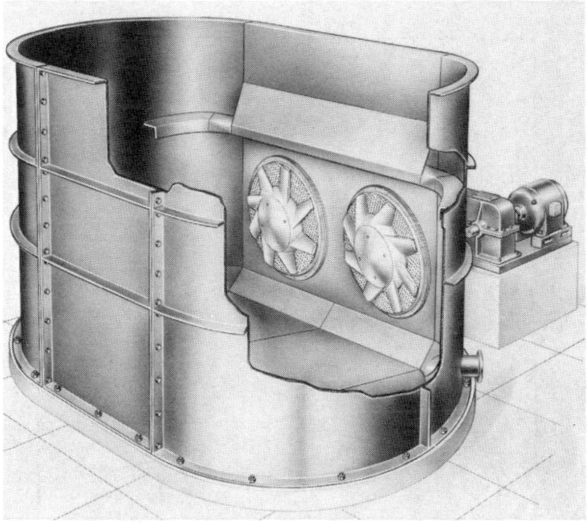

Figure 2. The Sydrapulper. Courtesy of Black Clawson Co.

relationship, these changes significantly increase the ability of the fibers to bond when dried from a water suspension and, therefore, enhance the strength of the sheet.

Optical micrographs of a softwood kraft pulp before and after several periods of extensive beating are given in Figure 3. In Figure 4, the tensile strength, bursting strength, and tearing resistance are shown as functions of beating time for a softwood kraft pulp. It generally is true that, within the commercial range, beating increases tensile strength, bursting strength, folding endurance, and sheet density, whereas it reduces tearing resistance.

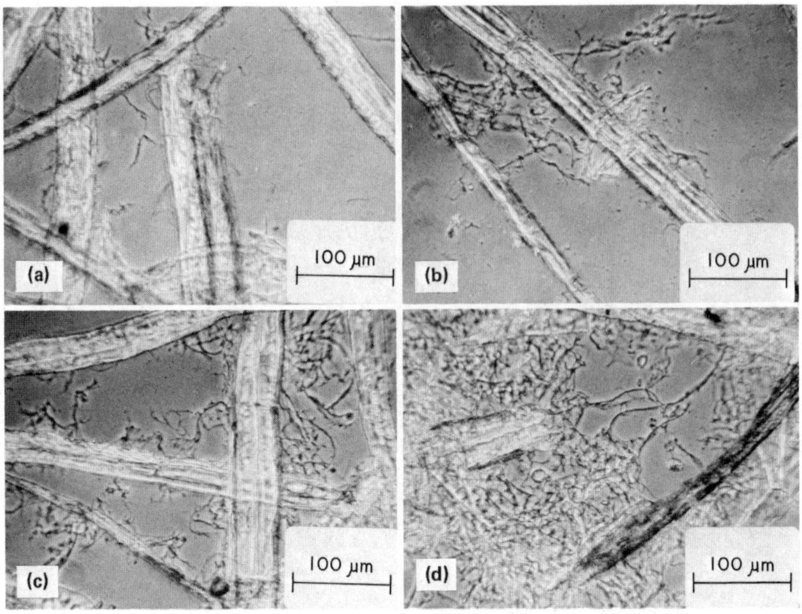

Figure 3. Optical micrographs of a softwood sulfite pulp (**a**) before beating and (**b–d**) after beating in a valley beater for different periods of time. Beating time increases from **b** to **d**.

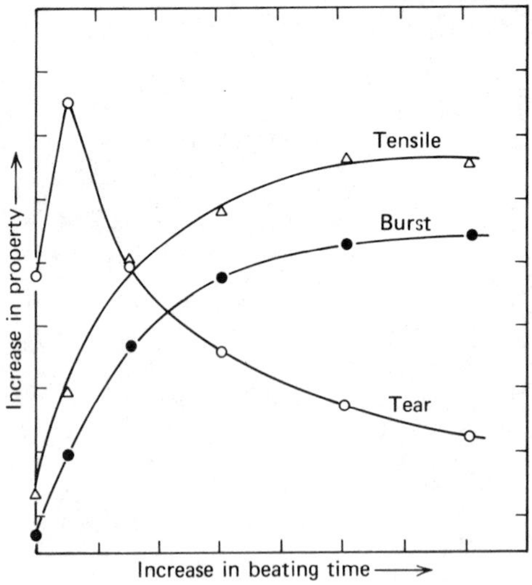

Figure 4. Typical beating curves for a softwood kraft pulp.

Batch systems have been replaced largely with continuous, pump-through equipment. Most batch systems are refinements of the hollander beater, which was developed in Holland in ca 1690. The hollander consists of an elongated tub with a central dividing partition, or midfeather which extends along the major axis to within a few meters of each end. A cylindrical beater roll is mounted on one side and knives are placed around the circumference parallel to the roll axis. The roll is mounted over a bedplate, which also contains a set of knives. Circulating stock passes between the roll and the bedplate, and the severity of beating is controlled by adjusting the load of one against the other. Most modifications of the original hollander represent attempts to increase capacity, improve stock circulation, and save power.

The first successful continuous refiner was the Jordan, which was developed in ca 1860. The Jordan refiner consists of a stationary conical shell which is fitted with knives or bars on the inside and with a conical plug which fits inside the shell and contains bars on its surface. The plug rotates, and the pressure between the bars on the shell and the plug is regulated by longitudinal movement of the plug. The pulp slurry is fed into the small end of the Jordan refiner and is discharged at the large end.

A variety of conical refiners have been developed that involve modifications of the Jordan refiner. The Hydrafiner, for example, has a short, low taper, high speed rotor, and wide bars (10). The stock is driven through the refiner by an impeller which is fitted to the small inlet end of the rotor shaft. The Claflin refiner has a very short, high taper plug; vanes, which fit on the wide end of the plug, draw the stock through the unit (11).

The disk refiner is a newer development and includes one or two rotary disks and two or four working surfaces, which are pressed together uniformly by hydraulic pressure and guided by high precision, heavy-duty bearing systems. Stock usually is fed through the center of one plate and leaves between the plates at their circumferences.

A considerable variation in the properties of the refined stock may be realized through control of the refiner operating variables. For example, high stock consistency, dull beater or refiner bars, or low pressure or excessively wide separation between the two sets of bars may yield a mild refining action resulting in accentuated fibrillation and swelling. Low stock consistency, sharp bars, and high pressure between elements can result in severe refining and severe fiber length reduction with subsequent negative effects on strength.

Filling and Loading. Materials, eg, mineral pigments for filling and loading, are added to the pulp slurry to make the papermaking furnish (see Papermaking additives). Pigments are used in varying amounts, depending upon the grade of paper, and may comprise 2–40 wt % of the final sheet. Fillers can improve brightness, opacity, softness, smoothness, and ink receptivity. They almost invariably reduce the degree of sizing and the strength of the sheet. The brightness, particle size, and refractive index of fillers influence the optical properties of the finished sheet, and the particle size and specific gravity are important in regard to the filler retention during sheet formation. All commercial fillers are essentially insoluble in water under the conditions of use.

Kaolin or China clay is used both as a filler material and as a coating pigment. It is a low cost, naturally occurring, hydrated aluminum silicate with widespread application. Titanium dioxide probably is the most desirable pigment for opacity improvement, and its use is increasing, particularly in fine papers. Two forms of titanium dioxide are used, ie, anatase and rutile. Calcium carbonate is used particularly in book and cigarette papers. It may be produced in the pulp mill as a by-product of causticization, or it may be obtained as a ground limestone or chalk. Calcium carbonate is not used in papers that are sized in an acid furnish because of its solubility and resultant foam problems. Several zinc pigments, eg, zinc sulfide and lithopone, ie, barium sulfate–zinc sulfide, also are used. Some properties of important fillers are listed in Table 2 (see Fillers).

The retention of pigments in the sheet during formation is important, particularly with titanium dioxide which is quite costly. Unless the white-water system is essentially closed, ie, there are no losses, high filler losses can result. Both hydrodynamic mechanisms and colloidal or coflocculation phenomena are significant in determining filler retention (6). Polymeric retention aids are used when titanium dioxide is used. Synthetic silicas and silicates are used to improve the optical efficiency of titanium pigments. Talc is used to reduce the deposition of pitchlike materials onto paper machinery.

Sizing. Sizing is the process of adding materials to the paper in order to render the sheet more resistant to penetration by liquids, particularly water. Unsized or waterleaf paper freely absorbs liquids. Writing and wrapping papers are typical sized sheets, as contrasted with blotting paper and facial tissue which usually are unsized. Rosin, various hydrocarbon and natural waxes (qv), starches, glues, casein, asphalt emulsions, synthetic resins, and cellulose derivatives are some of the materials which are used as sizing agents. The agents may be added directly to the stock as beater additives to produce internal or engine sizing, or the dry sheet may be passed through a size solution or over a roll which has been wetted with a size solution; such sheets are tub-sized or surface-sized.

Rosin, which is refined from pine trees or stumps, is one of the most widely used sizing agents. The extracted rosin is partially saponified with caustic soda and is processed to yield a thick paste of 70–80 wt % solids, of which as much as 30–40% is free, unsaponified rosin. Dry-rosin sizes and completely saponified rosins also are available. At the paper mill, the paste is diluted to ca 3 wt % solids with hot water and vigorous agitation. The solution is added to the stock (0.5–3.0 wt % size based on dry fiber) usually before but sometimes simultaneously with one to three times as much aluminum sulfate, which precipitates the rosin on the fibers as flocculated particles. The pH after the addition of the alum is critical and should be 4.5–5.5. In systems of higher pH, sodium aluminate also may be used to precipitate the rosin size. The exact mechanism of sizing with rosin and alum is controversial but good retention and distribution of the hydrophobic size precipitates on the fiber surface is necessary.

Use of acid-sensitive fillers, eg, calcium carbonate, in a rosin-sized furnish of pH 4.5–5.5 is problematic. Much attention has been focused upon the adverse influence of acid conditions during papermaking on the deterioration of paper with time. Effi-

Table 2. Properties of Paper Fillers

Filler	Specific gravity	Refractive index	Particle size, μm	Brightness, %
clay	2.5–2.8	1.55	0.5–1.0	80–85
titanium dioxide				
anatase	3.9	2.55	0.3	98–99
rutile	4.2	2.70	0.3	98–99
calcium carbonate	2.6–2.8	1.56	0.2–0.4	95–97
zinc sulfide	4.0	2.37	0.3	96–98
talc	2.8	1.57	1–10	70–90
synthetic silicates	2.1	1.55	0.1–4.0	93–94

cient internal sizing of papers and paperboard under alkaline papermaking conditions (pH 7.0–8.5) is achieved with synthetic sizes, eg, Aquapel (12), Hercon (13), and Fibran (14).

Coloring. The color of most paper and paperboards which are made from bleached pulps is achieved by the addition of dyes and other colored chemicals. White papers frequently are treated with small amounts of blue materials to achieve a whiter visual appearance. By far the largest proportion of dyes is added to the stock during stock preparation, although a limited amount of dry paper is colored by dip dyeing or by applying a dye solution during calendering. Water-soluble synthetic organic dyestuffs are the principal paper-coloring materials. Some coloring is with water-insoluble but water-dispersible pigments, eg, carbon black, vat colors, color lakes, and sulfur colors. The properties of basic, acid, and direct dyes are summarized in Table 3. Within each group there are wide variations and exceptions from the listed generalizations.

The kind of fiber and the degree to which it has been refined are important factors in paper dyeing. The undyed color of the pulp and the varying affinity for the same or different dyes, both from fiber to fiber within a pulp and between different pulps, are some of the variables which necessitate continual adjustment of dyeing techniques. The amount and kind of refining a pulp has received affects the pulp's optical properties and, therefore, the color effect of a given dye. Generally, refining deepens the shade from a given application of a water-soluble dye but does not change the amount of dye which is retained. Refining tends to increase the retention of pigments and other water-insoluble dyes, but it also may change the depth or hue by decreasing the pigment particle size. Many types of dyes are absorbed strongly by a wide range of fillers. The two-sidedness of sheets, which results from loss of fines and filler on the wire side, contributes to two-sidedness of the color, an effect that also may result from or be enhanced by contact with heated dryer surfaces. There is a tendency for pigment colors to be concentrated on the top side of the sheet. In a complex system containing fibers, fillers, size, and dye, much colloidal activity is possible, particularly when extraneous unknown ions or particles are present. An optimum order for the addition of the filler,

Table 3. Comparison of Basic, Acid, and Direct Dyes

	Basic dyes	Acid dyes	Direct dyes
cation	dye ion	Na^+, K^+, NH_4^+	Na^+
anion	Cl^-, SO_4^{2-}, NO_3^-	dye ion	dye ion
tinctorial strength	high	lower	lower
brilliancy	high	high	lower
lightfastness	poor	generally good	good
acid fastness	poor	poor	variable
alkali fastness	poor	poor	variable
waterbleed fastness	generally good	generally poor	generally good
solubility	good	high	lower
affinity	strong for unbleached, lignified fibers, therefore mottling occurs in mixed furnishes; no mordant necessary	none for cellulose; a mordant, eg, size and alum, is necessary	very strong for bleached or unbleached cellulose

size, alum, and color is very difficult to predict; however, addition of alum as the last in the sequence generally gives the best results.

Other Beater Additives. Beater adhesives are employed widely to enhance fiber-to-fiber bonding. Starches probably are used in the greatest tonnage. Natural gums (qv), eg, guar and locust bean, also are used as are modified celluloses, eg, the carboxymethyl and hydroxyethyl derivatives (see Cellulose derivatives). Urea–formaldehyde and melamine–formaldehyde polymers provide wet strength to the finished paper sheet (see Amino resins). Other natural and synthetic materials are used to alter the paper properties and to influence the behavior of the system during sheet forming and drying.

Sheet Forming, Pressing, and Drying

Sheet Forming. Continuous sheet forming and drying came into use in ca 1800. The equipment was of two types: the cylinder machine and the Fourdrinier machine. In the former, a wire-covered cylinder is mounted in a vat containing the fiber slurry. As the cylinder revolves, water drains inward through the screen and the paper web is formed on the outside. The wet web is removed at the top of the cylinder, passes through press rolls for water removal, and then passes into steam-heated, cylindrical drying drums. The Fourdrinier is more complex and basically consists of a long continuous wire screen which is supported by various devices that improve drainage. The fiber slurry, which is introduced at one end through a headbox and slice, loses water as it progresses down the wire, thereby forming the sheet. It then passes to presses and dryers as in the cylinder machine. A Fourdrinier paper machine is shown diagrammatically in Figure 5.

Continuous paper machines have undergone extensive mechanical developments in the past half century, although the principles employed have changed little. Cylinder machines still are operated and involve multiples of five to seven cylinders; they are used to produce heavy multi-ply boards. Fourdriniers are standard in the industry and are used to produce all grades of paper and paperboard. They vary from 1 to 10 meters in width and, including the press and dryer sections, may be more than 200 m long.

Subsequent to stock preparation and proper dilution, the paper furnish usually is fed to the paper machine through one or more screens or other devices to remove dirt and fiber bundles. It then enters a flow spreader which provides a uniform flowing stream and which is the width of the paper machine. The flow spreader discharges

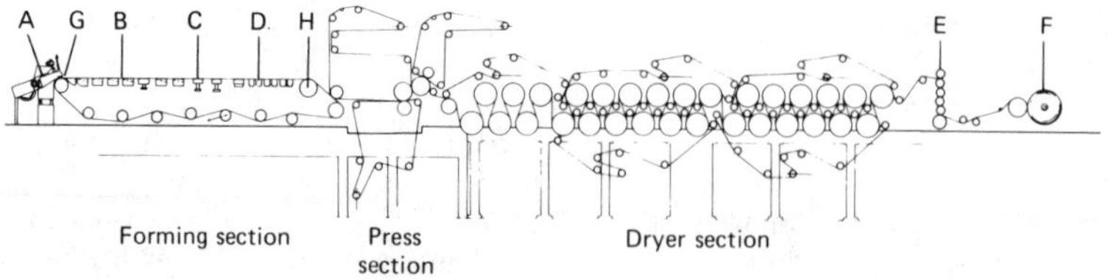

Figure 5. Fourdrinier paper machine with (A) headbox, (B) Fourdrinier wet end with foil boxes, (C) wet and (D) dry suction boxes, pickup and closed transfer of web through the press section, dryer section, (E) calender, and (F) reel. G and H are the breast roll and couch roll, respectively. Courtesy of Beloit Corporation.

the slurry into a headbox, where fiber flocculation is minimized by gentle agitation and where the proper pressure head is provided to cause the slurry to flow at the proper velocity through the slice and onto the moving Fourdrinier wire.

The Fourdrinier wire is mounted over the breast roll at the intake end and at the couch roll at the discharge end. Between the two rolls, it is supported for the most part by table rolls, foils, and suction boxes. The table rolls, which are driven by the wire, were originally devised solely to support the wire on a level table. A substantial vacuum is developed in the downstream nip between the table roll and the wire and promotes water drainage from the slurry on the wire. As speeds increase, however, the suction becomes too violent and deflects the wire, throwing stock into the air. A more controlled drainage action is accomplished by the use of foils (see Fig. 6). These are wing-shaped elements which support the wire and induce a vacuum at the downstream nip. Foil geometry can be varied to provide optimum conditions. After passing over the foils or table rolls, the wire and sheet pass over suction boxes, where more water is removed. Most machines also include a suction couch roll for additional water removal.

Machine speeds vary chiefly because of limitations imposed by the various products but also because of differences in production equipment. Heavy paperboards require a long drying time, and machine speeds are 50–250 m/min. Very dense papers, eg, glassine and greaseproof and condenser tissue, are difficult to dewater in the forming and press sections; speeds range from 20 to 300 m/min, depending on the product. Brown grades, eg, paper bags and linerboard, are produced at 200–800 m/min, depending on basis weight and the site of the paper machine. With the advent of the suction pick-up, which closes the draw between the forming and press sections, speeds of newsprint machines have doubled from 300 to 600 m/min. Closing the transfer of the sheet through the entire press section and increasing the dryness at the first open draw into the dryer section from 35–38% to 41–44% combined with careful designs of the web path through the dryer section has increased newsprint machine speeds to ca 1200 m/min. The majority of machines operate at 600–900 m/min. Drying capacity restraints and difficulties in reeling the product limit modern tissue machine speeds to 1500–1800 m/min. Most tissue machines operate at lower speeds. Novel designs for web handling, reeling and roll change will permit tissue machine speeds of up to ca 2000 m/min in continuous operation.

Twin-wire formers have replaced the Fourdrinier wet-ends on many machines, particularly for lightweight sheets, eg, tissue, towel, and newsprint. Twin-wire formers also are operated successfully on fine paper grades, corrugated media, and linerboard grades. In twin-wire formers, the water is drained from the slurry by pressure rather

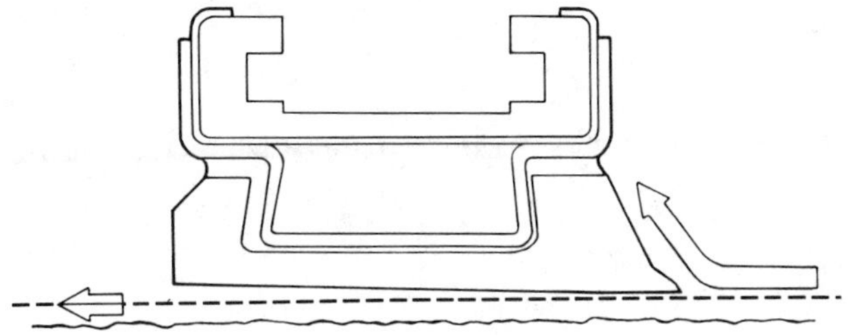

Figure 6. Dewatering foil. Water from the preceding unit is doctored off. The diverging wedge on the downstream side of the foil sucks water out of the slurry onto the wire.

than by vacuum. The two wires, with the slurry between, are wrapped around a cylinder or a set of supporting bars or foils. The tension in the outer wire results in a pressure which is transmitted through the slurry to the supporting structure. The pressurized slurry drains through one or both of the wires.

The Bel Baie II is an example of a twin-wire former (see Fig. 7). It is used extensively for newsprint and fine paper grades as well as for the lighter weight linerboards. Dewatering occurs through both wires; thus, a high drainage capacity is achieved. A typical roll-type twin-wire tissue former is shown in Figure 8. Drainage is single-sided and is limited to low basis weights, but the tissue former is sufficient for drainage at very high speeds, eg, ≥2100 m/min for thin tissue.

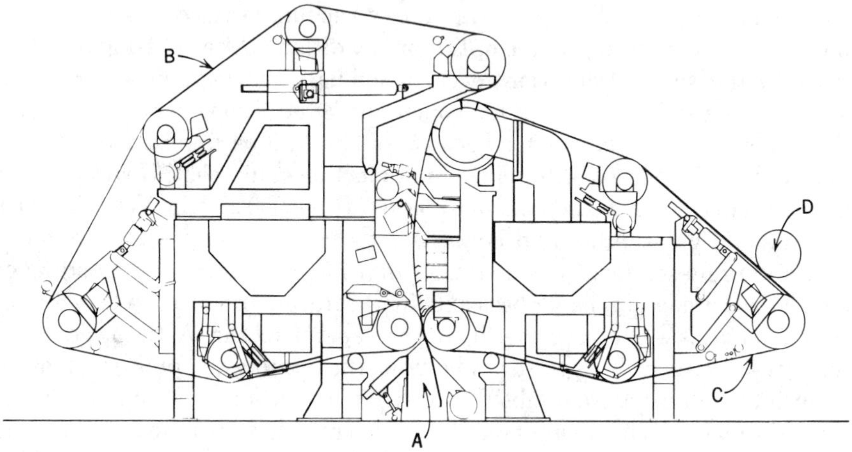

Figure 7. The Bel Baie II twin-wire former. Stock from (A) headbox is formed into sheet between the (B) number 1 wire and (C) number 2 wire. The web is removed from wire 2 by (D) a suction pick-up roll. Courtesy of Beloit Corporation.

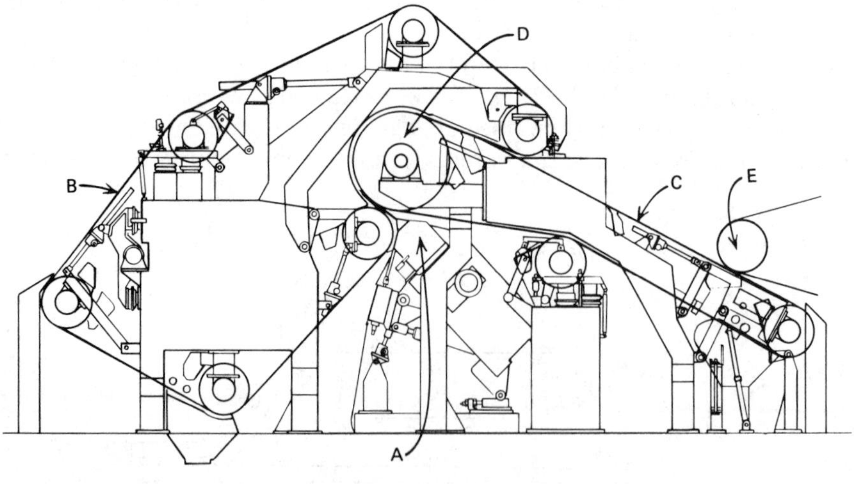

Figure 8. A roll-type twin-wire tissue former. Stock from (A) headbox passes between (B) wire number 1 and (C) wire number 2 around (D) a forming roll and the web is taken off the second wire by (E) a suction pickup roll. Courtesy of Beloit Corporation.

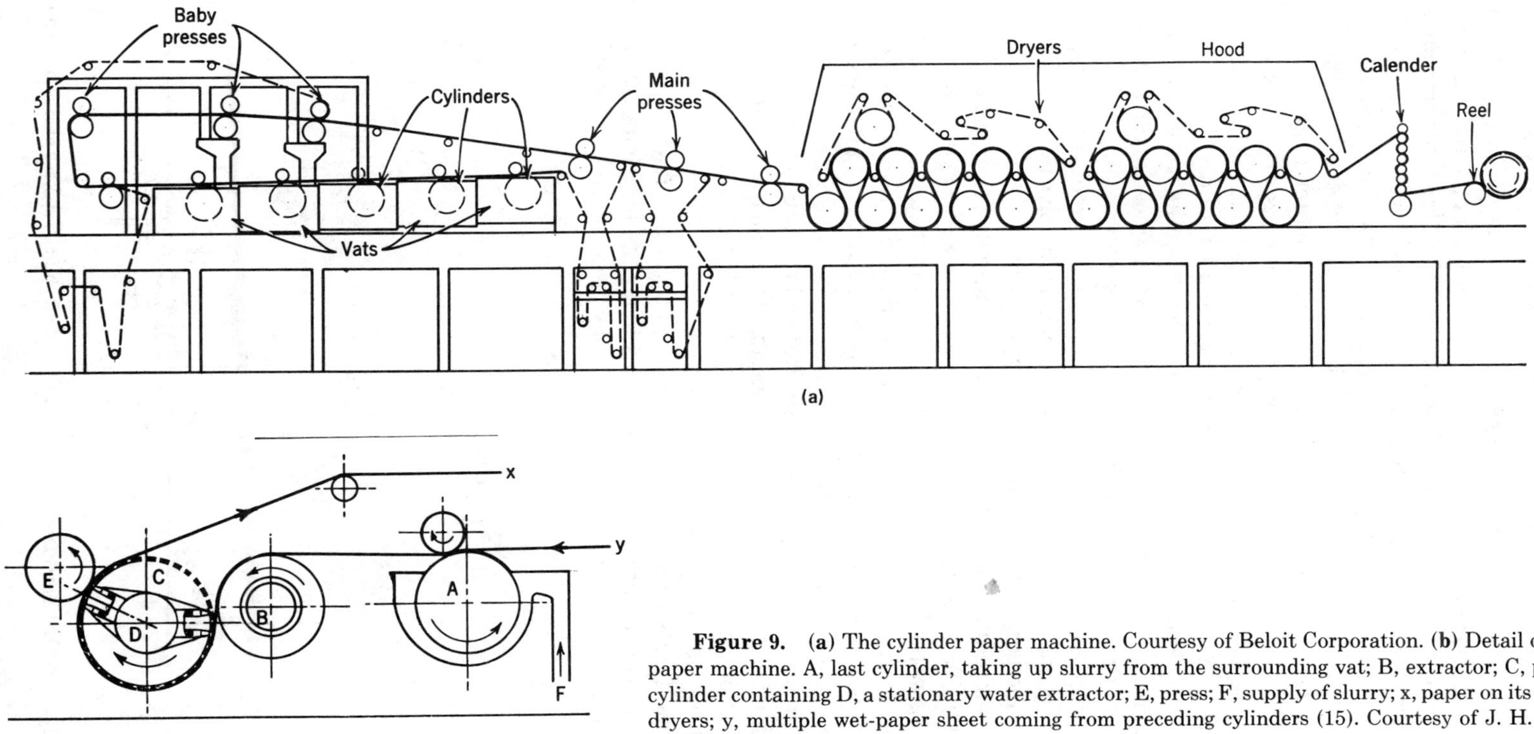

Figure 9. (**a**) The cylinder paper machine. Courtesy of Beloit Corporation. (**b**) Detail of cylinder paper machine. A, last cylinder, taking up slurry from the surrounding vat; B, extractor; C, perforated cylinder containing D, a stationary water extractor; E, press; F, supply of slurry; x, paper on its way to the dryers; y, multiple wet-paper sheet coming from preceding cylinders (15). Courtesy of J. H. de Bussy, Amsterdam, Netherlands.

Boxboard normally consists of 3–7 separate webs which are formed from different raw materials and are couched together. Because of the number of plies and the wide variations in speed requirements, there are a large number of different designs for board-forming sections. The classical forming section for a board machine is the vat machine (see Fig. 9). The incoming diluted slurry is introduced to the cylinder vat after final screening. The sheet is couched onto the underside of a long carrier felt or onto a previously formed paper layer which is carried by the felt. Because of the limited drainage capacities of the vats and the method of carrying the sheet, vat machines are limited to low speeds. Modern versions of the vat machine employ cast and drilled shells, suction boxes, and pressurized headboxes. These additions permit increase in the drainage capacity and facilitate water handling, so that the formers can be placed on top of the carrying fabric felt for higher speed operation.

Another development is the Inverform process and its more modern version, the Bel Bond in Figure 10. In the Inverform unit, several plies are formed on top of each other by consecutive, twin-wirelike forming units above a long carrying fabric. The Inverform process also is used for the forming of paper grades and is capable of moderately high speeds. Other versions of board machines involve mini-Fourdriniers and/or twin wires which are placed on top of a carrying fabric.

All the preceding forming units receive the incoming slurry at a low consistency, typically 100–300 kg water/kg solids, and the paper web leaving the couch typically contains 4 kg water/kg solids. The white water or drained water contains some fiber debris, clay filler, etc. The white water is reused for dilution of the incoming stock. In closed paper-machine systems, any excess white water is filtered, and the recovered solids are returned into the system.

Sheet Pressing. The sheet leaving the wet end contains approximately four parts of water per part of fiber; however, it is possible to remove additional water mechanically without adversely affecting sheet properties. This is achieved in rotary presses, of which there may be one or several on a given paper machine. The press rolls may be solid or perforated and, often, suction is applied through the interior. The sheet is passed through the presses on continuous felts (one for each press), which act as conveyors and porous receptors of water. They are essential to the effectiveness of

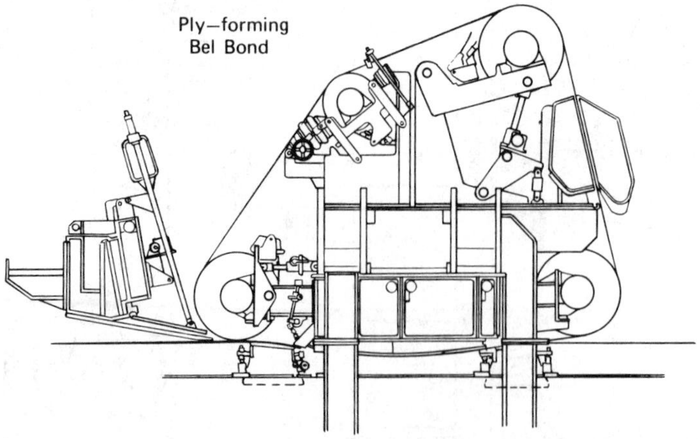

Figure 10. One Bel Bond unit of a multi-ply board machine. Courtesy of Beloit Corporation.

water removal. The water content of the sheet usually can be reduced by pressing to 1.9–1.2 parts of water per part of fiber without deleteriously affecting product quality.

Sheet Drying. At a water content of ca 1.2–1.9 parts of water per part of fiber, additional water removal by mechanical means is not feasible and evaporative drying must be employed. This is at best an efficient but costly process and often is the production bottleneck of papermaking. The dryer section most commonly consists of a series of steam-heated cylinders. Alternate sides of the wet paper are exposed to the hot surface as the sheet passes from cylinder to cylinder. In most cases, except for heavy board, the sheet is held closely against the surface of the dryers by fabrics of carefully controlled permeability to steam and air. Heat is transferred from the hot cylinder to the wet sheet, and water evaporates. The water vapor is removed by way of elaborate air systems. Most dryer sections are covered with hoods for collection and handling of the air, and heat recovery is practiced in cold climates. The final moisture content of the dry sheet usually is 4–10 wt %.

Other types of dryers are employed for special products or situations. The Yankee dryer is a steam-heated cylinder, 3.7–6.1 m in diameter, which dries the sheet from one side only. It is used extensively for tissues, particularly where creping is accomplished as the sheet leaves the dryer, and to produce machine-glazed papers where intimate contact with the polished dryer surface produces a high gloss finish on the contact side.

High velocity air drying, in which jets of hot air are directed against the sheet in a normal direction, is used in Yankee dryers and in combination with a percolation through-drying process. In the latter, hot air is sucked through the sheet, thereby effecting a high heat transfer to the sheet and efficient mass transfer of the water vapor from the sheet. The latter technology is commonly used for special, high quality tissue products. Infrared and other radiant drying techniques also are utilized in special cases.

Converting. Almost all paper is converted by undergoing further treatment after manufacture. Among the many converting operations are embossing, impregnating, saturating, laminating, and the forming of special shapes and sizes, eg, bags and boxes (see Uses).

Pigment Coatings. Pigment coatings are compositions of pigments and adhesives with small amounts of additives and are applied to one or both sides of a paper sheet. They generally are designed to mask or change the appearance of the base stock, improve opacity, impart a smooth and receptive surface for printing, or provide special properties for particular purposes. Pigment coatings are highly porous because the binder adhesive is insufficient to fill the void spaces among the pigment particles. This porous structure is responsible for many of the desired properties of the coated papers. For example, the high opacity of the coatings results from the scattering of light from the pigment–air and adhesive–air interfaces (see also Coatings, industrial; Coating processes; Pigments).

Pigment-coated printing papers usually are required to have high brightness to achieve contrast between the printed and unprinted areas. The coatings frequently have a glossy surface. There also is a demand for dull-coated papers and for dull papers upon which glossy ink films can be printed. The coated paper should be sufficiently smooth for printing to allow full contact between the inked image area of the plate or transfer blanket and the paper surface. Requirements for smoothness decrease from gravure printing to letterpress to offset lithography. Printing smoothness involves not only the smoothness and compressibility of the paper, but the amount of ink, the properties of the ink-transfer and impression materials, the printing pressure, and the printing process. Consequently, smoothness is best predicted by tests that simulate the printing conditions for which the paper is intended (see TAPPI UM 466 and UM 505).

Because only the coating layer is involved significantly in absorption of the small amount of ink fluid which is applied, tests that involve transudation of the sheet cannot be expected to provide pertinent information on ink absorbency. Comparison of ink absorbency of papers can be made according to TAPPI UM 553.

The printing process imposes tensile stress normal to the plane of the sheet; the stress depends upon the tack of the printing ink and the velocity of separation of the printing plate from the paper. The stress also tends to pick the paper, ie, remove material, unless the paper has adequate pick strength. Papers to be printed by offset lithography require high pick strength because of the very tacky inks which are employed, whereas letterpress papers need not be so demanding. In both cases, papers that are to be used for multicolor printing require higher pick strengths than those used for single color printing because of the range of tack required in the inks. For example, gravure papers do not require high surface strength because of the low viscosity inks which are used (see Printing processes).

To an increasing extent, pick strength is determined using printing tests that employ tack-graded inks or viscous test liquids at a controlled printing speed and pressure (see TAPPI T 499 and UM 507). The pick strength of a coating depends on the pigment and the type and amount of adhesive that are used as binders. The desired pick strength usually dictates a minimum amount of adhesive.

In offset printing, some water is transferred from the blanket to the paper. If the adhesive is water-soluble, the dampened coating is more susceptible to picking at subsequent impressions. Consequently, offset papers for multicolor printing must have water-resistant coatings. The water resistance is achieved by using insolubilizing agents in the adhesive (see TAPPI UM 513).

Application. Pigment coatings normally are applied to the base paper in the form of water suspensions and are referred to as coating colors. The total solids content, ie, pigment plus adhesive, may be 35–70%. After application, the coating must be dried by removal of water from the film. In some cases, a calendering operation serves to smooth the surface, control surface texture, and develop a glossy finish.

Paper may be coated either on equipment that is an integral part of the paper machine, ie, on-machine coating, or on separate converting equipment. High quality, coated paper is obtained by the on-machine procedure at high speeds. Many plants include both types of coating equipment and utilize each to its maximum advantage for paper and paperboard. The combination of techniques is of particular value where more than one coating must be applied to the sheet in order to obtain a product of desired quality.

In 1933, the first roll coater was installed as an integral part of a paper machine. These on-machine coaters produced a low cost coated paper which was used largely for magazines. Coating of paper at speeds greater than paper-machine manufacturing speeds is possible with the use of the various available blade coaters (16).

There are numerous roll coaters by which coatings are applied from rolls which travel at paper speed. Such machines apply coatings over a considerable range of thicknesses. Patterns often develop in the coated surface because of the way the film of coating color splits between the roll and paper surfaces. Therefore, these coaters are designed with metering and smoothing rolls to ensure optimum coating application.

With air doctor coaters, a coating color is applied with a roll and the excess is removed with an air doctor, ie, a long thin jet of air which acts as a doctor blade (see Fig. 11). A coating of uniform thickness is achieved and surface contours tend to follow those of the raw stock. The coating colors usually are more fluid than those used in other coating methods.

With blade coaters, the coating is smoothed and excess coating color is removed by a flexible blade. The blade is supported by the paper which tends to fill the depressions with a troweling action; thus, much less coating color is applied to the high points of the sheet. It is useful where the levelness of the coated surface is more important than uniform thickness of coating. Coating colors of high solids content are used with blade coaters. Since less water must be removed, the coatings can be dried at high speed. An inverted blade coater is illustrated in Figure 12.

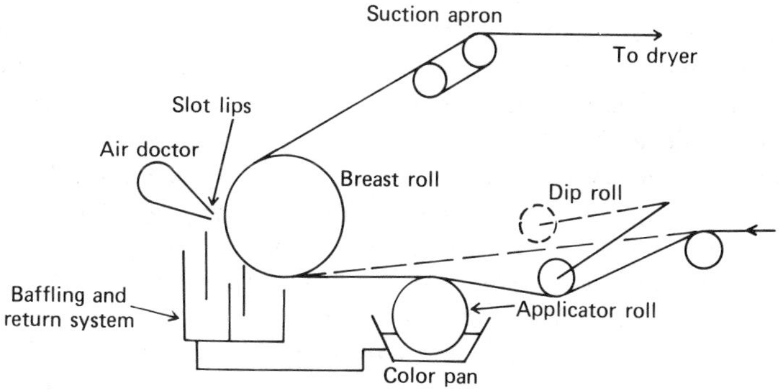

Figure 11. Air-knife coater.

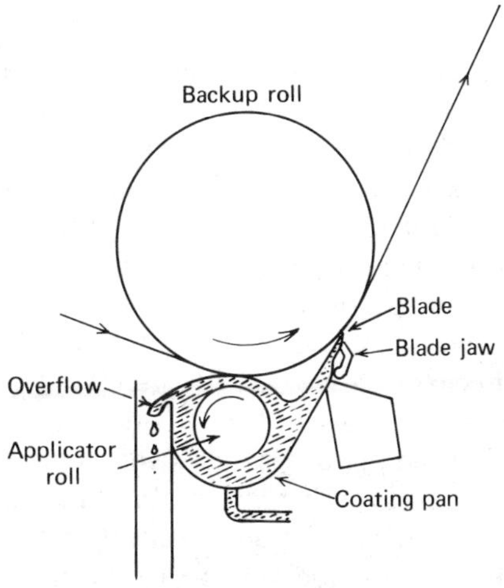

Figure 12. Inverted blade coater.

In cast coating, the coating color is pressed in contact with a highly polished metal drum immediately after application. The coating is dried in contact with the drum to give a smooth glossy surface, and subsequent calendering is unnecessary.

All pigment coatings must be dried to remove the water from the coating and the water that has penetrated the sheet. Drying methods include air or convection drying, contact or conduction drying, and radiant-energy drying. The speed of a coating operation often is restricted by the rate at which the water can be removed from the coating without blister formation or excessive migration of components to the hot surfaces.

Coated paper also is supercalendered to improve surface properties, eg, smoothness and gloss. The paper is passed successively through the nips of a stack of alternating hard steel and soft rolls. The pressure causes the soft roll to deform, which burnishes the surface of the sheet against the polished steel surface. This slipping action reforms the surface and produces a gloss, while smoothing or leveling the surface contour. The properties of the final surface are functions of the coating composition, number of rolls, pressure, temperature, and operation speed. The process also tends to densify the sheet and, thus, to reduce porosity (see Coating processes).

Pigments. Pigments comprise 70–90% of the dry solids in paper coatings (17). In nearly all cases, the individual particles of the pigments are less than 5 μm in equivalent spherical diameter and average less than 1 μm. Many pigments are less than 0.5 μm in size. These particles can fill the spaces between fibers on the sheet surface and form a nearly uniform surface mat. Pigments also control opacity, gloss, and the color of the raw stock. Refractive index, particle size, crystal structure, light scattering and absorption, and adhesive demand are important characteristics. Minimum amounts of adhesives are used to bind the pigments, because the adhesives are more expensive than the pigments and excessive amounts can adversely affect the light-scattering properties of the coatings. Clay requires 10–15% casein or 15–25% starch for adequate strength, and some synthetic pigments may require more.

Kaolin clays are hydrated aluminosilicates having the generalized formula $(Al_2(OH)_4(Si_2O_5))_2$. They occur as natural hexagonal plates with a diameter-to-thickness ratio of ca 10:1. Variations in the specific makeup of the clays influence the rheological, surface, chemical, and optical properties. Minor amounts of impurity minerals can adversely affect pigment properties (see Clays).

Commercial coating clays have been fractionated by centrifugation to remove grit and large particles and to obtain an optimum range of particle sizes. Special treatments improve the brightness and gloss properties and adjust the viscosity of the clay–water suspensions. Coating clays are produced in several grades, depending upon the properties desired. Brightness may vary from 75–85% for filler grades to greater than 90% for very special coating grades. The particles average ca 0.8 μm in equivalent spherical diameter but, because of the plate form, the kaolins produce glossy surfaces after calendering of the paper. The pigment produces many surfaces which scatter light and, thus, it contributes to opacity even though its refractive index is only 1.55, which is about the same as the cellulose fibers it must cover. Clays are the most common and most widely used pigments employed in paper coating.

The average particle size of coating-grade titanium dioxide is ca 0.3 μm. Because this size is optimum for maximum hiding power and because of its high refractive index, titanium dioxide pigment has a unique capacity to opacify and brighten coated paper. The pigment has a brightness greater than 95%. It is chemically inert and is easily dispersed in water. Although both anatase and rutile crystal forms are used, the rutile is preferred in coating because of its slightly higher refractive index. Partly because of its high cost, the amount of titanium dioxide in a coating seldom exceeds 25 wt % of the total pigment.

Calcium carbonate is available from natural sources, but the precipitated forms are most useful in coating because of their purity and brightness, ie, 90–95%. It has a particle size of ca 0.3 μm and a very low abrasive value. It is used with clays to produce improved brightness and to increase printability and ink receptivity. However, the pigment is not plate-shaped and it disrupts the orientation of the clay plates, reduces the surface gloss and maintains the porosity of the coating. It is a common component of matte-finish coatings.

Aluminum oxide dihydrate, $Al_2O_3.2H_2O$, is used in paper coating. This pigment has a brightness of ca 100%, is somewhat opacifying, and is used in combination with clays. It is platelike and contributes to coating gloss, but it requires large amounts of adhesive.

Satin white often is made in the mill from slaked lime and alum and is used in Europe. The exact composition of this calcium sulfoaluminate is variable and its properties may be inconsistent. The slurries are characterized by high viscosity, high adhesive demand, and high alkalinity. Satin white is characterized by high brightness and produces a high gloss coating with calender action.

Other pigments are used in coatings either in restricted amounts or for special purposes. The latter category includes the barium sulfate pigments, which are marked by high whiteness and density. They are ideal as bases for the sensitized layers in photographic print papers. Other minor pigments include silica for a nonskid or mat surface, calcium sulfate for brightening, and zinc oxide for electrophotographic reproduction papers.

Adhesives. The primary function of the adhesive in pigment coating is to bind the pigment particles together and to the raw stock. The strength must be sufficient to prevent picking of the coating from the paper by tacky printing inks. In addition, the type and proportion of the adhesive controls many of the characteristics of the finished paper, eg, surface strength, gloss, brightness, opacity, smoothness, ink receptivity, and firmness of surface. In many coating formulations, a combination of adhesives is used to maximize the properties of various types of binder systems. The amount of adhesive in a coating formulation varies from about 5 to 25%, based on the pigment. The actual amounts depend upon the properties desired in the coating, the specific use of the sheet, and the types of binders that are to be used (see Adhesives).

Animal glue (qv) was the first material used for bonding paper and as an adhesive in paper coating. Today, the use of glue is confined to specialty applications in paper converting.

In the late 1800s, when the demand for coated paper for the halftone printing process increased, casein rapidly replaced glue. Casein forms a hard, tough film when dry, and it can be waterproofed easily with formaldehyde. The flow properties of coating colors containing casein are well-suited to a wide range of coating methods. Casein is used in high quality coatings for offset printing where water resistance, smoothness, high gloss, and toughness of surface are required (see Milk products).

A binder, which is similar to casein, is obtained from soybeans and generally is referred to as soy or alpha protein. The properties of soy protein are very similar to those of casein and soy protein may be substituted for it in many types of coated papers requiring a casein-type binder (see Soybeans).

Starch (qv), which is obtained from corn, potatoes, tapioca, wheat, etc, is a polymer of glucose in which the units are arranged in the linear amylose and the branched amylopectin forms. Because of the complexity and size of the structure and the high viscosity of starch in solution, starches cannot be used directly in coating formulations. They must be modified in some manner to produce a lower viscosity system. Modification of starch by thermal conversion is the preferred process, especially in large plants where the coating may be achieved on the machine, eg, for newspapers. It is preferable to conduct the modification at the paper mill because of cost advantages and because the adhesives can be prepared at the required viscosities, thereby obviating a large inventory. Hypochlorite-oxidized starches are characterized by lower gelatination temperatures, clearer solutions, and lower viscosities at high concentration than their parent forms. Oxidized forms are available in a range of viscosity grades and are used in sizing operations as well as in pigment coating. The presence of side groups in hydroxyethylated starch prevents the gelling or retrogradation of the starch on cooling by separating the linear portions. Other methods of modifying starches, eg, acid and enzyme hydrolysis and dextrinization, are less important for adhesives that are used in pigment coatings.

Starches are quite hydrophilic and the granules swell in water to several times their volume. In the preparation of starches for coating, the granules are heated at 93°C to ensure complete breakdown. The starch then is mixed with the pigment while it is fluid. Normal practice is to coat the paper at an elevated temperature to help control viscosity.

Starches are used in many coated papers, especially in newspaper and magazine paper. If resistance to moisture is not required, starch is the principal adhesive which is employed. For uses where waterproofing is required, starch is often replaced by soy protein. Starches can be made more resistant to water by the use of aldehyde-donor cross-linking agents, eg, urea–formaldehyde and melamine–formaldehyde or glyoxal; however, these materials increase viscosity.

Exceptionally strong adhesives, eg, poly(vinyl alcohol) (PVA), can be used in coating formulations. The strength of poly(vinyl alcohol) in a coating is ca four times that of starch and three times that of casein. Although PVA is expensive, only small amounts are required and it is characterized by good optical properties. However, its high solution viscosity limits its use in coating formulations.

Various rubber latexes and other emulsions also are employed as adhesives. Such materials can be added to the coating without special preparation, as is required with the natural adhesives. Emulsions provide many properties that are superior to those of the aforementioned natural binders, eg, low viscosity which permits high solids content, easy handling, and low water content. The latter implies short drying times and applicability for high speed coating operations. Emulsions and latexes are distinguished by high gloss, good response to calendering, good ink "holdout" (nonblotting character), and good water resistance. The styrene–butadiene latex usually is supplied in a 60/40 ratio of styrene to butadiene. It is used with starch and primarily in publication-grade papers (see Elastomers, synthetic; Latex technology).

Acrylic-based emulsions are used mostly on paperboard. These materials are odorless, which is necessary in coatings which are to be used on paperboard for food packaging. The acrylics provide high gloss and good ink retention. However, use of these materials has been restricted because of their high cost. They are being replaced by poly(vinyl acetate), the adhesive strength of which is equivalent to the previously mentioned emulsions. Poly(vinyl acetate) also provides moisture and grease resistance.

Additives. Additives control coating behavior during application or they can be used to alter the properties of the finished product. A single chemical additive may be used for several purposes. Some additives are essential to the production of a salable product, and others may be added only to obviate problems of the coating operation.

A dispersing agent is used to transform pigments into a slurry form. The material, usually a polyphosphate, is adsorbed on the pigment, and causes the particles to repel one another, thereby reducing the coating viscosity. Proteins and casein also must be dispersed. The various emulsions contain stabilizers. Compatibility between these stabilizers and other coating components must be considered for all coating systems (see Dispersants; Emulsions).

Foam-control agents commonly are employed (see Defoamers). Adhesives that are good film formers tend to stabilize the foam which is present. Most antifoam or defoamer agents are surface-active materials. Materials, eg, pine oil, capryl and tridecyl alcohol, fuel oil, tributyl citrate and phosphate, and silicones, have been used.

Lubricants, plasticizers (qv), and flow modifiers include both soluble and insoluble soaps (qv), sulfated oils, wax emulsions, amine products, esters, etc. Certain materials, eg, urea and dicyandiamide, may reduce the viscosity of a coating color (see Cyanamides). Lubricants improve flow, coating smoothness, finish, printability, and antidusting effects. Humectants, eg, glycerol derivatives, are used in small amounts as plasticizers and they aid in the development of the finish on the sheet.

Materials that are used to increase the moisture resistance of a coating surface or to insolubilize the adhesive may not have film-forming properties. They may react with the hydrophilic groups in the adhesive or cross-link the polymer chains to prevent swelling with water and subsequent loss of binding strength. The decrease in water sensitivity can be achieved by mixing formaldehyde donors, eg, urea– or melamine–formaldehyde resins (see Amino resins), or glyoxal with the coating formulation. In some cases, the desired results may be obtained by exposing the surface to formaldehyde or by application of a zinc, aluminum, or other metal-salt solution during calendering. Some resistance to moisture always is obtained when a latex is included in the coating formulation.

Barrier Coatings. In packaging applications, a barrier may be needed against water, water vapor, oxygen, carbon dioxide, hydrogen sulfide, greases, fats and oils, odors, or miscellaneous chemicals. A water barrier can be formed by changing the wettability of the paper surface with sizing agents. A grease or oil barrier can be provided by hydrating the cellulose fibers to form a pinhole-free sheet or by coating the paper with a continuous film of a material which is resistant to the particular grease. Gas or vapor barriers are formed by coating the paper with a continuous film of a suitable material (see also Barrier polymers).

Paraffin wax is applied in a molten form; it resists water vapor and is colorless; it is free from odor, taste, or toxicity; and it is low in cost. It is applied by passing the paper through a molten bath or nip, removing the excess paraffin, and chilling. Modifiers, eg, microcrystalline wax, polyethylene, or ethylene–vinyl acetate copolymer, improve the durability and film strength, raise the softening point, and increase the gloss and heat-seal strength of the coating. Polyethylene is applied by extrusion. Polyethylene coatings are more durable and flexible than wax coatings. Polymer pellets are heated rapidly with minimum air contact, and the molten material is extruded through a die and immediately is laminated to the paper.

Solvent systems permit the formulation of highly sophisticated coatings which are comprised of a wide variety of polymers and various modifiers. Disadvantages include high solvent costs and the necessity for a solvent recovery system. The resins which are used include many of the same types that are used for coated fabrics or for industrial coatings, eg, cellulose derivatives, rubber derivatives, butadiene–styrene copolymers, vinyl copolymers, poly(vinylidene chloride), polyamides, polyesters, and alkyds. High solids content at minimum viscosity is available with emulsion or latex coatings. Poly(vinylidene chloride) provides excellent barrier properties (see Vinyl polymers; Vinylidene polymers).

Economic Aspects

The paper industry in the United States employs ca 6.6×10^5 people in 425 pulp mills, 730 paper and board mills, and ca 4500 converting plants. The United States produces ca 35% of the total world production, and has the highest per capita consumption of any country. The world production and consumption of paper and board for 1978 is defined in Table 4. The United States and Western Europe, which represent ca 13% of the world population, consumed 61% of production. Because of the relatively low consumption levels throughout most of the world, paper demand is growing at a high rate. Since 1960, growth in total demand averaged ca 5%/yr. Worldwide demand in the 1980s is forecast at 3.5% (18).

The total production of the U.S. paper industry in 1979 was ca 58.9×10^6 metric tons. This included 27.0×10^6 t of paper, 28.7×10^6 t of paperboard, and 3.2×10^6 t of construction paper and board. Some of the main grades were newsprint, 3.7×10^6 t; uncoated groundwood, 1.3×10^6 t, coated printing paper, 4.1×10^6 t; uncoated book paper 7.1×10^6 t; and tissue paper, 4.1×10^6 t. The main grades of paperboard were containerboard, 17.1×10^6 t, which is used largely for corrugated shipping containers; folding boxboard, 7.0×10^6 t; and foodboard, 3.6×10^6 t.

The apparent consumption of paper and paperboard products in the United States in 1979 is estimated at 63.8×10^6 t. The difference between production and consumption results largely from importation of newsprint to the United States from

Table 4. 1978 World Production and Consumption of Paper and Paperboard[a]

	10^6 Metric tons				Per capita
	Production	Imports	Exports	Apparent consumption	consumption, kg
North America	71.4	8.8	12.1	68.1	282
United States	57.8	8.5	2.6	63.6	291
Canada	13.6	0.3	9.4	4.5	193
Western Europe	40.6	13.6	15.6	38.6	110
Eastern Europe	15.6	1.8	1.7	15.7	40
Latin America	6.4	2.0	0.4	7.9	24
Africa	1.5	0.9	0.2	2.2	5
Asia and Oceania	32.1	3.9	1.2	34.8	14
Total	*239.0*	*39.8*	*43.2*	*235.4*	

[a] Ref. 18.

Canada. Current economic, statistical, and technological information on the U.S. pulp, paper, paperboard, and allied industries is given in ref. 4.

Analytical and Test Methods

Chemical Composition. Methods for paper analysis are reviewed and experimental details are provided in ref. 19. Chemical analyses which are used to characterize wood pulps for papermaking include determinations of α-, β-, and γ-cellulose, carbohydrates, lignin, carboxyl and carbonyl groups, copper number, and viscosity. The carbohydrate determination involves acid hydrolysis of the pulp, preparation of volatile derivatives, and separation of the individual monomeric sugars by gas chromatography (see TAPPI T 249 and ASTM D 1915). Results are used to compute percentages of cellulose and hemicellulose in the pulp. Lignin is that portion of the pulp that is insoluble in 72 wt % sulfuric acid (see TAPPI T 223 and ASTM D 1106). The α-, β-, and γ-cellulose test (TAPPI T 203) and other determinations which are based upon alkali solubility (TAPPI T 212 and T 235, ASTM D 1696) reflect empirically the hemicellulose content of the sample and the degradation of the cellulose. Also indicative of hydrolytic or oxidative degradation of the pulp are higher carboxyl (TAPPI T 237 and ASTM D 1926), carbonyl, and copper-number values and lower viscosity. Cupriethylenediamine solutions of cellulose are used for the viscosity test (TAPPI T 254 and ASTM D 539); results are related to the average degree of polymerization. Molecular weight distribution may be determined by separation of cellulose tricarbanilate derivatives by gel-permeation chromatography (gpc) (20).

Procedures are available for detecting and determining most of the noncellulosic constituents of papers. Rosin size is detected by the Raspail or Lieberman-Storch test; it may be determined by extracting the paper with acidified alcohol and isolating the ether-soluble portion of the alcohol extract (TAPPI T 408 and ASTM D 549). Starch is detected by the blue color produced with application of an iodine–potassium iodide solution. The intensity of the blue, which is measured with a spectrophotometer, provides the basis for a quantitative starch determination (TAPPI T 419 and ASTM D 591).

Kjeldahl nitrogen determinations are used to determine the wet-strength resins most commonly used in paper, eg, urea–formaldehyde and melamine–formaldehyde (TAPPI T 418 and ASTM D 982). Melamine may be determined by uv spectrophotometry (TAPPI T 493 and ASTM D 1597). Formaldehyde-containing wet-strength resins can be detected if a red-violet color appears after heating the paper in a solution of chromotropic and sulfuric acids (see Amino resins).

Acidity or alkalinity of paper is determined by measuring the pH of a cold- or hot-water extract (TAPPI T 435 and T 509, ASTM D 778). Alum is the most common source of acidity.

The amount of mineral filler or coating pigment is determined from the ash content of the paper (TAPPI T 413 and ASTM D 586). Factors may be necessary to correct for ash in the pulp and for pigment changes resulting from ashing. Distribution of filler through the thickness of the web is measured by removing increasingly greater amounts of paper by surface grinding and determining ash in the remaining paper. Pigments usually can be identified by x-ray diffraction analysis. Elemental analysis by emission spectrography or energy-dispersive x-ray analysis can aid in pigment identification.

Analysis of coatings is simplified if the coating can be removed from the paper in a water bath by ultrasonic cleaning. If fibers are not present, a carbohydrate determination can be used to identify gums and other carbohydrate polymers in the coating.

Latexes of synthetic resins are identified by ir spectrometry. Selective extraction with organic solvents is used to obtain purified fractions of the polymers for spectrometric identification. Polymeric films can be identified by the multiple internal reflectance ir technique, if the film is smooth enough to permit intimate contact with the reflectance plate. TAPPI and ASTM procedures have not been written for these instrumental methods, because the interpretation of spectra is not amenable to standardization.

Solvent extraction followed by gas-chromatographic analysis is used to determine paraffin wax; antioxidants (qv), ie, butylated hydroxyanisole and butylated hydroxytoluene; and other volatile materials. Trace amounts of chlorinated organic compounds, eg, polychlorinated biphenyls, can be determined by using a gas chromatograph with an electron-capture detector (21).

Fiber Analysis. Paper may be composed of one or several types of fibers, eg, animal, vegetable, mineral, and synthetic. Paper is generally composed of woody vegetable fibers obtained from coniferous (softwood) and deciduous (hardwood) trees. Qualitative and quantitative methods have been developed to determine the fibrous constituents in a sheet of paper (see TAPPI T 401). However, the recent proliferation in the number and types of pulping processes used have made the analysis of paper a much more complex problem. Comprehensive reviews of the methods are given in refs. 19 and 22.

A common method involves tearing a representative sample into small pieces, placing these in a beaker, and heating to a boil in 0.5% sodium hydroxide. The paper then is washed, neutralized with hydrochloric acid, and washed again. Disintegration is effected by vigorously shaking the flask. A desirable fiber concentration for suspension is ca 0.05%. A portion of this suspension is placed on a slide and examined microscopically. The fibers are stained in order to determine the pulping process and to produce contrast for the identification of the fibers. Graff's C stain is probably the best stain for general fiber analysis. Other stains which are used are Wilson's stain, Herzberg stain, and the Green-Yorston stain AZO. The colors which are developed by these stains vary according to the raw material used and the pulping process. The fibers should be examined in daylight or fluorescent lighting at a magnification of ca 100 diameters.

Analysis of certain papers requires special treatment before they can be disintegrated properly. Papers containing synthetics, tars, asphalt, rubber, viscose, or wet-strength resins must be analyzed individually (see TAPPI T 401 and ref. 19). Dyes or colors must be removed from highly colored papers before examination. The method of dye removal depends upon the type of dye.

Plant-fiber identification is described in TAPPI T 8 and T 10. In order to identify synthetic fibers, it usually is necessary to conduct solubility and physical properties tests in addition to light-microscopy observations.

Systematic sampling is required to obtain quantitative information on sample composition. Because different types of pulps contain varying numbers of fibers per unit weight, it is necessary to multiply the total number of each kind of fiber by a relative weight factor; thereby, the weight percentage that each fiber type contributes to the sample can be determined.

Environmental Issues and Plant Efficiency

Some Federal laws which affect the pulp and paper industry are the Clean Air Act, 1970, Clean Water Act, 1974, Resource and Recovery Act, 1976, Toxic Substances Control Act, 1977, Occupational Safety and Health Act, 1970 Federal Hazardous Substances Act, and Federal Insecticide, Fungicide and Rodenticide Act, 1972. The new requirements have resulted in increased lead time for expansion of basic production facilities and have affected the nature of industrial growth (18). It is thought that some of the regulations cannot be applied practically (23). Removal of the few percent of pollutants from paper-plant point sources could be prohibitively expensive (24).

Originally, the major concern was for general pollutants, eg, biochemical oxygen demand (BOD) and total suspended solids (TSS) in water and particulates, total reduced sulfur (TRS), and sulfur dioxide (SO_2) in air. Recently, there has been concern with those chemicals which are considered particularly toxic.

In general, most pollutants pass from air through rain and from solids through leaching into water. Therefore, water quality is of continuing high priority to the pulp and paper industry and to the public. With the growing concern over hazardous materials, the disposal of sludges as landfill will be questioned (25).

Pollution Control. Methods of pollution control in pulp and paper mills include in-mill control measures, end-of-the-pipe control measures, and monitoring and assessing environmental quality. Any change in the pulping or papermaking process or its control directly or indirectly affects environmental quality. Environmental control approaches cannot be segregated from production technology. Since ca 1965, air-pollution technology in the United States for the pulp and paper industry has undergone major advancements in the design of pollution-abatement systems for controlling gaseous and particulate emissions (see Air pollution control methods). In the kraft-pulping segment, sulfur gas and particulate emissions have been reduced within the production cycle.

Atmospheric emissions from the kraft process include gaseous and particulate matter. The principal gaseous emissions are malodorous, reduced sulfur compounds (TRS), eg, hydrogen sulfide, H_2S; methyl mercaptan, CH_3SH; dimethyl sulfide, CH_3SCH_3; and dimethyl disulfide, CH_3SSCH_3; and oxides of sulfur, SO_x, and nitrogen, NO_x. In addition, most pulp-mill flue-gas streams contain appreciable amounts of water vapor. The particulate matter emissions are primarily sodium sulfate, Na_2SO_4, and sodium carbonate, Na_2CO_3, from the recovery furnace and sodium compounds from the lime kiln and smelt tanks (26).

Odor is one of the principal air-pollution problems in a kraft pulp mill because the odorous gases are detectable at 1–10 ppb. The amount of odorous gases released per unit of production varies considerably between individual process units. The main sources for these include digester blow and relief gases, vacuum-washer hood and seal-tank vents, evaporation hot-well vents, the recovery furnace, dissolving tanks, black-liquor oxidation tank vents, the lime kiln, and some wastewater treatment operations.

Both SO_x and NO_x are emitted in varying quantities from specific sources in the kraft chemical-recovery system. The major source of SO_2 is the recovery furnace. Lesser quantities of SO_2 can be released from the lime kiln and smelt-dissolving tank. Nitrogen oxides, ie, NO and NO_2, are formed by oxidation of nitrogen-containing compounds at elevated temperatures, especially when auxiliary fuels, eg, natural gas and fuel oil, are added to the recovery furnace. When released into the atmosphere, these gases form acid rain.

Control devices for gases may include stripping, scrubbing, condensation, incineration, and adsorption. Control devices for particulate matter include cyclones, scrubbing, and electrostatic precipitators. Application of these individually or in combination has been quite effective in meeting the air-quality criteria.

The pulp and paper industry uses large amounts of water. In 1972 water-quality requirements have forced mills to install end-of-the-pipe secondary treatments, which principally are biological. Means other than installation of expensive end-of-the-pipe treatments are being evaluated in terms of meeting recent requirements. Mills are recycling and reusing more process water, which results in higher water temperatures and increased solids in the process water, thereby enhancing problems, eg, corrosion slime and other deposits. However, chemicals are used to control these problems. Dissolved solids may impart biochemical oxygen demand (BOD), color, and toxicity to mill effluents. The suspended solids impart turbidity and long-term BOD. Directly or indirectly, all of these may affect aquatic life.

The greatest bulk of water which is used in the paper mill is the white water. As a general rule, white water does not contain a serious amount of BOD. In most cases, receiving streams adsorb the demand without being adversely affected. However, in the manufacture of specialty papers, deleterious effects may occur. Such situations must be remedied by special mechanical or chemical treatments. White waters generally do not contain appreciable amounts of toxic materials; however, in the manufacture of asphalt-laminated paper, phenolic compounds may pass from the plant into streams. Excessive use of a slime-control agent might result in some being lost to the secondary treatment plant and river.

Modern practice is to maintain the white-water system as closed as possible, ie, as much water as is compatible with efficient machine operation is recycled. The loss of fibers and inert furnish components, particularly clay, has been greatly reduced. Fiber losses, however, still occur into the white water, and greater economy of operation might be achieved if these fibers could be recovered. Thus, it is common to design a fiber-recovery system into the white-water cycle. The three general types of save-all fiber recovery are based on filtration, flotation, and sedimentation. If these are operated efficiently, the net fiber loss can be less than 1%.

In order to conform to environmental quality guidelines, mills have installed a number of primary and secondary treatment systems to control effluents. The primary treatment is composed of settling basins and/or tanks, ie, clarifiers. These remove ca 85–100 wt % of solids, eg, fibers and clay. Primary sludges, which are removed by the primary clarifiers, cannot be reused by the mill in the same product. However, many mills use the sludges in lower grade product lines.

The secondary treatment generally consists of a biological treatment followed by secondary clarifying. Biological waste treatments include lagoons, aerated lagoons, activated sludge with air and oxygen, trickling filters, modified biological systems containing activated carbon, and combinations of these. The secondary treatments generally remove 90–95% BOD, most solids, most of the toxicity, but very little color. In some instances, color increases after the water has been treated. Some of the solids from the secondary clarifiers may be recycled but some is wasted. These solids or biological sludges are extremely hydrophilic and are hard to dewater.

As more process water is recycled to reduce overall water consumption and wastewater discharge volumes, more nonpathogenic microbial growth, ie, slime, occurs in the mill system. Slime formation prevents normal flow of stock suspensions, may make the furnish lumpy, prevents normal sheet formation, and interferes, in general, with papermaking. It has been a significant impediment to the goal of 100% closure of the water loop.

Fundamentally, the remedy for slime problems is to create conditions in the system that are inimical to the growth and propagation of slime-forming organisms. The efficiency of any form of slime control is greatly increased by ordinary good mill-cleaning procedures. The application of chlorine or chloramine with or without frequent cleaning is effective in many cases. Antiseptics and disinfectants reduce or inhibit slime formation (see Disinfectants and antiseptics). The use of these slimicides may result in an appreciable increase in the cost of producing paper. However, their use often reduces downtime that is caused by slime and, therefore, increases production which more than compensates for the initial cost of the slimicides.

White-water systems also often contain proteolytic microorganisms which attack the machine felts and reduce their useful life. Control of this problem may be accomplished by treating the felts with a slimicide followed by cleaning with a mild acid (see Industrial antimicrobial agents).

Sludge Handling and Disposal. Most waste-treatment processes generate solid wastes which must be disposed of. Two kinds of sludges are generated by pulp and paper mills: primary sludges contain fibers, clay filler materials, and other chemical additives and secondary sludges are largely biological in nature and are harder to handle and dewater. The disposal of sludges in landfills is being reevaluated and alternative disposal approaches are being developed (27–30).

The wide range of types of paper products results in a variety of sludges. Solid wastes result from several sources within the mill, eg, bark, sawdust, dirt, knots, pulpwood rejects, flyash, cinders, slag, and sludges. Sludges often are disposed of in combination with residuals from other sources. Approximately 300 kg of solid waste per ton of finished product is generated by the pulp and paper industry.

Solids content of wet sludges is 1–40 wt %. Ash content can vary from very little to over 50 wt % of the solids content. However, the solids or moisture content of a sludge is not enough for assessing the physical or engineering properties of that sludge. For example, sludges of 30–35 wt % solids from a paper recycling or deinking operation may be in a highly fluid state, whereas a low ash, high fiber pulp-mill sludge at 15–20 wt % solids may be quite dry and stable. The operating plan for a land-disposal site depends upon the fluid state of the sludge (29). Generally, sludge-handling processes include thickening, stabilization, conditioning, dewatering, incineration, and disposal (31).

Most sludges are disposed in landfills. Problems from this practice include possible leaching into surface and ground water, odors, methane generation, and other problems. Successful site-establishment processes are based on identifying all residual sources, including sludge, which originate from mills; characterizing the quantity and composition; assessing the storage, transportation, and material handling that are required to transport the residuals to a deposit site; and assessing the effects of various residuals on the environment of the site.

Water-Quality Assessment. Assessments of the effects of effluents on receiving streams until the mid-1970s was more subjective than objective (32–33). Since then, changes in attitude within the aquatic life-science field and the regulatory system have necessitated a restructuring of the design and foundations for effluent-impact assessments in receiving waters. For example, government regulations have proceeded from a system of stream-standards-based regulations to effluent-based regulations involving strict requirements on various pollution parameters. Pressure is being exerted to go back to the receiving water system as the ultimate test for new and more stringent discharge-control measures.

The increasing knowledge of the interrelationships between the various biological, chemical, and physical components of aquatic systems has provided significant restructuring of field assessment programs which are designed to analyze effluent impact. A single organism as an indicator of stream quality has been replaced by community compositional and structural analysis. Thus, total effluent effects on a broad scale can be realized. Other measured parameters are algal assays, fish surveys, sediment mapping, plume mapping, sediment oxygen demand, and socioeconomic impacts.

Uses

Paper and Paperboard Containers. ***Rigid Paper Containers.*** Rigid paper containers generally are constructed of paperboard or a combination of paperboard and paper.

Setup Boxes. The principal requirement of the setup box is stiffness. Because setup boxes are used for candy, stationery, etc, they need not be very strong. They constitute only a small percentage of total box production. Setup boxes usually are formed from a blank of single-ply, stiff paperboard (although pasted boards sometimes are used) by cutting the board from the outside almost entirely through its thickness along lines which are intended to be the edges of the final box. The board is folded along the precut lines, and the edges which are formed by the cut score, as well as those which are formed where the sides of the cut blank meet, normally are taped with paper to hold the box together and to reinforce the cut edges. A cover paper usually is glued to the outside surface of the box. Setup cartons are assembled in the manufacturer's plant before shipping.

Folding Cartons. Folding cartons differ from setup boxes in the type of paperboard used and the method of creasing. Because of its high stiffness, the paperboard is creased by means of a scoring or creasing rule. The board is crushed in the area of the crease and subsequently is folded along these predetermined crushed lines. Folding cartons are shipped flat and must be opened and set up for use.

The type of paperboard used by the carton industry is boxboard. Boxboard may be categorized, based on the raw material, as combination or solid boxboard. Combination boxboard, of which there are many grades, normally is made on a multicylinder paper machine using a substantial percentage of waste paper with virgin pulp. Solid boxboard usually is made on a Fourdrinier paper machine using only virgin pulp and it is bleached or coated.

Although folding cartons are made in many sizes and shapes, all are of three basic designs. Tray cartons are open on one face and are formed by folding a sheet of board to make the side panels. Covers are glued or locked in place. Top-opening cartons are similar to tray cartons, except that one side panel is extended to serve as the top. This is folded over to cover the open face. The cartons may be tucked, locked, or glued closed. End-opening cartons are essentially tubes, in which one or both ends are folded and sealed, locked, or tucked closed.

The folding carton primarily is a consumer's carton; as such, the requirements are greater for esthetics than for strength. However, ease of assembly, absence of cracking along edges, and stiffness are important.

The operations associated with the manufacture of folding cartons are printing, die cutting, and gluing. In general, most cartons are printed by letterpress, lithography, or rotogravure. Boxboard may be shipped as sheets or as a continuous web, as from a roll. The production of cartons from a continuous web is increasing. In the case of sheet stock, the sheets are fed by hand or by mechanical means to a printing press. One or more colors may be used. After printing, the sheets are fed to a die-cutting press where the carton blanks are cut and creased. The individual carton blanks then are passed through a machine which folds the die-cut blanks into tubes and glues the body closed. The formed tubes are packed flat for shipping. Examples of folding cartons are the toothpaste container, cereal box, butter carton, doughnut box, tack box, and milk carton.

Fiber Cans and Tubes. The basic material used for fiber tubes and cans is a bending board. The body of a fiber can usually is of paperboard and the ends usually are of metal, paperboard, or plastic. The construction of the body may be one of three general types: spiral-wound tubes and cans, convolutely wound tubes and cans, or laminated or lap-seam cans.

Spiral-wound bodies are made on a spiral winder, which consists of a stationary cylindrical mandrel. An endless belt is looped over one end of the mandrel and a traveling saw defines the other end. The raw stock is slit into narrow rolls which are fed into the winder from either side. Each web passes over a glue roll, around the mandrel, and under the endless belt at 25–85°. The pressure of the moving belt causes the formed tube to rotate on the mandrel, thus drawing in more material and causing the formed tube to move forward along the mandrel. The continuous spiral tube may be cut to length on the winder or rewound in multiple lengths and cut to size as a separate operation. The spirally wound tube bodies are restricted to those of circular cross section.

Convolute tubes or can bodies are made by winding two or more plies of board directly around a mandrel. The length of the tube may be the length or multiples of one can body. Convolute winding allows the formation of shapes other than round, eg, square or elliptical.

The lap-seam body is made similarly to the convolutely wound can, except that only one ply of board is used. The lap joint is glued by numerous different adhesives. Lap-seam cans can be made in various cross-sectional shapes.

The can bodies are made into cans by adding ends of either metal, plastic, or paperboard. Metal ends usually are applied using modified metal can-closing machines. Paperboard caps are drawn from special grades of board and may be glued to the can body.

Corrugated and Solid Fiber Boxes. Corrugated and solid fiber boxes are used primarily as shipping containers. Both types of containers are made from several layers of paperboard, normally referred to as combined board. Container board is the material from which the combined board is fabricated. Although both types of boxes serve the same general purpose, eg, in the handling, storage, and transportation of commodities, they differ markedly in their manufacture, structure, and performance.

Corrugated board is characterized by its cellular structure which imparts high compressive strength at a relatively low weight. This board usually consists of three layers: a corrugated layer with a liner glued to both sides. A relatively thin web of paperboard, ie, the corrugating medium, is passed between two fluted metal rolls to form the corrugation. The facings or liners of high strength container board or linerboard are glued to the tops of the flutes, thereby encasing the corrugated medium on both sides. The various steps take place sequentially: fluting of the central ply or corrugating medium; adhering the liner to one side of the fluted medium by means of aqueous adhesives; adding the other liner or facing by means of adhesives; curing the adhesive and bonding the second liner by passing the formed board over flat, steam-heated plates; and scoring to define flaps and cutting the box blanks to the desired size. Corrugated board, which is manufactured in this way, is referred to as double-faced corrugated board. Double-wall corrugated board is made by combining two fluted corrugating mediums with a central liner and then adding two outer liner facings. Triple-wall corrugated board is made by combining three corrugated mediums with two inner liners and two outer facings.

The blanks delivered at the end of the corrugator are passed to a printer–slotter, where they are printed using soft rubber dies and where the body scores and slots are introduced into the board. The body scores are similar to the flap scores except that they are parallel to the flutes of the corrugating medium and, thus, form the vertical edges of the box when it is assembled. The slotting permits the side flaps to fold over the end flaps. The flaps form the top and bottom of the box. After the printing and slotting operations, the blanks are formed into a flat tube by taping with paper or cloth tape, stitching with metal staples, or gluing the two ends of the box. The resultant flat tubes are bundled or palletized for shipment. The final box is set up by folding and gluing, taping, or stapling the top and bottom flaps.

Variations in construction are made by using different weight facings and different flutes. Conventional fluted rolls are designated A, B, C, and E flutes. They differ in height and in the number of flutes per length of board. The dimensions listed below are approximate, as they vary slightly from manufacturer to manufacturer:

Flute type	*Approximate height of flute, not including facings, cm*	*Number flutes per meter*
A flute	0.48	118 ± 10
B flute	0.24	164 ± 10
C flute	0.36	138 ± 10
E flute	0.12	308 ± 13

Cold corrugating is being developed and is expected to be a popular process (35).

Solid fiber combined board consists of numerous bonded plies of container board which form a solid board of high strength. It is much heavier in weight for a given thickness than corrugated board. Solid fiber combined board is made by passing two or more webs or plies of paperboard between a number of sets of press rolls. Adhesive is applied to each ply before it passes through the press nips. In general, solid fiber combined board is made of two to five plies, with three- and four-ply board being most common. The combined weight of the component plies, exclusive of adhesive, is 556–1758 g/m^2 (114–360 lb/1000 ft^2). In the combining operation it is customary to join the central plies first and then work outward so that the outer plies are applied last. It also is common to use a poor grade of paperboard, eg, chipboard, in the central plies and a strong linerboard as the outside facings or liners. The subsequent operations are similar to those described for corrugated boxes. As in the case of corrugated boxes, different constructions can be obtained by varying the components, number of plies, and caliper of the solid fiberboard.

The adhesives used in the manufacture of corrugated and solid fiber combined board usually are starch or silicate, except where water resistance is required, in which case starch–resin, modified silicate, and resin emulsions may be used.

Flexible Containers. *Paper Bags.* There are many types of paper bags, which differ in shape, style, and number of plies, eg, single-, double-, and multiwall bags are available. There are a number of bag-making machines which cut, fold, and glue bags from a continuous web of paper. A variety of papers are used, including bleached and unbleached ones. The type that is used depends on the requirements of the product to be packaged. Perhaps the most common are the brown kraft bags which include the grocery bag and the multiwall bag. Kraft bags are used for strength and frequently are of double thickness.

There are four types of single-wall bags. Flat bags are the simplest in construction and the least expensive to make. They have single lengthwise seam and the bottom simply is folded under and glued. Satchel-bottom bags provide a flat base when filled. Square-bottom bags, eg, grocery bags, have bottoms similar to flat bags, but have bellows folds at the sides to reduce the width of the closed bag without reducing capacity. Automatic self-opening bags combine the desirable features of the other types of bags. When filled, they form a neat, squared-up package with a stable base and a center or side seam.

Multiwall Shipping Sacks. The construction of multiwall-paper shipping sacks is dictated by the nature of the contents and the shipping and storage conditions. They are used primarily for the packaging of materials that need no protection against compressive forces. Their principal function is to contain the contents and to protect them from contamination. One of the primary requirements of multiwall sack paper is the ability to absorb energy without rupturing (see Packaging, industrial).

Shipping sacks are made from one to six plies of high quality paper and often in combination with special coatings, laminations, or films. In a multiwall sack, each ply or wall is fabricated as a tube and is arranged one within the other, so that each layer bears its share of any applied or induced stress. Better performance is obtained against shock or impact by the use of several plies of relatively lightweight papers than by the use of fewer plies of heavier-weight papers. However, the latter generally is considered to be more effective against externally applied point stresses, eg, protruding nails or broken floorboards. The average heavy-duty, multiwall sack is constructed of a number of plies of paper with basis weights of 65–114 g/m^2 (40–70 $lb/3000\ ft^2$). The most frequently used basis weights are 65, 81, and 98 g/m^2 (40, 50, and 60 $lb/3000\ ft^2$). Papers of heavier weight generally are used in single- and double-ply, pasted-type shipping sacks.

The shipment of many commodities may require special barriers on the sacks to impart resistance against liquids or vapors. Other treatments are used to provide grease resistance, acid resistance, and scuff resistance. Special coatings are used in sacks for packing commodities, eg, synthetic rubbers, asphalts, waxes, and resins, to prevent the contents from sticking to the paper.

BIBLIOGRAPHY

"Paper" in *ECT* 1st ed., Vol. 9, pp. 812–842, by H. F. Lewis, R. Shallcross, D. J. MacLaurin, T. A. Howells, W. A. Wink, B. L. Browning, I. H. Isenberg, R. C. McKee, and W. M. Van Horn, The Institute of Paper Chemistry; "Paper Coatings, Inorganic" in *ECT* 1st ed., Vol. 9, pp. 842–858, by G. Haywood, West Virginia Pulp and Paper Co.; "Paper Coatings, Organic" in *ECT* 1st ed., Vol. 9, pp. 858–867, by P. H. Yoder, Pyroxylin Products, Inc.; "Paper" in *ECT* 2nd ed., Vol. 14, pp. 494–532, by Roy P. Whitney, W. M. Van Horn, C. L. Carey, R. M. Leekley, T. A. Howells, R. C. McKee, W. A. Wink, I. H. Isenberg, and B. L. Browning, The Institute of Paper Chemistry.

1. J. N. McGovern, *Pulp Pap.* **52**(9), 112 (1978).
2. D. Hunter, *Papermaking*, 2nd ed., Alfred A. Knopf, Inc., New York, 1947.
3. D. C. Smith, *History of Papermaking in the United States (1691–1969)*, Lockwood Publishing Co., New York, 1970.
4. *FAO World Pulp and Paper Consumption Outlook—Phase II*, Industry Working Party for the Forestry Department of the Food and Agricultural Organization of the United Nations, Sept. 1977.
5. R. W. Mann, G. A. Baum, and C. C. Habeger, *Tappi* **63**(2), 163 (1980).
6. K. W. Britt, *Handbook of Pulp and Paper Technology*, 2nd ed., Van Nostrand Reinhold Publishing Corporation, New York, 1970.
7. J. P. Casey, *Pulp and Paper Chemistry and Chemical Technology*, 3rd ed., Vol. I, Wiley-Interscience, New York, 1980.
8. J. d'A. Clark, *Pulp Technology and Treatment for Paper*, Miller Freeman Publications, Inc., San Francisco, Calif., 1978.
9. R. G. MacDonald and J. N. Franklin, *Pulp and Paper Manufacture*, 2nd ed., Vol. 3, McGraw-Hill Book Co., New York, 1970.
10. U.S. Pat. 1,873,199 (Aug. 23, 1932), J. D. Haskell; U.S. Pat. 1,985,569 (Dec. 25, 1934), J. D. Haskell; U.S. Pat. 1,960,753 (May 29, 1934), G. P. Prathee.
11. U.S. Pat. 864,359 (Aug. 27, 1907), G. D. Claflin, Jr.
12. U.S. Pat. 2,627,477 (Feb. 3, 1953), W. F. Downey (to Hercules Powder Co.).
13. U.S. Pat. 3,923,745 (Dec. 2, 1975), D. H. Dumas (to Hercules, Inc.).
14. U.S. Pat. 3,102,064 (Aug. 27, 1963), O. B. Wurzburg and E. D. Mazzarella (to National Starch and Chemical Corp.).
15. J. F. van Oss and C. J. van Oss, *Warenkennis en Technologie*, Vol. V, p. 679.
16. *Pigmented Coating Processes for Paper and Board* (TAPPI Press Book No. 28), Technical Association of the Pulp and Paper Industry, Atlanta, Ga., 1962.
17. *Physical Chemistry of Pigments in Paper Coating* (TAPPI Press Book No. 38), Technical Association of the Pulp and Paper Industry, Atlanta, Ga., 1977.
18. J. E. Huber, *Kline Guide to the Paper Industry*, 4th ed., Charles H. Kline & Co., Fairfield, N.J., 1980.
19. B. L. Browning, *Analysis of Paper*, 2nd ed., Marcel Dekker, Inc., New York, 1977.
20. L. R. Schroeder and F. C. Haigh, *Tappi* **62**(10), 103 (1979).
21. S. J. V. Yound, C. Finsterwalder, and J. A. Burke, *J. Assoc. Off. Anal. Chem.* **56,** 957 (1973).
22. I. H. Isenberg, *Pulp and Paper Microscopy*, 3rd ed., The Institute of Paper Chemistry, Appleton, Wisc., 1967.
23. J. Quarles, *Federal Regulations of New Industrial Plants*, Morgan, Lewis and Brochius, Washington, D.C., 1979.
24. J. G. Strange, *The Paper Industry—A Clinical Study*, Graphic Communications Center, Inc., Appleton, Wisc., 1977.
25. H. S. Dugal, "Environmental Laws and Their Impact on Mill Processes," *paper presented at Third Symposium on Corrosion in Pulp and Paper Industry*, May 1980.
26. *Environmental Pollution Control, Pulp and Paper Industry, Part I—Air*, EPA 625/7-76-001, Oct. 1976.
27. *Process and Design Manual for Sludge Treatment and Disposal*, EPA 625/1-74-006, Oct. 1974.
28. D. Marshall, *South. Pulp Pap. Manuf.*, 19 (Dec. 1977).
29. J. J. Reinhardt and D. F. Kolberg, *Pulp Pap.*, 128 (Oct. 1978).
30. *Pap. Trade J.*, 34 (May 15, 1979).
31. *Process Design Manual for Sludge Treatment and Disposal*, EPA 625/1/74-006, Oct. 1974.
32. R. Patrick and D. M. H. Martin, *Biological Surveys and Biological Monitoring in Fresh Waters*, Academy of Natural Sciences, 1974.
33. J. M. Itellawell, *Biological Surveillance of Rivers*, Water Research Centre, Herts, UK, 1978.

G. A. Baum
E. W. Malcolm
D. Wahren
J. W. Swanson
D. B. Easty
J. D. Litvay
H. S. Dugal
The Institute of Paper Chemistry

PLASTIC BUILDING PRODUCTS

The use of plastics in building has grown rapidly in the last few years. Plastics were first used for decorative and nonstructural purposes, but because of increased knowledge of the long-term properties of plastics, particularly resistance to creep and environmental effects, some plastics are now available that maintain long-term structural integrity, such as piping that can contain moderate pressures for a long period of time.

The general advantages of plastics over other materials are that they provide a combination of properties and permit effects not possible with other materials; they provide lower costs, either of materials or in fabrication, or both; they give better performance in some critical respect; they are lighter in weight and have a greater strength-to-weight ratio; they provide a better appearance for a longer time and require less maintenance; and they may replace a scarce, often more expensive, material.

Preparation and Properties

The physical properties of the plastics that are of particular importance in building are the glass-transition or melt temperatures, ease of processing, as shown by the temperatures and pressures needed for molding, heat deflection temperature, tensile and impact strength, and elongation. For foams, the density, thermal conductivity, and fire resistance are important. The most important building plastics are described below and their physical properties are summarized in Table 1. Data on their costs are given in Table 2. Additional information on the uses of plastics in building products is given in Tables 3 and 4.

Low Density Polyethylene. Low density polyethylene (0.91–0.94 g/cm^3) melts at ca 115°C, is insoluble at room temperature, but dissolves in various solvents when molten. During polymerization, side reactions cause chain-branching. The degree and kinds of branching can be controlled to a considerable extent, and the properties of the polymer correspondingly modified. Branching decreases crystallinity and density, and increases the molecular weight distribution. A decrease in crystallinity decreases hardness, stiffness, melt temperature, and chemical resistance; it increases toughness, flexibility, and permeability. An increase in molecular weight distribution facilitates processing but reduces strength, toughness, and resistance to environmental stress-cracking. The average molecular weight also affects these properties. Increasing it increases the strength, toughness, and melt temperature, but decreases the ease of processing and the melt index (see Olefin polymers).

For film, toughness and flexibility are desired; for piping, high strength. The relation of branching to crystallinity, molecular weight, and molecular weight distribution are highly important in achieving the desired properties in the product.

Some of the specific physical properties of low density polyethylene are listed in Table 1. Its principal use in building products is for film, as a water barrier under below-grade floors, and as temporary enclosure material during construction. The film is made by extruding a thin-walled tube, which may be slit or wound up directly, or it may be fabricated by extrusion through a slot die and cast directly on to a cold roll, cooled, then wound up. The former method is used more widely. Another use for low density polyethylene is in piping, but this is much smaller, about 15,000 metric tons in 1979, compared with 106,000 t of film (see Film and sheeting materials; Piping systems).

Table 1. Physical Properties of Plastics[a]

Property	Plastics									
	HDPE	LDPE	PP	PVC	PS	ABS	Polyacrylic, glazing	Polycarbonate, glazing	Epoxy, mineral filled	Polyacetal
glass-transition temp, T_g, °C				75–105	85–105	90–120	90–105	105		
melting point, °C	135	115	168							
injection-molding temp, °C	150–260	150–230	200–290	150–215	225–250	200–275	165–260	290		195–250
injection-molding press, MPa[b]	70–140	55–200	70–140	70–270	70–200	55–170		70–140		70–140
tensile strength, MPa[b]	20–38	4–16	30–38	40–50	34–83	33–43	55–76	55–66	34–69	69
elongation, %	20–1300	90–800	200–700	40–80	1–2.5	5–70	2–7	100–130	16–21	25–75
Izod impact strength, J/m[c]	27–1070	no break	27–117	214–1070	13–21	27–267	16–21	640–690	16–21	75
heat-deflection temp, °C	60–90	45–50	50–60	75–80	80–110	100	70–100	130–135	125–200	125

	Laminates		
	Melamine	Phenolic, woodbase	Polyester, glass-filled
laminating temp, °C	145–165	150–160	RT–150
laminating press, MPa[b]	3.5–12	7–14	0–3.5
tensile strength, MPa[b]	69–172	110–220	69–172
Izod impact strength, J/m[c]	16–80	214–427	240–1500
heat-deflection temp, °C	105	90	100–200

	Foams			
	PU	Polyisocyanurate	PS board	PS expandable beads
density, kg/m^3	14–42	24–56	24–80	13–15
maximum service temp, °C	80–170	150	75–80	75–80
thermal conductivity, w/(m·K)	0.0125–0.034	0.012–0.02	0.023–0.034	0.03
fire resistance	HF-1[d]		HF-1	HF-1

[a] HDPE = high density polyethylene; LDPE = low density polyethylene; PP = polypropylene; PVC = poly(vinyl chloride); PS = polystyrene; ABS = poly(acrylonitrile–butadiene–styrene); PU = polyurethane.
[b] To convert MPa to psi, multiply by 145.
[c] To convert J/m to ft·lbf/in., divide by 53.38 (see ASTM D 256).
[d] HF-1 = UL Standard 94 for Foam Plastics.

Table 2. Prices of Plastics Used for Building Products, $/kg

Plastic	1974	1976	1978	1980	1981
ABS, medium impact		1.32	1.52	1.83	1.98
acrylic, glazing	1.00	1.19	1.30	1.50	1.68
epoxy		1.39	1.67	2.05	2.36
phenolic		0.88	1.03	1.12	1.23
polyacetal	1.32	1.72	2.07	2.47	2.95
polycarbonate, glazing	3.09	3.64	3.65	4.41	
polyester, reinforced			0.79	1.12	1.32
HDPE	0.51	0.62	0.70	0.92	0.93
LDPE, film	0.29	0.62	0.68	0.99	0.96
PP	0.52	0.57	0.73	0.88	1.01
PS	0.66	0.64	0.62	1.01	1.01
PS expandable beads	0.62	0.88		1.28	
PU foam			1.19	1.61	1.83
PVC	0.26	0.55	0.59	0.75	0.70
urea–formaldehyde for plywood, 65% NV	1.10	1.54	1.74	2.31	

Source: *Modern Plastics* and suppliers.

High Density Polyethylene. High density polyethylene (0.94–0.97 g/cm^3) is prepared commercially by either of two catalytic methods. In one, coordination catalysts may be prepared from an aluminum alkyl and titanium tetrachloride in heptane. The other method uses metal oxide catalysts supported on a carrier.

The high density homopolymer melts at about 135°C, is over 90% crystalline and quite linear, with more than 100 ethylene units per side chain. It is more rigid and harder than low density polyethylene and has a higher melting point, tensile strength, and heat-deflection temperature. The molecular weight distribution can be varied considerably, with consequent changes in properties. Typically, polymers of high density polyethylene are more difficult to process than those of low density polyethylene. The physical properties of high density polyethylene are listed in Table 1.

Additives are used extensively in compounding the resin. Antioxidants, uv stabilizers, especially carbon black, and fillers such as glass fibers, silica, or clay provide properties desirable for various purposes.

High density polyethylene is widely used for pipes and drains, especially in large-diameter corrugated forms. The corrugations provide stronger walls at less thickness, which reduces the materials cost of the pipe. About 239,000 metric tons was used for these purposes in 1979.

Polypropylene. Polypropylene and high density polyethylene are prepared in the same way, with a coordination catalyst and in the same equipment. The crystallinity of polypropylene gives it high tensile strength, stiffness, and hardness that are retained at high temperatures, thereby permitting articles to be sterilized. It also is free from environmental stress-cracking. However, the hydrogen atoms bonded to tertiary-carbons cause degradation by oxygen, uv light, and heat, but like high density polyethylene, polypropylene can be stabilized by antioxidants and uv absorbers (see Olefin polymers).

Polypropylene can be fabricated by almost any process used for plastics (see Plastics processing). The extrusion of pipe and injection molding of fittings present no unusual problem. The resin can be reinforced by glass fibers, mineral fillers, or other

Table 3. Uses in Building Products, According to Plastic, 10^3 Metric Tons

Plastic	1971	1972	1973	1974	1975	1976	1977	1978	1979
Poly(vinyl chloride)									
extruded foam moldings	3	23	26	22	18	21	21	21	23
flooring, calendered	131	156	142	89	72	155	75	62	86
coatings	52	58	69	67	59		70	85	82
lighting	4	5	5	6	6	6	6	7	8
panels and siding	27	32	39	44	40	42	45	61	91
pipe and conduit	265	405	525	526	425	616	800	934	1045
pipe fittings	34	39	44	31	27	42	40	51	56
rainware, soffits, fascias	13	14	16	14	13	14	15	7	9
swimming-pool liners	18	20	18	18	15	16	17	18	19
weatherstripping	14	18	16	15	12	13	13	16	17
windows, other profiles	23	25	26	22	20	21	22	26	27
wall coverings	46	58	54	58	45	47	48	57	64
Total	*630*	*853*	*980*	*912*	*752*	*993*	*1172*	*1345*	*1527*
Urea–formaldehyde									
bonded wood	199	232	262	256	171	274	307	346	350
laminated wood	22	24	24	19	13	14	15	17	18
plywood	31	40	40	39	28	31	37	40	40
Phenolic									
bonded wood	30	40	42	26	11	32	34	35	39
laminated wood	21	26	26	23	15	18	19	20	24
plywood	152	166	125	137	124	132	145	160	180
High density polyethylene									
pipe and conduit	42	121	152	150	110	113	184	229	239
Reinforced polyester									
bath and shower units	18	21	27	23	15				

glazing	7	8	8	8	7				
panels and siding	40	47	53	48	40	107	120	135	140
pipes, ducts, tanks	41	46	50	52	51				
Polyurethane foam									
insulation	34	46	65	70	80	75	87	110	133
ABS									
pipe and fittings	72	100	117	107	91	102	140	162	165
Polystyrene									
building and construction	65	72	122	66	47	50	60	57	50
foam, expandable beads	27	35	25	41	38	38	30	53	59
foam, expanded board	21	22	16	18	18	18	24	27	31
Low density polyethylene									
pipe and conduit	26	27	22	15	12	10	14	11	15
film, construction	46	68	70	98	51	51	108	80	106
Acrylic									
glazing	15.5	18.2	28.9	42.0	29.0	30	31	32	33
lighting	22.7	24.9	26.1	28.2	13.0	14	10	10	10
panels and siding	5.3	5.8	5.8	5.8	3.0	4	6	6	6
plumbing	3.6	4.2	10.0	25.0	12.0	14	11	11	13
Polycarbonate									
glazing	3	3	12	14	14	16	16	33	40
Epoxy									
flooring	3	4	5	5	3	5	6	8	8
Polypropylene									
pipe, fittings, and conduit	6	7	9	10	5	7	15	9	11
Polyacetal									
plumbing	3	3	4	4	4	6	8	10	10

Table 4. Use of Plastics, According to Building Products, 10^3 Metric Tons

Building product	1971	1972	1973	1974	1975	1976	1977	1978	1979
Decorative laminates									
phenolic	22	26	26	25	15	18	19	20	24
urea–formaldehyde	22	24	26	21	14	14	15	17	18
Flooring									
epoxy	3	5	5	5	3	5	6	8	8
vinyl, calendered	131	156	211	156	131	155	150	147	168
coatings	52	58							
urethane foam underlay			14	15	14	46	53	80	61
Glazing									
acrylic	16	18	29	42	39	30	34	32	33
polyester	7	8	8	8	7	11	13	15	18
polycarbonate	2	3	12	14	14	16	19	33	40
Insulation									
phenolic	88	108	112	108	74	96	100	125	130
polystyrene foam	28	34	22	23	23	20	24	57	93
polyurethane foam	34	46	55	70	80	75	87	110	133
Lighting fixtures									
acrylic	23	25	26	28	20	14	15	10	10
polycarbonate	1	1	2	3	2	2	2	5	5
polystyrene	17	19	22	24	11	12	12	12	11
vinyl	4	5	5	6	6	6	6	7	8
Panels and siding									
acrylic	6	6	6	6	5	4	6	6	6

vinyl	27	32	39	44	40	42	45	61	91
polyester	40	47	53	48	40	60	51	65	70
Pipe, fittings, conduit									
low density polyethylene	26	27	17	15	12	5	14	11	15
high density polyethylene	42	51		150	110	113	184	229	239
polypropylene	6	7	9	10	5	7	15	9	11
polystyrene	39	43	89	87	36	9	5	6	5
vinyl	299	444	561	549	452	648	840	985	1101
ABS	72	92	116	119	91	114	124	132	165
reinforced polyester	41	46	45	51	77	80	90	98	108
Windows and rainware									
vinyl	53	86	84	77	63	69	71	70	85
polyethylene	3	3	4	4	2	3	3	3	3
Plumbing, bath fixtures									
acrylic	4	4	10	25	21	14	16	11	13
polyacetal	3	3	4	4	4	6	8	9	10
reinforced polyester	18	21	27	23	15	37	40	55	63
Bonded wood									
phenolic	182	208	131	114	94	164	179	195	218
urea–formaldehyde	230	272	258	264	172	307	344	386	390
Vapor barriers									
polyethylene	46	68	96	100	70	74	108	80	78
vinyl	27	32	31	32	23	24	26	18	19
Wall coverings									
vinyl	46	58	54	58	45	47	48	57	64
Total, including items not listed	*1519*	*2073*	*2400*	*2257*	*1890*	*2411*	*2832*	*3311*	*3533*

types of fillers, and can be pigmented readily. Its usage in piping and conduit is small, about 11,000 t in 1979, and represents a very small fraction of total polypropylene production.

Poly(Vinyl Chloride). Poly(vinyl chloride) (PVC) for building products is prepared by either the bulk or the suspension polymerization process. In each process residual monomer is removed because it is carcinogenic. Oxygen must be avoided throughout the process (see Vinyl polymers).

The polymer is only slightly crystalline, mainly syndiotactic, but with so low a degree of order that only small crystallites are formed. It is fundamentally unstable to heat and light, and loses hydrogen chloride by an autocatalytic reaction. Zinc and iron salts also strongly catalyze the decomposition. Many stabilizers, especially combinations of acid acceptors and antioxidants, produce satisfactory results except at high temperatures. The high density, a result of the high chlorine content, is offset by the low cost; thus, the cost/volume ratio is quite attractive.

Vinyl chloride polymers are produced in two main types, homopolymers and copolymers, usually with vinyl acetate. Both types can be plasticized by a wide variety of plasticizers, usually esters. Rigid or unplasticized PVC is used extensively for pipe. The plasticized material is used largely in floor coverings. The homopolymer itself is inherently fire resistant, but addition of plasticizers, unless they are especially fire resistant, considerably reduces this characteristic (see Flame retardants).

Rigid, unplasticized PVC is stronger, with a higher tensile strength, than the polyolefins, including polystyrene and ABS resins. It is not as strong as oxygenated polymers, eg, polyacrylics, polycarbonates, polyacetals, and epoxy resins (qv), although it is similarly intermediate in elongation. Because of its instability to heat, rigid vinyl must be processed quickly, at low temperatures, and the pressures must be correspondingly high.

Poly(vinyl chloride) is the plastic used most widely in building products, and more than 1.5×10^6 t was used in 1979, nearly one-half of all the plastics used for this purpose. Most of the PVC, about 1.1×10^6 t, was used for pipe, fittings, and conduit, and the remainder, 1.68×10^5 t, in floor coverings. The use in siding is growing rapidly, about 9.1×10^4 t in 1979, a 50% increase over the use in 1978. Small amounts were used for lighting fixtures, wall coverings, and vapor barriers.

Polystyrene. Polystyrene is prepared by the polymerization of styrene, primarily by the suspension or bulk processes. Polystyrene is a linear polymer, atactic, amorphous, inert to acids and alkalies but attacked by chlorinated hydrocarbons (dry-cleaning fluids) and aromatic solvents. It is clear but yellows and crazes on outdoor exposure when attacked by uv light. It is brittle and does not accept plasticizers, although rubber can be compounded with it to raise the impact strength. Its principal use in building products is as a foamed plastic (see Foamed plastics). The foams are used for interior trim, door and window frames, cabinetry and, in the low density expanded form, for insulation (see Styrene plastics).

ABS Resins. Acrylonitrile–butadiene–styrene (ABS) resins have a wide variety of compositions, preparation conditions, and properties. Compositions generally run about 20–30% acrylonitrile, 20–30% butadiene and 40–60% styrene. The resins are typically tough and rigid, easy to extrude or mold, and have good thermal and abrasion resistance. They can be blended with other resins, especially with poly(vinyl chloride), and can be shaped by almost any plastics-fabrication process: injection molding, extrusion, or thermoforming. They are considered engineering plastics (qv), and are used in many automotive, marine, and communications applications. In building products, they are used for pipes, ducts, and structural foam. High impact grades used in piping have an Izod strength of 347–400 J/m (6.5–7.5 ft·lbf/in.) (see Acrylonitrile polymers).

The structural foams are made by means of a blowing agent or by mixing a gas into the hot melt, then injecting it into a mold where the gas expands to form a cellular product. The foams are used for large parts in which light weight and rigidity are desired. Extrusion and free expansion are also used.

Polyurethane. Polyurethane foams are prepared in several steps (see Urethane polymers). The preparation processes have many variations that lead to products of widely differing properties. Polyurethane foams can have quite low thermal conductivity (k) values, among the lowest of all types of thermal insulation, and have replaced polystyrene and glass fiber as insulation in refrigeration. The sprayed-on foam can be applied to walls, roofs, tanks, and pipes and between walls or surfacing materials directly. The slabs can be used as insulation in the usual ways. Typical values are listed in Table 1.

Polyisocyanurates. Polyisocyanurates are prepared by condensation of 4,4′-methylenebis(phenyl isocyanate) (MDI) with an acidic, basic, or organometallic catalyst (or a combination of them) to form a six-membered ring (see Cyanuric and isocyanuric acids). This structure is considerably more heat and fire resistant than urethanes, which dissociate at around 100–130°C; the isocyanurates are stable to 350–500°C, probably because there is no hydrogen atom on the ring. Polyisocyanurate foams are rather brittle; therefore, they are modified with an active-hydrogen compound to increase flexibility and resilience.

The preparation of polyisocyanurate foams is similar to that of polyurethanes. Building panels for walls and roof decks are made by pouring the mixed components into a closed mold in which the liquid foams. A solid skin forms against the surface of the mold, which produces a skin integral with the matrix foam. The density of the foam decreases from surface to center. These foams are being used, but reliable statistics are not yet available for them.

Phenolic Resins. Phenolic resins (qv) are prepared by the reaction of phenol with formaldehyde, through either the base-catalyzed one-stage or the acid-catalyzed two-stage process. The liquid intermediate may be used as an adhesive and bonding resin for plywood, particle board, fiberboard, insulation, and cores for laminates. The physical properties for typical phenolic laminates made with wood are listed in Table 1 (see Laminated wood-based composites).

Amino Resins. Amino resins (qv) include both urea– and melamine–formaldehyde condensation products. They are prepared similarly by the reaction of the amino groups in urea or melamine with formaldehyde to form the corresponding methylol derivatives. These are soluble in water or ethanol. For adhesive or bonding purposes to form plywood, particle board, and other wood products, a liquid resin is mixed with some acid catalyst and sprayed on the boards or granules, then cured and cross-linked under heat and pressure. The decorative plastic laminates widely used for countertops and cabinets are also based on melamine–formaldehyde resin (see Laminated and reinforced plastics). Several layers of phenolic-saturated kraft paper are placed in a press and a sheet of α-cellulose paper printed with the desired design and impregnated with melamine–formaldehyde resin is placed over them. Then a clear α-cellulose sheet, similarly impregnated with the resin, is placed on top to form a clear, protective surface over the decorative sheet. The assembly is cured under heat and pressure up to 138°C and 10 MPa (1400 psi). A similar process is used to make wall paneling, but as the surfaces need not be as resistant to abrasion and wear, laminates for wall panels are cured under lower pressure, about 2 MPa (275 psi). The amino resins are lighter in color and have better tensile strength and hardness than phenolic resins; their impact strength and heat and water resistance are less than those of phenolics. The melamine–formaldehyde resins are harder and have better heat and moisture resistance than the urea resins, but they are also more expensive. The physical properties of the melamine–formaldehyde laminates are listed in Table 1.

Polyester Resins. Reinforced polyester resins are based on unsaturated polyesters from glycols and dibasic acids, either or both of which contain reactive double bonds. The ratio of saturated to unsaturated components controls the degree of cross-linking and thus the rigidity of the product (see Polyesters, unsaturated). Typically, the glycols and acids are esterified until a viscous liquid results, to which an inhibitor is added to prevent premature gelation. Addition of the monomer, usually styrene, reduces the viscosity to an easily workable level. When the resin is to be used, fillers, eg, glass fibers, asbestos, or cotton, are mixed with it. The amount and kind of filler affect the strength, flexibility, and cost of the product. Alumina trihydrate acts as a fire retardant when needed. Thixotropic agents control the viscosity and prevent the mix from draining from sloping surfaces. Pigments are used to provide color, usually white, and uv absorbers for outdoor stability. To reduce smoke generation in burning resin and to improve its outdoor stability, styrene can be replaced by methyl methacrylate. A peroxide catalyst and a combination of a cobalt soap and a tertiary amine are added as initiators (qv). The reaction is exothermic, and gelation is usually rapid.

For sanitary ware, a gel coat containing no fiber reinforcement is applied first to the mold. It forms a smooth, strong, impervious, durable chemical and weather- and wear-resistant surface. The bulk of the resin, which may be reinforced with glass fiber, is applied by hand lay-up or by spray gun. The article is then cured at or near ambient conditions.

The physical properties of the reinforced polyester product made from chopped glass are listed in Table 1. The chemical resistance varies according to the composition but is generally good. Its principal uses in building products are for sanitary ware, eg, tub–shower units, and for panels, especially translucent or cement-filled types for roofing and walls of commercial or industrial buildings.

Polymethacrylates. Poly(methyl methacrylate) is the acrylic resin most used in building products, frequently as a blend or copolymer with other materials to improve its properties. The monomer is polymerized either by bulk or suspension processes. For glazing material, its greatest use, only the bulk process is used. Sheets are prepared either by casting between glass plates or by extrusion of pellets through a slit die. This second method is less expensive and more commonly used. Peroxide or azo initiators are used for the polymerization (see Methacrylic polymers).

The polymer is clear and colorless, and remains so on outdoor exposure; thus, it often is used for glazing and lighting. Because the resin is softer than glass, various coatings are applied to improve its abrasion resistance. It is strong, tough and, when used as glazing, does not shatter if broken. These qualities render it suitable for vandal-resistant window panes and outdoor light globes. It also is used in plumbing fixtures, simulated-marble compositions, lavatory bowls, vanity tops, countertops, and bathtub–shower units.

Polycarbonates. Polycarbonates (qv) are partly crystalline, with some disorder in the crystalline part and considerable order in the amorphous part. This disorder conveys high impact strength which, combined with its good transparency and outdoor exposure resistance, makes polycarbonates useful for vandal-resistant glazing and outdoor lighting. It is easily processed by extrusion and injection molding. Various uv and flame-retardant agents are often added (see Uv stabilizers).

Epoxy Resins. Epoxy resins (qv) or polyether resins are used as the binder for terrazzo flooring. The epoxy resin often is made from epichlorohydrin and Bisphenol A. An excess of epichlorohydrin is used to assure that the intermediate product contains terminal epoxide groups. This resin, usually a viscous liquid, is mixed with fillers, pigments, and a curing agent. The mix then is applied to the substrate, and cure is obtained in a few hours. The product is strong, tough, and resistant to chemicals and abrasion. It is used for industrial and other floors subject to hard wear. The use of epoxy resins for this purpose is only a small fraction of its total use.

Polyacetals. Polyacetals are of two types, homopolymer and copolymer (see Acetal resins). Both are based on formaldehyde, through acetals or trioxane, and are highly crystalline, strong, and rigid, with high melting points. They are made in a variety of grades of different melt indexes, and are processed easily by extrusion or injection molding. They can be reinforced with glass or fluorocarbon fibers and can be pigmented. Both have high resistance to creep and abrasion and low coefficients of friction. They are used as engineering resins and in building products for plumbing fittings such as ball cocks, faucets, pumps, and valves, which are subject to steady wear and must retain close dimensional tolerances. Polyacetals are flammable; however, because of the high oxygen content, they burn cleanly and produce no smoke. The volume of usage in building products is small, ca 10^4 t in 1979 (1–4).

Exterior Uses

Roofing. The roofing industry is currently pursuing two avenues of development for plastic roofing materials (see Roofing materials). One is the use of roofing tiles made by extrusion of rigid PVC with additives to stabilize it to uv light. The tiles are designed with internal ribs that add strength, and dual-wall construction to provide some thermal insulation.

The other approach to plastic roofing is the use of foam panels, usually of polyurethane, or for better fire resistance, of polyisocyanurate, sandwiched between surfacing materials. In some cases the foams are reinforced with glass fiber to improve dimensional stability. The surfaces may be asbestos roofing felt, reinforced panels, or expanded perlite boards, which give strength and additional fire resistance. The total thickness may be 5–14 cm. In other cases, foam boards made of polyurethane, polyisocyanurate, or polystyrene are used as part of a composite roof-decking assembly, to provide thermal insulation and sometimes a degree of water resistance. They usually are covered with a membrane of weather-resistance plastic, or more often by a layer of roofing felt and hot tar (5–9). A large portion of single-ply roofing systems for nonresidential buildings consists of elastomers of ethylene–propylene–diene monomer (EPDM) (10).

Siding. The resin most used for siding is poly(vinyl chloride) homopolymer, compounded with modifiers, stabilizers, and pigments. Modifiers are most often acrylic esters, followed by chlorinated polyethylene or ethylene-vinyl acetate, used at 6–8 phr (parts per hundred resin). The modifier increases the impact strength of the rigid PVC. Many other resin additives or copolymers are being studied to reduce the brittleness of the homopolymer, among them a copolymer of α-methyl styrene and an acrylic ester–styrene–acrylonitrile copolymer.

Heat stabilizers are usually methyl- or butyltin compounds, at 1.5–2.5 phr. Other organotin and barium–cadmium stabilizers are being developed that would permit the use of less TiO_2 in the composition. In light colors TiO_2 is a pigment but it also functions as a uv screen and, if less is used, stabilization must be increased by other means. A barium cadmium phosphite with an epoxy-plasticizer stabilizing system is said to be so effective that TiO_2 can be eliminated. However, the liquid components reduce the heat distortion temperature and impact strength so that more rigid modifiers may be needed. In further efforts to reduce costs, compounding is done by the extruder, thereby eliminating the pelletizing step by high intensity dry-blend mixing of powders that are fed directly to the extruders. This method also reduces the heat exposure and provides better quality with reduced amounts of stabilizers.

Siding is usually produced by extrusion of the PVC composition through a profile die. It also is prepared by extruding a flat sheet, embossing it, then postforming in a press. This process is relatively low cost at a high production rate, up to 270 kg/h, but the appearance obtained with extruded profiles is usually preferred. Extruded profiles are conventionally cooled with air, which requires a cooling length of 7.5–10.5 m. In some new processes the profile is held by means of a vacuum against a metal shape, through which water at a controlled temperature is passed. The profile is thus cooled at a controlled rate in a cooling length of 1–1.5 m, in a shorter time, and with less residual internal stress. The profile cooled in this manner shows about 1.5% shrinkage after 30 min at 30°C, as compared with 2–5% by conventional cooling. Output rates are 135–160 kg/h. The product also has a higher impact strength. In a new process more tolerant of formulation changes, some producers use twin-screw extruders to get high shear, lower temperatures with less power, and little or no backflow. The design of the dies and the control of stresses during cooling are vitally important to obtain a good product.

The postextrusion phase of the process is usually the most difficult to accelerate, and is the controlling limit on line speed, which often is no more than 4.5–9 m/min. The numerous stages include an embosser, vacuum-sizing/calibration, cooling, pulling, punching, cutting, and stacking. An output rate of 275 kg/h is considered very good; 240 kg/h is a present practice.

Residual stress in PVC is a factor in heat distortion of siding. When the sun shines on an installed strip of siding, the center of the strip becomes hotter than the edges or the covered portion. The uneven stress-relaxation and thermal expansion cause distortion, called oil-canning. The various temperatures have been measured and the resulting stresses estimated. To counteract these stresses, residual stresses have been set up in certain areas of the strip during production. Quenching the shaded portion, the butt, and the hanger creates compressive stresses; tensile stresses are produced by cooling the center slowly. When the siding is exposed to hot sunlight, the center shrinks and the shaded portions expand, thereby counteracting the oil-canning caused by thermal expansion; thus a good appearance is maintained.

Although PVC has several advantages as a siding, and affords good insulating properties, impact resistance, and low sound transmission, only in light colors does it offer adequate resistance to outdoor heat and light. Manufacturers who wish to produce PVC siding in dark or earth colors have had to consider the high cost of pigments and the fact that the TiO_2 used in light colors acts both as a pigment and stabilizer (see Pigments, inorganic). If the TiO_2 is eliminated, more stabilizer must be used; if it is retained, more pigment must be used. One way to overcome this problem is to use coextrusion of a dark-colored, thin-walled cap stock over a low cost, less pigmented substrate or core. The product is only slightly more expensive than the light-colored ones. Careful compounding is necessary to obtain adequate physical properties, including controlled cooling to avoid dimensional changes that lead to oil-canning. Another process uses a foamed rather than a solid core, and the coextruded film provides a smooth, dark-colored skin. The components may also be resins other than PVC. Acrylic films, which are inherently light stable, have been coextruded over PVC or chlorinated PVC substrates. Both methods raise the cost.

The usual inorganic pigments for dark colors may contain iron or other metal ions that can adversely affect the stability of PVC. Some organic pigments, although more expensive, contain only traces of heavy metals. Because inorganic pigments reflect more ir radiation, less heat is absorbed by the plastic, whereas organic pigments have a higher tinting strength and hiding power over TiO_2 (see Pigments, organic). Pricing according to color may be necessary. The inherent fire resistance of rigid PVC and its high flash-ignition temperature are definite advantages for its use in construction.

A new PVC siding is an extruded product of an internally ribbed, dual-wall profile. A conventional screw extruder, with a vacuum sizer cooled with water and air, forms a product about 20 cm wide, with 160-mm walls and 80-mm ribs on 25-cm centers. The total thickness is about 0.64 cm. The ribs provide rigidity and strength and the dual-wall construction adds thermal insulation. No backing or core is used.

Improvements desired in PVC siding are to increase the heat-distortion temperature and impact strength, to avoid the problems caused by thermal expansion. It also can be used to obtain darker earth colors at low cost, and to retain the original color and general appearance. The volume of use of poly(vinyl chloride) siding is shown in Table 3 (6–8,11–19).

Rainware. Plastic gutters and downspouts are made only of extruded rigid PVC. The development of light-stable dark-colored compositions will increase the markets for rainware as well as siding, although the market for the former will always be much smaller. Among the advantages of PVC for rainware are its low moisture absorption, resiliency (to absorb the impact of hail and of handling during installation without denting), uniformity and permanence of color, resistance to rot, rust, and corrosion, and its ease of fabrication and comparatively low cost (7,16).

Shutters and Exterior Trim. These have been made largely from thermoformed rigid cellular PVC. Foaming the PVC and reducing its density from about 1.4 to 0.4–0.5 g/cm^3 reduces the cost considerably although the composition itself is somewhat more expensive. Thermoforming may produce either cut sheet or roll-sheet type. The former uses stock up to 1.25 cm thick; the latter is limited to no more than 0.625 cm thickness. An extruder line may feed directly to a roll-sheet thermoformer line. Metal or wood reinforcements may be provided for structural support. Shutters and other exterior trim can also be extruded by conventional single-screw machines. Dies must be designed to allow the extrudate to expand to give a balanced flow to the sizing die, which holds and cools the extrudate to the desired dimensions, and allows for some residual shrinkage during final cooling. The product has a skin about 76–102-μm thick. For large profiles from PVC foam, the Celuka process can be used, whereby the hot and sizing dies are the same size, which causes an inward expansion into the hollow core. This product has a thicker skin and a variable cell structure than normal extrudate. All the problems and solutions that apply to siding, including capstock coextrusion for dark colors, apply to these products as well. Shutters have also been made from high impact polystyrene foam-coated with an acrylic ester, and from ABS capped with acrylic ester, to provide protection from uv light and improve outdoor weatherability (5–6,20–21).

Fascias, Soffits, Skirting, Panels and Curtain Walls. Use of glass-fiber reinforced polyester panels as panels and curtain walls for buildings has been studied extensively. Acrylic esters have been substituted for styrene in the polymer to reduce the smoke produced on burning, and fire-retardant agents have been added to increase char formation. Alumina trihydrate also is used to increase fire resistance. These panels can be made on continuous laminating machines. ABS plastics are covered by lamination with a pigmented acrylic ester to protect them from the weather, after which they are thermoformed to make soffits, fascias and external panels. Poly(vinyl chloride) is used widely for soffits, fascias, and skirting and is formulated and fabricated as for siding. These plastic products resist denting and chipping of the surface and do not blister, flake, peel, or need repainting (6–7).

Door and Window Frames and Sashes. The first successful application of rigid PVC for windows was as a cladding over a wood core to avoid the need for painting and repainting the sash and the exterior of the frame. Since then, extruded profiles of rigid PVC have been developed to give door and window frames and sashes that do not rot, corrode, or need paint and that provide good insulation and low air infiltration. One system extrudes profiles 1.8-mm thick with a series of closed channels for rigidity, thermal insulation, and low cost. The extrudate is cooled with a vacuum calibrator, then cooled further by forced air, sawn, cut, and the corners mitered. The pieces are put in holding jigs and holes are drilled for screws. Corners are secured with metal braces to resist the spring-loaded sash weight and to hold the screws. Joints are sealed with silicone polymers, then polycarbonate or metal hardware is attached, and double-pane glass installed. A 0.9×1.2-m window weighs about 11 kg and uses perhaps 8 kg of PVC. With a single-screw extruder, an output of 91 kg/h of lineals is considered very good, in view of the close tolerances that must be maintained to obtain a tight seal and smooth operation. Double-hung, side-slide, and picture windows can be made of extruded PVC.

All the difficulties in achieving dark colors for PVC siding apply also to door and window frames and sashes. Coextrusion of a PVC core capped with a vinyl or acrylic film from an adjacent extruder can be used. In addition, PVC–acrylic alloys can be used for dark colors. The modifiers, stabilizers, and pigments incorporated in siding are used in these products as well. To date, building codes restrict their use in new construction, so that their installation is primarily as replacements.

Poly(vinyl chloride) strips have been used as thermal breaks in aluminum windows, and aluminum doors and frames have been clad with PVC. A dual-wall, internally ribbed construction, like that used for siding, has been used for door and window frames and sashes in mild climates.

Foams have limited use for these purposes. Rigid cellular PVC is good as a thermal barrier but not for structural parts. Doors and frames of structural molded foam, eg, foamed high impact polystyrene, can be made by injection molding, with recesses for hinges, striker plates, and miter corners. Solid polystyrene and structural foam-molded polyurethane have been molded for door frames (6–8,11–12,17–19,21–26).

Glazing. Polyacrylates and polycarbonates are the resins most widely used for glazing (see Acrylic ester polymers; Methacrylic polymers). They are lighter in weight, easily formed, provide heat and sound insulation, but, as noted above, their chief advantage is their resistance to breakage and their failure to shatter when they break (see Insulation). Coatings of fluoropolymers often are used to improve abrasion resistance, and recently a cross-linked polysilicate, almost as hard as glass, has been used. Some grades of polyacrylates can be cemented with solvents instead of with adhesives that need curing. The surface can be rippled, or given a matte finish, or it can act as a sunscreen.

Polycarbonates are dimensionally more stable, have high impact strength and heat resistance, provide thermal insulation, and are fire resistant. Surface coatings of more abrasion-resistant films can be applied.

Polyarylates have been introduced for glazing where the amber tint and high temperature resistance, over 150°C, are advantageous. Their present cost limits their use to extreme service conditions (7–9,27–28).

Weather-Stripping and Sealants. The plastics used for these purposes are mainly the low cost oil-based or butyl rubber compounds; high cost silicones, polyurethanes, or polysulfides are used for specialty or extreme-service conditions. Acrylates adhere to metal, wood, glass, ceramic, masonry, concrete, and plastic as well as oily, greasy, wet, or dirty surfaces. They resist uv radiation, do not crack or discolor, and retain adhesion over many years. A new, lightly cross-linked product that has been produced remains flexible at low temperatures Solvent-based vehicles are used as sealants. Emulsion-based vehicles are used for caulking around window and door frames, between walls and roofs, in seams around plumbing fixtures, and as glazing compounds. They have a short tack time, good weather resistance, and they retain their adhesion and flexibility (7) (see Sealants).

Thermal Insulation. Foamed plastics (qv) are used as thermal insulation for all types of construction because of their low heat- and moisture-transmission values. Polystyrene is used either as foamed board or expandable beads. The foam may be faced with a structural surfacing material, eg, a kraft liner-board, to form a panel for insulating mobile homes. The foams can duplicate the appearance of wood and be used as trim. Foams can also be used as backing, for example, on aluminum siding, to provide heat and sound insulation. Foamed beads can be incorporated in concrete to reduce its density and provide some thermal insulation (see Insulation, thermal).

Poly(vinyl chloride) foams may be rigid or flexible; the former are used to replace wood for interior trim and some door frames, where the insulating properties are advantageous. They are dry-blended with a chemical blowing agent, fed to an extruder at 180–200°C and through a profile die, which provides up to 8 cm of land length. The die temperature should be cooler than the stock temperature to produce a good surface. The melt expands on leaving the die; then it is sized and cooled.

Urea–formaldehyde foam is low cost, easy to apply as spray between studding in walls, and is a good thermal insulator. It tends to shrink on curing, which may take 1–30 d, losing water by evaporation. In time it may emit toxic formaldehyde vapor; efforts are being made to develop compositions without this drawback.

Polyurethane foams for building purposes are normally of the rigid kind used for roof and wall insulation. Polyols containing halogen or phosphorus are used to increase the fire resistance of the foam.

Polyisocyanurate foams are superior to others in fire resistance. The initial step in their preparation is the trimerization of MDI to form a ring quite stable to heat. This polymer is mixed with blowing agents, as for polyurethanes, and expanded to obtain a low density closed-cell foam, stable up to 150°C. The process may be used for continuous buns or laminates. The product has dimensional stability and is much more fire resistant than polyurethane foams. In some fabrications, the foam is poured at a rate of 4.5–9 kg/min between facings held vertically in a jig, and then allowed to expand. The board can be removed from the jig after about one minute per 2.5 cm of board thickness. Because the pressure developed is not more than 0.14 MPa (20 psi), lightly built jigs can be used. The facings may differ in shape and material. Boards of 2.5–23-cm thickness and from 1.2 × 2.4 m–4.5 × 12 m have been made. The foam has uniform density from bottom to top and from side to side. It is used in wall and ceiling systems with a great variety of assembly and installation procedures. The wide choice of thickness, facing material, and foam density renders it suitable for many different construction designs. It is somewhat more expensive than polyurethane foam products (7,9,29–38).

Solar Heating. Plastics are used in both active and passive solar-heating systems, but more frequently in the latter type because of its lower-temperature (see Solar energy). However, even in the active systems there are many components being made of plastics. Among those under development are carbon-filled polypropylene pipes and poly(phenylene oxide) (PPO) plates and pipes. As covers, polycarbonate, polyacrylate, cellulose acetate–butyrate and glass-fiber reinforced polyester all transmit solar heat well and are resistant to uv light. Additional protection from uv radiation is provided by films of poly(vinyl fluoride), acrylate, or a new fluorocarbon film that is stable to sunlight. For frames to hold the assembly, high density polyethylene, ABS, polycarbonate, or rigid structural polyurethane foam can be used. Insulation behind the absorber plates can be of foamed plastic, especially foamed polyisocyanurate, which has good heat resistance. Phenolic-bonded glass fiber may vaporize and fog the inside of the cover, but for high temperature use, glass fiber is necessary. Polyurethane and urea–formaldehyde foams also can be used.

Focusing collectors are usually cast acrylic Fresnel lenses, or mirrors of aluminized polyester film, in frames of aluminum. These reflectors are enclosed in a bubble of poly(vinyl fluoride) film or under polycarbonate glazing, which may be covered with a fluorocarbon film to reduce the reflectivity. The absorbers for active systems are copper or aluminum since the temperatures are too high (325–370°C) for plastics. The frames, however, can be molded ABS, high density polyethylene or polyurethane, either solid or structural foam. Polybutylene or chlorinated PVC can be used for piping hot water, and tanks can be made of reinforced polyester or blow- or rotational-molded, high density polyethylene. All of these systems are under active development (7,9,39–40).

Interior Uses

Wall and Ceiling Panels. Most panels incorporating plastics use foamed plastic cores with various surfacing materials for inside and outside facings. These facings may be an integral part of the panel or applied separately. The foam may be of polystyrene, polyurethane, polyisocyanurate, phenolic or urea–formaldehyde. For some panels, expanded polystyrene beads are foamed in a large aluminum mold, with an accumulator feeding an injector with twenty or more nozzles. The foam has good moisture resistance and dimensional stability. Inserts and reinforcing materials, eg, glass-fiber or steel rods or wire mesh, have been used for greater strength. Door and window frames and conduits for services can be inserted during molding. After the foam is prepared, it is faced with a surfacing material for either interior or exterior exposures. Exterior facing may be wood, aluminum, vinyl, or plaster. Wall panels are used extensively for mobile homes. One panel uses polystyrene foam sheathing over the studs, covered by wood, aluminum, or vinyl siding. Polyurethane or urea–formaldehyde foam may be used between the studding. Gypsum board is used for the interior surface.

Roofing panels have been made from polyisocyanurate foams with asbestos roofing-felt facing, both foam and felt-reinforced with glass fiber.

Phenolic resins are used especially for decorative laminates for paneling. The substrate may be fiberboard or a core of expanded polystyrene beads. In one case the beads are coated with phenolic resin, then expanded in a mold to form a structural foam panel.

In another application, expanded polystyrene foam panels, 1.2- × 2.4-m are faced with a wire mesh and mounted in a metal channel bolted to a concrete slab. These panels are then sprayed on both sides with plaster, which is anchored to the wire mesh, and forms the interior and exterior surfaces. Roof and interior partitions provide low cost housing for mild climates (5,7–8,23,28–30,33,35–36,41–44).

Decorative Laminates. These laminates, composed of the layers described above (see Preparation and Properties, Amino Resins), are cured, then the sheets are trimmed and the backs sanded for bonding to a core of wood, particle board, fiberboard, or hardwood-surfaced plywood. The laminates are used for countertops, furniture tops, cabinetry, other furniture, and vertical paneling. The laminate can be applied to both sides of a core for paneling. Vertical surfaces that meet less wear may be covered with lower pressure laminates or may be faced with polyester or PVC coatings. Cores of expanded polystyrene beads or of honeycomb construction may also be covered with decorative laminates. The grades in use are general purpose for counter and table tops; a vertical-surface; postforming grade that is thinner and able to take curves down to 0.64-cm radius; fire-resistant with asbestos-cement cores; and heavy-duty grade with aluminum cores; and flooring grade for heavy-traffic areas (7,41) (see Laminated and reinforced plastics).

Moldings and Trim. These uses are the principal outlet for molded rigid PVC foam. Other outlets are trim for cabinets, especially kitchen cupboards and cabinets, furniture trim, and as a general substitute for wood in other than load-bearing conditions. It is stable to moisture and chemicals, can be finished easily, embossed to give a wood-grain appearance, and nailed or glued easily. The processing methods are similar to those for exterior PVC trim and siding but the formulation is simpler, as no exterior exposure is involved, and the temperature conditions are less severe (7,21,23,28).

Plastic Flooring. Plastic flooring is marketed as either tile or sheet flooring. Tile is supplied as pieces usually 30.5 × 30.5 cm with thickness of 0.16–0.32 cm and is usually homogeneous in composition. Sheet flooring, on the other hand, is produced in roll form 1.8-, 2.7-, 3.7- and 4.6-m wide (2- and 4-m in Europe), and generally consists of a plastic upper component on a fibrous backing.

Tile is based on vinyl chloride and vinyl acetate copolymers. A petroleum resin is usually employed as an extender and processing aid; conventional vinyl plasticizers and stabilizers also are incorporated. Reinforcing fibers and limestone constitute the remainder of the tile composition; the fibers contribute hot strength for processing and dimensional stability in the finished tile, limestone supplies bulk at an economical cost. Stable pigments are also incorporated. Since tile is installed on and below grade level, it is important that the finished product be resistant to the effects of moisture and alkali.

Tile is manufactured in several ways. In each method, a continuous sheet is formed and gauge refinement and planishing are carried out in subsequent calendering steps. Stresses that could lead to poor dimensional stability are avoided. The efforts to prevent stresses is governed by formulation, stock and roll temperatures, conveyor speeds, etc. After the final calendering, a resin-polymer-wax finish is applied to the surface of the sheet which is buffed before it moves to the punch press. Frame scrap and tile rejected because of defects are returned to the mixers and recycled.

Several techniques are used to introduce decorative elements into tile. Random straight-graining effects are obtained by introducing pigmented, filled vinyl chips and granules (mottle) into the tile composition at the appropriate point in the formation process. These flow to produce distinctive streaks as the sheet is formed. Less directional designs are produced by introducing grained vinyl chips, which form a continuous surface over a plain tile base. Random and registered designs are produced on a variety of tile bases by embossing and valley printing. Tile decorated by rotogravure printing is protected by a clear wear layer of plasticized PVC.

No-wax tile has become important in the residential market. Such products have a high-gloss surface coating that resists abrasion and soiling. When properly designed and maintained, no-wax tile retains its shiny appearance for an extended period of time, without application of floor polish.

Resilient sheet flooring, although based upon plasticized PVC, is manufactured from a variety of compositions by several processes. Generally, the type of decoration achieved depends upon the process and composition. Since compositions are specific to processes, resilient sheet flooring can be classified by process alone. Rotogravure printing is used for the largest volume of sheet flooring. This type of flooring, often called rotovinyl, is manufactured almost exclusively from PVC plastisols and organosols. These liquid dispersions of homopolymer and plasticizer, properly stabilized, can be applied as coatings that fuse into clear, tough films at temperatures of 180–200°C. Blowing agents can be incorporated to produce foams. Both clear coatings and foams are used in rotovinyl flooring. The typical structure consists of a fibrous organic or inorganic felt backing, a vinyl foam layer decorated with a rotogravure-printed design, and a clear vinyl surface layer. Many products of this type are embossed in register with portions of the design. Thickness varies from one product to another. Wear layers are 0.1–0.65 mm, depending on intended use, but the most common wear layer gauge is 0.25 mm. Rotovinyl flooring is produced in widths of 1.8, 2.7, 3.7, and 4.6 m. The fibrous backing is coated with a foamable plastisol containing a blowing agent. Clear plastisol is applied and the entire structure is heated enough to fuse the plastisols and cause expansion of the foamable gels. In this final step, embossing occurs either by chemical or mechanical means. The most widely used chemical technique involves a foaming inhibitor introduced in the appropriate inks.

In Europe, rotogravure flooring is made with a solid or foamed vinyl back (polyurethane froth also is used), and a glass mat with urea–formaldehyde resin binder is incorporated for dimensional stability. Most U.S. rotovinyls are defined as no-wax floors, although only a few have a shiny polyurethane surface coating of 25–50 μm which aids in gloss retention and maintenance. Other rotogravure-decorated flooring contains opaque, translucent, or transparent vinyl chips embedded in the clear plastisol wear layer for added decoration. Rotary screen printers are used in Europe with vinyl plastisols to produce sheet flooring similar to that described as cushioned roto-vinyl.

Stencil flooring is made 1.8-m wide on a similar fibrous organic or inorganic felt backing and consists of granules of a mixture of PVC homopolymer and copolymer formulated with plasticizers, stabilizers, filler, and pigments to achieve the proper consistency for this process. These granules are deposited on the backing through automated stencils, each of which delivers a portion of the total design. The sheet moves intermittently, allowing time for the simultaneous deposition of granules from as many as ten stationary stencils. The completed pattern is consolidated and embossed with heat and pressure in flatbed presses coordinated with the stencil line movement. This process is capable of a wide variety of designs that can extend through the thickness (ca 1 mm) of the deposited vinyl. A film of polyurethane is frequently applied to stencil sheet flooring to impart no-wax characteristics. Recently, a vinyl-backed product of this type was introduced, combining rotogravure decoration with the stencil design and polyurethane surface.

Another 1.8-m wide sheet floor is made by a continuous process, called roll-press, on a similar felt backing by deposition of filled and pigmented vinyl granules or chips with subsequent consolidation in the nip between a steel pressure roll and a back-up roll. Most products of this type employ shaped or variegated vinyl chips oriented edge-to-edge in a monolayer, with clear vinyl filling the spaces between them. The chips are formulated from homopolymers of PVC with as much as 65% limestone. Dry-blend, used as mortar between the chips, is made from suspension polymers of PVC in mixers that allow the absorption of plasticizers and other liquid components at relatively low temperatures so that free-flowing dry powders result. Wear layer thicknesses vary between 0.8 and 1.3 mm (30 and 50 mils). Some of these products are coated with polyurethane for gloss retention and easier maintenance.

The wide range of vinyl floors offers many choices for commercial and residential use, design, and color selection at prices of \$4.30–31.20/m^2 (\$0.40–2.90/ft^2). These floors resist staining, soiling, and physical damage.

Interior Hardware. Interior hardware has been molded from polyacetal for use in file-drawer rollers, casters, swivel-chair parts, door knobs, locks, latches, handles, and cranks and gears for casement windows. Polyacetal is strong, stable, moisture resistant, and has good lubricity. Polycarbonate is used for electrical hardware such as switches, as it is strong, tough, even at low temperatures, and has good fire resistance (7,45).

Pipe. Plastic pipe can be made for use under pressure, eg, potable water supply, gas pipelines, or pressurized sewers, or for nonpressure uses, eg, other sewers, drains-waste-vent (DWV), and other drainage. For some time, high density polyethylene was used more widely, with ABS preferred for DWV. More recently use of PVC in piping has grown rapidly. Pipe for potable water may be made from PVC, chlorinated PVC, or polybutylene. The last is also used for gas lines (see Pipelines). High molecular weight high density polyethylene is used for pressurized sewer lines, up to 0.7 MPa (100 psi). Most PVC is made by suspension polymerization; the rest by mass polymerization. A resin of molecular weight 75,000–90,000 is preferred; it is compounded with tin stabilizers, and a mixture of lubricants, and blended in a high speed mixer to obtain a uniform dry blend. Multiple-screw extruders, 8–13-cm dia, mix and push the compound through a die designed to give a streamlined, not turbulent, flow at a rate of 500–1600 kg/h. Impact modifiers of acrylic or ABS resins are added to reduce breakage and provide better performance under pressure (see Piping systems).

Chlorinated PVC is preferred in some cases, especially in mobile and site-constructed homes, for hot-water pipe because it is stable to 88°C at 0.7 MPa (100 psi). Polybutylene is also good for hot-water lines and has the advantage of flexibility. Chlorinated PVC can be solvent-welded, whereas polybutylene must be mechanically clamped or heat-welded. In some cases fittings of chlorinated PVC can be connected to polybutylene. High density polyethylene is extruded through single-screw extruders, which are lower in cost and have a higher output rate. ABS resin can be used for pressurized applications, eg, gas and water lines where adverse conditions are met. For extreme pressure conditions, plastic pipe may be filament-wound or overwrapped with reinforcing materials. A PVC pipe, for example, may be overwrapped with glass-fiber tape and impregnated with a polyester or an epoxy resin to provide the much greater bursting strength required for municipal water pipe, when relatively thin-walled PVC pipe is used.

For nonpressure conditions, a greater variety of resins and production processes is available. One of the largest uses is for DWV piping. The principal resins for that purpose are ABS and PVC modified with ABS to improve its impact resistance. ABS is extruded by single-screw machines at 200–240°C with no special problems. The pipe may be cut or sawed and is solvent-welded readily with an ABS cement. A new development is the advent of foamed ABS pipe which has lighter weight and lower cost. The foam has a closed cell structure with a solid skin. Two extruders are used, one for the surfaces and one for the core, which may be of a different and possibly lower cost resin. Both the skin thickness and the core density can be accurately controlled. An azodicarbonamide blowing agent is used as the foaming agent. The pipe has a density of 400–800 kg/m^3 (25–50 lbs/ft^3) and is more rigid and lower in cost than solid ABS pipe of the same diameter, but it also has lower tensile strength. PVC and HDPE also can be foam-extruded for the same purposes, with similar advantages and limitations. These pipes can be used for sewers and drains. Special methods have been developed for specific purposes, eg, helical wrapping, in which high density polyethylene, for example, is extruded to provide an internally ribbed profile that is machine wrapped around a collapsible mandrel to form pipe ≤6.1-m-dia, and ≤6.1-m-long. The wall thickness and strength can be increased by using multiple layers. The seams are overlapped and fused by an ir heater, welded by pressure applied to the joint area, and cooled at a controlled rate.

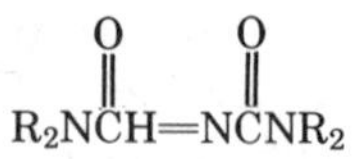

Plastic pipe of both PVC and high density polyethylene can be corrugated to increase its strength without increasing wall thickness. The bulk of the product is for irrigation pipe or for subsurface drainage. Since these are not building products, they are mentioned only because the use of corrugated pipe may increase in the construction industry.

Plumbing Fittings and Sanitary Ware. Plastics for fittings such as faucet handles, shower heads, and plumbing parts must have high strength and resistance to creep, abrasion, and dimensional change, especially when exposed to moisture. ABS resin is widely used for these purposes, and for soap dispensers, water filters, and other accessories. The article can be chrome-plated, or color can be molded in. Polyacetal also can be used for these articles as well as for valves, couplings, pumps, etc, where its natural lubricity is an advantage. Polycarbonate may be injection-molded to form handles and other parts which can be connected by a press fit instead of by threaded lock-nuts, as required for metal.

Sanitary ware, including tubs, showers, combined units, basins, and toilet tanks, may be made of thermoformed ABS or acrylic sheet, molded glass-fiber reinforced polyester, or cast acrylic resins. The glass-polyester type dominates the tub-shower market. It is now possible to install the units as a two-component system, assembled in place. Gel coats may be of thermoformed decorative acrylic skins. To reduce the smoke generated by fire, methyl methacrylate can be substituted for the styrene in the polyester, and alumina trihydrate used as a fire-retardant filler. Marble chips and dust can be included in the formulation to give a cultured-marble product for vanities, tubs, and panels. Thermoformed acrylic is usually given a glass-polyester backing which is sprayed on, or the acrylic and backing are press-molded together. A thermosetting acrylic filled with marble chips and dust is cast in an aluminum mold, cured to achieve cross-linking, then postcured to obtain a cultured marble basin, tub–shower unit, or paneling. It may have a polished or a matte finish, with a great variety of colors and patterns. It is possible also to make tub–shower units, toilet tanks, and basins from structural foam. The surface skins must be strong and impervious. They provide useful heat and sound insulating properties (4,7,20,29,45,60).

Electrical Applications. Plastics are used for electrical insulation, conduit and enclosures, lighting fixtures, and mechanical devices. The most widely used plastic for wire and cable insulation is flexible, plasticized PVC, which constitutes well over half the market in insulating wires for buildings, automobiles, appliances, and power and control lines. Polyethylene is a minor factor. For conduit and other enclosures, solid or foamed thermoformed ABS is widely used. Structural foam of ABS or polyethylene is used for boxes, basins, conduit, and various other enclosures. For switch and motor housings, thermoplastic poly(phenylene oxide), which has high impact strength, allowing thin walls, is used. With its low moisture absorption, it is a good electrical insulator. It is also corrosion, heat, and fire resistant. Polycarbonate is used for switches and outdoor electrical plugs because of its good low temperature impact strength and fire resistance. For lighting fixtures, polycarbonate is useful not only as a sheet glazing material for electrical lighting but also for molded housings and outdoor fixtures. It is strong, light weight, and has good weatherability (7,20) (see Insulation, electric).

Economic Aspects

Total sales volume for all plastics and for plastics in building products in 1979 are given in Table 5.

Table 5. Sales Volume, 1980

	Thousand metric tons	
	All plastics	Plastics in building products
polyethylene, low density (LDPE)	3,350	81
poly(vinyl chloride) (PVC)	2,469	1,240
polyethylene, high density (HDPE)	1,953	209
polystyrene (PS)	1,608	96
polypropylene (PP)	1,647	11
polyurethane, foam (PU)	744	170
phenolics	701	358
acrylonitrile–butadiene–styrene (ABS)	441	110
urea–melamine	589	447
reinforced polyester	440	240
Subtotal	*13,442*	*2,962*
acrylics	235	54
epoxy	144	12
polycarbonate	99	35
polyacetal	43	8
other	33	4
Total, all plastics	*13,996*	*3,075*

Source—*Modern Plastics*, January 1981.

Toxicity

The relative toxicity to laboratory mice of the fumes and smoke from burning plastics has been compared with that from natural materials, including wood (61).

Outlook

Recent economic trends indicate that building will become strongly responsive to the cost of energy. Insulation will become increasingly important, both for new and existing buildings. The materials and methods of construction chosen will reflect the total cost of energy involved in production, construction, and maintenance throughout the expected life of the building. Molded plastics can be expected to continue to replace scarcer and increasingly more expensive metals such as copper, zinc, and lead, because of both lower materials cost and simpler and less expensive fabrication. Also, larger units, eg, complete door and window frames and sashes, will be preassembled in factories and brought complete to the construction site, a development probably favorable to increased use of plastics. Building codes and labor arrangements will change to permit the increased use of plastics where their use and safety have been shown to be at least as satisfactory as those of materials being replaced.

The technical trend will follow to meet the demands of the market set by the previously described conditions. Improvements can be anticipated in the properties of the basic polymers through better control of polymerization conditions, providing better selection of order in the structure of the polymer, molecular weight and size distribution, fewer contaminants, and better additives, eg, stabilizers to heat, light and oxygen, fillers, and pigments. In addition, improvements can be expected in processing and finishing equipment, for greater output at lower cost and better quality of products.

BIBLIOGRAPHY

1. F. W. Billmeyer, *Textbook of Polymer Science*, 2nd ed., John Wiley & Sons, Inc., New York, 1971, Chapt. 8–10.
2. H. S. Kaufman and J. J. Falcetta, *Introduction to Polymer Science and Technology*, John Wiley & Sons, Inc., New York, 1977, Chapt. 2.
3. *Modern Plastics Encyclopedia*, Vol. 56, No. 10A, McGraw-Hill, Inc., New York, 1979–80, pp. 4–111, 132, 147, 292–300, 498–521, 528, 626.
4. N. C. Baldwin, *paper presented at 33rd Annual Technical Conference of Society of Plastic Engineers*, Vol. 21, Atlanta, Ga., 1975, p. 30.
5. R. Martino, *Mod. Plast.* **54**(1), 38 (1977).
6. R. Martino, *Mod. Plast.* **56**(6), 54 (1979).
7. J. A. Helgesen and co-workers, *Plast. Eng.* **34**(7), 22 (1978).
8. B. Miller, *Plast. World* **33**(10), 46 (1975).
9. D. V. Rosato, *Plast. World* **36**(8), 60 (1978).
10. *Chem. Week* **128,** 13 (Apr. 29, 1981).
11. R. Martino, *Mod. Plast.* **53**(6), 34 (1976).
12. L. F. Foro, *Plast. Des. Process.* **19**(2), 29 (1979).
13. M. Hartung, *Plast. Technol.* **25**(2), 67 (1979).
14. J. W. Summers and R. J. Brown, *Plast. Eng.* **35**(4), 34 (1979).
15. A. Stoloff, *Plast. Eng.* **35**(7), 29 (1979).
16. J. Changfoot, A. G. Dickson, F. Noel, and W. M. Stark, *paper presented at 34th Annual Technical Conference of the Society of Plastic Engineers*, Atlantic City, New Jersey, 1976, p. 42; J. W. Summers, p. 333.
17. J. T. Lutz, Jr., *paper presented at the 35th Annual Technical Conference of the Society of Plastic Engineers*, Montreal, Canada, 1977, p. 146; R. C. Durham, Jr., p. 151; J. D. Isner and J. W. Summers, p. 237; J. W. Summers, p. 240.
18. K. M. Tamski, *paper presented at the 36th Annual Conference of the Society of Plastic Engineers*, Washington, D.C., 1978, p. 20; R. C. Durham, Jr., p. 23; L. F. Fow, Jr., p. 29; T. W. Williams and J. W. Summers, p. 754; J. W. Summers, J. D. Isner, and E. B. Rabinovich, p. 757.
19. L. F. Fow, Jr., *paper presented at the Society Plastic Engineers Regional Technical Conference, Vinyl Plastics Division and Ontario Section*, Toronto, Canada, 1978, p. 17, L. C. Weaver, p. 45; D. S. Carr, B. Baum, and R. D. Deanin, p. 57; W. J. Reid, p. 70; J. H. Orem and J. K. Sears, p. 94.
20. R. Martino, *Mod. Plast.* **53**(8), 34 (1976).
21. T. J. Krauss and R. J. Straight, Jr., *Plast. World* **36**(7), 58 (1978).
22. R. Martino, *Mod. Plast.* **52**(6), 38 (1975).
23. *Mod. Plast.* **53**(1), 37 (1976).
24. R. Martino, *Mod. Plast.* **55**(5), 42 (1978).
25. R. R. McBride, *Mod. Plast.* **56**(6), 16 (1979).
26. R. C. Durham, Jr., *Plast. Eng.* **33**(6), 21 (1977).
27. *Mod. Plast.* **52**(5), 52 (1975).
28. *Mod. Plast.* **52**(10), 66 (1975).
29. R. Martino, *Mod. Plast.* **53**(2), 48 (1976).
30. A. S. Wood, *Mod. Plast.* **53**(4), 50 (1976).
31. R. Martino, *Mod. Plast.* **54**(4), 46 (1977).
32. *Mod. Plast.* **54**(7), 14 (1977).
33. R. Martino, *Mod. Plast.* **54**(10), 44 (1977).
34. *Mod. Plast.* **55**(12), 46 (1978).
35. *Mod. Plast.* **56**(3), 54 (1979).
36. *Mod. Plast.* **56**(10), 16 (1979).

37. R. S. Hallas, *Plast. Eng.* **31**(2), 36 (1975).
38. R. A. Kolakowski, H. G. Nadeau, and H. E. Reynolds, Jr., *35th Annual Technical Conference of Society of Plastic Engineers*, Vol. 23, Montreal, Canada, 1977, p. 171.
39. R. Martino, *Mod. Plast.* **53**(5), 52 (1976).
40. *Plastics Design Forum* **4**(5), 63 (1979).
41. *Mod. Plast.* **54**(9), 76 (1977).
42. *Mod. Plast.* **56**(9), 18 (1979).
43. *Plast. Des. Process.* **18**(1), 51 (1978).
44. R. R. Divis, *Plast. Eng.* **32**(2), 19 (1976).
45. A. S. Wood, *Mod. Plast.* **52**(3), 40 (1975).
46. A. S. Wood, *Mod. Plast.* **54**(5), 52 (1977).
47. *Mod. Plast.* **54**(8), 50 (1977).
48. R. Martino, *Mod. Plast.* **54**(12), 34 (1977).
49. R. Martino, *Mod. Plast.* **55**(1), 38 (1978).
50. L. J. Kovach and L. H. Barnett, *Mod. Plast.* **56**(9), 106 (1979).
51. R. Martino, *Mod. Plast.* **56**(11), 50 (1979).
52. M. H. Navitone, *Plast. Technol.* **22**(13), 50 (1976).
53. W. Neubert and W. A. Mack, *Plast. Eng.* **29**(8), 40 (1973).
54. R. B. Seymour, *Plast. Eng.* **29**(9), 40 (1973).
55. *Plast. Eng.* **29**(12), 26 (1973).
56. R. F. Heilmayr, *Plast. Eng.* **32**(1), 26 (1976).
57. E. C. Szamborski and R. A. Marcelli, *Plast. Eng.* **32**(11), 49 (1976).
58. D. V. Rosato, *Plast. World* **34**(4), 22 (1976).
59. *Plast. World* **37**(5), 10 (1979).
60. *Plast. Technol.* **21**(4), 9 (1975).
61. C. J. Hilado, H. J. Cumming, and C. J. Casey, *Mod. Plast.* **55**(4), 92 (1978).

LAWRENCE H. DUNLAP
Consultant

ROBERT DESCH
Armstrong World Industry

REFRACTORIES

Refractories are materials that resist the action of hot environments by containing heat energy and hot or molten materials (1). The type of refractories that are used in any particular application depends upon the critical requirements of the process. For example, processes that demand resistance to gaseous or liquid corrosion require low porosity, high physical strength, and abrasion resistance. Conditions that demand low thermal conductivity may require entirely different refractories. Indeed, combinations of several refractories are generally employed. There is no well established line of demarcation between those materials that are and those that are not refractory although the ability to withstand temperatures above 1100°C without softening has been cited as a practical requirement of industrial refractory materials (see also Ceramics).

Physical Forms

Refractories may be preformed (shaped) or formed and installed on site.

Brick. The standard dimensions of a refractory brick are 23 cm long by 11.4 cm wide and 6.4 cm thick (straight brick). Quantities of bricks are given in brick equivalents, that is, the number of standard 23-cm (9-in.) bricks with a volume equal to that of the particular installation. The actual shape and size of bricks depends upon the design of the vessel or structure in question and may vary considerably from the standard 23-cm straight brick. For example, bricks for basic oxygen furnaces (BOF vessels) may be in the shape of a key 65.6 cm long, 7.6 cm thick, and tapering in width from 15.2–10.2 cm. Numerous other shapes are available from manufacturers as standard items as well as custom made or special ordered shapes.

Bricks may be extruded or dry-pressed on mechanical or hydraulic presses. Formed shapes may be burned before use or, in the case of pitch, resin or chemically bonded brick, cured.

Setter Tile and Kiln Furniture. These products are formed in a similar manner to bricks and are used to support ware during firing operations. The wide variety of available shapes and sizes include flat slabs, posts, saggers, and car-top blocks.

Fusion-Cast Shapes. Refractory compositions are arc-melted and cast into shapes, eg, glass-tank flux blocks as large as 0.33 × 0.66 × 1.33 m. After casting and annealing, the blocks are accurately ground to ensure a precise fit.

Cast and Hand-Molded Refractories. Large shapes such as burner blocks, flux blocks, and intricate shapes such as glass feeder parts, saggers, and the like are produced by either slip or hydraulic cement casting or hand-molding techniques. Because these techniques are labor intensive, they are reserved for articles that cannot be satisfactorily formed in other ways.

Insulating Refractories. Insulating refractories in the form of brick are much lighter than conventional brick of the same composition by virtue of the brick porosity. The porosity may be introduced by means of lightweight grog or additives that create porosity by a foaming action or by evolution of combustion products upon burn-out. Refractory fiber made from molten oxides may be formed into bulk fiber, blankets, boards, or blocks. Such fibers are used as back-up thermal insulation and low heat-capacity linings in kilns and reheat furnaces (see Insulation, thermal; Refractory fibers).

Castables and Gunning Mixes. Castables consist of refractory grains to which a hydraulic binder is added. Upon mixing with water, the hydraulic agent reacts and binds the mass together. Gunning mixes are designed to be sprayed through a nozzle under water and air pressure. The mixture may be slurried before being shot through the gun or mixed with water at the nozzle. In addition to refractory grains, the mix may contain clay and nonclay additives to promote adherence to the furnace wall. Unlike casting mixes, gunning mixes may be used to make maintenance repairs without removing the furnace from service.

Plastic Refractories and Ramming Mixes. Plastic refractories are mixtures of refractory grains and plastic clays or plasticizers with water. Ramming mixes may or may not contain clay and are generally used with forms. The amount of water used with these products varies but is held to a minimum.

Mortars. These consist of finely ground refractory grain and plasticizers that can be thinly spread on brick during construction (see Cement). For air-setting mortars, sodium silicates or phosphates provide strength at room temperature. Heat-setting mortars contain no additives and develop strength only when a ceramic bond is formed at high temperatures.

Composite Refractories. Although many refractories may be considered composites, examples of composites analogous to metallurgical and organic composites are not common (see Composite materials). Recently, however, shaped and unshaped refractories containing metallic reinforcements have appeared on the market, as well as setter tile made with a sandwich construction, ie, one internal layer of deformation-resistant material such as silicon carbide bounded by two layers of oxidation-resistant material.

Refractory Coatings. Refractory coatings are applied either by painting or by spraying to a fine-grained refractory mix at room temperature. By heating, a dense sintered coating is formed. Other techniques include flame or plasma spraying. In the former, the powdered coating material is fed into a burner and sprayed at elevated temperatures. The pyroplastic grains form a dense monolithic coating when they impinge on the substrate. Plasma spraying is carried out in essentially the same manner, except that an electrically ionized gas plasma heats the coating powder to temperatures up to 16,700°C (see Refractory coatings).

Raw Materials

In the past, refractory raw materials were selected from a variety of available deposits and used essentially as mined minerals (2–7). Selective mining yielded materials of the desired properties and only in cases of expensive raw materials, such as magnesite, was a beneficiation process required. Today, however, high purity natural raw materials are increasingly in demand as well as synthetically prepared refractory grain made from combinations of high purity and beneficiated raw materials (see Tables 1 and 2). The material produced upon firing raw as-mined minerals or synthetic blends is called grain, clinker, co-clinker, or grog.

Silica. The most common refractory raw materials are ganister, which is a dense quartzite, and silica gravels (see Silica; Silicon compounds). The latter are generally purer than the former and are often further beneficiated by washing to give the raw material for superduty silica brick with no more than 0.5% impurities (alkalies, Al_2O_3, and TiO_2). Quartzite and gravel deposits are widespread throughout the world. The most important U.S. deposits are found in Pennsylvania, Ohio, Wisconsin, Alabama, Colorado, and Illinois. Synthetically produced electrofused silica is a thermally stable, shock-resistant, high purity raw material.

Fireclay. Fireclays consist mainly of the mineral kaolinite [*1318-74-7*], $Al_2O_3.2SiO_2.2H_2O$, with small amounts of other clay minerals, quartzite, iron oxide, titania, and alkali impurities. Clays can be used in the raw state or after being calcined. Raw clays may be coarsely sized or finely ground for incorporation in a refractory mix. Some high purity kaolins like those that occur in Georgia are slurried, classified, dried, and air-floated to achieve a consistent, high quality. The classified clays also may be blended and extruded or pelletized, and then calcined to produce burned synthetic kaolinitic grog, or coarsely crushed raw kaolinite may be burned to produce grog. Upon calcination or burning, kaolinite decomposes to mullite and a siliceous glass incorporating mineral impurities associated with the clay deposit (eg, quartzite, iron oxide, titania, and alkalies), and is consolidated into dense hard granular grog at high temperatures. Fireclay deposits are widely distributed; in the United States they occur in Pennsylvania, Missouri, and Kentucky.

High Alumina. The naturally occurring raw materials are bauxites, sillimanite [*12141-45-6*] group minerals, and diaspore clays (see Aluminum compounds). Other high alumina raw materials are made by beneficiation, blending, and other processing techniques.

Bauxites. Bauxites [*1318-16-7*] consist mainly of gibbsite [*14762-49-3*], $Al(OH)_3$, with varying amounts of kaolinite, and iron and titania impurities. Because the loss on ignition is high, bauxite must be calcined to high temperatures before use. During calcination, it is converted to a dense grain consisting mainly of corundum [*12252-63-0*], Al_2O_3, and mullite. Refractory-grade bauxite is relatively rare since a high iron content makes most bauxites unsuitable for refractory use. Commercially mined deposits are in South America, especially Guyana and Surinam, and the People's Republic of China. Other deposits occur in India and Central Africa but are not mined for refractory grades at present.

Sillimanite Minerals. This group includes sillimanite, andalusite [*12183-80-1*], and kyanite [*1302-76-7*], all of the formula $Al_2O_3.SiO_2$. Upon heating, a mixture of mullite, silica, and a siliceous glass is obtained. The specific gravity of sillimanite and andalusite is ca 3.2, and that of kyanite 3.5. Thus, kyanite expands upon conversion to mullite by ca 16–18 vol %, but sillimanite and andalusite expand only slightly. Kyanite is found in India and South Africa, and, in the United States, in Virginia and South Carolina. Large-grained kyanite is rare, and the largest size commercially available from U.S. sources is ca 500 μm (35 mesh). Refractory-grade sillimanite occurs in India and South Africa.

Diaspore Clays. These consist of a mixture of kaolin and the mineral diaspore [*14457-84-2*], $Al_2O_3.H_2O$. They were actively mined in Missouri, but these deposits are now largely depleted (see Clays).

Table 1. Composition of Refractory Raw Materials, %[a,b]

Name	Location	SiO_2	Al_2O_3	Fe_2O_3	TiO_2	CaO	MgO	Cr_2O_3	Alkalies	ZrO_2
silica raw materials										
ganister	Bwlehgwyn (N. Wales, UK)	97.4	0.73	0.78	0.1					
gravel	Sharon Conglomerate, Ohio	98.0	0.3	0.5	0.1					
clays										
flint clays	Pennsylvania	50.40	34.58	1.42	2.06	0.45	1.00		1.58	
	Missouri	43.80	38.29	0.60	2.33	0.07	0.20		0.39	
	Kentucky	44.94	35.17	1.56	3.12	0.11	0.18		1.45	
	People's Republic of China	51.20	46.62	0.87	0.87				0.15	
plastic clays	Pennsylvania	54.00	27.94	1.39	2.45	0.07	1.22		3.33	
	Missouri	55.12	28.65	1.66	1.55	0.11	0.10		2.54	
	Kentucky	56.42	26.92	2.05	1.95	0.18	0.51		1.75	
kaolin	Georgia	43.00	37.55	0.85	2.10	0.09	0.18		0.15	
	Florida	46.5	37.62	0.51	0.36	0.25	0.16		0.42	
fireclay	Stourbridge (Scotland)	68.1	27.2	1.95	1.1	0.72	0.35		1.28	
	Pfalz (FRG)	45.1	36.3	2.21	1.12	0.08	0.11		2.25	
plastic	Chasov-Yar (USSR)	51.6	33.3	0.9	1.37	0.53	0.57		3.28	
semikaolin	Suvorov (USSR)	46.1	33.9	2.14	1.52	0.41	0.23		0.44	
kaolin	Vladimirovka (USSR)	48.3	36.7	0.83	0.78	0.3	0.33		0.77	
high-alumina										
natural										
siliceous bauxite[c]										
ca 70% Al_2O_3	Eufaula, Alabama	25.9	70.1	1.13	2.9	0.05	0.03		0.13	
ca 60% Al_2O_3	Eufaula, Alabama	34.9	60.6	1.26	2.5	0.07	0.12		0.11	
ca 60% Al_2O_3	People's Republic of China	32.40	63.50	1.50	2.20				0.20	
ca 50% Al_2O_3	People's Republic of China	43.30	52.80	1.42	1.90				0.19	
South American bauxite[c]	Guyana	7.0	87.5	2.00	3.25	trace	trace		trace	
Chinese bauxite[c]	People's Republic of China	6.0	87.5	1.50	3.75				0.50	
kyanite[d]	Virginia	38.6	59–61	0.2–0.9	0.67	0.03	0.01		0.4	
sillimanite[d]	India		59–61							

synthetic										
fused alumina										
black		0.48	97.3	0.15	2.45	0.07	0.11		0.05	
gray		0.06	99.5+	0.15	0.06	trace	trace		0.07	
sintered alumina		0.06	99.5	0.06	trace	trace	trace		0.05	
sintered mullite	Georgia	27.90	68.00	1.33	2.61	0.06	0.04		0.06	
sintered magnesium aluminate	Japan	0.2	67.90	0.2		0.3	31.9			
fused mullite		22.0	77.7	0.12	0.05	trace	trace		0.35	
calcium aluminate cement										
low purity		8.4	42.0	10.7		37.0	1.2			
high purity		0.1	79.0	0.3	trace	18.0	0.1		0.5	
zirconium										
zircon		32.5	0.08	0.05	0.06	trace	trace	trace	trace	66.7
baddeleyite	Brazil			3–5						70.80
basic raw materials										
calcined magnesias										
natural magnesite	Austria	0.3	0.3	5.4		2.7	91.3			
natural magnesite	Greece	1.2	0.3	0.3		2.8	95.4			
natural magnesite	U.S.	1.00	0.36	0.65		3.02	94.97			
natural magnesite	People's Republic of China	0.40	0.13	1.73		1.31	96.36			
natural magnesite	Turkey	1.42	<0.02	<0.02		1.58	96.95			
seawater	Japan	0.37	0.04	0.01		1.23	98.32			
seawater	UK	0.70	0.20	0.10		2.30	96.66			
seawater	U.S.	0.80	0.12	0.14		2.22	96.70			
seawater	U.S.	0.80	0.15	0.22		2.65	96.12			
seawater	Ireland	0.83	0.18	0.17		1.07	97.75			
brine	Israel	0.14	0.02	0.04		0.77	99.45			
dolomite	UK	1.8	1.1	5.9		54.6	36.4			
dolomite, low flux	U.S.	0.7	0.3	0.9		57.7	40.4			
chrome ore	Africa	3.8	14.9	27.1		0.3	9.9	44.0		
chrome ore	Philippines	5.5	31.0	15.5		0.5	16.0	31.5		

[a] Refs. 2 and 5.
[b] Difference between total analysis and 100 = loss on ignition, %.
[c] Calcined.
[d] Raw.

Table 2. **Physical Properties of Refractory Raw Materials**[a]

Material	Pyrometer cone equivalent	Main crystalline phases	Specific gravity, g/cm³ Bulk	True	Apparent porosity, %
silica					
ganister		quartz		2.66	1.6
gravel		quartz		2.61	0.3
clays					
flint clays	32–33				
	34	kaolinite, illite, quartz			
	33	kaolinite, quartz, illite			
			2.55		
plastic clays	31	kaolinite, quartz, illite			
	30	kaolinite, quartz, illite			
	30	kaolinite, quartz, illite			
kaolin	33–34	kaolinite			
fireclay	33–34	kaolinite			
plastic	1720°C	kaolinite, illite			
semikaolin	1740°C				
kaolin	1750°C				
high-alumina					
natural					
siliceous bauxite[b]					
ca 70% Al_2O_3	38–39	mullite	2.85–2.95	3.1–3.2	4–8
ca 60% Al_2O_3		mullite	2.75–2.85	2.95–3.05	3–7
ca 60% Al_2O_3			2.70		
ca 50% Al_2O_3			2.65		
South American bauxite[b]		corundum, mullite	3.1	3.6–3.7	15–20
Chinese bauxite[b]			3.20		
kyanite[c]	36–37	kyanite		3.5–3.7	
sillimanite[c]		sillimanite		3.23	
synthetic					
fused alumina					
black	42	α-alumina	3.87	4.01	3.49
gray	42+	α-alumina	3.95	3.98	0.5–1.0
sintered alumina	42+	α-alumina	3.45–3.6	3.65–3.80	5.0
sintered mullite	39		2.85		
sintered magnesium aluminate		spinel, periclase	3.33		
fused mullite	39	mullite	3.1	3.45	0.1
calcium aluminate cement					
low purity		calcium monoaluminate			
high purity	34	α-alumina, calcium monoaluminate			
zirconium					
zircon		zircon		4.2–4.6	
baddeleyite (ZrO_2)		baddeleyite		5.5–6.5	
basic raw materials					
calcined magnesias					
natural magnesite		periclase	3.2		
natural magnesite		periclase	3.4		
natural magnesite			3.39		
natural magnesite		magnesite, dolomite, calcite	3.40		
natural magnesite			3.39		
seawater		periclase	3.44		
seawater		periclase	3.35		
seawater		periclase	3.40		

Table 2 (*continued*)

Material	Pyrometer cone equivalent	Main crystalline phases	Specific gravity, g/cm³ Bulk	True	Apparent porosity, %
seawater		periclase	3.44		
seawater		periclase	3.22		
brine		periclase	3.41		
dolomite		periclase + CaO			
dolomite, low flux		periclase + CaO			
chrome ore		chromite spinel	4.2		
chrome ore		chromite spinel	3.9		

[a] Refs. 2 and 5.
[b] Calcined.
[c] Raw.

Mullite. Although mullite is found in nature, for example, as inclusions in lava deposits on the island of Mull, Scotland, no commercial natural deposits are known. It is made by burning pure sillimanite minerals or sillimanite–alumina mixtures. Fused mullite of high purity is obtained by arc melting silica sand and calcined alumina. High purity sintered mullite is made from alumina and silica, but requires mineralizing agents and very high temperatures.

Alumina. A pure grade of alumina is obtained from bauxite (not necessarily refractory grade) by the Bayer process. In this process, the gibbsite from the bauxite is dissolved in a caustic soda solution and thus separated from the impurities. Alumina, calcined, sintered, or fused, is a stable and extremely versatile material used for a variety of heavy industrial, electronic, and technical applications.

Calcined alumina is a reactive powder which is used to make synthetic grain. It also may be used as a bonding or fine component in batched refractory mixes, as a raw material for molten cast refractories, or for refractory casting slips.

Sintered alumina, also known as tabular alumina, is formed by burning aggregates made from reactive calcined alumina to high temperatures to obtain a stable high purity corundum grain.

Fused alumina is obtained by fusing either calcined alumina or bauxite. Bauxite is beneficial during fusion as iron and silica are removed as ferrosilicon. A special grade of fused alumina is obtained by blowing air through a stream of molten alumina. Thus, small bubbles of alumina are formed, an excellent lightweight aggregate of high purity and refractoriness.

Calcium Aluminate Cements. Low purity calcium aluminate [*12042-78-3*] cements are obtained by sintering or fusing bauxite and lime in a rotary or shaft kiln. A high purity calcium aluminate cement, $2CaO.5Al_2O_3$, capable of withstanding service temperatures of 1750°C can be prepared by the reaction of high purity lime with calcined or hydrated alumina.

Zirconia. Zircon, the most widely occurring zirconium-bearing mineral, is dispersed in various igneous rocks and in zircon sands. The main deposits are in New South Wales, Australia; Travancore, India, and Florida in the United States. Zircon can be used as such in zircon refractories or as a raw material to produce zirconia. The zircon structure becomes unstable after about 1650°C, depending upon its purity, and decomposes into ZrO_2 and SiO_2 rather than melting (see Zirconium and zirconium compounds).

Zirconia occurs as the mineral baddeleyite [*12036-23-6*], for instance, in a deposit around Sao Paulo, Brazil. However, these baddeleyite deposits generally contain large amounts of impurities and only about 80% ZrO_2. High purity zirconia can be obtained from baddeleyite by leaching with concentrated sulfuric acid or chlorination at high temperatures. The zirconium sulfates and chlorides thus formed are readily separated from the impurities. Zirconia also is made from zircon by electric-arc melting under reducing conditions. Here, the silica is separated from the zirconia by adding iron to form ferrosilicon. The most remarkable property of zirconia is its volume instability. Upon heating, unstabilized ZrO_2 transforms from the low temperature monoclinic form to the tetragonal form at about 1000°C, resulting in a 9% volume contraction. By adding impurities with a cubic structure, such as magnesia, calcia, or yttria, zirconia can be transformed into a cubic crystal phase stable at all temperatures, although often only enough impurities are added for partial stabilization, ie, both cubic and either the tetragonal or monoclinic form exist.

Basic Raw Materials. ***Magnesite.*** Calcined or dead-burned magnesite [*13717-00-5*] is obtained by firing naturally occurring magnesium carbonate to 1540–2000°C. This treatment produces a dense product composed primarily of periclase [*1309-48-4*], MgO. Large sedimentary deposits of magnesite occur in Austria, Manchuria, Greece, the Ural Mountains, and in Washington, Nevada, and California. Calcia, silica, alumina, and iron-bearing phases occur as accessory minerals (see also Magnesium compounds).

Seawater contains approximately 1294 ppm Mg (seawater magnesite); higher concentrations occur in magnesium-rich brines (see Chemicals from brine). Treatment with hydrated lime, $Ca(OH)_2$ precipitates $Mg(OH)_2$:

$$CaO + H_2O + MgCl_2 \rightarrow Mg(OH)_2 + CaCl_2$$

The $Mg(OH)_2$ precipitate is filtered, dried, and calcined. For high quality refractory grain (98% MgO), an initial calcination is followed by mechanical compaction and final high temperature calcination to form a dense product. Large seawater or brine plants are located in the United States (Gulf of Mexico, California, Michigan, and New Jersey), Mexico, the United Kingdom, Ireland, Israel, Italy, Japan, and the USSR.

Magnesium hydroxide also occurs in sedimentary deposits as the mineral brucite [*1317-43-7*], eg, in Quebec and Nevada.

Dolomite. Dolomite [*17069-72-6*], $CaMg(CO_3)_2$, occurs in widespread deposits in many areas including southern Austria, the UK, the USSR, and the United States. Raw dolomite may be used for certain refractories, but in most instances it is calcined to form a grain consisting primarily of MgO (periclase) and CaO [*1305-78-8*]. Calcined dolomite absorbs H_2O and CO_2 from the atmosphere and eventually disintegrates. Fluxes such as SiO_2, Fe_2O_3, and Al_2O_3 increase hydration resistance but also sharply reduce the fusion point of the dolomite. High-purity, low-flux dolomites with less than

2% impurities are produced by high-temperature calcination of natural dolomites in Ohio and Pennsylvania (see Lime and limestone).

Forsterite. Pure forsterite is rare in nature; most natural magnesium orthosilicates form solid solutions of fayalite (Fe_2SiO_4) and forsterite. Forsterite refractories are usually made by calcining magnesium silicate rock such as dunite, serpentine, or olivine with sufficient magnesia added to convert all excess silica to fosterite and all sesquioxides to magnesia spinels.

Chrome Ore. Chromia-bearing spinel materials, although considered neutral, are generally used in combination with basic magnesite. Chrome ores consist essentially of a complex solid-solution series of spinels including hercynite, $FeO.Al_2O_3$; ferrous chromite, $FeO.Cr_2O$; magnesioferrite, $MgO.Fe_2O_3$; picrochromite, $MgO.Cr_2O_3$; spinel, $MgO.Al_2O_3$; and magnetite $FeO.Fe_2O_3$. Silicate phases such as serpentine, talc, and enstatite are commonly associated with the spinel grains. The principal deposits of chrome ore occur in Africa (Transvaal, South Africa, Zimbabwe), the USSR, Turkey, Greece, Cuba, and the Philippines; African and USSR chrome ores are high in iron content, and the Cuban and Philippine ores are higher in alumina.

Silicon Carbide. Silicon carbide is made by the electrofusion of silica sand and carbon. Silicon carbide is hard, abrasion resistant, and has a high thermal conductivity. It is relatively stable but has a tendency to oxidize above 1400°C. The silica thus formed affords some protection against further oxidation.

Beryllia and Thoria. These are specialty oxides for highly specialized applications that require electrical resistance and high thermal conductivity. Beryllia is highly toxic and must be used with care. Both are very expensive and are used only in small quantities.

Carbon and Graphite. Carbon and graphite [*7782-42- 5*] have been used alone to make refractory products although they are commonly used in conjunction with other refractory raw materials. Carbon blacks are commercially manufactured, whereas graphites have to be mined.

General Properties

Oxides. A number of simple and mixed refractory oxide materials are described in Table 3. At present, Al_2O_3 is the most widely used simple oxide; it has moderate thermal-shock resistance, good stability over a wide variety of atmospheres, and is a good electric insulator at high temperatures. The strength of ceramics is influenced by minor impurities and microstructural features. In general, polycrystalline alumina has reasonably good and nearly constant strength up to about 1000–1100°C; at higher temperatures, the strength drops to much less than one half of the room temperature value over a 400°C temperature increment. Single-crystal alumina is stronger than polycrystalline Al_2O_3 and actually increases in strength between 1000 and 1100°C. Fused silica glass has excellent thermal-shock properties but devitrifies on long heating above 1100°C and loses much of its shock resistance.

Beryllium and magnesium oxides are stable to very high temperatures in oxidizing environments. Above 1700°C MgO is highly volatile under reducing conditions and in vacuum, whereas BeO exhibits better resistance to volatilization but is readily

Table 3. Properties of Pure Refractory Materials[a]

Material	CAS Registry Number	Formula	Mp, °C	True specific gravity, g/cm^3	Mean specific heat		Thermal conductivity, W/(m·K)		Linear thermal expansion coefficient per °C × 10^6, from 20–1000°C
					J/(kg·K)[b]	Temp range, °C	at 500°C	at 1000°C	
aluminum oxide	[*1344-28-1*]	Al_2O_3	2015	3.97	795.5	25–1800	10.9	6.2	8.6
beryllium oxide	[*1304-56-9*]	BeO	2550	3.01	1004.8	25–1200	65.4	20.3	9.1
calcium oxide	[*1305-78-8*]	CaO	2600	3.32	753.6	25–1800	8.0	7.8	13.0
magnesium oxide	[*1309-48-4*]	MgO	2800	3.58	921.1	25–2100	13.9	7.0	14.2
silicon dioxide[c]	[*7631-86-9*]	SiO_2		2.20	753.6	25–2000	1.6	2.1	0.5
thorium oxide	[*1314-20-1*]	ThO_2	3300	10.01	251.2	25–1800	5.1	3.0	9.4
titanium oxide	[*13463-67-7*]	TiO_2	1840	4.24	711.8	25–1800	3.8	3.3	8.0
uranium oxide	[*1344-58-7*]	UO_2	2878	10.90	251.2	25–1500	5.1	3.4	
zirconium oxide[d]	[*1314-23-4*]	ZrO_2	2677	5.90	460.6	25–1100	2.1	2.3	
mullite	[*55964-99-3*]	$3Al_2O_3.2SiO_2$	1850[e]	3.16	628.0	25–1500	4.4	4.0	4.5
spinel	[*1302-67-6*]	$MgO.Al_2O_3$	2135	3.58	795.5		9.1	5.8	0.2
forsterite	[*15118-03-3*]	$2MgO.SiO_2$	1885	3.22	837.4		3.1	2.4	9.5
zircon	[*10101-52-7*]	$ZrO_2.SiO_2$	2340–2550[f]	4.60	544.3		4.3	4.1	4.0
carbon	[*7440-44-0*]	C		2.10	1046.7	25–1300	13.4	9.9	4.0
silicon carbide	[*409-21-2*]	SiC	3990[g]	3.21	795.5	25–1300	22.5	23.7	5.2

[a] Refs. 8–10.
[b] To convert J to cal, divide by 4.184.
[c] Silica glass.
[d] Cubic, stabilized with CaO.
[e] Congruent.
[f] Incongruent.
[g] Dissociates above 2450°C in reducing atmosphere and is readily oxidized above 1650°C.

volatized by water vapor above 1650°C. Beryllia has good electrical insulating properties and high thermal conductivities; however, its high toxicity restricts its use. Calcium oxide, and to a lesser extent uranium oxide, hydrate readily; UO_2 can be oxidized to lower melting U_3O_8. Zirconia in pure form is rarely used in ceramic bodies; however, stabilized or partially stabilized cubic ZrO_2 is currently the most useful simple oxide for operations above 1900°C. Thorium oxide exhibits good properties for high temperature operation but is very expensive. Since it is a fertile nuclear material, it is under the control of the U.S. NRC. Titanium oxide is readily reduced to lower oxides and cannot be used in neutral or reducing atmospheres.

Carbon, Carbides, and Nitrides. Carbon (graphite) is a good thermal and electrical conductor. It is not easily wetted by chemical action, which is an important consideration for corrosion resistance. As an important structural material at high temperature, pyrolytic graphite has shown a strength of 280 MPa (40,600 psi). It tends to oxidize at high temperatures, but can be used up to 2760°C for short periods in neutral or reducing conditions. The use of new composite materials made with carbon fibers is expected, especially in the field of aerospace structure.

When heated under oxidizing conditions, silicon carbide and silicon nitride, Si_3N_4 [*12033-89-5*], form protective layers of SiO_2 and can be used up to ca 1700°C. In reducing or neutral atmospheres, their useful range is much higher. Silicon carbide has very high thermal conductivity and can withstand thermal shock cycling without damage. It also is an electrical conductor and is used for electrical heating elements. Other carbides have relatively poor oxidation resistance, but under neutral or reducing conditions, they have potential usefulness as technical ceramics in aerospace application; eg, the carbides of B, Nb, Hf, Ta, Zr, Ti, V, Mo, and Cr. Ba, Be, Ca, and Sr carbides are hydrolyzed by water vapor.

Silicon nitride has good strength retention at high temperature and is the most oxidation resistant nitride. Boron nitride [*10043-11- 5*] has excellent thermal-shock resistance and is in many ways similar to graphite, except that it is not an electrical conductor.

Borides and Silicides. These materials do not show good resistance to oxidation; however, some silicides form SiO_2 coatings upon heating which retards further oxidation. Molybdenum disilicide [*1317-33-5*], $MoSi_2$, is used widely, primarily as an electrical heating element.

Metals. The highest-melting refractory metals are tungsten (3400°C), tantalum (2995°C), and molybdenum (2620°C); all show poor resistance to oxidation at high temperatures. Hafnium–tantalum alloys form a tightly adhering oxide layer that gives partial protection up to 2200°C; the layer continues to grow with extended use.

Phase Equilibria. Phase diagrams represent the chemical equilibria that exist among one, two, or three components of a system under the influence of temperature and pressure (11–12). Reference to a phase diagram permits the determination of the amount and composition of solid and liquid phases that coexist under certain specified conditions of temperature and pressure for a particular system. With such information, the occurrence of physical and chemical changes within a system or between systems at high temperatures can be predicted. Systems containing more than three components are difficult or impossible to present graphically; however, mathematical methods have been suggested (13–14).

Phase diagrams can be used to predict the reactions between refractories and various solid, liquid, and gaseous reactants. However, phase diagrams derived from phase equilibria of relatively simple pure compounds and real systems are highly complex and may contain a large number of minor impurities that significantly affect equilibria. Furthermore, equilibrium between the reacting phases may not be reached under actual service conditions, and the physical environment of a product may be more influential to its life than the chemical environment (15–19).

Physical Properties. The important physical properties of some refractories are listed in Tables 4–7. Brick bulk density depends mainly on the specific gravity of the constituents and the porosity. The latter is controlled by the porosity of the raw materials and the brick texture. Coarse, medium, and fine-sized material contribute to the degree of packing; usually the highest density possible is desired. Upon firing, the grains and matrix form glassy, direct, or solid-state ceramic bonds. Sintering is generally accompanied by shrinkage, unless new components are formed that may cause expansion. Differential volume change between the coarse and fine fraction caused by differential sintering rates and formation of addition phases may create stresses. Particle size distribution, forming method, and firing process contribute to texture, whereas permeability is related to porosity, which in turn is dependent upon texture.

Mechanical Properties. The physical properties of a particular refractory product depend upon its constituents and manner in which they were assembled. The physical properties may be varied to suit specific applications; for example, for thermal insulations highly porous products are employed, whereas dense products are used for slagging or abrasive conditions.

The strength (modulus of rupture) at room temperature is determined by the degree of bonding. Fine-grained refractories generally are stronger than coarse-grained types and those with a low porosity are stronger than those of high porosity. However, the room temperature strength of a refractory is not necessarily indicative of the strength at high temperature because the bond strength may be due to a glassy phase that softens upon heating.

Generally, high temperature strength is lower than room temperature strength. The former is a measure of the degree of solid-state bonding between refractory grains, whereas high temperature creep indicates the amount of associated liquid or glassy phases and their viscosity. The development of solid-state or direct-bonded basic brick requires high firing temperatures, and is impeded by glassy phases. By referring to phase diagrams, refractory compositions may be designed that avoid the development of such phases.

The modulus of elasticity is related to the strength and can be used as a nondestructive quality-control test on high cost special refractory shapes such as slide gate valves employed in the pouring of steel. Generally speaking, the strength is related directly to slide-gate performance.

Thermal Properties. Refractories, like most other solids, expand upon heating, but much less than most metals. The degree of expansion depends upon the chemical composition. A diagram of the thermal expansion of the most common refractories is shown in Figure 1.

Irreversible Thermal Expansion. Most refractory bricks are not chemically in equilibrium before use. During the prolonged heating in service, additional reactions occur that may cause the brick to shrink or expand. For instance, 70% Al_2O_3 bricks

Table 4. Composition of Alumina, Silica, and Zirconia Refractories, %[a]

Type	SiO_2	Al_2O_3	Fe_2O_3	TiO_2	CaO	MgO	Alkalies	ZrO_2
silica	95–96	0.2–1.5	0.8–1.0	0.3–0.1	3.3–2.7	<0.1	<0.2	
fireclay								
semisilica	70–79	17–27	0.6–2.0	0.8–1.6	0.1–0.4	0.1–0.4	0.2–0.4	
medium duty	56–70	25–36	1.8–3.4	1.3–2.7	0.2–0.4	0.4–0.6	1.0–2.7	
high duty	51–59	35–41	1.5–2.5	1.1–3.0	0.2–0.6	0.1–0.6	1.3–2.6	
super duty	50–54	40–46	0.8–2.3	1.1–2.5	0.1–0.5	0.4–0.6	0.1–1.4	
high alumina								
50% Al_2O_3	40–47	47–57	0.9–1.6	2.2–2.4	0.5–0.6	0.3–0.6	0.8–1.3	
60%	27–37	58–67	0.9–2.7	1.7–3.0	0.1–0.3	0.2–0.7	0.2–1.2	
70%	19–28	68–77	0.9–2.2	2.0–3.3	0.1–0.3	0.1–0.7	0.2–1.2	
80%	8–17	78–87	0.7–1.7	2.5–3.2	0.1–0.2	0.1–0.4	0.1–1.0	
85%[b]	6–10	82–86	1.0–2.5	2.0–3.0	trace–0.2	trace–0.4	0.1–0.3	
90%	3–10	88–96	0.1–1.4	0.1–2.6	0.1–0.2	0.1–0.3	trace–0.9	
100%	0.2–1.1	97–99	0.1–0.3	trace–0.3	0.1–0.3	0.1–0.3	0.1–0.3	
zircon	32	1	0.1	1.8	trace	0.1	0.02	64
molten cast								
Al_2O_3–ZrO_2–SiO_2	11–13	50–51	trace–1.0	trace–1.0	trace	trace	1.0–1.5	33–35
Al_2O_3–SiO_2	18–22	57–70	1–4	3.0–4.5	0–1	0–1	0.0–1.5	
Al_2O_3 high soda	0.04	93.8	0.1	0.05	0.1	0.1	5.6	
Al_2O_3 low soda	0.10	99.2	0.1	0.05	0.1	0.1	0.3	

[a] Refs. 2, 5, 20–23.
[b] Phosphate bonded.

Table 5. Typical Ranges of Physical Properties of Alumina, Silica, and Zirconia Refractory Brick[a]

Type	PCE[b]	Modulus of rupture MPa[c]	Deformation under 172 kPa[c] load, % linear change after 1½ hours		Linear reheat change[d]		Thermal spalling loss[e]		Bulk density, g/cm^3	Porosity, %
			Temperature, °C	Change, %	at °C	%	at °C	%		
silica		2.8–11.2	1650	0					1.60–1.80	20–30
fireclay										
semisilica	27–31	2.1–4.2	1450	0.1–2			1480	0–10	1.80–2.10	20–30
medium duty	29–31	7.0–11.2	1450	1–6			1600	2–6	2.11–2.20	17–21
high duty	31–33	2.8–21.0	1450	0.5–15	1600	0–1 S	1600	3–7	2.13–2.30	4–30
super duty	33–34	2.8–24.0	1450	0–9	1600	0–1 S	1650	2–8	2.28–2.48	5–22
high alumina										
50% Al_2O_3	34–35	7.0–11.2	1450	2–6	1600	0–1 E	1650	8–12	2.27–2.43	14–18
60%	36–37	7.0–11.2	1450	1–4	1600	0–5 E	1650	1–5	2.10–2.49	13–28
70%	37–38	7.0–11.2	1450	1–3	1600	1–7 E	1650	3–7	2.20–2.66	14–28
80%	38–39	7.0–12.6							2.50–2.90	14–29
85%[f]	38–39	21.0–35.0	1450	0.3–3	1600	0–2 E	1650	0–5	2.70–2.92	12–17
90%	40–41	14.0–35.0	1700	0–1	1700	0–1 E	1650	0–1	2.67–3.10	11–27
100%	41–42	12.6–21.0	1650	1–2	1700	0–0.5 S	1650	0–5	2.84–3.10	19–29
zircon			1600	2–5	1540	0			3.77	20
molten cast										
Al_2O_3–ZrO_2–SiO_2									3.70–3.74	0.8–3
Al_2O_3–SiO_2									3.00–3.20	1–3
Al_2O_3 high soda									2.89	
Al_2O_3 low soda									3.50	

[a] Refs. 2, 5, 20–23.
[b] Pyrometer cone equivalent, as determined by ASTM C 24.
[c] To convert MPa to psi, multiply by 145; for kPa, multiply by 0.145.
[d] S = Shrinkage; E = expansion.
[e] As determined by ASTM C 38.
[f] Phosphate bonded.

Table 6. Composition of Basic Refractory Bricks, %[a]

Type	SiO_2	Al_2O_3	Fe_2O_3	CaO	MgO	Cr_2O_3	Residual carbon
magnesite							
burned	0.4–4.5	0.1–1.0	0.1–2.3	0.6–3.8	91.0–98.0	0.0–0.9	
burned, tar impregnated	0.4–4.5	0.1–1.0	0.1–2.3	0.6–3.8	91.0–98.0	0.0–0.9	2.0–2.5
tar bonded, tempered	0.5–3.8	0.2–1.0	0.1–3.0	0.9–6.2	88.0–98.0	0.0–0.4	4.2–4.7
resin bonded	0.5–3.8	0.2–1.0	0.1–0.5	1.5–2.7	96.0–97.0	0.0–0.2	3.8–4.7
high carbon	0.7–2.3	0.1–0.7	0.2–0.9	0.9–3.0	94.0–98.0		7.0–20.0
magnesite–chrome							
burned	1.8–7.0	6.0–13.0	2.0–12.0	0.6–1.5	50.0–82.0	6.0–15.0	
direct bonded	1.0–2.6	3.0–16.0	3.0–10.0	0.6–1.1	50.0–80.0	7.0–20.0	
chemical bonded	2.0–8.0	5.0–14.0	2.0–8.0	1.0–2.0	50.0–80.0	6.0–15.0	
chrome							
burned	5.0–8.0	27.0–29.0	12.0–20.0	0.4–1.0	15.0–23.0	29.0–35.0	
chrome–magnesite							
burned	3.0–5.0	8.0–20.0	8.0–12.0	0.7–1.1	40.0–50.0	18.0–24.0	
dolomite							
burned	0.8–1.5	0.3–0.8	0.6–1.5	38.0–58.0	38.0–58.0		0.0–1.4
tar bonded	0.3–1.5	0.1–0.6	0.2–2.0	50.0–58.0	38.0–43.0	0.0–0.3	4.0–5.0

[a] Refs. 2, 5, 7, 20–24.

Table 7. Typical Ranges of Physical Properties of Basic Brick[a]

Type	Bulk density, g/cm^3	Apparent porosity, %	Modulus of rupture, MPa[b]			Refractoriness under load, shear temp, °C	Linear reheat change[d] at 1650°C, %
			at 20°C	at 1260°C	at 1480°C		
magnesite							
burned	2.8–3.0	15–19	7.0–24.5	3.5–18.5		1590–1760	0–0.4 S
burned, tar impregnated	3.0–3.2	(13–17)[c]	20.0–35.0	10.5–21.0	7.0–16.0		
tar bonded, tempered	3.0–3.1	3–7	7.0–11.0			1700+	
resin bonded	2.9–3.1	4–7	8.0–27.0				
high carbon	2.7–3.0	1–6	7.0–9.0				
magnesite–chrome							
burned	2.9–3.0	17–20	3.0–4.9	0.7–2.0		1500–1700	0–0.3 S
direct bonded	2.9–3.2	14–19	5.6–13.9	8.4–17.5	2.0–4.2	1700+	0.7 S–1.0 E
chemical bonded	3.0–3.2	18–20	7.0–14.0	0.7–2.8		1650–1760	1 S–5.0 S
chrome							
burned	3.1–3.3	18–20	6.0–14.0	0.4–1.1		1260–1370	0–0.6 S
chrome–magnesite							
burned	3.0–3.2	19–21	5.6–8.4	2.8–11.2		1650–1700	0–0.1 E
dolomite							
burned	2.7–3.1	6–19	7.0–32.8			1700+	0–0.2 S
tar bonded	2.8–3.0	(8–12)[c]	7.0–11.0				

[a] Refs. 2, 5, 7, 20–23, and 25.
[b] To convert MPa to psi, multiply by 145.
[c] Obtained on ignited sample.
[d] S = shrinkage; E = expansion.

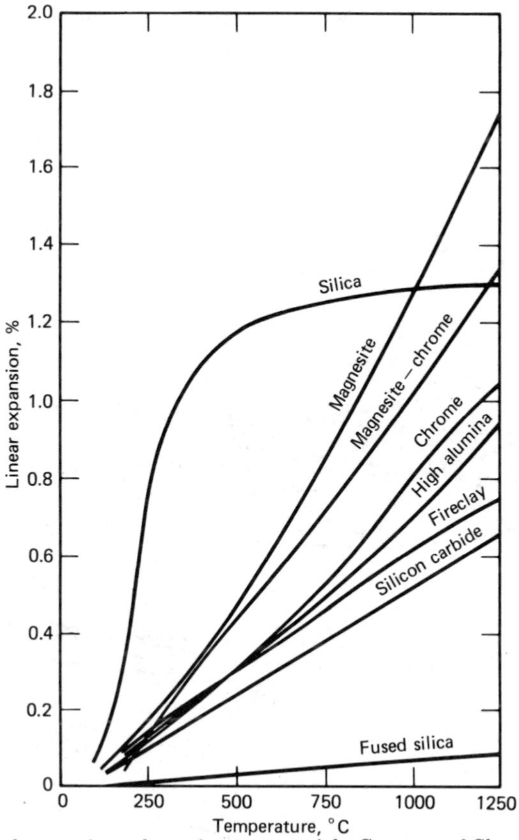

Figure 1. Thermal expansion values of some materials. Courtesy of *Chemical Engineering*.

consist of a mixture of corundum grains, mullite, and siliceous glass. In service, the siliceous glass reacts with corundum to form additional mullite, causing expansion. Mullite bricks that contain only mullite are volume-stable since equilibrium conditions have been reached during initial firing. Considerable expansion also may be caused by gas formation during heating, for instance, by the decomposition of sulfates. In the presence of a viscous siliceous glass, these gases cannot escape and expansion occurs. This mechanism explains the bloating behavior of common bloating ladle brick. The linear reheat changes of various refractories during an ASTM reheat test are shown in Tables 5 and 7.

In general, basic brick exhibits good volume stability at high temperatures. Some slight shrinkage may occur in chemically bonded or low fired brick; this shrinkage is most pronounced in high silicate compositions. In high fired, high purity brick containing chrome ore, some expansion on reheat is generated by periclase–spinel reactions, usually with very complex interdiffusion mechanisms, where the spinel constituents are dissolved in the periclase lattice at high temperature and the spinel phases precipitate from the periclase solid solution on cooling.

Thermal Conductivity. The refractory thermal conductivity depends on the chemical composition of the material and increases with decreasing porosity. The thermal conductivities of some common refractories are shown in Figure 2.

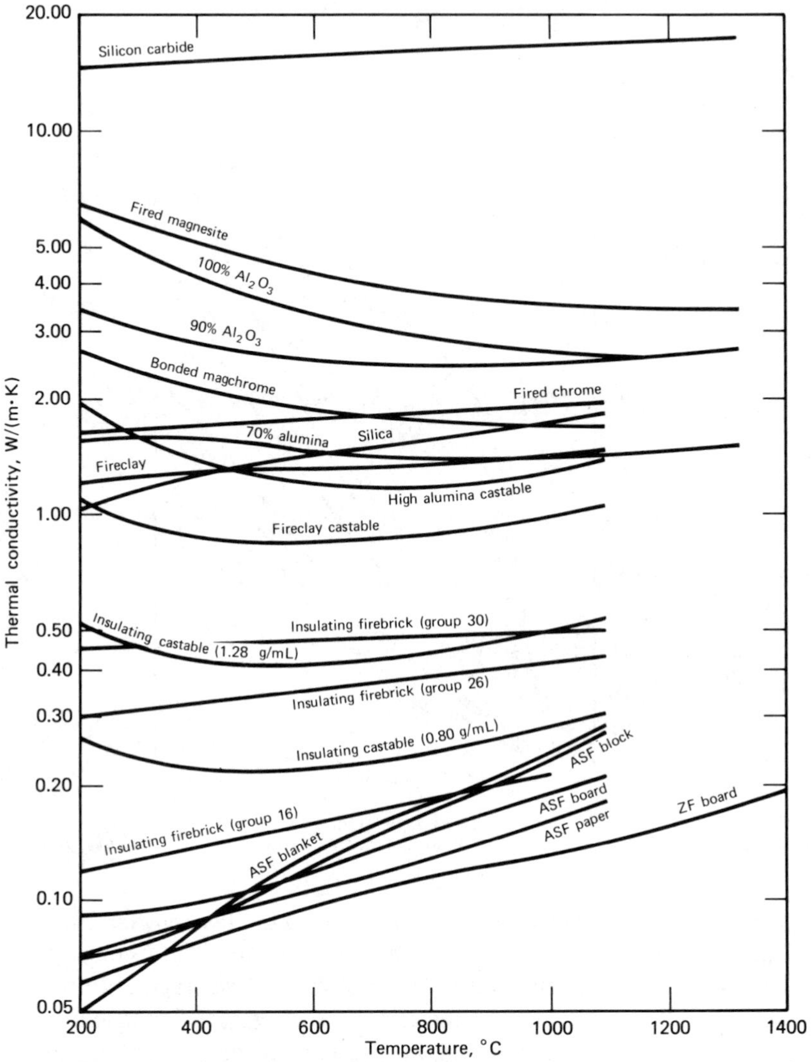

Figure 2. Thermal conductivity of typical refractories (5,25). ASF = aluminosilicate fiber; ZF = zirconia fiber. See Table 13 for group classification.

Specific Heat. In some applications, refractories are used for heat-exchange purposes on the regenerative principle, for instance, in blast-furnace stoves. High heat capacity is required in such applications (see Table 8).

Thermal Spalling. Refractories are brittle and stresses caused by sudden variations in temperatures can cause cracking and destruction. The susceptibility to thermal cracking and spalling depends upon certain characteristics of the raw material and the macrostructure of the particular refractory. Fireclay and high alumina refractories usually have a higher resistance to thermal shock than periclase refractories. Dense strong bodies withstand high stress and transmit it over large volumes; when failure occurs it is serious. Weak porous bodies, however, tend to crack before catastrophically large stresses are generated and thus are much less seriously damaged and generally remain intact.

Table 8. Mean Specific Heats of Refractory Brick and Minerals, Between 0°C and the Indicated Temperature, J/(kg·K)[a,b]

Temperature,	Brick								
°C	Fireclay	Silica	Magnesite	Chrome	Forsterite	Mullite	Cristobalite	Periclase	Corundum
0	249.5	218.5	268.9	219.8	232.9	237.9	213.3	268.9	221.1
93	257.3	243.0	283.1	227.5	258.5	248.2	236.6	293.5	253.4
204	266.4	272.8	300.0	235.3	279.3	276.7	263.8	310.3	276.7
316	274.1	296.1	312.9	243.1	297.4	288.3	309.0	324.5	292.2
427	284.5	307.7	324.5	250.8	310.3	296.1	324.5	332.3	303.9
538	293.5	318.1	333.6	257.3	318.1	301.3	333.6	338.8	312.9
649	302.6	325.8	340.1	259.9	323.3	306.4	341.4	345.2	320.7
760	311.6	333.6	346.5	268.9	328.4	310.3	346.5	349.1	325.8
871	320.7	336.2	353.0	274.1	333.6	312.9	350.4	354.3	332.3
982	327.1	341.4	359.5	279.3	338.8	316.8	353.0	358.2	336.2
1093	333.6	346.5	365.9	284.5	343.9	319.4	355.6	362.0	341.5
1204	338.8	351.7	372.4	287.0	349.1	322.0	358.2	364.6	345.2
1316	343.9	356.9	378.8	289.6	354.3	324.5	359.5	368.5	349.1
1427	347.8	360.7	384.0	292.2	359.5	327.1	360.7	372.4	353.0
1538						329.7	362.0	375.0	356.9
1642						331.0	363.3	377.6	360.7
1760						333.6		380.1	364.6

[a] Ref. 26.
[b] To convert J/(kg·K) to Btu/(lb·°F), multiply by 2.39×10^{-4}.

The resistance against thermal spalling of fireclay and high alumina brick is indicated in Table 5. No standard test has been adopted for basic brick. Refractories composed of 100% magnesia exhibit poor thermal-shock resistance, which is improved by addition of chrome ore.

Refractoriness. Most refractories are mixtures of different oxides, sometimes with significant quantities of impurities. Thus, they do not have sharp melting points but a softening range. Refractoriness is the resistance to physical deformation under the influence of temperature. It is determined by the pyrometric cone equivalent (PCE) test (see under Analytical and Test Methods).

Manufacture

Processing. Initial processing may include an extensive survey of the deposit, selective mining, stockpiling by grade, and beneficiation techniques such as weathering, grinding, washing, heavy-media separation, froth flotation, etc. Some materials can be used without further processing although many must be subjected to heat treatment. In the case of synthetic grain, the selected and beneficiated raw materials are blended in the desired proportions and formed into suitable shapes for calcination by briquetting, pelletizing, or extrusion. Slurries may be calcined; however, this practice is avoided in the interest of fuel economy. Originally, calcination referred specifically to the treatment of calcareous minerals to remove CO_2. Today, the term is often used to indicate heat treatment to sinter or burn (dead burn) the refractory grain to a stable dense material as well as to decompose minerals. Calcination may be carried out in rotary kilns, shaft kilns, multiple-hearth furnaces, or fluidized-bed reactors; the last two devices are reserved for relatively light calcining. The particular feed is dictated by the kiln type and the precalcination processes to which the raw materials or mixes may have been subjected. These hard, burned materials are called grain, clinker, or grog (term that is also used for ground firebrick). Low density or expanded aggregates can be made by burning clay or clay mixtures that evolve gas during burning and thereby expand or produce porosity. The burn-out material may be naturally present in the clay or may be mixed with the clay before burning.

Both raw and processed materials can be fused or melted in electric-arc furnaces. These materials can be melted and then cast into shapes, formed in the furnace itself as an ingot or formed into fibers. Partially fused ingots are crushed into grains. Since these ingots are not homogeneous, the grains have to be graded; the middle portion is purer than the outside material which is less well fused. Melts used for casting or fiber forming are homogeneous.

Crushing and Grinding. Some raw materials, such as hard clay and quartzite, must first be crushed to grains small enough for the grinding equipment. In general, a jaw, gyratory, or roll crusher is employed (see Size reduction).

Almost all raw materials require grinding after primary crushing. For coarse grinding, a dry pan or occasionally a wet pan is used. The dry pan is similar to a grist mill but has a perforated bottom through which the crushed material is continuously removed. The wet pan is similar, but has a solid bottom. For very fine grinding, a ring-roll, ball, or impact mill is employed.

Screening. To obtain a high density product, the mix is grain-sized. In a continuous screening operation, the ground raw materials are generally fed to vibrating high capacity screens that may be heated. Material that does not pass the screen is returned to the grinding system for further size reduction. A single screen with uniform openings produces a single product referred to as a straight grind. In gap-grain sizing, a series of two, three, or more vibrating screens are set in stacks yielding a coarse band, an intermediate band, and a fine band. These fractions are blended into the mix in the proper proportions to provide optimum properties. Gap-grain-sized material was originally used for the manufacture of basic brick; however, today it has been extended to many types of high alumina, fireclay, and silica brick. Coarse and medium-fine grain sizing is accomplished by the aforementioned methods, whereas fine-sized materials generated in rod mills, ball mills, ring-roll mills, and the like are classified by air separators (see Size separation).

Mixing. As in other ceramic processes, more than one type of raw material is often required for a refractory product. The purpose of mixing is to homogenize the various ingredients (see Mixing and blending). Although the specific steps and equipment involved in the mixing of batches for fireclay, high alumina, and basic refractories are somewhat different, the general principles are similar. Mixes that are to be dry-pressed contain 2–6% binding liquid, depending upon the plasticity of the raw material–bond system and the fineness of the mix. The ingredients may be blended in a pug mill, dry pan, or other type mixer and tempered with the bonding ingredients. Tempering, in the sense used here, denotes the kneading action produced on the mix, usually in a muller mixer. Mixes to be extruded or hand formed contain 10–20% liquid. These mixes may be prepared in a pug mill or wet pan.

The mixing of tar-, resin-, and chemically bonded brick material present special problems. Tar- and resin-bonded mixes are usually basic compositions. Because coal tar and petroleum pitches have ring-and-ball values of ca 100°C, ie, the softening point as determined by ASTM D 30, provisions for mixing the ingredients at high temperatures must be made. Typically, the grain is heated to ca 150°C and maintained at that temperature while the hot pitch is added. After a preset mixing time, the batch is transferred to the pressing equipment. For resin-bonded mixes, the grain does not have to be heated, although liquid resins may be heated to enhance their flow characteristics. Phenolic (phenol–formaldehyde), alkyd-oil–urethane, urea–formaldehyde, furan resins, and the like are used (see Phenolic resins; Alkyd resins; Amino resins). Resin-bonding systems are more expensive than pitches, but recent concerns regarding potential health risks of pitch fumes have renewed interest in resins.

Chemically bonded basic bricks are blended much the same as burned brick mixes except that a bonding agent, eg, magnesium sulfate or magnesium chloride is added to the mix as well as tempering water to form oxysulfates or oxychlorides.

Forming. Most refractory shapes are formed by mechanical equipment, but some very large or intricate shapes require hand molding in wooden, steel-lined molds with loose liners to permit easy removal.

A few fireclay refractories are produced by the stiff-mud process with an auger machine that pugs, de-airs, and continuously extrudes a clay column. A wire cutter cuts the clay into blanks which are then sized, shaped, and branded by a repress machine.

Refractory shapes are generally produced on a mechanical toggle press, screw press, or hydraulic press. In this operation, a mold cavity is filled with the damp mix. During the pressing cycle, some products such as fireclay brick are de-aired by applying a vacuum to the top and bottom press heads which contain a series of small openings;

de-airing promotes a denser product and reduces laminations. In some cases bricks are encased in steel or have internal steel plates which are placed in the mold box before charging; the plates become an integral part of the brick upon pressing. Plates may also be bonded to the brick after pressing. From one to four bricks may be pressed at a time, depending on the size. Pressures range from ca 17 (2500) to ca 98 MPa (14,000 psi) for plastic firebrick or nonplastic basic mixes, respectively. Pitch-bonded brick must be formed hot. Initial heating of the press may be required, but the residual heat from the mix is usually sufficient to maintain the necessary temperature. Another type is the impact- or jolt-mold press that consists of a mold box and press heads activated by pneumatic cylinders (eg, jack hammers). Some special shapes are produced by air-ramming which is similar to hand molding, except that reinforced steel molds are required and the damp mix is slowly fed to the mold while molders manually compact the material with pneumatic rammers. Special shapes can also be formed by slip casting and hot pressing.

Fusion-cast refractories are formed by first melting the material in an electric arc. The liquid is poured into a mold and allowed to cool. Various annealing processes and refining techniques ensure a uniform and structurally sound shape. The cast form is then cut or ground to size. Fused-cast refractory is very dense but may contain a system of closed pores and large, highly oriented grains may exist in a particular casting. The size and distribution of the pore and grain phases must be controlled.

Isostatic pressing gives a highly uniform product, although the production rate is somewhat low; it typically contains very small grains and little or no porosity. In this process, a rubber sock or bag of the desired shape is filled with the refractory mix. The sock is then subjected to extremely high pressure in a hydraulic pressure chamber.

Large and small shapes may be slip cast from both plastic and nonplastic mixes by the usual techniques. Precise shapes, such as glass feeder parts, are made in this way as well as large flux blocks. The process requires the formulation of a slip of suitably stable character to be poured into a plaster mold to be dewatered. After it solidifies, the mold is removed and dried further before firing.

Drying. The drying step for large shapes is critical; extremely large fireclay and silica shapes are sometimes allowed to dry on a temperature-controlled floor heated by steam or air ducts embedded in the concrete. Smaller shapes are generally dried in a tunnel dryer. The ware is placed on cars that enter the cold end and exit at the hot end. These dryers may be humidity-controlled and are heated by their own source or with waste heat from the burning operation. Microwave and infrared drying are being investigated.

Pitch-bonded and resin-bonded bricks are treated or cured in special ovens at temperatures higher than that used for drying other types of bricks. Pitch-bonded brick is cured at 230–320°C; this process is called tempering which is not to be confused with tempering during mixing. Tempering removes some of the volatiles from the pitch and eliminates or reduces thermoplastic or slumping behavior in service (see Drying).

Burning. Bricks are fired or burned in kilns to develop a ceramic bond within the refractory and attain certain desired properties. This step does not apply to chemically or organically bonded products. Preferred for this purpose is the continuous or tunnel kiln, a structure of narrow cross section and 61–183 m in length. The bricks are set on cars that move slowly through the kiln, which is divided into the preheating, burning, and cooling zones. The length of the zones controls the rates of heating and cooling, and the time at temperature (soak time). The temperature in the firing zone ranges from 1000 to 1700°C. Gas, oil, or coal may be used as a fuel.

Silica brick and large fireclay shapes are fired in circular down-draft kilns. These kilns vary in diameter and can accommodate up to 150,000 23-cm brick or their equivalent in other sizes. The complete burning cycle for a typical periodic kiln ranges from 21 to 27 days as compared with four to seven days for a tunnel kiln.

The shuttle kiln consists of a firing chamber with two or more kiln cars on which the bricks to be fired are set. While one load of brick is being fired, a second is being set. Somewhat similar is the bell top or top hat kiln which is raised and lowered above and over the kiln cars to be fired. These kilns are more expensive to operate than tunnel kilns but provide flexibility in burning conditions and production schedules.

Burned brick may be impregnated with tar or pitch to improve corrosion resistance. The heated bricks are placed in the impregnation unit which is sealed and evacuated to remove air from the pores of the refractory. The evacuated chamber is then filled with hot pitch and the vacuum is released. The treated product is allowed to drain free of excess pitch and is ready for shipment. Although this treatment is primarily used on basic refractories, it can be extended to other classes. The benefits derived from impregnation are lost if the pitch is burned off at high temperature; therefore, a reducing atmosphere is required such as is encountered in a basic oxygen steelmaking furnace.

Specialty Refractories. Bulk refractory products include gunning, ramming, or plastic mixes, granular materials, and hydraulic-setting castables, and mortars. These products are generally made from the same raw materials as their brick counterparts.

Granular materials are shipped raw or calcined and usually have been ground to a specified screen size or size distribution. The additives depend upon the application and service conditions. These materials are used in construction, repair, or maintenance of furnaces and vessels. Refractory mortars are used to lay brick of the same composition. They are manufactured wet premixed or dry.

Fireclay, ramming mixes, and high alumina and chrome ore plastics are manufactured similarly to brick; that is, the coarse refractory aggregate is added to a wet pan and combined with a small amount of raw clay for plasticity. Water is added to the batch; refractory plastics generally contain more water and raw clay than ramming mixes. After discharge from the wet pan, plastic mixes are formed by extrusion into 45-kg blocks and sliced into 9–12-kg slabs. The material is packaged damp and is ready for installation with pneumatic rammers. Ramming mixes are generally shredded into small lumps and placed damp in airtight steel drums ready for installation with pneumatic rammers. When installed, ramming mixes generally require forms to contain the material as it is rammed. Plastics, however, are soft and cohesive enough to be installed without forms.

Basic raw materials are susceptible to hydration and therefore specialty products are shipped dry and are mixed with water on site for gunning or ramming. Certain basic specialties are offered with organic vehicles such as oils and can be used without on-site mixing.

Information on manufacturing can be found in references 26–30.

Table 9. Distribution of U.S. Refractory Sales[a]

Industry	Percentage of total U.S. sales 1964	1977	1979
iron and steel	61.2	47.2	51.6
nonferrous metals	5.4	6.1	7.5
cement	2.0	3.6	4.9
glass	5.3	4.6	5.1
ceramics	5.0	8.7	9.7
chemical and petroleum	3.0	2.5	2.1
public utilities	1.1	0.7	0.9
export	5.3	6.5	7.4
all other and unspecified	11.7	20.1	10.9
Total, 10^8 $	*4.97*	*12.99*	*16.95*

[a] Ref. 31.

Economic Aspects

As can be seen from Table 9, the iron and steel industry is the principal consumer of refractories. The distribution, however, has changed since 1964; the decrease in refractories consumption coincides with the change from open-hearth steelmaking to BOF (Bessemer oxygen furnace) practice (see Steel).

The values of refractories shipped from 1967–1980 are given in Table 10. These data were obtained from the Bureau of Census current industry reports on refractories which have been issued on a quarterly basis since 1957. These reports provide reliable data on industry trends and give a fairly accurate economic picture.

The number of 23-cm brick equivalents of shaped refractories and the dollar sales values are given in Table 11. The dollar value for one brick equivalent of each type of refractory has increased by 76% for fireclay, 45% for high alumina, 305% for silica brick, and 173% for basic bricks in the period shown.

Table 10. Refractories Shipped, 1967–1980, 10^6 $[a,b]

Year	Constant dollar, total	Current dollar: Total	Clay refractories	Nonclay refractories
1967	524	524	225	299
1970	495	599	256	342
1973	572	780	327	453
1976	589	1084	464	621
1979	738	1727	712	1015
1980[c]	475	1211	512	698

[a] Ref. 31.

[b] The data were adjusted for price changes using the Producer Price Index for refractories as published by the Bureau of Labor Statistics. The base year is 1967.

[c] For nine months only.

Table 11. Shipments of Brick[a]

Year	Fireclay brick[b]		High alumina brick[c]		Silica brick		Basic brick[d]	
	1000[e]	10^6 $	1000[e]	10^6 $	1000[e]	10^6 $	1000[e]	10^6 $
1965	614	102	56	41	110	21	140	124
1975	507	142	129	137	41	32	126	279
1977	484	191	114	153	29	28	131	320

[a] Ref. 31.
[b] Includes regular fireclay, semisilica superduty fireclay, ladle brick, and IFB below 23.
[c] Includes high alumina, mullite, and extra-high alumina brick.
[d] Includes magnesite, magnesite–chrome, chrome, chrome–magnesite brick, and dolomitic brick.
[e] Brick equivalents.

ASTM Classifications

In addition to testing methods, ASTM publishes a list of classifications covering a wide variety of refractory types (32). The various brands from numerous producers and producing districts are grouped into classes with a nomenclature indicative of their chemical composition, heat resistance, and service properties.

Fireclay and High-Alumina Brick. ASTM designation C 27 covers fireclay and high alumina brick (see Table 12). High-alumina brick is classified according to alumina content, starting at 50% and continuing up to 99% Al_2O_3. Manufacturers are allowed ±2.5% of the nominal alumina content, except for 85 and 90% (±2.0%) and 99% (min 97%). An additional requirement for alumina bricks with Al_2O_3 content of 50, 60, 70, and 80%, are PCEs of 34, 35, 36, and 37, respectively.

Basic Bricks. Chrome brick, chrome–magnesite brick, magnesite–chrome brick, and magnesite brick are classified under ASTM C 455. The six classes of chrome–magnesite and magnesite–chrome brick start with 30% MgO and increase in 10% increments to 80% MgO; the minimum requirement for MgO is 5% less than the nominal percentage MgO specified for each class. For the three magnesite classes of 90, 95, and 98% MgO the minimum requirement is 86, 91, and 96% MgO, respectively. A chrome brick is manufactured entirely of chrome ore.

Insulating Brick. ASTM classifies insulating firebrick under C 155 by group; the group number indicates the service temperature multiplied by 100 degrees Fahrenheit (37.8°C) (see Table 13); eg, group 16 corresponds to a test temperature of ca 1600°F (871°C).

Mullite Refractories. Mullite refractories are classified under ASTM C 467. This brick must have an Al_2O_3 content between 56 and 79% and less than 5% impurities. Impurities are considered metal oxides other than those of aluminum and silicon. The hot-load subsidence is 5% max at 1593°C.

Silica Brick. Under ASTM C 416, types A and B silica bricks are classified according to chemical composition and strength. Silica brick must have an average modulus of rupture of 3.5 MPa (500 psi); they must have <1.50% Al_2O_3, no more than 0.20% TiO_2, <2.50% Fe_2O_3, and <4.00% CaO. Type A brick must have a flux factor equal to or <0.5. The flux factor is equal to percent alumina plus twice the percent of alkalies. Type B are all other silica brick covered by the standard chemical and strength specifications.

Table 12. ASTM C 27 Fireclay Brick Classification[a]

Classification	Type	PCE[b]	Panel spalling loss, max, %	Hot-load subsidence, max, %	Reheat shrinkage, at 1600°C, max, %	Cold modulus of rupture, MPa[c]	Other test requirements
superduty	regular	33	8 at 1650°C		1.0	4.14	
	spall-resistant	33	4 at 1650°C		1.0	4.14	
	slag-resistant	33				6.89	bulk density, min, 2.243 g/cm^3
high duty	regular	31½					
	spall-resistant	31½	10 at 1600°C			3.45	
	slag-resistant	31½				8.27	bulk density, min, 2.195 g/cm^3, or max porosity 15%
semisilica				1.5 at 1350°C		2.07	silica content, min, 72%
medium duty		29				3.45	
low duty		15				4.14	

[a] Ref. 33.
[b] Pyrometer cone equivalent.
[c] To convert MPa to psi, multiply by 145.

Table 13. ASTM C 155 Insulating Brick Classification[a]

Group identification	Reheat change of no more than 2% at test temperature, °C (°F)	Bulk density, g/cm³
group 16	815 (1550)	<0.545
group 20	1065 (1950)	<0.641
group 23	1230 (2250)	<0.768
group 26	1400 (2550)	<0.865
group 28	1510 (2750)	<0.961
group 30	1620 (2950)	<1.089
group 32	1730 (3150)	<1.522
group 33	1790 (3250)	<1.522

[a] Ref. 33.

Zircon Refractories. ASTM C 545 classifies zircon refractories in two types. Types A and B have the same chemical requirements of not less than 60% ZrO_2 and not less than 30% SiO_2. Type A (regular) must have a density of less than 3.85 g/cm^3 and type B (dense) more than 3.85 g/cm^3.

Castable Refractories. Hydraulic-setting refractory castables are classified under ASTM C 401 into three groups: normal strength with a modulus of rupture of 2.07 MPa (300 psi); high strength with a modulus of rupture of 4.14 MPa (600 psi); and insulating classified on the basis of bulk density. In addition, the normal and high strength groups are divided into classes A through G (see Table 14). The classification of insulating castables is given in Table 15.

Refractories used in steel-pouring pits are classified under ASTM C 435 (see Table 16).

Specifications

Among the many specifications covering refractory products, the best known are those published by ASTM. In addition, specifications are issued by the Federal Government and the armed forces. The former are generally preceded by the prefix HH and the latter by the prefix MIL. The ASTM refractory specifications always suggest a use, whereas Federal and military specifications are inconsistent in this respect.

Table 14. Classification of Normal and High Strength Refractories[a]

Class	No more than 5% shrinkage permitted after firing for 5 h at °C
A	1095
B	1260
C	1370
D	1480
E	1595
F	1705
G	1760

[a] Ref. 33.

Table 15. Classification of Insulating Castables[a]

Class	No more than 5% shrinkage permitted after firing for 5 h at °C	Density, g/cm³
N	925	0.88
O	1040	1.04
P	1150	1.20
Q	1260	1.44
R	1370	1.52
S	1480	1.52
T	1595	1.60
U	1650	1.68
V	1760	1.68

[a] Ref. 33.

Table 16. ASTM C 435 Classification For Steel-Pouring-Pit Refractories[a]

Class	Type	PCE[b]	Porosity, %	Reheat change above 1350°C, %[c]
nozzle	A	15–20	8 min	1.0, min
	B	20–29	8 min	1.0, min
	C	29 min	10 min	1.0, max
sleeve	A	15–20	10 min	1.0, min
	B	20–29	10 min	1.0, min
	C	29 min	10 min	1.0, max
laddle brick	A[d]	15 min	18 max	5.0, above 1290°C min
	B[d]	15 min	18 max	2.5, above 1350°C min
	C[d]	26 min	18 max	0.5, above 1500°C max

[a] Ref. 33.
[b] Pyrometer cone equivalent.
[c] Except ladle brick which is heated as stated; min refers to expansion, and max to shrinkage of diameter.
[d] Modulus of rupture of 4.83 MPa (700 psi) required.

No less important are specifications issued by industrial consumers. With the increasing emphasis on quality, this type of specification is becoming more and more prevalent in recent years. Where export is concerned, suppliers are faced with foreign specifications. Standard refractory samples are available from the Office of Standard Reference Materials, National Bureau of Standards, Washington, D.C., for chemical analysis and standardization, and for pyrometer cone equivalent (PCE) standardization from The Refractories Research Center, Ohio State University, Columbus, Ohio.

Analytical and Test Methods

The test methods applicable to refractories are found in the Annual Book of ASTM Standards, Part 17 (see Table 17). The chemical composition generally is reported on an oxide basis. ASTM designations C 571, C 572, C 573, C 574, C 575, and C 576 apply to the various types of refractories.

Table 17. ASTM Test Methods for Refractories[a]

Material	Test identification	Properties
burned brick	C 20, C 830	apparent porosity, water adsorption, bulk density
brick, various shapes	C 133, C 607, C 93	crushing strength, modulus of rupture
basic brick	C 456	hydration resistance
brick and tile	C 154	warpage
granules	C 357, C 493	bulk density
periclase grains	C 544	hydration
mortar	C 198	cold-bonding strength
air-setting plastics	C 491	modulus of rupture
castables	C 298	modulus of rupture
granular dead-burned dolomite	C 492	hydration
fireclay plastics	C 181	workability index
castables	C 417	thermal conductivity
plastics	C 438	thermal conductivity
general refractories	C 288	disintegration in CO atmosphere
	C 135	true specific gravity
	C 201	thermal conductivity
	C 92	sieve analysis and water content

[a] Ref. 33.

Refractoriness. Refractoriness is determined by several methods. The PCE test (ASTM C 24) measures the softening temperature of refractory materials. Inclined trigonal pyramids (cones) are formed from finely ground material, set on a base, and heated at a specific rate. The time and temperature (heat treatment) required to cause the cone to bend over and touch the base is compared to that for standard cones.

The standard ASTM PCE test is relative and used extensively only for alumina-silica refractories (see Table 5). However, the upper service limit is generally several hundred degrees below the nominal PCE temperature since some load is generally applied to the refractory during service. In addition, chemical reactions may occur that alter the composition of the hot face and, therefore, the softening point. The relationship between PCE numbers and temperature is described in ASTM C 24.

Another measure of refractoriness is the hot-compressive strength or hot-load test for refractory bricks or formed specialties. The specimen carries a static load from 69 kPa (10 psi) to 172 kPa (25 psi); it is heated at a specific rate to a specific temperature which is then held for 1½ h, or it is heated at a specific rate until it fails. The percent deformation or the temperature of failure is measured. The procedure is described in ASTM C 16. In a variation (ASTM C 546), the specimens are held at a specified temperature under a static load for 50 h. The results are reported as percent deformation or the time taken to achieve 13-mm deformation from the original length. A load of 345 kPa (50 psi) is applied to silicon carbide specimens and of 172 kPa (25 psi) to all other refractories.

Thermal Strength and Stability. Dimensional changes that occur upon reheating can be determined by ASTM C 605, C 436, C 210, C 179, or C 113. Specimens are selected and cut or formed to an appropriate size and measured before and after being heated at an appropriate temperature schedule to a specified temperature for 5 to 24 hours. The linear, diametral, or volume percentage change is noted. High temperature strength of refractory materials is determined on rectangular prisms 25 × 25 × 150 mm cut from the product being tested. The specimens are placed in a furnace, heated to a desired temperature, and the modulus of rupture is determined. A detailed description is given in ASTM C 583. Thermal spalling resistance is determined by ASTM C 439, C 180, C 122, C 107, and C 38. The last four methods apply to fireclay, high alumina bricks, and plastics using the apparatus described in C 38. The specimens are weighed and built into a panel (wall section) that was preheated for 24 h at a specific temperature. The panels are then subjected to twelve cycles of heating to 1400°C followed by cooling with a water-and-air spray. The panel is dismantled, the loose spalls are removed, and the weight loss is recorded as percent spalling loss. Thermal spalling of silica brick is determined on six specimens placed on a guarded hot plate. The specimens are heated and cooled at a specified rate. The heating rate is reported and any cracking that may have occurred is described.

Special Tests. Even though the American Society for Testing and Materials offers a wide range of test methods, there are other special tests that are imposed upon the manufacturer by consumers, the military, the Federal Government, and in some cases local or municipal governments. These tests are generally very specific and are oriented toward particular service conditions. In many instances, the producers develop special tests within their laboratories to solve customer problems. Many of these tests are adopted subsequently by ASTM.

Health and Safety Factors

Because industrial refractories are by their very nature stable materials, they usually do not constitute a physiological hazard. This statement does not apply however to unusual refractories that might contain heavy metals or radioactive oxides such as thoria, urania, and plutonia, or to binders or additives that may be toxic.

Inhalation of certain fine dusts may constitute a health hazard. For example, exposure to silica, asbestos, and beryllium oxide dusts over a period of time results in the potential risk of lung disease. OSHA regulations specify the allowable levels of exposure to ingestible and airborne particulate matter. Material Safety Data Sheets, OSHA form 20, available from manufacturers, provide information about hazards, precautions, and storage pertinent to specific refractory products.

Selection and Uses

Any manufacturing process requiring refractories depends upon proper selection and installation. When selecting refractories, the environmental conditions are evaluated first, then the functions to be served, and finally the expected length of service. All factors pertaining to the operation, service, design, and construction of equipment must be related to the physical and chemical properties of the various classes of refractories (34).

Service conditions that impare the effectiveness of refractories include chemical attack (ie, slags, fumes, gases, etc); operating conditions (ie, temperatures and cycling); and mechanical forces (ie, abrasion, erosion, and impact).

Design factors that influence the selection include type of equipment and its construction (ie, brick or monolithic material); refractory function (material containment, flow deflection, heat storage or release); heat environment (ie, exposure to constant or variable temperatures); refractory strength (ie, exposure to varying stress conditions); and thermal function (ie, insulation, dissipation, or transmission of heat).

The effects of processing conditions on refractories are given in Table 18.

Various standardized applications of different refractory types are listed in an excellent series of "Industrial Surveys of Refractory Service Conditions," compiled by Committee C-8 of the ASTM. These surveys cover the principal industrial applications of refractories and furnish a description of furnace operations and destructive influences, such as slagging, erosion, abrasion, spalling, and load deformation.

By far the most common industrial refractories are those composed of single or mixed oxides of single or mixed oxides of Al, Ca, Cr, Mg, Si, and Zr. These oxides exhibit relatively high degrees or stability under both reducing and oxidizing conditions. Carbon, graphite, and silicon carbide have been used both alone and in combination with the oxides. Refractories made from the above materials are used in ton-lot quantities in industrial applications. Other refractory oxides, nitrides, borides, and silicides are used in relatively small quantities for specialty applications in the nuclear, electronic, and aerospace industries.

The common industrial refractories are classified into acid, SiO_2 and ZrO_2, basic, CaO and MgO, and neutral, Al_2O_3 and Cr_2O_3. Oxides within each group are generally

Table 18. Effect of Processing Conditions on Refractories[a]

Service condition	Chemical resistance
oxidizing atmosphere	oxides and combinations of oxides (ie, silicates, fireclays) are unaffected; carbon and graphite oxidize; silicon carbide is fairly stable to 1650°C
steam or water vapor	can cause hydration of magnesite refractories at low temperatures and will oxidize carbon and graphite above 705°C
hydrogen	silica and silica-containing refractories are attacked above 1100°C; high-alumina, ZrO_2, MgO, and calcium-aluminate refractories show good resistance
sulfur and sulfates	above 870°C sulfur reacts with refractories containing silica; carbon and high purity oxides show good resistance; sulfates react to some degree; calcium-aluminate cement is more resistant than Portland cement
fuel ash	alkali and vanadium attack from ash can be severe on fireclay; high alumina resists
reducing atmosphere	most refractories are stable; however, iron oxide impurities, when reduced, can cause destruction, particularly if cycled
carbon monoxide	iron impurities can act as a catalyst to cause deposition of carbon in fireclay refractories; CO can oxidize graphite and SiC, and cause destructive changes in basic refractories
chlorine and fluorine	chlorine attacks silicates above 650°C; F attacks all refractory materials except graphite; basic refractories have poor resistance to both
acids	basic refractories have fair to poor resistance, fireclay and high alumina good resistance, except for HF; zircon, zirconia, and silicon carbide have good resistance; carbon and graphite do not react
alkalies	fireclay and high alumina perform well at low temperatures, magnesite refractories fair to good, chrome refractories poor, and graphite excellent

[a] Ref. 34.

compatible with each other whereas mixtures of acid and basic oxides often give low melting products; neutral oxides are generally compatible with both acidic and basic oxides.

Reactions between Refractories and Liquids. Molten metals are generally much less reactive than slags; therefore, the response of a refractory to chemical environments generally depends upon its slag resistance which, in turn, depends upon the compositions and properties of slag and refractory. Other factors include temperature, severity of thermal cycling or shock of the process, velocity and agitation of the slag in contact with the refractory, and the abrasion to which the refractory is subjected. Considering these factors, it is not surprising that similar refractories placed in similar furnaces can wear at vastly different rates if the operation practices are different.

As a general rule, acid slags (CaO/SiO_2 <1) require acid refractories, whereas basic slags (CaO/SiO_2 >1) require basic refractories. Fireclay and aluminosilicate refractories perform best for slags with a C/S ratio <1; however, acid slags containing considerable amounts of iron or manganese oxides, high alumina refractories are required. High-alumina refractories are also superior to fireclay refractories when the basicity-to-acidity ratio = 1. When the basicity of the slag >1, basic refractories should be employed, such as MgO, MgO–CaO, and $MgO.Cr_2O_3$. Magnesium oxide resists slags of a wide range of compositions; however, in actual practice all basic refractories contain some silica which usually occurs in various silicate phases at the grain boundaries. As slag attack proceeds, liquid phases migrate from the hot face to the cooler regions and attack of the grain-boundary silicate phase precedes solution of the periclase grain. For this reason, both the character of the bonding silicate phase and the porosity of the refractory are very important. Penetration of the slag may be impeded by materials such as carbon, pitch, or resins by lowering the porosity and changing the wetting characteristics.

The thermal conductivity of refractories in high wear areas of steelmaking vessels is increased with internal metal plates and by incorporating flake graphite with magnesite refractories. Slags penetrate refractories until their viscosity becomes too high for further migration. Refractories with high thermal conductivity can cause the slag to chill and reduce the penetrated volume.

Slag penetration alters the structure of the refractory by changing its porosity, density, and strength. If the altered refractories are subject to thermal cycling or if volume changes occur upon crystallization of the slag, stress concentrations build up immediately behind the densified zone and spalling or cracking may result. An example of structural alteration is the iron oxide bursting phenomenon in magnesite–chromite brick. Iron oxide contained in the chromite ore, or that which is allowed to penetrate the brick, can cause excessive expansion of the lattice because of the unequal diffusion of iron and chromium ions in magnesia chrome spinel, which leads to the production of pores in the iron-rich phase.

Reactions between Refractories and Gases. Reactions with gases can be quite destructive as the gases generally penetrate the pores of the refractory and destroy its structure. The refractory may either expand and crack because of the formation of new, low density compounds, or its refractoriness may be drastically reduced because of the formation of low melting compounds. An example is the disintegration of alumina silicates in blast furnaces caused by carbon monoxide. The deposition of carbon is catalyzed by iron in the bricks. The growth of the carbon deposit causes the brick to rupture and more surface is exposed. Therefore, a brick of low iron and alkali content with a dense, low permeability is preferred.

Reactions between Refractories. In Table 19, the compatibilities between various refractories are given for a range of temperatures. Generally, dissimilar refractories react vigorously with each other at high temperature. Phase diagrams are an excellent source of information concerning the reactivity between refractories.

Silica Refractories. This type consists mainly of silica in three crystalline forms: cristobalite [*14464-46-1*], tridymite [*1546-32-3*], and quartz [*14808-60-7*]. Quartzite sands and silica gravels are the main raw material ingredients, although lime and iron oxide are added to increase the mineral content. Uses include open-hearth roof linings, refractories for coke ovens, coreless induction foundry furnaces, and fused-silica technical ceramic products. Consumption of silica refractories has declined dramatically over the last 20 years, mainly because of the increasing use of oxygen in steelmaking.

Semisilica Refractories. Semisilica refractories consist essentially of silica, bonded by a glassy matrix. They contain 75–93% SiO_2 and may be made from siliceous clay or sand and fireclay. They tend to resist slag attack owing to a self-glazing tendency. Uses include shapes for open hearth stoves and checkers.

Fireclay Refractories. These products are made from clay minerals containing ca 17–45% Al_2O_3; pure kaolin has the highest alumina content. Fireclay refractories

Table 19. Approximate Initial Temperature, °C, at which Refractories React[a,b]

Refractory	Magnesite, 92% MgO	Magnesite–chrome, CB[b]	Chrome–magnesite, CB[c]	Chrome–magnesite, fired	Forsterite, stabilized	Chrome, fired	90% alumina	70% alumina
magnesite, 92% MgO		>1700	>1700	>1700	(1700)	1700	>1700	1600
magnesite–chrome, CB[c]	>1700		>1700	>1700	1650	>1700	1450	1450
chrome–magnesite								
CB[c]	>1700	>1700		>1700	1650	>1700	1650	1600
fired	>1700	>1700	>1700		1650	>1700	1650	1650
forsterite, stabilized	(1700)	1650	1650	1650		1650	1650	1600
chrome refractories, fired	1700	>1700	>1700	>1700	1650		1650	1600
alumina								
90%	>1700	1600	1600	1600	1650	1650		>1700
70%	1650	1600	1600	1500	1650	1600	>1700	
zircon	>1700	1600	1650	1600	1600	>1700	>1700	>1700
fireclay								
superduty	1400	(1700)	1650	1650	1600	1500	(1700)	(1700)
high-duty	1400	1650	1600	1600	1650	1500	(1700)	(1700)
semisilica	1500	1400	1400	1500	1500	1500	(1500)	(1500)
silicon carbide, clay-bonded	1500	1400	1400	1400	1600	1500	1650	1650
silica								
type B[d]	1500	1600	1600	1500	1650	1650	1650	1500
superduty	1500	1600	1600	1600	1650	1650	1650	1500

[a] Ref. 26.
[b] Max temperature tested = 1700°C.
[c] Chemically bonded (not fired); one or both materials not sufficiently refractory for test.
[d] Ref. 33.

are used in kilns, ladles, and heat-regenerators, acid-slag-resistant applications, boilers, blast-furnaces, and rotary kilns. They are generally inexpensive.

High Alumina Refractories. These refractories have alumina contents from 100% to just above 45%. The desired alumina content is obtained by adding bauxites, synthetic aluminosilicates, and synthetic aluminas to clay and other bonding agents. These refractories are used in kilns, ladles, and furnaces that operate at temperatures or under conditions for which fireclay refractories are not suited. Phosphate-bonded alumina bricks have exceptionally high strength and are employed in aluminum furnaces. High alumina and mullite are used in furnace roofs and petrochemical applications.

Chrome Refractories. Naturally occurring chrome ore composed mainly of chromite [*53293-42-8*], may be made into a brick or may be blended with fine calcined magnesite to obtain the desired chrome-to-magnesia ratio. When blended with magnesite, the products containing more chrome than magnesia are denoted chrome magnesite. These refractories are used in nonferrous metallurgical furnaces, rotary kiln linings, secondary refining vessels, such as argon–oxygen decarborizers (AODs) and glass-tank regenerators.

Magnesite Refractories. These refractories do not contain magnesite as their name implies, but rather periclase. The term magnesite refers to the ore from which the periclase was made, although magnesite brick can be made from synthetic periclase, ie, seawater MgO. Shaped magnesite refractories may be impregnated or bonded with pitch or resin to improve their resistance to slag attack. Chrome ore can be added to magnesite to produce magnesite–chrome refractories which are used in lining and maintenance of steelmaking and refining vessels and checkers.

Dolomite Refractories. These refractories contain dead-burned dolomite and possibly fluxes such as millscale, serpentine, or clay. Shaped refractories may be bonded or impregnated with pitch to improve slag resistance and inhibit hydration. Addition of magnesite gives magnesite–dolomite or magdol refractories. Dolomite refractories are primarily used in linings of BOF vessels and refining vessels, and in ladels and cement kilns.

Spinel Refractories. These refractories contain synthetic spinel, $MgAl_2O_4$. They exhibit good strength at high temperatures and thermal shock resistance. Chromium and magnesium oxides crystallize in the same structure and are referred to as chrome–magnesite spinels. Spinel refractories are used in cement kilns and steel-ladle linings.

Forsterite Refractories. These refractories are made from forsterite, Mg_2SiO_4. They resist alkali attack and have good volume stability, high temperature strength, and fair resistance to basic slags. Uses include nonferrous metal furnace roofs and glass-tank refractories not in contact with the melt, ie, checkers, ports, and uptakes.

Silicon Carbide Refractories. Silicon carbide has a wide range of refractory uses including chemical tanks and drains, kiln furniture, abrasion-resistant linings, blast-furnace linings, and nonferrous metallurgical crucibles and furnace linings.

Zirconia Refractories. The most common zirconia-containing refractories are made from zircon sand and are used mostly for glass-tank paver brick. Refractory blocks made from a composition of zircon and alumina, used to contain glass melts, are generally electromelted and then cast. They exhibit excellent corrosion resistance but are subject to thermal shock. Refractories made from pure ZrO_2 are extremely expensive and are reserved for extra high temperature service above 1900°C. Additives such as yttria, or CaO and MgO prevent deterioration during heating and cooling.

BIBLIOGRAPHY

"Refractories" in *ECT* 1st ed., Vol. 11, pp. 597–633, by L. J. Trostel and R. P. Heuer, General Refractories Co.; "Refractories" in *ECT* 2nd ed., Vol. 17, pp. 227–267, by W. T. Bakker, G. D. Mackenzie, G. A. Russell, Jr., and W. S. Treffner, General Refractories Co.

1. A. F. Greaves-Walker, *Bull. Am. Ceram. Soc.* **6,** 20, 213 (1941).
2. F. Singer and S. S. Singer, *Industrial Ceramics*, Chemical Publishing Company, Inc., New York, 1964.
3. R. D. Pehlke and co-eds., "Refractories" in *Basic Oxygen Furnace Steelmaking*, Vol. 4, The Iron and Steel Society of AIME, New York, 1977, Chapt. 11, pp. 1–58.
4. F. H. Norton, *Refractories*, 3rd ed., McGraw-Hill Company, Inc., New York, 1949.
5. J. J. Suec and co-eds., *Ceramic Data Book 1981 Suppliers' Catalog and Buyers' Directory*, Cahners Publishing Company, Denver, Col., 1981.
6. W. D. Kingery, *Introduction to Ceramics*, John Wiley & Sons, Inc., New York, 1960.
7. A. Alper, ed., *High Temperature Oxides*, (*1–4*) Vol. 5 of *Refractory Materials*, Academic Press, New York, 1970.
8. E. Ryshkewitch, *Oxide Ceramics*, Academic Press, Inc., New York, 1960.
9. H. Salmang, *Ceramics, Physical and Chemical Fundamentals*, Butterworth and Company, Ltd., London, 1961.
10. J. R. Hague, J. F. Lynch, A. Rudnick, F. C. Holden, and W. H. Duckworth, eds., *Refractory Ceramics for Aerospace: A Materials Selection Handbook*, The American Ceramic Society, Inc., Columbus, Ohio, 1964.
11. B. Phillips, *Res. Dev.* **18,** 22 (1967).
12. E. M. Levin, C. R. Robbins, and H. F. McMurdie, *Phase Diagrams for Ceramists*, The American Ceramic Society, Inc., Columbus, Ohio, 1964.
13. L. A. Dahl, "Rock Products, Sept. to Dec., 1938" in *PCA Research Bulletin*, Vol. 1, 1939.
14. L. A. Dahl, *J. Phys. Chem.* **52,** 698 (1948).
15. C. N. Fenner, *Am. J. Sci.* **36**(4), 383 (1913).
16. J. W. Greig, *Am. J. Sci.* **13**(5), 1 (1927).
17. A. Muan and E. F. Osborn, *Phase Equilibria Among Oxides in Steelmaking*, Addison-Wesley Publishing Company, Inc., Reading, Mass., 1965.
18. N. L. Bowen and J. F. Schairer, *Am. J. Sci.* **29**(5), 153 (1935).
19. P. Duwez, F. Odell, and F. H. Brown, Jr., *J. Am. Ceram. Soc.* **35**(5), 109 (1952).
20. H. E. McGonnon, ed., *The Marking, Shaping and Treating of Steel*, 9th ed., U.S. Steel Corporation, Pittsburgh, Pa., 1970.
21. *Refractories for Industry*, C-E Refractories, Valley Forge, Pa., 1976.
22. *Product Literature*, Kaiser Refractories, Oakland, Calif., 1977.
23. *Criterion I and Other Glass Furnace Refractories*, C-E Refractories, Valley Forge, Pa., 1977.
24. C. O. Fairchild and M. F. Peters, *J. Am. Ceram. Soc.* **9,** 700 (1926).
25. *Fibrous Ceramic Thermal Insulation for Ultra-High Temperature Use*, Zircar Products, Inc., New York, 1976.
26. *Modern Refractories Practice*, Harbison-Walker Refractories Company, Pittsburgh, Pa., 1961.
27. P. P. Budnikov, *The Technology of Ceramics and Refractories*, The MIT Press, Cambridge, Mass., 1964.
28. J. E. Neal and R. S. Clark, *Chem. Eng.*, 56 (May 4, 1981).
29. *Refractories*, General Refractories Company, Philadelphia, Pa., 1949.
30. A. A. Litvakovski, *Fused Cast Refractories*, Israel Program for Scientific Translations, Jerusalem, 1961.
31. *Refractories*, Current Industrial Reports, U.S. Bureau of Census, Industry Division, Annual Reports, Washington, D.C., 1965–1979.
32. *Manual of ASTM Standards on Refractory Materials*, 8th ed., American Society for Testing and Materials, Philadelphia, Pa., 1957.
33. "Refractories, Glass, Ceramic Materials; Carbon and Graphite Products" in *ASTM Annual Book of ASTM Standards Part 17*, 1981.
34. J. F. Burst and J. A. Spieckerman, *Chem. Eng.*, 85 (July 31, 1967).

General References

J. R. Rait, *Basic Refractories, Their Chemistry and Their Performance*, Ilife & Sons, Ltd., London, England, 1950.

J. H. Chesters, *Steel Plant Refractories, Testing Research and Development*, The United Steel Companies, Ltd., Sheffield, England, 1963.

J. R. Coxey, *Refractories*, The Pennsylvania State College, State College, Pa., 1950.

L. R. McCreight, H. W. Rauch, and W. H. Sulton, eds., *Ceramic and Graphite Fibers and Whiskers*, Vol. 1 of *Refractory Materials*, Academic Press, New York, 1970.

E. K. Storms, ed., *The Refractory Carbides*, Vol. 2 of *Refractory Materials*, Academic Press, New York, 1970.

A. M. Alper, ed., *Phase Diagrams, Materials Science and Technology (1–3)*, Vol. 6 of *Refractory Materials*, Academic Press, New York, 1970.

A. K. Kulkarni and V. K. Moorthy, ed., *Proceedings of the 1st Symposium on Material Science Research*, Series 2, Chemical Metallurgy Commission, Department of Atomic Energy, Bombay, India, 1970, pp. 208–218.

G. R. Belton, ed., *Proceedings of the International Conference of Metal Material Science, 1969*, Plenum Press, New York, 1970.

H. Bibring, G. Seibel, and M. Rabinouitch, eds., *2nd International Conference of Strength Metals Alloys*, Series 3, Conference Proceedings, ASM, Metals Park, Ohio, 1970, pp. 1178–1182.

G. H. Criss and A. R. Olsen, ed., *Proceedings of the 3rd National Incinerator Conference*, American Society of Mechanical Engineering, New York, 1968, pp. 53–68.

C. Brosset, ed., *Trans. Int. Ceram. Congr.* **10,** (1967).

R. M. Fulrath and J. A. Pask, eds., *Ceramic Microstructures, Proceedings of the 3rd International Material Symposium*, John Wiley & Sons, Inc., New York, 1968.

G. C. Kuczynski, ed., *Sintering Related Phenomena, Proceedings of the 2nd International Conference*, Gordon and Breach Science Publishers, New York, 1967.

H. H. Hausner, ed., *Fundamental Refractory Compounds*, Plenum Press, New York, 1968.

R. C. Bradt, D. P. H. Hasselman, and F. F. Lange, eds., *Mech. Ceramics, Proceedings of the Symposium*, Plenum Press, New York, 1974.

S. J. Lefond, *Industrial Mineral Rocks*, 4th ed., American Institute of Mechanical Engineers, New York, 1975.

C. S. Tedmon, Jr., *Corrosion Problems in Energy Conversion Generators*, Electrochemical Society, Princeton, N.J., 1974.

J. J. Burke, A. E. Gorum, and N. R. Katz, eds., *Ceramic High-Performance Applications, Proceedings of the 2nd Army Materials Technology Conference*, Brook Hill Publishing Company, Chestnut Hill, Mass., 1974.

R. C. Bradt and R. E. Tressler, eds., *Deformation of Ceramic Materials, Proceedings of the 1974 Symposium*, Plenum Press, New York, 1975.

R. F. S. Fleming, ed., *Proceedings of the Industrial Mineral International Congress, Met. Bull. Ltd. London* (1975).

Z. A. Foroulis and W. W. Smeltzer, eds., *Met. Slag-Gas React. Processes*, Electrochemical Society, Inc., Princeton, N.J., 1975.

S. Modry and M. Svata, eds., *Pore Structures Prop. Materials, Proceedings of the International Symposium*, Scademia Prague, Czechoslovakia, 1974.

F. V. Tooley, *Handbook of Glass Manufacturing*, Books Ind., Inc., New York, 1974.

10th International Congress of Glass, Ceramic Society of Japan, Tokyo, 1974.

N. Standish, ed., *Alkalis Blast Furnaces, Proceedings of the 1973 Symposium*, Department of Metallurgical Material Science, McMaster University, Hamilton, Ontario, Canada, 1973.

R. C. Marshall, ed., *Silicon Carbide, Proceedings of the 3rd International Conference, 1973*, University of South Carolina Press, Columbia, S.C., 1974.

B. Cockayne, ed., *Mod. Oxide Materials, Prep., Prop. Device Applications*, Academic Press, London, 1972.

R. R. M. Johnston, ed., *Corrosion Technology in the Seventies*, Technical Paper, 12th Annual Conference of the Australiasian Corrosion Association, Parkville, Victoria, Australia, 1972.

L. D. Pye, ed., *Introduction to Glass Science, Proceedings of a Tutorial Symposium*, Plenum Press, New York, 1972. Review of glass melting refractory corrosion.

High Temperature Material, Proceedings of the 3rd Symposium on Material Science Research, Department of Atomic Energy, Bombay, India, 1972. Papers cover a variety of oxide and nonoxide high temperature materials like spinel, alumina and silicon nitride.

J. J. Burke, ed., *Powder Metal High-Performance Applications, Proceedings of the 18th Sagamore Army Material Research Conference*, Syracuse University Press, Syracuse, N.Y., 1972. Review on silicon carbide silicon nitride ceramics.

Mechanical Behavior of Materials, Proceedings of the 1st International Conference, Society of Material Science, Kyoto, Japan, 1972.

J. D. Buckley, ed., *Advanced Materials, Composite Carbon. Pap. Symposium*, American Ceramics Society, Inc., Columbus, Ohio, 1972. Papers on composite materials including carbon, graphite of nitride composites.

J. I. Duffy, *Refractory Materials, Developments since 1977*, Chemical Technology Review, No. 178, Noyes Data Corporation, Park Ridge, N.J., 1980.

I. Ahmad and B. R. Noton, eds., *Advanced Fibers Compositions at Elevated Temperatures, Proceedings of the Symposium of the Metallurgical Society*, American Institute of Mechanical Engineers, Warrendale, Pa., 1980.

G. V. Samsonov and I. M. Vinitskii, *Handbook of Refractory Compounds*, Plenum Press, New York, 1980.

A. V. Levy, ed., *Proceedings of the Corrosion/Erosion Coal Conversion Systems Materials Conference*, National Association of Corrosion Engineers, Houston, Texas, 1979.

Basic Oxygen Steelmaking—New Technology Emerges?, Proceedings of the Conference, Metallurgical Society of London, 1979. Contains papers on refractories for conventional and new bottom blown vessels.

M. R. Louthan, Jr. and R. P. McNitt, eds., *Environmental Degradation of Engineering Materials, Proceedings of the Conference*, Virginia Polytechnical Institute, Blacksburg, Va., 1978.

19th International Refractories Colloquium, Institute Gesteinshuttenkunde RWTH Aachen, Aachen, Federal Republic of Germany, 1976. Contains a number of good papers covering a wide range of refractory materials, properties and problems.

Contin. Cast. Steel, Proceedings of the International Conference, Metallurgical Society of London, 1977.

3rd U.S.—U.S.S.R. Colloquium or Magnetohydrodynamic Electric Power Generation, National Technical Information Service, Springfield, Va., 1976.

P. Vincenzini, ed., *Advanced Ceramic Processes, Proceedings of the 3rd International Meeting of Modern Ceramics Technology*, National Research Council, Research Laboratory of Ceramics Technology, Faenza, Italy, 1978.

G. Y. Onoda and L. L. Hench, eds., *Ceramic Processing Before Firing*, John Wiley & Sons, Inc., New York, 1978.

V. I. Matkovich, ed., *Boron Refractory Borides*, Springer-Verlag, Berlin, Germany, 1977.

R. M. Fulrath and J. A. Pask, *Ceramic Microstructures, Proceedings of the 6th International Materials Symposium*, Westview Press, Boulder, Col., 1977.

H. D. LEIGH
C-E Basic, Inc.

REFRACTORY COATINGS

Refractory coatings denote those metallic, refractory-compound (ie, oxides, carbides, nitrides) and metal-ceramic coatings associated with high temperature service as contrasted to coatings used for decorative or corrosion-resistant applications. They also denote coatings of high melting materials that are used in other than high temperature applications. A coating may be defined as a near-surface region with properties that differ significantly from the bulk of the substrate (see Ceramics; Metallic coatings; Metal surface treatments).

The highest melting refractory metals are tungsten, tantalum, molybdenum, and niobium, although titanium, hafnium, zirconium, chromium, vanadium, platinum, rhodium, ruthenium, iridium, osmium, and rhenium may be included. Many of these metals do not resist air oxidation; hence, very few, if any, are used in their elemental form for high temperature protection. However, bulk alloys based on nickel, iron, and cobalt with alloying elements such as chromium, titanium, aluminum, vanadium, tantalum, molybdenum, silicon, and tungsten are used extensively in high temperature service. Some modern high temperature oxidation- and corrosion-resistant coatings have compositions similar to the high temperature bulk alloys and are applied by thermal spraying, evaporation, or sputtering. The protection mechanism for these high temperature alloy coatings is based on adherent impervious surface films of the Al_2O_3, SiO_2, CrO_2, or a spinel-type that grow on high temperature exposure to air.

Refractory coatings also include materials with high melting points, eg, silicides, borides, carbides, nitrides, or oxides, and combinations such as oxy-carbides, etc. In addition, mixtures of metals and refractory compounds (sometimes called metallides) of various microstructural configurations (ie, laminates, dispersed phases, etc) can also be classified as refractory coatings. Other terms used in the past to describe such multiphase coatings are multilayer coating, multicoating, composite coating, reinforced coating, and the like.

In some cases, a coating is a new material that is deposited onto the substrate by a variety of methods. It is then called a deposited or overlay coating. In other cases, the coating may be produced by altering the surface material to produce a surface layer composed of both the added and substrate materials. This is called a conversion coating, cementation coating, diffusion coating, or chemical-conversion coating when chemical changes in the surface are involved. Coatings may also be formed by altering the properties of the surface by melting and quenching, mechanical deformation, or other processes that change the properties without changing the composition.

Coating technology has developed extensively in the past 20 years. In this article, emphasis is given to new processes or processes that have undergone extensive development in the past decade.

All coating methods consist of three basic steps: synthesis or generation of the coating species or precursor at the source; transport from the source to the substrate; and nucleation and growth of the coating on the substrate. These steps can be completely independent of each other or may be superimposed on each other, depending on the coating process. A process in which the steps can be varied independently and controlled offers great flexibility, and a larger variety of materials can be deposited.

Numerous schemes can be devised to classify deposition processes. The scheme used here is based on the dimensions of the depositing species, ie, atoms and molecules, liquid droplets, bulk quantities, or the use of a surface-modification process (1–2) (see Table 1).

The coating has to adhere to the substrate. The bonding may be mechanical as a result of the interlocking between the asperities on the surface and the coating. Thus, the surface roughness, ie, the average distance between asperities, must be equal to or larger than the dimensions of the depositing particles. Consequently, plasma-sprayed coatings, where the material is deposited as droplets, do not adhere to a polished metal surface because the surface roughness is on a much smaller scale than the dimensions of the liquid droplets. The substrate surface has to be coarsened by techniques such as grit blasting that not only provide the required concave asperities, but also clean the surface (removal of oxide layers and scales) and provide high energy sites for a denser nucleation of coating crystallites and, to some extent, increase the real interfacial area.

In diffusion or chemical bonding, the substrate and the coating material interdiffuse at the interface. The latter depends on the equilbrium between the two materials at the effective temperature of deposition. If the two materials have no solid solubility, a sharp or abrupt interface forms. The bond strength depends on the affinity between the materials and increases with the degree of wetting of the substrate with the coated material. If the two materials exhibit extensive solid solubility, the interface is a solid solution. If intermetallic phases are indicated on the equilibrium diagram, they may also form at the interface as may gas-metal compounds, depending on the degree of contamination present on the substrate or arriving at the substrate from the environment.

Residual stresses, which are always present in coatings, arise from thermal-expansion mismatch between coating and substrate, or growth stresses caused by imperfections in the coating that are built-in during the process of film growth. Growth stresses are usually significant for coating processes carried out near room temperature. Failure or debonding of the coating may be caused by the residual stresses in the coating and substrate. The location of the failure may be in the substrate, at the interface, or in the coating, depending on plastic deformation and resistance to nucleation and growth of cracks, ie, fracture toughness. For example, a coating may fail at the interface because of crack propagation caused by a brittle intermetallic phase. A postfailure surface analysis of the substrate and the coating is recommended; the use of surface chemical-analysis techniques, eg, scanning electron microscopy (SEM) with energy-dispersive x-ray analysis (edax) capabilities, Auger spectroscopy, or esca analysis can determine the chemical composition at the failed interfaces and deduce its cause.

Coatings may be permeable either to the atmosphere or the substrate material. Diffusion of oxygen through a coating can result in gaseous products that may rupture the coating. Even if the gas can escape through the substrate, as with graphite, the substrate can be consumed and the bond weakened. Coatings permeable to the substrate material by outward diffusion can suffer similarly, as oxidation of the diffused species may occur at the atmosphere–coating interface.

The design of a coating that might equilibrate with its substrate during use is based on the phase diagram of the system. Extensive regions of solid solubility may indicate that rapid interdiffusion can be expected. For multicomponent coatings on

Table 1. Coating Methods[a]

Atomistic deposition	Particulate deposition	Bulk coatings	Surface modification
electrolytic environment	thermal spraying	wetting processes	chemical conversion
electroplating	plasma-spraying	painting	electrolytic
electroless plating	D-gun	dip coating	anodization (oxides)
fused-salt electrolysis	flame-spraying	electrostatic spraying	fused salts
chemical displacement	fusion coatings	printing	chemical–liquid
vacuum environment	thick-film ink	spin coating	chemical–vapor
vacuum evaporation	enameling	cladding	thermal
ion-beam deposition	electrophoretic	explosive	plasma
molecular-beam epitaxy	impact plating	roll bonding	leaching
plasma environment		overlaying	mechanical
sputter deposition		weld-coating	shot peening
activated reactive evaporation		liquid-phase epitaxy	thermal
plasma polymerization			surface enrichment
ion plating			diffusion from bulk
chemical-vapor environment			sputtering
chemical-vapor deposition			ion implantation
reduction			
decomposition			
plasma enhanced			
spray pyrolysis			

[a] Ref. 2.

multicomponent substrates, a large number of possibilities exists for formation of intermetallic compounds. Limited compound formation at the coating-substrate interface may be preferred if it decreases the permeability of the coating. Wetting or, more strictly speaking, high affinity between coating and substrate is conducive to high temperature service.

A reasonably close match of the thermal expansion of the coating and substrate over a wide temperature range to limit failure caused by residual stresses is desired for coatings. Since temperature gradients cause stress even in a well-matched system, the mechanical properties, strength, and ductility of the coating as well as the interfacial strength must be considered.

The simple case of protecting tungsten or molybdenum with platinum illustrates the above point. A wide range of solid solubility exists that contributes to interdiffusion between coating and base metal to form brittle intermetallic compounds. Both solutions and compounds oxidize preferentially, losing tungsten or molybdenum. Platinum is also somewhat permeable to oxygen, which permits subcoating oxidation. A thermal-expansion mismatch exists between platinum and these intermetallic compounds and adds another complication where thermal cycling is required. Silicides and beryllides exhibit sufficient conductivity to be electroplated and could well be used as diffusion barriers. However, the bond strengths and mechanical compatibility of two layers instead of one must be considered.

Atomic Deposition Processes

Electrodeposition. ***Aqueous Electrodeposition.*** The theory of electrodeposition is well known (see Electroplating). Of the numerous metals used in electrodeposition, only ten have been reduced to large-scale commercial practice. The most commonly plated metals are chromium, nickel, copper, zinc, rhodium, silver, cadmium, tin, and gold followed by the less frequently plated metals iron, cesium, platinum, and palladium and the infrequently plated metals iridium, ruthenium, and rhenium. Of these, only platinum, rhodium, iridium, and rhenium are refractory.

The electrodeposition of tungsten alloys of iron, nickel, and cobalt is commercially feasible but has remained largely experimental, although their properties should be of sufficient interest for engineering applications.

Cermets, ie, materials containing both ceramic and metal, eg, TiC–Ni and Al_2O_3–Cr, can be deposited from plating baths if the particulate matter is suspended by air agitation or stirring (see Ceramics; High temperature composites). Particle sizes from 1 to 50 μm may be deposited to concentrations over 20%. Chromium-based cermets with zirconium and tungsten borides, zirconium nitride, and molybdenum carbide can be plated on refractory metals and graphite. However, at 1650°C, protection is afforded only for a few minutes. Cermets also suffer from porosity; therefore, such coatings find application where exposure times are short and erosion conditions severe, and where they can be used to bridge the expansion mismatch between a metallic substrate and a ceramic coating.

Fused-Salt Electrodeposition. Molten-salt electrolysis has been used since Davy's and Faraday's pioneering experiments, but application as a method for coating metals has remained of academic interest. Fused-salt baths may be used to plate ruthenium, platinum, and iridium with improved coating soundness. Recent developments in this technology may lead to commercial plating of other refractory metals, such as zirconium, hafnium, vanadium, niobium, tantalum, molybdenum, and tungsten. Molten-salt electrolysis is based on the same principles as aqueous electrolysis processes, but the vehicle, ie, the bath, is a molten salt. The coating material deposits as a layer on the substrate.

Electroless Deposition. Electroless plating (qv) is defined as a controlled, autocatalytic chemical-reduction process for depositing metals. It resembles electroplating because it can be run continuously to build up a thick coating. Unlike displacement processes, it does not involve a chemical reaction with the substrate metal; and unlike the well-known silver-reduction process (Tollens reaction) for the silvering of optical glass, it is selective, ie, deposits form only on a catalytic surface. The metals, nickel, cobalt, platinum, palladium, and gold, and nickel–cobalt alloys can be deposited. Chromium, iron, and vanadium are often claimed, but electroless plating of active metals must be viewed with suspicion, since most such processes are of the displacement type.

Physical Vapor-Deposition Processes (PVD). The three physical vapor-deposition processes are evaporation, ion plating, and sputtering (see Film deposition techniques).

The materials that are deposited by PVD techniques include metals, semiconductors, alloys, intermetallic compounds, refractory compounds, ie, oxides, carbides, nitrides, borides, etc, and mixtures thereof. The source material must be pure and free of gases and inclusions; otherwise spitting may occur.

Metals and Elemental Semiconductors. Evaporation of single elements can be carried out from various evaporation sources subject to the restrictions with regard to melting point, container reactions, deposition rate, etc. A typical arrangement is shown in Figure 1 for electron-beam heating (3). This type of source is ideal for refractory coatings, since the material to be deposited is contained in a noncontaminating water-cooled copper crucible and the surface of the material can easily be heated to >3000°C.

Alloys. Alloys consist of two or more elements with different vapor pressures and hence different evaporation rates. As a result, the vapor phase and, therefore, the deposit constantly vary in compositions. This problem can be solved by multiple sources or a single rod-fed or wire-fed electron-beam source fed with the alloy. These solutions apply equally to evaporation or ion-plating processes.

Multiple Sources. Multiple sources offer a more versatile system. The number of sources evaporating simultaneously is equal to or less than the number of constituents in the alloy. The material evaporated from each source can be a metal, alloy, or compound. Thus, it is possible to synthesize a dispersion-strengthened alloy, eg, Ni–ThO_2. However, the evaporation rate from each source has to be monitored and controlled separately. The source-to-substrate distance would have to be sufficiently large (38 cm for 5-cm-dia sources) to blend the vapor streams prior to deposition, which decreases the deposition rate (see Fig. 2). Moreover, if the density of two vapors differs greatly, it may be difficult to obtain a uniform composition across the width of the substrate because of scattering of the lighter vapor atoms.

Evaporating each component sequentially produces a multilayered deposit that is homogenized by annealing. High deposition rates are difficult to obtain.

Single Rod-Fed Electron-Beam Source. The disadvantages of multiple sources for alloy deposition can be avoided by using a single wire-fed or rod-fed source (see Fig. 3) (3). A molten pool of limited depth is above the solid rod. If the equilibrium

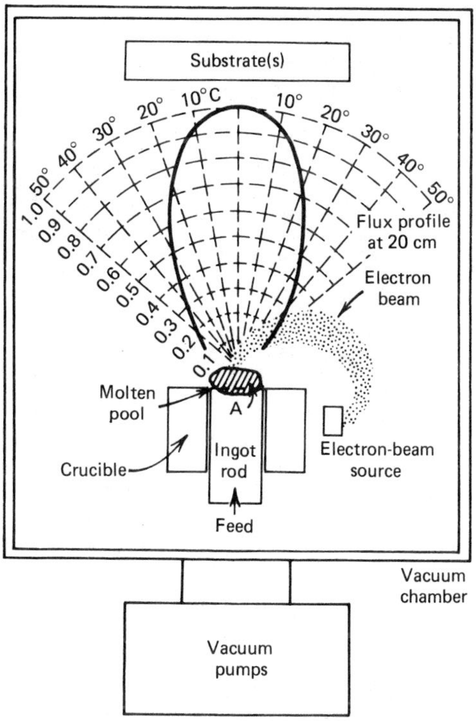

Figure 1. Vacuum-evaporation process with use of electron-beam heating.

vapor pressures of the components of an alloy A_1B_1 are in the ratio of 10:1 and the composition of the molten pool is A_1B_{10}, under steady-state conditions, the composition of the vapor is the same as that of the solid being fed into the molten pool. One can start the procedure with a pellet of appropriate composition A_1B_{10} on top of a rod A_1B_1 to form the molten pool initially, or one can start with a rod of alloy A_1B_1 and evaporate the molten pool until it reaches composition A_1B_{10}. The temperature and volume of the molten pool must be constant to obtain a constant vapor composition. A theoretical model has been developed and confirmed by experiment, and deposits of Ni–20 wt % Cr, Ti–6 wt % Al–4 wt % V, Ag–5 wt % Cu, Ag–10 wt % Cu, Ag–20 wt % Cu, Ag–30 wt % Cu, and Ni–Cr–yAl–zY alloy have been successfully prepared. This method can be used with a 5000-fold vapor-pressure difference between components. It cannot be used when one of the alloy constituents is a compound, eg, Ni–ThO_2.

Sputtering. Sputtering deposits alloys by means of an alloy target. The surface composition of the target changes in the inverse ratio of the sputtering yields of the individual elements, as in the alloy evaporation from a single rod-fed electron-beam source. Alternatively, the sputtering target can be made of strips of the components of the alloy with the respective surface areas inversely proportional to their sputtering yield (see also Film deposition techniques).

Refractory Compounds. Refractory compounds resemble oxides, carbides, nitrides, borides, and sulfides in that they have a very high melting point. In some cases, they form extensive defect structures, ie, they exist over a wide stoichiometric range. For example, in TiC, the C:Ti ratio can vary from 0.5 to 1.0, which demonstrates a wide range of vacant carbon-lattice sites.

In direction evaporation, the evaporant is the refractory compound itself, whereas

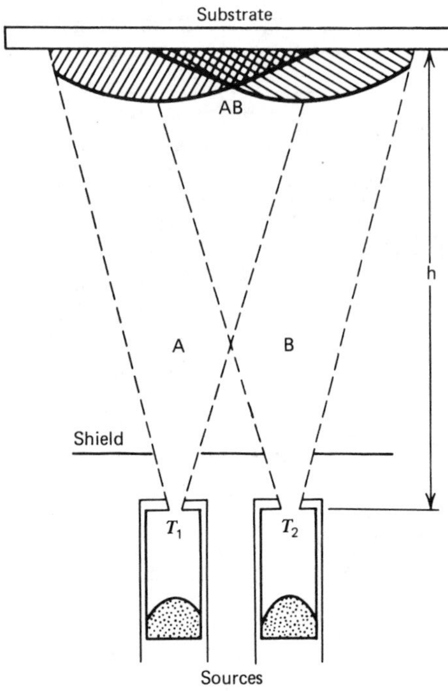

Figure 2. Two-source evaporation.

in reactive or activated reactive evaporation (ARE), a metal or a low-valency metal compound is evaporated in the presence of a partial pressure of a reactive gas to form a compound deposit, eg, Ti is evaporated in the presence of N_2 to form TiN, or Si or SiO is evaporated in the presence of O_2 to form SiO_2.

Direct Evaporation. Evaporation can occur with or without dissociation of the compound into fragments. The observed vapor species show that very few compounds evaporate without dissociation. Examples are MgF_2, B_2O_3, CaF_2, SiO, and other Group-IV divalent oxides (SiO homologues such as GeO and SnO).

In general, when a compound is evaporated or sputtered, the material is not transformed to the vapor state as compound molecules but as fragments thereof. Subsequently, the fragments recombine, most probably on the substrate, to reconstitute the compound. Therefore, the stoichiometry (anion:cation ratio) of the deposit depends on the deposition rate, the ratios of the various molecular fragments, the impingement of other gases present in the environment, the surface mobility of the fragments which, in turn, depends on their kinetic energy and substrate temperature, the mean residence time of the fragments of the substrate, the reaction rate of the fragments on the substrate to reconstitute the compound, and the impurities present on the substrate. For example, direct evaporation of Al_2O_3 results in a deposit deficient in oxygen, ie, that had the composition Al_2O_{3-x}. This O_2 deficiency could be compensated for by introducing O_2 at a low partial pressure into the environment.

Reactive Evaporation. In reactive evaporation (RE), metal or alloy vapors are produced in the presence of a partial pressure of reactive gas to form a compound either in the gas phase or on the substrate as a result of a reaction between the metal vapor and the gas atoms, eg,

$$2\,Ti + C_2H_2 \rightarrow 2\,TiC + H_2$$

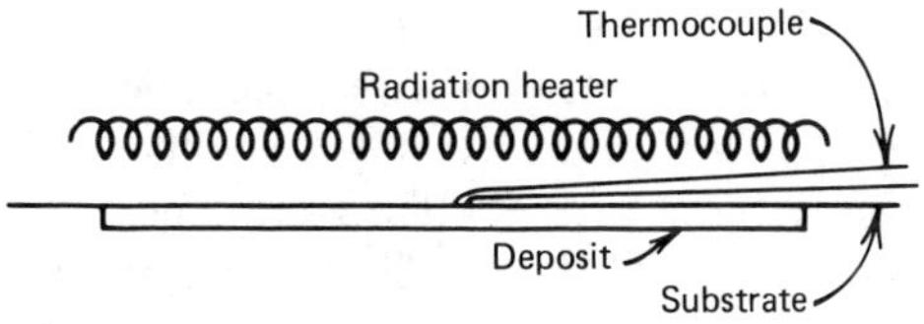

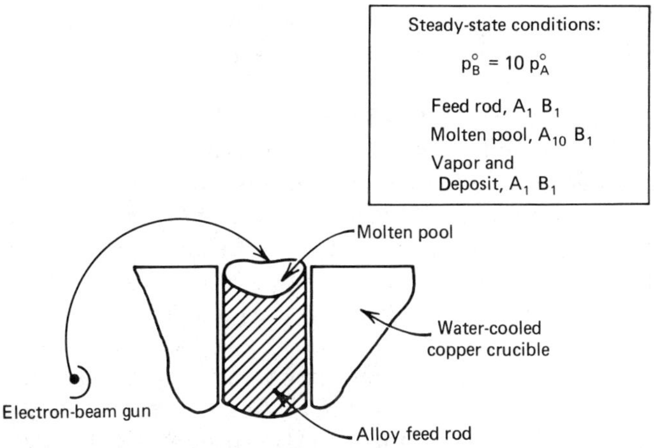

Figure 3. Alloy evaporation from a single rod-fed source. P_B° = the equilibrium vapor pressure of component B, P_A° = the equilibrium vapor pressure of component A.

If the metal and gas atoms are activated or ionized in the vapor phase, which activates the reaction, the process is called the activated reactive process (ARE), as illustrated in Figure 4 (4). The metal is heated and melted by a high acceleration-voltage electron beam. The melt has a thin plasma sheath on top from which low energy secondary electrons are pulled upward into the reaction zone by an electrode placed above the

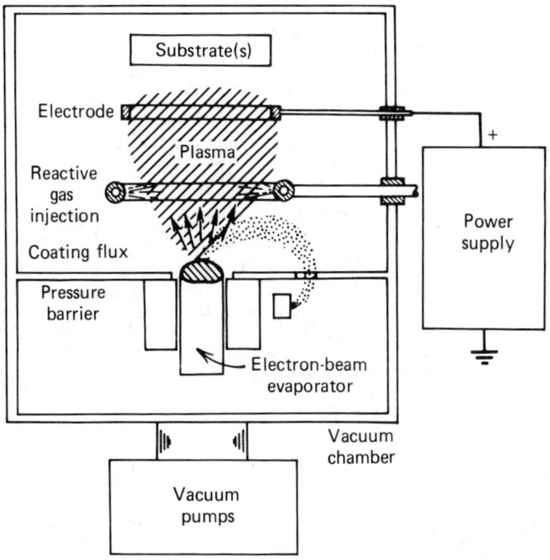

Figure 4. Activated reactive evaporation process (4).

pool; the electrode is biased to a low positive d-c potential (20–100 V). These low energy electrons have a high ionization cross-section, thus ionizing or activating the metal and gas atoms and increasing the reaction probability on collision. Titanium carbide was synthesized with this process by reaction of Ti metal vapor and C_2H_2 gas with a carbon:metal ratio approaching unity. Moreover, by varying the partial pressure of either reactant, the carbon:metal ratio of carbides could be varied at will. This process has been applied recently to the synthesis of the five different Ti oxides (5). With the ARE process (ie, with a plasma), as compared to the RE process (ie, without a plasma), a higher oxide formed for the same partial pressure of O_2, which demonstrates a better gas utilization in the presence of plasma (see Plasma technology, Supplement volume).

Reactive-Ion Plating. In reactive-ion plating (RIP), as in the reactive evaporation process, the metal atoms and reactive gases form a compound aided by the presence of a plasma. Since the partial pressures on the gases are much higher (>1.3 Pa or 10^{-2} mm Hg) than in the ARE process (13 mPa or 10^{-4} mm Hg), the deposits may be porous or sooty. In the simple diode ion-plating process, the plasma cannot be supported at a lower pressure. Therefore, an auxiliary electrode adjusted to a positive low voltage, as originally conceived for the ARE process, is used to initiate and sustain the plasma at a low pressure (ca 0.13 Pa or ca 10^{-3} mm Hg), as shown in Figure 5 (6).

Reactive Sputtering. Reactive sputtering is very similar to reactive evaporation and reactive-ion plating in that at least one coating species enters the system in the gas phase. Examples include sputtering Al in O_2 to form Al_2O_3, Ti in O_2 to form TiO_2, In–Sn in O_2 to form tin-doped In_2O_3, Nb in N_2 to form NbN, Cd in H_2S to form CdS, In in PH_3 to form InP, and Pb–Nb–Zr–Fe–Bi–La in O_2 to form a ferroelectric oxide.

By reactive sputtering, many complex compounds can be formed from relatively easy-to-fabricate metal targets; insulating compounds can be deposited with a d-c

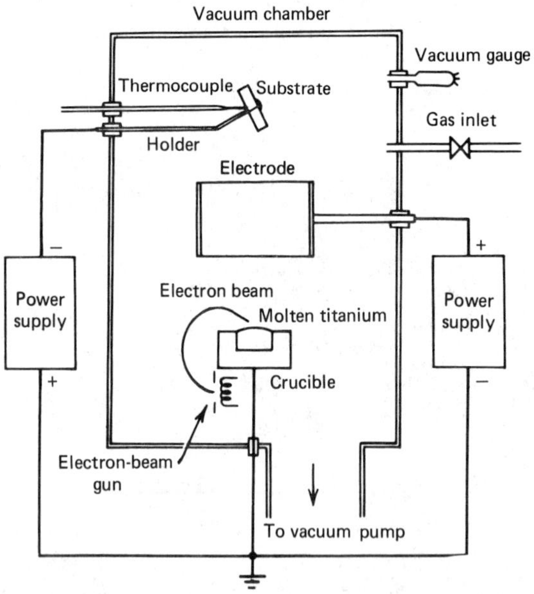

Figure 5. Reactive-ion plating with auxiliary electrode for low pressure operation in deposition of compounds (6). Courtesy of Kobayashi and Doi.

power supply; and graded compositions can be formed, as described in the preceding section. However, the process is complicated.

Chemical Vapor Deposition (CVD) and Plasma-Assisted Chemical Vapor Deposition (PACVD). In CVD, thin films or bulk coatings up to 2.5 cm in thickness are deposited by means of a chemical reaction between gaseous reactants passing over a substrate. Temperatures can be anywhere between 200 and 2200°C but are usually between 500 and 1100°C. The optimum for a given reaction often lies within a very narrow range, and the process needs to be tailored to the substrate and the intended application. The substrate's melting point and susceptibility to chemical attack by the reacting gases or by their side-products has to be considered. Coatings have application in a wide array of corrosion- and wear-resistant uses, but also in decorative layers, semiconductors, and magnetic and optical films (see Film deposition techniques).

In most cases, CVD reactions are activated thermally, but in some cases, notably in exothermic chemical-transport reactions, the substrate temperature is held below that of the feed material to obtain deposition. Other means of activation are available (7), eg, deposition at lower substrate temperatures is obtained by electric-discharge plasma activation. In some cases, unique materials are produced by plasma-assisted CVD (PACVD), such as amorphous silicon from silane where 10–35 mol % hydrogen remains bonded in the solid deposit. Except for the problem of large amounts of energy consumption in its formation, this material is of interest for thin-film solar cells. Passivating films of SiO_2 or SiO_2–Si_3N_4 deposited by PACVD are of interest in the semiconductor industry (see Semiconductors).

Plasma-assisted CVD processes use deposition temperatures lower than CVD processes; the desired deposition reaction is aided by the energy present in the plasma. An example of a PACVD process apparatus is shown in Figure 6 (8). The plasma greatly extends the utility of CVD processes, eg, the ability to deposit films on substrates that cannot withstand the temperature needed for the CVD process in reactions such as polymerization, anodization, nitriding, deposition of amorphous silicon, amorphous carbon (diamondlike carbon), etc. The films deposited by PACVD are generally more

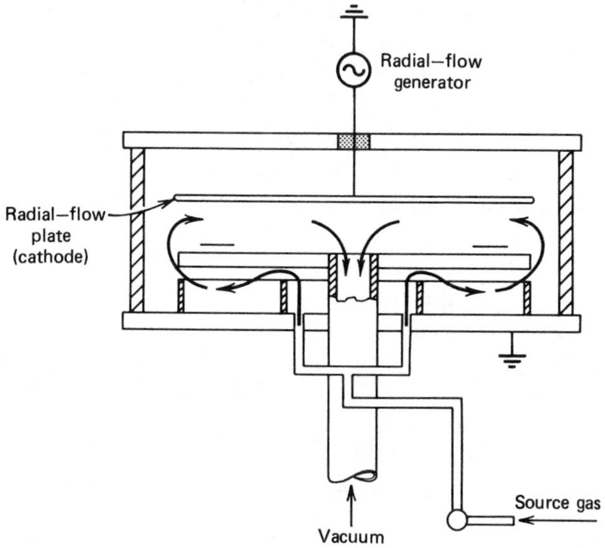

Figure 6. Radial-flow parallel-plate reactor used for the PACVD process (8).

complex than their analogues deposited by normal CVD methods. For example, silicon nitride films have a composition $Si_xN_yH_z$ and may contain as much as 15 to 30 at. % hydrogen as contrasted to the Si_3N_4 composition of the film deposited by the normal CVD process. Other advantages of PACVD are substrate surface etching and activation to produce good bonding at low deposition temperatures.

Ion Implantation. High energy ion implantation is a highly successful process for doping semiconductors to very precisely controlled concentrations (see Ion implantation). The part's surface is bombarded with high energy ions (about 50 keV), which results in a layer of implanted ions at an approximate depth of 8 nm. The potential for producing refractory coatings by implantation of specific metal ions or by reactions within the substrate to form refractory compounds exists but has not been exploited to date.

Particulate Deposition Processes

Melt Spraying. In melt spraying, a combustion flame, electric arc, or arc plasma or explosive-wave front heats particles of the refractory-coating material to a temperature sufficient to achieve sintering or cohesive solidification when the particles impinge on a substrate. Penetration of the substrate often accompanies the coalescence. The techniques include flame plating (D-gun), torch spraying, arc-plasma spraying, and induction-plasma spraying. These processes generate considerable noise, and acoustical protection is needed.

Flame plating employs oxyacetylene fuel. In this method, developed by the Linde Division of the Union Carbide Corporation, the oxyacetylene gas mixture is detonated by an electric spark at four detonations per second. The powders are fed under control into a chamber from which they are ejected when detonation occurs. The molten, 14–16-μm particles are sprayed at a velocity of 732 m/s at distances of 5.1–10.2 cm from the surface. The substrate is moved past the stationary gun.

Torch spraying is the most widely used melt-spraying process. In the power-feed method, powders of relatively uniform size (<44 μm or 325 mesh) are fed at a controlled rate into the flame. The torch, which can be held by hand, is aimed a few cm from the surface. The particles remain in the flame envelope until impingement. Particle velocity is typically 46 m/s, and the particles become at least partially molten. Upon impingement, the particles cool rapidly and solidify to form a relatively porous, but coherent, polycrystalline layer. In the rod-feed system, the flame impinges on the tip of a rod made of the material to be sprayed. As the rod becomes molten, droplets of material leave the rod with the flame. The rod is fed into the flame at a rate commensurate with melt removal. The torch is held at a distance of ca 8 cm from the object to be coated; particle velocities are ca 185 m/s.

Arc-plasma-spraying equipment has been commercially available since 1958. This process does not utilize a fully ionized gas (plasma). Uniform particles (<74 μm or <200 mesh) are fed into the jet that emerges through the nozzle of a high pressure d-c arc. The particles are melted and ejected at controllable velocities of 30–210 m/s. The composition of the atmosphere can be varied over a wide range, ie, wet, dry, oxidizing, reducing (≤80%), or inert. A 40–50 kW, 600-A d-c source supplies the average power requirements. The uv radiation generated during operations presents a safety hazard.

Induction-plasma spraying is still an experimental process. A graphite susceptor is inductively heated until the gas temperature creates a plasma. This plasma subsequently interrupts the inductive field and acts as the susceptor. Powders are fed into the plasma through a duct near the inlet of the plasma where they vaporize. Particle residence time can be varied over a greater range than in the arc-plasma process since gas velocity can be varied by adjusting the nozzle size. The flame temperature decreases rapidly with increasing gas velocity, and this affects the extent to which the particle melts. This apparatus is powered by a 50-kW, high frequency a-c source and cannot be held by hand. As in flame plating, the object is moved in the path of the gun.

Any refractory material that does not decompose or vaporize can be used for melt spraying. Particles do not coalesce within the spray. The temperature of the particles and the extent to which they melt depend on the flame temperature, which can be controlled by the fuel:oxidizer ratio or electrical input, gas flow rate, residence time of the particle in the heat zone, the particle-size distribution of the powders, and the melting point and thermal conductivity of the particle. Quenching rates are very high, and the time required for the molten particle to solidify after impingement is typically 10^{-4}–10^{-2} s.

A broad range of materials can be handled by plasma spraying. However, each material requires some optimization of conditions, such as modification of carrier gas, power, particle-size distribution, and substrate. In some melt-spray processes, a preferential loss of material may occur that leads to a change in stoichiometry. It can be compensated for by adding an appropriate additional gas to the gas stream. Thus, additions of oxygen can keep the reduction of TiO_2 and HfO_2 within acceptable limits, whereas methane reduces the loss of carbon from carbides. A wide selection of coating materials is available. Except for particle-velocity limitations, spray conditions are excellent because of the inert, clean carrier gases used. Inert atmospheres reduce the degradation or oxidation of the particles, preclude oxidation of the substrate, and diminish contamination of the sprayed material resulting from gas adsorption.

Torch spraying and arc-plasma spraying are the least expensive. Flame plating requires little substrate preparation, but is otherwise expensive. The adherence of melt-sprayed ceramic coatings onto metallic substrates varies widely from process to process and depends on procedure and substrate preparation (see Table 2). The high strength of flame-plated coatings is a consequence of the higher particle velocity that results in penetration of the steel surface. For other melt-spray processes, the substrate must be roughened to provide a suitable surface. Chemical bonding contributes to adhesion but the mechanical bond is preferred. Advantages and disadvantages of melt-spray processes are tabulated in Table 3.

Table 2. Tensile-Bond Test for Melt-Sprayed Al_2O_3

Method	Particle velocity, m/s	Bond strength, MPa[a]
torch spraying	46–83	3.4–13.8
flame plating	732	69
arc-plasma process	30–213	6.9–69
induction-plasma process	6–>300	na[b]

[a] To convert MPa to psi, multiply by 145.
[b] Na = not applicable.

Table 3. Melt-Spray Processes[a]

Advantages	Disadvantages
produces coatings on any substrate	undercuts, deep blind holes, and small internal diameters are difficult to coat
versatility in configuration size, shape, and structure	properties are generally inferior
good dimensional control, no shrinkage	nonstoichiometry may result
mixed compositions	residual stresses can limit coating thickness
fine grain size	
permits patching and repairing	
alternating layers	
graded coatings	
by masking, localized areas can be coated	

[a] Porosities are 0.5–20.0%.

Electrophoretic Processes. Electrophoretic coatings are obtained through the migration of charged particles when a potential is applied to electrodes immersed in a suitable suspension of the particles (see Electromigration). This process is particularly suited for applying uniform layers on complex bodies. Projecting edges become insulated and the current shifts to bare surfaces as the deposit builds up on them. Particles, not ions, are deposited. The early stages of coating are powdery, and densification is required to produce adherence. Electrophoresis can be used to apply metals and alloys, ceramics, and cermets. The three steps are preparation of the dispersion, deposition and conversion, and bonding. In the preparation phase, particle size is critical and diameters may be from 1 to 50 μm; however, too small a size leads to cracking after thick coatings are dried and consolidated.

Several suspension media with suitable viscosities, densities, conductivities, dielectric constants, and chemical stabilities are available. Binders improve the green strength of the coating. Deposition is effected rapidly with potentials of 50–1000 V dc. Simple d-c sources with no special filtering are sufficient. All inorganic materials are amenable to deposition and most are amenable to codeposition, even though they have different densities; for example, NiO (7.5 g/cm^3) and WC (15.7 g/cm^3) have been codeposited in the same ratio as they were dispersed. The practical limit of thickness is about 0.5 mm as set by shrinkage cracking. Conversion and bonding techniques vary with the type of coating. Hydrogen reduction, hydrostatic pressing, and sintering are used for most metal-base coatings. The oxides of iron, nickel, cobalt, molybdenum, and tungsten can be reduced in hydrogen at less than 600°C. For example, NiO–Cr green deposits are reduced in H_2 at 320°C to produce metallic nickel, with no particle sintering. After pressing at 69 MPa (10,000 psi), further densification is achieved by sintering to 90% of theoretical density at 1090°C. Vacuum sintering is preferred for titanium, zirconium, niobium, and tantalum. Glass and ceramic coatings can be either vitrified or fused.

Bulk Coating

In bulk-coating processes, bulk materials are joined to the substrate either by a surface-melt process or by attachment of the solid material. A recent example of the latter is the application of heat-resistant tiles of silica-type material to the aluminum-alloy skin of a space-shuttle vehicle; this enables the vehicle to withstand the reentry heat.

In cladding, one metal is coated with another by rolling or extruding the two metals in close contact to each other. Coherence of the two metals is induced by soldering, welding, or casting one in contact with the other prior to the rolling operation. By this mechanial process, steel can be coated with copper, nickel, or aluminum. The coextrusion process, which is used for lightweight rifle barrels made of a steel core and titanium alloy wrap, is a good example of the cladding process.

Another familiar commercial method is the immersion or hot-dipping process. The article to be coated is immersed in a molten metal bath. Usually little else is done to change the properties of the coating, which adheres to the surface upon removal of the article from the bath. For a successful coating, an alloying action must take place between the components to some extent. Zinc and tin coatings are applied to sheet steel by hot-dipping.

Surface coatings are also applied by welding processes, such as manual arc welding (oxy–acetylene, gas–tungsten), gas–metal arc welding, submerged-arc welding, spray welding, plasma-arc welding, and electro-slag welding (see Welding).

Thin coatings, 0.25–0.75 mm, are applied by spray welding. For heavier coatings in the range of 6 to 65 mm or more, other welding processes are used.

These welding processes are not suitable for the application of refractory coatings of reactive metals such as tungsten, tantalum, niobium, and chromium, although electro-slag welding is being studied for this purpose. On the other hand, such refractory-metal, alloy, and cermet coatings can be applied by electron-beam melting to form thick coatings, since this process is carried out in vacuum.

A new development in this area is the use of high power laser beams to surface-melt refractory-metal or cermet coatings onto substrates in a controlled atmosphere to prevent contamination of the reactive metals. This process is sometimes referred to as laser glazing (see Lasers).

Enameling meets decorative as well as protective requirements. Ceramic enamels are mainly based on alkali borosilicate glasses. The part to be enameled is dipped into or sprayed with a slip, ie, a water suspension of glass fragments called frit. The slip coating is dried and fused in a enameling furnace under careful heat control (see Enamels, porcelain or vitreous).

Troweling and painting methods are used to apply thick protective coatings of a refractory paste or cermet onto a variety of substrates for high temperature service. Fiberfrax (The Carborundum Co.) coating cements, composed of aluminosilicate fibers bonded with air-setting temperature-resistant inert binders, is commonly used as a coating for reducing the oxidation rate of graphite.

In the microelectronics industry, powdered metals and insulating materials that consist of nonnoble metals and oxides are deposited by screen printing in order to form coatings with high resistivities and low temperature coefficients of resistance. This technique may be useful in depositing oxide–metal refractory coatings.

Surface-Modification Processes

Cementation (Diffusion) Coatings. Cementation is defined as the introduction of one or more elements into the outer portion of a metal object by means of diffusion at elevated temperatures. Cementation was first used to convert iron to steel, and copper to brass, but today it is considered only as a method of surface treatment. The coating produced by cementation is formed by an alloying or chemical combination of the diffusing elements with the substrate material. Cementation coatings enjoy wide metallurgical application. A prime example is the case hardening of steel whereby a soft, ductile, low-carbon steel is heated to 810–900°C in a packing of carbonaceous material to produce a high carbon steel surface (see Steel). The coating formed in this manner is limited to use at temperatures of about 600°C or lower. Truly refractory coatings are alloyed onto molybdenum, tantalum, niobium, and tungsten by cementation. Such coatings provide short-term protection at 1650°C or higher, and long-term protection at 1370°C.

Cementation coatings are produced at temperatures well below the melting points of either the coating or substrate material by means of a vapor-transport mechanism. The coating material and the substrate form definite chemical compounds, such as silicides, aluminides, beryllides, or chromides. On the other hand, some cementation coatings can be solid solutions of indefinite composition. Oxidation-resistant coatings act as diffusion barriers for both the inward diffusion of oxygen and the outward diffusion of the substrate. The diffusion of the coating components into the substrate is generally the rate-controlling step. Cementation coatings are applied by pack cementation, activated or nonactivated slurry processes, and the fluidized-bed technique.

In pack cementation, the part to be coated is placed in a retort and surrounded with a powdered pack consisting of the coating component and an activator; the latter reacts with the coating component to form the carrier vapor, usually a halide or an inert diluent, to prevent the pack from sintering together and to permit vapor transport of the alloying component through the pack.

The slurry process requires less coating component. The latter is suspended in a vehicle, eg, lacquer or water, and is painted onto the substrate. The coated part is heated in an alumina retort containing a layer of activator at the bottom. The coating component forms a halide and is deposited onto and diffused into the substrate. Slurry processes can be either activated or nonactivated. In the latter case, development of the coating relies purely upon diffusion without the possible benefits of vapor deposition.

The fluidized-bed technique combines aspects of pack cementation and vapor deposition. A fluidized bed consists of a mass of finely divided solids contained in a column. The solids are brought into a fluidized state by the lifting action of a gas as it rises through the column. A vaporized halide may be carried by a gas into the bed, where it reacts with the fluidized coating powder to form the coating component halide, which then thermally decomposes and deposits on the substrate contained within the retort. Alternatively, the metal coating particles may be fluidized before entering the bed, whereby the halide gas permeates the fluidized particles to form the coating halide vapor. This vapor is carried into the bed of inert particles, where it thermally decomposes to deposit the coating (see Fluidization).

The chromizing of iron is described by three mechanisms:

Displacement	$Fe(alloy) + CrCl_2(g) \rightarrow FeCl_2(g) + Cr(alloy)$
Reduction	$H_2(g) + CrCl_2(g) \rightarrow 2\ HCl(g) + Cr(alloy)$
Thermal Decomposition	$CrCl_2(g) \rightarrow Cl_2(g) + Cr(alloy)$

The displacement mechanism involves placing the iron alloy packed in chromium powder, NH_4Cl, and Al_2O_3 in a sealed retort, which is heated to promote vapor-deposition and diffusion processes. The exact chemistry is not known, but the following steps probably occur:

$$NH_4Cl(g) \rightarrow NH_3(g) + HCl(g)$$
$$2\ NH_3(g) \rightarrow N_2(g) + 3\ H_2(g)$$
$$Cr(powder) + 2\ HCl(g) \rightarrow CrCl_2(g) + H_2(g)$$
$$CrCl_2(g) + Fe(alloy) \rightarrow FeCl_2(g) + Fe\text{–}Cr(alloy)$$

Several thermodynamic and kinetic requirements must be fulfilled for the reactions to proceed as shown. First, the vapor pressure of the activators or carriers must be sufficiently high at the coating temperature for the reaction to proceed. The activator, ie, the chloride, should have a boiling point slightly above the temperature of the coating to provide a reservoir for reaction without volatilizing too rapidly. The coating-component halide should have a boiling point below the coating temperature to saturate the pack. The volatility of the by-product of the coating reaction must be high; fast removal prevents the formation of a barrier to continued deposition of the coating element and also prevents contamination of the coating.

Another example is the silicidizing of tantalum, basically an oxidation–reduction reaction. The packing is sodium fluoride and silicon. After deposition, the coating diffuses continuously into the substrate, according to the following reactions:

$$6\ NaF(g) + 2\ Si(l) \rightarrow Si_2F_6(g) + 6\ Na(g)$$
$$Ta + 2\ Si_2F_6(g) \rightarrow 3\ SiF_4(g) + Si(TaSi_2)(alloy)$$
$$3\ SiF_4(g) + Si(l) \rightarrow 2\ Si_2F_6(g)$$

If the rules for volatilities and thermodynamics of the halides are followed, the reaction can be used for aluminizing, silicidizing, chromizing, and similar processing.

Silicide coatings of refractory metals may contain as much as three to five coating components other than silicon. A mixture of halide carriers is selected containing the best carrier for each component.

The outstanding characteristics of a fluidized bed are its high heat-transfer coefficient and its turbulence, which yield optimum temperature uniformity throughout the bed. These factors contribute to the successful treatment of large, complex objects, which might not be possible by other means.

Analogous to the carburization of steel is the nitrogen case-hardening of steel in which surfaces are hardened by heating the steel in the presence of nascent nitrogen. The nitrogen reacts with impurities or alloy constituents, eg, aluminum, chromium, vanadium, tungsten, etc, to form dispersed nitrides. Other examples of cementation coatings include the surface-hardening of steel by immersion in a molten sodium cyanide bath at 850–870°C to promote codiffusion of both carbon and nitrogen into the surface. Iron and steel are cemented with silicon at 800°C, and iron, nickel, and copper with tungsten at 800–1350°C. Similarly, iron and nickel are cemented with molybdenum by heating in ferromolybdenum. In the same temperature range, titanium coatings are obtained on ferrous metals by heating them for 1.5 h in a powdered mixture of sponge titanium containing 0.5–6.0% Fe at 800–1200°C. Boride coatings are produced by heating iron, cobalt, or nickel substrates in boron powder at 950°C in a vacuum of 67 mPa (5×10^{-4} mm Hg).

The most advanced cementation coatings are the intermetallic coatings, specifically silicides and aluminides, that protect refractory metals. The earliest and simplest of these is the molybdenum silicide coating developed for molybdenum substrates. It consists largely of $MoSi_2$ but is modified by additions of boron, manganese, titanium, chromium, beryllium, etc, singly or in combinations. Coating systems for niobium are more complex, but oxidation-resistant coatings result by diffusing silicon, chromium, aluminum, boron, titanium, and beryllium singly or in combinations into the substrate. Diffusion coatings for tantalum are based on the formation of $TaAl_3$, an oxidation-resistant material. Silicide coatings on tantalum also provide significant oxidation protection. Aluminum, boron, or manganese are often added as modifiers. Tungsten presents a more difficult problem, but various silicide coatings provide a measure of protection at temperatures as high as 1815°C. After 10 h at this temperature, WSi_3 is converted to the glassy $W_5Si_4O_2$ composition.

An activated-slurry process can be used to place impervious silicon carbide coatings on graphite. A slurry of silicon carbide, carbon, and appropriate organic binders is applied to the surface of the graphite by spraying, dipping, or painting. After a low temperature treatment to drive off the binder, the coated substance is heated in the presence of silicon vapor, which diffuses into the surface to form a SiC layer on the graphite. The coating composition can vary from self-bonded silicon carbide containing only a few percent of uncombined silicon to SiC crystals bonded with a continuous silicon matrix. Successful SiC coatings on complex graphite shapes have been obtained with isotropic graphite substrates possessing compatible thermal-expansion coefficients. Resistance to oxidation at 1400°C for 100 h or more has been accompanied by successful service in nuclear-reactor environments. Coatings of SiC on graphite fail under compressive loads or impact at levels that do not damage uncoated graphite. This occurs because the graphite is less brittle and deforms under load, whereas the thin, more rigid SiC coating does not.

Aluminide and silicide cementation coatings such as $TaAl_3$ on tantalum and $MoSi_2$ on molybdenum oxidize at slow rates and possess some inherent self-repair characteristics. Fine cracks that appear and are common to these coatings can be tolerated because stable, protective oxides form within the cracks and seal them. Thermal cycling, however, accelerates failure because of thermal-expansion mismatch that ultimately disrupts the protective oxide coating.

An important application is the aluminizing of air foils of gas-turbine engines made of high temperature Ni- or Co-base alloys. The aluminizing can be carried out either in a pack process or in an out-of-the-pack process.

Cementation coatings rely on diffusion to develop the desired surface alloy layer. Not only does the coating continue to diffuse into the substrate during service, thereby depleting the surface coating, but often the substrate material diffuses into the surface where it can be oxidized. Since the diffusion rate is temperature dependent, this may occur slowly at lower service temperatures.

The substrate has to be prepared for cementation. The surface must be clean and free of oxide. Corners and edges are particularly important in diffusion-type coatings; sharp edges are usually detrimental. Barrel finishing, ie, tumbling in a barrel with abrasive media, may result in the desired shape.

The quality of cementation coatings does not necessarily equal that produced by other techniques. There are cases, however, where a moderate degree of corrosion resistance is useful and where other requirements are best met by the application of cementation. With other processes, it may be difficult to coat porous surfaces or preserve the contour of machined surfaces. Cementation may then be the method of choice. Coatings for molybdenum, niobium, tantalum, and tungsten not easily obtained by other means can be achieved by this method. Coated-refractory-metal systems have been applied to leading edges, skins, and structural members of space vehicles, rocket combustion chambers, nozzle inserts, extension skirts, and to vanes or blades for advanced gas-turbine engines.

Metalliding. Metalliding, a General Electric Company process (9), is a high temperature electrolytic technique in which an anode and a cathode are suspended in a molten fluoride salt bath. As a direct current is passed from the anode to the cathode, the anode material diffuses into the surface of the cathode, which produces a uniform, pore-free alloy rather than the typical plate usually associated with electrolytic processes. The process is called metalliding since it encompasses the interaction, mostly in the solid state, of many metals and metalloids ranging from beryllium to uranium. It is operated at 500–1200°C in an inert atmosphere and a metal vessel; the coulombic yields are usually quantitative, and processing times are short; controlled uniform coatings from a few μm to many μm are obtained, many of which are unavailable by other techniques. Diffusion rates are high and the process can be run continuously. Boron and silicon anodes can be diffused into most metals of groups VB, VIB, VIIB, VIII, and IB of the periodic table.

The borides are extremely hard (9.8–29 GPa or 1000–3000 kgf/mm^2, Knoop) and, in the case of molybdenum, >39 GPa (4000 kgf/mm^2) (see Hardness). However, oxidation resistance is usually poor unless a subsequent coating is formed, such as siliciding or chromizing, which imparts oxidation resistance. Silicides are generally very oxidation resistant, but not as hard as borides. Silicide coatings formed on molybdenum (51 μm in 3 h) at 675°C have superior oxidation resistance. At these low temperatures, the molybdenum substrate does not embrittle and the coatings are quite flexible.

The metals that can be aluminized act similarly with boron, as do the metals in group IVB; resistance to oxidation is the principal benefit. Titanizing and zirconizing are extremely sensitive to oxygen impurities, and when the salts are competely free of oxides and blanketed by high purity argon, excellent diffusion coatings can be formed in LiF at 900–1100°C. The most promising area for applications appears to be with nickel- and iron- based alloys where intermetallic-compound formation gives rise to many unique coatings that are tough and oxidation resistant. Beryllide coatings can be formed on approximately forty metals ranging from titanium to uranium and with compositions such as $TiBe_{12}$, Ni_5Be_{21}, and UBe_{13}. They are very hard, usually oxidation resistant, and easily formed up to several mils in thickness.

Moving further to the right in the periodic table, the scope of the metalliding processes becomes much more limited. Iron, cobalt, and nickel are restricted to approximately twelve metals into which they can be diffused. Iron can be diffused into cobalt and nickel with very good results, but the reverse is unsuccessful. The diffusion of nickel into molybdenum, tungsten, and copper has been very successful. Germanium occupies a slightly higher position in the electromotive series than nickel and therefore diffuses into nickel; it is below cobalt and iron, however, as determined by voltage measurements in the salts, and cannot be diffused into either of these metals, even at high voltages and current densities. It is readily diffused into molybdenum, palladium, copper, platinum, gold, and Monel.

Microstructure of Coatings

The microstructure of bulk coatings resembles the normal microstructure of metals and alloys produced by melt solidification. The microstructure of particulate-deposited materials resembles a cross between rapidly solidified bulk materials with severe deformation and powder compacts produced by pressing and sintering. A special feature of particulate coatings is a significant degree of porosity (ca 2–20 vol %) that strongly affects the properties of the deposit.

The microstructure and imperfection content of coatings produced by atomistic deposition processes can be varied over a very wide range to produce structures and properties similar to or totally different from bulk-processed materials. In the latter case, the deposited materials may have high intrinsic stress, high point-defect concentration, extremely fine grain size, oriented microstructure, metastable phases, incorporated impurities, and macro- and microporosity, all of which may effect the physical, chemical, and mechanical properties of the coating.

Microstructure

PVD Condensates. Physical-vapor-deposition condensates can deposit as single-crystal films on certain crystal planes of single-crystal substrates, ie, by epitaxial growth or, in the more general case, the deposits are polycrystalline. In the case of films deposited by evaporation techniques, the main variables are the nature of the substrates; the temperature of the substrate during deposition; the rate of deposition; and the deposit thickness. Contrary to what might be expected, the deposit does not initially form a continuous film of one monolayer and grow. Instead, three-dimensional nuclei are formed on favored sites on the substrates, such as cleavage steps on a single-crystal substrate. These nuclei grow laterally and in thickness (growth state), ultimately impinging on each other to form a continuous film. The average thickness at which a continuous film forms depends on the nucleation density and the deposition temperature and rate; both influence the surface mobility of the adatom. This thickness varies from 1 nm for Ni condensed at 15 K to 100 nm for Au condensed at 600 K.

The microstructure and morphology of thick single-phase films have been extensively studied for a wide variety of metals, alloys, and refractory compounds. The structure model first proposed is shown in Figure 7; it was subsequently modified as shown in Figure 8 (10–11).

At low temperatures, the surface mobility of the adatoms is limited and the

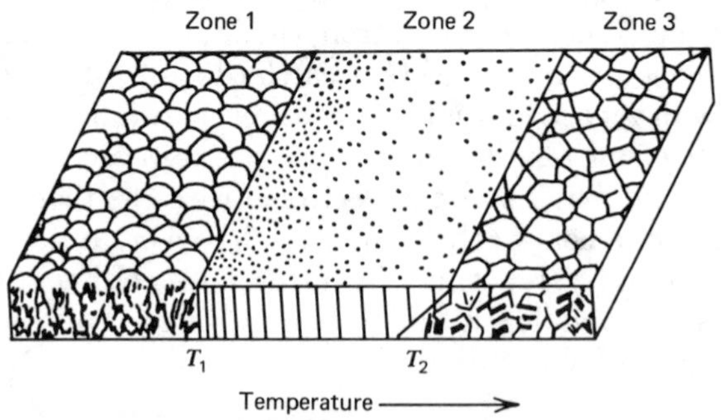

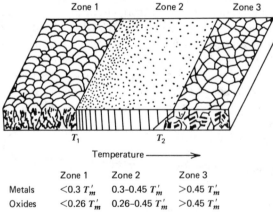

	Zone 1	Zone 2	Zone 3
Metals	$<0.3\ T'_m$	$0.3\text{–}0.45\ T'_m$	$>0.45\ T'_m$
Oxides	$<0.26\ T'_m$	$0.26\text{–}0.45\ T'_m$	$>0.45\ T'_m$

Figure 7. Structural zones in condensates at various substrate temperatures (10).

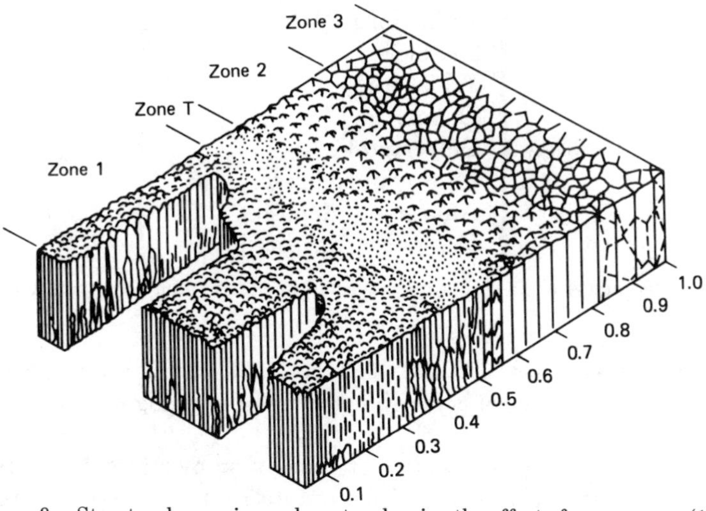

Figure 8. Structural zones in condensates showing the effect of gas pressure (11).

structure grows as tapered crystallites from a limited number of nuclei. It is not a full-density structure but contains longitudinal porosity on the order of a few tens of nm width between the tapered crystallites. It also contains numerous dislocations with a high level of residual stress. Such a structure has also been called botryoidal and corresponds to Zone 1 in Figures 7 and 8.

As the substrate temperature increases, the surface mobility increases and the structural morphology first transforms to that of Zone T, ie, tightly packed fibrous grains with weak grain boundaries, and then to a full-density columnar morphology corresponding to Zone 2 (see Fig. 8).

The size of the columnar grains increases as the condensation temperature increases. Finally, at still higher temperatures, the structure shows an equiaxed grain morphology, Zone 3. For pure metals and single-phase alloys, T_1 is the transition temperature between Zone 1 and Zone 2 and T_2 is the transition temperature between Zone 2 and Zone 3. According to the original model (10), T_1 is 0.3 T_m for metals and 0.22–0.26 T_m for oxides, whereas T_2 is 0.4–0.45 T_m for both (T_m is the melting point in K).

The modification shows that the transition temperature may vary significantly from those stated above and in general shift to higher temperatures as the gas pressure in the synthesis process increases. The transition from one zone to the next is not abrupt, but smooth. Hence, the transition temperatures should not be considered as absolute but as guidelines. Furthermore, not all zones are found in all types of deposit. For example, Zone T (Fig. 8) is not prominent in pure metals, but becomes more pronounced in complex alloys, compounds, or in deposits produced at higher gas pressures. Zone 3 is not seen very often in materials with high melting points.

CVD Coatings. As in PVD, the structure of the deposited material depends on the temperature and supersaturation, roughly as pictured in Figure 9 (12); however, in the case of CVD, the effective supersaturation (ie, the local effective concentration in the gas phase of the materials to be deposited, relative to its equilibrium concentration) depends not only on concentration, but on temperature, since the reaction is thermally activated. Since the effective supersaturation for thermally activated

Increasing supersaturation ↓		Increasing temperature ↑
	Epitaxial growth	
	Platelets	
	Whiskers	
	Dendrites	
	Polycrystals	
	Fine-grained polycrystals	
	Amorphous deposits	
	Gas-phase nucleated snow	

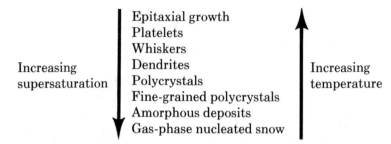

Figure 9. Morphological effects of supersaturation and temperature on vapor-deposited materials (12).

reactions increases with temperature, the opposing tendencies can lead in some cases to a reversal of the sequence of crystalline forms listed in Figure 9, as temperature is increased (12).

Growth of columnar grains is characteristic of many materials in certain ranges of conditions. This structure results from uninterrupted growth toward the source of supply. Where growth in one crystallographic direction is preferred over others, grains having that orientation engulf those of other orientations.

Electrodeposits. *Columnar* structures are characteristic of deposits from solutions (especially acid solutions) containing no additives, high metal-ion concentration solutions with high deposition rates, or from low metal-ion-concentration solutions at low deposition rates. They usually exhibit lower tensile strength, elongation, and hardness than other structures, but are generally more ductile. Such deposits are usually of highest purity (high density) and low electrical resistivity.

Fibrous structures represent a grain refinement of columnar structure. Stress-relieving additives, eg, saccharin or coumarin, promote such refinement, as do high deposition rates. These may be considered intermediate in properties between columnar and fine-grained structures.

Fine-grained deposits are usually obtained from complex-ion solutions, eg, cyanide, or with certain addition agents. These deposits are less pure, less dense, and exhibit higher electrical resistivities because of the presence of foreign material.

Banded structures are characteristic of some alloy deposits and of bright deposits resulting from brightening addition agents. Plating-current modifications (P.R., periodic reverse, IC, interrupted current, pulse) favor the conversion of normal structure from a solution to a banded structure. These deposits generally possess higher tensile strength, hardness, and internal stress and lower ductility than the other structures.

Grain size varies widely, from 10 to 5000 nm; the grain size of fine-grained or banded deposits is usually 10–100 nm. Some metals, notably copper, nickel, cobalt and gold, can be deposited in all four types of grain structure, depending on the solution composition and plating conditions.

Characterization and Testing

Evaluation for high temperature service has been less reliable and less standardized than evaluations for conventional service. This is partly the result of the difficulty of reproducing in the laboratory the severe service conditions capable of yielding acceptable correlations. Test conditions do not necessarily simulate the geometrical and environmental conditions of the service. Nevertheless, screening tests yield primary behavioral parameters that usually define the limits of operation (see Analytical methods). Various tests are summarized in Tables 4 and 5.

Selection Criteria

The selection of a particular deposition process depends on the material to be deposited and its availability; rate of deposition; limitations imposed by the substrate, eg, maximum deposition temperature; adhesion of deposit to substrate; throwing power; apparatus required; cost; and ecological considerations. Criteria for CVD, electrodeposition, and thermal spraying are given in Table 6 (13).

Applications

Coatings can be classified into six categories: chemically functional, mechanically functional, optically functional, electrically functional, biomedical, and decorative. In addition, there are some unique applications in the aerospace program, such as the

Table 4. Characterization of Refractory Coatings

Test	Factors evaluated
optical and electron microscopy metallographic and microscopic observation	substrate and coating structures; coating thickness; bond characteristics; detection of inclusions
physical electron microscope, SEM and TEM[a]	detection of injurious inclusions in substrate or coating
x-ray diffraction electron diffraction	effects of processing on coating composition; composition of coating at various depths within coating can be determined by controlled polishing followed by x ray
chemical microscopy edax aes[b] esca[c] sims[d]	chemical analysis of surface and subsurface layers; resolution can be as low as 5-nm segregation at imperfections
mechanical bend tests tensile tests	adhesion; ductility effect of coating or processing on base-material strength; ductility; elongation
microhardness transverse	cross-sectional hardness of coating and substrate; effect of processing on substrate; ductility
fatigue tests	effect of coating or coating process on substrate; fatigue properties; fatigue strength of coated-metal system
thermal chemical oxidation thermal cycling with sustained load in air	oxidation and thermal-shock resistance; effect of coating on ductility of base and of specimen creep on coating; integrity; thermal-shock and oxidation resistance
plasma-arc or oxyacetylene-torch test	coating emissivity; thermal shock and oxidation resistance; melting point

[a] Scanning electron microscopy; transmission electron microscopy.
[b] Auger electron spectroscopy.
[c] Electron spectroscopy for chemical analysis.
[d] Secondary-ion mass spectroscopy.

Table 5. Nondestructive Tests for Refractory Coatings

Method	Factors evaluated
evaporgraph[a]	discontinuities on flat panel surfaces[c]
electromagnetic inspection	pores, cracks, pits[c]
ultrasonic inspection	surface and subsurface flaws; unsuitable for thin skins[c]
fluorescent particle inspection[b]	surface cracks, pits, and similar coating defects
red-dye penetrant	surface flaws
radiographic inspection	small coating flaws in assembled structures; sensitivity is difficult to control
microscopy	flaws by observation of oxidation and weakness in bond strength resulting from thermal shock[d]

[a] Detection of residual moisture left behind in cracks, pores, or flaws.

[b] Finely divided, fluorescent-coated magnetic particles are attracted to and outline the pattern of any magnetic-leakage fields created by discontinuities.

[c] Unsuitable for corner or edge defects.

[d] Very reliable after an exposure test.

ablative coatings of pyrolytic carbon and graphite- and silica-based materials for protection of nose cones and the space shuttle during reentry (see Ablative materials). Another unique energy related application is the coating of low-Z elements such as TiC for the first wall of thermonuclear reactors to minimize contamination of the plasma.

Chemically Functional. Refractory coatings are used for corrosion-resistant high temperature service in gas turbine and diesel engines, components such as crucibles, thermocouple protection tubing, valve parts, etc.

Blades and vanes used in the hot-end of a gas turbine are subject to high stresses in a highly corrosive environment of oxygen, sulfur, and chlorine-containing gases. A single or monolithic material such as a high temperature alloy cannot provide protection against both. A bulk alloy designed for its mechanical properties provides the corrosion resistance by means of an overlay coating of an M–Cr–Al–Y alloy where M stands for Ni, Co, Fe or Ni + Co. In production, the coating is deposited by electron-beam evaporation; in the laboratory, by sputtering or plasma-spraying. These overlay coatings have several advantages over diffusion aluminide coatings. The latter lose their effectiveness at higher temperatures because of interactions with the substrate. The composition and properties of overlay coatings can be more easily tailored to the needs of specific applications. In addition, coatings of stabilized zirconia are used as thermal barriers in diesel engines and gas turbines, experimentally at this time, to raise operating temperature of the engine and protect it from corrosive fuels.

Boron nitride and titanium diboride coatings are used on graphite for the evaporation of liquid aluminum.

Mechanically Functional. Refractory coatings are used in engine parts, landing gears, soft-film lubricants, and cutting and forming tools (see Tool materials).

A large and rapidly growing application is the coating of cutting and forming tools and industrial knives with carbides, nitrides, oxides, or multiple layers of these materials. These coatings are deposited by CVD and PVD methods and increase the tool life by factors ranging from 2 to 10, depending on the operating conditions. Cermet coatings such as TiB_2–Ni are also deposited electrolytically to provide wear resistance.

Table 6. Characteristics of Deposition Processes

Characteristic	Evaporation	Ion plating	Sputtering	Chemical vapor deposition	Electro-deposition	Thermal spraying
mechanism of production of depositing species	thermal energy	thermal energy	momentum transfer	chemical reaction	deposition from solution	from flames or plasmas
deposition rate	can be up to 75,000 nm/min	can be up to 25,000 nm/min	low except for pure metals[a]	moderate, 20–2500 nm/min	low to high	very high
depositing specie	atoms and ions	atoms and ions	atoms and ions	atoms	ions	droplets
throwing power						
for complex shaped object	poor line-of-sight coverage except by gas scattering	good, but thickness distributions nonuniform	good, but thickness distribution nonuniform	good	good	none
into small blind holes	poor	poor	poor	limited	limited	very limited
deposition of						
metal	positive	positive	positive	positive	positive but limited	positive
alloy	positive	positive	positive	limited	limited	positive
refractory compound	positive	positive	positive	positive	limited	positive
energy of depositing species	ca 0.1–0.5 eV	1–100 eV	1–100 eV	high with PACVD	can be high	can be high
bombardment of substrate and deposit by inert gas ions	normally not	yes	possible, depending on geometry	possible	none	positive
growth interface perturbation	normally not	yes	yes	yes, by rubbing	none	none
substrate heating by external means	normally yes	yes	generally not	no	none	normally not

[a] For copper, 1000 nm/min.

These coatings are also employed as solid lubricants in engine components. Refractory materials such as MoS_2 and WSe_2 are lamellar compounds and provide very effective solid-state lubrication in spacecraft bearings and components used in radiation environments where conventional organic liquid lubricants are not stable (see Lubrication and lubricants).

Optically Functional. Laser optics, layer architectional glass panels (up to 3×4.3 m), lenses, TV-camera optical elements, and similar applications require optically functional coatings.

A large and expanding operation is coatings on architectural glass panels used in buildings to alter the transmission and reflection properties of glass in various wave-length ranges and thus conserve energy usage. Furthermore, attractive visual effects, such as a bronze appearance, can be achieved. These coatings consist of multiple layers of oxides and metals (the precise compositions are proprietary) and are deposited in large in-line sputtering and evaporation systems. Future applications will utilize the selective transmission of coatings of nitrides, carbides, and borides in solar-thermal applications.

Another growing application that overlaps the electrically functional area is the use of transparent conductive coatings or tin oxide, indium–tin oxide, and similar materials in photovoltaic solar cell and various optic electronic applications (see Photovoltaic cells). These coatings are deposited by PVD techniques as well as by spray-pyrolysis, which is a CVD process.

Electrically Functional. Refractory coatings are used in semiconductor devices, capacitors, resistors, magnetic tape, disk memories, superconductors, solar cells, and diffusion-barriers to impurity contamination from the substrate to the active layer.

Thin-film capacitors and resistors contain such dielectric materials as silicon monoxide and dioxide, tantalum oxide, silicon nitride, and the like. These coatings are deposited by a variety of atomic deposition techniques (PVD, CVD) and thick-film methods. In some cases coatings, eg, SiO, are formed by thermal oxidation. An important application is the deposition of passivating layers of SiO_2 or SiO_2–Si_3N_4 by plasma-assisted CVD techniques. A new area is insulation coatings for GaAs devices formed by plasma-assisted oxidation methods and the deposition of amorphous silicon by CVD techniques. The fabrication of tungsten-emitter elements for thermionic convertors by CVD is another application.

Biomedical. Heart-valve parts are fabricated from pyrolytic carbon, which is compatible with living tissue. Such parts are produced by high temperature pyrolysis of gases such as methane. Other potential biomedical applications are dental implants and other prostheses where a seal between the implant and the living biological surface is essential (see Prosthetic and biomedical devices).

Decorative. Titanium nitride has a golden color and is used extensively to coat steel and cemented carbide substrates for watch cases, watch bands, eyeglass frames, etc. It provides excellent scratch resistance as well as the desired esthetic appearance, and it replaces gold coatings used previously.

Economic Aspects

Diffusion aluminide and silicide coatings on external and internal surfaces for high temperature corrosion protection in parts such as gas-turbine blades is estimated at $\$40 \times 10^6$/yr in North America and about $\$50 \times 10^6$ worldwide.

Overlay coatings onto gas-turbine blades and vanes of M–Cr–Al–Y type alloys by electron-beam evaporation is estimated at $10 × 10^6 to coat 200,000 parts at an average cost of $50 per part.

Hard facing of various components in the aircraft gas-turbine engine and in industrial applications for textile machinery parts, oil and gas-machinery parts, paper-slitting knives, etc, is estimated at $400 × 10^6 in 1978 with an estimated growth rate of 7–10% annually. The mix is approximately 60% aerospace applications, 40% industrial applications. Additionally, repair coatings for gas-turbine blades and vanes is estimated at $300 × 10^6. These coatings are primarily deposited by plasma-spray and detonation-gun techniques.

Refractory compound coatings of carbides, nitrides, and oxides on cemented-carbide cutting tools, mainly by the CVD process, is estimated at $200 × 10^6 annually worldwide.

A current application is the coating of complex shaped high speed steel cutting tools such as drills, holes, gear-cutters, etc, with a titanium nitride coating deposited by the methods of plasma-aided reactive evaporation and ion plating. A very rough estimate of the add-on value of the coating is $(5–10) × 10^6 in 1981 but with a very rapid growth rate of 100–300% expected over the next few years. The same processes are also used to deposit gold-colored wear resistant titanium nitride coatings on watch bezels, watchbands and other decorative jewelry items. A very rough estimate is $5 × 10^6 annually. The last two applications are practiced principally in Japan, Europe, and the United States.

Finally, refractory coatings are extensively used in the semiconductor and optical-materials industry. The annual add-on value of such coatings exceeds $10^9.

BIBLIOGRAPHY

"Refractory Coatings" in *ECT* 2nd ed., Vol. 17, pp. 217–284, by Bruno R. Miccioli, The Carborundum Company.

1. H. H. Hausner, ed., *Coatings of High Temperature Materials*, Plenum Publishing Corp., New York, 1966.
2. R. F. Bunshah and D. M. Mattox, *Phys. Today* **33,** 50 (May 1980).
3. R. F. Bunshah in *New Trends in Materials Processing*, American Society of Metals, Metals Park, Ohio, 1974, pp. 200–269.
4. R. F. Bunshah and A. C. Raghuram, *J. Vac. Sci. Technol.* **9,** 1385 (1972).
5. W. Grossklaus and R. F. Bunshah, *Int. Vac. Sci. Technol.* **13,** 532 (1975).
6. M. Kobayashi and Y. Doi, *Thin Solid Films* **54,** 57 (1978).
7. K. K. Yee, *Int. Metal. Rev.* **1**(226), 19 (1978).
8. T. Bonifield, "Plasma Assisted Chemical Vapor Deposition" in R. F. Bunshah, ed., *Films and Coatings for Technology*, Noyes Data Corp., Park Ridge, N.J., 1982, Chapt. 9.
9. U.S. Pats. 3,024,175–3,024,177, 3,232,853 (Mar. 6, 1962), N. C. Cook (to General Electric Co.).
10. B. A. Movchan and A. V. Demchishin, *Fizika Metall.* **28,** 653 (1969).
11. J. A. Thornton, *J. Vac. Sci. Technol.* **12,** 830 (1975).
12. J. A. Blocher, "Chemical Vapor Deposition" in ref. 8, Chapt. 8.
13. R. F. Bunshah and co-workers in ref. 8.

General References

Ref. 8 is a general reference.

H. H. Hausner, ed., *Coatings of High Temperature Materials*, Plenum Publishing Corp., New York, 1966.

J. Huminik, Jr., ed., *High-Temperature Inorganic Coatings*, Reinhold Publishing Corp., New York, 1963.
D. H. Leeds, "Coatings on Refractory Metals" in J. E. Hove and W. C. Riley, eds., *Ceramics for Advanced Technologies*, John Wiley & Sons, Inc., New York, 1965, Chapt. 7.
F. A. Lowenheim, ed., *Modern Electroplating*, 2nd ed., John Wiley & Sons, Inc., New York, 1963.
C. F. Powell, J. H. Oxley, and J. M. Blocher, Jr., *Vapor Deposition*, John Wiley & Sons, Inc., New York, 1966.
R. F. Bunshah, ed., *Techniques of Metals Research*, Vol. I, Parts 1–3; Vol. VII, Part 1, John Wiley & Sons, Inc., New York, 1968.
L. Holland, *Vacuum Deposition of Thin Films*, Chapman & Hall, 1968.
L. I. Maissel and R. Glang, *Handbook of Thin Film Technology*, McGraw-Hill, Inc., New York, 1970.
B. Chapman and J. C. Anderson, eds., *Science and Technology of Surface Coatings*, Academic Press, Inc., New York, 1974.
J. L. Vossen and W. Kern, eds., *Thin Film Processes*, Academic Press, Inc., New York, 1978.
R. F. Bunshah, *Materials Coating Techniques*, Agard Lecture Series, No. 106, NATO, 1980.
A. R. Reinberg, "Plasma Deposition of Their Films," *Annual Rev. Mat. Sci. Technol.* **9,** 341 (1979).
R. W. Haskell and J. G. Byrne in H. Herman, ed., "Studies in Chemical Vapor Deposition," in *Treatise on Materials Science and Technology, VI*, Academic Press, Inc., New York.
R. Bakish, ed., "Electron and Ion Beam Science and Technology," a series of International Conferences.
R. Sard, H. Leidheiser, Jr., and F. Ogburn, eds., *Properties of Electrodeposits—Their Measurement and Significance*, Electrochemical Society, Princeton, N.J.
D. L. Hildenbrand and D. D. Cubicciotti, eds., *High Temperature Metal Halide Chemistry*, a 1977 symposium.
R. G. Frieser and C. J. Mogab, eds., *Plasma Processing*, a 1980 symposium.
K. K. Yee, "Protective Coatings for Metals by Chemical Vapor Deposition," *Int. Met. Rev.* **1,** 19 (1978).
K. C. Mittal, ed., *Adhesion Measurements of Thin Films, Thick Films and Bulk Coatings*, ASTM STP **640,** (1978).
J. I. Duffy, ed., *Electroless and Other Non-electrolytic Plating Techniques*, Noyes Data Corp., Park Ridge, N.J., 1980.
W. Kern and G. L. Schnable, "Low Pressure Chemical Vapor Deposition for Very Large Scale Integration Processing—A Review," *IEEE Trans.* **ED-26,** 647 (1979).
G. Wahl and R. Hoffman, "Chemical Engineering with CVD," *Rev. Int. Temp. Refract. Fr.* **17,** 7 (1980).
R. S. Holmes and R. G. Loasby, *Handbook of Thick Film Technology*, Electrochemical Publications, 1976.
"Heat Treating, Cleaning and Finishing, *ASTM Metals Handbook*, Vol. 2, American Society for Metals, 1964.
K. L. Mittal, ed., *Surface Contamination: Genesis, Detection and Control*, Vols. 1 and 2, Plenum Press, Inc., New York, 1979.

Journals

Thin Solid Films, Elsevier Sequoia S.A., Lausanne, Switzerland.
Journal Vac. Sci. Technol., American Vacuum Society.
Proceedings of International Conference on Chemical Vapor Deposition, Electrochemical Society.
J. Electrochem. Soc.

ROINTAN F. BUNSHAH
University of California, Los Angeles

REFRACTORY FIBERS

The term refractory fiber defines a wide range of amorphous and polycrystalline synthetic fibers used at temperatures generally above 1093°C (see also Fibers, chemical; Refractories). Chemically, these fibers can be separated into oxide and nonoxide fibers. The former include alumina–silica fibers and chemical modifications of the alumina–silica system, high silica fibers (>99% SiO_2), and polycrystalline zirconia, and alumina fibers. The diameters of these fibers are 0.5–10 μm (av ca 3 μm). Their length, as manufactured, ranges from 1 cm to continuous filaments, depending upon the chemical composition and manufacturing technique. Such fibers may contain up to ca 50 wt % unfiberized particles. Commonly referred to as shot, these particles are the result of melt fiberization usually associated with the manufacture of alumina–silica fibers. The presence of shot reduces the thermal efficiency of fibrous systems. Shot particles are not generated by the manufacturing techniques used for high silica and polycrystalline fibers, and consequently, these fibers usually contain <5 wt % unfiberized material. Refractory fibers are manufactured in the form of loose wool. From this state, they can be needled into flexible blanket form, combined with organic binders and pressed into flexible or rigid felts, fabricated into rope, textile, and paper forms, and vacuum-formed into a variety of intricate, rigid shapes.

The nonoxide fibers, silicon carbide, silicon nitride, boron nitride, carbon, or graphite [*7782-42-5*] have diameters of ca 0.5–50 μm. Generally, nonoxide fibers are much shorter than oxide fibers except for carbon, graphite, and boron fibers which are manufactured as continuous filaments. Carbon, graphite, and boron fibers are used for reinforcement in plastics in discontinuous form and for filament winding in continuous form. In addition to temperature resistance, these fibers have extremely high elastic modulus and tensile strength. Carbon and graphite fibers cannot be accurately classified as refractory fibers because they are oxidized above ca 400°C. This is also true of boron fibers which form liquid boron oxide at approximately 560°C. Most silicon carbide, silicon nitride, and boron nitride fibers are relatively short, ranging in length from single-crystal whiskers less than 1 mm to fibers as long as 5 cm. Diameters are ca 0.1–10 μm. These fibers are used mostly to reinforce composites of plastics, glass, metals, and ceramics. They also have limited application as insulation in and around rocket nozzles and in nuclear fusion technology requiring resistance to short-term temperatures above 2000°C. For inert atmosphere or vacuum applications, carbon-bonded carbon fiber composites provide effective thermal insulation up to 2500°C.

Until the early 1940s, natural asbestos was the principal source of insulating fibers (see Asbestos). In 1942, Owens-Corning Fiberglass Corporation developed a process for leaching and refiring E glass fibers to give a >99% pure silica fiber (1). This fiber was used widely in jet engines during World War II. Shortly after the war, the H.I. Thompson Company promoted insulating blankets for jet engines under the trade name Refrasil (2). The same material was manufactured in the UK by the Chemical and Insulating Company, Ltd. (3), and a similar material called Micro-Quartz was developed by Glass Fibers, Inc. (now Manville Corporation) from a glass composition specifically designed for the leaching process (4). Leached-glass fibers can be produced with extremely fine diameters using flame attenuation (see Glass).

Several techniques were developed by Engelhard Industries, Inc. in the early 1960s for manufacturing fused silica fibers by rod-drawing techniques (5). High thermal-efficiency insulating felts with very low densities are produced by rod-drawing. They are essentially free of shot, and were used as blankets for jet engines and space vehicles (see Ablative materials). High purity silica fiber was chosen for the tiles which form the first reusable thermal-protection system on the space shuttle Orbiter. Many of the leading edge and underside areas on the Orbiter are insulated with replaceable carbon-bonded carbon–fiber tile capable of protecting the metal skin beneath from the 1200–1800°C temperatures developed in those areas. In the early 1940s techniques for blowing molten kaolin [*1318-74-7*] clay ($Al_2O_3.2SO_2$) into fibers were also developed by Babcock & Wilcox Company (6) (see Clays). At about the same time, it was discovered that by blowing molten mullite [*1302-98-8*] ($3Al_2O_3.2SiO_2$) with air and steam jets, a portion of the melt is converted into glassy fibers (7).

The Johns-Manville Corporation became interested in alumina–silica fibers in 1946, establishing a cooperative development with Carborundum. Chemical modifications of the basic alumina–silica composition have resulted in three materials of interest. In the Carborundum Company process addition of ca 5 wt % zirconia to the basic 50 wt % alumina–50 wt % silica system produces a longer fiber for use at 1260°C. In 1965, Johns-Manville developed a chromia-modified, alumina–silica fiber with a temperature limit of 1427°C (8). More recently, the 3M Company has introduced a boria-modified alumina–silica fiber for continuous use at 1427°C. This fiber, trade named Nextel, is particularly suitable for fabrication into flexible high temperature textiles.

The next technological advance was the development of the precursor process by Union Carbide Corporation. An organic polymeric fiber is used as a precursor to absorb dissolved metal oxides. A subsequent heat treatment burns out the organic fiber, leaving a polycrystalline refractory metal oxide in the shape of the host polymeric material (9). With this technique Union Carbide developed the first commercially available 1650°C zirconia fiber, trade-named Zircar. At about the same time, Babcock & Wilcox, along with others, discovered that inorganic fibers could be made by spinning, blowing, or drawing a viscous, aqueous solution of metal salts into fibers followed by a heat treatment to convert the salts to the oxide form. The resulting fibers were polycrystalline and usually contained 5–10% porosity. This process allowed the fiberization of metallic oxides and combinations of metallic oxides, whose viscosity in the molten state would not allow fiberization by ordinary methods.

From this new technology, Babcock & Wilcox developed the first true mullite fiber, which was used experimentally in the development of tile insulation for the space shuttle Orbiter. In 1972, ICI introduced the first commercial fibers made by the sol process. Under the trade names Saffil alumina and Saffil zirconia, fibers were offered for continuous use at 1400 and 1600°C, respectively. The sol process eliminated the need for the relatively costly organic precursor fiber and thereby significantly reduced the cost of manufacturing relatively pure oxide fibers. The zirconia fiber, however, was dropped from production.

During the mid-1960s, ultralightweight high strength composites replaced traditional metallic structural members and other parts of the early space vehicles, and cast metals were reinforced with short fibers consisting of alumina, silicon carbide, silicon nitride, or boron nitride (see High temperature composites). Their strengths are 1.4–20 GPa ($(2\text{–}29) \times 10^5$ psi) with elastic moduli of 70–700 GPa (ca $(1\text{–}10) \times 10^7$

psi). Silicon carbide whiskers (single-crystal fibers) are of special interest because they offer not only high strength and stiffness but are useful up to 1800°C. Numerous companies worldwide became involved in carbide–fiber research. In many manufacturing processes these fibers are grown as reaction products of gaseous silicon monoxide and carbon monoxide at high temperatures. In some cases, carbon fibers were allowed to react with silicon monoxide. By 1968, both Thermokinetic Fibers, Inc. and Carborundum sold short silicon carbide fibers at ca $500/kg. A recent innovation is the production of high quality short silicon carbide fibers by the pyrolysis of rice hulls in an inert or ammonia atmosphere.

Boron nitride exhibits exceptional resistance to thermal shock and can be used in inert or nonoxidizing atmospheres to 1650°C (11). In a Carborundum process, boron oxide fibers are converted to polycrystalline boron nitride at elevated temperatures in an ammonia atmosphere. A 1971 Lockheed Aircraft patent describes a similar process in which a boron filament is heated first in air and then converted to the nitride in a nitrogen atmosphere (12).

Properties

Refractory-fiber insulating materials are generally used in applications above 1063°C. Table 1 gives the maximum long-term use temperatures in both oxidizing and nonoxidizing atmospheres. For short exposures, some of these fibers can be used at temperatures much closer to their melting temperature without degradation.

The most important properties of these fibers are thermal conductivity, resistance to thermal and physical degradation at elevated temperatures, and tensile strength and elastic modulus. The thermal conductivity, in W/(cm·K), is actually the inverse of the R factor that expresses the insulating value of building insulations (see Insulation, thermal). Fibrous insulations with increasingly lower thermal conductivities require reliable test methods, and currently ASTM C 201 and ASTM C 177 are the standard methods. In addition, ASTM C 892-78 defines maximum thermal conductivities and unfiberized contents. Thermal conductivity is affected by the bulk density of the material, the fiber diameter, the amount of unfiberized material (shot), and the mean temperature of application. Typical silica fibers have a mean fiber diameter of

Table 1. Maximum Use Temperatures of Refractory Fibers

Fiber type	CAS Registry Number	Mp, °C	Maximum use temperature, °C: Oxidizing atmosphere	Nonoxidizing atmosphere
Al_2O_3	*[1344-28-1]*	2040	1540	1600
ZrO_2	*[1314-23-4]*	2650	1650	1650
SiO_2	*[7631-86-9]*	1660	1060	1060
Al_2O_3–SiO_2	*[37287-16-4]*	1760	1300	1300
Al_2O_3–SiO_2–Cr_2O_3	*[65997-17-3]*	1760	1427	1427
Al_2O_3–SiO_2–B_2O_3	*[65997-17-3]*	1740	1427	1427
C	*[7440-44-0]*	3650	400	2500
B	*[7440-42-8]*	1260	560	1200
BN	*[10043-11-5]*	2980	700	1650
SiC	*[409-21-2]*	2690	1800	1800
Si_3N_4	*[12033-89-5]*	1900	1300	1800

1–2 μm and less than 5% unfiberized material. Alumina–silica fibers range in mean diameter from 2.5 to 5.5 μm and have 20–50 wt % shot, which significantly reduces their thermal efficiency (increases thermal conductivity). Alumina and zirconia fibers are manufactured with a uniform diameter of ca 3 μm and contain less than 5% unfiberized particles. With such a wide range of shot contents and fiber diameters, the thermal conductivity of equivalent bulk-density blankets or felts made from different fiber can differ significantly. A thermal conductivity-versus-mean-temperature plot is given for several fibers in Figure 1. The effect of density on the thermal conductivity of silica and alumina–silica fibers is shown in Figure 2. For reinforcing purposes, tensile strength measured at room temperature and stress-to-strain ratio measured as Youngs modulus are of primary importance (see Table 2). Alumina-silica fibers are not used as reinforcements, and consequently, strength and modulus measurements are not important.

At the relatively low densities of these materials, solid conduction of heat is negligible when considering total heat transfer (see Heat-exchange technology). Heat transfer is affected mainly by radiation, which at 1260°C, is reponsible for approximately 80% of all heat transfer. Heat transfer by convection is small, and gas conduction within the insulation is responsible for the balance of heat transfer through the material. Consequently, most efficient insulations are those that effectively block radiation. The temperature ratings given these fibers are generally the temperatures at which the linear shrinkage stabilizes within 24 h and does not exceed 5%. The 5% linear shrinkage value has become a design factor for furnace insulation and other high temperature applications.

Alumina–silica and leached-glass fibers have been converted from amorphous glass to their crystalline forms. The rate of crystallization, sometimes called devitrification, depends on time and temperature (see Crystallization). For some time, the degradation of mechanical strength of refractory fiber products was attributed to this crystallization. In alumina–silica fibers, crystallization is completed relatively quickly,

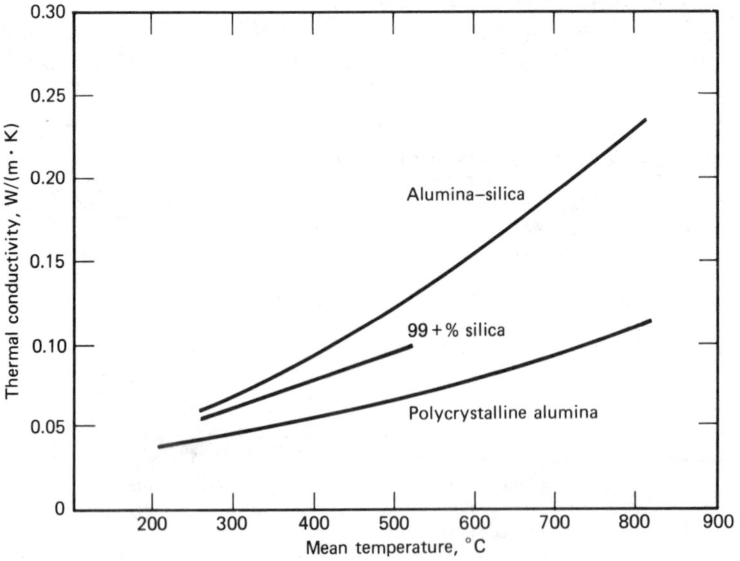

Figure 1. Thermal conductivity of refractory fiber insulations with 96 mg/cm³ density.

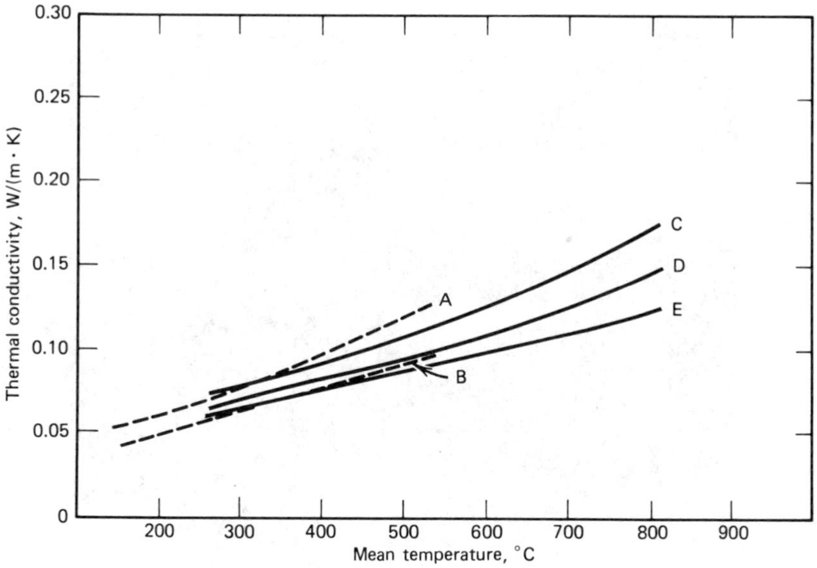

Figure 2. Effect of density on thermal conductivity. A, 48 mg/cm³ silica fiber; B, 96 mg/cm³ silica fiber; C, 128 mg/cm³ alumina–silica fiber; D, 192 mg/cm³ alumina–silica fiber; E, 384 mg/cm³ alumina–silica fiber.

within 24 h, at rated use temperatures; further degradation of physical properties with time is due to grain growth in the fiber (13). Sintering, the bonding together of these fibers at temperatures well below their softening point, is also responsible for linear shrinkage. At contact points, solid-state diffusion of molecules has a bridging effect whereby previously flexible material is transformed into a rigid structure. A scanning electron micrograph of alumina–silica–chromia fiber after 120 h exposure to 1427°C is shown in Figure 3. Both the formation of mullite and cristobalite crystals in the originally amorphous fiber and the sintering at fiber intersections can be clearly seen. The diffusion of oxides in the fibers toward the contact point has a slight shortening effect upon the fiber that ultimately contributes to the overall shrinkage of the product. In general, with increased sintering the resistance to abrasion or mechanical vibration decreases.

Table 2. Mechanical Properties of Oxide and Nonoxide Fibers

Fiber type	Density, g/cm³	Tensile strength, GPa[a]	Young's modulus, GPa[a]
SiO_2	2.19	5.9	72
Al_2O_3	3.15	2.1	170
ZrO_2	4.84	2.1	345
carbon	1.50	1.4	210
graphite[b]	1.66	1.8	700
BN	1.90	1.5	90
SiC[b]	3.21	2.0	480
Si_3N_4[b]	3.18	1.4	380

[a] To convert GPa to psi, multiply by 145,000.
[b] Single-crystal whiskers.

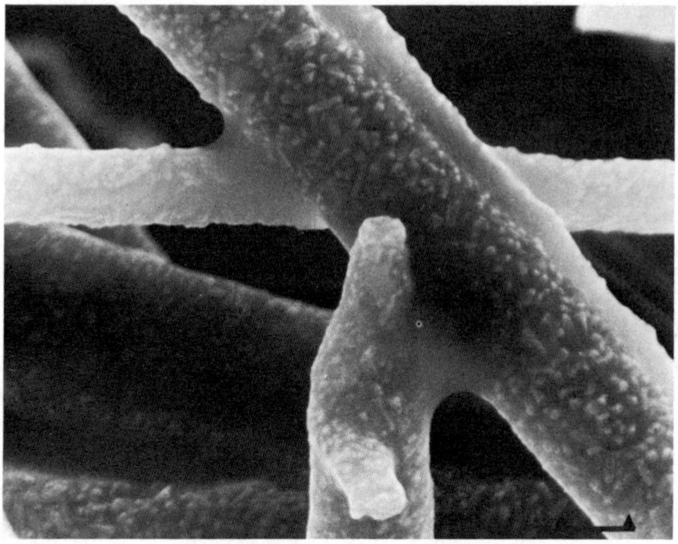

Figure 3. Alumina–silica–chromia fiber after 120 h at 1426°C showing crystallization and sintering at contact points. ×5000.

Alumina–Silica Fibers

Current refractory-fiber production consists mainly of melt-fiberized alumina–silica, and modified alumina–silica fibers. The 1260°C-alumina–silica fibers are produced by melting high purity alumina and silica or calcined kaolin in electric resistance furnaces. In either case, the resulting fiber contains ca 50 ± 2 wt % SiO_2 and Al_2O_3. The higher temperature grade, for use to 1427°C, is made by adding ca 3 wt % Cr_2O_3 to the basic composition or increasing the Al_2O_3 content to 55–60 wt %.

The raw materials are melted in a three-phase electric furnace that operates on the electrical resistance of the pool of molten material in which the electrodes are immersed. The initial melt is established by passing a current through graphite or coke granules that ultimately burn off, leaving the molten alumina–silica to serve as the conductive medium. Graphite is the common electrode material, although refractory metals, such as molybdenum and tungsten, are also used. The latter must be kept submerged in the molten pool to prevent oxidation. Some electrode loss occurs, but the refractory metal electrodes are generally not attacked by long-term use (14). The molten material is discharged through a temperature-controlled orifice at the bottom of the vessel (14–15), or the furnace is tilted to deliver the melt via a refractory trough to the fiberizing unit.

In the steam-blowing process, the liquid material is dropped in the path of a high velocity blast from a steam jet. Air jets are also used for fiberizing. Normally, the fiberizing pressures in the blowing processes are ca 700 kPa (ca 100 psi), but they can be much higher. The steam fiberizing of molten materials is an art long associated with the manufacture of mineral–wool insulations from low temperature melts. Ultrahigh-speed photography has revealed that the blast initially shreds the molten stream into tiny droplets. As each droplet picks up momentum from the velocity of the steam blast, it elongates into a teardrop shape and attenuates into a fiber with a spherical globule of molten material as its head, called shot. Several examples of fiber with the spherical shot particles still attached are shown in Figure 4. Each fiber is therefore the tail from a droplet of nonfibrous material. However, the amount of fiber in the crude product on a volume basis is very large compared with that of nonfibrous material.

In the melt-fiberization or spinning process, the molten material is dropped on the periphery of a vertically oriented, rotating disk. The molten material bonds to the high-speed rotating disk surface from which the melt droplets are ejected. The fiber is attenuated both by centrifugal action on the ejected droplet and the fact that the fiber tail frequently is still attached to the disk's surface. Generally, spinning produces a longer fiber than steam blowing, and converts more of melt to fiber, depending on the melt-delivery rate.

High Purity Silica Fibers

Refractory fibers of essentially >99 wt % silica are not made by conventional melt-fiberizing techniques. Thus, to achieve the high purity required for operation at 1093–1260°C, a leaching process is employed (16). A glass of ca 75 wt % SiO_2 and 25 wt % Na_2O is melted in a typical glass furnace at ca 1100°C. Filaments with a diameter of ca 0.3 mm are drawn from orifices in the furnace bottom. These filaments are then passed in front of a gas flame and attenuated to a diameter of approximately 1.5 μm. The loose fibers are subjected twice to an acid-leaching process to remove the Na_2O. The fiber is thoroughly rinsed, and dried at ca 315°C. The resulting fiber has a maximum of 0.01 wt % Na_2O and K_2O, 0.20 wt % Al_2O_3, and 0.04 wt % MgO and CaO.

The direct manufacture of pure silica fibers requires either a fused or extruded silica particulate rod of 5–25-mm dia. The fused silica rods with diameters of 6–7 mm are formed from molten high purity silica and mounted in groups of 20 on a motor-driven carriage. The rods are reduced to ca 2 mm in dia by being forced through a graphite guide and over a vertically oriented oxyhydrogen burner operating at 1800°C.

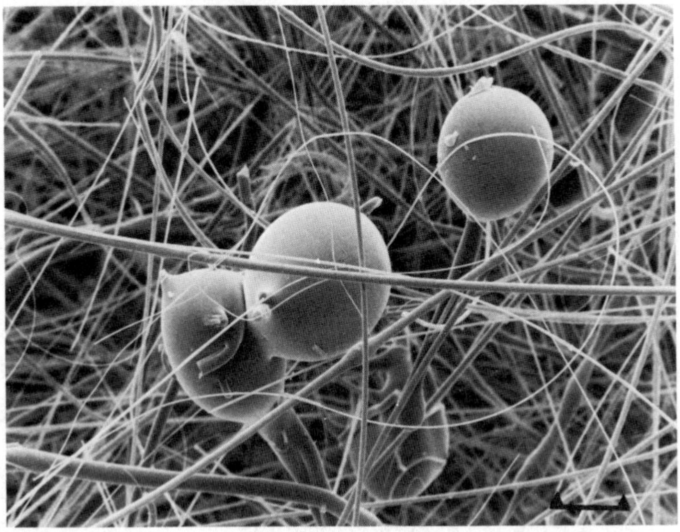

Figure 4. Refractory fiber (1260°C) and unfiberized shot particles. ×1000.

These relatively large fibers are passed through another graphite block containing 20 holes. As the fibers exit, they are hit by an axial oxyhydrogen jet attenuating the fibers toward a rotating drum which collects them. Fibers with a diameter of 4–10 μm and a silica content of 99.95 wt % are produced (17).

Chemically Produced Oxide Fibers

Ceramic oxide fibers are difficult to produce by common melt technology because oxides such as zirconia and alumina have very high melting points and low viscosities (see also Ceramics). In 1969, Union Carbide marketed the first 1650°C zirconia refractory fiber manufactured by a chemical technique termed the precursor process. The precursor can be any organic fiber containing extremely small crystallites of polymer chains held together in a matrix of amorphous polymer. When immersed in a solvent such as water, the fiber swells, thus expanding the spaces between the crystallites. Although a number of organic fibers have this swelling characteristic, eg, wool, cotton, and cellulose acetate, rayon is preferred because of its structural uniformity and high purity. After an initial swelling and dewatering by centrifuge, the rayon fiber is immersed in a 2 *M* aqueous solution of zirconyl chloride containing a small amount of yttrium salt. The excess solution is centrifuged, and the fibers are dried and heated to ca 400°C in an atmosphere containing <10 vol % oxygen. This treatment pyrolyzes the rayon to a carbonaceous residue; the zirconyl chloride is converted to microcrystalline zirconia fiber containing yttrium oxide for phase stabilization and to prevent embrittlement.

Imperial Chemical Industries produces 1600–1650°C fibers by the sol process (18). Both silica-stabilized alumina and calcia-stabilized zirconia fibers are marketed as Saffil. First a metal salt, such as aluminum oxychloride in the case of alumina fiber, is mixed with a medium molecular weight polymer such as 2 wt % poly(vinyl alcohol). This solution is then slowly evaporated in a rotary evaporator to a viscosity of ca 80 Pa·s (800 P). This solution is then extruded through a 100-μm spinneret; the fibers are collected on a drum where they are fired to a temperature of 800°C. This action burns the organic component away and a fine-grained aluminum oxide fiber is formed with a porosity of 5–10 vol % and a fiber diameter of 3–5 μm. These fibers are used as filter media because of their inherent porosity. For refractory application, they are heated to 1400–1500°C, just long enough to eliminate the porosity, which would result in 3–4% linear shrinkage in application. The same process is followed for zirconium oxide fibers, starting with zirconium oxychloride, zirconium acetate, and calcium oxide. The only truly continuous alumina–silica-type fiber is made by the 3M Company employing a similar solution process (19). In this case, a 10-μm dia alumina–silica–boria fiber is manufactured for use to 1427°C. Basic aluminum acetate is dissolved in water and the solution is mixed with an aqueous dispersion of colloidal silica and dimethylformamide. The resulting solution is concentrated in a Rotavapor flask and centrifuged. The solution is then extruded through a 75-μm spinneret at 100 kPa (1 atm). The resulting fibers are collected on a conveyor chain and passed through a furnace at 870°C converting the filaments to metallic oxides. Heating in another furnace at 1000°C produces a glassy aluminum borosilicate [*12794-54-5*] with the calculated composition $3Al_2O_3.B_2O_3.3SiO_2$.

Although fiber production by this technique is more expensive than melt fiberization, the sol process offers many other advantages. In melt fiberization, the effect upon viscosity and surface tension must be taken into account when adding even small amounts of other desirable oxides. In the sol process, however, the viscosity is controlled independently of the metals added and thus any number of metal salts can be easily added without adverse effect. The controlled addition of other metals can also serve as grain-growth inhibitors, sintering aids, phase stabilizers, or catalysts.

Nonoxide Refractory Fibers

The most important nonoxide refractory fibers are silicon carbide fibers and single-crystal whiskers. In the early work on silicon carbide, 1 part by volume SiO was mixed with 3 parts CO at 1300–1500°C in an inert or reducing carrier gas. The fibers formed on the colder parts of the reactor tube had 4–6-μm dia and were 50-mm long (20). In a similar but less complicated technique developed by Corning Glass Works, the SiO and CO gases are obtained by heating a mixture of carbon and silica in a molar ratio of 2:1 to 1300–1550°C in a hydrogen fluoride or hydrogen chloride atmosphere (21). The fibers grow more quickly than in an inert carrier gas and the process can be run continuously. These fibers are made up of beta silicon carbide crystals with a surface sheath of silica which prevents oxidation.

Since 1960, a number of processes for high quality silicon nitride fibers have been developed. These fibers are a by-product in the production of silicon nitride powder from silicon metal and nitrogen at high temperatures. Additions of a reducing agent greatly increased the fiber yield by converting silica first to silicon monoxide and then to silicon metal for the reaction with nitrogen (22).

$$SiO_2 + R \rightarrow SiO + RO$$

$$3\ SiO + R + 2\ N_2 \rightarrow 3\ RO + Si_3N_4$$

R is the reducing agent. Any silicate that forms thermally and chemically stable residual compounds as its SiO_2 content is reduced and provides a suitable source of silicon for the reaction. Alternate aluminum–silica and graphite plates are stacked in a graphite-lined alumina tube where the plates are separated by 2–4-cm graphite spacers. This tube is heated to 1400°C for 12 h in a nitrogen atmosphere in a silicon carbide-resistance furnace. After approximately 6 h the tube is cooled and the fibers are removed.

Boron nitride fibers are produced by nitriding boron filaments obtained by chemical vapor deposition of boron on heated tungsten wire. The tungsten wire is passed through a reactor containing boron trichloride in a carrier of hydrogen at 1100–1300°C. The boron trichloride is reduced and boron is deposited on the tungsten wire. The continuous boron filament exits the furnace and is wound upon a spool. In nonoxidizing atmospheres, boron filaments can be considered refractory fibers. When heated to 560°C, they develop a liquid boron oxide surface coating. For pure boron nitride fiber, the boron oxide filaments are further heated to 1000–1400°C in an ammonia atmosphere for ca 6 h. The following reactions are assumed to take place:

$$4\ B(s) + 3\ O_2(g) \rightarrow 2\ B_2O_3(l)$$

$$B_2O_3(l) + 2\ NH_3(g) \rightarrow 2\ BN(s) + 3\ H_2O(g)$$

Economic Aspects

Refractory fibers are 50–70% more efficient insulators than conventional brick linings at equivalent thicknesses. Although the initial cost of installing a fiber lining in a high temperature furnace is higher than that of installing brick lining, the difference can be quickly recovered in energy savings. As the cost of energy continues to increase, refractory-fiber furnace lining is becoming more and more attractive. In 1980, the average cost of 1260°C alumina–silica refractory fiber was ca \$3.75/kg, whereas the 1970 price was ca \$5/kg. The decrease in cost reflects the advancement in technology.

The 1981 selling price for the >99% silica fibers was approximately \$55/kg in both bulk and felted forms. This price has remained relatively constant over the past ten years. The more refractory Saffil alumina fiber sells for approximately \$33/kg. When first introduced in 1974, both the Saffil zirconia and alumina fibers sold for approximately \$24/kg. Production quantities are proprietary.

A substantial reinforced-composites market for continuous filaments has developed in the past few years. Carbon, graphite, and boron fibers are produced commercially at prices of \$40–400/kg, depending upon their properties. Silicon carbide whiskers have been produced commercially over the last 20 yr by several processes. Prices for these fibers have been based upon pilot plant quantities and have ranged from \$2000/kg to as low as \$150/kg in 1980. The price of silicon nitride fibers has been \$200–500/kg. Boron nitride fibers have not been produced commercially until about 1975. Their price has remained at ca \$50/kg since that time.

Health and Safety Factors

Synthetic or man-made refractory fibers do not appear to pose the health hazards of naturally occurring mineral fibers like asbestos. Because the diameter of most refractory fibers is <3.5 μm, they are considered respirable. Above 3.5 μm, fibers are not able to penetrate the functional components of the lung (23). Even though a portion of these fibers are respirable, they are considered a nuisance dust, because studies have shown they are not biologically active in living tissue. The Thermal Insulation Manufacturers Association is currently conducting animal inhalation studies on a number of man-made fibers including refractory fibers. Regardless of the results of these studies, proper respirators should be worn in environments of excessive exposure to refractory fibers.

Silica and alumina–silica refractory fibers, that have been in service above 1100°C, undergo partial conversion to cristobalite, a form of crystalline silica that can cause silicosis, a form of pneumoconiosis. The amount of cristobalite formed, the size of the individual crystallites, and the nature of the matrix in which they are embedded are time and temperature dependent. Under normal use conditions, refractory fibers are generally exposed to a temperature gradient. Consequently, it is most probable that only the fiber nearest the hot surface has an appreciable content of cristobalite. It is also possible that fiber containing devitrified cristobalite is more friable and therefore may generate a larger fraction of dust when it is removed from a high temperature furnace. Hence, removal of old furnace lining offers the greatest risk of exposure, and adequate protection against respiration should be provided. Fibers with diameters of ≥ 5 μm cause irritation to skin and mucous membranes. This is usually not a serious problem and can be avoided by wearing proper clothing and respirators. When heated in an oxidizing atmosphere, silicon carbide fibers form a surface coating of silica that converts to cristobalite. In addition, silicon carbide whiskers have been found to be biologically active in animal lung tissue and may cause pneumoconiosis. Adequate respiratory protection is advised when dealing with these fibers. The diameter of carbon and boron filaments places them above the respirable range, but they can cause skin irritation from penetration during handling. Silicon nitride and boron nitride fibers are not considered hazardous but may irritate the skin.

Uses

Blankets. Today, ca 60% of the refractory fiber production is used for 48–128-mg/cm^3 flexible needled blankets. The fiber is collected on a moving conveyor, run through compression rolls, and penetrated by barked needle boards. This needling has the effect of tying the fibers together; subsequent compression and heating increase the tensile strength of the product. Flexible needled blankets are commercially available in widths of 1.3 and 0.65 m, lengths $\leq$33 m, and thicknesses of 6 to 50 mm. They are primarily used as furnace-wall and roof insulations either as the exposed hot face or as back-up insulation behind refractory brick. Because of their lightweight flexible nature, blankets offer no structural support to the furnace wall and have to be anchored in place. Typically, the blankets are applied to furnace walls and roofs in overlapping layers by impaling them on metallic or ceramic studs fixed to the supporting metal framework. They can also be applied over existing brick walls using high temperature cement or mechanical anchoring systems providing a new, more insulating furnace interior (see Insulation, thermal; Furnaces). Because of their relatively low heat storage and thermal conductivity, they have replaced brick linings in most industrial kilns in order to reduce energy costs. Other applications for these blankets include insulation for automotive catalytic convertors and aircraft and space vehicle engines and a wide variety of uses in the steel-making industry.

Felts. Felts (qv) contain an organic binder, generally a phenolic resin or in some cases a latex material. As the fibers are collected after fiberization, they are mixed with a dry phenolic resin. The fiber is then passed through an oven and a constant pressure is applied by flight conveyors. Alternatively, the uncured fiber and resin are compressed between heated press plates to form felts with densities as high as 380 mg/cm^3. These high density felts are now used extensively in ingot-mold operations in steel foundries. Felts provide excellent expansion joints in high temperature applications because the fibers tend to expand after the organic binder has been burned out (see also Heat-resistant polymers).

Bulk Fibers. Bulk refractory fiber is used as a general-purpose high temperature filler for expansion joints, as stuffing wool, for furnace and oven construction, in steel mills and aluminum and brass foundries, in glass manufacturing operations, as a loose-fill insulation, and as a raw material for vacuum-formed shapes (see Fillers).

Vacuum-Formed Shapes. Approximately 20% of fiber production is converted to rigid shapes by a vacuum-forming process. The bulk fibers are mixed in aqueous suspension with clays, colloidal metal oxide particles, and organic binders. Molds with fine-mesh screen surfaces are used to accrete the solids into special shapes as the water is drawn through the screen by vacuum. These products are then dried at 100–200°C to obtain rigid shapes having densities generally between 200 and 300 mg/cm^3. In recent years, the more expensive 1650°C alumina and zirconia fibers have been mixed with small amounts of less expensive and lower temperature alumina–silica fibers in the vacuum-forming process to increase the use temperature. The uses of vacuum-formed refractory fiber products are extremely varied. The metal industry makes extensive use of vacuum-formed shapes from tap hole cones in furnaces and ladles to insulating risers in metal castings. The tiles of the space shuttle thermal protection system are made of 99.7 wt % silica fiber, trade named Q Fiber, made by a highly specialized vacuum-forming process. The largest application for vacuum-formed products is as lightweight board insulation for furnace linings. In the rigid vacuum-formed state, the fibers are less susceptible to abrasion from direct gas impingement emanating from burners or flue fans.

Modules. Lightweight refractory fiber modules are a recent innovation in furnace lining (24). The refractory fiber blanket is folded in an accordion fashion and then compressed and banded to form modules 0.3 by 0.3 m, and 0.2–0.3-m thick. Metallic hardware, attached to the back of each module as it is folded and compressed, is welded to the metal shells of furnace roofs and walls. The modules are snapped into place in a parquet fashion forming the complete insulation system for the furnace. This attachment technique offers distinct advantages over layered blanket construction by reducing installation time and labor and eliminating the need for metal or ceramic anchoring pins. The shrinkage effect is minimized because the resilient fibers expand somewhat when their compression banding is released during construction. This concept of furnace lining is gaining widespread acceptance in the industry because of the ease of installation and other technical advantages.

Fibers and Yarns. The continuous refractory oxide fibers, ie, silica and alumina–silica–boria can be twisted into yarns from which fabrics are woven. These fabrics are used extensively in heat-resistant clothing, flame curtains for furnace openings, thermocouple insulation, and electrical insulation. Coated with Teflon, these fibers are used for sewing threads for manufacturing speciality high temperature insulation shapes for air craft and space vehicles. The cloth is often used to encase other insulating fibers in flexible sheets of insulation. The spaces between the rigid tile on the space shuttle Orbiter are packed with this type of fiber formed into tape.

Nonoxide fibers in continuous form, such as carbon, graphite, and boron are primarily used in filament winding and the manufacture of high strength, high modulus fabrics. Lightweight, high strength pressure vessels are made by running these fibers through an epoxy impregnant and then winding them around plaster mandrels. After curing the resin, the plaster is dissolved leaving a thin-walled vessel. The drill stems used for the collection of lunar core samples were a combination of boron cloth with an epoxy–graphite filament. The carbon fibers in loose fill and cloth form are currently used in nonoxidizing processes as insulation up to 1800°C and higher temperatures for short duration. The shorter nonoxide fibers are primarily used as strength-enhancing reinforcements in resins, some ceramics, and metals. Silicon carbide and silicon nitride short fibers are dispersed in a number of resins that are then cast into various shapes. Applications for these cast parts are generally found in high technology areas such as the fabrication of specialty electrical parts, aircraft parts, and radomes (microwave windows). These fibers are also used as reinforcing inclusions in metals. Boron nitride is of particular value as a reinforcing fiber for cast aluminum parts because it is one of the few materials totally wet by molten aluminum. Silicon carbide and boron nitride fibers as well as short single-crystal alumina fibers are used to reinforce gold and silver castings.

The growth and commercialization of the nonoxide fiber industry parallels the growth of the high strength-composite industry. When these fibers can be produced in lengths of 2–10-cm in the \$10–20 per kilogram price range, a large >1600°C insulation market will open up.

BIBLIOGRAPHY

"Refractory Fibers" in *ECT* 2nd ed., Vol. 17, pp. 285–295, by T. R. Gould, Johns-Manville Corporation.

1. U.S. Pat. 2,461,841 (Feb. 15, 1949), M. E. Nordberg (to Corning Glass Works).
2. *Chem. Eng. News* **25**(18), 1290 (1947).
3. C. Z. Carroll-Porczynski, *Advanced Materials*, 1st ed., Astex Publishing Company, Ltd., Guildford, England, 1959.
4. U.S. Pat. 2,823,117 (Feb. 11, 1958), D. Labino (to Glass Fibers, Inc.).
5. U.S. Pat. 3,177,057 (Aug. 2, 1961), C. Potter and J. W. Lindenthal (to Engelhard Industries, Inc.).
6. U.S. Pat. 2,467,889 (Apr. 19, 1949), I. Harter, C. L. Horton, Jr., and L. D. Christie, Jr. (to Babcock & Wilcox Company).
7. *Ceram. Age* **78**(2), 37 (1962).
8. U.S. Pat. 3,449,137 (June 10, 1969), W. Ekdahl (to Johns-Manville Corporation).
9. U.S. Pat. 3,385,915 (May 28, 1968), B. Hamling (to Union Carbide Corporation).
10. H. W. Rauch, Sr., W. H. Sutton, and L. R. McCreight, *Ceramic Fibers and Fibrous Composite Materials*, Academic Press, New York, 1968, p. 24.
11. U.S. Pat. 3,429,722 (July 12, 1965), J. Economy and R. V. Anderson (to Carborundum Company).
12. U.S. Pat. 3,573,969 (Apr. 10, 1971), J. C. Millbrae and M. P. Gomez (to Lockheed Aircraft Corporation).
13. L. Olds, W. Miiller, and J. Pallo, *Am. Ceram. Soc. Bull.* **59**(7), 739 (1980).
14. U.S. Pat. 2,714,622 (Aug. 2, 1955), J. C. McMullen (to The Carborundum Company).
15. U.S. Pat. 3,066,504 (Dec. 4, 1962), F. J. Hartwig and F. H. Norton (to Babcock & Wilcox Company).
16. U.S. Pat. 4,200,485 (Apr. 29, 1980), G. Price and W. Kielmeyer (to Johns-Manville Corporation).
17. Brit. Pat. 507,951 (June 23, 1939).
18. Brit. Pat. 1,360,197 (July 17, 1974), M. Morton, J. Birchall, and J. Cassidy (to Imperial Chemical Industries, Ltd.).
19. U.S. Pat. 3,760,049 (Sept. 18, 1973), A. Borer and G. P. Krogseng (to Minnesota Mining and Manufacturing Company).
20. U.S. Pat. 3,246,950 (Apr. 19, 1966), B. A. Gruber (to Corning Glass Works).
21. U.S. Pat. 3,371,995 (March 5, 1968), W. W. Pultz (to Corning Glass Works).
22. U.S. Pat. 3,244,480 (Apr. 5, 1966), R. C. Johnson, J. K. Alley, and W. H. Warwick (to The United States of America).
23. J. Leineweber, *ASHRAE J.* **3**, 51 (1980).
24. U.S. Pat. 3,952,470 (Apr. 27, 1976), C. Byrd, Jr. (to J. T. Thorpe Company).

W. C. MIILLER
Manville Corporation

ROOFING MATERIALS

Roofs are a basic element of shelter from inclement weather. Natural or hewn caves, including those of snow or ice, are early evidence of human endeavors for protection from cold, wind, rain, and sun. Nomadic man, before the benefits of agriculture had been discovered and housing schemes were developed, depended upon the availability of natural materials to construct shelters. Portable shelters, eg, tents, probably appeared early in history. Later, more permanent structures were developed from stone and brick, with courses stepped inward to form the roof. Salient features depended strongly upon the availability of natural materials. The Babylonians used mud to form bricks and tiles that could be bonded with mortars or natural bitumen. Ancient buildings in Egypt were characterized by massive walls of stone with closely spaced columns to carry stone lintels to support a flat roof, often made of stone slabs.

As larger and larger structures were desired, building design became an important element in construction. Roofs evolved from a simple covering to roofing systems designed to perform a number of functions to separate indoor and outdoor environments. Selection of roof design depended not only upon factors of economy and comfort, but also on the availability of materials and structural and aesthetic factors (1). Thus, a modern design normally includes a structure to carry loads, insulation to control heat flow, a barrier to control air and vapor flow, and a roofing element to prevent water penetration (1–5).

The bituminous types are by far the most common roof coverings in the United States. The built-up roofing membrane of bituminous materials is estimated to cover over 90% of the commercial and industrial roof surfaces in North America. Asphalt-based materials predominate over coal-tar materials. Various types of roofing materials are employed for residential structures, both for new construction and reroofing, especially in the western area of the United States. Wood and miscellaneous materials account for 20% of residential roofing in the western region of the United States, but only for 5% in the northeast and central areas (see Wood); asphalt roofing accounts for the rest (6–7) (see Asphalt).

Continuous or Jointed Systems

Built-Up Roofings. Built-up roofing (BUR) is a continuous-membrane covering manufactured on site from alternate layers of bitumen, bitumen-saturated felts or asphalt-impregnated glass mats, saturated and coated felts, and surfacings. These membranes are usually applied with hot bitumens or by a cold process utilizing bituminous-solvent or water-emulsion cements (see Adhesives; Cements; Felts).

The deck may be nailable, eg, wood or light concrete, or not, eg, steel or dense concrete. The felts or mats may be organic (cellulose), asbestos, or fiber glass (see Cellulose; Asbestos; Glass). The ply adhesives may be hot-melt or cold-process bitumens (emulsion or cutback). The roof slope ranges from dead level (0–2.1 cm/m or 0–1.2°) to flat (2.1–12.5 cm/m or 1.2–7.1°) to steep (12.5–25 cm/m or 7.1–14°).

Specifications and Application Rates. Specifications as published in catalogs of roofing manufacturers or contractor associations appear complicated because of the many variations possible. Application methods depend on the type and slope of the deck, the types of insulation and roofing membrane, and the fastening method (see Table 1) (8–15). Built-up roofs are generally not applied to slopes steeper than 14° (25 cm/m) (14%) since application of hot asphalt to such roofs is difficult, whereas other forms of roofing, such as shingles, are easily applied.

Common BUR systems are installed in different ways: membrane adhered to deck without insulation; insulation adhered to deck with membrane applied to insulation; base sheet adhered to deck, insulation to base sheet, and top membrane to insulation; and membrane adhered to deck and insulation applied over the membrane, the so-called protected-membrane roof.

The components must be anchored as protection against wind uplift, slippage, and membrane movement.

Application rates for BUR membranes are given in Table 2 (16–19). Membrane strength is related to felt or glass-mat strengths and the number of plys. Continuous bonding is inadvisable if crack development or joint movement can be expected in the membrane substrate, since the elastic limit of the membrane will be exceeded and result in rupture failure. Nailing or spot adhesion provides relief from concentrated strains.

Properties of Ply Felts and Asphalt-Saturated and Coated Felts. The properties of asphalt-saturated and coated felts and impregnated fiber-glass mats are given in Table 3 (9,11,20–21). In addition, ASTM D 3672 (22) describes an asbestos-saturated or glass-asphalt-saturated and coated venting-base sheet. It has an embossed-granule design on the bottom surface to permit lateral relief of pressure that might be caused by moisture movement.

Table 1. Simplified Hot-Applied Membrane Specifications[a]

Type	ASTM specification
organic felt membrane[b]	D 226, Types I–IV
base sheet	D 2626 or D 3158
asbestos ply felts[b]	D 250, Types I–IV
organic base sheet	D 2626
asbestos base sheet	D 3378
fiber-glass mats[b]	D 2178, Types I–IV
base sheet	D 2178, Type V
surfacings[c]	
organic base sheet	D 371
fiber-glass-mat base sheet	D 3909
bitumens[d]	D 1683
emulsified asphalts	D 1227

[a] From refs. 8–15.
[b] Smooth surface or top pour of bitumen plus gravel or slag.
[c] Cap sheet covered with mineral granules.
[d] Usually with gravel or slag.

Table 4 describes properties of ASTM ply felts used for BUR. For additional details, the appropriate ASTM specifications should be consulted. Periodic revisions are made by ASTM committees to keep up with current practices.

Cold-Process Coatings. Cold-process BUR applications do not require heating of the bitumen on the job. Simple application and economical maintenance are primary considerations. Bitumens are liquefied either by dissolving in a solvent or by emulsifying in water. With both types, the base cutbacks or emulsions can be modified by addition of other components, commonly mineral fillers, to form specialized coatings. At ordinary temperatures, the solvent or water evaporates and adhesive bonds or weather-resistant surface coatings are formed.

The compositions of solvent-type asphalt coatings are illustrated in Figure 1. These coatings range from fluid cutbacks used for priming deck surfaces to viscous mineral-filled compositions used as flashing cements. The ASTM specifications in refs. 16–19 should be consulted for details (see Coatings; Coating processes). Lap cements used with asphalt-roll roofings are specified by ASTM D 3019 (23). This specification covers cements consisting of asphalt and petroleum solvent with or without asbestos fiber.

Emulsified asphalt used as a protective coating is specified by ASTM D 1227 (24). These emulsions are applied above freezing by brush, mop, or spray and bond to either damp or dry surfaces. Such application is not recommended for inclines $<4°$ (4%) to avoid the accumulation of water. However, curing by water evaporation can be slow, and these emulsions may remain water susceptible.

Protected-Membrane Roofs. Primitive roofs covered with earth and sod over sloping wood decks shingled with bark were of the protected-membrane type and gave excellent service. Grass and earth provided insulation and protected the shingled deck from inclement weather (25).

In modern construction, insulation is needed that is unaffected by water or that can be kept dry in some manner and that stays in place over the roof membrane. In the United States, extruded polystyrene-foam insulation boards are commonly employed (see Insulation, thermal). They are placed over the roof membrane, which is attached to the decks. The insulation may or may not be bonded to the membrane.

Table 2. Application Rates for BUR Membranes

Material	ASTM specification	L/m^2 [a]	kg/m^2 [b]
primer	D 41[c]	0.4	
bitumen ply	D 312		1.0
adhesive	D 450		
bitumen surfacing	D 312, D 450		2.4–2.9
gravel covering	D 1863		19.5
slag covering	D 1863		14.6
asphalt-liquid surfacings[d]		0.6–2.0	

[a] To convert L/m^2 to $gal/100\ ft^2$, multiply by 2.45.
[b] To convert kg/m^2 to $lb/100\ ft^2$, multiply by 20.5.
[c] Ref. 16.
[d] See refs. 16–19 for ASTM specifications.

Table 3. Properties of Asphalt-Saturated and Coated Felts for BUR

Property	Organic base sheet[a] ASTM D 2626		Asbestos base sheet[a] ASTM D 3378		Asphalt-impregnated fiber-glass mat[b], ASTM D 2178	Organic felt asphalt[c], ASTM D 3158
	Type I	Type II	Type I	Type II	Type V	
mass, min, kg/m^2 [d]	1.8	1.9	1.8	1.9	0.73	1.41
mass dry felt, min, g/m^2 [d]	254	341	415	537	49	254
mass saturant, min, g/m^2 [d]	356	478	166	215	454	356
mass coating and surfacing, min, g/m^2 [d]	878	878	878	878	461	585
breaking strength, kN/m [e]						
with grain	6.1	7.9	5.3	7.0	5.3	6.1
across grain	3.5	3.5	2.6	3.5	5.3	3.5
permeance, max, $pg/(Pa{\cdot}s{\cdot}m^2)$ [f]	1.7	1.7	1.7	1.7		
pliability radius, 5 pass (25°C)	12.7	12.7	12.7	12.7	1.27	12.7

[a] First ply or as vapor barrier under insulation; coated on both sides with asphalt and surfaced on top side with fine mineral granules; unrolls at or above 10°C (9,20).

[b] Combination base sheet; faced on one side with Kraft paper; unrolls at or above 10°C (11).

[c] Saturated and coated; perforated or not; perforations provide vents for gases liberated during applications; surfaced on top side with fine mineral granules; unrolls at or above 10°C (21).

[d] To convert g/m^2 to $lb/100\ ft^2$, multiply by 0.0205.

[e] To convert kN/m to lbf/in., divide by 0.175.

[f] To convert $pg/(Pa{\cdot}s{\cdot}m^2)$ to $grain/(h{\cdot}in.\ Hg{\cdot}ft^2)$, divide by 5.72.

Table 4. Properties of Roofing and Waterproofing Ply Felts[a] for BUR

Property	Organic felt, Asphalt-saturated[b,c], ASTM D 226, type I	II	III	Organic felt, Coal-tar saturated[c], ASTM D 227	Asbestos felt, asphalt-saturated[b,c], ASTM D 250, type I	II	Fiber-glass mat, asphalt impregnated[c,d], ASTM D 2178, type I	III	IV
nominal weight, g/m^2 [e]	732	1460	976	732	732	1460	356	474	342
felt or fiber-glass-mat mass, min, g/m^2 [e]	254	488	332	254	415	854	49	73	83
saturant mass, min, g/m^2 [e]	356	732	469	356	166	342	225	308	146
breaking strength, min, kN/m[f]									
with grain	5.25	7.00	6.13	5.25	3.50	7.00	2.63	3.85	7.71
across grain	2.63	5.25	2.98	2.63	1.75	3.50	2.63	3.85	7.71
pliability radius, pass at 25°C, mm	12.7	19.1	12.7	12.7	12.7	19.1	12.7	12.7	12.7
ash, %	10 max	10 max	10 max		70.0 min	70.0 min	75–88	75–88	75–88
loss on heating at 105°C, 5 h max, %	4	4	4	4.0	5	5	1.0	1.0	1.0

[a] All are nonsticking.
[b] Perforated or nonperforated.
[c] Unrolls above 10°C.
[d] Faced on one side with Kraft paper; combination base sheet.
[e] To convert g/m^2 to lb/100 ft^2, multiply by 0.0205.
[f] To convert kN/m to lbf/in., divide by 0.175.

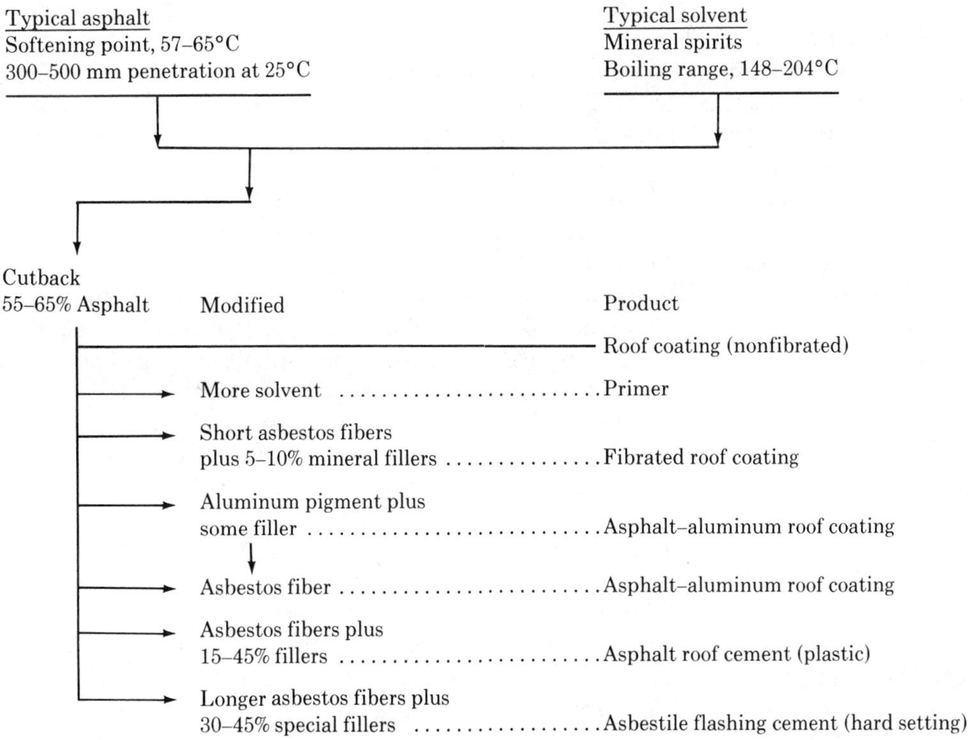

Figure 1. Production of asphaltic coatings, primer, and cements.

Gravel or slag at the rate of 48.8 kg/m^2 (1000 lb/100 ft^2) holds the insulation in place and offers protection from the sun. The insulation joints are open and drainage must be provided. Various other materials, eg, patio blocks, mortar, and concrete slabs, are also used as surfacings and ballast. The extra weight imposes more exacting requirements on construction. Although of relatively recent design, the protected-membrane roof is finding a place in roof systems (26–28).

Elastomeric Products Applied as Single- or Limited-Ply Roofing Membranes. New materials have proliferated during the last decade; their claimed advantages include light weight, high chemical and weather resistance, high elasticity, and ease of application and repair. These new roofing systems depend primarily on the application of single-ply-sheet membranes giving continuous coverage (see Film and sheeting materials). Materials applied as liquids prepared from a variety of base materials are often used in combination with glass-mat or plastic reinforcements (29–31) (see Plastic building products).

Although ASTM test methods are given in ref. 32, standards are currently not available to assess the merits of these materials. Elastic sheets are likely to have high extensibility. However, if the sheet is solidly attached and the substrate develops a crack, the latter is transmitted through the ply. The lapping of watertight and secure joints has presented a problem; excellent technology and careful application are necessary. Since usually only one ply is applied, a roof disaster can occur if the material or application is faulty. Until adequate field experience has been gained, particular material, back-up, and support from the supplier are essential.

Although estimates vary, the single-ply and liquid systems have probably captured

5–10% of the BUR market in terms of dollar volume; an increase of 50% is projected over the next 10 years (33).

In 1980, 36 manufacturers offered 42 sheet-membrane materials, 33% of which were poly(vinyl chloride), followed by modified asphalts and ethylene propylene diene monomer at 20% each; neoprene sheeting, chlorinated polyethylene, and polyisobutylene accounted for the rest. New materials and combinations of materials are continuously entering the market (34).

Single-ply loose-laid membranes are usually ballasted with smooth stones [22.0–48.8 kg/m^2 (450–1000 $lb/100\ ft^2$)].

Various liquid-applied single or two-component systems provide roofing membranes. Some are used with polymeric fabrics or glass reinforcements to obtain single- or multilayered membranes. Compositions are based on asphalt (solvent blends or emulsions), acrylic emulsions, urethane two-component systems, or chlorinated rubber (neoprene) with a top coat of chlorosulfonated polyethylene (Hypalon). Specification ASTM D 3468 describes solutions of neoprene and Hypalon for roofing and waterproofing applications (35) (see Elastomers, synthetic).

Other Types. Asbestos-cement corrugated sheets show excellent strength and durability properties in industrial and commercial roofing applications. Described by ASTM C 221, the sheets are lapped and usually applied over steel purlins (36). The usual span is 1.37 m; special products are suitable for spans up to ca 3 m.

Metal roofs are used on industrial and farm buildings and are usually made of corrugated aluminum or galvanized steel; a variety of other metals and coated metals is also applied, eg, ASTM A 755 specifies a zinc-coated steel sheet (37) and B 101 a lead-coated copper sheet (38). In the 1800s, standing-seam metal roofs of flat sheets, crimped at each edge, were popular in New England and the Midwest. Steel roofs with 92-cm wide panels, designed with trapezoidal ribs and a factory-applied finish, are used in some residential areas (39). Metal roofing systems also are promoted for plant reroofing (40).

Concrete-slab roofs that do not require a membrane save both on cost and maintenance (41–42). Cracks are prevented by posttensioning the roof slabs during installation.

Fabric roofs are either air-supported or tensioned. In the 1940s, air-supported structures were first used for radar-system enclosures. Later, a variety of military and recreational uses were developed. Both vinyl- and Teflon-coated fiber-glass fabrics have been used (43–44). A resin-covered glass-fiber fabric cover for a domed stadium is much less expensive than a steel-and-concrete roof (45).

Plastic sheets have been installed in roof-support systems. Because of their excellent strength properties, polycarbonate sheets have been reported to be highly successful in severe environments (46).

Shingle and Unit Construction

Shingling is highly effective for shedding water. Natural materials on hand have been used since ancient times, and even today straw-thatch constructions are employed in some countries.

Asphalt Shingles. For many years, organic-felt-base asphalt shingles have been the standard for residential roofing; specification ASTM D 225 was originally issued in 1925 (47). In general, the individual three-tab strip shingle is 91.4 by 30.5-cm wide.

Maximum exposure of 12.7 cm to the weather in application provides a head lap of not less than 18 cm to give double coverage. Other shapes and sizes are available, such as hexagonal tab strips. Seal-down shingles are treated with a factory-applied bonding adhesive after installation to prevent wind damage. Type II of ASTM D 225 has a thick butt, which is the portion exposed to the weather. The typical organic-felt shingle has a felt base saturated with an asphalt of a 60.0–68.3°C softening point and a penetration of 3–5 mm. The asphalt coating usually contains 60 wt % of a finely divided mineral stabilizer, eg, limestone or silica, which is applied to both sides of the saturated felt, with the heavier coating on the weather side. The coating asphalt usually has a softening point of 88–113°C and a penetration of 1.5 mm, min at 25°C. These coating properties also are specified for fiber-glass shingles by ASTM D 3462 (48). The mass of dry felt in organic shingles is specified at a minimum of 53.6 kg/m^2 (11 $lb/100\ ft^2$); a minimum of saturant, coating, and mineral matter is also required. The marketing unit is the sales square, ie, the area of shingles to cover 9.3 m^2 (100 ft^2) of roof. Sales squares are usually packaged in three bundles with a total weight of 107 kg (see Felts).

ASTM D 3462 specifies asphalt shingles with a glass-mat base (48). The glass mat is composed of fine fibers deposited in a nonwoven pattern that may be reinforced with glass yarns. A water-insoluble binding agent holds the fibers together. In shingle-manufacturing, one or more thicknesses of mat are impregnated with an asphalt that is usually compounded with a mineral stabilizer. Asbestos or glass fibers are sometimes added or used in place of the mineral stabilizer. The weather side of the shingle is surfaced with mineral granules and the reverse side with a pulverized mineral to prevent sticking. A self-sealing adhesive is specified, and requirements for wind resistance and tear strength are also given. The mass of the glass mat is specified at 65.9 g/m^2 (1.35 $lb/100\ ft^2$) min.

Other Shingles and Units. The preference for wood, slate, asbestos-cement, or tile roofing shingles and units varies in different parts of the United States. Nationwide, the average is ca 87% asphalt, 7% wood, and 6% miscellaneous (tile, concrete, and metal). The average in the western states is 65% asphalt, 20% wood, and 15% miscellaneous residential applications (6).

Specification ASTM C 222 describes and gives minimum requirements for asbestos-cement shingles (49); three grades of slate roofing units are specified by ASTM C 406, with expected service of 20–40, 40–75, and 75–100 yr (50). Tiles of fired clay with glazed surfaces are attractive and durable. Wood shingles and hand-split shakes have considerable esthetic appeal in certain areas (51). However, fire resistance has been a problem (52), and these materials have been banned in some localities unless a UL Class-C fire requirement is met (53).

Asphalt Roofing Components

The production of asphalt coatings, primer, and cements is illustrated in Figure 1.

Bitumens. Although native bitumens have been used since ancient times for waterproofing, in North America the use of saturated felts dates back to the 1850s. Coal tar and, to some extent, pine tar are the saturants. Coal tar, a by-product from gasworks, is also used as an adhesive in BURs (see Tar and Pitch). The petroleum industry, rapidly developed in the late 1800s, supplied residue asphalts for pavements.

Table 5. Properties of Roofing Asphalt, ASTM D 312[a]

Property	Type I[b]		Type II		Type III		Type IV	
	Min	Max	Min	Max	Min	Max	Min	Max
softening point, °C	57	66	70	80	85	96	99	107
penetration[c], mm								
at 0°C, 200 g, 60 s	0.3		0.6		0.6		0.6	
at 25°C, 100 g, 5 s	1.8	6.0	1.8	4.0	1.5	3.5	1.2	2.5
at 46°C, 50 g, 5 s	9.0	18.0		10.0		9.0		7.5
flash point, COC[d], °C	225		225		225		225	
ductility at 25°C, cm	10.0		3		2.5		1.5	
solubility in trichloro-ethylene, %	99		99		99		99	
slope								
%	up to 4.2		4.2–12.5		8.3–25		16.7–50	
cm per horizontal line	up to 3.8		3.8–11.4		7.6–22.8		15.2–45.6	

[a] Ref. 56.
[b] Self-healing, generally with slag or gravel surface.
[c] Of a certain weight, in g, for a period of time, in s.
[d] Cleveland open cup.

After the discovery of the air-blowing process, these residue asphalts became another source of bitumen for roofings. Roofing asphalts were developed that are highly resistant to flow after application and that retain sufficient fluidity in application (54–55).

Coal tars when used as ply adhesives are generally limited to slopes of ≤4° (4%) without nailing and ≤15° (15%) with back-nailing. The permissible slopes for asphalts without nailing range up to 50° (50%), depending on the asphalt grade specified. Properties of hot-applied bitumens for use in BUR as specified by ASTM are given in Tables 5 and 6 (56–58).

Table 6. Coal-Tar Bitumen Used in Roofing, Dampproofing, and Waterproofing, ASTM D 450[a]

Property	Type I[b]		Type II[c]		Type III[d]	
	Min	Max	Min	Max	Min	Max
softening point, °C	52	60	41	52	56	64
water, %		0		0		0
ash, %		0.5		0.5		0.5
flash point, COC, °C	120		120		120	
specific gravity, 25°/25°C	1.22	1.34	1.22	1.34	1.22	1.34
solubility in CS_2, %	72	85	72	85	72	85
distillate, up to 300°C, %		10		10		5[e]
specific gravity	1.03		1.03			
softening point of residue, °C		80		80		

[a] Ref. 57.
[b] For BUR with ASTM D 227 felts (58).
[c] For membrane waterproofing systems.
[d] For BUR, but less volatile than Type I.
[e] From 315–360°C.

Felts. Roofing felts consist of fiber mats impregnated or partially saturated and sometimes coated with bitumen. Usually, the felt is manufactured in sheets from wood-fiber pulp and some scrap-paper pulp. Earlier felts contained rag fibers thought to be necessary to impart durability. Extensive weathering tests and field experience demonstrated that Asplund wood-fiber felts were satisfactory for BUR and prepared roofings (59). Rag fibers are added only rarely. Organic felts are formed from a water slurry of pulp; the fibers are picked up on a rotating screened drum or by a traveling screen. The sheet is dried by vacuum suction and steam-heated drums and then calendered and wound into large rolls for subsequent trimming and slitting to the required width. The felt is saturated by dipping it into hot bitumen and passing it under and over a series of rolls. The saturated felts are cooled, marked with guide lines for application, and wound into rolls of convenient size. Small perforations let air and moisture escape during application. The saturant improves the bond to the bitumen used in the BUR process and to asphalts used in coated felts. Furthermore, the saturant reduces the rate of water absorption. Vapor permeability and water absorption are further reduced by asphalt coatings stabilized with mineral fillers (see Fillers). A heavier coating is placed on the weather side. A parting agent, eg, sand, mica, or talc, is spread on the coating to prevent sticking (see Abherents).

Asbestos felts are produced essentially in the same manner, except that the mix contains an 85 wt % min of asbestos fibers; the rest is glass or organic fibers. This increases the bulk of the asbestos and permits better saturation with bitumen. Asbestos felts are highly resistant to fungal attack and absorb water to a lesser extent than wood-fiber felts. On direct exposure, the asbestos felts show superior weather properties and excellent dimensional stability because of resistance to moisture and to other weather influences.

Glass felts are manufactured from fibrous glass prepared from a molten mixture of sand and other ingredients; the latter are drawn or blown into filaments. The filaments may be short or may be used in combination with continuous-strand filaments. They may be deposited dry or from a dilute water slurry on a screen conveyor. In either process, a binder is applied. Heat is applied to dry the mat and set the binder. Rolls of glass felt, as received at the roofing plant, are run directly to the coater to be impregnated with asphalt, spread with a mineral parting agent, cooled, marked with guide lines, and rolled to desired lengths. High dimensional stability and low moisture absorption are obtained with fiber-glass roofing felts.

For the preparation of mineral-surfaced roofings and shingles, mineral granules are applied to the heavier-coated side of the felt (see Fig. 2). Variations in manufacture also produce cap sheets, venting base sheets, and flashings.

Roofing Performance

Performance (effectiveness) of a roof covering is determined not only by environmental conditions, but also by the covering of the system. The design should be such that no component plays a critical part in the overall performance. Thus, each component should be lasting or durable in service to contribute to the overall roof performance (60).

Procedures for accelerated system testing are still in the development stage. They are largely based on durability tests of components, determination of system design, and the proper combination of the roof elements.

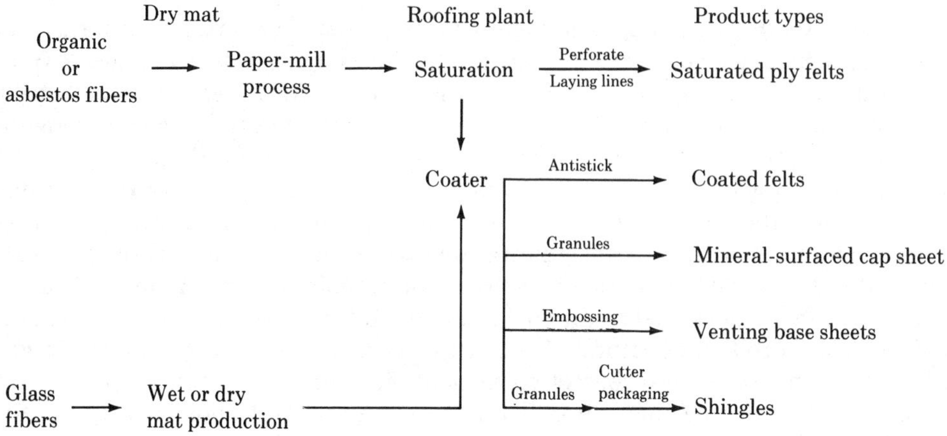

Figure 2. Manufacture of roof coverings, flow sheet.

Roof coverings must be protected from external abuse, eg, foot or other traffic. Penetration of BUR membranes by air-conditioning installations or other roof devices usually voids roof guarantees by the manufacturer, unless specific permission has been received.

Weathering. Bitumen films harden progressively owing to oxidation reactions and, to some extent, loss of plasticizing oils, especially with coal tars. Temperature changes may result in expansion and contraction of the roof covering. Other weather factors include moisture, ice, and hail. Wind affects shingles as well as BUR membrane performance.

In an accelerated weather study of fifteen coating-grade asphalts, a sixfold variation in durability was reported (61). Performance is correlated to properties and composition. Low asphaltene and high resin contents are desirable (62). As weathering progresses, the brittle-point temperature, taken by the Fraass procedure, increases (63). Blending of fractions or base stocks from different crudes improves performance more than antioxidants (64).

In a comprehensive review of theoretical and practical aspects of asphalt durability, six weathering factors were considered (65). A test apparatus for the accelerated weathering of organic materials and procedures was developed at the National Bureau of Standards; a carbon arc was used as the energy source (66). However, outdoor weather exposures are standard procedures for prepared roofings. The opaque granules and mineral stabilizers in the coating asphalts offer excellent protection against the destructive effects of radiant energy and, even in the accelerated weatherometer, an excessive number of daily cycles is required for failure.

Finely divided mineral stabilizers affect the weatherability of coating-grade roofing asphalts. Additions up to 60 wt % of mineral additives increase durability substantially (61). Platy minerals used as ground slates are the most effective. The nature of the asphalt, however, is the most important consideration (67–68). The results of weatherometer and outdoor exposure on combinations of four stabilizers and two asphalts are given in ref. 69.

A number of laboratory procedures that are more rapid than use of the accelerated weatherometer have been described for the determination of expected weatherability of coating asphalts (70–71). Recent work sponsored by the Asphalt Roofing Manufacturers Association describes a stepwise procedure to determine changes in the crude-asphalt source (72) (see Asphalt).

Thermal Effects. Temperature influences the oxidation rate, whereas temperature changes impart a mechanical stress to roof coverings. Daily temperature changes, sometimes quite abrupt, can result in a fatigue mechanism causing tensile failure. Repeated cycles of straining, especially at a high level, diminish the strength of roofing felts to the rupture point. Strain reduction by partial bonding of the membrane to the deck offers the biggest improvement in performance (73).

Data for thermal movement of various bitumens and felts and for composite membranes are given in refs. 74–75, which describe the development of a thermal-shock factor based on strength factors and the linear thermal-expansion coefficient.

Tensile and flexural fatigue tests on roof membranes were taken at 21 and −18°C, and performance criteria were recommended (76).

A study of four types of fluid-applied roofing membranes under cyclic conditions showed that they could not withstand movements of ≤ 1.0 mm over joints (77).

The limitations of present test methods for new roofing materials, such as prefabricated polymeric and elastomeric sheets and liquid-applied membranes, are described in ref. 78. For evaluation, both laboratory and field work are needed.

Water Effects. Water in its different forms (liquid, vapor, hail, and ice) profoundly influences the performance of roof coverings. Moisture migrations in roof insulation, eg, vapor that can accumulate and later liquefy, or water leakage through the roof covering, reduces insulating efficiency and leads to physical deterioration of the roofing material. Moisture can be detected by direct cut tests, electrical-resistance moisture meters, ir-scanning techniques, or nuclear moisture meters.

Flashing flaws in some roofs can be directly responsible for water leakage. The National Roof Contractors Association has developed criteria for curb and edge detail. The recommended flashings include those for stacks, pipes, and expansion joints (79). Factory Mutual Systems (FM) prescribe perimeter-flashing details dealing primarily with wind-uplift forces that can be encountered on built-up roofs (80). Roof-edge damage in windstorms can be extensive and would permit rain leakage and interior damage.

In roof coverings, organic felts are highly susceptible to expansion–contraction movements with moisture change. Shingles, with a low degree of asphalt saturation, especially if the felt has not been uniformly saturated, show warping, described as fish mouthing or clawing. Erratic movements in built-up roofing can be produced by moisture–thermal effects and may cause ridging and failure of the membrane by cracking (81).

The absorbed equilibrium moisture content of built-up roofing membranes varies with changing environmental conditions. In faulty designs, moisture accumulates during late fall and winter in amounts exceeding those that the system can accommodate in the summer. The weakening effects of moisture are substantially less on asbestos or glass with low equilibrium moisture contents (82).

Organic-felt laminates show more movement owing to humidity changes than to temperature changes. Asbestos felts respond moderately to both temperature and humidity and, as with the organic felts, the changes were larger in the longitudinal than in the transverse direction. Glass felts showed small and moderate dimensional changes in response to humidity and temperature changes, respectively; these changes are essentially nondirectional (83).

Hail damage occurs at a minimum impact energy (84). Icing is a function of the roof slope, roof-covering composition, and amount of sun exposure (85–86).

Fire and Wind Hazards. Weather resistance of roof coverings is not necessarily correlated to fire and wind resistance. Underwriter's Laboratories (UL) and the Factory Mutual Systems (FM) are nonprofit organizations that test and rate fire- and wind-hazard resistance.

Organic felt or fiber-glass-mat base shingles are commonly manufactured to meet minimum UL requirements which, in addition to minimum mass, require wind- and fire-resistance properties (87–89).

Fire Ratings. Above-deck fire hazards are rated by UL based on propagation of the flame along the roof covering and on the penetration of the fire into the deck or structure (90). Surface-burning characteristics are measured by the spread-of-flame test. The test deck, either of shingle or BUR construction, is set at a specific incline or slope and exposed to a standard gas flame or burning brand. Resistance Class A is considered effective against severe test exposure and affords a high degree of fire protection to roof decks; Class B affords a moderate degree of protection and Class C a light degree of protection. A list of systems and their classifications is published by UL. Metal deck assemblies are tested by UL with the tunnel test. The assembly, exposed to a gas flame, must protect the roofing membrane on top.

The FM uses a calorimeter fire-test chamber to evaluate the hazard of an under-deck fire. The deck is exposed to a gas flame and the rate of heat release is measured to relate to the rate of flame propagation. Another FM test assesses the damage to roof insulations exposed to radiant heat (91–92).

Wind Testing. Underwriters Laboratory employs two wind-resistance tests, one for shingles and another for BUR assemblies (93); ASTM D 3161 is a standard procedure for measuring the wind-resistance of asphalt shingles (89). In the latter and the UL shingle test, the conditioned test deck is placed at a specified slope and exposed to a wind velocity of 97 km/h for 2 h or until failure by tab or shingle lifting is shown.

Factory Mutual explains how to protect buildings from wind damage (94). Pressure coefficients are given for various structures and roofs. A laboratory uplift-pressure test checks roof assemblies and a field uplift test checks Class-I insulated steel-roof deck construction (93–94). An uplift pressure of 2.9 kPa (0.42 psi) must be withstood in either test under FM conditions to meet the Class-I requirements. The FM guide is revised periodically.

Economic Aspects

In the continental United States, 31 companies have 118 roofing manufacturing plants. As of 1977, 110 plants were in operation. The number of operating plants has remained relatively constant from 1954 through 1977, between 100 and 116 (95). They vary in size, with production in 53 plants ranging from 7,260–408,200 metric tons per year (96). In 1977, 40% of the plants were located in the south and produced 34% of the total production, exclusive of ply felts.

The prices of asphalt roofing products tripled between 1969 and 1978. These increases are attributed primarily to rising material costs, especially asphalt. The cost of felt components has also increased. Granules, parting agents, and stabilizers were estimated to constitute ca 16% of the total cost of materials in 1979. The cost of these materials does not have an appreciable effect on roofing products (97).

Table 7. Roll Roofing and Saturated Ply Felt, U.S. Shipments[a]

Year	Smooth roll, 10^6 sales squares		Mineral or grit, 10^6 sales squares		Saturated felt[b], 10^3 t
	Organic and inorganic	Inorganic	Organic and inorganic	Inorganic	
1970	21.2	na	13.2	na	834.3
1975	17.8	na	13.5	na	671.5
1977	17.4	na	12.3	na	826.7
1978	23.7	3.2	13.1	2.4	807.4
1979	22.1	6.7	15.0	3.3	857.8
1980	24.9	9.6	15.6	3.1	605.7

[a] From ref. 100.
[b] Asphalt or tarred.

A new trend is the replacement of organic felt by fiber-glass mats in shingles and roll roofings. By the mid-1980s, fiber-glass shingles are expected to account for half of the market (98).

The asphalt roofing industry is not seasonal like the housing industry. Supply and demand are well balanced, primarily since there is a large reroofing segment that comprises 50–70% of the market, depending on new construction activity. Although a roof is an indispensible part of a building, it represents less than 1% of its cost. Wood shingles are considerably more expensive than asphalt shingles. Asphalt roofing production has shown a 2% annual growth rate from 1969 to 1977 and is expected to continue at this rate into the mid-1980s (99).

Sales trends for strip shingles, roll roofings, and saturated felts are shown in Tables 7 and 8 (100).

Health and Safety Factors

OSHA enforces standards established by NIOSH to protect workers against safety and health hazards (101). In addition to Federal requirements, state and local agencies may have more stringent health and safety regulations. Health-hazard evaluation reports for various plants are issued after inspection by NIOSH (102–103). According to the EPA, the impact of OSHA regulations on the roofing industry seems to be minimal (104).

Asphalt derived from petroleum differs from coal tar, which is a condensation

Table 8. Strip Shingles, U.S. Shipments, 10^6 Sales Squares[a]

Year	Organic base		Fiber-glass base	Total strip shingles
	Self-sealing	Standard		
1970	26.2	20.0	na	46.2
1975	55.6	1.5	na	57.1
1977	60.2	2.4	na	62.6
1978	67.0	6.3	6.2	79.5
1979	64.3	9.3	9.3	82.9
1980	47.5	2.2	11.3	61.0

[a] From ref. 100.

by-product obtained from the carbonization of coal. The predominant emissions from asphalts are paraffinic or cycloparaffinic in character, whereas coal-tar emissions are predominantly aromatic in character. Low temperatures for handling are suggested as a precaution (105). The Asphalt Institute has issued reports describing emissions from asphalt-paving plants (106) and asphalt built-up roofing operations (107). The emissions from paving asphalts correspond to those expected from saturants used in the asphalt roofing industry. No significant air pollution or health problems were reported in paving operations. In order to reduce worker exposure to emissions from roofing kettles, it was suggested that the kettles operate at the lowest possible temperature and that the lid remain closed.

BIBLIOGRAPHY

"Roofing Materials" in *ECT* 2nd ed., Vol. 17, pp. 459–475, by G. W. Berry, Johns-Manville Corporation.

1. T. Eastwood, *Proc. Geol. Assoc. London* **62,** 6 (1951).
2. "Roof" in *Encyclopedia Britannica*, Vol. 19, Chicago, Ill., 1962, pp. 527–533.
3. M. C. Baker, *Roofs*, Multiscience Publications Ltd., Montreal, Can., 1980, pp. 1–6.
4. S. D. Probert and T. J. Thirst, *Appl. Energy* **6,** 79 (1980).
5. H. W. Busching, *Int. J. Hous. Sci. Appl.* **3,** 21 (1979).
6. Unpublished reports, Asphalt Roofing Manufacturers Association, Washington, D.C.
7. Ref. 3, p. 34.
8. *Asphalt-Saturated Organic Felt Used in Roofing and Waterproofing*, ASTM D 226.
9. *Asphalt-Saturated and Coated Asbestos Felt Base Sheet Used in Roofing*, ASTM D 3378.
10. *Asphalt-Saturated Asbestos Felt Used In Roofing and Waterproofing*, ASTM D 250.
11. *Asphalt-Impregnated Glass (Felt) Mat Used in Roofing and Waterproofing*, ASTM D 2178.
12. *Wide-Selvage Asphalt Roll Roofing (Organic Felts) Surfaced with Mineral Granules*, ASTM D 371.
13. *Asphalt Roll Roofing (Glass Mat) Surfaced with Mineral Granules*, ASTM D 3909.
14. *Emulsified Asphalt Used as a Protective Coating for Roofing*, ASTM D 1227.
15. *Mineral Aggregate Used in Built-Up Roofs*, ASTM D 1863.
16. *Asphalt Primer Used in Roofing, Dampproofing, and Waterproofing*, ASTM D 41.
17. *Asphalt Roof Coatings*, ASTM D 2823.
18. *Aluminum-Pigmented Asphalt Roof Coatings*, ASTM D 2824.
19. *Asphalt Roof Cement*, ASTM D 2822.
20. *Asphalt-Saturated and Coated Organic Felt Base Sheet Used in Roofing*, ASTM D 2626.
21. *Asphalt-Saturated and Coated Organic Felt Used in Roofing*, ASTM D 3158.
22. *Venting Asphalt-Saturated and Coated Inorganic Felt Base Sheet Used in Roofing*, ASTM D 3672.
23. *Lap Cement Used with Asphalt Roll Roofing*, ASTM D 3019.
24. *Emulsified Asphalt Used as a Protective Coating for Roofing*, ASTM D 1227.
25. Ref. 3, p. 194.
26. C. P. Hedlin in D. E. Brotherson, ed., *Moisture Content in Protected Membrane Roof Insulations*, ASTM Special Technical Publication 603, ASTM, Philadelphia, Pa., 1976.
27. K. A. Epstein and L. E. Putnam, *J. Thermal Insul.* **1,** 149 (Oct. 1977).
28. Ref. 3, p. 197.
29. W. J. Rossiter, Jr., and R. G. Mathey, *Nat. Bur. Stand. U.S. Tech. Note 972*, (July, 1978).
30. D. A. Davis and M. P. Krenick in ref. 26, p. 30.
31. Ref. 3, p. 55.
32. *Elastomeric and Plastomeric Roofing and Waterproofing Materials*, ASTM D 3105.
33. "Single Ply Penetrates Roofing Market" in *Roofing, Siding, Insulation*, **55,** 57, 92, 108 (Aug. 1978).
34. "Elasto/Plastic Products," unpublished memorandum, National Roofing Contractors Association, Dec. 1980.

35. *Liquid-Applied Neoprene and Chlorosulfonated Polyethylene Used in Roofing and Waterproofing*, ASTM D 3468.
36. *Corrugated Asbestos-Cement Sheets*, ASTM C 221.
37. *General Requirements for Steel Sheet Zinc-Coated (Galvanized) by the Hot-Dip Process and Coil Coated for Roofing and Siding*, ASTM A 755.
38. *Lead-Coated Copper Sheets*, ASTM B 101.
39. "Steel Roofs Gain Foothold," *Denver Post*, (Dec. 7, 1980).
40. R. C. Baldwin, *Plant Eng.* **33,** 118 (1979).
41. B. Starkovich, *Concr. Int.* **2,** 65 (May 1980).
42. I. Martin, *Concr. Int.* **1,** 59 (Jan. 1979).
43. A. Morrison, *Civ. Eng. ASCE* **50,** 60 (Aug. 1980).
44. B. Mead, *Architectural Fabric Roofs*, ACS Preprints, 179th Meeting, Houston, Tex., Mar. 1980.
45. *Chem. Week*, 60 (Sept. 2, 1981).
46. *Plast. World* **39,** 30 (Feb. 1981).
47. *Asphalt Shingles (Organic Felts) Surfaced with Mineral Granules*, ASTM D 225.
48. *Asphalt Shingles Made from Glass Mat and Surfaced with Mineral Granules*, ASTM D 3462.
49. *Asbestos-Cement Roofing Shingles*, ASTM C 222.
50. *Roofing Slate*, ASTM C 406.
51. Sweet's Division, McGraw-Hill Information Services, New York (issued annually).
52. C. A. Holmes, *Evaluation of Fire-Retardant Treatment for Wood Shingles*, Forest Service Research Paper, FSRP-FPL-158 (May 1971).
53. *Houston Chronicle and Houston Post* (July 31, 1979); *Los Angeles Times* (Nov. 25, 1980).
54. L. W. Corbett, "Manufacture of Petroleum Asphalt," in A. J. Hoiberg, ed., *Bituminous Materials*, Vol. 2, 1965, reprint ed., Krieger Publishing Co., Huntington, N.Y., 1979, p. 81.
55. E. O. Rhodes in ref. 54, Vol. 3, p. 1.
56. *Asphalt Used in Roofings*, ASTM D 312.
57. *Coal-Tar Bitumen Used in Roofing, Dampproofing, and Waterproofing*, ASTM D 450.
58. *Coal-Tar-Saturated Organic Felt Used in Roofing and Waterproofing*, ASTM D 227.
59. S. H. Greenfield, *Nat. Bur. Stand. U.S. Tech. Note 477*, (Mar. 1969).
60. A. J. Hoiberg, *Nat. Bur. Stand. U.S. Spec. Publ. 361*, *1*, Vol. 1, 1972, pp. 777–787; *NBS Special Publication 361*, 1; *Proc. Joint RELEM-ASTM-CIB Symposium*, 777 (May 1972).
61. S. H. Greenfeld, *Nat. Bur. Stand. J. Res.* **64C,** 299 (Oct.–Dec. 1960).
62. *Ibid.*, p. 297.
63. P. M. Jones, *I.E.C. Prod. Res. Dev.* **4,** 57 (Mar. 1965).
64. B. D. Beitchman, *Nat. Bur. Stand. J. Res.* **64C,** 13 (Jan.–Mar. 1960).
65. J. R. Wright in ref. 54, p. 249.
66. *Accelerated Weathering Test of Bituminous Materials*, ASTM D 529.
67. S. H. Greenfield, "Effect of Mineral Additives on the Durability of Coating-Grade Roofing Asphalts, *Nat. Bur. Stand. Bldg. Mat. Struct. Rep. 147* (Sept. 1956).
68. S. H. Greenfeld, *Natural Weathering of Mineral Stabilized Asphalt Coatings*, NBS Bldg. Science Ser. No. 24, Oct. 1969, 15 pp.
69. D. A. Davis and E. J. Bastian, Jr. in Sereda and Litvan, eds., *Durability of Building Materials and Components*, ASTM Special Technical Publication 691, ASTM, Philadelphia, Pa., 1980, p. 767.
70. C. D. Smith, C. C. Schuetz, and R. S. Hodgson, *I.E.C. Prod. Res. Dev.* **5,** 153 (1966).
71. S. H. Greenfeld and J. R. Wright, *Mat. Res. Stand.*, 738 (Sept. 1962).
72. A. J. Hoiberg, *I.E.C. Prod. Res. Dev.* **19,** 450 (1980).
73. K. G. Martin, "Rupture of Built-Up Roofing Components," in *Engineering Properties of Roofing Systems*, ASTM Special Technical Publication 409, ASTM, Philadelphia, Pa., 1967.
74. W. C. Cullen, *Nat. Bur. Stand. U.S. Monogr. 89* (Mar. 1965).
75. W. C. Cullen and T. H. Boone, *Thermal-Shock Resistance for Built-Up Membranes*, NBS Bldg. Science Ser. 9, Aug. 1967.
76. G. F. Sushinsky and R. G. Mathey, *Fatigue Tests of Bituminous Membrane Roofing Specimens*, Report No. NBS-TN-863, Apr. 1975.
77. M. Koike in ref. 73, p. 26.
78. J. O. Laaly and P. J. Sereda in ref. 69, p. 757.
79. *Criteria for NRCA Roof Curb Approval*, National Roofing Contractors Assoc., Oak Park, Ill.
80. *Perimeter Flashing*, Factory Mutual Engineering Corp., Norwood, Mass.
81. E. C. Shuman in ref. 73, p. 41.

82. K. Tator and S. J. Alexander, in ref. 73, p. 187.
83. E. G. Long, in ref. 73, p. 71.
84. S. H. Greenfeld, *Hail Resistance of Roofing Products*, NBS Bldg. Science Ser., BSS-23, Aug. 1969.
85. J. W. Lane, S. J. Marshall, and R. H. Mumis, *Cold Req. Res. Eng. Lab., Rep. 79-17*, (July, 1979).
86. B. J. Dempsey in ref. 69, p. 779.
87. *Building Materials Directory*, Underwriters Laboratories, Oak Park, Ill. (issued annually).
88. *Fire Tests of Roof Coverings*, ASTM E 108.
89. *Wind Resistance of Asphalt Shingles*, ASTM D 3161.
90. *Tests for Fire Resistance of Roof Covering Materials*, UL 790, Underwriters Laboratories, Oak Park, Ill., 1978.
91. *Approval Guide*, Factory Mutual System, Norwood, Mass.
92. *Approval Standard for Class I Insulated Steel Deck Roofs*, Factory Mutual Research Corp., Norwood, Mass.
93. *Wind Resistance of Prepared Roof-Covering Materials*, UL 997, Underwriters Laboratories, Oak Park, Ill., 1973.
94. *Wind Forces on Buildings and Other Structures*, FM System, Norwood, Mass.
95. "*Asphalt Roofing Manufacturing Industry—Background Information for Proposed Standards*," U.S. EPA, Research Triangle Park, N.C., 1980, pp. 8–21.
96. *Ibid.*, pp. 3–6.
97. *Ibid.*, pp. 8–38.
98. *Ibid.*, pp. 8–129.
99. *Ibid.*, pp. 8–126.
100. From asphalt and tar roofing and side products. U.S. Department of Commerce, Bureau of Census, M4-29A Series.
101. "General Industry Standards," U.S. Dept. of Labor, "OSHA Safety and Health Standards" (29 CFR 1910), OSHA 2206, Washington, D.C. (Nov. 7, 1978).
102. A. G. Apol and M. Okawa, *Health Hazard Evaluation Determination Report No. 76-55-443*, NIOSH, National Technical Information Service, Springfield, Va. (Nov. 1977).
103. M. T. Okawa and A. G. Apol, *Health Hazard Evaluation Determination Report No. 77-56-467*, NIOSH, National Technical Information Service, Springfield, Va. (Feb. 1978).
104. Ref. 95, pp. 8–122.
105. V. P. Puzinauskas and L. W. Corbett, *Differences between Petroleum Asphalt, Coal-Tar Pitch and Road Tar*, RR-78-1, Asphalt Institute, College Park, Md., 1978.
106. V. P. Puzinauskas and L. W. Corbett, *Report on Emissions from Asphalt Hot Mixes*, RR-75-1A, Asphalt Institute, College Park, Md., 1975.
107. V. P. Puzinauskas, *Emissions from Asphalt Roofing Kettles*, RR-79-2, Asphalt Institute, College Park, Md., 1979.

General References

Refs. 1–3 are general references.

ARNOLD J. HOIBERG
ERNEST G. LONG
Manville Corporation

SEALANTS

The Bureau of Census of the U.S. Department of Commerce includes sealants with adhesives (qv) in its SIC 2891 classification. According to this source, the adhesive and sealants industry employed almost 16,000 workers in 1977, and the total value of shipments was ca $\$2 \times 10^9$. The annual sales of caulking compounds and sealants was ca $\$400 \times 10^6$. The annual consumption of adhesives and sealants in the United States is 4×10^6 metric tons, of which ca 2×10^6 is sealants (1).

Although there is some overlap in application, adhesives and sealants usually have different functions. The former are selected for their ability to bind two materials, whereas the latter are selected as load-bearing elastic jointing materials that exclude dust, dirt, moisture, and chemicals that contain a liquid or gas. Sealants are also used to reduce noise and vibrations and to insulate or serve as space fillers (see Insulation, acoustic; Chemical grouts).

The term caulking compound is derived from the verb to caulk, which describes the forcing of tarred oakum between the planks of a ship. The original caulking compounds were nonload-bearing materials based on mixtures of bitumens and asbestos (qv). The sealant is a more modern term and describes compositions based on synthetic resins as well as asphaltic and oil-based caulking compounds (2–4). The formulations for several hundred commercially available adhesives and sealants have been described (5).

These composites require little if any energy for conversion to solids and usually conserve energy when applied. The principal resins used as sealants are polysulfides, silicones, polyurethanes, acrylics, neoprene [*31727-55-6*], butyl rubber [*9006-49-9*], and sulfochlorinated polyethylene [*68037-39-8*] (6).

Polysulfides

Poly(ethylene sulfide) [*24936-67-2*] was one of the first synthetic elastomers and was produced by the condensation of α,ω-dichloroalkanes with sodium polysulfide (7) (see Polymers containing sulfur). The principal polysulfide is prepared from bis(chloroethyl) formal. The utility of these solvent-resistant solid elastomers was increased when they were reduced to lower molecular weight liquid mercaptans by heating with sodium sulfite and sodium hydrosulfite. The liquid products can be oxidized by metallic oxides, eg, lead dioxide, or by epoxy resins *in situ* at ambient temperatures to produce flexible caulking compounds (8). It is customary to add fillers (qv), eg, carbon black (qv); calcium carbonate or clay; plasticizers (qv), eg, dibutyl phthalate; retarders, eg, stearic acid; accelerators, eg, amino compounds; and adhesion promotors, eg, epoxy resins (qv) (9) (see Rubber chemicals). Polysulfide sealants are commercially designated as LP-2, LP-32, LP-3, LP-33, etc. These liquid prepolymers have different average molecular weights and different cross-link densities.

A typical formulation for a two-package polysulfide sealant in parts by weight is as follows:

Composition	*Package 1*	*Package 2*
polysulfide	100	
liquid epoxy resin		90
solid epoxy resin		60
titanium dioxide	100	
carbon black	5	
urea–formaldehyde resin		20
methyl ethyl ketone	45	
plasticizer	15	

These materials, like most two-package sealants, caulking compounds, and cements, are thoroughly mixed before application by trowel, knife, or caulking gun. The prepolymer hardens in place at ambient temperatures to a semielastic solid.

Polysulfide sealants are the most widely used sealants for airplane fuel tanks, curtain-wall construction, glazing, marine decks, binders for solid-rocket propellants, and sealants for joints in airport runways, highways, and canals (10). The principal advantages of polysulfide sealants are ease of application, good adhesion, good resistance to weathering and solvents, negligible shrinkage, and good moisture-vapor transmission (MVT) (11).

Silicones

Silicone sealants based on polydimethylsiloxanes were developed in the early 1940s, and the two-package, room-temperature-vulcanizing (RTV), elastomeric silicone sealants with silanol ends were introduced in the 1950s (12). One-package silicone sealants, which cure by exposure to moisture in the air, were introduced in the 1960s (13). The first one-package silicone sealants were characterized by a tacky surface. This deficiency was eliminated by the addition of metal salts of carboxylic acids. These one-component RTV systems account for the main share of the silicone-sealant market. These products are used in adhesives, encapsulation, impregnation, moldmaking, and sealants.

Silicone sealants are based on mixtures of fillers, eg, silica, silicone polymers, cross-linking components, and catalysts. The polymer has a siloxane backbone, ie, —Si—O—Si, with alkyl and alkoxy or acetoxy pendant groups. The latter are readily hydrolyzed to silanol groups (SiOH), which form larger chains by condensation and loss of alcohol or acetic acid.

Trifunctional silanes, eg, trimethoxylmethylsilane or triacetoxymethylsilane, are used as cross-linking agents. Metal carboxylates such as stannous octanoate and titanates, eg, tetraisopropyl titanate, are used as catalysts (14).

The components of the two-package silicone sealants or caulking compositions are mixed thoroughly just before use, and the mixtures are usually applied with a caulking gun or knife. The one-component systems are cured by exposure in moist air. The cured products are used as flexible space fillers or sealants.

Silicone sealants that were introduced in 1967 have low shrinkage characteristics and can be applied and used over a wide temperature range (−40–65°C). The fluorosilicones (polyfluoroalkylsiloxanes), which retain their flexibility at even lower temperatures (−50°C), are used as sealants for aircraft, automobiles, and light bulbs as well as for sealing oil-well heads in Alaskan oil fields (15).

Silicone sealants with cyanoalkyl pendant groups have superior solvent resistance and are used as oil rings on fuel lines of jet aircraft (16–17). Room-temperature-vulcanizing silicones have been used for formed-in-place gaskets and potting compounds (18–19) (see Silicon compounds, silicones).

Butyl Rubber

Polyisobutylene [*9003-27-4*] produced by the low temperature cationic polymerization of isobutylene has been used to a small extent as a one-package sealant since the 1930s. However, butyl rubber, which is a copolymer of isobutylene and isoprene, is more readily available and can be readily cross-linked by agents, eg, *p*-quinone dioxime and an oxidizer (see Elastomers, synthetic).

Commercial butyl sealant compositions usually contain fillers, such as carbon black and zinc oxide, or silica; tackifiers, such as pentaerythritol esters of rosin; and solvents, eg, cyclohexane. A typical butyl rubber caulking compound has the following composition:

Composition	*Parts by weight*
butyl rubber	175
mineral spirits	270
petroleum resins	34
pentaerythritol esters of rosin	8
bentone clay derivative	23
finely divided silica	364
fiber	91
titanium dioxide	45

Both polyisobutylene and butyl rubber sealants harden by evaporation of the solvent. Butyl rubber compositions may be used as hot-melt sealants. These compositions may be admixed with ethyl acrylate copolymers and used as a gunnable hot melt for sealing dual-pane windows. Butyl rubber sealants are used for joint-and-panel sealing as well as for glazing. Butyl rubber is also available as a tape sealant, which is pressed into the cavities to be filled by sealants.

Polyurethanes

Polyurethanes, which were developed in the 1940s for use as fibers, elastomers, plastic foams, and adhesives, are also useful as sealants. These sealants are usually two-package systems, but one-package systems are also available. The two-package systems are based on a diisocyanate, eg, tolyl diisocyanate (TDI), and a poly(ethylene glycol). These systems also contain antioxidants, pigments, fillers, alkylenetriol cross-linking agents, and organometallic catalysts, eg, dibutyltin dilaurate (see Urethane polymers).

These two-component systems are used on a large scale in reaction-injection molding (RIM), in which the components are allowed to react in a large mold (20). Castable polyurethane sealants are characterized by a wide range of hardness and flexibility. They have good sag resistance, good thixotropy, and good adhesion to the substrate (21–22). These sealant compositions are also used as dental fillings (23) (see Dental materials). The surfaces to be sealed must be dry to prevent foam formation from the reaction of water and the diisocyanates. Polyurethane sealants also are being used in the assembly of several makes of automobiles.

These compositions, like most sealants, are not energy-intensive. Their hardness and flexibility may be controlled by the selection of appropriate diols and fillers. The life expectancy of polyurethane sealants is ca 10 yr, in contrast to a life expectancy of 15 yr for polysulfides and butyl sealants (24). Polyurethane sealants have been used successfully in lip-type launch seals in submarine missile systems (25). Over 75,000 metric tons of polyurethane-type sealants are used annually.

Acrylics

Emulsions and solutions of poly(methyl methacrylate) or its copolymers have been used as caulking materials and sealants. The solvent-type acrylics are similar to high solids (90 wt %), acrylic coatings. A typical formula for a water-based acrylic sealant is as follows:

Composition	*Parts by weight*
acrylic latex (50 wt % solids)	41.9
wetting agent	1.7
plasticizer	9.5
ethylene glycol	32
calcined China clay	42.7
titanium dioxide	1.7
mineral spirits	0.3

Two-package systems consist of acrylic monomers, fillers, and initiators. They polymerize at RT and are also used as polymer concrete (26). These compositions have been used for repairing potholes and cracks in concrete highways and for bone cement (27). One-package acrylic sealants harden by evaporation of water and solvent. The two-package systems are essentially solvent-free but precautions must be taken to avoid inhalation of acrylic monomers. Automated equipment is available for mixing the components of both systems (see Acrylic ester polymers).

Polychloroprene

Polychloroprene (neoprene) was commercialized by E. I. du Pont in the 1930s. The preferred sealant composition is a two-package system consisting of the polymers in one container and accelerators, eg, tertiary amine, in the other (28). A typical polychloroprene sealant has the following composition:

Composition	*Parts by weight*
Neoprene AG	110
magnesia	4
zinc oxide	5
heat-reactive resin	45
methyl ethyl ketone, naphtha, or toluene	600

The components are mixed in an intensive mixer. Some two-package systems contain litharge (PbO) and cure at ambient temperatures. The one-package systems set by evaporation of solvent and cure at elevated temperatures. Appropriate precautions must be taken to avoid inhalation of the flammable solvent used in the one-package system (see Elastomers, synthetic).

Chlorosulfonated Polyethylene

Hypalon is the trade name of the product obtained by chlorination and sulfonation of polyethylene, and it is used as a sealant (28). It is customary to blend at least two polymers with different degrees of substitution (DS) with a plasticizer, eg, chlorinated naphthalene; fillers, eg, carbon black; and curing agents, eg, litharge. Compounded mixtures of sulfochlorinated polyethylene and neoprene are also used as sealants for sealing masonry. A typical formulation for a sealant of this type is as follows:

Composition	*Parts by weight*
blend of sulfochlorinated polyethylene	17.5
chlorinated paraffins	17.5
asbestos and other fillers	20.0
titanium dioxide	14.0
curing agents, eg, litharge	7.5
dibutyl phthalate plasticizer	19.0
solvents, eg, isopropyl alcohol, etc	4.5

These compositions are cured by cross-linking at ambient temperatures to produce flexible sealants with good adhesion (see Elastomers, synthetic).

Bitumens

Bitumens or asphaltic materials have been used as sealants for many centuries (see Asphalt). Bituminous hot-melt compositions usually contain scrap rubber or neoprene, which acts as a flexibilizing agent. These dark-colored sealants are used for joints in highways and buildings. Their tendency to cold-flow is reduced by the incorporation of epoxy resins. Sealants produced from asphalt emulsions have been used to prevent the escape of radon gas from storage bins of uranium-mill tailings.

Latex Caulking Compositions

Filled polymeric emulsions, which are related to latex paints, have been used as tub caulks and spackling compounds. A typical caulking composition consists of 45 wt % poly(vinyl acetate), filler, plasticizer, and water. Since these systems set by evaporation, their shrinkage is greater than hot-melt or prepolymer sealants (see Paint; Latex technology).

Oil-Based Caulks and Sealants

Oil-based caulks or putty have been used for centuries as glazing sealants. These 100% solids compositions are much more rigid than other sealants, but the rigidity is often overcome by the addition of elastomers, eg, neoprene. A typical formula for an oil-based caulking composition is as follows:

Composition	*Parts by weight*
bodied vegetable oil	100
kettle-bodied vegetable oil	70
polyisobutylene	100
cobalt carboxylate	0.20
calcium carbonate	483
asbestos (long fiber)	23
asbestos (short fiber)	27
titanium dioxide	17

Hot-Melt Sealants

In addition to butyl rubber and bitumens, several other polymers are used as hot-melt sealants. Among these are the copolymers of ethylene and vinyl acetate (EVA), atactic polypropylene, and mixtures of paraffin wax and polyolefins (see Olefin polymers). Over 250,000 t of hot-melt sealants is used annually. Silica- or carbon-black-filled sulfur cements are also used as hot-melt sealants. Because they must be heated before use, hot-melt sealants are more energy-intensive than many room-temperature-curing sealants.

Poly(vinyl Chloride) Plastisols

Poly(vinyl chloride) (PVC) plastisols have been used widely as gap fillers and sealants in automobiles. These plastisols are dispersions of finely divided PVC in liquid plasticizers, eg, dioctyl phthalate (DOP). These liquid sealants set to form flexible compositions when heated, ie, they are activated at 125–200°C. Filled plastisols or plastigels may be used like putty. Plastigels have less tendency to leak through large crevices. However, the liquid plasticizer may bleed from either of these PVC sealants and cause softening of painted surfaces (see Vinyl polymers).

Polyesters

Fibrous-glass-reinforced unsaturated polyesters have been used as structural materials since the early 1940s. Silica- and calcium carbonate-filled polyester grouting compositions called cultured marble, plastic concrete, and Vitroplast were developed in the early 1950s (29). These strong, lightweight plastic cements have been used for making sewer pipe (30). Plastic concrete, trade name Polysil, is used for making utility poles, bathroom fixtures, and manholes (31). Polyester compositions have been used to fill crevices and for automobile body repair.

A typical two-package, unsaturated, polyester sealant consists of a filler, eg, silica, calcium carbonate, or alumina trihydrate (ATH), and an initiator, eg, benzoyl peroxide (BPO). The liquid compound consists of an ethylene glycol maleate dissolved in styrene monomer and a tertiary amine, eg, *N*,*N*-dimethylaniline. Since the styrene monomer is flammable and has a relatively high vapor pressure, adequate ventilation must be provided to protect the applicators.

Epoxy Resins

Epoxy resins (qv) were first synthesized by the reactions of bisphenol A and epichlorohydrin in the early 1940s by Castans in Switzerland (32). Most of the annual production of 400,000 t of epoxy resin is used as *in situ* polymerized plastics, which are used as adhesives (6.5%), flooring (6.0%), protective coatings (45%), and laminates (16.6%) (33) (see also Embedding).

It is customary to cure or cross-link epoxy prepolymers with polyamines, eg, diethylenetriamine. Epoxy resins have been used to seal bridge surfaces and to repair concrete structures. Filled epoxy resins have been used as jointing materials for brick and as monolithic heavy-duty coatings. These compositions are resistant to many nonoxidizing acids, salts, and alkalines (34–35).

A typical formulation for a corrosion-resistant epoxy resin sealant is as follows:

Composition	*Parts by weight*
liquid epoxy resin prepolymer	100
liquid poly(ethylene sulfide)	50
dimethylaminopropylamine	10

Phenolic Resins

In situ cured silica-filled phenolic resin composites were introduced under the trade name Asplit in Germany in the 1930s (36) (see Phenolic resins). The two-package system consists of a silica or carbon filler containing an acid setting agent, eg, *p*-toluenesulfonic acid (PTA), and an A-stage resole phenolic resin. The filler and liquid resin must be thoroughly mixed and the mixture must be applied to bricks or tile while it is a fluid. Phenolic grouting compositions and other resinous sealants that are cured by strong acids cannot be used as grouts for concrete or other alkaline substrates.

Urea Resins

The disadvantage of the dark color associated with the phenolic resin cement was overcome by the use of urea–formaldehyde prepolymers. Kaolin-filled, *in situ* cast urea resinous composites had been used for sealing cracks in underground rocks and for the fabrication of irrigation pipes. The acid setting agents used with urea resins are usually inorganic acids, which are less apt to cause dermatitis than the *p*-toluenesulfonic acid (PTA) used with phenolic resin compositions. However, these resin compositions can be used for filling voids if the surfaces of the substrate are coated with a protective coating (see Amino resins).

Furan Resin

The lack of resistance of phenolic resin cements to alkali was overcome by the development of comparable cements based on furan resins (37). These two-package resinous cements have been used extensively as mortars, grouts, and setting beds for brick and tile (38). Most of these high solids composites are essentially solvent-free. These jet-black materials are chemurgic products (see Chemurgy). The furfuryl alcohol used to produce furan resins is obtained by the reduction of furfural. Furfural is obtained by acid-degradation of pentosans, eg, those present in corn cobs and oat hulls (see Furan derivatives).

Economic Aspects

The use of sealants is growing at an annual rate of 6%, and the total sales should exceed $\$10^9$ in 1985 (1). The annual U.S. sales for various uses are shown in Table 1.

There are over 350 compounders and producers of sealants in the United States. Over 60% of the total value of sealants used is supplied by 15 producers. General Electric, Dow Corning, and Thiokol are the only vertically integrated producers of sealants; General Electric, DAP, and Dow Corning are the leading sealant suppliers.

The modern automobile contains 40 kg of adhesives, sealants, and sound deadeners. Poly(vinyl chloride) and asphalt have been the principal products used for these applications, but asphalt is being replaced by more complex polymers. Considerable quantities of sound deadeners are being used in the floor and rear-seat areas of diesel automobiles. The emphasis on the reduction of energy costs has resulted in an increase in the use of insulated glass in residential construction. Polysulfides account for 90% of the sealants used in this application, but these products are being displaced by butyl rubber hot melts.

Health and Safety Factors

Although the death rate of sealant workers is no greater than that of other workers, appropriate precautions must be taken to assure the health and safety of those working with the production or application of sealants. Care should be taken to prevent exposure to solvents, plasticizers, and fillers used with polysulfide sealants. The solvents should not be allowed to evaporate in a closed area, and potable water or foods should

Table 1. Sales Volume for Sealants, $\$10^6$

Use	1979	1985
building construction	135	283
do-it-yourself market	95	201
insulating glass	56	122
glazing	42	88
membranes	37	84

not contact the set sealants for long periods. Adequate precaution should be taken during production and application if asbestiform fillers are used in the formulation. Polysulfide sealants produce hydrogen sulfide when heated and, therefore, should not be used at elevated temperatures.

Silicone sealants are usually considered to be nontoxic. However, the acetoxysilanes produce acetic acid in the setting reaction, and precaution must be taken to counteract any deleterious effects of this weak acid.

Butyl rubber melt sealants are nontoxic, but room-temperature-setting butyl rubber sealants contain solvents that are volatile and flammable. Thus, provisions must be made for exhausting these solvents when the sealants are applied in enclosed areas.

The diisocyanates present in two-package polyurethane systems are toxic, and adequate ventilation must be provided. The cured polyurethanes are considered to be nontoxic unless heated above their decomposition temperature.

Sealants based on acrylic emulsions are nontoxic. Monomers and solvents in the two-package systems are toxic and flammable and, therefore, adequate precautions should be taken to remove the fumes from working areas. Additional precautions should be taken when acrylic sealant systems containing asbestos fillers are used.

Sealants based on chlorosulfonated polyethylene may contain chlorinated aromatic and phthalate plasticizers, asbestos fillers, volatile solvents, and lead curing agents. Appropriate precautions must be taken to avoid contact with the asbestos filler or plasticizers during application of these sealants. Solvents must be exhausted from the working area and foodstuffs should not be placed in contact with the cured sealant.

Hot-melt asphaltic compositions should not be allowed to come in contact with the skin. The solvent in bituminous mastics must be exhausted from the working area. Both the solvent and the set bitumens are flammable.

Since many oil-based caulks contain asbestos, appropriate precautions must be taken during production and application. Since most hot melts are flammable, adequate precaution must be taken to prevent them from igniting and from coming in contact with the skin of the applicator.

Initiators or catalysts used to cure polyester resins are unstable compounds that may explode by impact. Hence, initiators, eg, benzoyl peroxide, should be used in small quantities and preferably in a liquid medium. Styrene is said to be carcinogenic; it is also flammable. Therefore, adequate ventilation must be used to maintain concentrations of styrene monomer in air of less than 10 ppm.

Because many epoxy resin curing agents are vesicants, the skin of the applicator should be protected to prevent dermatitis. Since the fumes of these amines may cause asthma, precautions should be taken to prevent the inhalation of these fumes. These problems can be reduced by automation and good housekeeping.

Formaldehyde which evolves from phenolic and urea resins is said to be carcinogenic. The acid curing agent can cause dermatitis if it comes in contact with the skin. If the process is not automated, adequate precaution should be taken to prevent inhalation of formaldehyde fumes and to prevent contact between the applicator's skin and the acid curing agent. The acid setting agent in furan cements can also cause dermatitis if the applicator's skin is not protected by salves or clothing.

BIBLIOGRAPHY

1. *Sealants*, Skeist Laboratories, Inc., Livingston, N.J.
2. A. Damusis, *Sealants*, Reinhold Publishing Co., New York, 1967.
3. R. B. Seymour, ed., *Plastic Mortars, Sealants, and Caulking Compounds*, ACS Symposium Series 113, Washington, D.C., 1979.
4. H. H. Buchter, *Industrial Sealing Technology*, John Wiley & Sons, Inc., New York, 1979.
5. E. W. Fleck, *Adhesive and Sealant Compounds and Their Formulations*, Noyes Data Corp., Park Ridge, N.J., 1978.
6. S. S. Maharajan and N. D. Ghatge, *Paintindia* **29**(6), 37 (1979).
7. Brit. Pat. 302,220 (Dec. 13, 1927), J. C. Patrick and N. M. Mnookin (to Thiokol Corp.).
8. U.S. Pat. 2,466,963 (Apr. 12, 1949), J. C. Patrick and H. R. Ferguson (to Thiokol Corp.).
9. E. R. Bertozzi, *Rubber Chem. Technol.* **41,** 114 (1968).
10. U.S. Pat. 2,910,922 (Nov. 3, 1959), F. P. Horning (to Thiokol Corp.).
11. M. A. Schuman and A. D. Yazujian in ref. 3, Chapt. 11.
12. U.S. Pat. 2,571,039 (Oct. 9, 1951), J. F. Hyde (to Dow Corning, Inc.).
13. U.S. Pat. 3,105,061 (Sept. 24, 1963), L. B. Bruner (to Dow Corning, Inc.).
14. J. M. Klosowski and G. A. L. Gent in ref. 3, Chapt. 10.
15. O. R. Pierce and Y. K. Kim, *J. Elastoplast.* **3,** 82 (1971).
16. U.S. Pat. 3,513,508 (Sept. 29, 1970), G. K. Goldman and L. Morris (to Dow Corning, Inc.).
17. W. J. Ratelle, *Mach. Des.* **49,** 133 (1971).
18. T. J. Gair and W. W. Wadsworth, *SAE Tech. Pap. 720129* (1972).
19. A. Grey, *Elastomerics* **109,** 23 (1977).
20. W. M. Haines, *Elastomerics* **110**(9), 26 (1978).
21. G. Eagle, *Plast. Eng.* **36**(7), 29 (1980).
22. L. J. Lee, *Rubber Chem. Technol.* **53,** 542 (1980).
23. G. G. Stecher, *New Dental Materials*, Noyes Data Corp., Park Ridge, N.J., 1980.
24. R. B. Seymour in ref. 3, Chapt. 8.
25. J. P. Meier, G. F. Rudd, A. J. Molna, V. D. Jerson, and M. A. Mendelsohm, *Resins for Aerospace*, ACS Symposium Series 132, Washington, D.C., 1980, Chapts. 14–15.
26. R. Martino, *Mod. Plast.* **57**(3), 46 (1980).
27. S. W. Shalaby, *Polymer News* **5,** 176 (1979).
28. L. S. Bake in I. Skeist, ed., *Handbook of Adhesives*, R. E. Kreiger Publishing Co., Huntington, N.Y., 1973, Chapt. 20.
29. R. B. Seymour in ref. 3, Chapt. 5.
30. B. J. Schrock and G. Rangel, *Paper presented at 30th Anniv. Tech. Conf. Reinforced Plastics/Composites Institute SPI*, 1975.
31. G. Miller, *Plast. Des. Process.* **19**(1), 30 (1979).
32. I. Skeist, *Epoxy Resins*, Van Nostrand Reinhold Co., New York, 1958.
33. J. M. Sosa in ref. 3, Chapt. 3.
34. R. B. Seymour and R. H. Steiner, *Chem. Eng. Progr.* **49,** 220 (1953).
35. F. T. Watson, *Symposium on Engineering Applications of Epoxy Resins at 57th Annual Meeting of AICHE*, Boston, Mass., Dec. 6, 1964.
36. R. B. Seymour in ref. 3, Chapt. 1.
37. U.S. Pat. 2,226,049 (Dec. 26, 1944), C. R. Payne and R. B. Seymour (to Atlas Mineral and Chemical Products Co.).
38. R. H. Leitheiser, M. E. Londrigan, and C. A. Rude in ref. 3, Chapt. 2.

RAYMOND B. SEYMOUR
The University of Southern Mississippi

A

B

C

D

E

G

H

I

J

K

L

M

N

O

P

Q

R

S

T

U

V

W